T0181959

Exact Solutions of Einstein's Field Equations

A revised edition of the now classic text, *Exact Solutions of Einstein's Field Equations* gives a unique survey of the known solutions of Einstein's field equations for vacuum, Einstein–Maxwell, pure radiation and perfect fluid sources. It starts by introducing the foundations of differential geometry and Riemannian geometry and the methods used to characterize, find or construct solutions. The solutions are then considered, ordered by their symmetry group, their algebraic structure (Petrov type) or other invariant properties such as special subspaces or tensor fields and embedding properties.

This edition has been expanded and updated to include the new developments in the field since the publication of the first edition. It contains five completely new chapters, covering topics such as generation methods and their application, colliding waves, classification of metrics by invariants and inhomogeneous cosmologies. It is an important source and guide for graduates and researchers in relativity, theoretical physics, astrophysics and mathematics. Parts of the book can also be used for preparing lectures and as an introductory text on some mathematical aspects of general relativity.

The late HANS STEPHANI gained his Diploma, Ph.D. and Habilitation at the Friedrich-Schiller-Universität Jena. He became Professor of Theoretical Physics in 1992, before retiring in 2000. He lectured in theoretical physics from 1964 and published numerous papers and articles on relativity and optics. He was also the author of four books.

DIETRICH KRAMER was Professor of Theoretical Physics at the Friedrich-Schiller-Universität Jena, and retired in 2003. He graduated from this university, where he also finished his Ph.D. (1966) and Habilitation (1970). His research concerns classical relativity. The majority of his publications are devoted to exact solutions in general relativity.

MALCOLM MACCALLUM is Professor of Applied Mathematics at the School of Mathematical Sciences, Queen Mary, University of London, where he was also Vice-Principal for Science and Engineering for three years. He graduated from Kings College, Cambridge and went on to complete his M.A. and Ph.D. there. His research covers general relativity and computer algebra, especially tensor manipulators and differential equations. He has published many papers, review articles and books. In 2009–12 he will be Director of the Heilbronn Institute for Mathematical Research at Bristol.

CORNELIUS HOENSELAERS gained his Diploma at Technische Universit at Karlsruhe, his D.Sc. at Hiroshima Daigaku and his Habilitation at Ludwig-Maximilian Universität München. He is Reader in Relativity Theory at Loughborough University. He has specialized in exact solutions in general relativity and other non-linear partial differential equations, and published a large number of papers, review articles and books.

EDUARD HERLT was wissenschaftlicher Mitarbeiter at the Theoretisch Physikalisches Institut der Friedrich-Schiller-Universität Jena and retired in 2008. Having studied physics as an undergraduate at Jena, he went on to complete his Ph.D. there as well as his Habilitation. He has had numerous publications including one previous book.

Exact Solutions of Einstein's Field Equations

CAMBRIDGE MONOGRAPHS ON MATHEMATICAL PHYSICS

General Editors: P. V. Landshoff, D. R. Nelson, S. Weinberg

S. J. Aarseth *Gravitational N-Body Simulations: Tools and Algorithms*
J. Ambjørn, B. Durhuus and T. Jonsson *Quantum Geometry: A Statistical Field Theory Approach*
A. M. Anile *Relativistic Fluids and Magneto-fluids: With Applications in Astrophysics and Plasma Physics*
J. A. de Azcárraga and J. M. Izquierdo *Lie Groups, Lie Algebras, Cohomology and Some Applications in Physics*†
O. Babelon, D. Bernard and M. Talon *Introduction to Classical Integrable Systems*†
F. Bastianelli and P. van Nieuwenhuizen *Path Integrals and Anomalies in Curved Space*
V. Belinski and E. Verdaguer *Gravitational Solitons*
J. Bernstein *Kinetic Theory in the Expanding Universe*
G. F. Bertsch and R. A. Broglia *Oscillations in Finite Quantum Systems*
N. D. Birrell and P. C. W. Davies *Quantum Fields in Curved Space*†
K. Bolejko, A. Krasiski, C. Hellaby and M-N. Célérier *Structures in the Universe by Exact Methods: Formation, Evolution, Interactions*
D. M. Brink *Semi-Classical Methods for Nucleus-Nucleus Scattering*†
M. Burgess *Classical Covariant Fields*
E. A. Calzetta and B.-L. B. Hu *Nonequilibrium Quantum Field Theory*
S. Carlip *Quantum Gravity in 2+1 Dimensions*†
P. Cartier and C. DeWitt-Morette *Functional Integration: Action and Symmetries*
J. C. Collins *Renormalization: An Introduction to Renormalization, the Renormalization Group and the Operator-Product Expansion*†
P. D. B. Collins *An Introduction to Regge Theory and High Energy Physics*†
M. Creutz *Quarks, Gluons and Lattices*†
P. D. D'Eath *Supersymmetric Quantum Cosmology*
F. de Felice and C. J. S Clarke *Relativity on Curved Manifolds*
B. DeWitt *Supermanifolds, 2nd edition*†
P. G. O Freund *Introduction to Supersymmetry*†
Y. Frishman and J. Sonnenschein *Non-Perturbative Field Theory: From Two Dimensional Conformal Field Theory to QCD in Four Dimensions*
J. A. Fuchs *Affine Lie Algebras and Quantum Groups: An Introduction, with Applications in Conformal Field Theory*†
J. Fuchs and C. Schweigert *Symmetries, Lie Algebras and Representations: A Graduate Course for Physicists*†
Y. Fujii and K. Maeda *The Scalar-Tensor Theory of Gravitation*
J. A. H. Futterman, F. A. Handler, R. A. Matzner *Scattering from Black Holes*†
A. S. Galperin, E. A. Ivanov, V. I. Orievetsky and E. S. Sokatchev *Harmonic Superspace*
R. Gambini and J. Pullin *Loops, Knots, Gauge Theories and Quantum Gravity*†
T. Gannon *Moonshine beyond the Monster: The Bridge Connecting Algebra, Modular Forms and Physics*
M. Göckeler and T. Schücker *Differential Geometry, Gauge Theories and Gravity*†
C. Gómez, M. Ruiz-Altaba and G. Sierra *Quantum Groups in Two-Dimensional Physics*
M. B. Green, J. H. Schwarz and E. Witten *Superstring Theory Volume 1: Introduction*†
M. B. Green, J. H. Schwarz and E. Witten *Superstring Theory Volume 2: Loop Amplitudes, Anomalies and Phenomenology*†
V. N. Gribov *The Theory of Complex Angular Momenta: Gribov Lectures on Theoretical Physics*
J. B. Griffiths and J. Podolský *Exact Space-Times in Einstein's General Relativity*
S. W. Hawking and G. F. R. Ellis *The Large Scale Structure of Space-Time*†
F. Iachello and A. Arima *The Interacting Boson Model*†
F. Iachello and P. van Isacker *The Interacting Boson-Fermion Model*
C. Itzykson and J. M. Drouffe *Statistical Field Theory Volume 1: From Brownian Motion to Renormalization and Lattice Gauge Theory*†
C. Itzykson and J. M. Drouffe *Statistical Field Theory Volume 2: Strong Coupling, Monte Carlo Methods, Conformal Field Theory and Random Systems*†
C. V. Johnson *D-Branes*†
P. S. Joshi *Gravitational Collapse and Spacetime Singularities*

J. I. Kapusta and C. Gale *Finite-Temperature Field Theory: Principles and Applications, 2nd edition*
V. E. Korepin, N. M. Bogoliubov and A. G. Izergin *Quantum Inverse Scattering Method and Correlation Functions*†
M. Le Bellac *Thermal Field Theory*†
Y. Makeenko *Methods of Contemporary Gauge Theory*
N. Manton and P. Sutcliffe *Topological Solitons*†
N. H. March *Liquid Metals: Concepts and Theory*
I. Montvay and G. Münster *Quantum Fields on a Lattice*†
L. O'Raifeartaigh *Group Structure of Gauge Theories*†
T. Ortín *Gravity and Strings*†
A. M. Ozorio de Almeida *Hamiltonian Systems: Chaos and Quantization*†
L. Parker and D. J. Toms *Quantum Field Theory in Curved Spacetime: Quantized Fields and Gravity*
R. Penrose and W. Rindler *Spinors and Space-Time Volume 1: Two-Spinor Calculus and Relativistic Fields*†
R. Penrose and W. Rindler *Spinors and Space-Time Volume 2: Spinor and Twistor Methods in Space-Time Geometry*†
S. Pokorski *Gauge Field Theories, 2nd edition*†
J. Polchinski *String Theory Volume 1: An Introduction to the Bosonic String*
J. Polchinski *String Theory Volume 2: Superstring Theory and Beyond*
V. N. Popov *Functional Integrals and Collective Excitations*†
R. J. Rivers *Path Integral Methods in Quantum Field Theory*†
R. G. Roberts *The Structure of the Proton: Deep Inelastic Scattering*†
C. Rovelli *Quantum Gravity*†
W. C. Saslaw *Gravitational Physics of Stellar and Galactic Systems*†
R. N. Sen *Causality, Measurement Theory and the Differentiable Structure of Space-Time*
M. Shifman and A. Yung *Supersymmetric Solitons*
H. Stephani, D. Kramer, M. MacCallum, C. Hoenselaers and E. Herlt *Exact Solutions of Einstein's Field Equations, 2nd edition*
J. Stewart *Advanced General Relativity*†
T. Thiemann *Modern Canonical Quantum General Relativity*
D. J. Toms *The Schwinger Action Principle and Effective Action*
A. Vilenkin and E. P. S. Shellard *Cosmic Strings and Other Topological Defects*†
R. S. Ward and R. O. Wells, Jr *Twistor Geometry and Field Theory*†
J. R. Wilson and G. J. Mathews *Relativistic Numerical Hydrodynamics*

†Issued as a paperback

Exact Solutions of Einstein's Field Equations

Second Edition

HANS STEPHANI
Friedrich-Schiller-Universität, Jena

DIETRICH KRAMER
Friedrich-Schiller-Universität, Jena

MALCOLM MACCALLUM
Queen Mary, University of London

CORNELIUS HOENSELAERS
Loughborough University

EDUARD HERLT
Friedrich-Schiller-Universität, Jena

CAMBRIDGE
UNIVERSITY PRESS

University Printing House, Cambridge CB2 8BS, United Kingdom

Cambridge University Press is part of the University of Cambridge.

It furthers the University's mission by disseminating knowledge in the pursuit of education, learning and research at the highest international levels of excellence.

www.cambridge.org
Information on this title: www.cambridge.org/9780521467025

© H. Stephani, D. Kramer, M. MacCallum, C. Hoenselaers and E. Herlt 2009

This publication is in copyright. Subject to statutory exception and to the provisions of relevant collective licensing agreements, no reproduction of any part may take place without the written permission of Cambridge University Press.

First published 1981
Second edition 2003
Reprinted 2006
First paperback edition 2009

A catalogue record for this publication is available from the British Library

ISBN 978-0-521-46136-8 Hardback
ISBN 978-0-521-46702-5 Paperback

Cambridge University Press has no responsibility for the persistence or accuracy of URLs for external or third-party internet websites referred to in this publication, and does not guarantee that any content on such websites is, or will remain, accurate or appropriate.

Contents

Preface

When, in 1975, two of the authors (D.K. and H.S.) proposed to change their field of research back to the subject of exact solutions of Einstein's field equations, they of course felt it necessary to make a careful study of the papers published in the meantime, so as to avoid duplication of known results. A fairly comprehensive review or book on the exact solutions would have been a great help, but no such book was available. This prompted them to ask 'Why not use the preparatory work we have to do in any case to write such a book?' After some discussion, they agreed to go ahead with this idea, and then they looked for coauthors. They succeeded in finding two.

The first was E.H., a member of the Jena relativity group, who had been engaged before in exact solutions and was also inclined to return to them.

The second, M.M., became involved by responding to the existing authors' appeal for information and then (during a visit by H.S. to London) agreeing to look over the English text. Eventually he agreed to write some parts of the book.

The quartet's original optimism somewhat diminished when references to over 2000 papers had been collected and the magnitude of the task became all too clear. How could we extract even the most important information from this mound of literature? How could we avoid constant rewriting to incorporate new information, which would have made the job akin to the proverbial painting of the Forth bridge? How could we decide which topics to include and which to omit? How could we check the calculations, put the results together in a readable form and still finish in reasonable time?

We did not feel that we had solved any of these questions in a completely convincing manner. However, we did manage to produce an outcome, which was the first edition of this book, Kramer *et al.* (1980).

In the years since then so many new exact solutions have been published that the first edition can no longer be used as a reliable guide to the subject. The authors therefore decided to prepare a new edition. Although they knew from experience the amount of work to be expected, it took them longer than they thought and feared. We looked at over 4000 new papers (the cut-off date for the systematic search for papers is the end of 1999). In particular so much research had been done in the field of generation techniques and their applications that the original chapter had to be almost completely replaced, and C.H. was asked to collaborate on this, and agreed.

Compared with the first edition, the general arrangement of the material has not been changed. But we have added five new chapters, thus reflecting the developments of the last two decades (Chapters 9, 10, 23, 25 and 36), and some of the old chapters have been substantially rewritten. Unfortunately, the sheer number of known exact solutions has forced us to give up the idea of presenting them all in some detail; instead, in many cases we only give the appropriate references.

As with the first edition, the labour of reading those papers conceivably relevant to each chapter or section, and then drafting the related manuscript, was divided. Roughly, D.K., M.M. and C.H. were responsible for most of the introductory Part I, M.M., D.K. and H.S. dealt with groups (Part II), H.S., D.K. and E.H. with algebraically special solutions (Part III) and H.S. and C.H. with Part IV (special methods) and Part V (tables). Each draft was then criticized by the other authors, so that its writer could not be held wholly responsible for any errors or omissions. Since we hope to maintain up-to-date information, we shall be glad to hear from any reader who detects such errors or omissions; we shall be pleased to answer as best we can any requests for further information. M.M. wishes to record that any infelicities remaining in the English arose because the generally good standard of his colleagues' English lulled him into a false sense of security.

This book could not have been written, of course, without the efforts of the many scientists whose work is recorded here, and especially the many contemporaries who sent preprints, references and advice or informed us of mistakes or omissions in the first edition of this book. More immediately we have gratefully to acknowledge the help of the students in Jena, and in particular of S. Falkenberg, who installed our electronic files, of A. Koutras, who wrote many of the old chapters in LaTeX and simultaneously checked many of the solutions, and of the financial support of the Max-Planck-Group in Jena and the Friedrich-Schiller-Universität Jena. Last but not least, we have to thank our wives, families and colleagues for tolerating our incessant brooding and discussions and our obsession with the book.

Hans Stephani
Jena
Dietrich Kramer
Jena
Malcolm MacCallum
London
Cornelius Hoenselaers
Loughborough
Eduard Herlt
Jena

List of tables

Notation

All symbols are explained in the text. Here we list only some important conventions which are frequently used throughout the book.

Complex conjugates and constants

Complex conjugation is denoted by a bar over the symbol. The abbreviation const is used for 'constant'.

Indices

Small Latin indices run, in an n-dimensional Riemannian space V_n, from 1 to n, and in space-time V_4 from 1 to 4. When a general basis $\{\boldsymbol{e}_a\}$ or its dual $\{\boldsymbol{\omega}^a\}$ is in use, indices from the first part of the alphabet $(a, b, \ldots, h)$ will normally be tetrad indices and $i, j, \ldots$ are reserved for a coordinate basis $\{\partial/\partial x^i\}$ or its dual $\{\mathrm{d}x^i\}$. For a vector $\boldsymbol{v}$ and a l-form $\boldsymbol{\sigma}$ we write $\boldsymbol{v} = v^a \boldsymbol{e}_a = v^i \partial/\partial x^i$, $\boldsymbol{\sigma} = \sigma_a \boldsymbol{\omega}^a = \sigma_i \mathrm{d}x^i$. Small Greek indices run from 1 to 3, if not otherwise stated. Capital Latin indices are either spinor indices $(A, B = 1, 2)$ or indices in group space $(A, B = 1, \ldots, r)$, or they label the coordinates in a Riemannian 2-space V_2 $(M, N = 1, 2)$.

Symmetrization and antisymmetrization of index pairs are indicated by round and square brackets respectively; thus

$$v_{(ab)} \equiv \tfrac{1}{2}(v_{ab} + v_{ba}), \quad v_{[ab]} \equiv \tfrac{1}{2}(v_{ab} - v_{ba}).$$

The Kronecker delta, δ^a_b, has the value 1 if $a = b$ and zero otherwise.

Metric and tetrads

Line element in terms of dual basis $\{\boldsymbol{\omega}_a\}$: $\mathrm{d}s^2 = g_{ab}\boldsymbol{\omega}^a\boldsymbol{\omega}^b$.
Signature of space-time metric: $(+ + + -)$.

Commutation coefficients: $D^c{}_{ab}$; $[\boldsymbol{e}_a, \boldsymbol{e}_b] = D^c{}_{ab}\boldsymbol{e}_c$.
(Complex) null tetrad: $\{\boldsymbol{e}_a\} = (\boldsymbol{m}, \overline{\boldsymbol{m}}, \boldsymbol{l}, \boldsymbol{k})$, $g_{ab} = 2m_{(a}\overline{m}_{b)} - 2k_{(a}l_{b)}$,

$$\mathrm{d}s^2 = 2\omega^1\omega^2 - 2\omega^3\omega^4.$$

Orthonormal basis: $\{\boldsymbol{E}_a\}$.
Projection tensor: $h_{ab} \equiv g_{ab} + u_a u_b$, $u_a u^a = -1$.

Bivectors

Levi-Civita tensor in four dimensions: ε_{abcd}; $\varepsilon_{abcd}m^a\overline{m}^b l^c k^d = i$.
in two dimensions: $\varepsilon_{ab} = -\varepsilon_{ba}$, $\varepsilon_{12} = 1$
Dual bivector: $\widetilde{X}_{ab} \equiv \frac{1}{2}\varepsilon_{abcd}X^{cd}$.
(Complex) self-dual bivector: $X^*_{ab} \equiv X_{ab} + i\widetilde{X}_{ab}$.
Basis of self-dual bivectors: $U_{ab} \equiv 2\overline{m}_{[a}l_{b]}$, $V_{ab} \equiv 2k_{[a}m_{b]}$,
$W_{ab} \equiv 2m_{[a}\overline{m}_{b]} - 2k_{[a}l_{b]}$.

Derivatives

Partial derivative: comma in front of index or coordinate, e.g.

$$f_{,i} \equiv \partial f/\partial x^i \equiv \partial_i f, \quad f_{,\zeta} \equiv \partial f/\partial\zeta.$$

Directional derivative: denoted by stroke or comma, $f_{|a} \equiv f_{,a} \equiv \boldsymbol{e}_a(f)$; if followed by a numerical (tetrad) index, we prefer the stroke, e.g. $f_{|4} = f_{,i}k^i$. Directional derivatives with respect to the null tetrad $(\boldsymbol{m}, \overline{\boldsymbol{m}}, \boldsymbol{l}, \boldsymbol{k})$ are symbolized by $\delta f = f_{|1}$, $\overline{\delta} f = f_{|2}$, $\Delta f = f_{|3}$, $Df = f_{|4}$.
Covariant derivative: ∇; in component calculus, semicolon. (Sometimes other symbols are used to indicate that in V_4 a metric different from g_{ab} is used, e.g. $h_{ab||c} = 0, \gamma_{ab:c} = 0$.)
Lie derivative of a tensor $\boldsymbol{T}$ with respect to a vector $\boldsymbol{v}$: $\mathcal{L}_{\boldsymbol{v}}\boldsymbol{T}$.
Exterior derivative: d.

When a dot is used to denote a derivative without definition, e.g. $\dot{Q}$, it means differentiation with respect to the time coordinate in use; a prime used similarly, e.g. Q', refers either to the unique essential space coordinate in the problem or to the single argument of a function.

Connection and curvature

Connection coefficients: $\Gamma^a{}_{bc}$, $v^a{}_{;c} = v^a{}_{,c} + \Gamma^a{}_{bc}v^b$.
Connection 1-forms: $\boldsymbol{\Gamma}^a{}_b \equiv \Gamma^a{}_{bc}\boldsymbol{\omega}^c$, $\mathrm{d}\boldsymbol{\omega}^a = -\boldsymbol{\Gamma}^a{}_b \wedge \boldsymbol{\omega}^b$.
Riemann tensor: $R^d{}_{abc}$, $2v_{a;[bc]} = v_d R^d{}_{abc}$.

Curvature 2-forms: $\boldsymbol{\Theta}^a{}_b \equiv \frac{1}{2}R^a{}_{bcd}\boldsymbol{\omega}^c \wedge \boldsymbol{\omega}^d = \mathrm{d}\boldsymbol{\Gamma}^a{}_b + \boldsymbol{\Gamma}^a{}_c \wedge \boldsymbol{\Gamma}^c{}_b$.
Ricci tensor, Einstein tensor, and scalar curvature:

$$R_{ab} \equiv R^c{}_{acb}, \quad G_{ab} \equiv R_{ab} - \tfrac{1}{2}Rg_{ab}, \quad R \equiv R^a{}_a.$$

Weyl tensor in V_4:

$$C_{abcd} \equiv R_{abcd} + \tfrac{1}{3}Rg_{a[c}g_{d]b} - g_{a[c}R_{d]b} + g_{b[c}R_{d]a}.$$

Null tetrad components of the Weyl tensor:

$$\Psi_0 \equiv C_{abcd}k^a m^b k^c m^d, \quad \Psi_1 \equiv C_{abcd}k^a l^b k^c m^d,$$

$$\Psi_2 \equiv C_{abcd}k^a m^b \overline{m}^c l^d,$$

$$\Psi_3 \equiv C_{abcd}k^a l^b \overline{m}^c l^d, \quad \Psi_4 \equiv C_{abcd}\overline{m}^a l^b \overline{m}^c l^d.$$

Metric of a 2-space of constant curvature:

$$\mathrm{d}\sigma^2 = \mathrm{d}x^2 \pm \Sigma^2(x, \varepsilon)\mathrm{d}y^2,$$

$$\Sigma(x, \varepsilon) = \sin x,\ x,\ \sinh x \ \text{ resp. when } \varepsilon = 1, 0 \text{ or } -1.$$

Gaussian curvature: K.

Physical fields

Energy-momentum tensor: T_{ab}, $T_{ab}u^a u^b \geq 0$ if $u_a u^a = -1$.
Electromagnetic field: Maxwell tensor F_{ab}, $T_{ab} = F^{*c}_a \overline{F}^*_{bc}/2$.
Null tetrad components of F_{ab}:

$$\Phi_0 \equiv F_{ab}k^a m^b, \quad \Phi_1 \equiv \tfrac{1}{2}F_{ab}(k^a l^b + \overline{m}^a m^b), \quad \Phi_2 = F_{ab}\overline{m}^a l^b.$$

Perfect fluid: pressure p, energy density μ, 4-velocity $\boldsymbol{u}$,

$$T_{ab} = (\mu + p)u_a u_b + pg_{ab}.$$

Cosmological constant: Λ.
Gravitational constant: κ_0.
Einstein's field equations: $R_{ab} - \frac{1}{2}Rg_{ab} + \Lambda g_{ab} = \kappa_0 T_{ab}$.

Symmetries

Group of motions (r-dim.), G_r; isotropy group (s-dim.), I_s; homothety group (q-dim.), H_q.
Killing vectors: $\boldsymbol{\xi}$, $\boldsymbol{\eta}$, $\boldsymbol{\zeta}$, or $\boldsymbol{\xi}_A$, $A = 1, \ldots, r$
Killing equation: $(\mathcal{L}_\xi \boldsymbol{g})_{ab} = \xi_{a;b} + \xi_{b;a} = 0$.
Structure constants: $C^C{}_{AB}$; $[\boldsymbol{\xi}_A, \boldsymbol{\xi}_B] = C^C{}_{AB}\boldsymbol{\xi}_C$.
Orbits (m-dim.) of G_r or H_q: S_m (spacelike), T_m (timelike), N_m (null).

1
Introduction

1.1 What are exact solutions, and why study them?

The theories of modern physics generally involve a mathematical model, defined by a certain set of differential equations, and supplemented by a set of rules for translating the mathematical results into meaningful statements about the physical world. In the case of theories of gravitation, it is generally accepted that the most successful is Einstein's theory of general relativity. Here the differential equations consist of purely geometric requirements imposed by the idea that space and time can be represented by a Riemannian (Lorentzian) manifold, together with the description of the interaction of matter and gravitation contained in Einstein's famous field equations

$$R_{ab} - \tfrac{1}{2}Rg_{ab} + \Lambda g_{ab} = \kappa_0 T_{ab}. \tag{1.1}$$

(The full definitions of the quantities used here appear later in the book.) This book will be concerned only with Einstein's theory. We do not, of course, set out to discuss all aspects of general relativity. For the basic problem of understanding the fundamental concepts we refer the reader to other texts.

For any physical theory, there is first the purely mathematical problem of analysing, as far as possible, the set of differential equations and of finding as many exact solutions, or as complete a general solution, as possible. Next comes the mathematical and physical interpretation of the solutions thus obtained; in the case of general relativity this requires global analysis and topological methods rather than just the purely local solution of the differential equations. In the case of gravity theories, because they deal with the most universal of physical interactions, one has an additional class of problems concerning the influence of the gravitational field on other fields and matter; these are often studied by working within a fixed gravitational field, usually an exact solution.

This book deals primarily with the solutions of the Einstein equations, (1.1), and only tangentially with the other subjects. The strongest reason for excluding the omitted topics is that each would fill (and some do fill) another book; we do, of course, give some references to the relevant literature. Unfortunately, one cannot say that the study of exact solutions has always maintained good contact with work on more directly physical problems. Back in 1975, Kinnersley wrote "Most of the known exact solutions describe situations which are frankly unphysical, and these do have a tendency to distract attention from the more useful ones. But the situation is also partially the fault of those of us who work in this field. We toss in null currents, macroscopic neutrino fields and tachyons for the sake of greater 'generality'; we seem to take delight at the invention of confusing anti-intuitive notation; and when all is done we leave our newborn metric wobbling on its vierbein without any visible means of interpretation." Not much has changed since then.

In defence of work on exact solutions, it may be pointed out that certain solutions have played very important roles in the discussion of physical problems. Obvious examples are the Schwarzschild and Kerr solutions for black holes, the Friedmann solutions for cosmology, and the plane wave solutions which resolved some of the controversies about the existence of gravitational radiation. It should also be noted that because general relativity is a highly non-linear theory, it is not always easy to understand what qualitative features solutions might possess, and here the exact solutions, including many such as the Taub–NUT solutions which may be thought unphysical, have proved an invaluable guide. Though the fact is not always appreciated, the non-linearities also mean that perturbation schemes in general relativity can run into hidden dangers (see e.g. Ehlers *et al.* (1976)). Exact solutions which can be compared with approximate or numerical results are very useful in checking the validity of approximation techniques and programs, see Centrella *et al.* (1986).

In addition to the above reasons for devoting this book to the classification and construction of exact solutions, one may note that although much is known, it is often not generally known, because of the plethora of journals, languages and mathematical notations in which it has appeared. We hope that one beneficial effect of our efforts will be to save colleagues from wasting their time rediscovering known results; in particular we hope our attempt to characterize the known solutions invariantly will help readers to identify any new examples that arise.

One surprise for the reader may lie in the enormous number of known exact solutions. Those who do not work in the field often suppose that the intractability of the full Einstein equations means that very few solutions are known. In a certain sense this is true: we know relatively few exact solutions for real physical problems. In most solutions, for example, there is no complete description of the relation of the field to sources. Problems which are without an exact solution include the two-body problem, the realistic description of our inhomogeneous universe, the gravitational field of a stationary rotating star and the generation and propagation of gravitational radiation from a realistic bounded source. There are, on the other hand, some problems where the known exact solutions may be the unique answer, for instance, the Kerr and Schwarzschild solutions for the final collapsed state of massive bodies.

Any metric whatsoever is a 'solution' of (1.1) if no restriction is imposed on the energy-momentum tensor, since (1.1) then becomes just a definition of T_{ab}; so we must first make some assumptions about T_{ab}. Beyond this we may proceed, for example, by imposing symmetry conditions on the metric, by restricting the algebraic structure of the Riemann tensor, by adding field equations for the matter variables or by imposing initial and boundary conditions. The exact solutions known have all been obtained by making some such restrictions. We have used the term 'exact solution' without a definition, and we do not intend to provide one. Clearly a metric would be called an exact solution if its components could be given, in suitable coordinates, in terms of the well-known analytic functions (polynomials, trigonometric funstions, hyperbolic functions and so on). It is then hard to find grounds for excluding functions defined only by (linear) differential equations. Thus 'exact solution' has a less clear meaning than one might like, although it conveys the impression that in some sense the properties of the metric are fully known; no generally-agreed precise definition exists. We have proceeded rather on the basis that what we chose to include was, by definition, an exact solution.

1.2 The development of the subject

In the first few years (or decades) of research in general relativity, only a rather small number of exact solutions were discussed. These mostly arose from highly idealized physical problems, and had very high symmetry. As examples, one may cite the well-known spherically-symmetric solutions of Schwarzschild, Reissner and Nordström, Tolman and Friedmann (this last using the spatially homogeneous metric form now associated with the names of Robertson and Walker), the axisymmetric static electromagnetic and vacuum solutions of Weyl, and the plane wave metrics. Although such a limited range of solutions was studied, we must, in fairness, point out that it includes nearly all the exact solutions which are of importance in physical applications: perhaps the only one of comparable importance which was discovered after World War II is the Kerr solution.

In the early period there were comparatively few people actively working on general relativity, and it seems to us that the general belief at that time was that exact solutions would be of little importance, except perhaps as cosmological and stellar models, because of the extreme weakness of the relativistic corrections to Newtonian gravity. Of course, a wide variety of physical problems were attacked, but in a large number of cases they were treated only by some approximation scheme, especially the weak-field, slow-motion approximation.

Moreover, many of the techniques now in common use were either unknown or at least unknown to most relativists. The first to become popular was the use of groups of motions, especially in the construction of cosmologies more general than Friedmann's. The next, which was in part motivated by the study of gravitational radiation, was the algebraic classification of the Weyl tensor into Petrov types and the understanding of the properties of algebraically special metrics. Both these developments led in a natural way to the use of invariantly-defined tetrad bases, rather than coordinate components. The null tetrad methods, and some ideas from the theory of group representations and algebraic geometry, gave rise to the spinor techniques, and equivalent methods, now usually employed in the form given by Newman and Penrose. The most recent of these major developments was the advent of the generating techniques, which were just being developed at the time of our first edition (Kramer *et al.* 1980), and which we now describe fully.

Using these methods, it was possible to obtain many new solutions, and this growth is still continuing.

1.3 The contents and arrangement of this book

Naturally, we begin by introducing differential geometry (Chapter 2) and Riemannian geometry (Chapter 3). We do not provide a formal textbook of these subjects; our aim is to give just the notation, computational methods and (usually without proof) standard results we need for later chapters. After this point, the way ahead becomes more debatable.

There are (at least) four schemes for classification of the known exact solutions which could be regarded as having more or less equal importance; these four are the algebraic classification of conformal curvature (Petrov types), the algebraic classification of the Ricci tensor (Plebański or Segre types) and the physical characterization of the energy-momentum tensor, the existence and structure of preferred vector fields, and the groups of symmetry 'admitted by' (i.e. which exist for) the metric (isometries and homotheties). We have devoted a chapter (respectively, Chapters 4, 5, 6 and 8) to each of these, introducing the terminology and methods used later and some general theorems. Among these chapters we have interpolated one (Chapter 7) which gives the Newman–Penrose formalism; its position is due to the fact that this formalism can be applied immediately to elucidating some of the relationships between the considerations in the preceding three chapters. With more solutions being known, unwitting rediscoveries happened more frequently; so methods of invariant characterization became important which we discuss in Chapter 9. We close Part I with a presentation of the generation methods which became so fruitful in the 1980s. This is again one of the subjects which, ideally, warrants a book of its own and thus we had to be very selective in the choice and manner of the material presented.

The four-dimensional presentation of the solutions which would arise from the classification schemes outlined above may be acceptable to relativists but is impractical for authors. We could have worked through each classification in turn, but this would have been lengthy and repetitive (as it is, the reader will find certain solutions recurring in various disguises). We have therefore chosen to give pride of place to the two schemes which seem to have had the widest use in the discovery and construction of new solutions, namely symmetry groups (Part II of the book) and Petrov types (Part III). The other main classifications have been used in subdividing the various classes of solutions discussed in Parts II and III, and they are covered by the tables in Part V. The application of the generation techniques and some other ways of classifying and constructing exact solutions are presented in Part IV.

The specification of the energy-momentum tensor played a very important role because we decided at an early stage that it would be impossible to provide a comprehensive survey of all energy-momentum tensors that have ever been considered. We therefore restricted ourselves to the following energy-momentum tensors: vacuum, electromagnetic fields, pure radiation, dust and perfect fluids. (The term 'pure radiation' is used here for an energy-momentum tensor representing a situation in which all the energy is transported in one direction with the speed of light: such tensors are also referred to in the literature as null fields, null fluids and null dust.) Combinations of these, and matching of solutions with equal or different energy-momentum tensors (e.g. the Schwarzschild vacuoli in a Friedmann universe) are in general not considered, and the cosmological constant Λ, although sometimes introduced, is not treated systematically throughout.

These limitations on the scope of our work may be disappointing to some, especially those working on solutions containing charged perfect fluids, scalar, Dirac and neutrino fields, or solid elastic bodies. They were made not only because some limits on the task we set ourselves were necessary, but also because most of the known solutions are for the energy-momentum tensors listed and because it is possible to give a fairly full systematic treatment for these cases. One may also note that unless additional field equations for the additional variables are introduced, it is easier to find solutions for more complex energy-momentum tensor forms than for simpler ones: indeed in extreme cases there may be no equations to solve at all, the Einstein equations instead becoming merely definitions of the energy-momentum from a metric ansatz. Ultimately, of course, the choice is a matter of taste.

The arrangement within Part II is outlined more fully in §11.1. Here we remark only that we treated first non-null and then null group orbits (as defined in Chapter 8), arranging each in order of decreasing dimension of the orbit and thereafter (usually) in decreasing order of dimension of the group. Certain special cases of physical or mathematical interest were separated out of this orderly progression and given chapters of their own, for example, spatially-homogeneous cosmologies, spherically-symmetric solutions, colliding plane waves and the inhomogeneous fluid solutions with symmetries. Within each chapter we tried to give first the differential geometric results (i.e. general forms of the metric and curvature) and then the actual solutions for each type of energy-momentum in turn; this arrangement is followed in Parts III and IV also.

In Part III we have given a rather detailed account of the well-developed theory that is available for algebraically special solutions for vacuum, electromagnetic and pure radiation fields. Only a few classes, mostly very special cases, of algebraically special perfect-fluid solutions have been thoroughly discussed in the literature: a short review of these classes is given in Chapter 33. Quite a few of the algebraically special solutions also admit groups of motions. Where this is known (and, as far as we are aware, it has not been systematically studied for all cases), it is of course indicated in the text and in the tables.

Part IV, the last of the parts treating solutions in detail, covers solutions found by the generation techniques developed by various authors since 1980 (although most of these rely on the existence of a group of motions, and in some sense therefore belong in Part II). There are many such techniques in use and they could not all be discussed in full: our choice of what to present in detail and what to mention only as a reference simply reflects our personal tastes and experiences. This part also gives some discussion of the classification of space-times with special vector and tensor fields and solutions found by embedding or the study of metrics with special subspaces.

The weight of material, even with all the limitations described above, made it necessary to omit many proofs and details and give only the necessary references.

1.4 Using this book as a catalogue

This book has not been written simply as a catalogue. Nevertheless, we intended that it should be possible for the book to be used for this purpose. In arranging the information here, we have assumed that a reader who wishes to find (or, at least, search for) a solution will know the original author (if the reader is aware the solution is not new) or know some of its invariant properties.

If the original author[1] is known, the reader should turn to the alphabetically-organized reference list. He or she should then be able to identify the relevant paper(s) of that author, since the titles, and, of course, journals and dates, are given in full. Following each reference is a list of all the places in the book where it is cited.

A reader who knows the (maximal) group of motions can find the relevant chapter in Part II by consulting the contents list or the tables. If the reader knows the Petrov type, he or she can again consult the contents list or the tables by Petrov type; if only the energy-momentum tensor is known, the reader can still consult the relevant tables. If none of this information is known, he or she can turn to Part IV, if one of the special methods described there has been used. If still in doubt, the whole book will have to be read.

If the solution is known (and not accidentally omitted) it will in many cases be given in full, possibly only in the sense of appearing contained in a more general form for a whole class of solutions: some solutions of great complexity or (to us) lesser importance have been given only in the sense of a reference to the literature. Each solution may, of course, be found in a great variety of coordinate forms and be characterized invariantly in several ways. We have tried to eliminate duplications, i.e. to identify those solutions that are really the same but appear in the literature as separate, and we give cross-references between sections where possible. The solutions are usually given in coordinates adapted to some invariant properties, and it should therefore be feasible (if non-trivial) for the reader to transform to any other coordinate system he or she has discovered (see also Chapter 9). The many solutions obtained by generating techniques are for the most part only tabulated and not given explicitly,

[1] There is a potential problem here if the paper known to the reader is an unwitting re-discovery, since for brevity we do not cite such works.

since it is in principle possible to generate infinitely many such solutions by complicated but direct calculations.

Solutions that are neither given nor quoted are either unknown to us or accidentally omitted, and in either case the authors would be interested to hear about them. (We should perhaps note here that not all papers containing frequently-rediscovered solutions have been cited: in such a case only the earliest papers, and those rediscoveries with some special importance, have been given. Moreover, if a general class of solutions is known, rediscoveries of special cases belonging to this class have been mentioned only occasionally. We have also not in general commented, except by omission, on papers where we detected errors, though in a few cases where a paper contains some correct and some wrong results we have indicated that.)

We have checked most of the solutions given in the book. This was done by machine and by hand, but sometimes we may have simply repeated the authors' errors. It is not explicitly stated where we did not check solutions.

In addition to references within the text, cited by author and year, we have sometimes put at the ends of sections some references to parallel methods, or to generalizations, or to applications. We would draw the reader's attention to some books of similar character which have appeared since the first edition of this book was published and which complement and supplement this one. Krasiński (1997) has extensively surveyed those solutions which contain as special cases the Robertson–Walker cosmologies (for which see Chapter 14), without the restrictions on energy-momentum content which we impose. Griffiths (1991) gives an extensive study of the colliding wave solutions discussed here in Chapter 25, Wainwright and Ellis (1997) similarly discusses spatially-homogeneous and some other cosmologies (see Chapters 14 and 23), Bičák (2000) discusses selected exact solutions and their history, and Belinski and Verdaguer (2001) reviews solitonic solutions obtainable by the methods of Chapter 34, especially §34.4: these books deal with physical and interpretational issues for which we do not have space.

Thanks are due to many colleagues for comments on and corrections to the first edition: we acknowledge in particular the remarks of J.E. Åman, A. Barnes, W.B. Bonnor, J. Carot, R. Debever, K.L. Duggal, J.B. Griffiths, G.S. Hall, R.S. Harness, R.T. Jantzen, G.D. Kerr, A. Koutras, J.K. Kowalczyński, A. Krasiński, K. Lake, D. Lorenz, M. Mars, J.D. McCrea, C.B.G. McIntosh, G.C. McVittie, G. Neugebauer, F.M. Paiva, M.D. Roberts, J.M.M. Senovilla, S.T.C. Siklos, B.O.J. Tupper, C. Uggla, R. Vera, J.A. Wainwright, Th. Wolf and M. Wyman.

Part I
General methods

2
Differential geometry without a metric

2.1 Introduction

The concept of a tensor is often based on the law of transformation of the components under coordinate transformations, so that coordinates are explicitly used from the beginning. This calculus provides adequate methods for many situations, but other techniques are sometimes more effective. In the modern literature on exact solutions coordinatefree geometric concepts, such as forms and exterior differentiation, are frequently used: the underlying mathematical structure often becomes more evident when expressed in coordinatefree terms.

Hence this chapter will present a brief survey of some of the basic ideas of differential geometry. Most of these are independent of the introduction of a metric, although, of course, this is of fundamental importance in the space-times of general relativity; the discussion of manifolds with metrics will therefore be deferred until the next chapter. Here we shall introduce vectors, tensors of arbitrary rank, p-forms, exterior differentiation and Lie differentiation, all of which follow naturally from the definition of a differentiable manifold. We then consider an additional structure, a covariant derivative, and its associated curvature; even this does not necessarily involve a metric. The absence of any metric will, however, mean that it will not be possible to convert 1-forms to vectors, or vice versa.

Since we are primarily concerned with specific applications, we shall emphasize the rules of manipulation and calculation in differential geometry. We do not attempt to provide a substitute for standard texts on the subject, e.g. Eisenhart (1927), Schouten (1954), Flanders (1963), Sternberg (1964), Kobayashi and Nomizu (1969), Schutz (1980), Nakahara (1990) and Choquet-Bruhat *et al.* (1991) to which the reader is referred for fuller information and for the proofs of many of the theorems. Useful introductions can also be found in many modern texts on relativity.

For the benefit of those familiar with the traditional approach to tensor calculus, certain formulae are displayed both in coordinatefree form and in the usual component formalism.

2.2 Differentiable manifolds

Differentiable manifolds are the most basic structures in differential geometry. Intuitively, an (n-dimensional) manifold is a space $\mathcal{M}$ such that any point $p \in \mathcal{M}$ has a neighbourhood $\mathcal{U} \subset \mathcal{M}$ which is homeomorphic to the interior of the (n-dimensional) unit ball. To give a mathematically precise definition of a differentiable manifold we need to introduce some additional terminology.

A *chart* $(\mathcal{U}, \Phi)$ in $\mathcal{M}$ consists of a subset $\mathcal{U}$ of $\mathcal{M}$ together with a one-to-one map Φ from $\mathcal{U}$ onto the n-dimensional Euclidean space E^n or an open subset of E^n; Φ assigns to every point $p \in \mathcal{U}$ an n-tuple of real variables, the *local coordinates* $(x^1, \dots, x^n)$. As an aid in later calculations, we shall sometimes use pairs of complex conjugate coordinates instead of pairs of real coordinates, but we shall not consider generalizations to complex manifolds (for which see e.g. Flaherty (1980) and Penrose and Rindler (1984, 1986)).

Two charts $(\mathcal{U}, \Phi)$, $(\mathcal{U}', \Phi')$ are said to be *compatible* if the combined map $\Phi' \circ \Phi^{-1}$ on the image $\Phi(\mathcal{U} \cup \mathcal{U}')$ of the overlap of $\mathcal{U}$ and $\mathcal{U}'$ is a homeomorphism (i.e. continuous, one-to-one, and having a continuous inverse): see Fig. 2.1.

An *atlas* on $\mathcal{M}$ is a collection of compatible charts $(\mathcal{U}_\alpha, \Phi_\alpha)$ such that every point of $\mathcal{M}$ lies in at least one chart neighbourhood $\mathcal{U}_\alpha$. In most cases, it is impossible to cover the manifold with a single chart (an example which cannot be so covered is the n-dimensional sphere, $n > 0$).

An n-dimensional (topological) *manifold* consists of a space $\mathcal{M}$ together with an atlas on $\mathcal{M}$. It is a (C^k or analytic) *differentiable manifold* $\mathcal{M}$ if the maps $\Phi' \circ \Phi^{-1}$ relating different charts are not just continuous but differentiable (respectively, C^k or analytic). Then the coordinates are related by n differentiable (C^k, analytic) functions, with non-vanishing

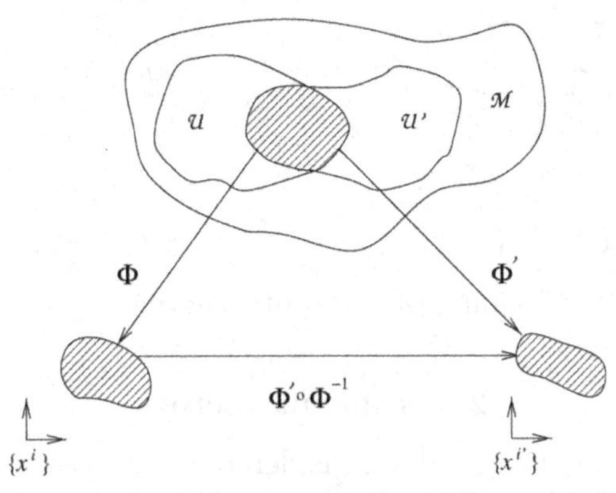

Fig. 2.1. Two compatible charts of a differentiable manifold

Jacobian at each point of the overlap:

$$x^{i'} = x^{i'}(x^j), \qquad \det(\partial x^{i'}/\partial x^j) \neq 0. \tag{2.1}$$

Definitions of manifolds often include additional topological restrictions, such as paracompactness and Hausdorffness, and these are indeed essential for the rigorous proof of some of the results we state, as is the precise degree of smoothness, i.e. the value of k. For brevity, we shall omit any consideration of these questions, which are of course fully discussed in the literature cited earlier.

A differentiable manifold $\mathcal{M}$ is called *orientable* if there exists an atlas such that the Jacobian (2.1) is positive throughout the overlap of any pair of charts.

If $\mathcal{M}$ and $\mathcal{N}$ are manifolds, of dimensions m and n, respectively, the $(m+n)$-dimensional product $\mathcal{M} \times \mathcal{N}$ can be defined in a natural way.

A *map* $\Phi : \mathcal{M} \to \mathcal{N}$ is said to be *differentiable* if the coordinates $(y^1, \ldots, y^n)$ on $\mathcal{V} \subset \mathcal{N}$ are differentiable functions of the coordinates $(x^1, \ldots, x^n)$ of the corresponding points in $\mathcal{U} \subset \mathcal{M}$ where Φ maps (a part of) the neighbourhood $\mathcal{U}$ into the neighbourhood $\mathcal{V}$. If $\Phi(\mathcal{M}) \neq \mathcal{N}$, $\Phi(\mathcal{M})$ is called a *submanifold* of $\mathcal{N}$: submanifolds $\mathcal{P} \subset \mathcal{N}$ of dimension $p < n$ can also be defined by the existence of charts $(\mathcal{V}, \Psi)$ in $\mathcal{N}$ such that $\mathcal{P} \cap \mathcal{V} \subset \mathbb{R}^p \times 0$ where the 0 is the zero of $\mathbb{R}^{n-p}$. A submanifold of dimension $n-1$ will be called a *hypersurface.*

A smooth *curve* $\gamma(t)$ in $\mathcal{M}$ is defined by a differentiable map of an interval of the real line into $\mathcal{M}$, $\gamma(t) : -\varepsilon < t < \varepsilon \to \mathcal{M}$ (or sometimes by a similar map of a closed interval $[\varepsilon, \varepsilon]$). A differentiable map $\Phi : \mathcal{M} \to \mathcal{N}$ and its action on a curve are illustrated in Fig. 2.2.

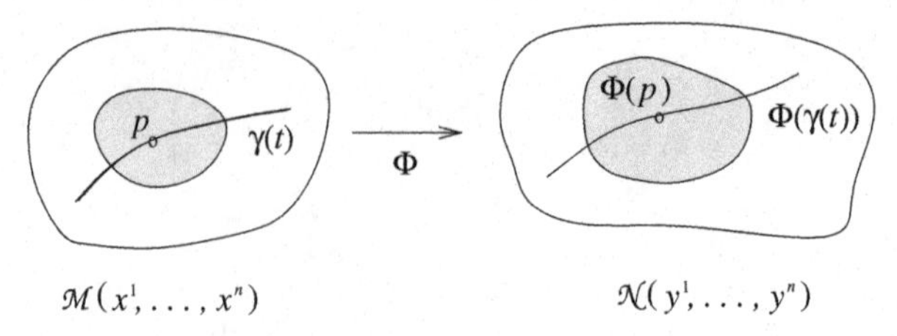

Fig. 2.2. The map of a smooth curve $\gamma(t)$ to $\Phi(\gamma(t))$

2.3 Tangent vectors

In general a vector cannot be considered as an arrow connecting two points of the manifold. To get a consistent generalization of the concept of vectors in E^n, one identifies vectors on $\mathcal{M}$ with tangent vectors. A *tangent vector* $\boldsymbol{v}$ at p is an *operator* (*linear functional*) which assigns to each differentiable function f on $\mathcal{M}$ a real number $\boldsymbol{v}(f)$. This operator satisfies the axioms

$$\begin{aligned} &\text{(i)} \quad \boldsymbol{v}(f+h) = \boldsymbol{v}(f) + \boldsymbol{v}(h), \\ &\text{(ii)} \quad \boldsymbol{v}(fh) = h\boldsymbol{v}(f) + f\boldsymbol{v}(h), \\ &\text{(iii)} \quad \boldsymbol{v}(cf) = c\boldsymbol{v}(f), \qquad c = \text{const.} \end{aligned} \tag{2.2}$$

It follows from these axioms that $\boldsymbol{v}(c) = 0$ for any constant function c. The definition (2.2) is independent of the choice of coordinates. A tangent vector is just a *directional derivative* along a curve $\gamma(t)$ through p: expanding any function f in a Taylor series at p, and using the axioms (2.2), one can easily show that any tangent vector $\boldsymbol{v}$ at p can be written as

$$\boldsymbol{v} = v^i\, \partial/\partial x^i. \tag{2.3}$$

The real coefficients v^i are the *components* of $\boldsymbol{v}$ at p with respect to the local coordinate system $(x^1, \ldots, x^n)$ in a neighbourhood of p. According to (2.3), the directional derivatives along the coordinate lines at p form a basis of an n-dimensional vector space the elements of which are the tangent vectors at p. This space is called the tangent space T_p. The basis $\{\partial/\partial x^i\}$ is called a *coordinate basis* or *holonomic frame*.

A *general basis* $\{\boldsymbol{e}_a\}$ is formed by n linearly independent vectors $\boldsymbol{e}_a$; any vector $\boldsymbol{v} \in T_p$ is a linear combination of these basis vectors, i.e.

$$\boldsymbol{v} = v^a \boldsymbol{e}_a. \tag{2.4}$$

The action of a basis vector $\boldsymbol{e}_a$ on a function f is denoted by the symbol $f_{|a} \equiv \boldsymbol{e}_a(f)$. In a coordinate basis we use a comma in place of a solidus, $f_{,i} \equiv \partial f/\partial x^i$. A non-singular linear transformation of the basis $\{\boldsymbol{e}_a\}$

induces a change of the components v^a of the vector $\boldsymbol{v}$,

$$\boldsymbol{e}_{a'} = L_{a'}{}^b \boldsymbol{e}_b, \qquad v^{a'} = L^{a'}{}_b v^b, \qquad L^{a'}{}_b L_{a'}{}^c = \delta^c_b. \tag{2.5}$$

A coordinate basis $\{\partial/\partial x^i\}$ represents a special choice of $\{\boldsymbol{e}_a\}$. In the older literature on general relativity the components with respect to coordinate bases were preferred for actual computations. However, for many purposes it is more convenient to use a general basis, often called a frame or n-bein (in four dimensions, a tetrad or vierbein), though when there is a metric, as in Chapter 3, these terms may be reserved for the cases with constant lengths. Well-known examples are the Petrov classification (Chapter 4) and the Newman–Penrose formalism (Chapter 7).

The set of all tangent spaces at points p in $\mathcal{M}$ forms the *tangent bundle* $T(\mathcal{M})$ of $\mathcal{M}$. To make this a differentiable manifold, the charts on $T(\mathcal{M})$ can be defined by extending the charts (U, Φ) of $\mathcal{M}$ to charts $(U \times \mathbb{R}^n, \Phi \times Id)$ where Id is the identity map on $\mathbb{R}^n$, i.e. we can use the components (2.3) to extend coordinates x^i on $\mathcal{M}$ to coordinates (x^i, v^j) on $T(\mathcal{M})$. The tangent bundle thus has dimension $2n$. If M is a C^k manifold, $T(\mathcal{M})$ is C^{k-1}.

We can construct a *vector field* $\boldsymbol{v}(p)$ on $\mathcal{M}$ by assigning to each point $p \in \mathcal{M}$ a tangent vector $\boldsymbol{v} \in T_p$ so that the components v^i are differentiable functions of the local coordinates. Thus a vector field can be regarded as a smooth map $\mathcal{M} \to T(\mathcal{M})$ such that each point $p \to \boldsymbol{v}(p)$, and is then referred to as a *section* of the tangent bundle.

From the identification of vectors with directional derivatives one concludes that in general the result of the successive application of two vectors to a function depends on the order in which the operators are applied. The *commutator* $[\boldsymbol{u}, \boldsymbol{v}]$ of two vector fields $\boldsymbol{u}$ and $\boldsymbol{v}$ is defined by $[\boldsymbol{u}, \boldsymbol{v}](f) = \boldsymbol{u}(\boldsymbol{v}(f)) - \boldsymbol{v}(\boldsymbol{u}(f))$. For a given basis $\{\boldsymbol{e}_a\}$, the commutators

$$[\boldsymbol{e}_a, \boldsymbol{e}_b] = D^c{}_{ab}\boldsymbol{e}_c, \qquad D^c{}_{ab} = -D^c{}_{ba}, \tag{2.6}$$

define the commutator coefficients $D^c{}_{ab}$, which obviously vanish for a coordinate basis: $[\partial/\partial x^i, \partial/\partial x^j] = 0$. Commutators satisfy the *Jacobi identity*

$$[\boldsymbol{u}, [\boldsymbol{v}, \boldsymbol{w}]] + [\boldsymbol{v}, [\boldsymbol{w}, \boldsymbol{u}]] + [\boldsymbol{w}, [\boldsymbol{u}, \boldsymbol{v}]] = 0 \tag{2.7}$$

for arbitrary $\boldsymbol{u}$, $\boldsymbol{v}$, $\boldsymbol{w}$, from which one infers, for constant $D^c{}_{ab}$, the identity

$$D^f{}_{d[a} D^d{}_{bc]} = 0. \tag{2.8}$$

2.4 One-forms

By definition, a *1-form* (Pfaffian form) $\boldsymbol{\sigma}$ maps a vector $\boldsymbol{v}$ into a real number, the *contraction*, denoted by the symbol $\langle \boldsymbol{\sigma}, \boldsymbol{v} \rangle$ or $\boldsymbol{v} \lrcorner \boldsymbol{\sigma}$, and this mapping is linear:

$$\langle \boldsymbol{\sigma}, a\boldsymbol{u} + b\boldsymbol{v} \rangle = a\langle \boldsymbol{\sigma}, \boldsymbol{u} \rangle + b\langle \boldsymbol{\sigma}, \boldsymbol{v} \rangle \tag{2.9}$$

for real a, b, and $\boldsymbol{u}$, $\boldsymbol{v} \in T_p$. Linear combinations of 1-forms $\boldsymbol{\sigma}$, $\boldsymbol{\tau}$ are defined by the rule

$$\langle a\boldsymbol{\sigma} + b\boldsymbol{\tau}, \boldsymbol{v} \rangle = a\langle \boldsymbol{\sigma}, \boldsymbol{v} \rangle + b\langle \boldsymbol{\tau}, \boldsymbol{v} \rangle \tag{2.10}$$

for real a, b. The n linearly independent 1-forms $\boldsymbol{\omega}^a$ which are uniquely determined by

$$\langle \boldsymbol{\omega}^a, \boldsymbol{e}_b \rangle = \delta^a_b \tag{2.11}$$

form a basis $\{\boldsymbol{\omega}^a\}$ of the *dual space* T^*_p of the tangent space T_p. This basis $\{\boldsymbol{\omega}^a\}$ is said to be dual to the basis $\{\boldsymbol{e}_b\}$ of T_p. Any 1-form $\boldsymbol{\sigma} \in T^*_p$ is a linear combination of the basis 1-forms $\boldsymbol{\omega}^a$;

$$\boldsymbol{\sigma} = \sigma_a \boldsymbol{\omega}^a. \tag{2.12}$$

For any $\boldsymbol{\sigma} \in T^*_p$, $\boldsymbol{v} \in T_p$ the contraction $\langle \boldsymbol{\sigma}, \boldsymbol{v} \rangle$ can be expressed in terms of the components σ_a, v^a of $\boldsymbol{\sigma}$, $\boldsymbol{v}$ with respect to the bases $\{\boldsymbol{\omega}^a\}$, $\{\boldsymbol{e}_a\}$ by

$$\langle \boldsymbol{\sigma}, \boldsymbol{v} \rangle = \sigma_a v^a. \tag{2.13}$$

The *differential* $\mathrm{d}f$ of an arbitrary function f is a 1-form defined by the property

$$\langle \mathrm{d}f, \boldsymbol{v} \rangle = \boldsymbol{v}(f) \equiv v^a f_{|a}. \tag{2.14}$$

Specializing this definition to the functions $f = x^1, \dots, x^n$ one obtains the relation

$$\langle \mathrm{d}x^i, \partial/\partial x^j \rangle = \delta^i_j, \tag{2.15}$$

indicating that the basis $\{\mathrm{d}x^i\}$ of T^*_p is dual to the coordinate basis $\{\partial/\partial x^i\}$ of T_p. Any 1-form $\boldsymbol{\sigma} \in T^*_p$ can be written with respect to the basis $\{\mathrm{d}x^i\}$ as

$$\boldsymbol{\sigma} = \sigma_i \mathrm{d}x^i. \tag{2.16}$$

In local coordinates, the differential $\mathrm{d}f$ has the usual form

$$\mathrm{d}f = f_{|a}\boldsymbol{\omega}^a = f_{,i}\mathrm{d}x^i. \tag{2.17}$$

From 1-forms at points in $\mathcal{M}$ we can build the 1-form bundle $T^*(\mathcal{M})$ of $\mathcal{M}$, also called the *cotangent bundle* of $\mathcal{M}$, and define *fields* of 1-*forms* on $\mathcal{M}$, analogously to the constructions for vectors, the components σ_i of a 1-form field being differentiable functions of the local coordinates. In tensor calculus the components σ_i are often called 'components of a covariant vector'.

2.5 Tensors

A tensor $\boldsymbol{T}$ of type (r, s), and of order $(r + s)$, at p is an element of the product space

$$T_p(r, s) = \underbrace{T_p \otimes \cdots \otimes T_p}_{r \text{ factors}} \otimes \underbrace{T_p^* \otimes \cdots \otimes T_p^*}_{s \text{ factors}}$$

and maps any ordered set of r 1-forms and s vectors,

$$(\boldsymbol{\sigma}^1, \ldots, \boldsymbol{\sigma}^r; \boldsymbol{v}_1, \ldots, \boldsymbol{v}_s), \tag{2.18}$$

at p into a real number. In particular, the tensor $\boldsymbol{u}_1 \otimes \cdots \otimes \boldsymbol{u}_r \otimes \boldsymbol{\tau}^1 \otimes \cdots \otimes \boldsymbol{\tau}^s$ maps the ordered set (2.18) into the product of contractions, $\langle \boldsymbol{\sigma}^1, \boldsymbol{u}_1 \rangle \cdots \langle \boldsymbol{\sigma}^r, \boldsymbol{u}_r \rangle \langle \boldsymbol{\tau}^1, \boldsymbol{v}_1 \rangle \cdots \langle \boldsymbol{\tau}^s, \boldsymbol{v}_s \rangle$. The map is multilinear, i.e. linear in each argument. In terms of the bases $\{\boldsymbol{e}_a\}$, $\{\boldsymbol{\omega}^b\}$ an arbitrary tensor $\boldsymbol{T}$ of type (r, s) can be expressed as a sum of tensor products

$$\boldsymbol{T} = T^{a_1 \cdots a_r}{}_{b_1 \cdots b_s} \boldsymbol{e}_{a_1} \otimes \cdots \otimes \boldsymbol{e}_{a_r} \otimes \boldsymbol{\omega}^{b_1} \cdots \otimes \boldsymbol{\omega}^{b_s}, \tag{2.19}$$

where all indices run from 1 to n. The coefficients $T^{a_1 \cdots a_r}{}_{b_1 \cdots b_s}$ with covariant indices $b_1 \cdots b_s$ and contravariant indices $a_1 \cdots a_r$ are the *components* of $\boldsymbol{T}$ with respect to the bases $\{\boldsymbol{e}_a\}$, $\{\boldsymbol{\omega}^b\}$. For a general tensor, the factors in the individual tensor product terms in (2.19) may not be interchanged.

Non-singular linear transformations of the bases,

$$\boldsymbol{e}_{a'} = L_{a'}{}^a \boldsymbol{e}_a, \qquad \boldsymbol{\omega}^{a'} = L^{a'}{}_a \boldsymbol{\omega}^a, \qquad L^{a'}{}_b L_{a'}{}^c = \delta^c_b, \tag{2.20}$$

change the components of the tensor $\boldsymbol{T}$ according to the transformation law

$$T^{a'_1 \cdots a'_r}{}_{b'_1 \cdots b'_s} = L^{a'_1}{}_{a_1} \cdots L^{a'_r}{}_{a_r} L_{b'_1}{}^{b_1} \cdots L_{b'_s}{}^{b_s} T^{a_1 \cdots a_r}{}_{b_1 \cdots b_s}. \tag{2.21}$$

For transformations connecting two coordinate bases $\{\partial/\partial x^a\}$, $\{\partial/\partial x^{a'}\}$, the $(n \times n)$ matrices $L^{a'}{}_a$, $L_{a'}{}^a$ take the special forms $L^{a'}{}_a = \partial x^{a'}/\partial x^a$, $L_{a'}{}^a = \partial x^a/\partial x^{a'}$.

The following algebraic operations are independent of the basis used in (2.19): addition of tensors of the same type, multiplication by a real number, tensor product of two tensors, contraction on any pair of one contravariant and one covariant index, and formation of the (anti)symmetric part of a tensor.

Maps of tensors. The map Φ (Fig. 2.2) sending $p \in \mathcal{M}$ to $\Phi(p) \in \mathcal{N}$ induces in a natural way a map Φ^* of the real-valued functions f defined on $\mathcal{N}$ to functions on $\mathcal{M}$,

$$\Phi^* f(p) = f(\Phi(p)). \tag{2.22}$$

Moreover, induced maps of vectors and 1-forms

$$\begin{aligned} \Phi_* : \boldsymbol{v} \in T_p \quad &\rightarrow \quad \Phi_* \boldsymbol{v} \in T_{\Phi(p)}, \\ \Phi^* : \boldsymbol{\sigma} \in T^*_{\Phi(p)} \quad &\rightarrow \quad \Phi^* \boldsymbol{\sigma} \in T^*_p, \end{aligned} \tag{2.23}$$

are defined by the postulates:

(i) the image of a vector satisfies

$$\Phi_* \boldsymbol{v}(f)|_{\Phi(p)} = \boldsymbol{v}(\Phi^* f)|_p, \tag{2.24a}$$

$\Phi_* \boldsymbol{v}$ being the tangent vector to the image curve $\Phi(\gamma(t))$ at $\Phi(p)$, if $\boldsymbol{v}$ is the tangent vector to $\gamma(t)$ at p (see Fig. 2.2);

(ii) the maps (2.23) preserve the contractions,

$$\langle \Phi^* \boldsymbol{\sigma}, \boldsymbol{v} \rangle|_p = \langle \boldsymbol{\sigma}, \Phi_* \boldsymbol{v} \rangle|_{\Phi(p)}. \tag{2.24b}$$

It follows immediately from (2.24a) that, for any $\boldsymbol{u}$ and $\boldsymbol{v}$,

$$[\Phi_* \boldsymbol{u}, \Phi_* \boldsymbol{v}] = \Phi_* [\boldsymbol{u}, \boldsymbol{v}]. \tag{2.25}$$

Let us denote local coordinates in corresponding neighbourhoods of p and $\Phi(p)$ by $(x^1, \ldots, x^m)$ and $(y^1, \ldots, y^n)$ respectively. The map of a 1-form $\boldsymbol{\sigma}$ is given simply by coordinate substitution,

$$\begin{gathered} \Phi^* : \boldsymbol{\sigma} = \sigma_i(y) \mathrm{d}y^i \rightarrow \Phi^* \boldsymbol{\sigma} = \sigma_i(y(x)) \left(\partial y^i / \partial x^k \right) \mathrm{d}x^k = \widetilde{\boldsymbol{\sigma}}_k(x) \mathrm{d}x^k, \\ i = 1, \ldots, n, \quad k = 1, \ldots, m. \end{gathered} \tag{2.26}$$

These maps can immediately be extended to tensors of arbitrary type (r, s) provided that the inverse Φ^{-1} exists, i.e. that Φ is a one-to-one map. In this case, (2.24b) can be rewritten in the form

$$\langle \Phi^* \boldsymbol{\sigma}, \Phi_*^{-1} \boldsymbol{v} \rangle|_p = \langle \boldsymbol{\sigma}, \boldsymbol{v} \rangle|_{\Phi(p)}. \tag{2.27}$$

Note that Φ^* maps tensors on $\mathcal{N}$ to tensors on $\mathcal{M}$, starting from a map Φ of $\mathcal{M}$ to $\mathcal{N}$. Although (2.26) looks like a coordinate transformation, it defines *new* tensors, $\Phi^* \boldsymbol{\sigma}$ etc. In contrast, under the transformation (2.20) of the basis of a given manifold $\mathcal{M}$ any tensor remains the same object; only its components are changed. Tensors are invariantly defined.

Up to now we have considered tensors at a given point p. The generalization to *tensor bundles* and *tensor fields* is straightforward. As special cases we have defined fields of 1-forms and vector fields on $\mathcal{M}$ at the ends of §§2.3 and 2.4. In the next few sections we shall introduce various derivatives of tensor fields. For brevity, we shall call tensor fields simply tensors.

2.6 Exterior products and p-forms

Let $\boldsymbol{\alpha}^1, \ldots, \boldsymbol{\alpha}^p$ denote p 1-forms. We define an algebraic operation $\wedge$, the *exterior product* or *wedge product* (up to a factor) by the axioms: the exterior product $\boldsymbol{\alpha}^1 \wedge \boldsymbol{\alpha}^2 \wedge \cdots \wedge \boldsymbol{\alpha}^p$

(i) is linear in each variable, and

(ii) vanishes if any two factors coincide.

From these axioms it follows that the exterior product changes sign if any two factors are interchanged, i.e. it is completely antisymmetric. From the basis 1-forms $\boldsymbol{\omega}^1, \ldots, \boldsymbol{\omega}^n$ we obtain $\binom{n}{p}$ independent *p-forms*

$$\boldsymbol{\omega}^{a_1} \wedge \cdots \wedge \boldsymbol{\omega}^{a_p}, \qquad 1 \leq a_1 < a_2 < \cdots < a_p \leq n, \qquad p \leq n. \tag{2.28}$$

Axiom (ii) implies that these exterior products vanish for $p > n$.

A general p-form $\underset{(p)}{\boldsymbol{\alpha}}$ is a linear combination of the p-forms (2.28),

$$\underset{(p)}{\boldsymbol{\alpha}} = \alpha_{a_1 \cdots a_p} \boldsymbol{\omega}^{a_1} \wedge \cdots \wedge \boldsymbol{\omega}^{a_p}, \tag{2.29}$$

where all the indices run from 1 to n, the restriction in (2.28) for the indices having been dropped. If $\{\boldsymbol{\omega}^a\}$ is a dual coordinate basis $\{\mathrm{d}x^i\}$ this expansion has the form

$$\underset{(p)}{\boldsymbol{\alpha}} = \alpha_{i_i \cdots i_p} \mathrm{d}x^{i_1} \wedge \cdots \wedge \mathrm{d}x^{i_p}. \tag{2.30}$$

The exterior product can be extended to forms of arbitrary degree by the rule that

$$(\boldsymbol{\alpha}^1 \wedge \cdots \wedge \boldsymbol{\alpha}^p) \wedge (\boldsymbol{\beta}^1 \wedge \cdots \wedge \boldsymbol{\beta}^q) = \boldsymbol{\alpha}^1 \wedge \cdots \wedge \boldsymbol{\alpha}^p \wedge \boldsymbol{\beta}^1 \wedge \cdots \wedge \boldsymbol{\beta}^q. \tag{2.31}$$

Exterior multiplication is associative and distributive. However, the commutative law is slightly changed:

$$\underset{(p)}{\boldsymbol{\alpha}} \wedge \underset{(q)}{\boldsymbol{\beta}} = (-1)^{pq} \underset{(q)}{\boldsymbol{\beta}} \wedge \underset{(p)}{\boldsymbol{\alpha}}. \tag{2.32}$$

This property can easily be derived from the axioms defining exterior products of 1-forms and from expansions like (2.29).

In analogy with the contraction of a vector $\boldsymbol{v}$ and a 1-form $\boldsymbol{\sigma}$, which gives a function, we define the contraction of a vector $\boldsymbol{v}$ and a p-form to give a $(p-1)$-form,

$$\boldsymbol{v} \lrcorner \underset{(p)}{\boldsymbol{\alpha}} = \underset{(p-1)}{\boldsymbol{\beta}}, \tag{2.33}$$

which we assume to be linear in both $\boldsymbol{v}$ and $\underset{(p)}{\boldsymbol{\alpha}}$. This implies that the exterior product $\wedge$ is just the antisymmetrization of the tensor product $\otimes$, so that the p-forms are precisely the antisymmetric tensors of type $(0, p)$ (antisymmetric covariant tensors). The factor of proportionality is fixed by the rule that if $\underset{(p)}{\boldsymbol{\alpha}}$ and $\underset{(q)}{\boldsymbol{\beta}}$ have the components $\alpha_{a_1\cdots a_p}$ and $\beta_{b_1\cdots b_q}$, respectively, then their exterior product has the components

$$\Big(\underset{(p)}{\boldsymbol{\alpha}} \wedge \underset{(q)}{\boldsymbol{\beta}}\Big)_{a_1\cdots a_p b_1\cdots b_q} = \alpha_{[a_1\cdots a_p}\beta_{b_1\cdots b_q]}, \tag{2.34}$$

which leads, for example, to $\boldsymbol{\omega}^1 \wedge \boldsymbol{\omega}^2 = (\boldsymbol{\omega}^1 \otimes \boldsymbol{\omega}^2 - \boldsymbol{\omega}^2 \otimes \boldsymbol{\omega}^1)/2$. The component form of (2.33) is then

$$(\boldsymbol{v} \lrcorner \underset{(p)}{\boldsymbol{\alpha}})_{a_2\cdots a_p} = v^b \alpha_{b a_2\cdots a_p}. \tag{2.35}$$

A p-form $\underset{(p)}{\boldsymbol{\alpha}}$ is said to be *simple* if it admits a representation as an exterior product of p linearly independent 1-forms,

$$\underset{(p)}{\boldsymbol{\alpha}} = \boldsymbol{\alpha}^1 \wedge \boldsymbol{\alpha}^2 \wedge \cdots \wedge \boldsymbol{\alpha}^p. \tag{2.36}$$

2.7 The exterior derivative

In §2.4 we defined the differential $\mathrm{d}f$ of a function f by the equation (2.14). The operator d generates a 1-form $\mathrm{d}f$ from a 0-form f by

$$\mathrm{d} : f \to \mathrm{d}f = f_{,i}\mathrm{d}x^i. \tag{2.37}$$

We generalize this differentiation to apply to any p-form. The *exterior derivative* d maps a p-form into a $(p+1)$-form and is completely determined by the axioms:

$$\text{(i)} \qquad \mathrm{d}(\boldsymbol{\alpha} + \boldsymbol{\beta}) = \mathrm{d}\boldsymbol{\alpha} + \mathrm{d}\boldsymbol{\beta}, \tag{2.38a}$$

$$\text{(ii)} \qquad \mathrm{d}\Big(\underset{(p)}{\boldsymbol{\alpha}} \wedge \underset{(q)}{\boldsymbol{\beta}}\Big) = \mathrm{d}\underset{(p)}{\boldsymbol{\alpha}} \wedge \underset{(q)}{\boldsymbol{\beta}} + (-1)^p \underset{(p)}{\boldsymbol{\alpha}} \wedge \mathrm{d}\underset{(q)}{\boldsymbol{\beta}}, \tag{2.38b}$$

$$\text{(iii)} \qquad \mathrm{d}f = f_{,i}\mathrm{d}x^i, \tag{2.38c}$$

$$\text{(iv)} \qquad \mathrm{d}(\mathrm{d}f) = 0. \tag{2.38d}$$

Because of axiom (2.38a) it is sufficient to verify the existence and uniqueness of the exterior derivative for the p-form $f\mathrm{d}x^{i_1} \wedge \cdots \wedge \mathrm{d}x^{i_p}$. One can prove (see e.g. Flanders (1963)) that

$$\mathrm{d}(f\mathrm{d}x^{i_1} \wedge \cdots \wedge \mathrm{d}x^{i_p}) = \mathrm{d}f \wedge \mathrm{d}x^{i_1} \wedge \cdots \wedge \mathrm{d}x^{i_p}. \tag{2.39}$$

and that all the axioms (2.38) are then satisfied.

From a general p-form (2.30) we obtain the $(p+1)$-form

$$\mathrm{d}\underset{(p)}{\boldsymbol{\alpha}} = \alpha_{i_1\cdots i_p,j}\mathrm{d}x^j \wedge \mathrm{d}x^{i_1} \wedge \cdots \wedge \mathrm{d}x^{i_p} \tag{2.40}$$

by exterior differentiation. The (completely antisymmetric) components of $\mathrm{d}\underset{(p)}{\boldsymbol{\alpha}}$ involve only partial derivatives of the components $\alpha_{i_1\cdots i_p}$. We remark that axiom (2.38d) is just the equality of the mixed second partial derivatives of f,

$$\mathrm{d}(\mathrm{d}f) = \mathrm{d}(f_{,i}\mathrm{d}x^i) = f_{,i,j}\mathrm{d}x^j \wedge \mathrm{d}x^i = 0. \tag{2.41}$$

From (2.40) we see that

$$\mathrm{d}(\mathrm{d}\boldsymbol{\alpha}) = 0 \tag{2.42}$$

for any p-form $\boldsymbol{\alpha}$.

The following theorems, for proofs of which we refer the reader to the literature (e.g. for Theorem 2.2 see Flanders (1963)), hold locally, i.e. in a neighbourhood of a point p.

Theorem 2.1 (Poincaré's theorem). *If* $\boldsymbol{\alpha}$ *is a* p*-form* $(p \geq 1)$ *and* $\mathrm{d}\boldsymbol{\alpha} = 0$, *then there is a* $(p-1)$*-form* $\boldsymbol{\beta}$ *such that* $\boldsymbol{\alpha} = \mathrm{d}\boldsymbol{\beta}$. *In components,*

$$\alpha_{[i_1\cdots i_p,j]} = 0 \quad \Leftrightarrow \quad \alpha_{i_1\cdots i_p} = \beta_{[i_1\cdots i_{p-1},i_p]}. \tag{2.43}$$

Theorem 2.2 (Frobenius's theorem). *Let* $\boldsymbol{\sigma}^1, \ldots, \boldsymbol{\sigma}^r$ *be* r *1-forms linearly independent at a point* $p \in \mathcal{M}$. *Suppose there are 1-forms* $\boldsymbol{\tau}^A{}_B$ $(A, B = 1, \ldots, r)$ *satisfying* $\mathrm{d}\boldsymbol{\sigma}^A = \boldsymbol{\tau}^A{}_B \wedge \boldsymbol{\sigma}^B$. *Then in a neighbourhood of* p *there are functions* $f^A{}_B$, h^A *such that* $\boldsymbol{\sigma}^A = f^A{}_B \mathrm{d}h^B$.

Other formulations of Frobenius's theorem. Introducing the r-form $\boldsymbol{\Sigma} \equiv \boldsymbol{\sigma}^1 \wedge \cdots \wedge \boldsymbol{\sigma}^r$, we can replace the condition $\mathrm{d}\boldsymbol{\sigma}^A = \boldsymbol{\tau}^A{}_B \wedge \boldsymbol{\sigma}^B$ by either of the two equivalent conditions:

(i) $\mathrm{d}\boldsymbol{\sigma}^A \wedge \boldsymbol{\Sigma} = 0$,

(ii) there exists a 1-form $\boldsymbol{\lambda}$ such that $\mathrm{d}\boldsymbol{\Sigma} = \boldsymbol{\lambda} \wedge \boldsymbol{\Sigma}$.

In the case of a single 1-form $\boldsymbol{\sigma}$ we have the result

$$\boldsymbol{\sigma} \wedge \mathrm{d}\boldsymbol{\sigma} = 0 \Leftrightarrow \boldsymbol{\sigma} = f\mathrm{d}h, \tag{2.44}$$

or in components,

$$\sigma_{[a,b}\sigma_{c]} = 0 \Leftrightarrow \sigma_a = f h_{,a}. \tag{2.45}$$

The surfaces $h =$ constant are called the integral surfaces of the equation $\boldsymbol{\sigma} = 0$, f^{-1} being the integrating factor.

Frobenius's theorem is important in the construction of exact solutions because it allows us to introduce local coordinates f, h adapted to given normal 1-forms (see e.g. §27.1.1).

The *rank* q of a 2-form $\boldsymbol{\alpha}$ is defined by

$$\underbrace{\boldsymbol{\alpha}\wedge\cdots\wedge\boldsymbol{\alpha}}_{q\text{ factors}}\neq 0,\qquad \underbrace{\boldsymbol{\alpha}\wedge\cdots\wedge\boldsymbol{\alpha}}_{(q+1)\text{ factors}}=0,\qquad 2q\leq n. \tag{2.46}$$

Using this definition we can generalize the statement (2.44) to

Theorem 2.3 (Darboux's theorem). *Let $\boldsymbol{\sigma}$ be a 1-form and let the 2-form $\mathrm{d}\boldsymbol{\sigma}$ have rank q. Then we can find local coordinates $x^1,\ldots,x^q$, $\xi^1,\ldots,\xi^{n-q}$ such that*

$$\text{if }\ \boldsymbol{\sigma}\wedge\underbrace{\mathrm{d}\boldsymbol{\sigma}\wedge\cdots\wedge\mathrm{d}\boldsymbol{\sigma}}_{q\text{ factors}}\begin{cases}=0: \boldsymbol{\sigma}=x^1\mathrm{d}\xi^1+\cdots+x^q\mathrm{d}\xi^q,\\ \neq 0: \boldsymbol{\sigma}=x^1\mathrm{d}\xi^1+\cdots+x^q\mathrm{d}\xi^q+\mathrm{d}\xi^{q+1}.\end{cases} \tag{2.47}$$

(For a proof, see Sternberg (1964).)

This theorem gives the possible normal forms of a 1-form $\boldsymbol{\sigma}$. Specializing Darboux's theorem to a four-dimensional manifold one obtains the following classification of a 1-form $\boldsymbol{\sigma}$ in terms of its components:

$$\begin{aligned}
&q=0: \sigma_{[a,b]}=0: && && \sigma_a=\xi_{,a}\\
&q=1: \sigma_{[a,b]}\neq 0, && \sigma_{[a,b}\sigma_{c,d]}=0,\ \sigma_{[a,b}\sigma_{c]}=0: && \sigma_a=x\xi_{,a}\\
&\qquad\ \ \sigma_{[a,b]}\neq 0, && \sigma_{[a,b}\sigma_{c,d]}=0,\ \sigma_{[a,b}\sigma_{c]}\neq 0: && \sigma_a=x\xi_{,a}+\eta_{,a}\\
&q=2: \sigma_{[a,b}\sigma_{c,d]}\neq 0 && && \sigma_a=x\xi_{,a}+y\eta_{,a}.
\end{aligned} \tag{2.48}$$

The real functions denoted by x, y, ξ, η are independent. The second subcase is just Frobenius's theorem applied to a single 1-form $\boldsymbol{\sigma}$.

Now we give a theorem concerning 2-forms.

Theorem 2.4 *For any 2-form $\boldsymbol{\alpha}$ of rank q there exists a basis $\{\boldsymbol{\omega}^a\}$ such that*

$$\boldsymbol{\alpha}=(\boldsymbol{\omega}^1\wedge\boldsymbol{\omega}^2)+(\boldsymbol{\omega}^3\wedge\boldsymbol{\omega}^4)+\cdots+(\boldsymbol{\omega}^{2q-1}\wedge\boldsymbol{\omega}^{2q}). \tag{2.49}$$

If $\mathrm{d}\boldsymbol{\alpha}=0$, then we can introduce local coordinates $x^1,\ldots,x^q$, $\xi^1,\ldots,\xi^{n-q}$ such that

$$\boldsymbol{\alpha}=\mathrm{d}x^1\wedge\mathrm{d}\xi^1+\cdots+\mathrm{d}x^q\wedge\mathrm{d}\xi^q. \tag{2.50}$$

(For a proof, see Sternberg (1964).)

To conclude this series of theorems, we consider a map $\Phi:\mathcal{M}\to\mathcal{N}$ between two manifolds, as in (2.23), and show by induction

Theorem 2.5 *For the exterior derivative* $\mathrm{d}\boldsymbol{\alpha}$ *of a p-form* $\boldsymbol{\alpha}$ *we have*

$$\mathrm{d}(\Phi^*\boldsymbol{\alpha}) = \Phi^*(\mathrm{d}\boldsymbol{\alpha}). \tag{2.51}$$

Proof: Let us denote local coordinates in corresponding neighbourhoods of $p \in \mathcal{M}$ and $\Phi(p) \in \mathcal{N}$ by $(x^1, \dots, x^m)$ and $(y^1, \dots, y^n)$ respectively. Obviously, (2.51) is true for a 0-form f:

$$\mathrm{d}(\Phi^* f) = \frac{\partial(\Phi^* f)}{\partial x^k}\mathrm{d}x^k = \frac{\partial f(y(x))}{\partial y^i}\frac{\partial y^i}{\partial x^k}\mathrm{d}x^k = \Phi^*(\mathrm{d}f). \tag{2.52}$$

Suppose the relation is valid for the $(p-1)$-form $\boldsymbol{\beta}$ and let $\boldsymbol{\alpha} = f\mathrm{d}\boldsymbol{\beta}$. (This is sufficiently general.) Then,

$$\mathrm{d}(\Phi^*\boldsymbol{\alpha}) = \mathrm{d}[(\Phi^* f)\mathrm{d}(\Phi^*\boldsymbol{\beta})] = \mathrm{d}(\Phi^* f) \wedge \mathrm{d}(\Phi^*\boldsymbol{\beta}) = \Phi^*(\mathrm{d}\boldsymbol{\alpha}). \tag{2.53}$$

We do not consider integration on manifolds, except to note that the operator d of exterior derivation has been defined so that *Stokes's theorem* can be written in the simple form

$$\int_{\partial\mathcal{V}} \boldsymbol{\alpha} = \int_{\mathcal{V}} \mathrm{d}\boldsymbol{\alpha}, \tag{2.54}$$

where $\boldsymbol{\alpha}$ is any $(k-1)$-form and $\partial\mathcal{V}$ denotes the oriented boundary of a k-dimensional manifold with boundary $\mathcal{V}$. (An n-dimensional manifold with boundary is defined by charts which map their neighbourhoods $\mathcal{U}$ into the half space H^n defined by $x^n \geq 0$ rather than into E^n, the *boundary* then being the set of points mapped to $x^n = 0$.)

2.8 The Lie derivative

For each point $p \in \mathcal{M}$, a vector field $\boldsymbol{v}$ on $\mathcal{M}$ determines a unique curve $\gamma_p(t)$ such that $\gamma_p(0) = p$ and $\boldsymbol{v}$ is the tangent vector to the curve. The family of these curves is called the congruence associated with the vector field. Along a curve $\gamma_p(t)$ the local coordinates $(y^1, \dots, y^n)$ are the solutions of the system of ordinary differential equations

$$\frac{\mathrm{d}y^i}{\mathrm{d}t} = v^i(y^1(t), \dots, y^n(t)) \tag{2.55}$$

with the initial values $y^i(0) = x^i(p)$.

To introduce a new type of differentiation we consider the map Φ_t dragging each point p, with coordinates x^i, along the curve $\gamma_p(t)$ through p into the image point $q = \Phi_t(p)$ with coordinates $y^i(t)$. For sufficiently small values of the parameter t the map Φ_t is a one-to-one map which induces a map $\Phi_t^* \boldsymbol{T}$ of any tensor $\boldsymbol{T}$, called *Lie transport*. The *Lie derivative* of $\boldsymbol{T}$ with respect to $\boldsymbol{v}$ is defined by

$$\mathcal{L}_{\boldsymbol{v}} \boldsymbol{T} \equiv \lim_{t \to 0} \frac{1}{t} (\Phi_t^* \boldsymbol{T} - \boldsymbol{T}). \tag{2.56}$$

The tensors $\boldsymbol{T}$ and $\Phi_t^* \boldsymbol{T}$ are of the same type (r, s) and are both evaluated at the same point p. Therefore, the Lie derivative (2.56) is also a tensor of type (r, s) at p. The Lie derivative vanishes if the tensors $\boldsymbol{T}$ and $\Phi_t^* \boldsymbol{T}$ coincide. In this case the tensor field $\boldsymbol{T}$ remains in a sense the 'same' under Lie transport along the integral curves of the vector field $\boldsymbol{v}$. However, the components of $\boldsymbol{T}$ with respect to the coordinate basis $\{\partial / \partial x^i\}$ may vary along the curves. Using coordinate bases $\{\partial / \partial x^i\}$ and $\{\partial / \partial y^i\}$, we compute the *components* of the Lie derivative. The relations

$$\left. \frac{\partial y^i}{\partial x^k} \right|_{t=0} = \delta^i_k, \qquad \left. \frac{\mathrm{d} y^i}{\mathrm{d} t} \right|_{t=0} = v^i, \qquad \left. \frac{\mathrm{d} x^i}{\mathrm{d} t} \right|_{t=0} = -v^i \tag{2.57}$$

will be used. We start with the Lie derivatives of functions, 1-forms, and vectors:

function f:
$$\mathcal{L}_{\boldsymbol{v}} f = v^i f_{,i} \quad (= \boldsymbol{v}(f)). \tag{2.58}$$

Proof:

$$\Phi_t^* f|_p = f(y(x,t)), \quad \mathcal{L}_{\boldsymbol{v}} f|_p = \left. \frac{\partial f}{\partial y^i} \frac{\mathrm{d} y^i}{\mathrm{d} t} \right|_p .$$

1-form $\boldsymbol{\sigma}$:
$$\mathcal{L}_{\boldsymbol{v}} \boldsymbol{\sigma} = (v^m \sigma_{i,m} + \sigma_m v^m{}_{,i}) \mathrm{d} x^i. \tag{2.59}$$

Proof:

$$\Phi_t^* \boldsymbol{\sigma}|_p = \sigma_j(y(x,t)) \frac{\partial y^j}{\partial x^i} \mathrm{d} x^i,$$

$$\mathcal{L}_{\boldsymbol{v}} \boldsymbol{\sigma}|_p = \left[\frac{\partial \sigma_j}{\partial y^m} \frac{\mathrm{d} y^m}{\mathrm{d} t} \frac{\partial y^j}{\partial x^i} + \sigma_j \frac{\partial}{\partial x^i} \left(\frac{\mathrm{d} y^j}{\mathrm{d} t} \right) \right]_{t=0} \mathrm{d} x^i .$$

vector $\boldsymbol{u}$:
$$\mathcal{L}_{\boldsymbol{v}} \boldsymbol{u} = (v^m u^i{}_{,m} - u^m v^i{}_{,m}) \frac{\partial}{\partial x^i}. \tag{2.60}$$

Proof:

$$\Phi_t^* \boldsymbol{u}|_p = u^j\,(y(x,t))\, \frac{\partial x^i}{\partial y^j} \frac{\partial}{\partial x^i},$$

$$\mathcal{L}_{\boldsymbol{v}} \boldsymbol{u}|_p = \left[\frac{\partial u^j}{\partial y^m} \frac{\mathrm{d} y^m}{\mathrm{d} t} \frac{\partial x^i}{\partial y^j} + u^j \frac{\partial}{\partial y^j} \left(\frac{\mathrm{d} x^i}{\mathrm{d} t} \right) \right]_{t=0} \frac{\partial}{\partial x^i} .$$

The Lie derivative of $\boldsymbol{u}$ with respect to $\boldsymbol{v}$ is equal to the commutator $[\boldsymbol{v}, \boldsymbol{u}]$,

$$\mathcal{L}_{\boldsymbol{v}}\boldsymbol{u} = [\boldsymbol{v}, \boldsymbol{u}] = v^m \frac{\partial}{\partial x^m}\left(u^i \frac{\partial}{\partial x^i}\right) - u^m \frac{\partial}{\partial x^m}\left(v^i \frac{\partial}{\partial x^i}\right). \tag{2.61}$$

Two *commuting* vector fields generate a family of two-dimensional submanifolds of $\mathcal{M}$ on which the parameters of the integral curves of both vector fields can be taken as coordinates.

From the Leibniz product rule and (2.59), (2.60) one obtains the components of the Lie derivative of an arbitrary tensor,

$$(\mathcal{L}_{\boldsymbol{v}}\boldsymbol{T})^{ij\cdots}{}_{kl\cdots} = v^m T^{ij\cdots}{}_{kl\cdots,m} - T^{mj\cdots}{}_{kl\cdots} v^i{}_{,m} - T^{im\cdots}{}_{kl\cdots} v^j{}_{,m} - \cdots$$
$$+ T^{ij\cdots}{}_{ml\cdots} v^m{}_{,k} + T^{ij\cdots}{}_{km\cdots} v^m{}_{,l} + \cdots \quad . \tag{2.62}$$

Equation (2.61) and the Jacobi identity (2.7) imply that when applied to vectors

$$\mathcal{L}_{\mathbf{u}}\mathcal{L}_{\mathbf{v}} - \mathcal{L}_{\mathbf{v}}\mathcal{L}_{\mathbf{u}} = \mathcal{L}_{[\mathbf{u},\, \mathbf{v}]}, \tag{2.63}$$

and (2.58) and the Leibniz rule then imply that this is true for any tensor.

From (2.51), (2.56), it follows that the Lie derivative applied to forms commutes with the exterior derivative:

$$\mathrm{d}(\mathcal{L}_{\boldsymbol{v}}\boldsymbol{\alpha}) = \mathcal{L}_{\boldsymbol{v}}(\mathrm{d}\boldsymbol{\alpha}) \tag{2.64}$$

for any p-form $\boldsymbol{\alpha}$. Of course, this rule can also be verified by using (2.40), (2.62) in terms of components.

As will be seen later, the Lie derivative plays an important role in describing symmetries of gravitational fields and other physical fields.

The exterior derivative and the Lie derivative are operations defined on a differentiable manifold without imposing additional structures. Both operations are generalizations of the partial derivative. The exterior derivative is a limited generalization acting only on forms. The Lie derivative depends on the vector $\boldsymbol{v}$ not only at p, but also at neighbouring points. To introduce invariantly defined derivatives which have neither of these defects we have to impose a new structure on $\mathcal{M}$, and we proceed to do so in the following section.

2.9 The covariant derivative

The covariant derivative $\nabla_{\boldsymbol{v}}$ in the direction of the vector $\boldsymbol{v}$ at p maps an arbitrary tensor into a tensor of the same type. If $\boldsymbol{v}$ is unspecified, the covariant derivative ∇ generates a tensor of type $(r, s+1)$ from a tensor of type (r, s). In particular, for a vector $\boldsymbol{u}$ we have the expansion

$$\nabla\boldsymbol{u} = u^a{}_{;b}\boldsymbol{e}_a \otimes \boldsymbol{\omega}^b \tag{2.65}$$

with components $u^a{}_{;b}$ as yet unspecified. The directional covariant derivative is given by the vector

$$\nabla_{\boldsymbol{v}} \boldsymbol{u} = (u^a{}_{;b} v^b) \boldsymbol{e}_a. \tag{2.66}$$

The covariant derivative of the basis vector $\boldsymbol{e}_a$ in the direction of the basis vector $\boldsymbol{e}_b$ can be expanded in terms of basis vectors:

$$\nabla_b \boldsymbol{e}_a = \Gamma^c{}_{ab} \boldsymbol{e}_c, \qquad \Gamma^c{}_{ab} = \langle \boldsymbol{\omega}^c, \nabla_b \boldsymbol{e}_a \rangle. \tag{2.67}$$

For consistency of (2.67) with the Leibniz rule applied to (2.11), the covariant derivative of a dual basis $\{\boldsymbol{\omega}^a\}$ is given by

$$\nabla_b \boldsymbol{\omega}^a = -\Gamma^a{}_{cb} \boldsymbol{\omega}^c. \tag{2.68}$$

The coefficients $\Gamma^c{}_{ab}$, called the *connection coefficients*, relate the bases at different points of $\mathcal{M}$, and they have to be imposed as an extra structure on $\mathcal{M}$. We restrict ourselves to covariant derivatives satisfying

$$\nabla_{\boldsymbol{u}} \boldsymbol{v} - \nabla_{\boldsymbol{v}} \boldsymbol{u} = [\boldsymbol{u}, \boldsymbol{v}] \tag{2.69}$$

for two arbitrary vectors $\boldsymbol{u}$ and $\boldsymbol{v}$. This relation is equivalent to the equation

$$2\Gamma^c{}_{[ab]} = -D^c{}_{ab}, \tag{2.70}$$

where the commutation coefficients are defined by (2.6). In a coordinate basis, the connection coefficients $\Gamma^c{}_{ab}$ have a symmetric index pair (ab). Therefore a covariant derivative satisfying (2.69) is called *symmetric* (or *torsionfree*).

Using the symmetry axiom (2.70), we may replace the partial derivatives in (2.40) and (2.62) for the components of, respectively, the exterior derivative and the Lie derivative by covariant derivatives, so that the commas can be replaced by semicolons.

Once the connection coefficients are prescribed, the components $u^a{}_{;c}$ of the covariant derivative of $\boldsymbol{u}$ in the direction of the basis vector $\boldsymbol{e}_c$ are completely determined,

$$\nabla_c \boldsymbol{u} = \nabla_c (u^a \boldsymbol{e}_a) = (u^a{}_{|c} + \Gamma^a{}_{dc} u^d) \boldsymbol{e}_a = u^a{}_{;c} \boldsymbol{e}_a, \tag{2.71a}$$

and the components of the covariant derivative $\nabla \boldsymbol{T}$ of a tensor (2.19) are

$$\begin{aligned} T^{a_1 \cdots a_r}{}_{b_1 \cdots b_s;c} = (T^{a_1 \cdots a_r}{}_{b_1 \cdots b_s})_{|c} &+ \Gamma^{a_1}{}_{dc} T^{d \cdots a_r}{}_{b_1 \cdots b_s} + \cdots + \Gamma^{a_r}{}_{dc} T^{a_1 \cdots d}{}_{b_1 \cdots b_s} \\ &- \Gamma^d{}_{b_1 c} T^{a_1 \cdots a_r}{}_{d \cdots b_s} - \cdots - \Gamma^d{}_{b_s c} T^{a_1 \cdots a_r}{}_{b_1 \cdots d}, \end{aligned} \tag{2.71b}$$

where the symbol $f_{|a} \equiv \boldsymbol{e}_a(f) = f_{,i} e_a{}^i$ has been used. Note that (2.71b) is valid for a general basis $\{\boldsymbol{e}_a\}$.

Since the bases $\{\boldsymbol{e}_a\}$, $\{\boldsymbol{\omega}^a\}$ are linear combinations of coordinate bases,

$$\boldsymbol{e}_a = e_a{}^i \partial/\partial x^i, \qquad \boldsymbol{\omega}^a = \omega^a{}_i \mathrm{d}x^i, \tag{2.72}$$

the connection coefficients (2.67) can be written with respect to these bases as

$$\Gamma^c{}_{ab} = \omega^c{}_k e_a{}^k{}_{;i} e_b{}^i = -e_a{}^k \omega^c{}_{k;i} e_b{}^i . \tag{2.73}$$

These are also referred to as the Ricci rotation coefficients. For the exterior derivative of the basis 1-forms we get

$$\mathrm{d}\boldsymbol{\omega}^a = \omega^a{}_{i,j} \mathrm{d}x^j \wedge \mathrm{d}x^i = \omega^a{}_{i;j} \mathrm{d}x^j \wedge \mathrm{d}x^i = \Gamma^a{}_{bc} \boldsymbol{\omega}^b \wedge \boldsymbol{\omega}^c. \tag{2.74}$$

Introducing the *connection 1-forms*

$$\boldsymbol{\Gamma}^a{}_b \equiv \Gamma^a{}_{bc} \boldsymbol{\omega}^c, \tag{2.75}$$

we can write (2.74) in the form

$$\mathrm{d}\boldsymbol{\omega}^a = -\boldsymbol{\Gamma}^a{}_b \wedge \boldsymbol{\omega}^b \tag{2.76}$$

due to Cartan (the first Cartan equation). For a given basis, the *antisymmetric part* $\Gamma^a{}_{[bc]}$ of the connection coefficients can be computed from this.

The definition of the covariant derivative is equivalent to a definition of parallellism; the relation is that if $\boldsymbol{w}(q)$ is the vector at q parallel to $\boldsymbol{u}(p)$ at p, and $\boldsymbol{v}(p)$ is the tangent vector at p to a curve γ from $p = \gamma(0)$ to $q = \gamma(\epsilon)$, the covariant derivative $\nabla_{\boldsymbol{v}}\boldsymbol{u}$ is the limit of $[\boldsymbol{w}(q) - \boldsymbol{u}(q)]/\epsilon$ as $\epsilon \to 0$. Thus, a tensor $\boldsymbol{T}$ is said to be *parallelly-transported* along the curve with tangent vector $\boldsymbol{v}$ if $\nabla_{\boldsymbol{v}}\boldsymbol{T} = 0$. An *autoparallel* curve is one whose tangent vector is parallel to itself along the curve.

2.10 The curvature tensor

The *curvature tensor* (*Riemann tensor*), $\boldsymbol{R} = R^a{}_{bcd} \boldsymbol{e}_a \otimes \boldsymbol{\omega}^b \otimes \boldsymbol{\omega}^c \otimes \boldsymbol{\omega}^d$ is a tensor of type (1, 3) mapping the ordered set $(\boldsymbol{\sigma};\ \boldsymbol{w},\boldsymbol{u},\boldsymbol{v})$ of a 1-form $\boldsymbol{\sigma}$ and three vectors $\boldsymbol{w},\boldsymbol{u},\boldsymbol{v}$ into the real number

$$\begin{aligned}
\sigma_a w^b u^c v^d R^a{}_{bcd} &= \langle \boldsymbol{\sigma}, (\nabla_{\boldsymbol{u}}\nabla_{\boldsymbol{v}} - \nabla_{\boldsymbol{v}}\nabla_{\boldsymbol{u}} - \nabla_{[\boldsymbol{u},\boldsymbol{v}]})\boldsymbol{w}\rangle \\
&= \sigma_a[(w^a{}_{;c}v^c)_{;d}u^d - (w^a{}_{;c}u^c)_{;d}v^d - w^a{}_{;c}(u^d v^c{}_{;d} - v^d u^c{}_{;d})] \\
&= \sigma_a(w^a{}_{;cd} - w^a{}_{;dc})v^c u^d.
\end{aligned} \tag{2.77}$$

As the components σ_a, v^c, u^d can be chosen arbitrarily we arrive at the *Ricci identity*

$$w^a{}_{;cd} - w^a{}_{;dc} = w^b R^a{}_{bdc}. \tag{2.78}$$

The general rules (2.71) for the components of the covariant derivative of a tensor imply the formula

$$R^a{}_{bcd} = \Gamma^a{}_{bd|c} - \Gamma^a{}_{bc|d} + \Gamma^e{}_{bd}\Gamma^a{}_{ec} - \Gamma^e{}_{bc}\Gamma^a{}_{ed} - D^e{}_{cd}\Gamma^a{}_{be}. \tag{2.79}$$

In a coordinate basis, the last term vanishes. The components (2.79) of the curvature tensor satisfy the symmetry relations

$$R^a{}_{bcd} = -R^a{}_{bdc}, \qquad R^a{}_{[bcd]} = 0. \tag{2.80}$$

The covariant derivatives of the curvature tensor obey the *Bianchi identities*

$$R^a{}_{b[cd;e]} = 0. \tag{2.81}$$

By contraction we obtain the identities

$$R^a{}_{bcd;a} + 2R_{b[c;d]} = 0, \tag{2.82}$$

where the components R_{bd} of the *Ricci tensor* are defined by

$$R_{bd} \equiv R^a{}_{bad}. \tag{2.83}$$

If a vector is parallelly transported round a closed curve, the initial and final vectors will in general not be equal: this phenomenon is called *holonomy.* For infinitesimally small curves the holonomy is given by an integral of the curvature tensor over an area enclosed by the curve, and conversely this gives an alternative way to define curvature.

A compact and efficient method for calculating the components (2.79) with respect to a general basis is provided by Cartan's procedure. Defining the *curvature 2-forms* $\boldsymbol{\Theta}^a{}_b$ by

$$\boldsymbol{\Theta}^a{}_b \equiv \tfrac{1}{2}R^a{}_{bcd}\boldsymbol{\omega}^c \wedge \boldsymbol{\omega}^d, \tag{2.84}$$

equation (2.79) is completely equivalent to the second Cartan equation

$$\mathrm{d}\boldsymbol{\Gamma}^a{}_b + \boldsymbol{\Gamma}^a{}_c \wedge \boldsymbol{\Gamma}^c{}_b = \boldsymbol{\Theta}^a{}_b, \tag{2.85}$$

which gives an algorithm for the calculation of the curvature from the connection. We collect the relations between the various quantities in Fig. 2.3. In this notation the Bianchi identities (2.81) are the components of

$$\mathrm{d}^2\boldsymbol{\Gamma}^a{}_b = \mathrm{d}\boldsymbol{\Theta}^a{}_b - \boldsymbol{\Theta}^a{}_c \wedge \boldsymbol{\Gamma}^c{}_b + \boldsymbol{\Gamma}^a{}_c \wedge \boldsymbol{\Theta}^c{}_b = 0. \tag{2.86}$$

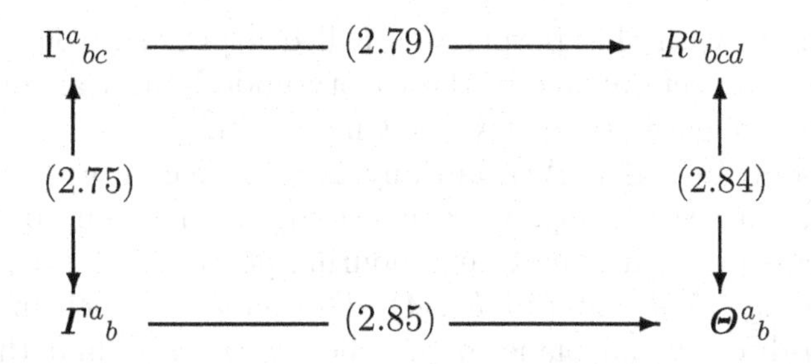

Fig. 2.3. How to get the curvature from the connection

2.11 Fibre bundles

The tensor bundles introduced in §2.4 are examples of fibre bundles. A simple picture of a fibre bundle $\mathcal{E}$ over a manifold $\mathcal{M}$ can be obtained by imagining a copy of another manifold $\mathcal{F}$, the fibre, to be attached to each point of $\mathcal{M}$ in such a way that for suitable regions where coordinates x^j are given on $\mathcal{M}$ and ξ^a on the fibre, the fibre bundle has coordinates (x^j, ξ^a), i.e. $\mathcal{E}$ is locally $\mathcal{M} \times \mathcal{F}$; $\mathcal{M}$ is called the base manifold. On regions where two coordinate systems in $\mathcal{M}$ apply, the corresponding coordinate transformations of $\mathcal{E}$ are given by the usual coordinate transformations in $\mathcal{M}$ together with fibre transformations $\xi^{b'} = \xi^{b'}(\xi^a)$ for each $p \in \mathcal{M}$; these fibre transformations belong to a group of transformations of $\mathcal{F}$ called the structure group of the bundle. (For the tangent bundle the structure group just consists of linear transformations of $\mathbb{R}^m$.) The structure group is often a Lie group G, as defined in Chapter 8. (For fuller details, see e.g. Steenrod (1951), Crampin and Pirani (1986).)

The map $\pi : \mathcal{E} \to \mathcal{M}; (p,\, f) \mapsto p$ is called the *projection* onto the base manifold $\mathcal{M}$. A map $\sigma : \mathcal{M} \to \mathcal{E}$ which for each point p in $\mathcal{M}$ gives a unique point $\sigma(p) \in \pi^{-1}(p)$ is called a section, consistently with the use of this term above. All the maps involved in the definition must of course be suitably smooth. If $\mathcal{F}$ is a vector space, the bundle is called a vector bundle.

The *frame bundle* $F(\mathcal{M})$ for a manifold $\mathcal{M}$ has as the fibre $\mathcal{F}$ at p the set of all possible bases of $T_p(\mathcal{M})$, so the structure group is the group of non-singular linear transformations $GL(n, \mathbb{R})$. Similarly $T_p^*(\mathcal{M})$ defines the *coframe bundle*. Various restricted (co)frame bundles can be obtained by restricting the set of allowed (co)frames, the most common case, in space-times, being restriction to one of the special classes of tetrads, orthonormal or null (see Chapter 3). In frame bundles the structure group and the fibre can be identified, $\mathcal{F} \cong G$, i.e. each choice of frame can be obtained by applying a unique element of the structure group to a basic choice of frame: a bundle with this property is called a principal bundle. Frame bundles play an essential role in the theory underlying Cartan's method for testing equivalence of metrics (see Chapter 9).

Given a point in $p \in F(\mathcal{M})$ and any tangent vector $\boldsymbol{v}$ at $\pi(p) \in \mathcal{M}$, tangent to a curve $\gamma(t)$ say, we can use the connection to define the parallelly-transported frame at neighbouring points of $\gamma(t)$, i.e. to define a corresponding *lifted curve* in $F(\mathcal{M})$. The set of all such lifted curves defines an n-dimensional plane at p, called *horizontal*, and the tangent to the lifted curve at p defines a vector in $T(F\mathcal{M}))$, the *horizontal lift* of $\boldsymbol{v}$. The distribution of horizontal subspaces in fact completely defines the connection. The basis corresponding to p can be lifted to give a uniquely defined basis of horizontal vectors at p. One-forms on $\mathcal{M}$ can be lifted to the horizontals in $F(\mathcal{M})$ by π^*; at a point p, they have a uniquely-defined basis given by lifting the basis of 1-forms dual to the basis defined by p.

Now consider a general curve through $p \in F(\mathcal{M})$ with tangent vector $\boldsymbol{V}$, and a frame $\{\boldsymbol{e}_a\}$. The change of frame along the horizontal part of $\boldsymbol{V}$, defined by $\pi_*(\boldsymbol{V})$, is given by the usual connection, while the change due to the vertical part of $\boldsymbol{V}$, i.e. the part tangent to the fibre, is given by (2.5) with $L_a{}^b = \widehat{\Gamma}^b{}_{ac}V^c$, where $\widehat{\Gamma}^b{}_{ac}$ depends on the parametrization of $GL(n,\mathbb{R})$. The quantities $\widehat{\Gamma}^b{}_{ac}$ can be added to $\Gamma^b{}_{ac}$ to define a connection $\mathbf{\Gamma}^a{}_b$ on the bundle; the formulae for covariant derivatives given in §2.9, in particular (2.76), can then all be extended to $F(\mathcal{M})$. Moreover, given the connection (and the structure group) the 1-form fields $\boldsymbol{\omega}^a$ and $\boldsymbol{\Gamma}^a{}_b$ are a uniquely defined basis on $F(\mathcal{M})$. Similar remarks apply to the various restricted (co)frame bundles.

From this connection, one can define a curvature in $F(\mathcal{M})$. Direct calculation shows (see e.g. Araujo *et al.* (1992)) that the non-zero components of the curvature of $F(\mathcal{M})$ at p are just given by those of the usual curvature at $\pi(p)$ in the frame $\{\boldsymbol{e}_a\}$ which p represents, and (2.85) still applies although $\boldsymbol{\Gamma}^a{}_b$ and $\boldsymbol{\Theta}^a{}_b$ now refer to the connection and curvature on $F(\mathcal{M})$. Part of the reason is that since the action of the structure group on a fibre maps the horizontal subspaces to one another in a unique way, transport in the vertical direction has no holonomy and correspondingly components of the curvature in the vertical direction are zero. One should note that the curvature components are invariantly-defined scalars on $F(\mathcal{M})$ since they are known for a given point in $\mathcal{M}$ and frame.

The exterior derivative of the components of curvature on $F(\mathcal{M})$ obeys

$$\mathrm{d}R_{abcd} = R_{abcd;e}\boldsymbol{\omega}^e + R_{ebcd}\boldsymbol{\Gamma}^e{}_a + R_{aecd}\boldsymbol{\Gamma}^e{}_b + R_{abed}\boldsymbol{\Gamma}^e{}_c + R_{abce}\boldsymbol{\Gamma}^e{}_d\,, \quad (2.87)$$

where $R_{abcd;e}$ is evaluated, in the tetrad given by p, at $\pi(p) \in \mathcal{M}$, and similar equations hold for higher derivatives.

Jet bundles $J^{(n)}(X, U)$ are another type of fibre bundle of interest in exact solutions (see Chapter 10). They are used in describing systems of differential equations. Here the base manifold X is the space of independent variables and the fibre is the set U of values of the dependent variables and their first n partial derivatives with respect to the X. A (partial) differential equation, or a system of such equations, specifies a submanifold in a jet bundle.

3
Some topics in Riemannian geometry

3.1 Introduction

In Chapter 2 we treated differential geometry without a metric. In order to define covariant derivatives we imposed an extra structure on the differential manifold $\mathcal{M}$, the connection. Adding a further structure, the metric g_{ab}, and postulating $g_{ab;c} = 0$, we arrive at Riemannian geometry.

General relativity is based on the concept of *space-time*, which is a four-dimensional differentiable (C^∞, Hausdorff) manifold $\mathcal{M}$ endowed with a *Lorentzian metric* g_{ab} which can be transformed to

$$g_{ab} = \eta_{ab} \equiv \operatorname{diag}(1, 1, 1, -1) \tag{3.1}$$

at any point of $\mathcal{M}$, i.e. space-time is a normal-hyperbolic Riemannian space V_4. In what follows, a knowledge of fundamental facts about Riemannian geometry as given in most textbooks on general relativity is presumed; we give here only some notation and results used in the remainder of this book. For further details the reader is referred to standard texts on Riemannian geometry, e.g. Eisenhart (1949) and Schouten (1954).

3.2 The metric tensor and tetrads

We introduce as a new structure a symmetric tensor of type (0,2), called the *metric tensor* $\boldsymbol{g}$, which endows each vector space T_p with a scalar product (inner product)

$$\boldsymbol{e}_a \cdot \boldsymbol{e}_b = g_{ab}. \tag{3.2}$$

The tensor $\boldsymbol{g}$, sometimes called the line element $\mathrm{d}s^2$, is written

$$\boldsymbol{g} = \mathrm{d}s^2 = g_{ab}\boldsymbol{\omega}^a\boldsymbol{\omega}^b. \tag{3.3}$$

The scalar product of two vectors $\boldsymbol{v}$, $\boldsymbol{w}$ is given by

$$\boldsymbol{v}\cdot\boldsymbol{w} = g_{ab}v^a w^b. \tag{3.4}$$

Two vectors $\boldsymbol{v}$, $\boldsymbol{w}$ are *orthogonal* if their scalar product vanishes. A non-zero vector $\boldsymbol{v}$ is said to be *spacelike*, *timelike* or *null*, respectively, when the product $\boldsymbol{v}\cdot\boldsymbol{v} = g_{ab}v^a v^b$ is positive, negative or zero. In a coordinate basis, we write the line element $\mathrm{d}s^2$ as

$$\mathrm{d}s^2 = g_{ij}\mathrm{d}x^i\mathrm{d}x^j. \tag{3.5}$$

The contravariant components, g^{ab}, form the matrix inverse to g_{ab}. Raising and lowering the indices of the tensor components has to be performed in the usual manner:

$$v_a = g_{ab}v^b, \qquad v^a = g^{ab}v_b. \tag{3.6}$$

In this sense, the vector $v^a\boldsymbol{e}_a$ and the 1-form $v_a\boldsymbol{\omega}^a$ represent the same geometric object.

In space-time, an *orthonormal basis* or *orthonormal tetrad* or *Lorentz frame* $\{\boldsymbol{E}_a\}$ consists of three spacelike vectors $\boldsymbol{E}_\alpha$ and one timelike vector $\boldsymbol{E}_4 \equiv \boldsymbol{t}$, such that

$$\begin{aligned} &\{\boldsymbol{E}_a\} = \{\boldsymbol{E}_\alpha, \boldsymbol{t}\} = \{\boldsymbol{x}, \boldsymbol{y}, \boldsymbol{z}, \boldsymbol{t}\}, \qquad g_{ab} = x_a x_b + y_a y_b + z_a z_b - t_a t_b \\ \Leftrightarrow \qquad &\boldsymbol{E}_\alpha\cdot\boldsymbol{E}_\beta = \delta_{\alpha\beta}, \qquad \boldsymbol{t}\cdot\boldsymbol{t} = -1, \qquad \boldsymbol{E}_\alpha\cdot\boldsymbol{t} = 0. \end{aligned} \tag{3.7}$$

If one has a fluid with four-velocity parallel to $\boldsymbol{t}$, the symbol $\boldsymbol{t}$ is often replaced by $\boldsymbol{u}$.

Complex null tetrads play an important role. A complex null tetrad consists of two real null vectors $\boldsymbol{k}$, $\boldsymbol{l}$ and two complex conjugate null vectors $\boldsymbol{m}$, $\overline{\boldsymbol{m}}$:

$$\begin{aligned} &\{\boldsymbol{e}_a\} = (\boldsymbol{m}, \overline{\boldsymbol{m}}, \boldsymbol{l}, \boldsymbol{k}), \\ &g_{ab} = 2m_{(a}\overline{m}_{b)} - 2k_{(a}l_{b)} = \begin{pmatrix} 0 & 1 & 0 & 0 \\ 1 & 0 & 0 & 0 \\ 0 & 0 & 0 & -1 \\ 0 & 0 & -1 & 0 \end{pmatrix} \end{aligned} \tag{3.8}$$

(g_{ab} are the components of $\boldsymbol{g}$ with respect to the complex null tetrad), i.e. the scalar products of the tetrad vectors vanish apart from

$$k^a l_a = -1, \qquad m^a\overline{m}_a = 1. \tag{3.9}$$

In terms of a coordinate basis, a complex null tetrad $\{\boldsymbol{e}_a\}$ and its dual $\{\boldsymbol{\omega}^a\}$ will take the form

$$\begin{aligned}
&\boldsymbol{e}_1 = m^i\frac{\partial}{\partial x^i},\ \boldsymbol{e}_2 = \overline{m}^i\frac{\partial}{\partial x^i},\ \boldsymbol{e}_3 = l^i\frac{\partial}{\partial x^i},\quad \boldsymbol{e}_4 = k^i\frac{\partial}{\partial x^i};\\
&\boldsymbol{\omega}^1 = \overline{m}_i\,\mathrm{d}x^i,\ \boldsymbol{\omega}^2 = m_i\,\mathrm{d}x^i,\ \boldsymbol{\omega}^3 = -k_i\,\mathrm{d}x^i,\ \boldsymbol{\omega}^4 = -l_i\,\mathrm{d}x^i.
\end{aligned}\tag{3.10}$$

The explicit expressions for the directional derivatives $f_{|a}$ of a function f with respect to the complex null tetrad (3.8) are

$$f_{|1} = f_{,i}m^i,\qquad f_{|2} = f_{,i}\overline{m}^i,\qquad f_{|3} = f_{,i}l^i,\qquad f_{|4} = f_{,i}k^i.\tag{3.11}$$

An orthonormal tetrad (3.7) and a complex null tetrad (3.8) may be related by

$$\begin{aligned}
\sqrt{2}\,\boldsymbol{m} &= \boldsymbol{E}_1 - \mathrm{i}\boldsymbol{E}_2,\ \ \sqrt{2}\,\overline{\boldsymbol{m}} = \boldsymbol{E}_1 + \mathrm{i}\boldsymbol{E}_2,\\
\sqrt{2}\,\boldsymbol{l} &= \boldsymbol{E}_4 - \boldsymbol{E}_3,\quad \sqrt{2}\,\boldsymbol{k} = \boldsymbol{E}_4 + \boldsymbol{E}_3.
\end{aligned}\tag{3.12}$$

In flat space-time, (3.12) implies the relations

$$\zeta = \tfrac{1}{\sqrt{2}}(x+\mathrm{i}y),\ \ \overline{\zeta} = \tfrac{1}{\sqrt{2}}(x-\mathrm{i}y),\ \ u = \tfrac{1}{\sqrt{2}}(t-z),\ \ v = \tfrac{1}{\sqrt{2}}(t+z),\tag{3.13}$$

between the null coordinates ζ, $\overline{\zeta}$, u, v (adapted to the basis vectors $\boldsymbol{m} = \partial_\zeta$, $\overline{\boldsymbol{m}} = \partial_{\overline{\zeta}}$, $\boldsymbol{l} = \partial_u$, $\boldsymbol{k} = \partial_v$) and the Minkowski coordinates x, y, z, t (adapted to the basis vectors $\boldsymbol{E}_1 = \partial_x$, $\boldsymbol{E}_2 = \partial_y$, $\boldsymbol{E}_1 = \partial_z$, $\boldsymbol{E}_4 = \partial_t$). Here we have adopted the convention $\partial_\zeta = \partial/\partial\zeta$ etc. In some approaches the coordinates ζ, $\overline{\zeta}$, u and v are considered as four independent *complex* variables, thus leading to a complexification of Einstein's field equations and the task of regaining real cuts from its complex solutions. Sometimes also the mixed form of a *real null tetrad* or *half null tetrad* $(\boldsymbol{x}, \boldsymbol{y}, \boldsymbol{l}, \boldsymbol{k})$ is used.

Lorentz transformations give rise to the following changes of the basis (3.8):

null rotations (*$\boldsymbol{l}$ fixed*),

$$\boldsymbol{l}' = \boldsymbol{l},\ \boldsymbol{m}' = \boldsymbol{m} + E\boldsymbol{l},\ \ \boldsymbol{k}' = \boldsymbol{k} + E\overline{\boldsymbol{m}} + \overline{E}\boldsymbol{m} + E\overline{E}\boldsymbol{l},\ E\ \text{complex};\tag{3.14}$$

null rotations (*$\boldsymbol{k}$ fixed*),

$$\boldsymbol{k}' = \boldsymbol{k},\ \boldsymbol{m}' = \boldsymbol{m} + B\boldsymbol{k},\ \ \boldsymbol{l}' = \boldsymbol{l} + B\overline{\boldsymbol{m}} + \overline{B}\boldsymbol{m} + B\overline{B}\boldsymbol{l},\ B\ \text{complex};\tag{3.15}$$

spatial rotations in the $\boldsymbol{m}$–$\overline{\boldsymbol{m}}$-plane,

$$\boldsymbol{m}' = \mathrm{e}^{\mathrm{i}\Theta}\boldsymbol{m},\qquad \Theta\ \text{real};\tag{3.16}$$

special Lorentz transformations (boosts) in the $\boldsymbol{k}$–$\boldsymbol{l}$-plane,

$$\boldsymbol{k}' = A\boldsymbol{k}, \qquad \boldsymbol{l}' = A^{-1}\boldsymbol{l}, \qquad A > 0. \tag{3.17}$$

The transformations (3.14)–(3.17) contain six real parameters. The transformations preserving the $\boldsymbol{k}$-direction are

$$\boldsymbol{k}' = A\boldsymbol{k}, \quad \boldsymbol{m}' = \mathrm{e}^{\mathrm{i}\Theta}(\boldsymbol{m}+B\boldsymbol{k}), \quad \boldsymbol{l}' = A^{-1}(\boldsymbol{l}+B\overline{\boldsymbol{m}}+\overline{B}\boldsymbol{m}+B\overline{B}\boldsymbol{k}). \tag{3.18}$$

Symmetric connection coefficients (2.67) are uniquely determined by adding the *metric condition*

$$\nabla\boldsymbol{g} = 0 \quad \Leftrightarrow \quad g_{ab;c} = 0 = g_{ab|c} - 2\Gamma_{(ab)c}, \quad \Gamma_{abc} \equiv g_{ad}\Gamma^{d}{}_{bc}. \tag{3.19}$$

Combining the metric condition (3.19) and the symmetry condition (2.70) one obtains the general formula

$$\Gamma_{abc} = \tfrac{1}{2}(g_{ab|c}+g_{ac|b}-g_{bc|a}+D_{cab}+D_{bac}-D_{abc}), \quad D_{abc} \equiv g_{ad}D^{d}{}_{bc}, \tag{3.20}$$

expressing the connection coefficients in terms of the metric tensor and the commutation coefficients. We mention two cases of special interest:

coordinate basis (*holonomic frame*): $D_{ijk} = 0$: $\quad \Gamma_{i[jk]} = 0,$
$\Gamma^{i}{}_{jk} \equiv \{{}^{\,i}_{jk}\}$ (= Christoffel symbols);

constant metric (*rigid frame*): $g_{ab|c} = 0$: $\quad \Gamma_{(ab)c} = 0.$

In a holonomic frame, the connection coefficients Γ_{abc} are symmetric in the index pair (bc), while in a rigid frame they are antisymmetric in the index pair (ab).

When using tetrad methods for a rigid frame, it is common to calculate simultaneously the components of the basis vectors for the essential coordinates of the metric under study and the connection (and hence commutation) coefficients, which involve only those coordinates. In order to obtain a coordinate form for the metric, one then has to integrate (2.6) successively for the remaining unknown coordinate components of the tetrad basis. The coordinates $\{y^{\alpha}\}$ not appearing in the commutation coefficients can be chosen so that the operators $\partial_{y^{\alpha}}$ are the 'constants of integration' arising in this process. The relevant integrability conditions are the Jacobi identities: for further discussion, in the null tetrad context, see §7.3.

In §2.9 autoparallels were introduced. One can also consider *geodesic* curves, those for which the total length, measured using the metric, is extremal. The tangent vector $\boldsymbol{v}$ of a geodesic satisfies

$$\nabla_{\boldsymbol{v}}\boldsymbol{v} = f\boldsymbol{v}; \qquad \text{in components: } v^{b}v^{a}{}_{;b} = fv^{a}. \tag{3.21}$$

By a suitable scaling of the parameter of the geodesic curve we can make the function f in (3.21) vanish, so that the curve is also autoparallel. Such a parameter is called an *affine parameter* τ and in terms of such a parameter the geodesic equation becomes

$$\frac{Dv^i}{\mathrm{d}\tau} \equiv v^k v^i{}_{;k} = 0. \tag{3.22}$$

3.3 Calculation of curvature from the metric

The components of the curvature tensor are given by (2.79). In a coordinate basis, one may simply substitute the Christoffel symbols into this expression. However, Cartan's method for the calculation of curvature is more compact and efficient in many applications. It immediately yields the tetrad components. The algorithm is divided into two steps.

(i) *Calculation of the connection 1-forms* (2.75), $\boldsymbol{\Gamma}^a{}_b = \Gamma^a{}_{bc}\boldsymbol{\omega}^c$, from the first Cartan equation (2.76) and the metric condition (3.19),

$$\mathrm{d}\boldsymbol{\omega}^a = -\boldsymbol{\Gamma}^a{}_b \wedge \boldsymbol{\omega}^b, \qquad \mathrm{d}g_{ab} = \boldsymbol{\Gamma}_{ab} + \boldsymbol{\Gamma}_{ba}, \tag{3.23}$$

which determine $\boldsymbol{\Gamma}_{ab}$ uniquely. In a rigid frame ($\mathrm{d}g_{ab} = 0$), at most six independent connection 1-forms survive.

(ii) *Calculation of the curvature 2-forms* $\boldsymbol{\Theta}^a{}_b$ (2.84) from the second Cartan equation (2.85),

$$\boldsymbol{\Theta}^a{}_b = \mathrm{d}\boldsymbol{\Gamma}^a{}_b + \boldsymbol{\Gamma}^a{}_c \wedge \boldsymbol{\Gamma}^c{}_b = \tfrac{1}{2}R^a{}_{bcd}\boldsymbol{\omega}^c \wedge \boldsymbol{\omega}^d. \tag{3.24}$$

This calculus gives the *components* $R^a{}_{bcd}$ *with respect to a general basis* $\{\boldsymbol{e}_a\}$.

For the complex null tetrad $\{\boldsymbol{e}_a\} = (\boldsymbol{m}, \overline{\boldsymbol{m}}, \boldsymbol{l}, \boldsymbol{k})$, the second Cartan equation (3.24) takes the form of three complex equations,

$$\mathrm{d}\boldsymbol{\Gamma}_{41} + \boldsymbol{\Gamma}_{41} \wedge (\boldsymbol{\Gamma}_{21} + \boldsymbol{\Gamma}_{43}) = \tfrac{1}{2}R_{41cd}\boldsymbol{\omega}^c \wedge \boldsymbol{\omega}^d, \tag{3.25a}$$

$$\mathrm{d}\boldsymbol{\Gamma}_{32} - \boldsymbol{\Gamma}_{32} \wedge (\boldsymbol{\Gamma}_{21} + \boldsymbol{\Gamma}_{43}) = \tfrac{1}{2}R_{32cd}\boldsymbol{\omega}^c \wedge \boldsymbol{\omega}^d, \tag{3.25b}$$

$$\mathrm{d}(\boldsymbol{\Gamma}_{21} + \boldsymbol{\Gamma}_{43}) + 2\boldsymbol{\Gamma}_{32} \wedge \boldsymbol{\Gamma}_{41} = \tfrac{1}{2}(R_{21cd} + R_{43cd})\boldsymbol{\omega}^c \wedge \boldsymbol{\omega}^d, \tag{3.25c}$$

where the indices refer to the basis vectors $\boldsymbol{e}_1 = \boldsymbol{m}$, $\boldsymbol{e}_2 = \overline{\boldsymbol{m}}$, $\boldsymbol{e}_3 = \boldsymbol{l}$, and $\boldsymbol{e}_4 = \boldsymbol{k}$; $\boldsymbol{\Gamma}_{41} = \overline{\boldsymbol{\Gamma}}_{42} = \boldsymbol{\Gamma}_{ab}k^a m^b$, $\boldsymbol{\Gamma}_{21} = \overline{\boldsymbol{\Gamma}}_{12} = \boldsymbol{\Gamma}_{ab}\overline{m}^a m^b$, etc. (Exchanging the indices 1 and 2 implies complex conjugation.)

The metric can be used to lower the first index on the curvature and one then has symmetries additional to (2.80), the full set being

$$R_{abcd} = -R_{bacd} = -R_{abdc} = R_{cdab}, \qquad R_{a[bcd]} = 0. \tag{3.26}$$

3.4 Bivectors

Bivectors are antisymmetric tensors of second order, or 2-forms,

$$\boldsymbol{X} = X_{ab}\boldsymbol{\omega}^a \wedge \boldsymbol{\omega}^b. \tag{3.27}$$

A *simple bivector*, $X_{ab} = u_{[a}v_{b]}$, represents a 2-surface element spanned by the two tangent vectors $\boldsymbol{u} = u^a\boldsymbol{e}_a$ and $\boldsymbol{v} = v^a\boldsymbol{e}_a$. This surface element is spacelike, timelike or null according to whether $X_{ab}X^{ab}$ is positive, negative or zero, respectively.

Taking a particular orientation of (a neighbourhood in) $\mathcal{M}$, we define the Levi-Civita 4-form $\boldsymbol{\varepsilon}$ to be $-4!\sqrt{-g}\,\boldsymbol{\omega}^1 \wedge \boldsymbol{\omega}^2 \wedge \boldsymbol{\omega}^3 \wedge \boldsymbol{\omega}^4$, where g is the determinant of the matrix g_{ab} of metric tensor components with respect to a positively-oriented basis $\{\boldsymbol{e}_a\}$. Its components are written ε_{abcd}, and, if the positively-oriented basis is an orthonormal tetrad $\{\boldsymbol{E}_a\}$ as in (3.7), are defined by

$$\varepsilon_{1234} = -1. \tag{3.28}$$

This amounts to a choice of orientation of the four-dimensional manifold in which the basis $\{\boldsymbol{E}_a\}$ represents a Lorentz frame with $\boldsymbol{E}_4$ pointing toward the future and with a right-handed spatial triad as $\{\boldsymbol{E}_\alpha\}$. If that basis is related to the complex null tetrad (3.8) by the formula (3.12), then (3.28) can be written as

$$\varepsilon_{abcd}m^a\overline{m}^b l^c k^d = \mathrm{i}. \tag{3.29}$$

The corresponding three-dimensional tensor obtained by contraction with a timelike unit vector $\boldsymbol{u} = \boldsymbol{E}_4$, giving components $\varepsilon_{abcd}u^d$, will be denoted $\varepsilon_{\alpha\beta\gamma}$, where as usual α, β, $\gamma = 1, 2, 3$.

With the aid of the Levi-Civita 4-form we define the dual bivector $\widetilde{\boldsymbol{X}}$, in index notation, by

$$\widetilde{X}_{ab} \equiv \tfrac{1}{2}\varepsilon_{abcd}X^{cd}. \tag{3.30}$$

To avoid confusion we emphasize that the concepts of dual basis and dual bivector have entirely distinct meanings. Repeated application of the duality operation (3.30) gives

$$(\widetilde{X}_{ab})^{\widetilde{}} = -X_{ab}. \tag{3.31}$$

A bivector is called *null* (or *singular*) if

$$X_{ab}X^{ab} = 0 = X^{ab}\widetilde{X}_{ab} \tag{3.32}$$

holds. Two bivectors $\boldsymbol{X}$ and $\boldsymbol{Y}$ satisfy the identities

$$X_{ac}Y_b{}^c - \widetilde{X}_{bc}\widetilde{Y}_a{}^c = \tfrac{1}{2}g_{ab}X_{cd}Y^{cd}, \quad \widetilde{X}_{ab}Y^{ab} = X_{ab}\widetilde{Y}^{ab}, \tag{3.33}$$

which can be verified from the well-known formula

$$\varepsilon_{abcd}\varepsilon^{fghd} = -6\delta^f_{[a}\delta^g_b\delta^h_{c]}. \tag{3.34}$$

The complex bivector defined by

$$X^*_{ab} \equiv X_{ab} + \mathrm{i}\widetilde{X}_{ab} \tag{3.35}$$

is self-dual, i.e. it fulfils the condition

$$(X^*_{ab})^{\sim} = -\mathrm{i}X^*_{ab}. \tag{3.36}$$

A self-dual bivector is completely determined by a timelike unit vector $\boldsymbol{u}$ and the projection

$$X_a \equiv X^*_{ab}u^b, \qquad X_a u^a = 0, \qquad u_c u^c = -1, \tag{3.37}$$

according to the equation

$$X^*_{ab} = 2u_{[a}X_{b]} + \mathrm{i}\varepsilon_{abcd}u^c X^d = 2(u_{[a}X_{b]})^*. \tag{3.38}$$

As consequences of this important relation we get

$$X^*_{ab}X^{*ab} = -4X_a X^a, \qquad X^*_{\alpha\beta} = \mathrm{i}\varepsilon_{\alpha\beta\gamma}X^\gamma. \tag{3.39}$$

A general *self-dual* bivector can be expanded in terms of the basis $\boldsymbol{Z}^\mu = (\boldsymbol{U}, \boldsymbol{V}, \boldsymbol{W})$ constructed from the complex null tetrad (3.8) by

$$\begin{aligned}
\boldsymbol{Z}^1 &\equiv \boldsymbol{U} = 2\overline{\boldsymbol{m}} \wedge \boldsymbol{l} : && U_{ab} = -l_a\overline{m}_b + l_b\overline{m}_a,\\
\boldsymbol{Z}^2 &\equiv \boldsymbol{V} = 2\boldsymbol{k} \wedge \boldsymbol{m} : && V_{ab} = k_a m_b - k_b m_a,\\
\boldsymbol{Z}^3 &\equiv \boldsymbol{W} = 2(\boldsymbol{m} \wedge \overline{\boldsymbol{m}} - \boldsymbol{k} \wedge \boldsymbol{l}) : && W_{ab} = m_a\overline{m}_b - m_b\overline{m}_a - k_a l_b + k_b l_a.
\end{aligned} \tag{3.40}$$

All contractions vanish except

$$U_{ab}V^{ab} = 2, \qquad W_{ab}W^{ab} = -4. \tag{3.41}$$

With the aid of (3.29) we can verify that the bivectors (3.40) are self-dual: $\widetilde{Z}^\alpha_{ab} = -\mathrm{i}Z^\alpha_{ab}$. The complex conjugate bivectors $\overline{\boldsymbol{U}}$, $\overline{\boldsymbol{V}}$, $\overline{\boldsymbol{W}}$ form a basis $\{\overline{\boldsymbol{Z}}^\alpha\}$ of the space of anti-self-dual bivectors, i.e. those obeying $\widetilde{\overline{Z}}{}^\alpha_{ab} = \mathrm{i}\overline{Z}^\alpha_{ab}$.

The null rotations (3.14), (3.15) induce the following transformations of the bivectors (3.40):

$\boldsymbol{l}$ fixed:

$$U'_{ab} = U_{ab},\ V'_{ab} = V_{ab} - EW_{ab} + E^2U_{ab},\ W'_{ab} = W_{ab} - 2EU_{ab}, \tag{3.42a}$$

k fixed:

$$V'_{ab} = V_{ab},\ U'_{ab} = U_{ab} - \overline{B}W_{ab} + \overline{B}^2 V_{ab},\ W'_{ab} = W_{ab} - 2\overline{B}V_{ab}. \quad (3.42b)$$

A general bivector can be expanded in terms of the basis $\{\boldsymbol{Z}^\alpha, \overline{\boldsymbol{Z}}^\alpha\}$:

$$X_{ab} = c_\alpha Z^\alpha_{ab} + d_\alpha \overline{Z}^\alpha_{ab}. \quad (3.43)$$

Finally, we mention the relation

$$\overline{Z}^\alpha{}_{a[b} Z^\beta{}_{c]}{}^a = 0. \quad (3.44)$$

Ref.: For bivectors and their application, see also Debever (1966), Zund and Brown (1971), and Israel (1970). For the connection between bivectors and the complex 3-space used in Chapter 4 see Cahen *et al.* (1967).

3.5 Decomposition of the curvature tensor

The curvature tensor, with components (2.79) with respect to a basis $\{\boldsymbol{e}_a\}$, can be uniquely decomposed into parts which are irreducible representations of the full Lorentz group,

$$R_{abcd} = C_{abcd} + E_{abcd} + G_{abcd}, \quad (3.45)$$

where the following abbreviations have been used:

$$E_{abcd} \equiv \tfrac{1}{2}(g_{ac}S_{bd} + g_{bd}S_{ac} - g_{ad}S_{bc} - g_{bc}S_{ad}), \quad (3.46)$$

$$G_{abcd} \equiv \tfrac{1}{12}R(g_{ac}g_{bd} - g_{ad}g_{bc}) \equiv \tfrac{1}{12}Rg_{abcd}, \quad (3.47)$$

$$S_{ab} \equiv R_{ab} - \tfrac{1}{4}Rg_{ab}, \qquad R \equiv R^a{}_a. \quad (3.48)$$

R and S_{ab} respectively denote the trace and the traceless part of the Ricci tensor R_{ab} defined by (2.83).

The decomposition (3.45) defines *Weyl's conformal tensor* C_{abcd} (see §3.7 for the relation of this tensor to conformal transformations). It and the other parts in the decomposition (3.45) have the same symmetries (3.26) as the Riemann tensor. Moreover, we have the relations

$$C^a{}_{bad} = 0, \qquad E^a{}_{bad} = S_{bd}, \qquad G^a{}_{bad} = \tfrac{1}{4}g_{bd}R. \quad (3.49)$$

The Weyl tensor is completely traceless, i.e. the contraction with respect to each pair of indices vanishes, and it has ten independent components. A space-time with zero Weyl tensor is said to be *conformally flat*.

A decomposition slightly different from (3.45)–(3.48) is

$$R^{ab}{}_{cd} = C^{ab}{}_{cd} - \tfrac{1}{3}R\delta^a_{[c}\delta^b_{d]} + 2\delta^{[a}_{[c}R^{b]}{}_{d]}. \tag{3.50}$$

Because the tensors C_{abcd}, E_{abcd}, G_{abcd} have two pairs of bivector indices we can introduce the notions of the left dual and the right dual, e.g.

$$ {}^{\sim}C_{abcd} \equiv \tfrac{1}{2}\varepsilon_{abef}C^{ef}{}_{cd}, \qquad C^{\sim}_{abcd} \equiv \tfrac{1}{2}\varepsilon_{cdef}C_{ab}{}^{ef}. \tag{3.51}$$

It turns out that these dual tensors obey the relations

$$ {}^{\sim}C_{abcd} \equiv C^{\sim}_{abcd}, \qquad {}^{\sim}E_{abcd} \equiv -E^{\sim}_{abcd}, \qquad {}^{\sim}G_{abcd} \equiv G^{\sim}_{abcd}. \tag{3.52}$$

For algebraic classification (see Chapters 4, 5 and 9) it is convenient to introduce the complex tensors

$$C^*_{abcd} \equiv C_{abcd} + \mathrm{i}C^{\sim}_{abcd}, \qquad {}^{\sim}C^*_{abcd} = -\mathrm{i}C^*_{abcd}, \tag{3.53}$$

$$E^*_{abcd} \equiv E_{abcd} + \mathrm{i}E^{\sim}_{abcd}, \qquad {}^{\sim}E^*_{abcd} = +\mathrm{i}E^*_{abcd}, \tag{3.54}$$

$$G^*_{abcd} \equiv G_{abcd} + \mathrm{i}G^{\sim}_{abcd}, \qquad {}^{\sim}G^*_{abcd} = -\mathrm{i}G^*_{abcd}. \tag{3.55}$$

The 'unit tensor' defined by

$$I_{abcd} \equiv \tfrac{1}{4}(g_{abcd} + \mathrm{i}\varepsilon_{abcd}) = \tfrac{1}{2}(V_{ab}U_{cd} + U_{ab}V_{cd}) - \tfrac{1}{4}W_{ab}W_{cd} \tag{3.56}$$

(so that $I_{abcd}Z^{\alpha cd} = Z^{\alpha}_{ab}$, cp. (3.40), and $G^*_{abcd} = (R/3)I_{abcd}$) is self-dual with respect to both pairs of bivector indices. Therefore I_{abcd} admits the double expansion, given in (3.56), in terms of the basis $\{\boldsymbol{Z}^{\alpha}\}$. The decompositions

$$C^*_{abcd} = c_{\alpha\beta}Z^{\alpha}_{ab}Z^{\beta}_{cd}, \qquad E^*_{abcd} = e_{\alpha\beta}\overline{Z}^{\alpha}_{ab}Z^{\beta}_{cd}, \tag{3.57}$$

are valid for the tensors defined by (3.53), (3.54).

Because of the tracelessness of C^*_{abcd} we have the explicit expansion

$$\begin{aligned}\tfrac{1}{2}C^*_{abcd} = {} & \Psi_0 U_{ab}U_{cd} + \Psi_1(U_{ab}W_{cd} + W_{ab}U_{cd}) + \Psi_2(V_{ab}U_{cd} + U_{ab}V_{cd} \\ & + W_{ab}W_{cd}) + \Psi_3(V_{ab}W_{cd} + W_{ab}V_{cd}) + \Psi_4 V_{ab}V_{cd},\end{aligned} \tag{3.58}$$

the five complex coefficients $\Psi_0, \ldots, \Psi_4$, being defined by

$$\begin{aligned} &\Psi_0 \equiv C_{abcd}k^a m^b k^c m^d, && \Psi_3 \equiv C_{abcd}k^a l^b \overline{m}^c l^d, \\ &\Psi_1 \equiv C_{abcd}k^a l^b k^c m^d, && \Psi_4 \equiv C_{abcd}\overline{m}^a l^b \overline{m}^c l^d, \\ &\Psi_2 \equiv C_{abcd}k^a m^b \overline{m}^c l^d = \tfrac{1}{2}C_{abcd}k^a l^b(k^c l^d - m^c \overline{m}^d). && \end{aligned} \tag{3.59}$$

In these definitions, $C^*_{abcd}/2$ may be substituted for C_{abcd}. The various terms in (3.58) admit the following physical interpretation (Szekeres 1965): the Ψ_4-term represents a transverse wave in the $\boldsymbol{k}$-direction, the Ψ_3-term a longitudinal wave component, and the Ψ_2-term a 'Coulomb' component; the Ψ_0- and Ψ_1-terms represent transverse and longitudinal wave components in the $\boldsymbol{l}$-direction.

With the aid of (3.42) we find the transformation laws of $\Psi_0, \ldots, \Psi_4$ under the null rotations (3.14), (3.15):

$$\begin{aligned} \boldsymbol{l}\text{ fixed: } \Psi'_4 &= \Psi_4, \quad \Psi'_3 = \Psi_3 + E\Psi_4, \\ \Psi'_2 &= \Psi_2 + 2E\Psi_3 + E^2\Psi_4, \\ \Psi'_1 &= \Psi_1 + 3E\Psi_2 + 3E^2\Psi_3 + E^3\Psi_4, \\ \Psi'_0 &= \Psi_0 + 4E\Psi_1 + 6E^2\Psi_2 + 4E^3\Psi_3 + E^4\Psi_4. \end{aligned} \tag{3.60}$$

$$\begin{aligned} \boldsymbol{k}\text{ fixed: } \Psi'_0 &= \Psi_0, \quad \Psi'_1 = \Psi_1 + \overline{B}\Psi_0, \\ \Psi'_2 &= \Psi_2 + 2\overline{B}\Psi_1 + \overline{B}^2\Psi_0, \\ \Psi'_3 &= \Psi_3 + 3\overline{B}\Psi_2 + 3\overline{B}^2\Psi_1 + \overline{B}^3\Psi_0, \\ \Psi'_4 &= \Psi_4 + 4\overline{B}\Psi_3 + 6\overline{B}^2\Psi_2 + 4\overline{B}^3\Psi_1 + \overline{B}^4\Psi_0. \end{aligned} \tag{3.61}$$

Generalizing (3.37), (3.38), we can express C^*_{abcd} in terms of the complex tensor

$$-Q_{ab} \equiv C^*_{abcd}u^b u^d \equiv E_{ac} + \mathrm{i}B_{ac}, \qquad u_c u^c = -1, \tag{3.62}$$

according to the formula

$$\begin{aligned} -\tfrac{1}{2}C^*_{abcd} &= 4u_{[a}Q_{b][d}u_{c]} + g_{a[c}Q_{d]b} - g_{b[c}Q_{d]a} \\ &\quad + \mathrm{i}\varepsilon_{abef}u^e u_{[c}Q_{d]}{}^f + \mathrm{i}\varepsilon_{cdef}u^e u_{[a}Q_{b]}{}^f. \end{aligned} \tag{3.63}$$

E_{ac} and B_{ac} respectively denote the 'electric' and 'magnetic' parts of the Weyl tensor for the given four-velocity u^a (Matte 1953). The components Q_{ab} satisfy the relations

$$Q^a{}_a = 0, \qquad Q_{ab} = Q_{ba}, \qquad Q_{ab}u^b = 0, \tag{3.64}$$

and can be considered as a symmetric complex (3×3) matrix $\boldsymbol{Q}$ with zero trace. Using (3.40), (3.58) and (3.62), and expressing the 3×3 matrix with respect to the orthonormal basis given by (3.12),

$$\boldsymbol{Q} = \begin{pmatrix} \Psi_2 - \frac{1}{2}(\Psi_0 + \Psi_4) & \frac{1}{2}\mathrm{i}(\Psi_4 - \Psi_0) & \Psi_1 - \Psi_3 \\ \frac{1}{2}\mathrm{i}(\Psi_4 - \Psi_0) & \Psi_2 + \frac{1}{2}(\Psi_0 + \Psi_4) & \mathrm{i}(\Psi_1 + \Psi_3) \\ \Psi_1 - \Psi_3 & \mathrm{i}(\Psi_1 + \Psi_3) & -2\Psi_2 \end{pmatrix}. \tag{3.65}$$

The matrix $\boldsymbol{Q}$ determines ten real numbers corresponding to the ten independent components of the Weyl tensor.

3.6 Spinors

Spinor formalism provides a very compact and elegant framework for numerous calculations in general relativity, e.g. algebraic classification of the Weyl tensor (Chapter 4) and the Newman-Penrose technique (Chapter 7).

It can be shown that the (connected) group $SL(2,\mathbb{C})$ of linear transformations in two complex dimensions, with determinant of modulus 1, has a two-to-one homomorphism onto the group $L_+^\uparrow$. The space on which $SL(2,\mathbb{C})$ acts is called *spinor space*, and its elements are (one-index) *spinors* with components φ^A. Spinor indices like A obviously range over 1 and 2, or, commonly, 0 and 1. Every proper Lorentz transformation defines an element of $SL(2,\mathbb{C})$ up to overall sign. Since the defining property of $L_+^\uparrow$ (within all linear transformations in four dimensions) is that it is the (connected) group preserving the Minkowski metric, and since $SL(2,\mathbb{C})$ is defined (within all linear transformations of two complex dimensions) as the (connected) group that preserves determinants, we expect that the determinant-forming 2-form in spin space, with components

$$\varepsilon_{AB} = \begin{pmatrix} 0 & 1 \\ -1 & 0 \end{pmatrix} = \varepsilon^{AB}, \tag{3.66}$$

will play the role of the metric. Spinor indices are raised and lowered according to the rule

$$\varphi^A = \varepsilon^{AB}\varphi_B \quad \Leftrightarrow \quad \varphi_A = \varphi^B\varepsilon_{BA}. \tag{3.67}$$

Note that $\varphi_A\varepsilon^{AB} \neq \varepsilon^{BA}\varphi_A$. The scalar product of two spinors (with components φ^A and ψ^A) is then defined by

$$\varepsilon_{AB}\varphi^A\psi^B = \varphi_A\psi^A = -\varphi^A\psi_A. \tag{3.68}$$

If φ^B transforms under $S^A{}_B \in SL(2,\mathbb{C})$, the complex conjugate spinor $\overline{\varphi}^{\dot{B}}$ must, for consistency, transform under the complex conjugate $\overline{S}^{\dot{A}}{}_{\dot{B}}$, and similarly φ_A transforms under the inverse of $S^A{}_B$. Dotted indices are used to indicate that the complex conjugate transformations are to be applied. The order of dotted and undotted indices is clearly irrelevant. One can obviously build multi-index spinors, in just the same way that tensors are developed from vectors.

It is now natural to seek a correspondence between the vectors $\boldsymbol{v}$ of Minkowski space and spinors. To do so we shall need not one-index spinors, but two-index spinors $v^{A\dot{B}}$, because the sign ambiguity arising from the map of $SL(2,\mathbb{C})$ to $L_+^\uparrow$ must be removed and it must be possible to relate the length of a vector in the Minkowski metric (quadratic in the components v^a) to a determinant (also quadratic in the entries of a (2×2) matrix). Such a map will be given by any set of *spin tensors* $\sigma_{aA\dot{B}}$ satisfying

$$\sigma_{aA\dot{B}}\sigma^{aC\dot{D}} = -\delta_A^C\delta_{\dot{B}}^{\dot{D}} \quad\Leftrightarrow\quad \sigma_{aA\dot{B}}\sigma^{bA\dot{B}} = -\delta_a^b. \tag{3.69}$$

Then v^a corresponds to $v^{A\dot{B}}$ by

$$v^a = -\sigma^a{}_{A\dot{B}}v^{A\dot{B}} \quad\Leftrightarrow\quad v^{A\dot{B}} = \sigma_a{}^{A\dot{B}}v^a. \tag{3.70}$$

(Note that the formulae here exhibit some sign changes as compared with Penrose (1960), due to a change of convention about the signature of the space-time metric.) The spin tensors will be Hermitian,

$$\sigma_a{}^{A\dot{B}} = \overline{\sigma_a{}^{B\dot{A}}} \equiv \overline{\sigma}_a{}^{A\dot{B}}. \tag{3.71}$$

Given a null vector, its spinor counterpart must be an outer product $\zeta_A\overline{\eta}_{\dot{B}}$ since the matrix $v_{A\dot{B}}$ must have determinant zero and thus be of rank 1. Given a null tetrad $(\boldsymbol{m},\overline{\boldsymbol{m}},\boldsymbol{l},\boldsymbol{k})$ and a pair of basis spinors o^A, ι^A such that $o_A\iota^A = 1$ (a *dyad*), one can choose the map $\sigma_a{}^{A\dot{B}}$ so that in the orthonormal tetrad associated with $(\boldsymbol{m},\overline{\boldsymbol{m}},\boldsymbol{l},\boldsymbol{k})$ by (3.12), and in the spin basis consisting of o^A, ι^A themselves (so $o^A = (1,0)$ and $\iota^A = (0,1)$), one has

$$\begin{aligned} \sigma_1{}^{A\dot{B}} &= \frac{1}{\sqrt{2}}\begin{pmatrix} 0 & 1\\ 1 & 0\end{pmatrix}, \quad \sigma_2{}^{A\dot{B}} = \frac{1}{\sqrt{2}}\begin{pmatrix} 0 & \mathrm{i}\\ -\mathrm{i} & 0\end{pmatrix},\\ \sigma_3{}^{A\dot{B}} &= \frac{1}{\sqrt{2}}\begin{pmatrix} 1 & 0\\ 0 & -1\end{pmatrix}, \quad \sigma_4{}^{A\dot{B}} = \frac{1}{\sqrt{2}}\begin{pmatrix} 1 & 0\\ 0 & 1\end{pmatrix}. \end{aligned} \tag{3.72}$$

Then

$$m^a \leftrightarrow o^A\overline{\iota}^{\dot{B}}, \quad \overline{m}^a \leftrightarrow \iota^A\overline{o}^{\dot{B}}, \quad l^a \leftrightarrow \iota^A\overline{\iota}^{\dot{B}}, \quad k^a \leftrightarrow o^A\overline{o}^{\dot{B}}. \tag{3.73}$$

One can check from (3.69), (3.70) that this is consistent with the normalization of $(\boldsymbol{m},\overline{\boldsymbol{m}},\boldsymbol{l},\boldsymbol{k})$. Conversely (3.73) could be used to define a dyad so that (3.72) arises.

The null rotations (3.14), (3.15) correspond to the transformations

$$o'^A = o^A + E\iota^A, \qquad \iota'^A = \iota^A + Bo^A. \tag{3.74}$$

In Table 3.1 we give some examples of spinor equivalents of tensors, constructed according to the relation (3.70). The spinor form of the decomposition (3.45) of the curvature tensor is obtained from the spinor

Table 3.1. Examples of spinor equivalents, defined as in (3.70)

Indices a, b, c, d correspond to index pairs $A\dot{W}$, $B\dot{X}$, $C\dot{Y}$, $D\dot{Z}$ respectively.

Tensor	Spinor equivalent
Metric g_{ab}:	$\varepsilon_{AB}\varepsilon_{\dot{W}\dot{X}}$; ε_{AB} from (3.66)
Levi-Civita tensor ε_{abcd}:	$\varepsilon_{A\dot{W}B\dot{X}}{}^{C\dot{Y}D\dot{Z}} \equiv \mathrm{i}(\delta_A{}^C\delta_B{}^D\delta_{\dot{W}}{}^{\dot{Z}}\delta_{\dot{X}}{}^{\dot{Y}} - \delta_A{}^D\delta_B{}^C\delta_{\dot{W}}{}^{\dot{Y}}\delta_{\dot{X}}{}^{\dot{Z}})$
Null vector n_a:	$\zeta_A\overline{\eta}_{\dot{W}}$
Real null vector, $\overline{n}_a = n_a$	$\pm\zeta_A\overline{\zeta}_{\dot{W}}$
Bivector X_{ab}:	$\varepsilon_{AB}\overline{\zeta}_{\dot{W}\dot{X}} + \varepsilon_{\dot{W}\dot{X}}\eta_{AB}$, $\eta_{[AB]} = 0 = \overline{\zeta}_{[\dot{A}\dot{B}]}$
Real bivector F_{ab}:	$\varepsilon_{AB}\overline{\Phi}_{\dot{W}\dot{X}} + \varepsilon_{\dot{W}\dot{X}}\Phi_{AB}$, $\Phi_{[AB]} = 0$
Dual bivector $\widetilde{F}_{ab}$:	$\mathrm{i}(\varepsilon_{AB}\overline{\Phi}_{\dot{W}\dot{X}} - \varepsilon_{\dot{W}\dot{X}}\Phi_{AB})$
Complex self-dual bivector F^*_{ab}:	$2\Phi_{AB}\varepsilon_{\dot{W}\dot{X}}$
V_{ab}, U_{ab}, W_{ab}:	$o_Ao_B\varepsilon_{\dot{W}\dot{X}}$, $\iota_A\iota_B\varepsilon_{\dot{W}\dot{X}}$, $-2o_{(A}\iota_{B)}\varepsilon_{\dot{W}\dot{X}}$
Curvature tensor R_{abcd}:	$\chi_{ABCD}\varepsilon_{\dot{W}\dot{X}}\varepsilon_{\dot{Y}\dot{Z}} + \varepsilon_{AB}\varepsilon_{CD}\overline{\chi}_{\dot{W}\dot{X}\dot{Y}\dot{Z}} + \Phi_{AB\dot{Y}\dot{Z}}\varepsilon_{CD}\varepsilon_{\dot{W}\dot{X}} + \varepsilon_{AB}\varepsilon_{\dot{Y}\dot{Z}}\overline{\Phi}_{\dot{W}\dot{X}CD}$, $\Phi_{AB\dot{C}\dot{D}} = \Phi_{(AB)(\dot{C}\dot{D})} = \overline{\Phi_{CD\dot{A}\dot{B}}} = \overline{\Phi}_{AB\dot{C}\dot{D}}$
Weyl tensor C_{abcd}:	$\Psi_{ABCD}\varepsilon_{\dot{W}\dot{X}}\varepsilon_{\dot{Y}\dot{Z}} + \varepsilon_{AB}\varepsilon_{CD}\overline{\Psi}_{\dot{W}\dot{X}\dot{Y}\dot{Z}}$, $\Psi_{ABCD} = \chi_{(ABCD)}$
C^*_{abcd}:	$2\Psi_{ABCD}\varepsilon_{\dot{W}\dot{X}}\varepsilon_{\dot{Y}\dot{Z}}$
Traceless Ricci tensor S_{ab}:	$2\Phi_{AB\dot{W}\dot{X}}$

equivalent of R_{abcd} by using the relation

$$\chi_{ABCD} = \Psi_{ABCD} + \tfrac{1}{12}R(\varepsilon_{AC}\varepsilon_{BD} + \varepsilon_{AD}\varepsilon_{BC}). \qquad (3.75)$$

The reason why spinors are frequently used in general relativity is that the spinor formalism simplifies some relations involving null vectors and bivectors. For example, Table 3.1 shows that the Weyl tensor has a completely symmetric spinor equivalent Ψ_{ABCD} while the corresponding tensorial symmetry relations (3.26), (3.49) are much more complicated. In addition, the definitions (3.59) of the complex tetrad components $\Psi_0, \ldots, \Psi_4$

are very symmetric in the spinor calculus:

$$\begin{aligned}
&\Psi_0 = \Psi_{ABCD}o^A o^B o^C o^D, \ \Psi_1 = \Psi_{ABCD}o^A o^B o^C \iota^D, \\
&\Psi_2 = \Psi_{ABCD}o^A o^B \iota^C \iota^D, \ \Psi_3 = \Psi_{ABCD}o^A \iota^B \iota^C \iota^D, \\
&\Psi_4 = \Psi_{ABCD}\iota^A \iota^B \iota^C \iota^D.
\end{aligned} \tag{3.76}$$

Up to now we have been concerned only with algebraic relations. The covariant derivatives also have their spinor equivalents:

$$\nabla_{A\dot{B}} = \sigma^a{}_{A\dot{B}}\nabla_a \quad \Leftrightarrow \quad \nabla_a = -\sigma_a{}^{A\dot{B}}\nabla_{A\dot{B}}. \tag{3.77}$$

The Bianchi identities (2.81), written as

$$R_{ab[cd;e]} = 0 \quad \Leftrightarrow \quad R_{\widetilde{ab}cd}{}^{;d} = 0 \tag{3.78}$$

(Lanczos 1962), when transcribed into spinor language become

$$\nabla^D{}_{\dot{E}}\chi_{ABCD} = \nabla_C{}^{\dot{F}}\Phi_{AB\dot{E}\dot{F}}. \tag{3.79}$$

For vacuum fields ($R_{ab} = 0$), in virtue of (3.75), these equations take the simpler form

$$\nabla^D{}_{\dot{E}}\Psi_{ABCD} = 0. \tag{3.80}$$

The Weyl tensor C_{abcd} can be written in terms of the derivatives of a third order tensor L_{abc}, the Lanczos potential (Lanczos 1962). The spinor equivalent of this relation reads

$$\Psi_{ABCD} = \nabla_D{}^{\dot{S}}L_{ABC\dot{S}}. \tag{3.81}$$

In vacuum, L_{abc} satisfies a wave equation, see Illge (1988), Dolan and Kim (1994) and Edgar and Höglund (1997) for references.

The directional derivatives along the null tetrad $(\boldsymbol{m}, \overline{\boldsymbol{m}}, \boldsymbol{l}, \boldsymbol{k})$ are denoted by the symbols

$$\begin{aligned}
D &\equiv k^a\nabla_a = -o^A\bar{o}^{\dot{B}}\nabla_{A\dot{B}}, & \Delta &\equiv l^a\nabla_a = -\iota^A\bar{\iota}^{\dot{B}}\nabla_{A\dot{B}}, \\
\delta &\equiv m^a\nabla_a = -o^A\bar{\iota}^{\dot{B}}\nabla_{A\dot{B}}, & \bar{\delta} &\equiv \overline{m}^a\nabla_a = -\iota^A\bar{o}^{\dot{B}}\nabla_{A\dot{B}}.
\end{aligned} \tag{3.82}$$

Ref.: For spinors see Penrose and Rindler (1984, 1986) and Bichteler (1964).

3.7 Conformal transformations

A special type of map of metric spaces is given by dilatation (or contraction) of all lengths by a common factor which varies from point to point,

$$\widehat{g}_{ab} = \mathrm{e}^{2U} g_{ab}, \qquad \widehat{g}^{ab} = \mathrm{e}^{-2U} g^{ab}, \qquad U = U(x^n). \tag{3.83}$$

The connection coefficients and the covariant derivative of a 1-form $\boldsymbol{\sigma}$ are transformed to

$$\begin{aligned} \widehat{\Gamma}^c{}_{ab} &= \Gamma^c{}_{ab} + 2\delta^c_{(a} U_{,b)} - g_{ab} U^{,c}, \quad U^{,c} \equiv g^{cd} U_{,d}, \\ \widehat{\nabla}_a \widehat{\sigma}_b &= \nabla_a \sigma_b - U_{,b} \sigma_a - U_{,a} \sigma_b + g_{ab} U^{,c} \sigma_c, \quad \widehat{\sigma}_a = \sigma_a. \end{aligned} \tag{3.84}$$

The curvature tensors of the two spaces with metrics $\widehat{g}_{ab}$ and g_{ab} are connected by the relation

$$\begin{aligned} \mathrm{e}^{2U} \widehat{R}^{da}{}_{bc} &= R^{da}{}_{bc} + 4Y^{[a}_{[b}\delta^{d]}_{c]}, \\ Y^a{}_b &\equiv U^{,a}{}_{;b} - U^{,a} U_{,b} + \tfrac{1}{2}\delta^a{}_b U_{,e} U^{,e} , \end{aligned} \tag{3.85}$$

which holds for n-dimensional Riemannian spaces $\widehat{V}_n$ and V_n (the covariant derivative is taken with respect to g_{ab}). From (3.85) one obtains the equation

$$\widehat{R}_{ab} = R_{ab} + (2-n) Y_{ab} - g_{ab} Y_c{}^c \tag{3.86}$$

for the Ricci tensors in $\widehat{V}_n$ and V_n. In three dimensions, this equation takes the form

$$\widehat{R}_{\alpha\beta} = R_{\alpha\beta} - U_{,\alpha;\beta} + U_{,\alpha} U_{,\beta} - g_{\alpha\beta}(U_{,\gamma}{}^{;\gamma} + U_{,\gamma} U^{,\gamma}). \tag{3.87}$$

The application of (3.86) to a (flat) space V_2 yields

$$\mathrm{d}\widehat{s}^2 = \mathrm{e}^{2U}(\mathrm{d}x^2 + \mathrm{d}y^2): \quad \widehat{R}_{AB} = K \widehat{g}_{AB}, \quad K = -\mathrm{e}^{-2U} U^{,A}{}_{,A}. \tag{3.88}$$

A space is called conformally flat if it can be related, by a conformal transformation, to flat space. A space V_2 is always conformally flat. A space V_3 is conformally flat if and only if the Cotton tensor

$$C^a{}_{bc} \equiv 2(R^a{}_{[b} - \tfrac{1}{4} R \delta^a{}_{[b})_{;c]} \tag{3.89}$$

vanishes, see Schouten (1954); York (1971) defined a related conformally invariant tensor density. A space V_n, $n > 3$, is conformally flat if and only if the conformal tensor

$$\begin{aligned} C_{abcd} \equiv\ & R_{abcd} + R(g_{ac} g_{bd} - g_{ad} g_{bc})/(n-1)(n-2) \\ & - (g_{ac} R_{bd} - g_{bc} R_{ad} + g_{bd} R_{ac} - g_{ad} R_{bc})/(n-2) \end{aligned} \tag{3.90}$$

(for V_4, see (3.50)) vanishes. The conformal tensor components $C^a{}_{bcd}$ are unchanged by (3.83).

The special case $U = \text{const}$ of (3.83), i.e. $\hat{g}_{ab} = k^2 g_{ab}$ with constant k, corresponds to a general symmetry of Einstein's field equations (1.1), see §10.3: applied to an already known solution, it may lead to a (trivial) new solution, related to the old by a homothety, and often described by the same metric form with rescaled values of parameters such as mass.

Other than such trivial homothetic relations, attempts to generate one solution from another by conformal transformation are restricted for vacuum and Einstein spaces by:

Theorem 3.1 (Brinkmann's theorem). *If two distinct Einstein space-times are properly conformally related, then they are either (a) both vacuum pp-waves (see Chapter 24) or (b) both conformally flat, one being flat and the other a de Sitter space-time* (Brinkmann 1925).

Daftardar-Gejji (1998) has generalized this theorem to the cases where the two Einstein tensors are equal and where they differ by a cosmological constant term. In the former case, both spaces are (not necessarily vacuum) *pp*-waves; in the latter, for perfect fluids with $\mu \neq 0$, both spaces are Robertson–Walker (see Chapter 14) with equations of state $\mu + 3p = 0$ or $\mu = p$. Further results for other energy-momentum tensors are summarized in §10.11.2.

3.8 Discontinuities and junction conditions

In the preceding discussion we have not pointed out differentiability requirements: in practice exact solutions are almost always given in a form which is analytic or at least C^∞. However, physical models often require two or more such regions to be joined across a hypersurface of discontinuity, for example at the boundary between a star and interstellar space. This book does not attempt a systematic discussion of such possible solutions, but in this section we briefly introduce the conditions which must be imposed at such a boundary. We shall assume an elementary knowledge of distribution theory.

In general a jump discontinuity in the metric would, by (3.20), lead to a δ-function in the connection and thence, by (2.79), to products of δ-functions in the curvature, which are not among the usually allowed distributions. So attention is usually restricted to the case where the connection has at worst a jump discontinuity and the curvature at worst a δ-function (see e.g. Taub (1980): for methods that go beyond this restriction see e.g. Grosser *et al.* (2000) and references therein). To meet this requirement the space-time manifold must be at least C^1 and piecewise C^3. If there is a δ-function, it models a thin shell or surface layer of matter or an impulsive gravitational wave.

We consider manifolds V^+ and V^- with respective metrics g^+ and g^- and bounding hypersurfaces Σ^+ and Σ^- which are to be identified as a single hypersurface Σ in space-time. In either V^+ or V^-, there is a map $\Phi : \Sigma \to V$, so one can define the first fundamental form on Σ by the pullback $\Phi^*(\boldsymbol{g})$ (see §2.5) of the metric. The first requirement at a boundary is that the first fundamental forms calculated on the two sides are the same. This enables us to pass from a coordinate-free description to one in terms of coordinates: Mars and Senovilla (1993b), generalizing work of Clarke and Dray (1987), showed that if V^+, V^- and Φ are C^3, g^+ and g^- are C^2 and the first fundamental forms induced on Σ are the same, then there is a C^1 atlas of charts covering the whole space-time and a continuous metric on the whole space-time which coincides with g^+ in V^+ and with g^- in V^-. In the subsequent discussion, use of such an atlas is assumed.

In general Σ may be null in some regions and non-null elsewhere. Older literature treats only cases where Σ has the same character everywhere. The null case is more awkward because the first fundamental form is degenerate and the vector normal to Σ is tangent to Σ. Here we follow Mars and Senovilla (1993b) and treat all cases simultaneously by using a vector $\boldsymbol{n}$ normal to Σ (taken in the direction from V^- to V^+) and a suitably smooth 'rigging' vector field $\boldsymbol{l}$ on Σ with the property $\boldsymbol{n} \cdot \boldsymbol{l} = 1$; for non-null surfaces $\boldsymbol{n}$ and $\boldsymbol{l}$ (or $-\boldsymbol{l}$) can both be taken to be the unit normal. Define the tensor $\mathcal{H} = \Phi^*(\nabla \boldsymbol{l})$, i.e.

$$\mathcal{H}_{ab} = P^c{}_a P^d{}_b l_{d;c}, \tag{3.91}$$

where $P^a{}_b = \delta^a{}_b - l^a n_b$ projects into Σ. The second fundamental form K is given by the same formula (3.91) with $\boldsymbol{n}$ in place of $\boldsymbol{l}$. Then if there is a discontinuity $[\mathcal{H}_{ab}] = (\mathcal{H}_{ab})_{|V_+} - (\mathcal{H}_{ab})_{|V_-}$, the Riemann tensor has a δ-function singularity with coefficient

$$\mathcal{Q}^a{}_{bcd} = 2\{n^a([\mathcal{H}_{b[c}]n_{d]}) - n_b([\mathcal{H}^a{}_{[c}]n_{d]})\}, \tag{3.92}$$

which for the non-null case can easily be written in terms of the second fundamental form $[K_{ab}] = (\boldsymbol{n} \cdot \boldsymbol{n})[\mathcal{H}_{ab}]$. The resulting coefficient for the δ-function part of the Einstein tensor (the energy-momentum of the surface layer) is

$$\tau_{bc} = 2n^a[\mathcal{H}_{a(b}]n_{c)} - (\boldsymbol{n} \cdot \boldsymbol{n})[\mathcal{H}_{bc}] - [\mathcal{H}^a{}_a]n_b n_c - \tfrac{1}{2} g_{bc|\Sigma} H, \tag{3.93}$$

where $H = 2[\mathcal{H}_{ab}]n^a n^b - 2(\boldsymbol{n} \cdot \boldsymbol{n})[\mathcal{H}^a{}_a]$; this can again readily be expressed, in the non-null case, in terms of K_{ab}.

The Einstein tensor has no δ-function part (no surface layer) if and only if $\boldsymbol{n}$ is non-null and $[\mathcal{H}_{ab}] = 0$, or $\boldsymbol{n}$ is null, $n^a[\mathcal{H}_{ab}] = 0$ and $[\mathcal{H}^a{}_a] = 0$.

If $[\mathcal{H}_{ab}] = 0$ the possible (step function) discontinuities in the Riemann tensor satisfy $n^a[G_{ab}] = 0 = n_a[C^a{}_{bcd}]P^c{}_e P^d{}_f$ and can be characterized by specific Riemann tensor components (Mars and Senovilla 1993b).

In the non-null case, the required conditions for a matching without a δ-function part in terms of the equality of first and second fundamental forms were given first by Darmois (1927) (a formulation in terms of the connection in C^1 coordinates had earlier been given by Sen (1924)). They guarantee the existence of coordinates, e.g. Gaussian normal coordinates on both sides of Σ, in which the metric and its first derivative are continuous, which is the form of junction condition given by Lichnerowicz (1955), and similarly the existence of coordinates in which the conditions of O'Brien and Synge (1952) are true. The coordinate forms (Lichnerowicz or O'Brien and Synge) imply the Darmois form. In that sense the formulations are equivalent (Bonnor and Vickers 1981). However, if other coordinates are used the O'Brien and Synge conditions, for example, give additional, and physically unnecessary, restrictions.

For the null case, the corresponding restriction of the above results has been developed in Taub (1980), Clarke and Dray (1987), Barrabés (1989) and Barrabés and Israel (1991).

Junction conditions are hard to use in exact solutions except when the hypersurface Σ shares a symmetry with the space-time. Most of the applications have been to cases with spherical, cylindrical or plane symmetry. The best known example in the non-null case is the Einstein and Straus (1945) or 'Swiss cheese' model, in which the Schwarzschild solution (15.19) is matched to a Friedmann Robertson–Walker solution (14.6). For some examples in the null case see Chapter 25 on colliding plane waves.

Note that if two space-times M_1 and M_2 are each divided into two regions, giving V_1^+, V_1^-, V_2^+ and V_2^-, and if V_1^+ is matched with V_2^-, then the same conditions will match V_2^+ with V_1^-.

The conditions stated above concern the gravitational field, and thus, indirectly, the total energy-momentum (see Chapter 5). However, in non-vacuum space-times, the matter content will have its own field equations leading to additional boundary conditions which also have to be imposed.

4
The Petrov classification

There are two approaches to the classification of the Weyl tensor – the eigenvalue problem for the matrix $\boldsymbol{Q}$ and the principal null directions – which are completely equivalent. This classification enables one to divide the gravitational fields in an invariant way into distinct types: the Petrov (1954) types (we use this term although other authors obtained analogous results, see Géhéniau (1957), Pirani (1957), Debever (1959, 1964), Bel (1959) and Penrose (1960)). The connection between the Penrose and Debever approaches has been discussed by Adler and Sheffield (1973) and Ludwig (1969). For introductions to the subject, see also Synge (1964) and Pirani (1965).

4.1 The eigenvalue problem

We are interested in invariant characterizations of a gravitational field, independent of any special coordinate system. For this purpose we investigate the algebraic structure of the tensors C_{abcd} and E_{abcd} introduced in (3.45)–(3.48). The classification of S_{ab} (which is equivalent to E_{abcd}) will be treated in Chapter 5. Here we consider the classification of the Weyl tensor C_{abcd} (*Petrov classification*).

The starting point is the *eigenvalue equation*

$$\tfrac{1}{2}C_{abcd}X^{cd} = \lambda X^{ab} \tag{4.1}$$

with eigenbivectors X^{ab} and eigenvalues λ. With each solution (X_{ab}, λ) of this eigenvalue equation is associated its complex conjugate solution $(\overline{X}_{ab}, \overline{\lambda})$. Without loss of generality we can rewrite (4.1) in the form

$$\tfrac{1}{4}C^{*}_{abcd}X^{*cd} = \lambda X^{*ab}\ . \tag{4.2}$$

We note that an analogous equation with E_{abcd} in place of C_{abcd} would be inconsistent because of the property (3.54).

Multiplying the eigenvalue equation (4.2) by a timelike unit vector u^a, and taking into account the definitions (3.37), (3.62), and the expressions (3.38), (3.63), we reduce the eigenvalue problem to the simple form

$$Q_{ab}X^b = \lambda X_a, \tag{4.3}$$

which is completely equivalent to the original formulation. In the three-dimensional vector notation suggested by (3.64) we write this as

$$\boldsymbol{Q}\boldsymbol{r} = \lambda \boldsymbol{r}, \tag{4.4}$$

i.e. we now have to determine the eigenvectors $\boldsymbol{r}$ and the eigenvalues λ of the *complex* symmetric and traceless (3×3) matrix $\boldsymbol{Q}$: from the four-dimensional Lorentz frame we have passed to a three-dimensional complex space with Euclidean metric.

The group $SO(3, \mathbb{C})$ of proper orthogonal transformations in this complex 3-space is isomorphic to the group $L_+^\uparrow$ of proper orthochronous Lorentz transformations. The transformation matrices of these two groups,

$$\begin{aligned} SO(3, \mathbb{C}) : X_{\beta'} &= A^{\alpha}{}_{\beta'} X_\alpha, & A^{\alpha}{}_{\gamma'} A_{\beta}{}^{\gamma'} &= \delta^{\alpha}_{\beta}, \\ L_+^\uparrow : X^*_{a'b'} &= \Lambda^{c}{}_{a'} \Lambda^{d}{}_{b'} X^*_{cd}, & \Lambda^{a}{}_{c'} \Lambda_{b}{}^{c'} &= \delta^{a}_{b}, \end{aligned} \tag{4.5}$$

are related by the formula

$$A^{\alpha}{}_{\beta'} = \Lambda^{\alpha}{}_{\beta'} \Lambda^{4}{}_{4'} - \Lambda^{4}{}_{\beta'} \Lambda^{\alpha}{}_{4'} + \mathrm{i}\varepsilon^{\alpha}{}_{\gamma\delta} \Lambda^{\gamma}{}_{\beta'} \Lambda^{\delta}{}_{4'}, \tag{4.6}$$

which follows from (3.39): each Lorentz transformation induces a unique orthogonal transformation in the complex 3-space. The isomorphism is explicitly verified in Synge (1964).

The eigenvalue problem (4.4) leads to the characteristic equation $\det(\boldsymbol{Q} - \lambda \boldsymbol{I}) = 0$, and determines the orders $[m_1, \ldots, m_k]$ of the elementary divisors $(\lambda - \lambda_1)^{m_1}, \ldots, (\lambda - \lambda_k)^{m_k}$, $m_1 + \cdots + m_k = 3$, belonging to the eigenvalues $\lambda_1, \ldots, \lambda_k$.

4.2 The Petrov types

The distinct algebraic structures studied by Petrov (1954) (see also Petrov (1966)) are characterized by the elementary divisors and multiplicities of the eigenvalues discussed above; the results are displayed in Table 4.1. The algebraic type of the matrix $\boldsymbol{Q}$ provides an invariant characterization of the gravitational field at a given point p; these characteristics

Table 4.1. The Petrov types

Round brackets indicate that the corresponding eigenvalues coincide, e.g. [(11) 1] means: simple elementary divisors, and $\lambda_1 = \lambda_2 \neq \lambda_3$.

Petrov types	Orders of the elementary divisors $[m_1, \ldots, m_k]$	Matrix criterion
I	[111]	$(\boldsymbol{Q}-\lambda_1\boldsymbol{I})(\boldsymbol{Q}-\lambda_2\boldsymbol{I})(\boldsymbol{Q}-\lambda_3\boldsymbol{I})=0$
D	[(11) 1]	$(\boldsymbol{Q}+\frac{1}{2}\lambda\boldsymbol{I})(\boldsymbol{Q}-\lambda\boldsymbol{I})=0$
II	[2 1]	$(\boldsymbol{Q}+\frac{1}{2}\lambda\boldsymbol{I})^2(\boldsymbol{Q}-\lambda\boldsymbol{I})=0$
N	[(2 1)]	$\boldsymbol{Q}^2=0$
III	[3]	$\boldsymbol{Q}^3=0$
O		$\boldsymbol{Q}=0$

are independent of the coordinate system and of the choice of the tetrad at p.

Table 4.1 also gives matrix criteria for the distinct Petrov types. At a given point p, the field is of the Petrov type corresponding, in Table 4.1, to the most restrictive of the criteria which the matrix $\boldsymbol{Q}$ satisfies; for instance

$$\text{Type } III \quad \Leftrightarrow \quad \boldsymbol{Q}^3=0, \;\; \boldsymbol{Q}^2\neq 0. \tag{4.7}$$

A real (or purely imaginary) matrix $\boldsymbol{Q}$ has simple elementary divisors (Petrov types I, D or O). No vacuum solution with a purely magnetic Weyl tensor is known (McIntosh *et al.* 1994).

In order to determine the Petrov type of a given metric we can calculate the complex matrix $\boldsymbol{Q}$ with respect to an arbitrary orthonormal basis $\{\boldsymbol{E}_a\}$ and use the invariant criteria listed in Table 4.1 (see also §9.3).

Gravitational fields of Petrov types D, II, N, III, and O are said to be *algebraically special.*

Lorentz rotations, or, equivalently, elements of the group $SO(3,\mathbb{C})$, can be applied to find simple *normal forms* for the various types (Table 4.2). The normal forms of $\boldsymbol{Q}$ and C_{abcd} are uniquely associated.

Apart from reflections, which are not considered here, the basis $\{\boldsymbol{E}_a\}$ (see (3.7)) of the normal form of a *non-degenerate* Petrov type (I, distinct λ_α; II, $\lambda\neq 0$; or III) is uniquely determined; for the non-degenerate types there is no subgroup of $L_+^\uparrow$ preserving the corresponding normal

Table 4.2. Normal forms of the Weyl tensor, and Petrov types

Normal forms of the matrix $\boldsymbol{Q}$	Eigenvalues λ_α of $\boldsymbol{Q}$ and corresponding eigenbivectors $\boldsymbol{r}_\alpha$	Eigenbivectors X^*_{ab} of C^*_{abcd}	Coefficients $\Psi_0, \ldots, \Psi_4$ in the expansions (3.58) and normal forms of C^*_{abcd}	Petrov types
$\boldsymbol{Q} = \begin{pmatrix} \lambda_1 & & \\ & \lambda_2 & \\ & & \lambda_3 \end{pmatrix}$ $\lambda_1 + \lambda_2 + \lambda_3 = 0$	$\lambda_1 : \boldsymbol{r}_1 = (1, 0, 0)$ $\lambda_2 : \boldsymbol{r}_2 = (0, 1, 0)$ $\lambda_3 : \boldsymbol{r}_3 = (0, 0, 1)$	$X^*_{(1)ab} = V_{ab} - U_{ab}$ $X^*_{(2)ab} = \mathrm{i}(V_{ab} + U_{ab})$ $X^*_{(3)ab} = W_{ab}$	$\Psi_0 = \Psi_4 = (\lambda_2 - \lambda_1)/2$ $\Psi_1 = \Psi_3 = 0, \Psi_2 = -\lambda_3/2;$ $C^*_{abcd} = -\sum_{\alpha=1}^{3} \lambda_\alpha X^*_{(\alpha)ab} X^*_{(\alpha)cd}$	*I*, *D* (for$\lambda_1 = \lambda_2$)
$\boldsymbol{Q} = \begin{pmatrix} 1 - \frac{\lambda}{2} & -\mathrm{i} & 0 \\ -\mathrm{i} & -\frac{\lambda}{2} - 1 & 0 \\ 0 & 0 & \lambda \end{pmatrix}$	$\lambda_1 = \lambda_2 = -\frac{\lambda}{2} :$ $\boldsymbol{r}_1 = (\frac{1}{2}, -\frac{\mathrm{i}}{2}, 0)$ $\lambda_3 = \lambda :$ $\boldsymbol{r}_2 = (0, 0, 1)$	$X^*_{(1)ab} = V_{ab}$ $X^*_{(2)ab} = W_{ab}$	$\Psi_0 = \Psi_1 = \Psi_3 = 0,$ $\Psi_2 = -\frac{\lambda}{2}, \Psi_4 = -2;$ $C^*_{abcd} = -4V_{ab}V_{cd} - \lambda(V_{ab}U_{cd}$ $+U_{ab}V_{cd} + W_{ab}W_{cd})$	*II*, *N* (for$\lambda = 0$)
$\boldsymbol{Q} = \begin{pmatrix} 0 & 0 & \mathrm{i} \\ 0 & 0 & 1 \\ \mathrm{i} & 1 & 0 \end{pmatrix}$	$\lambda_1 = \lambda_2 = \lambda_3 = 0 :$ $\boldsymbol{r} = (\frac{1}{2}, -\frac{\mathrm{i}}{2}, 0)$	$X^*_{ab} = V_{ab}$	$\Psi_0 = \Psi_1 = \Psi_2 = \Psi_4 = 0,$ $\Psi_3 = -\mathrm{i};$ $C^*_{abcd} = -2\mathrm{i}(V_{ab}W_{cd} + W_{ab}V_{cd})$	*III*

forms given in Table 4.2. We call the uniquely determined basis $\{\boldsymbol{E}_a\}$ the *Weyl principal tetrad.*

In the case of Petrov type I we have spacelike and timelike 2-planes ('blades') associated with the complex self-dual eigenbivectors

$$\begin{aligned} \boldsymbol{V} - \boldsymbol{U} &= 2(\boldsymbol{E}_{[4}\boldsymbol{E}_{1]} + \mathrm{i}\boldsymbol{E}_{[2}\boldsymbol{E}_{3]}), \\ \mathrm{i}(\boldsymbol{V} + \boldsymbol{U}) &= 2(\boldsymbol{E}_{[4}\boldsymbol{E}_{2]} + \mathrm{i}\boldsymbol{E}_{[3}\boldsymbol{E}_{1]}), \\ \boldsymbol{W} &= 2(\boldsymbol{E}_{[4}\boldsymbol{E}_{3]} + \mathrm{i}\boldsymbol{E}_{[1}\boldsymbol{E}_{2]}), \end{aligned} \tag{4.8}$$

and the intersections of these 2-planes determine the principal tetrad.

The principal tetrad is only partially determined by the metric for the two *degenerate* Petrov types D (I, $\lambda_1 = \lambda_2$), and N (II, $\lambda = 0$). It is not difficult to find the remaining subgroups of $L_+^\uparrow$ which preserve the normal forms

$$C^*_{abcd} = -\tfrac{1}{2}\lambda(g_{abcd} + i\varepsilon_{abcd}) - \tfrac{3}{2}\lambda W_{ab}W_{cd} \quad \text{for type } D, \tag{4.9}$$

$$C^*_{abcd} = -4V_{ab}V_{cd} \quad \text{for type } N. \tag{4.10}$$

In type D metrics this invariance group consists of special Lorentz transformations in the $\boldsymbol{E}_3$–$\boldsymbol{E}_4$-plane and spatial rotations in the $\boldsymbol{E}_1$–$\boldsymbol{E}_2$-plane. In terms of the complex null tetrad, these transformations are given by (3.16), (3.17). In type N metrics the invariance group with $V'_{ab} = V_{ab}$ is just the two-parameter subgroup (3.15).

The spinor form (§3.6) of the eigenvalue equation (4.2) with the eigenbivector $X^*_{ab} \leftrightarrow \eta_{AB}\varepsilon_{\dot{C}\dot{D}}$ reads

$$\Psi_{ABCD}\eta^{CD} = \lambda\eta_{AB}. \tag{4.11}$$

The invariants

$$\begin{aligned} I &\equiv \tfrac{1}{2}\Psi_{ABCD}\Psi^{ABCD} = \tfrac{1}{2}(\lambda_1^2 + \lambda_2^2 + \lambda_3^2), \\ J &\equiv \tfrac{1}{6}\Psi_{ABCD}\Psi^{CDEF}\Psi_{EF}{}^{AB} = \tfrac{1}{6}(\lambda_1^3 + \lambda_2^3 + \lambda_3^3) = \tfrac{1}{2}\lambda_1\lambda_2\lambda_3, \end{aligned} \tag{4.12}$$

of the Weyl tensor are useful in Petrov classification, since the eigenvalues satisfy $\lambda^3 - I\lambda - 2J = 0$; algebraically special fields (all Petrov types except type I) satisfy the relation

$$I^3 = 27J^2 \tag{4.13}$$

and, in particular, for the types III, N and O both invariants (4.12) vanish, $I = J = 0$.

In terms of the electric and the magnetic parts (3.62) of the Weyl tensor, the invariant I can be written as

$$I = \tfrac{1}{2}(E^{ab}E_{ab} - B^{ab}B_{ab}) + \mathrm{i}\, E^{ab}B_{ab} \qquad (4.14)$$

(Matte 1953).

Next we give two theorems valid for vacuum fields ($R_{ab} = 0$).

Theorem 4.1 *A type I vacuum solution for which one of the eigenvalues λ_α of the Weyl tensor vanishes over an open region (so that $J = 0$ in that region) must be flat space-time* (Brans 1975).

Theorem 4.2 *Vacuum fields satisfying the equation $R_{abcd;e}{}^{e} = \alpha R_{abcd}$ are either type N ($\alpha = 0$) or type D ($\alpha \neq 0$)* (Zakharov 1965, 1970).

The *proof* of Theorem 4.2 follows immediately from $R_{ab} = 0$ and the identity (Zakharov 1972)

$$R_{abcd;m}{}^{m} = R^{m}{}_{nab}R^{n}{}_{mcd} + 2(R^{m}{}_{adn}R^{n}{}_{cbm} - R^{m}{}_{bdn}R^{n}{}_{cam}) \qquad (4.15)$$

written down with respect to a principal tetrad (cp. the normal forms in Table 4.2).

4.3 Principal null directions and determination of the Petrov types

For Petrov types *II* and *III*, the real and imaginary parts of the eigenbivector $V_{ab} = 2k_{[a}m_{b]}$ represent 2-spaces containing the real null vector $\boldsymbol{k}$. The normal forms (Table 4.2) for these types are adapted to this preferred null vector, which is significant in what follows.

The classification based on the distinct possible solutions of the eigenvalue problem (4.4) is equivalent to the characterization of the Weyl (conformal) tensor in terms of principal null directions $\boldsymbol{k}$ with the property (Penrose 1960)

$$k_{[e}C_{a]bc[d}k_{f]}k^{b}k^{c} = 0 \quad \Leftrightarrow \quad \Psi_0 \equiv C_{abcd}k^{a}m^{b}k^{c}m^{d} = 0. \qquad (4.16)$$

There are at most four such null vectors; to determine them, we apply the inverse of the null rotation (3.14) to an arbitrary complex null tetrad $(\boldsymbol{m}', \overline{\boldsymbol{m}}', \boldsymbol{l}', \boldsymbol{k}')$ defined by (3.8). By this means the null vector $\boldsymbol{k}'$ can be transformed into any other real null vector except $\boldsymbol{l}'$. The coefficients $\Psi_0, \ldots, \Psi_4$ defined by (3.59) then undergo the transformations (3.60), in particular,

$$\Psi_0 = \Psi_0{}' - 4E\Psi_1{}' + 6E^2\Psi_2{}' - 4E^3\Psi_3{}' + E^4\Psi_4{}', \qquad (4.17)$$

and so, from the condition (4.16), we obtain an algebraic equation of at most fourth order for the complex number E:

$$\Psi_0{}' - 4E\Psi_1{}' + 6E^2\Psi_2{}' - 4E^3\Psi_3{}' + E^4\Psi_4{}' = 0. \tag{4.18}$$

The invariants (4.12) are given by

$$I \equiv \Psi_0\Psi_4 - 4\Psi_1\Psi_3 + 3\Psi_2^2, \\ J \equiv \begin{vmatrix} \Psi_4 & \Psi_3 & \Psi_2 \\ \Psi_3 & \Psi_2 & \Psi_1 \\ \Psi_2 & \Psi_1 & \Psi_0 \end{vmatrix}. \tag{4.19}$$

Starting with the normal forms for the various Petrov types given in Table 4.2 and in (4.9), (4.10), we can calculate the roots E given in Table 4.3. Conversely, by solving (4.18) for E for a given Weyl tensor, we can determine its Petrov type.

The four distinct principal null directions of a Petrov type I space need not span a four-dimensional space (Trümper 1965). If there exists an observer (with four-velocity u^a) who sees the Weyl tensor as purely electric or purely magnetic, cp. §3.5, then (McIntosh *et al.* 1994) the principal null directions are linearly dependent, so (McIntosh and Arianrhod 1990a) the invariant I defined in (4.12) is real and $M = I^3/J^2 - 6$ is non-negative (possibly infinite). In the pure electric case I is positive and J is real, while in the pure magnetic case I is negative and J is imaginary. One may note that by conformal transformation one can, from a given metric with pure electric or magnetic Weyl tensor, construct others, with different Ricci tensors.

In types D and III there is an additional principal null direction $\boldsymbol{l}$ not obtainable with the aid of the null rotations (3.14); $\boldsymbol{l}$ fulfils the condition

$$l_{[e}C_{a]bc[d}l_{f]}l^b l^c = 0 \iff \Psi_4 \equiv C_{abcd}l^a\overline{m}^b l^c \overline{m}^d = 0. \tag{4.20}$$

In an arbitrarily given tetrad we can determine the Petrov type by determining the roots of the quartic algebraic equation (4.18). If the order of this equation is $(4-m)$, then there are $(4-m)$ principal null directions $\boldsymbol{k}$, and $\boldsymbol{l}$ represents an m-fold principal null direction. The Petrov type can be obtained immediately, by inspection of Table 4.3, once the multiplicities of the roots of (4.18) are known. An equivalent method for determining the Petrov type is based on the eigenvalue equation (4.4). One can use the invariant criteria for the matrix $\boldsymbol{Q}$ which are listed in Table 4.1, $\boldsymbol{Q}$ being calculated with respect to an arbitrary orthonormal basis $\{\boldsymbol{E}_a\}$. For further discussion and references to efficient methods of computation see Chapter 9.

Table 4.3. The roots of the algebraic equation (4.18) and their multiplicities

The corresponding multiplicities of the principal null directions are symbolically depicted on the right of this table.

Type	Roots E	Multiplicities
I	$\dfrac{\sqrt{\lambda_2 + 2\lambda_1} \pm \sqrt{\lambda_1 + 2\lambda_2}}{\sqrt{\lambda_1 - \lambda_2}}$	(1,1,1,1)
D	$0, \infty$	(2,2)
II	$0, \pm \mathrm{i}\sqrt{\frac{3}{2}\lambda}$	(2,1,1)
III	$0, \infty$	(3,1)
N	0	(4)

A Weyl tensor is said to be *algebraically special* if it admits at least one *multiple* principal null direction (the multiplicity of a null direction is equal to the multiplicity of the corresponding root of the algebraic equation (4.18)).

One can show the validity of the following equations (Jordan *et al.* 1961)

$$k_{[e}C_{a]bc[d}k_{f]}k^b k^c = 0 \Longleftrightarrow \qquad \Psi_0 = 0, \qquad \Psi_1 \neq 0, \quad (4.21)$$

$$C_{abc[d}k_{f]}k^b k^c = 0 \Longleftrightarrow \qquad \Psi_0 = \Psi_1 = 0, \qquad \Psi_2 \neq 0, \quad (4.22)$$

$$C_{abc[d}k_{f]}k^c = 0 \Longleftrightarrow \quad \Psi_0 = \Psi_1 = \Psi_2 = 0, \quad \Psi_3 \neq 0, \quad (4.23)$$

$$C_{abcd}k^c = 0 \Longleftrightarrow \Psi_0 = \Psi_1 = \Psi_2 = \Psi_3 = 0, \ \Psi_4 \neq 0, \quad (4.24)$$

for principal null directions $\boldsymbol{k}$ of multiplicity 1, 2, 3 and 4 respectively. An equivalent formulation of the criterion (4.22) is

$$k^a k^c C^*_{abcd} = \lambda k_b k_d, \qquad \lambda \neq 0. \quad (4.25)$$

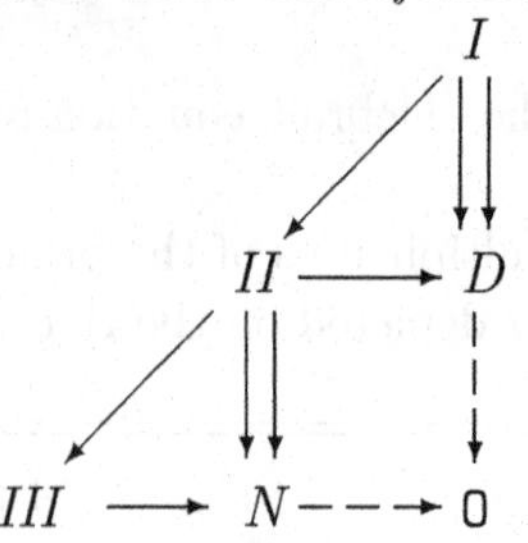

Fig. 4.1. Penrose diagram

Any two of the following conditions for a null direction $\boldsymbol{k}$ imply the third (Hall 1973)

$$C_{abc[d}k_{f]}k^bk^c = 0, \qquad R_{abc[d}k_{f]}k^bk^c = 0, \qquad R_{a[b}k_{c]}k^a = 0. \tag{4.26}$$

The last condition in (4.26) means that $\boldsymbol{k}$ is a Ricci eigendirection (§5.1).

Type D is characterized by the existence of *two double* principal null directions, $\boldsymbol{k}$ and $\boldsymbol{l}$,

$$\begin{aligned} C_{abc[d}k_{f]}k^bk^c = 0 &\Longleftrightarrow \Psi_0 = \Psi_1 = 0,\ \Psi_2 \neq 0, \\ C_{abc[d}l_{f]}l^bl^c = 0 &\Longleftrightarrow \Psi_4 = \Psi_3 = 0,\ \Psi_2 \neq 0. \end{aligned} \tag{4.27}$$

Type O (zero Weyl tensor) does not single out any null directions.

In the *Penrose diagram* (Fig. 4.1) the arrows point in the direction of increasing multiplicity of the principal null directions; every arrow indicates one additional degeneration.

The classification in terms of principal null directions can be formulated in terms of spinors (Penrose 1960). The completely symmetric spinor Ψ_{ABCD} can be written as a symmetrized product of one-index spinors, which are uniquely determined apart from factors. The proof is an application of the fundamental theorem of algebra: any polynomial may be factorized over $\mathbb{C}$ into linear forms,

$$\Psi_{ABCD}\zeta^A\zeta^B\zeta^C\zeta^D = (o_A\zeta^A)(\beta_B\zeta^B)(\gamma_C\zeta^C)(\iota_D\zeta^D). \tag{4.28}$$

The Petrov types are then characterized by the criteria:

$$\begin{aligned} \text{Type}\quad I &: \Psi_{ABCD} \sim o_{(A}\beta_B\gamma_C\iota_{D)}, \\ D &: \Psi_{ABCD} \sim o_{(A}o_B\iota_C\iota_{D)}, \\ II &: \Psi_{ABCD} \sim o_{(A}o_B\gamma_C\iota_{D)}, \\ III &: \Psi_{ABCD} \sim o_{(A}o_Bo_C\iota_{D)}, \\ N &: \Psi_{ABCD} \sim o_{(A}o_Bo_Co_{D)} \end{aligned} \tag{4.29}$$

$(k^a \longleftrightarrow o^A\bar{o}^{\dot{B}},\ \ l^a \longleftrightarrow \iota^A\bar{\iota}^{\dot{B}})$.

5

Classification of the Ricci tensor and the energy-momentum tensor

5.1 The algebraic types of the Ricci tensor

In §3.5 we decomposed the curvature tensor into irreducible parts. The invariant classification of the Weyl tensor was treated in Chapter 4. Now we consider the algebraic classification of the remaining part, the traceless Ricci tensor S_{ab}.

In a Riemannian space, every second-order symmetric tensor defines a linear mapping which takes a vector $\boldsymbol{v}$ into another vector $\boldsymbol{w}$. To classify S_{ab}, it is natural to examine the eigenvalue equation

$$S^a{}_b v^b = \widehat{\lambda} v^a. \tag{5.1}$$

Because a term proportional to $g^a{}_b$ merely shifts all eigenvalues by the same amount, we may as well consider the eigenvalue equation for the Ricci tensor $R^a{}_b$

$$R^a{}_b v^b = \lambda v^a; \qquad \lambda = \widehat{\lambda} + \tfrac{1}{4}R. \tag{5.2}$$

In a positive definite metric, a real symmetric matrix can always be diagonalized by a real orthogonal transformation. However, the Lorentz metric g_{ab} leads to a more complicated algebraic structure; the elementary divisors can be non-simple, and the eigenvalues can be complex. The eigenvalue equation (5.2) determines the orders $m_1, \ldots, m_k$ of the elementary divisors belonging to the various eigenvalues. The *Segre notation* (1884), which also appears in Weiler (1874), gives just these orders, and round brackets indicate that the corresponding eigenvalues coincide. If two eigenvalues are complex conjugates they are symbolized by Z and $\overline{Z}$.

The *Plebański notation* (1964) indicates whether the space spanned by the eigenvectors belonging to a certain real eigenvalue is timelike (T), null (N) or spacelike (S). The multiplicity of the eigenvalue is written in front of this symbol. Finally the orders of the corresponding factors in the minimal polynomial for the Ricci tensor (considered as a matrix) are added as indices enclosed in brackets.

Table 5.1 gives a complete list of all possible types for the Ricci tensor $R^a{}_b$ of a space-time in both Segre and Plebański notation. In accordance with the conventions established in §§3.1 and 3.2 we have arranged the eigenvalues in such an order that those whose corresponding eigenvectors are null or timelike (or complex) appear last, and we have used a comma to separate these from the eigenvalues with spacelike eigenvectors. The table also gives the possible physical interpretations of the types discussed in detail in §5.2. Further refinements of the classification given are possible (see e.g. Ludwig and Scanlan (1971)). A limiting diagram for the Segre types can be found in Paiva *et al.* (1998).

One can establish the list of possibilities by the following sequence of results (Churchill 1932, Hall 1976a). We first define an *invariant 2-plane* (at any point) as a two-dimensional subspace of the tangent space which is mapped to itself by $R^a{}_b$; any vector $\boldsymbol{v}$ lying in an invariant 2-plane is mapped by $R^a{}_b$ to a vector $\boldsymbol{w}$ in the same plane.

(1) There is always an invariant 2-plane.

Proof: Either there are two real eigenvectors, or (at least) one complex eigenvector. In the former case the two eigenvectors give the required plane; in the latter case, the real and imaginary parts of the eigenvector do the same.

(2) The 2-plane orthogonal to that given in (1) is also invariant.

Proof: If the 2-plane of (1) is timelike or spacelike, taking an orthonormal basis with $\boldsymbol{E}_1$, $\boldsymbol{E}_2$, lying in the 2-plane breaks $R^a{}_b$ into block diagonal form, and so shows the orthogonal 2-plane is invariant. If the 2-plane of (1) is null, let $\boldsymbol{k}$, $\boldsymbol{x}$ span it. Then in an expansion of R_{ab} using a null tetrad $(\boldsymbol{m}, \overline{\boldsymbol{m}}, \boldsymbol{l}, \boldsymbol{k})$ with $\sqrt{2}\boldsymbol{m} = \boldsymbol{x} + \mathrm{i}\boldsymbol{y}$ only terms in $k_a k_b$, $k_{(a}l_{b)}$, $k_{(a}x_{b)}$, $k_{(a}y_{b)}$, $x_a x_b$ and $y_a y_b$ survive. Thence $\boldsymbol{k}$, $\boldsymbol{y}$ span another invariant 2-plane.

(3) If $R^a{}_b$ has an invariant timelike (or spacelike) 2-plane, it has two distinct spacelike eigenvectors.

Proof: By (2) above the spacelike and timelike cases are the same and $R^a{}_b$ takes block diagonal form in an orthonormal tetrad. The (2×2) matrix acting on the spacelike 2-plane can be diagonalized by spatial rotation in the usual way for symmetric matrices.

(4) If $R^a{}_b$ has an invariant null plane, it has a null eigenvector.

Proof: In the proof of (2), we see that $\boldsymbol{k}$ is a null eigenvector.

Following Hall (1976a), one can now enumerate the cases listed in Table 5.1 by taking a null tetrad basis and systematically considering first

Table 5.1. The algebraic types of the Ricci tensor (for explanation, see the text)

	Segre notation	Plebański notation	Physical interpretations
A1	$[111,1]$	$[S_1 - S_2 - S_3 - T]_{(1111)}$	
	$[11(1,1)]$	$[S_1 - S_2 - 2T]_{(111)}$	
	$[(11)1,1]$	$[2S_1 - S_2 - T]_{(111)}$	
	$[(11)(1,1)]$	$[2S - 2T]_{(11)}$	Non-null Maxwell field
	$[1(11,1)]$	$[S - 3T]_{(11)}$	
	$[(111),1]$	$[3S - T]_{(11)}$	Perfect fluid
	$[(111,1)]$	$[4T]_{(1)}$	Λ-term
A2	$[11,Z\overline{Z}]$	$[S_1 - S_2 - Z - \overline{Z}]_{(1111)}$	
	$[(11),Z\overline{Z}]$	$[2S - Z - \overline{Z}]_{(111)}$	
A3	$[11,2]$	$[S_1 - S - 2N]_{(112)}$	
	$[1(1,2)]$	$[S - 3N]_{(12)}$	
	$[(11),2]$	$[2S - 2N]_{(12)}$	
	$[(11,2)]$	$[4N]_{(2)}$	Null Maxwell field, pure rad.
B	$[1,3]$	$[S - 3N]_{(13)}$	
	$[(1,3)]$	$[4N]_{(3)}$	

the cases where $R^a{}_b$ has a null eigenvector, and then the cases where there is no null eigenvector. In the table the different cases are divided into classes according to whether or not there is a timelike (and hence also a spacelike) invariant 2-plane. The distinct types are

A: *Timelike invariant 2-plane*

A1: Two real orthogonal eigenvectors exist in this plane

A2: No real eigenvectors exist in this plane

A3: One double null real eigenvector exists in this plane.

B: *Null invariant 2-plane*: one triple null real eigenvector exists.

In each case one can transform R_{ab} to a canonical form:

$$\text{A1:}\quad R_{ab} = \lambda_1 x_a x_b + \lambda_2 y_a y_b + \lambda_3 z_a z_b - \lambda_4 u_a u_b, \tag{5.3a}$$

$$\text{A2:}\quad R_{ab} = \lambda_1 x_a x_b + \lambda_2 y_a y_b + \lambda_3 k_{(a} l_{b)} + \lambda_4 (k_a k_b - l_a l_b), \tag{5.3b}$$

$$\text{A3:}\quad R_{ab} = \lambda_1 x_a x_b + \lambda_2 y_a y_b - 2\lambda_3 k_{(a} l_{b)} \pm k_a k_b, \tag{5.3c}$$

$$\text{B:}\quad R_{ab} = \lambda_1 x_a x_b + \lambda_2 y_a y_b + \lambda_3 k_{(a} l_{b)} + k_{(a} x_{b)}. \tag{5.3d}$$

Table 5.2. Invariance groups of the Ricci tensor types

Invariance group	Ricci tensor types
None	$[111,1], [11,Z\overline{Z}], [11,2], [1,3]$
Spatial rotations (3.16)	$[(11)1,1], [(11),Z\overline{Z}], [(11),2]$
Boosts (3.17)	$[11(1,1)]$
Boosts (3.17) and rotations (3.16)	$[(11)(1,1)]$
$SO(3)$ rotations	$[(111),1]$
$SO(2,1)$: three-dimensional Lorentz group	$[1(11,1)]$
One-parameter group of null rotations	$[1(1,2)], [(1,3)]$
Null rotations (3.15) and rotations (3.16)	$[(11,2)]$
Full Lorentz group	$[(111,1)]$, $R^a{}_b = 0$

Here $(\boldsymbol{x}, \boldsymbol{y}, \boldsymbol{z}, \boldsymbol{u})$ is an orthonormal tetrad, and $(\boldsymbol{x}, \boldsymbol{y}, \boldsymbol{k}, \boldsymbol{l})$ a real (or half-) null tetrad. In each case the associated orthonormal tetrad is called the *Ricci principal tetrad.*

The Ricci tensor types are called *degenerate* when there is more than one elementary divisor with the same eigenvalue; in the Segre notation these degeneracies are indicated by round brackets (in Table 5.1).

If the Ricci tensor is non-degenerate and the elementary divisors are simple, the type is said to be *algebraically general.* Otherwise it is called *algebraically special.* These ideas are analogous to those for Petrov types (Chapter 4). As we see in the next section, the physically most important types are algebraically special.

The Ricci principal tetrads of the non-degenerate types (where the eigenvalues of different elementary divisors are distinct) are uniquely determined (cp. §4.2), but in other cases some freedom is allowed. We list the possibilities in Table 5.2.

Ref.: For other approaches to the classification of symmetric tensors see Crade and Hall (1982), Penrose and Rindler (1986) and Chapter 9.

5.2 The energy-momentum tensor

The *Einstein field equations* (1.1)

$$R_{ab} - \tfrac{1}{2}Rg_{ab} + \Lambda g_{ab} = \kappa_0 T_{ab} \tag{5.4}$$

(κ_0 being Einstein's gravitational constant and Λ the cosmological constant) connect the Ricci tensor R_{ab} with the energy-momentum tensor

T_{ab}. The Bianchi identities (2.81) imply the important relation

$$\kappa_0 T^{ab}{}_{;b} = (R^{ab} - \tfrac{1}{2}Rg^{ab})_{;b} = 0. \tag{5.5}$$

As well as *vacuum fields* (empty spaces)

$$R_{ab} = 0, \tag{5.6}$$

with zero T_{ab} and Λ, we shall consider solutions of the field equations (5.4) for the following physically relevant energy-momentum tensors:

(i) *electromagnetic field* (*Maxwell field*):

$$T_{ab} = F_{ac}F_b{}^c - \tfrac{1}{4}g_{ab}F_{cd}F^{cd} = \tfrac{1}{2}(F_{ac}F_b{}^c + \widetilde{F}_{ac}\widetilde{F}_b{}^c) = \tfrac{1}{2}F^{*c}_a\overline{F}^*_{bc},$$
$$F^*_{ab} \equiv F_{ab} + \mathrm{i}\widetilde{F}_{ab}, \quad \widetilde{F}_{ab} \equiv \tfrac{1}{2}\varepsilon_{abcd}F^{cd}, \quad F^{*ab}{}_{;b} = 0, \tag{5.7}$$

(ii) *pure radiation field* (*null dust*):

$$T_{ab} = \Phi^2 k_a k_b, \qquad k_a k^a = 0, \tag{5.8}$$

(iii) *perfect fluid*:

$$T_{ab} = (\mu + p)u_a u_b + pg_{ab}, \quad u_a u^a = -1. \tag{5.9}$$

In the perfect fluid case we normally assume $\mu + p \neq 0$, $\mu > 0$. In the particular case where $T_{ab} = 0$ and $\Lambda \neq 0$, or where T_{ab} is of perfect fluid type (5.9) but with $\mu + p = 0$, we shall say the Ricci tensor is of *Λ-term type*. Thus the perfect fluid solutions formally include Λ-term cases, the *Einstein spaces* $R_{ab} = \Lambda g_{ab}$. They also include the combination of a perfect fluid and a Λ-term. The latter can be incorporated in the fluid quantities by substituting $(p - \Lambda/\kappa_0)$ for p and $(\mu + \Lambda/\kappa_0)$ for μ; of course this substitution may violate the condition $\mu > 0$. Note that no invariant direction for u^a is determined by a Λ-term; the kinematic quantities defined in §6.2 have no invariant meaning in this case. In the other non-vacuum cases the *cosmological constant* Λ is usually set equal to zero; occasionally solutions including Λ are listed.

In general we do not consider superpositions of these energy-momentum tensors.

By virtue of the field equations (5.4), T_{ab} has the same algebraic type as R_{ab}. We shall now determine these types for the energy-momentum tensors (5.7)–(5.9).

(i) The complex self-dual *electromagnetic field* tensor F^*_{ab} can be expanded in terms of the basis ($\boldsymbol{U}$, $\boldsymbol{V}$, $\boldsymbol{W}$) (see §3.4) as

$$\tfrac{1}{2}F^*_{ab} = \Phi_0 U_{ab} + \Phi_1 W_{ab} + \Phi_2 V_{ab}, \tag{5.10}$$

Φ_0, Φ_1, Φ_2 being complex functions. There is an invariant, $F^*_{ab}F^{*ab} = 16(\Phi_0\Phi_2 - \Phi_1^2)$. If it is non-zero, the electromagnetic field is said to be *non-null* (or non-singular), while if the invariant is zero, the electromagnetic field is said to be *null* (or singular). In either case one can, with the aid of a tetrad rotation, set $\Phi_0 = 0$.

A *non-null electromagnetic field* and the corresponding energy-momentum tensor (5.7) can be transformed into

$$F^*_{ab} = 2\Phi_1 W_{ab} = 4\Phi_1(m_{[a}\overline{m}_{b]} - k_{[a}l_{b]}), \tag{5.11}$$

$$T_{ab} = 4\Phi_1\overline{\Phi}_1(m_{(a}\overline{m}_{b)} + k_{(a}l_{b)}) \tag{5.12}$$

(the null tetrad is adapted to the two null eigendirections of the Maxwell field). In terms of an orthonormal tetrad related to the null tetrad by (3.12), expression (5.12) can be rewritten in the canonical form (5.3*a*),

$$T_{ab} = \Phi^2(x_a x_b + y_a y_b - z_a z_b + u_a u_b), \qquad \Phi^2 = 2\Phi_1\overline{\Phi}_1. \tag{5.13}$$

In the principal tetrad ($\boldsymbol{x}$, $\boldsymbol{y}$, $\boldsymbol{z}$, $\boldsymbol{u}$) so defined, the electric and magnetic fields (E_a and B_a) are parallel to each other,

$$E_a + \mathrm{i}B_a \equiv F^*_{ab}u^b = (E + \mathrm{i}B)z_a = 2\Phi_1 z_a. \tag{5.14}$$

From the canonical form (5.13) one infers that for gravitational fields produced by a non-null electromagnetic field the Ricci tensor has the type [(11)(1,1)] with $\lambda_1 = \lambda_2 = -\lambda_3 = -\lambda_4 = 2\kappa_0\Phi_1\overline{\Phi}_1 = \kappa_0(E^2 + B^2)/2$; the double roots have equal magnitude and opposite sign. The Ricci tensor obeys the relation

$$(R^b{}_a - \lambda g^b_a)(R^c{}_b + \lambda g^c_b) = 0 \tag{5.15}$$

(compare the similar equations for the $\boldsymbol{Q}$-matrix in Table 4.1).

A *null electromagnetic field* and the corresponding energy-momentum tensor (5.7) can be transformed into

$$F^*_{ab} = 2\Phi_2 V_{ab} = 4\Phi_2 k_{[a}m_{b]}, \tag{5.16}$$

$$T_{ab} = \Phi^2 k_a k_b, \qquad \Phi^2 = 2\Phi_2\overline{\Phi}_2. \tag{5.17}$$

Obviously, the Ricci tensor is [(11,2)] with eigenvalue zero.

Of course, having the correct structure (5.12) or (5.17) of the energy-momentum tensor does not guarantee that the corresponding F^*_{ab} satisfies the Maxwell equations.

(ii) The energy-momentum tensor (5.8) of a *pure radiation field* has the same algebraic type as (5.17). However, it need not arise from a Maxwell field, and when such a Maxwell field exists it need not be unique (see e.g. (13.47)). The same form of energy-momentum tensor also arises from other types of directed massless radiation, for example, massless scalar fields or neutrino fields. It may be considered as representing the incoherent superposition of waves with random phases and polarizations but the same propagation direction. For any particular solution with $T_{ab} = \Phi^2 k_a k_b$ one may investigate whether or not there is a null electromagnetic field satisfying the (sourcefree) Maxwell equations or whether one could actually solve the equations for the other possibly underlying physical fields.

(iii) The energy-momentum tensor (5.9) of *perfect fluid type* has the algebraic type [(111),1]; three eigenvalues coincide. For *dust* solutions ($p = 0$) the triple eigenvalue of T_{ab} is equal to zero and the eigenvalues of R_{ab} are $\lambda_1 = \lambda_2 = \lambda_3 = -\lambda_4 = \kappa_0\mu/2$. Energy-momentum tensors of Λ-term type are clearly of algebraic type [(111,1)].

Energy-momentum tensors of the perfect fluid type also arise from scalar fields ψ with a timelike gradient $\psi_{,a}$, and in particular massless scalar fields give '*stiff matter*' $\mu = p$. It may happen that a perfect fluid admits an interpretation as a viscous fluid (King and Ellis 1973) or a magnetohydrodynamic field (and an electromagnetic field as a viscous fluid), see e.g. Tupper (1981, 1983), Raychaudhuri and Saha (1981) and Hall and Negm (1986).

The energy-momentum tensors (5.7)–(5.9) (and the associated Ricci tensors) have very simple algebraic types, namely [(11)(1,1)], [(11,2)] and [(111),1].

5.3 The energy conditions

A physically reasonable energy-momentum tensor has to obey the *dominant energy condition*: the local energy density as measured by an observer with 4-velocity $\boldsymbol{u}$ is non-negative and the local energy flow vector $\boldsymbol{q}$ is non-spacelike,

$$T_{ab}u^a u^b \geq 0, \tag{5.18a}$$

$$q^a q_a \leq 0, \quad q^a \equiv T^a{}_b u^b. \tag{5.18b}$$

For discussion of energy conditions, see Plebański (1964) and Hawking and Ellis (1973). The dominant energy condition (5.18) should hold for *all* timelike (unit) vectors $\boldsymbol{u}$ and, by continuity, these inequalities must still be true if we replace $\boldsymbol{u}$ by a null vector $\boldsymbol{k}$.

For type [111,1] (and its degeneracies), T_{ab} can be diagonalized, so that $T_{ab} = \mathrm{diag}(p_1, p_2, p_3, \mu)$, and (5.18) is then satisfied if

$$\mu \geq 0, \qquad -\mu \leq p_\alpha \leq \mu \qquad (\alpha = 1, 2, 3). \tag{5.19}$$

These inequalities hold for a non-null electromagnetic field (see (5.13)) and impose reasonable restrictions on the energy density μ and pressure p ($p = p_1 = p_2 = p_3$) of a perfect fluid. The dominant energy condition (5.18) is also satisfied by the energy-momentum tensors of pure radiation fields and null electromagnetic fields.

Types [11,$Z\overline{Z}$] and [1,3] (and their degeneracies) in Table 5.1, i.e. types A2 and B, violate even the weak energy condition (5.18a). Therefore these types are not physically significant.

5.4 The Rainich conditions

Locally, a gravitational field originates in a *non-null electromagnetic field* (outside matter and charges), or is a *Rainich geometry*, if and only if the space-time metric and its derivatives satisfy the *Rainich conditions* (Rainich 1925, Misner and Wheeler 1957). These conditions naturally split into two parts:

$$\begin{aligned}\text{algebraic}: \quad & R^a{}_b R^b{}_c = \tfrac{1}{4}\delta^a_c R_{bd}R^{bd} \neq 0, \quad R^a{}_a = 0, \\ & u^a u_a < 0 \quad \Rightarrow \quad R_{ab}u^a u^b > 0,\end{aligned} \tag{5.20}$$

$$\text{analytic}: \quad \alpha_{a,b} - \alpha_{b,a} = 0, \quad \alpha_a \equiv (R_{mn}R^{mn})^{-1}\varepsilon_{abcd}R^b{}_e R^{ed;c}. \tag{5.21}$$

To prove this assertion of Rainich's 'already unified theory' one first has to find the so-called *'extremal' field* f_{ab} which satisfies

$$f_{ab}\widetilde{f}^{ab} = 0, \qquad f_{ab}f^{ab} < 0, \qquad \widetilde{f}_{ab} \equiv \tfrac{1}{2}\varepsilon_{abcd}f^{cd}, \tag{5.22}$$

and then obtain from f_{ab} a solution F_{ab} of the Maxwell equations with the aid of a duality rotation

$$F_{ab} = f_{ab}\cos\alpha - \widetilde{f}_{ab}\sin\alpha, \quad \widetilde{F}_{ab} = f_{ab}\sin\alpha + \widetilde{f}_{ab}\cos\alpha \tag{5.23}$$

at each point p of V_4.

The Einstein–Maxwell system of simultaneous equations (outside the charge and current distribution) reads

$$R_{ab} = \tfrac{1}{2}\kappa_0(F_{ac}F_b{}^c + \widetilde{F}_{ac}\widetilde{F}_b{}^c) \quad \Leftrightarrow \quad R_{ab} = \tfrac{1}{2}\kappa_0(f_{ac}f_b{}^c + \widetilde{f}_{ac}\widetilde{f}_b{}^c), \tag{5.24}$$

$$F^{ab}{}_{;b} = 0 = \widetilde{F}^{ab}{}_{;b} \quad \Leftrightarrow \quad f^{ab}{}_{;b} - \alpha_{,b}\widetilde{f}^{ab} = 0 = \widetilde{f}^{ab}{}_{;b} + \alpha_{,b}f^{ab}. \tag{5.25}$$

The Einstein equations (5.24) can be rewritten, using (3.46) and (5.20), in the form

$$E_{abcd} = \tfrac{1}{2}(g_{ac}R_{bd} - g_{bc}R_{ad} + g_{bd}R_{ac} - g_{ad}R_{bc}) = \tfrac{1}{2}\kappa_0(f_{ab}f_{cd} + \widetilde{f}_{ab}\widetilde{f}_{cd}). \tag{5.26}$$

Together with (5.22), they imply the formula

$$\kappa_0 f_{ab} f_{cd} = E_{abcd} - (R_{mn}R^{mn})^{-1/2} E_{abef} E_{cd}{}^{ef}, \tag{5.27}$$

which enables one to find f_{ab} (up to an overall sign) provided that (5.20) is satisfied. The explicit determination of f_{ab} can most easily be carried out in a tetrad system in which $R_a{}^b$ has the diagonal form

$$R_a{}^b = \mathrm{diag}(\lambda, \lambda, -\lambda, -\lambda), \qquad \lambda > 0, \tag{5.28}$$

in accordance with the algebraic Rainich conditions (5.20).

As the next step one has to determine the scalar field α in the duality rotation (5.23) so that F_{ab} is a solution of the Maxwell equations (5.25), from which one obtains

$$\alpha_{,b} = 2(f_{mn}f^{mn})^{-1}(\tilde{f}_{ba} f^{ac}{}_{;c} + f_{ab}\tilde{f}^{ac}{}_{;c}). \tag{5.29}$$

(The identity (3.33) for arbitrary bivectors has been applied to f_{ab} and $\tilde{f}_{ab}$.) In order to express the gradient $\alpha_{,a}$ in terms of geometrical quantities, one uses the relations

$$\begin{aligned} E_{\widetilde{abcd}} &= \tfrac{1}{2}\varepsilon_{cdef}(\delta^e_a R^f{}_b - \delta^e_b R^f{}_a) = \tfrac{1}{2}\kappa_0(f_{ab}\tilde{f}_{cd} - \tilde{f}_{ab} f_{cd}), \\ E^{abcd}{}_{;d} &= \tfrac{1}{2}(R^{ac;b} - R^{bc;a}) = \tfrac{1}{2}\kappa_0(f^{ab}f^{cd} + \tilde{f}^{ab}\tilde{f}^{cd})_{;d}, \end{aligned} \tag{5.30}$$

which follow from (5.26). The resulting formula

$$\alpha_{,a} = -8\,(\kappa_0 f_{mn}f^{mn})^{-2}\, E_{\widetilde{abcd}} E^{cdbf}{}_{;f} = (R_{mn}R^{mn})^{-1}\varepsilon_{abcd}R^b{}_e R^{ed;c} = \alpha_a \tag{5.31}$$

enables one to find the *'complexion'* α (up to an additive constant) provided that (5.21) is satisfied. Hence, a Rainich geometry determines the associated electromagnetic field F_{ab} uniquely up to a *constant* duality rotation. This method, implemented in the computer algebra package CLASSI (Åman 2002), was used to check Maxwell fields for Einstein–Maxwell solutions in this book.

The problem of determining a *null electromagnetic field* from the geometry has not yet been completely solved (Jordan and Kundt 1961, Ludwig 1970).

5.5 Perfect fluids

In order to describe a perfect fluid completely, the energy-momentum tensor (5.9) of perfect fluid type has to be supplemented by an equation of state expressing, say, the density ρ of rest mass as a function of the energy density μ and the pressure p (taking a picture of a fluid composed of conserved microscopic particles). Conservation of the number of particles can be formulated as

$$(\rho u^a)_{;a} = 0. \tag{5.32}$$

The conservation law $T^{ab}{}_{;b} = 0$ and the thermodynamic relation

$$\mathrm{d}h = \mathrm{d}p/\rho + T\mathrm{d}s, \qquad h \equiv (\mu + p)/\rho, \tag{5.33}$$

where h and s denote respectively the specific enthalpy and entropy, lead to the equation

$$\Omega_{ab}u^b = -Ts_{,a}, \qquad \Omega_{ab} \equiv (hu_a)_{;b} - (hu_b)_{;a}. \tag{5.34}$$

From (5.34) we obtain $s_{,a}u^a = 0$ (isentropic motion).

For constant specific entropy s, (5.33) reads

$$\mathrm{d}\mu = (\mu + p)\mathrm{d}\rho/\rho, \tag{5.35}$$

i.e. both μ and p depend only on ρ, or, p is a function of $\mu : p = p(\mu)$.

Two approaches are commonly used to deal with perfect fluid solutions: either a relation $p = p(\mu)$ is prescribed, or μ and p are evaluated from the field equations, there being, in general, no relation $p = p(\mu)$ between them. We shall not investigate the question of whether or not solutions of the latter type allow a thermodynamic interpretation in accordance with (5.33) and so have physical significance; for a discussion of this problem and further references see e.g. Krasiński *et al.* (1997) and Israel and Stewart (1980). On the other hand one can construct physical situations, such as the presence of viscous fluids, in which it becomes imperative to consider more general Ricci tensor types than those treated in this book.

In those cases where an equation of state of a perfect fluid is prescribed before the field equations are solved, a *'barotropic equation of state'* $p = p(\mu)$ of the simple form

$$p = (\gamma - 1)\mu, \tag{5.36}$$

where γ is a constant, has frequently been taken; we shall refer to such cases as 'γ-law perfect fluids'. Cases regarded as of particular interest are the *'dust'* case

$$p = 0, \qquad \gamma = 1, \tag{5.37}$$

the *'incoherent radiation'* case

$$p = \mu/3, \qquad \gamma = 4/3, \tag{5.38}$$

so called because it represents the superposition of waves of a massless field (e.g. the electromagnetic field) with random propagation direction, and the *'stiff matter'* case

$$p = \mu, \qquad \gamma = 2. \tag{5.39}$$

Because (5.39) leads to a sound speed equal to the velocity of light, the characteristics of its governing equations are the same as those of the gravitational field, and consequently such solutions can often be derived from vacuum solutions as described in Theorem 10.2.

The question of whether a *thermodynamic perfect fluid* can be determined from the geometry in a manner analogous to the Rainich treatment of the Maxwell field (§5.4) has also been addressed. The problem divides into two questions: whether the Ricci tensor R_{ab} has the correct algebraic structure, and whether the necessary thermodynamic relations above can be satisfied, see Coll and Ferrando (1989) and Quevedo and Sussman (1995) for further references.

6
Vector fields

6.1 Vector fields and their invariant classification

Vector fields in four-dimensional Riemannian spaces are frequently characterized by the properties of their first covariant derivatives and the invariants of the fields which can be built from these derivatives. The methods being standard, we give only the definitions and some simple applications for further reference. The physical meaning and the interpretation of the invariants in question can be found in the literature (Ehlers 1961, Jordan *et al.* 1961).

A vector field $\boldsymbol{v}(x^i)$ is said to be *hypersurface-orthogonal* or non-rotating or normal, if it is proportional to a gradient,

$$v_a = \lambda f_{,a}, \tag{6.1}$$

i.e. if and only if the rotation ω^a,

$$\omega^a := \varepsilon^{abcd} v_{b;c} v_d, \tag{6.2}$$

vanishes. (A factor $\frac{1}{2}$ might be expected on the right-hand side of (6.2), by analogy with (3.30) or (6.18), and is often used in defining the vorticity of a fluid, cp. §6.2.1, but would complicate other formulae later in this book.)

A vector field $\boldsymbol{v}(x^i)$ is said to be *geodesic* if it is proportional to the tangent vector field $\boldsymbol{t}$ of a congruence of geodesics (see (3.22)),

$$v_a = \lambda t_a, \qquad Dt_a/\mathrm{d}\tau = 0, \tag{6.3}$$

τ being an affine parameter. Equation (6.3) is equivalent to

$$v_{[a} v_{b];c} v^c = 0. \tag{6.4}$$

A vector field $\boldsymbol{v}$ is said to be *recurrent* (or parallel), if its covariant derivative is proportional to itself,

$$v_{a;b} = v_a K_b, \tag{6.5}$$

$\boldsymbol{K} = K^i\partial/\partial x^i$ being the recurrence vector. As can be checked, a recurrent vector is geodesic and non-rotating and its components can therefore be written as $v_a = \lambda f_{,a}$. Together with (6.5) this gives

$$f_{,a;b} = f_{,a}(K_b - \lambda_{,b}/\lambda). \tag{6.6}$$

If $\boldsymbol{v}$ is non-null ($f^{,a}f_{,a} \neq 0$), then (6.6) implies

$$K_a = \frac{\lambda_{,a}}{\lambda} + \frac{f^{,b}f_{,b;a}}{f^{,c}f_{,c}} = \frac{1}{2}(\ln \lambda^2 f^{,b}f_{,b})_{,a}, \tag{6.7}$$

$\boldsymbol{K}$ is a gradient, and a function $\alpha = (\lambda^2 f^{,b}f_{,b})^{-1/2}$ can be found such that

$$\alpha v_a = w_a, \qquad w_{a;b} = 0. \tag{6.8}$$

Consequently, a non-null recurrent vector field is proportional to a (covariantly) constant vector field.

If $\boldsymbol{v}$ is null, we can infer from (6.6) that the null vector $\boldsymbol{k}$ ($k_a \equiv f_{,a}$) obeys

$$k_{a;b} = \beta k_a k_b, \tag{6.9}$$

because $f_{,a;b}$ is symmetric in a and b, and the right-hand side must be symmetric too. This null vector is proportional to a constant vector only if $\beta_{,[a}k_{b]} = 0$. Space-times admitting a recurrent null vector are algebraically special (Debever and Cahen 1961, Öktem 1976).

A vector field $\boldsymbol{\xi}$ is a *conformal motion* if it satisfies

$$\mathcal{L}_{\boldsymbol{\xi}} g_{ab} \equiv \xi_{a;b} + \xi_{b;a} = 2\phi(x^k)g_{ab}. \tag{6.10}$$

If ϕ is constant, $\boldsymbol{\xi}$ is a *homothetic vector*, and if $\phi = 0$ a *Killing vector* satisfying Killing's equation

$$\xi_{a;b} + \xi_{b;a} = 0. \tag{6.11}$$

Because of their considerable importance, we devote an extra chapter to Killing and homothetic vectors and groups of motion (Chapter 8); conformal motions are discussed in Chapter 35.

Two vector fields $\boldsymbol{v}$, $\boldsymbol{w}$ are said to be *surface-forming* if the Lie derivative of one vector with respect to the other lies in the plane defined by $\boldsymbol{v}$ and $\boldsymbol{w}$,

$$v^a w^b{}_{;a} - w^a v^b{}_{;a} = (\mathcal{L}_{\boldsymbol{v}}\boldsymbol{w})^b = -(\mathcal{L}_{\boldsymbol{w}}\boldsymbol{v})^b = \lambda v^b + \mu w^b, \tag{6.12}$$

or, equivalently,

$$\varepsilon_{abcd}v^b w^c(\mathcal{L}_{\boldsymbol{v}}\boldsymbol{w})^d = 0. \tag{6.13}$$

6.1.1 Timelike unit vector fields

The covariant derivative of a timelike unit vector field $\boldsymbol{u}(x^i)$, $u^a u_a = -1$, can be decomposed as follows:

$$
\begin{aligned}
u_{a;b} &= -\dot{u}_a u_b + \omega_{ab} + \sigma_{ab} + \Theta h_{ab}/3, \\
\dot{u}_a &:= u_{a;b}u^b = Du_a/d\tau, \qquad \dot{u}_a u^a = 0, \\
\omega_{ab} &:= u_{[a;b]} + \dot{u}_{[a}u_{b]}, \qquad \omega_{ab}u^b = 0, \\
\sigma_{ab} &:= u_{(a;b)} + \dot{u}_{(a}u_{b)} - \Theta h_{ab}/3, \qquad \sigma_{ab}u^b = 0, \\
h_{ab} &:= g_{ab} + u_a u_b, \qquad h_{ab}u^b = 0, \\
\Theta &:= u^a{}_{;a}.
\end{aligned}
\tag{6.14}
$$

Physically, the timelike vector field $\boldsymbol{u}$ is often taken to be the 4-velocity of a fluid, and the quantities $\dot{u}_a$, Θ, ω_{ab} and σ_{ab} are accordingly called acceleration, expansion, rotation (or vorticity) and shear, respectively.

When we study the conformal properties of space-time, it is interesting to know how these invariants change under a conformal transformation $\widehat{g}_{ab} = \mathrm{e}^{2U} g_{ab}$, cp. (3.83). The world lines $x^i(\tau)$ being the same, the 4-velocities are related to each other by $\hat{u}^a = \mathrm{e}^{-U} u^a$, $\hat{u}_b = \mathrm{e}^{U} u_b$, and we obtain

$$
\begin{aligned}
\widehat{\sigma}_{ab} &= \mathrm{e}^U \sigma_{ab}, \qquad \widehat{\Theta} = \mathrm{e}^{-U}(\Theta + 3u^a U_{,a}), \\
\widehat{\omega}_{ab} &= \mathrm{e}^U \omega_{ab}, \qquad \widehat{\dot{u}}_a = \dot{u}_a + u_a u_b U^{,b} + U_{,a}\ ,
\end{aligned}
\tag{6.15}
$$

for the invariant parts of their derivatives.

6.1.2 Geodesic null vector fields

To get the decomposition of the covariant derivative of a (affinely parametrized) geodesic null vector field $\boldsymbol{k}(x^i)$ fulfilling

$$k_{a;b}k^b = 0, \tag{6.16}$$

one usually introduces the complex null tetrad $(\boldsymbol{m}, \overline{\boldsymbol{m}}, \boldsymbol{l}, \boldsymbol{k})$ defined in (3.8). The result is

$$
\begin{aligned}
k_{a;b} &= 2\,\mathrm{Re}\left[(\Theta + \mathrm{i}\omega)\,\overline{m}_a m_b - \sigma \overline{m}_a \overline{m}_b\right] + v_a k_b + k_a w_b, \\
v^a k_a &= 0 = w^a k_a, \\
\Theta &:= \tfrac{1}{2}k^a{}_{;a}, \qquad \omega^2 := \tfrac{1}{2}k_{[a;b]}k^{a;b}, \\
\sigma\overline{\sigma} &:= \tfrac{1}{2}k_{(a;b)}k^{a;b} - \tfrac{1}{4}\left(k^a{}_{;a}\right)^2.
\end{aligned}
\tag{6.17}
$$

$\boldsymbol{k}$ is given, but the vectors $\boldsymbol{m}$, $\overline{\boldsymbol{m}}$, $\boldsymbol{l}$ are not uniquely defined by it. Nevertheless, the quantities Θ, ω^2 and $\sigma\overline{\sigma}$ are invariants of $\boldsymbol{k}$ and independent of the choice of the null tetrad.

As a consequence of (3.29) and of (6.17), ω may also be defined by

$$\tfrac{1}{2}\varepsilon^{abcd}k_b k_{c;d} = \omega k^a. \tag{6.18}$$

As can be seen from (6.17), $k_{a;b}$ obeys $\varepsilon^{abcd}k_{a;b}k_{c;d} = 0$ (the rank of the matrix $k_{a;b}$ is maximally two); this and the identity $\varepsilon^{abcd}k_{c;da} = 0$ allow one to conclude from (6.18) that

$$(\omega k^a)_{;a} = \omega_{,a}k^a + 2\Theta\omega = 0. \tag{6.19}$$

The null geodesic congruence $\boldsymbol{k}$ is *hypersurface-orthogonal* if and only if ω vanishes. If an arbitrary null congruence is normal, it is also geodesic.

Physically, geodesic null vector fields can be interpreted as the tangent vectors of optical rays. Accordingly, the quantities Θ, ω and σ are called *expansion* (or divergence), *twist* (or rotation) and *shear*, respectively. The decomposition (6.17) implies

$$\begin{aligned} \rho &:= -(\Theta + \mathrm{i}\omega) = -k_{a;b}m^a\overline{m}^b, \\ \sigma &:= -k_{a;b}m^a m^b, \qquad \kappa := -k_{a;b}m^a k^b = 0. \end{aligned} \tag{6.20}$$

The expressions ρ, σ and κ are three of the twelve (complex) Newman–Penrose spin coefficients, see §7.1. Note that we have changed the sign of σ in (6.17) and (6.20) from that of Ehlers and Kundt (1962) to achieve conformity with the usual definition of the spin coefficients.

If we carry out a conformal transformation $\widehat{g}_{ab} = \mathrm{e}^{2U}g_{ab}$, see (3.83), but retain the condition $k_{a;b}k^b = 0$ for the transformed vector, then the null vector field has to be transformed by $\widehat{k}_a = k_a$; this implies

$$\widehat{k}_{a;b} = k_{a;b} - k_a U_{,b} - k_b U_{,a} + g_{ab}k^c U_{,c} \tag{6.21}$$

and

$$\widehat{\omega} = \mathrm{e}^{-2U}\omega, \qquad \widehat{\Theta} = \mathrm{e}^{-2U}(\Theta + k^a U_{,a}). \tag{6.22}$$

The importance of the invariant classification of vector fields in the context of exact solutions is twofold. If a solution defines a preferred vector field (velocity field, eigenvector field of the Weyl tensor, etc.), a classification of the vector field is also a classification of the solution in question. On the other hand, the existence of a vector field a_n with some special properties (shearfree, rotationfree, etc.) imposes conditions on the metric via

$$2a_{a;[bc]} = a_d R^d{}_{abc}. \tag{6.23}$$

We shall discuss this in detail in the following section.

6.2 Vector fields and the curvature tensor

6.2.1 Timelike unit vector fields

The main idea of this section is to evaluate the Ricci identity (6.23),

$$u_{a;bc} - u_{a;cb} = R^d{}_{abc}u_d, \tag{6.24}$$

using the decomposition (6.14),

$$u_{a;b} = -\dot{u}_a u_b + \omega_{ab} + \sigma_{ab} + \Theta h_{ab}/3, \tag{6.25}$$

and see what limitations on the metric (curvature tensor) follow from the existence of a vector field $\boldsymbol{u}$ with special derivative properties.

Writing down (6.24) in detail, one easily gets

$$\begin{aligned}\tfrac{1}{2}R^d{}_{abc}u_d = &-\dot{u}_a(\omega_{bc} - \dot{u}_{[b}u_{c]}) - \dot{u}_{a;[c}u_{b]} + \omega_{a[b;c]} + \sigma_{a[b;c]} + \tfrac{1}{3}\Theta_{,[c}h_{b]a}\\ &+ \tfrac{1}{3}\Theta(u_a\omega_{bc} - u_a\dot{u}_{[b}u_{c]} + \omega_{a[c}u_{b]} + \sigma_{a[c}u_{b]} + \tfrac{1}{3}\Theta h_{a[c}u_{b]}),\end{aligned} \tag{6.26}$$

and from this, by contraction and/or multiplication with u^b,

$$\begin{aligned}R^a{}_b u_a = &\ \dot{u}^a\omega_{ab} - \dot{u}^a{}_{;a}u_b - \dot{u}^a\sigma_{ab} + \omega^a{}_{b;a}\\ &+ \sigma^a{}_{b;a} - \tfrac{2}{3}\Theta_{,b} + \tfrac{1}{3}\dot{\Theta}u_b + \tfrac{1}{3}\Theta^2 u_b,\end{aligned} \tag{6.27}$$

$$\begin{aligned}R^d{}_{abc}u_d u^b = &\ \dot{u}_a\dot{u}_c - \omega_{ab}\omega^b{}_c - \tfrac{1}{3}\left(\dot{\Theta} + \tfrac{1}{3}\Theta^2\right)h_{ca}\\ &- \sigma_{ab}\sigma^b{}_c - \tfrac{2}{3}\Theta\sigma_{ac} + h^d_a h^e_c(\dot{u}_{(d;e)} - \dot{\sigma}_{de}),\end{aligned} \tag{6.28}$$

$$R^a{}_b u_a u^b = \dot{u}^a{}_{;a} + \omega_{ab}\omega^{ab} - \sigma_{ab}\sigma^{ab} - \dot{\Theta} - \Theta^2/3. \tag{6.29}$$

Equation (6.29) is often called the Raychaudhuri equation (Raychaudhuri 1955).

Formulae (6.27)–(6.29) may be considered as equations governing the temporal variation of u^a, σ_{ab} and Θ. In the context of exact solutions, however, we prefer to interpret them as equations determining some components of the curvature tensor if the properties of the vector field are prescribed or known. The implications of (6.26)–(6.29) for the self-dual Weyl tensor C^*_{abcd} can be calculated using the definition (3.50) and the previous equations. The result is

$$\begin{aligned}C^*_{abcd}u^a u^c = &\ \dot{u}_b\dot{u}_d - \omega_{bc}\omega^c{}_d - \sigma_{bc}\sigma^c{}_d - \tfrac{2}{3}\Theta\sigma_{bd} + h^e_b h^f_d\left[\dot{u}_{(e;f)} - \dot{\sigma}_{ef}\right]\\ &- \tfrac{1}{3}h_{bd}(\dot{u}^a{}_{;a} + \omega_{ae}\omega^{ae} - \sigma_{ae}\sigma^{ae}) + \tfrac{1}{2}h^e_b h^f_d R_{ef}\\ &- \tfrac{1}{6}h_{bd}h^{ef}R_{ef} - \mathrm{i}h^g{}_{(b}\varepsilon_{d)c}{}^{ef}u^c\left[\omega_{ge;f} + \sigma_{ge;f} - \dot{u}_g\omega_{ef}\right].\end{aligned} \tag{6.30}$$

The most remarkable property of this formula is the simplicity of the imaginary part of the tensor $Q_{ab} \equiv -C^*_{acbd}u^c u^d = E_{bd} + \mathrm{i}\, B_{bd}$. As shown in Chapter 4, the Petrov classification is the classification of this tensor Q_{ab}, and if Q_{ab} is real (i.e. the Weyl tensor is purely electric, $B_{ab} = 0$) then the space-time is of Petrov type *I*, *D* or *O*. The following theorems hold:

Theorem 6.1 *If a space-time admits a timelike unit vector field* $\boldsymbol{u}$ *satisfying*

$$B_{bd} \equiv h^g{}_{(b}\varepsilon_{d)c}{}^{ef} u^c \left[\omega_{ge;f} + \sigma_{ge;f} - \dot{u}_g \omega_{ef}\right] = 0, \tag{6.31}$$

then it is of Petrov type I, D or O (Trümper 1965).

Theorem 6.2 *If there is a shearfree perfect fluid* ($\sigma_{ab} = 0$) *with four-velocity* $\boldsymbol{u}$ *such that the Weyl tensor is purely electric, then the four-velocity is hypersurface-orthogonal* ($\omega_{ab} = 0$), *or* ω_a *is a Weyl eigenvector with eigenvalue* $-(\mu + p)/3$ (Glass 1975, Barnes 1984). *The rotating solutions are given by* (21.61) (Collins 1984, Senovilla 1987b).

Theorem 6.3 *If a space-time has a purely electric Weyl tensor and admits an irrotational timelike unit vector field* $\boldsymbol{u}$ *with non-vanishing shear, then the eigenframes of the shear and the Weyl tensor coincide. For Petrov type I, all the eigenvectors of this frame are hypersurface orthogonal* (Barnes and Rowlingson 1989).

Simple examples of vector fields satisfying (6.31) are those which are hypersurface-orthogonal and shearfree ($\omega_{ab} = 0 = \sigma_{ab}$). All vacuum solutions admitting such a vector field are static (Barnes 1973b). Shearfree and non-rotating perfect fluids were also discussed by Barnes (1973b); those of type *I* are necessarily static.

Diverging vacuum type *D* solutions (Chapters 28 and 29) are twistfree exactly if the Weyl tensor is purely electric (McIntosh *et al.* 1994).

For *static* metrics, characterized by the existence of a timelike non-rotating Killing vector $\boldsymbol{\xi}$, the vector field $\boldsymbol{u} = \boldsymbol{\xi}/\sqrt{-\xi_a \xi^a}$ obeys $u_{a;b} = -\dot{u}_a u_b$ because of the Killing equation (6.11). Consequently, all static metrics are of type *I*, *D* or *O*.

Rigid motions of a (test) body correspond to vector fields $\boldsymbol{u}$ satisfying $\Theta = 0 = \sigma_{ab}$, i.e. $\mathcal{L}_{\boldsymbol{u}} h_{ab} = 0$. Einstein spaces admitting rigid motions are either flat, or of constant curvature, or they are degenerate static metrics of class *B* (Wahlquist and Estabrook (1966); see also Mason and Pooe (1987)).

The energy-momentum tensor being specified, (6.26)–(6.31) together with the Bianchi identities can be used to derive results relevant to exact solutions; for example, if a perfect fluid with non-constant pressure p is geodesic, then it is irrotational (Synge 1937). Another such result is:

Theorem 6.4 *For shearfree dust,* $\Theta(\omega_{ab}\omega^{ab}) = 0$ (Ellis 1967).

Much effort has been put into trying to prove the analogous result for perfect fluids, i.e. to show that all shearfree perfect fluids with an equation of state $p = p(\mu)$, $p + \mu \neq 0$, are non-rotating ($\omega_{ab} = 0$) or non-expanding ($\Theta = 0$) or both. All known solutions support this conjecture, but so far it has been shown to be true only for subcases, e.g. if Θ and μ or Θ and ω are functionally dependent (Lang and Collins 1988, White and Collins 1984, Sopuerta 1998a), for a purely electric Weyl tensor (Collins 1984), for a purely magnetic Weyl tensor or an equation of state $\mu + 3p = \text{const}$ (Lang 1993), for vanishing acceleration (Senovilla *et al.* 1998), for Petrov type N (Carminati 1987), and for Petrov type III (for which necessarily $\Theta = 0$ and $\omega_{ab} \neq 0$) (Carminati and Cyganowski 1997). For the history of this conjecture and a more complete list of its confirmations, see Collins (1986), Lang (1993), Sopuerta (1998a) and Van den Bergh (1999).

Dust solutions which are non-rotating and have a purely electric Weyl tensor are called *silent universes.* Lists of those universes, and discussions of the conjecture that there are no spatially inhomogeneous silent universes of Petrov type I, can be found in Sopuerta (1997), van Elst *et al.* (1997) and Mars (1999). A corresponding discussion of solutions with a purely magnetic Weyl tensor is given in Maartens *et al.* (1998).

Ref.: For additional relations between the Weyl tensor and kinematic properties of a 4-velocity see also Lesame *et al.* (1996) and Collins and Wainwright (1983). For an application in covariant characterization of space-times see van Elst and Ellis (1996).

6.2.2 Null vector fields

One can perform a detailed evaluation of

$$k_{a;bc} - k_{a;cb} = R^{d}{}_{abc}k_d, \tag{6.32}$$

similar to the one which was carried out for timelike vector fields in the preceding section. This is best done by means of the Newman–Penrose formalism, which we shall introduce in the next chapter. We only mention here one conclusion which can be drawn *without* specifying the tetrad $(\boldsymbol{m}, \overline{\boldsymbol{m}}, \boldsymbol{k}, \boldsymbol{l})$: from (6.32) and (6.17) we get (for geodesic null vector fields)

$$\Theta_{,a}k^a - \omega^2 + \Theta^2 + \sigma\overline{\sigma} = -\tfrac{1}{2}R_{ab}k^a k^b, \tag{6.33}$$

and, using (6.19), we can write this in the form

$$(\Theta + \mathrm{i}\omega)_{,a}k^a + (\Theta + \mathrm{i}\omega)^2 + \sigma\overline{\sigma} = -\tfrac{1}{2}R_{ab}k^a k^b. \tag{6.34}$$

Ref.: For spacelike congruences see e.g. Tsamparlis and Mason (1983).

7

The Newman–Penrose and related formalisms

7.1 The spin coefficients and their transformation laws

The null tetrad formalism due to Newman and Penrose (1962) has proved very useful in the construction of exact solutions, in particular for studying algebraically special gravitational fields (for some of the earliest examples, see Kinnersley (1969b), Talbot (1969), and Lind (1974)). Despite the fact that we have to solve a considerably larger number of equations than arise when we use coordinates directly, this formalism has great advantages. All differential equations are of first order. Gauge transformations of the tetrad can be used to simplify the field equations. One can extract invariant properties of the gravitational field without using a coordinate basis. We give here an outline of this important approach to general relativity; see also Frolov (1977), Penrose and Rindler (1984, 1986) and Stewart (1990).

Using the complex null tetrad $\{\boldsymbol{e}_a\} = (\boldsymbol{m}, \overline{\boldsymbol{m}}, \boldsymbol{l}, \boldsymbol{k})$, and recalling the definition (2.67),

$$\nabla_b \boldsymbol{e}_a = \Gamma^c{}_{ab} \boldsymbol{e}_c, \tag{7.1}$$

of the connection coefficients $\Gamma^c{}_{ab}$, we can define the so-called *spin coefficients*, 12 independent complex linear combinations of the connection coefficients. Explicitly, the spin coefficients are defined in tensor and spinor notation as follows:

$$-\kappa \equiv \Gamma_{144} = k_{a;b} m^a k^b = m^a D k_a = o^A \bar{o}^{\dot{B}} o^C \nabla_{A\dot{B}} o_C, \tag{7.2a}$$

$$-\rho \equiv \Gamma_{142} = k_{a;b} m^a \overline{m}^b = m^a \bar{\delta} k_a = \iota^A \bar{o}^{\dot{B}} o^C \nabla_{A\dot{B}} o_C, \tag{7.2b}$$

$$-\sigma \equiv \Gamma_{141} = k_{a;b} m^a m^b = m^a \delta k_a = o^A \bar{\iota}^{\dot{B}} o^C \nabla_{A\dot{B}} o_C, \tag{7.2c}$$

$$-\tau \equiv \Gamma_{143} = k_{a;b} m^a l^b = m^a \Delta k_a = \iota^A \bar{\iota}^{\dot{B}} o^C \nabla_{A\dot{B}} o_C, \tag{7.2d}$$

$$\nu \equiv \Gamma_{233} = l_{a;b}\overline{m}^a l^b = \overline{m}^a \Delta l_a = -\iota^A \bar{\iota}^{\dot{B}} \iota^C \nabla_{A\dot{B}} \iota_C, \tag{7.2e}$$

$$\mu \equiv \Gamma_{231} = l_{a;b}\overline{m}^a m^b = \overline{m}^a \delta l_a = -o^A \bar{\iota}^{\dot{B}} \iota^C \nabla_{A\dot{B}} \iota_C, \tag{7.2f}$$

$$\lambda \equiv \Gamma_{232} = l_{a;b}\overline{m}^a \overline{m}^b = \overline{m}^a \bar{\delta} l_a = -\iota^A \bar{o}^{\dot{B}} \iota^C \nabla_{A\dot{B}} \iota_C, \tag{7.2g}$$

$$\pi \equiv \Gamma_{234} = l_{a;b}\overline{m}^a k^b = \overline{m}^a D l_a = -o^A \bar{o}^{\dot{B}} \iota^C \nabla_{A\dot{B}} \iota_C, \tag{7.2h}$$

$$-\varepsilon \equiv \tfrac{1}{2}(\Gamma_{344} - \Gamma_{214}) = \tfrac{1}{2}(k_{a;b} l^a k^b - m_{a;b}\overline{m}^a k^b)$$

$$= \tfrac{1}{2}(l^a D k_a - \overline{m}^a D m_a) = o^A \bar{o}^{\dot{B}} \iota^C \nabla_{A\dot{B}} o_C, \tag{7.2i}$$

$$-\beta \equiv \tfrac{1}{2}(\Gamma_{341} - \Gamma_{211}) = \tfrac{1}{2}(k_{a;b} l^a m^b - m_{a;b}\overline{m}^a m^b)$$

$$= \tfrac{1}{2}(l^a \delta k_a - \overline{m}^a \delta m_a) = o^A \bar{\iota}^{\dot{B}} \iota^C \nabla_{A\dot{B}} o_C, \tag{7.2j}$$

$$\gamma \equiv \tfrac{1}{2}(\Gamma_{433} - \Gamma_{123}) = \tfrac{1}{2}(l_{a;b} k^a l^b - \overline{m}_{a;b} m^a l^b)$$

$$= \tfrac{1}{2}(k^a \Delta l_a - m^a \Delta \overline{m}_a) = -\iota^A \bar{\iota}^{\dot{B}} o^C \nabla_{A\dot{B}} \iota_C, \tag{7.2k}$$

$$\alpha \equiv \tfrac{1}{2}(\Gamma_{432} - \Gamma_{122}) = \tfrac{1}{2}(l_{a;b} k^a \overline{m}^b - \overline{m}_{a;b} m^a \overline{m}^b)$$

$$= \tfrac{1}{2}(k^a \bar{\delta} l_a - m^a \bar{\delta} \overline{m}_a) = -\iota^A \bar{o}^{\dot{B}} o^C \nabla_{A\dot{B}} \iota_C, \tag{7.2l}$$

where we have used the notation (3.82), i.e.

$$D \equiv k^a \nabla_a = -o^A \bar{o}^{\dot{B}} \nabla_{A\dot{B}}, \qquad \Delta \equiv l^a \nabla_a = -\iota^A \bar{\iota}^{\dot{B}} \nabla_{A\dot{B}},$$
$$\delta \equiv m^a \nabla_a = -o^A \bar{\iota}^{\dot{B}} \nabla_{A\dot{B}}, \qquad \bar{\delta} \equiv \overline{m}^a \nabla_a = -\iota^A \bar{o}^{\dot{B}} \nabla_{A\dot{B}}, \tag{7.3}$$

for the directional derivatives $D, \Delta, \delta, \bar{\delta}$. Some of these spin coefficients have already been introduced in (6.20). From the spinor expressions for the spin coefficients and from the relation

$$o_A \iota^A = 1 \quad \Longrightarrow \quad \iota^C \nabla_{A\dot{B}} o_C = o^C \nabla_{A\dot{B}} \iota_C \tag{7.4}$$

it follows that all connection coefficients can be expressed in terms of the 12 complex spin coefficients (7.2). One can also give the spin coefficients as partial derivatives, e.g. $\kappa = k_{[a,b]} m^a k^b$. These formulae can be obtained from the commutator relations (see (7.6) below) and are given e.g. in Cocke (1989).

Since the metric components in the complex null basis are constant, the *commutators* introduced previously in (2.6) and (2.70),

$$[\boldsymbol{e}_a, \boldsymbol{e}_b] = D^c{}_{ab}\boldsymbol{e}_c, \qquad D^c{}_{ab} = -2\Gamma^c{}_{[ab]}, \tag{7.5}$$

are given explicitly in the present notation as follows:

$$(\Delta D - D\Delta) = (\gamma+\bar{\gamma})D + (\varepsilon+\bar{\varepsilon})\Delta - (\tau+\bar{\pi})\bar{\delta} - (\bar{\tau}+\pi)\delta, \tag{7.6a}$$

$$(\delta D - D\delta) = (\bar{\alpha}+\beta-\bar{\pi})D + \kappa\Delta - \sigma\bar{\delta} - (\bar{\rho}+\varepsilon-\bar{\varepsilon})\delta, \tag{7.6b}$$

$$(\delta\Delta - \Delta\delta) = -\bar{\nu}D + (\tau-\bar{\alpha}-\beta)\Delta + \bar{\lambda}\bar{\delta} + (\mu-\gamma+\bar{\gamma})\delta, \tag{7.6c}$$

$$(\bar{\delta}\delta - \delta\bar{\delta}) = (\bar{\mu}-\mu)D + (\bar{\rho}-\rho)\Delta - (\bar{\alpha}-\beta)\bar{\delta} - (\bar{\beta}-\alpha)\delta. \tag{7.6d}$$

These relations and the rigid frame conditions are equivalent to (7.2), as noted by Papapetrou (1971a, 1971b). The application of the commutator relations (7.6a)–(7.6d) to scalar functions (e.g. the space-time coordinates x^i) may yield information useful in solving the field equations.

In many applications, the null direction $\boldsymbol{k}$ is fixed; under Lorentz transformations (3.18) preserving this direction, i.e. under

$$\boldsymbol{k}' = \boldsymbol{k}, \quad \boldsymbol{m}' = \boldsymbol{m} + B\boldsymbol{k}, \quad \boldsymbol{l}' = \boldsymbol{l} + B\overline{\boldsymbol{m}} + \overline{B}\boldsymbol{m} + B\overline{B}\boldsymbol{k}, \tag{7.7a}$$

and

$$\boldsymbol{k}' = A\boldsymbol{k}, \quad \boldsymbol{m}' = \mathrm{e}^{\mathrm{i}\Theta}\boldsymbol{m}, \quad \boldsymbol{l}' = A^{-1}\boldsymbol{l}, \tag{7.7b}$$

the spin coefficients have the following transformation laws (Carmeli and Kaye 1976, Stewart 1990):

$$\begin{aligned}
&\kappa' = \kappa, && \tau' = \tau + \overline{B}\sigma + B\rho + \overline{B}B\kappa,\\
&\rho' = \rho + \overline{B}\kappa, && \alpha' = \alpha + \overline{B}(\varepsilon+\rho) + \overline{B}^2\kappa,\\
&\sigma' = \sigma + B\kappa, && \beta' = \beta + \overline{B}\sigma + B\varepsilon + \overline{B}B\kappa,\\
&\varepsilon' = \varepsilon + \overline{B}\kappa, && \pi' = \pi + 2\overline{B}\varepsilon + \overline{B}^2\kappa + D\overline{B},\\
&\gamma' = \gamma + B\alpha + \overline{B}(\tau+\beta) + \overline{B}B(\rho+\varepsilon) + \overline{B}^2\sigma + \overline{B}^2B\kappa,\\
&\lambda' = \lambda + \overline{B}(\pi+2\alpha) + \overline{B}^2(\rho+2\varepsilon) + \overline{B}^3\kappa + \overline{B}D\overline{B} + \bar{\delta}\,\overline{B},\\
&\mu' = \mu + 2\overline{B}\beta + B\pi + \overline{B}^2\sigma + 2B\overline{B}\varepsilon + B\overline{B}^2\kappa + BD\overline{B} + \delta\overline{B},\\
&\nu' = \nu + \overline{B}(2\gamma+\mu) + B\lambda + \overline{B}^2(\tau+2\beta) + B\overline{B}(\pi+2\alpha) + \overline{B}^3\sigma\\
&\qquad + B\overline{B}^2(\rho+2\varepsilon) + B\overline{B}^3\kappa + \Delta\overline{B} + \overline{B}\delta\overline{B} + B\bar{\delta}\,\overline{B} + B\overline{B}D\overline{B},
\end{aligned} \tag{7.7c}$$

and

$$
\begin{aligned}
\kappa' &= A^2 e^{i\Theta}\kappa, & \nu' &= A^{-2}e^{-i\Theta}\nu, & \varepsilon' &= A\left[\varepsilon + \tfrac{1}{2}D(\ln A + i\Theta)\right],\\
\rho' &= A\rho, & \mu' &= A^{-1}\mu, & \beta' &= e^{i\Theta}\left[\beta + \tfrac{1}{2}\delta(\ln A + i\Theta)\right],\\
\sigma' &= Ae^{2i\Theta}\sigma, & \lambda' &= A^{-1}e^{-2i\Theta}\lambda, & \gamma' &= A^{-1}\left[\gamma + \tfrac{1}{2}\Delta(\ln A + i\Theta)\right],\\
\tau' &= e^{i\Theta}\tau, & \pi' &= e^{-i\Theta}\pi, & \alpha' &= e^{-i\Theta}\left[\alpha + \tfrac{1}{2}\overline{\delta}(\ln A + i\Theta)\right].
\end{aligned}
\tag{7.7d}
$$

To get the transformation law for a null rotation (3.14) about $\boldsymbol{l}$,

$$\boldsymbol{l}' = \boldsymbol{l}, \quad \boldsymbol{m}' = E\boldsymbol{m}, \quad \boldsymbol{k}' = \boldsymbol{k} + E\overline{\boldsymbol{m}} + \overline{E}\boldsymbol{m} + E\overline{E}\boldsymbol{l}, \tag{7.8}$$

one can take (7.7c), replace B by $\overline{E}$, and make the interchange $(\kappa, \sigma, \rho, \tau, \beta, \varepsilon) \longleftrightarrow (\nu, -\lambda, -\mu, -\pi, -\alpha, -\gamma)$.

7.2 The Ricci equations

To derive the curvature in terms of the spin coefficients, it is convenient to work with the connection forms

$$
\begin{aligned}
\boldsymbol{\Gamma}_{14} &\equiv m^a k^b \Gamma_{abc}\boldsymbol{\omega}^c = -\sigma\boldsymbol{\omega}^1 - \rho\boldsymbol{\omega}^2 - \tau\boldsymbol{\omega}^3 - \kappa\boldsymbol{\omega}^4,\\
\boldsymbol{\Gamma}_{23} &\equiv \overline{m}^a l^b \Gamma_{abc}\boldsymbol{\omega}^c = \mu\boldsymbol{\omega}^1 + \lambda\boldsymbol{\omega}^2 + \nu\boldsymbol{\omega}^3 + \pi\boldsymbol{\omega}^4,\\
\tfrac{1}{2}(\boldsymbol{\Gamma}_{12} + \boldsymbol{\Gamma}_{34}) &\equiv \tfrac{1}{2}(m^a\overline{m}^b + l^a k^b)\Gamma_{abc}\boldsymbol{\omega}^c = -\beta\boldsymbol{\omega}^1 - \alpha\boldsymbol{\omega}^2 - \gamma\boldsymbol{\omega}^3 - \varepsilon\boldsymbol{\omega}^4,
\end{aligned}
\tag{7.9}
$$

introduced in §3.3. We now take the second Cartan equation in the form (3.25). On the right-hand side of these equations we insert the decomposition (3.45) of the curvature tensor, using the following abbreviations for the tetrad components of the traceless Ricci tensor ($S_{ab} \equiv R_{ab} - g_{ab}R/4$) and the Weyl tensor:

$$\Phi_{00} \equiv \tfrac{1}{2}S_{ab}k^a k^b = \Phi_{AB\dot{C}\dot{D}}o^A o^B \overline{o}^{\dot{C}} \overline{o}^{\dot{D}} = \overline{\Phi}_{00} = \tfrac{1}{2}R_{44}, \tag{7.10}$$

$$\Phi_{01} \equiv \tfrac{1}{2}S_{ab}k^a m^b = \Phi_{AB\dot{C}\dot{D}}o^A o^B \overline{o}^{\dot{C}} \overline{\iota}^{\dot{D}} = \overline{\Phi}_{10} = \tfrac{1}{2}R_{41}, \tag{7.11}$$

$$\Phi_{02} \equiv \tfrac{1}{2}S_{ab}m^a m^b = \Phi_{AB\dot{C}\dot{D}}o^A o^B \overline{\iota}^{\dot{C}} \overline{\iota}^{\dot{D}} = \overline{\Phi}_{20} = \tfrac{1}{2}R_{11}, \tag{7.12}$$

$$
\begin{aligned}
\Phi_{11} \equiv \tfrac{1}{4}S_{ab}(k^a l^b + m^a\overline{m}^b) &= \Phi_{AB\dot{C}\dot{D}}o^A \iota^B \overline{o}^{\dot{C}} \overline{\iota}^{\dot{D}}\\
&= \overline{\Phi}_{11} = (R_{43} + R_{12})/4,
\end{aligned}
\tag{7.13}
$$

$$\Phi_{12} \equiv \tfrac{1}{2}S_{ab}l^a m^b = \Phi_{AB\dot{C}\dot{D}}o^A \iota^B \overline{\iota}^{\dot{C}} \overline{\iota}^{\dot{D}} = \overline{\Phi}_{21} = \tfrac{1}{2}R_{31}, \tag{7.14}$$

$$\Phi_{22} \equiv \tfrac{1}{2}S_{ab}l^a l^b = \Phi_{AB\dot{C}\dot{D}}\iota^A\iota^B\bar{\iota}^{\dot{C}}\bar{\iota}^{\dot{D}} = \overline{\Phi}_{22} = \tfrac{1}{2}R_{33}; \tag{7.15}$$

$$\Psi_0 \equiv C_{abcd}k^a m^b k^c m^d = \Psi_{ABCD}o^A o^B o^C o^D, \tag{7.16}$$

$$\Psi_1 \equiv C_{abcd}k^a l^b k^c m^d = \Psi_{ABCD}o^A o^B o^C \iota^D, \tag{7.17}$$

$$\Psi_2 \equiv -C_{abcd}k^a m^b l^c \overline{m}^d = \Psi_{ABCD}o^A o^B \iota^C \iota^D, \tag{7.18}$$

$$\Psi_3 \equiv C_{abcd}l^a k^b l^c \overline{m}^d = \Psi_{ABCD}o^A \iota^B \iota^C \iota^D, \tag{7.19}$$

$$\Psi_4 \equiv C_{abcd}l^a \overline{m}^b l^c \overline{m}^d = \Psi_{ABCD}\iota^A \iota^B \iota^C \iota^D \tag{7.20}$$

(the definitions (7.16)–(7.20) agree with (3.59), (3.76)). On the left-hand side of equations (3.25) we calculate the exterior derivatives $\mathrm{d}\boldsymbol{\Gamma}_{ab}$, using the notation (7.3) for the directional derivatives. We thus arrive at the Ricci identities, often called the *Newman–Penrose equations:*

$$D\rho - \bar{\delta}\kappa = \rho^2 + \sigma\bar{\sigma} + (\varepsilon + \bar{\varepsilon})\rho - \bar{\kappa}\tau - \kappa(3\alpha + \bar{\beta} - \pi) + \Phi_{00}, \tag{7.21a}$$

$$D\sigma - \delta\kappa = (\rho + \bar{\rho})\sigma + (3\varepsilon - \bar{\varepsilon})\sigma - (\tau - \bar{\pi} + \bar{\alpha} + 3\beta)\kappa + \Psi_0, \tag{7.21b}$$

$$D\tau - \Delta\kappa = (\tau + \bar{\pi})\rho + (\bar{\tau} + \pi)\sigma + (\varepsilon - \bar{\varepsilon})\tau - (3\gamma + \bar{\gamma})\kappa + \Psi_1 + \Phi_{01}, \tag{7.21c}$$

$$D\alpha - \bar{\delta}\varepsilon = (\rho + \bar{\varepsilon} - 2\varepsilon)\alpha + \beta\bar{\sigma} - \bar{\beta}\varepsilon - \kappa\lambda - \bar{\kappa}\gamma + (\varepsilon + \rho)\pi + \Phi_{10}, \tag{7.21d}$$

$$D\beta - \delta\varepsilon = (\alpha + \pi)\sigma + (\bar{\rho} - \bar{\varepsilon})\beta - (\mu + \gamma)\kappa - (\bar{\alpha} - \bar{\pi})\varepsilon + \Psi_1, \tag{7.21e}$$

$$D\gamma - \Delta\varepsilon = (\tau + \bar{\pi})\alpha + (\bar{\tau} + \pi)\beta - (\varepsilon + \bar{\varepsilon})\gamma - (\gamma + \bar{\gamma})\varepsilon + \tau\pi - \nu\kappa + \Psi_2 + \Phi_{11} - R/24, \tag{7.21f}$$

$$D\lambda - \bar{\delta}\pi = \rho\lambda + \bar{\sigma}\mu + \pi^2 + (\alpha - \bar{\beta})\pi - \nu\bar{\kappa} - (3\varepsilon - \bar{\varepsilon})\lambda + \Phi_{20}, \tag{7.21g}$$

$$D\mu - \delta\pi = \bar{\rho}\mu + \sigma\lambda + \pi\bar{\pi} - (\varepsilon + \bar{\varepsilon})\mu - \pi(\bar{\alpha} - \beta) - \nu\kappa + \Psi_2 + R/12, \tag{7.21h}$$

$$D\nu - \Delta\pi = (\pi + \bar{\tau})\mu + (\bar{\pi} + \tau)\lambda + (\gamma - \bar{\gamma})\pi - (3\varepsilon + \bar{\varepsilon})\nu + \Psi_3 + \Phi_{21}, \tag{7.21i}$$

$$\Delta\lambda - \bar{\delta}\nu = -(\mu + \bar{\mu})\lambda - (3\gamma - \bar{\gamma})\lambda + (3\alpha + \bar{\beta} + \pi - \bar{\tau})\nu - \Psi_4, \tag{7.21j}$$

$$\delta\rho - \bar{\delta}\sigma = \rho(\bar{\alpha} + \beta) - \sigma(3\alpha - \bar{\beta}) + (\rho - \bar{\rho})\tau + (\mu - \bar{\mu})\kappa - \Psi_1 + \Phi_{01}, \tag{7.21k}$$

$$\delta\alpha - \bar{\delta}\beta = \mu\rho - \lambda\sigma + \alpha\bar{\alpha} + \beta\bar{\beta} - 2\alpha\beta + \gamma(\rho - \bar{\rho}) + \varepsilon(\mu - \bar{\mu})$$
$$-\Psi_2 + \Phi_{11} + R/24, \quad (7.21l)$$

$$\delta\lambda - \bar{\delta}\mu = (\rho - \bar{\rho})\nu + (\mu - \bar{\mu})\pi + \mu(\alpha + \bar{\beta}) + \lambda(\bar{\alpha} - 3\beta)$$
$$-\Psi_3 + \Phi_{21}, \quad (7.21m)$$

$$\delta\nu - \Delta\mu = \mu^2 + \lambda\bar{\lambda} + (\gamma + \bar{\gamma})\mu - \bar{\nu}\pi + (\tau - 3\beta - \bar{\alpha})\nu + \Phi_{22}, \quad (7.21n)$$

$$\delta\gamma - \Delta\beta = (\tau - \bar{\alpha} - \beta)\gamma + \mu\tau - \sigma\nu - \varepsilon\bar{\nu} - \beta(\gamma - \bar{\gamma} - \mu)$$
$$+\alpha\bar{\lambda} + \Phi_{12}, \quad (7.21o)$$

$$\delta\tau - \Delta\sigma = \mu\sigma + \bar{\lambda}\rho + (\tau + \beta - \bar{\alpha})\tau - (3\gamma - \bar{\gamma})\sigma - \kappa\bar{\nu} + \Phi_{02}, \quad (7.21p)$$

$$\Delta\rho - \bar{\delta}\tau = -(\rho\bar{\mu} + \sigma\lambda) + (\bar{\beta} - \alpha - \bar{\tau})\tau + (\gamma + \bar{\gamma})\rho + \nu\kappa$$
$$-\Psi_2 - R/12, \quad (7.21q)$$

$$\Delta\alpha - \bar{\delta}\gamma = (\rho + \varepsilon)\nu - (\tau + \beta)\lambda + (\bar{\gamma} - \bar{\mu})\alpha + (\bar{\beta} - \bar{\tau})\gamma - \Psi_3. \quad (7.21r)$$

The definitions of $\{\boldsymbol{e}_a\}$, ∇_a, $\Gamma^a{}_{bc}$ and $R^a{}_{bcd}$ coincide with those of the original paper (Newman and Penrose 1962). Whenever the metric is employed to move indices we have to remember the change of signature (for sign conventions see Ernst (1978a)).

The Einstein field equations have not been used so far: they give conditions on, or relations between, the quantities $\Phi_{AB'}$ and Λ in the above equations.

In the Newman–Penrose formalism, the Maxwell equations read

$$D\Phi_1 - \bar{\delta}\Phi_0 = (\pi - 2\alpha)\Phi_0 + 2\rho\Phi_1 - \kappa\Phi_2, \quad (7.22)$$

$$D\Phi_2 - \bar{\delta}\Phi_1 = -\lambda\Phi_0 + 2\pi\Phi_1 + (\rho - 2\varepsilon)\Phi_2, \quad (7.23)$$

$$\delta\Phi_1 - \Delta\Phi_0 = (\mu - 2\gamma)\Phi_0 + 2\tau\Phi_1 - \sigma\Phi_2, \quad (7.24)$$

$$\delta\Phi_2 - \Delta\Phi_1 = -\nu\Phi_0 + 2\mu\Phi_1 + (\tau - 2\beta)\Phi_2, \quad (7.25)$$

where the notation

$$\Phi_0 \equiv F_{ab}k^a m^b = \tfrac{1}{4}F^*_{ab}V^{ab} = \Phi_{AB}o^A o^B, \quad (7.26)$$

$$\Phi_1 \equiv \tfrac{1}{2}F_{ab}(k^a l^b + \overline{m}^a m^b) = -\tfrac{1}{8}F^*_{ab}W^{ab} = \Phi_{AB}o^A \iota^B, \quad (7.27)$$

$$\Phi_2 \equiv F_{ab}\overline{m}^a l^b = \tfrac{1}{4}F^*_{ab}U^{ab} = \Phi_{AB}\iota^A \iota^B, \quad (7.28)$$

is used for the tetrad components of the electromagnetic field tensor (cp. (5.10)). The Ricci tensor components (7.10)–(7.15) of Einstein–Maxwell fields are given by

$$\Phi_{\alpha\beta} = \kappa_0 \Phi_\alpha \overline{\Phi}_\beta, \qquad \alpha, \beta = 0, 1, 2. \tag{7.29}$$

Ref.: Hall *et al.* (1987) developed an analogous formalism for (2+1)-dimensional space-times, Stoeger *et al.* (1992) for a tetrad incorporating a time-like unit vector, and Ramos and Vickers (1996b) a calculus based on a single null direction which generalizes the GHP-formalism to be treated in §7.4.

7.3 The Bianchi identities

The *Bianchi identities* (3.78) (better referred to as equations in this context),

$$R_{ab[cd;e]} = 0, \tag{7.30}$$

are the remaining set of equations to be satisfied. Written in terms of the tetrad components and the directional derivatives, they have the form

$$R_{ab[cd|f]} = -2R_{abe[c}\Gamma^e{}_{df]} + \Gamma^e{}_{a[c}R_{df]eb} - \Gamma^e{}_{b[c}R_{df]ea}. \tag{7.31}$$

In full, they read (Pirani 1965, p. 350):

$$\begin{aligned}\bar\delta\Psi_0 - D\Psi_1 + D\Phi_{01} - \delta\Phi_{00} &= (4\alpha - \pi)\Psi_0 - 2(2\rho + \varepsilon)\Psi_1 + 3\kappa\Psi_2 \\ &+ (\bar\pi - 2\bar\alpha - 2\beta)\Phi_{00} + 2(\varepsilon + \bar\rho)\Phi_{01} + 2\sigma\Phi_{10} - 2\kappa\Phi_{11} - \bar\kappa\Phi_{02},\end{aligned} \tag{7.32a}$$

$$\begin{aligned}\Delta\Psi_0 - \delta\Psi_1 + D\Phi_{02} - \delta\Phi_{01} &= (4\gamma - \mu)\Psi_0 - 2(2\tau + \beta)\Psi_1 + 3\sigma\Psi_2 \\ &+ (2\varepsilon - 2\bar\varepsilon + \bar\rho)\Phi_{02} + 2(\bar\pi - \beta)\Phi_{01} + 2\sigma\Phi_{11} - 2\kappa\Phi_{12} - \bar\lambda\Phi_{00},\end{aligned} \tag{7.32b}$$

$$\begin{aligned}\bar\delta\Psi_3 - D\Psi_4 + \bar\delta\Phi_{21} - \Delta\Phi_{20} &= (4\varepsilon - \rho)\Psi_4 - 2(2\pi + \alpha)\Psi_3 + 3\lambda\Psi_2 \\ &+ (2\gamma - 2\bar\gamma + \bar\mu)\Phi_{20} + 2(\bar\tau - \alpha)\Phi_{21} + 2\lambda\Phi_{11} - 2\nu\Phi_{10} - \bar\sigma\Phi_{22},\end{aligned} \tag{7.32c}$$

$$\begin{aligned}\Delta\Psi_3 - \delta\Psi_4 + \bar\delta\Phi_{22} - \Delta\Phi_{21} &= (4\beta - \tau)\Psi_4 - 2(2\mu + \gamma)\Psi_3 + 3\nu\Psi_2 \\ &+ (\bar\tau - 2\bar\beta - 2\alpha)\Phi_{22} + 2(\gamma + \bar\mu)\Phi_{21} + 2\lambda\Phi_{12} - 2\nu\Phi_{11} - \bar\nu\Phi_{20},\end{aligned} \tag{7.32d}$$

$$\begin{aligned}&D\Psi_2 - \bar\delta\Psi_1 + \Delta\Phi_{00} - \bar\delta\Phi_{01} + \tfrac{1}{12}DR = \\ &\quad - \lambda\Psi_0 + 2(\pi - \alpha)\Psi_1 + 3\rho\Psi_2 - 2\kappa\Psi_3 \\ &\quad + (2\gamma + 2\bar\gamma - \bar\mu)\Phi_{00} - 2(\bar\tau + \alpha)\Phi_{01} - 2\tau\Phi_{10} + 2\rho\Phi_{11} + \bar\sigma\Phi_{02},\end{aligned} \tag{7.32e}$$

$$\begin{aligned}&\Delta\Psi_2 - \delta\Psi_3 + D\Phi_{22} - \delta\Phi_{21} + \tfrac{1}{12}\Delta R = \\ &\quad \sigma\Psi_4 + 2(\beta - \tau)\Psi_3 - 3\mu\Psi_2 + 2\nu\Psi_1 \\ &\quad + (\bar\rho - 2\varepsilon - 2\bar\varepsilon)\Phi_{22} + 2(\bar\pi + \beta)\Phi_{21} + 2\pi\Phi_{12} - 2\mu\Phi_{11} - \bar\lambda\Phi_{20},\end{aligned} \tag{7.32f}$$

$$D\Psi_3 - \bar{\delta}\Psi_2 - D\Phi_{21} + \delta\Phi_{20} - \tfrac{1}{12}\bar{\delta}R =$$
$$- \kappa\Psi_4 + 2(\rho - \varepsilon)\Psi_3 + 3\pi\Psi_2 - 2\lambda\Psi_1$$
$$+ (2\bar{\alpha} - 2\beta - \bar{\pi})\Phi_{20} - 2(\bar{\rho} - \varepsilon)\Phi_{21} - 2\pi\Phi_{11} + 2\mu\Phi_{10} + \bar{\kappa}\Phi_{22}, \quad (7.32g)$$

$$\Delta\Psi_1 - \delta\Psi_2 - \Delta\Phi_{01} + \bar{\delta}\Phi_{02} - \tfrac{1}{12}\delta R =$$
$$\nu\Psi_0 + 2(\gamma - \mu)\Psi_1 - 3\tau\Psi_2 + 2\sigma\Psi_3$$
$$+ (\bar{\tau} - 2\bar{\beta} + 2\alpha)\Phi_{02} + 2(\bar{\mu} - \gamma)\Phi_{01} + 2\tau\Phi_{11} - 2\rho\Phi_{12} - \bar{\nu}\Phi_{00}, \quad (7.32h)$$

$$D\Phi_{11} - \delta\Phi_{10} - \bar{\delta}\Phi_{01} + \Delta\Phi_{00} + \tfrac{1}{8}DR =$$
$$(2\gamma - \mu + 2\bar{\gamma} - \bar{\mu})\Phi_{00} + (\pi - 2\alpha - 2\bar{\tau})\Phi_{01} + (\bar{\pi} - 2\bar{\alpha} - 2\tau)\Phi_{10}$$
$$+ 2(\rho + \bar{\rho})\Phi_{11} + \bar{\sigma}\Phi_{02} + \sigma\Phi_{20} - \bar{\kappa}\Phi_{12} - \kappa\Phi_{21}, \quad (7.32i)$$

$$D\Phi_{12} - \delta\Phi_{11} - \bar{\delta}\Phi_{02} + \Delta\Phi_{01} + \tfrac{1}{8}\delta R =$$
$$(-2\alpha + 2\bar{\beta} + \pi - \bar{\tau})\Phi_{02} + (\bar{\rho} + 2\rho - 2\bar{\varepsilon})\Phi_{12} + 2(\bar{\pi} - \tau)\Phi_{11}$$
$$+ (2\gamma - 2\bar{\mu} - \mu)\Phi_{01} + \bar{\nu}\Phi_{00} - \bar{\lambda}\Phi_{10} + \sigma\Phi_{21} - \kappa\Phi_{22}, \quad (7.32j)$$

$$D\Phi_{22} - \delta\Phi_{21} - \bar{\delta}\Phi_{12} + \Delta\Phi_{11} + \tfrac{1}{8}\Delta R =$$
$$(\rho + \bar{\rho} - 2\varepsilon - 2\bar{\varepsilon})\Phi_{22} + (2\bar{\beta} + 2\pi - \bar{\tau})\Phi_{12} + (2\beta + 2\bar{\pi} - \tau)\Phi_{21}$$
$$- 2(\mu + \bar{\mu})\Phi_{11} + \nu\Phi_{01} + \bar{\nu}\Phi_{10} - \bar{\lambda}\Phi_{20} - \lambda\Phi_{02}. \quad (7.32k)$$

The consistency, completeness and integrability of the Newman–Penrose formalism has been considered in a number of papers, e.g. Papapetrou (1971a, 1971b) and Edgar (1980, 1992). As given here, the equations are a set of differential equations for the tetrad components ${e_a}^i$ with respect to a coordinate basis $\{\partial/\partial x^i\}$, the spin coefficients (7.2) and the Riemann tensor components (7.10)–(7.20), the corresponding equations being respectively either the commutator relations (7.6) together with the rigid frame condition $\mathrm{d}g_{ab} = 0$ or the definitions (7.2), the Ricci (Newman–Penrose) equations (7.21) and the Bianchi equations (7.32). It is (implicitly) assumed that the connection coefficients and Riemann tensor components not mentioned explicitly in (7.2) and (7.10)–(7.20) can be found from the symmetry relations $\Gamma_{(ab)c} = 0$ and (3.26).

There is redundancy between these equations in the sense that some of them (or combinations of some of them) are integrability conditions for others. The underlying reason is that (7.6) and (7.21) are versions of the Cartan structure equations (2.76) and (2.85), which have as integrability conditions the first and second Bianchi identities, i.e. $\mathrm{d}^2\boldsymbol{\omega}^a = 0$

(or equivalently the Jacobi identities (2.7) or the symmetry given by the second part of (2.80)) and $\mathrm{d}^2\boldsymbol{\Gamma}^a{}_b = 0$ or (2.81); however, the situation is complicated by the fact that the Cartan equations have been contracted with tetrad vectors, substitutions have been made from (7.6) in (7.21), and so on. Details of the resulting relations, which show that the general system is complete and consistent, are given by Edgar (1980, 1992), not only for the Newman–Penrose formalism but for general sets of tetrads with constant metrics and for the GHP formalism discussed in §7.4.

In the system as defined above, the integrability conditions of the second Bianchi equations (2.86), $\mathrm{d}^2\boldsymbol{\Theta}^a{}_b = 0$, which could be called the third Bianchi identities, are identically satisfied. The analysis above could, however, be extended by introducing derivatives of Riemann tensor components as extra variables so that the third Bianchi identities become additional equations, and so on.

One may note that the 36 real equations formed by the real and imaginary parts of (7.21) can be combined into the 16 real Jacobi (first Bianchi) equations for the commutators (7.6) and 20 real equations giving the Riemann tensor components (7.10)–(7.20) as differential expressions in the spin coefficients. Thus if functions $\Gamma^a{}_{bc}(x^i)$ for the spin coefficients are given which satisfy the 16 Jacobi relations, the integrability conditions of the commutator equations (7.6) are satisfied and one can integrate the commutator relations for the $e_a{}^i$ and thence the line-element, without considering the remaining 20 real components of (7.21), or (7.32). (Here we assume that there are no analytic or topological obstructions to the integrability.) However, calculation of the derivatives of the spin coefficients implies at least partial knowledge of the $e_a{}^i$ already.

It is less simple to characterize useful subsystems which involve R_{abcd} and guarantee integrability of the remaining equations, because to satisfy either the remaining 20 real parts of the Ricci equations or the Bianchi equations we need both $\Gamma^a{}_{bc}(x^i)$ and $R_{abcd}(x^i)$ (and information on how to calculate their tetrad derivatives). In principle for any given situation the completeness of a given subset could be studied by specialization of the results of Edgar (1980, 1992).

In practice, additional constraints are always imposed. For example, the consistency of the general vacuum case was studied by Papapetrou (1971a, 1971b). A number of papers have studied the system for Petrov type I vacua arising from the further specialization to a tetrad in which $\Psi_0 = \Psi_4$ and $\Psi_1 = \Psi_3 = 0$, i.e. the tetrad of the canonical form in Table 4.2, valid in a region where the Petrov type does not degenerate and the special tetrad can be chosen in a smooth manner. In this case the tetrad conditions and their derivatives have to be added to the Newman–Penrose system, and the third Bianchi equations are also needed. The resulting additional equations in Ψ_0 and Ψ_2 and their derivatives (Brans 1977, Edgar 1979) are called the post-Bianchi or Brans–Edgar equations (they were first derived as conditions on a spin-2 field in space-time by Bell and Szekeres (1972)). The consistency conditions, i.e. fourth Bianchi identities, were explicitly checked, some of them turning out to depend algebraically on the Ricci equations in this case, but consistency is guaranteed a priori by the fact that the Newman–Penrose system could in principle be solved in a general tetrad and then the special tetrad could have been found pointwise, algebraically.

Individual investigations, such as many of those quoted in later chapters, use various combinations of assumptions and may therefore require a separate consistency analysis (which may be negative: see for example Theorem 33.2); the assumptions may also imply that only a particular subset of the general set of equations is needed, which might for example not explicitly include the commutator equations, these instead following as consistency conditions (Edgar 1992).

Ref.: For further discussion of the post-Bianchi equations see Ludwig (1996), Bonanos (1996) and references therein.

7.4 The GHP calculus

Geroch *et al.* (1973) developed a modified calculus, the GHP-*formalism*, adapted to physical situations in which a pair of real null directions is naturally picked out at each space-time point. (Further modified and extended versions of the formalism were given by Held (1974a, 1975) and Ludwig (1988).) This version of the spin coefficient method leads to even simpler formulae than the standard Newman–Penrose technique.

In spinor notation, the most general transformation preserving the two preferred null directions and the dyad normalization $o_A\iota^A = 1$ is given by

$$o^A \longrightarrow C o^A, \qquad \iota^A \longrightarrow C^{-1}\iota^A, \qquad C \text{ complex.} \tag{7.33}$$

The corresponding two-parameter subgroup of the Lorentz group (boost and spatial rotations) affects the complex null tetrad $(\boldsymbol{m}, \overline{\boldsymbol{m}}, \boldsymbol{l}, \boldsymbol{k})$ as follows (cp. (7.7*b*)):

$$\boldsymbol{k} \longrightarrow A\boldsymbol{k}, \;\; \boldsymbol{l} \longrightarrow A^{-1}\boldsymbol{l}, \;\; \boldsymbol{m} \longrightarrow \mathrm{e}^{\mathrm{i}\Theta}\boldsymbol{m}; \;\; A = C\bar{C}, \;\; \mathrm{e}^{\mathrm{i}\Theta} = C\bar{C}^{-1}. \tag{7.34}$$

A scalar η which undergoes the transformation

$$\eta \longrightarrow C^p \bar{C}^q \eta \tag{7.35}$$

is called a weighted *scalar of type* (p, q), or a spin- and boost-weighted scalar of type $(\frac{1}{2}(p-q), \frac{1}{2}(p+q))$. The components of the Weyl and Ricci tensors, and the spin coefficients κ, λ, μ, ν, π, ρ, σ, τ, have the types

$$\begin{array}{lllll}\Psi_0:(4,0), & \Psi_1:(2,0), & \Psi_2:(0,0), & \Psi_3:(-2,0), & \Psi_4:(-4,0),\\ \Phi_{00}:(2,2), & \Phi_{01}:(2,0), & \Phi_{10}:(0,2), & \Phi_{02}:(2,-2), & \Phi_{20}:(-2,2),\\ \Phi_{11}:(0,0), & \Phi_{12}:(0,-2), & \Phi_{21}:(-2,0), & \Phi_{22}:(-2,-2), & \\ \kappa:(3,1), & \lambda:(-3,1), & \mu:(-1,-1), & \nu:(-3,-1), & \\ \pi:(-1,1), & \rho:(1,1), & \sigma:(3,-1), & \tau:(1,-1). & \end{array} \tag{7.36}$$

In the GHP calculus the replacement

$$\boldsymbol{k} \longleftrightarrow \boldsymbol{l}, \qquad \boldsymbol{m} \longleftrightarrow \overline{\boldsymbol{m}}, \tag{7.37}$$

is indicated by a prime on the symbols, for instance

$$\kappa' \equiv -\nu, \quad \sigma' \equiv -\lambda, \quad \rho' \equiv -\mu, \quad \tau' \equiv -\pi, \quad \beta' \equiv -\alpha, \quad \varepsilon' \equiv -\gamma. \tag{7.38}$$

There is also the star operation given by

$$\boldsymbol{k} \longleftrightarrow \boldsymbol{m}, \qquad \boldsymbol{l} \longleftrightarrow \overline{\boldsymbol{m}}, \tag{7.39}$$

which leads to

$$\begin{array}{llllll}\kappa^* \equiv \sigma, & \lambda^* \equiv \nu, & \mu^* \equiv \pi, & \rho^* \equiv \tau, & \beta^* \equiv \varepsilon, & \alpha^* \equiv \gamma,\\ \bar\kappa^* \equiv -\bar\lambda, & \bar\mu^* \equiv -\bar\tau, & \bar\rho^* \equiv -\bar\pi, & \bar\sigma^* \equiv -\bar\nu, & \bar\beta^* \equiv -\bar\gamma, & \bar\alpha^* \equiv -\bar\varepsilon.\end{array} \tag{7.40}$$

Applied to a type (p, q) scalar η, the operations given by the substitutions (7.37) and (7.39) and by complex conjugation respectively produce scalars η' of type $(-p, -q)$, η^* of type $(p, -q)$ and $\bar\eta$ of type (q, p). The prime and star conventions considerably reduce the notational effort, and are helpful in checking formulae; e.g. the Newman–Penrose equation (7.21n) is simply the primed version of (7.21a), etc.

The spin coefficients β, β', ε, ε' transform, under the tetrad change (7.33)–(7.34), according to inhomogeneous laws (7.7d) containing derivatives of C (resp. A and Θ). Therefore, these spin coefficients do not appear directly in the modified equations. However, they enter the new derivative operators acting on weighted scalars η of type (p, q):

$$\begin{array}{ll}\text{þ}\eta \equiv (D - p\varepsilon - q\bar\varepsilon)\eta, & \text{þ}'\eta \equiv (\Delta + p\varepsilon' + q\bar\varepsilon')\eta,\\ \eth\eta \equiv (\delta - p\beta + q\bar\beta')\eta, & \eth'\eta \equiv (\bar\delta + p\beta' - q\bar\beta)\eta.\end{array} \tag{7.41}$$

One can understand this transformation of operators as an example of 'absorption of torsion' (Gardner 1989, Olver 1995).

The operators þ and ð ('thorn' and 'edth') respectively map a scalar of type (p, q) into scalars of types $(p + 1, q + 1)$ and $(p + 1, q - 1)$. In consequence of the definitions (7.38) and (7.41), the commutators (7.6), Ricci (Newman–Penrose) equations (7.21), and Bianchi equations (7.32) get new explicit forms. They contain only scalars and derivative operators of good weight, and split into two sets of equations, one being the primed version of the other. It is important to note that the new commutators include weight-dependent terms, which, as can be seen from (7.41), contain some of the information formerly in the Newman–Penrose equations (7.21), so that to extract the full information contained in them one must apply them to quantities of non-zero weight, as well as some of zero weight.

To use this formalism for exact solutions, one attempts to choose the tetrad invariantly, any remaining gauge freedom being used to simplify formulae, and then find as many zero-weighted independent combinations of the variables as possible (Edgar and Ludwig 1997b). These can then be used as invariantly-defined coordinates. By completing the tables for the action of the operators (7.41) on these coordinates and on some suitable object of non-zero spin and boost weight, adding new variables and tables where needed (cp. Kerr (1998)), one may be able to carry out a complete integration of the field equations, the advantage of the method being that the first part of the integration is a coordinate-free procedure, and the final coordinates are invariantly chosen.

Examples of elegant discussions on these lines are given by Held for a class of type D vacuum metrics (1974a) and algebraically special metrics in general (1975), and by Edgar and Ludwig (1997b) for conformally flat pure radiation metrics. The GHP technique is also useful in dealing with fields in algebraically special background metrics, especially the Kerr metric, see e.g. Breuer (1975).

Ref.: For a presentation of the GHP-formalism see also Penrose and Rindler (1984).

7.5 Geodesic null congruences

In §6.1 we dealt with geodesic null congruences, whose tangent vector fields $\boldsymbol{k}$ satisfy

$$\kappa \equiv -k_{a;b}m^a k^b = 0, \tag{7.42}$$

and introduced the complex divergence ρ and the complex shear σ,

$$\rho \equiv -k_{a;b}m^a\overline{m}^b = -(\Theta + \mathrm{i}\omega), \qquad \sigma \equiv -k_{a;b}m^a m^b, \tag{7.43}$$

(see (6.17) and (6.20)), in agreement with the definitions (7.2*b*), (7.2*c*), of these spin coefficients. If $\sigma = 0$, the congruence will be called *shearfree.*

Here we consider the possible simplifications of the Newman–Penrose equations when the null vector $\boldsymbol{k}$ is geodesic. Null vector fields which are both geodesic and shearfree will be the subject of the next section.

Choosing $\boldsymbol{k}$ so that the geodesics are affinely parametrized, we have

$$k_{a;b}k^b = 0 \quad \Leftrightarrow \quad \kappa = 0, \qquad \varepsilon + \bar{\varepsilon} = 0. \tag{7.44}$$

Then (7.21a) is equivalent to the propagation equation (6.34),

$$D\rho = \rho^2 + \sigma\bar{\sigma} + \Phi_{00}. \tag{7.45}$$

If $\boldsymbol{k}$ is a double principal null direction of the Weyl tensor ($\Psi_0 = \Psi_1 = 0$), then for vacuum fields the Bianchi identity (7.32e) takes the simple form

$$D\Psi_2 = 3\rho\Psi_2. \tag{7.46}$$

If the null tetrad $\{\boldsymbol{e}_a\}$ is parallelly propagated along the geodesic null congruence $\boldsymbol{k}$, we obtain

$$\kappa = \varepsilon = \pi = 0, \tag{7.47}$$

i.e. three (complex) spin coefficients are zero. This choice of tetrad is very convenient for certain calculations; the left-hand sides of (7.21a)–(7.21i) become directional derivatives $D\rho, \ldots, D\nu$ of the remaining spin coefficients and thus in a coordinate system with $\boldsymbol{k} = \partial_r$, these equations determine the r-dependence of the spin coefficients.

A simple geodesic principal null direction $\boldsymbol{k}$ of a vacuum field is non-twisting, $k_{[a;b}k_{c]} = 0$ (Kammerer 1966).

7.6 The Goldberg–Sachs theorem and its generalizations

The Goldberg–Sachs theorem is very useful in constructing algebraically special solutions. It exhibits a close connection between certain geometrical properties of a null congruence and Petrov type. The original paper (Goldberg and Sachs 1962) presents the proof of two theorems:

Theorem 7.1 *If a gravitational field contains a shearfree geodesic null congruence* $\boldsymbol{k}$ ($\kappa = 0 = \sigma$) *and if*

$$R_{ab}k^a k^b = R_{ab}k^a m^b = R_{ab}m^a m^b = 0, \tag{7.48}$$

then the field is algebraically special, and $\boldsymbol{k}$ *is a degenerate eigendirection;*

$$C_{abc[d}k_{e]}k^b k^c = 0 \quad \Leftrightarrow \quad \Psi_0 = 0 = \Psi_1. \tag{7.49}$$

Remark: The conditions (7.48) are invariant with respect to the null rotations (7.7a).

Theorem 7.2 *If a vacuum metric ($R_{ab} = 0$) is algebraically special, then the multiple principal null vector is tangent to a shearfree geodesic null congruence.*

Combining these two theorems one obtains the well-known form of the Goldberg–Sachs theorem:

Theorem 7.3 (Goldberg–Sachs theorem). *A vacuum metric is algebraically special if and only if it contains a shearfree geodesic null congruence,*

$$\kappa = 0 = \sigma \quad \Leftrightarrow \quad \Psi_0 = 0 = \Psi_1. \tag{7.50}$$

All statements in (7.50) remain unchanged under conformal transformations (3.83). This remark leads to an obvious generalization of the Goldberg–Sachs theorem, namely to any gravitational field that is conformal to a vacuum field (Robinson and Schild 1963). The Goldberg–Sachs theorem was proved in Newman and Penrose (1962) using the formalism outlined in §§7.1–7.3. From the Bianchi equations (7.32a)–(7.32k) it is easily seen that with the assumption $\Psi_0 = 0 = \Psi_1$ we obtain $\kappa = 0 = \sigma$. The converse is more difficult to prove. In the special case $\rho = 0$ we obtain $\Psi_0 = 0 = \Psi_1$ from equations (7.21b), (7.21k). If $\rho \neq 0$ one can always set $\alpha + \bar{\beta} = 0$ by tetrad rotations and from (7.21b), (7.21d), (7.21e) and (7.44) we arrive at $\Psi_0 = 0$, $\Psi_1 = \rho\pi$. Various steps using the commutators and Bianchi equations lead to the final result $\pi = 0 = \Psi_1$.

For empty space-times which are algebraically special on a given *submanifold* $\mathcal{S}$, which is either a spacelike hypersurface or a timelike world line, the vector field tangent to a principal null direction of the curvature tensor and pointing, on $\mathcal{S}$, in the repeated principal null direction is geodesic and shearfree on $\mathcal{S}$ (Collinson 1967).

We give (without proof) an interesting theorem (Mariot 1954, Robinson 1961) which allows a reformulation of the Goldberg–Sachs theorem.

Theorem 7.4 (Mariot–Robinson theorem). *An arbitrary space-time V_4 admits a geodesic shearfree null congruence if and only if V_4 admits an electromagnetic null field ('test field') satisfying the Maxwell equations in V_4:*

$$\kappa = 0 = \sigma \quad \Leftrightarrow \quad F^*_{ab}k^b = 0, \qquad F^{*ab}{}_{;b} = 0. \tag{7.51}$$

Conditions (7.48) show that only a part of the vacuum field equations is needed to prove Theorem 7.1. Thus we have the

Corollary. *The Weyl tensor of Einstein–Maxwell fields with an electromagnetic null field is algebraically special.*

Proof: The Ricci tensor obeys the conditions (7.48) and the Maxwell equations demand that $\boldsymbol{k}$ is geodesic and shearfree (Theorem 7.4).

A generalization of the Goldberg–Sachs theorem, not restricted to vacuum solutions, is due to Kundt and Thompson (1962) and Robinson and Schild (1963):

Theorem 7.5 (Kundt–Thompson theorem). *Any two of the following imply the third:*

(A) the Weyl tensor is algebraically special, $\mathbf{k}$ *being the repeated null vector.*

(B) $\mathbf{k}$ *is shearfree and geodesic* ($\sigma = \kappa = 0$).

(C) $V^{ab}C_{abcd}{}^{;d}V^{ce} = 0$ *for Petrov type II or D,*
$V^{ab}C_{abcd}{}^{;d} = 0$ *for Petrov type III,*
$U^{ab}C_{abcd}{}^{;d}V^{ce} = 0$ *for Petrov type N.*

Proof: By elementary calculation we derive the following equations (Szekeres 1966b):

$$\text{Type } II, D\ (\Psi_0 = \Psi_1 = 0):\quad V^{ab}C_{abcd}{}^{;d}V^{ce} = 6\Psi_2(\sigma k^e - \kappa m^e), \tag{7.52a}$$

$$\text{Type } III\ (\Psi_0 = \Psi_1 = \Psi_2 = 0):\quad V^{ab}C_{abcd}{}^{;d} = 4\Psi_3(\sigma k_c - \kappa m_c), \tag{7.52b}$$

$$\text{Type } N\ (\Psi_0 = \Psi_1 = \Psi_2 = \Psi_3 = 0):$$
$$U^{ab}C_{abcd}{}^{;d}V^{ce} = 2\Psi_4(\sigma k^e - \kappa m^e), \tag{7.52c}$$

from the decomposition (3.58). Here we have used the definitions (3.40) of the complex self-dual bivectors U_{ab}, V_{ab}, W_{ab}, and the normalization (3.9) of the complex null tetrad. From (7.52) it is clear that (A), (B) $\Rightarrow$ (C) and (A), (C) $\Rightarrow$ (B). The proof that (B), (C) implies (A) is less trivial and is omitted here. Condition (C) of the Kundt–Thompson theorem may be replaced (Bell and Szekeres 1972) by the condition (C′): there exists a null type solution ($\Phi_{AB\cdots M} = \Phi o_A o_B \cdots o_M$) of the zero rest-mass free field equation $\nabla^{A\dot{X}}\Phi_{AB\cdots M} = 0$ for some spin value $s > 1$.

For vacuum fields, condition (C) of Theorem 7.5 is automatically true and (A) $\Leftrightarrow$ (B) is just the Goldberg–Sachs theorem (Theorem 7.3). This follows from the Bianchi identities (3.78) written in the form

$$C_{abcd}{}^{;d} = R_{c[a;b]} - \tfrac{1}{6}g_{c[a}R_{,b]} \tag{7.53}$$

(this form can be obtained by using the relations (3.50) and (3.52)).

Suppose we have an algebraically special Einstein–Maxwell field with a non-null electromagnetic field such that one of the eigendirections of the

Maxwell tensor is *aligned* with the multiple principal null direction of the Weyl tensor. Then it follows from the Bianchi identities (7.32*a*), (7.32*b*), and expressions (7.29), (5.12) that

$$(2\kappa_0\Phi_1\overline{\Phi}_1 + 3\Psi_2)\sigma = 0, \qquad (-2\kappa_0\Phi_1\overline{\Phi}_1 + 3\Psi_2)\kappa = 0 \tag{7.54}$$

(Kundt and Trümper 1962). If $\Psi_2 = 0$ (Petrov type *III* and more special types), then $\kappa = \sigma = 0$. If $\Psi_2 \neq 0$ one obtains $\kappa\sigma = 0$, i.e. either κ or σ must vanish. Relation (7.52*a*) also leads to (7.54).

Unfortunately, the Kundt–Thompson theorem does not directly specify the most general matter distribution which would allow one to conclude that (A) ⇒ (B). For instance, the assumption that the Ricci tensor is of pure radiation type,

$$R_{ab} = \kappa_0\Phi^2 k_a k_b = \kappa_0 T_{ab}, \tag{7.55}$$

does not guarantee condition (C) for fields of Petrov type N; in general the shear of $\boldsymbol{k}$ does not vanish, cp. §26.1. However, if T_{ab} in (7.55) is the energy-momentum tensor of an electromagnetic (null) field, the congruence $\boldsymbol{k}$ is necessarily shearfree because of the above corollary to Theorem 7.4.

8
Continuous groups of transformations; isometry and homothety groups

In this chapter we shall summarize those elements of the theory of continuous groups of transformations which we require for the following chapters. As far as we know, the most extensive treatment of this subject is to be found in Eisenhart (1933), while more recent applications to general relativity can be found in the works of Petrov (1966) and Defrise (1969), for example. General treatments of Lie groups and transformation groups in coordinatefree terms can be found in, for example, Cohn (1957), Warner (1971) and Brickell and Clark (1970), but none of these cover the whole of the material contained in Eisenhart's treatise.

Einstein's equations have as the possible generators of similarity solutions either isometries or homotheties (see §10.2.3). Hence we treat these types of symmetry here, the other types of symmetry, which are more general in the sense of imposing weaker conditions, but are more special in the sense of occurring rarely in exact solutions, being discussed in Chapter 35. Isometries have been widely used in constructing solutions, as the results described in Part II show. Many of the solutions found also admit proper homotheties (homotheties which are not isometries), and these are listed in Tables 11.2–11.4, but only since the 1980s have homotheties been used explicitly in the construction of solutions.

8.1 Lie groups and Lie algebras

We begin by introducing the concepts of Lie groups and Lie algebras, and the relation between them. A *Lie group* G is (i) a group (in the usual sense of algebra), with elements q_0, q_1, $q_2, \ldots$ say, q_0 being the identity element, and (ii) a differentiable manifold (§2.2) such that the map $\Phi : G \times G \to G$ given by the algebraic product $(q_1, q_2) \to q_1 q_2$ is analytic. Coordinates in (a neighbourhood of the identity q_0 of) G are called group *parameters*, and the analytic functions describing Φ in such coordinates are called *composition functions.*

We shall not discuss the differentiability conditions required to ensure that a continuous group (and, in later sections, the action of a continuous transformation group) is analytic; for these see e.g. Cohn (1957). In our applications such conditions always hold. Moreover, as in the rest of the book, many of the results stated apply only locally, i.e. in a neighbourhood: for example, in what follows, we may say that a manifold is invariant under a group, whereas all that we really require is that a neighbourhood (on which we solve the Einstein equations) is isometric to a neighbourhood in a space with the stated symmetry. In a few places where the distinction is crucial we shall remind the reader of it by inserting 'locally'. Schmidt (1971) and Hall (1989) give some results on the relation of local and global groups of transformations.

Our aim is to study transformation groups. An abstract Lie group naturally has associated with it two transformation groups. One of them consists of the *left translations*, the left translation associated with $q \in G$ being the map L_q of G to G such that

$$q' \longrightarrow qq'; \tag{8.1a}$$

the other consists of *right translations* R_q defined similarly by

$$q' \longrightarrow q'q. \tag{8.1b}$$

Each of these has associated with it a set of vector fields related to one-dimensional subgroups of transformations in the same way as $\boldsymbol{v}$ is related to Φ_t in §2.8.

As is easily seen, right translations commute with left translations,

$$R_q L_{q'} = L_{q'} R_q. \tag{8.2}$$

If we follow the convention that maps are written on the left (e.g. $L_q(q') = qq'$), then the left translation group is isomorphic to G and is called the *parameter group*, while the right translation group is algebraically dual to G. (If maps are written on the right, 'right' and 'left' must be interchanged in all subsequent statements.) The vector fields related to left translations turn out to be right-invariant vector fields, which we now study.

A *right-invariant vector field* $\boldsymbol{v}$ on G is defined to be one satisfying

$$(R_q)_*\boldsymbol{v} = \boldsymbol{v} \tag{8.3}$$

(for definitions, see (2.23), (2.24)). The value $\boldsymbol{v}(q)$ of such a vector field at a point q gives, and is given by, its value $\boldsymbol{v}(q_0)$ at the group identity:

$$\boldsymbol{v}(q) = (R_q)_*\, \boldsymbol{v}(q_0), \qquad \boldsymbol{v}(q_0) = (R_{q^{-1}})_*\, \boldsymbol{v}(q). \tag{8.4}$$

Equations (8.4) show that the group G has the same dimension r at all points, and that the set of all right-invariant vector fields and the tangent space T_{q_0} to G at q_0 are isomorphic vector spaces. An r-dimensional group is denoted by G_r and said to be of r parameters.

The transformations Φ_t generated by a right-invariant vector field $\boldsymbol{v}$, in the way described in §2.8, clearly commute with right translations. If $\Phi_t q_0 = q(t)$, we find

$$\Phi_t q' = \Phi_t R_{q'} q_0 = R_{q'} \Phi_t q_0 = R_{q'} q(t) = q(t) q', \tag{8.5}$$

so that $\Phi_t = L_{q(t)}$; the right-invariant vector fields represent infinitesimal left translations. From (2.25), the commutator of two right-invariant vector fields is also right-invariant, so that if we take a basis $\{\boldsymbol{\xi}_A, A = 1, \ldots, r\}$ of the space of right-invariant vector fields we must have

$$[\boldsymbol{\xi}_A, \boldsymbol{\xi}_B] = C^C{}_{AB} \boldsymbol{\xi}_C, \qquad C^C{}_{AB} = -C^C{}_{BA}. \tag{8.6}$$

The coefficients $C^C{}_{AB}$ are known as the *structure constants* of the group. A *Lie algebra* is defined to be a (finite-dimensional) vector space in which a bilinear operation $[\boldsymbol{u}, \boldsymbol{v}]$, obeying $[\boldsymbol{u}, \boldsymbol{v}] = -[\boldsymbol{v}, \boldsymbol{u}]$ and the Jacobi identity (2.7), is defined. Thus we have proved

Theorem 8.1 *A Lie group defines a unique Lie algebra.*

It is possible to show that the converse also holds.

Theorem 8.2 *Every Lie algebra defines a unique (simply-connected) Lie group.*

For a proof, see e.g. Cohn (1957). The elements of the Lie algebra, or a basis of them, are said to *generate* the group. Noting that the Jacobi identity (2.7) holds for (8.6) if and only if

$$C^E{}_{[AB} C^F{}_{C]E} = 0, \tag{8.7}$$

we can rewrite Theorem 8.2 as

Theorem 8.3 (Lie's third fundamental theorem). *Any set of constants $C^A{}_{BC}$ satisfying $C^A{}_{BC} = C^A{}_{[BC]}$ and (8.7) are the structure constants of a group.*

Theorem 8.2 does not imply that a given Lie algebra arises from only one Lie group. For example the Lorentz group $L_+^\uparrow$ and the group SL(2,$\mathbb{C}$) (see §3.6) have the same Lie algebra. It is true, however, that all connected Lie groups with a given Lie algebra are homomorphic images of the one specified in Theorem 8.2.

All the above work can be repeated interchanging left and right. We shall denote a basis of the Lie algebra of left-invariant fields by $\{\boldsymbol{\eta}_A,\ A = 1,\dots,r\}$. For all A and B, (8.2) implies

$$[\boldsymbol{\xi}_A, \boldsymbol{\eta}_B] = 0. \tag{8.8}$$

Clearly there must be some position-dependent matrix ${M_A}^B(q)$ such that $\boldsymbol{\eta}_A = {M_A}^B \boldsymbol{\xi}_B$, with inverse ${(M^{-1})^A}_B$. Equations (8.8) and (8.6) show that

$$[\boldsymbol{\eta}_A, \boldsymbol{\eta}_B] = -{M_A}^C {M_B}^D {C^E}_{CD} {(M^{-1})^F}_E \boldsymbol{\eta}_F, \tag{8.9}$$

so that the structure constants ${D^A}_{BE}$ of the basis $\{\boldsymbol{\eta}_A\}$ are related to the ${C^A}_{BE}$. Choosing $\boldsymbol{\eta}_A = -\boldsymbol{\xi}_A$ at q_0 shows that the Lie algebras, and hence the Lie groups, of the left and right translations are isomorphic. However, it is more usual to take $\boldsymbol{\eta}_A = \boldsymbol{\xi}_A$, leading to

$${D^A}_{BC} = -{C^A}_{BC}. \tag{8.10}$$

The commutators $[\boldsymbol{u}, \boldsymbol{v}]$ of the right-invariant vector fields are the infinitesimal generators of the commutator subgroup of G (i.e. that formed from all products of the form $q_1 q_2 (q_1)^{-1} (q_2)^{-1}$). This is also known as the (first) *derived group*, and its Lie algebra, which is spanned by ${C^A}_{BC}\boldsymbol{\xi}_A$, is the *derived algebra*. A group is said to be *Abelian* if every pair of elements commutes: for Lie groups, this is true if and only if all the structure constants are zero. A subgroup H of a group G is said to be *normal* or *invariant* if $qhq^{-1} \in H$ for any $h \in H$ and $q \in G$; for Lie groups this is true if and only if the generators $\boldsymbol{\zeta}_i$ $(i = 1,\dots,p)$ of H obey

$$[\boldsymbol{\xi}_A, \boldsymbol{\zeta}_i] = {C^j}_{Ai} \boldsymbol{\zeta}_j \tag{8.11}$$

for all A and i. A group is said to be *simple* if it has no invariant subgroup other than the group itself and the identity, and *semisimple* if it similarly has no invariant Abelian subgroup. The derived group is always invariant. If the sequence $\{G_{r_i}\}$, where G_{r_i} is the derived group of $G_{r_{i-1}}$ and $G_{r_0} = G_r$, satisfies $r > r_1 > \cdots > r_k = 0$, then the group G_r is said to be *solvable* (or *integrable*); this can be tested by calculating the dimensions of the successive derived algebras.

Any subalgebra of the Lie algebra of a Lie group generates a Lie subgroup, and a subalgebra with basis $\{\boldsymbol{\zeta}_i\}$ satisfying (8.11), known as an *ideal*, generates an invariant subgroup.

It is possible to define *canonical coordinates* on a Lie group G in such a way that a given basis $\{\boldsymbol{\xi}_A\}$ has $\boldsymbol{\xi}_A = \partial/\partial x^A$ at q_0; actually, this can be done in more than one way (Cohn 1957).

8.2 Enumeration of distinct group structures

Linear transformations of the basis $\{\boldsymbol{\xi}_a\}$ transform the $C^A{}_{BC}$ of (8.6) as a tensor. To find distinct algebras we need sets of constants $C^A{}_{BC}$ which cannot be related by such a linear transformation. The enumerations therefore naturally use properties, such as the dimension of the derived algebra, which are invariant under these transformations. Methods of enumerating all complex Lie algebras are wellknown, being useful in pure mathematics and quantum physics, but the enumeration of the *real* Lie algebras, although its foundations have also long been known, is not so widely studied. We give here the distinct structures for groups G_2 and G_3, and some information on classification of the G_4. We omit the full list of G_4 because many cases do not arise in exact solutions. MacCallum (1999) gives an enumeration and compares it with previous classifications such as those of Petrov (1966) and Patera and Winternitz (1977).

In a G_2 there is only one (non-trivial) commutator; hence all G_2 are solvable. If the G_2 is Abelian, it is called type G_2I. If it is non-Abelian, one can choose $\boldsymbol{\xi}_1$ in the derived algebra, and scale $\boldsymbol{\xi}_2$ so that

$$[\boldsymbol{\xi}_1, \boldsymbol{\xi}_2] = \boldsymbol{\xi}_1; \tag{8.12}$$

this case is called type G_2II.

The G_3 were originally enumerated by Bianchi (1898). There are nine types, Bianchi *I* to Bianchi *IX*, two of which, *VI* and *VII*, are one-parameter families of distinct group structures. Complex transformations relate types *VIII* and *IX*, and types *VI* and *VII*. Bianchi's method began, like that above for the G_2, by considering the dimension of the derived algebra, but we shall obtain the result in a different way (Schücking, unpublished, 1957, Estabrook *et al.* 1968, Ellis and MacCallum 1969). Taking any completely skew tensor ε^{ABC} on the Lie algebra we write

$$\tfrac{1}{2}C^D{}_{BC}\varepsilon^{BCE} = N^{DE} + \varepsilon^{DEF}A_F, \quad A_D \equiv \tfrac{1}{2}C^B{}_{DB}, \quad N^{DE} = N^{(DE)}, \tag{8.13}$$

so that, with ε_{DEF} obeying $\varepsilon^{ABC}\varepsilon_{DEF} = 6\delta^A_{[D}\delta^B_E\delta^C_{F]}$ as usual,

$$C^D{}_{BC} = \varepsilon_{BCF}N^{DF} + 2\delta^D_{[C}A_{B]}. \tag{8.14}$$

The Jacobi identity (8.7) reduces to

$$N^{DE}A_E = 0. \tag{8.15}$$

N^{DE} is defined up to an overall factor (since ε^{ABC} is). Its invariant properties are its rank and the modulus of its signature. In types *VI* and *VII* there is a further invariant h, defined by

$$(1-h)C^A{}_{BA}C^D{}_{CD} = -2hC^A{}_{DB}C^D{}_{AC}, \tag{8.16}$$

Table 8.1. Enumeration of, and canonical structure constants for, the Bianchi types

Class	G_3A						G_3B				
Type	I	II	VI_0	VII_0	VIII	IX	V	IV	III	VI_h	VII_h
Rank (N^{DE})	0	1	2	2	3	3	0	1	2	2	2
$\lvert$Signature $(N^{DE})\rvert$	0	1	0	2	1	3	0	1	0	0	2
A	0	0	0	0	0	0	1	1	1	$\sqrt{-h}$	$\sqrt{h}$
N_1	0	1	0	0	-1	1	0	0	0	0	0
N_2	0	0	-1	1	1	1	0	0	-1	-1	1
N_3	0	0	1	1	1	1	0	1	1	1	1
Dimensions of canonical basis freedom	9	6	4	4	3	3	6	4	4	4	4

which supplies the one parameter required to subdivide these Bianchi types: in type *VI*, $h < 0$, and in type *VII*, $h > 0$. The relation between h and Bianchi's parameters q for types *VI* and *VII* is given by $h = -(1+q)^2/(1-q)^2$ and $h = q^2/(4-q^2)$ respectively. Bianchi type *III* is the same as VI_h with $h = -1$.

There are two main classes of G_3, Class G_3A ($A_E = 0$) and Class G_3B ($A_E \neq 0$). In all cases, by rotation and rescaling of the basis $\{\boldsymbol{\xi}_A\}$, one can set $N^{DE} = \mathrm{diag}(N_1, N_2, N_3)$, $A_E = (A, 0, 0)$, with N_1, N_2, N_3 equal to 0 or ± 1 as appropriate, and $A = \sqrt{hN_2N_3}$ (for Bianchi types *VI*, *VII*, and *III*). Thus one obtains Table 8.1 which lists all types and canonical forms of the structure constants. All types are solvable, except *VIII* and *IX* which are semisimple. The canonical form does not uniquely specify the basis. The dimension of the subgroup of the linear transformations which preserves the canonical form is shown in Table 8.1 (for a proof see Siklos (1976a)).

The G_4 can similarly be divided into two classes by whether $A_E = \frac{1}{2}C^B{}_{EB} = 0$ or not. In the first case one has (Farnsworth and Kerr 1966):

Theorem 8.4 *If* $A_E = 0$, *then either (i) the structure constants of the* G_4 *can be written in the form*

$$C^A{}_{BC} = \Theta^A{}_{[B}P_{C]} \tag{8.17}$$

or (ii) if no form (8.17) *exists, there is a non-zero vector* L^A *such that*

$$C^A{}_{BC}L^B = 0 \tag{8.18}$$

As a corollary of this theorem, we have

Theorem 8.5 (Egorov). *Every G_4 contains a G_3 (locally);* (see Petrov (1966), p. 180).

Proof. If $A_E \neq 0$, the Jacobi identities yield $A_B C^B{}_{CD} = 0$, showing that the derived algebra is three-dimensional (at most). If $A_E = 0$ the forms (8.17), (8.18) show clearly that the derived algebra is again at most three-dimensional. In all cases the derived algebra (together, if necessary, with enough linearly independent vectors to make the dimension three) generates a G_3.

A slightly different proof was found by Kantowski (see Collins (1977a)). Patera and Winternitz (1977) have explicitly calculated all subgroups G_2 and G_3 of the real G_4.

Another result due to Egorov (see Petrov (1966), p. 180) is

Theorem 8.6 *Every G_5 contains a subgroup G_4.*

8.3 Transformation groups

Let $\mathcal{M}$ be a differentiable (analytic) manifold and G a Lie group of r parameters. An *action* of G on $\mathcal{M}$ is an (analytic) map $\mu : G \times \mathcal{M} \to \mathcal{M}; (q, p) \to \tau_q p$. Each element q of G is associated with a transformation $\tau_q : \mathcal{M} \to \mathcal{M}$. It is assumed that the identity q_0 of G is associated with the identity map $I : p \to p$ of $\mathcal{M}$, and that

$$\tau_q \tau_{q'} p = \tau_{qq'} p \tag{8.19}$$

so that the transformations τ_q form a group isomorphic with G. The group is said to be *effective* (and the parameters *essential*) if $\tau_q = I$ implies $q = q_0$; only such groups need be considered.

The *orbit* (or *trajectory*, or *minimum invariant variety*) of G through a given p in $\mathcal{M}$ is defined to be $\mathcal{O}_p = \{p' : p' \in \mathcal{M}$ and $p' = \tau_q p$ for some $q \in G\}$. It is a submanifold of $\mathcal{M}$. The group G is said to be *transitive* on its orbits, and to be either *transitive* on $\mathcal{M}$ (when $\mathcal{O}_p = \mathcal{M}$) or *intransitive* ($\mathcal{O}_p \neq \mathcal{M}$). It is *simply-transitive* on an orbit if $\tau_q p = \tau_{q'} p$ implies $q = q'$; otherwise it is *multiply-transitive*. A group may be simply-transitive on general orbits but multiply-transitive on some special orbit(s). The set of q in G such that $\tau_q p = p$ forms a subgroup of G called the *stability group* $S(p)$ of p. If $p' \in \mathcal{O}_p$, so that there is a q in G such that $\tau_q p = p'$, and if $q' \in S(p)$, then $\tau_q \tau_{q'} \tau_{q^{-1}} p' = p'$ and hence $qq'q^{-1} \in S(p')$. Thus $S(p)$ and $S(p')$ are conjugate subgroups of G, and have the same dimension, s say; for brevity, one often refers to the stability subgroup S_s of an orbit.

For each orbit, a map $\mu_p : G \to \mathcal{O}_p; q \to \tau_q p$ can be defined. The map $(\mu_p)_*$ then maps the right-invariant vector fields on G to vector fields tangent to $\mathcal{O}_p$. It can be shown that the choice of base point p in $\mathcal{O}_p$ does not affect $(\mu_p)_*$. Hence, using a map $(\mu_p)_*$ in each $\mathcal{O}_p$, we can define a Lie algebra of vector fields on $\mathcal{M}$ by taking the image of the Lie algebra of G. At the risk of some confusion we use $\{\boldsymbol{\xi}_A\}$ to denote a basis of either Lie algebra. The two algebras are isomorphic because G is assumed to be effective, and so $(\mu_p)_*\boldsymbol{v} = \boldsymbol{0}$ for all p only if $\boldsymbol{v} = \boldsymbol{0}$.

The stability group of p is generated by those $\boldsymbol{v}$ such that $(\mu_p)_*\boldsymbol{v} = \boldsymbol{0}$ at p; this is clearly the kernel of the map $(\mu_p)_*$ at q_0. Denoting the dimension of $\mathcal{O}_p$ by d we thus have

$$r = d + s. \tag{8.20}$$

The classical theorems on continuous transformation groups can be expressed as

Theorem 8.7 (Lie's first fundamental theorem). *An action $\mu : G \times \mathcal{M} \to \mathcal{M}$ of a continuous (Lie) group of transformations defines and is defined by a linear map of the right-invariant vector fields on G_r onto an r-dimensional set of (smooth) vector fields on $\mathcal{M}$.*

Theorem 8.8 (Lie's second fundamental theorem). *A set of r (smooth) linearly independent vector fields $\{\boldsymbol{\xi}_A\}$ on $\mathcal{M}$ obeying* (8.6) *defines and is defined by a continuous (Lie) group of transformations on $\mathcal{M}$.*

A single generator $\boldsymbol{\xi}$ of a transformation group G_r gives rise to a one-parameter subgroup Φ_x (see §2.8) of G_r, and by choosing one point p in each orbit of this group as $x = 0$ we can find a coordinate x in $\mathcal{M}$ such that $\boldsymbol{\xi} = \partial_x$ (the term *trajectory* is sometimes reserved for such one-dimensional orbits). If there are m commuting generators $\{\boldsymbol{\xi}_A\}$ (forming an Abelian subgroup), all non-zero at p, then one can thus find m coordinates $(x^1, \ldots, x^m)$ such that $\boldsymbol{\xi}_A = \partial/\partial x^A$ $(A = 1, \ldots, m)$.

8.4 Groups of motions

Manifolds with structure, such as Riemannian manifolds V_n, may admit (continuous) groups of transformations preserving this structure. In a V_n, the map Φ_t corresponding (as in §2.8) to a conformal motion obeying (6.10) has the property $(\Phi_t\boldsymbol{g})_{ab} = \mathrm{e}^{2U} g_{ab}$, where U is the integral of the ϕ in (6.10) along a curve, i.e. it preserves the metric up to a factor. This is a conformal transformation (§3.7), whence the name *conformal motion.* It is a *homothety* if ϕ is constant, and a *motion* (or *isometry*), whose generator obeys Killing's equation (6.11) and which preserves the metric, if $\phi = 0$. Here we shall consider motions. Homothety groups are discussed further in §8.7 and more general symmetries in §35.4.

The set of all solutions of Killing's equation (6.11), i.e.

$$\xi_{a;b} + \xi_{b;a} = 0, \tag{8.21}$$

can easily be seen to form a Lie algebra (8.6) and hence, by Theorem 8.8, to generate a Lie group of transformations, called a *group of motions* or *isometry group*. If we use the coordinate x adapted to a Killing vector $\boldsymbol{\xi}$, so that $\boldsymbol{\xi} = \partial_x$, then g_{ab} is independent of x.

Some special terminology is used for groups of motions. The stability group of p in a group G_r of motions is called the *isotropy group* I_s of p. We use the term *generalized orthogonal group* for the set of linear transformations of the tangent space at p which preserve scalar products formed with the metric $\boldsymbol{g}$ by (3.4). By Lie dragging, (2.60), I_s gives rise to the *linear isotropy group* $\widehat{I}$ of p, acting in the tangent space to the orbit $\mathcal{O}_p$ at p, which is a subgroup of the generalized orthogonal group. For a space-time, the generalized orthogonal group is the Lorentz group. The orbits (with dimension d) of a group of motions may be spacelike, null, or timelike submanifolds, and these are denoted by S_d, N_d and T_d, respectively. If we use V_d it denotes either an S_d or a T_d. A space V_d on which a group of motions acts transitively is called *homogeneous*.

The quantity $K_{ab} = \xi_{a;b} = \xi_{[a,b]}$, the *Killing bivector*, can be interpreted in terms of a Lorentz transformation between a tetrad Lie-dragged along an integral curve of $\boldsymbol{\xi}_A$ and one parallelly transported along the same curve (a rotation if K_{ab} is simple and spacelike, a boost if K_{ab} is simple and timelike and a null rotation if K_{ab} is null) (Kobayashi and Nomizu 1969, Hall 1988a). Properties of the Killing bivector can be related to those of the curvature (see e.g. Catenacci *et al.* (1980)): for example, a non-flat vacuum with a null Killing bivector must be algebraically special.

We now consider the question of the dimension r of the (maximal) group of motions admitted by a given Riemannian manifold. A useful step is provided by the following result.

Theorem 8.9 *If a Killing vector field* $\boldsymbol{\xi}$ *has* $\xi^a = 0$ *and* $\xi_{a;b} = 0$ *at a point* p, *then* $\boldsymbol{\xi} \equiv \mathbf{0}$.

Proof. Locally, any point p' may be joined to p by a geodesic, with tangent vector $\boldsymbol{v}$ at p, say. Then $\boldsymbol{\xi}$ fixes p and $\boldsymbol{v}$ (by (2.60)), and preserves the affine parameter distance along the geodesic with tangent vector $\boldsymbol{v}$ at p. It thus fixes p'. Thus $\boldsymbol{\xi} = \mathbf{0}$ at any point p'.

Under appropriate smoothness conditions this can also be proved by considering the linear differential equation (8.22).

From this result we see that (i) the isotropy and linear isotropy groups of p are isomorphic, and (ii) a Killing vector field will be completely specified given the $n(n+1)/2$ values of ξ_a and $\xi_{a;b}$ at a point p. Thus we have only to check if there are further restrictions on these values. In general all such restrictions for systems of partial differential equations are obtained by repeated differentiation (see e.g. Eisenhart (1933)). In the case of the Killing equation, the first differentiation of (8.21) gives

$$(\mathcal{L}_{\boldsymbol{\xi}}\boldsymbol{\Gamma})^a{}_{bc} = 0 \Leftrightarrow \xi_{a;bc} = R_{abcd}\xi^d, \tag{8.22}$$

which, together with (8.21), gives a system of first-order differential equations for the quantities ξ_a, $\xi_{b;c}$. Vectors satisfying (only) (8.22) are called affine collineations, cp. §35.4. The integrability conditions given by further differentiation are exactly the equations

$$\mathcal{L}_{\boldsymbol{\xi}}\boldsymbol{R} = 0, \quad \mathcal{L}_{\boldsymbol{\xi}}(\nabla_{a_1}\cdots\nabla_{a_N}\boldsymbol{R}) = 0, \quad N = 1, 2, \ldots \tag{8.23}$$

for the successive covariant derivatives of the Riemann tensor $\boldsymbol{R}$. Each of these gives an equation linear in ξ_a and $\xi_{b;c}$ (as we see from §2.8). Since there can be at most $n(n+1)/2$ independent conditions, we see there must exist an integer Q such that the conditions (8.23) for $N > Q$ depend algebraically on those for $N \leq Q$ at any point. From this argument, and similar considerations for the isotropy group ($\xi_a = 0$) and the group of conformal motions, one obtains the following results.

Theorem 8.10 *If the rank of the linear algebraic equations* (8.23) *for* ξ_a *and* $\xi_{a;b}$ *is* q, *then the maximal group* G_r *of motions of the* V_n *has* $r = \frac{1}{2}n(n+1) - q$ *parameters.*

Theorem 8.11 *For a* V_n *admitting a group* G_r *of motions, the rank of the linear algebraic equations* (8.23) *for* $\xi_{a;b}$ *with* $\xi_c = 0$ *is* p *if and only if there is an isotropy subgroup* I_s, $s = \frac{1}{2}n(n-1) - p$ (Defrise 1969).

Theorem 8.12 *The maximal order of a group* G_r *of conformal transformations in a* V_n *is* $r = \frac{1}{2}n(n+1)(n+2)$ (see e.g. Eisenhart 1949).

To find the Killing vectors of a metric, or, alternatively, to find the restrictions on the metric and curvature of a space admitting a group G_r of motions with given r, one can use (8.23). Petrov (1966) largely worked by this method. For a given metric, one can often obtain results equivalent to (8.23) by remarking that any invariantly-defined geometric object (e.g. a principal null direction of the Weyl tensor, the velocity vector of a perfect fluid, the bivector fields defined by the eigenblades of a type D Weyl tensor) must be invariant under the isometries; the coordinates are usually adapted to some such invariant structure, and this facilitates the calculation.

In particular, scalar invariants of the Riemann tensor and its derivatives (see §9.1) must be invariant under isometries. Kerr (1963b) proved that in a four-dimensional Einstein space, the number of functionally independent scalar invariants is $4 - d$, where d is the dimension of the orbits of the maximal group of motions.

Ref.: Swift *et al.* (1986) considered classes of solutions where the isometries are specializations of diffeomorphisms preserving the whole class.

8.5 Spaces of constant curvature

A two-dimensional Riemannian space has only one independent component, R_{1212} say, of its curvature tensor. The tensor g_{abcd}, defined as in (3.47), has the same index symmetries as R_{abcd}, and is non-zero (being, in two dimensions, essentially the determinant of g_{ab}). Thus in two dimensions

$$R_{abcd} = K(g_{ac}g_{bd} - g_{ad}g_{bc}). \tag{8.24}$$

K is called the *Gaussian curvature.*

In a Riemannian space of more than two dimensions one can, at any point p, form a two-dimensional submanifold by taking all geodesics through p whose initial tangent vector is of the form $\alpha\boldsymbol{v} + \beta\boldsymbol{w}$, where α, β are real and $\boldsymbol{v}$ and $\boldsymbol{w}$ are fixed vectors at p. Equation (8.24) then defines the *sectional curvature* K of this two-dimensional manifold, assuming it is non-null, and it can be shown that

$$K = \frac{R_{abcd}v^a w^b v^c w^d}{(g_{ac}g_{bd} - g_{ad}g_{bc})v^a w^b v^c w^d}. \tag{8.25}$$

The space V_n is said to be of *constant curvature* if K in (8.25) is independent of p and of $\boldsymbol{v}$ and $\boldsymbol{w}$. Then (8.25) leads to

$$\begin{aligned} &Q_{abcd} + Q_{adcb} + Q_{cbad} + Q_{cdab} = 0,\\ &Q_{abcd} \equiv R_{abcd} - K(g_{ac}g_{bd} - g_{ad}g_{bc}), \end{aligned} \tag{8.26}$$

and the Riemann tensor symmetries yield (8.24) for the Riemann tensor of the V_n, with constant K.

If we take a space of constant curvature, conditions (8.23) are all identically satisfied, so by Theorem 8.10 there is a group of motions of $\frac{1}{2}n(n+1)$ parameters.

If a space V_n admits an isotropy group of $\frac{1}{2}n(n-1)$ parameters, it is the whole of the relevant generalized orthogonal group (see Theorem 8.11). In this case (8.25) is independent of the choice of $\boldsymbol{v}$ and $\boldsymbol{w}$ and, as above, we obtain (8.26) and (8.24). The Bianchi identities $R_{ab[cd;e]} = 0$, contracted on b and d, yield $(n-2)(K_{,a}g_{ce} - K_{,c}g_{ae}) = 0$, and contracting again on a and e gives $(n-2)(n-1)K_{,c} = 0$. Thus, if $n \geq 3$, K is constant and the space admits a G_r $(r = \frac{1}{2}n(n+1))$ of motions. Conversely if a space V_n admits a G_r $(r = \frac{1}{2}n(n+1))$ of motions then by (8.20) it admits an I_s $(s = \frac{1}{2}n(n-1))$ of isotropies and is thus of constant curvature if $n \geq 3$. If $n = 2$, a G_2 (or G_3) of motions must be transitive and then $\mathcal{L}_{\boldsymbol{\xi}}\boldsymbol{R} = 0$ for the Riemann tensor leads to $K =$constant.

Collecting together these arguments we find we have proved

Theorem 8.13 *A Riemannian space is of constant curvature if and only if it (locally) admits a group G_r of motions with $r = \frac{1}{2}n(n+1)$.*

Theorem 8.14 *A Riemannian space V_n $(n \geq 3)$ is of constant curvature if and only if it (locally) admits an isotropy group I_s of $s = \frac{1}{2}n(n-1)$ parameters at each point.*

Theorem 8.15 *A two-dimensional Riemannian space admitting a G_2 of motions admits a G_3 of motions.*

Substituting (8.24) into the definition (3.50) of the Weyl tensor we find $C^a{}_{bcd} = 0$, and so (3.85) can be solved to find the factor e^{2U} in (3.83) relating the metric to that of a flat space of the same dimension and signature, $\mathring{g}_{ab} = \mathrm{diag}(\varepsilon_1, \ldots, \varepsilon_n)$, where $\varepsilon_1, \ldots, \varepsilon_n = \pm 1$ as appropriate. Equations (3.85) are satisfied if

$$2(\mathrm{e}^{-U})_{,ab} = K\mathring{g}_{ab}, \quad (\mathrm{e}^{-U})_{,a}(\mathrm{e}^{-U})^{,a} = K(\mathrm{e}^{-U} - 1), \tag{8.27}$$

the solution of which can be transformed to

$$\mathrm{e}^{-U} = 1 + \tfrac{1}{4}K\mathring{g}_{ab}x^a x^b. \tag{8.28}$$

Hence the metric of a space V_n of constant curvature can always be written as

$$\mathrm{d}s^2 = \frac{\mathrm{d}x_a \mathrm{d}x^a}{\left(1 + \frac{1}{4}Kx_b x^b\right)^2} \tag{8.29}$$

(indices raised and lowered with $\mathring{g}_{ab}$), for any value of K or signature of V_n, and any two metrics of the same constant curvature and signature must be locally equivalent.

A space V_n of non-zero constant curvature, $K \neq 0$, can be considered as a hypersurface

$$Z_a Z^a + k(Z^{n+1})^2 = kY^2, \quad K = kY^{-2}, \quad k = \pm 1, \tag{8.30}$$

in an $(n+1)$-dimensional pseudo-Euclidean space with metric

$$\mathrm{d}s^2 = \mathrm{d}Z_a \mathrm{d}Z^a + k(\mathrm{d}Z^{n+1})^2. \tag{8.31}$$

For each parametrization of Z^a and Z^{n+1} in terms of coordinates (e.g. angular coordinates) in V_n, in accordance with the surface equation (8.30), the metric of V_n can be obtained from (8.31), and the relation

$$x^a = \frac{2Z^a}{1 + (1 - KZ_bZ^b)^{1/2}} \tag{8.32}$$

yields the transformation from the x^a in (8.29) to the new coordinates in V_n.

In the remainder of this section we consider some special cases which play an important role in general relativity. The metric

$$\mathrm{d}s^2 = \frac{\mathrm{d}x^2 + \mathrm{d}y^2 + \mathrm{d}z^2 - \mathrm{d}t^2}{\left[1 + \frac{1}{4}K(x^2 + y^2 + z^2 - t^2)\right]^2} \tag{8.33}$$

of a space-time V_4 of constant curvature (*de Sitter space* if $K > 0$ or *anti de Sitter* space if $K < 0$) can be given in the equivalent form

$$\mathrm{d}s^2 = \frac{\mathrm{d}r^2}{1 - Kr^2} + r^2(\mathrm{d}\vartheta^2 + \sin^2\vartheta\,\mathrm{d}\varphi^2) - (1 - Kr^2)\mathrm{d}t^2. \tag{8.34}$$

In this metric, K can be related to a Λ-term (see §5.2) by $\Lambda = 3K$.

Gravitational fields often admit subspaces of constant curvature. On a single subspace, K is of course constant, but it may have differing values on different subspaces (see Chapter 36).

The metric

$$\mathrm{d}s^2 = \frac{\mathrm{d}x^2 + \mathrm{d}y^2 + \mathrm{d}z^2}{\left[1 + \frac{1}{4}K(x^2 + y^2 + z^2)\right]^2} \tag{8.35}$$

of a three-dimensional positive-definite space (e.g. a spacelike hypersurface in a space-time), the Killing vectors of which are given by (12.25), can be transformed (cp. (37.13)–(37.14)) to the form

$$\mathrm{d}s^2 = a^2\left[\mathrm{d}r^2 + \Sigma^2(r,k)(\mathrm{d}\vartheta^2 + \sin^2\vartheta\,\mathrm{d}\varphi^2)\right], \quad K = ka^{-2}, \tag{8.36}$$

$$\Sigma(r,k) = \sin r,\ r \text{ or } \sinh r, \text{respectively, when } k = 1,\ 0 \text{ or } -1. \tag{8.37}$$

The metrics of 2-spaces of constant curvature have six distinct types

$$\mathrm{d}s^2 = Y^2\left[(\mathrm{d}x^1)^2 \pm \Sigma^2(x^1,k)(\mathrm{d}x^2)^2\right], \quad K = kY^{-2}, \tag{8.38}$$

with $\Sigma(x^1,k)$ as in (8.37). In the case of a metric with signature zero, and $k = -1$, the parametrization

$$Z^1 = Y\sin x^1 \sinh x^2, \quad Z^2 = Y\sin x^1\cosh x^2, \quad Z^3 = Y\cos x^1, \tag{8.39}$$

leads (Barnes 1973a) to the form

$$d\sigma^2 = Y^2[-(dx^1)^2 + \sin^2 x^1 (dx^2)^2]. \tag{8.40}$$

The specific case of spacelike surfaces S_2 frequently occurs. For these

$$d\sigma^2 = \frac{2d\zeta d\bar{\zeta}}{\left(1+\frac{1}{2}K\zeta\bar{\zeta}\right)^2}, \quad \zeta = \frac{1}{\sqrt{2}}(x^1 + ix^2), \tag{8.41}$$

where x^1, x^2 are as in (8.29). For $k = 1$, $\zeta = \sqrt{2/K}\cot(\vartheta/2)\exp(i\varphi)$ gives the usual form for the sphere of radius Y

$$d\sigma^2 = Y^2(d\vartheta^2 + \sin^2\vartheta\, d\varphi^2). \tag{8.42}$$

For $k = -1$, the transformation $z = (1+z')/(1-z')$, $z' = \sqrt{-\frac{1}{2}K}\zeta$, leads to

$$d\sigma^2 = 4Y^2 \frac{dz\, d\bar{z}}{(z+\bar{z})^2}. \tag{8.43}$$

Besides (8.41)–(8.43), other coordinate systems are frequently used in the literature, e.g. for S_2, $k = -1$,

$$d\sigma^2 = Y^2(d\vartheta^2 + \cosh^2\vartheta\, d\varphi^2), \tag{8.44}$$

$$d\sigma^2 = Y^2(dx^2 + e^{2x}dy^2). \tag{8.45}$$

All the results given above follow from well-known classical methods and theorems and are described in many texts, e.g. Eisenhart (1933), Petrov (1966), Plebański (1967), Weinberg (1972).

8.6 Orbits of isometry groups

From the previous section we know a great deal about orbits V_n of groups of motions G_r with $r = \frac{1}{2}n(n+1)$. In the present section we shall discuss orbits of smaller groups of motions. We first note the following well-known theorems (see e.g. Eisenhart (1933)).

Theorem 8.16 *If the orbits of a group of motions are hypersurfaces then their normals are geodesics, and if the hypersurfaces are non-null they are geodesically parallel, i.e. taking an affine parameter along the normal geodesics as the coordinate x^n, the metric has the form*

$$ds^2 = g_{\mu\nu}dx^\mu x^\nu + \varepsilon(dx^n)^2, \quad \mu, \nu = 1,\ldots,(n-1), \quad \varepsilon = \pm 1. \tag{8.46}$$

Theorem 8.17 (Fubini's theorem). *A Riemannian manifold V_n of dimension $n \geq 2$ cannot have a maximal group of motions of $\frac{1}{2}n(n+1) - 1$ parameters.*

8.6.1 Simply-transitive groups

If a group is simply-transitive, the map $\mu_p : G \to \mathcal{O}_p$ used in §8.3 is an isomorphism and thus $(\mu_p)_*$ can be used to map the left-invariant vector fields. Taking a basis of these, $\{\boldsymbol{e}_A = \boldsymbol{\eta}_A; A = 1, \ldots, r\}$, in each orbit, the metric in the orbit is

$$\mathrm{d}\sigma^2 = (\boldsymbol{\eta}_A \cdot \boldsymbol{\eta}_B)\boldsymbol{\omega}^A\boldsymbol{\omega}^B = g_{AB}\boldsymbol{\omega}^A\boldsymbol{\omega}^B, \tag{8.47}$$

where the $\boldsymbol{\omega}^A$ are dual, in $\mathcal{O}_p$, to the $\boldsymbol{\eta}_A$. For any Killing vector $\boldsymbol{\xi}$, (8.21) and (8.8) show that $\mathcal{L}_{\boldsymbol{\xi}} g_{AB} = 0$, so that the g_{AB} are constant in the orbit.

If a simply-transitive group G_r is intransitive on the V_n, one can choose a basis $\{\boldsymbol{e}_A = \boldsymbol{\eta}_A; A = 1, \ldots, r\}$ separately in each orbit, and complete a basis $\{\boldsymbol{e}_a; a = 1, \ldots, n\}$ of the tangent space at one point p in each orbit by adding $(n-r)$ arbitrary vectors, in a suitably smooth way. If vector fields $\{\boldsymbol{e}_a; a = r+1, \ldots, n\}$ are then defined throughout $\mathcal{O}_p$ by using the $(\tau_q)_*$ on the $\{\boldsymbol{e}_a\}$ at p we find, using the dual basis $\{\boldsymbol{\omega}^a\}$, that

$$\mathrm{d}s^2 = g_{ab}\boldsymbol{\omega}^a\boldsymbol{\omega}^b \tag{8.48}$$

with g_{ab} *constant in each orbit.* This can be done in such a way that $\boldsymbol{\omega}^a = \boldsymbol{\omega}^A$ for $A = 1, \ldots, r$: for a non-null $\mathcal{O}_p$ one can use vectors orthogonal to $\mathcal{O}_p$ to complete the basis; this cannot be done for a null $\mathcal{O}_p$ because the null normal lies in the orbit but a related prescription can be given.

The vector fields $\boldsymbol{\eta}_A$ generate a group of transformations on each orbit, called the *reciprocal group*; it will not necessarily have any of the symmetry properties of the transformation group G_r, i.e. in the present case it will not in general consist of isometries.

It is often convenient to choose an orthonormal basis of reciprocal group generators in each orbit. Their Lie algebra cannot then be completely reduced to canonical form because only the generalized orthogonal group of linear transformations, and not the general linear group, is available. For a simply-transitive G_3 on S_3 we can reduce the commutators of such orthonormal reciprocal group generators $\{\boldsymbol{E}_\alpha; \alpha = 1, 2, 3\}$ to the form

$$[\boldsymbol{E}_\alpha, \boldsymbol{E}_\beta] = \gamma^\delta{}_{\alpha\beta}\boldsymbol{E}_\delta, \qquad \tfrac{1}{2}\gamma^\delta{}_{\alpha\beta}\varepsilon^{\alpha\beta\varphi} = n^{(\delta\varphi)} + \varepsilon^{\delta\varphi\nu}a_\nu, \tag{8.49}$$

$$n^{(\delta\varphi)} = \mathrm{diag}(n_1, n_2, n_3) \qquad \text{and} \qquad a^\nu = (a, 0, 0), \tag{8.50}$$

where $\varepsilon^{\alpha\beta\gamma}$ is the natural skew tensor defined by $g_{\alpha\beta}$ (up to sign). In the case of a G_4 simply-transitive on space-time the orthonormal reciprocal group generators have commutators

$$[\boldsymbol{E}_a, \boldsymbol{E}_b] = D^c{}_{ab}\boldsymbol{E}_c, \tag{8.51}$$

where, from (8.10) and Theorem 8.4, $D^c{}_{ca} \neq 0$ or (8.17) or (8.18) holds.

Most steps in this argument do not use the fact that the $\{\boldsymbol{\xi}_A\}$ are motions, and so, by a very similar argument, one can prove (Hoenselaers 1988), in n dimensions,

Theorem 8.18 *An n-dimensional simply-transitive group of affine collineations with generators $\{\boldsymbol{\xi}_A\}$ exists if and only if there is a frame $\{\boldsymbol{\eta}_A\}$ the Ricci rotation coefficients of which are constant; the $\boldsymbol{\eta}_A$ generate the reciprocal group and the bases are related by* (8.8).

In the case of a (non-null) orbit of a simply-transitive group the curvature tensor of the orbit is easily calculated from (3.20) and (2.79), remembering that here, using (8.47), the g_{AB} and $D^C{}_{AB}$ are constants. One gets

$$\Gamma_{CAB} = \tfrac{1}{2}(D_{ACB} + D_{BCA} - D_{CAB}), \tag{8.52}$$

$$R^D{}_{ABC} = \Gamma^E{}_{AC}\Gamma^D{}_{EB} - \Gamma^E{}_{AB}\Gamma^D{}_{EC} - D^E{}_{BC}\Gamma^D{}_{AE}, \tag{8.53}$$

$$\begin{aligned} R_{AB} &= -\tfrac{1}{2}D^E{}_{DA}D^D{}_{EB} - \tfrac{1}{2}D^E{}_{DA}D_E{}^D{}_B + \tfrac{1}{4}D_{ADE}D_B{}^{DE} \\ &\quad - \tfrac{1}{2}(D^D{}_{DE})(D_{AB}{}^E + D_{BA}{}^E), \end{aligned} \tag{8.54}$$

where indices are to be raised and lowered by g^{AB} and g_{AB}. One can, by the choice giving (8.10), use the group structure constants directly. For space-time metrics with groups G_3 simply-transitive on hypersurfaces (e.g. spatially-homogeneous cosmologies), it will be convenient to have expressions for $\boldsymbol{\xi}_A$, $\boldsymbol{\eta}_A$ and $\boldsymbol{\omega}^A$ obeying (8.10), (8.47) and Table 8.1 in terms of the canonical coordinates mentioned at the end of §8.1. These are given as Table 8.2. Note that such coordinates can still be chosen in many different ways, owing to the initial basis freedom listed in Table 8.1 and the freedom of choice of p for μ_p in the orbit (§8.3).

8.6.2 Multiply-transitive groups

Here we consider only non-null orbits. Schmidt (1968) has shown how to calculate all possible Lie algebras for a given isotropy group and dimension and signature of orbit, and how to find the curvature of the resulting orbits.

The method is as follows. The isotropy subgroup I_s of a chosen point p must be a (known) subgroup of the generalized orthogonal group and hence the commutators of its generators, $\{\boldsymbol{Y}_i; i = 1, \ldots, s\}$, are known. The basis of generators of the complete group of motions can be completed by adding d non-zero Killing vectors $\{\boldsymbol{\xi}_\alpha; \alpha = 1, \ldots, d\}$ which may be chosen at p in a way adapted to the isotropy group (e.g. if I_s consists

Table 8.2. Killing vectors and reciprocal group generators by Bianchi type

Expressions are given in canonical coordinates: for full explanation, see text.

	I	*II*	*IV*	*V*	*VI* (including *III*)	*VII*
$\boldsymbol{\xi}_A$	∂_x	∂_x	$\partial_x - y\partial_y - (y+z)\partial_z$	$\partial_x - y\partial_y - z\partial_z$	$\partial_x + (z - Ay)\partial_y + (y - Az)\partial_z$	$\partial_x + (z - Ay)\partial_y - (y + Az)\partial_z$
	∂_y	∂_y	∂_y	∂_y	∂_y	∂_y
	∂_z	$\partial_z + y\partial_x$	∂_z	∂_z	∂_z	∂_z
$\boldsymbol{\eta}_A$	∂_x	∂_x	∂_x	∂_x	∂_x	∂_x
	∂_y	$\partial_y + z\partial_x$	$\mathrm{e}^{-x}(\partial_y - x\partial_z)$	$\mathrm{e}^{-x}\partial_y$	$\mathrm{e}^{-Ax}(\cosh x\,\partial_y + \sinh x\,\partial_z)$	$\mathrm{e}^{-Ax}(\cos x\,\partial_y - \sin x\,\partial_z)$
	∂_z	∂_z	$\mathrm{e}^{-x}\partial_z$	$\mathrm{e}^{-x}\partial_z$	$\mathrm{e}^{-Ax}(\sinh x\,\partial_y + \cosh x\,\partial_z)$	$\mathrm{e}^{-Ax}(\sin x\,\partial_y + \cos x\,\partial_z)$
$\boldsymbol{\omega}^A$	$\mathrm{d}x$	$\mathrm{d}x - z\,\mathrm{d}y$	$\mathrm{d}x$	$\mathrm{d}x$	$\mathrm{d}x$	$\mathrm{d}x$
	$\mathrm{d}y$	$\mathrm{d}y$	$\mathrm{e}^{x}\mathrm{d}y$	$\mathrm{e}^{x}\mathrm{d}y$	$\mathrm{e}^{Ax}(\cosh x\,\mathrm{d}y - \sinh x\,\mathrm{d}z)$	$\mathrm{e}^{Ax}(\cos x\,\mathrm{d}y - \sin x\,\mathrm{d}z)$
	$\mathrm{d}z$	$\mathrm{d}z$	$\mathrm{e}^{x}(\mathrm{d}z + x\,\mathrm{d}y)$	$\mathrm{e}^{x}\mathrm{d}z$	$\mathrm{e}^{Ax}(-\sinh x\,\mathrm{d}y + \cosh x\,\mathrm{d}z)$	$\mathrm{e}^{Ax}(\sin x\,\mathrm{d}y + \cos x\,\mathrm{d}z)$

	VIII	*IX*
$\boldsymbol{\xi}_A$	$\operatorname{sech} y \cosh z\,\partial_x + \sinh z\,\partial_y - \tanh y \cosh z\,\partial_z$	$\sec y \cos z\,\partial_x + \sin z\,\partial_y - \tan y \cos z\,\partial_z$
	$\operatorname{sech} y \sinh z\,\partial_x + \cosh z\,\partial_y - \tanh y \sinh z\,\partial_z$	$-\sec y \sin z\,\partial_x + \cos z\,\partial_y + \tan y \sin z\,\partial_z$
	∂_z	∂_z
$\boldsymbol{\eta}_A$	∂_x	∂_x
	$-\sin x \tanh y\,\partial_x + \cos x\,\partial_y - \sin x \operatorname{sech} y\,\partial_z$	$\sin x \tan y\,\partial_x + \cos x\,\partial_y - \sin x \sec y\,\partial_z$
	$\cos x \tanh y\,\partial_x + \sin x\,\partial_y + \cos x \operatorname{sech} y\,\partial_z$	$-\cos x \tan y\,\partial_x + \sin x\,\partial_y + \cos x \sec y\,\partial_z$
$\boldsymbol{\omega}^A$	$\mathrm{d}x - \sinh y\,\mathrm{d}z$	$\mathrm{d}x + \sin y\,\mathrm{d}z$
	$\cos x\,\mathrm{d}y - \sin x \cosh y\,\mathrm{d}z$	$\cos x\,\mathrm{d}y - \sin x \cos y\,\mathrm{d}z$
	$\sin x\,\mathrm{d}y + \cos x \cosh y\,\mathrm{d}z$	$\sin x\,\mathrm{d}y + \cos x \cos y\,\mathrm{d}z$

of null rotations fixing a vector $\boldsymbol{k}$, choose $\boldsymbol{\xi}_1 = \boldsymbol{k}$). The action of the (linear) isotropy group on vectors at p is known (from that of the generalized orthogonal group) and hence the commutators $[\boldsymbol{\xi}_\alpha, \boldsymbol{Y}_i]$ are known up to terms in $\boldsymbol{Y}_j$, i.e. the structure constants $C^\beta{}_{\alpha i}$ are known. The unknown structure constants $C^j{}_{i\alpha}$, $C^j{}_{\alpha\beta}$, $C^\gamma{}_{\alpha\beta}$ must satisfy the Jacobi identities (8.7) and all possibilities can then be enumerated. One can add to Schmidt's remarks that if the group has a simply-transitive subgroup, this latter must have a basis $\{\boldsymbol{Z}_\alpha; \alpha = 1, \ldots, d\}$ agreeing with $\{\boldsymbol{\xi}_\alpha\}$ at p. Therefore one must have

$$\boldsymbol{Z}_\alpha = \boldsymbol{\xi}_\alpha + A_\alpha{}^i \boldsymbol{Y}_i, \tag{8.55}$$

where the $A_\alpha{}^i$ are constants. One can easily evaluate the commutators $[\boldsymbol{Z}_\alpha, \boldsymbol{Z}_\beta]$, and the condition that these should be spanned by the $\boldsymbol{Z}_\alpha$ (so that a subalgebra is generated) gives restrictions on the $A_\alpha{}^i$. All possible simply-transitive subgroups can thus be determined.

Using the basis vector fields $\{\boldsymbol{\xi}_\alpha\}$ in a neighbourhood of p, it is possible to determine the curvature of $\mathcal{O}_p$ at p as follows (Schmidt 1971). The connection coefficients are given by

$$\nabla_{\boldsymbol{\xi}_\alpha} \boldsymbol{\xi}_\beta = \Gamma^\gamma{}_{\beta\alpha} \boldsymbol{\xi}_\gamma. \tag{8.56}$$

The commutator gives

$$-2\Gamma^\gamma{}_{[\alpha\beta]} = C^\gamma{}_{\alpha\beta} \tag{8.57}$$

at p. Note that $[\boldsymbol{\xi}_\alpha, \boldsymbol{\xi}_\gamma] \cdot \boldsymbol{\xi}_\beta = C_{\beta\alpha\gamma}$ need only hold at p, since $\boldsymbol{Y}_i \cdot \boldsymbol{\xi}_\beta$ need not and in general will not be zero elsewhere. The symmetric part $\Gamma^\gamma{}_{(\alpha\beta)}$ of the connection can be found using the Killing equations which yield $\Gamma_{\beta\alpha\gamma} + \Gamma_{\gamma\alpha\beta} = 0$ and thus

$$2\Gamma_{\gamma(\alpha\beta)} = C_{\beta\alpha\gamma} + C_{\alpha\beta\gamma}, \tag{8.58}$$

whence

$$2\nabla_{\boldsymbol{\xi}_\alpha} \boldsymbol{\xi}_\beta = [\boldsymbol{\xi}_\alpha, \boldsymbol{\xi}_\beta] + \{[\boldsymbol{\xi}_\alpha, \boldsymbol{\xi}_\gamma] \cdot \boldsymbol{\xi}_\beta + [\boldsymbol{\xi}_\beta, \boldsymbol{\xi}_\gamma] \cdot \boldsymbol{\xi}_\alpha\} g^{\gamma\delta} \boldsymbol{\xi}_\delta. \tag{8.59}$$

Now to compute the next derivative, and hence the Riemann tensor, by (2.77), we need $\nabla_{\boldsymbol{\xi}_\alpha} \boldsymbol{g}$ and $\nabla_{\boldsymbol{\xi}_\alpha} \boldsymbol{Y}_i$. The first of these involves only $\nabla_{\boldsymbol{\xi}_\alpha} \boldsymbol{\xi}_\beta$ and $\nabla_{\boldsymbol{\xi}_\alpha} \boldsymbol{\xi}_\gamma$ since $\mathcal{L}_{\boldsymbol{\xi}_\alpha} \boldsymbol{g} = 0$; these are already known at p. Also $\nabla_{\boldsymbol{\xi}_\alpha} \boldsymbol{Y}_i = \nabla_{\boldsymbol{Y}_i} \boldsymbol{\xi}_\alpha + [\boldsymbol{\xi}_\alpha, \boldsymbol{Y}_i]$ and since $\boldsymbol{Y}_i = \boldsymbol{0}$ at p, this simplifies, at p, to $\nabla_{\boldsymbol{\xi}_\alpha} \boldsymbol{Y}_i = [\boldsymbol{\xi}_\alpha, \boldsymbol{Y}_i]$ which is known. Thus the components of the Riemann tensor of the orbit can be evaluated at p in the basis $\{\boldsymbol{\xi}_\alpha\}$.

We now give some examples of this method with applications in the sequel. First we determine the possible isometry groups of two-dimensional positive-definite spaces of constant curvature. The isotropy $\boldsymbol{Y}$ is a spatial rotation, and $\boldsymbol{\xi}_1$ and $\boldsymbol{\xi}_2$ can be chosen at p so that

$$[\boldsymbol{\xi}_1, \boldsymbol{Y}] = \boldsymbol{\xi}_2 + \alpha\boldsymbol{Y}, \quad [\boldsymbol{\xi}_2, \boldsymbol{Y}] = \boldsymbol{\xi}_1 + \beta\boldsymbol{Y}. \tag{8.60a}$$

A linear transformation $\boldsymbol{\xi}_1 \to \boldsymbol{\xi}_1 + \beta\boldsymbol{Y}$, $\boldsymbol{\xi}_2 \to \boldsymbol{\xi}_2 + \alpha\boldsymbol{Y}$ eliminates α and β. Using the finite rotation $\Phi : (\boldsymbol{\xi}_1, \boldsymbol{\xi}_2) \to (-\boldsymbol{\xi}_1, -\boldsymbol{\xi}_2)$, $\Phi[\boldsymbol{\xi}_1, \boldsymbol{\xi}_2] = [\boldsymbol{\xi}_1, \boldsymbol{\xi}_2]$ implies

$$[\boldsymbol{\xi}_1, \boldsymbol{\xi}_2] = K\boldsymbol{Y}. \tag{8.60b}$$

By the method outlined above one can compute the Riemann tensor and show that K is its constant curvature. Clearly the isometry group is a G_3 of Bianchi type *IX*, *VII*$_0$, or *VIII* respectively when K is positive, zero or negative. On changing the basis to $\boldsymbol{Z}_1 = \boldsymbol{\xi}_1 + \alpha\boldsymbol{Y}$, $\boldsymbol{Z}_2 = \boldsymbol{\xi}_2 + \beta\boldsymbol{Y}$ we find

$$[\boldsymbol{Z}_1, \boldsymbol{Z}_2] = -\alpha\boldsymbol{Z}_1 - \beta\boldsymbol{Z}_2 + (K + \alpha^2 + \beta^2)\boldsymbol{Y}. \tag{8.61}$$

Thus there is a simply-transitive subgroup of type G_2I if $K = 0$ (given by $\alpha = \beta = 0$), and a one-parameter family of simply-transitive subgroups of type G_2II (conjugate to one another within the G_3) if $K < 0$, given by $\alpha = |K| \sin\varphi$, $\beta = |K| \cos\varphi$ for arbitrary angle φ. The second of these results does not appear to be widely known. If $K > 0$ there are no simply-transitive subgroups; this is equivalent to the statement that the rotation group of three-dimensional space (whose orbits are the spheres centred at the origin) has no two-dimensional subgroup. The three-dimensional 'Lorentz group' (Bianchi *VIII*), however, has simply-transitive G_2 subgroups; they are generated by the combinations of a null rotation and a boost.

It is quite useful to calculate the Killing vectors for the S_2 of constant curvature. For the form (8.41) they have components given by

$$\xi^\zeta = \tfrac{1}{2}\gamma K\zeta^2 + \mathrm{i}a\zeta + \bar{\gamma} \tag{8.62}$$

where a is real and γ complex.

A second application of Schmidt's method is to prove that the maximal isotropy group of a space-time cannot consist of a (non-trivial) combination of a boost and a spatial rotation. If it did, the full isometry group must be G_5 on V_4 (because a G_3 on V_2 or G_4 on V_3 would have isotropies acting only in two- or three-dimensional subspaces of the tangent space). Using a basis at p in which $\boldsymbol{\xi}_1$ and $\boldsymbol{\xi}_2$ are spacelike unit vectors in the rotation plane, and $\boldsymbol{\xi}_3$ and $\boldsymbol{\xi}_4$ null vectors in the boosted plane such that $\boldsymbol{\xi}_3 \cdot \boldsymbol{\xi}_4 = -1$, Schmidt's calculation shows that the four Killing vectors $(\boldsymbol{\xi}_1, \boldsymbol{\xi}_2, \boldsymbol{\xi}_3, \boldsymbol{\xi}_4)$ form an Abelian group and thus the space is flat. Hence its isotropy group is really an I_6.

The result of Theorem 8.16 is a special case of a phenomenon known as *orthogonal transitivity.* This occurs when the orbits of a group of motions are submanifolds of a V_n which have orthogonal surfaces and is of particular interest for space-times admitting a group G_2I (see Chapter 17 et seq.). Schmidt proved a number of further theorems on this matter, including

Theorem 8.19 *If a group G_r of motions of $r = \frac{1}{2}d(d+1)$, $(d > 1)$, parameters has orbits of dimension d the orbits admit orthogonal surfaces* (Schmidt 1967).

8.7 Homothety groups

The equation defining a homothetic (or 'homothetic Killing') vector,

$$(\mathcal{L}_{\boldsymbol{\xi}} g)_{ab} = \xi_{a;b} + \xi_{b;a} = 2\Phi g_{ab}, \quad \Phi = \text{const}, \tag{8.63}$$

implies related equations for other geometrically defined tensors, e.g.

$$(\mathcal{L}_{\boldsymbol{\xi}} R)^a{}_{bcd} = 0, \quad \mathcal{L}_{\boldsymbol{\xi}} R = -2\Phi R, \tag{8.64}$$

and conversely these equations give a sequence of integrability conditions in the same way as (8.23). Note that from (8.64) Einstein spaces with $\Lambda \neq 0$ cannot admit proper homothetic motions. In classifying homotheties, the 'homothetic bivector' $\xi_{[a,b]}$ is of importance; for example, if it vanishes and $\boldsymbol{\xi}$ is null, the space-time must be algebraically special (McIntosh and van Leeuwen 1982).

Now consider a homothety group (i.e. a Lie group each of whose elements is a homothety). A basis of its generators will obey

$$(\mathcal{L}_{\boldsymbol{\xi}_A} g)_{ab} = \Phi_A g_{ab}, \tag{8.65}$$

where each Φ_A is a constant, possibly 0, and in general $\Phi_A \neq \Phi_B$ if $A \neq B$. Since the Lie derivative is linear (over $\mathbb{R}$) in the vector field used, the generators $\boldsymbol{w} = C^A \boldsymbol{\xi}_A$ satisfy $\mathcal{L}_{\boldsymbol{w}} g_{ab} = (\sigma^c w_c) g_{ab}$ for some 1-form $\boldsymbol{\sigma}$. Generators satisfying $\sigma^c w_c = 0$ are isometries and so a space-time admitting a group H_r of homothetic motions necessarily admits a group G_{r-1} of motions. From (2.63) the commutator of any two homotheties or isometries must be an isometry, so the G_{r-1} is an invariant subgroup of the H_r. The structure constants of the basis (8.65) must satisfy

$$C^A{}_{BC} \Phi_A = 0, \tag{8.66}$$

(Yano 1955) and so $\mathrm{d}\boldsymbol{\sigma} = 0$. The generators of the H_n can thus be chosen so that only one of them is a proper homothety (i.e. a homothetic motion which is not an isometry, so $\Phi \neq 0$ in (8.63)), and this generator can itself be scaled so that $\Phi = 1$.

Using these properties, Eardley (1974) has enumerated all possible H_3, giving a refinement of the classification of Table 8.2 (see also McIntosh (1979)), and Koutras (1992b) has enumerated the possible H_4, by refinement of the classification in MacCallum (1999).

As a special case of Theorem 35.12, a sufficiently smooth space-time admitting a group of homothetic motions is locally conformal, except where the homothety has a fixed point, to a space-time admitting the same group as a group of motions, unless it is a *pp*-wave or conformally flat. Although the related space-time will have an energy-momentum tensor of a different and perhaps unphysical type, this may still provide a way to find solutions with proper homothetic motions since the relation between the two metrics must be as given in §3.7; see Kerr (1998).

In a region where a proper homothety has no fixed point, one can consider a coordinate z such that $\boldsymbol{\xi} = \partial_z$. Then one can write the metric as $\mathrm{d}s^2 = \mathrm{e}^{2\Phi z}\mathrm{d}s_0^2$ where $\mathrm{d}s_0^2$ is independent of z. A change of origin of z can then be used to remove an overall scale parameter in the metric. Moreover, invariants such as the eigenvalues of the Ricci tensor will have a simple exponential behaviour along integral curves of $\boldsymbol{\xi}$, which may imply the existence of a singularity (Collins and Lang 1986, Hall 1988b).

9
Invariants and the characterization of geometries

When discussing solutions, we should often like to be able to decide, in an invariant manner, whether two metrics, each given in some specific coordinate system, are identical or not, or whether a given metric is new or not. For such purposes it is useful to have an invariantly-defined and unique complete characterization of each metric. Such a characterization can be attempted using scalar polynomial invariants, whose definition and construction are discussed in §9.1. However, it turns out that those invariants do not characterize space-times uniquely.

A method which does provide a unique coordinate-independent characterization, using Cartan invariants, is described in §9.2. This enables one to compare metrics given in differing coordinate systems, which distinguishes the results from those on uniqueness of the metric given the coordinate components for curvature and its derivatives (for which see e.g. Ihrig (1975), Hall and Kay (1988)). That uniqueness is related to the structure of the holonomy group, defined for each point p as the group of linear transformations of the tangent space at p generated by the holonomy (see §2.10) for different closed curves, or of the infinitesimal holonomy group, which is generated by the curvature and its derivatives but is equal to the holonomy group at almost all points in simply-connected smooth manifolds. These groups are subgroups of the Lorentz group and their properties can also be related to classification of curvature and the existence of constant tensor fields (Goldberg and Kerr 1961, Beiglböck 1964, Ihrig 1975, Hall 1991).

To apply the method using Cartan invariants in practice, one may proceed by finding the Petrov type (Chapter 4) and Segre type (Chapter 5) of the space-time, and §9.3 discusses the methods for doing so. The remainder of this chapter concerns applications of these ideas.

All the considerations here concern purely local equivalence. If two metrics are given which cover disjoint regions of a single analytic space-time, the fact that each is in the analytic continuation of the other cannot be detected by these methods. Conversely, topological identifications, or non-trivial homotopies of the metric as a section of the bundle of symmetric tensors (called 'kinks'; see e.g. Whiston (1981)), may give locally equivalent but globally inequivalent metrics. A further limitation is that we assume that all classifying quantities, discrete or continuous, are respectively constant or sufficiently smooth in the (open) neighbourhoods considered. For classifications like the Petrov type this follows from the smoothness of the invariants whose vanishing or otherwise characterizes the type; for smooth metrics such types change only on submanifolds of lower dimension.

In principle a space-time for which a Cauchy problem is well posed, and all its properties, can also be characterized by Cauchy data on a suitable hypersurface, since such data completely determine (a neighbourhood in) the space-time (see e.g. Friedrich and Rendall (2000)).

The method described in §9.2 and various of the methods described in §9.3 have been implemented in computer algebra programs. We do not describe such programs here as this information would rapidly become outdated: instead we refer interested readers to the reviews of Hartley (1996) and MacCallum (1996). However, we do discuss some efficiency considerations in §9.3. In preparing this book, we used the system CLASSI (Åman 2002), based on SHEEP and REDUCE, as described in MacCallum and Skea (1994).

9.1 Scalar invariants and covariants

Scalars constructed from the metric and its derivatives must be functions of the metric itself and the Riemann tensor and its covariant derivatives (Christoffel 1869). In a manifold $\mathcal{M}$ of n dimensions, at most n such scalars can be functionally independent, i.e. independent functions on $\mathcal{M}$. (The term 'functional independence' may also, confusingly, be used for functional independence over the bundle of symmetric tensors or some jet bundle thereof.) The number of *algebraically* independent scalar invariants, i.e. invariants not satisfying any polynomial relation (called a *syzygy*) is rather larger: it can be calculated, e.g. by considering Taylor expansions of the metric and of the possible coordinate transformations (Siklos 1976a). The result (Thomas 1934) is that in a general V_n the number of algebraically independent scalars constructible from the metric and its derivatives up to the pth order (the Riemann tensor and its derivatives up to the $(p-2)$th) is 0 for $p=0$ or $p=1$ and

$$N(n,p) = \frac{n[n+1][(n+p)!]}{2n!\,p!} - \frac{(n+p+1)!}{(n-1)!\,(p+1)!} + n, \qquad (9.1)$$

for $p \geq 2$, except for $N(2,2) = 1$. Thus in a general space-time the Riemann tensor has $N(4,2) = 14$ algebraically independent scalar invariants. In particular cases, including all exact solutions here, the number is reduced.

It is natural to try to express the algebraically independent scalar invariants as polynomials in the curvature and its derivatives, i.e. *scalar polynomial invariants*, and to aim to find such a set which contains the maximum number of independent scalar invariants even in special cases. One may also attempt to find a set of such invariants $\{I_1, I_2, \ldots, I_n\}$ that is complete in the sense that any other such scalar can be written as a polynomial (or, in some definitions, a rational function, i.e. a ratio of polynomials) in the I_j but no invariant in the set can be so expressed in terms of the others. It has been shown that any complete set of scalar polynomial invariants of the Riemann tensor, and any set which always contains a maximal set of independent scalars, contains redundant elements. Such sets contain more than 14 scalars, in four dimensions, and these are related by syzygies (but the syzygies cannot be solved for one of the invariants as a *polynomial* in the others). Hence all attempts to satisfy the above aims for space-times using 14 explicit scalar polynomials in the Riemann tensor failed (these attempts are reviewed in Zakhary and McIntosh (1997)).

The smallest known set of scalar polynomial invariants for the Riemann tensor which will always contain a maximal set of algebraically independent scalars consists of 17 polynomials (Zakhary and McIntosh 1997), though 16 suffice for perfect fluids and Einstein-Maxwell fields (Carminati and McLenaghan 1991). The origin of many of the syzygies can be understood in terms of the vanishing of any object skewed over $(n+1)$ indices in n dimensions (Harvey 1995, Bonanos 1998, Edgar 1999). The smallest set known to be complete has 38 scalars (Sneddon 1999).

For the Weyl tensor, there are four scalar invariants given by the real and imaginary parts of I and J defined by (4.12). The Ricci tensor defines the invariant Ricci scalar and the three eigenvalues of (5.1) which satisfy

$$\lambda^4 - \tfrac{1}{2}I_6\lambda^2 - \tfrac{1}{3}I_7\lambda + \tfrac{1}{8}(I_6^2 - 2I_8) = 0, \qquad (9.2)$$

where the scalar polynomial invariants I_k are given by

$$\begin{aligned} I_6 &= S^a{}_b S^b{}_a &&= 4\Phi_{AB\dot{X}\dot{Y}}\Phi^{AB\dot{X}\dot{Y}}, \\ I_7 &= S^a{}_b S^b{}_c S^c{}_a &&= 8\Phi_{AB\dot{X}\dot{Y}}\Phi^{A}{}_{E}{}^{\dot{X}}{}_{\dot{Z}}\Phi^{EB\dot{Z}\dot{Y}}, \\ I_8 &= S^a{}_b S^b{}_c S^c{}_d S^d{}_a &&= 16\Phi_{AB\dot{X}\dot{Y}}\Phi^{AC\dot{X}\dot{Z}}\Phi_{CD\dot{Z}\dot{W}}\Phi^{DB\dot{W}\dot{Y}}. \end{aligned} \tag{9.3}$$

The further invariants of the Riemann tensor needed to make up the set of 17 mentioned above, or the larger complete set, will not be given in detail here. They involve contractions between the Weyl and Ricci curvatures and are known as *mixed invariants*; they give information on the relative alignment of frames defined by the Weyl and Ricci tensors.

Less attention has been given, except in the context of possible Lagrangians in quantum gravity (cp. Fulling *et al.* (1992)), to the explicit construction of scalar polynomial invariants or algebraically invariant scalars of higher differential order in general metrics.

Even if algebraically independent or complete, no set of scalar polynomial invariants serves to characterize all space-times uniquely, though they are often sufficient to prove inequivalence, may be useful in proving equivalence (Cartan 1946, Eleuterio and Mendes 1982), and are important in investigating singularities and other properties. One can see this easily by noting that homogeneous plane waves and flat space both have all scalar polynomial invariants, of all orders, equal to zero (Jordan *et al.* 1960). Non-vacuum solutions, and metrics without symmetries, can also have this property (see e.g. Bueken and Vanhecke (1997), Skea (2000)). Algebraically special solutions with a cosmological constant provide examples of inequivalent metrics with equal but non-zero scalar polynomial invariants of all orders (Siklos 1985, Bičák and Pravda 1998, Pravda 1999). These ambiguities, and the consequent inadequacy of scalar polynomial invariants as classifying quantities, are associated with the indefiniteness of the metric and the non-compactness of the Lorentz group (Schmidt 1998).

Thus to find a set of invariants defined by the Riemann tensor and its derivatives that will always suffice to characterize a space-time, one must follow a different approach. A starting point is that 14 invariants of the Riemann tensor can be considered to arise from the four independent real eigenvalues of the Ricci tensor and the four real quantities in the canonical forms of the Weyl tensor (Table 4.2), together with six parameters specifying the Lorentz transformation between the Weyl and Ricci principal tetrads (Ehlers and Kundt 1962). More generally one can define tetrads invariantly from the Riemann tensor and its derivatives and then use the remaining non-zero components of those tensors in the chosen tetrads, which are scalar contractions between the curvature or its derivatives and the tetrad vectors, as the scalar invariants required. These are the so-called *Cartan invariants* or *Cartan scalars.* That they are not equivalent to the scalar polynomial invariants can be seen by considering plane waves again; in that case the surviving Cartan invariant of the Riemann tensor is $\Psi_4 = 1$, whereas all the scalar polynomial invariants are zero. We describe in the next section how the Cartan invariants can be used to locally characterize space-times.

One may also consider *covariants* of the Riemann tensor and its derivatives. In general, for a given tensor or spinor and a scalar polynomial invariant obtained by contracting it with arbitrary vectors or spinors, a covariant is defined to be any other scalar polynomial in the same variables; for example, a covariant of $\phi_{AB\dot{A}}\zeta^A\zeta^B\bar{\zeta}^{\dot{A}}$ is a scalar expression in the coefficients $\phi_{AB\dot{A}}$ and the variables ζ^B, $\bar{\zeta}^{\dot{A}}$. Covariants differ from invariants in that they do not depend solely on the (metric and the) tensor or spinor itself. The degree of a covariant is the degree in the tensor or spinor, e.g. in $\phi_{AB\dot{A}}$. Covariants and invariants together are called *concomitants.* For example, the covariants

$$
\begin{aligned}
Q &= \Psi_{AB}{}^{EF}\Psi_{CDEF}\zeta^A\zeta^B\zeta^C\zeta^D, \\
R &= \Psi_{ABC}{}^{K}\Psi_{DE}{}^{LM}\Psi_{FKLM}\zeta^A\zeta^B\zeta^C\zeta^D\zeta^E\zeta^F,
\end{aligned}
\tag{9.4}
$$

of the expression (4.28) involving the Weyl spinor are used in some methods of Petrov classification; see §9.3.

9.2 The Cartan equivalence method for space-times

The method to be outlined here is a specialized form of a more general method, due to Cartan, applicable to the equivalence of sets of differential forms on manifolds under appropriate transformation groups (Gardner 1989, Olver 1995). For sufficiently smooth metrics, it gives sets of scalars providing a unique local characterization, and thus leads to a procedure for comparing metrics.

To relate two apparently different metrics, we need to consider coordinate or basis transformations, and therefore to consider a frame bundle as defined in §2.11. The basis of the method is that if the metrics are equivalent, the frame bundles they define are identical (locally). Moreover, the frame bundles possess uniquely-defined bases of 1-form fields, $\{\omega^a, \boldsymbol{\Gamma}^a{}_b\}$, as described in §2.11, which would therefore also be identical for equivalent metrics. The same is more generally true for any (sufficiently smooth) manifolds equipped with uniquely-defined 1-form bases

$\{\sigma^I\}$. For two such manifolds to be identified, the exterior derivatives $\mathrm{d}\sigma^I = C^I{}_{JK}\sigma^J \wedge \sigma^K$ of these basis 1-forms, and hence the $C^I{}_{JK}$, must agree. In the case of space-times, the Cartan structure equations (2.76) and (2.85) show that this implies the components of the curvature on the frame bundle must be equatable.

This condition is necessary, but not sufficient. Cartan showed that a sufficient condition is obtained by repeatedly taking (exterior) derivatives starting with $\mathrm{d}C^I{}_{JK}$ until no new functionally independent quantity arises; if at any step of differentiation no such quantity arises, the process terminates because then any further derivatives depend on those already known. The relations between the independent invariants and the dependent ones must be the same in (neighbourhoods in) both manifolds for equivalence. The number k of functionally independent quantities, called the *rank*, is assumed constant in a neighbourhood whose points are then called *regular*. Since k is at most the dimension m of the manifold, the process necessarily terminates in a finite number of steps. If $k < m$, this is due to the presence of symmetries. The final step, testing whether or not the relations obtained by equating corresponding quantities on the two manifolds have a solution, is formally undecidable (unless the functions that arise lie in some simple class such as rational functions) but in practice usually turns out to be feasible.

For space-times, (2.87) shows that the repeated differentiation is equivalent to repeatedly taking covariant derivatives of the Riemann tensor, viewed as functions on $F(\mathcal{M})$. Hence a metric can be uniquely characterized by the Riemann tensor and a finite number of its covariant derivatives, regarded as functions on $F(\mathcal{M})$. We use $\mathcal{R}^q$ to denote the set $\{R_{abcd}, R_{abcd;f}, \ldots, R_{abcd;f_1f_2\cdots f_q}\}$ of the components of the Riemann tensor and its derivatives up to the qth. If p is the last derivative at which a new functionally independent quantity arises, called the *order*, we need to calculate $\mathcal{R}^{p+1}$ (unless $p = \dim F(\mathcal{M})$). If there are k elements in a maximal set of functionally independent invariants on $F(\mathcal{M})$, let the invariants be denoted by I^α, $\alpha = 1, \ldots, k$, and their *index basis*, i.e. the set of indices of the corresponding components of $\mathcal{R}^p$, by $\mathcal{A}$.

The first work on the equivalence problem for Riemannian manifolds was due to Christoffel (1869), using the full (coordinate) frame bundle, which has dimension $n(n+1)$. It was the context of the invention of the Christoffel symbols (for a fuller history, see Ehlers (1981)) but dealt only with the case of metrics without symmetry, and implied that for space-times the twentieth derivatives of the Riemann tensor might be required. This is computationally impractical, however.

Cartan (1946) made an important reduction by using frames with constant metric components (e.g. complex null frames), in which case the frame bundle has dimension $n(n+1)/2$ so that the maximum order of differentiation of the Riemann tensor for space-times is at most 10. Cartan also showed how the method applied to metrics with isometries, the argument being completed by Sternberg (1964); for $k < n(n+1)/2$ there is an isometry group of dimension $n(n+1)/2 - k$. The result as stated in Ehlers (1981), following Sternberg (1964), applies to frame bundles with the generalized orthogonal group transitive on the fibres. For orthonormal frames on space-times it gives, using the notation introduced above:

Theorem 9.1 *Let $\mathcal{M}$ and $\overline{\mathcal{M}}$ be space-times of differentiability class C^{13}, x be a regular point of $\mathcal{M}$ and E be a frame at x, and similarly for $\overline{\mathcal{M}}$. Then there is an isometry which maps (x, E) to $(\overline{x}, \overline{E})$ if and only if $\mathcal{R}^{p+1}$ for $\mathcal{M}$ is such that:*
(i) $\mathcal{A}$ indexes quantities $\overline{I}^{\alpha}$ which are functionally independent in $F(\overline{\mathcal{M}})$,
(ii) $I^{\alpha}(x, E) = \overline{I}^{\alpha}(\overline{x}, \overline{E})$ for $\alpha = 1, \ldots, k$, and
(iii) the functions giving all other components of $\mathcal{R}^{p+1}$ in terms of the I^{α} and $\overline{I}^{\alpha}$ are the same for $\mathcal{M}$ and $\overline{\mathcal{M}}$.

The differentiability class required, $n(n+1)/2+3$ in n dimensions, can be reduced since, as we show below, $p < n(n+1)/2-1 = 9$ in space-times. The size of the set of quantities to be compared in *(iii)* can be reduced since $\mathcal{R}^{p+1}$ can be replaced by a minimal set of its elements from which the others follow algebraically by use of the Ricci and Bianchi identities, e.g. the set specified by MacCallum and Åman (1986); see §9.3.2. This can be further reduced by using canonical forms as described below, or, for example, by giving, on a region of $F(\mathcal{M})$, the Riemann tensor components and the derivatives $I^{\alpha}{}_{|J}$ as functions of the I^{α} (Bradley and Karlhede 1990). Data sufficient to define the $\mathcal{R}^{p+1}$ can be given in various ways other than just giving the metric itself; see §9.4.

The practical application of Theorem 9.1 was considered by Brans (1965), who initially proposed a scheme using canonical forms chosen by lexicography of bases but later (1977) considered canonical forms of the Weyl tensor at the first step, similar to the scheme introduced by Karlhede and implemented by Åman and others (Karlhede and Åman 1979, Karlhede 1980b, MacCallum and Skea 1994) which we now discuss.

In this method the idea is to reduce the frame bundle to the smallest possible dimension at each step by casting the curvature and derivatives into a canonical form and only permitting those frame changes which preserve the canonical form. (Note that the horizontal and vertical dimensions are not treated in the same way, since the independent functions on the fibres are removed by restricting the choices of frame.) Because there are many vacuum solutions and few conformally flat solutions, the method is usually implemented by first putting the Weyl tensor into the appropriate one of the forms given in Table 4.2 and then using any residual freedom in the tetrad to put the Ricci tensor into canonical form, if possible (in the exceptional case of conformally flat space-times, the first step is to bring the Ricci tensor to one of the forms discussed in Chapter 5, as, e.g., in Bradley (1986)). Finding such a canonical tetrad, though formally decidable in four dimensions (as it can be done by finding roots of quartics), can be somewhat intractable in practice. The curvature components in this tetrad are the first set of the invariants required. The next step is to calculate the first derivatives of the curvature and use them to further fix the tetrad if necessary, and so on. Details of the canonical forms used in practice can be found from MacCallum and Skea (1994) and references therein ; see also Pollney *et al.* (2000). It is the remaining non-zero components that are referred to as the Cartan scalars.

The resulting procedure is as follows:

1. Set the order of differentiation q to 0.
2. Calculate the derivatives of the Riemann tensor up to the qth.
3. Find the canonical form of the Riemann tensor and its derivatives.
4. Fix the frame as far as possible by this canonical form, and note the residual frame freedom (the group of allowed transformations is the linear isotropy group $\hat{I}_q$). The dimension of $\hat{I}_q$ is the dimension of the remaining vertical part of the frame bundle.
5. Find the number t_q of independent functions of space-time position in the components of the Riemann tensor and its derivatives in canonical form. This tells us the remaining horizontal freedom.
6. If the isotropy group and number of independent functions are the same as at the previous step, let $p + 1 = q$ and stop; if they differ (or if $q = 0$) increment q by 1 and go to step 2.

The space-time is then characterized by the canonical form used, the successive isotropy groups and independent function counts and the values of the non-zero Cartan invariants. Since there are t_p essential space-time coordinates, clearly the remaining $4 - t_p$ are ignorable, so the isotropy group of the space-time will have dimension $s = \dim \hat{I}_p$ and the isometry group has dimension $r = s + 4 - t_p$ (see e.g. Karlhede (1980b)). To compare two space-times one can first compare the discrete properties such as the sequence of isotropy groups, and only if those all match does one have to check whether the set of equations obtained by equating corresponding Cartan invariants has a solution. This final step is not algorithmic, and could in principle be unsolvable, but in practice is not usually the difficult step. Note that *while the discrete properties can prove inequivalence of manifolds, they are insufficient to prove equivalence in general.*

Since the method gives the dimension of the isometry group, it can be used to find those subcases within a class of solutions which have (additional) isometries (e.g. Seixas (1992b)). In principle, it could be used to identify metrics given separately one of which is a subcase of another (as Schwarzschild is of Kerr), but this depends on inspection of the values of invariants and is probably better tackled by taking limits of families (see §9.5) and comparing them with other solutions.

The system CLASSI, mentioned in §9.1, uses the Newman–Penrose formalism (Chapter 7) in implementing this procedure, but the calculations can be carried out in other sets of frames. Conformally flat perfect fluids have been investigated in orthonormal frames (Seixas 1992b) and the formalism of Ramos and Vickers (1996b) based on a single null congruence has also been applied (see e.g. Ramos (1998)).

Since the continuous isotropy group for non-zero Weyl and/or tracefree Ricci tensors is at most an I_3 (see §4.2 and Table 5.2), the maximum dimension of the reduced tetrad frame bundle required in this procedure is 7. (One should note, however, that a discrete isotropy group consisting of frame changes that interchange principal null directions may need to be considered.) The exceptional case where both tensors vanish has constant curvature and is detected at the first step of differentiation. Thus no more than the seventh derivatives of the curvature could be needed; for Petrov types *I*–*III* this reduces by a similar argument to at most five (Karlhede 1980b). More detailed consideration leads to the results in Table 9.1 (Skea, unpublished). The cases where the two bounds given disagree are those which are not yet fully understood: these include two cases where exhaustive inspection of possible metrics gives a bound smaller than has been proved in any other way (Petrov type D vacua and conformally flat metrics of Segre type [(11)(1,1)]).

One can consider partial equivalences, where only $\mathcal{R}^q$, $q \leq p$, or parts thereof, are equated: these give families of space-times related in some way. Brinkmann's theorem (Theorem 3.1) is an example: see also e.g. Collinson and Vaz (1982), Lor and Rozoy (1991), Bueken and Vanhecke (1997). Cartan scalars can similarly be used to characterize or subclassify families of metrics (see e.g. Edgar (1986)).

9.3 Calculating the Cartan scalars

9.3.1 Determination of the Petrov and Segre types

The Petrov and Segre classifications both depend on the handling of quartics which may have multiple roots.

For the *Petrov type* of (the Weyl tensor of) a gravitational field at a given point p, we have (see Table 4.3) to determine the multiplicities of

Table 9.1. Maximum number of derivatives required to characterize a metric locally

The column '$p \geq$' gives the value from examples analysed so far; if marked A this is obtained by exhaustive analysis of all possible metrics. '$p \leq$' gives the upper bound from theoretical arguments.

Petrov type	Ricci tensor	$p \geq$	$p \leq$	References
I	All	1	5	
II	Vacuum	2	4	Paiva, unpublished
	Others	1	5	
III	Vacuum	2	5	
	Others	1	5	
D	Vacuum	2 A	3	Collins *et al.* (1991), Åman (1984)
	Others	2	6	Collins and d'Inverno (1993)
N	Vacuum	3	5	Ramos and Vickers (1996a)
	Others	7	7	
O	[(111),1]	3 A	3	Bradley (1986), Seixas (1992b)
	[1(11,1)]	1 A	7	
	[(112)]	4 A	4	Koutras (1992c), Skea (1997 and unpublished)
	[(11)(1,1)]	1 A	5	Paiva and Skea, unpublished
	Others	1	5	Paiva and Skea, unpublished

the roots of the quartic algebraic equation (4.18),

$$\Psi_0 - 4E\Psi_1 + 6E^2\Psi_2 - 4E^3\Psi_3 + E^4\Psi_4 = 0, \tag{9.5}$$

where the coefficients $\Psi_0, \ldots, \Psi_4$ defined by (3.59) can be calculated with respect to an *arbitrary* complex null tetrad at p. If the degree of the algebraic equation (9.5) is $(4-m)$, then there are $(4-m)$ principal null directions $\boldsymbol{k}$, and $\boldsymbol{l}$ represents an m-fold principal null direction.

The classical algorithm for determining the multiplicities from the coefficients $\Psi_0, \ldots, \Psi_4$, based on considering the discriminant of the quartic, is displayed in Fig. 9.1 (d'Inverno and Russell-Clark 1971). Provided that $\Psi_4 \neq 0$, the additional definitions used in the flow diagram are

$$\begin{aligned} K &\equiv \Psi_1\Psi_4^2 - 3\Psi_4\Psi_3\Psi_2 + 2\Psi_3^3 \\ L &\equiv \Psi_2\Psi_4 - \Psi_3^2, \quad N \equiv 12L^2 - \Psi_4^2 I \end{aligned} \tag{9.6}$$

(K and L are coefficients in the covariants (9.4)). If $\Psi_4 = 0$, but $\Psi_0 \neq 0$, one has to interchange Ψ_0 with Ψ_4 and Ψ_1 with Ψ_3 in these definitions,

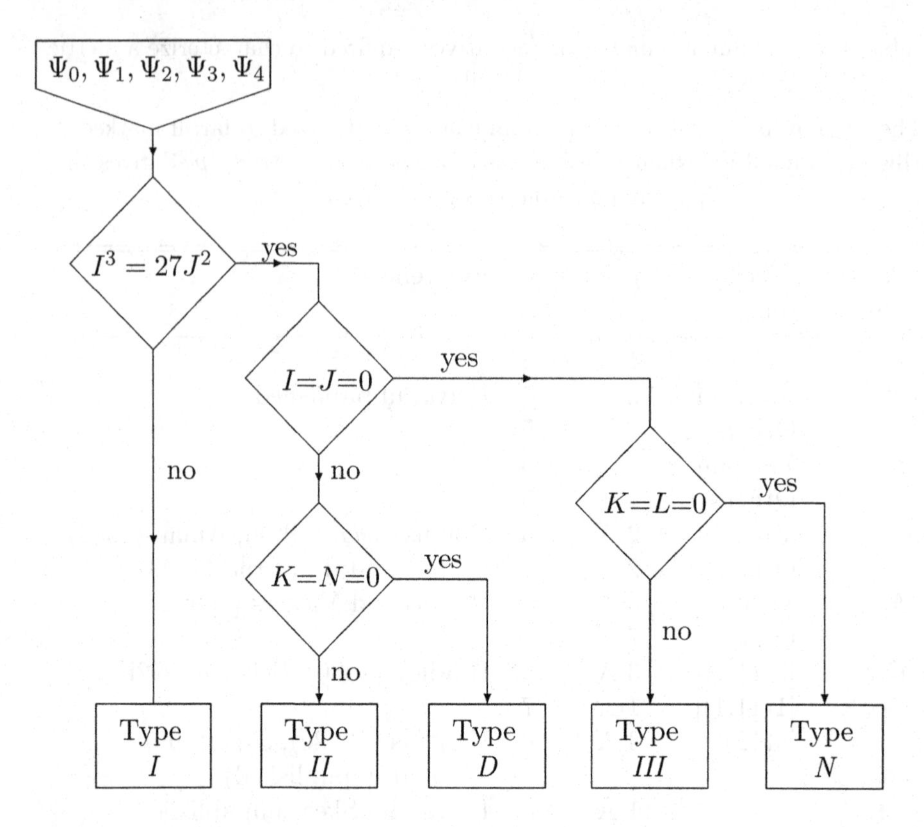

Fig. 9.1. Flow diagram for determining the Petrov type by the classical method

and the algorithm proceeds as before. For $\Psi_0 = \Psi_4 = 0$, the multiplicities of the roots of (9.5) are very simple to determine.

If both the invariants I and J vanish, then at least three of the principal null directions coincide. If not, then from the diagram and the definitions (9.6) it follows that gravitational fields with a repeated principal null direction $\boldsymbol{k}$ ($\Psi_0 = \Psi_1 = 0$) are type D if and only if the remaining tetrad components of the Weyl tensor satisfy the condition

$$3\Psi_2\Psi_4 = 2\Psi_3^2. \tag{9.7}$$

An *equivalent method* for determining the Petrov type is based on the eigenvalue equation (4.4). One can use the invariant criteria for the matrix $\boldsymbol{Q}$ which are listed in Table 4.1, $\boldsymbol{Q}$ being calculated with respect to an arbitrary orthonormal basis $\{\boldsymbol{E}_a\}$.

These algorithms are far from optimal computationally. The first improvements were those of Fitch (1971), which used ideas similar to those later (re-)introduced and extended by others (Hon 1975, Åman *et al.* 1984,

1991, Letniowski and McLenaghan 1988). The main sources of improvements are as follows.

First, cases can be separated according to whether or not one or more of the Newman–Penrose components Ψ_A of the Weyl tensor vanish. In many cases this immediately gives the Petrov type. The basic idea was introduced independently by Hon (1975) and Åman *et al.* (1984). Secondly, and relatedly, one arranges the tests so that any calculation of the full discriminant (a sixth degree polynomial, a sextic, in the components Ψ_A) is put off until it becomes unavoidable (Hon 1975, Åman *et al.* 1984). Thirdly, one can build up more complicated expressions in the components Ψ_A by a series of binary operations followed by simplification. This will in general save both time and computer memory in the calculations of quantities of high degree (Åman *et al.* 1984). Finally, one must look very closely at the tests required to determine the Petrov type in those remaining cases where a number of types are possible. The algorithms in successive treatments differ essentially only at this point. A deeper understanding of the origin of the special tests arises from formulating them as consequences of the Euclidean algorithm applied to finding simultaneous roots of the quartic (9.5) and its derivative (Åman *et al.* 1991). An alternative understanding and procedure comes from considering the covariants (Zund 1986, Penrose and Rindler 1986, Zakhary 1994); this may in practice involve longer calculations (Piper 1997). The ranking of the different versions of the improved methods now depends on rather subjective estimates of whether the expense of a test is justified by its probability of success.

The calculation of the *Segre type* (Joly and MacCallum 1990, Seixas 1991, Paiva *et al.* 1998) makes use of the same methods for quartics as the Petrov classification. It has two parts. The first part classifies the Plebański spinor (the coefficient in a degree-two covariant of the Ricci spinor)

$$\Phi_{(AB}{}^{\dot{C}\dot{D}}\Phi_{CD)\dot{C}\dot{D}} \tag{9.8}$$

into types algebraically equivalent to the Petrov types, called Plebański–Petrov types, by the same method as for Petrov classification. Unfortunately, different Ricci tensor types give the same Plebański–Petrov type, so a second stage is usually required, in which particular methods for separating the subcases are used. These are based on considering (9.2), by similar methods. The process aims only at separation into the different Segre types, and does not consider refinements, e.g. those discussed by Penrose and Rindler (1986). Further improvements to the computation have been suggested by Paiva and Skea (unpublished, 1998) and Zakhary (1994).

9.3.2 The remaining steps

Having thus dealt with step 2 for $q = 0$ of the procedure given in the previous section, we now have to bring the Riemann tensor into canonical form by Lorentz transformations. Although this is in principle algorithmic (since quartics have an algorithmic solution, and the quartic for Petrov classification has as its roots the parameters of the required null rotations), it is in practice often the most difficult step, especially in the Petrov type I case. In algebraically special cases, we are helped by the fact that the method used for Petrov classification can be related to the Euclidean algorithm for simultaneous roots of polynomials, so that when there are equal roots they are readily found.

The fourth and fifth steps are straightforward in principle. Functional dependence is tested by finding the rank of the Jacobians between the possibly independent functions, and the isotropy can be found from the canonical form.

Now we reach step 1 again for $q = 1$. For $q \geq 1$ it is very important in calculations to cut down the number of quantities to be computed by taking only a minimal set of derivatives of the Riemann tensor, which can be done because the derivatives obey the Ricci and Bianchi identities. An explicit minimal set of derivatives was found by MacCallum and Åman (1986), generalizing the treatment of the electrovac case in Penrose (1960). In terms of the Newman–Penrose quantities Ψ_{ABCD}, $\Phi_{AB\dot{A}\dot{B}}$ and Λ it can be described as follows. One has to take, in the general case, the following components for $q \geq 0$ (where the term 'totally symmetrized' means that a spinor is to be symmetrized over all its free dotted indices and over all its free undotted indices):

1. The totally symmetrized qth derivatives of Λ.
2. The totally symmetrized qth derivatives of Ψ_{ABCD}.
3. The totally symmetrized qth derivatives of $\Phi_{AB\dot{A}\dot{B}}$.
4. For $q \geq 1$, the totally symmetrized $(q-1)$th derivatives of $\Xi_{ABC\dot{D}} = \nabla^{D}{}_{\dot{D}}\Psi_{DABC}$ which is one side of one of the Bianchi identities.
5. For $q \geq 2$, the d'Alembertian $\nabla^{A\dot{A}}\nabla_{A\dot{A}}$ applied to all the quantities calculated for the derivatives of order $q-2$.

Steps 2 and 3 of the procedure for $q \geq 1$ are handled in existing programs in a manner which is capable of refinements that would probably be useful in creating more precise classifications. In fact, in most cases all that is tested is whether any invariance at the $q = 0$ stage persists, and if it does not, new canonical forms are not found. Steps 4 and 5 continue to be in principle straightforward, if more time-consuming (because of the increasing numbers of components to be tested).

Ref.: For a covariant approach to location of the frame in Petrov type I see Ferrando and Sáez (1997).

9.4 Extensions and applications of the Cartan method

The procedure also supplies information from which the Lie algebraic structure of the isometry group can be obtained (Karlhede and MacCallum 1982, Araujo *et al.* 1992, MacCallum and Skea 1994): the application of similar ideas to three-dimensional manifolds has been considered by Bona and Coll (1992, 1994). An extension of the arguments about isometries, by considering ratios of Cartan scalars, enables one to test for the presence of homotheties (Koutras 1992b, Koutras and Skea 1998). Note the close relation to methods based on extensions of the GHP technique (§7.4) in which invariantly-defined scalars are sought (e.g. Ludwig and Edgar (2000)).

Although the Cartan invariants give a complete local characterization of the space-time, the detailed relationships between these scalars and physical properties are unknown at present.

Another particular question of some interest is whether a metric not given in diagonal form can be expressed in coordinates in which it is diagonal. The solution to this question (Tod 1992) can be formulated in a manner similar to but not identical with the classification by Cartan scalars.

Brans (1965) pointed out that one could find a metric from an invariant characterization by Cartan scalars. This idea has been developed and applied by Karlhede (1980a), Karlhede and Lindström (1983), Bradley (1986), Bradley and Karlhede (1990), Bradley and Marklund (1996) and Marklund (1997). One has to assume part of the structure of the frame bundle, or a specified reduction thereof. Then one takes a manifold $\mathcal{N}$ of suitable dimension, a constant matrix η_{ij} and a set of $C^I{}_{JK}$ on $\mathcal{N}$ satisfying the Cartan structure equations (2.76) and (2.85) (note that the indexing implies some information on the fibration of $\mathcal{N}$, and that the usual index symmetries are implicitly assumed). Specifying a set of I^α among the Riemann tensor and its derivatives which are to be independent and will be used as coordinates, one has to find suitable $I^\alpha{}_{|L}$ as functions of I^β. The quantities found either have to be from a minimal set which defines the rest of $\mathcal{R}^{p+1}$, using the Bianchi and Ricci identities, or (Bradley and Karlhede 1990, Bradley and Marklund 1996) have to satisfy those identities as integrability conditions. Note that the Ricci identities for the curvature are Bianchi identities $\mathrm{d}^2 I^\alpha = 0$ for the scalars on the frame bundle representing the components, due to (2.87).

To obtain the metric from these values if the matrix of $I^\alpha{}_{|L}$ is invertible (i.e. if the number of functionally independent I^α equals the dimension of $\mathcal{N}$), one obtains the basis 1-forms from $\boldsymbol{\omega}^K = I^K{}_\beta \mathrm{d}I^\alpha$, where the components $I^K{}_\beta$ form the inverse of $I^\alpha{}_{|L}$, and the metric is then given by $\eta_{ij}\boldsymbol{\omega}^i \wedge \boldsymbol{\omega}^j$ (Karlhede 1980a). If the matrix $(I^\alpha{}_{|L})$ is not invertible there is a continuous symmetry: if the rank of $(I^\alpha{}_{|L})$ (the number of index sets in $\mathcal{A}$) is k, one has to give $(\dim \mathcal{N}) - k$ 1-forms $\boldsymbol{\omega}^P$, introducing additional coordinates on $\mathcal{N}$, subject to $\mathrm{d}^2\boldsymbol{\omega}^P = 0$ and then one can find $\boldsymbol{\omega}^A = I^A{}_\alpha(\mathrm{d}I^\alpha - I^\alpha{}_{|P}\boldsymbol{\omega}^P)$ for $1 \leq A \leq k$ using the inverse of $I^\beta{}_A$ (Bradley and Karlhede 1990, Bradley and Marklund 1996). When using a reduced frame bundle with an assumed symmetry group, one can give values in a fixed frame together with the generators of the rotations and appropriately specialize the method just outlined. This method has mainly been applied to recovering already known solutions (Karlhede and Lindström 1983, Bradley and Karlhede 1990, Bradley 1986) but some new solutions have been found also (see Bradley and Marklund (1996), Marklund (1997), Marklund and Bradley (1999)).

A considerable number of papers report applications of the above Cartan equivalence procedure (see e.g. the references in MacCallum and Skea (1994)), some of them cited in Table 9.1. Special adaptations of the methods can be made for classes of metrics with special properties. For example, the spatially homogeneous Bianchi metrics can be dealt with in terms of the automorphism group variables described in Chapter 13 (Araujo and Skea 1988a) and the subset of metrics with two commuting Killing vectors which are computationally tractable can be widened using the factorization method of §20.7 (Seixas 1992a), although such metrics can be made arbitrarily complicated and intractable by repeated application of the generating techniques (Chapter 34).

These methods can also be applied to classification of the potential spaces arising in Hamiltonian descriptions, e.g. of those for space-times with a G_3 on S_3 (Chapters 13 and 14) or H_3 on T_3 containing a G_2I on S_2 (Uggla *et al.* 1995a). The potential spaces may themselves have symmetries (see §34.1 and Uggla *et al.* (1995a)). The general Cartan method has been used to classify the null bundle of space-time (Nurowski *et al.* 1999), leading to a classification procedure for non-conformally-flat Einstein spaces, and null hypersurfaces (Nurowski and Robinson 2000).

9.5 Limits of families of space-times

It is possible in principle that one could find new solutions of Einstein's equations as limits of known solutions. In practice, since the limits are generally simpler than the space-times in the parent families, this rarely happens: the limits are usually already known. The general situation for the case where the limit is a non-singular (region of) space-time was investigated first by Geroch (1969) and more recently by Paiva *et al.* (1993). There is also the possibility of singular limits which can be interpreted as shocks or impulsive waves (cp. Chapters 24 and 25), cp. §3.8.

The idea is that one has a family of metrics sufficiently smoothly dependent on some parameter λ, and wishes to study the limit as $\lambda \to 0$. Geroch (1969) defined hereditary properties to be those which the limiting solution must inherit from the general family. For example, if the general family all have an isometry group G_r, the limit will have an isometry group G_t where $t \geq r$. From the methods of Petrov classification described in Chapter 4 and §9.3, and the fact that invariants will be continuous functions of λ, one can see that 'being of a Petrov type at least as special as ... ' is a hereditary property. The possible limiting processes are as implied by Fig. 4.1. The corresponding specialization diagram for Segre type has been given by Paiva *et al.* (1998); essentially the same information arises from considering perturbing the Weyl or Ricci tensor of the limiting type, in the sense of Arnol'd (Ellis and McCarthy 1987, Guzman S. *et al.* 1991).

Since the coordinate transformations giving the form of the line element from which the limit is to be derived may themselves be λ-dependent, and in a way that may be singular as $\lambda \to 0$, most of the examples have been derived in an ad hoc manner. Examples of limits found in this way are given in §§20.6 and 21.1.2, but such methods do not provide a systematic way of finding all possibilities. Paiva *et al.* (1993) pointed out that the possibilities can be enumerated without directly seeking the coordinate transformations, by using the Cartan invariants defined in §9.2. This approach arises because the Cartan invariants give a unique local characterization of the geometry, and will be continuous functions of λ which, for regular limits, will have finite values as $\lambda \to 0$ if the limiting space-time is non-singular. Thus one can enumerate the limits by studying the possible limiting values of the Cartan invariants: in particular any λ-independent equation relating Cartan invariants must be a hereditary property. Once the limiting spaces are known one can then find appropriate coordinate transformations.

This can lead to cases overlooked in applications of other methods. For example, Paiva *et al.* (1993) rediscussed the limits of the family of Schwarzschild metrics (15.19) as $1/m \to 0$ and found five possible limits, rather than the two (flat space and the axisymmetric Kasner metric (13.53) with $p_1 = p_2 = 2/3$, $p_3 = -1/3$) found by Geroch (1969). The extra cases are one inhomogeneous and two homogeneous forms of vacuum plane wave solutions. For $m \to 0$ one gets only flat space. The method has also been used to show that the Levi-Civita metric (22.7) appears as a limit of the Zipoy–Voorhees metrics (20.11) (Herrera *et al.* 1999).

As yet, this method has limitations. It does not so far give a complete algorithm for all possible limits, due to complications like the singular relation of canonical frames for different Petrov types and the absence of a complete classification of the derivatives for all cases; it assumes that the limit is a four-dimensional space-time (limits which become singular or shock hypersurfaces, representing shells or impulsive waves that can be boundaries of regions of other space-times, are not covered by this technique though examples have been given in the literature, e.g. Aichelburg and Sexl (1971), Ferrari and Pendenza (1990), Hotta and Tanaka (1993), Balasin and Nachbagauer (1996)); it does not give the coordinate transformation for coordinates which are ignorable in the limiting space-time; and it does not ensure that the limit is global.

10
Generation techniques

10.1 Introduction

Einstein's equations are a system of non-linear partial differential equations and thus rather difficult to solve. A book on exact solutions is thus – among other things – a catalogue of tricks invented to solve such a complicated system. Many of these tricks are not special to Einstein's equations, but can be used on any similar system; in fact, several of them come from outside relativity.

In this chapter we want to give a concise survey of some of these methods. We have selected those which could help in understanding the general approach to finding solutions, or those which have been frequently used in the search for solutions, or those which have been proved to be extremely powerful. Here, we shall make only cursory references to general relativity; the applications are discussed in Chapter 34.

10.2 Lie symmetries of Einstein's equations

10.2.1 Point transformations and their generators

A Lie point symmetry of a system

$$H_A(x^n, u^\alpha, u^\alpha_{,n}, u^\alpha_{,nm}, \dots) = 0 \qquad (10.1)$$

of partial differential equations in the independent variables x^n, the dependent variables $u^\alpha(x^n)$ and their derivatives is a mapping (point transformation)

$$\tilde{x}^n = \tilde{x}^n(x^i, u^\beta; \varepsilon), \quad \tilde{u}^\alpha = \tilde{u}^\alpha(x^i, u^\beta; \varepsilon) \qquad (10.2)$$

in the space of independent and dependent variables which maps solutions into solutions. These mappings form a group. They may depend on arbitrary functions or arbitrary parameters; in the following we shall assume that at least one arbitrary parameter ε exists.

It is immediately clear from the above definition that symmetries can be used to generate solutions: if one has found a solution, then the application of a symmetry may lead to a new solution.

To apply a symmetry to a differential equation, one has to extend (prolong) the mapping (10.2) to the derivatives. This is done in the obvious way by defining

$$\widetilde{u^\alpha_{,n}} = \tilde{u}^\alpha_{,n} = \frac{\partial \tilde{u}^\alpha}{\partial \tilde{x}^n}, \quad \widetilde{u^\alpha_{,nm}} = \tilde{u}^\alpha_{,nm} = \frac{\partial^2 \tilde{u}^\alpha}{\partial \tilde{x}^n \partial \tilde{x}^m}, \quad \text{etc.} \tag{10.3}$$

The corresponding infinitesimal transformations are

$$\begin{aligned} \tilde{x}^n &= x^n + \varepsilon \xi^n(x^i, u^\beta) + \cdots, & \xi^n &= \partial \tilde{x}^n / \partial \varepsilon|_{\varepsilon=0}, \\ \tilde{u}^\alpha &= u^\alpha + \varepsilon \eta^\alpha(x^i, u^\beta) + \cdots, & \eta^\alpha &= \partial \tilde{u}^\alpha / \partial \varepsilon|_{\varepsilon=0}, \\ \tilde{u}^\alpha_{,n} &= u^\alpha_{,n} + \varepsilon \eta^\alpha_n(x^i, u^\beta, u^\beta_{,i}) + \cdots, & & \ldots, \\ \tilde{u}^\alpha_{,nm} &= u^\alpha_{,nm} + \varepsilon \eta^\alpha_{nm}(x^i, u^\beta, u^\beta_{,i}, u^\beta_{,ik}) + \cdots, & & \ldots, \end{aligned} \tag{10.4}$$

where the $\eta^\alpha_n, \eta^\alpha_{nm}, \ldots$ are given in terms of the functions ξ^n and η^α by

$$\eta^\alpha_n = \frac{\mathrm{D}\eta^\alpha}{\mathrm{D}x^n} - u^\alpha_{,i} \frac{\mathrm{D}\xi^i}{\mathrm{D}x^n}, \quad \eta^\alpha_{nm} = \frac{\mathrm{D}\eta^\alpha_n}{\mathrm{D}x^m} - u^\alpha_{,ni} \frac{\mathrm{D}\xi^i}{\mathrm{D}x^n}, \quad \ldots, \tag{10.5}$$

and the operator $\mathrm{D}/\mathrm{D}x^n$ is defined as

$$\frac{\mathrm{D}}{\mathrm{D}x^n} = \frac{\partial}{\partial x^n} + u^\alpha_{,n} \frac{\partial}{\partial u^\alpha} + u^\alpha_{,nm} \frac{\partial}{\partial u^\alpha_{,m}} + \cdots. \tag{10.6}$$

Note that the η^α_n etc. are *not* simply the partial derivatives of the η^α etc!

From the infinitesimal transformations (10.4) the *generator* $\mathbf{X}$ can be read off as

$$\mathbf{X} = \xi^n \frac{\partial}{\partial x^n} + \eta^\alpha \frac{\partial}{\partial u^\alpha} + \eta^\alpha_n \frac{\partial}{\partial u^\alpha_{,n}} + \eta^\alpha_{nm} \frac{\partial}{\partial u^\alpha_{,nm}} + \cdots. \tag{10.7}$$

By performing an appropriate point transformation in the ξ^i–u^α-space (taking the integral curves of $\mathbf{X}$ as coordinate lines) the generator can in principle always be transformed to its normal form

$$\mathbf{X} = \frac{\partial}{\partial s}. \tag{10.8}$$

If the point transformations depend on r parameters ε_A, then there exist r generators $\mathbf{X}_A$ that form the Lie algebra of the corresponding group G_r.

10.2.2 How to find the Lie point symmetries of a given differential equation

If a point transformation (10.2) is a symmetry of a system $H_A = 0$, then, as well as $u^\alpha(x^i)$, $\tilde{u}^\alpha(\tilde{x}^i)$ is also a solution of that system, i.e.

$$H_A(\tilde{x}^i, \tilde{u}^\alpha, \tilde{u}^\alpha_{,n}, \tilde{u}^\alpha_{,nm}, \ldots) = 0 \tag{10.9}$$

has to be a consequence of (10.1). Inserting the infinitesimal transformation (10.4) into this equation, one obtains the necessary condition

$$\mathbf{X}H_A \equiv 0 \ (\mathrm{mod}\ H_A = 0). \tag{10.10}$$

Here the identity sign $\equiv$ indicates that this condition on the coefficients of $\mathbf{X}$ has to be satisfied identically in x^i, u^α, *and all the derivatives*, since it has to be true for every solution, which means that at each point arbitrary values can be assigned to all these variables. These conditions are also sufficient if e.g. the differential equations are written so that they are linear in the highest derivatives.

Written in full, the symmetry conditions are a system of *linear* partial differential equations for the components ξ^i, η^α of the generator. It may, or may not, have non-trivial solutions, which may depend on arbitrary parameters or functions. For ordinary differential equations, which are included here as a special case, the generators cannot depend on arbitrary functions. In most cases occurring in practice, the system (10.10) can be solved and the symmetries can thus be determined. For a review of the available computer programs see e.g. Hereman (1996).

For Einstein's vacuum field equations, the symmetries have the form (Ibragimov 1985)

$$\mathbf{X} = \xi^i(x^n)\partial/\partial x^i - (\xi^k_{,m}g_{kn} + \xi^k_{,n}g_{mk} - 2ag_{nm})\partial/\partial g_{nm}, \tag{10.11}$$

where the $\xi^i(x^n)$ are arbitrary functions, and a is a constant.

For a perfect fluid, the four-velocity components u^i and pressure p and mass-density μ have to be added to the list of dependent functions. The symmetries for this case are

$$\begin{aligned}\mathbf{X} = \xi^i(x^k)\partial/\partial x^i &- (\xi^k_{,i}g_{kn} + \xi^k_{,n}g_{ik} - 2ag_{in})\partial/\partial g_{in} \\ &+(\xi^i_{,k}u^k + au^i)\partial/\partial u^i + 2ap\,\partial/\partial p + 2a\mu\,\partial/\partial\mu.\end{aligned} \tag{10.12}$$

Equations (10.11)–(10.12) show that the finite Lie symmetries of Einstein's equations are (only) diffeomorphisms ($\tilde{x}^i = x^i + \varepsilon\xi^i(x^n) + \cdots$, together with the appropriate change of vectors and tensors), and scalings $\tilde{g}_{in} = \mathrm{e}^{2\varepsilon a}g_{ik}$. If one is dealing only with a subset of the solutions of

Einstein's equations, e.g. with the class of algebraically special or some spherically-symmetric solutions, new types of Lie symmetries could in principle arise; in practice, in most cases one only recovers the generator given above or its subcases.

10.2.3 How to use Lie point symmetries: similarity reduction

A first way to use a symmetry is to generate new solutions from known ones. In the case of the symmetries of Einstein's equations this idea is less fruitful. Although a coordinate transformation of a solution leads again to a solution, the two are trivially related. Similarly, the constant $\mathrm{e}^{2\varepsilon a}$ by which the two line elements

$$\mathrm{d}\tilde{s}^2 = \mathrm{e}^{2\varepsilon a}\mathrm{d}s^2 \tag{10.13}$$

differ can often be incorporated into the coordinates (as for flat space-time) or leads only to new values of a parameter already existing in the metric (as for the exterior Schwarzschild solution (15.19), where it can be absorbed into the mass m by the transformation $\tilde{r} = \mathrm{e}^{\varepsilon a}$, $\tilde{t} = \mathrm{e}^{\varepsilon a}t$, $\tilde{m} = \mathrm{e}^{\varepsilon a}m$).

A second way to use a symmetry is *similarity reduction*. For (systems of) partial differential equations the existence of a symmetry shows that it makes sense, and does not lead to immediate contradictions, to ask for fixed points of the transformations, i.e. solutions for which the solution surface

$$u^\alpha - u^\alpha(x^i) = 0 \tag{10.14}$$

does not change under the action of the symmetry; these solutions are also called *similarity solutions*. Applying the generator $\mathbf{X}$ to (10.14), one obtains

$$\eta^\alpha - u^\alpha_{,k}\xi^k = 0, \tag{10.15}$$

or, with (10.11), for Einstein's equations,

$$\mathcal{L}_\xi g_{nm} = g_{nm,i}\xi^i + g_{ni}\xi^i_{,m} + g_{mi}\xi^i_{,n} = 2a g_{nm}. \tag{10.16}$$

The similarity solutions of Einstein's equations are those which admit a homothetic vector (for $a \neq 0$) or a Killing vector (for $a = 0$). This connection with the symmetries explains the outstanding rôle Killing vectors and homothetic vectors play compared with other vector fields.

If one transforms the generator to its normal form (10.8) and takes s as a coordinate, say x^1, then (10.15) implies that the solutions are independent of x^1 and depend only on those coordinates y^k for which $\mathbf{X}y^k = 0$: a similarity reduction (of partial differential equations) is a reduction of the number of independent variables. In the case of Einstein's equations, one usually does not achieve a completely normal form of the generator: one often takes $\mathbf{X} = \partial/\partial x^1 - 2ag_{nm}\partial/\partial g_{nm}$ or $\mathbf{X} = x^1\partial/\partial x^1 - 2ag_{nm}\partial/\partial g_{nm}$, so that (10.16) amounts to either of the two forms

$$g_{nm}(x^1, x^2, x^3, x^4) = \mathrm{e}^{2ax^1}\hat{g}_{nm}(x^2, x^3, x^4), \tag{10.17}$$

$$g_{nm}(x^1, x^2, x^3, x^4) = (x^1)^{2a}\hat{g}_{nm}(x^2, x^3, x^4) \tag{10.18}$$

(power-law behaviour). If the coordinate system is already fixed, e.g. in time-dependent spherically-symmetric fields, then the generator may have the typical form $\mathbf{X} = r\partial/\partial r + t\partial/\partial t$; the solutions depend only on the similarity variable $z = r/t$.

For an ordinary differential equation such as

$$y'' = \omega(x, y, y'), \tag{10.19}$$

where the condition (10.10) for the components ξ, η of the generator $\mathbf{X}$ reads

$$\begin{aligned}\omega(\eta_{,y} - 2\xi_{,x} - 3y'\xi_{,y}) - \omega_{,y'}[\eta_{,x} + y'\,(\eta_{,y} - \xi_{,x}) - y'^2\xi_{,y}] - \omega_{,x}\xi \\ -\omega_{,y}\eta + \eta_{,xx} + y'\,(2\eta_{,xy} - \xi_{,xx}) + y'^2\,(\eta_{,yy} - 2\xi_{,xy}) - y'^3\xi_{,yy} = 0,\end{aligned} \tag{10.20}$$

the existence of a symmetry implies the possibility of a reduction *in order*: if one takes x as the coordinate s in which $\mathbf{X}$ has its normal form, then because of the symmetry condition (10.10) ω cannot depend on x: the differential equation (10.19) is in fact a first-order equation for $y'(y)$.

Of course, the above procedure can be applied repeatedly: for partial differential equations this can lead to solutions depending on fewer and fewer variables (for the Einstein equations: solutions with more than one homothetic or Killing vector), ending up with ordinary differential equations, and for ordinary differential equations with more than one first integral. The details depend very much on the structure of the group of symmetries and its Lie algebra. If for an ordinary second-order differential equation two or more symmetries exist, these symmetries can be used to construct its general solution.

The numerous applications of the similarity reduction to Einstein's equations are dealt with in the relevant chapters of this book.

Ref.: For textbooks on symmetries of differential equations see e.g. Ovsiannikov (1982), Ibragimov (1985), Olver (1986), Bluman and Kumei (1989), and Stephani (1989).

10.3 Symmetries more general than Lie symmetries

10.3.1 Contact and Lie–Bäcklund symmetries

Lie symmetries can be generalized by admitting derivatives in the components of the generator.

For *contact symmetries* first-order derivatives are admitted,

$$\mathbf{X} = \xi^n(x^i, u^\beta, u^\beta_{,k})\frac{\partial}{\partial x^n} + \eta^\alpha(\ldots)\frac{\partial}{\partial u^\alpha} + \eta^\alpha_n(x^i, u^\beta, u^\beta_{,k})\frac{\partial}{\partial u^\alpha_n} + \cdots. \quad (10.21)$$

Although the extension law (10.5) applies, no higher derivatives of u^β are admitted in η^α_n and its extensions. This imposes severe restrictions: contact transformations and symmetries exist only for partial differential equations with only one dependent variable u, or for ordinary differential equations. In the context of Einstein's equations they are rarely used.

For *Lie–Bäcklund symmetries* the components ξ^n and η^α depend on derivatives of u^α up to an arbitrary finite order; because of the extension law (10.5), this order grows unboundedly when going to higher and higher extensions of η^α. An analysis of Einstein's vacuum equations (Torre and Anderson 1993, Capovilla 1994, Anderson and Torre 1996) shows that all their Lie–Bäcklund symmetries are in fact disguised Lie symmetries: they have again the form (10.11), but the arbitrary functions $\xi^i(x^n)$ are replaced by arbitrary functions $\xi^i(x^n, g_{pq}, g_{pq,n}, \ldots)$ which means that some extra conditions are imposed on these functions.

10.3.2 Generalized and potential symmetries

Roughly speaking, if a class of solutions depends on arbitrary parameters (or functions), a change of these parameters (or functions) always maps solutions into solutions and is, therefore, a symmetry. It may be difficult to find the explicit form of this symmetry, but it is obvious that symmetries are more common than one may guess at first thought. To incorporate such general symmetries, one has to enlarge the set of admitted transformations. One way to do this is the following.

Formally one can go a step further than with the Lie–Bäcklund symmetries and admit derivatives up to an arbitrary or even infinite order in the components ξ^n and η^α of the generator. Sometimes these symmetries are also called Lie–Bäcklund symmetries. The main trick for avoiding these infinitely many derivatives is to introduce new variables (potentials) in terms of which the symmetry is again a Lie point symmetry. Since for Lie–Bäcklund symmetries one can gauge the symmetry generator to be of the form

$$\mathbf{X} = f_{ab}\,\partial/\partial g_{ab} + \cdots, \quad (10.22a)$$

the symmetry condition $\mathbf{X}R_{ab} = 0$ for the vacuum equations amounts to

$$f^n{}_{a;bn} + f^n{}_{b;an} - f_{ab}{}^{;n}{}_{;n} - f^n{}_{n;ab} = 0. \qquad (10.22b)$$

Its solution f_{ab} has to satisfy (10.22b) for *every* vacuum metric g_{ab}.

For the full set of Einstein's vacuum equations no such symmetry (except the trivial one $f_{ab} = \mathcal{L}_\xi g_{ab} - 2ag_{ab}$) is known. But if one assumes the existence of a Killing vector ξ^i, then for the enlarged set

$$R_{ab} = 0, \quad \mathcal{L}_\xi g_{ab} = \xi_{a;b} + \xi_{b;a} = 0 \qquad (10.23)$$

such non-trivial symmetries can be found.

Following Kramer and Neugebauer (1968b) and Geroch (1972), for a non-null Killing vector ($\xi^a\xi_a = \lambda \neq 0$) one can introduce the functions ω, α_a and β_b by

$$\begin{aligned} \omega_{,a} &= \varepsilon_{abcd}\xi^{d;c}\xi^b, \qquad \alpha_{b,a} - \alpha_{a,b} = \varepsilon_{abcd}\xi^{d;c}, \\ \beta_{b,a} - \beta_{a,b} &= 4\lambda\xi_{b;a} + 2\omega\varepsilon_{abcd}\xi^{d;c}, \end{aligned} \qquad (10.24)$$

and gauge them by

$$\xi^a\alpha_a = \omega, \quad \xi^a\beta_a = \omega^2 + \lambda^2. \qquad (10.25)$$

The integrability conditions are satisfied because of (10.23). In terms of these quantities, the generator of the symmetry is given by

$$\begin{aligned} \mathbf{X}g_{ab} &= f_{ab} = -2\omega g_{ab} + 2(\xi_a\alpha_b + \xi_b\alpha_a), \\ \mathbf{X}\alpha_a &= 2\omega\alpha_a - \beta_a, \quad \mathbf{X}\beta_a = 2(\omega^2 + \lambda^2)\alpha_a. \end{aligned} \qquad (10.26)$$

If one calculates the action of this generator on λ, ω, and the 3-metric $h_{ab} = g_{ab} - \xi_a\xi_b/\lambda$, then one obtains

$$\mathbf{X}\lambda = 2\omega\lambda, \quad \mathbf{X}\omega = \omega^2 - \lambda^2, \quad \mathbf{X}h_{ab} = -2\omega h_{ab}. \qquad (10.27)$$

The components of the generator (10.26) are non-local functions of the metric and the Killing vector's components, but in terms of the 'potentials' ω, λ and h_{ab} (10.27) describes the action of a simple Lie symmetry. The finite symmetry transformations can be found by solving the system

$$\partial\tilde{\lambda}/\partial\varepsilon = 2\tilde{\omega}\tilde{\lambda}, \quad \partial\tilde{\omega}/\partial\varepsilon = \tilde{\omega}^2 - \tilde{\lambda}^2, \quad \partial\tilde{h}_{ab}/\partial\varepsilon = -2\tilde{\omega}\tilde{h}_{ab}. \qquad (10.28)$$

The result is the equations

$$\begin{aligned} \tilde{\lambda} &= \lambda/\left[1 - 2\varepsilon\omega + \varepsilon^2(\lambda^2 + \omega^2)\right], \quad \tilde{g}_{ab} = \lambda h_{ab}/\tilde{\lambda} + \tilde{\xi}_a\tilde{\xi}_b/\tilde{\lambda}, \\ \tilde{\omega} &= \tilde{\lambda}\left[\omega - \varepsilon(\lambda^2 + \omega^2)\right]/\lambda, \qquad \tilde{\xi}_a = \xi_a + \tilde{\lambda}(2\varepsilon\alpha_a - \varepsilon^2\beta_a). \end{aligned} \qquad (10.29)$$

They show how to find the new metric $\tilde{g}_{ab}$ if the old metric g_{ab} (and $\lambda, \omega, \alpha_a$, and β_a constructed from it) is known. This is discussed from another point of view in §34.1.

This idea can be generalized to include some special perfect fluid cases which we now give (see e.g. Stephani (1988)). R_{ab} and T_{ab} are separately invariant under the symmetry; the transformation of the metric is again given by (10.29). If the Killing vector ξ_a is parallel to the four-velocity and the equation of state is $\mu + 3p = 0$, then the fluid variables obey

$$\mathbf{X}\mu = 2\omega\mu, \quad \tilde{\mu} = \mu/[1 - 2\varepsilon\omega + \varepsilon^2(\lambda^2 + \omega^2)], \tag{10.30}$$

$$\mathbf{X}u^a = -\omega u^a, \quad \tilde{u}^a = u^a[1 - 2\varepsilon\omega + \varepsilon^2(\lambda^2 + \omega^2)]^{1/2}. \tag{10.31}$$

If the Killing vector is orthogonal to the four-velocity and the equation of state is $\mu = p$, then (10.30) is still valid, but instead of (10.31) we have

$$\mathbf{X}u_a = -\omega u_a, \quad \tilde{u}_a = u_a[1 - 2\varepsilon\omega + \varepsilon^2(\lambda^2 + \omega^2)]^{1/2}. \tag{10.32}$$

For a null Killing vector ξ_a, one knows that λ and the twist ω vanish; it then follows from (10.24)–(10.25) that α_a and β_a are parallel to ξ_a. In generalizing (10.26), one may start from

$$\mathbf{X}g_{ab} = f_{ab} = N(x^i)\xi_a\xi_b. \tag{10.33}$$

The symmetry conditions (10.22*b*) then yield

$$N_{,a}\xi^a = 0, \quad N^{,r}{}_{;r}\xi_i\xi_k + 4N\xi^m\xi^r R_{mirk} - 2N^{,r}(\xi_{r;k}\xi_i + \xi_{r;i}\xi_k) = 0, \tag{10.34}$$

and the finite symmetry transformation is given by

$$\tilde{g}_{ab} = g_{ab} + \varepsilon N(x^i)\xi_a\xi_b, \tag{10.35}$$

where $N(x^i)$ is a solution of (10.34). This transformation can be used to generate the vacuum *pp*-waves from flat space-time and all other vacuum solutions with a null Killing vector from a special type D metric, see §24.4, equations (24.37*b*) and (24.39).

Also the method of constructing a stiff perfect fluid solution from the vacuum (Theorem 10.2) or a perfect fluid (which applies if there is an Abelian G_2) can be understood in terms of a symmetry (Stephani 1988). In terms of the metric (17.4), the finite symmetry transformation reads

$$\widetilde{M} = M + \widetilde{\Omega}, \quad \tilde{\sigma} = \sigma + \varepsilon\tau, \tag{10.36a}$$

where τ satisfies $\tau_{;a}^{\ ,a} = 0$, and $\widetilde{\Omega}$ is to be determined from

$$\widetilde{\Omega}_{,a} = 2W(W_{,b}W^{,b})^{-1}[2\varepsilon\,(\tau^{,n}W_{,n}\sigma_{,a} + \sigma^{,n}W_{,n}\tau_{,a} - \tau^{,n}\sigma_{,n}W_{,a}) + \varepsilon^2\,(\tau^{,n}W_{,n}\tau_{,a} - \tau^{,n}\tau_{,nb}W_{,a})]. \tag{10.36b}$$

Other examples of a generalized symmetry are the Kerr–Schild transformations (Chapter 32) and the linear superposition of Einstein–Maxwell fields discussed by Xanthopoulos (1986).

10.4 Prolongation

The method of prolongation structures was invented by Harrison and Estabrook (1971), Wahlquist and Estabrook (1975) and Estabrook and Wahlquist (1976), cp. also Estabrook (1982). It consists of a systematic way to introduce new functions, called pseudopotentials, and use them to derive new solutions from known ones. It uses the calculus of differential forms, cp. §§2.4–2.8, and is particularly well adapted to equations with two independent variables.

For the purpose of the following two subsections, a partial differential equation of order n is a subvariety of a finite jet bundle $J^{(n)}(X, U)$, i.e. a space parametrized by the independent variables and all derivatives of the dependent variables up to order n, where X is the space of independent variables and U the space of dependent variables in the original equation.

10.4.1 Integral manifolds of differential forms

Solutions of differential equations lie in and foliate the integral manifolds of a certain set of differential forms (Cartan 1945). A subspace $\mathcal{V}_l$ of dimension l immersed in a space $\mathcal{V}_k$ of dimension k can be described by giving $k-l$ variables as functions of l others, thus showing the distinction between dependent and independent variables. In this section we shall define the so-called integral manifolds of a given set of differential forms, see e.g. Estabrook (1976, 1982). Consider, for instance, that we are given in a k-dimensional space a set of l forms ω^A together with their exterior derivatives $\mathrm{d}\omega^A$; here, we take without loss of generality the ω^A to be 1-forms and $A = 1, \ldots, l$. We then find a set of linearly independent vector fields $\mathbf{V}_i, i = 1, \ldots, k$ such that they lie in a family of k-dimensional subspaces – this requires $[\mathbf{V}_a, \mathbf{V}_b] = f^c_{ab}\mathbf{V}_c$ – and such that they annul any ω^A, viz.

$$\mathbf{V}_i \lrcorner \omega^A = 0. \tag{10.37}$$

By operating with $\mathcal{L}_{\mathbf{V}_j}$ we find that also

$$\mathbf{V}_i \lrcorner \mathbf{V}_j \lrcorner \mathrm{d}\omega^A = 0. \tag{10.38}$$

The subspace spanned by the $\mathbf{V}_i$ is called an integral manifold of the set $\{\omega^A, \mathrm{d}\omega^A\}$. The definition given here can easily be extended to include forms of higher rank and their exterior derivatives.

Describing a subspace as a map $\Phi : \mathcal{V}_l \to \mathcal{V}_k$ we have seen in Chapter 2 that forms transform under the inverse map Φ^*; they can be "pulled back" by Φ^* onto $\mathcal{V}_l$ where we denote them by $\tilde{\omega}^A$ and $\mathrm{d}\tilde{\omega}^A$. Since the contraction of ω^A with any vector in $\mathcal{V}_l$ and $\mathrm{d}\omega^A$ with any pair of vectors in $\mathcal{V}_l$ both vanish, the condition determining integral submanifolds can be stated as $\tilde{\omega}^A = 0$ and $\mathrm{d}\tilde{\omega}^A = 0$. For a general set of forms the integral manifold is defined by requiring them to vanish when pulled back onto it.

Since the forms ω^A, $\mathrm{d}\omega^A$, etc. only enter the definition of an integral manifold through linear homogeneous equations, any algebraically equivalent set can be used. Indeed it is the entire ideal – defined as in e.g. Cartan (1945) – of forms generated by the given forms, here ω^A and $\mathrm{d}\omega^A$, which vanishes on an integral manifold. 1-forms like $f_A\omega^A$, 2-forms like $\psi_A \wedge \omega^A + f_A\mathrm{d}\omega^A$ (with arbitrary functions f_A and 1-forms ψ_A) etc. obviously all vanish when pulled back to the integral manifold. An ideal of forms is called closed if it contains the exterior derivatives of all forms in the set; we shall consider only closed ideals.

Introducing coordinates y^a in the immersing space $\mathcal{V}_k$ and x^α in the subspace $\mathcal{V}_l$ we get with $\omega^A = \omega^A_a \mathrm{d}y^a$ and $\mathrm{d}\omega^A = \omega^A_{a,b}\mathrm{d}y^b \wedge \mathrm{d}y^a$

$$\tilde{\omega}^A = \omega^A_a \frac{\partial y^a}{\partial x^\alpha}\mathrm{d}x^\alpha, \quad \mathrm{d}\tilde{\omega}^A = \omega^A_{a,b}\frac{\partial y^b}{\partial x^\beta}\frac{\partial y^a}{\partial x^\alpha}\mathrm{d}x^\beta \wedge \mathrm{d}x^\alpha. \tag{10.39}$$

Setting these expressions equal to zero gives a coupled set of partial differential equations for k unknown functions y^a in terms of l independent variables x^α.

How can the integral manifold of a given set of forms be constructed, at least in principle? To this end we consider some general point p with coordinates y^a and a vector $\mathbf{V}_1$ such that

$$\mathbf{V}_1 \lrcorner\, \omega^A = 0 \tag{10.40}$$

at p. These are homogeneous linear algebraic equations for the components of $\mathbf{V}_1$ in terms of the components of ω^A. Denote the rank of this system by s_0; thus $l_1 = k - s_0$ components of $\mathbf{V}_1$ can be chosen arbitrarily. We assume the rank of (10.40) to be maximal at the point p and in a neighbourhood thereof. We now construct a one-dimensional integral manifold $\mathcal{V}_1 = \{y^a \mid y^a = f^a(s)\}$ by integrating

$$\mathrm{d}f^a/\mathrm{d}s = \mathbf{V}_1^a(s). \tag{10.41}$$

In doing so we have made l_1 choices of functions of one variable. At each point of $\mathcal{V}_1$ we now choose a vector $\mathbf{V}_2$ such that

$$\mathbf{V}_2 \lrcorner\, \omega^A = 0, \qquad \mathbf{V}_1 \lrcorner\, \mathbf{V}_2 \lrcorner\, \mathrm{d}\omega^A = 0. \tag{10.42}$$

If there were other 2-forms given apart from the $\mathrm{d}\omega^A$ we would include them in the last equation. The rank of this set of linear equations is not less than s_0; we denote it by s_0+s_1. Now we can choose $l_2 = n-(s_0+s_1)$ components of $\mathbf{V}_2$ arbitrarily. N.b., $\mathbf{V}_1$ is a solution of the above equations and thus is to be included in the $n-(s_0+s_1)$ arbitrary choices. If it were the only solution, we would not be able to construct a $\mathcal{V}_2$; thus we assume $l_2 > 1$. We now construct an integral manifold $\mathcal{V}_2 = \{y^a \mid y^a = f^a(s,t)\}$ by integrating

$$\mathrm{d}f^a/\mathrm{d}s = \mathbf{V}_1^a(s,t), \quad \mathrm{d}f^a/\mathrm{d}t = \mathbf{V}_2^a(s,t), \quad [\mathbf{V}_1, \mathbf{V}_2] = 0. \tag{10.43}$$

The last equation guarantees that $\mathbf{V}_1$ and $\mathbf{V}_2$ are indeed surface-forming. Moreover, in the process of integration we have chosen l_2 arbitrary functions of two variables.

The construction continues. At each point of $\mathcal{V}_2$ we find a vector $\mathbf{V}_3$ such that

$$\mathbf{V}_3 \lrcorner \omega^A = 0, \quad \mathbf{V}_1 \lrcorner \mathbf{V}_3 \lrcorner \mathrm{d}\omega^A = 0, \quad \mathbf{V}_2 \lrcorner \mathbf{V}_3 \lrcorner \mathrm{d}\omega^A = 0. \tag{10.44}$$

We would include any independent 3-forms at this stage. The algebraic rank of this system is $s_0+s_1+s_2$ and $l_3 = n-(s_0+s_1+s_2)$ components of $\mathbf{V}_3$ can be chosen arbitrarily. If $l_3 > 2$ an integral manifold $\mathcal{V}_3$ can be constructed containing l_3 free functions of three variables. $\mathbf{V}_1$ and $\mathbf{V}_2$ are again constructed along the way by $[\mathbf{V}_1, \mathbf{V}_2] = 0$, $[\mathbf{V}_1, \mathbf{V}_3] = 0$, $[\mathbf{V}_2, \mathbf{V}_3] = 0$.

The integers s_i are the so-called Cartan characters and they are numerical invariants of the ideal of forms. Since $l_i \le l_{i-1}$ and we require $l_i > i-1$ the process must eventually terminate and there must be a maximal dimension for the integral manifolds. Let the largest value of i be denoted g and called the genus. Then we have $l_g = n-(s_0+s_1+\cdots+s_{g-1}) > g-1$ and $l_{g+1} = n-(s_0+s_1+\cdots+s_g) \le g$. If $l_g > g$ the integration scheme remains undetermined at the last step: the equivalent set of partial differential equations (10.38) for $n-g$ functions of g variables will be such that arbitrary functions of g variables enter the solution.

The most interesting case is $l_g = g$. In this case there is no freedom in the last step of the construction; $\mathcal{V}_g$ is determined by data given at its boundary, i.e. at $\mathcal{V}_{g-1}$. The above inequalities require the relation

$$n-g = s_0+s_1+\cdots+s_g. \tag{10.45}$$

This is an important criterion for a closed differential ideal to represent a proper set of partial differential equations. If this equation holds, the solution will not contain an arbitrary function of g variables.

10.4.2 Isovectors, similarity solutions and conservation laws

In what follows we shall assume that we are given a closed ideal of differential forms. A characteristic vector is defined by the property that contracting it with any form of the ideal gives again a form in the ideal; it corresponds to a 'conditional symmetry'. For instance, let the set in question consist of 1-forms ω^A and 2-forms α^a. A characteristic vector satisfies

$$\mathbf{Y} \lrcorner \omega^A = 0, \qquad \mathbf{Y} \lrcorner \alpha^a = \phi^a_A\, \omega^A = 0 \text{ (mod } \textit{ideal}) : \tag{10.46}$$

ϕ^a_A are undetermined functions. If there is another characteristic vector, $\mathbf{Z}$, we find by taking the Lie derivative of (10.46) with respect to $\mathbf{Z}$ that $[\mathbf{Y}, \mathbf{Z}]$ is also a characteristic vector. Thus the characteristics are surface-forming. We note that adjoining a characteristic to each point of a regular integral manifold gives again a regular integral manifold. The maximal integral manifold thus must contain all characteristics.

A generalization of a characteristic vector is an isovector. Now it is the Lie derivative of any form in the ideal which has to be in the ideal. Thus, for the example considered above, $\mathbf{X}$ is an isovector if

$$\mathcal{L}_{\mathbf{X}}\omega^A = 0 \text{ (mod } \textit{ideal}), \qquad \mathcal{L}_{\mathbf{X}}\alpha^a = 0 \text{ (mod } \textit{ideal}). \tag{10.47}$$

$\mathbf{X}$ is a solution of an overdetermined set of linear partial differential equations; any constant linear combination of isovectors is again an isovector. Moreover, if $\mathbf{X}$ and $\mathbf{Y}$ are isovectors then $[\mathbf{X}, \mathbf{Y}]$ is again an isovector. The set of all isovectors generates the isogroup. Isovectors correspond to Lie–Bäcklund symmetries, cp. §10.3.1.

An ideal of forms remains invariant under the infinitesimal mappings generated by its isovectors. Thus the isovectors also map integral manifolds into integral manifolds and the finite transformations, found in analogy to (10.87), generate new solutions of the underlying differential equations.

Given an isovector which is not a characteristic vector a number of new forms can be found by contracting it with the forms in the ideal. Adjoining these to the original ideal one finds a larger closed ideal of forms the integral manifold of which is a subset of the integral manifold of the original ideal. These are the most general similarity solutions. The integration to find similarity solutions involves at least one less independent variable since the larger ideal has a characteristic vector.

Suppose that, in a given closed ideal of genus g, we have found an exact k-form, $k = 1, \ldots, g$. Denote this exact form by $\mathrm{d}\vartheta$, where ϑ is a $(k-1)$-form determined up to the exterior derivative of a $(k-2)$-form. From

Stokes's theorem we have

$$\int_{\mathcal{V}} \mathrm{d}\vartheta = \int_{\partial\mathcal{V}} \vartheta \tag{10.48}$$

for any k-volume $\mathcal{V}$ with the closed boundary $\mathrm{d}\mathcal{V}$. If $\mathcal{V}$ lies in an integral manifold, the pull-back $\mathrm{d}\tilde{\vartheta}$ vanishes and we get $\int_{\partial\mathcal{V}} \tilde{\vartheta} = 0$ over any closed $(k-1)$-manifold in an integral manifold. From the point of view of g independent variables, this is an integral conservation law which is non-trivial if $\tilde{\vartheta} \neq 0$.

10.4.3 Prolongation structures

Conservation laws as discussed in the previous subsection with $k = 2$ allow a useful extension of the ideal. When such a conservation law can be found, we add the 1-form $\mathrm{d}y - \vartheta$ to the ideal and at the same time introduce a new variable, y, called a potential. Since the exterior derivative of this 1-form is already in the ideal, the extended ideal still has the same genus and independent variables. To find such a conservation law one is integrating an overdetermined set of linear partial differential equations for auxiliary functions Θ_i defined as the components of ϑ in the local coordinate base, i.e. $\vartheta = \Theta_i \mathrm{d}x^i$. The Θ_i have to satisfy $\mathrm{d}\vartheta = 0$ (mod *ideal*). Once a potential has been found the process can be iterated and the new potentials depend on the original variables and the previously found potentials.

The generalization is to let Θ_i depend on y itself. This leads to the concept of *pseudopotentials*: a set of an a priori unspecified number of functions y^α such that the exterior derivatives of the 1-forms $\mathrm{d}y^\alpha - \Theta_i^\alpha\, \mathrm{d}x^i$ (where Θ_i^α depends on the original variables as well as all the y^α) are in the prolonged ideal, i.e. in the ideal one obtains by adjoining the above 1-forms to the original ideal. Once pseudopotentials have been found one tries to find a Bäcklund transformation by writing a new solution of the equations as a function of a known solution and its pseudopotentials. The method is particularly useful if the genus of the system is 2 and we shall discuss it in more detail below (see also Guo *et al.* (1983a), for reviews see Harrison (1984) and Gaffet (1988)).

The starting point is the differential equations (10.1) but now written as a system of first-order equations with two independent variables which we shall, for the purpose of this section, call x and t. The equations read

$$H_A(x, t, u^a_{,x}, u^a_{,t}) = 0. \tag{10.49}$$

According to the methods outlined above, we now consider a manifold $\mathcal{U}$ with coordinates $\{x, t, u^a\}$ and a set of 2-forms ω^A defined on this manifold such that the pull-back of ω^A to the submanifold labelled by $\{x, t\}$ yields (10.49). It should be noted that this process is by no means unique; in many cases there are several spaces with corresponding ideals which are effective in the sense that they describe the original differential equation (Finley and McIver 1995). The resulting prolongation structure will depend on the choice of the ideal. For the ideal – which we here take to be generated by ω^A – to be closed we require

$$\mathrm{d}\omega^A = 0 \ (\mathrm{mod}\ \omega^A). \tag{10.50}$$

Then we introduce the 1-forms

$$\vartheta^\alpha = \mathrm{d}y^\alpha - F^\alpha(x, t, u^a, y^\alpha)\,\mathrm{d}x - G^\alpha(x, t, u^a, y^\alpha)\,\mathrm{d}t. \tag{10.51}$$

The prolonged ideal is now $\{\omega^A, \vartheta^\alpha\}$ and the requirement that the exterior derivative of forms in the ideal is contained in the ideal reads

$$\mathrm{d}\vartheta^\alpha = 0\,(\mathrm{mod}\ \omega^A, \vartheta^\alpha) = -\mathrm{d}F^\alpha\wedge\mathrm{d}x - \mathrm{d}G^\alpha\wedge\mathrm{d}t = f^\alpha_A\,\omega^A + \eta^\alpha_\beta\wedge\vartheta^\beta. \tag{10.52}$$

By comparing the coefficients of the various independent 2-forms in (10.52) the following information is obtained. First, we determine the dependence of F^α and G^α on the variables in $\mathcal{U}$. Secondly, we are able to express the previously unknown multipliers f^α_A and η^α_β in terms of derivatives of F^α and G^α. While the continuation of the prolongation method does not require explicit knowledge of these multipliers, it is nonetheless quite important that the final F^α and G^α are such that $f^\alpha_A \neq 0$. After all, they contain the information enabling the prolongation structure to 'remember' the original ideal and the equation from where it came. Thirdly, the $\mathrm{d}x\wedge\mathrm{d}t$ generate a commutator equation for F^α and G^α as vector fields vertical to $\mathcal{U}$. More specifically we get

$$\begin{aligned}
-\mathrm{d}\vartheta^\alpha &= \frac{\partial F^\alpha}{\partial y^\beta}\mathrm{d}y^\beta \wedge \mathrm{d}x + \frac{\partial F^\alpha}{\partial u^a}\mathrm{d}u^a \wedge \mathrm{d}x - \frac{\partial F^\alpha}{\partial t}\mathrm{d}x \wedge \mathrm{d}t + \frac{\partial G^\alpha}{\partial y^\mu}\mathrm{d}y^\mu \wedge \mathrm{d}t \\
&\quad + \frac{\partial G^\alpha}{\partial u^a}\mathrm{d}u^a \wedge \mathrm{d}t + \frac{\partial G^\alpha}{\partial t}\mathrm{d}x \wedge \mathrm{d}t \\
&= \left(\frac{\partial G^\alpha}{\partial y^\mu}F^\mu - \frac{\partial F^\alpha}{\partial y^\mu}G^\mu + \frac{\partial G^\alpha}{\partial x} - \frac{\partial F^\alpha}{\partial t}\right)\mathrm{d}x \wedge \mathrm{d}t + \frac{\partial F^\alpha}{\partial u^a}\mathrm{d}u^a \wedge \mathrm{d}x \\
&\quad + \frac{\partial G^\alpha}{\partial u^a}\mathrm{d}u^a \wedge \mathrm{d}t + \frac{\partial F^\alpha}{\partial y^\mu}\vartheta^\mu \wedge \mathrm{d}x + \frac{\partial G^\alpha}{\partial y^\mu}\vartheta^\mu \wedge \mathrm{d}t.
\end{aligned} \tag{10.53}$$

The first two terms in the last expression form the commutator of F^α and G^α with respect to the y^α; we thus define

$$[F + \partial_x, G + \partial_t]^\alpha = \frac{\partial G^\alpha}{\partial y^\mu}F^\mu - \frac{\partial F^\alpha}{\partial y^\mu}G^\mu + \frac{\partial G^\alpha}{\partial x} - \frac{\partial F^\alpha}{\partial t}. \tag{10.54}$$

The terms with $du^a \wedge dx$ and $du^a \wedge dt$ in (10.53) are eliminated as much as possible using ω^A and the resulting equations always have the general forms

$$[F+\partial_x, G+\partial_t]^\alpha = A_a \frac{\partial F^\alpha}{\partial u^a} + B_a \frac{\partial G^\alpha}{\partial u^a}, \qquad (10.55a)$$

$$0 = C_a \frac{\partial F^\alpha}{\partial u^a} + D_a \frac{\partial G^\alpha}{\partial u^a}. \qquad (10.55b)$$

A_a, B_a, C_a and D_a are functions of $\{x, t, u^a\}$ the precise form of which depends on the system under consideration. These equations are used to determine the dependence of F^α and G^α on $\{x, t, u^a\}$. N.b., this process involves solving first-order linear partial differential equations and is thus not algorithmic; in practical applications, however, one rarely encounters problems here. Having integrated (10.55) with respect to $\{x, t, u^a\}$ we get the solution in the form

$$F^\alpha = F^A X^\alpha_A, \qquad G^\alpha = G^A X^\alpha_A, \qquad (10.56)$$

where F^A and G^A are functions of $\{x, t, u^a\}$ and the vertical vector fields X^α_A depend on y^α only. If F^α and G^α are inserted into (10.55a) and the resulting expression sorted with respect to functions of $\{x, t, u^a\}$ one obtains relations for the X^α_A which typically look like

$$[X_A, X_B] = C^M_{AB}\, X_M, \qquad (10.57a)$$

$$0 = [X_M, X_N] + [X_U, X_V] + X_W \qquad (10.57b)$$

(in the remainder of this section we shall omit the superscript α on X_A: it is understood that the X_A are vector fields). The first of these equations just prescribes the commutator of two vector fields whereas equations of the type (10.57b) may be absent in a particular situation. In any case, not all commutators will be given. For instance, the commutator of two vectors appearing only in F^α cannot show up in (10.57).

One then constructs the free algebra generated from (10.57a) by checking the Jacobi identities and introducing new vectors for as yet unknown commutators subject to the restrictions imposed by (10.57b). This algebra is called the prolongation algebra. In most cases the Jacobi identities will at a certain stage determine all commutators and the algebra is finite-dimensional. In those cases where this happens the original equation cannot be reconstructed from the algebra, i.e. some of the multipliers f^α_A in (10.52) vanish (e.g. Kramer 1988b). In general no Bäcklund transformations exist; there is, however one rather curious counterexample known (Hoenselaers 1986).

Assume now that the process of introducing new vectors appears to be open-ended. The main problem of the procedure then is to identify the algebra. In some cases it suffices to make the ansatz that a vector to be introduced is a linear combination of the previous ones (Wahlquist and Estabrook 1975, Hoenselaers 1985b). The choice has to be such that the Jacobi identities determine the commutators and one is left with a finite algebra the structure constants of which depend on at least one parameter. Of course, the ansatz has to leave the functions f^α_A in (10.52) non-zero. This process requires some familiarity with algebras and we shall not go into the details; rather we shall briefly describe the results (for a suitable text on algebras see e.g. Kac (1990)). In every case where the prolongation method has been successful the resulting algebra could be recognized as, essentially, the loop algebra of some semisimple algebra or, sometimes, the semidirect product of such an algebra with a finite-dimensional one (Wahlquist and Estabrook 1975).

It has been mentioned above that one uses the pseudopotentials to derive Bäcklund transformations. It is thus essential to have a representation for the algebra one has found. Even though infinite-dimensional algebras have been found as prolongation algebras and in some cases have also been identified – Finley (1996) has found the contragradient algebra known as K_2 for a certain subcase of the Robinson–Trautman equations – the lack of a representation means that one cannot proceed further.

Let us thus assume that the prolongation algebra is the loop algebra of some semisimple algebra. In this case we have a natural grading of the algebra by introducing a spectral parameter λ (Kac 1990). For practical purposes we use a matrix representation of the semisimple algebra and a matrix generating function $\Phi(\lambda)$ for the pseudopotentials. The pull-back of (10.51) onto a solution manifold, i.e. $\bar{\vartheta}^\alpha = 0$, can be rewritten as

$$\mathrm{d}\Phi(\lambda) = H(\lambda)\Phi(\lambda) \tag{10.58}$$

with

$$H(\lambda) = F^A X_{Aa}\lambda^a \mathrm{d}x + G^A X_{Aa}\lambda^a \mathrm{d}t. \tag{10.59}$$

In this expression F^A and G^A are the coefficients from (10.56), X_{Aa} are matrices from the representation of the semisimple algebra – they are consequently tracefree and so is $H(\lambda)$ – and the summation over a extends only from finite, possibly negative, values up to finite values. The integrability condition for (10.58) reads

$$\mathrm{d}H(\lambda) = H(\lambda) \wedge H(\lambda). \tag{10.60}$$

This equation is to hold identically in λ and yields the original equations (10.49). It is the matrix version of (10.52) with non-vanishing multipliers f^α_A. The linear problem (10.58) is the starting point for many solution generating techniques. Loosely speaking, they centre around the construction of a new $\Phi_{new}(\lambda)$ from a given $\Phi(\lambda)$ and then reading off $H_{new}(\lambda)$ from (10.58). The fields which are of interest are coded in $H(\lambda)$ and the construction has to be such that the algebraic properties of $H(\lambda)$ are preserved. By construction, $H_{new}(\lambda)$ will satisfy the original equations which are coded in (10.60). Methods for doing this will be developed below.

Often it is also useful, and indeed shorter, to try to construct a new solution directly from a given solution and the pseudopotentials constructed from it by writing in (10.49) $\widetilde{u} = f(u, y)$ with the ys being solutions of (10.51) with $\mathrm{d}\vartheta^\alpha = 0$. The vectors F and G are given by the prolongation structure, and $f(u, y)$ has to be determined by inserting $\widetilde{u}$ into the original equation.

Some generalization to more than three independent variables is possible, see Morris (1976).

10.5 Solutions of the linearized equations

We assume that we have a one-parameter, ε, say, family of solutions and we denote the derivative with respect to ε by an overdot. The idea is to find the general solution to the linearized equations and then exponentiate this solution. Taking a derivative of (10.58) with respect to ε yields

$$\mathrm{d}\dot{\Phi}(\lambda) = \dot{H}(\lambda)\Phi(\lambda) + H(\lambda)\dot{\Phi}(\lambda). \tag{10.61}$$

To find a solution of this equation we introduce an additional parameter σ, say, and use the ansatz $\dot{\Phi}(\lambda) = \dot{\Phi}\,(\lambda, \sigma) = \mu(\lambda, \sigma)P(\sigma)\Phi(\lambda)$ with a matrix $P(\sigma)$ and a scalar function $\mu(\lambda, \sigma)$. By inserting this ansatz into (10.61) and choosing $\mu(\lambda, \sigma) = \lambda \cdot (\lambda - \sigma)^{-1}$, it can be shown (Hoenselaers 1993) that $P(\sigma)$ satisfies

$$dP(\sigma) = H(\sigma)P(\sigma) - P(\sigma)H(\sigma). \tag{10.62}$$

Using (10.58), this equation can be integrated immediately yielding $P(\sigma) = \Phi(\sigma)\alpha\Phi^{-1}(\sigma)$ with a constant matrix α. However, the resulting $\dot{\Phi}$ would not be defined at $\lambda = \sigma$. This can be remedied by adding a term proportional to $\Phi(\lambda)\alpha$, which is a solution of (10.61), to $\dot{\Phi}(\lambda)$ to obtain

$$\dot{\Phi}(\lambda) = \lambda(\lambda - \sigma)^{-1}[\Phi(\sigma)\alpha\Phi^{-1}(\sigma)\Phi(\lambda) - \Phi(\lambda)\alpha]. \tag{10.63}$$

Here, α enters as an additional parameter. To find the finite transformation one has to integrate

$$\begin{aligned}\partial_\varepsilon\Phi(\lambda, \sigma, \alpha, \varepsilon) = \lambda(\lambda - \sigma)^{-1}[&\Phi(\sigma, \sigma, \alpha, \varepsilon)\alpha\Phi^{-1}(\sigma, \sigma, \alpha, \varepsilon)\Phi(\lambda, \sigma, \alpha, \varepsilon) \\ &- \Phi(\lambda, \sigma, \alpha, \varepsilon)\alpha],\end{aligned} \tag{10.64}$$

where we have listed all the arguments on which Φ depends. It is the first argument of Φ which is relevant for (10.60) and (10.61).

We now consider the function

$$G(\lambda,\mu) = (\lambda-\mu)^{-1}[-\lambda\mathbf{1} + \mu\Phi^{-1}(\lambda)\Phi(\mu)] = \sum N_{nm}\lambda^n\mu^m \qquad (10.65)$$

and its expansion coefficients N_{nm}. Under an infinitesimal transformation (10.63) it transforms as

$$\begin{aligned}\dot{G}(\lambda,\mu) &= \alpha(\lambda-\sigma)^{-1}[G(\lambda,\mu) - G(\sigma,\mu)] \\ &\quad -(\mu-\sigma)^{-1}[\mu G(\lambda,\mu) - \sigma G(\lambda,\sigma)]\alpha - G(\lambda,\sigma)\,\alpha G(\sigma,\mu)\,.\end{aligned} \qquad (10.66)$$

The transformation of the coefficients N_{nm} depends continuously on the parameter σ. For each power of σ there is thus an infinitesimal transformation $T(\alpha_k)$ parametrized by a matrix α_k which maps the N_{nm} such that they obey

$$\dot{N}_{nm} = \alpha_k N_{n+k,m} - N_{n,m+k}\alpha_k - \sum_{s=1}^{k} N_{ns}\alpha_k N_{k-s,m}. \qquad (10.67)$$

The commutator of two such transformations is

$$[T(\alpha_k), T(\alpha_l)] = T(\alpha_{k+l}) \ , \ \alpha_{k+l} = [\alpha_k, \alpha_l]\,. \qquad (10.68)$$

This is precisely the structure which Kinnersley and Chitre (1977, 1978a) have found for the Geroch (1972) 'group' in general relativity. Using the methods of Schmidt (1984) it can be shown that this algebra of infinitesimal transformations yields a Banach Lie group, see also the review article by Breitenlohner and Maison (1987).

10.6 Bäcklund transformations

We return to (10.58) and seek a new solution by the ansatz (Neugebauer and Meinel 1984)

$$\Phi(\lambda) = P(\lambda)\,\Phi_0(\lambda)\,, \qquad (10.69)$$

where $\Phi_0(\lambda)$ is a solution of (10.58) for some given $H_0(\lambda)$, and, here, $P(\lambda)$ is a finite polynomial in λ of degree m, i.e.

$$P(\lambda) = \sum_{k=0}^{m} P_k\lambda^k \ , \ \det(P_0) \neq 0 \ , \ \det(P_m) \neq 0 \qquad (10.70)$$

(cp. Meinel *et al.* (1991), Chapter 4.3). The new $H(\lambda)$ is given by

$$H(\lambda) = [\mathrm{d}P(\lambda) + P(\lambda)\,H_0(\lambda)]\,P(\lambda)^{-1}\,. \qquad (10.71)$$

To preserve the expansion of $H(\lambda)$ as finite Taylor resp. Laurent series in λ we want to choose $P(\lambda)$ such that $H(\lambda)$ has the same pole structure as $H_0(\lambda)$. First we note that at $\lambda = 0$ and at $\lambda = \infty$ the poles of $H(\lambda)$ are of the same order as the ones of $H_0(\lambda)$. It remains to analyse the behaviour of $H(\lambda)$ at the zeros of $\det(P(\lambda))$. We have

$$\det(P(\lambda)) = \det(P_m) \prod_{k=1}^{m \cdot n} (\lambda - \lambda_k), \tag{10.72}$$

where n is the size of the matrix. The zeros of the determinant, which we assume for the moment to be disjunct, are simple poles of the inverse matrix, and with

$$P(\lambda)^{-1} = \widetilde{P}(\lambda) / \det(P(\lambda)) \tag{10.73}$$

we have

$$H(\lambda) = [\det P(\lambda)]^{-1} [\mathrm{d}P(\lambda) + P(\lambda) H_0(\lambda)] \widetilde{P}(\lambda). \tag{10.74}$$

Hence using (10.69) we require

$$[\mathrm{d}P(\lambda_k) + P(\lambda_k) H_0(\lambda_k)] \widetilde{P}(\lambda_k) = \frac{\mathrm{d}\Phi(\lambda_k) \widetilde{\Phi}(\lambda_k)}{\det(\Phi_0(\lambda_k))} = 0. \tag{10.75}$$

As $\mathrm{d}\det(\Phi_0(\lambda)) = \mathrm{Tr}\, H(\lambda) = 0$, we have without loss of generality $\det(\Phi_0(\lambda)) \neq 0$. Moreover, $\det(\Phi(\lambda_k)) = 0$ and consequently there exist non-trivial eigenvectors $\mathbf{p}^{(k)}$ such that

$$\Phi(\lambda_k) \mathbf{p}^{(k)} = 0, \quad k = 1, \ldots, n \cdot m. \tag{10.76}$$

Here, $\mathbf{p}^{(k)}$ is a column vector of length n. For constant $\mathbf{p}^{(k)}$ we also have $\mathrm{d}\Phi(\lambda_k) \mathbf{p}^{(k)} = 0$ and thus

$$\mathrm{d}\Phi(\lambda_k) \widetilde{\Phi}(\lambda_k) = 0. \tag{10.77}$$

This result follows from the fact that if two square matrices A and B have a common non-trivial eigenvector with eigenvalue zero, i.e. $Ap = Bp = 0$, then $A\widetilde{B} = 0$. The matrices $\mathrm{d}\Phi(\lambda_k)$ and $\widetilde{\Phi}(\lambda_k)$ are such a pair of matrices. The matrix $P(\lambda)$ is to be constructed algebraically from the equation

$$P(\lambda_k) \Phi_0(\lambda_k) \mathbf{p}^{(k)} = 0, \quad k = 1, \ldots, n \cdot m. \tag{10.78}$$

With a normalization condition, e.g. $P_m = \mathbf{1}$, this is an inhomogeneous algebraic system for the other expansion coefficients. Finally, $H(\lambda)$ is calculated via (10.71). N.b., the constants λ_k and the constant components of $\mathbf{p}^{(k)}$ enter as parameters into the new solution. In a concrete case the choice of parameters may be restricted by the requirement of preserving not only the pole structure of $H(\lambda)$ but also its 'inner structure', i.e. the algebraic structure of its expansion coefficients. This may necessitate that the λ_k and the components of $\mathbf{p}^{(k)}$ are complex.

10.7 Riemann–Hilbert problems

We again seek a new solution of (10.58) by the ansatz (10.69), viz.

$$\mathrm{d}\Phi(\lambda) = H(\lambda)\Phi(\lambda)\ , \ \ \Phi(\lambda) = P(\lambda)\,\Phi_0(\lambda)\,, \qquad (10.79)$$

and assume without loss of generality $\det(\Phi) \neq 0$ and $\Phi(0) = 1$. The parameter λ is extended into the complex plane. The requirement that $\Phi(\lambda)$ has the same analyticity properties as $\Phi_0(\lambda)$ implies that $P(\lambda)$ has to be a holomorphic function in a neighbourhood of $\lambda = 0$. However, $P(\lambda)$ cannot be holomorphic in the whole complex plane without being constant. To get a non-trivial transformation one has to use different matrices in different neighbourhoods, say of the origin and $\lambda = \infty$, and match them on the intersection of those neighbourhoods.

Let $\mathcal{L}$ be a contour enclosing the origin in the complex λ-plane, $\mathcal{L}_+$ its interior and $\mathcal{L}_-$ its exterior. The Riemann–Hilbert problem consists of finding matrices $P_+(\lambda)$ and $P_-(\lambda)$ such that $P_+(\lambda)$ is holomorphic in $\mathcal{L}_+$ and continuous with non-vanishing determinant in $\mathcal{L}_+ \cup \mathcal{L}$ and $P_-(\lambda)$ is holomorphic in $\mathcal{L}_-$ and continuous with non-vanishing determinant in $\mathcal{L}_- \cup \mathcal{L}$. On the contour $\mathcal{L}$ we have the condition

$$\begin{aligned}&[\mathrm{d}P_+(\lambda) + P_+(\lambda)\,H_0(\lambda)]P_+(\lambda)^{-1}\\ &\qquad = [\mathrm{d}P_-(\lambda) + P_-(\lambda)\,H_0(\lambda)]P_-(\lambda)^{-1} = H(\lambda) \quad \forall\lambda \in \mathcal{L}. \qquad (10.80)\end{aligned}$$

This is equivalent to

$$P_-(\lambda) = P_+(\lambda)\,\Phi_0(\lambda)\,u(\lambda)\,\Phi_0(\lambda)^{-1} \quad \forall\lambda \in \mathcal{L}, \qquad (10.81)$$

where $u(\lambda)$ is a constant matrix depending on λ only.

The solution of this problem leads to integral equations of the Cauchy type which are discussed in e.g. Muskhelishvili (1953) and Ablowitz and Fokas (1997), where also methods for solving them are given. For the relations of Riemann–Hilbert problems to Kac–Moody algebras cp. Chau and Ge (1989) and Li and Hou (1989).

10.8 Harmonic maps

Let $\mathcal{M}$ be a m-dimensional Riemannian space with coordinates x^a and metric $\gamma^{ab}(x)$ and let $\mathcal{N}$ be an n-dimensional space with coordinates φ^A and metric $G_{AB}(\varphi)$; $\mathcal{M}$ is called the background space and $\mathcal{N}$ is the potential space. A map $\mathcal{M} \to \mathcal{N}$ is called harmonic if it is such that $\varphi^A(x^a)$ satisfies the Euler–Lagrange equations of a variational principle

$$\delta L/\delta\varphi^A = 0 \qquad (10.82)$$

with a Lagrangian of the form (Neugebauer and Kramer 1969, Neugebauer 1969)

$$L = \sqrt{\gamma} G_{AB}(\varphi^C)\gamma^{ab}\varphi^A_{,a}\varphi^B_{,b}, \tag{10.83}$$

where $\gamma = \det(\gamma_{ab})$. Moreover, we assume $\det(G_{AB}) \neq 0$. The field equations are

$$\varphi^{A;a}{}_{;a} + \Gamma^A{}_{BC}\varphi^B_{,a}\varphi^{C,a} = 0, \tag{10.84}$$

where $\Gamma^A{}_{BC}$ are the Christoffel symbols formed with G_{AB}. We shall see in Chapter 34 that the equations for electrovac space-times with one non-null Killing vector are just of this form.

The field equations are invariant under infinitesimal transformations of the form

$$\varphi^A \to \varphi^A + \varepsilon X^A(\varphi), \tag{10.85}$$

if X^A is an affine collineation, i.e. if it satisfies

$$X_{A;BC} = R_{ABCD}\, X^D, \tag{10.86}$$

where R_{ABCD} is the Riemann tensor of G_{AB}. The finite transformation can be found by integrating

$$\mathrm{d}\varphi^A/\mathrm{d}\,\varepsilon = X^A(\varphi). \tag{10.87}$$

If $\varphi^A|_{\varepsilon=0}$ is a solution of the field equations (10.84) then a solution of (10.87), i.e. a symmetry, yields a one-parameter family of solutions.

In the particular case where X^A is a Killing vector, i.e. $X_{(A;B)} = 0$, the infinitesimal transformation (10.85) leaves the Lagrangian (10.83) invariant and gives rise to a conserved quantity, viz.

$$(X_A\, \varphi^{A,a})_{;a} = 0. \tag{10.88}$$

This is Noether's theorem in the present context. Moreover, the tensor

$$T_{ab} = K_{AB}\varphi^A_{,a}\varphi^B_{,b} - \tfrac{1}{2}\gamma_{ab}K_{AB}\varphi^A_{,n}\varphi^{B,n} \tag{10.89}$$

is divergencefree if and only if K_{AB} is covariantly constant, i.e.

$$T_a{}^b{}_{;b} = 0 \qquad \Longleftrightarrow \qquad K_{AB;C} = 0. \tag{10.90}$$

In particular, if $K_{AB} = G_{AB}$, T_{ab} is the canonical energy-momentum tensor.

Besides the study of the affine symmetries of the metric G_{AB} associated with (10.83) other geometrical investigations of the potential space (geodesics, subspaces) are important in simplifying the field equations (Neugebauer and Kramer 1969, Hoenselaers 1978d). Let us assume that the fields φ^A depend only on a single potential λ. The ansatz $\varphi^A = \varphi^A(\lambda)$ reduces the field equations (10.84) to

$$\frac{\mathrm{d}\varphi^A}{\mathrm{d}\lambda}\lambda_{,a}{}^{;a} + \left(\frac{\mathrm{d}^2\varphi^A}{\mathrm{d}\lambda^2} + \Gamma^A_{BC}\frac{\mathrm{d}\varphi^B}{\mathrm{d}\lambda}\frac{\mathrm{d}\varphi^C}{\mathrm{d}\lambda}\right)\lambda_{,a}\lambda^{,a} = 0. \tag{10.91}$$

The potential λ is determined up to a transformation $\lambda \to \lambda'(\lambda)$. It follows from (10.91) that $\lambda_{,a}^{;a} = f(\lambda)\lambda_{,a}\lambda^{,a}$ and thus the above transformation can be used to achieve $\lambda_{,a}^{;a} = 0$. For $\lambda_{,a}\lambda^{,a} \neq 0$ the field equations reduce to

$$\lambda_{,a}^{;a} = 0, \qquad \mathrm{D}^2\varphi^A/\mathrm{d}\lambda^2 = 0. \tag{10.92}$$

Here, φ^A are solutions of the geodesic equation in the potential space and λ is a solution of the Laplace–Beltrami equation in the background space. If, however, λ not only solves the Laplace–Beltrami equation but also is such that its gradient is a null vector, i.e. $\lambda_{,a}\lambda^{,a} = 0$ (this presupposes that the background space is of indefinite signature), then it can be seen from (10.91) that there are no restrictions on $\varphi^A(\lambda)$; these functions can then be chosen arbitrarily.

The case in which the background space is two-dimensional and flat is of particular interest. In this case (see e.g. Hoenselaers (1988)) one can show that the canonical energy-momentum tensor associated with the Lagrangian (10.83), viz.

$$T_{ab} = G_{AB}\varphi^A_{,a}\varphi^B_{,b} - \tfrac{1}{2}\gamma_{ab}G_{AB}\varphi^A_{,n}\varphi^{B,n} \equiv 0, \tag{10.93}$$

actually vanishes identically. As this tensor is symmetric and tracefree, these are two equations of first order in addition to the field equations (10.84).

In what follows we assume in addition to the above that the potential space admits a hypersurface-orthogonal Killing vector, viz.

$$X_{(A;B)} = 0, \qquad X_{[A;B}X_{C]} = 0. \tag{10.94}$$

One can introduce a function κ such that $X_A = \partial_\kappa$. The Lagrangian (10.83) can be rewritten as

$$L = G'_{AB}\varphi^A_{,a}\varphi^{B,a} + f(\varphi^C)\kappa_{,a}\kappa^{,a} \qquad (A, B = 1, ..., N-1), \tag{10.95}$$

where neither G'_{AB} nor f depend on κ. Variation with respect to κ yields

$$(f\kappa^{,a})_{,a} = 0, \tag{10.96}$$

which is the integrability condition for a function ψ defined by

$$\psi_{,a} = f\varepsilon_{ab}\kappa^{,b}, \quad \varepsilon_{ab} = \begin{pmatrix} 0 & -1 \\ 1 & 0 \end{pmatrix}. \tag{10.97}$$

The equations for the other fields are obtained by the variational derivative of the Lagrangian (10.95) with respect to the φ^A. Replacing κ by ψ in the field equations is tantamount to the Legendre transformation

$$L' = L - \kappa_{,a}\partial L/\partial\kappa_{,a} = G'_{AB}\varphi^A_{,a}\varphi^{B,a} - \psi_{,a}\psi^{,a}/f(\varphi^C). \qquad (10.98)$$

Clearly, L' admits at least one Killing vector, ∂_ψ, and the transformation associated with it is $\psi \to \psi +$ const which leaves κ unaffected. If, however, L' admits more than one Killing vector it can be used to generate new solutions. Moreover, if this other Killing vector is hypersurface-orthogonal one can iterate the process and at each step generate new solutions.

For a two-dimensional background space with real coordinates x and y, say, we can introduce derivative operators $\partial = \partial_x + i\partial_y$, $\partial^* = \partial_x - i\partial_y$, respectively $\partial = \partial_x + \partial_y$, $\partial^* = \partial_x - \partial_y$, depending on whether the signature is 2 or 0. Moreover, we use an n-bein in the potential space such that

$$G_{AB} = e^\alpha_A\, e_{B\alpha}, \qquad G_{\alpha\beta} = e^A_\alpha\, e_{A\beta}, \qquad (10.99)$$

where we do not require the frame metric $G_{\alpha\beta}$ to be constant or diagonal. Projecting the derivatives of the fields φ^A onto the frame vectors we define

$$M^\alpha = e^\alpha_A\, \partial\varphi^A, \qquad M^{*\alpha} = e^\alpha_A\, \partial^*\varphi^A. \qquad (10.100)$$

The field equations (10.84) together with the integrability conditions of these equations become

$$\partial^* M^\alpha + \Gamma^\alpha{}_{\beta\gamma} M^\beta M^{*\gamma} = 0, \qquad \partial M^\alpha + \Gamma^\alpha{}_{\beta\gamma} M^{*\beta} M^\gamma = 0, \qquad (10.101)$$

where $\Gamma^\alpha{}_{\beta\gamma}$ are the Ricci rotation coefficients, $\Gamma^\alpha{}_{\beta\gamma} = e^{\alpha A}\, e_{\beta A;B}\, e^B_\gamma$. The advantage of this formulation of the field equations is that the nonlinearities become purely quadratic if it is possible to choose the frame vectors such that the rotation coefficients are constant. According to Theorem 8.18 this is possible if G_{AB} admits a simply-transitive n-dimensional group of affine collineations.

10.9 Variational Bäcklund transformations

In this section we assume that the equations in question are again Euler–Lagrange equations of the form (10.82) with a general Lagrangian $L(\varphi^A, \varphi^A_{,a}, \varphi^A_{,ab}, \ldots)$ which is supposed to be polynomial in the derivatives of the fields. The field equations to be derived from this Lagrangian contain derivatives of order $2n$ if L is of order n. If L can be factorized, i.e.

$$L = L_1(\varphi^A, \varphi^A_{,a}, \varphi^A_{,ab}, \ldots)\; L_2(\varphi^A, \varphi^A_{,a}, \varphi^A_{,ab}, \ldots) + \text{divergence terms}, \qquad (10.102)$$

where L_1 and L_2 are again polynomial in the derivatives of φ^A, then a solution of the differential equations

$$L_1(\varphi^A, \varphi^A_{,a}, \varphi^A_{,ab}, \ldots) = 0, \qquad L_2(\varphi^A, \varphi^A_{,a}, \varphi^A_{,ab}, \ldots) = 0 \qquad (10.103)$$

is also a solution of the field equations derived from L (Rund 1976). The term 'variational Bäcklund transformation' comes from the use of (10.103) for deriving Bäcklund transformations for the sine-Gordon equation and other non-linear partial differential equations. If one has one equation to be derived from a Lagrangian $L(\varphi, \varphi_{,a}, \ldots)$ for one function and wants to derive an auto-Bäcklund transformation one uses the total Lagrangian $L_{tot} = L(\varphi, \varphi_{,a}, \ldots) - L(\psi, \psi_{,a}, \ldots)$ and tries to factorize it. If successful, (10.103) provides relations between φ and ψ which are of lower order than the field equations and constitute the well-known Bäcklund transformations for various equations.

10.10 Hirota's method

Hirota's direct method introduces derivative operators D_x defined by

$$D^n_x(a * b) = (\partial_x - \partial_{x'})^n \, [a(x)\ b(x')] \,|_{x=x'} \qquad (10.104)$$

(Hirota 1976). Thus $D_x(a * b) = b\partial_x a - a\partial_x b$ and $D^2_x(a * b) = b\partial^2_x a - 2\partial_x a\ \partial_x b + a\partial^2_x b$ etc. This is particularly effective if the solution to the equation(s) in question can be written as a ratio f/g, say, and additional equations can be introduced such that the resulting system can be written using D operators only. It has been applied to many equations and the standard n-soliton solutions can be found easily.

10.11 Generation methods including perfect fluids

For perfect fluids, the known generation methods are less powerful than in the vacuum case. In this section we want to present some of the more general ones. Those using only properties of a particular type of equation, e.g. a linear differential equation or a Riccati equation, for which one known solution may be used to get a second solution, will be treated in the relevant chapters.

10.11.1 Methods using the existence of Killing vectors

If a perfect fluid solution admits a (spacelike) Killing vector $\boldsymbol{\xi}$ perpendicular to the four-velocity and has an equation of state $\mu = p$, or admits a (timelike) Killing vector $\boldsymbol{\xi}$ parallel to the four-velocity (rigid motion), and has an equation of state $\mu + 3p = 0$, then a one-parameter family of solutions can be generated from this solution. The method relies on the existence of a twist potential $\omega_{,a} = \varepsilon_{abcd}\xi^{d;c}\xi^{b}$; for details see §10.3.2 above and Stephani (1988). Since the equation of state $\mu+3p = 0$ is rather unphysical, only very few examples have been discussed for this case, see e.g. Rácz and Zsigrai (1996). For stiff matter $\mu = p$, two more generation methods are available which we shall discuss now.

If the four-velocity is irrotational, it can be written in terms of a scalar function $\sigma(z,t)$ as

$$u_n = \sigma_{,n}(-\sigma^{,a}\sigma_{,a})^{-1/2}, \tag{10.105}$$

and Einstein's field equations for a stiff fluid read

$$R_{ab} = 2\sigma_{,a}\sigma_{,b}, \quad \kappa_0 p = \kappa_0 \mu = -\sigma^{,n}\sigma_{,n} \tag{10.106}$$

(Tabensky and Taub 1973). If the metric admits a spacelike hypersurface-orthogonal Killing vector $\xi = \partial_x$, it can be written as

$$\mathrm{d}s^2 = \mathrm{e}^{2U}\mathrm{d}x^2 + \mathrm{e}^{-2U}\gamma_{\alpha\beta}\mathrm{d}x^\alpha\mathrm{d}x^\beta, \quad \alpha,\beta = 1,\,2,\,3. \tag{10.107}$$

With respect to (10.107), the field equations read

$$\overset{3}{R}_{\alpha\beta} = 2U_{,\alpha}U_{,\beta} + 2\sigma_{,\alpha}\sigma_{,b}, \quad U_{,\alpha}^{;\alpha} = 0 \tag{10.108}$$

(with $\xi^n\sigma_{,n} = 0 = \xi^n U_{,n}$), cp. Theorem 18.1. Four-velocity ($\sigma_{,a}$) and the gradient of U enter in a very symmetric way, and the following theorem (Krori and Nandy 1984) can easily be read off:

Theorem 10.1 *If* $\mathrm{d}s^2 = \mathrm{e}^{2V}\mathrm{d}x^2 + \mathrm{e}^{-2V}\gamma_{\alpha\beta}\mathrm{d}x^\alpha\mathrm{d}x^\beta$ *is a vacuum solution with* $V_{,\alpha}V^\alpha < 0$, *then* (10.107) *with* $U = (1-\lambda)V$, $\lambda =$ *const, is a stiff fluid solution with* $\sigma = (2\lambda - \lambda^2)^{1/2}V$.

Unfortunately there are not many non-static solutions with only one Killing vector (so that in the applications (Krori and Nandy 1984, Baillie and Madsen 1985) of this theorem some seed metrics admit even a G_3). Most of the possible seed metrics admit a G_2, and then a more powerful method is available.

If the metric admits two commuting spacelike Killing vectors ∂_x and ∂_y (orthogonally transitive), the line element can be written as

$$\mathrm{d}s^2 = \mathrm{e}^M(\mathrm{d}z^2 - \mathrm{d}t^2) + W[\mathrm{e}^{-\Psi}\mathrm{d}y^2 + \mathrm{e}^{\Psi}(\mathrm{d}x + A\mathrm{d}y)^2], \quad W > 0, \tag{10.109}$$

cp. §17.1.2. If the four-velocity is orthogonal to the group orbits, it is necessarily irrotational, so that (10.105) is satisfied. The Bianchi identities then imply that σ satisfies the same equation (17.13) as ψ does for $A = 0$, i.e.

$$\sigma^{,a}{}_{;a} = 0 = (W\sigma_{,z})_{,z} - (W\sigma_{,t})_{,t} \tag{10.110}$$

(Letelier 1975, Ray 1976). Taking now a vacuum solution of the form (10.109) and adding a function Ω to M,

$$\widetilde{M} = M_{vac} + \Omega, \tag{10.111}$$

one sees that the resulting Ricci tensor $\widetilde{R}_{ab}$ is exactly of the form

$$\widetilde{R}_{ab} = 2\sigma_{,a}\sigma_{,b}, \tag{10.112}$$

provided the function Ω satisfies the equation

$$\Omega_{,a} = 2W\sigma^{,b}(2W_{,b}\sigma_{,a} - \sigma_{,b}W_{,a})/(W^{,n}W_{,n}). \tag{10.113}$$

Note that only the metric function W enters this equation (M drops out).

Theorem 10.2 *If the metric* (10.109) – *with* $W^{,a}W_{,a} \neq 0$ – *satisfies the vacuum field equations* $R_{ab} = 0$*, then the perfect fluid field equations for a stiff fluid* $\widetilde{R}_{ab} = 2\sigma_{,a}\sigma_{,b}$ *are satisfied by the metric* $\mathrm{d}\widetilde{s}^2$ *which differs from* (17.4) *by the substitution* $\widetilde{M} = M + \Omega$*, where* σ *and* Ω *are solutions of* (10.110) *and* (10.113)*, respectively* (Letelier and Tabensky (1975) for $A = 0 = \Psi$, Wainwright *et al.* (1979), Belinskii (1979)).

For $W^{,n}W_{,n} = 0$, $W \neq$ const, by an obvious generalization one can generate from a vacuum solution with $W = W(z+t)$ a pure radiation solution with $\sigma = \sigma(z+t)$ by adding an $\Omega = \omega_0(z-t) + \omega_1(z+t)$ with $\omega_1' = W(\sigma')^2/2W'$.

The technique described by the above theorem can be – and has been – applied to many solutions: the class of vacuum metrics (17.4) belongs to those metrics with an orthogonally transitive Abelian G_2 on S_2 where soliton techniques can be used to generate *vacuum* solutions, see above and Chapter 34. Moreover, since the metric function W does not change under these generation techniques, the same functions σ and Ω can be used for all vacuum solutions obtained from the same vacuum seed. A different way of looking at the class of solutions covered by Theorem 10.2 is to start from a stiff fluid solution, perform a transformation to a vacuum solution, apply a soliton-generation technique and then go back to the stiff fluid: one then may speak of a solution describing solitons travelling in the background of a (particular) stiff fluid of higher symmetry. Of course, since σ and Ω need not be changed, it suffices to immediately transform only the vacuum part of the metric.

We cannot give in detail all solutions found by using these ideas, but only give the references (see also §23.3). The generation techniques have been applied to:

flat space: Belinskii and Fargion (1980a), Jantzen (1980a),

FRW-cosmologies: Belinskii (1979), Jantzen (1980a), Kitchingham (1986), Alencar and Letelier (1986), Gleiser *et al.* (1988a), Griffiths (1993b), Alekseev and Griffiths (1995);

vacuum solutions with a G_3 (different Bianchi types): Wainwright *et al.* (1979), Wainwright and Marshman (1979), Letelier (1979), Jantzen (1980a), Kitchingham (1984, 1986), Alekseev and Griffiths (1995);

other vacuum metrics: Wainwright *et al.* (1979), Carmeli *et al.* (1981, 1983), Tseitlin (1985), Kitchingham (1986), Oliver and Verdaguer (1989), Davidson (1993a, 1998), Fernandez-Jambrina (1997).

We end this subsection by mentioning two methods which work for *static* metrics and use a discrete symmetry of the field equations. The first is the Ehlers transformation described in Theorem 21.1 which relates a rigidly rotating stationary dust solution to each static vacuum solution. The second is the Buchdahl transformation (Buchdahl 1956). For a static perfect fluid with metric

$$ds^2 = e^{-2U}\gamma_{\mu\nu}dx^\mu dx^\nu - e^{2U}dt^2, \tag{10.114}$$

the field equations are equivalent to the system

$$\widehat{R}_{ab} = 2U_{,a}U_{,b} - 2\kappa_0 p e^{-2U}\gamma_{ab}, \quad U^{,a}{}_{;a} = -\tfrac{1}{2}\kappa_0(\mu+3p)e^{-2U}, \tag{10.115}$$

cp. Theorem 18.1. If we make the substitution

$$\widetilde{U} = -U, \quad \widetilde{p} = e^{-4U}p, \quad \widetilde{\mu} = -\,e^{-4U}(\mu+6p), \tag{10.116}$$

we get a 'reciprocal' static perfect fluid solution. The resulting mass density $\widetilde{\mu}$ will be physical (positive) only in those regions of the seed solution (choice of constants of integration!) where $\mu + 6p$ is negative (Stewart 1982).

10.11.2 Conformal transformations

While few space-times admit proper conformal motions (see §35.4), it may still be possible to obtain new solutions from old by conformal transformation. Conformally flat solutions are discussed in full in Chapter 37, so we restrict attention here to cases where two space-times which are not conformally flat are conformally related. If one applies a conformal transformation $g_{ab} = e^{-2\Phi}\widehat{g}_{ab}$ to a vacuum metric $\widehat{R}_{ab} = 0$, then the resulting Ricci tensor

$$R_{ab} = 2\Phi_{,a;b} - 2\Phi_{,a}\Phi_{,b} - g_{ab}\left(\Phi^{,n}{}_{;n} + 2\Phi^{,n}\Phi_{,n}\right) \tag{10.117}$$

may be of the perfect fluid type, in which case

$$2\Phi_{,a;b} - 2\Phi_{,a}\Phi_{,b} + g_{ab}\Phi^{,n}\Phi_{,n} = \kappa_0(\mu + p)u_a u_b + \tfrac{1}{3}\kappa_0 p g_{ab} \qquad (10.118)$$

holds, cp. §3.7 (the covariant derivatives are understood with respect to g_{ab}). The integrability conditions for this system of differential equations for Φ have been studied by Kozameh *et al.* (1985) and in a series of papers by Van den Bergh. Obviously, all conformally flat perfect fluids (see Theorem 37.17) are contained here. The general solution of (10.117) is not known, but (for $\mu+p \neq 0$) the following results have been obtained.

(1) If the fluid and vacuum solutions share a G_r, $r \geq 2$, or if the fluid has an equation of state $p = p(\mu)$ and the conformal factor obeys $\Phi_{,a}u^a = 0$, the solutions are shearfree or irrotational (Van den Bergh 1988b).

(2) Shearfree solutions are conformally flat and thus covered by Theorem 37.17, or of Petrov type D and locally rotationally symmetric and thus known (Van den Bergh 1986b, 1986e, Ellis 1967, Stewart and Ellis 1968).

(3) Irrotational fluids are conformally flat (if $\Phi_{,a}$ is parallel to u_a), or shearfree and non-expanding with a G_3 on V_2 or conformally related to the Bianchi type VI_0 vacuum solution (13.58) (if $\Phi_{,a}$ is orthogonal to u_a), or expansionfree and admit a G_3 on V_2 (in the general case) (Van den Bergh 1986d).

(4) The vacuum solutions cannot be of Petrov type N, and $\Phi_{,n}$ cannot be proportional to a Killing vector (Van den Bergh 1986c, 1987).

The only null Einstein–Maxwell fields obtainable in a similar manner are *pp*-waves (Van den Bergh 1986a). The conformally Ricci-flat pure radiation solutions are either *pp*-waves or contained in a Petrov type N solution in Kundt's class, (31.38) with $W^o = 0$ and (31.39) replaced by

$$(\zeta + \bar{\zeta})\left(\frac{H_0}{\zeta + \bar{\zeta}}\right)_{,\zeta\bar{\zeta}} = (\zeta + \mathrm{i}a)\left[\frac{H_o}{(\zeta + \mathrm{i}a)(\bar{\zeta} - \mathrm{i}a)}\right]_{,\zeta} + \text{c.c.} \qquad (10.119)$$

(Wils 1989a).

The Einstein spaces conformal to *pp*-waves have been given by Siklos (1985) within a more general class including a pure radiation field; by Theorem 3.1, the *pp*-waves are necessarily non-vacuum.

All these results are somewhat disappointing in that conformal transformation did not really lead to new solutions. The same is true for attempts to use five-dimensional backgrounds for a generation method (which leads to equations very similar to (10.117) for the scalar field connected with g_{55}).

Part II
Solutions with groups of motions

11
Classification of solutions with isometries or homotheties

11.1 The possible space-times with isometries

In specifying the symmetry properties of a metric one has to state the dimension of the maximal group of motions or homotheties, its algebraic structure, and the nature and dimension of its orbits. For this purpose we shall, as in §8.4, use the following notation: the symbols S, T and N will denote, respectively, spacelike, timelike and null orbits, and will be followed by a subscript giving the dimension. If an isometry group is transitive on the whole manifold V_4, the space-time will be said to be *homogeneous.* If an isometry group is transitive on S_3, T_3 or N_3, the space-time will be called *hypersurface-homogeneous* (or, respectively, spatially-homogeneous, time-homogeneous, or null-homogeneous).

Petrov (1966) and his colleagues were the first to give a systematic treatment of metrics with isometries, and we therefore inevitably recover many of Petrov's results in the following chapters.

It turns out that if the orbits are null, the construction of the metric and the understanding of its properties have to be achieved by a rather different method from that used when the orbits are non-null. Accordingly we give first the discussion of non-null orbits (Chapters 12–22) and later the discussion of null orbits (Chapter 24). Within this broad division we proceed in order of decreasing dimension of the orbits. A further subdivision occurs because the group may be multiply- or simply-transitive on its orbits. We shall in general treat the higher-dimensional (multiply-transitive) case first. However, it frequently happens that the multiply-transitive group contains a subgroup simply-transitive on the same orbits, and in this case it may be advantageous to make use of the simply-transitive subgroup.

All these subdivisions give a rather long list of inequivalent possible structures. As far as possible we shall try to group these together when looking for exact solutions, and for each case we shall first discuss the form of the metric and only subsequently insert this into the Einstein field equations. As usual we shall restrict ourselves to vacuum, electromagnetic, pure radiation and perfect fluid energy-momentum tensors and only occasionally mention combinations such as Einstein–Maxwell fields with a cosmological constant.

Even a short search of the literature reveals that certain types of space-time symmetry have attracted much more attention than others, either because of their physical importance, or because they are mathematically tractable and interesting. We have therefore devoted separate chapters to some of these special cases, these chapters following the appropriate more general chapters.

One particular complication is that the maximal group of motions of a space-time may contain a considerable number of inequivalent subgroups. The metric may therefore be rediscovered as a special case of a space-time invariant under one of these subgroups, often in a form which makes it difficult to recognize unless the methods of Chapter 9 are applied. We shall try to indicate such possibilities.

Isometries necessarily leave the energy-momentum tensor unchanged. For perfect fluids, this implies that the quantities appearing in (5.9) have the same symmetries as the metric (Hoenselaers 1978c). For pure radiation, invariance of (5.8) shows that $\mathcal{L}_\xi k_a \propto k_a$, and one can scale k_a so that $\mathcal{L}_\xi k_a = 0 = \mathcal{L}_\xi \Phi$ (this is possible simultaneously for all generators of a group since k_a must be invariant under any isotropy in the group). The analogous result for Einstein–Maxwell fields is not true, however. Instead we have

Theorem 11.1 *If the metric of a sourcefree Einstein–Maxwell field admits a Killing vector field* $\boldsymbol{\xi}$, *then*

$$\mathcal{L}_\xi F_{ab} = \Psi \widetilde{F}_{ab}, \tag{11.1}$$

where Ψ *is a constant for a non-null electromagnetic field, and* $\Psi_{,[a}k_{b]} = 0$ *for a null field with repeated principal null direction* $\boldsymbol{k}$ (Ray and Thompson 1975, Coll 1975). *When* $\Psi = 0$, *the Maxwell field is said to inherit the symmetry.*

Proof. In the non-null case the Rainich formulation (§5.4) is used. The extremal field f_{ab} and the gradient $\alpha_{,a}$ of the complexion are determined by the metric. Hence $\mathcal{L}_\xi f_{ab} = 0 = \mathcal{L}_\xi \alpha_{,a}$, and (5.23) then yields (11.1) with $\Psi = \xi^a \alpha_{,a}$. For null Maxwell fields, the form (5.17) allows us, as with pure radiation, to take $\mathcal{L}_\xi k_a = 0 = \mathcal{L}_\xi \Phi$. Then, writing the real and imaginary parts of (5.16) as $F_{ab} = 2k_{[a}E_{b]}$ and $\widetilde{F}_{ab} = 2k_{[a}H_{b]}$, invariance of T_{ab} and of $k^a E_a = 0$ yields $E^a \mathcal{L}_\xi E_a = k^a \mathcal{L}_\xi E_a = 0$, so $\mathcal{L}_\xi E_a$ lies in the k_a–H_a-plane and (11.1) follows. The condition $\Psi_{,[a}k_{b]} = 0$ follows immediately from (2.64) and the Maxwell equations: it implies that either Ψ is constant or $\boldsymbol{k}$ is hypersurface-orthogonal.

To investigate possible non-inheritance we first note (Henneaux 1984) that whether F_{ab} is null or not, Theorem 11.1 shows that any isometry $T \in G_r$ acts on F_{ab} by $T^* F_{ab} = F^0_{ab} \cos\beta - \widetilde{F}^0_{ab} \sin\beta$, where F^0_{ab} is the value at some arbitrarily-chosen initial point in the orbit, in analogy with (5.23). Thus Theorem 11.1 implies that in all cases there is a homomorphism from G_r to a one-parameter group of duality rotations, the kernel of which consists of inherited symmetries. Consequently the derived algebra must be inherited, and if Ψ_A is the factor in (11.1) for a generator $\boldsymbol{\xi}_A$ then

$$\Psi_A C^A{}_{BC} = 0. \tag{11.2}$$

This and the previous results restrict possible non-inheritance. For example, in space-times admitting groups of Bianchi type *IX* (including spherical symmetry; Michalski and Wainwright (1975), Wainwright and Yaremovicz (1976b)) the Maxwell field must inherit those symmetries, and the same is true for space-times with two hypersurface-orthogonal commuting non-null Killing vectors and a non-null Maxwell field (Michalski and Wainwright 1975). Consideration of the action of an isotropy on the invariant $F^*_{ab}F^{*ab}$ shows that non-inheritance of an isotropy is only possible for a null field (Henneaux 1984). Further related results are discussed by Goenner (1984), Henneaux (1984) and Carigi and Herrera (1986).

A number of solutions are known in which the Maxwell field does not inherit the symmetry. For non-null fields the non-inherited Killing vector may be hypersurface-orthogonal (see §22.2), or twisting, as in (12.21). Examples of null non-inheriting Maxwell fields, which are also pure radiation solutions, are given by (13.46) and the *pp*-waves described in §24.5 (Wainwright and Yaremovicz 1976b, cp. (12.37); see also Lukács and Perjés (1976)), and in §15.4.

11.2 Isotropy and the curvature tensor

At fixed points of a group of motions, only suitable energy-momentum and Petrov types are possible. The isometry concerned is then usually given the same name as the local isotropy it induces at the fixed points, so one may have a rotation, i.e. a Killing vector field whose zeros form a fixed axis of rotation at which the isotropy can be represented by (3.16), and similarly a boost, with fixed points with isotropy (3.17), or a null rotation whose fixed points have isotropy (3.14) (or (3.15)). The implications for the curvature at the fixed points are the same whether the maximal group is simply-transitive except on a submanifold, or is multiply-transitive, which is the case we now discuss.

Following the remarks in §8.4 we see that if a space-time admits a group G_r of motions transitive on orbits of dimension $d(< r)$, each point has an isotropy group of dimension $r - d$ (see (8.20)), which must be (isomorphic to) a subgroup of the Lorentz group.

Space-times with such isotropy have been particularly intensively studied. The term *locally isotropic* has been introduced for the cases where every point p has a non-trivial continuous isotropy group (Cahen and Defrise 1968); when this group consists of spatial rotations the space-time is called *locally rotationally symmetric* or LRS (Ellis 1967). Some general remarks about the possible cases are given in this section. It can be shown that local isotropy of the Riemann tensor and a few of its derivatives (in some important cases, just the first) is sufficient to ensure the existence of a group of motions (see Chapter 9). Such a group must clearly be continuous, since every point in some neighbourhood has an isotropy subgroup of at least one parameter, and is at least a G_3 (acting on 2-surfaces). Weakening the assumptions to invariance of the Weyl tensor and the quantities in (6.14) only adds Szekeres solutions (see §33.3.2) to the LRS cases (Mustapha *et al.* 2000).

Schmidt (1968) gave a calculation of all possible Lie algebras of isometry groups acting transitively on a V_4 with local isotropy, and Defrise (1969) determined all distinct locally isotropic metric forms. MacCallum (1980) gave the Lie algebras and other information for locally isotropic metrics with non-null homogeneous hypersurfaces. Cahen and Defrise (1968) found all locally isotropic vacuum type D metrics, while Ellis (1967) and Stewart and Ellis (1968) studied LRS metrics with perfect fluid and electromagnetic field; the approaches used were rather similar, in that tetrads were defined up to the isotropy by properties of the curvature tensor, and the invariance under the isotropy was then imposed.

In §§4.2 and 5.1 (see Table 5.2) the maximal linear isotropy group was determined for each Petrov and Segre type. We recall that only in Petrov types D, N or O was any continuous isotropy possible. In these cases the permitted isotropies were, respectively, the group generated by spatial rotations (3.16) and boosts (3.17), the group of null rotations (3.15), and the Lorentz group; subgroups of these are, of course, permitted.

One can determine every possible subgroup of the Lorentz group; the details are given in many texts. For our purpose, some of these possibilities are irrelevant, because only certain isotropies are consistent with given Weyl and Ricci tensors. To begin with we can combine the information of §§4.2, 5.1 and 8.5 to obtain

Theorem 11.2 *The only Einstein spaces ($R_{ab} = \Lambda g_{ab}$) with a group of motions G_r, $r \geq 7$, are the spaces of constant curvature* (8.33) *which admit a G_{10}.*

Further we can see that there are no space-times with a G_r ($r \geq 7$) containing an electromagnetic field with $C = 0$ in (11.1), and that the only possible metrics with exactly a G_7 are a perfect fluid solution with an I_3 of spatial rotations at each point, and a pure radiation solution with an I_3 generated by null rotations (3.15) and spatial rotations (3.16). Both these spaces must be conformally flat. Both in fact exist, being the Einstein static universe and special plane waves (see Chapter 12). The same energy-momentum tensors would be required for the cases where the maximal group is a G_6 with the same isotropy groups I_3.

Now let us briefly list the less-highly-symmetric cases with isotropy permitted by the algebra of the curvature tensor.

For vacuum, one might have locally isotropic *homogeneous* Petrov type D or type N solutions with an I_2 or I_1. Actually the only case will be shown in Chapter 12 to be the plane waves of Petrov type N, with a G_6, and an I_2 of null rotations at each point. One has inhomogeneous spaces with a multiply-transitive G_3 or G_4 (see Chapters 13 and 15), or with null orbits (see Chapter 24) and a G_5 (which is impossible for non-null orbits by Theorem 8.17).

For the Λ-term, the situation is essentially the same as for vacuum, except that there is a Petrov type D solution with a G_6 and a metric of the form (12.8).

With a non-null Einstein–Maxwell field the locally isotropic metrics are either conformally flat or of Petrov type D, and in the latter case the invariant planes of R_{ab}, C^*_{abcd} and F^*_{ab} all coincide. The only such homogeneous space-time is found in Chapter 12 to admit a G_6; it is the Bertotti–Robinson metric, and is conformally flat. The solutions with a G_4 on V_3 or G_3 on V_2 are discussed in Chapters 13 and 15.

For a null electromagnetic field, a locally isotropic solution must be conformally flat or Petrov type N, the invariant planes of C^*_{abcd} (if non-zero) and F^*_{ab} again agree, and they define the same principal null direction as R_{ab}. The only homogeneous cases turn out to be special plane waves (see Chapter 12). Cases with lesser symmetry could have a G_5 or G_4 on

N_3 or a G_3 on N_2, in principle (but see Chapter 24). The only conformally flat cases are the special plane waves, see §37.5.

The situation for pure radiation is essentially the same as for null electromagnetic fields (disregarding remarks about F^*_{ab}).

Finally we come to the case of perfect fluid solutions. The conformally flat perfect fluid solutions (§37.5) admitting an isotropy are the Friedmann–Robertson–Walker universes (§14.2) and the interior Schwarzschild metric (§16.1). The locally isotropic non-conformally-flat perfect fluids admit a I_1 of spatial rotations. Such solutions exist for a G_5 on V_4 (the Gödel solution), G_4 on V_3 and G_3 on V_2, and were discussed in detail by Ellis (1967) and Stewart and Ellis (1968). Note that no solution with an I_2 of spatial rotations arises, essentially because, as noted in §8.6, the rotation group has no two-dimensional subgroup.

These locally rotationally symmetric solutions were divided by Ellis (1967) into three classes, assuming the metric has an invariantly-defined orthonormal basis ($\boldsymbol{e}_1$, $\boldsymbol{e}_4$) in the plane orthogonal to the plane of the rotational symmetry, which is spanned by ($\boldsymbol{e}_2$, $\boldsymbol{e}_3$). Then one can show that $\omega k = 0$, where $D^4{}_{23} = \omega$, $D^1{}_{23} = k$ in (2.6) (Ellis 1967, Stewart and Ellis 1968). Ellis thus divided the cases into three classes. In Class *I* $\omega \neq 0$, so that there is a timelike congruence with rotation, and the 2-planes in which the isotropy acts are non-integrable; these are stationary or static metrics admitting a G_4 on T_3. In Class *II* $\omega = k = 0$, the 2-planes in which the isotropy acts are integrable, and the timelike congruence is non-rotating; the metrics admit a G_3 on S_2. In Class *III* $k \neq 0$, so the timelike congruence is non-rotating but the 2-planes are non-integrable; the metrics are spatially-homogeneous, admitting a G_4 on S_3.

The arrangement of the subsequent chapters of Part II is given by Table 11.1.

11.3 The possible space-times with proper homothetic motions

Relatively few solutions have been found by assuming in advance the existence of a proper homothetic motion. However, a great many solutions in fact have proper homothetic motions whose existence accounts for the comparatively simple forms of their line elements and consequently for their discovery. A number of these appear in Chapters 13, 14 and 23, and others are summarized in §11.4.

The relation between homothety and similarity solutions of the Einstein equations gives rise to the alternative name 'self-similar' for solutions with homothetic motions, more precisely 'similarity of the first kind', similarity of the zeroth and second kinds being defined like the kinematic similarity of (35.64) (Carter and Henriksen 1989). There are arguments for believing

Table 11.1. Metrics with isometries listed by orbit and group action, and where to find them

Equations (3.15) are null rotations, (3.16) spatial rotations, and (3.17) boosts.

Orbit	Maximal group	Isotropy subgroup	Relevant chapters and sections
V_4	G_7	I_3 of space rotations	12.4
		(3.15) and (3.16)	12.5
	G_6	(3.15)	12.2, 12.5
		(3.16) and (3.17)	12.3, 12.5
	G_5	(3.15), real B	12.5
		(3.16)	12.4
	G_4	None	12
S_3	G_6	I_3 of space rotations	13.1, 14
	G_4	(3.16)	13, 14, 15.4, 15.7
	G_3	None	13, 14
T_3	G_6	3-dim. Lorentz group	13
	G_4	(3.16) and (3.17)	13, 15.4, 15.7, 16.1
	G_3	None	13, 22.2
S_2	G_3	(3.16)	15, 16
	G_2	None	17, 22, 23, 25
T_2	G_3	(3.17)	15
	G_2	None	17, 19–21
S_1	G_1	None	17.3, 23.4
T_1	G_1	None	17.3, 18
N_3, N_2, N_1	G_r ($1 \leq r \leq 6$)	If any, (3.15) and/or (3.16)	24

that self-similar solutions will in general represent asymptotic states of more general solutions, for example the behaviour of cosmologies near a big-bang or at late stages of expansion (Coley 1997b, Carr and Coley 1999), and this has been proved in, for example, some classes of the hypersurface-homogeneous cosmologies of Chapter 14 (Wainwright and Ellis 1997). However, self-similar solutions cannot in general be asymptotically flat or spatially compact (Eardley *et al.* 1986).

At a fixed point of a proper homothety all scalar invariants polynomial in the curvature vanish, because they would have to map to themselves under the homothety, but this would multiply them by a non-trivial factor.

Thence a space-time admitting a proper homothetic motion with a fixed point p must be of Petrov type III, N or O and Segre type $[(2,11)]$ or $[(3,1)]$ at p. Such fixed points are either isolated or form null geodesics; in the former case, the Ricci tensor can only be non-zero if it has the unphysical Segre type $[(3\ 1)]$ at least on a three-dimensional submanifold (or vacuum, for which in Petrov type N see da Costa and Vaz (1992)), and in the latter case, the space-time is either flat or a generalized plane wave (Hall 1988a). These restrictions apply to all points if the H_r and its G_{r-1} subgroup have the same orbits; this can only happen if the orbits are either V_4 or N_3 (Hall and Steele 1990).

This further limits the groups of proper homothetic motions H_r for large r, which must anyway be included among the possible spaces with a group G_{r-1} of motions listed in §11.2; for reviews of space-times with H_r for $r \geq 3$ see Hall and Steele (1990) and Carot and Sintes (1997). The possible Segre types include some not considered in this book. Flat space has an H_{11} but there are no space-times with an H_r for $r = 9$ or 10. The only solutions with an H_8 (and G_7) are (12.37). Other homogenous plane waves (12.12) admit an H_6 and may admit an H_7 on V_4, while inhomogeneous plane waves may admit an H_7 on N_3. The Robertson-Walker metrics, with a G_6 on S_3, may have an H_7 on V_4. For an H_6, only the generalized plane waves, in which the orbits are N_3, are possible, while an H_5 must act on a V_4 in which the G_4 acts on three-dimensional orbits (Hall and Steele 1990). Solutions with an H_5 must therefore be locally rotationally symmetric or similarly locally boost or null rotation symmetric, and hence of Petrov types D or N (or conformally flat). An H_4 on V_3 or V_4, with an isometry group G_3 on, respectively, a three- or two-dimensional orbit, is possible but the case of an H_4 and G_3 on N_3 is excluded. Examples of spaces explicitly found by these assumptions appear in Chapters 14 and 23; see also Tables 11.2–11.4. There are no multiply-transitive H_3. An Abelian orthogonally-transitive proper H_2 acting on non-null orbits in a non-flat vacuum is impossible (Kolassis and Ludwig 1996).

In a non-flat vacuum, a proper homothety must be non-null: if it has a non-null homothetic bivector it must be shearfree and expanding but not hypersurface-orthogonal or geodesic; if the homothetic bivector is null the Petrov type is III or N (McIntosh 1976a).

As with isometries, homotheties may imply inheritance of symmetry by the matter content. For a perfect fluid, homothety implies that

$$\mathcal{L}_\xi \mu = -2\Phi\mu, \quad \mathcal{L}_\xi p = -2\Phi p, \quad \mathcal{L}_\xi u_a = \Phi u_a, \tag{11.3}$$

and the symmetry is always inherited. These equations imply that if there is a barotropic equation of state $p = p(\mu)$, it must be (5.36), i.e. $p = (\gamma - 1)\mu$ for some constant γ (Cahill and Taub 1971, Collins 1977b, Wainwright 1985), and the homothety cannot then be orthogonal to the

Table 11.2. Solutions with proper homothety groups H_r, $r > 4$

The H_4 act on V_4 except in (24.46) where the action is on N_3. References are to where the homothety (not the solution) was first given. The proper homothety generator is only defined up to addition of a Killing vector. The energy-momentum notation is defined in the main text.

H_r	Metric	T_{ab}	Homothety	Reference(s)
H_8	(12.37) [in (12.12)]	E	$2v\partial_v + \zeta\partial_\zeta + \bar{\zeta}\partial_{\bar{\zeta}}$	Eardley (1974)
H_7	(12.12), (37.104)	E	$2v\partial_v + \zeta\partial_\zeta + \bar{\zeta}\partial_{\bar{\zeta}}$	Eardley (1974)
	(14.8b)	F	$t\partial_t + (1 - 2/3\gamma)r\partial_r$	Eardley (1974)
	§14.2: (12.9), $\varepsilon \neq 0$, $\gamma = \frac{2}{3}$	F	$t\partial_t$	
H_6	(24.46)	E	$2v\partial_v + \zeta\partial_\zeta + \bar{\zeta}\partial_{\bar{\zeta}}$	Eardley (1974)
H_5	LRS subcase of (13.51)≡(22.7), cp. (13.53) and Table 18.2, including	V	$x^4\partial_{x^4} + \sum_{i=1}^{3}(1 + a_4 - a_i)x^i\partial_{x^i}$	Godfrey (1972)
	(15.12), $k = 0$, $e^{2\nu} = 1/x^3$ ($BIII$)	V	$4x^4\partial_{x^4} + 2x^3\partial_{x^3} + x^1\partial_{x^1}$	Koutras (1992b)
	(15.17), $k = 0$, $H = -2m/r$ ($AIII$)	V	$4u\partial_u + 2r\partial_r + x^1\partial_{x^1}$	Halford and Kerr (1980)
	(13.79)	F	$w\partial_w$	Daishev (1984)
	(13.85), $a = 0$	F	$4x\partial_x + (2q + 1)(y\partial_y + z\partial_z + t\partial_t)$	Wainwright (1985)
	(14.19)	F	$t\partial_t + (1 - c)x\partial_x$	Sintes (1996)
	(14.20)	F	(13.52), $a_4 = 0$, $a_i = p_i$	Hsu and Wainwright (1986)
	(14.28b), LRS case	F	(13.52)	Koutras (1992b)
	(15.17), $k = 0$, $H = e^2/r^2$	E	$3u\partial_u + r\partial_r + x^1\partial_{x^1}$	Carminati and McIntosh (1980)
	(15.50), $L = 1$: see §15.6.2	F	$r\partial_r + at\partial_t$	
	(15.65) with $A = a/r^n$, $B = br^n$	F	$t\partial_t + r\partial_r$	Ponce de León (1988)
	(15.76a) with $6/7 < \gamma < 1$	F	$(\gamma - 2)[2(\gamma - 1)z\partial_z + (2 - 3\gamma)t\partial_t]$	Sintes (1996)
	and $G = (2 - \gamma)z^3/(7\gamma - 6)$	F	$+(\gamma^2 + 4\gamma - 4)(x\partial x + y\partial y)$	
	Table 16.2: $\mu = p/(\gamma - 1) \propto 1/r^2$	F	$r\partial_r + (2 - \gamma)t/\gamma\,\partial_t$	Henriksen and Wesson (1978)

Table 11.3. Solutions with proper homothety groups H_4 on V_4, described using the conventions of Table 11.2

Metric	T_{ab}	Homothety	Reference(s)
(13.51)≡(22.7), cp. (13.53)	V	(13.52)	Godfrey (1972)
(13.72)	E	$2t\partial_t + y\partial_y + z\partial_z$	Koutras (1992b)
(13.73)	E	$2r\partial_r + y\partial_y + z\partial_z$	Koutras (1992b)
(13.76)	E	$(1-k^2)v\partial_v + t\partial_t + y\partial_y - k^2\phi\partial_\phi$	McIntosh (1979)
(13.87)	F	$r\partial_r + s(\phi\partial_\phi + t\partial_t)$	Hermann (1983)
(14.28*b*)	F	(13.52)	Hsu and Wainwright (1986)
(14.30)	F	$t\partial_t + 2qx\partial_x + (1-3q)y\partial_y + (1-q)z\partial_z$	Hewitt (1991b)
(14.31)	F	$3\partial_t + 6x\partial_x + 4y\partial_y + 2z\partial_z$	Hewitt *et al.* (2001)
(14.33)	F	$t\partial_t + (1-p_2)y\partial_y + (1-p_3)z\partial_z$	Hsu and Wainwright (1986)
(14.41)	F	$5t\partial_t + 4y\partial_y + 2z\partial_z$	Hsu and Wainwright (1986)
(14.42)	F	$t\partial_t + (1+q-s)y\partial_y + (1-q-s)z\partial_z$	Rosquist and Jantzen (1985)
(15.17), $k=0$, $H = \mathrm{const}\, u^q/r$	R	$4u\partial_u + 2(q+1)r\partial_r + (1-q)x^1\partial_{x^1}$	Koutras (1992b)
(20.32) with $\Omega = \log\rho$	V	$(\phi - 2t)\partial_\phi + t\partial_t + 4(\rho\partial_\rho + z\partial_z)$	Steele (1991)
(20.32) with $\Omega = z$	V	$3\partial_\phi - t\partial_t + 4(\rho\partial_\rho + z\partial_z)$	Steele (1991)
(22.27*a*), $a=0$, $\beta=\chi$, $\alpha = 4(\gamma-1)(q\rho + r)$	F	$\partial_\rho + 4q(\gamma-1)(\phi\partial_\phi + z\partial_z)$ $+ ((2\gamma-4)q + 2b)t\partial_t$	Debever and Kamran (1982)
(24.40), $H = \ln(\zeta\bar{\zeta})$, $\zeta = x + \mathrm{i}y$	V	$x\partial_x + y\partial_y + u\partial_u + (v-2u)\partial_v$	Steele (1991)
(28.16), cp. (13.64)	V	$r\partial_r + u\partial_u$	McIntosh (1976b)
(33.30)	F	$x\partial_x + y\partial_y - u\partial_u$, $y = \mathrm{Im}\,\zeta$	Koutras (1992b)

fluid four-velocity $\boldsymbol{u}$ unless $\mu = p$ (McIntosh 1976a), or parallel to $\boldsymbol{u}$ unless $\gamma = 2/3$ (Wainwright 1985).

Fluid solutions with an H_r have been studied in several works. Robertson–Walker metrics with an H_7 and solutions with an H_4 on V_4 or an H_5 on V_4 containing a G_4 on V_3 are covered by Chapters 13 and 14, solutions with an H_4 on V_3 (and G_3 on V_2) by Chapters 15 and 16, and those with a maximal H_3 by Chapter 23. See also the tables in §11.4.

The analogue of (11.1) for homothetic motions in Einstein–Maxwell solutions reads (Wainwright and Yaremovicz 1976a, 1976b)

$$\mathcal{L}_{\xi_A} F_{ab} = \Psi_A \widetilde{F}_{ab} + \Phi_A F_{ab}, \tag{11.4}$$

and (11.2), $\Psi_A C^A{}_{BC} = 0$, again holds.

For a non-null Maxwell field, if there is a non-null proper homothety $\boldsymbol{\xi}$, it cannot be hypersurface-orthogonal (McIntosh 1979); if a geodesic shearfree principal null direction of the Maxwell field coincides with one for a non-null homothetic bivector, the solution is algebraically special; and if both principal null directions coincide, they cannot be geodesic and the homothety cannot be hypersurface-orthogonal (Faridi 1990). For the null case, $\Psi_{,[a}k_{b]} = 0$, where k_b is the repeated principal null direction of the Maxwell field, and Ψ is therefore constant if k_b is twisting (Wainwright and Yaremovicz 1976b). An example of a non-inherited homothety is given by (13.76).

A homothetic motion is often apparent in a metric's power-law form, which indicates that a homogeneity transformation of the type $x'^i = k^{n_i} x^i$ for each i, with some constants k, n_i, maps the metric to a multiple of itself: in that case $\boldsymbol{\xi} = \sum_i n_i x_i \partial_{x^i}$ is a homothetic vector. For example the proper homothety (13.56) of the Kasner metric (13.53) is of this type. Homothety is also readily recognized if one or more of the x^i is replaced by $\exp(y^i)$. The detection of homothety by coordinate independent methods is discussed in Koutras (1992b), Koutras and Skea (1998) and Chapter 9. It is rather common for a proper homothetic vector field to be timelike in some regions and spacelike in others.

11.4 Summary of solutions with homotheties

Whereas solutions with G_r are treated systematically in the following chapters, solutions among them and elsewhere which admit an H_r are not. Hence, for reference, Tables 11.2–11.4 list all solutions given explicitly in this book and known to admit a homothety group H_r for $r \geq 3$. In the tables the abbreviations for the energy-momentum are as follows: V denotes vacuum, E Einstein–Maxwell, F perfect fluid, and R radiation

Table 11.4. Solutions with proper homothety groups H_r on V_3, described using the conventions of Table 11.2

H_r	Metric	T_{ab}	Homothety	Reference(s)
H_4	(15.39)	F	$t\partial_t + r\partial_r$	Henriksen and Wesson (1978), Bona (1988a)
	(15.72), $A = \mathrm{e}^{2r}$		∂_r	Bogoyavlensky and Moschetti (1982)
	(15.75), $a = 0$	F	$r\partial_r$	Collins and Lang (1987)
	(15.81), $a = 0$	F	$\sum_a x^a\partial_{x^a}$	Collins and Lang (1987)
	(15.81), $b = 0$	F	$r\partial_r - D(x\partial_x + y\partial_y)$	
	(15.82)	F	∂_z	Lorenz (1983c)
	(15.83), $c = 0$	F	$t\partial_t + (D-1)(y\partial_y + z\partial_z)$	Collins and Lang (1987)
	(15.84)	F	$2\gamma t\partial_t + (3\gamma - 2)(y\partial_y + z\partial_z)$	Goode (1980)
	(15.85)	F	$\partial_t + \partial_x - (y\partial_y + z\partial_z)$	Carot and Sintes (1997)
	(15.86), $N = z$	F	$z\partial_z$	
	(15.90)	F	$x^n\partial_n$	
	See §16.2.2	F	$t\partial_t + r\partial_r$	Dyer *et al.* (1987), Havas (1992)
	(22.17) with $c = 0$	E	$4(r\partial_r + z\partial_z) + 3\theta\partial_\theta - t\partial_t$	Koutras (1992b)
	(37.57)	F	$r\partial_r$	Collins and Lang (1987)
H_3	(17.14)	V	$\partial_y - \partial_t$	Phan (1993)
	(18.66), $a = -1$, $N = \mathrm{e}^z$	F	∂_z	Bogoyavlensky and Moschetti (1982)
	(19.21), $A = 0$, $U = \alpha\ln\rho + z$, $k = \alpha^2\ln\rho + 2\alpha z - \rho^2/2$	V	$\partial_z + 2\alpha\varphi\partial_\varphi + 2(\alpha-1)t\partial_t$	Godfrey (1972)
	(19.21), $A = 0$, $\mathrm{e}^U = \rho^\alpha(r+z)^\beta$, $\mathrm{e}^k = \rho^{\alpha^2}(r+z)^{2\beta(\alpha+\beta)}/r^{2\beta^2}$	V	$\rho\partial_\rho + z\partial_z + \alpha(2\beta+\alpha)\varphi\partial_\varphi + (\alpha-1)(2\beta+\alpha-1)t\partial_t$	Godfrey (1972)
	(21.61), $\mu = p$	F	∂_z	Hermann (1983)
	(21.72)	F	$\partial_z - \sigma[\varphi\partial_\varphi + t\partial_t)$	Hermann (1983)
	(23.1)	F	$\partial_x - 2ncz\partial_y - ab(y\partial_y + z\partial_z)$	McIntosh (1978a)

Table 11.4 contd.

Metric	Type	Vector field	Reference
(23.2)	F	$\partial_x - [(m^2 - n^2 - 1 + 2(a^2 - b^2))$ $(y\partial_y + z\partial_z)/2] - n(y\partial_y - z\partial_z)$	McIntosh (1978a)
(23.3)	F	$\partial_x + (nm - \alpha\beta)(y\partial_y + z\partial_z) - n(y\partial_y - z\partial_z)$	McIntosh (1978a)
(23.4)	F	$A[\partial_w + (\frac{3}{4} + b^2 - a^2 - \frac{1}{2}m^2)(y\partial_y + z\partial_z)]$ $+ z\partial_y - y\partial_z$	McIntosh (1978a)
(23.6), $F = \mathrm{e}^{-u}$	F	$2\partial_x + (2z + y)\partial_y + z\partial_z$	McIntosh (1978a)
(23.6), $F = u^{-2p}$	F	$t\partial_t + x\partial_x - \frac{1}{2}z\partial_z + (q - p - \frac{1}{4})(y\partial_y + z\partial_z)$	Collins (1991)
(23.7)	F	$\partial_x + 4y\partial_y - z\partial_z$	Wils (1991)
(23.8)	F	$\partial_x + 4(2 - a)y\partial_y - 2(1 + a)z\partial_z$	Wils (1991)
(23.11)	F	$\partial_t + \partial_x + y\partial_y$	Carot and Sintes (1997)
(23.14), $T = \mathrm{e}^{at}$ [§23.2]	F	$(1 - n^2)\partial_t + ma[(1 - n)y\partial_y + (1 + n)z\partial_z]$	
(23.14), $P = 1$, $T = t$, $G = x^c$	F	$2t\partial_t + 2x\partial_x + (2m(1 - c) + 1 - n)y\partial_y$ $+ (2m(1 - c) + 1 + n)z\partial_z$	A special case is in Carot and Sintes (1997)
(23.23)	F	$2\partial_t + b(y\partial_y + 3z\partial_z$	
(23.32), $\varepsilon = 0$	F	∂_t	Senovilla and Vera (1998)
(23.36)	F	$\partial_x + (c - \frac{1}{2}a)(y\partial_y + z\partial_z) + b\ell(y\partial_y - z\partial_z)$	Senovilla and Vera (2001)
(23.38), $(i)/(ii)$, $c_2 = 0$, $\lambda = 4b/(2b + 2q + 1)$	F	$[t + (1 - \lambda)x]\partial_t + [x + (1 - \lambda)t]\partial_x$ $+ [b(\lambda - 2) + 1]y\partial_y + [(b - 1)\lambda + 1]z\partial_z$	Senovilla and Vera (1997)
(23.46)	F	$2\partial_t + (2q - b)(y\partial_y + z\partial_z) - (y\partial_y - z\partial_z)$	Koutras (see Haager and Mars (1998))
(23.51), $\delta = 0$	F	$6t\partial_t + 6x\partial_x + (5c + 16)y\partial_y + 8z\partial_z$	Wils (1991), Koutras (1992b)
(28.17)	V	$2r\partial_r + u\partial_u + \zeta\partial_\zeta + \bar\zeta\partial_{\bar\zeta}$	Halford and Kerr (1980), Lun *et al.* (1988)
(28.25), C-metric	V	$r\partial_r + 2u\partial_u + 2\zeta\partial_\zeta + 2\bar\zeta\partial_{\bar\zeta}$	Godfrey (1972)
(28.26)	V	$r\partial_r + 2u\partial_u - \zeta\partial_\zeta - \bar\zeta\partial_{\bar\zeta}$	Halford and Kerr (1980), Lun *et al.* (1988)
(28.72)	R	$r\partial_r + u\partial_u$	Koutras (1992b)
(29.46)–(29.47), $l = 0$	V	$r\partial_r + 2u\partial_u - \zeta\partial_\zeta - \bar\zeta\partial_{\bar\zeta}$	Halford and Kerr (1980), Lun *et al.* (1988)
(35.77)	F	$2t\partial_t + 2x\partial_x + (1 + a)(\partial_y + \partial_z) + b(\partial_y - \partial_z)$	Mars and Wolf (1997)

solutions. Some references to solutions known to admit an H_r, $r \geq 3$, which are not given explicitly, appear in §§15.7, 16.2.2, 20.2, 21.1.5, 22.2–22.4, 23.2, 30.7.4 and 33.1. Hsu and Wainwright (1986) and Koutras (1992b) proved that (14.28*b*), (14.33) and (14.41) contain all Petrov type *I* perfect fluid solutions in which the fluid flow is orthogonal to the orbits of the G_3.

For vacuum, Kerr and Debney (1970), McIntosh and Steele (1991) and Kerr (1998) have shown that certain sets of solutions with H_r, $r \geq 3$, are exhaustive. The algebraically special cases with diverging rays and an H_r, for $5 \geq r \geq 1$, were studied by Halford and Kerr (1980), Halford (1980) and Lun *et al.* (1988). Not all cases could be solved explicitly. Most are previously known solutions.

Robinson–Trautman solutions admitting an H_r are reviewed in Singleton (1990); homotheties of Petrov type *N* vacuum solutions in Kundt's class are discussed in Salazar I. *et al.* (1983); and reductions of the field equations for twisting algebraically special vacua with an H_2 or H_1 are discussed in §§29.2 and 29.3; see also Stephani and Herlt (1985) and McIntosh *et al.* (1987).

There are a number of solutions known to admit an H_2, which we now cite. There are 16 explicitly known algebraically special vacuum metrics with a maximal H_2: six are twisting solutions in the class discussed in §29.2.5, arising from special choices of the arbitrary functions (Halford 1980, Lun *et al.* 1988); five are Robinson–Trautman solutions of type *N* (Singleton 1990); two are Robinson–Trautman solutions of type *III* (Lun *et al.* 1988, Singleton 1990 and §17.3); two are type *N* solutions in Kundt's class (Salazar I. *et al.* 1983); and the other is the Hauser solution (29.72).

Of the 17 non-degenerate Harrison solutions of Petrov type *I* (see §17.3), 13 admit homotheties, i.e. an H_2, at least for certain subcases (Koutras 1992b). A further solution with an H_2 is given by (18.48) and others are mentioned in §29.2. Godfrey (1972) found solutions with H_3 and H_4 in Weyl's static axisymmetric vacuum class (§20.2), and Kolassis (1996) has shown that these include all Petrov type *I* vacuum solutions admitting an H_2 which is orthogonally transitive.

The null Einstein–Maxwell solution (28.56*b*) with $m_0 = 0$, $\epsilon = 1$, admits a homothety (Koutras 1992b). Pure radiation solutions with an H_2 are mentioned in §30.7.4, and some conformally flat solutions with homothety in §37.5.1.

Perfect fluid solutions with a maximal H_2 arise in §17.3. Koutras (1992b) has shown that certain subcases of the Petrov type *N* perfect fluid solutions of Oleson's Class *I*, (33.48), admit an H_2 or H_1, and that one subcase of Class *II* also has an H_1.

12
Homogeneous space-times

12.1 The possible metrics

A homogeneous space-time is one which admits a transitive group of motions. It is quite easy to write down all possible metrics for the case where the group is or contains a simply-transitive G_4; see §8.6 and below. Difficulties may arise when there is a multiply-transitive group G_r, $r > 4$, *not* containing a simply-transitive subgroup, and we shall consider such possibilities first. In such space-times, there is an isotropy group at each point. From the remarks in §11.2 we see that there are only a limited number of cases to consider, and we take each possible isotropy group in turn.

For G_r, $r \geq 8$, we have only the metrics (8.33) with constant curvature admitting an I_6 and a G_{10}.

If the space-time admits a G_6 or G_7, and its isotropy group contains the two-parameter group of *null rotations* (3.15), but its metric is not of constant curvature, then it is either of Petrov type N, in which case we can find a complex null tetrad such that (4.10) holds, or it is conformally flat, with a pure radiation energy-momentum tensor, and we can choose a null tetrad such that (5.8) holds with $\Phi^2 = 1$. In either case the tetrad is fixed up to null rotations (together with a spatial rotation in the latter case). The covariant derivative of $\boldsymbol{k}$ in this tetrad must be invariant under the null rotations, which immediately gives

$$\kappa = \rho = \sigma = \varepsilon + \bar{\varepsilon} = \tau + \bar{\alpha} + \beta = 0. \tag{12.1}$$

Since τ and σ are invariantly defined for the tetrad described, (7.21p) yields

$$\tau(\tau + \beta - \bar{\alpha}) = 0, \tag{12.2}$$

and thus either $\tau = 0$, in which case $\boldsymbol{k}$ is (proportional to) a covariantly constant vector and we arrive at homogeneous plane waves (§24.5), or

$$\tau + \beta = 0 = \alpha. \tag{12.3}$$

In the latter case (7.21o) yields $\gamma = 0$ and (7.21q) shows that

$$\Lambda = -6\tau\bar{\tau} \neq 0, \tag{12.4}$$

so we must have Petrov type N. It is convenient to alter the tetrad choice so that $\Psi_4 = \Lambda/2$; (7.32d) then shows that $\tau = \pm\bar{\tau}$, and, correspondingly, $\Phi_{22} = \mp 5\Lambda/6$, so for positive energy we need $\Lambda < 0$. A position-dependent null rotation of the tetrad is still permitted, and may be used to set

$$\bar{\pi} = -\tau, \qquad \lambda = \mu = 0, \qquad \nu = -\tau. \tag{12.5}$$

Since the resulting tetrad has constant spin coefficients, it generates a transformation group whose reciprocal group must be a simply-transitive isometry group. With the choice (12.5) the commutators enable one to introduce coordinates so that the metric is

$$\mathrm{d}s^2 = \frac{3}{|\Lambda|y^2}\left[\mathrm{d}y^2 + \mathrm{d}z^2 - \mathrm{d}v\left(\mathrm{d}u - \frac{\Lambda \mathrm{d}v}{|\Lambda|y^2}\right)\right]. \tag{12.6}$$

This is a pure radiation solution of Petrov type N with a cosmological constant and a G_6. It was first given by Defrise (1969).

In the case with an additional spatial rotation symmetry and a G_7, the symmetry implies that for the null tetrad fixed by $\Phi_{22} = 1$, $\tau = 0$, and thus only a special homogeneous plane wave is possible. From §24.5, the plane wave solutions are

$$\mathrm{d}s^2 = 2\mathrm{d}\zeta\,\mathrm{d}\bar{\zeta} - 2\mathrm{d}u\,\mathrm{d}v - 2[\overline{A}(u)\bar{\zeta}^2 + A(u)\zeta^2 + B(u)\zeta\bar{\zeta}]\mathrm{d}u^2 \tag{12.7}$$

with $R_{ab} = B(u)k_a k_b$, and for them to admit a G_6 or G_7 we require special forms for $A(u)$ and $B(u)$ (see Table 24.2 and §12.5).

The other possible cases with a maximal G_6 are those with an isotropy group composed of *boosts* (3.17) and *rotations* (3.16). A short calculation by Schmidt's method (§8.6) reveals that the metric must be that of the product of two 2-spaces of constant curvature, i.e.

$$\mathrm{d}s^2 = A^2[\mathrm{d}x^2 + \Sigma^2(x,k)\mathrm{d}y^2] + B^2[\mathrm{d}z^2 - \Sigma^2(z,k')\mathrm{d}t^2], \tag{12.8}$$

where A and B are constants. This space is symmetric, cp. (35.29).

A metric with a G_7 and an isotropy group I_3 consisting of *rotations* must contain a preferred timelike vector field $\boldsymbol{u}$. The isotropy of the covariant derivative of $\boldsymbol{u}$ shows that $\boldsymbol{u}$ is hypersurface-orthogonal and shearfree. The hypersurfaces to which $\boldsymbol{u}$ is orthogonal have constant curvature (by Theorem 8.14), and Theorem 8.16 and (8.36) give the metric as

$$\mathrm{d}s^2 = a^2(t)[\mathrm{d}r^2 + \Sigma^2(r,k)(\mathrm{d}\vartheta^2 + \sin^2\vartheta\,\mathrm{d}\varphi^2)] - \mathrm{d}t^2, \tag{12.9}$$

which is the well-known Robertson–Walker metric form (Robertson 1929, 1935, 1936, Walker 1936). For a homogeneous space-time we shall require $a(t)$ to be constant, since it is clearly an invariant (assuming the I_3 is the maximal isotropy group), or, in the case $k = 0$, invariant up to a constant scale.

Now we have to consider the possibility of a maximal *isotropy group* I_1. For the case of an I_1 of spatial rotations, Schmidt's calculations (1968) show that there is a simply-transitive subgroup G_4 except in the case where the full group of motions is a G_6 and the metric is (12.8). The same holds for the case of an I_1 of Lorentz transformations (3.17). Finally we have the case of an I_1 of null rotations. Here a calculation by Schmidt's method again shows that there is a simply-transitive G_4 subgroup in all cases. We have now exhausted the possible multiply-transitive groups.

The existence of a *simply-transitive group* G_4 enables one to make the solution of the field equations into a purely algebraic problem. To do so one simply chooses a set of reciprocal group generators which form an orthonormal or a complex null tetrad; the connection coefficients (§3.3) or spin coefficients (§7.1) will be constants, and the curvature is easily calculated (cp. §8.6, (8.52)–(8.54)), in terms of the structure constants of the simply-transitive isometry group; such computations are given by e.g. Sengier-Diels (1974b, 1974a) and Fee (1979).

The classification of groups G_4 following Theorem 8.4 (§8.2) permits the further simplification of aligning one of the tetrad vectors with the distinguished vector of the class (A_E, P_C or L^B), except in the trivial case $\boldsymbol{A} = \boldsymbol{L} = \boldsymbol{P} = \boldsymbol{0}$ when the space-time is flat. The resulting Ricci tensors have been given in detail by Hiromoto and Ozsváth (1978). (Essentially the same results can be achieved by taking the reciprocal group generators so that $C^A{}_{BC}$ is in canonical form and algebraically determining the g_{AB}.)

The space-times thus found tend to have a rather large number of different invariant characterizations, and so may be recovered in various ways. As a consequence of Theorem 8.5 all the space-times with a simply-transitive G_4 have a G_3 transitive on hypersurfaces, and so, in principle, recur in Chapters 13, 14 and 24.

Computation of the permissible homogeneous spaces for a given energy-momentum tensor by the systematic treatment of the various possibilities listed above is very laborious, and more elegant proofs are available in some cases, see §§12.2 and 12.3.

12.2 Homogeneous vacuum and null Einstein-Maxwell space-times

Consider a homogeneous null electromagnetic field. Taking a complex null tetrad such that $\Phi_2 = 1$, $\Phi_0 = \Phi_1 = 0$, (7.22)–(7.25) yield

$$\kappa = \sigma = \rho - 2\varepsilon = \tau - 2\beta = 0 \tag{12.10}$$

and (6.34) then shows that $\rho = 0$ (since ρ must be constant, being an invariant). From the Goldberg–Sachs theorem, $\Psi_0 = \Psi_1 = 0$. Now (7.21p) and (7.21q) give $\Psi_2 = 0$, $\tau(\tau+\beta-\bar{\alpha}) = 0$, (7.21$l$) gives $\alpha - \bar{\beta} = 0$ and thus $\tau = 0$. This leads to plane waves (§24.5), since $\boldsymbol{k}$ must be proportional to a covariantly constant vector.

The homogeneous vacuum spaces were given by Petrov (1962) (cp. Hiromoto and Ozsváth (1978)). Non-flat homogeneous vacua with a multiply-transitive group must be type D or N. Taking a geodesic shear-free $\boldsymbol{k}$, the Bianchi identities in the type D case give

$$\kappa = \sigma = \lambda = \nu = \rho = \mu = \tau = \pi = 0, \tag{12.11}$$

and (7.21q) then gives $\Psi_2 = 0$. In the type N case, τ is an invariant, and (7.21p) and (7.21q) yield $\tau = 0$, again giving plane waves. Thus we have

Theorem 12.1 *The plane waves*

$$\mathrm{d}s^2 = 2\mathrm{d}\zeta\mathrm{d}\bar{\zeta} - 2\mathrm{e}^{\varepsilon u}\mathrm{d}u\,\mathrm{d}v - 2\mathrm{d}u^2[2a\,\mathcal{R}e(\zeta^2\mathrm{e}^{-2\mathrm{i}\gamma u}) + b^2\zeta\bar{\zeta}] \tag{12.12}$$

represent all homogeneous null Einstein–Maxwell fields (with $\mathcal{L}_{\boldsymbol{\xi}}F_{ab} = 0$), and all non-flat vacuum homogeneous solutions with a multiply-transitive group.

In (12.12) a, b, γ are real constants, $\varepsilon = 0$ or 1, the Petrov type is N if $a \neq 0$ or O if $a = 0$, and the space-time is empty if $b = 0$. If $a = 0 = \varepsilon$ one can set $b = 1$, and if $a \neq 0 = \varepsilon$ one can set $2a = 1$. The latter case, with $\gamma = 1$, $\zeta = (x + \mathrm{i}y)/\sqrt{2}$, gives an interesting special solution

$$\mathrm{d}s^2 = \mathrm{d}x^2 + \mathrm{d}y^2 - 2\mathrm{d}u\,\mathrm{d}v - 2[(x^2 - y^2)\cos(2u) - 2xy\sin(2u)]\mathrm{d}u^2, \tag{12.13}$$

the 'anti-Mach' metric of Ozsváth and Schücking (1962). It is geodesically complete and without curvature singularities.

The metrics (12.12) admit null Maxwell fields with non-zero components given by $\sqrt{\kappa_0}F_{u\zeta} = b\mathrm{e}^{\mathrm{i}f(u)}$, where $f(u)$ is an arbitrary function; hence the Maxwell field may but need not share all the space-time's symmetries (cp. §11.1) and in general is invariant only under the subgroup G_5 of the group of motions which acts in surfaces $u = \text{const}$ (for the special cases (12.37) this was noted by Pasqua (1975)).

The group G_6 acting on (12.12) (see §24.5) may contain a simply-transitive G_4 and subgroups G_3 of Bianchi types IV, VI_h or VII_h acting on hypersurfaces (see Siklos (1976a, 1981) and §13.3.2). The isometry group structure of the vacuum metrics was studied by Klekowska and Osinovski (1971). The vacuum solutions admit an H_7 and the conformally flat Einstein–Maxwell fields an H_8.

By the methods described in §12.1, one can show that

Theorem 12.2 *The only vacuum solution admitting a simply-transitive G_4 as its maximal group of motions is given by*

$$k^2 \mathrm{d}s^2 = \mathrm{d}x^2 + \mathrm{e}^{-2x}\mathrm{d}y^2 + \mathrm{e}^{x}[\cos\sqrt{3}x(\mathrm{d}z^2 - \mathrm{d}t^2) - 2\sin\sqrt{3}x\,\mathrm{d}z\,\mathrm{d}t], \quad (12.14)$$

where k is an arbitrary constant (Petrov 1962).

The Killing vectors of (12.14) are

$$\partial_t, \quad \partial_z, \quad \partial_y, \quad \partial_x + y\partial_y + \tfrac{1}{2}(\sqrt{3}t - z)\partial_z - \tfrac{1}{2}(t + \sqrt{3}z)\partial_t, \quad (12.15)$$

and the group obeys (8.17) with spacelike $\boldsymbol{P}$. There are subgroups G_3 of Bianchi types I and VII_h acting in timelike hypersurfaces. The solution (12.14) is Petrov type I, and the eigenvalues of the Riemann tensor are the roots of $\lambda^3 = -k^3$. It is, after a complex coordinate transformation, a special case of (13.51) with $a_1 = -1$ and $(a_2, a_3, a_4) = (-1, \tfrac{1}{2}(1+\mathrm{i}\sqrt{3}), \tfrac{1}{2}(1+\mathrm{i}\sqrt{3}))$ (cp. Debever (1965)) and thus of a cylindrically symmetric vacuum metric (Bonnor 1979a; see §22.2).

12.3 Homogeneous non-null electromagnetic fields

Theorem 12.3 *The only Einstein–Maxwell field that is homogeneous and has a homogeneous non-null Maxwell field is*

$$\mathrm{d}s^2 = k^2(\mathrm{d}\vartheta^2 + \sin^2\vartheta\,\mathrm{d}\varphi^2 + \mathrm{d}x^2 - \sinh^2 x\,\mathrm{d}t^2). \quad (12.16)$$

Proof: (Kramer 1978) From the Rainich conditions (5.21), and the invariance of α, we can obtain

$$\begin{aligned}&(\overline{m}^c m_{a;c} - m^c \overline{m}_{a;c})k^a = (\overline{m}^c m_{a;c} - m^c \overline{m}_{a;c})l^a = 0,\\ &(k^c l_{a;c} - l^c k_{a;c})m^a = 0,\end{aligned} \quad (12.17)$$

by inserting (5.12) in (5.31). (Note that $\Phi_1\overline{\Phi}_1$ must be a constant for homogeneous fields.) Equations (12.17) show that there are two families of orthogonal 2-surfaces, and the space-time is of the form (12.8), the 2-surfaces having equal and opposite curvatures.

The group G_6 admitted by (12.16) contains no simply-transitive G_4, in agreement with Ozsváth's conclusion (1965a) that (12.12) includes all electromagnetic solutions with simply-transitive G_4 obeying $F_{ab} \neq 0$, $\mathcal{L}_{\boldsymbol{\xi}} F_{ab} = 0$, $\Lambda = 0$. The metric (12.16) does admit subgroups transitive on S_3, N_3 and T_3, and has therefore been rediscovered as a spherically-symmetric and spatially-homogeneous solution (Lovelock 1967, Dolan 1968). It was first given by Levi-Civita (1917b), but is more usually associated with the names of Bertotti (1959), who gave the more general case with $\Lambda \neq 0$, and Robinson (1959).

Some alternative forms of the line element are

$$\mathrm{d}s^2 = \frac{k^2}{r^2}(\mathrm{d}r^2 + r^2[\mathrm{d}\vartheta^2 + \sin^2\vartheta\,\mathrm{d}\varphi^2] - \mathrm{d}\tau^2), \tag{12.18}$$

$$\mathrm{d}s^2 = (1-\lambda y^2)\mathrm{d}x^2 + (1-\lambda y^2)^{-1}\mathrm{d}y^2 + (1+\lambda z^2)^{-1}\mathrm{d}z^2 - (1+\lambda z^2)\mathrm{d}t^2, \tag{12.19}$$

where $\lambda k^2 = 1$. In the metric form (12.19) the electromagnetic field is

$$\sqrt{\kappa_0}F_{12} = \sqrt{2\lambda}\sin\beta, \qquad \sqrt{\kappa_0}F_{43} = \sqrt{2\lambda}\cos\beta, \qquad \beta = \text{const.} \tag{12.20}$$

The solution is conformally flat; it is the only conformally flat non-null solution of the (sourcefree) Einstein–Maxwell equations (cp. Theorem 37.18). By taking the more general form (12.8) with unequal curvatures for the two 2-spaces one can obtain similar homogeneous Einstein–Maxwell solutions with a cosmological constant, see, e.g., Bertotti (1959), Cahen and Defrise (1968) and §35.2.

McLenaghan and Tariq (1975) and Tupper (1976) presented a homogeneous metric obeying the Einstein–Maxwell equations whose non-null Maxwell field does not share the space-time symmetry, showing that the condition that the Maxwell field shares the space-time symmetry is essential in Theorem 12.3 (cp. §11.1). It can be written as

$$\mathrm{d}s^2 = a^2x^{-2}(\mathrm{d}x^2+\mathrm{d}y^2) + x^2\mathrm{d}\varphi^2 - (\mathrm{d}t - 2y\,\mathrm{d}\varphi)^2, \qquad a = \text{const.} \tag{12.21}$$

The parameter a just gives an overall constant scale: the case $a^2 = 2$ had been given earlier by Tariq and Tupper (1975). This metric admits a simply-transitive group G_4 with Killing vectors ∂_t, ∂_φ, $2\varphi\,\partial_t + \partial_y$ and $\boldsymbol{\xi} \equiv x\partial_x + y\partial_y - \varphi\partial_\varphi$; there is no hypersurface orthogonal Killing vector. It is of Petrov type I and its Maxwell and Weyl tensors have no common null eigendirection (non-aligned case); it is characterized by the existence of a tetrad parallelly propagated along the two geodesic non-expanding null congruences of the non-null Maxwell field. In (12.21), the Maxwell field, for which $\mathcal{L}_{\boldsymbol{\xi}} F_{ab} \neq 0$, is

$$\sqrt{\kappa_0} F_{tx} = \sqrt{\kappa_0} F_{xy}/2y = \cos(2\log x)/x, \quad \sqrt{\kappa_0} F_{y\varphi} = -\sin(2\log x). \tag{12.22}$$

12.4 Homogeneous perfect fluid solutions

Pure Λ-term solutions, which could be regarded as fluids with $\mu+p=0$, are treated in the next section, so we assume here that $\mu+p\neq 0$. With this restriction, the following result was proved by Ozsváth (1965c) and Farnsworth and Kerr (1966), by the method outlined in §12.1. Their results are stated for dust, with a cosmological constant, but as all invariants are constants, only $\kappa_0\mu+\Lambda$ and $\kappa_0 p-\Lambda$ are fixed and the solutions can be reinterpreted as perfect fluid solutions with $\Lambda=0$ (see §5.2) or with some given relation $p=p(\mu)$ and a cosmological constant.

Theorem 12.4 *The homogeneous perfect fluid solutions are* (12.24) *and* (12.26)–(12.33) *below. The only such solution with a maximal G_5 is the Gödel solution* (12.26).

The resulting list of solutions is as follows (note that a perfect fluid solution cannot have a transitive maximal G_6, see §11.2).

(i) The only fluid solution with a G_7 is (12.9) with constant a. The field equations give

$$\kappa_0\mu+\Lambda = -3(\kappa_0 p-\Lambda) = 3\varepsilon/a^2. \tag{12.23}$$

To satisfy the dominant energy condition (5.19) we require $\varepsilon=1$ ($\Lambda\geq 0$ implies $\mu+3p\geq 0$ and $\Lambda\geq\varepsilon/a^2$ gives $p\geq 0$). This is Einstein's static universe (Einstein 1917). Some alternative metric forms are

$$\mathrm{d}s^2 = \left(1+\frac{r^2}{4a^2}\right)^{-2}\mathrm{d}x_\alpha\mathrm{d}x^\alpha - \mathrm{d}t^2, \quad r^2 = x_\alpha x^\alpha, \tag{12.24a}$$

$$\mathrm{d}s^2 = a^2[\mathrm{d}\chi^2+\sin^2\chi(\mathrm{d}\vartheta^2+\sin^2\vartheta\,\mathrm{d}\varphi^2)-\mathrm{d}t^2], \tag{12.24b}$$

$$\mathrm{d}s^2 = \frac{\mathrm{d}r^2}{(1-r^2/a^2)}+r^2(\mathrm{d}\vartheta^2+\sin^2\vartheta\,\mathrm{d}\varphi^2)-\mathrm{d}t^2. \tag{12.24c}$$

The G_7 includes a simply-transitive G_4, groups G_4 and G_3 transitive on both S_3 and T_3, and a G_6 on S_3 with generators

$$\boldsymbol{\xi}_\alpha = \left(1-\frac{r^2}{4a^2}\right)\partial_\alpha+\frac{1}{2a^2}x_\alpha(x^\beta\partial_\beta), \qquad \boldsymbol{\eta}^\alpha = \varepsilon^{\alpha\beta\gamma}x_\beta\partial_\gamma\,, \tag{12.25a}$$

in the coordinates of (12.24a) and

$$[\overset{\pm}{\boldsymbol{\eta}}_\alpha, \overset{\pm}{\boldsymbol{\eta}}_\beta] = \mp\frac{2}{K}\varepsilon_{\alpha\beta\gamma}\overset{\pm}{\boldsymbol{\eta}}^\gamma, \quad \overset{\pm}{\boldsymbol{\eta}}_\alpha \equiv \boldsymbol{\xi}_\alpha \pm \frac{1}{K}\boldsymbol{\eta}_\alpha, \tag{12.25b}$$

as the non-vanishing commutators. The G_6 is $SO(4) \cong SO(3) \times SO(3)$, arising from the embedding of the S_3 in R^4 (see §8.5). The seventh Killing vector is ∂_t. The fluid velocity is covariantly constant.

(ii) The solutions with a G_5 arise as special cases of those with a transitive G_4. In fact there is only one such solution, the Gödel (1949) solution

$$\mathrm{d}s^2 = a^2[\mathrm{d}x^2 + \mathrm{d}y^2 + \tfrac{1}{2}\mathrm{e}^{2x}\mathrm{d}z^2 - (\mathrm{d}t + \mathrm{e}^x\mathrm{d}z)^2]. \tag{12.26}$$

This solution has $\kappa_0\mu + \Lambda = \kappa_0 p - \Lambda = 1/2a^2$. Admitting an Abelian G_3 on T_3, it can be interpreted as a stationary cylindrically-symmetric solution (see Chapter 22). There are also groups of Bianchi type III on both S_3 and T_3, and type VIII on T_3.

(iii) Solutions whose G_4 has $A_E = 0$ (see §8.2). These all obey (8.18) and so are of case (i). If $\boldsymbol{L}$ is null we get only the Einstein static and Gödel solutions.

If $\boldsymbol{L}$ is timelike, one obtains the Farnsworth–Kerr class I solution

$$\mathrm{d}s^2 = a^2\Big[(1-k)(\boldsymbol{\omega}^1)^2 + (1+k)(\boldsymbol{\omega}^2)^2 + 2(\boldsymbol{\omega}^3)^2 - (\mathrm{d}t + \sqrt{1-2k^2}\boldsymbol{\omega}^3)^2\Big], \tag{12.27}$$

where the $\boldsymbol{\omega}^A$ are those for type IX in Table 8.2 and $2|k| < 1$. This is a rotating solution, extensively discussed by Ozsváth and Schücking (1969). $k = 0$ in (12.27) gives the Einstein static solution. (12.27) admits a G_3 of Bianchi type IX on S_3.

If $\boldsymbol{L}$ is spacelike, there are two metrics, Farnsworth–Kerr classes II and III, given by

$$\mathrm{d}s^2 = a^2\Big[(1-k)(\boldsymbol{\omega}^1)^2 + (1+k)(\boldsymbol{\omega}^2)^2 + (\mathrm{d}u + \sqrt{1-2k^2}\boldsymbol{\omega}^3)^2 - 2(\boldsymbol{\omega}^3)^2\Big], \tag{12.28}$$

$$\mathrm{d}s^2 = a^2[(1-s)(\boldsymbol{\omega}^1)^2 + (1+s)(\boldsymbol{\omega}^2)^2 + \mathrm{d}u^2 - 2(\boldsymbol{\omega}^3)^2], \tag{12.29}$$

with the $\boldsymbol{\omega}^A$ of Bianchi type VIII (Table 8.2); a, k and s are constants obeying $1 < 4k^2 \leq 2$ and $|s| < 1$. In (12.29), $\kappa_0 p - \Lambda = \kappa_0\mu + \Lambda = 1/2a^2$ and in (12.27) and (12.28) $\kappa_0(\mu + p) = 2(4k^2-1)$ $(\kappa_0 p - \Lambda) = |4k^2 - 1|/4a^2(1-k^2)$. The metrics (12.28) and

(12.29) were investigated in detail by Ozsváth (1970); see also Rosquist (1980). They both admit groups G_3 of Bianchi type *VIII* on T_3 and type *III* on S_3. The limit $|k| = 1/\sqrt{2}$ of (12.28) is included in (12.29), while the limit $|k| = 1/2$ is a pure radiation solution of Petrov type I with a Λ-term. (12.29) contains the Gödel solution (12.26) as the special case $s = 0$.

(iv) Solutions whose G_4 has $A_E \neq 0$. Following Ozsváth (1965b) we use a parameter s^2, $\frac{1}{2} \le s^2 \le 2$. If $\beta^2 \equiv 1 + 2s^2(1-s^2)(3-s^2)$, then we have

$$\beta^2 > 0 : \mathrm{d}s^2 = a^2[4(A\mathrm{e}^{Az}\mathrm{d}t + B\mathrm{e}^{Bz}\mathrm{d}x)^2/b^2 + (\mathrm{e}^{Fz}\mathrm{d}y)^2 + \mathrm{d}z^2 - (\mathrm{e}^{Az}\mathrm{d}t + \mathrm{e}^{Bz}\mathrm{d}x)^2], \tag{12.30}$$

$$\beta^2 = 0 : \mathrm{d}s^2 = a^2[\mathrm{e}^z\mathrm{d}x^2 + \mathrm{e}^{2Fz}\mathrm{d}y^2 + \mathrm{d}z^2 - \tfrac{1}{4}(b - 1/b)^2\mathrm{e}^z(\mathrm{d}t + z\,\mathrm{d}x)^2], \tag{12.31}$$

$$\beta^2 = -4k^2 < 0 : \mathrm{d}s^2 = a^2[(\mathrm{e}^{Fz}\mathrm{d}y)^2 + \mathrm{d}z^2 + \mathrm{e}^z((\cos kz - 2k\sin kz)\mathrm{d}t + (2k\cos kz + \sin kz)\mathrm{d}x)^2/b^2 - \mathrm{e}^z(\cos kz\,\mathrm{d}t + \sin kz\,\mathrm{d}x)^2], \tag{12.32}$$

where $A = \frac{1}{2}(1-\beta)$, $B = \frac{1}{2}(1+\beta)$, $F = 1-s^2$, $b = \sqrt{2}s(3-s^2)$, and a is constant (a correction found by Koutras (private communication) has been incorporated in (12.31)). These metrics all have

$$\kappa_0 p - \Lambda = (2-s^2)/2a^2; \quad \kappa_0(\mu + p) = (2s^2-1)(2-s^2)/a^2. \tag{12.33}$$

The special cases $s^2 = 2$, $s^2 = 1$, $s^2 = \frac{1}{2}$ give, respectively, the Petrov vacuum solution (12.14), the Gödel solution (12.26), and the homogeneous Petrov type N Einstein space (12.34). All these metrics have an Abelian G_3 on T_3 and (12.30), (12.31), have G_3 of Bianchi type VI_h on S_3; (12.32) has a G_3 of type VII_h on T_3, (12.31) has a group G_3 of type IV on T_3 and (12.30) a group of type VI on T_3.

Gödel's solution and (12.29) are of Petrov type D, but the other metrics above are in general of Petrov type I. In Gödel's solution the four-velocity of the fluid is a Killing vector but not hypersurface-orthogonal; the vorticity $\boldsymbol{\omega} = \partial_y$ is covariantly constant. Gödel's solution has interesting global properties and has been widely used to illustrate possible cosmological effects of rotation (Hawking and Ellis 1973, Ryan and Shepley 1975).

12.5 Other homogeneous solutions

All homogeneous solutions with a (non-zero) Λ-term have been found. Plane waves cannot have a Λ-term, so the type N solutions have a maximal G_5. The type D solutions are either of the form (12.8), in which case they must be composed of two 2-spaces of equal curvature, or have at most a G_5. By the usual arguments, all these spaces have simply-transitive G_4, and G_3 transitive on hypersurfaces. They were given by Kaigorodov (1962) and have been investigated and rediscovered by several other authors (e.g. Cahen (1964), Siklos (1981), Ozsváth (1987)).

Theorem 12.5 *The homogeneous Einstein spaces with* $\Lambda \neq 0$ *are:* (8.33) *with* $K \neq 0$; (12.8) *with* $3\Psi_2 = -\Lambda = k/A^2 = k'/B^2 \neq 0$ (Nariai 1950);

$$ds^2 = 3dz^2/|\Lambda| + \varepsilon e^z dx^2 + e^{-2z}(dy^2 + 2du\, dx) \tag{12.34}$$

where $\varepsilon = \pm 1$ *and* $\Lambda < 0$; *and*

$$ds^2 = 3dz^2/|\Lambda| + e^{4z}dx^2 + 4e^z dx\, dy + 2e^{-2z}(dy^2 + du\, dx) \tag{12.35}$$

with $\Lambda < 0$.

Solutions (8.33) are conformally flat and admit a G_{10}; (12.8) is of Petrov type D and admits a G_6 and its Einstein space specialization admits groups G_3III acting on T_3 and, if $\Lambda < 0$, on S_3; (12.34) is of Petrov type N with a maximal G_5 and groups G_3I, II and VI_h, $h = -4/9$, $-49/9$ and -16, on T_3 and $G_3VI_{-1/9}$ on S_3; and (12.35) is of Petrov type III with a maximal G_4 and G_3I and G_3VI_h, $h = -1/9$ or $-49/9$, acting on T_3 (for the groups, see MacCallum and Siklos (1992)). Thus there are no homogeneous Einstein spaces with $\Lambda \neq 0$ of Petrov types I or II.

Theorem 12.6 *The only homogeneous pure radiation solutions are of Petrov type* N *and are given by* (12.12) *with* $b \neq 0$, *and*

$$ds^2 = dx^2 + dy^2 + 2du\, dv - 2e^{2\rho x}du^2, \tag{12.36}$$

where ρ *is a constant* (Wils 1989a, Steele 1990).

The plane waves (12.12) have a G_6 or G_7; the cases with a G_7 have $A(u) = 0$, $B(u) = b/u^2$ or $B = b$, where b is a constant, and can be transformed to the form (Petrov 1966)

$$ds^2 = C^2(u)(dx^2 + dy^2) - 2\, du\, dv, \qquad \ddot{C} + 2BC = 0, \tag{12.37}$$

cp. §24.5 and the metrics (15.18), (24.51). The metric (12.36) was given by Sippel and Goenner (1986), has a maximal G_5, and does not admit

Table 12.1. Homogeneous solutions

Source	Maximal group			
	G_4	G_5	G_6	G_7
Vacuum	A Petrov (12.14)	$\nexists$	A Plane waves (12.12), $b=0$	$\nexists$
E-M non-null	$\nexists$	$\nexists$	A Bertotti–Robinson (12.16)	$\nexists$
E-M null	$\nexists$	$\nexists$	A Plane waves (12.12), $b \neq 0$	$\nexists$
Perfect fluid	A Ozsváth (12.27)–(12.32)	A Gödel (12.26)	$\nexists$	A Einstein (12.24)
Λ-term	A (12.35)	A (12.34)	A (12.8), special case	$\nexists$
Pure radiation	$\nexists$	A (12.36)	A Plane waves (12.12), $b \neq 0$	A Plane waves (12.37)

a Maxwell field. Thus (12.12) are the only homogeneous geometries with non-inheriting null Maxwell fields; in general they also have inheriting null Maxwell fields, but (12.37) does not.

Other energy-momentum tensors have been considered. For example, Ozsváth (1965a) found a non-null Maxwell field with $\Lambda \neq 0$ and the unique null Maxwell field with $\Lambda \neq 0$. The latter can also be considered as the member $k = 2$ of the family (Siklos 1985)

$$ds^2 = \frac{3}{|\Lambda| y^2}\left(dy^2 + dz^2 - du\,dv - \varepsilon y^{2k} dv^2\right) \tag{12.38}$$

of pure radiation solutions of Petrov type N with $\Lambda < 0$, where $\varepsilon = \pm 1$ (and $k(2k-3)\varepsilon > 0$ for positive energy) admitting a G_5 and containing (12.6) and (12.34) as the special cases $k = -1$ and $k = 3/2$; the family (12.38) are conformal to (non-flat) *pp*-waves (§24.5). Homogeneous pure radiation solutions with $\Lambda \neq 0$ and a maximal G_r, $r \leq 5$, and homogeneous solutions with non-inheriting non-null Maxwell fields, have not been exhaustively treated.

12.6 Summary

The results in this chapter are summarized in Table 12.1. There it is assumed that $\mathcal{L}_\xi F_{ab} = 0$ for electromagnetic solutions (the known solutions given above which do not satisfy this condition, other than the reinterpretation of (12.12), are the metrics (12.21) with Maxwell field (12.22), and (12.37)). The spaces of constant curvature, with a G_{10}, are omitted from the table. The solutions with a maximal G_7 shown are the only possible ones. The symbol A means that all solutions are known; $\nexists$ indicates non-existence.

13
Hypersurface-homogeneous space-times

13.1 The possible metrics

This chapter is concerned with metrics admitting a group of motions transitive on S_3 or T_3. Some solutions, such as the well-known Taub–NUT (Newman, Unti, Tamburino) metrics (13.49), cover regions of both types, joined across a null hypersurface which is a special group orbit (metrics admitting a G_r whose general orbits are N_3 are considered in Chapter 24). As in the case of the homogeneous space-times (Chapter 12) we first consider the cases with multiply-transitive groups. From Theorems 8.10 and 8.17 we see that only G_6 and G_4 are possible.

13.1.1 Metrics with a G_6 on V_3

From §12.1, the space-times with a G_6 on S_3 have the metric (12.9); this always admits G_3 transitive on hypersurfaces $t =$const and the various cases are thus included in (13.1)–(13.3) and (13.20) below. The relevant G_3 types are V and VII_h if $k = -1$, I and VII_0 if $k = 0$, and IX if $k = 1$.

Of the energy-momentum tensors considered in this book, the space-times with a G_6 on T_3 permit only vacuum and Λ-term Ricci tensors (see Chapter 5). Thus they will give only the spaces of constant curvature, with a complete G_{10}, which also arise with G_6 on S_3 and those energy-momentum types. Metrics with maximal G_6 on S_3 are non-empty and have an energy-momentum of perfect fluid type: see §14.2.

13.1.2 Metrics with a G_4 on V_3

The metric forms with a G_4 on S_3 or T_3 are easily determined by using Schmidt's method (§8.6) to find the possible G_4, followed by use of Theorems 8.16 and 8.19 to determine the complete metric (MacCallum

1980). The (one-dimensional) isotropy subgroup of the G_4 may consist of spatial rotations, boosts or null rotations, and we refer to these cases as spatial rotation isotropy, also known as local rotational symmetry (LRS), boost isotropy and null rotation isotropy respectively: the isotropy has the same nature everywhere in an orbit and thus throughout a neighbourhood. From §4.2 the metrics are of Petrov type D (or conformally flat) for the LRS and boost isotropic cases, and type N or conformally flat for null rotation isotropy, and they are all among the metrics found by Cahen and Defrise (1968) and Defrise (1969) except for the most general form of one of the Petrov type N metrics. We list them below: an almost complete list was given by Petrov (1966).

Spatial rotation isotropy

We give first the LRS metrics with a G_4 (Ellis 1967, Stewart and Ellis 1968). The possible cases are, with $\epsilon = \pm 1$, $k = \pm 1$ or 0 and the Σ of (8.37),

$$\mathrm{d}s^2 = \epsilon[-\mathrm{d}t^2 + A^2(t)\mathrm{d}x^2] + B^2(t)[\mathrm{d}y^2 + \Sigma^2(y,k)\mathrm{d}z^2], \tag{13.1}$$

$$\mathrm{d}s^2 = \epsilon[-\mathrm{d}t^2 + A^2(t)(\boldsymbol{\sigma}^1)^2] + B^2(t)[\mathrm{d}y^2 + \Sigma^2(y,k)\mathrm{d}z^2], \tag{13.2}$$

$$\mathrm{d}s^2 = \epsilon[-\mathrm{d}t^2 + A^2(t)\mathrm{d}x^2] + B^2(t)\mathrm{e}^{2x}(\mathrm{d}y^2 + \mathrm{d}z^2), \tag{13.3}$$

where in (13.2) we have

$$\boldsymbol{\sigma}^1 = \mathrm{d}x + \cos y\,\mathrm{d}z \qquad \text{if } k = 1, \tag{13.4}$$

$$\boldsymbol{\sigma}^1 = \mathrm{d}x + \tfrac{1}{2}y^2\mathrm{d}z \qquad \text{if } k = 0, \tag{13.5}$$

$$\boldsymbol{\sigma}^1 = \mathrm{d}x + \cosh y\,\mathrm{d}z \quad \text{if } k = -1. \tag{13.6}$$

The metric (13.2) with $\epsilon = -1$ is in Ellis class I (as defined in §11.2); (13.1) and (13.3) are in class II; and (13.2) with $\epsilon = 1$ is in class III.

Except for (13.1) with $k = 1$, these metrics can all be written as

$$\mathrm{d}s^2 = \epsilon[-\mathrm{d}t^2 + A^2(t)(\boldsymbol{\omega}^1)^2] + B^2(t)[(\boldsymbol{\omega}^2)^2 + (\boldsymbol{\omega}^3)^2], \tag{13.7}$$

where the $\boldsymbol{\omega}^\alpha$ are dual to a basis of reciprocal group generators of a G_3. The possible G_3 types are: for (13.1), I or VII_0 if $k = 0$, and III if $k = -1$; for (13.2), IX if $k = 1$, II if $k = 0$, and VIII or III if $k = -1$; and for (13.3), V or VII_h. If the $\boldsymbol{\omega}^\alpha$ of Table 8.2 are substituted in (13.7), the results in general differ from (13.1)–(13.6) only by coordinate transformations; however, for the G_3III one must, before making the substitution, transform to a new basis $\boldsymbol{\omega}'^\alpha$, e.g. for (13.1), $k = -1$, one requires

$$\boldsymbol{\omega}'^2 = \boldsymbol{\omega}^1, \qquad \boldsymbol{\omega}'^3 = \boldsymbol{\omega}^2 - \boldsymbol{\omega}^3, \qquad \boldsymbol{\omega}'^1 = \boldsymbol{\omega}^2 + \boldsymbol{\omega}^3. \tag{13.8}$$

The exceptional case, (13.1) with $k = 1$, admits no simply-transitive G_3, and, with $\epsilon = 1$, gives the only spatially-homogeneous solutions with this property (Kantowski and Sachs 1966, Kantowski 1966; cp. Collins 1977a); with $\epsilon = -1$ it gives the spherically-symmetric static metrics (for which see Chapters 15 and 16). The metrics (13.1)–(13.3) can also be derived by considering the extensions of Lie algebras of groups G_3 on S_2 or S_3, see Kantowski (1966), Shikin (1972).

The metrics (13.1) and (13.2) can jointly be written as

$$\mathrm{d}s^2 = Y^2(w)\frac{2\mathrm{d}\zeta\mathrm{d}\bar{\zeta}}{[1+\frac{1}{2}k\zeta\bar{\zeta}]^2} + \frac{\mathrm{d}w^2}{f(w)} - f(w)\left[\mathrm{d}t + \mathrm{i}l\frac{\zeta\mathrm{d}\bar{\zeta} - \bar{\zeta}\mathrm{d}\zeta}{1+\frac{1}{2}k\zeta\bar{\zeta}}\right]^2, \tag{13.9}$$

where $l \neq 0$ corresponds to (13.2), and $f(w)$ may have either sign, $\epsilon \equiv -f(w)/|f(w)|$. The Killing vectors of (13.9) are

$$\begin{aligned} \boldsymbol{\xi}_1 &= \mathrm{i}[1 - \tfrac{1}{2}k\zeta^2]\partial_\zeta - \mathrm{i}[1 - \tfrac{1}{2}k\bar{\zeta}^2]\partial_{\bar{\zeta}} + l(\zeta + \bar{\zeta})\partial_t, \quad \boldsymbol{\xi}_3 = \mathrm{i}(\zeta\partial_\zeta - \bar{\zeta}\partial_{\bar{\zeta}}), \\ \boldsymbol{\xi}_2 &= [1 + \tfrac{1}{2}k\zeta^2]\partial_\zeta + [1 + \tfrac{1}{2}k\bar{\zeta}^2]\partial_{\bar{\zeta}} + \mathrm{i}l(\zeta - \bar{\zeta})\partial_t, \quad \boldsymbol{\xi}_4 = \partial_t. \end{aligned} \tag{13.10}$$

A fifth Killing vector, making the space-time homogeneous, cannot occur for the energy-momentum tensors we consider; for such metrics see Ozsváth (1966). In the basis

$$\begin{aligned} &\boldsymbol{\omega}^1 = Y\frac{\mathrm{d}\zeta}{[1+\frac{1}{2}k\zeta\bar{\zeta}]}, \qquad \boldsymbol{\omega}^4 = X\left[\mathrm{d}t + \mathrm{i}l\frac{\zeta\mathrm{d}\bar{\zeta} - \bar{\zeta}\mathrm{d}\zeta}{1+\frac{1}{2}k\zeta\bar{\zeta}}\right], \\ &\boldsymbol{\omega}^2 = \bar{\boldsymbol{\omega}}^1, \qquad \boldsymbol{\omega}^3 = \mathrm{d}w/X, \qquad f = -\epsilon X^2, \quad \epsilon = \pm 1, \end{aligned} \tag{13.11}$$

the Ricci tensor of (13.9) has, as its only non-zero tetrad components,

$$\begin{aligned} R_{12} &= \frac{k}{Y^2} + \frac{2l^2 f}{Y^4} - \frac{f'Y'}{Y} - \frac{fY'^2}{Y^2} - \frac{fY''}{Y}, \\ R_{33} &= \epsilon\left(\frac{f'Y'}{Y} + \frac{2fY''}{Y} + \frac{f''}{2}\right), \qquad R_{14} = -\epsilon\left(\frac{f'Y'}{Y} + \frac{2l^2 f}{Y^4} + \frac{f''}{2}\right), \end{aligned} \tag{13.12}$$

and is of type [(11)1, 1] or its specializations. The Ricci tensor for (13.3), in the basis $(A\boldsymbol{\omega}^1, B\boldsymbol{\omega}^2, B\boldsymbol{\omega}^3, \mathrm{d}t)$, has non-zero tetrad components,

$$\begin{aligned} &R_{44} = -\frac{\ddot{A}}{A} - \frac{2\ddot{B}}{B}, \quad R_{14} = \frac{2}{A}\left(\frac{\dot{A}}{A} - \frac{\dot{B}}{B}\right), \quad R_{11} = \frac{\ddot{A}}{A} + \frac{2\dot{A}\dot{B}}{AB} - \frac{2}{A^2}, \\ &R_{22} = R_{33} = \epsilon\left(\frac{\ddot{B}}{B} + \frac{\dot{B}^2}{B^2} + \frac{\dot{A}\dot{B}}{AB} - \frac{2}{A^2}\right), \end{aligned} \tag{13.13}$$

whose Segre type is [(11)1,1] or [(11), 2], or their specializations. No G_5 on V_4 is possible in (13.3), but G_6 on V_3 occurs if $R_{14} = 0$, so for $\epsilon = 1$ we then recover (12.9).

Boost isotropy

The metrics with a G_4 on T_3 and boost isotropy are

$$ds^2 = dw^2 + A^2(w)dx^2 + B^2(w)[dy^2 - \Sigma^2(y,k)dt^2], \tag{13.14}$$

$$ds^2 = dw^2 + A^2(w)(\boldsymbol{\omega}^1)^2 - 2B^2(w)\boldsymbol{\omega}^2\boldsymbol{\omega}^3, \tag{13.15}$$

where (13.15) covers four cases given by taking the $\boldsymbol{\omega}^\alpha$ of Table 8.2 for G_3V or G_3II, or $\boldsymbol{\omega}'^\alpha$ related to those given for G_3VIII by

$$\boldsymbol{\omega}'^1 = \boldsymbol{\omega}^1, \qquad \boldsymbol{\omega}'^2 = \boldsymbol{\omega}^2 + \boldsymbol{\omega}^3, \qquad \boldsymbol{\omega}'^3 = \pm(\boldsymbol{\omega}^2 - \boldsymbol{\omega}^3). \tag{13.16}$$

The possible simply-transitive G_3 are then: for (13.14), G_3III if $|k| = 1$ and G_3I or G_3VI_0 if $k = 0$; for (13.15) with (13.16), G_3VIII and G_3III; (13.15) with $\boldsymbol{\omega}^\alpha$ of G_3II, G_3II; and (13.15) with $\boldsymbol{\omega}^\alpha$ of G_3V, G_3V or G_3VI_h or G_3III. With the exception of the last case (which admits a normal null Killing vector, see §24.4), these metrics can be combined as

$$ds^2 = Y^2(w)\frac{2\,du\,dv}{[1+\frac{1}{2}kuv]^2} + \frac{dw^2}{f(w)} + f(w)\left(dx + l\frac{u\,dv - v\,du}{1+\frac{1}{2}kuv}\right)^2 \tag{13.17}$$

with $f(w) > 0$. Equation (13.17) can be derived from (13.9) with $f(w) > 0$ by the complex substitution

$$\zeta \to u, \qquad \bar{\zeta} \to v, \qquad t \to ix, \tag{13.18}$$

which, when applied to (13.10) and (13.12), also yields the Killing vectors and Ricci tensor; the latter must be of type [1 1(1, 1)] or its specializations.

Null rotation isotropy

The (Petrov type N) metrics with a G_4 on T_3 and a null rotation isotropy are either special cases of the metrics with a G_3 on N_2 including a null Killing vector (for which see Petrov (1966), Defrise (1969), Barnes (1979) and §24.4) or

$$ds^2 = dw^2 + A^2(w)[dy^2 - 2e^y dv(du + B(w)e^y dv)], \tag{13.19}$$

which also has a null Killing vector, ∂_u. They all have a Ricci tensor of type [1 (1,2)] or a specialization thereof, so only vacuum, null Einstein–Maxwell and pure radiation energy-momentum tensors are possible.

Summarizing the results on the multiply-transitive groups G_4 we have:

Theorem 13.1 *Apart from the cases with null (sub)orbits, all metrics with a* G_4 *on* V_3 *are covered by the metrics* (13.3), (13.9) *and* (13.17).

The metrics with null (sub)orbits are discussed in Chapter 24.

13.1.3 Metrics with a G_3 on V_3

Finally, there are the space-times with a maximal G_3 on S_3 or T_3, with metrics given by the $\boldsymbol{\omega}^\alpha$ of Table 8.2 and

$$ds^2 = -dt^2 + g_{\alpha\beta}(t)\boldsymbol{\omega}^\alpha\boldsymbol{\omega}^\beta, \qquad \det(g_{\alpha\beta}) > 0, \tag{13.20}$$

$$ds^2 = dt^2 + g_{\alpha\beta}(t)\boldsymbol{\omega}^\alpha\boldsymbol{\omega}^\beta, \qquad \det(g_{\alpha\beta}) < 0. \tag{13.21}$$

From the detailed results above we have:

Theorem 13.2 *All metrics admitting a* G_r, $r \geq 3$, *acting on* S_3 *or* T_3, *are included in* (13.20), (13.21), *and* (13.1) *with* $k = 1$.

The spatially-homogeneous cosmologies, i.e. solutions with a perfect fluid matter content and a group transitive on S_3, are discussed in Chapter 14. A number of the other metric forms just given recur elsewhere in the book, and the solutions of their field equations are accordingly also not discussed in full here. The metrics concerned are as follows. Metrics (13.1) and (13.3) admit a G_3 on S_2, while their counterparts contained in (13.14) and (13.15) admit G_3 on T_2; these metrics are therefore treated in Chapter 15. In particular, metrics (13.1) with $k = 1$ or 0 and $\epsilon = -1$ are spherically and plane symmetric static metrics, and are treated in both Chapters 15 and 16. All solutions with a group of motions transitive on T_3 are stationary or static. Metrics of the form

$$ds^2 = dw^2 + A^2(w)dx^2 + B^2(w)dy^2 - C^2(w)dt^2 \tag{13.22}$$

with a G_3I on T_3 are often interpreted as cylindrically-symmetric static or stationary metrics (assuming a combination of x and y to be an angular coordinate) and these are treated in Chapter 22; this includes some metrics with plane symmetry. Metrics (13.14) with $k = 0$ and (13.15) with the $\boldsymbol{\omega}^\alpha$ of G_3II and G_3V admit groups on null orbits and are covered in Chapter 24, as are all the metrics with a null rotation isotropy. Some of the cases of (13.21) also admit groups on null orbits; for example, nearly all Kellner's metrics with a G_3I on T_3 contain a normal null Killing vector (Kellner 1975), see e.g. (20.32), the $BIII$ metric of Table 18.2, and (33.30). The tables at the end of this chapter summarize the solutions given in this and the next chapter, and give references to occurrences of solutions with G_3 on V_3 elsewhere in this book.

It should be noted that hypersurface-homogeneous perfect fluids are always locally barotropic, provided $\mu(x)$ and $p(x)$ have non-zero x derivatives, where x labels the homogeneous hypersurfaces.

13.2 Formulations of the field equations

The Einstein equations for hypersurface-homogeneous space-times reduce to a system of ordinary differential equations. At least for the spatially-homogeneous case, they form a well-posed Cauchy problem (Taub 1951) and, although they have not been completely integrated, their qualitative properties have been discussed in many papers: see Wainwright and Ellis (1997) for an extensive survey of the results, and e.g. Ryan and Shepley (1975), MacCallum (1973, 1979a), Bogoyavlenskii (1980), and Rosquist and Jantzen (1988) for useful earlier reviews. Methods from dynamical systems theory which proved fruitful in elucidating these properties have also led to ways of restricting the general case to more readily solvable subcases and thence to new exact solutions (see e.g. Uggla *et al.* (1995b) for a summary). Nearly all of these methods were developed initially for use in the spatially-homogeneous case ('cosmologies', for brevity), and we shall therefore describe the methods in this context although they can be adapted to the G_3 on T_3 and H_3 on V_3 cases also (e.g. for an orthonormal tetrad method for G_3 on T_3 see Harness (1982)).

The number of degrees of freedom, i.e. the number of essential arbitrary constants required in a general cosmology for each Bianchi type, has been studied by Siklos (1976a) (cp. MacCallum (1979b), Wainwright and Ellis (1997)). Table 13.1 summarizes the results for vacua and perfect fluids. The fluid here may be 'tilted', i.e. the velocity $\boldsymbol{u}$ need not coincide with the normal $\boldsymbol{n}$ to the hypersurfaces of homogeneity. In general such fluid cosmologies can have four more parameters than the corresponding vacua but Bianchi types I, where $\boldsymbol{u} = \boldsymbol{n}$, II and $VI_{-1/9}$ are special cases in which the constraints on the Cauchy problem arising from the $G^4{}_\alpha$ field equations are not linearly independent. In the last case, type $VI_{-1/9}$, the effect is an extra degree of freedom in the vacuum solutions, and solutions where this is activated are denoted type $VI^*_{-1/9}$.

The residual set of ordinary differential equations to be solved can be formulated in various ways (see e.g. MacCallum (1973), Wainwright and Ellis (1997)). One can use a time-independent basis as in (13.20) and parametrize the components $g_{\alpha\beta}$ in some suitable way. This is called the *metric approach.* One may then choose spatial coordinates and a new time coordinate τ so that the dt of (13.20) is replaced by $\boldsymbol{\sigma} = N(\tau)\mathrm{d}\tau + N_\alpha(\tau)\boldsymbol{\omega}^\alpha$, N being the *lapse* and N_α the *shift.* Such a change of basis could be interpreted as introducing rotation, if $\boldsymbol{\sigma} \wedge \mathrm{d}\boldsymbol{\sigma} \neq 0$, but this

Table 13.1. The number of essential parameters, by Bianchi type, in general solutions for vacuum and for perfect fluids with given equation of state

The number for a non-tilted fluid, including Λ-term, is one more than for vacuum. Type *III* is included in VI_h. h itself is regarded as fixed in a solution, i.e. is not counted as a parameter. The number is reduced if there is extra symmetry or a condition such as (13.25) is imposed.

Energy-momentum	Bianchi type						
	I	II	VI_0 & VII_0	VIII & IX	V	IV, VII_h & $VI_{h\neq -1/9}$	$VI_{-1/9}$
Vacuum	1	2	3	4	1	3	4
Perfect fluid	2	5	7	8	5	7	7

has no physical meaning unless the direction $\boldsymbol{\sigma}$ is invariantly defined in some way (see Jantzen (1986), who notes that one may also meaningfully speak of rotation when eigenvectors of the shear of dt rotate relative to a Fermi–Walker propagated frame).

In the metric approach, a first step in simplifying the equations is to factorize $g_{\alpha\beta}$ as $A^{\gamma}{}_{\alpha}(t)A^{\phi}{}_{\beta}(t)\hat{g}_{\gamma\phi}(t)$, where the linear transformation $\widetilde{\boldsymbol{\omega}}^{\gamma} = A^{\gamma}{}_{\alpha}\boldsymbol{\omega}^{\alpha}$ is an automorphism of the Lie algebra, i.e. preserves the forms of the commutators given in §8.2. This idea was introduced by Collins and Hawking (1973) and developed by several authors (see e.g. Siklos (1980), Jantzen (1984), Rosquist and Jantzen (1988)). It clarifies what the true degrees of freedom are, and reduces the system of equations. In general the $A^{\gamma}{}_{\alpha}$ are chosen to make $\hat{g}_{\gamma\phi}(t)$ diagonal, where possible, and in some cases they can be used to reduce its independent components even further. The time-dependent variables are then the remaining diagonal components of $\hat{g}_{\gamma\phi}(t)$ and a suitable parametrization of the possible matrices $A^{\gamma}{}_{\alpha}$ for the particular Bianchi type (e.g. as in Harvey (1979) or Roque and Ellis (1985)).

To parametrize the remaining metric components this approach can be coupled with a special case of *Misner's* (1968) *parametrization*, which we write in the form

$$S^6 \equiv \mathrm{e}^{6\lambda} \equiv \det(g_{\alpha\beta}), \qquad g_{\alpha\gamma} = S^2(\exp 2\boldsymbol{\beta})_{\alpha\gamma}, \tag{13.23}$$

where $\boldsymbol{\beta}$ is a symmetric tracefree matrix function of t. If $g_{\alpha\beta}$ is diagonal, one may write

$$\beta_{\alpha\gamma} = \mathrm{diag}\left(\beta_1, -\tfrac{1}{2}(\beta_1 - \sqrt{3}\beta_2), -\tfrac{1}{2}(\beta_1 + \sqrt{3}\beta_2)\right). \tag{13.24}$$

One may now take λ or S to be a new time variable.

The metric formalism is simplified if the automorphism variables can be taken to be constants, i.e. the $g_{\alpha\beta}$ in (13.20) can be taken to be a diagonal matrix. MacCallum *et al.* (1970) showed that if $\boldsymbol{t} = \partial_t$ is a Ricci eigenvector, then in Class G_3A, except for types G_3I and G_3II, $g_{\alpha\beta}$ can be taken to be diagonal, and in Class G_3B, except for type $VI^*_{-1/9}$, the vector a^α in (8.50) is an eigenvector of $t^\alpha{}_{;\beta}$ and of the Ricci tensor. The excluded cases require additional assumptions to reach these conclusions, because of the linear dependence of the $G^4{}_\alpha$ mentioned above. For realistic matter content, the G_3B metrics can only be diagonal if either they are LRS, or the $n_{\alpha\beta}$ of (8.50) obey

$$n^\alpha{}_\alpha = 0 \tag{13.25}$$

(MacCallum 1972). The restriction imposed by (13.25) can apply only in Bianchi types *I*, *III*, *V*, *VI*, and *VIII*, and it turns out to give useful new restrictions only in types *III* and *VI* (Ellis and MacCallum 1969). With (13.25), it can be convenient to alter the canonical form of the group generators to read

$$[\boldsymbol{\xi}_2, \boldsymbol{\xi}_3] = 0, \qquad [\boldsymbol{\xi}_3, \boldsymbol{\xi}_1] = (1 - A)\boldsymbol{\xi}_3, \qquad [\boldsymbol{\xi}_1, \boldsymbol{\xi}_2] = (1 + A)\boldsymbol{\xi}_2, \tag{13.26}$$

where $h = -A^2$. The $\boldsymbol{\omega}^\alpha$ analogous to those of Table 8.2 are

$$\boldsymbol{\omega}^1 = \mathrm{d}x, \qquad \boldsymbol{\omega}^2 = \mathrm{e}^{(A+1)x}\mathrm{d}y, \qquad \boldsymbol{\omega}^3 = \mathrm{e}^{(A-1)x}\mathrm{d}z. \tag{13.27}$$

The main alternative to the metric approach is the *orthonormal tetrad approach* using a tetrad basis (3.7) chosen as in (8.49). The two are closely related when variables are chosen as just described above (see Jantzen and Uggla (1998)), and in both approaches a suitable choice of lapse (or, to include the G_3 on T_3 case, 'slicing gauge'), or directly of independent variable, may decouple and simplify the equations; for details see e.g. Jantzen (1988), Uggla *et al.* (1995b). A power-law lapse, i.e. a product of powers of the dependent variables, an idea introduced in Bonanos (1971), may be useful, as may an 'intrinsic slicing' (e.g. making a product of invariantly-defined metric components the independent variable).

A *Lagrangian* or *Hamiltonian formulation* (using only functions of t) proved to be possible only for Class G_3A in general (MacCallum and Taub 1972, Sneddon 1976); it can also be achieved for the metrics with $n^\alpha{}_\alpha = 0$. Distinct physical problems can give rise to equivalent Hamiltonians (Uggla *et al.* 1995a). To understand the dynamics it may be useful to work with the Jacobi form in which the kinetic part contains all the dependence on variables (Uggla *et al.* 1990). The form of the Hamiltonian for the remaining time-dependent variables in a metric approach may enable one to determine a good choice of lapse. This happens in particular when the potential space described by the Hamiltonian admits a Killing tensor (see §35.3) or Killing vectors (see Rosquist and Uggla (1991), Uggla *et al.* (1995b) and §10.8). Indeed Uggla *et al.* (1995b) say 'We are not aware of any solvable case in the literature on hypersurface-homogeneous models which cannot be explained by the existence of rank two Killing tensor and Killing vector symmetries'.

'Regularizing' the equations, which further reduces, and usually compactifies, the phase space of the system and assists in understanding the trajectories within that space, can be done in both the metric and orthonormal tetrad formalisms by introducing normalized variables, in which an overall time-dependent scaling factor is removed (see e.g. Bogoyavlenskii (1980), Rosquist and Uggla (1991), Wainwright and Ellis (1997)). The variational formalism led naturally to choices which were bounded and compactified the phase space using a single coordinate patch (see e.g. Rosquist and Jantzen (1988) and references therein). Then applying methods from dynamical systems, in the simplest cases the *phase plane methods* first used by Shikin (1967) and Collins (1971), provided an approach leading to a number of new solutions. Within a phase plane, when the equations can be written, using an overdot to denote a time derivative, in the form

$$\dot{x} = P(x, y), \qquad \dot{y} = Q(x, y), \tag{13.28}$$

with polynomial right-hand sides P and Q, one may seek 'Darboux polynomials' $f(x, y)$ obeying $\boldsymbol{D}f = gf$, where $\boldsymbol{D} = P\partial_x + Q\partial_y$ and $g(x, y)$ is a polynomial (for the literature on this general method see Man and MacCallum (1997)): $f = 0$ then gives a solution (Hewitt 1991a).

These methods have mainly been applied to the spatially-homogeneous case and to perfect fluids with the γ-law equation of state, $p = (\gamma - 1)\mu$, including vacuum and Λ-term solutions as special cases, and many of the new solutions of the last two decades have been discovered using them. The results show that the known exact solutions of this type arise either as fixed points in the reduced phase space or as orbits joining such points. The former type have homothety groups H_r, $r \geq 4$ and components $g_{\alpha\beta}$ which are powers of t, and are listed in Wainwright and Ellis (1997). The metrics can be expressed relative to orbits of those H_3 other than the G_3 which are present (Jantzen and Rosquist 1986). The methods could be and to some extent have been applied to solutions with G_3 on T_3 and to other energy-momentum contents but most known solutions with a Maxwell field have been obtained by other means.

An especially simple situation arises in Bianchi type I metrics, and in those G_3V metrics where ∂_t is a Ricci eigenvector, if the Ricci tensor is of the perfect fluid type [(111), 1], because the field equations then yield

$$3\dot{S}^2 = \Sigma^2 S^{-4} + \kappa_0\mu S^2 + \Lambda S^2 - 3k, \tag{13.29a}$$

$$S\dot{\mu} + 3\dot{S}(\mu + p) = 0, \quad S^3\dot{\beta}_{\alpha\gamma} \equiv \Sigma_{\alpha\gamma}, \quad \dot{\Sigma}_{\alpha\gamma} = 0, \tag{13.29b}$$

using (13.23), where $2\Sigma^2 \equiv \Sigma_{\alpha\beta}\Sigma^{\alpha\beta}$ and $k = 0$ for G_3I, $k = -1$ for G_3V. One may now take

$$\beta_{\alpha\gamma} = \left(\int \frac{2\Sigma \mathrm{d}t}{\sqrt{3}S^3}\right) \mathrm{diag}\left(\cos\psi,\ \cos(\psi + \tfrac{2}{3}\pi),\ \cos(\psi + \tfrac{4}{3}\pi)\right). \tag{13.30}$$

In G_3V the remaining field equation implies $\psi = \pi/2$. The metrics can then be written (cp. Heckmann and Schücking (1958)) as

$$\mathrm{d}s^2 = -\mathrm{d}t^2 + S(t)^2\left(F^{2\cos\psi}\mathrm{d}x^2 + F^{2\cos(\psi+2\pi/3)}\mathrm{d}y^2 + F^{2\cos(\psi-2\pi/3)}\mathrm{d}z^2\right), \tag{13.31}$$

$$\mathrm{d}s^2 = -\mathrm{d}t^2 + S(t)^2\left(\mathrm{d}x^2 + e^{2x}\mathrm{d}y^2/F^{\sqrt{3}} + F^{\sqrt{3}}e^{2x}\mathrm{d}z^2\right), \tag{13.32}$$

for Bianchi types I and V respectively, where

$$F = \exp\left(2\Sigma \int \mathrm{d}t/\sqrt{3}S^3\right). \tag{13.33}$$

In G_3I we may take $\Sigma = \sqrt{3}$. If we can solve the first of (13.29b) for $\mu(S)$, then, by taking S as the time variable and replacing $\mathrm{d}t^2$ in the metric by $\mathrm{d}S^2/\dot{S}^2$ with $\dot{S}^2$ taken from (13.29a), we have found an exact solution up to the quadrature (13.33). Similarly, Fiser *et al.* (1992) used an 'intrinsic time' variable S^δ but in the general case took $\delta = 1$. However, for physical applications it may be important to know S as a function of t.

Another case of this type arises in the metric (13.1), $\epsilon = 1$, where, for a Ricci tensor of type [(111), 1], the field equations are

$$\frac{2\ddot{B}}{B} + \frac{\dot{B}^2}{B^2} + \frac{k}{B^2} = \frac{\ddot{B}}{B} + \frac{\ddot{A}}{A} + \frac{\dot{A}}{A}\frac{\dot{B}}{B} = \Lambda - \kappa_0 p, \tag{13.34a}$$

$$\frac{2\dot{A}\dot{B}}{AB} + \frac{\dot{B}^2}{B^2} + \frac{k}{B^2} = \Lambda + \kappa_0\mu. \tag{13.34b}$$

These equations reduce to a series of quadratures if one starts with a choice of an arbitrary function $A(t)$ and finds successively $B(t)$, $p(t)$ and $\mu(t)$; there are special cases of embedding class one which are included in (37.50) and (34.59). The equations also give a series of quadratures if one imposes the γ-law equation of state (with constant γ); for $k = 0$, $\gamma \neq 2$, one obtains (Stewart and Ellis 1968)

$$t = \int (B^{3(2-\gamma)/2} + c)^{(\gamma-1)/(2-\gamma)} B^{1/2} \mathrm{d}B,$$
$$A = (B^{3(2-\gamma)/2} + c)^{1/(2-\gamma)} B^{-1/2}, \quad \kappa_0\mu = \kappa_0 p/(\gamma - 1) = 3/(AB^2)^{\gamma}, \tag{13.35}$$

where c is a constant; $c = 0$ is the conformally flat Robertson–Walker subcase. For cases where the integration can be done explicitly see §14.3.

A quite different approach leading to solutions of the Einstein–Maxwell equations that are homogeneous on S_3 or T_3 arises from an ansatz introduced by Tariq and Tupper (1975) and generalized by Barnes (1978). The latter ansatz is that the principal null tetrad of the Maxwell bivector, $(\boldsymbol{m}, \overline{\boldsymbol{m}}, \boldsymbol{l}, \boldsymbol{k})$, is *weakly parallelly-propagated* along $\boldsymbol{k}$ and $\boldsymbol{l}$, i.e. obeys a series of equations such as $k_{a;b}l^b = Bk_a$, or equivalently that there is a non-null Maxwell field obeying

$$F^*_{ab;c}l^c = fF^*_{ab}, \qquad F^*_{ab;c}k^c = gF^*_{ab}, \tag{13.36}$$

where f and g are not both zero. Equations (13.36) imply that

$$\kappa = \nu = \pi = \tau = 0. \tag{13.37a}$$

The original ansatz was that the tetrad itself was parallelly propagated along $\boldsymbol{k}$ and $\boldsymbol{l}$, so one would also have

$$\varepsilon = \gamma = 0. \tag{13.37b}$$

This *parallelly-propagated* case differs from assuming $f = g = 0$ in (13.36): that would lead instead to $\rho = \mu = 0$ and thence only to the Bertotti–Robinson solution (12.16) (Tupper, private communication).

There are dual ansätze for propagation along $\boldsymbol{m}$ and $\overline{\boldsymbol{m}}$ which give rise, correspondingly, to the conditions

$$\sigma = \lambda = \rho = \mu = 0, \tag{13.38a}$$

$$\alpha = \beta = 0. \tag{13.38b}$$

By studying the consistency of the remaining Einstein–Maxwell equations, it can be shown that (13.37a) implies, assuming $\rho^2 \neq \bar{\rho}^2$, $\mu^2 \neq \bar{\mu}^2$ and $\lambda \neq 0$, that it is possible to choose a tetrad such that

$$\rho = e\mu, \qquad \sigma = e\lambda, \qquad \varepsilon = e\gamma, \qquad \alpha = \beta = 0,$$
$$\Psi_0 = \Psi_4, \qquad \Psi_1 = \Psi_3 = 0, \tag{13.39}$$

and such that, for any Newman-Penrose quantity x,

$$\delta x = \bar{\delta} x = (D + e\Delta)x = 0, \tag{13.40}$$

where $e = \pm 1$. The excluded cases lead to doubly-aligned Petrov D solutions or to metrics in Kundt's class (see §30.6 and Chapter 31). The similar conditions arising from (13.38a) are

$$\tau = \pi, \quad \alpha = \beta, \quad \kappa = e\nu, \quad \Psi_0 = e\Psi_4, \quad \varepsilon = \gamma = \Psi_1 = \Psi_3 = 0, \tag{13.41}$$

$$Dx = \Delta x = (\delta + \bar{\delta})x = 0. \tag{13.42}$$

Equations (13.40) and (13.42) show that an isometry group acts in the hypersurfaces spanned by $(\boldsymbol{m}, \overline{\boldsymbol{m}}, \boldsymbol{l} + e\boldsymbol{k})$ and $(\boldsymbol{k}, \boldsymbol{l}, \boldsymbol{m} + \overline{\boldsymbol{m}})$ respectively. The solutions all admit a G_3 on S_3 or T_3. In the further integration by Barnes (1978), who gives an extensive list of solutions, it is assumed that in the spacelike case, (13.38), the matrices $\boldsymbol{A}$ and $\boldsymbol{B}$ arising in

$$\delta \begin{pmatrix} \boldsymbol{k} \\ \boldsymbol{l} \end{pmatrix} = (\boldsymbol{A} + i\boldsymbol{B}) \begin{pmatrix} \boldsymbol{k} \\ \boldsymbol{l} \end{pmatrix} \tag{13.43}$$

commute, and a similar assumption is made in the timelike case (13.37). All known solutions in the class obey this extra restriction.

Equations analogous to (13.40) and (13.42) arise in the treatment (Siklos 1981, MacCallum and Siklos 1992) of algebraically special hypersurface-homogeneous Einstein spaces. If the normal $\boldsymbol{n}$ to the homogeneous hypersurfaces can be written as $\sqrt{2}\boldsymbol{n} = \mathrm{e}^{-\eta}\boldsymbol{k} + \mathrm{e}^{\eta}\boldsymbol{l}$, the homogeneity implies (Siklos 1981) that

$$\delta x = \bar{\delta} x = (\mathrm{e}^{-2\eta} D - \Delta)x = 0. \tag{13.44}$$

Applying the commutators (7.6a)–(7.6d) to a coordinate t constant on the group orbits gives relations between the spin coefficients, and the assumption that $\boldsymbol{k}$ is a repeated principal null direction gives $\kappa = \sigma = 0$. The remaining tetrad freedom can be used to set $\varepsilon = 0$. Then the field equations for each possible case can be integrated. In the other case (MacCallum and Siklos 1992), where the homogeneous hypersurfaces contain the repeated principal null direction, one can similarly take $\boldsymbol{n} = \boldsymbol{m} + \overline{\boldsymbol{m}}$ and arrive at

$$(\delta - \bar{\delta})x = Dx = \Delta x = 0, \tag{13.45}$$

with $\rho = \lambda + \mu = \alpha + \beta = \kappa = \sigma = \varepsilon - \bar{\varepsilon} = 0$.

13.3 Vacuum, Λ-term and Einstein–Maxwell solutions

13.3.1 Solutions with multiply-transitive groups

The case G_6 on V_3 is covered by the Robertson–Walker line element (12.9). No Einstein–Maxwell fields exist. The vacuum and Λ-term solutions are the spaces of constant curvature (8.33); the case $\Lambda > 0$ appears, with different choices of the spatial hypersurfaces, as a solution with each k, and the $\Lambda = 0$ case occurs with $k = 0$ and $k = -1$.

According to Theorem 13.1, the metrics with a G_4 on S_3 or T_3, apart from those admitting a null rotation symmetry, are (13.3), (13.9) and (13.17). In the Einstein–Maxwell case they permit only non-null electromagnetic fields whose principal null directions are aligned with those of the (Petrov type D) Weyl tensor, assuming that the Maxwell field shares the space-time symmetry. (If that assumption is false, the Maxwell field is null and the metric is a pure radiation solution; an example using (13.3) with $\epsilon = 1$ is:

$$t = \int A\,\mathrm{d}\tau, \quad A = Ce^{3\tau/2}/\sinh^{1/4}(2\tau), \quad B = \sinh^{1/2}(2\tau), \tag{13.46}$$

(Ftaclas and Cohen 1978), where C is a constant. The general Maxwell field for (13.46) is given by the components

$$\begin{aligned} F_{xy} = -F_{\tau y} &= \sqrt{3/\kappa_0}e^{x-\tau}\cos(k(x-\tau)), \\ -F_{xz} = F_{\tau z} &= \sqrt{3/\kappa_0}e^{x-\tau}\sin(k(x-\tau)), \end{aligned} \tag{13.47}$$

where k is a constant (Henneaux 1984).)

For the metric (13.3), the condition $R_{14} = 0$, true for the cases being considered here, gives only Robertson–Walker solutions.

For the metrics (13.9) and (13.17), the solutions with constant Y are just the Bertotti–Robinson solution and its generalization to non-zero Λ (cp. §§12.3 and 12.5). If Y in (13.9) is not constant, one finds that

$$\begin{aligned} f(w) &= (w^2+l^2)^{-1}\left[k(w^2-l^2) - 2mw + e^2 - \Lambda\left(\tfrac{1}{3}w^4 + 2l^2w^2 - l^4\right)\right], \\ Y^2 &= w^2 + l^2, \quad \sqrt{\kappa_0}\Phi_1 = e/\sqrt{2}Y^2, \end{aligned} \tag{13.48}$$

is the general solution (Cahen and Defrise 1968), where Φ_1 defines the non-null Maxwell field as in (5.11), up to a constant complexion factor. The solution for (13.17) is the same with the signs of the k and e^2 terms changed. These solutions contain numerous well-known and frequently rediscovered particular cases, e.g. special Kasner metrics (13.51), the Schwarzschild and Reissner–Nordström metrics (§15.4), plane symmetric cosmologies (§15.7), the Taub–NUT metrics, and even de Sitter space (8.34), cp. also the metrics (15.12), (15.27), (22.13) and (31.59).

The solutions (13.48) have expanding principal null congruences of the Weyl tensor. For $l = 0 = e = \Lambda$ one obtains the 'A-metrics' of Table 18.2 (including Kasner metrics if $m = 0$ also). The $l \neq 0$ vacuum cases are the Taub–NUT metrics (Taub 1951, Newman *et al.* 1963); for $k = 1$, using

$\zeta = \sqrt{2}\cot(\vartheta/2)\exp(\mathrm{i}\varphi)$ and $r = w$, one obtains the metric

$$\begin{aligned} \mathrm{d}s^2 &= (r^2+l^2)(\mathrm{d}\vartheta^2 + \sin^2\vartheta\,\mathrm{d}\varphi^2) - f(r)(\mathrm{d}t + 2l\cos\vartheta\,\mathrm{d}\varphi)^2 + f^{-1}(r)\mathrm{d}r^2, \\ f(r) &= (r^2+l^2)^{-1}(r^2 - 2mr - l^2), \qquad m,\ l \text{ constants}, \end{aligned} \tag{13.49}$$

which has the Killing vectors (cp. Table 8.2)

$$\begin{aligned} \boldsymbol{\xi}_1 &= \sin\varphi\,\partial_\vartheta + \cos\varphi\left(\cot\vartheta\,\partial_\varphi - 2l\frac{1}{\sin\vartheta}\partial_t\right), \quad \boldsymbol{\xi}_3 = \partial_\varphi, \\ \boldsymbol{\xi}_2 &= \cos\varphi\,\partial_\vartheta - \sin\varphi\left(\cot\vartheta\,\partial_\varphi - 2l\frac{1}{\sin\vartheta}\partial_t\right), \quad \boldsymbol{\xi}_4 = \partial_t. \end{aligned} \tag{13.50}$$

The Taub–NUT metrics are stationary in the region $f(w) > 0$ (see §20.3) and spatially-homogeneous in the region $f(w) < 0$. The two solutions form part of a single manifold, being joined across a null hypersurface; this manifold has interesting topological properties (see Misner and Taub (1968), Ryan and Shepley (1975), Siklos (1976b)). The 'charged' generalization of the Taub metric ($k = 1$, $\Lambda = 0 \neq el$ in (13.48)) was first given by Brill (1964).

The vacuum solutions for (13.17) have non-expanding principal null congruences of the Weyl tensor, and are the only such Petrov type D solutions (Kinnersley 1969b, 1975); they thus belong to Kundt's class (Chapter 31). For $l = 0$, they give the 'B-metrics' of Table 18.2, which again include some special Kasner metrics.

13.3.2 Vacuum spaces with a G_3 on V_3

A number of the vacuum solutions with G_r, $r > 3$, given above can be rediscovered as solutions with a G_3 on V_3, namely:
(a) flat space (for G_3 on an S_3 with expanding normals this can happen only with LRS S_3 admitting groups G_3 of Bianchi types I and VII_0, III, or V and VII_h);
(b) plane waves (§12.2) with G_3IV and/or G_3VI_h or G_3VII_h (see Siklos (1981), Araujo and Skea (1988a) for details): these have been frequently rediscovered, e.g. Lifshitz and Khalatnikov (1963), who give the only linearly-polarized case, Collins (1971), Doroshkevich *et al.* (1973), Harvey and Tsoubelis (1977);
(c) the vacuum solutions contained in (13.48), which in the various cases may admit G_3 of types I, II, III, VIII or IX (see §13.1); and
(d) the Petrov solution (12.14) with G_3I and G_3VII_h on T_3.
The reader should also consult the chapters referred to in §13.1.

Apart from the solutions with a null Killing vector given in Tables 24.1 and 24.2, and the vacuum Petrov type III case with a G_3VI contained in

(13.64) below, which admits an H_4, the known vacuum solutions with a G_3 on V_3 as their maximal isometry group are algebraically general. They give solutions with homogeneity on S_3 and T_3 related by substitutions of the form $t \leftrightarrow ix$.

The first example of this arises in the well-known general solutions for a diagonal vacuum metric with a G_3I on S_3 or T_3 and an H_4 on V_4. These are named after Kasner (1921), who wrote them as a metric of signature +4, although the signs of the metric coefficients can in fact be chosen arbitrarily, and stated he had found all similar solutions: his metric was

$$\begin{gathered} \mathrm{d}s^2 = (x^4)^{2a_1}(\mathrm{d}x^1)^2 + (x^4)^{2a_2}(\mathrm{d}x^2)^2 + (x^4)^{2a_3}(\mathrm{d}x^3)^2 + (x^4)^{2a_4}(\mathrm{d}x^4)^2, \\ a_1 + a_2 + a_3 = a_4 + 1, \quad {a_1}^2 + {a_2}^2 + {a_3}^2 = (a_4+1)^2. \end{gathered} \tag{13.51}$$

The homothety generator is (Godfrey 1972)

$$\boldsymbol{X} = x^4\partial_{x^4} + \sum_{i=1}^{3}(a_4 + 1 - a_i)x^i\partial_{x^i}. \tag{13.52}$$

For the S_3 case one can take $g_{44} < 0$, $x^4 = t$, $a_4 = 0$, $a_\alpha = p_\alpha$ to get the well-known form,

$$\begin{gathered} \mathrm{d}s^2 = t^{2p_1}\mathrm{d}x^2 + t^{2p_2}\mathrm{d}y^2 + t^{2p_3}\mathrm{d}z^2 - \mathrm{d}t^2, \\ p_1 + p_2 + p_3 = 1 = {p_1}^2 + {p_2}^2 + {p_3}^2; \qquad p_1,\ p_2,\ p_3 \ \text{constants}, \end{gathered} \tag{13.53}$$

which can easily be found directly from (13.29), (13.30). Using (13.30),

$$p_\alpha = \tfrac{1}{3}[1 + 2\cos(\psi + \tfrac{2}{3}(\alpha - 1)\pi)], \quad \alpha = 1, 2, 3. \tag{13.54}$$

Another form of the metric which is frequently cited is obtained by taking $t = (1 + kt')$ for some constant k and was given by Narlikar and Karmarkar (1946). It should be noted that for $p_1 = 1$, $p_2 = p_3 = 0$, (13.53) is flat space-time. The non-flat plane symmetric case ($p_2 = p_3 = 2/3$) was given by Taub (1951), cp. (15.29); other rediscoveries were listed by Harvey (1990). Special cases appear included in (15.12) and (15.17) and elsewhere. These Kasner solutions play an important role in the discussion of certain cosmological questions (see the reviews cited in §13.1).

The solution analogous to (13.53) but with a G_3I on T_3 and real p_i can be transformed to the Levi-Civita cylindrically symmetric vacuum solution (22.7) and includes the *AIII* and *BIII* metrics of Table 18.2. Space-times with G_3I on T_3 can also be obtained by taking the appropriate signature in (13.51), allowing $a_4 = -1$, and making a complex coordinate transformation to find real, but not diagonal, metrics of a 'windmill' character which include (12.14) as a special case (McIntosh 1992), as well as (22.6) with imaginary n.

The G_3I solutions in which the metric on the T_3 cannot be diagonalized, even in complexified variables, have null Killing vectors (see §24.4). The possible solutions are the *pp*-wave with $f = \ln\zeta$ in Table 24.2, and the special case of (24.37a) which is (20.32) with $\Omega \propto \ln\rho$.

The general vacuum solution with a G_3II on S_3 (Taub 1951) and its T_3 counterpart are, using the $\boldsymbol{\omega}^\alpha$ of Table 8.2,

$$\begin{aligned} &\mathrm{d}s^2 = \epsilon X^{-2}(\boldsymbol{\omega}^1)^2 + X^2[\mathrm{e}^{2A\tau}(\boldsymbol{\omega}^2)^2 + \mathrm{e}^{2B\tau}(\boldsymbol{\omega}^3)^2 - \epsilon\mathrm{e}^{2(A+B)\tau}\mathrm{d}\tau^2], \\ &kX^2 = \cosh k\tau, \quad 4AB = k^2, \quad \epsilon = \pm 1, \quad k > 0, A, B \quad \text{const.} \end{aligned} \tag{13.55}$$

This can be put into a Kasner-like form, with constants p_i and b:

$$\begin{aligned} &\mathrm{d}s^2 = \epsilon t^{2p_1}G^{-2}(\mathrm{d}x + 4p_1bz\mathrm{d}y)^2 + G^2[t^{2p_2}\mathrm{d}y^2 + t^{2p_3}\mathrm{d}z^2 - \epsilon\mathrm{d}t^2], \\ &G^2 = 1 + b^2t^{4p_1}, \qquad p_1 + p_2 + p_3 = 1 = {p_1}^2 + {p_2}^2 + {p_3}^2. \end{aligned} \tag{13.56}$$

The solution (24.37a) with $M = ay + b\ln x$ (see Table 24.1) has a G_3II on T_3 and a null Killing vector, and includes (20.32) with $\Omega \propto z$.

From Table 24.2, there are *pp*-waves with maximal G_3 of types III and V on T_3.

The vacuum solution with a G_3VI_h, $n^\alpha{}_\alpha = 0$, acting on T_3 or S_3,

$$\begin{aligned} &\mathrm{d}s^2 = k^2\sinh 2u[\epsilon W^n(-\mathrm{d}u^2 + \mathrm{d}x^2) + W\mathrm{e}^{2(1+n)x}\mathrm{d}y^2 + W^{-1}\mathrm{e}^{2(1-n)x}\mathrm{d}z^2], \\ &W = (\sinh 2u)^n(\tanh u)^m, \quad m^2 = 3 + n^2, \quad n^2h = -1, \end{aligned} \tag{13.57}$$

where $\epsilon = \pm 1$ and k is constant, was found as the general solution of this type, excluding $VI^*_{-1/9}$, in which the G_3 acts on S_3 (Ellis and MacCallum 1969, MacCallum 1971). The cases $mn > 0$ and $mn < 0$ are inequivalent. The $n = 1$ (G_3III) case is LRS and included in (13.48).

Taking a limit of (13.57) by transforming to $\tilde{x} = nx$, $\tilde{u} = nu$ and letting $n \to \infty$, one can derive the vacuum solution

$$\mathrm{d}s^2 = k^2\left\{\epsilon u^{-1/2}\mathrm{e}^{u^2}\left[(\boldsymbol{\omega}^1)^2 - \mathrm{d}u^2\right] + 2u\left[(\boldsymbol{\omega}^2)^2 + (\boldsymbol{\omega}^3)^2\right]\right\} \tag{13.58}$$

with G_3VI_0, $n^\alpha{}_\alpha = 0$, on S_3 (Ellis and MacCallum 1969) or T_3, in a basis (13.27). For $n = 0$, (13.57) gives the general G_3V on S_3 vacuum solution (Joseph 1966) and its counterpart for T_3,

$$\begin{aligned} \mathrm{d}s^2 = k^2\sinh 2A\tau\Big[&\epsilon((\boldsymbol{\omega}^1)^2 - \mathrm{d}\tau^2) + (\tanh A\tau)^{\sqrt{3}}(\boldsymbol{\omega}^2)^2 \\ &+ (\tanh A\tau)^{-\sqrt{3}}(\boldsymbol{\omega}^3)^2\Big], \end{aligned} \tag{13.59}$$

using $\omega^1 = \mathrm{d}x$, $\omega^2 = e^{Ax}\mathrm{d}y$ and $\omega^3 = e^{Ax}\mathrm{d}z$. A solution with a G_3V on T_3 is given by $M = \mathrm{J}_0(2ax)\exp(-2ay)$ in (24.37a), where J_0 is a Bessel function: see Table 24.1.

The solution (23.6) with $F = u^{-2p}$, $p = q = 13/16$ is a vacuum solution with a G_3VI_{-4} group whose existence was pointed out by Collins (1991).

Lorenz-Petzold (1984) found vacuum solutions with G_3VI_0 and VII_0 on S_3 up to a quadrature in the form

$$\mathrm{d}s^2 = \mathrm{e}^{2g}(-\mathrm{d}u^2/4u + (\omega^1)^2) + q\sqrt{uw}(\omega^2)^2 + q\sqrt{u/w}(\omega^3)^2, \qquad (13.60a)$$

with ω^α from Table 8.2: here q is a constant, g is given by

$$8g' = -1/u + u(w'^2/w^2) + w + 1/w - 2\epsilon, \qquad (13.60b)$$

where for G_3VI_0 we take $\epsilon = -1$, and for VII_0, $\epsilon = 1$, and $w(u)$ is a Painlevé transcendental function of the third type solving

$$w'' = w'^2/w - [w' + \tfrac{1}{2}(w^2 - 1)]/u. \qquad (13.60c)$$

Barnes (1978) found vacuum solutions with G_3VI_0 and VII_0 on T_3; with ϵ as above, Σ as in (8.37) and $2\Sigma'(2x,\epsilon) = \mathrm{d}\Sigma(2x,\epsilon)/\mathrm{d}x$, these are

$$\mathrm{d}s^2 = f^4[\Sigma(2x,\epsilon)(\mathrm{d}u^2 - \epsilon\,\mathrm{d}v^2) - 2\Sigma'(2x,\epsilon)\mathrm{d}u\,\mathrm{d}v] + P^2\mathrm{d}x^2 + U^2\mathrm{d}z^2;$$
$$P^{-1} = \sqrt{2}kf\exp(1/2z), \quad U = P/(z^2f^4), \quad f^8 = (\epsilon(z-4)/z). \qquad (13.61)$$

The only known type VII vacuum solutions other than (13.60), (13.61), and the algebraically special ones listed earlier, are a solution due to Lukash (1974), with G_3VII_h, $h = A^2 = 4/11$, on S_3 and its timelike counterpart, i.e., with ω^α from Table 8.2 and $w = Ax$

$$\mathrm{d}s^2 = k^2\left[\epsilon G^2(-\mathrm{d}\xi^2 + \mathrm{d}w^2) + \sinh(2\xi)\left(f(\cos\Phi\,\omega^2 + \sin\Phi\,\omega^3)^2 + (-\sin\Phi\,\omega^2 + \cos\Phi\,\omega^3)^2/f\right)\right], \qquad (13.62)$$

$$G^2 = \mathrm{e}^{11\xi/4}\sinh^{-3/8}2\xi, \quad \Phi = \sqrt{11}\xi/2, \quad f^2\tanh\xi = 1.$$

Apart from the spaces with higher symmetries, no vacuum solutions with G_3 of types III, IV, VIII or IX are known.

13.3.3 Einstein spaces with a G_3 on V_3

As in the vacuum case, a number of Einstein spaces with extra symmetry can be rediscovered as solutions with a G_3 on V_3, namely: the homogeneous Petrov type N solution with a Λ-term (12.34), and the Petrov type *III* solution (12.35), both of which belong to Kundt's class (Chapter 31) and admit a G_3VI_h, $h = -1/9$; the Einstein spaces contained in (13.48), which in the various cases may admit G_3 of types *I*, *II*, *III*, *VIII* or *IX* (see §13.1); and the Bertotti–Robinson-like solutions (12.8), with G_3 of types *VIII*, VI_0, VI_h and *III* (in the non-flat cases). All these solutions are *algebraically special.*

Equations (13.53) can be generalized to non-zero Λ as

$$ds^2 = -dt^2 + G(t)^{2/3}\left\{\sum_{i=1}^{3} \exp[2(p_i - 1/3)U(t)]d(x^i)^2\right\}, \quad (13.63a)$$

$$\dot{U} = 1/G, \quad \dot{G}^2 + 3\Lambda G^2 = 1 = \sum_1^3 p_i = \sum_1^3 p_i^2, \quad (13.63b)$$

cp. also (15.31) and (22.8). The $\Lambda > 0$ case was given by Kasner (1925) and both cases by Saunders (1967). G and U can be given explicitly for both signs of Λ. Except for special parameter values, this solution is algebraically general. The algebraically special G_3I solutions analogous to the *pp*-wave and (24.37*a*) forms for vacuum were given by Kellner (1975), who showed that the only Einstein spaces with a G_3I on T_3 without a null Killing vector are a stationary cylindrically-symmetric Kasner vacuum metric and its generalization to non-zero Λ.

Lorenz (1983b) has given an implicit form for an Einstein space with a G_3VI_0 ($n^\alpha{}_\alpha = 0$) on S_3.

The remaining known solutions complete the class of hypersurface-homogeneous algebraically special Einstein spaces. By the method of Siklos (1981) (see §13.2), one obtains four more such Einstein spaces with repeated principal null directions not lying in the group orbits, each admitting a maximal $G_3VI^*_{-1/9}$.

The first is a Robinson–Trautman solution of Petrov type *III*,

$$ds^2 = -2e^{-z}du\,dy + u^2(dz^2 + e^{4z}dx^2)/2 + 2ue^{-z}dy\,dz$$
$$+ 2\left(6 + \tfrac{1}{6}\Lambda u^2\right)e^{-2z}dy^2. \quad (13.64)$$

This is the $\Lambda \neq 0$ generalization (Theorem 28.7) of (28.16), which is itself the only algebraically special vacuum space-time with diverging rays and a maximal G_3 (Kerr and Debney 1970), admits an H_4 (see Table 11.2) and is also associated with the names of Collinson and French (1967).

The second is a non-diverging Petrov type *II* solution with metric

$$ds^2 = \tfrac{1}{2}e^{2z}dx^2 + \tfrac{1}{8}b^{-2}dz^2 + 4ue^{-2z}dy\,dz + \tfrac{2}{3}b^{-1}e^{-z}dy\,dx$$
$$-2e^{-2z}du\,dz - (8b^2u^2 - 1/18b^2)e^{-4z}dy^2, \quad (13.65)$$

where $\Lambda = -8b^2$. It has no (non-trivial) vacuum limit.

The Killing vectors for (13.64) and (13.65) are ∂_x, ∂_y and respectively

$$\partial_z - 2x\partial_x + y\partial_y \quad \text{and} \quad \partial_z - x\partial_x + 2y\partial_y. \tag{13.66}$$

The other two solutions are twisting solutions, one of Petrov type N and the other of Petrov type III, which can be jointly written as

$$\begin{aligned} k\,\mathrm{d}s^2 = -2\left(x^a \mathrm{d}u - \frac{\mathrm{d}y}{(a+1)x}\right)\Big[\mathrm{d}t + a(t\,\mathrm{d}x + \mathrm{d}y)/x \\ + f(t)\left(x^a \mathrm{d}u - \frac{\mathrm{d}y}{(a+1)x}\right)\Big] + (\mathrm{d}x^2 + \mathrm{d}y^2)(t^2+1)/2x^2. \end{aligned} \tag{13.67}$$

$$\text{Type } N: \; a = 2, \quad k\Lambda = -3, \quad f(t) = (t^2-1)/2.$$

$$\text{Type } III: \; a = 1/2, \; 78k\Lambda = -32, \; f(t) = (13t^2+17)/32.$$

The type N solution was found by Leroy (1970), and the type III solution by MacCallum and Siklos (1980). Both admit the Killing vectors

$$\partial_u, \quad \partial_y, \quad x\partial_x + y\partial_y - au\partial_u. \tag{13.68}$$

The algebraically special hypersurface-homogeneous Einstein spaces in which the repeated principal null direction lies in the surfaces of transitivity of the group belong to Kundt's class (see Chapter 31): the rather long list, and earlier literature, was discussed in detail by MacCallum and Siklos (1992). Apart from solutions with a G_r, $r \geq 4$, given in Chapter 12 which contain a simply-transitive G_3, and the *pp*-wave solutions (see §24.5), there are, for each Λ (or each $\Lambda < 0$): three-parameter families of Petrov type II solutions with groups of Bianchi types VI_h, VII_h and VIII and two-parameter families of Petrov type II and Bianchi types III and VI_0; two-parameter families of Petrov type III and Bianchi types VIII and VI_h; Petrov type N solutions of Bianchi types VIII, III, VI_h, VI_0, and with a G_4 containing no simply-transitive G_3. In some of these families special cases of other Bianchi types or with extra symmetry arise.

13.3.4 Einstein–Maxwell solutions with a G_3 on V_3

The plane wave and Bertotti–Robinson-like Einstein–Maxwell metrics admit subgroups G_3 of the full group of symmetry, like their vacuum and Λ-term analogues (for details of the plane wave cases, see Araujo and Skea (1988b)). The other hypersurface-homogeneous metrics which admit multiply-transitive groups are given in §13.3.1. Again we remind the reader to consult also the chapters referred to at the end of §13.1. For example, some cylindrically-symmetric solutions with G_3I and G_3II on T_3 are included in §22.2, and solutions with a null Killing vector are given in Tables 24.1 and 24.2.

In this section we give the Maxwell fields as values for the Φ_1 of (5.11), which is uniquely defined up to a constant duality rotation factor e^{ip} (see §5.4): the choice of this rotation may, however, have physical significance in interpreting the solution. If the electromagnetic field in an Einstein–Maxwell metric admitting a maximal G_3 inherits the symmetry and is purely magnetic or purely electric, only particular non-zero components and complexions may be possible (Hughston and Jacobs 1970).

The Einstein–Maxwell solutions admitting a G_3I on S_3 are

$$ds^2 = A^{-2}dx^2 + A^2[t^2dz^2 + t^{2m^2}(dy^2 - dt^2)], \quad k_1k_2m^2 > 0, \tag{13.69a}$$

$$A = (k_1t^m + k_2t^{-m}), \quad \Phi_1 = \sqrt{(2k_1k_2/\kappa_0)}\, m/(A^2t^{1+m^2}), \tag{13.69b}$$

k_1, k_2 and m^2 being real constants. They were given by Datta (1965) and are related by complex transformations to (22.11)–(22.14), e.g. $t \leftrightarrow i\rho$ for (22.11); see §22.2 for discussion of these analogues with G_3I on T_3, which can interpreted as cylindrically-symmetric static or stationary metrics. The solutions (13.69) are related to the vacuum G_3II solution (13.55) (Collins 1972) and include solutions of interest as cosmologies with a magnetic field; solutions of this type were given by Rosen (1962) and Jacobs (1969), and can be generalized to include fluids (see §§14.3, 14.4). The special case $m = 1$ admits a G_4, being an LRS plane symmetric solution included in (15.27) and (13.48).

Studying solutions with a G_2I in the formalism of Chapter 19, Wils (1989b) found the Einstein–Maxwell spaces with a G_3II on T_3 or S_3 given by

$$ds^2 = -\epsilon f(dv + wu\,dy)^2 + r^2dy^2/f + \delta e^{2\gamma}(dr^2 + \epsilon du^2)/f, \tag{13.70}$$

where δ and ϵ are ± 1 (excluding $-\delta = 1 = \epsilon$), the Φ of (18.31) is qu, q and w are constants, γ is obtained from a line integral and f obeys an ordinary differential equation of the third Painlevé type. Here $\sqrt{2\kappa_0}\Phi_1 = qe^{-\gamma}$. For the pairs $(\epsilon, \delta) = (-1, 1)$ and $(1, 1)$ the equations are analogous to the cases discussed in §22.2 which lead to a Painlevé equation but the solutions differ: the pair $(\epsilon, \delta) = (-1, -1)$ gives a solution of cosmological type. The special cases analogous to (22.12) are

$$ds^2 = r^{4/3}dy^2 + \delta r^{-4/9}\exp(\tfrac{9}{8}w^2r^{4/3})(dr^2 + \epsilon du^2) - \epsilon r^{2/3}(dv + wu\,dy)^2,$$

$$\sqrt{\kappa_0}\Phi_1 = \tfrac{1}{4}Wr^{-2/9}\exp(-\tfrac{9}{16}w^2r^{4/3}). \tag{13.71}$$

The remaining Einstein–Maxwell hypersurface-homogeneous solutions known to us all arise from the ansätze (13.37) or (13.38). The case (13.37) leads to (12.21) when $\rho + \bar{\rho} = 0$. It also, in the case of twistfree rays ($\rho - \bar{\rho} = 0$), leads to the solution

$$ds^2 = \tfrac{4}{3}t^2 dx^2 + t(e^{-2x}dy^2 + e^{2x}dz^2) - dt^2, \quad \sqrt{\kappa_0}\Phi_1 = 1/2t, \tag{13.72}$$

with a G_3VI_0 ($n^\alpha{}_\alpha = 0$) on S_3 (Tariq and Tupper 1975) and a homothety $\boldsymbol{X} = 2t\partial_t + y\partial_y + z\partial_z$ (Koutras 1992b). The dual ansatz leads (Barnes 1977) to the solution

$$ds^2 = dr^2 + \tfrac{4}{3}r^2(\omega^1)^2 + r\left((\omega^2)^2 - (\omega^3)^2\right), \quad \sqrt{\kappa_0}\Phi_1 = 1/2r, \tag{13.73}$$

with the ω^α of G_3VII_0 from Table 8.2 and a homothety $\boldsymbol{X} = 2r\partial_r + y\partial_y + z\partial_z$ (Koutras 1992b). The solutions (13.72) and (13.73) form two real slices of a complex manifold (Barnes 1977). A number of solutions satisfying the weaker ansatz (13.37a) have been found, both twistfree and twisting (Barnes 1978): for the twistfree case only the vacuum solution (13.58) with G_3VI_0 on S_3 or T_3 has been found; the twisting solutions admit a G_3I or G_3II on S_3 or T_3. Among these, the only solutions not discussed earlier are the electromagnetic G_3II solutions given in the general case by

$$\begin{gathered} ds^2 = P^2dx^2 + R^2dy^2 + 2\epsilon Q^2(dv + 2fx\,dy)^2 - \epsilon U^2dt^2, \\ \epsilon = \pm 1, \quad a, b, c, f \quad \text{const}, \quad a > 0, \quad a^2 + b^2 = c^2 + 1, \end{gathered} \tag{13.74}$$

where

$$\begin{aligned} &Q^2 = [2ct + b(1+t^2)]/(1+t^2), \\ &QP = F^{-(a+1)/2a}, \quad QR = F^{-(a-1)/2a}, \quad \sqrt{2}f(1+t^2)Q^3FU = 1, \\ &F = \begin{cases} \exp\left(\dfrac{2a}{\sqrt{1-a^2}}\arctan\dfrac{bt+c}{\sqrt{1-a^2}}\right), & a < 1, \\ \exp\left(\dfrac{-2a}{b(1+t)}\right), & \text{if } a = 1, \\ \left(\dfrac{bt+c-\sqrt{a^2-1}}{bt+c+\sqrt{a^2-1}}\right)^{a/\sqrt{a^2-1}}, & a > 1, \end{cases} \\ &\sqrt{\kappa_0}\Phi_1 = \Phi\frac{(1-t^2-2it)}{(1+t^2)}, \quad \Phi \text{ real}, \quad -\epsilon\kappa_0\Phi^2 = 2bf^2F^2Q^4, \quad b\epsilon < 0. \end{aligned} \tag{13.75}$$

For real $Q(t)$ for some t, we need $b > 0$ or $a > 1$. The case (13.74)–(13.75) with $\epsilon = 1$, $a > 1$, was given earlier by Ruban (1971), as a generalization of a form of the solution (13.48) with $k = 0$. The $a = 0$ limit is LRS and included in (13.48); the $b = 0$ limit is the vacuum solution (13.55). If

$c = 0$, $\epsilon = -1$, and the solution can be written as

$$\begin{aligned} ds^2 &= t^{-2k^2}(dt^2 + dy^2) + t^2 d\varphi^2 - (dv + 2ky\, d\varphi)^2, \\ \sqrt{\kappa_0}\Phi_1 &= 2kt^{k^2-1+2ik}, \qquad k \text{ const.} \end{aligned} \tag{13.76}$$

It admits, in addition to the G_3II on T_3, a twisting homothety $\boldsymbol{\xi} \equiv (1-k^2)v\partial_v + t\partial_t + y\partial_y - k^2\varphi\partial_\varphi$ not inherited by the Maxwell field (McIntosh 1979). The special case $k = 1$ of (13.76) is (12.21), but in taking this limit one has to introduce a parameter a for the overall scale which in (13.76) can be absorbed by a change of variables, due to the homothety.

Similarly, the dual ansatz (13.38a), with $\tau + \bar{\pi} = 0$, gives the vacuum metric (13.61) with G_3VI_0 or G_3VII_0 on T_3, while the case $\tau + \bar{\pi} \neq 0$ gives solutions with G_3I and G_3II on T_3, together with (12.14). Using the Σ of (8.37), the G_3II electromagnetic solutions are

$$\begin{aligned} ds^2 = F^{-1}P^{-2}\Sigma(B,\epsilon)(du^2 - \epsilon\, dv^2) - 2F^{-1}P^{-2}\Sigma'(B,\epsilon)du\, dv \\ + P^2(dx - 2fu\, dv)^2 + U^2 dz^2, \quad \epsilon = \pm 1, \end{aligned}$$

$$W^2 = 2cz + b(1+z^2), \quad F = \left(\frac{bz + c - \sqrt{a^2+\epsilon}}{bz + c + \sqrt{a^2+\epsilon}}\right)^{a/\sqrt{a^2+\epsilon}}, \tag{13.77}$$

$$P^2 = W^2/(1+z^2), \quad aB = \ln F, \quad U^{-1} = fPFW^2,$$

$$\sqrt{\kappa_0}\Phi_1 = \sqrt{-b/2}fP^2F(1 - z^2 - 2iz)/(1+z^2),$$

where a, b, c, f and p are constants obeying $a^2 + b^2 = c^2 - \epsilon$.

13.4 Perfect fluid solutions homogeneous on T_3

Apart from the stationary or static solutions with spherical, plane or cylindrical symmetry (see Chapters 15, 16 and 22) and the examples provided by the solutions given in §12.4, few perfect fluid solutions with a G_3 on T_3 are known. (The more numerous known S_3-homogeneous perfect fluid cosmologies are treated in Chapter 14. Tilted fluid solutions containing both T_3 and S_3 surfaces of homogeneity are also given there, because they usually include as a special case a non-tilted solution with S_3 surfaces.) Most authors have concentrated on fluids obeying the γ-law or some other pre-determined equation of state: solutions not obeying such an assumption can be somewhat easier to find, as follows.

From (13.12), any pair of functions f and Y in (13.9) obeying

$$\tfrac{1}{2}f'' + \frac{k}{Y^2} + f\left[(3+\epsilon)\frac{l^2}{Y^4} - \epsilon\frac{Y''}{Y} - \frac{Y'^2}{Y^2}\right] = 0, \quad \epsilon \equiv \frac{-f(w)}{|f(w)|}, \tag{13.78}$$

gives an LRS perfect fluid solution. Thus one can put any function $Y(w)$ or $f(w)$ in (13.78) and solve for the other of Y and f to get such a solution. Herlt (1988) provides examples. If a solution (Y, f) is already known, one can regard using the same f or Y and solving for the most general resulting pair (Y, f) as generating a new perfect fluid solution from a known one. Yet another variation is to assume a relationship $g(Y, f) = 0$ between Y and f. When using such methods, there is no guarantee that the new solutions obey physical equations of state, and one also has to be careful to check that the constants of integration introduced cannot be eliminated by coordinate transformations. One can apply similar methods to the field equations for other perfect fluid solutions determined by two functions, of t say, which appear in the equation(s) for isotropy of pressure: such a situation arises in several cases with a maximal G_3 (see §14.4). Solutions obtained by such methods will not usually be given in full below. For brevity we refer to these methods as 'pressure isotropy' methods.

Of the solutions with G_4 on T_3 only the LRS cases need be considered, because the four-velocity must be invariant under the isotropy; such solutions are algebraically special (cp. Chapter 33) and of Ellis Class I or II. For G_4 on T_3 we must take $f(w) > 0$ in (13.9), $\epsilon = -1$, and a four-velocity $\boldsymbol{u} = u\partial_t$. The Ellis Class II perfect fluid solutions with a G_4 on T_3 and G_3 on S_2 are included in §§15.5 and 15.7 and Chapter 16.

Perfect fluid solutions for (13.9) in which $w\partial_w$ is a homothety generator are given by

$$f = w, \quad Y^2 = aw, \quad 2ka + 2(3+\epsilon)l^2 - a^2 = 0. \tag{13.79}$$

For $\epsilon = 1$ these have $\mu = p = (a-k)/2\kappa_0 aw$ (cp. §11.3) and if $l \neq 0$ (Ellis Class I) admit a G_3 on T_3 of type II, or III and VIII, or IX, while if $l = 0$ only $k = a/2 = 1$ (spherical symmetry; cp. §16.1) is possible. The solutions (13.79) were given for $k \leq 0$ by Hermann (1983), and for all k by Daishev (1984).

The (stationary) dust solutions of Ellis' Class I are completely known. They have $\omega \neq 0 = \sigma = \Theta$. Apart from the Gödel solution (12.26), they have $f = 1$, $\omega = l/Y^2$ and

$$\Lambda > 0,\ Y^2 = a\cos\beta w + b\sin\beta w + k/2\Lambda,\ \Lambda(a^2+b^2) = l^2 + k^2/4\Lambda; \tag{13.80}$$

$$\Lambda = 0,\ Y^2 = kw^2 + 2aw + b, \qquad a^2 = l^2 + kb\ ; \tag{13.81}$$

$$\Lambda < 0,\ Y^2 = ae^{\beta w} + be^{-\beta w} + k/2\Lambda, \qquad 4\Lambda ab = l^2 + k^2/4\Lambda, \tag{13.82}$$

(Ellis 1967), where $\beta^2 = 4|\Lambda|$ when $\Lambda \neq 0$, and $\kappa_0\mu = 2\Lambda + 4l^2/Y^4$.

Some solutions of Ellis Class I containing Maxwell fields and (in some cases γ-law) perfect fluids are given up to quadratures in Stewart and

Ellis (1968). Equation (33.34) gives a solution with $\mu + 3p = \text{const}$. By the methods outlined in §9.4, Bradley and Marklund (1996) and Marklund (1997) obtained a number of Ellis Class I perfect fluid solutions not obeying the γ-law, e.g. the metric

$$\mathrm{d}s^2 = -\tfrac{1}{2}(4c-1)r^2(\mathrm{d}t - 2\cos\vartheta\,\mathrm{d}\varphi)^2 + \mathrm{d}r^2/f + r^2(\mathrm{d}\vartheta^2 + \sin^2\vartheta\,\mathrm{d}\varphi^2), \tag{13.83}$$

where $f = c - \frac{1}{6}p_0 r^2$, which has an equation of state $\mu = p + p_0$, or can be interpreted as a stiff fluid together with a cosmological constant.

We now survey metrics with a maximal G_3 on T_3.

Krasiński (1998a, 1998b) gave a comprehensive survey of the Bianchi types, metric forms and remaining field equations for hypersurface-homogeneous rotating dust solutions. A number of known exact solutions were recovered, and some new classes given, up to differential equations.

Some fluid solutions with a G_3I on T_3 can be found in §22.2. Studying solutions admitting flat slices, Wolf (1986b) found (36.22) and (36.23), which admit a G_3I on T_3, are of Petrov type I, and contain shearing and rotating fluid with, in general, an equation of state not of the γ-law form. Other examples with a G_3I on T_3 are provided by the following metrics: (18.65) (Barnes 1972); (33.30), which admits an H_4 and is of Petrov type II; and (36.34).

In his study of shearfree dust, Ellis (1967) found a solution for dust with a Λ-term and a G_3II on T_3,

$$\begin{aligned}
\mathrm{d}s^2 &= -(\mathrm{d}t + 2ay\,\mathrm{d}z)^2 + \mathrm{d}x^2 + Y^2F^{-2}\mathrm{d}y^2 + Y^2F^2\mathrm{d}z^2,\\
Y^2 &= (c\sin 2\sqrt{\Lambda}x)/\sqrt{\Lambda} \qquad (\Lambda > 0),\\
Y^2 &= 2cx \qquad (\Lambda = 0), \quad \kappa_0\mu = 4a^2/Y^4 + 2\Lambda,\\
Y^2 &= c\sinh 2\sqrt{|\Lambda|}x/\sqrt{|\Lambda|} \quad (\Lambda < 0),
\end{aligned} \tag{13.84}$$

where a, b, and $c = \sqrt{a^2+b^2}$ are constants and $F = \exp\left(b\int Y^{-2}\mathrm{d}x\right)$ can be integrated explicitly.

As part of a qualitative study of the dynamics of γ-law perfect fluid solutions admitting a G_3II on T_3 in which the distinguished direction in the group orbits is timelike, Nilsson and Uggla (1997b) found the solutions

$$\begin{aligned}
\mathrm{d}s^2 &= -\sqrt{x(x+a)}(\mathrm{d}t + cy\,\mathrm{d}z)^2 + [x(x+a)]^q\mathrm{d}x^2\\
&\quad + [x(x+a)]^s([x/(x+a)]^\delta\,\mathrm{d}y^2 + [x/(x+a)]^{-\delta}\mathrm{d}z^2),\\
q &= (4-3\gamma)/4(\gamma-1), \quad s = (3\gamma-2)/8(\gamma-1),\\
\delta^2 &= (11\gamma-10)(3\gamma-2)/64(\gamma-1)^2, \quad c^2 = (5\gamma^2-4)/16(\gamma-1)^2,\\
\kappa_0\mu &= (7\gamma-6)/16(\gamma-1)^2(x(x+a))^{q+1},
\end{aligned} \tag{13.85}$$

with arbitrary constant a. If $a \neq 0$ it can be taken to be 1. The case with $a = 0$, which was found by Wainwright (1985), is LRS, can be regarded as a stationary axisymmetric metric, and admits an H_5, $4x\partial_x + (2q + 1)(y\partial_y + z\partial_z + t\partial_t)$ being a homothety generator; its $\gamma = 1$ limit is the case $\Lambda = 0 = b$ of (13.84) and the $\gamma = 2$ solution is one case of (13.79).

Analogously to the use of (13.78), the perfect fluid equations for a metric with a G_3II on T_3 and the ω_1 of Table 8.2 spacelike have been reduced to a single second-order equation by Misra and Narain (1971); as a limit these contain the Einstein static solution. Similarly Van den Bergh (1988c) reduced a diagonal metric with a G_3V on T_3 (a special case of (23.22)) to a single second-order differential equation. The metric (14.46) has a G_3VI or V on T_3 for some x and t. Another fluid solution not obeying the γ-law is, in comoving coordinates,

$$\begin{aligned} ds^2 &= f^2[b(-dt^2 + dx^2) + e^{b(t+x)-2x}dy^2 + e^{-b(t-x)-2x}dz^2], \\ \kappa_0\mu &= (-2ff'' + f'^2 - 2bff' + 4ff' - b^2f^2 + 3bf^2 - 3f^2)/bf^4, \qquad (13.86) \\ \kappa_0 p &= 2ff'' - f'^2 + bff' - 2ff' - b^2f^2 + f^2)/bf^4, \ f = c/(1 - e^{(2-b)x/2}), \end{aligned}$$

where b and c are constants and $'$ denotes d/dx (Sintes *et al.* 1998). This has a G_3VI on T_3 and a conformal Killing vector $\partial_t + \partial_x + (1-b)y\partial_y + z\partial_z$; there is a three-parameter conformal group acting on null orbits.

In a study of stationary axisymmetric fluids, González-Romero (1994) gave solutions for G_3II, G_3VI_0 and G_3VII_0 on T_3 up to an ordinary differential equation and quadratures: these generalize or complement the known rotating solutions (see §21.2). An explicit perfect fluid solution with an H_4 on V_4 and G_3VII_0 on T_3 was found by Hermann (1983); it can be written, using a tetrad aligned with the fluid velocity, as

$$\begin{aligned} ds^2 &= r^{2(1-s)}\left[-(2+s)(\sin bz\, d\varphi + \cos bz\, dt)^2 + s(\cos bz\, d\varphi - \sin bz\, dt)^2\right] \\ &\quad + dr^2 + r^2dz^2, \qquad b^2 = s(s+2)(1-s)/(1+s), \qquad (13.87) \\ \kappa_0\mu &= (2 - 2s^2)/r^2, \quad \kappa_0 p = 2(1-s)^2/r^2, \end{aligned}$$

where $0 < s < 1$ is a constant, so $\gamma = 2/(1+s)$.

13.5 Summary of all metrics with G_r on V_3

To help the reader in identifying any metric with a G_r on V_3 which he or she may have found, we append here three tables. Table 13.2 lists the places where line elements with larger maximal symmetry may be found, Table 13.3 summarizes the actual solutions with a maximal G_4 on V_3 for each energy-momentum type considered in this book, and Table 13.4 similarly lists the solutions with a maximal G_3 on V_3.

Table 13.2. Subgroups G_3 on V_3 occurring in metrics with multiply-transitive groups

The spaces (8.33) of constant curvature are omitted here.

Bianchi	Maximal group				
type of G_3	G_7 or G_6 on V_4	G_5 or G_4 on V_4	G_6 on S_3	G_4 on S_3 or T_3, LRS	G_4 on T_3, non-LRS
I	(12.24) (12.8)	(12.14), (12.26), (12.30)–(12.32), (12.34)–(12.36), (12.38)	(14.2)	(13.1), $k=0$	(13.14)
II		(12.21), (12.34), (12.36), (12.38)		(13.2), $k=0$	(13.15)
III	(12.8)	(12.21), (12.26), (12.28), (12.29) (12.36)		(13.1), $k=-1$ (13.2), $k=-1$	(13.14), (13.15)
IV	§12.2	(12.31)			
V			(14.2)	(13.3)	(13.15)
VI	§12.2 (12.8)	(12.30), (12.31) (12.34)–(12.36) (12.38)			(13.14), (13.15)
VII	§12.2 (12.8)	(12.14) (12.32)	(14.2)	(13.1), $k=0$ (13.3)	
VIII	(12.8)	(12.26), (12.28), (12.29), (12.36)		(13.2), $k=-1$	(13.15)
IX	(12.24)	(12.27)	(14.2)	(13.2), $k=1$	

Table 13.3. Solutions given in this book with a maximal G_4 on V_3

Energy-	Metric			
	(13.9)		(13.3)	(13.17)
momentum	(13.1)	(13.2)		
Dust	(13.80)–(13.82) (14.15)	(14.20)	(14.22)	∄
Perfect fluid	§§15.7, 16.1 (13.35), (14.14) (14.16)–(14.19) (13.79), (33.34)	(13.83), (13.79) (14.20) (14.21) (14.23), (14.25)	 (14.24), (14.44) with $m=n=0$	∄
Vacuum and Einstein–Maxwell (with $\Lambda\neq 0$)	(13.48) (includes, e.g. (15.21)–(15.32), *A* of Table 18.2)		(13.46)	(13.48), $e^2\to -e^2$ (includes (15.12), *B* of Table 18.2)

Table 13.4. Solutions given explicitly in this book with a maximal G_3 on V_3

A means all such solutions are known. Note that type *III* is included in type VI_h here. For brevity, some solutions are given by reference to another table.

Bianchi type	Vacuum	Λ-term	Einstein–Maxwell	Dust	Perfect fluid
I	A (13.51), (13.53) (20.32) $\Omega \propto \ln \rho$, (22.5)–(22.7) Tables 24.1, 24.2	(13.63) (22.8)	(13.69) (22.11)–(22.18)	A for S_3 (14.26) (22.19)	γ-law: A for S_3 (13.31), (14.26)–(14.28) (14.37)–(14.38), $\beta = 0$ (18.65), (22.27)–(22.39) (33.30), (36.22), (36.23), (36.34)
II	A for S_3 (13.55), (20.32) $\Omega \propto z$, Table 24.1		(13.70)–(13.71) (13.74)–(13.77)	(13.84) (14.29), $\gamma = 1$	(13.82), (13.85) (14.29)–(14.31) (23.1) $b = 0$
V	A for S_3 (13.59), Table 24.1				(13.32), (14.32)
VI_0	(13.58), (13.60) (13.61)		(13.72)	(14.33), $A = 0$, $\gamma = 1$	(14.33), $A = 0$ (14.42), (23.3) $nm = ab$
VI_h	(13.57), Table 24.2 (28.16)	(13.64)–(13.67)		(14.33), $\gamma = 1$ (14.37)–(14.39), $\gamma = 1$	(13.86), (14.33)–(14.41) (14.46), (23.2)/(14.44), (23.6) $p = q$, (23.36), $c = 0 \neq \nu$
VII_0	(13.60), (13.61)		(13.73)		(13.87), §14.4
VII_h	(13.62)				(23.4)/(14.45)

14
Spatially-homogeneous perfect fluid cosmologies

14.1 Introduction

In this chapter we give solutions containing a perfect fluid (other than the Λ-term, treated in §13.3) and admitting an isometry group transitive on spacelike orbits S_3. By Theorem 13.2 the relevant metrics are all included in (13.1) with $\varepsilon = -1$, $k = 1$, and (13.20).

The properties of these metrics and their implications as cosmological models are beyond the scope of this book, and we refer the reader to standard texts, which deal principally with the Robertson–Walker metrics (12.9) (e.g. Weinberg (1972), Peacock (1999), Bergstrom and Goobar (1999), Liddle and Lyth (2000)), and to the reviews cited in §13.2. Solutions containing both fluid and magnetic field are of cosmological interest, and exact solutions have been given by many authors, e.g. Doroshkevich (1965), Shikin (1966), Thorne (1967) and Jacobs (1969). Details of these solutions are omitted here, but they frequently contain, as special cases, solutions for fluid without a Maxwell field. Similarly, they and the fluid solutions may contain as special cases the Einstein–Maxwell and vacuum fields given in Chapter 13.

There is an especially close connection between vacuum or Einstein–Maxwell solutions and corresponding solutions with a stiff perfect fluid (equation of state $p = \mu$) or equivalently a massless scalar field. If they admit an orthogonally-transitive G_2I on S_2, such solutions can be generated by the procedure given by Wainwright *et al.* (1979) (see Theorem 10.2), and many of the known stiff fluid solutions are obtainable in this way; spatially-homogeneous metrics with groups of Bianchi types I to VII and LRS metrics of types VIII and IX may or must admit such a G_2I on S_2. Another procedure adapted to these G_3 (and corresponding H_3) was given by Jantzen (1980b).

Many of the known solutions assume a γ-law equation of state (5.36), i.e. $p = (\gamma - 1)\mu$, or its specializations with $\gamma = 1$, 4/3, 2; these solutions were surveyed in Chapter 9 of Wainwright and Ellis (1997). If the four-velocity is aligned with the ∂_t of (13.1) or (13.20), then, using the Bianchi identities (5.5) and the definition (13.23), we find

$$\kappa_0\mu = MS^{-3\gamma}, \qquad \dot{M} = 0. \tag{14.1}$$

We shall also use the notation $m{=}M/3$. Note that S and hence M may only be fixed up to a constant scale factor. In each section we treat such solutions first and follow them by the perfect fluids not obeying the γ-law. Perfect fluids with bulk viscosity have been investigated by a number of authors (see e.g. Murphy (1973), Golda *et al.* (1983) and references therein) because they may give solutions without singularities.

The general dynamics of the tilted solutions have been studied by King and Ellis (1973) and Ellis and King (1974). Tilt complicates the equations, in part because the diagonalization theorems discussed in §13.2 do not apply; consequently, for example, Bradley (1988) has shown, using the methods of Chapter 9, that dust power-law solutions, which exist for zero tilt, do not exist in the tilted case. Thus rather few exact tilted solutions are known: they are included in sections §§14.3–14.4.

Since the equations for spatially-homogeneous cosmologies are already reduced to ordinary differential equations, it may make little sense to give here solutions in which some such equations remain unsolved: however, we have included such solutions where they are of sufficient interest for physical or mathematical reasons.

14.2 Robertson–Walker cosmologies

Here the metric is (12.9), i.e.

$$\mathrm{d}s^2 = -\mathrm{d}t^2 + a^2(t)[\mathrm{d}r^2 + \Sigma^2(r,k)(\mathrm{d}\vartheta^2 + \sin^2\vartheta\,\mathrm{d}\varphi^2)]. \tag{14.2}$$

The field equations, which necessitate an energy-momentum tensor of perfect fluid type, are (assuming $\dot{a} \neq 0$, since a constant $a(t)$ gives only the Einstein static solution and flat space)

$$3\dot{a}^2 = \kappa_0\mu a^2 + \Lambda a^2 - 3k, \tag{14.3}$$

$$\dot{\mu} + 3(\mu + p)\dot{a}/a = 0. \tag{14.4}$$

If $\mu(a)$ satisfies (14.4) for suitable p, then, using (14.3), one can replace $\mathrm{d}t^2$ in (14.2) by $\mathrm{d}a^2/[\kappa_0\mu(a)a^2 + \Lambda a^2 - 3k]$, giving an exact solution with a as the time variable. This is similar to the situation arising from (13.29).

The problem of integrating (14.3) and (14.4) has frequently been attacked. Apart from the choice $a(t)$, the most commonly used new time variable is

$$\Psi = \int \mathrm{d}t/a. \tag{14.5}$$

The form using this coordinate has become known as the conformal form.

The general behaviour for fluids obeying (14.1), with $S \equiv a$, has been investigated by Harrison (1967), who gave many references to the earlier literature. Apart from the de Sitter metrics (3.29) and the Einstein static universe (12.24), some of the best known solutions are the dust solutions ($\gamma = 1$) with $\Lambda = 0$ (Friedmann 1922, 1924, Einstein and de Sitter 1932), which are

$$k = 1, \quad a = m \sin^2 \Psi/2, \quad 2t = m(\Psi - \sin \Psi), \tag{14.6a}$$

$$k = 0, \quad a = (\tfrac{3}{2}\sqrt{m}t)^{2/3}, \tag{14.6b}$$

$$k = -1, \quad a = m \sinh^2 \Psi/2, \quad 2t = m(\sinh \Psi - \Psi), \tag{14.6c}$$

and the Tolman (1934b) radiation ($\gamma = 4/3$) solution

$$\Lambda = k = 0, \qquad a = (2\sqrt{m}t)^{1/2}. \tag{14.7}$$

The general solutions for $\Lambda k = 0$ with (14.1) are known. For $k = 0$,

$$\Lambda > 0, \quad a^{3\gamma} = M \sinh^2(\tfrac{1}{2}\sqrt{3\Lambda}\gamma t)/\Lambda, \tag{14.8a}$$

$$\Lambda = 0, \quad a^{3\gamma} = \left(\tfrac{3}{2}\gamma\sqrt{m}t\right)^2, \tag{14.8b}$$

$$\Lambda < 0, \quad a^{3\gamma} = -M \sin^2(\tfrac{1}{2}\sqrt{-3\Lambda}\gamma t)/\Lambda. \tag{14.8c}$$

The metrics (14.8b) have a homothety $3\gamma t\partial_t + (3\gamma - 2)r\partial_r$.

For $\Lambda = 0 \neq k$, if $\gamma \neq 2/3$,

$$k = 1, \quad a^{3\gamma-2} = m \sin^2 \eta, \quad |3\gamma - 2|t = 2\int a\,\mathrm{d}\eta, \tag{14.9a}$$

$$k = -1, \quad a^{3\gamma-2} = m \sinh^2 \eta, \quad |3\gamma - 2|t = 2\int a\,\mathrm{d}\eta. \tag{14.9b}$$

These parametric solutions, which include (14.6a) and (14.6c) and a model due to Tolman, were given by Harrison (1967). The special case $\gamma = 2/3$ has solutions $a = t/b$, where $\kappa_0\mu = 3(1 + kb^2)/t^2$, which admit a homothety $t\partial_t$. Using the manifestly conformally flat form of the metric, Tauber (1967) found an explicit elementary function form for $k = -1$,

$\Lambda = 0$ and general γ; for $k = 1$ hypergeometric functions were needed in general but reduced to elementary functions for $\gamma = 1$, $4/3$, and 2. These solutions were also given, with a useful compendium of other cases, by Vajk (1969); the equations can be regarded as describing two fluids (McIntosh 1972) and solved from that point of view.

For $\Lambda k \neq 0$ and $\gamma = 1$ or $4/3$, general solutions can be given in terms of elliptic functions (Lemaître 1933, Edwards 1972, Kharbediya 1976). There are a number of special solutions in terms of elementary functions, at least in implicit forms, see e.g. Lemaître (1927) and Harrison (1967), though some of these are for $\gamma < 1$. For example, for $k = 1$, $\gamma = 4/3$, $\Lambda > 0$, we have (Harrison 1967)

$$\frac{2\Lambda}{3}a^2 = 1 - \cosh(2\sqrt{\Lambda}t/\sqrt{3}) + \left(\frac{\Lambda}{\Lambda_c}\right)^{1/2} \sinh(2\sqrt{\Lambda}t/\sqrt{3}), \qquad (14.10)$$

where $\Lambda_c = 3/4m$ is the value of Λ required in the Einstein static solution (12.23) for the same m.

It is of cosmological interest to consider a perfect fluid composed of a sum of dust and incoherent radiation with the same four-velocity. Solutions with such a mixture, with or without Λ, and with or without interactions between the components, have been given by a number of authors (see e.g. McIntosh (1968), Sapar (1970), May (1975), Coquereaux and Grossmann (1982), Dąbrowski and Stelmach (1986) and references therein). If $\Lambda = 0$ and m and N are constants giving non-interacting dust and radiation densities, one gets

$$\kappa_0\mu = \frac{3m}{a^3} + \frac{3N}{a^4}, \qquad \kappa_0 p = \frac{N}{a^4}, \qquad (14.11)$$

$$t = \frac{m}{2}\sin^{-1}\left(\frac{2a - m}{(m^2 + 4N)^{1/2}}\right) - (ma + N - a^2)^{1/2}, \qquad k = 1, \quad (14.12a)$$

$$t = \frac{2(ma - 2N)(ma + N)^{1/2}}{3m^2}, \qquad k = 0, \quad (14.12b)$$

$$t = x - \tfrac{1}{2}m\ln\left(a + \tfrac{1}{2}m + x\right), \quad x \equiv (ma + N + a^2)^{1/2}, \quad k = -1. \quad (14.12c)$$

Expressions (14.12) are awkward to use in applications, for which the form with time variable a may be better, and for practical calculation parametric forms of the relations (14.12) have been developed.

Other combinations of fluids, formed like (14.11), can be integrated to give $a(t)$, at least in terms of elliptic, hypergeometric or other special functions; for details see e.g. Vajk (1969), McIntosh (1972), McIntosh and Foyster (1972) and Sistero (1972). Since the 1980s many solutions for scalar fields with various potentials, whose energy-momenta do not in general obey a γ-law, have been given, stimulated by the interest in inflationary cosmology: we have made no attempt to survey these. For such models, unless the potential has strong physical motivation, it may be simpler to postulate the desired behaviour of $a(t)$ and use the field equations to define the behaviour required for the matter (see Ellis and Madsen (1991)). One may note also that although the total energy-momentum in (14.2) must have the perfect fluid form, this can be composed of several physical fields which separately do not have that form (cp. remarks in §5.2).

14.3 Cosmologies with a G_4 on S_3

Within this class, metrics of the form (13.1) with $\epsilon = 1$, i.e.

$$\mathrm{d}s^2 = -\mathrm{d}t^2 + A^2(t)\mathrm{d}x^2 + B^2(t)\left[\mathrm{d}y^2 + \Sigma^2(y,k)\mathrm{d}z^2\right], \qquad (14.13)$$

have been extensively investigated as cosmological models. The field equations for perfect fluids are (13.34). The known exact γ-law solutions were collected by Vajk and Eltgroth (1970): see also Lorenz (1983a).

For the case $k = 0$ these (LRS) solutions are special cases of the solutions (14.26)–(14.28) with a G_3I on S_3, in which the ψ of (13.30) is 0 or π. Particular solutions were found by Doroshkevich (1965) (for $\gamma = 1$, 4/3 and 2) and Jacobs (1968). The general solutions for $1 < \gamma < 2$ can be given in the form (14.27) or (13.35), or as

$$\begin{aligned}\mathrm{d}s^2 = {} & (\cosh u)^{4/(2-\gamma)}(\sinh u)^{-4/3(2-\gamma)}\mathrm{d}x^2 + (\sinh u)^{8/3(2-\gamma)}(\mathrm{d}y^2 + y^2\,\mathrm{d}z^2) \\ & -16(\cosh u \sinh u)^{4/(2-\gamma)}\mathrm{d}u^2/3M(2-\gamma)^2 \end{aligned} \qquad (14.14)$$

or the same with cosh and sinh interchanged (Vajk and Eltgroth 1970).

For the cases $k = \pm 1$, the solutions for $\gamma = 1$, 4/3 and 2 can be given as follows. For $\gamma = 1 = k$,

$$\begin{aligned} A\cos\Psi &= M(\Psi\sin\Psi + \cos\Psi) + K\sin\Psi, \\ B &= b\cos^2\Psi, \quad \mathrm{d}t = 2B\mathrm{d}\Psi; \quad b, K \text{ const}; \end{aligned} \qquad (14.15a)$$

while for $\gamma = 1 = -k$ we have three cases,

$$A = C/F, \quad B = bF^2, \quad \mathrm{d}t = 2B\mathrm{d}\Psi,$$

$$F = \sinh\Psi, \quad C = M(\Psi\cosh\Psi - \sinh\Psi) + K\cosh\Psi, \qquad (14.15b)$$

$$F = \cosh\Psi, \quad C = M(\Psi\sinh\Psi - \cosh\Psi) + K\sinh\Psi, \qquad (14.15c)$$

$$F = e^{\Psi}, \quad C = e^{\Psi}(M\Psi + K), \qquad (14.15d)$$

where b and K are constants. The solutions (14.15a) were found by Kantowski (1966), Kantowski and Sachs (1966), Shikin (1966) and Thorne (1967), (14.15b)–(14.15c) by Kantowski (1966) (cp. Kantowski and Sachs (1966)), (14.15c) with an added magnetic field by Shikin (1966) and Thorne (1967), and (14.15d) by Shukla and Patel (1977).

For $\gamma = 4/3$ (Kantowski 1966),

$$A\sqrt{B} = u^{3/2}, \quad B = a + mu - u^3/9k, \quad \mathrm{d}t^2 = uB\,\mathrm{d}u^2, \tag{14.16}$$

which for $k = -1$ is the special case $n = \frac{1}{2}$ of (14.37). For $k = 1$ there is also a special solution (Kantowski 1966) with $M = 1$ which can be written

$$A\sqrt{B} = 1, \quad B = u, \quad \mathrm{d}t^2 = 3u\,\mathrm{d}u^2/4(c-u), \quad c \text{ const.} \tag{14.17}$$

For $\gamma = 2$ (Kantowski 1966),

$$\lambda^2\mathrm{d}s^2 = f^2 g^{2c}(-\mathrm{d}t^2 + \mathrm{d}y^2 + \Sigma^2(y,k)\mathrm{d}z^2) + g^{-2c}\mathrm{d}x^2,$$

$$f = \sin(t), \qquad g = \tan t/2, \qquad \text{if } k = 1, \tag{14.18a}$$

$$f = \sinh(t), \quad g = \tanh t/2, \quad \text{if } k = -1, \tag{14.18b}$$

$$c^2 = 1 - M\lambda^4, \quad c,\ \lambda \text{ const},$$

which for $k = -1$ is the special case $n = 1$, $b = 0$, of (23.2).

Special perfect fluid solutions for (13.1) with $k = -1$, an H_5 and a γ-law equation of state, but with $p < 0$, are given by

$$A = t^c,\ B = t/b,\ b^2 = 1 - c^2,\ \kappa_0\mu = c(c+2)/t^2,\ \kappa_0 p = -c^2/t^2, \tag{14.19}$$

which is a special case of (14.33). The subcase $c = 1$ is among the spatially-homogeneous solutions analogous to (13.79), but these have the rather unphysical equation of state $\mu + 3p = 0$ (see §11.3). If a Λ-term is added to a γ-law fluid in (13.1), elliptic functions are needed in general but some cases have elementary function solutions (see e.g. Kantowski (1966), Lorenz (1982b), Gron and Eriksen (1987)).

The field equations for γ-law perfect fluid solutions, $\gamma \neq 0$, of the LRS G_3II metric, (13.2) with $k = 0$, have been reduced to a series of quadratures and a second-order differential equation by Maartens and Nel (1978) (but note some corrections in Lorenz (1981)). Some explicit solutions were given by Collins and Stewart (1971) (cp. Collins (1971)); they arise from the non-LRS G_3II solutions (14.29) in the next section and can be written

$$\lambda^2\mathrm{d}s^2 = -\mathrm{d}t^2 + t^{2p_1}(\mathrm{d}x + nz\,\mathrm{d}y/2\gamma)^2 + t^{2p_2}(\mathrm{d}y^2 + \mathrm{d}z^2), \tag{14.20}$$

with $p_1 = (2-\gamma)/2\gamma$, $p_2 = (2+\gamma)/4\gamma$, $n^2 = (2-\gamma)(3\gamma-2)$ and $\lambda^2 = 4\gamma^2 M/(6-\gamma)$. Solution (14.20) has a homothety of the form (13.52) with $a_4 = 0$, $a_i = p_i$.

The $\gamma = 2$ perfect fluid solutions for (13.2) were found, with in addition a magnetic field, by Ruban (1971) (see also Ruban (1977, 1978)) and for $k = 1$ by Batakis and Cohen (1972), with the fluid considered as a scalar field (cp. Theorem 10.2). One has

$$\begin{aligned} &\mathrm{d}t = AB^2\mathrm{d}\eta, \quad A^2 = a/\cosh(a\eta), \quad 2AB = c/F(\tfrac{1}{2}c(\eta+\eta_0)), \\ &c^2 = a^2 + 4M, \quad F(u) = \cosh u,\ \mathrm{e}^u,\ \sinh u, \text{ respectively for } k = 1,\ 0,\ -1, \end{aligned} \tag{14.21}$$

where a, c and η_0 are constants. The $k = 0$ case of (14.21) is contained in the more general G_3II solution due to Ruban (1971), which is (23.1) with $b = 0$.

The Ricci tensor components (13.13) for the metric (13.3) show that if it contains a perfect fluid with 4-velocity ∂_t it is Robertson–Walker. The equations for tilted dust were reduced by Farnsworth (1967) to the form

$$\begin{aligned} &\mathrm{d}s^2 = a^2(b\dot{g}-g)^2\mathrm{d}x^2 + (g\mathrm{e}^{-x})^2(\mathrm{d}y^2+\mathrm{d}z^2) - \mathrm{d}t^2, \\ &g = g(t+bx), \quad \kappa_0\mu = 6\left(\ddot{g} - \tfrac{1}{3}\Lambda g\right)/(b\dot{g}-g), \\ &2g\ddot{g} + \dot{g}^2 - \Lambda g^2 - 1/a^2 = 0; \quad a, b \ \text{const}, \end{aligned} \tag{14.22}$$

where $\dot{g}(u) = \mathrm{d}g(u)/\mathrm{d}u$. Farnsworth also gave, as a special case, a Tolman–Bondi solution (see §15.5).

The (tilted) $\gamma = 2$ solution for (13.3) is included, as the case $m = n = 0$, in the solution (23.2) due to Wainwright *et al.* (1979); the case $a = \sqrt{2}$ was found by Maartens and Nel (1978).

The remaining known perfect fluid solutions for (13.1)–(13.3) do not obey the γ-law (5.36). Some were found by assuming other equations of state, e.g. the solution for the LRS G_3I metric (13.1), $k = 0$, with $\mu + 3p = \text{const}$ was found by Vishwakarma *et al.* (1999).

From (13.12) the metric (13.9), $f(w) < 0$, permits a perfect fluid matter content only if the four-velocity is $\boldsymbol{u} = u\partial_w$; then any pair of functions f and Y obeying (13.78) with $\epsilon = 1$ gives a perfect fluid solution. Several authors (e.g. Gaete and Hojman (1990), Hajj-Boutros and Sfeila (1989)) have used such pressure isotropy methods, as described in §13.4, to obtain new solutions, either from known solutions or ansätze on the metric functions.

For (13.1), Hajj-Boutros generated solutions in this way (1985) from (14.14) and (1986) from (14.18), Ram (1989d) from $A = at + b$, and

Bayin and Krisch (1986) from various choices of B in the case $k = -1$. Solutions with $B = bt$ for the Kantowski–Sachs case, (13.1) with $k = 1$, are included in (15.65); see also Shukla and Patel (1977). For the LRS G_3I metric (13.1), $k = 0$, new solutions have been obtained starting from Robertson–Walker metrics with $k = 0$ (Hajj-Boutros and Sfeila 1987), and the relevant special cases of (13.48), (14.26) and (14.28b) (Lorenz-Petzold 1987b, Ram 1989a, Singh and Ram 1995). Other starting points have been $\mathrm{d}t = A\mathrm{d}\tau$, $A = \sin\tau$ (Kitamura 1995a), $B = \sqrt{t}$ (Singh and Singh 1968, Ram 1989c), a condition arising from embedding considerations, $B = t$ (Singh and Abdussattar 1974) and $B = T$, $bT\mathrm{e}^{a/T}\mathrm{d}T = \mathrm{d}t$ (Singh and Abdussattar 1973). The assumption $A^2 \propto B$ was used by Assad and Damião Soares (1983) to obtain solutions for (13.1) with $k \neq 0$, Novikov (1964) found a special $k = 1$ solution illustrating the T-region concept, Biech and Das (1990) studied $A = \mathrm{e}^B$ with time variable $\mathrm{d}\tau = \mathrm{e}^{-B}\mathrm{d}t$ for $k = 1$, and Bradley and Sviestins (1984) studied a number of special cases for $f(t) = A/B$ in the case $k = -1$.

Another approach to (13.1) was given by Senovilla (1987a), who took a generalized Kerr-Schild ansatz and showed that it led from one perfect fluid solution to another (see §32.5, (32.101)–(32.103)).

For (13.2) perfect fluid solutions have been found by other ansätze. The first one found, by Collins *et al.* (1980), was for the case $k = -1$ (with G_3 of types III and VIII) and can be written as

$$\mathrm{d}s^2 = c^2\left\{\sinh^{10}\xi\left[-\mathrm{d}\xi^2 + \tfrac{1}{9}(\mathrm{d}y^2 + \sinh^2 y\,\mathrm{d}z^2)\right] + \tfrac{8}{81}\sinh^8\xi[\mathrm{d}x + \cosh y\,\mathrm{d}z]^2\right\}, \tag{14.23}$$

$$\kappa_0\mu = (56\sinh^2\xi + 63)/c^2\sinh^{12}\xi, \quad \kappa_0 p = -(16\sinh^2\xi + 9)/c^2\sinh^{12}\xi,$$

where c is an arbitrary constant. It was obtained (in another form) by complexifying the coordinates in a stationary axisymmetric solution. The hypersurface normal $\boldsymbol{n}$ has the properties that Θ and σ (as defined in §6.1) are proportional, and σ_{ab} has a repeated eigenvalue: Collins *et al.* (1980) generalized the metric form and shear eigenvector properties of (14.23) to give ansätze which led to new solutions, at least up to quadratures of an arbitrary function of t, in the other cases of (13.2) (LRS solutions with G_3 of types II and IX, $k = 0$ or 1), and in Bianchi type VI_0 and II metrics without rotational symmetry. This has been studied further by Banerjee and Santos (1984) for (13.2), $k = 0$. One of the Collins *et al.* (1980) solutions for (13.2) with $k = 0$ admits an H_4 (Koutras 1992b).

In examining shearfree cases, Collins and Wainwright (1983) found solutions for tilted LRS G_3V perfect fluids, up to an ordinary differential equation, which in comoving coordinates take the form

$$\begin{aligned} &\mathrm{d}s^2 = c^2[-U'^2\mathrm{d}t^2/a^2 + \mathrm{d}x^2 + \mathrm{e}^{-2x}(\mathrm{d}y^2+\mathrm{d}z^2)]/U^2, \quad U''+U'+U^2=0, \\ &\kappa_0\mu = (3a^2+2U^3+3U^2+6UU'+3U'^2)/c^2, \quad \kappa_0(\mu+p) = 2U^4/c^2U', \end{aligned} \tag{14.24}$$

where a and c are constants, $U=U(t+x)$, and $U'=\mathrm{d}U(w)/\mathrm{d}w$.

Starting from hypersurfaces of the Taub solution, Lozanovski and Aarons (1999) constructed a perfect fluid solution for (13.2) with $k=1$ which has a pure magnetic Weyl tensor (in the sense of §3.5) in the frame of the fluid,

$$\begin{aligned} k^2\mathrm{d}s^2 &= T[-\mathrm{d}T^2/2(T+\ell) + (T+\ell)(\boldsymbol{\sigma}^1)^2 + T(\mathrm{d}y^2+\sin^2 x\,\mathrm{d}z^2)], \\ \kappa_0\mu &= (27T+15\ell)/4k^2T^3, \quad p = (3\ell-9T)/4k^2T^3, \end{aligned} \tag{14.25}$$

where k and ℓ are constants.

A number of Ellis Class *III* perfect fluid solutions not obeying the γ-law were given by Marklund (1997), using the methods of Chapter 9 in the same way as (13.83) was derived.

14.4 Cosmologies with a G_3 on S_3

Once again, because of the occurrence of G_3 as subgroups of larger symmetry groups, we have to refer to §§12.4, 13.1, 14.2 and 14.3. We now list the other known solutions, taking first those obeying $p=(\gamma-1)\mu$. These may include as special cases LRS solutions admitting a G_4 but not explicitly given in §14.3.

Following the remark after (13.32), (14.1) can be substituted in (13.29a) which can then be used in (13.31), i.e.

$$\mathrm{d}s^2 = -\mathrm{d}t^2 + S(t)^2\left(F^{2\cos\psi}\mathrm{d}x^2 + F^{2\cos(\psi+2\pi/3)}\mathrm{d}y^2 + F^{2\cos(\psi-2\pi/3)}\mathrm{d}z^2\right),$$

changing the time variable to S, to give the general solution for perfect fluid in the G_3I case, up to a quadrature $F=\exp(2\int\Sigma\mathrm{d}t/\sqrt{3}S^3)$. However, it is often possible to perform this quadrature and/or to give the solution $S(t)$ explicitly in terms of the original proper time t of (13.20).

For dust, with $\Lambda\neq 0$, the general solutions (Saunders 1967), using the metric (13.31), are

$$\begin{aligned} &\Lambda>0, \quad S^3 = a\sinh\omega t + M(\cosh\omega t-1)/2\Lambda, \quad S^2F = (\cosh\omega t-1)^{2/3}, \\ &\Lambda=0, \quad S^3 = \tfrac{3}{4}\left(Mt^2+4t\right), \quad S^2F = t^{4/3}, \\ &\Lambda<0, \quad S^3 = a\sin\omega t + M(\cos\omega t-1)/2\Lambda, \quad S^2F = (1-\cos\omega t)^{2/3}, \end{aligned} \tag{14.26}$$

where $\omega^2 = 3|\Lambda|$, $\Lambda a^2 = 3$. There is a special solution $M^2 = 12\Lambda > 0$, $S^3 = a(\mathrm{e}^{\omega t} - 1)$, $F = (1 - \mathrm{e}^{-\omega t})^{2/3}$ (Lorenz 1982a). With $M = 0 \neq \Lambda$, (14.26) gives (13.63). The case $\Lambda = 0$ was earlier given by Raychaudhuri (1958) and Heckmann and Schücking (1958). The LRS cases ($\psi = 0$ or π in (13.30)) have been rediscovered several times.

Choosing a time variable $y \propto (S^3)\dot{}$, Jacobs (1968) was able to integrate G_3I γ-law perfect fluids with a general constant γ $(1 \neq \gamma \neq 2)$ for $F(y)$ and so give an exact solution. This general family can be written, with a different choice of new time variable τ, as (Wainwright 1984, Wainwright and Ellis 1997)

$$\mathrm{d}s^2 = -G^{2(\gamma-1)}\mathrm{d}\tau^2 + \tau^{2p_1}G^{2q_1}\mathrm{d}x^2 + \tau^{2p_2}G^{2q_2}\mathrm{d}y^2 + \tau^{2p_3}G^{2q_3}\mathrm{d}z^2, \tag{14.27a}$$

$$G^{2-\gamma} = \sqrt{3}\Sigma + \tfrac{9}{4}m\tau^{2-\gamma}, \quad q_\alpha = \tfrac{2}{3} - p_\alpha, \tag{14.27b}$$

where p_α, Σ and m, which contain two essential parameters, are as in (13.54), (13.29) and (14.1). A closed form in terms of S can be given for $\gamma = 4/3$ (Jacobs 1968; see also Shikin (1968)), and solutions as power series in S for $\gamma = 1 + n/(n+1)$ and $\gamma = 1 + (2n+1)/(2n+3)$ with integer n (Jacobs 1968). The limits $\Sigma = 0$ and $m = 0$ are respectively the Robertson–Walker and vacuum solutions. Solutions for G_3I with combinations of fluids with $\gamma = 1$, $4/3$, $5/3$ and 2 can be expressed in terms of elliptic functions (Jacobs 1968, Ellis and MacCallum 1969).

The solutions with $\gamma = 2$, including $\Lambda \neq 0$, are (Ellis and MacCallum 1969)

$$\Lambda > 0,\ S^3 = \sqrt{(3+M)/\Lambda}\,\sinh\omega t, \quad F = (\tanh(\omega t/2))^b, \tag{14.28a}$$

$$\Lambda = 0,\ S^3 = \sqrt{3(3+M)t}, \quad F = t^b, \tag{14.28b}$$

$$\Lambda < 0,\ S^3 = \sqrt{(3+M)/|\Lambda|}\,\sin\omega t, \quad F = (\tan(\omega t/2))^b, \tag{14.28c}$$

where $b \equiv 2/\sqrt{3(3+M)}$ and $\omega^2 = 3|\Lambda|$. The case (14.28b), which was found by Jacobs (1968) and Shikin (1968), can be expressed in the form (13.53), the only change being that $\sum_1^3(p_i)^2 < 1$, and it has a homothety of the form (13.52) (Koutras 1992b). Solution (14.28b) with $b\cos\psi = -1/3$ is of Petrov type D (Allnutt 1980), in addition to the LRS subcases.

The G_3I solutions with electromagnetic fields, given by Jacobs (1969), can be used to solve the G_3II fluid case (Collins 1972). Collins (1971) found a special G_3II solution, which can be written as

$$\begin{aligned}\mathrm{d}s^2 = &-G^{2(\gamma-1)}\mathrm{d}\tau^2 + \tau^{2p_1}G^{2p_1}(\mathrm{d}x + nz\mathrm{d}y)^2 v + \tau^{2p_2}G^{2p_3}\mathrm{d}y^2 \\ &+\tau^{2p_3}G^{2p_2}\mathrm{d}z^2, \qquad G^{2-\gamma} = a + 4M\tau^{2-\gamma}/(6-\gamma)\end{aligned} \tag{14.29}$$

(Wainwright and Ellis 1997), where a is a constant appearing in the shear, M gives μ as in (14.1), the ψ of (13.54) is given by $8\cos\psi = 2-3\gamma$, and $(6-\gamma)n^2 = (3\gamma-2)(2-\gamma)M$: here $\frac{2}{3} < \gamma < 2$. Special cases of (14.29) were integrated in other forms by Ruban (1978).

Hewitt (1991b) gave a tilted Bianchi II γ-law solution, $\frac{10}{7} < \gamma < 2$:

$$\begin{aligned} ds^2 &= -dt^2 + t^{2-4q}dx^2 + t^{6q}dy^2 + t^{2q}[dz + (4Wt^{2q}/q + 2nx)dy/\gamma]^2, \\ q &= (2-\gamma)/2\gamma, \quad n^2 = (2-\gamma)(3\gamma-4)(5\gamma-4)/(17\gamma-18), \\ w^2 &= (2-\gamma)(7\gamma-10)(11\gamma-10)/64(17\gamma-18), \quad \kappa_0\mu = 2(2-\gamma)/\gamma^2t^2. \end{aligned} \tag{14.30}$$

This solution admits an H_4, with a homothety $t\partial_t + 2qx\partial_x + (1-3q)y\partial_y + (1-q)z\partial_z$, and is the unique such solution with an orthogonally-transitive G_2. The fluid is accelerated but not rotating, and is tilted relative to the $t =$ const surfaces.

A family of tilted Bianchi II γ-law solutions with $\gamma = 14/9$ was found by Hewitt *et al.* (2001). It has the form

$$\begin{aligned} ds^2 &= -e^{14t/3}dt^2 + [e^{t/3}(dx + nz\,dy) + c(e^t dy + be^{5t/3}dz)]^2 \\ &\quad +(e^t dy + 2be^{5t/3}dz)^2 + e^{10t/3}dz^2, \\ c^2 &= 4(4b^2+1)(8-3b^2)/19, \quad n^2 = 4(2b^2+1)(17-8b^2)/57, \\ \kappa_0\mu &= 2(1-b^2)e^{-14t/3}, \quad v^2 = 4(4b^2+1)(2b^2+1)/19cn, \quad 0 < b < 1, \end{aligned} \tag{14.31}$$

where, using components in the orthonormal tetrad basis implied by expression (14.31) for the metric, $v = u^3/u^0$ gives the fluid velocity. These solutions also admit an H_4, with a homothety $3\partial_t + 6x\partial_x + 4y\partial_y + 2z\partial_z$.

The general non-tilted G_3II $\gamma = 2$ solution was found, with an added magnetic field, by Ruban (1971). In the form due to Wainwright *et al.* (1979) derived from the vacuum solution (13.55), it is given by (23.1) with $b = 0$. The tilted G_3II solution with $\gamma = 2$ has been reduced to an ordinary differential equation defining the third Painlevé transcendent, followed by quadratures (Maartens and Nel 1978). The G_3VI_0 and VII_0 vacuum solutions (13.60) similarly give a $\gamma = 2$ solution if (13.60b) is modified by adding $4M/q^2u$ on the right hand side (Lorenz-Petzold 1984).

For Bianchi type V, (13.29), (13.32) and (14.1) provide the general solution for non-tilted γ-law fluids up to a quadrature, using a time variable S, as remarked in §13.2. Limits giving vacuum, Robertson–Walker and G_3I metrics exist. Fiser *et al.* (1992) noted that for suitable time variables S^δ and values of γ the remaining quadrature gives an elementary function or elliptic integrals: the γ concerned are $\gamma = 1, \frac{10}{9}, \frac{4}{3}, \frac{14}{9}, \frac{5}{3}$ and 2. Formulae in terms of the t of (13.20) could be given in terms of elliptic functions for $\gamma = 2$ or $\frac{4}{3}$. For $\gamma = \frac{4}{3}$ one can write the solution as (Ruban 1977)

$$ds^2 = c^2 S^2 \left[-dt^2 + dx^2 + e^{2x}(F^{\sqrt{3}}dy^2 + F^{-\sqrt{3}}dz^2)\right],$$
$$S^2 = \sinh t(\alpha \cosh t + m \sinh t), \quad S^2 F = \sinh^2 t, \quad \kappa_0 \mu = 3m/c^2 S^4. \tag{14.32}$$

The metric (13.59) with $\boldsymbol{\omega}^1 = dx$, $\boldsymbol{\omega}^2 = e^{Bx}dy$ and $\boldsymbol{\omega}^3 = e^{Bx}dz$, $A \neq B$, is a perfect fluid solution with $\mu + 3p = 0$ (Koutras, private communication).

Solutions (23.36) with $c = 0 \neq b\ell$ give a generalization of (14.32) with a tilted fluid, $1 + 1/\sqrt{10} < \gamma < 2$ and a group of type G_3VI_h with $h = -a^2/4b^2\ell^2$, or type V when $b\ell = 0$.

The only other explicit solutions known to us with G_3 on S_3 containing a γ-law perfect fluid, $\gamma \neq 2$, are special cases of the Bianchi type VI_h metrics with $n^\alpha{}_\alpha = 0$. They include metrics of Bianchi types III and V and as limits metrics of types VI_0 and I.

Collins (1971) found special G_3VI_h solutions with $n^\alpha{}_\alpha = 0$ which are included in (14.35) below and, using (13.27), can be put in the form

$$ds^2 = -dt^2 + 4\gamma^2 t^2 (\boldsymbol{\omega}^1)^2/q^2 + t^{b+c}(\boldsymbol{\omega}^2)^2 + t^{b-c}(\boldsymbol{\omega}^3)^2,$$
$$b = (2/\gamma - 1), \quad c = A\,(3 - 2/\gamma), \quad q^2 = (2-\gamma)(3\gamma - 2), \tag{14.33}$$
$$\kappa_0 \mu = ((2-\gamma) - A^2(3\gamma - 2))/\gamma^2 t^2,$$

where $\mu > 0 \Leftrightarrow (3\gamma - 2)(1 - 3h) < 4$. These solutions have a homothetic motion $\boldsymbol{H} = t\partial_t + (1 - (b+c)/2)y\partial_y + (1 - (b-c)/2)z\partial_z$. A class of perfect fluid G_3VI_0 space-times with $n^\alpha{}_\alpha = 0$, also given by Collins (1971) and admitting an H_4, is obtained by specializing (14.33) to $A = 0$; they include the dust solution found by Ellis and MacCallum (1969). This class and its dust subcase have been rediscovered by other authors.

There are several solutions for G_3VI_h ($n^\alpha{}_\alpha = 0$) which can be expressed in the form

$$ds^2 = c^2 \left[-F^{2a_4}G^{2b_4}dt^2 + F^{2a_1}G^{2b_1}dx^2 + F^{2a_2}G^{2b_2}e^{2c_2 x}dy^2 + F^{2a_3}G^{2b_3}e^{2c_3 x}dz^2\right], \tag{14.34}$$

where c, a_i, b_i and c_i are constants and F and G are functions of t; (14.34) can be generalized to a form including the G_3I and G_3II solutions (14.27) and (14.29) (Wainwright 1984).

A perfect fluid solution of the form (14.34) not obeying the γ-law was given by Wainwright (1983). It has

$$\begin{gathered}a_1 = 1,\ a_2 = c_2 = u + v,\ a_3 = c_3 = u - v,\ a_4 = 0,\ b_4 = b_1 = u + w - 1,\\ b_2 = w - v, \quad b_3 = w + v, \quad \kappa_0\mu = Ma(1 + 2u) + Q(v^2 + 3w^2),\\ F = t, \quad G^{2w} = a + mt^{2(w-u)}, \quad \kappa_0 p = Ma(1 - 2w) + Q(v^2 - w^2),\\ M \equiv 2m(w - u)G^{2(1-u-3w)}/k^2t^{2(1-w+u)}, \quad Q \equiv Mm(u + w)t^{2(w-u)}/2w^2,\end{gathered} \tag{14.35}$$

where a, m, u, v and w are constants obeying $u = u^2 + v^2$, $u < w < u + 1$. This reduces to a plane wave if $m = 0$ and (14.33) if $a = 0$: in general one can set $a = 1 = k$.

To describe the known γ-law perfect fluid solutions in the form (14.34) we use

$$n\sqrt{1 - 3h} = 1, \quad n = -\cos\psi, \quad q_\alpha = \tfrac{2}{3} - p_\alpha, \tag{14.36}$$

where p_α is given by (13.54).

For the Collins (1971) G_3VI_h solution with $4n = 3\gamma - 2$, $\frac{2}{3} < \gamma < 2$,

$$\begin{gathered}a_4 = b_4 = (4n - 1)/4(1 - n), \quad 4(1 - n)a_\alpha = 3p_\alpha, \quad 4(1 - n)b_\alpha = 3q_\alpha,\\ F = t, \qquad 2(1 - n)c_{2,3} = \beta(\sqrt{1 - n^2} \pm \sqrt{3}n),\\ G = \alpha + mt + \beta^2(\frac{1 - n}{1 + n})t^{(1+n)/(1-n)}, \quad \kappa_0\mu = \frac{4m}{3c^2(2 - \gamma)^2(Gt)^{\gamma/(2-\gamma)}}.\end{gathered} \tag{14.37}$$

Here $\beta = 0$ gives G_3I solutions, which are Robertson–Walker if $\alpha = 0$, $m = 0$ gives the vacuum solution, (13.57) with $\epsilon = 1$, and $n = 1/2$ gives the LRS G_3III radiation solution, (14.16) with $k = -1$. Only two of α, β, c and m are essential in general.

For $8n = 3\gamma + 2$, $0 < \gamma < 2$, Uggla and Rosquist (1990) found, with the a_i and b_i of (14.37) in terms of n, and the c_i interchanged,

$$G = F_{,t}\left(\alpha + \beta^2 \int \frac{F^r}{F_{,t}^2}\mathrm{d}t\right), \quad \kappa_0\mu = \frac{16}{3c^2(2 - \gamma)^2(FG)^{\gamma/(2-\gamma)}}, \tag{14.38}$$

where $r = 2n/(1 - n)$ and the three possible forms of F are $F = \cosh t$, e^t, and $\sinh t$. Again $\beta = 0$ gives G_3I solutions, which are Robertson–Walker in the case $F = \mathrm{e}^t$. Otherwise we can set $\beta = 1$.

For $2n = 4 - 3\gamma$, $\frac{2}{3} < \gamma < \frac{4}{3}$, Uggla and Rosquist (1990) found solutions with the F, G and r of (14.38) but with

$$\begin{gathered}a_4 = a_1, \quad 2(1 - n)a_\alpha = 3q_\alpha, \quad b_4 = b_1, \quad 2(1 + n)b_\alpha = 3p_\alpha,\\ \kappa_0\mu = \frac{4\beta^2(F^{1/3(\gamma-2)}G^{1/(3\gamma-2)})^{3\gamma-8}}{(2 - \gamma)(3\gamma - 2)c^2}, \quad c_{2,3} = \frac{\sqrt{1 - n^2} \mp \sqrt{3}n}{\sqrt{1 - n^2}},\end{gathered} \tag{14.39}$$

and for $\frac{4}{3} < \gamma < 2$ the same with $n \to -n$. If $\beta = 0$ we obtain the vacuum solution, (13.57), and if $\gamma = 1$ the LRS G_3III dust solution, (14.15b), while the limit $n = 0$ gives the G_3V radiation solution (14.32).

As the last case with (14.34), we have (Uggla 1990),with $n = \frac{4}{5}$, $\gamma = \frac{6}{5}$,

$$\begin{aligned} &4a_4 = 1,\ 4a_\alpha = 5p_\alpha,\ 4b_4 = 11,\ 4b_\alpha = 15q_\alpha,\ 2\sqrt{2}\,c_{2,3} = \sqrt{3} \mp 4, \\ &F = \sinh t + \sin(t+\alpha),\quad G = \sinh t - \sin(t+\alpha),\quad \kappa_0\mu = \frac{25}{8c^2\sqrt{F^3G^9}}. \end{aligned} \tag{14.40}$$

Wainwright (1984) gave a $G_3VI^*_{-1/9}$ solution for $\gamma = 10/9$,

$$\begin{aligned} \mathrm{d}s^2 &= -\mathrm{d}t^2 + t^2\mathrm{d}x^2 + t^{2/5}\left[\exp(-\sqrt{6}rx/5)\mathrm{d}y + bt^{4/5}\mathrm{d}x\right]^2 \\ &\quad + t^{6/5}\exp(4\sqrt{6}rx/5)\mathrm{d}z^2,\quad 4(b^2+1) = 9r^2,\quad 2/3 < r < 1, \end{aligned} \tag{14.41}$$

with a homothety $\boldsymbol{H} = 5t\partial_t + 4y\partial_y + 2z\partial_z$. The density is $\kappa_0\mu = 27(1 - r^2)/25t^2$. In the limit $r \to 2/3$ (14.41) reduces to (14.33) with $\gamma = 10/9$, and for $r \to 1$ to the vacuum case of (13.64).

A class of tilted γ-law perfect fluid Bianchi VI_0 solutions was given by Rosquist and Jantzen (1985), using (13.27) with $A = 0$, as

$$\begin{aligned} \mathrm{d}s^2 &= -\mathrm{d}t^2 + (kt\boldsymbol{\omega}^1)^2 + (t^{(-q+s)}\boldsymbol{\omega}^2 + mkt\boldsymbol{\omega}^1)^2 + (t^{(q+s)}\boldsymbol{\omega}^3)^2, \\ s &= (2-\gamma)/2\gamma,\quad 4\gamma(36-35\gamma)q^2 + 4(5\gamma-6)(\gamma-2)q + (5\gamma-6)^2 = 0, \\ k^2 &= -(3s+3q-1)/(s+3q-1)(3s^2+(6q-1)s-q^2-q), \\ m^2 &= -32q^2s/(s-q-1)^2(3s+3q-1). \end{aligned} \tag{14.42}$$

Writing $q_\pm = (6-5\gamma)[2-\gamma \pm 2\sqrt{(9\gamma-1)(\gamma-1)}\,]/2\gamma(35\gamma-36)$ for the values of q, the fluid has corresponding density and orthonormal tetrad velocity components given by

$$\begin{aligned} \kappa_0\mu &= \frac{2[\mp(5\gamma-6)\sqrt{(9\gamma-1)(\gamma-1)} - 15(\gamma-2)(\gamma-1)]}{\gamma^2(35\gamma-36)t^2}, \\ u^1 &= k(q-3s+1)u^4,\quad u^2 = m(1+q-s)u^1/4q, \\ u^3 &= 0,\quad \gamma\kappa_0\mu k^2t^2(q-3s+1)(u^4)^2 = -2q. \end{aligned} \tag{14.43}$$

The pair (γ, q) has to give $k^2 \geq 0$; for $q = q_-$, $1 < \gamma < 2$ but for $q = q_+$, $6/5 < \gamma < 1.7169\ldots$. The case $\gamma = 4/3$ was given by Rosquist (1983). These solutions admit a homothety generated by $t\partial_t + (1+q-s)y\partial_y + (1-q-s)z\partial_z$. They have vorticity as well as expansion and shear and give the only such explicitly known solutions. In principle solutions with other G_3 or H_3 on S_3 were also given in Rosquist and Jantzen (1985), but the explicit parameter values required were not found.

The matter is also rotating, shearing and expanding in the G_3VII_0 solution discussed by Demiański and Grishchuk (1972), who imposed the condition that the orbits of the group are flat, but this is only known up to a differential equation. Some other solutions with flat slices, of Petrov type I, and with shear and rotation, similar to (36.22), are given in Wolf (1986b).

A number of stiff fluid solutions obtained by using Theorem 10.2 are known as special cases of solutions with maximal groups H_3 and G_2 (for which see Chapter 23). The G_3VI_h stiff fluid solution corresponding to (13.57) is obtained by setting

$$m^2 = 3 + n^2 + 2(b^2 - a^2) \tag{14.44}$$

in (23.2) (Wainwright *et al.* 1979). It includes LRS and non-LRS G_3III ($n^2 = 1$) and G_3V ($n = 0$) solutions, in particular (14.18b): the G_3V solutions can be generated from (13.59). The non-tilting case was also found, in another form, by Collins (1971), and by Ruban (1977, 1978).

Setting $nm = ab$ in (23.3) gives a tilted fluid with a G_3VI_0 (Wainwright *et al.* 1979), which is a limit of (23.2) with (14.44) as (13.58) is of (13.57). The non-tilted case is the G_3VI_0 ($n^\alpha{}_\alpha = 0$) solution found by Ellis and MacCallum (1969).

Similarly for Bianchi type VII_h, one needs

$$m^2 - \tfrac{11}{4} + 2(a^2 - b^2) = 0 \tag{14.45}$$

in (23.4). The non-tilted case was found by Barrow (1978).

Putting $p = q$ in (23.6) with $F = u^{-2p}$ gives a tilted stiff fluid solution with a G_3VI_{-4} on $u/t =$ const (Collins 1991), and the case $a = \frac{1}{2}$ of the stiff fluid solution (23.8) admits a $G_3VI_{-1/9}$.

The Petrov type D fluid solution with a group G_3VI_h on surfaces $\mathrm{e}^{ax}/t =$ const given in comoving coordinates by Allnutt (1980) as

$$\mathrm{d}s^2 = dx^2 + \mathrm{e}^{-2ax}(t^{1+n}\mathrm{d}y^2 + t^{1-n}\mathrm{d}z^2 - \mathrm{d}t^2) \tag{14.46}$$

can be interpreted as a stiff fluid with density $\kappa_0\mu = (1-n^2)\mathrm{e}^{2ax}/4t^2$ and a cosmological constant $\Lambda = -3a^2$. For $n = 0$, where the group is a G_3V, this is (15.75) with $b = k = 0$, $a > 0$, $c = 1$. The group orbits are S_3 or T_3 depending on the sign of $t^2a^2 - \mathrm{e}^{2ax}$. This and some analogous G_3VI_h solutions have an inheriting conformal Killing vector (Czapor and Coley 1995, Vera 1998a).

As mentioned in §10.11.2, a G_3VI_0 solution for perfect fluid not obeying the γ-law was found by Van den Bergh (1986d), as a conformal transform of the vacuum metric (13.58): the required conformal factor multiplying (13.58) is $\int u^{1/4}e^{(4u^2-1)/8}du$.

Fluid solutions not obeying the γ-law (5.36) can be obtained by the pressure isotropy methods outlined in §13.4. Ram (1990) applied these ideas to $k=-1$ Robertson–Walker metrics to obtain G_3V solutions, and for G_3VI_0, $n^\alpha{}_\alpha=0$, considered $A=at+b$ (1988). Lorenz-Petzold (1987a) generated G_3VI solutions from (14.33); for $\gamma=1$, Ram (1989b) iterated the process.

Fluid solutions of type G_3I with equation of state $\mu=a\sqrt{p}+p$ for constant a and a Λ-term were given by Kellner (1975). Further G_3V solutions with σ/Θ constant and G_3VI_0 solutions with purely magnetic Weyl tensor were obtained by Roy and Tiwari (1982). Sklavenites (1992a) gave some G_3 solutions containing orthogonally-transitive G_2I on S_2 (see Chapter 23), which include as special cases some of the γ-law solutions above. Metric (23.26) contains some special G_3VI solutions. As in the case of the Robertson–Walker metrics, a number of authors, e.g. Aguirregabiria *et al.* (1993b), have considered various scalar field models in the metric forms of this section.

15

Groups G_3 on non-null orbits V_2. Spherical and plane symmetry

15.1 Metric, Killing vectors, and Ricci tensor

A Riemannian space V_q admitting a group G_r, $r = q(q+1)/2$, is a space of constant curvature (§8.5). Hence the orbits V_2 of a group G_3 of motions must have constant Gaussian curvature K, and the 2-metric $\mathrm{d}\sigma^2$ of the spacelike (S_2) or timelike (T_2) orbits can be written in the form (8.38):

$$\mathrm{d}\sigma^2 = Y^2[(\mathrm{d}x^1)^2 \pm \Sigma^2(x^1, k)(\mathrm{d}x^2)^2] \tag{15.1}$$

with $k = KY^2$, and, as in §8.5,

$$\Sigma(x^1, k) = (\sin x^1,\ x^1,\ \sinh x^1) \quad \text{for} \quad k = (1,\ 0,\ -1). \tag{15.2}$$

In (15.1) and in the following formulae the upper and lower signs refer to spacelike and timelike orbits, respectively. The function Y in (15.1) is independent of the coordinates x^1 and x^2 in the orbits V_2. However, Y in general depends on coordinates x^3 and x^4 because the orbits V_2 are subspaces ($x^3 = \text{const}$, $x^4 = \text{const}$) of the space-time V_4.

From Theorem 8.19 it follows that the orbits V_2 admit *orthogonal surfaces* in V_4. By performing a coordinate transformation in the 2-spaces orthogonal to the orbits we can put the space-time metric into diagonal form (Goenner and Stachel 1970)

$$\begin{gathered}\mathrm{d}s^2 = Y^2[(\mathrm{d}x^1)^2 \pm \Sigma^2(x^1, k)(\mathrm{d}x^2)^2] + \mathrm{e}^{2\lambda}(\mathrm{d}x^3)^2 \mp \mathrm{e}^{2\nu}(\mathrm{d}x^4)^2, \\ Y = Y(x^3, x^4), \qquad \lambda = \lambda(x^3, x^4), \qquad \nu = \nu(x^3, x^4).\end{gathered} \tag{15.3}$$

Note that for timelike orbits T_2 the coordinate x^2 is timelike, while x^4 is spacelike. For spacelike orbits S_2, it is sometimes more convenient to use the space-time metric

$$\mathrm{d}s^2 = Y^2(u,v)[(\mathrm{d}x^1)^2 + \Sigma^2(x^1,k)(\mathrm{d}x^2)^2] - 2G(u,v)\mathrm{d}u\,\mathrm{d}v \tag{15.4}$$

(cp. (15.18) and (15.24) below).

In the S_2 case, we have to distinguish between $Y_{,m}Y^{,m} > 0$ (R-region), $Y_{,m}Y^{,m} < 0$ (T-region) (see McVittie and Wiltshire (1975) and references cited therein) and $Y_{,m}Y^{,m} = 0$. The case $Y_{,m}Y^{,m} < 0$ cannot occur for timelike orbits T_2, because of the Lorentzian signature of the metric.

Coordinate transformations which preserve the form (15.3) of the metric can be used to reduce the number of functions in the line element (15.3). For instance, for $Y_{,a}Y^{,a} \neq 0$ we can always set $Y = x^3 \mathrm{e}^{\lambda}$ (isotropic coordinates), and for $Y_{,m}Y^{,m} > 0$ we can choose $Y = x^3$ (canonical coordinates).

In the coordinate system (15.3), the Killing vectors $\boldsymbol{\xi}_A$ are given by

$$\begin{aligned} \boldsymbol{\xi}_1 &= \cos x^2\,\partial_1 - \sin x^2\,\Sigma_{,1}\Sigma^{-1}\partial_2, \qquad \boldsymbol{\xi}_2 = \partial_2, \\ \boldsymbol{\xi}_3 &= \sin x^2\,\partial_1 + \cos x^2\,\Sigma_{,1}\Sigma^{-1}\partial_2 \end{aligned} \tag{15.5a}$$

for spacelike orbits and

$$\begin{aligned} \boldsymbol{\xi}_1 &= \cosh x^2\,\partial_1 - \sinh x^2\,\Sigma_{,1}\Sigma^{-1}\partial_2, \qquad \boldsymbol{\xi}_2 = \partial_2, \\ \boldsymbol{\xi}_3 &= -\sinh x^2\,\partial_1 + \cosh x^2\,\Sigma_{,1}\Sigma^{-1}\partial_2 \end{aligned} \tag{15.5b}$$

for timelike orbits. The group types are: for S_2, *IX* ($k=1$), *VII*$_0$ ($k=0$), *VIII* ($k=-1$), and for T_2, *VIII* ($k=1$), *VI*$_0$ ($k=0$), and *VIII* ($k=-1$). These Bianchi types belong to class G_3A (§8.2). The spacelike and timelike metrics on V_2, and their corresponding groups, are related by complex transformations.

The existence of a higher-dimensional group of motions, G_r, $r > 3$, imposes further restrictions on the functions ν, λ and Y in the metric (15.3). The de Sitter universe (8.34), the hypersurface-homogeneous space-times (13.1)–(13.3) with a spatial rotation isotropy, the Friedmann models (§14.2), the Kantowski–Sachs solutions (14.15), and the static spherically-symmetric perfect fluid solutions (§16.1) are solutions admitting a G_r, $r > 3$, with a subgroup G_3 on V_2. For certain Ricci tensor types, a group G_3 on V_2 *implies* a G_4 on V_3 (§15.4).

For the metric (15.3), the non-zero components G_a^b of the Einstein tensor are (prime and dot denote differentiation with respect to the coordinates x^3 and x^4 respectively)

$$G_4^4 = -\frac{k}{Y^2} + \frac{2}{Y}\mathrm{e}^{-2\lambda}\left(Y'' - Y'\lambda' + \frac{Y'^2}{2Y}\right) \mp \frac{2}{Y}\mathrm{e}^{-2\nu}\left(\dot{Y}\dot{\lambda} + \frac{\dot{Y}^2}{2Y}\right), \tag{15.6a}$$

$$G^3_3 = -\frac{k}{Y^2} \mp \frac{2}{Y}\mathrm{e}^{-2\nu}\left(\ddot{Y} - \dot{Y}\dot{\nu} + \frac{\dot{Y}^2}{2Y}\right) + \frac{2}{Y}\mathrm{e}^{-2\lambda}\left(Y'\nu' + \frac{Y'^2}{2Y}\right), \quad (15.6b)$$

$$G^1_1 = G^2_2 = \mathrm{e}^{-2\lambda}[\nu'' + \nu'^2 - \nu'\lambda' + Y''/Y + Y'\nu'/Y - Y'\lambda'/Y] \mp \mathrm{e}^{-2\nu}[\ddot{\lambda} + \dot{\lambda}^2 - \dot{\lambda}\dot{\nu} + \ddot{Y}/Y + \dot{Y}\dot{\lambda}/Y - \dot{Y}\dot{\nu}/Y], \quad (15.6c)$$

$$G^4_3 = \pm 2e^{-2\nu}(\dot{Y}' - \dot{Y}\nu' - Y'\dot{\lambda})/Y. \quad (15.6d)$$

The Bianchi identities imply the existence of a *mass function*

$$m(r,t) = \tfrac{1}{2}Y(k - Y_{,a}Y^{,a}) \quad (15.7a)$$

for which

$$m' = \tfrac{1}{2}Y^2(\dot{Y}G^4_3 - Y'G^4_4), \quad \dot{m} = \tfrac{1}{2}Y^2(Y'G^3_4 - \dot{Y}G^3_3) \quad (15.7b)$$

holds (Lemaître 1949);this is the same m which occurs e.g. in (15.12) and (16.5).

Ref.: For a classification of spherically-symmetric metrics, see Takeno (1966), and of plane-symmetric metrics, Hergesell (1985).

15.2 Some implications of the existence of an isotropy group I_1

The group G_3 on V_2 implies an isotropy group I_1 (§11.2) and this, in turn, implies a (one-dimensional) linear isotropy subgroup of the Lorentz group $L^{\uparrow}_{+}$ in the tangent space T_p. Therefore the principal tetrad (§4.2) cannot be determined uniquely; only degenerate Petrov types (N, D, O) are possible. As G_3 acts on *non-null* orbits, I_1 describes spatial rotations (boosts) for spacelike (timelike) orbits. For Petrov type N, the invariance subgroup of $L^{\uparrow}_{+}$ consists of null rotations (3.15). Therefore type N cannot occur and we have proved

Theorem 15.1 *Space-times admitting a group G_3 of motions acting on non-null orbits V_2 are of Petrov type D or O.*

Similarly, the existence of an isotropy group I_1 leads to

Theorem 15.2 *The Ricci tensor of a space-time admitting a group G_3 on V_2 has at least two equal eigenvalues.*

Any invariant timelike vector field in V_4 (with zero Lie derivative with respect to the Killing vectors (15.5)) necessarily lies in the 2-spaces orthogonal to the group orbits S_2, because otherwise the preferred vector field would not be invariant under the isotropies. Hence we can formulate

Theorem 15.3 *Perfect fluid (with $T_{ik} \neq \Lambda g_{ik}$) and dust solutions cannot admit a group G_3 on timelike orbits T_2.*

In the coordinate system (15.3) (upper signs) the 4-velocity $\boldsymbol{u}$ of a perfect fluid has the form $\boldsymbol{u} = u^3\partial_3 + u^4\partial_4$ and from the invariance of $\boldsymbol{u}$ under isometries one infers that the components u^3, u^4 cannot depend on x^1 and x^2. Therefore $\boldsymbol{u}$ is hypersurface-orthogonal and there is a transformation of x^3 and x^4 which takes $\boldsymbol{u}$ into

$$\boldsymbol{u} = \mathrm{e}^{-\nu}\partial_4 \tag{15.8}$$

(comoving coordinates) and, simultaneously, preserves the form of the metric (15.3).

15.3 Spherical and plane symmetry

The group G_3 on V_2 contains two special cases of particular physical interest: spherical and plane symmetry.

In the first four decades of research on general relativity the majority of exact solutions were obtained by solving the field equations under the assumption of spherical symmetry. The exterior and interior Schwarzschild solutions and the Friedmann model of relativistic cosmology are well-known examples.

Originally, problems with spherical and plane symmetry were treated more or less intuitively. In the modern literature the group theoretical approach is preferred and spherical and plane symmetry are invariantly defined by the

Definition: *A space-time V_4 is said to be spherically- (plane-) symmetric if it admits a group G_3IX (G_3VII_0) of motions acting on spacelike 2-spaces S_2 and if the non-metric fields inherit the same symmetry.*

Each orbit S_2 has constant positive (zero) Gaussian curvature, $k = 1$ ($k = 0$). The isotropy group I_1 represents a spatial rotation in the tangent space of S_2. According to Theorem 15.1, spherically- and plane-symmetric space-times are of Petrov type D or O.

In the expressions (15.6) for the components of the Einstein tensor we have to choose the upper signs, and $k = 1$ or $k = 0$. For *spherical symmetry* ($k = 1$), we can specialize the metric (15.3) to

$$\mathrm{d}s^2 = Y^2(r,t)(\mathrm{d}\vartheta^2 + \sin^2\vartheta\,\mathrm{d}\varphi^2) + \mathrm{e}^{2\lambda(r,t)}\mathrm{d}r^2 - \mathrm{e}^{2\nu(r,t)}\mathrm{d}t^2. \tag{15.9}$$

For *plane symmetry* ($k = 0$), we often use Cartesian coordinates in the orbits:

$$\begin{aligned} \mathrm{d}s^2 &= Y^2(z,t)(\mathrm{d}x^2 + \mathrm{d}y^2) + \mathrm{e}^{2\lambda(z,t)}\mathrm{d}z^2 - \mathrm{e}^{2\nu(z,t)}\mathrm{d}t^2, \\ \boldsymbol{\xi}_1 &= \partial_x, \qquad \boldsymbol{\xi}_2 = \partial_y, \qquad \boldsymbol{\xi}_3 = x\partial_y - y\partial_x. \end{aligned} \tag{15.10}$$

For vacuum, Einstein–Maxwell and pure radiation fields, and for dust, the general solutions admitting a G_3 on V_2 are known (§§15.4, 15.5). Perfect fluid solutions in general are discussed in §15.6, and plane-symmetric perfect fluids in §15.7. Chapter 16 gives a survey of the spherically-symmetric perfect fluids.

15.4 Vacuum, Einstein–Maxwell and pure radiation fields

For vacuum and Einstein–Maxwell fields (including the cosmological constant Λ), and pure radiation fields, the algebraic types of the energy-momentum tensor are [(111, 1)], [(11) (1,1)], and [(11, 2)], see §5.2.

The eigenvalues λ_α of the Einstein tensor are

$$\begin{aligned} &\lambda_1 = \lambda_2 = G^1_1 = G^2_2, \\ &\lambda_{3,4} = \tfrac{1}{2}(G^3_3 + G^4_4) \pm \Delta^{1/2}, \ \Delta \equiv \tfrac{1}{4}(G^3_3 - G^4_4)^2 \mp (G^4_3)^2. \end{aligned} \tag{15.11}$$

The symmetry implies that a double eigenvalue ($\lambda_1 = \lambda_2$) exists (see Theorem 15.2). For the algebraic types under consideration, there are *two* double eigenvalues (which might coincide); $\lambda_3 = \lambda_4$ implies $\Delta = 0$.

15.4.1 Timelike orbits

For timelike orbits T_2, Table 5.2 shows immediately that the algebraic type is [11(1, 1)]. Since the type [(11, 2)] is impossible we have

Theorem 15.4 *Einstein–Maxwell fields with an electromagnetic null field, and pure radiation fields, cannot admit a G_3 on T_2.*

For vacuum and non-null Einstein–Maxwell fields, we have to distinguish between $Y_{,a} \neq 0$ and $Y_{,a} = 0$. If $Y_{,a} \neq 0$, one can put $Y = x^3$ (canonical coordinates), because of $\underline{Y_{,a}Y^{,a}} > 0$, and the evaluation of $G^3_3 = G^4_4$ and $G^4_3 = 0$ in the metric (15.3) (lower signs) gives $\lambda' + \nu' = 0$ and $\dot\lambda = 0$. In ν an additive term dependent on x^4 can be made zero by transforming x^4. Thus we have $\nu = -\lambda$, $\dot\lambda = 0$. In the coordinates of (15.3), the general solution of the Einstein–Maxwell equations (including Λ) reads (cp. (13.48))

$$\begin{aligned} &e^{2\nu} = k - 2m/x^3 - e^2/(x^3)^2 - \Lambda(x^3)^2/3 > 0, \\ &\lambda = -\nu, \quad Y = x^3, \quad m, e = \text{const}, \quad k = 0, \pm 1, \end{aligned} \tag{15.12}$$

the only non-vanishing tetrad component of the (non-null) electromagnetic field tensor being (up to a constant duality rotation, see §5.4)

$$\sqrt{2\kappa_0}\, F_{34} = 2e/(x^3)^2. \tag{15.13}$$

Table 15.1. The vacuum, Einstein–Maxwell and pure radiation solutions with G_3 on S_2 ($Y_{,a}Y^{,a} > 0$)

Segre type	$2H(u,r)$	Solution for $k = 1$
[(111,1)]	$k - 2m/r - \frac{1}{3}\Lambda r^2$	$\Lambda = 0$: Schwarzschild (1916a) $\Lambda \neq 0$: Kottler (1918)
[(11)(1,1)]	$k - 2m/r + e^2/r^2 - \frac{1}{3}\Lambda r^2$	$\Lambda = 0$: Reissner (1916), Nordström (1918)
[(11,2)]	$k - 2m(u)/r - \frac{1}{3}\Lambda r^2$	$\Lambda = 0$: Vaidya (1943)

These type D Einstein–Maxwell fields belong to Kundt's class (Chapter 31).

For $Y = Y_0 =$const, two double eigenvalues exist:

$$\lambda_1 = \lambda_2 = K_\perp, \qquad \lambda_3 = \lambda_4 = -kY_0^{-2}, \tag{15.14}$$

where $K_\perp$ denotes the Gaussian curvature of the 2-spaces with

$$\mathrm{d}\sigma_\perp^2 = (\mathrm{e}^\lambda \mathrm{d}x^3)^2 + (\mathrm{e}^\nu \mathrm{d}x^4)^2 \tag{15.15}$$

orthogonal to the orbits T_2. There is only one Einstein–Maxwell field (for $K_\perp = kY_0^{-2} < 0$), namely the Bertotti–Robinson solution (12.16).

15.4.2 Spacelike orbits

For spacelike orbits S_2, and $Y_{,a}Y^{,a} > 0$, $\lambda_3 = \lambda_4$ implies that

$$\Delta = 0 \quad \leftrightarrow \quad \partial(\mathrm{e}^{\nu+\lambda})/\partial r + \partial(\mathrm{e}^{2\lambda})/\partial t = 0 \tag{15.16}$$

($Y = x^3 = r$, $x^4 = t$) is satisfied (Plebański and Stachel 1968, Goenner and Stachel 1970). Then, introducing a null coordinate u according to $\mathrm{d}u = \mathrm{e}^{\nu+\lambda}\mathrm{d}t - \mathrm{e}^{2\lambda}\mathrm{d}r$, one can transform the line element (15.3) (upper signs) into the simpler form

$$\mathrm{d}s^2 = r^2\mathrm{d}\sigma^2 - 2\mathrm{d}u\,\mathrm{d}r - 2H(u,r)\mathrm{d}u^2, \quad 2H = \mathrm{e}^{-2\lambda} \tag{15.17}$$

($\mathrm{d}\sigma^2$ as in (15.1)). For the Ricci tensor types under consideration, the corresponding metrics can be completely determined. The results are listed in Table 15.1. For vacuum solutions in double-null coordinates $\mathrm{d}s^2 = r^2(u,v)\mathrm{d}\sigma^2 - 2f(u,v)\mathrm{d}u\,\mathrm{d}v$, see Curry and Lake (1991).

In the coordinate frame (15.3), the general solution of the Einstein–Maxwell equations (including Λ) differs from its counterpart

(15.12)–(15.13) in the case of timelike orbits only by a sign ($e^2 \to -e^2$ in the expression for $\mathrm{e}^{2\nu}$), see §13.3.

The assumption $Y_{,a}Y^{,a} > 0$ enabled us to put $Y = x^3$. For $Y_{,a}Y^{,a} < 0$, an analogous treatment with $Y = x^4$ leads again to this solution, but now with $x^3 = r$ and $x^4 = t$ interchanged, and with $(k - 2m/t + e^2/t^2 - \Lambda t^2/3) < 0$. The cases $Y_{,a}Y^{,a} > 0$ and $Y_{,a}Y^{,a} < 0$ correspond to the R- and T-regions of the same solutions. There is an additional timelike or spacelike Killing vector $\boldsymbol{\xi} = \partial_t$ (resp. $\boldsymbol{\xi} = \partial_r$) which is hypersurface-orthogonal and commutes with the three generators of G_3 on V_2.

For plane symmetry ($k = 0$), $2H = -2m(u)/r$ can also represent a null Einstein–Maxwell field, cp. (28.43).

The case of spacelike orbits with $Y_{,a}Y^{,a} = 0$ must be treated separately (Foyster and McIntosh 1972). It is advisable to start with the coordinate system (15.4). It turns out that no vacuum solutions exist. For $\Lambda \neq 0$, the metric is of the form (35.35) (Nariai 1951). The only Einstein–Maxwell fields are the Bertotti–Robinson solution (12.16) ($Y = \mathrm{const}$), and the special pp-wave (§§24.5 and 12.5)

$$\mathrm{d}s^2 = Y^2(u)(\mathrm{d}x^2 + \mathrm{d}y^2) - 2\mathrm{d}u\,\mathrm{d}r \tag{15.18}$$

(which is flat for $Y_{,uu} = 0$).

15.4.3 Generalized Birkhoff theorem

From the results obtained hitherto in this section we see that all the Einstein–Maxwell fields (including Λ-terms) admitting a group G_3 on V_2 have (at least) one additional Killing vector.

Theorem 15.5 *Metrics with a group G_3 of motions on non-null orbits V_2 and with Ricci tensors of types* [(11)(1,1)] *and* [(111,1)] *admit a group G_4, provided that $Y_{,a} \neq 0$* (Cahen and Debever 1965, Barnes 1973a, Goenner 1970, Bona 1988b).

Obviously, this theorem generalizes Birkhoff's theorem (Birkhoff 1923): the only vacuum solution with spherical symmetry is the *Schwarzschild solution* (Reissner 1916, Droste 1916-17)

$$\mathrm{d}s^2 = r^2(\mathrm{d}\vartheta^2 + \sin^2\vartheta\,\mathrm{d}\varphi^2) + (1 - 2m/r)^{-1}\mathrm{d}r^2 - (1 - 2m/r)\mathrm{d}t^2. \tag{15.19}$$

Note that the additional Killing vector $\boldsymbol{\xi} = \partial_t$ of the Schwarzschild solution is spacelike in the T-region ($r < 2m$). The original formulation of Birkhoff's theorem (the only vacuum solution with spherical symmetry is *static*) was criticized by Petrov (1963a, 1963b). Petrov's contribution to Birkhoff's theorem is discussed in Bergmann *et al.* (1965).

For the Ricci tensor type [(11,2)] the group G_3 on V_2 does not imply the existence of a G_4: the Vaidya metric (Table 15.1)

$$ds^2 = r^2(d\vartheta^2 + \sin^2\vartheta\, d\varphi^2) - 2du dr - (1 - 2m(u)/r)du^2, \tag{15.20}$$

$m(u)$ being an arbitrary function of the null coordinate u, has G_3 on V_2 as the maximal group of motions (unless $m =$ const).

15.4.4 Spherically- and plane-symmetric fields

The spherically-symmetric Einstein–Maxwell field with $\Lambda = 0$ is the Reissner–Nordström solution

$$ds^2 = r^2(d\vartheta^2 + \sin^2\vartheta\, d\varphi^2) + (1 - 2m/r + e^2/r^2)^{-1}dr^2 - (1 - 2m/r + e^2/r^2)dt^2, \tag{15.21}$$

which describes the exterior field of a spherically-symmetric charged body (its form in isotropic coordinates can be found e.g. in Prasanna (1968)). For $e = 0$, we obtain the Schwarzschild solution (15.19). We give it here in various other coordinate systems which are frequently used:
ISOTROPIC COORDINATES:

$$ds^2 = [1 + m/2\overline{r}]^4[d\overline{x}^2 + d\overline{y}^2 + d\overline{z}^2] - [1 - m/2\overline{r}]^2 dt^2/[1 + m/2\overline{r}]^2, \quad r = \overline{r}[1 + m/2\overline{r}]^2 \tag{15.22}$$

(for isotropic coordinates covering also $r < 2m$, see Buchdahl 1985).
(OUTGOING) EDDINGTON–FINKELSTEIN COORDINATES (Eddington 1924, Finkelstein 1958) (t', r), which, with $u = t' - r$, give

$$\begin{aligned} ds^2 &= r^2(d\vartheta^2 + \sin^2\vartheta\, d\varphi^2) - 2du\, dr - (1 - 2m/r)du^2, \\ u &= t - \textstyle\int(1 - 2m/r)^{-1}dr = t - r - 2m\ln(r - 2m). \end{aligned} \tag{15.23}$$

KRUSKAL–SZEKERES COORDINATES (Kruskal 1960, Szekeres 1960):

$$\begin{aligned} &ds^2 = r^2(d\vartheta^2 + \sin^2\vartheta\, d\varphi^2) - 32m^3r^{-1}e^{-r/2m}du\, dv, \\ &u = -(r/2m - 1)^{1/2}e^{r/4m}e^{-t/4m}, \quad v = (r/2m - 1)^{1/2}e^{r/4m}e^{t/4m}. \end{aligned} \tag{15.24}$$

LEMAITRE–NOVIKOV COORDINATES:

$$\begin{aligned} &ds^2 = Y^2(d\vartheta^2 + \sin^2\vartheta\, d\varphi^2) + [1 - \varepsilon f^2(r)]^{-1}(Y'dr)^2 - d\tau^2, \\ &\dot{Y}^2 - 2m/Y = -\varepsilon f^2(r) \end{aligned} \tag{15.25}$$

($\varepsilon = 0$: Lemaître (1933); $\varepsilon = 1$, $f^2 = (1 + r^2)^{-1}$: Novikov (1963)).

ISRAEL COORDINATES (Israel 1966):

$$ds^2 = 4m^2\left(4\mathrm{d}x[\mathrm{d}y + y^2\mathrm{d}x/(1+xy)] + [1+xy]^2\left(\mathrm{d}\vartheta^2 + \sin^2\vartheta\,\mathrm{d}\varphi^2\right)\right),$$
$$r = 2m(1+xy), \quad t = 2m\left(1 + xy + \ln|y/x|\right). \tag{15.26}$$

The plane-symmetric Einstein–Maxwell field (with $\Lambda = 0$) either has $Y_{,a}Y^{,a} = 0$ and is then given by (12.16) or (15.18), see the discussion above, or it has $Y_{,a}Y^{,a} \neq 0$ and

$$ds^2 = r^2(\mathrm{d}x^2 + \mathrm{d}y^2) - 2\mathrm{d}u\,\mathrm{d}r - \left[e^2(u)/r^2 - 2m(u)/r\right]\mathrm{d}u^2 \tag{15.27}$$

(Kar (1926), McVittie (1929), see also (28.43)–(28.44)). To include a cosmological constant Λ, a term $-\Lambda r^2/3$ has to be added in the coefficient of $\mathrm{d}u^2$ (Theorem 28.7).

For $e \neq 0$, both e and m are constant, and the metric is either static (r being a spacelike coordinate, $-2m/r + e^2/r^2 > 0$, $Y_{,a}Y^{,a} > 0$), or spatially homogeneous ($-2m/r + e^2/r^2 < 0$, $Y_{,a}Y^{,a} < 0$), cp. (13.48). By a transformation of the r-coordinate (15.27) can be transformed into the form given by Patnaik (1970) and Letelier and Tabensky (1974):

$$ds^2 = Y^2(z)(\mathrm{d}x^2 + \mathrm{d}y^2) + \tfrac{1}{2}Y'(z)(\mathrm{d}z^2 - \mathrm{d}t^2), \tag{15.28a}$$

where $Y(z)$ is determined implicitly by the equation

$$(Y - A)^2 + 2A^2\ln(Y + A) = -Cz, \quad A, C \text{ const.} \tag{15.28b}$$

In the metric (15.28*a*), the (non-null) electromagnetic field is given by

$$F_{12} = C_1, \quad F_{34} = \tfrac{1}{2}C_2Y'Y^{-2}, \quad A \equiv \tfrac{1}{2}\kappa_0(C_1^2 + C_2^2)/C. \tag{15.28c}$$

For $e = 0$, m is an arbitrary function of u (and the Maxwell field is a null field). Note that in this case and in (15.18) the Maxwell field does not share the plane symmetry (Kuang *et al.* 1987).

The plane-symmetric vacuum solution with $Y_{,a}Y^{,a} > 0$ is the static metric

$$ds^2 = z^{-1/2}(\mathrm{d}z^2 - \mathrm{d}t^2) + z(\mathrm{d}x^2 + \mathrm{d}y^2), \quad z > 0. \tag{15.29}$$

(Taub 1951), see also (13.51) and (22.7). The case $Y_{,a}Y^{,a} < 0$ leads to the Kasner metric (13.53) with $p_1 = p_2 = 2/3$, $p_3 = -1/3$,

$$ds^2 = t^{-1/2}(\mathrm{d}z^2 - \mathrm{d}t^2) + t(\mathrm{d}x^2 + \mathrm{d}y^2), \quad t > 0. \tag{15.30}$$

The plane-symmetric vacuum solution with a Λ-term can be written in the form (Novotný and Horský 1974)

$$\begin{aligned} &\mathrm{d}s^2 = \sin^{4/3}(az)(\mathrm{d}x^2 + \mathrm{d}y^2) + \mathrm{d}z^2 - \cos^2(az)\sin^{-2/3}(az)\mathrm{d}t^2, \\ &a \equiv \sqrt{3\Lambda}/2, \quad \Lambda > 0. \end{aligned} \tag{15.31}$$

This solution belongs to the class given by Carter (1968b).

The plane-symmetric pure radiation field

$$\mathrm{d}s^2 = z^2(\mathrm{d}x^2 + \mathrm{d}y^2) - 2\mathrm{d}u\,\mathrm{d}z + 2m(u)z^{-1}\mathrm{d}u^2 \tag{15.32}$$

is the subcase $e = 0$ of the metric (15.27) with a different interpretation.

15.5 Dust solutions

For dust, the group orbits cannot be timelike (Theorem 15.3). We take the comoving system of reference (15.8), and, as $T^{ab}{}_{;b} = (\mu u^a u^b)_{;a} = 0$ implies $\nu' = 0$, we start with

$$\mathrm{d}s^2 = Y^2(r,t)[\mathrm{d}\vartheta^2 + \Sigma^2(\vartheta,k)\mathrm{d}\varphi^2] + \mathrm{e}^{2\lambda(r,t)}\mathrm{d}r^2 - \mathrm{e}^{2\nu(t)}\mathrm{d}t^2, \tag{15.33}$$

$\Sigma(\vartheta, k)$ being defined as in (15.2). Of the components (15.6) of the Einstein tensor (to be taken with the upper signs), only $G^4_4 = -\kappa_0\mu$ is non-zero.

For $Y' \neq 0$, we choose $\nu = 0$ and integrate the field equation $G^4_3 = 0$ by

$$\mathrm{e}^{2\lambda} = Y'^2/[k - \varepsilon f^2(r)], \quad \varepsilon = 0, \pm 1, \tag{15.34}$$

$f(r)$ is an arbitrary function, and ε is to be chosen such that $\mathrm{e}^{2\lambda}$ becomes positive. With (15.34), we obtain a first integral of $G^3_3 = 0$ by

$$\dot{Y}^2 - 2m(r)/Y = -\varepsilon f^2(r). \tag{15.35}$$

$G^1_1 = G^2_2 = 0$ are satisfied identically, and $G^4_4 = -\kappa_0\mu$ yields

$$\kappa_0\mu(r,t) = 2m'/Y'Y^2. \tag{15.36}$$

The differential equation (15.35) can be completely integrated. The solution for $\varepsilon = 0$ is

$$t - t_0(r) = \pm\tfrac{2}{3}Y^{3/2}[2m(r)]^{-1/2}, \quad \varepsilon = 0, \tag{15.37a}$$

and for $\varepsilon \neq 0$

$$\begin{aligned} t - t_0(r) &= \pm h(\eta)m(r)f^{-3}(r), \quad Y = h'(\eta)m(r)f^{-2}(r), \\ h(\eta) &= \{\eta - \sin\eta, \sinh\eta - \eta\} \quad \text{for} \quad \varepsilon = \{+1, -1\}. \end{aligned} \tag{15.37b}$$

Theorem 15.6 *The general dust solution admitting a G_3 on V_2 with $Y' \neq 0$ is given in a comoving system of reference by*

$$ds^2 = Y^2(r,t)[d\vartheta^2 + \Sigma^2(\vartheta,k)d\varphi^2] + Y'^2 dr^2[k - \varepsilon f^2(r)]^{-1} - dt^2, \quad (15.38)$$

where Y is given by (15.37), and m, f and t_0 are arbitrary functions of r (Lemaître 1933, Tolman 1934a, Datt 1938, Bondi 1947).

Note that the radial coordinate is definded only up to a scale transformation.

Originally, dust solutions were considered only for the case of spherical symmetry ($k = +1$), and $Y' \neq 0$ was tacitly assumed. Special solutions contained here are: (i) the Schwarzschild solution ($m =$ const; neither t nor r are uniquely defined; the choice $\varepsilon = 0$, $t_0 = r$ leads to the Lemaître form (15.25)), (ii) the Friedmann dust universes (14.6) and (iii) the solutions with $m = r$, $-\varepsilon f^2 = a =$ const and Y given by

$$t = br + \int^Y \left[\frac{x}{2r - ax}\right]^{1/2} dx, \quad b = \text{const}, \quad (15.39)$$

admitting the homothetic vector $\boldsymbol{\xi} = t\partial_t + r\partial_r$ (Henriksen and Wesson 1978, Bona 1988a). There are spherically-symmetric solutions ($k = 1$) which recollapse ($\varepsilon = 1$) but have infinite spatial sections $T =$ const (Bonnor 1985).

For $Y' = 0$, $\dot{Y} \neq 0$, we choose $Y = t$ ($Y =$ const leads to $\mu = 0$). All these solutions have the property $Y_{,a}Y^{,a} < 0$. Now $G^3_3 = 0$ is integrated by

$$e^{-2\nu} = at^{-1} - k \quad (15.40)$$

and, writing $e^\lambda = Ve^{-\nu}$, $G^1_1 = G^2_2 = 0$ gives the differential equation

$$\ddot{V} + \dot{V}(t^{-1} - 3\dot{\nu}) = 0 \quad (15.41)$$

for $V(r,t)$, which is solved by

$$V(r,t) = B(r)\int t^{1/2}(a - kt)^{-3/2}dt + A(r). \quad (15.42)$$

If $B(r)$ is zero, we regain the vacuum case. Assuming $B(r) \neq 0$, we can transform B to unity by a scale transformation of r. Introducing $x = e^\nu$ as a new variable of integration in (15.42), the final result is the

Theorem 15.7 *The general dust solution admitting a G_3 on a V_2, with $Y' = 0$, is given in comoving coordinates by*

$$ds^2 = t^2[d\vartheta^2 + \Sigma^2(\vartheta,k)d\varphi^2] + e^{2\lambda(r,t)}dr^2 - e^{2\nu(t)}dt^2, \quad e^{2\nu} = t/(a - kt),$$
$$e^\lambda = e^{-\nu}\left[\int^{e^\nu} \frac{2x^2dx}{1 + kx^2} + A(r)\right], \quad \kappa_0\mu(r,t) = 2(\dot{\lambda} + \dot{\nu})/te^{2\nu}. \quad (15.43)$$

These solutions are generalizations of the Kantowski–Sachs ($k \neq 0$) and Bianchi type I ($k = 0$) solutions given in Chapter 14 (in rescaled timelike coordinates) and specialize to them for $A = \text{const}$ (Ellis 1967).

Ref.: For dust solutions including a cosmological constant, see Lemaître (1933). For a discussion of voids in a Tolman model, see e.g. Sato (1984) and Bonnor and Chamorro (1990) and the references given there.

15.6 Perfect fluid solutions with plane, spherical or pseudospherical symmetry

Most of the perfect fluid solutions with a G_3 on a spacelike orbit S_2 have been found by specializing at the very beginning to spherical or plane symmetry. Here we want to discuss some general properties and treat those approaches which cover all three subcases at one go. The plane-symmetric case will then be discussed in the following §15.7, and the spherically-symmetric case in Chapter 16 .

15.6.1 Some basic properties

In dealing with perfect fluid solutions, most authors prefer a comoving system of reference, i.e. they start from

$$\begin{aligned} ds^2 &= Y^2(r,t)\left[(dx^1)^2 + \Sigma^2(x^1,k)(dx^2)^2\right] + e^{2\lambda(r,t)}dr^2 - e^{2\nu(r,t)}dt^2, \\ u^i &= (0,\,0,\,0,\,e^{-\nu}). \end{aligned} \tag{15.44}$$

For the field equations we then have to take (15.6) with the upper sign, and $G^4_4 = -\kappa_0\mu, G^1_1 = G^2_2 = G^3_3 = \kappa_0 p$ as the only non-zero components of G^b_a. For the integration procedure, the two equations $G^4_3 = 0$, i.e.

$$\dot{Y}' - \dot{Y}\nu' - Y'\dot{\lambda} = 0, \tag{15.45}$$

and $G^1_1 - G^3_3 = 0$ (isotropy of pressure) are the most important; the other two equations may be considered as defining μ and p.

Two simple consequences of $T^{ab}{}_{;b} = 0$ are the relations

$$p' = -(\mu + p)\nu', \quad \dot{\mu} = -(\mu + p)(\dot{\lambda} + 2\dot{Y}/Y). \tag{15.46}$$

Perfect fluid solutions can be classified according to their kinematical properties, i.e. the 4-velocity's rotation, acceleration, expansion, and shear, cp. §6.1. Here the symmetry implies that $\omega_{ab} = 0$, so the velocity field must be hypersurface-orthogonal, and in the coordinate system

(15.44) the other quantities in question are given by

$$\begin{aligned} \dot{u}_i &= (0,0,\nu',0), \quad \Theta = \mathrm{e}^{-\nu}(\dot{\lambda} + 2\dot{Y}/Y), \\ \sigma^1_1 &= \sigma^2_2 = -\tfrac{1}{2}\sigma^3_3 = \tfrac{1}{3}\mathrm{e}^{-\nu}(\dot{Y}/Y - \dot{\lambda}). \end{aligned} \tag{15.47}$$

In particular the static and the shearfree cases have been studied in some detail (see e.g. Barnes (1973b)), and we shall present the results in the following subsections.

15.6.2 *Static solutions*

For static solutions one may regard the two field equations $G^4_4 = -\kappa_0\mu$ and $G^2_2 = \kappa_0 p$ as the definitions of μ and p. The field equation $G^4_3 = 0$ is satisfied identically, only the condition of isotropy (of pressure), $G^1_1 = G^3_3$, remains to be solved. In coordinates

$$\mathrm{d}s^2 = L^{-2}(r)\left\{R^2(r)\left[(\mathrm{d}x^1)^2 + \Sigma^2(x^1,k)(\mathrm{d}x^2)^2\right] + \mathrm{d}r^2 - G^2(r)\mathrm{d}t^2\right\} \tag{15.48}$$

this condition reads

$$R(-2GL'R' + 2GL''R + LG'R' - G''LR) = GL(RR'' + k - R'^2). \tag{15.49}$$

$R(r)$ is an arbitrary function which can be fixed conveniently by a gauge transformation $\overline{r} = \overline{r}(r)$. We do that by demanding that the right-hand side of (15.49) vanishes, and also introduce a new independent variable $x(r)$ by $x''/x' = R'/R$, i.e. we take

$$\begin{array}{lll} R = r, & x = r^2 & k = 1 \\ R = 1, & x = r \quad \text{for} & k = 0 \\ R = \cosh r, & x = \sinh r & k = -1. \end{array} \tag{15.50}$$

In these variables, the condition of isotropy reads

$$2GL_{,xx} = LG_{,xx} \tag{15.51}$$

(see Kustaanheimo and Qvist (1948) for the spherically-symmetric case). All three subclasses $k = 0, \pm 1$ are governed by the same differential equation. Obviously one can prescribe one of the metric functions and then get the second by solving a linear differential equation.

The metrics (15.50) admit a homothety $\mathbf{H} = r\partial_r + at\partial_t$ when $L = 1$, $R = A^2r^2$, $G = r^{2-2a}$, $a^2 = 2 - k/A^2$.

15.6.3 *Solutions without shear and expansion*

Because of (15.47), vanishing shear and expansion implies $\dot{\lambda} = \dot{Y} = 0$; ν is the only metric function that may depend on t. But as $\dot{\nu}$ enters the field equations (see (15.6b) and (15.6c)) only as a coefficient of $\dot{Y}$ and $\dot{\lambda}$, and $\ddot{\nu}$ does not appear at all, no time derivative is contained in the field equations, which are the same as in the static case. These equations show that the energy density μ is a function of r alone, whereas the pressure p may depend on t if ν' does.

The solutions in question are therefore either static (all static solutions are shear- and expansion-free) or can be generated from static solutions as follows. Take any static solution

$$\mathrm{d}s^2 = Y(r)^2\left[(\mathrm{d}x^1)^2 + \Sigma^2(x^1,k)(\mathrm{d}x^2)^2\right] + \mathrm{e}^{2\lambda(r)}\mathrm{d}r^2 - \mathrm{e}^{2\nu}\mathrm{d}t^2 \qquad (15.52)$$

and replace $\nu(r)$ by the general time-dependent solution $\nu(r,t)$ of the condition of isotropy $G^1_1 = G^3_3$. Introducing $N \equiv \mathrm{e}^{\nu}$, this condition reads

$$N''Y^2 - N'(YY' + Y^2\lambda') + N(k\mathrm{e}^{2\lambda} - Y'^2 - YY'\lambda' + YY'') = 0. \qquad (15.53)$$

The functions $\lambda(r)$ and $Y(r)$ being taken from the static solution, this is a linear differential equation for $N \equiv \mathrm{e}^{\nu}$, with coefficients independent of t, and so its general solution can be written as

$$\mathrm{e}^{\nu} = N = f_1(t)N_1(r) + f_2(t)N_2(r). \qquad (15.54)$$

The functions f_1 and f_2 are disposable, and N_1 and N_2 are any two linearly independent solutions of (15.53). Equations (15.52) and (15.54) give all non-static, expansion- and shear-free solutions (see Kustaanheimo and Qvist (1948) and Leibovitz (1971) for the spherically-symmetric case).

15.6.4 Expanding solutions without shear

For $\sigma_{ab} = 0$, but $\Theta \neq 0$ it follows from (15.47) that $\dot{\lambda}$ must be non-zero. From $\dot{\lambda} = \dot{Y}/Y$ and (15.45) we infer that

$$Y = R(r)\mathrm{e}^{\lambda}, \quad \mathrm{e}^{\nu} = \dot{\lambda}\mathrm{e}^{-f(t)}, \quad \Theta = 3\mathrm{e}^{f(t)}, \qquad (15.55)$$

where $R(r)$ is an arbitrary function which can be incorporated into e^{λ}, or fixed conveniently, by a gauge transformation $\overline{r} = \overline{r}(r)$. The condition of isotropy $G^1_1 - G^3_3 = 0$ can then be written as

$$\frac{\partial}{\partial t}\left[\mathrm{e}^{\lambda}(-k - RR'' + R'^2 - R^2\lambda'' + RR'\lambda' + R^2\lambda'^2)\right] = 0 \qquad (15.56)$$

and is integrated by

$$-k - RR'' + R'^2 - R^2\lambda'' + RR'\lambda' + R^2\lambda'^2 = \mathrm{e}^{-\lambda}\varphi(r)R^{-2}(r) \qquad (15.57)$$

(Wyman 1946, Narlikar 1947, Kustaanheimo and Qvist 1948). We choose the function $R(r)$ and introduce a new independent variable $x(r)$ as in the static case in (15.50). Written in terms of the function $L(x,t) = \mathrm{e}^{-\lambda}$, equation (15.57) then leads to the differential equation

$$L_{,xx} = F(x)L^2, \quad F(x) = \varphi(r)(x_{,r})^{-2}, \tag{15.58}$$

where $F(x)$ is an arbitrary function which is zero if space-time is conformally flat. All three subclasses ($k = 0, \pm 1$) are governed by the same differential equation (Krasiński 1989, Mészáros 1985) which was originally found by Kustaanheimo and Qvist (1948) for the spherically-symmetric case. In terms of L, the line element reads

$$\mathrm{d}s^2 = L^{-2}(x,t)\left\{R^2(r)\left[(\mathrm{d}x^1)^2 + \Sigma^2(x^1,k)(\mathrm{d}x^2)^2\right] + \mathrm{d}r^2 - \dot{L}^2\mathrm{e}^{-2f(t)}\mathrm{d}t^2\right\}, \tag{15.59}$$

with R and x taken from (15.50). To get a solution, one has to prescribe $f(t)$ – which fixes the t-coordinate – and $F(x)$, and find a solution $L(x,t)$ of (15.58), which in general will contain two arbitrary functions of time which enter via the constants of integration (note that $\dot{L}$ must be nonzero). Energy density and pressure can then be computed from

$$\begin{aligned} \kappa_0\mu &= 3\mathrm{e}^{2f} - \mathrm{e}^{-2\lambda}\left[2\lambda'' + \lambda'^2 + 4\lambda' R'/R + 3R''/R\right], \\ \kappa_0 p\dot{\lambda} &= \mathrm{e}^{-3\lambda}\partial_t\left[\mathrm{e}^{\lambda}\left(\lambda'^2 + 2\lambda' R'/R + R''/R\right) - \mathrm{e}^{3\lambda+2f}\right]. \end{aligned} \tag{15.60}$$

The differential equation (15.58) has been widely discussed in the context of the spherically-symmetric solutions, and we refer the reader to §16.2.2 for a survey of its solutions. Only a few plane- or pseudospherically-symmetric counterparts of the many spherically-symmetric solutions have been discussed in detail, e.g. McVittie's solution (16.46) by Hogan (1990). If an equation of state $p = p(\mu)$ is assumed, then the metric is either spherically-symmetric (with $F = 0, 1$, cp. §16.2.2) or plane-symmetric (with $F = -x^2$, $p = p(t + \ln x)$, and a fourth Killing vector), see Collins and Wainwright (1983).

15.6.5 Solutions with nonvanishing shear

Solutions with shear but without acceleration

For $\dot{u}^\alpha = 0$, in the comoving system of reference (15.44) ν is a function only of t, and so is p. The field equation (15.6d) therefore implies

$$\dot{Y}' = \dot{\lambda}Y'. \tag{15.61}$$

If $Y = \text{const} = a$, then we have $\kappa_0\mu = -\kappa_0 p = k/a^2$, and if we choose $\nu = 0$, the resulting metric is

$$ds^2 = a^2\left[(dx^1)^2 + \Sigma(x^1,k)(dx^2)^2\right] + \Sigma^2(t/a,-k)dr^2 - dt^2, \quad (15.62)$$

which is a special case of (12.8) with a G_6 on V_4.

If $Y' = 0$ ($Y \neq$ const), then we can choose $Y = t$, and the remaining field equations read

$$\kappa_0\mu t^2 = 2\dot{\lambda}te^{-2\nu} + k + e^{-2\nu}, \quad \kappa_0 p t^2 = 2\dot{\nu}te^{-2\nu} - k - e^{-2\nu}, (15.63a)$$

$$\kappa_0 p = -e^{-2\nu}[\ddot{\lambda} + \dot{\lambda}^2 - \dot{\lambda}\dot{\nu} + (\dot{\lambda} - \dot{\nu})/t]. \quad (15.63b)$$

These equations closely resemble the static case, cp. §16.1 and equations (16.2). To solve the field equations, one should realise that although λ' is in general non-zero, no derivatives with respect to r appear in this system. So one can e.g. prescribe the function $\nu(t)$ and then determine $\lambda(r,t) = \ln W$ from the condition of isotropy

$$\ddot{W} - \dot{W}(\dot{\nu} - 1/t) + W(\dot{\nu}t - 1 - ke^{2\nu})/t^2 = 0 \quad (15.64)$$

by taking any solution $W = c_1W_1(t) + c_2W_2(t)$ of this ordinary linear differential equation for W, and allowing c_1 and c_2 to become arbitrary functions of r. One could also start from any solution with $\lambda = \lambda(t)$, i.e. any solution of the Kantowski–Sachs class (§14.3), take the function $\nu(t)$ from this solution and generalize λ as indicated above (Herlt 1996).

An example is the solution

$$ds^2 = t^2 d\Omega^2 + \left[A(r)t^n + B(r)t^{-n}\right]^2 dr^2 + (1-n^2)dt^2, \quad (15.65)$$

which generalizes the Kantowski–Sachs type solution of McVittie and Wiltshire (1975), with $e^\lambda = t^n$, and includes the similarity solution found by Ponce de León (1988) as the special case $A(r) = ar^{-n}$, $B(r) = br^n$.

If $Y' \neq 0$, then we can choose $\nu = 0$ and integrate (15.61) by

$$e^{2\lambda} = Y'^2/[k - \varepsilon f^2(r)], \quad \varepsilon = 0, \pm 1. \quad (15.66)$$

From the field equation (16.21b) we obtain

$$\kappa_0 p(t) Y^2 = -2Y\ddot{Y} - \dot{Y}^2 - \varepsilon f^2(r). \quad (15.67)$$

Equation (15.6c), with $G^1_1 = \kappa_0 p$, follows from (15.67) by differentiation with respect to r, and μ can be computed from

$$\kappa_0\mu = -3\kappa_0 p(t) - 2\ddot{Y}'/Y' - 4\ddot{Y}/Y. \quad (15.68)$$

In (15.67), $p(t)$ can be prescribed, but to avoid zero shear, $\dot{Y}'Y = \dot{Y}Y'$ is forbidden, i.e. Y'/Y has to depend on the time t.

Substituting $Y = Z^{2/3}$ into (15.67) leads to

$$\ddot{Z} + \tfrac{3}{4}\kappa_0 p(t) Z + \varepsilon f^2(r) Z^{-1/3} = 0. \tag{15.69}$$

For p = const, (15.69) can easily be solved by quadratures. For non-constant p, but $\varepsilon = 0$ (which because of (15.66) is possible only for $k = 1$, i.e. in the spherically-symmetric case), one is led to

$$\ddot{Z} + \tfrac{3}{4}\kappa_0 p(t) Z = 0 \tag{15.70}$$

(Bona *et al.* 1987b). By an appropriate choice of $p(t)$, solutions to (15.70) can be constructed (choosing the 'constants of integration' as arbitrary functions of r; note that Z'/Z has to be a function of time to avoid zero shear). Alternatively (Leibovitz 1971), one can prescribe any function $h(t)$, take Z as

$$Z = A(r)h(t) + B(r)h(t)\int^t h^{-2}\mathrm{d}t \tag{15.71}$$

and calculate the pressure from (15.70).

Except for the cases treated above, no solutions of (15.69) have been found, see Herlt (1996) and Soh and Mahomed (1999).

Solutions with shear and acceleration

Metrics of the form

$$\mathrm{d}s^2 = A(r,t)\left[\mathrm{d}r^2 + B(t)\left((\mathrm{d}x^1)^2 + \Sigma^2(x^1,k)(\mathrm{d}x^2)^2\right) - \mathrm{d}t^2\right] \tag{15.72}$$

(in comoving coordinates (15.44)) have been considered by Herrera and Ponce de León (1985) and Kitamura (1994) in the spherically-symmetric case and by Bogoyavlensky and Moschetti (1982) for $A = \mathrm{e}^{2r}$ in the pseudospherically-symmetric case; also the search for perfect fluids with a conformal Killing vector orthogonal to the four-velocity and the orbits of the rotational group for which the fluid inherits the conformal symmetry led to these metrics, see §35.4.4 (Coley and Tupper 1990b, Kitamura 1994, 1995a, 1995b, Coley and Czapor 1992). For (15.72), the field equation $G^4_3 = 0$ gives $2\dot{A}'A = 3\dot{A}A'$, which is integrated by

$$A(r,t) = [H(r) + F(t)]^{-2}. \tag{15.73}$$

For $A' \neq 0$, the condition of isotropy $G^1_1 = G^3_3$ yields the system of ordinary differential equations

$$\begin{aligned} &\ddot{B} - 4c_1 B + 2k = 0, \quad H'' - c_1 H = c_2, \\ &\dot{B}\dot{F} - 2c_1 BF + 2c_2 B = 0, \end{aligned} \tag{15.74}$$

which can easily be integrated; μ and p can then be calculated from the resulting metric. The solutions (37.57) are included here as a subcase. For $A' = 0$, the acceleration is zero, the metric (15.72) admits the additional Killing vector ∂_r and thus belongs to the Kantowski–Sachs class, see §14.3. Here one can prescribe e.g. $B(t)$ and then determine $F(t)$ from $\ddot{B}F - 2\dot{B}\dot{F} + 2kF = 0$.

In a search for solutions with a generalized similarity Collins and Lang (1987) found a class of solutions given (in comoving coordinates) by

$$\mathrm{d}s^2 = r^2t^2\left[(\mathrm{d}x^1)^2 \pm \Sigma^2(x^1,k)(\mathrm{d}x^2)^2\right] + \frac{\mathrm{d}r^2}{ar^2-b} - \frac{r^2t^2\mathrm{d}t^2}{c-kt^2-bt^4}, \qquad (15.75)$$
$$\kappa_0 p = \kappa_0\mu + 6a.$$

The solutions (15.75) contain the spherically-symmetric metrics (16.66), see §16.2.3 for further details, and the plane-symmetric solutions found by Hajj-Boutros and Léauté (1985).

All solutions of embedding class one, Petrov type D, with acceleration, given in §37.4.3, admit a G_3 on S_2.

For a solution in noncomoving coordinates, see (16.77).

15.7 Plane-symmetric perfect fluid solutions

15.7.1 Static solutions

Using (15.51), many static solutions could be obtained from the vast number of known static spherically-symmetric solutions given in §16.1, but so far this method has not been exploited.

The plane-symmetric static perfect fluids *with a prescribed equation of state* $\mu = \mu(p)$ are given by (Taub 1956)

$$\mathrm{d}s^2 = z^2(\mathrm{d}x^2+\mathrm{d}y^2) + zF^{-1}(z)\mathrm{d}z^2 - \mathrm{e}^{2\nu}\mathrm{d}t^2, \qquad (15.76a)$$

$$2zp'/[(\mu(p)+p] = 1 - \kappa_0 pz^3/F = -2z\nu', \quad F' = -\kappa_0\mu(p)z^2, \qquad (15.76b)$$

($' = \mathrm{d}/\mathrm{d}z$, $-F$ is the $2m$ of (15.7a)). For a given function $\mu = \mu(p)$, the differential equations (15.76b) determine $F = F(z)$ and $p = p(z)$ and from $p(z)$ then $\nu(z)$. Equations (15.76b) lead to the condition

$$\frac{p'}{p} = \frac{z^2+G'}{G}\frac{G+z^3}{z^3-G}, \quad G(z) \equiv -F(z)/\kappa_0 p(z) \qquad (15.77)$$

(Hojman and Santamarina 1984). So one may prescribe $G(z)$ and then obtain p and ν as line integrals. For an equation of state $p = (\gamma-1)\mu$ the function G has to be of the form $G = Az^{-\gamma/(2\gamma-2)} + (2-\gamma)z^3/(7\gamma-6)$ (Collins 1985).

The solution for $\mu = p$ is contained in (15.80) as $\sigma = z$. A solution for $p = \mu/3$ was obtained by Teixeira *et al.* (1977b). In this case the functions p and F are given by

$$p = p_0 z^2(1 - z^5)^2, \quad F = \kappa_0 p_0 (1 - z^5)^3/5. \tag{15.78}$$

The solution for $\mu = \text{const}$ was given by Taub (1956) as

$$\begin{aligned} &\mathrm{d}s^2 = \sin^{4/3} bw(\mathrm{d}x^2 + \mathrm{d}y^2) + \mathrm{d}w^2 - \mathrm{e}^{2\nu}\mathrm{d}t^2, \\ &\mathrm{e}^{\nu} = 1 - \frac{\cos bw}{2\sin^{1/3} bw}\int_{bw_0}^{bw} \frac{\mathrm{d}v}{\sin^{2/3} v}, \quad \kappa_0\mu = \tfrac{4}{3}b^2, \quad p = \mu(\mathrm{e}^{-\nu} - 1), \end{aligned} \tag{15.79}$$

and by Horský (1975) in terms of hypergeometric functions; it contains (15.31) ($\kappa_0\mu = -\kappa_0 p = \Lambda$) as a special case. Some other special solutions have been given by Davidson (1987, 1989a). For metrics with a proper conformal Killing vector see (35.76).

Static plane-symmetric perfect fluids occur also as subcases of the static cylindrically-symmetric solutions, see §22.2 and the references given there. They can be constructed by solving (22.23) with $a_1 = 0$, (22.24) with $a_0 = 0$, or (22.26) with $y = z$. Similarly, the general solution for an equation of state $p = (\gamma - 1)\mu$ is contained in (22.27) for $a = 0$, see also Bronnikov and Kovalchuk (1979).

15.7.2 *Non-static solutions*

Besides the classes of solutions described in §15.6, several other classes have been found by making special assumptions for the metric or the equation of state.

For $p = \mu$, Tabensky and Taub (1973) reduced the field equation to a single linear differential equation

$$\begin{aligned} &\mathrm{d}s^2 = t^{-1/2}\mathrm{e}^{\Omega}(\mathrm{d}z^2 - \mathrm{d}t^2) + t(\mathrm{d}x^2 + \mathrm{d}y^2), \quad t > 0, \\ &\Omega = 2\textstyle\int t[(\sigma_{,t}^2 + \sigma_{,z}^2)\mathrm{d}t + 2\sigma_{,t}\sigma_{,z}\mathrm{d}z], \quad \sigma_{,tt} + t^{-1}\sigma_{,t} - \sigma_{,zz} = 0, \\ &\kappa_0 p = \kappa_0\mu = t^{1/2}\mathrm{e}^{-\Omega}(\sigma_{,t}^2 - \sigma_{,z}^2), \quad (\kappa_0 p)^{1/2}u_i = \sqrt{2}\sigma_{,i}, \end{aligned} \tag{15.80}$$

(see also Theorem 10.2). For $\sigma = \text{const}$, we regain the vacuum solution (15.30), and – taking the negative root in $\sqrt{t}$ so that z is the timelike coordinate – for $\sigma = az$, $\Omega = a^2t^2$ the static solutions. Tabensky and Taub (1973) also gave the special solution $\sigma = \alpha \ln t + \beta \arccos(z/t)$, $\Omega = 2(\alpha^2 + \beta^2)\ln t + 2\beta^2 \ln(1 - z^2/t^2) + 4\alpha\beta \arccos(z/t)$. Contained here are the solutions (Collins and Lang 1987)

$$\begin{aligned} ds^2 &= r^2t^2(dx^2+dy^2) + \frac{t^{2D}r^{4-2D}}{ar^4+b}dr^2 - \frac{t^{2D+2}r^{2-2D}}{at^4+c}dt^2, \\ & b(2D-3)=0, \quad a,b,c,D = \text{ const} \end{aligned} \tag{15.81}$$

(the special case $a=0$ is due to Goode (1980); this case and $b=0$ admit a homothety), the special case of (15.86) mentioned below, and the metric

$$\begin{aligned} ds^2 &= e^{2bz}[U(dz^2-dt^2)+\sinh(2bt)(dx^2+dy^2)], \\ U &= c[\sinh(2bt)]^{m^2/b^2}[\tanh(bt)]^{m\sqrt{m^2+b^2}/b^2}, \quad c,b,m = \text{ const} \end{aligned} \tag{15.82}$$

(Lorenz 1983c).

Collins and Lang (1987) also found the perfect fluids (in comoving coordinates)

$$\begin{aligned} ds^2 &= r^2t^2(dx^2+dy^2) + \frac{t^{2D}r^{2D}dr^2}{ar^{4D}+br^N} - \frac{t^{2D+2}r^{2-2D}}{c+t^4}dt^2, \\ & N = 4(D^2-D-1)/(D-2), \quad a,b,c,D = \text{ const.} \end{aligned} \tag{15.83}$$

They have an equation of state – which then is of the form $p=(D-2)\mu/(D+2)$ – only if $c=0$ (and there is a homothety) or if $D=-1/2$.

Solutions with an equation of state $p=(\gamma-1)\mu$ were found by Goode (1980). In comoving coordinates, they can be written as

$$\begin{aligned} ds^2 &= M(z)^{4(\gamma-1)/(2-\gamma)}(t^2dz^2-dt^2) \\ &\quad + t^{(2-\gamma)/\gamma}M(z)^{-4(\gamma-1)/3\gamma-2)}(dx^2+dy^2), \\ M(z) &= A\cosh az + B\sinh az, \ a=(2-\gamma)(3\gamma-2)/4\gamma(\gamma-1), \ A^2\neq B^2. \end{aligned} \tag{15.84}$$

They admit a homothetic vector $H = x\partial_x + y\partial_y + 2t\gamma/(3\gamma-2)\partial_t$, see also Carot and Sintes (1997).

Non-barotropic solutions (in comoving coordinates), admitting a homothety, have been found by Carot and Sintes (1997) as

$$ds^2 = \exp[b\,e^{-2(t-z)}]e^{2z}(a^2dz^2-dt^2)+e^{2(t+z)}(dx^2+dy^2), \quad a,b = \text{const.} \tag{15.85}$$

All solutions of the form

$$\begin{aligned} ds^2 &= dz^2 + N^2(z)\left[B(t)(dx^2+dy^2)-dt^2\right], \\ & NN''-N'^2 = -ka^2, \quad \ddot{B} = 4ka^2B, \\ & \kappa_0(p-\mu) = 6N''/N, \quad \kappa_0(p+\mu) = (B_{,t}^2-4ka^2B^2)/(2N^2B^2), \end{aligned} \tag{15.86}$$

(comoving coordinates) admit a conformal vector $\zeta = N(z)\partial_z$, which is homothetic for $N = z$. Contained here as $k = 0, N = z,\ B = \sinh 2t$, is a metric due to Bray (1983) (with an equation of state $\mu = p$), and as $k = 1,\ N = \sinh az,\ B = \sinh 2at$, a solution due to Tariq and Tupper (1992).

The metrics

$$
\begin{aligned}
&\mathrm{d}s^2 = \mathrm{d}z^2 - \mathrm{d}t^2 + P(z)Q(t)[\mathrm{d}x^2 + \mathrm{d}y^2],\\
&\qquad \kappa_0(p-\mu) = P''/P - \ddot{Q}/Q, \quad \kappa_0(p+\mu) = (\dot{Q}^2/Q^2 - P'^2/P^2)/2,\\
&\quad \text{with } P'' = ka^2P, \quad \ddot{Q} = -ka^2Q, \quad k = 0,\ \pm 1, \qquad (15.87)\\
&\qquad \text{or } P(z) = \exp(bz^2 + c_1 z), \quad Q(t) = \exp(-bt^2 + c_2 t).
\end{aligned}
$$

are also perfect fluids (in *non*-comoving coordinates). Contained here as $P(z) = \exp z$, $Q(t) = \cos t$ is a metric found by Bray (1983).

Incompressible fluids (μ = const) have been investigated in Taub (1956). Davidson (1988) found the special solution (in comoving coordinates)

$$
\mathrm{d}s^2 = tz^{m(m-1)}\left[\mathrm{d}x^2 + \mathrm{d}y^2\right] - 2\mathrm{d}z\,\mathrm{d}t - t^{-1}(z + z^m t^{1/(m-1)})\mathrm{d}t^2. \quad (15.88)
$$

Götz (1988) considered metrics of the form

$$
\mathrm{d}s^2 = U(z)V(t)\left[\mathrm{d}x^2 + \mathrm{d}y^2\right] + V(t)^{1-\alpha}\mathrm{d}z^2 - U(z)^{\alpha}\mathrm{d}t^2 \qquad (15.89)
$$

with an equation of state $p = (\gamma - 1)\mu$; the equations for U and V decouple and can be solved by quadratures.

Following Taub (1972), Shikin (1979) determined implicitly all solutions of the form

$$
\mathrm{d}s^2 = Y^2(z/t)[\mathrm{d}x^2 + \mathrm{d}y^2] + X^2(z/t)\mathrm{d}z^2 - T^2(z/t)\mathrm{d}t^2 \qquad (15.90)
$$

(in comoving coordinates, and with an equation of state $p = (\gamma - 1)\mu$) in terms of quadratures; they admit an additional homothetic vector $\boldsymbol{\xi} = x^n\partial_n$.

Plane-symmetric solutions (in non-comoving coordinates) are contained in the metrics (36.20a) possessing flat slices. Plane-symmetric solutions of embedding class one have been constructed by Gupta and Sharma (1996a, 1996b). Metrics of plane symmetry also occur as subcases of perfect fluid solutions admitting a G_2 on S_2 treated in Chapter 23.

16
Spherically-symmetric perfect fluid solutions

Contrary to what may be the common belief, only a minority of spherically-symmetric solutions is known. Most of the known solutions are static or shearfree, and only very few of them satisfy fundamental physical demands such as a plausible equation of state or the absence of singularities.

16.1 Static solutions

16.1.1 Field equations and first integrals

Static spherically symmetric perfect fluid solutions are hypersurface-homogeneous space-times, cp. Chapter 13. They have been widely discussed as models of stars in mechanical and thermodynamical equilibrium. One often takes *Schwarzschild* (or *canonical*) *coordinates* defined by

$$ds^2 = r^2 d\Omega^2 + e^{2\lambda(r)} dr^2 - e^{2\nu(r)} dt^2, \quad d\Omega^2 \equiv d\vartheta^2 + \sin^2\vartheta\, d\varphi^2, \quad (16.1)$$

and the field equations then read

$$\kappa_0 \mu r^2 = -G^4_4 r^2 = [r(1 - e^{-2\lambda})]', \quad (16.2a)$$

$$\kappa_0 p r^2 = G^3_3 r^2 = -1 + e^{-2\lambda}(1 + 2r\nu'), \quad (16.2b)$$

$$\kappa_0 p = G^1_1 = G^2_2 = e^{-2\lambda}[\nu'' + \nu'^2 - \nu'\lambda' + (\nu' - \lambda')/r]. \quad (16.2c)$$

These field equations should be supplemented by an equation of state

$$f(\mu, p) = 0. \quad (16.3)$$

From the four equations (16.2)–(16.3), the four unknown functions μ, p, λ and ν can be determined. Physically, and to get a realistic stellar model, one should start with a reasonable equation of state and impose some regularity conditions, see e.g. Glass and Goldman (1978). In practice, to get analytic expressions for the solutions, the field equations are often solved by making an ad hoc assumption for one of the metric functions or for the energy density, the equation of state being computed from the resulting line element.

The field equations can be cast into various mathematical forms, each of them admitting different tricks for finding solutions; e.g. one may try to transform one of the field equations into a linear differential equation, so that by choosing its coefficients in a suitable way solutions can be obtained. Often the starting point for constructing exact solutions is the condition of isotropy (of pressure) $G^1_1 = G^3_3$, which in full reads

$$\nu'' + \nu'^2 - \nu'\lambda' - (\nu' + \lambda')/r + (\mathrm{e}^{2\lambda} - 1)/r^2 = 0. \tag{16.4}$$

Once a solution (ν, λ) of this equation has been found, one can compute μ and p from (16.2).

An obvious first integral of (16.2a) is

$$\mathrm{e}^{-2\lambda} = 1 - 2m(r)/r, \quad 2m(r) \equiv \kappa_0 \int^r \mu(r) r^2 \mathrm{d}r, \tag{16.5}$$

cp. (15.7). Inserting this into (16.2c) one obtains

$$2r(r - 2m)\nu' = \kappa_0 r^3 p + 2m. \tag{16.6}$$

Eliminating ν' by means of $(\mu + p)\nu' = -p'$, which immediately follows from $T^{ab}_{\ ;a} = 0$, one gets

$$2r(r - m)p' = -(\mu + p)(\kappa_0 r^3 p + 2m). \tag{16.7}$$

Equations (16.5)–(16.7) can be useful if $\lambda(r)$ or $\mu(r)$ or an equation of state is prescribed.

Buchdahl (1959) introduced new variables

$$x = r^2, \quad \zeta = \mathrm{e}^{\nu}, \quad w = mr^{-3} \leftrightarrow \mathrm{e}^{-2\lambda} = 1 - 2xw. \tag{16.8}$$

Equations (16.4)–(16.7) then yield

$$\kappa_0 \mu = 6w + 4xw_{,x}, \quad \kappa_0 p = -2w + (4 - 8xw)\zeta_{,x}/\zeta, \tag{16.9}$$

and the condition of isotropy reads

$$(2 - 4xw)\zeta_{,xx} - (2w + 2xw_{,x})\zeta_{,x} - w_{,x}\zeta = 0. \tag{16.10}$$

The last equation is a differential equation *linear* in *both* ζ and w, and an analytic expression for one of them may be found if the other is prescribed suitably. Moreover, if (by some other method) a solution (ζ, w) is known, possibly new solutions $(\hat{\zeta}, \hat{w})$ can be generated by

$$\hat{\zeta} = \zeta, \quad \hat{w} = w + C(\zeta + 2x\zeta_{,x})^{-2} \exp\left[4 \int \zeta_{,x}(\zeta + 2x\zeta_{,x})^{-1} \mathrm{d}x\right] \tag{16.11}$$

(Heintzmann 1969), or by

$$\hat{w} = w, \quad \hat{\zeta} = C\zeta \int \zeta^{-2}(2 - 4wx)^{-1/2}\,\mathrm{d}x. \tag{16.12}$$

Once a solution (ζ, w) of (16.10) is known, λ, μ and p can be computed from (16.9).

Introducing a new function α by $\mathrm{e}^{2\lambda} = (1 + r\nu')^2/\alpha$, Fodor (2000) transformed the condition of isotropy (16.4) into a linear equation for α,

$$r(1 + r\nu')\alpha' + 2[(1 - r\nu')^2 - 2]\alpha + 2(1 + r\nu')^2 = 0 \tag{16.13}$$

(see also Burlankov (1993)), which for given ν can be solved by quadratures (or, for prescribed α, gives a quadratic equation for $r\nu'$).

Sometimes *isotropic coordinates*

$$\mathrm{d}s^2 = \mathrm{e}^{2\lambda}(r^2\mathrm{d}\Omega^2 + \mathrm{d}r^2) - \mathrm{e}^{2\nu}\mathrm{d}t^2 \tag{16.14}$$

prove useful. In these coordinates, the condition of isotropy of pressure reads

$$\lambda'' + \nu'' + \nu'^2 - \lambda'^2 - 2\lambda'\nu' - (\lambda' + \nu')/r = 0, \tag{16.15}$$

which is a Riccati equation in either λ' or ν'. It can also be written as

$$LG_{,xx} = 2GL_{,xx}\,, \quad L \equiv \mathrm{e}^{-\lambda}, \quad G \equiv L\mathrm{e}^{\nu}, \quad x \equiv r^2 \tag{16.16}$$

(Kustaanheimo and Qvist 1948), cp. §15.6.2. This equation is linear in both L and G, and solutions can be easily found by prescribing one of these two functions appropriately.

The condition of isotropy (16.15) is invariant under the substitution

$$\hat{\nu} = -\nu, \quad \hat{\lambda} = \lambda + 2\nu \tag{16.17}$$

(Buchdahl (1956), cp. §10.11). This substitution can be used to generate static perfect fluid solutions from known ones.

Also the general form (15.9) of the line element (Buchdahl 1967, Simon 1994, Roy and Rao 1972) or coordinates with $\nu(r) = r$ (Roy and Rao 1972) may make the analytic expression for the solution simple and/or lead to physically interesting solutions.

16.1.2 Solutions

The best known of the spherically-symmetric static perfect fluid solutions is the interior Schwarzschild solution (Schwarzschild 1916b)

$$\kappa_0\mu = 3R^{-2} = \text{const}, \quad \kappa_0 p = \frac{3b\sqrt{1-r^2/R^2}-a}{R^2(a-b\sqrt{1-r^2/R^2})},$$
$$\mathrm{d}s^2 = r^2\mathrm{d}\Omega^2 + \mathrm{d}r^2/(1-r^2/R^2) - \left(a-b\sqrt{1-r^2/R^2}\right)^2\mathrm{d}t^2. \tag{16.18}$$

Solutions with $\mu = \text{const}$, but possessing a singularity at $r = 0$ ($\exp[-2\lambda] = 1 - c_1r^2 + c_2/r$), have been discussed by Volkoff (1939) and Wyman (1949). Solutions with a simple equation of state have been found in various cases, e.g. for $\mu + 3p = \text{const}$ (Whittaker 1968), for $p = \mu +$ const (Buchdahl and Land 1968), for $\mu = 3p$ (Hajj-Boutros 1989), and for $\mu = (1+a)\sqrt{p} - ap$ (Buchdahl 1967). Most of these equations of state are not very realistic. But if one takes e.g. polytropic fluid spheres $p = a\mu^{1+1/n}$ (Klein 1953, Tooper 1964, Buchdahl 1964) or a mixture of an ideal gas and radiation (Suhonen 1968), one soon has to use numerical methods.

Many classes of explicit static solutions are known, most of them being unphysical. Solutions which have a singularity at $r = 0$ may nevertheless be used for outer layers of composite spheres. Tables 16.1 and 16.2 give the key assumptions (and references for further details) of many of the known solutions; we have selected those for which μ or the metric functions are particularly simple. Solutions where $\exp(-\lambda+\nu)$ is simple have been discussed by Whitman (1983). For reviews of known classes of solutions and a discussion of their properties see Finch (1987), Finch and Skea (1989) and Delgaty and Lake (1998), where more complicated metrics can also be found.

Using a Hamiltonian formulation of the field equations, Rosquist (1994) found the solution

$$\mathrm{d}s^2 = Z^{-1}\mathrm{d}R^2 + W^2 d\Omega^2 - Z\mathrm{d}t^2, \; Z = V/W, \; W = A+B, \; V = A-B,$$
$$A = \cosh c \sin[\omega_-(R-R_-)]/\omega_-, \; B = \sinh c \sin[(R-R_+)]/\omega_+, \tag{16.19}$$
$$p = a(Z^2 - 2\delta Z + 1), \; \mu = a(-5Z^2 + 6\delta Z - 1), \; \omega_\pm = \sqrt{2a\kappa_0(\delta \pm 1)}.$$

Solutions admitting a homothetic vector $\alpha t\partial_t + r\partial_r$ (Henriksen and Wesson 1978) are contained in the case $\mu = ar^{-2}$ of Table 16.2.

Table 16.1. Key assumptions of some static spherically-symmetric perfect fluid solutions in *isotropic coordinates*

$$ds^2 = e^{2\lambda}(dr^2 + r^2 d\Omega^2) - e^{2\nu}dt^2;\ a, b, c, \alpha, \beta = \text{const.}$$

$e^{-2\lambda}$	ar^b	Narlikar *et al.* (1943)
	$(a + br^2)^\alpha$	Nariai (1950), Tolman (1939)
	$(ar^{1+\alpha} + br^{1-\alpha})^2$	Narlikar *et al.* (1943) Nariai (1950)
	$cr^2(1 + ar^b)^4/(1 - ar^b)^4$	Kuchowicz (1972b)
	$r^{a+2}/(br^a + c)^2$	Kuchowicz (1972b)
	$a\left(\dfrac{r^2 + b - \sqrt{3}/2}{r^2 + b + \sqrt{3}/2}\right)^{\sqrt{3}}$	Burlankov (1993)
	$r^2(a + b\ln r)^2$	Nariai (1950)
	$r^2(\ln br)^4/(1 + c\ln br)^4$	Kuchowicz (1972b)
	$a\exp(br^2)$	Kuchowicz (1972a)
	$a\cos(b + cr^2)$	Nariai (1950)
$e^{2\nu}$	ar^b	Kuchowicz (1971a, 1972a)
	$(a + br^2)^\alpha$	Nariai (1950), Tolman (1939) Kuchowicz (1972a, 1973) Bayin (1978)
	$c(1 + ar^2)^2(1 - br^2)^{-2}$	Stewart (1982)
	$(r^2 + a)^c(r^2 + b)^{-c}$ with restr. on a, b, c	Glass and Goldman (1978) Goldman (1978)
	$a\left(\dfrac{1 - b\delta}{1 + b\delta}\right)^2, \delta = \left(\dfrac{1 + \alpha r^2}{1 + \beta r^2}\right)^{\frac{1}{2}}$	Pant and Sah (1985) $\beta = 0$: Buchdahl (1964)
	ae^{br}	Kuchowicz (1972a)
	$a\exp(br^\alpha)$	Kuchowicz (1972a), Bayin (1978)
	$\dfrac{\cosh(a + br^2) - 1}{\cosh(a + br^2) + 1}$	Goldman (1978)

16.2 Non-static solutions

16.2.1 The basic equations

As in § 15.1, we take a comoving frame of reference

$$ds^2 = Y^2(r,t)d\Omega^2 + e^{2\lambda(r,t)}dr^2 - e^{2\nu(r,t)}dt^2, \quad u^i = (0, 0, 0, e^{-\nu}). \tag{16.20}$$

Table 16.2. Key assumptions of some static spherically-symmetric perfect fluid solutions in *canonical coordinates*

$ds^2 = r^2 d\Omega^2 + e^{2\lambda(r)} dr^2 - e^{2\nu(r)} dt^2$; $a, b, m, \alpha =$ const, possibly complex.

μ	ar^b	Wyman (1949), Kuchowicz (1966)
	$a - br^2$	Tolman (1939)
$e^{2\nu}$	ar^b	Tolman (1939), Kuchowicz (1968c)
	$1 + a/r$	Kuchowicz (1968a)
	$a + br^2$	Tolman (1939), Kuchowicz (1968b)
	$a(r+b)^2$	Kuchowicz (1967, 1968c)
	$(a + br^n)^2$, $n = 3, \pm 1, -2$	Heintzmann (1969)
	$a(1 + br^2)^n$	Heintzmann (1969), Korkina (1981) Durgapal (1982) Durgapal *et al.* (1984)
	$(1 + br^{2n})^m$	Paklin (1994)
	$(ar^{1-\alpha} - br^{1+\alpha})^2$	Tolman (1939), Wyman (1949) Kuchowicz (1968b)
	$\left(c_1 r^{2(a+b)} + c_2 r^{2(a-b)}\right)^\alpha$, with restr. on a, b, α	Kuchowicz (1970), Leibovitz (1969) Pant and Sah (1982), Pant (1994)
	$\left(\dfrac{a + (3k^2 + 2k - 1)r^2}{a + (3k^2 - 2k - 1)r^2}\right)^k$	Orlyanski (1997)
	$a(5 + br^2)^2(2 - br^2)$	Heintzmann (1969)
	$r^2(a + b \ln r)^2$	Kuchowicz (1968b)
	ae^{br^2}	Kuchowicz (1968b), Leibovitz (1969)
$e^{-2\lambda}$	a	Tolman (1939), Kuchowicz (1968b)
	ar^2	Patwardhan and Vaidya (1943)
	ar^b	Kuchowicz (1968b, 1971b)
	$a + br$	Kuchowicz (1968c)
	$a + br^2$	Kuchowicz (1968b) Buchdahl and Land (1968)
	$a - 2r^{-2}$	Bayin (1978)
	$a + br^\alpha$ with restr. on a, b, α	Tolman (1939), Wyman (1949) Kuchowicz (1968a, 1968b)
	$1 + ar^2 + br^4$	Tolman (1939), Mehra (1966) Patwardhan and Vaidya (1943)
	$(1 + ar^2)/(1 + cr^2)$	Buchdahl (1959, 1984)
	$1 - \dfrac{8ar^2(3 + ar^2)}{7(1 + ar^2)}$	Durgapal and Fuloria (1985)
	$a - 2 \ln r$	Kuchowicz (1968b)

In this coordinate system, the field equations read (cf. (15.6))

$$\kappa_0\mu = \frac{1}{Y^2} - \frac{2}{Y}\mathrm{e}^{-2\lambda}\left(Y'' - Y'\lambda' + \frac{Y'^2}{2Y}\right) + \frac{2}{Y}\mathrm{e}^{-2\nu}\left(\dot{Y}\dot{\lambda} + \frac{\dot{Y}^2}{2Y}\right), \tag{16.21a}$$

$$\kappa_0 p = -\frac{1}{Y^2} + \frac{2}{Y}\mathrm{e}^{-2\lambda}\left(Y'\nu' + \frac{Y'^2}{2Y}\right) - \frac{2}{Y}\mathrm{e}^{-2\nu}\left(\ddot{Y} - \dot{Y}\dot{\nu} + \frac{\dot{Y}^2}{2Y}\right), \tag{16.21b}$$

$$\begin{aligned}\kappa_0 pY = {}& \mathrm{e}^{-2\lambda}\left[\left(\nu'' + \nu'^2 - \nu'\lambda'\right)Y + Y'' + Y'(\nu' - \lambda')\right]\\ & -\mathrm{e}^{-2\nu}\left[\left(\ddot{\lambda} + \dot{\lambda}^2 - \dot{\lambda}\dot{\nu}\right)Y + \ddot{Y} + \dot{Y}(\dot{\lambda} - \dot{\nu})\right],\end{aligned} \tag{16.21c}$$

$$0 = \dot{Y}' - \dot{Y}\nu' - Y'\dot{\lambda}. \tag{16.21d}$$

Many of the known solutions have vanishing shear. In this case, (15.47) implies the relation $\dot{Y}/Y = \dot{\lambda}$, whose integral is $Y = \mathrm{e}^{\lambda}g(r)$. Thus, by a coordinate transformation $\hat{r} = \hat{r}(r)$, we can transform (16.20) into

$$\mathrm{d}s^2 = \mathrm{e}^{2\lambda(r,t)}(r^2\mathrm{d}\Omega^2 + \mathrm{d}r^2) - \mathrm{e}^{2\nu(r,t)}\mathrm{d}t^2, \tag{16.22}$$

i.e. we can introduce a coordinate system which is *simultaneously* comoving and isotropic. In (16.22), the r-coordinate is defined up to a transformation (inversion)

$$\hat{r} = 1/r, \quad \mathrm{e}^{2\hat{\lambda}} = \mathrm{e}^{2\lambda}r^4. \tag{16.23}$$

If the shear does not vanish, isotropic coordinates (16.22) can again be introduced, but they cannot be comoving (and (16.21) no longer hold).

16.2.2 Expanding solutions without shear

Some basic properties

Solutions without shear and expansion are either static or can easily be generated from static solutions, see §15.6.3.

For expanding solutions without shear it was shown in §15.6.4 that one can introduce coordinates

$$\mathrm{d}s^2 = \mathrm{e}^{2\lambda(r,t)}(r^2\mathrm{d}\Omega^2 + \mathrm{d}r^2) - \dot{\lambda}^2\mathrm{e}^{-2f(t)}\mathrm{d}t^2 \tag{16.24}$$

and reduce the field equations to the ordinary differential equation

$$\mathrm{e}^{\lambda}(\lambda'' - \lambda'^2 - \lambda'/r) = -\varphi(r), \tag{16.25}$$

where $\varphi(r)$ is an arbitrary function. Introducing the variables

$$L \equiv \mathrm{e}^{-\lambda}, \quad x \equiv r^2, \tag{16.26}$$

one can write this equation as

$$L_{,xx} = F(x)L^2, \quad F(x) \equiv \varphi(r)/4r^2. \tag{16.27}$$

Because of $\Psi_2 = 4xL^3F/3$, the function $F(x)$ characterizes the only non-vanishing Weyl-tensor component Ψ_2 (Barnes 1973b).

As explained in §15.6.4, one has to prescribe the functions $f(t)$ and $F(r^2)$ and to find a solution $L(r^2, t)$ of (16.27), which in general will contain two arbitrary functions of time which enter via the constants of integration (remember that $\dot{\lambda} = -\dot{L}/L$ must be non-zero). Energy density and pressure can then be computed from

$$\kappa_0\mu = 3\mathrm{e}^{2f} - \mathrm{e}^{-2\lambda}(2\lambda'' + \lambda'^2 + 4\lambda'/r), \tag{16.28a}$$

$$\kappa_0 p\dot{\lambda} = \mathrm{e}^{-3\lambda}\partial_t[\mathrm{e}^{\lambda}(\lambda'^2 + 2\lambda'/r) - \mathrm{e}^{3\lambda+2f}]\,, \tag{16.28b}$$

the expansion being given by $\Theta(t) = 3\mathrm{e}^{f(t)}$.

Known classes of solutions of $L_{,xx} = F(x)L^2$

The history of the spherically-symmetric and shearfree solutions is long, and rich in rediscoveries. To the authors' knowledge, all known solutions can be found in McVittie (1933), Kustaanheimo and Qvist (1948) and Wyman (1976), where in the later papers the results of the foregoing ones are always contained as special cases.

Three different approaches to finding solutions to the field equations can be distinguished.

The first approach was to make an ad hoc ansatz for the metric functions or for the function $F(x)$. Many solutions have been found this way, but since the authors did not characterize them invariantly, the same solutions were discovered again and again. We want to mention here McVittie (1933, 1984), but for the rest refer the reader to the papers by Srivastava (1987, 1992) and Sussman (1987, 1988a), where many of these solutions are given and their interrelation is discussed. The properties of some of these solutions have been discussed by Knutsen, see e.g. Knutsen (1986).

The second approach was to ask for which functions $F(x)$ the equation admits one (or two) Lie point symmetries or Noether symmetries, and to use these symmetries for the integration procedure. This approach was successfully initiated by Kustaanheimo and Qvist (1948), but pursued further only much later, see e.g. Stephani (1983b), Stephani and Wolf (1996) and Soh and Mahomed (1999).

The third approach was Wyman's (1976) search for solutions of $L_{,xx} = F(x)L^2$ which have the Painlevé property. All known solutions belong to this class.

We shall now characterize the known solutions and classify them by classifying $F(x)$. Following Wyman (1976), we first observe that solutions of $L_{,xx} = F(x)L^2$ can be obtained from solutions of

$$\mathrm{d}^2\tilde{L}/\mathrm{d}\tilde{x}^2 = \tilde{F}(\tilde{x})\,\tilde{L}^2 + p\tilde{x} + q \tag{16.29}$$

via a mapping

$$\frac{\mathrm{d}x}{\mathrm{d}\tilde{x}} = \tilde{\psi}(\tilde{x})^{-2}, \quad L(x) = \tilde{\psi}(\tilde{x})^{-1}[\tilde{L}(\tilde{x}) - \tilde{H}(\tilde{x})], \quad F(x) = \tilde{\psi}^5(\tilde{x})\,\tilde{F}(\tilde{x}), \tag{16.30}$$

if the functions $\tilde{\psi}$ and $\tilde{H}$ satisfy

$$\frac{\mathrm{d}^2\tilde{H}}{\mathrm{d}\tilde{x}^2} = \tilde{F}(\tilde{x})\,\tilde{H}^2(\tilde{x}) + p\tilde{x} + q\,, \quad \frac{\mathrm{d}^2\tilde{\psi}(\tilde{x})}{\mathrm{d}\tilde{x}^2} = 2\,\tilde{H}(\tilde{x})\,\tilde{F}(\tilde{x})\,\tilde{\psi}(\tilde{x}) \tag{16.31}$$

(the group-theoretical background of this transformation has been discussed by Herlt and Stephani (1992)).

For $p = 0 = q$, these mappings leave the *form* of the differential equation invariant, as $\mathrm{d}^2L/\mathrm{d}x^2 = F(x)L^2$ is transformed (mapped) into $\mathrm{d}^2\tilde{L}/\mathrm{d}\tilde{x}^2 = \tilde{F}(\tilde{x})\,\tilde{L}^2$. We may therefore refer to these transformations as gauge transformations of the function F, as opposed to symmetry transformations which leave the function F fixed.

The general solution of the differential equation (16.29) is known in the following three cases: for

$$\mathrm{d}^2\tilde{L}/\mathrm{d}\tilde{x}^2 = \tilde{L}^2 + p\tilde{x} + q, \quad p, q = \text{const}, \tag{16.32}$$

where it defines a Painlevé transcendent (and admits no Lie point symmetry if $p \neq 0$), and for its subcases

$$\mathrm{d}^2\tilde{L}/\mathrm{d}\tilde{x}^2 = \tilde{L}^2 + q \tag{16.33}$$

(with exactly one Lie point symmetry if $q \neq 0$) and

$$\mathrm{d}^2\tilde{L}/\mathrm{d}\tilde{x}^2 = \tilde{L}^2 \tag{16.34}$$

(with two Lie point symmetries), in which case it leads to elliptic integrals. Because of the existence of the gauge transformations (16.30)–(16.31), the solutions are also known if $\mathrm{d}^2L/\mathrm{d}x^2 = F(x)L^2$ can be mapped onto one of these three cases. All known solutions can be obtained by applying these transformations to the solutions of (16.32)–(16.34).

To decide whether for a given $F(x)$ such a mapping to (16.32), (16.33) or (16.34) exists, one has to check whether $F(x)$ satisfies

$$F^{-2/5}\Big(2F^{-2/5}\left\{F^{-2/5}[F^{-3/5}(F^{-1/5})'']'\right\}' + F^{-6/5}[(F^{-1/5})'']^2\Big)' = -4p, \tag{16.35}$$

$$2F^{-2/5}\left\{F^{-2/5}[F^{-3/5}(F^{-1/5})'']'\right\}' + F^{-6/5}[(F^{-1/5})'']^2 = -4q \quad (16.36)$$

or

$$2F^{-2/5}\left\{F^{-2/5}[F^{-3/5}(F^{-1/5})'']'\right\}' + F^{-6/5}[(F^{-1/5})'']^2 = 0. \quad (16.37)$$

To find all functions $F(x)$ which belong to one of these three classes, one can either apply the transformation (16.30)–(16.31), with $\tilde{F} = 1$, to each of these three cases (Wyman 1976), or one can solve the relevant conditions for the existence of one (or two) Lie point symmetries in the cases where symmetries exist (Stephani and Wolf 1996). Except for the case (16.35) of the Painlevé transcendent, there is an algorithmic procedure for constructing $F(x)$. The general solution of $\mathrm{d}^2L/\mathrm{d}x^2 = F(x)L^2$ for these cases is either also supplied by the transformation procedure, or by the standard methods of the symmetry approach (§ 10.2).

To summarize: The general solution of $\mathrm{d}^2L/\mathrm{d}x^2 = F(x)L^2$ is known for all functions $F(x)$ that satisfy (16.35) and therefore do not admit a symmetry, or that satisfy (16.36) and therefore admit one (special) symmetry, or that satisfy (16.37) or equivalently admit two symmetries. In each of the last two cases, all functions $F(x)$ can be constructed.

Examples of solvable classes $F(x)$

The cases most studied in the literature are those which can be generated from (16.33) or (16.34); they all lead to solutions $L(x)$ which can be expressed in terms of elliptic functions or their subcases. Examples are:

$F(x) = (ax^2 + 2bx + c)^{-5/2}$	Kustaanheimo and Qvist (1948)	
$F(x) = a(x - b)^{-15/7}$	Wyman (1976)	(16.38)
$F(x) = a(x - b)^{-15/7}(x - c)^{-20/7}$	Srivastava (1987)	

The Kustaanheimo–Qvist class of solutions

A rather large class of solutions of (16.27) was found by Kustaanheimo and Qvist (1948); many of the physically most interesting solutions fall into this class. They chose

$$F(x) = \pm\,(ax^2 + 2bx + c)^{-5/2}. \quad (16.39)$$

This $F(x)$ satisfies (16.36) and admits one symmetry. The perhaps surprising power $-5/2$ can be understood by observing that this form of $F(x)$ is invariant under the reflection (16.23) and its consequence $\hat{F}(x) = x^{-5}F(x^{-1})$.

Table 16.3. Some subclasses of the class $F = (ax^2+2bx+c)^{-5/2}$ of solutions

		F	A, B
McVittie solution (1933)		$\left[x(x+4R^2)\right]^{-5/2}$	$A = 0$
$\mu = \mu(t)$		$(2bx)^{-5/2}, b \neq 0$	$6A = b(3\mathrm{e}^{2f} - \kappa_0\mu)$
Kustaanheimo (1947)		0	$12AB = 3\mathrm{e}^{2f} - \kappa_0\mu$
Equation of state $p = p(\mu)$ Wyman (1946)	$\mu = \mu(t)$ (Friedmann 1922)	0	$B = \varepsilon A$, $\varepsilon = 0, \pm 1$
	$\mu = \mu(r, t)$	1	$A =$ const, $B = t$ $\mathrm{e}^{-2f} = -4At$

With (16.39), $\mathrm{d}^2L/\mathrm{d}x^2 = F(x)L^2$ for $F \neq 0$ leads to

$$(ax^2+2bx+c)^2u_{,xx} + 2(ax+b)(ax^2+2bx+c)u_{,x} + (ac-b^2)u = u^2, \tag{16.40}$$

$$u \equiv \pm\,(ax^2+2bx+c)^{-1/2}L, \tag{16.41}$$

which is integrated by

$$\int \frac{\mathrm{d}u}{\sqrt{\frac{2}{3}u^3 + (b^2-ac)u^2 + A(t)}} = \int \frac{\mathrm{d}x}{ax^2+2bx+c} + B(t). \tag{16.42}$$

The function $u = u(x) = u(r^2)$ – which may be expressed in terms of elliptic functions – gives $\mathrm{e}^{-\lambda} = L$ via (16.41), and choosing $f(t)$, we can compute the full metric (16.24). For $F = 0$, (16.27) gives

$$\mathrm{e}^{-\lambda} = L = A(t)r^2 + B(t). \tag{16.43}$$

Table 16.3 lists some subcases which correspond to special choices of the functions $A(t)$, $B(t)$ and/or of the real constants a, b, and c. We shall discuss these subcases in the following paragraphs.

The subclass $A(t) = 0$

For $A(t) = 0$, $a \neq 0$, $ac - b^2 \neq 0$, and with the notation

$$ax^2 + 2bx + c = a(x-x_1)(x-x_2), \quad B(t) = 2\ln C(t)/a(x_1-x_2), \tag{16.44}$$

we obtain from (16.42) the solution

$$\mathrm{e}^{-\lambda(r,t)} = \tfrac{3}{2}a^{5/2}(x_1-x_2)^2C(t)(r^2-x_1)\left[1-\sqrt{\frac{r^2-x_1}{r^2-x_2}}C(t)\right]^{-2} \tag{16.45}$$

(Kustaanheimo and Qvist 1948). It contains, for $x_1 = 0, x_2 = -4R^2$, the metric of McVittie (1933)

$$\begin{aligned} \mathrm{d}s^2 &= (1+f)^4 \mathrm{e}^{g(t)} \,[1 + r^2/4R^2\,]^{-2}(r^2\mathrm{d}\Omega^2 + \mathrm{d}r^2) - (1-f)^2\mathrm{d}t^2/(1+f)^2, \\ 2f &= m\mathrm{e}^{-g(t)/2}\,[1 + r^2/4R^2\,]^{1/2}/r, \quad R = \text{const}, \end{aligned} \tag{16.46}$$

which has been interpreted as a mass in a Robertson–Walker universe (for a (global) interpretation of the shearfree perfect fluids see Sussman (1988b)).

For $A(t) = 0$ and $a = 0$, one obtains the solution

$$\begin{aligned} \mathrm{d}s^2 &= S^2(t)(1+h)^4(r^2\mathrm{d}\Omega^2 + \mathrm{d}r^2) - (1-h)^2(1+h^2)^{-2}\mathrm{d}t^2, \\ h &\equiv S^{-1}(t)(\alpha r^2 + \beta)^{-1/2}. \end{aligned} \tag{16.47}$$

Solutions with a homogeneous distribution of matter $\mu = \mu(t)$

If we assume a homogeneous distribution of matter, $\mu = \mu(t)$, then we can differentiate the field equation (16.28a) with respect to r and eliminate λ'' and λ''' by means of (16.25). We obtain $3\varphi + r\varphi' = 0$, i.e. because of (16.27)

$$F(x) = (2bx)^{-5/2}. \tag{16.48}$$

To satisfy (16.28a), which requires that $\mathrm{e}^{-2\lambda}(2\lambda'' + \lambda'^2 + 4\lambda'/r)$ depends only on t, we have to make an appropriate choice of $A(t)$ and $B(t)$ in the general formula (16.42). The final result is: all shearfree, spherically-symmetric, expanding perfect fluid solutions with the energy density μ depending only on t are given by

$$\mathrm{d}s^2 = \mathrm{e}^{2\lambda(r,t)}(r^2\mathrm{d}\Omega^2 + \mathrm{d}r^2) - \dot{\lambda}^2\mathrm{e}^{-2f(t)}\mathrm{d}t^2 \tag{16.49}$$

with

$$\mathrm{e}^{-\lambda} = A(t) + B(t)r^2, \quad 12AB = 3\mathrm{e}^{2f} - \kappa_0\mu \quad (b = 0) \tag{16.50}$$

or

$$\begin{aligned} &\mathrm{e}^{-\lambda} = \sqrt{2bur}, \quad b = \text{const} \neq 0, \\ &\int \left[2u^3/3 + b^2u^2 + b(3\mathrm{e}^{2f} - \kappa_0\mu)/6\right]^{-1/2} \mathrm{d}u = (\ln r)/b + B(t) \end{aligned} \tag{16.51}$$

(Kustaanheimo 1947). In general, the pressure p will depend on both t and r; the subcase $p = p(t)$ is contained in the solutions considered in the following paragraphs.

Solutions with an equation of state $p = p(\mu)$

Solutions which obey an equation of state have been discussed by Wyman (1946), and special cases by Taub (1968). Because of (15.46), (15.47) and $\nu = \ln\dot{\lambda} - f(t)$, vanishing shear implies

$$p' = -(\mu + p)\dot{\lambda}'/\dot{\lambda}, \quad \dot{\mu} = -3(\mu + p)\dot{\lambda}. \tag{16.52}$$

If $\mu + p$ vanishes, then μ and p are constant, and the solution in question is the static vacuum solution of Kottler, see Theorem 15.5 and Table 15.1. If $\mu + p$ is different from zero, but μ' vanishes ($\mu = \mu(t), p = p(t)$), then we have $\mu' = p' = \nu' = \dot{\lambda}' = 0$. We choose the time coordinate t so that ν is zero, and infer from (16.24) that

$$\lambda = \lambda_1(r) + \lambda_2(t), \quad \dot{\lambda} = \dot{\lambda}_2 = \mathrm{e}^f \tag{16.53}$$

holds. This special time dependence of λ is compatible with (16.24) only if $\varphi = 0 = F$. So the solutions with $\mu = \mu(t), p = p(t)$ are the subcase

$$\mathrm{e}^{-\lambda} = A(t)[1 + \varepsilon r^2/4], \quad \varepsilon = 0, \pm 1, \quad \mathrm{e}^\nu = 1 \tag{16.54}$$

of (16.50). These are exactly the Friedmann-like universes (§14.2).

If neither $\mu + p$ nor μ' vanishes, then $\dot{\lambda}'$ is not zero and can be eliminated from (16.52), which leads to $\dot{\mu}\mu' = (\mu + p)\dot{\mu}'$. This is integrated by

$$\ln\dot{\mu} = \ln M(\mu) + \ln\dot{\alpha}(t), \quad \ln M(\mu) \equiv \int [\mu + p(\mu)]^{-1}\mathrm{d}\mu, \tag{16.55}$$

and in a further step by

$$H(\mu) = \alpha(t) + \beta(t), \quad H(\mu) \equiv \int M^{-1}(\mu)\mathrm{d}\mu. \tag{16.56}$$

Choosing the time coordinate so that $\alpha = t$ ($\dot{\alpha} = 0$ is prohibited by $\mu + p \neq 0, \dot{\lambda} \neq 0 \Rightarrow \dot{\mu} \neq 0$, cp. (16.52)), we see that because of (16.56) and (16.52) the functions μ and λ must have the special t dependence

$$\mu(r,t) = \mu(v), \quad \lambda(r,t) = \lambda_1(v) + \lambda_2(r^2), \quad v \equiv t + G(r^2). \tag{16.57}$$

To determine the functions λ_1, λ_2, G and the function F occurring in the first integral (16.27), we insert (16.57) into (16.27). Writing

$$\mathrm{e}^{-\lambda} = L = u(v)l(x), \quad x = r^2, \tag{16.58}$$

we obtain

$$\dot{u}(G_{,xx}l + 2G_{,x}l_{,x}) + ul_{,xx} + \ddot{u}lG_{,x}^2 = u^2l^2F. \tag{16.59}$$

In this equation, only u and its derivatives depend on t. Since by assumption, $\dot{u}, l$ and $G_{,x}$ are non-zero, either $u, \dot{u}^2$ and $\ddot{u}$ are proportional to each other (which is impossible), or at least one of the coefficients of these functions vanishes. From this reasoning, we conclude that

$$G_{,xx}l + 2G_{,x}l_{,x} = 0, \quad l_{,xx} = 0, \quad \ddot{u} = \text{const} \cdot u^2 = l^2 F u^2 G_{,x}^{-2} \qquad (16.60)$$

holds. One can show that the two cases $l = ax+b$ and $l = b$ are equivalent (i.e. connected by a transformation (16.23)), and that $F = 0$ gives $\mu = \mu(t)$. So we need only consider the case $F = 1$, $l = \text{const}$. For this case we now have to choose the arbitrary functions of integration so that μ and λ have the functional form (16.57). This is done by specializing (16.42) to

$$\begin{gathered} ds^2 = e^{2\lambda}(dr^2 + r^2 d\Omega^2) - \dot{\lambda}^2 (a_0 - 4At)^{-1} dt^2, \\ \int (2u^3/3 + A)^{-1/2}\, du = t + r^2, \quad A = \text{const}, \quad e^{-\lambda} = u(t + r^2) \end{gathered} \qquad (16.61)$$

(Wyman 1946). One can easily check by computing μ and p from (16.28) that $\mu = \mu(t + r^2)$ and $p = p(t + r^2)$ hold.

Solutions with a homothetic vector or a conformal Killing vector

If a shearfree spherically-symmetric perfect fluid admits a homothetic vector $\boldsymbol{\xi} = r\partial_r + t\partial_t = 2x\partial_x + t\partial_t$, then it can be shown (see Dyer *et al.* (1987)) that the function F has to be of the form $F = cx^n$ (note that a rescaling of the x-coordinate by a transformation $\hat{x} = x^\alpha$ is not possible since the line element (16.24) is not invariant under this transformation). Since for this case the differential equation (16.22)

$$L_{,xx} = cx^n L^2 \qquad (16.62)$$

admits the Lie point symmetry $X = 2x\partial_x - (n+2)L\partial_L$, there always exist solutions of the form $\hat{L} = t^{-2(n+2)} L(x/t^2)$ which together with $e^f = t^{2n+5}$ lead to metrics admitting a homothetic vector (i.e. the constants of integration entering into the solution of (16.62) can be chosen as functions of time appropriately). For the cases $n = -5, -5/2, -15/7$, in which (16.62) can be solved in terms of elliptic functions, the explicit metrics were constructed by Havas (1992).

Conformal Killing vectors have been found so far only in the (trivial) case of the conformally flat metrics $F = 0$ (Sussman 1989).

16.2.3 Solutions with non-vanishing shear

As shown in the preceding subsection, solutions without shear are fairly well known. In contrast, only a few classes of solutions with shear have been found and discussed so far. For a physical approach to the basic quantities and equations, see Misner and Sharp (1964).

A negative result concerning solutions with shear is due to Thompson and Whitrow (1967) and Misra and Srivastava (1973): if in the comoving frame of reference (16.20) the mass density μ is a function only of t, and if the metric is regular at $r = 0$ (i.e. $Y = 0$, $Y' = \mathrm{e}^{\lambda}$), then the four-velocity is necessarily shearfree.

For solutions *with shear but without acceleration,* see §15.6.5.

Solutions with shear but without expansion

Because of (15.47), solutions with zero expansion but non-zero shear have to obey $\dot{\lambda} = -2\dot{Y}/Y$, $\dot{Y} \neq 0$. Together with the field equation (16.21d), this leads to

$$\mathrm{d}s^2 = Y^2\mathrm{d}\Omega^2 + Y^{-4}\mathrm{d}r^2 - Y^4\dot{Y}^2 f^2(t)\mathrm{d}t^2. \tag{16.63}$$

Equation (15.46) shows that the mass density μ is a function only of r, so that the field equations (16.21a)–(16.21b) read

$$2Y^5Y'' + 5Y^4Y'^2 + 3Y^{-4}f^{-2} + \kappa_0\mu(r)Y^2 - 1 = 0, \tag{16.64a}$$

$$\kappa_0 p = -Y^{-2} + 5Y'^2Y^2 + 2\dot{Y}'Y'\dot{Y}^{-1}Y^3 + 3Y^{-6}f^{-2} + 2\dot{f}f^{-3}\dot{Y}^{-1}Y^{-5}. \tag{16.64b}$$

The condition of isotropy is satisfied if (16.64) are. In (16.64a), $\mu(r)$ and $f(t)$ can be prescribed, and p can be computed from (16.64b) once Y is known. For constant μ, (16.64a) is integrated by

$$Y'^2Y^5 - Y^{-3}f^{-2}(t) + \kappa_0\mu Y^3/3 - Y = A(t), \tag{16.65}$$

which can be solved by quadratures (Skripkin 1960).

Solutions with shear, acceleration and expansion

A large variety of spherically-symmetric perfect fluid solutions is to be expected in this most general class, but only a few special cases have been treated so far. Some of them have already been given above, see §15.6.5.

Solutions of the form

$$\begin{aligned} \mathrm{d}s^2 &= r^2[\varepsilon/2 + h(t)]\mathrm{d}\Omega^2 + (\varepsilon + cr^2)^{-1}\mathrm{d}r^2 - r^2\mathrm{d}t^2/4, \\ h(t) &= \begin{cases} A\sin t + B\cos t \\ -t^2/4 + 2At + B \\ A\exp(t) + B\exp(-t) \end{cases} \quad \text{for } \varepsilon = \begin{cases} -1 \\ 0 \\ +1 \end{cases} \end{aligned} \tag{16.66}$$

were found by Leibovitz (1971), Lake (1983) and Van den Bergh and Wils (1985a), and further discussed by Collins and Lang (1987) and Maharaj *et al.* (1993). They have an equation of state $\kappa_0 p = \kappa_0\mu + 6c$; the case $c = 0$ is also contained in (37.57). They admit a hypersurface-orthogonal conformal Killing vector (Koutras, private communication).

Marklund and Bradley (1999) gave the solutions

$$\mathrm{d}s^2 = \frac{t}{r^2}\mathrm{d}\Omega^2 + \frac{\mathrm{d}r^2}{a - br^2} - \frac{r^2\mathrm{d}t^2}{4t^2(ct^2 - t + a)^2}, \quad \kappa_0 p = \kappa_0\mu + 6b. \tag{16.67}$$

Some authors pursued the idea that a metric may be simple when written in non-comoving coordinates

$$\mathrm{d}s^2 = \mathrm{e}^{2\lambda(r,t)}[f^2(r)\mathrm{d}\Omega^2 + \mathrm{d}r^2] - \mathrm{e}^{2\nu(r,t)}\mathrm{d}t^2, \quad u^n = (0, 0, u^3, u^4). \tag{16.68}$$

The field equations then have the form

$$\begin{aligned} G^1_1 = G^2_2 &= \kappa_0 p, & G^3_3 &= \kappa_0(\mu + p)u^3u_3 + \kappa_0 p, \\ G^3_4 &= \kappa_0(\mu + p)u^3u_4, & G^4_4 &= \kappa_0(\mu + p)u^4u_4 + \kappa_0 p. \end{aligned} \tag{16.69}$$

When solutions to the condition of isotropy

$$\mathrm{e}^{2\lambda}(G^3_4)^2 + \mathrm{e}^{2\nu}(G^3_3 - G^1_1)(G^4_4 - G^1_1) = 0 \tag{16.70}$$

have been found, then μ, p and the components u^3 and u^4 of the four-velocity can be computed from (16.69). Narlikar and Moghe (1935) gave some classes of solutions (in isotropic coordinates, (16.68) with $f = r$, and with $G^3_4 \neq 0$), which – when corrected – read

$$\begin{aligned} &\nu = \lambda + \ln(r/t) - \ln a, \quad \lambda(r/t) = g(x), \quad x = r/t, \\ &\quad (a^2 + 1)g'' - (a^2 - 1)(g'^2 + g'/x) - a^2/2x^2 = 0, \end{aligned} \tag{16.71}$$

$$\nu = -\lambda, \quad \lambda = 1/hr + \ln h + c_1 + (c_2 - a^2t)/ah, \quad h = at + b, \tag{16.72}$$

$$\mathrm{e}^\nu = a(\ln r + t + b), \quad \lambda = t, \tag{16.73}$$

$$\nu = 0, \quad \mathrm{e}^\lambda = a\mathrm{e}^{c_1 t}/r^2 + b\mathrm{e}^{c_2 t}. \tag{16.74}$$

McVittie and Wiltshire (1977) found the following classes of solutions:

$$\mathrm{d}s^2 = A(\mathrm{d}r^2 + r^2\mathrm{d}\Omega^2) - A^2\mathrm{d}t^2, \quad A = (ar^2 + bt)^{2/3}, \tag{16.75}$$

$$\begin{aligned} &\mathrm{d}s^2 = \exp\left[2\alpha(r) + 2\psi(t)\right](\mathrm{d}r^2 + \mathrm{d}\Omega^2 - \mathrm{d}t^2), \\ &\quad \alpha_{,rr} - (a + 1)\alpha^2_{,r} + 1/2 = 0, \quad \psi_{,tt} - (a + 1)\psi^2_{,t}/a - 1/2 = 0, \end{aligned} \tag{16.76}$$

$$\begin{aligned} &\mathrm{d}s^2 = \mathrm{e}^{2bt}S^2(1 + aS)\left[\mathrm{d}r^2 + \Sigma^2(r, k)\mathrm{d}\Omega^2\right] - (1 + aS)^{4/3}\mathrm{d}t^2, \\ &\quad S = \Sigma^{-2}(r/2, k)t\mathrm{e}^{-2bt}, \end{aligned} \tag{16.77}$$

(with $\Sigma(r,k)$ as in (15.3)), see also Bonnor and Knutsen (1993) and Knutsen (1995).

Vaidya (1968) studied (non-comoving) metrics of the form

$$ds^2 = r^2 d\Omega^2 + dr^2/[1 - \kappa_0 \mu(r,t) r^2/3] - e^{2\nu} dt^2, \tag{16.78}$$

but no new explicit solutions were found.

Spherically-symmetric solutions are also contained in the solutions (36.20) possessing flat slices.

17

Groups G_2 and G_1 on non-null orbits

17.1 Groups G_2 on non-null orbits

17.1.1 Subdivisions of the groups G_2

The groups G_2 can be divided into several subclasses depending on the properties of the (appropriately chosen) two Killing vectors which (*a*) do commute or not, (*b*) are orthogonally transitive or not, (*c*) are hypersurface-orthogonal or not. We shall discuss these alternatives now in turn.

(*a*) The normal forms of the space-time metrics in the commuting (G_2I) and in the non-commuting case (G_2II) (§8.2) are given by (Petrov 1966, p. 150)

$$G_2I: \quad g_{ij} = g_{ij}(x^3, x^4), \qquad \boldsymbol{\xi} = \partial_1, \qquad \boldsymbol{\eta} = \partial_2, \tag{17.1}$$

$$G_2II: \quad g_{ij} = \begin{pmatrix} e^{-2x^2}a_{11} & e^{-x^2}a_{12} & e^{-x^2}a_{13} & 0 \\ e^{-x^2}a_{12} & a_{22} & a_{23} & 0 \\ e^{-x^2}a_{13} & a_{23} & a_{33} & 0 \\ 0 & 0 & 0 & e_4 \end{pmatrix}, \tag{17.2}$$

$$a_{ij} = a_{ij}(x^3, x^4), \quad e_4 = \pm 1, \quad \boldsymbol{\xi} = \partial_1, \quad \boldsymbol{\eta} = x^1\partial_1 + \partial_2.$$

The 2-surfaces of transitivity (group orbits) spanned by the two Killing vectors $\boldsymbol{\xi}$ and $\boldsymbol{\eta}$ are spacelike or timelike respectively when the square of the simple bivector $\xi_{[a}\eta_{b]}$ is positive or negative.

The general field equations, both for G_2I and G_2II, for space-times admitting a group G_2 of motions are very complicated to solve and no exact solutions have been obtained for either of the metrics (17.1), (17.2) without additional simplifications. The further restrictions imposed may be degeneracy of the Weyl tensor, or special properties of the Killing vector fields (see below), or an additional homothetic vector. The symmetry groups of the known algebraically special solutions have only partially been investigated, and we cannot completely answer the question of which of the algebraically special solutions (Part III) admit a G_2 or G_1 on non-null orbits. Some remarks concerning the link between groups of motions and Petrov types are contained in Chapter 38.

(*b*) A restriction which is often imposed is the existence of 2-surfaces orthogonal to the group orbits (*orthogonally transitive group*). The Killing vectors then obey the relations

$$\xi_{[a;b}\xi_c\eta_{d]} = 0 = \eta_{[a;b}\eta_c\xi_{d]} \tag{17.3}$$

(see (6.13) and §19.2). An invertible Riemann–Maxwell structure in general implies the existence of an Abelian G_2 which is orthogonally transitive (Debever *et al.* 1979; see also Duggal 1978). Solutions which are not orthogonally transitive are rarely considered; they occur, however, when three Killing vectors are present and a second pair satisfies (17.3), see §§22.2 and 23.3.3 for examples.

(*c*) Hypersurface-orthogonal Killing vectors satisfy the more stringent condition $\xi_{[a;b}\xi_{c]} = 0 = \eta_{[a;b}\eta_{c]}$ which, of course, implies (17.3). They usually occur as subcases of the general case. Note that for two non-null orthogonally transitive Killing vectors none or both are hypersurface-orthogonal (cp. (17.4)).

17.1.2 Groups G_2I on non-null orbits

There are several physically interesting classes of solutions admitting an Abelian group G_2 on non-null orbits; because of their importance and the large amount of relevant material, they will be divided into separate chapters. The stationary axisymmetric fields (Chapters 19–21 and 34) have *timelike* group orbits T_2. The classes with *spacelike* group orbits S_2 (time-dependent *cylindrically-symmetric* fields and their stationary subclasses, *colliding plane waves* with their typical dependence on the retarded/advanced time, and *inhomogeneous perfect fluid solutions*) will be treated in Chapters 22, 23, 25 and 34. In the boost-rotation-symmetric space-times (§17.2), the group orbits have different characters (timelike or spacelike) in different regions. In this chapter we restrict ourselves to some general facts and to the boost-rotation-symmetric space-times.

If an orthogonally transitive group G_2I acts on non-null orbits V_2, the space-time metric can be written in the form

$$ds^2 = e^M(dz^2 + \varepsilon dt^2) + W[e^{-\Psi}dy^2 - \varepsilon e^{\Psi}(dx + Ady)^2], \quad W > 0, \tag{17.4}$$

where all functions are independent of x and y. The orbits are spacelike for $\varepsilon = -1$, and timelike for $\varepsilon = +1$. The function W is invariantly defined by $W^2 := -2\varepsilon\xi_{[a}\eta_{b]}\xi^a\eta^b$. An equivalent form of the metric is

$$\mathrm{d}s^2 = \mathrm{e}^M(\mathrm{d}z^2 + \varepsilon\mathrm{d}t^2) + l(z,t)\mathrm{d}y^2 + 2m(z,t)\mathrm{d}y\mathrm{d}x - \varepsilon n(z,t)\mathrm{d}x^2. \quad (17.5)$$

In general, the x–y part of the metric is described by three independent functions l, m, n. If, however, these three functions obey a linear relation

$$al + 2bm - c\varepsilon n = 0 \quad (17.6)$$

(with constant coefficients a, b, c), then the metric can be simplified by means of a linear transformation

$$x = \alpha\widetilde{x} + \beta\widetilde{y}, \quad y = \gamma\widetilde{x} + \delta\widetilde{y}, \quad \alpha\delta - \beta\gamma \neq 0, \quad (17.7)$$

leading to

$$\begin{aligned} \widetilde{m} &= \gamma\delta l + (\alpha\delta + \gamma\beta)m - \alpha\beta\varepsilon n, \quad -\varepsilon\widetilde{n} = \gamma^2 l + 2\alpha\gamma m - \alpha^2\varepsilon n, \\ \widetilde{l} - \varepsilon\widetilde{n} &= (\delta^2 + \gamma^2)l + 2(\alpha\gamma + \beta\delta)m - (\alpha^2 + \beta^2)\varepsilon n. \end{aligned} \quad (17.8)$$

Comparing (17.6) and (17.8), one sees that three cases occur, depending on the relation of $\Delta = ac - b^2$ to the corresponding discriminants for the three equations of (17.8), see MacCallum (1998) and the references given there. If $\Delta < 0$, then $\widetilde{m}$ (or, equivalently, A in (17.4)) can be made zero: the metric admits two hypersurface-orthogonal Killing vectors. If $\Delta = 0$, then $\varepsilon\widetilde{n}$ can be transformed to zero (and to have the correct signature of space-time, ε is -1): the metric admits a null Killing vector. If $\Delta > 0$, then $\widetilde{l} - \varepsilon\widetilde{n}$ can be made zero; this case corresponds to a pair of complex conjugate hypersurface-orthogonal Killing vectors, and the metric can be written as

$$\mathrm{d}s^2 = \mathrm{e}^M(\mathrm{d}z^2 + \mathrm{d}t^2) + W\left[(\cos\Psi\mathrm{d}y + \sin\Psi\mathrm{d}x)^2 - (\cos\Psi\mathrm{d}x - \sin\Psi\mathrm{d}y)^2\right]. \quad (17.9)$$

For $A = 0 = \Psi$ one regains from (17.4) the plane-symmetric line element (15.10).

For the Ricci tensor of the metric (17.4), see e.g. Wainwright *et al.* (1979). For later use we only give the relation

$$R_1{}^1 + R_2{}^2 = -\mathrm{e}^{-M}W^{-1}(W_{,33} + \varepsilon W_{,44}) = -\kappa_0(T_3{}^3 + T_4{}^4), \quad (17.10)$$

where the coordinates are labelled as $x^a = (x, y, z, t)$.

In some cases (vacuum fields, Einstein–Maxwell fields with $F^*_{ab}\xi^a\eta^b = 0$, perfect fluids with $p = \mu$ for $\varepsilon = -1$), the energy-momentum tensor obeys the condition

$$T_3{}^3 + T_4{}^4 = 0. \quad (17.11)$$

Then (17.10) shows that for $\varepsilon = -1$ we obtain the general solution

$$W = f(u) + g(v), \quad \sqrt{2}u = t - z, \quad \sqrt{2}v = t + z, \tag{17.12}$$

with arbitrary functions $f(u)$ and $g(v)$, and for $\varepsilon = +1$ W is analytic.

In the vacuum case with $A = 0$ in (17.4) (diagonal form of the metric), the function Ψ obeys the linear differential equation

$$(W\Psi_{,3})_{,3} + \varepsilon(W\Psi_{,4})_{,4} = 0 \tag{17.13}$$

(see also §25.2), and the rest of the field equations determine M in terms of a line integral.

In most applications (e.g. for stationary axisymmetric gravitational fields or colliding plane waves) the condition (17.3) for orthogonal transitivity follows from other physical assumptions. The more general case when the metric with two commuting Killing vectors is not reducible to block-diagonal form has been investigated by Gaffet (1990). In that paper the field equations and a corresponding Lagrangian were derived, reductions gave rise to several cases of integrability, either by quadratures, or by elliptic functions, and a non-orthogonally transitive generalization of Weyl's static metrics (§20.2) was obtained.

Non-orthogonally transitive metrics with a preferred null direction (e.g. a pure radiation source) have been discussed by Kolassis and Santos (1987).

17.1.3 G_2II on non-null orbits

The case of the non-Abelian group G_2II has been considered in the literature only occasionally. A reason for that may be that the Lorentz group does not possess a G_2II as a subgroup so that a G_2II will not occur within asymptotically flat solutions, in agreement with the result derived by Carter (1970).

Kolassis (1989) formulated the necessary and sufficient conditions for a space-time to admit a G_2I or a G_2II in the context of the modified spin coefficient formalism of Geroch *et al.* (1973) (§7.4).

Aliev and Leznov (1992a, 1992b) have rewritten Einstein's vacuum equations in the form of a covariant gauge theory in two dimensions. A special solution of the field equations is the metric AII in Table 18.2 (all metrics with pseudospherical symmetry admit a G_2II, cp. §8.6.2); a second ansatz reduces the problem to an ordinary differential equation, and a third ansatz leads to a solution with an additional timelike Killing vector.

An example of an algebraically special vacuum solution is provided by (38.6). Perfect fluid solutions with a G_2II are considered in Chapter 23.

Phan (1993) found a Petrov type I vacuum solution admitting a G_2II with Killing vectors ∂_x and $x\partial_x - \partial_y$ and a homothetic vector $\partial_y - \partial_t$. It reads

$$\mathrm{d}s^2 = 2\mathrm{e}^{2y}\mathrm{d}x^2 + \frac{8z}{3}\mathrm{e}^{-2t}\mathrm{d}y^2 + \frac{\mathrm{e}^{-2t}}{z}\left[\left(\frac{\mathrm{d}z}{2zA} + \frac{\mathrm{d}y}{B}\right)^2 - (\mathrm{d}t - H\mathrm{d}y)^2\right],$$

$$A(z) = \sqrt{[(2z-1)^2+2]/8z}, \quad B(z) = \sqrt{3/4z}, \tag{17.14}$$

$$H(z) = 1 + 2A(z)/B(z).$$

Orthogonal transitivity and spacelike orbits of the group G_2II reduce the metric (17.2) to the simpler form

$$\mathrm{d}s^2 = \mathrm{e}^M(\mathrm{d}z^2 - \mathrm{d}t^2) + W[\mathrm{e}^\Psi(\mathrm{e}^{-y}\mathrm{d}x + A\mathrm{d}y)^2 + \mathrm{e}^{-\Psi}\mathrm{d}y^2], \tag{17.15}$$

where the metric functions M, Ψ, A and W $(W > 0)$ depend only on z and t. The line element (17.15) differs from (17.4) only by the factor e^{-y}. Bugalho (1987) has shown that the vacuum field equations imply that Ψ and A in (17.2) are necessarily constants and that the resulting solutions are pseudospherically-symmetric, i.e. they admit at least one additional Killing vector. The same is true for perfect fluids for which $\mu + p > 0$ with the four-velocity orthogonal to the group orbit.

17.2 Boost-rotation-symmetric space-times

In Minkowski space-time, the generators of rotations around the z-axis and of boosts along the z-axis commute; hence these two symmetries form an Abelian group G_2I of motions. Accordingly, curved space-times are called *boost-rotation-symmetric* if they admit, in addition to the hypersurface-orthogonal Killing vector describing axial symmetry, a second Killing vector which becomes a boost in the flat-space limit. The corresponding symmetries are compatible with gravitational radiation and asymptotic flatness. In axially-symmetric space-times, the only allowable second symmetry that does not exclude radiation is boost symmetry. The proof (Bičák and Schmidt (1984), for corrections see Bičák and Pravdová (1998)) uses the asymptotic form of the Killing equations near future null infinity $\mathscr{I}^+$ in Bondi coordinates.

Asymptotically flat radiative space-times with boost-rotation symmetry were systematically investigated by Bičák and Schmidt (1989) (this review article and the papers by Schmidt (1996) and Bičák (1997) may also be consulted for global aspects of the boost-rotation symmetry). In accordance with the metric (17.4) and the field equation $W_{,uv} = 0$, see (17.10), the canonical form of the metric considered by these authors in the context of boost-rotation-symmetric space-times is given by

$$ds^2 = -e^{\lambda} du\, dv + \tfrac{1}{4}(v-u)^2 e^{-\mu} d\varphi^2 + \tfrac{1}{4}(v+u)^2 e^{\mu} d\chi^2. \qquad (17.16)$$

The metric functions $\mu = \mu(u,v)$ and $\lambda = \lambda(u,v)$, which are defined for $0 < v_0 \le v < \infty$ and $u_0 \le u \le u_1$, $u < v$, $u \ne -v$, have to satisfy the conditions (for fixed u)

$$\lim_{v\to\infty} \lambda(u,v) = \lambda_0(u), \quad \lim_{v\to\infty} \mu(u,v) = \kappa = \text{const.} \qquad (17.17)$$

The vacuum field equations reduce to a flat-space wave equation for μ,

$$\left[\partial_u\partial_v + (v^2-u^2)^{-2}(v\partial_u - u\partial_v)\right]\mu = 0. \qquad (17.18)$$

Once μ is given, λ can be determined, by a line integral, from

$$\partial_v\lambda = \frac{(v^2-u^2)}{4v}(\partial_v\mu)^2 - \frac{u}{v}\partial_v\mu, \quad \partial_u\lambda = -\frac{(v^2-u^2)}{4u}(\partial_u\mu)^2 - \frac{v}{u}\partial_u\mu. \qquad (17.19)$$

The wave equation (17.18) is just the integrability condition of (17.19). By means of suitable solutions of the wave equation (17.18) one can construct a large class of solutions with boost-rotation symmetry, the C-metric being a member of this class.

The solution given by Bonnor and Swaminarayan (1965) is another famous example of a boost-rotation-symmetric space-time. It describes the gravitational field of two independent pairs of particles that are uniformly accelerated along the axis of symmetry in opposite directions and that are symmetrically located with respect to the plane $z = 0$.

The Bonnor–Swaminarayan solution is given by the metric

$$\begin{aligned}
ds^2 &= e^{2\lambda}\frac{d\rho^2 + dz^2}{2r} + e^{-2\widetilde{U}}\frac{\rho^2}{r+z}d\varphi^2 - e^{2\widetilde{U}}(r+z)dt^2, \quad r^2 = \rho^2 + z^2,\\
\widetilde{U} &= -m_1/r_1 - m_2/r_2 + \text{const}, \quad {r_A}^2 = \rho^2 + (z-b_A)^2, \quad A = 1,2,\\
2\lambda &= -\frac{{m_1}^2\rho^2}{{r_1}^4} - \frac{{m_2}^2\rho^2}{{r_2}^4} + 4m_1m_2\frac{\rho^2 + (z-b_1)(z-b_2)}{r_1r_2(b_1-b_2)^2}\\
&\quad + 2m_1r/b_1r_1 + 2m_2r/b_2r_2 + \text{const}, \quad b_A, m_A = \text{const(real)}.
\end{aligned} \qquad (17.20)$$

By means of the inverse of the transformation

$$Z = \sqrt{r+z}\cosh t, \quad R = \sqrt{r-z}, \quad T = \sqrt{r+z}\sinh t, \quad \Phi = \varphi, \qquad (17.21)$$

the metric (17.20) may be transformed into

$$\begin{aligned}
ds^2 &= e^{2\lambda}dR^2 + R^2 e^{-2\widetilde{U}}d\Phi^2\\
&\quad + (Z^2 - T^2)^{-1}[e^{2\lambda}(ZdZ - TdT)^2 - e^{2\widetilde{U}}(ZdT - TdZ)^2],
\end{aligned} \qquad (17.22)$$

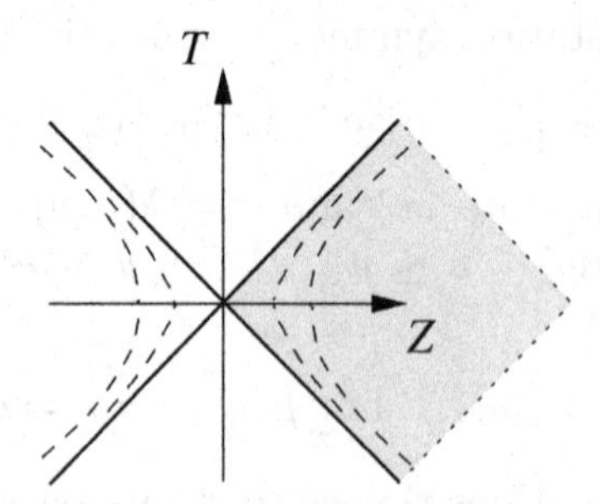

Fig. 17.1. Accelerated particles. The world lines of particles are dashed

with $\widetilde{U}$ and λ from (17.20). The Killing vectors $\boldsymbol{\xi} = T\partial_Z + Z\partial_T$ and $\boldsymbol{\eta} = \partial_\Phi$ generate an Abelian group G_2; $\boldsymbol{\xi}$ is timelike for $Z^2 > T^2$, but spacelike for $Z^2 < T^2$.

For $\lambda = 0 = \widetilde{U}$, the metric (17.22) reduces to flat space-time,

$$\mathrm{d}s^2 = \mathrm{d}R^2 + R^2\mathrm{d}\Phi^2 + \mathrm{d}Z^2 - \mathrm{d}T^2. \tag{17.23}$$

The part of Minkowski space for which $Z^2 > T^2, Z > 0$ (the hatched region in Fig. 17.1) can be transformed with the aid of (17.21) into Weyl's canonical coordinates ((19.21) with $A = 0$)

$$\mathrm{d}s^2 = \mathrm{e}^{-2U}[\mathrm{e}^{2k}(\mathrm{d}\rho^2 + \mathrm{d}z^2) + \rho^2\mathrm{d}\varphi^2] - \mathrm{e}^{2U}\mathrm{d}t^2. \tag{17.24}$$

The metric coefficients U and k in (17.24) given by

$$\mathrm{e}^{2U} = \mathrm{e}^{2U_0} = r + z, \quad \mathrm{e}^{2k} = (r + z)/2r, \quad r = (\rho^2 + z^2)^{1/2}. \tag{17.25}$$

lead to flat space-time. The metric (17.20) belongs to the Weyl class and the function U in (17.24), which for vacuum fields satisfies the linear flat-space potential equation $\Delta U = 0$, is the linear superposition of the potential $U_0 = \frac{1}{2}\ln(r + z)$ for flat space-time and the potential $\widetilde{U}$ given in (17.20) for the field of two point particles.

In the region $Z^2 > T^2, Z < 0$, two additional (mirror) particles appear. The additive constants in U and λ may be chosen so that two particles of positive mass are moving freely, whereas the other two particles are either connected by stress singularities on the axis or the singularities extend from these two particles to infinity. By appending an appropriate mass at infinity (Ernst 1976) these singularities can be removed. This has been done by Bičák *et al.* (1983a, 1983b) for the solution (17.20) and for a general freely falling pair of particles.

The world lines of the point particles are $Z = \pm(T^2 + 2b_A)^{1/2}, R = 0$. The point sources moving with uniform acceleration do not remain in a finite region of space for all times. Consequently, the solution (17.20) cannot serve as a model describing the emission of gravitational waves from an *isolated source*. Nevertheless, the Bonnor–Swaminarayan solution is radiative.

Similarly, the C-metric (§18.6) can be interpreted as the field of two black holes uniformly accelerated in opposite directions (Kinnersley and Walker 1970). Bonnor (1983) transformed the line element given for the C-metric in Table 18.2 into Weyl's form (17.24), extended the manifold and arrived at the interpretation of the gravitational field of two particles uniformly accelerated by a spring between them. Cornish and Uttley (1995) presented a simplified version of Bonnor's approach. There is a close analogy of the C-metric with the Born solution for accelerated charges in electrodynamics.

Ernst (1978b) applied his procedure summarized by (20.21), (20.22) to derive the *generalized* C-metric

$$\begin{aligned} ds^2 = r^2 \Big[& e^{\lambda(F+G)} \exp(-\lambda^2 r^4 AB)(dx^2/A + dy^2/B) + e^{\lambda(G-F)} A dz^2 \\ & - e^{\lambda(F-G)} B dt^2 \Big], \quad F = Ar^2 + mx, \quad G = Br^2 + my, \qquad (17.26) \\ A = 1 - x^2 - mx^3, \quad & B = y^2 - my^3 - 1, \quad r = (x+y)^{-1}, \quad \lambda, m = \text{const.} \end{aligned}$$

In this boost-rotation-symmetric solution the nodal singularity of the original C-metric can be eliminated by an appropriate choice of the additional parameter λ which is related to the external gravitational field that causes the acceleration of the particles (Dray and Walker 1980).

It can be shown (Valiente Kroon 2000) that all the solutions in §29.2.6 are boost-rotation-symmetric as the asymptotic form of at least two of their Killing vectors is the same as those of the C-metric.

Bičák and Pravda (1999) investigated some properties of the twisting generalization of the C-metric which can be interpreted as a radiative space-time with accelerating and *rotating* black holes.

17.3 Group G_1 on non-null orbits

Stationary gravitational fields (Chapter 18) admit a timelike Killing vector. The reduction formulae for the Ricci tensor derived in §18.2 for a timelike Killing vector also hold in the case of a spacelike Killing vector.

For vacuum and some restricted classes of perfect fluids, the generation procedure outlined in §10.3.2 can be used to obtain a one-parameter family of solutions from any seed solution admitting a G_1, see e.g. Garfinkle *et al.* (1997) for applications.

In contrast to the case of a null Killing vector (§24.4), it is very difficult to solve Einstein's field equations without any further assumptions. In this section we consider special assumptions which lead to solutions admitting a group G_1 (on non-null orbits). Such solutions will also appear in Chapters 26–33, e.g. as subcases of algebraically special metrics.

Harrison (1959) started with the metrical ansatz ('linked pairs' form)

$$g_{ij} = e_i \delta_{ij} A_i^2(x^1, x^4) B_i^2(x^3, x^4) \quad \text{(no summation)}, \qquad e_1 = e_3 = 1, \quad e_2 = \pm 1, \quad e_2 e_4 = -1, \tag{17.27}$$

and obtained a series of exact vacuum solutions admitting a group G_1. The spacelike (or timelike) Killing vector $\boldsymbol{\xi} = \partial_2$ (or $\boldsymbol{\xi} = \partial_4$) is hypersurface-orthogonal. The separation of variables in the particular form (17.27) either leads to solutions in closed form or reduces the problem to an ordinary differential equation. We give one example from each of these cases for illustration.

Solution in closed form:

$$\mathrm{d}s^2 = \sum_{i=1}^{4} e_i a_i^2 [(x^4)^2 + x^1]^{n_i} (x^3)^{2k_i} (\mathrm{d}x^i)^2, \tag{17.28}$$

i	1	2	3	4
n_i	$1+\sqrt{2}$	$-\sqrt{2}$	$2+\sqrt{2}$	$1+\sqrt{2}$
k_i	$(1+2\sqrt{2})/7$	$(2-3\sqrt{2})/7$	0	1
a_i	$c = \text{const}$	1	$(3-\sqrt{2})/7$	1

Solution in non-closed form:

$$\mathrm{d}s^2 = \sum_{i=1}^{4} e_i \left[\frac{x^1}{\sqrt{3}} - \frac{(x^4)^2}{12}\right]^{n_i} \left[\exp \int \frac{v \mathrm{d}z}{z}\right]^{k_i} (-z)^{l_i} \left(\frac{v^2-1}{z-1}\right)^{m_i} (\mathrm{d}x^i)^2,$$

$$\frac{\mathrm{d}v}{\mathrm{d}z} = \frac{v^2-1}{4z}\left(\frac{2v}{z-1} + \frac{4}{\sqrt{3}}\right), \quad z = x^3, \tag{17.29}$$

i	1	2	3	4
n_i	$1+\sqrt{3}$	$-\sqrt{3}$	$3+\sqrt{3}$	$2+\sqrt{3}$
k_i	$1+2/\sqrt{3}$	-1	$1+2/\sqrt{3}$	$1+2/\sqrt{3}$
l_i	$-(1+1/\sqrt{3})$	$1/\sqrt{3}$	$-(2+1/\sqrt{3})$	$-1/\sqrt{3}$
m_i	0	0	0	1

The complex transformation $x^2 \to ix^4$, $x^4 \to ix^2$ takes (17.28) and (17.29) into static solutions. However, not every Harrison solution with a spacelike Killing vector has a (real) static counterpart.

Because of their analytical complexity, the metrics given by Harrison have been checked by computer (d'Inverno and Russell-Clark 1971). Altogether there are 17 vacuum solutions – including (17.28), (17.29) – which are of non-degenerate Petrov type I and do not admit a $G_r, r > 1$ (Collinson 1964). The ansatz (17.27) also leads to solutions, mostly of type D, which belong to the Weyl class of static fields (§20.2) or to the class of Einstein–Rosen waves (§22.3).

By means of a separation ansatz which is similar to (but not identical with) ansatz (17.27), Harris and Zund (1978) obtained a class of (type I) vacuum solutions admitting (at least) a G_1. These solutions have been investigated by MacCallum (1990).

Vandyck (1985) considered time-dependent generalizations of Weyl's class (§20.2) in the sense that the metric functions in

$$ds^2 = e^{2K}\left[(dx^1)^2 + (dx^2)^2\right] + W^2 e^{-2U} d\varphi^2 - e^{2U} dt^2 \tag{17.30}$$

are allowed to depend on the three variables x^1, x^2 and t, and corresponding generalizations of the Einstein–Rosen class (§22.3), and proved that all vacuum solutions of that kind necessarily admit an Abelian G_2.

Fackerell and Kerr (1991) reformulated the vacuum equations with one spacelike Killing vector. The vacuum field equations imply that there is a Ricci collineation (§35.4) in the 3-space Σ_3 of Killing trajectories (§18.1) and the metric of Σ_3 can be represented in terms of potentials. In particular, the case where the Ricci collineation vector generates a proper homothetic motion in Σ_3 leads to Petrov type III solutions of Robinson–Trautman type, cp. (28.15). Solutions admitting a homothetic vector are also considered by Lun *et al.* (1988) and Singleton (1990).

If a vacuum space-time admits a Killing vector $\boldsymbol{\xi}$ with a null bivector $\xi_{a;b}$, then space-time is algebraically special and of Kundt's class (Chapter 31), Petrov type III being excluded and type N leading to *pp*-waves; if a vacuum space-time admits a geodesic Killing vector $\boldsymbol{\xi}$, then $\boldsymbol{\xi}$ is the null eigenvector of an algebraically special metric (Debney 1971).

Hoenselaers (1978a) found a class of vacuum solutions with a non-null Killing vector which are algebraically special, see also Hoenselaers and Skea (1989); the corrected form of the line element can be found in Hoenselaers (1992). Taking the norm F and the twist potential ω of the Killing vector (see Chapter 18) as coordinates and assuming $\omega_{,n}\omega^{,n} = 0 = F_{,n}\omega^{,n}$, the resulting metrics are

$$\mathrm{d}s^2 = F\left(\mathrm{d}x + P\mathrm{d}\omega + \omega\frac{\mathrm{d}F}{F}\right)^2 + \frac{1}{F}\left[2\mathrm{d}\omega(\mathrm{d}r - M\mathrm{d}\omega - NQ\frac{\mathrm{d}F}{F}) + \frac{\mathrm{d}F^2}{Q^2}\right]$$

$$M = \tfrac{1}{2}ABQr^2 - Cr + M_0, \quad Q = (A\sqrt{F} + B/\sqrt{F})^2, \tag{17.31a}$$

$$N = -rFQ_{,F}, \quad P = \ln F + \frac{1 + 2AF/B}{ABF^2 + B^2F} - \frac{2A}{B^3}\log\left(A + B/F\right).$$

$A(\omega)$ and $B(\omega)$ are arbitrary functions, and $C(F,\omega)$ and $M_0(F,\omega)$ can be obtained by performing the quadratures

$$\begin{aligned} Q\partial_F C &= \tfrac{1}{2}\partial_F(Q_{,\omega}) - Q_{,\omega}Q_{,F}/Q, \quad F\partial_F\left(QM_0\right) = h, \\ FQ\partial_F h &= -\partial_\omega(Q_{,\omega}/Q) + (Q_{,\omega}/Q)^2 - CQ_{,\omega}/Q + F^{-2}/2, \end{aligned} \tag{17.31b}$$

ω being a null coordinate, and the metrics (17.31) belong to Kundt's class (Chapter 31).

It should be remarked that the Hauser type N twisting vacuum solution given in §29.3 also admits only one Killing vector (and one homothetic Killing vector).

For perfect fluid solutions with a maximal G_1 see §23.4.

18
Stationary gravitational fields

Stationary gravitational fields are characterized by the existence of a timelike Killing vector, i.e. one can choose coordinates so that the metric is independent of a timelike coordinate. A stationary space-time is said to be *static* if the Killing vector is hypersurface-orthogonal.

In §§18.1–18.3 we derive the field equations from a projection formalism using differential geometric concepts. Some methods outlined for stationary fields, e.g. the projection formalism (§18.1) or geodesic eigenrays (§18.5) also apply, with slight changes, to the case with a spacelike Killing vector.

Only in a few cases are exact stationary solutions without an additional symmetry known. They are given in §§18.5, 18.7 and 17.3. The stationary fields admitting a second Killing vector describing axial symmetry will be treated in the subsequent chapters.

18.1 The projection formalism

A stationary space-time invariantly determines a differentiable 3-manifold Σ_3 defined by the smooth map (§2.2) $\Psi : \mathcal{M} \to \Sigma_3$ (Geroch 1971), where $\mathcal{M}$ is the space-time V_4 and $\Psi = \Psi(p)$ denotes the trajectory of the timelike Killing vector $\boldsymbol{\xi}$ passing through the point p of V_4. The elements of Σ_3 are the orbits of the one-dimensional group of motions generated by $\boldsymbol{\xi}$. The 3-space Σ_3 is called the quotient space V_4/G_1.

Only in the case of static gravitational fields is there a natural way of introducing subspaces V_3 (orthogonal to the Killing trajectories). The quotient space Σ_3 provides a generalization applicable to stationary, as well as static, space-times; it must be regarded as the image of a map rather than a hypersurface in V_4.

Geroch (1971) has shown in detail that there is a one-to-one correspondence between tensor fields on Σ_3 and tensor fields $T_{a...}^{...b}$ on V_4 satisfying

$$\xi^a T_{a...}^{...b} = 0, \quad \xi_b T_{a...}^{...b} = 0, \quad \mathcal{L}_\xi T_{a...}^{...b} = 0. \tag{18.1}$$

Tensors on V_4 subject to these conditions are called tensors on Σ_3; the algebra of the space-time tensors satisfying (18.1) is completely and uniquely mapped to the tensor algebra on Σ_3. Examples of tensors on Σ_3 are the projection tensor h_{ab} (the metric tensor on Σ_3) and the Levi–Civita tensor ε_{abc} on Σ_3, where

$$h_{ab} = g_{ab} - \xi_a\xi_b/F, \quad \varepsilon_{abc} = \varepsilon_{dabc}\xi^d/\sqrt{-F}, \quad F \equiv \xi_a\xi^a. \tag{18.2}$$

The derivative on Σ_3 defined by

$$T_{a...||e}^{...b} \equiv h_a^c \ldots h_d^b h_e^f T_{c...;f}^{...d} \tag{18.3}$$

satisfies all the axioms for the covariant derivative associated with the metric h_{ab}. In particular, it is symmetric (torsionfree) and the metric tensor h_{ab} is covariantly constant, $h_{ab||c} = 0$.

The Riemann curvature tensor on Σ_3 can be calculated from the identity (cp. (2.78))

$$v_{a||bc} - v_{a||cb} = v^d \overset{3}{R}_{dabc}, \tag{18.4}$$

$\boldsymbol{v}$ being an arbitrary vector field on Σ_3. The curvature tensors on Σ_3 and V_4 are related by the equation

$$\overset{3}{R}_{abcd} = h^p_{[a}h^q_{b]}h^r_{[c}h^s_{d]}\left\{R_{pqrs} + 2(\xi_{q;p}\xi_{s;r} + \xi_{r;p}\xi_{s;q})/F\right\} \tag{18.5}$$

(Lichnerowicz 1955, Jordan *et al.* 1960). In order to simplify our expressions we define the *twist vector* $\boldsymbol{\omega}$:

$$\omega^a = \varepsilon^{abcd}\xi_b\xi_{c;d}, \quad \omega^a\xi_a = 0, \quad \mathcal{L}_{\boldsymbol{\xi}}\boldsymbol{\omega} = 0. \tag{18.6}$$

Applying (3.38) for complex self-dual bivectors to $\xi^*_{a;b} = \xi_{a;b} + \mathrm{i}\,\tilde{\xi}_{a;b}$, we obtain the simple relation

$$2\xi_{a;b} = F^{-1}(\varepsilon_{abcd}\xi^c\omega^d + 2\xi_{[a}F_{,b]}) \tag{18.7}$$

for the covariant derivative of the Killing vector with respect to the space-time metric g_{ab}, so that such terms can be eliminated from (18.5).

For practical calculations, it is convenient to take a *coordinate system* adapted to the congruence $\boldsymbol{\xi} = \partial_t$:

$$\begin{aligned} \mathrm{d}s^2 &= h_{\mu\nu}\mathrm{d}x^\mu\mathrm{d}x^\nu + F(\mathrm{d}t + A_\mu\mathrm{d}x^\mu)^2, \\ g_{\mu\nu} &= h_{\mu\nu} + F^{-1}\xi_\mu\xi_\nu, \quad g_{\mu 4} = \xi_\mu = FA_\mu, \quad g_{44} = \xi_4 = F. \end{aligned} \tag{18.8}$$

18.2 The Ricci tensor on Σ_3

First we introduce a complex vector $\boldsymbol{\Gamma}$ on Σ_3,

$$\Gamma_a = -2\xi^c \xi^*_{c;a} = -F_{,a} + \mathrm{i}\omega_a, \quad \Gamma^a \xi_a = 0, \quad \mathcal{L}_{\boldsymbol{\xi}} \boldsymbol{\Gamma} = 0. \tag{18.9}$$

With the aid of (8.22) and the symmetry relations (2.80) of the curvature tensor it can easily be verified that the equations

$$\xi^{*\ ;b}_{a;b} = -R_{ad}\xi^d, \quad \xi^*_{a;b} = -F^{-1}(\xi_{[a}\Gamma_{b]})^* \tag{18.10}$$

hold. Taking the divergence of the complex vector $\boldsymbol{\Gamma}$ gives

$$\Gamma^a{}_{;a} = \Gamma^a{}_{||a} + \tfrac{1}{2}F^{-1}F_a\Gamma^a = -F^{-1}\Gamma^a\Gamma_a + 2\xi^a\xi^b R_{ab}. \tag{18.11}$$

The real and imaginary parts of this equation are

$$F^{,a}{}_{||a} = \tfrac{1}{2}F^{-1}F_{,a}F^{,a} - F^{-1}\omega_a\omega^a - 2\xi^a\xi^b R_{ab}, \tag{18.12}$$

$$\omega^a{}_{||a} = \tfrac{3}{2}F^{-1}F_{,a}\omega^a. \tag{18.13}$$

Equation (18.12) expresses the component $\xi^a\xi^b R_{ab}$ of the Ricci tensor in V_4 in terms of tensors and their covariant derivatives on Σ_3. In order to derive an analogous formula for the components $h^a_c\xi^b R_{ab}$, we calculate the curl of $\boldsymbol{\omega}$:

$$\begin{aligned} \omega_{[b||a]} &= \varepsilon_{smnr}(\xi^m\xi^{n;r})_{;c}h^c_{[a}h^s_{b]} = 2\xi^m\xi^d R^{\sim}_{dcms}h^c_{[a}h^s_{b]} \\ &= -2\xi^d R^{\sim}_{d[ab]m}\xi^m = \varepsilon_{abmn}\xi^m R^n_d\xi^d. \end{aligned} \tag{18.14}$$

The result of this short calculation is the formula

$$(-F)^{-1/2}\varepsilon^{abc}\omega_{c||b} = 2h^a_b R^b_c\xi^c. \tag{18.15}$$

From this equation we conclude that the vanishing of the components $h^a_b R^b_c\xi^c$ implies that, at least locally, the twist vector $\boldsymbol{\omega}$ is a gradient, $\omega_a = \omega_{,a}$ (see Theorem 2.1).

Finally, we can derive the formula

$$\overset{3}{R}_{ab} = \tfrac{1}{2}F^{-1}F_{,a||b} - \tfrac{1}{4}F^{-2}F_{,a}F_{,b} + \tfrac{1}{2}F^{-2}(\omega_a\omega_b - h_{ab}\omega_c\omega^c) + h^m_a h^n_b R_{mn} \tag{18.16}$$

by a straightforward calculation in which we insert (18.7) into (18.5), contract (18.5) for the curvature tensor with the projection tensor and apply some of the previous relations of the present section.

Equations (18.12), (18.15) and (18.16) express the Ricci tensor of a stationary space-time in terms of tensors and their derivatives on the 3-space Σ_3. All the equations are written in a four-dimensionally covariant manner and refer to a general basis $\{\boldsymbol{e}_a\}$ of V_4. Nevertheless, they are in fact three-dimensional relations. If we take a basis such that

$$\{\boldsymbol{e}_a\} = \{\boldsymbol{e}_\alpha, \boldsymbol{\xi}\}, \quad \boldsymbol{e}_\alpha \cdot \boldsymbol{\xi} = 0, \tag{18.17}$$

$\boldsymbol{\xi}$ being the Killing vector, then (18.15) and (18.16) give just the tetrad components with respect to the 3-basis $\{\boldsymbol{e}_\alpha\}$ in Σ_3.

For a specified Ricci tensor, $R_{ab} = \kappa_0(T_{ab} - Tg_{ab}/2)$, the Einstein field equations are equivalent to the system of differential equations (18.12), (18.13), (18.15), (18.16).

Given a four-dimensional manifold $\mathcal{M}$, a vector field $\boldsymbol{\xi}$ on $\mathcal{M}$ (with prescribed *contravariant* components), and a solution (h_{ab}, F, ω^a) of these field equations, we can find the corresponding stationary *metric*

$$g_{ab} = h_{ab} + F^{-1}\xi_a\xi_b, \tag{18.18}$$

for which $\boldsymbol{\xi}$ is the timelike Killing vector field.

That the quantities ξ_a (covariant components!) can be calculated from (h_{ab}, F, ω^a) is proved as follows. The 1-form $F^{-1}\xi_a \mathrm{d}x^a$ can be determined up to a gradient,

$$F^{-1}\xi_a \to F^{-1}\xi_a + \chi_{,a}, \quad \chi_{,a}\xi^a = 0, \tag{18.19}$$

from (18.7),

$$2(F^{-1}\xi_{[a})_{;b]} = F^{-2}\varepsilon_{abcd}\xi^c\omega^d, \tag{18.20}$$

if the exterior derivative of the bivector $(F^{-1}\xi_{[a})_{;b]}$ vanishes, i.e. if

$$\xi^d(\omega^a{}_{;a} - 2F^{-1}F_{,a}\omega^a) + \omega^a\xi^d{}_{;a} - \xi^a\omega^d{}_{;a} = 0. \tag{18.21}$$

Since $\mathcal{L}_{\boldsymbol{\xi}}\boldsymbol{\omega}$ is zero, (18.21) gives, as the integrability condition for (18.20), precisely (18.13).

If (18.13) is satisfied, the quantities ξ_μ (covariant components of $\boldsymbol{\xi}$) in the line element (18.8) can be obtained from (18.20), which, in this special coordinate system, takes the form

$$2A_{[\mu;\nu]} = (-F)^{-3/2}\varepsilon_{\mu\nu\rho}\omega^\rho \tag{18.22}$$

(metric $h_{\mu\nu}$). The remaining freedom in the choice of the gauge function in (18.19) is irrelevant; it corresponds simply to a transformation of the time coordinate, $t \to t + \chi(x^\mu)$.

18.3 Conformal transformation of Σ_3 and the field equations

Equation (18.16) containing second derivatives of F can be simplified considerably by applying a conformal transformation of Σ_3:

$$\Sigma_3 \to \widehat{\Sigma}_3 : \gamma_{ab} \equiv \hat{h}_{ab} = -Fh_{ab}. \tag{18.23}$$

According to (3.87), the Ricci tensor $\overset{3}{R}_{ab}$ of Σ_3 is connected with the Ricci tensor $\widehat{R}_{ab}$ of $\widehat{\Sigma}_3$ by the formula

$$\widehat{R}_{ab} = \overset{3}{R}_{ab} - \tfrac{1}{2}F_{,a||b}/F + \tfrac{3}{4}F_{,a}F_{,b}/F^2 - \tfrac{1}{2}h_{ab}(F_{,c}{}^{||c} - \tfrac{1}{2}F^{-1}F_{,c}F^{,c})/F. \quad (18.24)$$

We insert this equation into (18.16) and denote covariant derivatives with respect to γ_{ab} by a colon in front of the index. Then we arrive at

Theorem 18.1 *The complete set of Einstein equations for stationary fields takes the form*

$$\widehat{R}_{ab} = \tfrac{1}{2}F^{-2}(F_{,a}F_{,b} + \omega_a\omega_b) + \kappa_0(h^c_a h^d_b - F^{-2}\gamma_{ab}\xi^c\xi^d)(T_{cd} - \tfrac{1}{2}Tg_{cd}), \quad (18.25)$$

$$F_{,a}{}^{:a} = F^{-1}\gamma^{ab}(F_{,a}F_{,b} - \omega_a\omega_b) + 2\kappa_0 F^{-1}\xi^a\xi^b(T_{ab} - \tfrac{1}{2}Tg_{ab}), \quad (18.26)$$

$$\omega_a{}^{:a} = 2F^{-1}\gamma^{ab}F_{,a}\omega_b, \quad (18.27)$$

$$\hat{\varepsilon}^{abc}\omega_{c,b} = -2F^{-1}\kappa_0 h^a_b T^b_c \xi^c. \quad (18.28)$$

Rácz (1997) gave these equations also for the case of a spacelike Killing vector and studied their interrelations.

18.4 Vacuum and Einstein–Maxwell equations for stationary fields

For stationary *vacuum fields*, (18.28) implies the existence of a twist potential ω with $\omega_a = \omega_{,a}$ (Papapetrou 1963). The real functions F and ω can be combined to form a complex scalar potential (cp. (18.9))

$$\Gamma = -F + \mathrm{i}\omega. \quad (18.29)$$

Scalar potentials play an important role in a procedure for generating solutions. This question is discussed in Chapters 10 and 34. In terms of Γ, (18.25)–(18.27) reduce to

$$\widehat{R}_{ab} = \tfrac{1}{2}F^{-2}\Gamma_{,(a}\overline{\Gamma}_{,b)}, \quad \Gamma_{,a}{}^{:a} + F^{-1}\gamma^{ab}\Gamma_{,a}\Gamma_{,b} = 0. \quad (18.30)$$

Theorems on the existence and uniqueness of asymptotically flat stationary vacuum solutions are proven in Reula (1989).

For stationary *Einstein–Maxwell fields*, one can introduce a complex potential Φ by

$$\tfrac{1}{2}\sqrt{2\kappa_0}\,\xi^a F^*_{ab} = \Phi_{,b}, \quad \Phi_{,a}\xi^a = 0, \quad F^{*ab}{}_{;b} = 0, \quad (18.31)$$

provided that the Lie derivative of the Maxwell tensor vanishes. With the energy-momentum tensor (5.7),

$$T_{ab} = \tfrac{1}{2} F^{*c}_{a} \overline{F^{*}_{bc}}, \quad \sqrt{2\kappa_0}\, F^{*}_{ab} = 4F^{-1}(\xi_{[a}\Phi_{,b]})^{*}, \tag{18.32}$$

the right-hand side of the field equation (18.28) takes the form

$$\begin{aligned} \kappa_0 h^{a}_{b} \xi^{c} F^{*bd}\, \overline{F^{*}_{cd}} &= \sqrt{2\kappa_0}\, h^{a}_{b} F^{*bd} \overline{\Phi}_{,d} \\ &= 2\mathrm{i} F^{-1} \varepsilon^{adbc} \xi_b \Phi_{,c} \overline{\Phi}_{,d} = 2\mathrm{i} F \hat{\varepsilon}^{abc} (\overline{\Phi} \Phi_{,c})_{,b}. \end{aligned} \tag{18.33}$$

Thus we conclude from (18.28) that the combination $\omega_c + 2\mathrm{i}\overline{\Phi}\Phi_{,c}$ is a gradient (Harrison 1968). It is convenient to introduce a complex scalar potential, the Ernst potential $\mathcal{E}$, by the equation (Ernst 1968b)

$$\mathcal{E}_{,a} \equiv -F_{,a} + \mathrm{i}\omega_a - 2\overline{\Phi}\Phi_{,a}, \quad \mathrm{Re}\,\mathcal{E} = -F - \overline{\Phi}\Phi. \tag{18.34}$$

The existence of the potential $\mathcal{E}$ is guaranteed, in a similar manner to that of the potential Φ in (18.31), by the equations

$$\xi^{a} K^{*}_{ab} = \mathcal{E}_{,b}, \ \mathcal{E}_{,a}\xi^{a} = 0, \ K^{*ab}{}_{;b} = 0, \ K^{*}_{ab} = -2\xi^{*}_{a;b} - \sqrt{2\kappa_0}\, \overline{\Phi} F^{*}_{ab}. \tag{18.35}$$

Now we insert the expressions

$$h^{c}_{a} h^{d}_{b} R_{cd} = 2F^{-1}\left[\Phi_{,(a}\overline{\Phi}_{,b)} - \tfrac{1}{2}\gamma_{ab}\Phi^{,c}\overline{\Phi}_{,c}\right], \quad \xi^{a}\xi^{b} R_{ab} = \Phi_{,c}\overline{\Phi}^{,c}, \tag{18.36}$$

into the field equations and substitute for ω_a with the aid of (18.34).

Theorem 18.2 (Harrison 1968, Neugebauer and Kramer 1969) *The field equations for stationary Einstein–Maxwell fields outside the sources reduce to the following system of equations, referred to the metric γ_{ab}:*

$$\widehat{R}_{ab} = \tfrac{1}{2} F^{-2} (\mathcal{E}_{,(a} + 2\overline{\Phi}\Phi_{,(a})(\overline{\mathcal{E}}_{,b)} + 2\Phi\overline{\Phi}_{,b)}) + 2F^{-1}\Phi_{,(a}\overline{\Phi}_{,b)}, \tag{18.37}$$

$$\mathcal{E}_{,a}{}^{;a} + F^{-1}\gamma^{ab}\mathcal{E}_{,a}(\mathcal{E}_{,b} + 2\overline{\Phi}\Phi_{,b}) = 0, \tag{18.38}$$

$$\Phi_{,a}{}^{;a} + F^{-1}\gamma^{ab}\Phi_{,a}(\mathcal{E}_{,b} + 2\overline{\Phi}\Phi_{,b}) = 0. \tag{18.39}$$

The sourcefree Maxwell equations are equivalent to (18.39), the field equations (18.25)–(18.27) have been written in terms of the complex potential $\mathcal{E}$ in the form (18.37)–(18.38), and the field equation (18.28) is automatically satisfied by introducing the complex potential $\mathcal{E}$.

By specialization of the potentials $\mathcal{E}$ and Φ we can describe the different physical situations given in Table 18.1.

The potentials $\mathcal{E}$ and Φ can also be introduced when electromagnetic and material *currents*, which are everywhere parallel to the Killing vector,

$$j_{[a}\xi_{b]} = 0 = u_{[a}\xi_{b]}, \tag{18.40}$$

Table 18.1. The complex potentials $\mathcal{E}$ and Φ for some physical problems

Physical problem	$\mathcal{E}$	Φ
Stationary Einstein–Maxwell fields	Complex	Complex
Electrostatic Einstein–Maxwell fields	Real	Real
Magnetostatic Einstein–Maxwell fields	Real	Imaginary
Stationary vacuum fields	Complex	0
Static vacuum fields	Real	0
Conformastationary Einstein–Maxwell fields (§18.7)	0	Complex

are present. In this case, the (generalized Poisson) equations for $\mathcal{E}$ and Φ read

$$\begin{aligned} &\mathcal{E}_{,a}{}^{:a} + F^{-1}\mathcal{E}_{,a}\Gamma^a = \xi_a T^a, \qquad \Phi_{,a}{}^{:a} + F^{-1}\Phi_{,a}\Gamma^a = \tfrac{1}{2}\sqrt{2\kappa_0}\xi_a j^a, \\ &j^a = \sigma u^a, \qquad T^a = -\kappa_0\xi^a[3p + \mu + \sqrt{2}(-\kappa_0 F)^{-1/2}\sigma\overline{\Phi}\,]. \end{aligned} \tag{18.41}$$

Theorem 18.3 *Stationary asymptotically flat and asymptotically source-free Einstein–Maxwell fields are static provided that the currents satisfy the condition* (18.40) *everywhere.*

The proof uses (18.41) and Stokes's theorem, see Carter (1972).

18.5 Geodesic eigenrays

Assuming the existence of a timelike Killing vector, the Einstein field equations have been written as equations in a three-dimensional space. It is also possible to develop a (three-dimensional) triad formalism and a corresponding spinor technique (Perjés 1970) in the three-dimensional Riemannian space Σ_3 of the Killing trajectories.

In a stationary space-time, a null congruence $\boldsymbol{k}$ normalized by $k_a\xi^a = 1$ determines a spacelike unit vector $\boldsymbol{n}$ by

$$n^a = k^a - F^{-1}\xi^a, \quad n^a n_a = 1, \quad n^a\xi_a = 0, \quad \mathcal{L}_{\boldsymbol{\xi}}\boldsymbol{n} = 0. \tag{18.42}$$

One can introduce a triad $\{\boldsymbol{e}_\alpha\} = (\boldsymbol{n}, \boldsymbol{m}, \overline{\boldsymbol{m}})$ orthogonal to $\boldsymbol{\xi}$. Applying the conformal transformation (18.23), we rewrite the geodesic condition on $\boldsymbol{k}$ in the form of an equation for $\boldsymbol{n}$ over $\widehat{\Sigma}_3$ (metric γ_{ab}),

$$k^b k_{a;b} = 0 \quad\Leftrightarrow\quad F n^b n_{a:b} + F_{,a} - n_a n^b F_{,b} + \hat{\varepsilon}_{abc}\omega^b n^c = 0. \tag{18.43}$$

An *eigenray* $\boldsymbol{n}$ is defined by the equation

$$F_{,a} - n_a n^b F_{,b} + \hat{\varepsilon}_{abc}\omega^b n^c = 0. \tag{18.44}$$

In the case of static gravitational fields (§18.6), there are two different eigenrays given by

$$n_a = \pm(F_{,b}F^{,b})^{-1/2}F_{,a}. \qquad (18.45)$$

From (18.43) we infer that an eigenray $\boldsymbol{n}$ is geodesic in $\widehat{\Sigma}_3$ if and only if the null congruence $\boldsymbol{k}$ is geodesic in V_4,

$$k^b k_{a;b} = 0 \quad \Leftrightarrow \quad n^b n_{a:b} = 0 \quad \Leftrightarrow \quad \kappa = n_{a:b}m^a n^b = 0. \qquad (18.46)$$

The shear σ of an eigenray is defined by

$$\sigma = n_{a:b}m^a m^b, \quad n_a m^a = 0, \quad m_a \overline{m}^a = 1. \qquad (18.47)$$

One can show that the existence of geodesic and shearfree eigenrays ($\kappa = \sigma = 0$) implies that space-time admits a geodesic and shearfree null congruence. All stationary vacuum solutions with geodesic *shearfree* eigenrays are given in Horváth *et al.* (1997). The stationary vacuum solutions admitting geodesic (but not necessarily shearfree) eigenrays are completely known (Kóta and Perjés 1972). The geodesic eigenray conditions (18.44), (18.46) allow the integration of the field equations in a three-dimensional version of the Newman–Penrose formalism. The three resulting vacuum metrics with geodesic *shearing* eigenrays (Kóta–Perjés solutions) are:

$$\mathrm{d}s^2 = f^{-1}\Big[f^0(r^{1-a}\mathrm{d}x^2 + r^{1+a}\mathrm{d}y^2) + \mathrm{d}r^2\Big] - f\left[\mathrm{d}t + \sqrt{2}Qx\,\mathrm{d}y + f^{-1}\mathrm{d}r\right]^2,$$
$$f = f^0(r^a + Q^2r^{-a})^{-1}, \quad f^0 = P(x+Qy), \quad a = 1/\sqrt{2}, \qquad (18.48)$$

$$\mathrm{d}s^2 = f^{-1}\Big[f^0(r^{1-a}\mathrm{d}x^2 + r^{1+a}\mathrm{d}y^2) + \mathrm{d}r^2\Big] - f\left[\mathrm{d}t - ax^2y\,\mathrm{d}y + f^{-1}\mathrm{d}r\right]^2,$$
$$f = f^0(y^2r^a + x^2r^{-a})^{-1}, \quad f^0 = P(x+Qy), \quad a = 1/\sqrt{2}, \qquad (18.49)$$

$$\mathrm{d}s^2 = \frac{r^{2p}+B^2}{2r^p}\left(2\lambda^{-1}\mathrm{d}r^2 + r^{1-q}\mathrm{d}x^2 + r^{1+q}\mathrm{d}y^2\right) - \frac{\lambda r^p(\mathrm{d}t - pBx\;\mathrm{d}y)^2}{r^{2p}+B^2}, \qquad (18.50)$$

where P, Q, B, λ, p and q are real constants with $p^2+q^2=1$. The last solution (18.50) admits two Killing vectors, ∂_t and ∂_y, and belongs to Papapetrou's class (§20.3). For $Q=0$, the metric (18.48) admits the homothetic vector $\boldsymbol{H} = r\partial_r + \frac{1}{3}(a+1)(x\partial_x + 2t\partial_t) + \frac{1}{3}(1-2a)y\partial_y$. The given vacuum solutions with shearing geodesic eigenrays arise also as solutions of the vacuum Kerr–Schild problem, cp. §32.2.

All vacuum solutions with $\kappa\sigma = 0$ are known and it was proved by Horváth and Lukács (1992) that stationary vacuum solutions with $\kappa = c_1$, $\sigma = c_2$ do not exist unless both constants c_1 and c_2 are zero.

Lukács (1973, 1983) gave all vacuum metrics with a *spacelike* Killing vector and shearing geodesic eigenrays. These metrics are very similar to (18.48) and (18.49). The stationary Einstein–Maxwell fields for which the shearing geodesic eigenrays of the Maxwell and gravitational fields coincide are given by Lukács and Perjés (1973). In this case the potentials $\mathcal{E}$ and Φ can be gauged so that $\mathcal{E} = q\Phi$, $q = \text{const}$ (Lukács 1985). For dust solutions, see Lukács (1973).

18.6 Static fields

18.6.1 Definitions

A stationary solution is called *static* if the timelike Killing vector is hypersurface-orthogonal,

$$\xi_{(a;b)} = 0, \quad \xi_{[a}\xi_{b;c]} = 0, \quad \xi^a\xi_a < 0. \tag{18.51}$$

We mention an *equivalent characterization*: in a static space-time there is a vector field $\boldsymbol{u}$ with the properties (Ehlers and Kundt 1962)

$$u_{a;b} = -\dot{u}_a u_b, \quad \ddot{u}_{[a}u_{b]} = 0, \quad u^a u_a = -1 \tag{18.52}$$

(a dot means $\nabla_{\boldsymbol{u}}$). Locally static space-times in the sense of this definition can be *globally* stationary (Stachel 1982). From (6.25), (6.26) we obtain the equation

$$\dot{u}_a\dot{u}_c + \dot{u}_{a;c} + \ddot{u}_a u_c = u^b u^d R_{dabc}, \tag{18.53}$$

the antisymmetric part of which tells us that $\dot{\boldsymbol{u}}$ is a gradient, $\dot{u}_a = U_{,a}$. Therefore, $\boldsymbol{\xi} = \mathrm{e}^U\boldsymbol{u}$ is a hypersurface-orthogonal Killing vector field.

In §6.2 we have shown that the Weyl tensor of a static space-time is of Petrov type *I*, *D* or *O*. The vector $\boldsymbol{u}$ satisfying (18.52) is a principal vector of the Weyl tensor and an eigenvector of the Ricci tensor. The tensor Q_{ac} defined in (3.62) is purely real, and the traceless symmetric matrix $\boldsymbol{Q}$ can be transformed to a diagonal form with *real eigenvalues* with the aid of orthogonal tetrad transformations preserving the principal vector $\boldsymbol{u}$.

Equations (18.25)–(18.27) reduce to

$$\begin{aligned} \widehat{R}_{ab} &= 2U_{,a}U_{,b} + h_a^c h_b^d R_{cd} - \mathrm{e}^{-4U}\gamma_{ab}\xi^c\xi^d R_{cd}, \\ U_{,a}{}^{;a} &= \mathrm{e}^{-4U}\xi^a\xi^b R_{ab}, \quad F = -\,\mathrm{e}^{2U}. \end{aligned} \tag{18.54}$$

In particular, the *vacuum field equations* have the remarkably simple form

$$\widehat{R}_{ab} = 2U_{,a}U_{,b}. \tag{18.55}$$

The potential equation $U_{,a}{}^{;a} = 0$ follows from (18.55) in virtue of the contracted Bianchi identities for $\widehat{R}_{ab}$. It can easily be verified that $\widehat{R} = 0$ implies $R_{abcd} = 0$.

The spacelike hypersurfaces (space sections) orthogonal to the Killing vector are totally geodesic, i.e. geodesics in the space sections are simultaneously geodesics of the space-time. In general, in a static space-time there exists only one Killing vector satisfying (18.51); the space sections are uniquely determined by the metric.

In a coordinate frame with $\boldsymbol{\xi} = \partial_t$ the line element has the structure

$$ds^2 = d\sigma^2 - e^{2U}dt^2, \quad d\sigma^2 = h_{\mu\nu}dx^\mu dx^\nu = e^{-2U}\gamma_{\mu\nu}dx^\mu dx^\nu, \tag{18.56}$$

the metric functions being independent of the time coordinate t. The preferred coordinate system (18.56) is unique up to purely spatial transformations, $x^{\nu'} = x^{\nu'}(x^\mu)$, and linear transformations of t with constant coefficients, $t' = at + b$.

18.6.2 Vacuum solutions

All *degenerate* (type D) static vacuum fields are known. They are given in Table 18.2, together with the simple eigenvalue λ of the Weyl tensor. All these metrics admit at least an Abelian G_2 and belong to the class of solutions investigated by Weyl (§20.2).

The degenerate static vacuum solutions were originally found by Levi-Civita (1917a). The invariant classification into the subclasses of Table 18.2 is given in Ehlers and Kundt (1962). Classes A and B are connected by the complex substitution $t \to i\varphi$, $\varphi \to it$. The fields of classes A and B admit an isometry group G_4 and an isotropy group I_1. The 'C-metric' admits an Abelian group G_2 of motions. The spacelike (class A) or timelike (class B) surfaces determined by the eigenbivectors of the curvature tensor (§4.2) have constant Gaussian curvature. AI is the Schwarzschild solution (15.19). Classes A and B are included in (13.48) and its timelike counterpart.

The Harrison metrics (§17.3) include some static *non-degenerate* (type I) vacuum solutions.

18.6.3 Electrostatic and magnetostatic Einstein–Maxwell fields

Restriction to *electrostatic* fields,

$$\Phi = \chi, \quad \mathcal{E} = e^{2U} - \chi^2 \tag{18.57}$$

(χ is the electrostatic potential), or *magnetostatic* fields,

$$\Phi = i\psi, \quad \mathcal{E} = e^{2U} - \psi^2 \tag{18.58}$$

(ψ is the magnetostatic potential), simplifies the differential equations (18.37)–(18.39) for stationary Einstein–Maxwell fields outside the sources.

Table 18.2. The degenerate static vacuum solutions

Class		Metric	Eigenvalue λ
A	AI	$\mathrm{d}s^2 = r^2(\mathrm{d}\theta^2 + \sin^2\theta\mathrm{d}\varphi^2) + (1-b/r)^{-1}\mathrm{d}r^2$ $-(1-b/r)\mathrm{d}t^2$	$-br^{-3}$
	AII	$\mathrm{d}s^2 = z^2(\mathrm{d}r^2 + \sinh^2 r\mathrm{d}\varphi^2) + (b/z-1)^{-1}\mathrm{d}z^2$ $-(b/z-1)\mathrm{d}t^2$	bz^{-3}
	$AIII$	$\mathrm{d}s^2 = z^2(\mathrm{d}r^2 + r^2\mathrm{d}\varphi^2) + z\mathrm{d}z^2 - z^{-1}\mathrm{d}t^2$	z^{-3}
B	BI	$\mathrm{d}s^2 = (1-b/r)^{-1}\mathrm{d}r^2 + (1-b/r)\mathrm{d}\varphi^2$ $+r^2(\mathrm{d}\theta^2 - \sin^2\theta\mathrm{d}t^2)$	$-bz^{-3}$
	BII	$\mathrm{d}s^2 = (b/z-1)^{-1}\mathrm{d}z^2 + (b/z-1)\mathrm{d}\varphi^2$ $+z^2(\mathrm{d}r^2 - \sinh^2 r\mathrm{d}t^2)$	br^{-3}
	$BIII$	$\mathrm{d}s^2 = z\mathrm{d}z^2 + z^{-1}\mathrm{d}\varphi^2 + z^2(\mathrm{d}r^2 - r^2\mathrm{d}t^2)$	z^{-3}
C ('C-metric')		$\mathrm{d}s^2 = (x+y)^{-2}[\mathrm{d}x^2/f(x) - \mathrm{d}y^2/f(-y)$ $+f(x)\mathrm{d}\varphi^2 + f(-y)\mathrm{d}t^2]$ $f(x) = \pm(x^3 + ax + b) > 0$	$\pm(x+y)^3$

In the electrostatic case, the equations read

$$\widehat{R}_{ab} = 2(U_{,a}U_{,b} - \mathrm{e}^{-2U}\chi_{,a}\chi_{,b}), \tag{18.59a}$$

$$U_{,a}{}^{:a} = \mathrm{e}^{-2U}\gamma^{ab}\chi_{,a}\chi_{,b}, \qquad \chi_{,a}{}^{:a} = 2\gamma^{ab}U_{,a}\chi_{,b}. \tag{18.59b}$$

The equations governing magnetostatic fields follow from (18.59) by the substitution $\chi \to \psi$.

The simplest way to find solutions to these differential equations is the assumption of a relationship $U = U(\chi)$ between the potentials U and χ. This ansatz and (18.59b) imply

$$\mathrm{e}^{2U} = 1 - 2c\chi + \chi^2, \quad c = \text{const}, \tag{18.60}$$

or, in parametric representation:

$$\begin{aligned} c^2 > 1: \ & \chi = -(c^2-1)^{1/2}\coth Y + c, \ \mathrm{e}^{2U} = (c^2-1)\sinh^{-2}Y, \\ c^2 < 1: \ & \chi = -(1-c^2)^{1/2}\cot Y + c, \ \ \mathrm{e}^{2U} = (1-c^2)\sin^{-2}Y, \\ c^2 = 1: \ & \chi = -Y^{-1} + 1, \qquad\qquad\quad \mathrm{e}^{2U} = Y^{-2}. \end{aligned} \tag{18.61}$$

From (18.61) it follows that the field equations (18.59) reduce to $\widehat{R}_{ab} = \pm Y_{,a}Y_{,b}$ for $c^2 \neq 1$, and to $\widehat{R}_{ab} = 0$, $Y_{,a}{}^{:a} = 0$ for $c^2 = 1$. The class

$c^2 = 1$ of static fields (without spatial symmetry) is a special case of the conformastationary Einstein–Maxwell fields given in §18.7.

Das (1979) considered the field equations when both electro- and magnetostatic fields are present and proved that in a static space-time the electric and magnetic field vectors must be parallel to each other. Electrovac solutions which admit a vector field $\boldsymbol{u}$ satisfying the condition

$$h_a^c h_b^d \dot{u}_{c;d} = 0 \tag{18.62}$$

are treated in Srinivasa Rao and Gopala Rao (1980). Inserting (18.60) into the field equations with charged perfect fluid sources (see (18.41)) one obtains (Gautreau and Hoffman 1973)

$$c\sigma = \tfrac{1}{2}\sqrt{2\kappa_0}\,\varepsilon; \tag{18.63}$$

the charge density σ divided by the active gravitational mass density $\varepsilon = (3p+\mu)\mathrm{e}^U + \sigma\chi$ is a constant.

We end this section with

Theorem 18.4 *There are no static Einstein–Maxwell fields with an electromagnetic null field* (Banerjee 1970).

18.6.4 Perfect fluid solutions

Barnes (1972) determined *all static degenerate* (type D or O) perfect fluid solutions by a method closely analogous to that for vacuum fields. The metrics admitting an isotropy group I_1 are already contained in Chapters 13, 15 and 16. The metrics without an isotropy group are

$$\mathrm{d}s^2 = \frac{1}{(x+y)^2}\left\{\frac{\mathrm{d}x^2}{f(x)} + \frac{\mathrm{d}y^2}{h(y)} + f(x)\mathrm{d}\varphi^2 - h(y)\left[A\int h^{-3/2}\mathrm{d}y + B\right]^2\mathrm{d}t^2\right\},$$
$$f(x) = \pm x^3 + ax + b, \quad h(x) = -f(-x) - \kappa_0\mu/3, \quad \mu = \text{const}; \tag{18.64}$$

$$\mathrm{d}s^2 = (n+mx)^{-2}\left\{F^{-1}(x)\mathrm{d}x^2 + F(x)\mathrm{d}\varphi^2 + \mathrm{d}z^2 - x^2\mathrm{d}t^2\right\},$$
$$F(x) = a(n^2\ln x + 2mnx + m^2x^2/2) + b,\ m = \pm 1, 0,\ x > 0; \tag{18.65}$$

$$\mathrm{d}s^2 = N^{-2}(z)\left\{G^{-1}(x)\mathrm{d}x^2 + G(x)\mathrm{d}\varphi^2 + \mathrm{d}z^2 - x^2\mathrm{d}t^2\right\},$$
$$G(x) = ax^2 + b\ln x + c, \quad x > 0, \quad N(z)'' = -aN(z); \tag{18.66}$$

(a, b, c, m, n, A, B real constants). Metrics (18.64) and (18.66) admit an Abelian group G_2; metric (18.65) admits a G_3. For vanishing pressure and energy density, (18.64) goes over into the 'C-metric' of Table 18.2.

The class of *conformastat* (in the sense of (18.67) below) perfect fluid solutions was studied by Melnick and Tabensky (1975) and later reconsidered by Basu *et al.* (1990). The conformastat form (18.67) requires no assumption about symmetry. However, further investigation reveals that all conformastat perfect fluid solutions admit a G_4 and are spherically-symmetric solutions equivalent to metrics discussed in §16.2 or are their plane- or pseudospherically-symmetric counterparts (Bonnor and MacCallum 1982, Karlhede and MacCallum 1982).

The conjecture that all static stellar models must be spherically-ymmetric was proved only for perfect fluids with uniform mass density μ (Lindblom(1980, 1981)).

Internal symmetries of the field equations can be used to generate perfect fluid solutions with the equation of state $p=\mu$ or $\mu+3p=0$ according to (10.29)–(10.32).

18.7 The conformastationary solutions

Conformastat metrics (see Synge 1960, p. 339) form the subclass of static fields which admits conformally flat 3-spaces orthogonal to the Killing vector $\boldsymbol{\xi}=\partial_t$. The conformastat line element can be written in the form

$$\mathrm{d}s^2=\Psi^4(x,y,z)(\mathrm{d}x^2+\mathrm{d}y^2+\mathrm{d}z^2)-\mathrm{e}^{2U(x,y,z)}\mathrm{d}t^2. \qquad (18.67)$$

The curvature scalar of the conformally flat spaces $t=$ const is

$$\overset{3}{R}=-8\Psi^{-5}\Delta\Psi. \qquad (18.68)$$

Generalizing the notion of conformastat metrics (18.67) one can consider *stationary* space-times with a conformally flat 3-space Σ_3 introduced in §18.1; in the coordinate system (18.8) the 3-metric is assumed to be conformally flat, $h_{\mu\nu}=\Psi^4\delta_{\mu\nu}$ (cp. (18.67)). We call space-times with this property *conformastationary.*

18.7.1 Conformastationary vacuum solutions

With the aid of the complex-triad method (§18.5) Lukács *et al.* (1983) attacked the problem of finding the conformastationary vacuum solutions and determined the metrics in the special case when the real and imaginary parts of the Ernst potential $\mathcal{E}$ depend on each other,

$$\mathcal{E}_{,[a}\overline{\mathcal{E}}_{,b]}=0 \quad\Leftrightarrow\quad \mathcal{E}=\mathcal{E}(\overline{\mathcal{E}}). \qquad (18.69)$$

The solutions obtained are the NUT-solution (20.28) and two closely related metrics; the static limits of these solutions are the class A metrics in Table 18.2. In the generic case one has

Theorem 18.5 *All conformastationary vacuum solutions which do not satisfy* (18.69) *have axial symmetry* (Perjés 1986a).

For stationary axisymmetric metrics, Perjés (1986b) solved the condition for conformal flatness of the 3-spaces Σ_3 (the Cotton–York tensor $C^{\alpha\beta}$ as defined in (3.89) has to vanish). The calculations are facilitated by using the Ernst potential $\mathcal{E}$ and its complex conjugate (provided that they are independent) as two of the local space coordinates (Ernst coordinates) (Perjés (1986c, 1988)). The final result can be formulated as

Theorem 18.6 *Conformastationary vacuum space-times are always characterized by the relation* (18.69) (Perjés 1986b).

Hence the famous Kerr solution is not conformastationary; instead it is characterized by the property that the Simon tensor defined by

$$S_{\alpha}^{\beta} = f^{-2}\varepsilon^{\mu\nu\beta}\left(\mathcal{E}_{,\alpha;\mu}\mathcal{E}_{,\nu} - h_{\alpha\mu}h^{\rho\sigma}\mathcal{E}_{,\rho;[\sigma}\mathcal{E}_{,\nu]}\right), \quad f = \mathrm{Re}\,\mathcal{E} \tag{18.70}$$

(Simon 1984) is zero, $S_{\alpha}^{\beta} = 0$. (In the definition of the Simon tensor S_{α}^{β}, the metric operations refer to the metric $\gamma_{\mu\nu}$ of $\widehat{\Sigma}_3$, cp. (18.8).)

Starting with this condition, Perjés (1985b) integrated (in Ernst coordinates) the stationary vacuum equations and found that, apart from some exceptional cases, the Kerr–NUT solution is the only zero Simon tensor solution. The Simon tensor S_{α}^{β} is a complex generalization of the conformal tensor $C^{\alpha\beta}$ defined in (3.89). For static fields ($\mathcal{E}$ real), the conditions $S_{\alpha}^{\beta} = 0$ and $C^{\alpha\beta} = 0$ are equivalent; for proper stationary fields they differ. Krisch (1988) classified the vacuum solutions with zero Simon tensor in terms of a Frenet tetrad.

18.7.2 Conformastationary Einstein–Maxwell fields

An interesting class of stationary Einstein–Maxwell fields without spatial symmetry is characterized by the 3-space $\widehat{\Sigma}_3$ being flat,

$$\mathrm{d}s^2 = \mathrm{e}^{-2U}(\mathrm{d}x^2 + \mathrm{d}y^2 + \mathrm{d}z^2) - \mathrm{e}^{2U}(\mathrm{d}t + A_{\mu}\mathrm{d}x^{\mu})^2. \tag{18.71}$$

This conformastationary metric results from (18.8) with $h_{\mu\nu} = \mathrm{e}^{-2U}\delta_{\mu\nu}$. The field equations (18.37)–(18.39) show that the particular choice

$$\widehat{R}_{ab} = 0, \qquad \mathcal{E} = 0, \qquad \Phi^{-1}{}_{,a}{}^{:a} = 0 \tag{18.72}$$

is possible. In the class characterized by (18.72) the functions U and A_{μ} in the line element (18.71) can be determined from a solution $V = \Phi^{-1}$

of the potential equation $\Delta V = 0$ in the flat 3-space $\widehat{\Sigma}_3$ via the relations (18.34) and (18.22), i.e. in three-dimensional vector notation,

$$e^{2U} = (V\overline{V})^{-1}, \quad \text{curl}\,\boldsymbol{A} = \mathrm{i}(V\,\text{grad}\,\overline{V} - \overline{V}\,\text{grad}\,V), \quad \Delta V = 0. \quad (18.73)$$

This class of solutions was discovered by Neugebauer (1969), Perjés (1971), and Israel and Wilson (1972). The fields are static if $\text{curl}\,\boldsymbol{A} = 0$. In particular, purely electric fields ($\overline{\Phi} = \Phi$) are static. Papapetrou (1947) and Majumdar (1947) have given this special class of static Einstein–Maxwell fields without spatial symmetry.

The asymptotic form of the electromagnetic potentials and the space-time metric shows that the source of the class under consideration satisfies $|Q| = \sqrt{\kappa_0/2}M$, $\mu = \pm\sqrt{\kappa_0/2}J$ (M = mass, Q = charge, μ = magnetic moment, J = angular momentum). The conformastationary solutions are the exterior fields of charged spinning sources in equilibrium under their mutual electromagnetic and gravitational forces.

The linearity of the differential equation for V allows a *superposition* of solutions. An example of an exterior field of N isolated sources will be given in (21.29).

If the geometry is regular outside the sources, the condition

$$\int_S (\overline{V}\,\text{grad}\,V - V\,\text{grad}\,\overline{V})\mathrm{d}\boldsymbol{f} = 0 \quad (18.74)$$

is satisfied for every exterior closed 2-surface S. From this regularity condition one can derive restrictions on the source parameters which guarantee that no stresses between the spinning sources occur (Israel and Spanos 1973). For a discussion of the regularity conditions in stationary fields, and further references, see Ward (1976).

Linet (1987) considered in the metric

$$\mathrm{d}s^2 = e^{-2U}\left(\mathrm{d}z^2 + \mathrm{d}\rho^2 + B^2\rho^2\mathrm{d}\varphi^2\right) - e^{2U}\mathrm{d}t^2, \qquad 0 < B \leq 1, \quad (18.75)$$

the superposition of a charged black hole and a cosmic string.

18.8 Multipole moments

Stationary, asymptotically flat gravitational fields should admit a multipole expansion which provides an invariant tool for classifying and interpreting them. In this section we shall define the multipole moments for stationary, asymptotically flat *vacuum* fields. These space-times can be described by the line element

$$\mathrm{d}s^2 = -F^{-1}\gamma_{\alpha\beta}\mathrm{d}x^\alpha\mathrm{d}x^\beta + F(\mathrm{d}t + A_\alpha\mathrm{d}x^\alpha)^2, \quad (18.76)$$

cp. (18.8), (18.23). The 3-metric $\gamma_{\alpha\beta}$ refers to the 3-space $\widehat{\Sigma}_3$ of the Killing trajectories. It has been shown in §18.4 that the stationary vacuum space-times are characterized by $\gamma_{\alpha\beta}$ and two real scalars Q_A, $A=1,2$, which can be considered as appropriate combinations of the real and imaginary parts of the Ernst potential.

A stationary vacuum field ($\widehat{\Sigma}_3$, $\gamma_{\alpha\beta}$, Q_A) is said to be *asymptotically flat* if there is a manifold $\widetilde{\Sigma}_3 = \widehat{\Sigma}_3 \cup \Lambda$ consisting of $\widehat{\Sigma}_3$ plus one additional point Λ such that

$$\widetilde{\gamma}_{\alpha\beta} = \Omega^2 \gamma_{\alpha\beta}, \qquad \widetilde{Q}_A = \Omega^{-\frac{1}{2}} Q_A \tag{18.77}$$

are the conformally related metric and potentials in $\widetilde{\Sigma}_3$ and the conformal factor satisfies the conditions

$$\Omega|_\Lambda = 0 = \widetilde{\mathrm{D}}_\alpha \Omega|_\Lambda, \qquad \widetilde{\mathrm{D}}_\alpha \widetilde{\mathrm{D}}_\beta \Omega\,|_\Lambda = 2\widetilde{\gamma}_{\alpha\beta}|_\Lambda \tag{18.78}$$

at Λ. The covariant derivative $\widetilde{D}_\alpha$ refers to $\widetilde{\gamma}_{\alpha\beta}$.

The transformation (18.77), (18.78) effects a conformal compactification mapping spatial 'infinity' to the single point Λ which can be taken as the origin. Roughly speaking, the conformal factor Ω behaves as $\widetilde{r}^2 \sim r^{-2}$, where r means the radial coordinate in the space $\widehat{\Sigma}_3$ at large distance from the sources and $\widetilde{r}$ is the distance from Λ in the space $\widetilde{\Sigma}_3$. The potentials Q_A are assumed to decrease with r^{-1} and thus $\widetilde{Q}_A\,|_\Lambda$ exists. Possible candidates are the potentials $(F+1)$ and ω defined in (18.9).

Using the construction of conformal compactification in flat space, one notices that a change of origin in the physical space leads to a change of Ω; i.e. $\Omega \to \Omega\, f$ with $f\,|_\Lambda = 1$. In flat space one defines the multipole moments as the symmetric tracefree part of the partial derivatives (of arbitrary order) of $\widetilde{Q}_A$ evaluated at Λ. An infinitesimal shift of origin by a vector ε_α gives $\Omega \to \Omega(1 + x^\alpha \varepsilon_\alpha + \cdots)$ and the multipole moments $m_{A\alpha_1\ldots\alpha_n}$ transform as

$$m_{A\alpha_1\ldots\alpha_n} \to \mathcal{C}\,(m_{A\alpha_1\ldots\alpha_{n-1}}\varepsilon_{\alpha_n}) + m_{A\alpha_1\ldots\alpha_n}, \tag{18.79}$$

where the symbol $\mathcal{C}$ denotes the operation of taking the symmetric and tracefree part of the subsequent expression with respect to the tensor indices $\alpha_1 \ldots \alpha_n$. To emulate this behaviour in the case when $\widetilde{\Sigma}_3$ is not flat one wants

$$m_{A\alpha_1\ldots\alpha_n} \to \mathcal{C}\,(m_{A\alpha_1\ldots\alpha_{n-1}} \widetilde{D}_{\alpha_n} f\,|_\Lambda) + m_{A\alpha_1\ldots\alpha_n} \tag{18.80}$$

under infinitesimal conformal transformations. It is now clear that the covariant derivatives of $\widetilde{Q}_A$ will not have the desired behaviour. However, it can be shown (Geroch 1970a, 1970b, Hansen 1974) that the fields

$P^{(n+1)}_{A\alpha_1\ldots\alpha_{n+1}}$ recursively determined by

$$P^{(0)}_A = \widetilde{Q}_A, \qquad P^{(1)}_{A\alpha} = \widetilde{\mathrm{D}}_\alpha \widetilde{Q}_A, \tag{18.81}$$

$$P^{(n+1)}_{A\alpha_1\ldots\alpha_{n+1}} = \mathcal{C}\left[\widetilde{\mathrm{D}}_{\alpha_{n+1}} P^{(n)}_{A\alpha_1\ldots\alpha_n} - \tfrac{1}{2}n(2n-1)\widetilde{R}_{\alpha_1\alpha_2} P^{(n-1)}_{A\alpha_3\ldots\alpha_{n+1}}\right]$$

define two sets of (Geroch–Hansen) *multipole moments* $m^{(n)}_{A\alpha_1\ldots\alpha_n}$ according to

$$m^{(n)}_{A\alpha_1\ldots\alpha_n} = P^{(n)}_{A\alpha_1\ldots\alpha_n}|_\Lambda, \quad A = 1, 2. \tag{18.82}$$

These covariantly defined multipole moments behave as required under infinitesimal conformal transformations $\Omega \to \Omega(1 + x^\alpha \varepsilon_\alpha)$.

Continuing the Kundu (1981) approach, Beig and Simon (1980, 1981) have verified a conjecture by Geroch and proved

Theorem 18.7 *A given asymptotically flat stationary (vacuum) space-time is uniquely characterized by its multipole moments.*

The main step in the proof is the result that for any solution $(\gamma_{\alpha\beta}, Q_A)$ there exists a chart defined in some neighbourhood of Λ in $\widetilde{\Sigma}_3$ such that $\widetilde{\gamma}_{\alpha\beta}, \widetilde{Q}_A$ are analytic.

The formalism summarized in this section has been applied to stationary axisymmetric vacuum solutions by Fodor *et al.* (1989). In that paper the quantities Q_A are identified with the real resp. imaginary parts of the complex potential $\xi = (1-\Gamma)/(1+\Gamma)$, $\Gamma = -F + \mathrm{i}\omega$, which correspond to the mass resp. angular momentum multipoles. The construction can be extended to electrovac fields by considering, in addition to $\mathcal{E}$, the electromagnetic potential Φ.

Other references for studies on multipole moments are e.g. Chaudhuri and Das (1996), Gürsel (1983), Suen (1986).

19

Stationary axisymmetric fields: basic concepts and field equations

19.1 The Killing vectors

We now consider physical systems which, in addition to being stationary (with Killing vector $\boldsymbol{\xi}$) possess a further symmetry: axial symmetry (Killing vector $\boldsymbol{\eta}$).

Independent of the field equations and of the presence of a timelike Killing vector, axial symmetry is defined as an isometric $SO(2)$ mapping of space-time such that the set of fixed points forms a (regular) two-dimensional surface W_2 which is usually called the *axis of rotation.* Mars and Senovilla (1993a) collected some properties of axisymmetric gravitational fields. In particular, it turns out that W_2 is timelike and that the Killing vector field $\boldsymbol{\eta}$ describing axial symmetry must be *spacelike* in a neighbourhood of the axis and zero only at points q on the axis. Outside W_2 the Killing trajectories are *closed* (compact) curves. The Killing vector $\boldsymbol{\eta} = x^1\partial/\partial_2 - x^2\partial/\partial_1 = \partial_\varphi$ vanishes on the rotation axis ($x^1 = 0 = x^2$). The tensor field

$$\mathcal{H}_{ab} = (\nabla_a\eta^c)(\nabla_b\eta_c) \tag{19.1}$$

is at any point of W_2 the projection tensor to the space orthogonal to W_2.

To ensure Lorentzian geometry ('elementary flatness') in the vicinity of the rotation axis, the length of an orbit which passes through a point p in some neighborhood of $q \in W_2$ should be, at first relevant order, 2π times the distance from p to the axis. This can be achieved (by scaling the group parameter φ along the trajectories of $\boldsymbol{\eta}$ to have the standard periodicity 2π) if the norm X of the Killing vector $\boldsymbol{\eta}$ is proportional to the square of that distance. In this case (for points $q \in W_2$ and p with coordinates q^m and x^m respectively), the expansions

$$\begin{aligned} \eta^m &= (x^n - q^n)\nabla_n \eta^m \,|_q + \mathcal{O}(2), \\ X \equiv \eta^m \eta_m &= (x^m - q^m)(x^n - q^n)\mathcal{H}_{mn} \,|_q + \mathcal{O}(3) \end{aligned} \tag{19.2}$$

hold and therefore the *regularity condition*

$$X_{,a}X^{,a}/4X \to 1 \tag{19.3}$$

in the limit at the rotation axis is satisfied. Otherwise, if (19.3) is violated, there are conical singularities (rods or struts) on the axis.

The behaviour of the metric near the axis was carefully studied by Carlson and Safko (1980) using pseudo-Cartesian coordinates and their transformation to polar coordinates perpendicular to the symmetry axis. Carot (2000) generalized these results and proved some other implications of axial symmetry. Remarkably, a group of motions G_2 containing an axial symmetry must be Abelian. The 'extended' regularity conditions (Van den Bergh and Wils 1985b) do not necessarily hold even when (19.3) is satisfied. A *cyclic* Killing vector has closed orbits but – contrary to the axisymmetric case – fixed points and a regular axis are not assumed to exist, cp. Szabados (1987). The possible Bianchi types of a G_3 containing a cyclic symmetry are determined in Barnes (2000).

In this chapter we demand that there are two commuting Killing vectors $\boldsymbol{\xi}$ and $\boldsymbol{\eta}$,

$$\xi^a{}_{;b}\eta^b - \eta^a{}_{;b}\xi^b = 0, \quad \xi^a\xi_a < 0, \quad \eta^a\eta_a > 0, \tag{19.4}$$

i.e. they generate an Abelian group G_2I, see §17.1. Carter (1970) has shown that the cases of the greatest physical importance, *asymptotically flat* stationary axisymmetric gravitational fields, necessarily admit an Abelian group G_2, so that for them (19.4) does not impose an additional restriction. If the cyclic Killing vector $\boldsymbol{\eta}$ has a non-null orbit, than any G_2 containing $\boldsymbol{\eta}$ is Abelian (Ernotte 1980).

The Killing vectors $\boldsymbol{\xi}$ and $\boldsymbol{\eta}$ are *uniquely* determined if we demand that (i) $\boldsymbol{\eta}$ has compact trajectories (in general other subgroups G_1 will have non-compact trajectories), (ii) $\boldsymbol{\eta}$ is normalized by (19.3) and (iii) space-time is asymptotically flat and $\boldsymbol{\xi}$ is normalized so that $F = \xi^a\xi_a \to -1$ holds in the asymptotic region.

19.2 Orthogonal surfaces

The Killing trajectories form two-dimensional orbits T_2; the simple bivector

$$v_{ab} = 2\xi_{[a}\eta_{b]}, \quad v_{ab}v^{ab} < 0, \tag{19.5}$$

is surface-forming (see (19.4) and (6.12)). The surface element orthogonal to the group orbits is spanned by the dual bivector

$$\tilde{v}_{ab} = \tfrac{1}{2}\varepsilon_{abcd}v^{cd} = 2m_{[a}\overline{m}_{b]}, \quad m_a\xi^a = 0 = m_a\eta^a. \tag{19.6}$$

The simple bivector $\tilde{v}_{ab}$ is surface-forming and the metric is said to be *orthogonally transitive* if and only if the condition

$$\begin{aligned} v_{ab}\tilde{v}^{bc}{}_{;c} = 2\xi_{[a}\eta_{b]}(\overline{m}^c m^b{}_{;c} - m^c\overline{m}^b{}_{;c}) = 0 \\ \Leftrightarrow \quad \varepsilon_{abcd}\eta^a\xi^b\xi^{c;d} = 0 = \varepsilon_{abcd}\xi^a\eta^b\eta^{c;d} \end{aligned} \tag{19.7}$$

is satisfied (cp. (6.12)). In general, the Killing vectors of an arbitrary stationary axisymmetric space-time do not obey (19.7), but only solutions obeying (19.7) are explicitly known.

The condition (19.7) can be formulated as a restriction on the algebraic form of the Ricci tensor:

Theorem 19.1 *Stationary axisymmetric fields admit 2-spaces orthogonal to the group orbits if and only if the conditions*

$$\xi^d R_{d[a}\xi_b\eta_{c]} = 0 = \eta^d R_{d[a}\xi_b\eta_{c]} \tag{19.8}$$

are satisfied (Kundt and Trümper 1966).

Proof (see e.g. also Carter 1972): For a Killing vector $\boldsymbol{\xi}$ the identity (8.22),

$$\xi_{a;bc} = R_{abcd}\xi^d, \tag{19.9}$$

holds. From (19.9) (and an analogous relation for $\boldsymbol{\eta}$) we get

$$(\xi_{[a;b}\xi_{c]})^{;c} = \tfrac{2}{3}\xi^d R_{d[a}\xi_{b]}, \quad (\eta_{[a;b}\eta_{c]})^{;c} = \tfrac{2}{3}\eta^d R_{d[a}\eta_{b]}. \tag{19.10}$$

With the aid of these equations, together with the cyclic symmetry relation of the curvature tensor and the commutativity of $\boldsymbol{\xi}$ and $\boldsymbol{\eta}$, we obtain

$$(\xi_{[a;b}\xi_c\eta_{d]})^{;d} = -\tfrac{1}{2}\xi^d R_{d[a}\xi_b\eta_{c]}, \quad (\eta_{[a;b}\eta_c\xi_{d]})^{;d} = -\tfrac{1}{2}\eta^d R_{d[a}\eta_b\xi_{c]}. \tag{19.11}$$

Using the fact that a completely antisymmetric tensor is proportional to the ε-tensor, we convert (19.11) into the equivalent dual equations

$$\begin{aligned} (\varepsilon^{abcd}\xi_{a;b}\xi_c\eta_d)^{;e} &= -2\varepsilon^{abce}\xi^d R_{da}\xi_b\eta_c, \\ (\varepsilon^{abcd}\eta_{a;b}\eta_c\xi_d)^{;e} &= -2\varepsilon^{abce}\eta^d R_{da}\eta_b\xi_c. \end{aligned} \tag{19.12}$$

From these equations we conclude that the conditions (19.8) are a necessary consequence of the original criterion (19.7). Conversely, if the conditions (19.8) are satisfied, the twist scalars $\varepsilon_{abcd}\eta^a\xi^b\xi^{c;d}$ and $\varepsilon_{abcd}\xi^a\eta^b\eta^{c;d}$

are constants. Both twist scalars vanish on the rotation axis ($\eta = 0$). Hence they vanish in a connected domain which intersects the rotation axis if the conditions (19.8) hold in that domain. Then the geometry admits 2-spaces orthogonal to the orbits of the Abelian group of motions G_2.

For vacuum fields ($R_{ab} = 0$), the existence of orthogonal 2-spaces was first demonstrated by Papapetrou (1966) in his pioneering work on this subject.

Conditions (19.8) hold for a wide class of energy-momentum tensors. For example, decomposition of the metric into orthogonal 2-spaces is possible for perfect fluid solutions, and for Einstein–Maxwell fields, provided that the four-velocity of the fluid, and the electromagnetic 4-current density vector, respectively, satisfy the so-called circularity condition

$$u_{[a}\xi_b\eta_{c]} = 0 = j_{[a}\xi_b\eta_{c]}, \tag{19.13}$$

i.e. the trajectories of $\boldsymbol{u}$ and $\boldsymbol{j}$ lie in the surfaces of transitivity of the group G_2. In this terminology, conditions (19.8) mean that $R^a_b\xi^b$ and $R^a_b\eta^b$ are circular vector fields. The circularity condition (19.13) is a natural generalization of the condition (18.40) (four-velocity and current parallel to the Killing vector) to the case of stationary axisymmetric fields. Obviously, for perfect fluids the condition $u_{[a}\xi_b\eta_{c]} = 0$ implies (19.8).

Now we show that $j_{[a}\xi_b\eta_{c]} = 0$ for Einstein–Maxwell fields also implies (19.8). The Maxwell equations

$$F^{*ab}{}_{;b} = j^a, \tag{19.14}$$

and the assumption that the Lie derivatives of F^*_{ab} with respect to both the commuting Killing vectors $\boldsymbol{\xi}$ and $\boldsymbol{\eta}$ vanish, lead to

$$(F^*_{ab}\xi^a\eta^b)_{;d} = \mathrm{i}\varepsilon_{abcd}\xi^a\eta^b j^c = 0. \tag{19.15}$$

$F^*_{ab}\xi^a\eta^b$ is constant, and it is zero on the axis of rotation. Hence,

$$F^*_{ab}\xi^a\eta^b = 0 = \xi_{[a}\eta_b F^*_{cd]}. \tag{19.16}$$

(In flat space-time electrodynamics this relation says that the azimuthal components of the electric and magnetic field vectors are zero.) Relation (19.16) in turn implies (19.8).

If in an Einstein–Maxwell geometry, outside the sources, two Killing vectors $\boldsymbol{\xi}$, $\boldsymbol{\eta}$ satisfy the conditions (19.4), (19.5), (19.7), then the Maxwell field shares the space-time symmetry: $\mathcal{L}_{\boldsymbol{\xi}}F_{ab} = 0 = \mathcal{L}_{\boldsymbol{\eta}}F_{ab}$ (Michalski and Wainwright 1975).

Gourgoulhon and Bonazzola (1993) considered the more general case when the conditions (19.8) do not hold, e.g. for toroidal magnetic fields. The field equations can be written in an entirely two-dimensional covariant form.

19.3 The metric and the projection formalism

Theorem 19.1 is of fundamental importance in the construction of exact solutions of the Einstein equations. If we adapt the coordinate system to the Killing vectors, the line element of a stationary axisymmetric field admitting 2-spaces orthogonal to the Killing vectors $\boldsymbol{\xi} = \partial_t$ and $\boldsymbol{\eta} = \partial_\varphi$ can be written in the form (Lewis 1932, Papapetrou 1966)

$$\mathrm{d}s^2 = \mathrm{e}^{-2U}(\gamma_{MN}\mathrm{d}x^M\mathrm{d}x^N + W^2\mathrm{d}\varphi^2) - \mathrm{e}^{2U}(\mathrm{d}t + A\mathrm{d}\varphi)^2, \qquad (19.17)$$

where the metric functions U, γ_{MN}, W, A depend only on the coordinates $x^M = (x^1, x^2)$ which label the points on the 2-surfaces S_2 orthogonal to the orbits. The form (19.17) is preserved under the transformations

$$x^{M'} = x^{M'}(x^N), \quad t' = at + b\varphi, \quad \varphi' = c\varphi + dt, \qquad (19.18)$$

$(a, \ldots, d$ constants). The functions W, U, A behave like scalars under $x^{M'} = x^{M'}(x^N)$. The metric (19.17) has block-diagonal form and exhibits the reflection symmetry $(t, \varphi) \to (-t, -\varphi)$.

Equation (19.17) is the specialization of (18.8) (including the conformal transformation (18.23)) to the stationary axisymmetric fields obeying the conditions (19.7). In the coordinate system (19.17), the scalar products of the Killing vectors are

$$\begin{aligned} \xi^a\xi_a &= -\mathrm{e}^{2U}, & \eta^a\eta_a &= \mathrm{e}^{-2U}W^2 - \mathrm{e}^{2U}A^2, \\ \xi^a\eta_a &= -\mathrm{e}^{2U}A, & 2\xi_{[a}\eta_{b]}\xi^a\eta^b &= W^2. \end{aligned} \qquad (19.19)$$

If a linear combination of the two Killing vectors leads to the relation $A = \pm W\mathrm{e}^{-2U}$, the space-time admits a null Killing vector. This special case will be considered in Chapter 24. Vacuum solutions with this property belong to the class $S = S(A)$ (§20.4), provided that W is not constant.

In general, $\boldsymbol{\xi}$ and $\boldsymbol{\eta}$ are not hypersurface-orthogonal. However, it can easily be verified (Bardeen 1970) that the timelike vector $\boldsymbol{\zeta} = \boldsymbol{\xi} - \xi^a\eta_a(\eta^b\eta_b)^{-1}\boldsymbol{\eta}$ is orthogonal to the hypersurfaces $t =$ const. In general, it is not a Killing vector.

Without loss of generality we can introduce *isotropic coordinates* in V_2,

$$\gamma_{MN} = \mathrm{e}^{2k}\delta_{MN}. \qquad (19.20)$$

Moreover, if W is a non-constant potential function, it can be transformed to $W = x^1$ with the aid of a coordinate transformation $\zeta' = f(\zeta)$, $\sqrt{2}\zeta = x^1 + \mathrm{i}x^2$, preserving the isotropic coordinate condition. The case $W =$ const is not of interest here because it is not compatible with the regularity condition (19.3); vacuum metrics of this type are plane waves (24.35) with $F = 0$, $f = f(\zeta)$.

The isotropic coordinates with $W = x^1 = \rho$ are called *Weyl's canonical coordinates* ($x^1 = \rho$, $x^2 = z$; $\sqrt{2}\zeta = \rho + \mathrm{i}z$). In these coordinates, the space-time metric (19.17) is given by

$$\mathrm{d}s^2 = \mathrm{e}^{-2U}[\mathrm{e}^{2k}(\mathrm{d}\rho^2 + \mathrm{d}z^2) + \rho^2\mathrm{d}\varphi^2] - \mathrm{e}^{2U}(\mathrm{d}t + A\mathrm{d}\varphi)^2. \qquad (19.21)$$

For stationary axisymmetric fields, one can develop a projection formalism (Geroch 1972) similar to that for stationary fields given in §18.1. We introduce the notation

$$\lambda_{AB} \equiv \xi_A{}^a\xi_{Ba}, \quad A, B = 1, 2, \qquad (19.22)$$

for the scalar products of the Killing vectors $\boldsymbol{\xi}_1 = \boldsymbol{\xi}$, $\boldsymbol{\xi}_2 = \boldsymbol{\eta}$, and identify the projection tensor

$$H_{ab} \equiv g_{ab} + W^{-2}\lambda^{AB}\xi_{Aa}\xi_{Bb}, \quad W^2 \equiv -\tfrac{1}{2}\lambda^{AB}\lambda_{AB}, \qquad (19.23)$$

with the metric tensor on a differentiable manifold Σ_2. Capital Latin indices are raised and lowered with the aid of the alternating symbols (3.66), as in (3.67). Covariant derivatives (denoted by D_a) and the Riemann curvature tensor on Σ_2 are defined in analogy to (18.3) and (18.5) ($h_{ab} \to H_{ab}$).

If 2-spaces V_2 orthogonal to the group orbits exist, Σ_2 can be identified with these 2-spaces and the two scalars $C_A = \varepsilon^{MN}\varepsilon^{abcd}\xi_{Ma}\xi_{Nb}\xi_{Ac;d}$ vanish. Then the Ricci tensor R_{ab} in V_4 can be written in terms of tensors and metric operations on V_2 as follows,

$$\overset{2}{R}_{ab} = (2W)^{-2}\lambda^{AB}{}_{,a}\lambda_{AB,b} + W^{-1}D_aW_{,b} + H^c_aH^d_bR_{cd}, \qquad (19.24)$$

$$D^a(W^{-1}\lambda_{AB,a}) = \tfrac{1}{2}W^{-3}\lambda_{AB}H^{ab}\lambda^{CD}{}_{,a}\lambda_{CD,b} - 2W^{-1}R_{ab}\xi^a_A\xi^b_B, \qquad (19.25)$$

$$H^c_a\xi^d_AR_{cd} = 0. \qquad (19.26)$$

The contracted Bianchi identities (2.82) in the projection formalism are given by Whelan and Romano (1999) and read (for $C_A = 0$)

$$D_a(W\xi^b_AH^{ac}G_{bc}) = 0, \quad D_b(WH^c_aH^{bd}G_{cd}) = \tfrac{1}{2}W\xi^c_A\xi^d_BG_{cd}(W^{-2}\lambda^{AB})_{,a}, \qquad (19.27)$$

where G_{ab} is the Einstein tensor.

Kitamura (1978) gave an invariant characterization of the stationary axisymmetric metric (19.17) in terms of a tetrad and its first covariant derivatives. This characterization is independent of the field equations.

19.4 The field equations for stationary axisymmetric Einstein–Maxwell fields

We now reduce the field equations (18.37)–(18.39) for stationary Einstein–Maxwell fields outside the sources to the axisymmetric case. Assuming that the electromagnetic field satisfies the condition (19.16), $F^*_{ab}\xi^a\eta^b = 0$, we can start with the line element (19.17) (see Theorem 19.1), and the potentials $\mathcal{E}$ and Φ depend only on the spatial coordinates in the 2-spaces orthogonal to the group orbits. Exceptional cases (e.g. the field of an axial current) where condition (19.16) is violated are given in §22.2.

All field equations can be written covariantly with respect to the 2-metric γ_{MN} in (19.17). After the substitution $-F = \mathrm{Re}\,\mathcal{E} + \Phi\overline{\Phi}$ (see (18.34)), the field equations (18.38) and (18.39) read (a semicolon denotes covariant derivative with respect to γ_{MN})

$$(\mathrm{Re}\,\mathcal{E} + \Phi\overline{\Phi})W^{-1}(W\mathcal{E}_{,M})^{;M} = \mathcal{E}_{,M}(\mathcal{E}^{,M} + 2\overline{\Phi}\Phi^{,M}), \qquad (19.28)$$

$$(\mathrm{Re}\,\mathcal{E} + \Phi\overline{\Phi})W^{-1}(W\Phi_{,M})^{;M} = \Phi_{,M}(\mathcal{E}^{,M} + 2\overline{\Phi}\Phi^{,M}). \qquad (19.29)$$

The field equations (18.37) reduce to

$$W_{,M}{}^{;M} = 0, \qquad (19.30)$$

$$K\gamma_{MN} - \frac{W_{,M;N}}{W} = \frac{(\mathcal{E}_{,(M} + 2\overline{\Phi}\Phi_{,(M})(\overline{\mathcal{E}}_{,N)} + 2\Phi\overline{\Phi}_{,N)})}{2(\mathrm{Re}\mathcal{E} + \overline{\Phi}\Phi)^2} - 2\frac{\overline{\Phi}_{,(M}\Phi_{,N)}}{\mathrm{Re}\mathcal{E} + \overline{\Phi}\Phi}, \qquad (19.31)$$

where $K = -2\mathrm{e}^{-2k}k_{,\zeta\overline{\zeta}}$ denotes the Gaussian curvature of the 2-space $\widehat{V}_2$ with metric γ_{MN}. According to (18.22) and (18.34), the metric coefficients A and $F = -\mathrm{e}^{2U}$ in the line element (19.17) can be determined from the complex potentials $\mathcal{E}$ and Φ via

$$A_{,M} = W\mathrm{e}^{-4U}\varepsilon_{MN}\omega^N, \quad \omega_N = \mathrm{Im}\,\mathcal{E}_{,N} - \mathrm{i}(\overline{\Phi}\Phi_{,N} - \Phi\overline{\Phi}_{,N}), \qquad (19.32)$$

$$\mathrm{e}^{2U} = \mathrm{Re}\,\mathcal{E} + \overline{\Phi}\Phi, \qquad (19.33)$$

where ε_{MN} is the Levi-Civita tensor in $\widehat{V}_2$.

Because of (19.30) we can introduce Weyl's canonical coordinates, ρ $(= W)$ and z. Then the functions U, k and A in the metric (19.21) are determined by (19.33) and

$$k_{,\zeta} = \sqrt{2}\rho\left[\frac{(\mathcal{E}_{,\zeta} + 2\overline{\Phi}\Phi_{,\zeta})(\overline{\mathcal{E}}_{,\zeta} + 2\Phi\overline{\Phi}_{,\zeta})}{4(\mathrm{Re}\,\mathcal{E} + \overline{\Phi}\Phi)^2} - \frac{\overline{\Phi}_{\zeta}\Phi_{\zeta}}{\mathrm{Re}\,\mathcal{E} + \overline{\Phi}\Phi}\right], \qquad (19.34)$$

$$A_{,\zeta} = \rho\frac{\mathrm{i}(\mathrm{Im}\,\mathcal{E})_{,\zeta} + \overline{\Phi}\Phi_{,\zeta} - \Phi\overline{\Phi}_{,\zeta}}{(\mathrm{Re}\,\mathcal{E} + \overline{\Phi}\Phi)^2}, \quad \sqrt{2}\partial_\zeta = \partial_\rho - \mathrm{i}\partial_z. \qquad (19.35)$$

Because of the field equations (19.28), (19.29), the integrability conditions of (19.34) and (19.35) are satisfied identically, and k and A can be calculated simply by line integration. Two of the three equations (19.31) are just equations (19.34) for determining k. The remaining field equation of (19.31) has, in Weyl's canonical coordinates, the form

$$k_{,MM} + \frac{(\mathcal{E}_{,M} + 2\overline{\Phi}\Phi_{,M})(\overline{\mathcal{E}}_{,M} + 2\Phi\overline{\Phi}_{,M})}{4(\mathrm{Re}\,\mathcal{E} + \overline{\Phi}\Phi)^2} - \frac{\overline{\Phi}_{,M}\Phi_{,M}}{\mathrm{Re}\,\mathcal{E} + \overline{\Phi}\Phi} = 0. \qquad (19.36)$$

It is identically satisfied because of the other field equations.

Theorem 19.2 *The Einstein–Maxwell equations for stationary axisymmetric fields can be completely reduced to the simultaneous system* (19.28), (19.29) *(with $W = \rho$) of elliptic differential equations of the second order for the two complex potentials $\mathcal{E}$ and Φ. Every solution of these equations determines a metric* (19.21) *via* (19.33)–(19.35).

For exterior electro*static* fields ($\Phi = \overline{\Phi} = \chi$), the field equations (19.28)–(19.29) read

$$W^{-1}(WU_{,M})^{;M} = \mathrm{e}^{-2U}\chi_{,M}\chi^{,M}, \quad (W\mathrm{e}^{-2U}\chi_{,M})^{;M} = 0. \qquad (19.37)$$

In Weyl's canonical coordinates, (19.37) can be replaced by the equations

$$\begin{aligned} \Delta U &= \rho^{-1}(\rho U_{,A})_{,A} = U_{,A}U_{,A} + k_{,AA}, \\ (\Delta U)^2 &= (2U_{,1}U_{,2} - k_{,2}/\rho)^2 + ({U_{,1}}^2 - {U_{,2}}^2 - k_{,1}/\rho)^2 \end{aligned} \qquad (19.38)$$

for the metric functions U and k. Two special classes of solutions are those with $\Delta U = 0$ (Weyl's *vacuum* solutions, §20.2), and $k = 0$ (Papapetrou–Majumdar solutions (§18.7) with axial symmetry).

19.5 Various forms of the field equations for stationary axisymmetric vacuum fields

All stationary axisymmetric vacuum fields ($\Phi = 0$, $\mathcal{E} = \Gamma$) are covered by the metric (19.17). The field equation (19.28), specialized to the vacuum case, is the *Ernst equation* (Ernst 1968a)

$$(\Gamma + \overline{\Gamma})W^{-1}(W\Gamma_{,M})^{;M} = 2\Gamma_{,M}\Gamma^{,M}. \qquad (19.39)$$

Splitting Γ into its real and imaginary parts, we have to solve a simultaneous system of two elliptic differential equations of the second order,

$$W^{-1}(WU_{,M})^{;M} = -\tfrac{1}{2}\mathrm{e}^{-4U}\omega_{,M}\omega^{,M}, \quad (W\mathrm{e}^{-4U}\omega_{,M})^{;M} = 0. \qquad (19.40)$$

These equations are covariant with respect to transformations in the 2-space $\widehat{V}_2$ with the metric γ_{MN}. In Weyl's canonical coordinates (ρ, z), (19.39) reads

$$(\Gamma+\overline{\Gamma})(\Gamma_{,\rho\rho}+\rho^{-1}\Gamma_{,\rho}+\Gamma_{,zz}) = 2(\Gamma_{,\rho}{}^2+\Gamma_{,z}{}^2). \tag{19.41}$$

Once a solution $\Gamma = \mathrm{e}^{2U}+\mathrm{i}\omega$ of this differential equation is given, we find the full metric (19.21) with the aid of line integrals for k and A,

$$k_{,\zeta} = \sqrt{2}\,\rho\frac{\Gamma_{,\zeta}\overline{\Gamma}_{,\zeta}}{(\Gamma+\overline{\Gamma})^2}, \quad A_{,\zeta} = 2\rho\frac{(\Gamma-\overline{\Gamma})_{,\zeta}}{(\Gamma+\overline{\Gamma})^2}, \quad \sqrt{2}\partial_\zeta = \partial_\rho - \mathrm{i}\partial_z, \tag{19.42}$$

the integrability conditions being automatically satisfied. Thus the problem has been reduced essentially to (19.41) for the complex potential Γ (see Theorem 19.2). At first glance, the simple structure of this non-linear differential equation is encouraging, but actually only some special solutions and restricted classes of solutions had been found before the arrival of the generation methods treated in Chapter 34. At that time reformulations of the field equations were of some help for finding solutions and for preparing the way for the powerful generation techniques. Therefore we list here some of these formulations.

(i) Introducing a new function $S \equiv -U+\frac{1}{2}\ln W$, we obtain from (19.40) the system of differential equations

$$W^{-1}(WS_{,M})^{;M} = \tfrac{1}{2}\mathrm{e}^{-4S}A_{,M}A^{,M}, \quad (W\mathrm{e}^{-4S}A_{,M})^{;M} = 0 \tag{19.43}$$

for the unknown functions S and A. Up to a sign (19.43) and (19.40) have exactly the same form.

(ii) We retain Weyl's canonical coordinates ($\sqrt{2}\zeta = \rho+\mathrm{i}z$), but go over to new variables M and N defined by

$$M = 2(\zeta+\bar{\zeta})(\Gamma+\overline{\Gamma})^{-1}(\Gamma+\overline{\Gamma})_{,\zeta}, \quad N = 2(\zeta+\bar{\zeta})(\Gamma+\overline{\Gamma})^{-1}(\Gamma-\overline{\Gamma})_{,\zeta}. \tag{19.44}$$

Inserting these expressions into the field equation (19.41), and taking into account the integrability condition, we obtain a simultaneous system of partial differential equations of the *first* order for the two complex functions M and N,

$$2(\zeta+\bar{\zeta})M_{,\bar{\zeta}} = M-\overline{M}-\overline{N}N, \quad 2(\zeta+\bar{\zeta})N_{,\bar{\zeta}} = N+\overline{N}-\overline{N}M. \tag{19.45}$$

Once a solution of these equations is known, we get the potential Γ by a line integral.

(iii) The Ernst equation (19.39) can be reformulated by introducing the new potential ξ by

$$\xi \equiv (1-\mathcal{E})/(1+\mathcal{E}). \tag{19.46}$$

ξ satisfies the differential equation

$$(\xi\bar{\xi} - 1)W^{-1}(W\xi_{,M})^{;M} = 2\bar{\xi}\xi_{,M}\xi^{,M}. \tag{19.47}$$

This last version of the field equations has proved to be especially useful in constructing new solutions (see §20.5).

(iv) The function k in the metric (19.21) satisfies a differential equation of the fourth order, which can be derived from the fact that the right-hand side of (19.31) (with $\Phi = 0$) multiplied by $(-2)\mathrm{d}x^M \mathrm{d}x^N$ is the metric of a space of constant negative curvature. From the left-hand side of (19.31) we infer that the line element $\rho^{-1}(\rho k_{,A})_{,A}(\mathrm{d}\rho^2 + \mathrm{d}z^2) - 2k_{,A}\mathrm{d}x^A\mathrm{d}\rho/\rho$ is associated with this space of constant curvature so that we obtain a differential equation for k alone:

$$\begin{aligned} 2D(A_{,zz} + C_{,\rho\rho} - 2B_{,\rho z}) - D_{,z}A_{,z} - D_{,\rho}C_{,\rho} - BA_{,\rho}C_{,z} + BA_{,z}C_{,\rho} \\ +2CA_{,\rho}B_{,z} + 2AC_{,z}B_{,\rho} - 4BB_{,\rho}B_{,z} = 4D^2, \end{aligned} \tag{19.48}$$

$$A \equiv k_{,\rho}/\rho - 2k_{,AA}, \quad C \equiv -k_{,\rho}/\rho - 2k_{,AA}, \quad B \equiv 2k_{,z}/\rho, \quad D \equiv AC - B^2.$$

From a given (non-constant) solution k of (19.48), the potential ξ can be constructed up to a phase factor, $\xi \to e^{i\alpha}\xi$ (ignoring pure coordinate transformations). Independently, Cosgrove (1978a), Herlt (1978a) and Cox and Kinnersley (1979) have also reduced the field equations to a fourth-order differential equation for a real superpotential, see also Tomimatsu (1981) and Lorencz and Sebestyén (1986).

(v) Perjés (1985a), see also Theorems 18.5 and 18.6, introduced the Ernst potential and its complex conjugate as new space coordinates. Other forms of the field equations are given, e.g., in Chandrasekhar (1978) and Kramer and Neugebauer (1968b).

Comparison of (19.40) with (19.43), and with the very similar equations (19.37) for electrostatic fields, leads to

Theorem 19.3 *Given a stationary axisymmetric vacuum solution (U, ω), the substitution*

$$S' = -U' + \tfrac{1}{2}\ln W = U, \quad A' = i\omega, \tag{19.49}$$

yields another vacuum solution (U', A') (Kramer and Neugebauer 1968b), *and the substitution*

$$U' = 2U, \quad \chi = i\omega, \quad k' = 4k \tag{19.50}$$

generates from a stationary vacuum solution (U, ω, k) a static solution (U', χ, k') of the Einstein–Maxwell equations (Bonnor 1961).

The new solutions will be *real* if one can analytically continue the parameters in the solutions to compensate for the imaginary unit in the substitutions of this theorem.

Kordas (1995) and Meinel and Neugebauer (1995) have shown that an asymptotically flat solution of the stationary axisymmetric vacuum equations is reflection-symmetric with respect to the equatorial plane $z = 0$ if and only if the Ernst potential Γ_+ on the upper part of the symmetry axis ($\rho = 0, z > 0$) satisfies the relation

$$\Gamma(z)\bar{\Gamma}(-z) = 1. \tag{19.51}$$

Applying the Newman-Penrose formalism (Chapter 7) to stationary axisymmetric vacuum solutions, and choosing the real null directions $\boldsymbol{k}$ and $\boldsymbol{l}$ as linear combinations of the Killing vectors $\boldsymbol{\xi}$ and $\boldsymbol{\eta}$, we obtain the result: solutions of Petrov type *III* cannot occur. Type *II* solutions are members of the class $A^2 = W^2 \mathrm{e}^{-4U}$ (§20.4) (Collinson and Dodd 1969). The Petrov classification of the vacuum space-times with orthogonally transitive Killing vectors has also been studied by Morisetti *et al.* (1980).

19.6 Field equations for rotating fluids

Two different approaches using non-holonomic frames have been developed for stationarily rotating perfect fluids with axial symmetry.

Chinea and González-Romero (1992) wrote the perfect fluid equations as a set of differential-form equations. The 1-forms $\theta^0 \equiv u$ and θ^1 lie in the group orbits, whereas the remaining 1-forms θ^2 and θ^3 span the 2-space orthogonal to the orbits. The authors introduced the Hodge dual operation (denoted by $*$) in the subspace generated by θ^2 and θ^3, and the tilde operation $\widetilde{\theta}^2 = \theta^2, \widetilde{\theta}^3 = -\theta^3$. From the zero torsion conditions, their integrability condition $R^a_{[bcd]} = 0$, and the Einstein equations one gets the following total set of field equations for the 1-forms $u = \theta^0$, θ^1, $b = \mathrm{e}^{-Q}\,\theta^2$ (or $*b = \mathrm{e}^{-Q}\,\theta^3$), the 1-forms a, w, s which are linear combinations of θ^2 and θ^3, and the functions Q, p (pressure) and μ (mass density) which are independent of t and φ.

$$\begin{gathered}
\mathrm{d}u = a \wedge u + w \wedge \theta^1, \quad \mathrm{d} * (w - s) + 2a \wedge *w + 2(a - b) \wedge *s = 0, \\
\mathrm{d}a = w \wedge s, \quad \mathrm{d}w = -(b - 2a) \wedge w, \quad \mathrm{d}s = (b - 2a) \wedge s, \quad \mathrm{d}b = 0, \\
\mathrm{d}\theta^1 = (b - a) \wedge \theta^1 + s \wedge u, \quad \mathrm{d} * b + b \wedge *b = 2\kappa_0 p\, \mathrm{e}^{2Q}\, b \wedge *b, \\
2\mathrm{d} * a + 2b \wedge *a + w \wedge *w - s \wedge *s = \kappa_0(\mu + 3p)\, \mathrm{e}^{2Q}\, b \wedge *b, \\
\mathrm{d}Q \wedge b + b \wedge \widetilde{a} - a \wedge \widetilde{a} + \tfrac{1}{4}(s - w) \wedge (\widetilde{s} - \widetilde{w}) = 0, \quad \mathrm{d}p + (\mu + p)a = 0, \\
\mathrm{d}Q \wedge *b + \tfrac{1}{2}\mathrm{d} * b - \tfrac{1}{2}b \wedge *b + b \wedge *\widetilde{a} - a \wedge *\widetilde{a} + \tfrac{1}{4}(s - w) \wedge *(\widetilde{s} - \widetilde{w}) = 0.
\end{gathered} \tag{19.52}$$

The kinematical quantities a, s and w are related to the fluid's acceleration, shear and rotation, respectively. The advantage of this approach is the possibility of investigating particular cases by imposing simplifying ansätze. In particular, $w = 0$, $b \wedge s = 0$ leads to the irrotational solution given by Chinea and González-Romero (1992). The differential-form approach summarized here can also be applied to construct stationary axisymmetric solutions of the Einstein–Maxwell equations (Fernandez-Jambrina and Chinea 1994), see §21.1.

Marklund and Perjés (1997) used a complex non-holonomic triad in the 3-space of the timelike Killing trajectories and introduced a complex tensor which is symmetric and tracefree both in vacuum and in the presence of matter. This tensor was constructed from the Simon tensor (18.70) by adding an appropriate matter term. In the limit of a static space-time this tensor becomes real and in vacuo it equals the Simon tensor. The triad formulation of the field equations for stationary axisymmetric perfect fluids simplifies considerably when the generalized Simon tensor vanishes. The method covers the general case of differential rotation.

Using a tetrad approach Fodor *et al.* (1999) proved that all rigidly rotating stationary axisymmetric perfect fluid solutions with a purely magnetic Weyl tensor ($E_{mn} = 0$) and with circular motion are locally rotationally symmetric.

20
Stationary axisymmetric vacuum solutions

20.1 Introduction

In this chapter we survey some simple axisymmetric vacuum solutions. In general these solutions had been already derived before the powerful generation methods outlined in Chapter 10 were known. Stationary axisymmetric vacuum solutions obtained by these methods will be given in Chapter 34. The stationary cylindrically-symmetric vacuum field is treated in §22.2.

For review articles devoted to the solutions in the following sections we refer the reader to Quevedo (1990) and Islam (1985). For a review of interpretations which have been given to some of these solutions, see Bonnor (1992).

20.2 Static axisymmetric vacuum solutions (Weyl's class)

If we assume that one of the Killing vectors in the metric

$$\mathrm{d}s^2 = \mathrm{e}^{-2U}(\gamma_{MN}\mathrm{d}x^M\mathrm{d}x^N + W^2\mathrm{d}\varphi^2) - \mathrm{e}^{2U}(\mathrm{d}t + A\mathrm{d}\varphi)^2, \tag{20.1}$$

say $\boldsymbol{\xi} = \partial_t$, is hypersurface-orthogonal, then A can be put equal to zero and the second Killing vector is also hypersurface-orthogonal. The *static* axisymmetric vacuum solutions (Weyl 1917) are invariantly characterized by the existence of two commuting hypersurface-orthogonal Killing vectors $\boldsymbol{\xi}$, $\boldsymbol{\eta}$ satisfying

$$\begin{aligned} &\xi_{(a;b)} = 0, \quad \xi_{[a;b}\xi_{c]} = 0, \ \xi^a\xi_a < 0, \\ &\eta_{(a;b)} = 0, \quad \eta_{[a;b}\eta_{c]} = 0, \ \eta^a\eta_a > 0, \end{aligned} \quad \xi^c\eta^a{}_{;c} - \eta^c\xi^a{}_{;c} = 0. \tag{20.2}$$

Because in this case the twist potential ω vanishes, $\Gamma = -\,\mathrm{e}^{2U}$ is real, and the Ernst equation (19.41) is simply the potential equation $\Delta U = 0$ for a real function U independent of the azimuthal coordinate φ. In Weyl's canonical coordinates (19.21), the *metric* and the *field equations* for static axisymmetric vacuum solutions read

$$\begin{aligned} ds^2 &= e^{-2U}[e^{2k}(d\rho^2 + dz^2) + \rho^2 d\varphi^2] - e^{2U}dt^2, \\ \Delta U &= \rho^{-1}(\rho U_{,M})_{,M} = 0, \quad k_{,\rho} = \rho(U_{,\rho}^2 - U_{,z}^2), \quad k_{,z} = 2\rho U_{,\rho}U_{,z}. \end{aligned} \tag{20.3}$$

The function k can be calculated by means of a line integral. Although $\Delta U = 0$ is a linear differential equation, the equations for k manifest the non-linearity of the Einstein field equations. For static axisymmetric vacuum solutions the *regularity condition* (19.3) on the axis of symmetry means $k = 0$ in the limit $\rho \to 0$.

Two solutions of Weyl's class are completely equivalent if the functions U and k of one solution differ only by additive constants from those of the other solution.

The potential equation $\Delta U = 0$ may be solved using various coordinates in the Euclidean 3-space. In spherical coordinates (r,ϑ), the asymptotically flat solutions are

$$\begin{aligned} U &= \sum_{n=0}^{\infty} a_n r^{-(n+1)} P_n(\cos\vartheta), \\ k &= -\sum_{l,m=0}^{\infty} \frac{a_l a_m (l+1)(m+1)}{(l+m+2)r^{l+m+2}}(P_l P_m - P_{l+1}P_{m+1}), \end{aligned} \tag{20.4}$$

$P_n = P_n(\cos\vartheta)$ being the Legendre polynomials. The simplest case is the solution

$$U = -m/r, \quad 2k = -m^2 \sin^2\vartheta / r^2 \tag{20.5}$$

(Chazy 1924, Curzon 1924). Though the potential U is spherically-symmetric, the solution (20.5) is not.

Prolate spheroidal coordinates (x, y) are connected with Weyl's canonical coordinates (ρ, z) by the relations (Zipoy 1966)

$$\begin{aligned} \rho &= \sigma(x^2-1)^{1/2}(1-y^2)^{1/2}, \quad z = \sigma xy, \quad \sigma = \text{const}, \\ 2\sigma x &= r_+ + r_-, \quad 2\sigma y = r_+ - r_-, \quad r_\pm^2 = \rho^2 + (z \pm \sigma)^2, \end{aligned} \tag{20.6}$$

$$d\rho^2 + dz^2 = \sigma^2(x^2 - y^2)\left[(x^2-1)^{-1}dx^2 + (1-y^2)^{-1}dy^2\right]. \tag{20.7}$$

The surfaces $x = \text{const}$ and $y = \text{const}$ are orthogonal families of, respectively, ellipsoids and hyperboloids. The general form of the asymptotically flat solutions of Weyl's class which are, outside the sources, regular at the symmetry axis can be derived from the potential

$$U = \sum_{n=0}^{\infty} q_n Q_n(x) P_n(y) \tag{20.8}$$

(Quevedo 1989), $Q_n(x)$ being the Legendre functions of the second kind. In particular, the Erez and Rosen (1959) metric is obtained in the case in which the constants q_n vanish except $q_0 = 1$ and q_2,

$$2U = \ln\left(\frac{x-1}{x+1}\right) + q_2(3y^2-1)\left[\tfrac{1}{4}(3x^2-1)\ln\left(\frac{x-1}{x+1}\right) + \tfrac{3}{2}x\right]. \quad (20.9)$$

For the solution due to Gutsunaev and Manko (1985), the potential U expressed in terms of prolate spheroidal coordinates has the form

$$U = \frac{1}{2}\ln\left(\frac{x-1}{x+1}\right) + \sum_{n=0}^{\infty} a_{n+1}\left[\frac{P_n^+}{(x+y)^{n+1}} - \frac{P_n^-}{(x-y)^{n+1}}\right], \quad (20.10)$$

where a_n are arbitrary real parameters, and $P_n^\pm$ are the Legendre polynomials of the arguments $(xy \pm 1)/(x \pm y)$. The metric function k is given in Denisova *et al.* (1994).

The special case of the Weyl solution with

$$e^{2U} = (x-1)^\delta (x+1)^{-\delta} \quad (20.11)$$

has been investigated by Zipoy (1966) and Voorhees (1970). For $\delta = 1$, (20.11) represents the Schwarzschild solution; it is expressed in terms of Weyl's canonical coordinates (ρ, z) by

$$U = \frac{1}{2}\ln\left(\frac{r_+ + r_- - 2m}{r_+ + r_- + 2m}\right), \quad k = \frac{1}{2}\ln\left(\frac{(r_+ + r_-)^2 - 4m^2}{4r_+ r_-}\right),$$
$$r_\pm^2 = \rho^2 + (z \pm m)^2 \quad (20.12)$$

(put $\sigma = m$ in (20.6)). For $\delta = 2$, (20.11) represents the Darmois (1927) solution with the potential

$$U = \ln\left(\frac{r_1 + r_2 - m}{r_1 + r_2 + m}\right), \quad r_{1,2}^2 = \rho^2 + (z \pm m/2)^2 \quad (20.13)$$

(put $\sigma = m/2$ in (20.6)), m being again the mass parameter. For a discussion of this solution, see Fernandez-Jambrina (1994).

The linear superposition of N collinear particles with masses m_A and positions b_A on the z-axis yields the value (Israel and Khan 1964, for two particles see Bach and Weyl 1922)

$$\lim_{\rho\to 0} k = \frac{1}{4}\sum_{A,B=1}^{N} \ln\left[\frac{(r_{A+}r_{B-} + l_{A+}l_{B-})(r_{A-}r_{B+} + l_{A-}l_{B+})}{(r_{A-}r_{B-} + l_{A-}l_{B-})(r_{A+}r_{B+} + l_{A+}l_{B+})}\right],$$
$$l_{A\pm} = z - b_A \pm m_A, \quad r_{A\pm} = |l_{A\pm}| \quad (20.14)$$

for k on the axis $\rho = 0$. The regularity condition $\lim_{\rho\to 0} k = 0$ cannot be satisfied everywhere on the axis outside the sources; the stringlike singularities (or struts) prevent the masses moving towards each other by their mutual gravitational attraction. In Einstein's theory, a two-body system in static equilibrium is impossible without such singularities – a very satisfactory feature of this nonlinear theory.

We notice that a Weyl solution is flat if the potential U is (up to an additive constant) equal to 0, $\ln\rho$ or $\frac{1}{2}\ln(\sqrt{\rho^2+z^2}+z)$. In the last case the Killing vectors are adapted to a boost and a rotation, cf. §17.2.

Waylen (1982) gives the general closed-form solution to $\Delta U = 0$, which includes all Weyl fields with finite values of U on the axis $\rho = 0$, by the integral

$$U(\rho, z) = \frac{1}{\pi}\int_0^{\pi} f(u)\mathrm{d}\Theta\,, \qquad u = z + \mathrm{i}\rho\cos\Theta\,, \tag{20.15}$$

where ρ and z denote Weyl coordinates and f is an arbitrary function of its argument.

A class of vacuum space-times with two commuting Killing vectors $(\partial_\sigma, \partial_\tau)$ was given by Plebański (1980), see also Plebański and García D. (1982). The metric has the diagonal form

$$\mathrm{d}s^2 = \mathcal{A}(q+p)^{1-(b+c)^2}(q-p)^{1-(b-c)^2}\left(\frac{\mathrm{d}q^2}{q^2-1} + \frac{\mathrm{d}p^2}{1-p^2}\right) + \mathcal{B}\mathrm{d}\sigma^2 - \mathcal{C}\mathrm{d}\tau^2,$$

$$\begin{aligned}
\mathcal{A} &= m^2(q+1)^{\alpha^2-1/4}(q-1)^{\beta^2-1/4}(1+p)^{\gamma^2-1/4}(1-p)^{\delta^2-1/4},\\
\mathcal{B} &= m^2(q+1)^{1/2+\alpha}(q-1)^{1/2+\beta}(1+p)^{1/2+\gamma}(1-p)^{1/2+\delta},\\
\mathcal{C} &= m^2(q+1)^{1/2-\alpha}(q-1)^{1/2-\beta}(1+p)^{1/2-\gamma}(1-p)^{1/2-\delta},\\
\alpha &= a+b, \quad \beta = a-b, \quad \gamma = a+c, \quad \delta = a-c,
\end{aligned} \tag{20.16}$$

where a, b, c and m are real constants, and the coordinates (p, q) are restricted to the range $-1 < p < 1 < q < \infty$. This class is in general of non-degenerate Petrov type I and contains as special cases well-known solutions such as the Kasner solution (13.51).

The degenerate static vacuum solutions of classes A and B in Table 18.2 can be transformed to Weyl coordinates and are interpreted by Martins (1996). The C-metric can also be expressed as a Weyl metric and is given by Bonnor (1990) in Bondi coordinates. For the physical interpretation of the C-metric and its generalization, see §17.2. Bonnor and Martins (1991) interpret a particular degenerate Weyl metric as the gravitational field of a (semi-)infinite line mass. The solution for which U is the Newtonian potential of a ring is due to Bach and Weyl (1922).

Because of the linear equation for U one can superpose Weyl solutions. This superposition has been explicitly done e.g. for a particle and a ring or for a gravitational monopole endowed with a quadrupole term similar to (20.9); see Letelier and Oliveira (1988), Gleiser and Pullin (1989), Chakrabarti (1988) and Hernandez P. and Martín (1994). In particular, Szekeres (1968) considered superpositions of monopole–quadrupole particles and could remove the conical singularity between the particles. The required quadrupole moments, however, cannot be constructed from a distribution of positive masses.

The exact superposition of a central static black hole with a surrounding finite thin disk of counterrotating particles has been studied by Lemos and Letelier (1994). Morgan and Morgan (1969) found a disk metric which in terms of oblate ellipsoidal coordinates (ξ, η) connected to (ρ, z) by

$$\rho^2 = a^2(1+\xi^2)(1-\eta^2), \quad z = a\xi\eta, \quad |\eta| \le 1, \ 0 \le \xi < \infty \tag{20.17}$$

$(a = \text{const})$ reads

$$\begin{aligned} U &= -M\left(\operatorname{arccot}\xi + \tfrac{1}{4}[(3\xi^2+1)\operatorname{arccot}\xi - 3\xi](3\eta^2-1)\right)/a, \\ k &= \tfrac{9}{4}M^2\rho^2a^{-4}\left[(\rho/a)^2B^2(\xi) - (1+\eta^2)A^2(\xi) - 2\xi(1-\eta^2)A(\xi)B(\xi)\right], \end{aligned} \tag{20.18}$$

with the notation

$$A(\xi) = \xi \operatorname{arccot}\xi - 1, \quad B(\xi) = [\xi/(1+\xi) - \operatorname{arccot}\xi]/2. \tag{20.19}$$

The disk $(\xi = 0, \ |\eta| \le 1)$ has an outer edge and is characterized by its total mass M and radius a.

Letelier and Oliveira (1998b) found a much simpler form of the Morgan–Morgan solution in terms of Weyl coordinates by the complex substitution $m \to \mathrm{i}\beta$, β real, of the parameter m in (20.12) so that the disk metric is essentially given by

$$U = C\ln\left|\frac{\mathrm{Re}R - \mathrm{i}\beta}{\mathrm{Re}R + \mathrm{i}\beta}\right|, \quad k = -2C^2\ln\left|\frac{(\mathrm{Re}R)^2 + \beta^2}{|R|^2}\right|, \quad R^2 = \rho^2 + (z - \mathrm{i}\beta)^2. \tag{20.20}$$

Properties of counterrotating relativistic disks have been studied by Bičák *et al.* (1993).

According to a procedure due to Ernst (1978b), from a given Weyl metric (U_0, k_0) one can generate another Weyl metric (U, k) as follows:

$$2U = 2U_0 + c(F+G), \quad 2k = 2k_0 + 2cF - c^2\rho^2, \tag{20.21}$$

where c is a constant and the real functions F and G are determined from the original metric by the relations

$$\nabla F = 2\mathrm{i}\nabla U_0, \qquad \nabla G = 2\mathrm{i}\nabla(-U_0 + \ln\rho), \qquad \nabla = \partial_\rho + \mathrm{i}\partial_z . \tag{20.22}$$

This procedure has been applied to studying a black hole in an external gravitational field. The corresponding metric (in Schwarzschild coordinates) is given by (Kerns and Wild 1982a)

$$\begin{aligned} ds^2 = \exp\left[2c(r-3m)\cos\theta - c^2(r^2-2mr)\sin^2\theta\right]\left[\frac{dr^2}{r^2-2mr} + r^2 d\theta^2\right] \\ + r^2\sin^2\theta \exp[-2c(m-r)\cos\theta]d\varphi^2 \\ - (1-2m/r)\exp[2c(m-r)\cos\theta]dt^2 . \end{aligned} \tag{20.23}$$

For the application of Ernst's method to the C-metric, see §17.2. The Weyl solution for a ring immersed in a homogeneous field contains a discontinuity extending from the ring to its center (Hoenselaers 1995). The solution resulting from the Zipoy–Voorhees class of metrics (20.11) as seed has been derived by Kerns and Wild (1982b). Solutions admitting an additional homothetic vector have been discussed by McIntosh (1992).

Finally we remark that some authors, e.g. Peters (1979), Xanthopoulos (1983a), Stewart *et al.* (1987), have dealt with special Weyl metrics describing black holes with toroidal topology. In contrast to the isolated black holes, the (asymptotically flat) gravitational field cannot satisfy the vacuum equations everywhere outside the horizon. The vacuum region around the black hole does not extend to infinity; there is necessarily matter or some other field surrounding the local toroidal black hole.

20.3 The class of solutions $U = U(\omega)$ (Papapetrou's class)

The assumption $U = U(\omega)$ considerably simplifies the problem of finding stationary axisymmetric vacuum solutions. From (19.50) one learns that

$$e^{4U} = -\omega^2 + C_1\omega + C_2, \tag{20.24}$$

and that there is a function $V = V(\omega)$ satisfying the potential equation in flat 3-space,

$$V = \int e^{-4U} d\omega = \int \frac{d\omega}{-\omega^2 + C_1\omega + C_2}, \qquad \Delta V = 0. \tag{20.25}$$

The right-hand side of (20.24) can assume positive values only if $s^2 = C_2 + C_1^2/4 > 0$. Given a solution V of $\Delta V = 0$, one obtains the associated complex potential $\Gamma = e^{2U} + i\,\omega$ from

$$\Gamma = s\left[\text{sech}(sV) + i\tanh(sV)\right]. \tag{20.26}$$

The class $U = U(\omega)$ was discovered by Papapetrou (1953) and given originally in the equivalent form

$$e^{-2U} = \alpha\cosh\Omega_{,z} + \beta\sinh\Omega_{,z}, \quad A = (\alpha^2-\beta^2)^{1/2}\rho\Omega_{,\rho}, \quad \Delta\Omega = 0 \tag{20.27}$$

(α, β constants), in which the metric function A can be obtained from the potential function Ω by pure differentiation. In general, this class of solutions is Petrov type I. The solutions (20.27) are asymptotically well behaved only if a mass term $\sim r^{-1}$ does not occur in e^{2U}. This is an obvious defect of these solutions; whereas the isolated sources have angular momentum, they are without mass.

One well-known member of the Papapetrou class $U = U(\omega)$ is the NUT solution (13.49). In terms of Weyl's canonical coordinates the NUT solution is given by

$$e^{2U} = \frac{(r_+ + r_-)^2 - 4(m^2 + l^2)}{(r_+ + r_- + 2m)^2 + 4l^2}, \quad A = \frac{l}{\sqrt{m^2 + l^2}}(r_+ - r_-),$$
$$e^{2k} = \frac{(r_+ + r_-)^2 - 4(m^2 + l^2)}{4r_+ r_-}, \quad r_\pm^2 = \rho^2 + \left(z \pm \sqrt{m^2 + l^2}\right)^2 \tag{20.28}$$

(Gautreau and Hoffman 1972). For $l = 0$, these metric functions go over into the corresponding expressions (20.12) of the Schwarzschild solution. Thus the NUT solution can be considered as an exterior field of a *rotating* source.

Halilsoy (1992) constructed rotating generalizations ($A \neq 0$ in (20.1)) from the static Chazy–Curzon and Zipoy–Voorhees solutions (20.5) and (20.11), respectively. In particular, the metric functions of the rotating Chazy–Curzon solution read (in Weyl's canonical coordinates with $r^2 = \rho^2 + z^2$ and $p^2 + q^2 = 1$)

$$e^{-2U} = \cosh(2m/r) - p\sinh(2m/r), \quad 2k = -m^2\rho^2/r^4, \quad A = 2qmz/r \tag{20.29}$$

and have been discussed in Halilsoy and Gürtuĝ (1994).

20.4 The class of solutions $S = S(A)$

The substitution (19.49) generates from the Papapetrou class $U = U(\omega)$ another class of stationary vacuum solutions characterized by the functional relationship

$$e^{4S} = \rho^2 e^{-4U} = A^2 + C_1 A + C_2. \tag{20.30}$$

The function $\Psi = \int e^{-4S} dA$ satisfies the potential equation $\Delta\Psi = 0$. One has to consider three distinct cases:

$$\rho^2 e^{-4U} = \begin{cases} A^2 - 1, & h < 0, \\ A^2 + 1, & h > 0, \quad h \equiv 4C_2 - C_1^2, \\ A^2, & h = 0 \end{cases} \tag{20.31}$$

(this classification is related to that given after (17.8)). The subclass $h < 0$ is just Weyl's class: with the aid of a real linear transformation $t' = at+b$, $\varphi' = c\varphi+dt$, the function A can be made equal to zero. The subclass $h > 0$ was discovered by Lewis (1932) and can be transformed to static solutions ($A = 0$) by *complex* linear transformations of the coordinates φ and t (Hoffman 1969a). The last subclass $h = 0$, due to van Stockum (1937), has a line element of the simple form

$$ds^2 = \rho^{-1/2}(d\rho^2 + dz^2) - 2\rho d\varphi dt + \rho\Omega dt^2, \quad \Delta\Omega = 0. \tag{20.32}$$

These solutions are of Petrov type *II*. They admit at most a group G_3, and ∂_φ is a null Killing vector (§24.4).

The cylindrically-symmetric stationary vacuum solutions form a subcase of the class $S = S(A)$ and are given in §22.2.

20.5 The Kerr solution and the Tomimatsu–Sato class

The Kerr and Tomimatsu–Sato solutions possibly describe exterior gravitational fields of stationary rotating axisymmetric isolated sources. However, no satisfactory interior solutions are known. We refer the reader to the review article by Sato (1982) on the Kerr–Tomimatsu–Sato class of vacuum solutions.

The Kerr solution was found by a systematic study of algebraically special vacuum solutions. From its original form (32.47) (Kerr 1963a) the metric can be transformed to Boyer–Lindquist coordinates (r, ϑ) which are related to Weyl's canonical coordinates (ρ, z) and to prolate spheroidal coordinates (x, y) by

$$\rho = \sqrt{r^2 - 2mr + a^2}\sin\vartheta, \quad z = (r - m)\cos\vartheta, \tag{20.33a}$$

$$\sigma x = r - m, \quad y = \cos\vartheta, \quad \sigma = \text{ const} \tag{20.33b}$$

(Boyer and Lindquist 1967). In these coordinates, the Kerr solution reads

$$\begin{aligned} ds^2 = {} & \left(1 - \frac{2mr}{r^2 + a^2\cos^2\vartheta}\right)^{-1} \left[(r^2 - 2mr + a^2)\sin^2\vartheta d\varphi^2 \right. \\ & \left. + \left(r^2 - 2mr + a^2\cos^2\vartheta\right)\left(d\vartheta^2 + \frac{dr^2}{r^2 - 2mr + a^2}\right)\right] \\ & - \left(1 - \frac{2mr}{r^2 + a^2\cos^2\vartheta}\right)\left(dt + \frac{2mar\sin^2\vartheta d\varphi}{r^2 - 2mr + a^2\cos^2\vartheta}\right)^2. \end{aligned} \tag{20.34}$$

Special cases are: $a = 0$ (Schwarzschild solution) and $a = m$ ('extreme' Kerr solution). The form (20.34) exhibits the existence of 2-surfaces orthogonal to the trajectories of the two Killing vectors ∂_t and ∂_φ. The Kerr solution is of Petrov type D and admits a non-trivial Killing tensor (35.3). Starting from (20.34) Ruiz (1986) obtained harmonic coordinates for the Kerr solution. The Kerr solution is characterized by the vanishing of the Simon tensor (18.70) (Simon 1984).

Carter (1968b) generalized the Kerr solution (20.34) to include the Λ-term. Demiański and Newman (1966) constructed a solution which contains the Kerr and NUT metrics as special cases. For further references see Table 21.1.

For the Kerr solution, the complex potential ξ defined in (19.46) takes the very simple form (Ernst 1968a)

$$\xi^{-1} = px - \mathrm{i}qy, \quad p^2 + q^2 = 1, \tag{20.35}$$

if one uses prolate spheroidal coordinates. The full metric can be obtained from ξ as

$$\begin{aligned} \mathrm{e}^{2U} &= \frac{p^2x^2 + q^2y^2 - 1}{(px+1)^2 + q^2y^2}, \quad \mathrm{e}^{2k} = \frac{p^2x^2 + q^2y^2 - 1}{p^2(x^2 - y^2)}, \\ A &= \frac{2mq}{p^2x^2 + q^2y^2 - 1}(1 - y^2)(px + 1) \end{aligned} \tag{20.36}$$

($mq = a$, $mp = \sigma$). The potential ξ in (20.35) is a special solution of the differential equation (19.47),

$$\begin{aligned} \left(\xi\bar{\xi} - 1\right)\left\{\left[(x^2 - 1)\xi_{,x}\right]_{,x} + \left[(1 - y^2)\xi_{,y}\right]_{,y}\right\} \\ = 2\bar{\xi}\left[(x^2 - 1)\xi_{,x}^2 + (1 - y^2)\xi_{,y}^2\right]. \end{aligned} \tag{20.37}$$

Exploiting the very symmetric form of this equation in the variables x and y, Tomimatsu and Sato (1972, 1973) succeeded in constructing a series of new solutions (TS solutions) containing an integer distortion parameter δ. The potential ξ of these solutions is a quotient $\xi = \beta/\alpha$, α and β being polynomials in the coordinates x and y. For $\delta = 1, 2, 3$, these polynomials are ($p^2 + q^2 = 1$)

$$\begin{aligned} \delta = 1: \quad & \alpha = px - \mathrm{i}qy, \quad \beta = 1 \quad \text{(Kerr solution)}. \\ \delta = 2: \quad & \alpha = p^2(x^4 - 1) - 2\mathrm{i}pqxy(x^2 - y^2) + q^2(y^4 - 1), \\ & \beta = 2px(x^2 - 1) - 2\mathrm{i}qy(1 - y^2). \\ \delta = 3: \quad & \alpha = p(x^2 - 1)^3(x^3 + 3x) + \mathrm{i}q(1 - y^2)^3(y^3 + 3y) \\ & \quad -pq^2(x^2 - y^2)^3(x^3 + 3xy^2) - \mathrm{i}p^2q(x^2 - y^2)^3(y^3 + 3x^2y), \\ & \beta = p^2(x^2 - 1)^3(3x^2 + 1) - q^2(1 - y^2)^3(3y^2 + 1) \\ & \quad -12\mathrm{i}pqxy(x^2 - y^2)(x^2 - 1)(1 - y^2). \end{aligned} \tag{20.38}$$

The constant σ in the relation (20.6) between the coordinates (ρ, z) and (x, y), the angular momentum J and the quadrupole moment Q are given by

$$\sigma = mp/\delta, \quad J = m^2 q, \quad Q = m^2[q^2 + (\delta^2 - 1)p^2/3\delta^2], \tag{20.39}$$

m being the mass parameter of the TS family. The TS solutions are asymptotically flat. They stimulated further investigation of the internal symmetries of the field equations which later led to the generation methods (Chapters 10 and 34). The corresponding Weyl solutions ($q = 0$) are the solutions (20.11) for integer parameter δ.

Hori (1996a, 1996b) derived generalized TS solutions which are defined for any real value of the parameter δ and coincide with the TS class for any integral value of δ. Manko and Moreno (1997) considered the complex continuation $p \to \mathrm{i}p$, $\sigma \to \mathrm{i}\sigma$ of the parameters in the TS metrics.

A class closely related to the TS solutions was given by Chandrasekhar (1978) and investigated by Xu (1987). A common property, which characterises the Kerr–Tomimatsu–Sato metrics (written in prolate spheroidal coordinates x and y), is the validity of the relation

$$\Gamma_{,x}\bar{\Gamma}_{,y} + \bar{\Gamma}_{,x}\Gamma_{,y} = 0 \tag{20.40}$$

for the complex Ernst potential Γ.

Some properties of the Kerr–Tomimatsu–Sato family of spinning mass solutions are studied and a closed form of the TS metrics (with an arbitrary positive integer parameter δ) is given by Yamazaki (1982). The calculation of the Weyl tensor invariants for the TS $\delta = 3$ in Hoenselaers (1979b) showed that curvature singularities occur on rings in the equatorial plane. Hoenselaers and Ernst (1983) wrote the TS metrics in a modified form and investigated their behaviour near the poles $(x = 1, \; y = \pm 1)$.

The TS solution with integer δ ($\delta = N$) can be obtained via a limiting process from the non-linear superposition of N Kerr–NUT solutions with common symmetry axis (Tomimatsu and Sato 1981).

20.6 Other solutions

Kinnersley and Chitre (1978b) generated an extended $\delta = 2$ Tomimatsu–Sato solution. The complex potential $\xi = \beta/\alpha$ is a rational function of the prolate spheroidal coordinates (x, y) and is explicitly given by

$$\begin{aligned} \alpha &= p^2(x^4 - 1) - 2\mathrm{i}pqxy(x^2 - y^2) + q^2(y^4 - 1) - 2\mathrm{i}\lambda(x^2 + y^2 - 2x^2y^2) \\ &\quad + (\lambda^2 - \mu^2)(x^2 - y^2)^2 - 2\mathrm{i}\mu xy(x^2 + y^2 - 2), \quad p^2 + q^2 = 1, \\ \beta &= 2px(x^2 - 1) - 2\mathrm{i}qy(1 - y^2) - 2\mathrm{i}(x^2 - y^2)[x(p\lambda + \mathrm{i}q\mu) - y(p\mu + \mathrm{i}q\lambda)]. \end{aligned} \tag{20.41}$$

This solution contains the two additional real parameters λ and μ . The full metric for the Kinnersley–Chitre solution has been given by Yamazaki (1980c). In general it is not invariant under reflection at the equatorial plane. Hoenselaers (1982a) considered a limit of the Kinnersley–Chitre solution and Yamazaki (1980b, 1980a) generalized the Kinnersley–Chitre solution to include an arbitrary positive distortion parameter δ.

The function k in the Kerr metric (20.36) depends only on the coordinate $\eta = (x^2-1)/(1-y^2)$. The fact that $k = k(\eta)$ is a property of the whole class of TS solutions leads to a generalization of this class to arbitrary continuous real parameter δ. Cosgrove (1978a, 1978b) studied the ansatz

$$\begin{aligned} &(1-\nu^2)k_{,\nu} = 2h, \quad 2\eta(1+\eta)k_{,\eta} = l(\eta), \\ &\eta = (x^2-1)/(1-y^2), \qquad \nu = y/x, \quad h = \text{const} \end{aligned} \tag{20.42}$$

for the function k obeying the fourth-order differential equation (19.48) and obtained the ordinary differential equation (see also Dale (1978))

$$\eta^2(1+\eta)^2 l''^2 = 4[\eta l'^2 - ll' - h^2][-(1+\eta)l' + l - \delta^2] \tag{20.43}$$

(δ constant). For $h = 0$, and with the boundary condition $l(\eta) = \delta^2 p^{-2} + O(\eta^{-1})$ as $\eta \to \infty$, equation (20.43) defines asymptotically flat solutions which are regular on the axis outside a finite region. This class contains three parameters, $\sigma = mp/\delta$, q and δ ($p^2 + q^2 = 1$), related to mass m, angular momentum J and quadrupole moment Q according to (20.39), and goes over into the TS class for integer values of δ.

For $h \neq 0$, Cosgrove (1978b) obtained special solutions in closed form. Using the parametrization $\delta^2 = n^2 + 2bn + 2b^2$, $h = b(n+b)$, one finds e.g. for $n = 1$ that

$$\begin{aligned} \mathrm{e}^{2k} &= C^2(x^2-1)^{b^2}(1-y^2)^{b^2}(x-y)^{-(2b+1)^2}(x+y)^{-1} \\ &\qquad\qquad \times[p^2(x^2-1)^{2b+1} - q^2(1-y^2)^{2b+1}], \\ \Gamma &= \left[\frac{(x-1)(1+y)}{(x+1)(1-y)}\right]^b \\ &\quad \times\frac{p^2(x^2-1)^{2b+1} - q^2(1-y^2)^{2b+1} - 2\mathrm{i}pq(x-1)^{2b}(1+y)^{2b}(x+y)}{p^2(x^2-1)^{2b}(x+1)^2 + q^2(1-y^2)^{2b}(1-y)^2} \end{aligned} \tag{20.44}$$

This solution is a $h \neq 0$ generalization of the Kerr solution and is, in general, not asymptotically flat.

Limiting procedures give rise to new solutions. There are several ways of performing the limit $q \to 1$ in the TS solutions. Assuming that the product px remains finite, one always obtains the extreme Kerr metric ($m = a$)

regardless of which value of δ one starts with. Kinnersley and Kelley (1974) took the limit such that px^{2S-1} remains finite and introduced this quantity as a new coordinate. In this way they obtained a new class of solutions,

$$\xi = \frac{(\Theta_+\Theta_-)^{S-1}[\Theta_+{}^S - \Theta_-{}^S] + \mathrm{i}(r/2m)^{2S-1}[\Theta_+{}^{S-1} + \Theta_-{}^{S-1}]}{(r/2m)^{2S-1}[\Theta_+{}^{S-1} - \Theta_-{}^{S-1}] - \mathrm{i}(\Theta_+\Theta_-)^{S-1}[\Theta_+{}^S + \Theta_-{}^S]} \tag{20.45}$$

($\Theta_\pm = 1 \pm \cos\vartheta$, and r, ϑ are spherical coordinates), for all real values of the parameter S. These solutions are not asymptotically flat, except for $S = 1$.

Other limiting procedures lead to the closed-form solution determined by the complex potential (Cosgrove 1978b)

$$\Gamma = r^{2+2c}\frac{\eta^c[p^2\eta^{2c+3} - q^2 + \mathrm{i}pq(2c+3)(1+\eta)\eta^{c+1}]}{(1+\eta)^{2c+2}[p^2\eta^{2c+1} - q^2 - \mathrm{i}pq(2c+1)(1+\eta)\eta^c]} \tag{20.46}$$

with $\eta = \Theta_+/\Theta_-$, and to a rotating version of the Chazy–Curzon solution (20.5) (Cosgrove 1977).

The Ernst equation (19.39) may be separated in some coordinate systems. The ansatz (Ernst 1977)

$$\Gamma = r^k Y_k(\cos\vartheta) \tag{20.47}$$

(no summation) leads to the ordinary differential equation

$$\tfrac{1}{2}(Y_k + \overline{Y}_k)[\sin\vartheta)^{-1}(\sin\vartheta Y_{k,\vartheta})_{,\vartheta} + k(k+1)Y_k] = k^2 Y_k{}^2 + (Y_{k,\vartheta})^2 \tag{20.48}$$

for the complex function $Y_k = Y_k(\cos\vartheta)$. For $k = 2$, the solution is the special case $c = 0$ of (20.46). Tseitlin (1985) integrated the Ernst equation with an ansatz of the form

$$W_{,\zeta}\Gamma_{,\overline{\zeta}} + W_{,\overline{\zeta}}\Gamma_{,\zeta} = 0. \tag{20.49}$$

Marek (1968) found stationary axisymmetric vacuum solutions under the assumption

$$a_{,\zeta} = \rho e^{-2U}h(\rho), \qquad \sqrt{2}\partial_\zeta = \partial_\rho - i\partial_z, \tag{20.50}$$

which corresponds to a separation of the Ernst equation (19.39) in Weyl's canonical coordinates,

$$\Gamma = \mathrm{e}^{mz}R_m(\rho). \tag{20.51}$$

This product ansatz leads to an ordinary differential equation defining Painlevé transcendents. Dodd and Morris (1983) derived the general solution by solving an associated Riemann–Hilbert problem. Particular cases are considered in Léauté and Marcilhacy (1979). For a special range of the parameters the problem reduces to the differential equation

$$H_{,\rho\rho} + \frac{1}{\rho} H_{,\rho} - \left[\frac{C^2}{\rho^2 (1 + \cosh H)^2} + m^2 \right] \sinh H = 0 \tag{20.52}$$

for the real function H from which the Ernst potential Γ can be obtained. Non-trivial solutions of the types (20.47) and (20.51) do not have the desired asymptotic behaviour $\Gamma \to 1$ for $r \to \infty$. Halilsoy and El-Said (1993) and Léauté and Marcilhacy (1982) discussed the separation of the potential $\xi = (1 - \Gamma)/(1 + \Gamma)$ in the coordinates ρ and z. It turns out that the metric function k in (19.21) is independent of z.

Persides and Xanthopoulos (1988) systematically investigated the separation of the field equations and derived from Painlevé transcendents two families of solutions which, away from the symmetry axis, become asymptotically flat. The relation of the Ernst equation with various types of Painlevé equations has also been studied by Calvert and Woodhouse (1996).

Explicit solutions of the Ernst equation can be obtained by exploiting the Lie point symmetries (§10.2), see Fischer (1980) and Pryse (1993).

The procedure developed by Herlt (1978b, 1979) for constructing static axisymmetric Einstein–Maxwell fields (see §21.1) was used by Das (1983) to generate *real* stationary axisymmetric vacuum fields from *complex* solutions Ω of the (modified) Laplace equation (21.5a). The example given by Das is equivalent to the extreme Kerr solution ($a = m$) (Bonanos and Kyriakopoulos 1987).

20.7 Solutions with factor structure

The metric functions in the Kerr and Tomimatsu–Sato solutions are rational functions of the prolate spheroidal coordinates x and y. It is advantageous to introduce coordinates such that $x = \cosh u$, $y = \cos v$. It can then be seen that the numerator of e^{2U} for the Kerr metric, (20.36), factorizes into $p \sinh u \pm q \sin v$, and the other metric functions can also be factorized. Similar properties hold for the Tomimatsu–Sato and Kinnersley–Chitre solutions. Therefore we summarize in this section some facts on solutions of the Ernst equation with a factor structure (for more information see Hoenselaers (1997)).

We use coordinates u and v related to Weyl's canonical coordinates by any pair of the relations

$$\rho = \sin v \begin{pmatrix} \sinh u \\ \exp u \\ \cosh u \end{pmatrix}, \qquad z = \cos v \begin{pmatrix} \cosh u \\ \exp u \\ \sinh u \end{pmatrix}. \tag{20.53}$$

The coordinates (u, v) correspond to prolate spheroidal, spherical or oblate spheroidal coordinates. We write the metric as

$$\mathrm{d}s^2 = B\mathrm{e}^{2k}(\mathrm{d}u^2 + \mathrm{d}v^2)/A + (G\,\mathrm{d}\varphi^2 + 2C\,\mathrm{d}\varphi\,dt - A\,\mathrm{d}t^2)/B, \quad (20.54)$$

where A, B, C and G are *assumed* to be polynomials in $\sin v$, $\cos v$ and in $\sinh u$, $\cosh u$, $\exp u$. In what follows we shall use prolate spheroidal coordinates; it can easily be seen that everything remains valid in the other cases.

In this case A, B, C and G are polynomials in $\cos v$ and $\cosh u$ only because due to the field equations (19.31) being elliptic the solution must be analytic and, if it exists on the axis $\rho = 0$, can depend only on z and ρ^2. The condition on the determinant of the $\varphi - t$-part, the last of the equations (19.19), reads

$$AG + C^2 = \rho^2\,B^2. \quad (20.55)$$

This can be rewritten as

$$AG = (\rho B + C)(\rho B - C). \quad (20.56)$$

As neither A nor G contains a factor $(\rho B + C)$ it follows (Kerr and Wilson 1989) that A and G factorize as well as $(\rho B + C)$ and $(\rho B - C)$. This finally implies that the metric can be written as

$$\mathrm{d}s^2 = B\mathrm{e}^{2k}(\mathrm{d}x^2 + \mathrm{d}y^2)/A + [{\lambda_2}^2(\mu\,\mathrm{d}\varphi + \nu\,\mathrm{d}t)^2 - {\lambda_1}^2(\sigma\,\mathrm{d}\varphi + \tau\,\mathrm{d}t)^2]/B,$$
$$B = \mu\tau - \nu\sigma, \qquad A = {\lambda_1}^2\tau^2 - {\lambda_2}^2\nu^2, \qquad \rho = \lambda_1\,\lambda_2. \quad (20.57)$$

We have either $\lambda_1 = \sinh u$, $\lambda_2 = \sin v$ or $\lambda_1 = 1$, $\lambda_2 = \sinh u\ \sin v$. The quantities μ, ν, σ and τ are polynomials in $\sinh u$, $\cosh u$, $\sin v$ and $\cos v$.

Assuming Ernst's ξ potential (19.46),

$$\xi = \beta/\alpha, \qquad \Gamma = (\alpha - \beta)/(\alpha + \beta), \quad (20.58)$$

to be rational in the trigonometric resp. hyperbolic functions we get the relations

$$A = \alpha\bar{\alpha} - \beta\bar{\beta} = (\lambda_1\tau + \lambda_2\nu)(\lambda_1\tau - \lambda_2\nu) = A_+A_-,$$
$$B = (\alpha + \beta)(\bar{\alpha} + \bar{\beta}), \quad C = {\lambda_2}^2\mu\nu - {\lambda_1}^2\sigma\tau, \quad \rho B \pm C = A_\pm N_\mp. \quad (20.59)$$

Introducing the differential operator

$$\partial_\pm = \partial_u \pm i\partial_v \quad (20.60)$$

one can show (Hoenselaers and Perjés 1990) that there are polynomials K_+, K_- and L_+, L_- such that

$$\alpha\partial_\pm\beta - \beta\partial_\pm\alpha = K_\pm A_\pm\,, \quad \bar{\alpha}\partial_\pm\alpha - \bar{\beta}\partial_\pm\beta = L_\pm A_\pm\,. \tag{20.61}$$

Moreover, it can be shown that

$$\partial_\pm A_\mp = L_\pm\,. \tag{20.62}$$

From (20.61) we get for the derivatives of α and β

$$A_\mp\partial_\pm\alpha = \bar{\beta}K_\pm + \alpha L_\pm, \qquad A_\mp\partial_\pm\beta = \bar{\alpha}K_\pm + \beta L_\pm. \tag{20.63}$$

Using the polynomials $K_\pm$, $L_\pm$ and $A_\pm$ the field equation (19.47) reads

$$A_\pm\partial_\mp K_\pm + \tfrac{1}{2}(A_+K_+\partial_-\rho + A_-K_-\partial_+\rho)/\rho - L_\mp K_\pm = 0. \tag{20.64}$$

This equation comprises the Ernst equation and part of the integrability conditions of (20.61). The remaining integrability conditions are written as

$$A_\pm\partial_+\partial_-A_\pm - \partial_+A_\pm\partial_-A_\pm + K_\mp\bar{K}_\mp = 0. \tag{20.65}$$

It can be shown that the conformal factor e^{2k} in (20.57) is given by

$$e^{2k} = c\, A(\partial_+\rho\,\partial_-\rho)^{-(n/2-1)}, \tag{20.66}$$

where n is the degree of A in $\sinh u$ – it is necessarily even – and c is a constant. From (19.42) one derives the existence of polynomials $P_\pm$ which satisfy

$$\begin{aligned} A_\mp(\alpha+\beta)\partial_\pm\rho + 2\rho(\bar{\alpha}+\bar{\beta})K_\pm &= P_\pm A_\pm \\ A_\mp\partial_\pm N_\mp - N_\mp\partial_\pm A_\mp &= (\bar{\alpha}+\bar{\beta})P_\pm. \end{aligned} \tag{20.67}$$

As an example we give the polynomials $A_\pm$ and $K_\pm$ for the Kinnersley–Chitre solution (20.41),

$$\begin{aligned} A_\pm &= (p^2+l^2-m^2)\sinh^4 u + (q^2+l^2-m^2)\sin^4 v \\ &\quad \mp 2pq\sinh u\sin v\,(\sinh^2 u + \sin^2 v) + 2(l^2-m^2)\sinh^2 u\sin^2 v \\ &\quad \pm 4l\cosh u\sinh u\cos v\sin v \mp 2m\sinh u\sin v\,(\cosh^2 u + \cos^2 v), \\ K_\pm &= -2p\sinh^3 u \pm 2q\sin^3 v \\ &\quad + 2\left[\cosh^2 u + \cos^2 v\right][(ipl+qm)\sinh u \pm (pm+iql)\sin v] \\ &\quad + 4\cosh u\cos v\,[(ql-ipm)\sinh u \mp (pl+iqm)\sin v] \end{aligned} \tag{20.68}$$

$(p, q, m, l = \text{const},\ p^2+q^2=1)$.

It should be noted that the TS solutions are characterized by $K_\pm =$ real.

21

Non-empty stationary axisymmetric solutions

In this chapter we continue the survey of stationary axisymmetric solutions. In §21.1 we give primarily those Einstein–Maxwell fields which were already known before the arrival of the powerful solution-generating techniques treated in Chapters 10 and 34. References to Einstein–Maxwell fields generated by these methods can be found in Chapter 34. Known perfect fluid solutions are listed in §21.2. The stationary *cylindrically-symmetric* fields are contained in §22.2.

21.1 Einstein–Maxwell fields

21.1.1 Electrostatic and magnetostatic solutions

The metric can be cast into the form (19.21), with $A = 0$,

$$ds^2 = e^{-2U}[e^{2k}(d\rho^2 + dz^2) + \rho^2 d\varphi^2] - e^{2U}dt^2, \qquad (21.1)$$

and the Einstein–Maxwell equations reduce to (19.37). Substituting the magnetostatic potential ψ for the potential χ of an electrostatic solution one obtains its magnetostatic counterpart (Bonnor 1954). Thus we need not consider the electro- and the magnetostatic solutions separately.

Assuming the functional dependence $e^{2U} = 1 - 2c\chi + \chi^2$ (cp. (18.60)) of the potentials U and χ plus axial symmetry, Weyl (1917) has given a class of solutions (*Weyl's electrovac class*). (The sources of these solutions have a constant specific charge density, see §18.6.3). For axisymmetric electrostatic Einstein–Maxwell fields the formula (19.34) yields

$$k_{,\zeta} = (\zeta + \bar{\zeta})(U_{,\zeta}^{2} - e^{-2U}\chi_{,\zeta}^{2}), \qquad (21.2)$$

so that for the Weyl solutions the function k can easily be constructed from a solution Y of the potential equation $\Delta Y = 0$, according to

$$k_{,\zeta} = \pm(\zeta + \bar{\zeta})Y_{,\zeta}^{2} \quad (c^2 \neq 1); \quad k_{,\zeta} = 0 \quad (c^2 = 1). \qquad (21.3)$$

The Weyl solutions with $c^2 = 1$ are just the Papapetrou–Majumdar solutions (see §18.7) with axial symmetry. In the sense of (19.50), Weyl's electrovac class corresponds to Papapetrou's class (§20.3). The spherically-symmetric solution in Weyl's electrovac class is the well-known Reissner–Nordström solution (15.21). In Weyl's canonical coordinates, it is given by

$$\begin{aligned} \mathrm{d}s^2 &= \frac{(R+m)^2}{r_+r_-}(\mathrm{d}\rho^2 + \mathrm{d}z^2) + \frac{(R+m)^2}{R^2-d^2}\rho^2\mathrm{d}\varphi^2 - \frac{R^2-d^2}{(R+m)^2}\mathrm{d}t^2, \\ \chi &= \frac{e}{R+m}, \quad R \equiv \tfrac{1}{2}(r_+ + r_-),\ {r_\pm}^2 \equiv \rho^2 + (z \pm d)^2,\ d \equiv (m^2 - e^2)^{1/2} \end{aligned} \tag{21.4}$$

(see e.g. Gautreau *et al.* 1972). The two independent parameters in (21.4) denote the mass (m) and the charge (e) (in geometrical units). Because of the *linear* equation $\Delta Y = 0$ one can superpose, say, two charged Curzon solutions. In this two-body configuration the mass attraction and the repulsion of charges balance to form an equilibrium state without conical singularities on the axis outside the particles provided that the masses m_1 and m_2 and the charges e_1 and e_2 of the two constituents satisfy the Newtonian relation $m_1m_2 = e_1e_2$ (Cooperstock and de la Cruz 1979). The superposition of N Reissner–Nordström solutions on the symmetry axis is discussed in Azuma and Koikawa (1994).

Herlt (1978b) found a new class of solutions which differs from Weyl's class and contains a subclass of asymptotically flat solutions. From every real solution Ω of the linear differential equation

$$\Omega_{,\rho\rho} - \rho^{-1}\Omega_{,\rho} + \Omega_{,zz} = 0 \tag{21.5a}$$

the gravitational and electrostatic potentials are calculated according to the relations

$$\begin{aligned} e^{2U} &= (\Omega^{-1} + G)^2, \quad \chi = \Omega^{-1} - G, \\ G &= \Omega_{,\rho}\left[\rho(\Omega_{,\rho}^2 + \Omega_{,z}^2) - \Omega\Omega_{,\rho}\right]^{-1}. \end{aligned} \tag{21.5b}$$

The metric function k can be obtained from (21.2). The solutions of *Herlt's class* (21.5) are in general of non-degenerate Petrov type I. This class was constructed by an application of generation techniques (Chapter 34) to the complex van Stockum class (20.32). Some special solutions of the class (21.5) are considered by Carminati (1981). To obtain the Bonnor (1966) solution for a mass endowed with a magnetic dipole, the function Ω has to be chosen as

$$\Omega = A(e^{\alpha}r_1 + e^{-\alpha}r_2 + 2m) \tag{21.6}$$

(r_1, r_2 as in (20.13); A, α = real constants). A generalization of this solution to the exterior field of a rotating charged source was given by Neugebauer and Kramer (1969). The asymptotically flat vacuum solution in the class (21.5) is the Darmois solution (20.13). Herlt (1979) generalized his class to include asymptotically flat electrovac solutions which go over to the Schwarzschild solution for vanishing electrostatic field. Within the Herlt framework the superposition of two separated Reissner–Nordström sources leads to an equilibrium condition (existence and regularity of the axis) which depends on their distance (Carminati and Cooperstock 1992).

An asymptotically flat three-parameter static electrovac solution with mass, electric charge and dipole moment was given by Bonnor (1979b). In terms of prolate spheroidal coordinates (x, y) this metric reads

$$\mathrm{d}s^2 = \frac{A^4B^4G^2}{(x^2-y^2)^3}\left(\frac{\mathrm{d}x^2}{x^2-1}+\frac{\mathrm{d}y^2}{1-y^2}\right)+\frac{(x^2-1)(1-y^2)\mathrm{d}\varphi^2}{G^2}-G^2\mathrm{d}t^2, \tag{21.7}$$

where

$$A = cy-x+b, \quad B = cy+x+a, \quad c^2 = 1+ab, \quad G = 1-a/B-b/A \tag{21.8}$$

and the electrostatic potential χ is given by

$$\chi = a/B - b/A\ . \tag{21.9}$$

The three independent parameters are a, b and the constant σ hidden in the definition (20.6). The gravitational and electrostatic potentials are functionally non-related. This is also true for the solutions generated by Das (1980a). Kóta *et al.* (1982) considered the electrovac class of stationary solutions with non-geodesic shearfree eigenrays. Following Bonnor's theorem, see (19.50), Das (1980b) and Lukács (1992) constructed the static electrovac counterparts of the stationary Kinnersley–Chitre vacuum solution (20.41) and the Tomimatsu–Sato solutions (20.38), respectively.

Carminati and Cooperstock (1983) attacked the static axisymmetric electrovac problem by adapting the (orthogonal) independent coordinates to the electrostatic equipotential surfaces.

The static gravitational field of a mass endowed with a magnetic dipole moment can be given by the metric (21.1) written in terms of prolate spheroidal coordinates (x, y) as defined in (20.6). In the asymptotically flat solution due to Gutsunaev and Manko (1987), the metric functions and the magnetostatic potential read

$$\begin{aligned} \mathrm{e}^{2U} &= \frac{x-1}{x+1}\left(\frac{[x^2-y^2+\alpha^2(x^2-1)]^2+4\alpha^2x^2(1-y^2)}{[x^2-y^2+\alpha^2(x^2-1)]^2-4\alpha^2y^2(x^2-1)}\right), \\ \mathrm{e}^{2k} &= \frac{x^2-1}{x^2-y^2}\,\frac{\{[x^2-y^2+\alpha^2(x^2-1)]^2+4\alpha^2x^2(1-y^2)\}^4}{(1+\alpha^2)^8(x^2-y^2)^8}, \\ \psi &= \frac{4\sigma\alpha^3(1-y^2)[2(1+\alpha^2)x^3+(1-3\alpha^2)x^2+y^2+\alpha^2]}{(1+\alpha^2)\{[x^2-y^2+\alpha^2(x^2-1)]^2+4\alpha^2x^2(1-y^2)\}}. \end{aligned} \tag{21.10}$$

For $\alpha = 0$ one gets the Schwarzschild solution. This solution was obtained by means of the 'method of variation of constants' (Gutsunaev and Manko 1988). For generalizations and other examples of this method, see e.g. Castejon-Amenedo and Manko (1990a), Manko (1990), Manko and Khakimov (1991), Krori and Goswami (1992), Manko and Novikov (1992), Gutsunaev and Elsgolts (1993) and the references given there.

The Schwarzschild black hole can be immersed in the Bertotti–Robinson space-time (12.16) describing a spatially homogeneous magnetic (or electric) background field; Alekseev and García D. (1996) give a discussion of the corresponding static electrovac solution and explain how it was constructed by means of the methods treated in Chapter 34.

21.1.2 Type D solutions: A general metric and its limits

The general Einstein–Maxwell type D solution (with a non-null double aligned electromagnetic field, cp. §§26.1–26.2, and including the cosmological constant Λ) is known (Debever *et al.* 1982, 1984, García D. 1984). It admits (at least) a group G_2 with commuting Killing vectors which acts on null or non-null orbits (Debever and McLenaghan 1981). Its various subcases have been widely discussed in the literature; they can all be derived by limiting procedures from the line element (Plebański and Demiański (1976), see also Debever (1971) and García D. and Macias (1998))

$$\begin{aligned} \mathrm{d}s^2 = (1-pq)^{-2}\Big[&(p^2+q^2)\mathrm{d}p^2/X + X(\mathrm{d}\tau + q^2\mathrm{d}\sigma)^2/(p^2+q^2) \\ &+ (p^2+q^2)\mathrm{d}q^2/Y - Y(\mathrm{d}\tau - p^2\mathrm{d}\sigma)^2/(p^2+q^2)\Big], \\ X = X(p) &= (-g^2+\gamma-\Lambda/6) + 2lp - \varepsilon p^2 + 2mp^3 - (e^2+\gamma+\Lambda/6)p^4, \\ Y = Y(q) &= (e^2+\gamma-\Lambda/6) - 2mq + \varepsilon q^2 - 2lq^3 + (g^2-\gamma-\Lambda/6)q^4. \end{aligned} \tag{21.11}$$

To get a Lorentzian signature, X must be positive. The orbits of the Killing vectors ∂_τ and ∂_σ are spacelike for $Y<0$ and timelike for $Y>0$. In the latter case, the vector ∂_τ is timelike in certain ranges of the coordinates, and σ may be interpreted as an azimuthal coordinate: the solutions are stationary and axisymmetric or belong to the class of boost-rotation-symmetric space-times (§17.2). $Y=0$ corresponds to null orbits; in that case, one should first perform a coordinate transformation $\mathrm{d}u = \mathrm{d}\tau + q^2\mathrm{d}q/Y$, $\mathrm{d}\sigma = -\,\mathrm{d}v - \mathrm{d}q/Y$ and then let Y go to zero, thus arriving at the metrics (24.21)–(24.22).

Besides the cosmological constant Λ, (21.11) contains six real parameters: m and l are the mass and the NUT parameter (see (20.28)); γ and

ε are related to the angular momentum per unit mass, a, and the acceleration b; e and g are the electric and magnetic charges. With respect to a complex null tetrad associated with the preferred null directions

$$Y^{-1}(q^2\partial_\tau - \partial_\sigma) \pm \partial_q, \tag{21.12}$$

the only non-vanishing tetrad component of the Weyl tensor is

$$\Psi_2 = -(m + \mathrm{i}\,l)\left(\frac{1-pq}{q+\mathrm{i}p}\right)^3 + (e^2+g^2)\left(\frac{1-pq}{q+\mathrm{i}p}\right)^3 \frac{1+pq}{q-\mathrm{i}p}. \tag{21.13}$$

The space-time is flat if $m = l = 0$, $e = g = 0$, $\Lambda = 0$. The complex potential Φ (formed with respect to the Killing vector ∂_τ) and the complex invariant of the non-null electromagnetic field are given by

$$\Phi = \frac{e+\mathrm{i}g}{q+\mathrm{i}p}, \quad \kappa_0 F^*_{ab}F^{*ab} = \frac{8}{F}\Phi_{,a}\Phi^{,a} = -8(e+\mathrm{i}g)^2\left(\frac{1-pq}{q+\mathrm{i}p}\right)^4. \tag{21.14}$$

Some well-known classes of solutions can be obtained from the general metric (21.11) by appropriate *limiting procedures* (§9.5). We give two examples of such 'contractions' consisting of both a coordinate transformation and a simultaneous redefinition of constants.

Case I: We consider the scale transformation

$$\begin{aligned} &p \to n^{-1}p, \quad q \to n^{-1}q, \quad \sigma \to n^3\sigma, \quad \tau \to n\tau, \\ &m + \mathrm{i}l \to n^{-3}(m+\mathrm{i}l), \quad e + \mathrm{i}g \to n^{-2}(e+\mathrm{i}g), \quad \varepsilon \to n^{-2}\varepsilon, \\ &\gamma - \Lambda/6 \to n^{-4}\gamma, \quad \Lambda \to \Lambda, \end{aligned} \tag{21.15}$$

perform the limit $n \to \infty$ and obtain from (21.11) the metric

$$\begin{aligned} \mathrm{d}s^2 = {} & (p^2+q^2)\mathrm{d}p^2/X + X(\mathrm{d}\tau + q^2\mathrm{d}\sigma)^2/(p^2+q^2) \\ & + (p^2+q^2)\mathrm{d}q^2/Y - Y(\mathrm{d}\tau - p^2\mathrm{d}\sigma)^2/(p^2+q^2), \\ X = {} & \gamma - g^2 + 2lp - \varepsilon p^2 - \Lambda p^4/3, \quad Y = \gamma + e^2 - 2mq + \varepsilon q^2 - \Lambda q^4/3. \end{aligned} \tag{21.16}$$

This metric has been studied in detail by Plebański (1975). It includes the family of Einstein–Maxwell fields found by Carter (1968a, 1968b), who investigated space-times with an Abelian group of motions G_2 in which the Hamilton–Jacobi equation is separable. In addition to (21.16) (with $g = 0$), Carter gives other families which are also limiting cases of the general form (21.11). We remark that, for instance, the Bertotti–Robinson solution (12.16) can be obtained from (21.11) by a contraction procedure (Plebański 1975). Another solution which appears as a limiting case of the general metric (21.11) is the Reissner–Nordström black hole embedded in a uniform external electric field which asymptotically approaches the Bertotti–Robinson solution (Halilsoy and Al-Badawi 1998).

Besides Λ, the metric (21.16) contains five real parameters (ε can be reduced to $\varepsilon = 0, \pm 1$ by a scale transformation (21.15) with finite n). For $\Lambda = 0$, $\varepsilon = 1$ we obtain from (21.16) the metric

$$\begin{aligned} \mathrm{d}s^2 &= (p^2+q^2)(\mathrm{d}\theta^2 + \mathrm{d}q^2/Y) + a^2\sin^2\theta(\mathrm{d}\tau + q^2\mathrm{d}\sigma)^2/(p^2+q^2) \\ &\quad -Y(\mathrm{d}\tau - p^2\mathrm{d}\sigma)^2/(p^2+q^2), \qquad (21.17) \\ p &= l - a\cos\theta,\ \ a^2 = \gamma - g^2 + l^2,\ \ Y = a^2 + e^2 + g^2 - l^2 - 2mq + q^2, \end{aligned}$$

which in turn contains as an important special case ($l = g = 0$) the well-known Kerr–Newman solution, which will be dealt with separately in §21.1.3. The coordinate transformation

$$q = r, \quad p = -a\cos\theta, \quad \sigma = -\varphi/a, \quad \tau = t + a\varphi \qquad (21.18)$$

brings the metric (21.17) into the form (21.24) and the multiple principal null directions (21.12) take the form given in (21.23) (up to normalization).

Case II: We make the substitution

$$\begin{aligned} &p \to n^{-1}p, \quad q \to -nq^{-1}, \quad \sigma \to n^{-1}\sigma, \quad \tau \to n^{-1}\tau, \\ &l \to nl, \quad \varepsilon \to n^2\varepsilon, \quad m \to n^3 m, \quad e + ig \to n^2(e+ig), \qquad (21.19) \\ &\gamma \to \gamma + n^4 g^2, \quad \Lambda \to \Lambda, \end{aligned}$$

in the metric (21.11). Taking the limit $n \to \infty$, we obtain the line element

$$\begin{aligned} \mathrm{d}s^2 &= (p+q)^{-2}[\mathrm{d}p^2/X + \mathrm{d}q^2/Y + X\mathrm{d}\sigma^2 - Y\mathrm{d}\tau^2], \\ X &= (\gamma - \Lambda/6) + 2lp - \varepsilon p^2 + 2mp^3 - (e^2+g^2)p^4, \qquad (21.20) \\ Y &= -(\gamma + \Lambda/6) + 2lq + \varepsilon q^2 + 2mq^3 + (e^2+g^2)q^4. \end{aligned}$$

These space-times are generalizations of the degenerate static vacuum field denoted the C-metric in Table 18.2. We now consider the subcase $\Lambda = 0$, $\varepsilon = 1$, $g = 0$. The parameter l can be put equal to zero by a simple coordinate transformation. Then the transformation ($\gamma = b^2$)

$$q = r^{-1} + bx, \quad p = -bx, \quad \sigma = z/b, \quad \tau = u - \int Q^{-1}\mathrm{d}q, \qquad (21.21)$$

turns the metric (21.20) into the form

$$\begin{aligned} \mathrm{d}s^2 &= r^2(G^{-1}\mathrm{d}x^2 + G\mathrm{d}z^2) - 2\mathrm{d}u\,\mathrm{d}r - 2br^2\mathrm{d}u\,\mathrm{d}x - 2H\mathrm{d}u^2, \\ G &= 1 - x^2 - 2mbx^3 - e^2b^2x^4, \qquad (21.22) \\ 2H &= 1 - 2m/r + e^2/r^2 - b^2r^2G + brG' + 6mbx + 6e^2b^2x^2 - 4be^2x/r. \end{aligned}$$

Table 21.1. Stationary axisymmetric Einstein–Maxwell fields

Only the parameters marked by a cross (×) are different from zero in the corresponding solution.

m	l	a	b	e	g	Λ	References
×	×	×	×	×	×	×	Debever (1971), Plebański and Demiański (1976)
×	×	×	×	×	×		Kinnersley (1969b)
×	×	×		×	×	×	Plebański (1975)
×	×	×		×	×		Demiański and Newman (1966)
×	×	×		×		×	
×	×		×	×		×	Carter (1968a, 1968b) (applies to these three rows)
×	×		×	×			
×	×			×			Brill (1964)
×		×		×			Newman *et al.* (1965)
				×	×		Bertotti (1959), Robinson (1959)
×				×			Reissner (1916), Nordström (1918)

This is the gravitational field generated by two uniformly accelerated charged mass points, the parameter b being the acceleration parameter (Kinnersley and Walker 1970, Walker and Kinnersley 1972, Plebański and Demiański 1976). Note that b may be put equal to zero in (21.22) but not in (21.20), because the transformation (21.21) involves b explicitly. Putting $b = 0$ in (21.22) we get the Reissner–Nordström solution in terms of the retarded time coordinate u (cp. (15.17) and Table 15.1). The singularity between the sources is removed in a more general metric (Ernst 1976), which contains an additional parameter describing an electric field which causes the uniform acceleration of the charges.

A particular limit of the general metric (21.11) can be considered as a superposition of the Schwarzschild and Bertotti–Robinson solutions (Halilsoy 1993a).

In Table 21.1, taken from Plebański and Demiański (1976), special cases of the metric (21.11) which had been given earlier in the literature are listed.

21.1.3 The Kerr–Newman solution

The Kerr–Newman solution is a special case of the type D solutions discussed in §21.1.2. Newman *et al.* (1965) found it by applying a complex substitution (see §21.1.4) to the preferred complex null tetrad of the Kerr solution. They obtained the new complex null tetrad

$$k^i = (1, 0, a/\Delta, (r^2+a^2)/\Delta),\ \ l^i = \tfrac{1}{2}(r^2+a^2\cos^2\theta)^{-1}(-\Delta, 0, a, r^2+a^2),$$
$$m^i = 2^{-1/2}(0, 1, \mathrm{i}/\sin\theta, \mathrm{i}a\sin\theta)/(r+\mathrm{i}a\cos\theta),\ \ \Delta \equiv r^2+a^2+e^2-2mr \tag{21.23}$$

($x^1 = r$, $x^2 = \theta$, $x^3 = \varphi$, $x^4 = t$), which represents a solution of the Einstein–Maxwell equations. The corresponding metric is given by

$$\begin{aligned} \mathrm{d}s^2 = {} & (r^2+a^2\cos^2\theta)\left(\frac{\mathrm{d}r^2}{\Delta}+\mathrm{d}\theta^2\right) - \left(1-\frac{2mr-e^2}{r^2+a^2\cos^2\theta}\right)\mathrm{d}t^2 \\ & + \sin^2\theta\left[r^2+a^2+\frac{a^2\sin^2\theta}{r^2+a^2\cos^2\theta}(2mr-e^2)\right]\mathrm{d}\varphi^2 \\ & - \frac{2a\sin^2\theta}{r^2+a^2\cos^2\theta}(2mr-e^2)\mathrm{d}\varphi\mathrm{d}t. \end{aligned} \tag{21.24}$$

For $e = 0$, it goes over into the Kerr metric (20.34). The Kerr–Newman solution (21.24) may describe the exterior gravitational field of a rotating charged source and contains three real parameters: m (mass), e (charge) and a (angular momentum per unit mass).

The only non-vanishing components of the Weyl and Maxwell tensors, with respect to the null tetrad (21.23), are

$$\Psi_2 = -\frac{m(r+\mathrm{i}a\cos\theta)-e^2}{(r-\mathrm{i}a\cos\theta)^3(r+\mathrm{i}a\cos\theta)},\quad \sqrt{\frac{\kappa_0}{2}}\Phi_1 = \frac{e}{2(r-\mathrm{i}a\cos\theta)^2}. \tag{21.25}$$

In the metric (21.24) the complex scalar potentials Φ and $\mathcal{E}$, with respect to the Killing vector $\boldsymbol{\xi} = \partial_t$, are given by

$$\Phi = \frac{e}{r-\mathrm{i}a\cos\theta},\quad \mathcal{E} = 1-\frac{2m}{r-\mathrm{i}a\cos\theta}. \tag{21.26}$$

$\mathcal{E}$ is a linear function of Φ. As can be seen from Φ, the magnetic dipole moment vanishes if the source is either non-rotating ($a = 0$, Reissner–Nordström solution) or uncharged ($e = 0$, Kerr solution).

Ernst and Wild (1976) applied the Harrison transformation (34.12)

$$\mathcal{E}' = \Lambda^{-1}\mathcal{E},\ \ \Phi' = \Lambda^{-1}(\Phi - B\mathcal{E}),\ \ \Lambda = 1-2B\Phi+B^2\mathcal{E} \tag{21.27}$$

(B being a real parameter) to obtain the solution for a Kerr–Newman black hole immersed in a homogeneous magnetic field. Note that the Ernst potentials in (21.27) are formed with respect to the Killing vector $\boldsymbol{\eta} = \partial_\varphi$. For a discussion of the Ernst–Wild solution and further references, see Aliev and Gal'tsov (1989a) and the review article Aliev and Gal'tsov (1989b). Another generalization of the Kerr–Newman solution (with quadrupole deformation of the mass) was derived by Denisova *et al.* (1991).

The Kerr–Newman solution with $|e| = m$ is conformastationary (see §18.7) and can be derived from the solution

$$V = \left(1 - \frac{m}{r - \mathrm{i}a\cos\theta}\right)^{-1} = 1 + \frac{m}{R}, \quad R^2 = \rho^2 + (z - \mathrm{i}a)^2,$$
$$\rho = \sin\theta\sqrt{(r-m)^2 + a^2}, \quad z = (r-m)\cos\theta, \tag{21.28}$$

of the potential equation $\Delta V = 0$, see (18.73). The *superposition* of N collinear Kerr–Newman sources for which the gravitational attraction and the electrostatic repulsion are balanced ($|e_A| = m_A$, $A = 1, \ldots, N$) and for which the spins are parallel or antiparallel along the axis of symmetry leads to the expression

$$V = 1 + \sum_{A=1}^{N} \frac{m_A}{R_A}, \quad {R_A}^2 \equiv \rho^2 + (z - l_A)^2. \tag{21.29}$$

The real and imaginary parts of l_A give respectively the position of the source A on the z-axis and the angular momentum per unit mass. The explicit metric for $N = 2$ has been obtained by Parker *et al.* (1973) and Kobiske and Parker (1974). The regularity condition (18.74) implies $\mathrm{Im}\,[m_1 + m_1\overline{m}_2(l_1 - \overline{l}_2)^{-1}] = 0 = \mathrm{Im}\,(m_1 + m_2)$. For two equal sources with oppositely directed spins ($m_1 = m_2$ real, $l_1 = -l_2$), no singularities along the axis between the particles occur. In general, this superposition (21.29) of Kerr–Newman solutions with $|e| = m$ gives rise to naked singularities (Hartle and Hawking 1972); black hole metrics necessarily belong to the *static* subclass (l_A real) discovered by Papapetrou (1947) and Majumdar (1947), see §18.7.2. This subclass can be generalized to include in addition the cosmological constant Λ (Kastor and Traschen 1993).

The solution given by Bonnor (2000) adapts to cosmology the Papapetrou–Majumdar metric. The line element of Bonnor's solution reads

$$\mathrm{d}s^2 = X^{-1}R(t)^2[\mathrm{d}r^2 + \Sigma^2(\mathrm{d}\vartheta^2 + \sin^2\vartheta\mathrm{d}\varphi^2] - X\mathrm{d}t^2,$$
$$X = (1 + U/R(t))^{-2}, \qquad U = U(r, \vartheta, \varphi), \tag{21.30}$$

where $\Sigma = \sin r$, r or $\sinh r$ for positive, zero or negative spatial curvature of the background FRW cosmology. It is a generalization of the solution derived in Kastor and Traschen (1993). Assuming that charge is convected with matter, one obtains outside the charges the *linear* field equation $\Delta U = 0$, where Δ denotes the Laplacian with respect to the 3-metric $\mathrm{d}r^2 + \Sigma^2 \mathrm{d}\Omega^2$ in (21.30). Bonnor (2000) discusses the non-radiative character of this Einstein–Maxwell field (with charged matter).

Examples of *axisymmetric* conformastationary (§18.7.2) Einstein–Maxwell fields are given by Chatterjee and Banerji (1979). Electrovac generalizations of the Tomimatsu–Sato solutions are derived in Ernst (1973), see also Panov (1979d).

21.1.4 Complexification and the Newman–Janis 'complex trick'

A few new metrics were discovered by a formal procedure which can roughly be described as follows: a given metric is first complexified and then a complex coordinate transformation is performed in such a way that the result is a new real metric.

Such a 'complex trick' was introduced by Newman and Janis (1965). It was subsequently justified by Talbot (1969) for algebraically special vacuum metrics (27.27) for which $P_{,u} = L_{,u} = \Sigma_{,u} = 0$ and $m + \mathrm{i}M = cu + a + ib$ for constants a, b and c; thus it applies to some Robinson–Trautman metrics and to the metrics discussed in §29.2.1. This can be extended to electrovacuum (see also Quevedo (1992)) and to perfect fluids.

Newman and Janis (1965) used the method to obtain the Kerr solution from the Schwarzschild solution. Newman *et al.* (1965) derived a new solution of the Einstein–Maxwell equations from the Reissner–Nordström metric

$$\mathrm{d}s^2 = r^2(\mathrm{d}\vartheta^2 + \sin^2\vartheta\,\mathrm{d}\varphi^2) - 2\mathrm{d}u\,\mathrm{d}r - (1 - 2m/r + e^2/r^2)\mathrm{d}u^2. \quad (21.31)$$

The radial coordinate $x^1 = r$ and the retarded time $x^4 = u$ are allowed to take complex values and the null tetrad is formally replaced by the expressions

$$\begin{aligned} &\boldsymbol{k} = \partial_r, \quad \boldsymbol{l} = \partial_u - \tfrac{1}{2}M\,\partial_r, \quad \boldsymbol{m} = (\partial_\vartheta + i\,\mathrm{cosec}\,\vartheta\,\partial_\varphi)/\sqrt{2}\bar{r}, \\ &M = M(r,\bar{r}) = 1 - m/r - m/\bar{r} + e^2/r\bar{r}. \end{aligned} \quad (21.32)$$

For real values of the coordinate r, (21.32) is a null tetrad for the metric (21.31). After the *complex* coordinate transformation

$$r' = r + \mathrm{i}a\cos\vartheta, \quad u' = u - \mathrm{i}a\cos\vartheta, \quad (21.33)$$

one obtains from (21.32) the new tetrad components

$$\begin{aligned} &\boldsymbol{k}' = \partial_r, \quad \boldsymbol{l}' = \partial_u - \frac{1}{2}\left(1 - \frac{2mr' - e^2}{r'^2 + a^2\cos^2\vartheta}\right)\partial_r, \\ &\boldsymbol{m}' = [\partial_\vartheta + \mathrm{i}\,\mathrm{cosec}\,\vartheta\,\partial_\varphi + \mathrm{i}a\sin\vartheta(\partial_u - \partial_r)]/\sqrt{2}(r' + \mathrm{i}a\cos\vartheta). \end{aligned} \quad (21.34)$$

For real values of the coordinates r', u', the associated metric is the Kerr–Newman metric (21.24).

Demiański (1972) found the most general vacuum solution which results from the complexified null tetrad (21.32), $M = \bar{M} = M(r, \bar{r})$ still being unspecified, when the complex coordinate transformation $r' = r + \mathrm{i}F(\vartheta, \varphi)$, $u' = u + \mathrm{i}G(\vartheta, \varphi)$, $\vartheta' = \vartheta$, $\varphi' = \varphi$ (F and G being real functions of their real arguments ϑ and φ) is performed and the new coordinates r', U', are restricted to being real. The resulting solution is given by (29.62): in general it is of Petrov type II.

Twisting type D vacuum metrics (§29.5) may be obtained from corresponding non-twisting metrics by the Newman–Janis method (Basey 1975).

The original generation of the Kerr solution from Schwarzschild (Newman and Janis 1965) is obtained by putting $e = 0$ in (21.31) and (21.32). The Kerr solution can, however, be considered to be a complexification of the Schwarzschild solution in quite another sense. Some Kerr–Schild vacuum metrics (§32.2) can be computed from a (complex) generating potential γ which simultaneously satisfies the two equations $\Delta\gamma = 0$, $(\nabla\gamma)^2 = \gamma^4$ in flat 3-space. The solution $\gamma = 1/r = 1/\sqrt{x^2 + y^2 + z^2}$ of these equations yields the Schwarzschild solution; the Kerr solution results from $\gamma = 1/r$ by an imaginary translation of the origin, $z \to z - \mathrm{i}a$ (Schiffer *et al.* 1973). For a corresponding treatment of Einstein–Maxwell fields, see Finkelstein (1975), Collins (1976).

21.1.5 Other solutions

Solutions of the Ernst equations (19.28)–(19.29) when the complex potentials $\mathcal{E}$ and Φ depend only on the similarity variable z/ρ (in Weyl coordinates) are given in Kaliappan and Lakshmanan (1981). The solutions which have been found by the method of separation of variables depend on the fifth Painlevé transcendent (Léauté and Marcilhacy 1984, Halilsoy 1985).

Starting with the Kerr–Newman solution as seed metric, Hiscock (1981) and Aliev *et al.* (1980) constructed the Einstein–Maxwell field of a rotating charged black hole in a strong magnetic field. For the generation method used in these papers, and for other applications of these methods in General Relativity, we refer the reader to Chapters 10 and 34.

With the aid of their differential form approach for stationary axisymmetric Einstein–Maxwell fields, Fernandez-Jambrina and Chinea (1994) derived a solution with the metric

$$\mathrm{d}s^2 = \rho^{-2/3}\left[\rho^2\mathrm{d}\varphi^2 + \mathrm{e}^{2k}(\mathrm{d}\rho^2 + \mathrm{d}z^2)\right] - \rho^{2/3}(\mathrm{d}t - A\mathrm{d}\varphi)^2, \qquad (21.35)$$

where A satisfies the linear equation

$$A_{,\rho\rho} + (1/3\rho)A_{,\rho} + A_{,zz} = 0 \tag{21.36}$$

and k is determined by a line integral from A.

Perjés (1993) found a class of Einstein–Maxwell fields without a functional relationship $\Phi = \Phi(\mathcal{E})$ between the complex Ernst potentials. The field equations lead to the key equation

$$\mu(u)\left[(1+\sigma^2)\alpha + (1+u^2)\alpha^{-1}\right]_{,\sigma\sigma} + \lambda(\sigma)\alpha_{,uu} = 0 \tag{21.37}$$

for $\alpha = \alpha(u,\sigma)$, where the functions $\mu = \mu(u)$ and $\lambda = \lambda(\sigma)$ are arbitrary. In terms of the real space-time coordinates u and σ the Ernst potentials are given by

$$\mathcal{E} = \frac{1-(1+\mathrm{i}\sigma)\alpha}{1+(1+\mathrm{i}\sigma)\alpha}, \qquad \Phi = \frac{u}{1+(1+\mathrm{i}\sigma)\alpha}. \tag{21.38}$$

For one branch of solutions one is led to an ordinary third-order differential equation. The metric and the electromagnetic field corresponding to the particular solution

$$\alpha = a\sqrt{\frac{1+u^2}{1+\sigma^2}}, \quad \lambda = \frac{1+1/a^2}{1+\sigma^2}, \quad \mu = -(1+u^2)^{-2}, \quad a = \text{ const} \tag{21.39}$$

of (21.37) are calculated in Perjés and Kramer (1996).

Das and Chaudhuri (1993) generated stationary axisymmetric Einstein–Maxwell fields from solutions of the Laplace equation as seed.

All Einstein–Maxwell fields considered in this chapter have a non-null electromagnetic field. In the null case, and for pure radiation fields, the stationary axisymmetric solutions admit a null Killing vector (§24.4) (Gürses 1977).

21.2 Perfect fluid solutions

21.2.1 Line element and general properties

In most rotating perfect fluid solutions the four-velocity of the fluid obeys the circularity condition (19.13) so that the four-velocity is a linear combination of the two Killing vectors:

$$\begin{aligned} &u^{[a}\xi^b_A\xi^{c]}_B = 0, \quad u^a = (-H)^{-1/2}(\xi^a + \Omega\eta^a) = (-H)^{-1/2}S^A\xi^a_A, \\ &H \equiv \lambda_{AB}S^AS^B, \quad \lambda_{AB} \equiv \xi^a_A\xi_{Ba}, \quad S^A \equiv (1,\Omega). \end{aligned} \tag{21.40}$$

Then there exist 2-surfaces orthogonal to the group orbits (§19.2) and the metric can be written as

$$\begin{aligned} ds^2 &= e^{-2U}\left[e^{2k}(d\rho^2+dz^2)+W^2d\varphi^2\right]-e^{2U}(dt+Ad\varphi)^2, \\ H &= W^2\Omega^2e^{-2U}-(1+A\Omega)^2e^{2U}. \end{aligned} \tag{21.41}$$

The velocity field is necessarily expansionfree, $u^a{}_{;a}=0$. If the angular velocity Ω of the system (with respect to infinity) is a constant, then the system rotates rigidly, i.e. the fluid is in shearfree and expansionfree motion,

$$\sigma = 0 = \Theta \quad \leftrightarrow \quad u_{(a;b)}+u_{(a}\dot{u}_{b)}=0. \tag{21.42}$$

The general case ($\Omega_{,a}\neq 0$) is that of differential rotation. The fluid's rotation $\omega^a=\varepsilon^{abcd}u_{b;c}u_d$ vanishes if $\Omega = A/(W^2e^{-4U}-A^2)$ (up to linear φ–t-transformations, or a redefinition $t\leftrightarrow\varphi$ which leads to $\Omega=-1/A$ and a negative e^{2U}). For *static* axisymmetric solutions see also §18.6.4.

In the metric (21.41), the *regularity condition* (19.3) takes the form

$$\lim_{\rho\to 0}\,[\rho^{-1}e^{U-k}(e^{-2U}W^2-e^{2U}A^2)^{1/2}]=1. \tag{21.43}$$

A perfect fluid solution *not* satisfying the circularity condition is provided by (33.13). Mars and Senovilla (1998) investigated in detail the problem of matching a given interior perfect fluid solution and an exterior vacuum solution which turns out to be unique.

For a review of perfect fluid solutions see also Senovilla (1993).

21.2.2 The general dust solution

All stationary axisymmetric (rotating) dust metrics satisfying condition (21.40) are known up to quadratures.

For *differentially rotating dust* ($\Omega_{,a}\neq 0$), we proceed as follows (Winicour 1975). We substitute $R_{ab}=\kappa_0\mu(u_au_b+\frac{1}{2}g_{ab})$ into (19.25),

$$\begin{aligned} D^a(W^{-1}\lambda_{AB,a}) &= \kappa_0\mu W^{-1}(\lambda_{AB}+2W^2H^{-1}S_AS_B) \\ &\quad +\tfrac{1}{2}W^{-3}\lambda_{AB}\lambda^{CD,a}\lambda_{CD,a}, \quad 2W^2\equiv -\lambda^{AB}\lambda_{AB}, \end{aligned} \tag{21.44}$$

and take into account the relation

$$T^{ab}{}_{;b}=0 \quad \rightarrow \quad S^AS^B\lambda_{AB,a}=0. \tag{21.45}$$

The last equation, if rewritten in the form

$$H_{,a}=2\eta\Omega_{,a}, \quad \eta\equiv\lambda_{12}+\Omega\lambda_{22}, \tag{21.46}$$

shows that for differential rotation both η and Ω are functions of H. The arbitrary function $\eta = \eta(H)$ determines an auxiliary function $\beta = \beta(H)$ defined by

$$\beta_{,a} \equiv H_{,a}/(H\eta). \tag{21.47}$$

From the field equations (21.44) we obtain the relations

$$W^{-1}[(\beta W^2)_{,a} + (H/\eta)(\eta^2/H)_{,a}] = \varepsilon_{ab}\gamma^{,b} \;\rightarrow\; \Delta\gamma = 0, \tag{21.48}$$

$$D^a W_{,a} = 0, \tag{21.49}$$

and an expression for the mass density μ. Because of the two-dimensional Laplace equation (21.49) for W we can choose $W = \rho$ in the metric (21.41). The choice $W = 1$ is also possible, but does not allow us to interpret φ as an azimuthal coordinate. The remaining field equations obtained from (19.24)–(19.26) either determine the conformal factor e^{2k} in the line element (21.41) or are satisfied in consequence of (21.44)–(21.45).

In order to construct dust metrics, we have to choose (i) a function $\eta = \eta(H)$ and (ii) an axisymmetric solution γ of the potential equation $\Delta\gamma = 0$ in flat 3-space. Once $\gamma = \gamma(\rho, z)$ and $\eta = \eta(H)$ are given, one obtains $\beta = \beta(H)$ from (21.47), the function $\beta W^2 + 2\eta - \int \eta H^{-1}\mathrm{d}H$ and consequently $H = H(\rho, z)$ from (21.48), and finally the angular velocity Ω from (21.46). Hence, the scalar products of the Killing vectors are completely known:

$$\begin{aligned}
\lambda_{11} &= \xi^a\xi_a = g_{44} = H^{-1}[(H - \eta\Omega)^2 - \Omega^2\rho^2] = -e^{2U}, \\
\lambda_{12} &= \xi^a\eta_a = g_{34} = H^{-1}\Omega(\rho^2 - \eta^2) + \eta = -Ae^{2U}, \\
\lambda_{22} &= \eta^a\eta_a = g_{33} = -H^{-1}(\rho^2 - \eta^2) = e^{-2U}\rho^2 - A^2e^{2U}.
\end{aligned} \tag{21.50}$$

The conformal factor e^{2k} can be determined by means of a line integral and the mass density is given by the formula

$$4\kappa_0\mu = \eta^{-2}[H^2\rho^{-2}(\eta^2/H)_{,a}(\eta^2/H)^{,a} - \rho^2H^{-2}H_{,a}H^{,a}]. \tag{21.51}$$

The general class of dust solutions described above contains the cylindrically-symmetric dust solutions (§22.2).

For *rigidly rotating dust* ($\Omega = \Omega_0 = \text{const}$), one can either start from the special case $H = -(1 + p^2\eta^2)$, $\Omega = \Omega_0 - p^2\eta$ of the above class and then take the limit $p \rightarrow 0$ (Vishveshwara and Winicour 1977), or directly attack the field equations (21.54). With $\sqrt{2}\zeta = \rho + \mathrm{i}\,z$, $W = \rho$, $U = 0$, they read (Lanczos 1924, van Stockum 1937)

$$\begin{aligned}
&\Delta A = 0 = 2(\zeta + \bar\zeta)A_{,\zeta\bar\zeta} - A_{,\zeta} - A_{,\bar\zeta}, \\
&2k_{,\zeta} = -(\zeta + \bar\zeta)^{-1}(A_{,\zeta})^2, \quad \kappa_0\mu = 2\rho^{-2}A_{,\zeta}A_{,\bar\zeta}e^{-2k} = 2u_{[a;b]}u^{a;b}.
\end{aligned} \tag{21.52}$$

The mass density is positive definite. A special member of this class was analysed by Bonnor (1977) with the surprising result that, contrary to Newtonian mechanics, a density gradient in the z-direction (axis of rotation) occurs. (For $r \to \infty$, the mass density tends very rapidly to zero.) A closed form solution of (21.52) has been discussed by Islam (1983a).

All stationary rigidly rotating dust metrics with geodesic and/or shear-free eigenrays (§18.5) are given in Lukács (1974). They are axially-symmetric and belong to the Lanczos–van Stockum class (21.52). For a solution whose source can be interpreted as a rigidly rotating disk of dust (Neugebauer and Meinel 1995) see Chapter 34.

Comparing (21.52) with (20.3) for Weyl's vacuum solutions, we conclude that there exists a one-to-one correspondence between static axisymmetric vacuum solutions and stationary axisymmetric dust solutions. This statement is a special case of the more general

Theorem 21.1 *To every static vacuum solution (metric g_{ij}, Killing vector ξ_k) a (rigidly rotating) stationary dust solution ($\hat{g}_{ij}, u_i, \mu$) can be assigned by*

$$\begin{aligned} \hat{g}_{ij} &= -\xi_m\xi^m g_{ij} + \xi_i\xi_j - u_i u_j, \quad u_i\xi^i = -1, \quad \mathrm{e}^{2U} = -\xi^i\xi_i, \\ u_{[i;j]} &= \varepsilon_{ijkl}U^{,k}\xi^l, \quad u_{(i;j)} = 0, \quad \kappa_0\mu = 4U_{,i}U^{,i}. \end{aligned} \tag{21.53}$$

(Ehlers 1962).

21.2.3 Rigidly rotating perfect fluid solutions

For rigid rotation (Ω = const in (21.40)), there exists a (timelike) Killing vector (linear combination of $\boldsymbol{\xi}$ and $\boldsymbol{\eta}$ with constant coefficients) parallel to the four-velocity of the fluid. We can identify this Killing vector with $\boldsymbol{\xi} = \partial_t$ (comoving system, with $\Omega = 0$).

In a comoving coordinate system (21.41), the field equations (18.25)–(18.28) for the energy-momentum tensor of a rigidly rotating perfect fluid read (with $\sqrt{2}\zeta = \rho + \mathrm{i}\,z$)

$$W_{,\zeta\bar{\zeta}} = \kappa_0 p W \mathrm{e}^{2k-2U}, \tag{21.54a}$$

$$\begin{aligned} &U_{,\zeta\bar{\zeta}} + (2W)^{-1}(U_{,\zeta}W_{,\bar{\zeta}} + U_{,\bar{\zeta}}W_{,\zeta}) + \tfrac{1}{2}W^{-2}\mathrm{e}^{4U}A_{,\zeta}A_{,\bar{\zeta}} \\ &\quad = \kappa_0(\mu + 3p)\mathrm{e}^{2k-2U}/4, \end{aligned} \tag{21.54b}$$

$$A_{,\zeta\bar{\zeta}} - (2W)^{-1}(A_{,\zeta}W_{,\bar{\zeta}} + A_{,\bar{\zeta}}W_{,\zeta}) + 2(A_{,\zeta}U_{,\bar{\zeta}} + A_{,\bar{\zeta}}U_{,\zeta}) = 0, \tag{21.54c}$$

$$2W_{,\zeta}k_{,\zeta} = W_{,\zeta\zeta} + 2W(U_{,\zeta})^2 - (2W)^{-1}\mathrm{e}^{4U}(A_{,\zeta})^2, \tag{21.54d}$$

$$k_{,\zeta\bar{\zeta}} + U_{,\zeta}U_{,\bar{\zeta}} + (2W)^{-2}\mathrm{e}^{4U}A_{,\zeta}A_{,\bar{\zeta}} = \kappa_0 p \mathrm{e}^{2k-2U}/2. \tag{21.54e}$$

The conservation law

$$T^{ab}{}_{;a} = 0 \quad \rightarrow \quad p_{,\zeta} + (\mu + p)U_{,\zeta} = 0 \tag{21.55}$$

is a consequence of the field equations (21.54). Conversely, (21.54*e*) follows from (21.55) (Trümper 1967). For a different form of the field equations see Bonanos and Sklavenites (1985).

Many of the known rigidly rotating solutions have an equation of state (Senovilla 1993)

$$\mu + 3p = \text{const} = \mu_0, \tag{21.56}$$

and most of them are subcases of the Wahlquist (1968, 1992) solution. This solution is given in terms of generalized oblate-spheroidal coordinates (ξ, η, φ) by

$$\begin{aligned} ds^2 &= r_0^2(\xi^2+\eta^2)\left[\frac{d\xi^2}{(1-k^2\xi^2)h_1} + \frac{d\eta^2}{(1+k^2\eta^2)h_2} + \frac{c^2 h_1 h_2}{h_1 - h_2} d\varphi^2\right] \\ &\quad - e^{2U}(dt + A d\varphi)^2, \\ e^{2U} &= \frac{h_1 - h_2}{\xi^2+\eta^2}, \quad A = c r_0 \left(\frac{\xi^2 h_2 + \eta^2 h_1}{h_1 - h_2} - \eta_0^2\right), \\ h_1 &= h_1(\xi) = 1 + \xi^2 - 2m r_0^{-1}\xi(1-k^2\xi^2)^{1/2} \\ &\quad + \xi b^{-2}[\xi - k^{-1}(1-k^2\xi^2)^{1/2}\arcsin(k\xi)], \\ h_2 &= h_2(\eta) = 1 - \eta^2 - 2a r_0^{-1}\eta(1+k^2\eta^2)^{1/2} \\ &\quad - \eta b^{-2}[\eta - k^{-1}(1+k^2\eta^2)^{1/2}\operatorname{arcsinh}(k\eta)], \\ h_2(\eta_0) &= 0, \quad c^{-1} = \tfrac{1}{2}(1+k^2\eta_0^2)^{1/2} dh_2/d\eta|_{\eta=\eta_0}, \\ p &= \tfrac{1}{2}\mu_0(1-b^2e^{2U}), \quad \mu = \tfrac{1}{2}\mu_0(3b^2e^{2U}-1), \quad \kappa_0\mu_0 = 2k^2(br_0)^{-2}. \end{aligned} \tag{21.57}$$

The real constants r_0, k, b, m, a are arbitrary and the constants η_0, c are adjusted so that the solution behaves properly on the axis, i.e. so that the metric satisfies the regularity condition (21.43). For $m = a = 0$, the solution (21.57) is a singularityfree interior solution for a rigidly rotating fluid body bounded by a finite surface of zero pressure. For $m = a = 0$, $\xi = r/r_0$, $\eta = \cos\theta$ and in the limit $r_0 \rightarrow 0$ we obtain from (21.57) a spherically-symmetric solution (§16.1.2) with $\mu + 3p = $ const. For different forms of the Wahlquist solution, see Senovilla (1993) and Rácz and Zsigrai (1996).

Wahlquist's solution (21.57) is of Petrov type D, the four-velocity not being spanned by the two null eigenvectors. Conversely it can be shown (Senovilla 1987c, 1993) that the general Petrov type D solution with an equation of state $\mu + 3p$ = const is the Wahlquist solution or one of its limits. Solutions known to be among these limits are: (i) the solution

$$
\begin{aligned}
ds^2 &= M^2[(1 - a^2 \sin^2 \alpha/R^2)^{-1} d\alpha^2 + \sin^2 \alpha \, d\beta^2] \\
&\quad -2(du + a \sin^2 \alpha \, d\beta) d\tau + (1 + 2\widehat{m}N)(du + a \sin^2 \alpha \, d\beta)^2, \\
M^2 &= (R^2 - a^2) \sin^2(r/R) + a^2 \cos^2 \alpha, \\
N &= R \sin(r/R) \cos(r/R)/M^2, \quad u = \tau - r, \quad \widehat{m}, a, R \text{ const},
\end{aligned}
\tag{21.58}
$$

(with $\mu + 3p = 0$) given by Vaidya (1977), cp. Herlt and Hermann (1980); (ii) the metrics (33.34) and (33.35) due to Kramer (1984c); (iii) the solutions with a vanishing Simon tensor discussed by Kramer (1985, 1986a) and Papacostas (1988); (iv) the Carter metrics studied by García D. and Hauser (1988); (v) metrics with $\mu + 3p = 0$ found in the search for solutions with conformal motions by Mars and Senovilla (1994), see also Rácz and Zsigrai (1996); (vi) the generalized Kerr–Schild solution (32.104) by Patel and Vaidya (1983) and (vii) the cylindrically-symmetric rigidly rotating perfect fluid solution given by Davidson (2000). Not all of these cases are different!

Solutions with the equation of state $\mu + 3p = 0$ but Petrov type I have been found by Bonanos and Kyriakopoulos (1994) as

$$
\begin{aligned}
ds^2 &= \tfrac{1}{2} x(1 - \alpha)^2 e^{-y} \{ dx^2/(ax^2 - b^2) + dy^2 [y + \alpha(e^{-y} - 1)] \} \\
&\quad + x [y + \alpha(e^{-y} - 1)] d\varphi^2/a - x^{-1}(dt - by d\varphi/a)^2,
\end{aligned}
\tag{21.59}
$$

and by Kyriakopoulos (1992) as

$$
\begin{aligned}
ds^2 &= e^{-2ax} \left[dx^2 + dy^2 + F^2 (d\varphi - be^{2ax} dt)^2 \right] - e^{2ax} dt^2, \\
F &= \frac{\sqrt{1+c}}{b} \frac{6\wp(y; g_2, g_3) + 2a^2 - 3a^2 c}{6\wp(y; g_2, g_3) + 2a^2 + 3a^2 c},
\end{aligned}
\tag{21.60}
$$

where $\wp$ is a Weierstrass elliptic function with invariants $g_2 = a^2(a^2/3 - b^2)$ and $g_3 = -a^4(b^2 + a^2/9)/3$.

Solutions which are of Petrov type D, with the four-velocity spanned by the two multiple eigenvectors, divide into three classes (Senovilla 1987b): (i) the metric is LRS and admits a G_4; (ii) the metric is static and admits an Abelian G_2; (iii) the metric is not static, admits an Abelian G_2, is shearfree, the magnetic part of the Weyl tensor vanishes, the equation of state is $\mu = p +$ const and the line element is given by

$$\mathrm{d}s^2 = N^{-2}(z)\left[\mathrm{d}z^2 + G^{-1}(x)\mathrm{d}x^2 + G(x)\mathrm{d}\varphi^2 - x^2(\mathrm{d}t + nx^{-2}\mathrm{d}\varphi)^2\right],$$

$$G(x) = \varepsilon x^2 + b\ln x + c + n^2/x^2, \quad N''(z) = \varepsilon N(z), \tag{21.61}$$

$$\kappa_0(p-\mu) = 6(N'^2 - \varepsilon N^2), \;\; \kappa_0(\mu+p) = bN^2(z)/x^2, \;\; \varepsilon = 1,\ 0,\ -1.$$

This last metric generalizes the static solution (18.66) to which it reduces for $n = 0$. The solutions with $\mu = p$ have flat three-dimensional slices (see (4.9) of Wolf (1986b)) and admit the homothetic vector $\mathbf{H} = \partial_z$ (see also Hermann (1983)).

Static solutions admitting an additional homothetic vector have been found by Kolassis and Griffiths (1996).

Using the field equations as given by Bonanos and Sklavenites (1985), a solution of Petrov type I, with a vanishing magnetic part of the Weyl tensor, and an equation of state $\mu = p +$ const has been found by Sklavenites (1985) and Kyriakopoulos (1999).

Starting with special assumptions for the metric functions, rigidly rotating solutions have been found by Kyriakopoulos (1987, 1988) and Sklavenites (1992b). The solutions (36.34)–(36.35) are also rigidly rotating.

Besides the two Killing vectors, a rigidly rotating perfect fluid may admit an additional proper conformal vector (Kramer 1990, Kramer and Carot 1991). For the only known non-static solution with this property the three symmetries commute, and the metric is given by (21.61).

Known rigidly rotating axisymmetric perfect fluid solutions with more than two Killing vectors belong to the locally rotationally-symmetric space-times (§§13.4, 14.3) or to the homogeneous space-times (§12.4) or they are cylindrically-symmetric (§22.2). For instance, if in generalizing Papapetrou's class of vacuum solutions (§20.3) one assumes $\omega = \omega(U)$, $\omega_{,\zeta} = -\mathrm{i}W^{-1}\mathrm{e}^{4U}A_{,\zeta}$ and $\mu + 3p \neq 0$ (Herlt 1972), then one arrives at a locally rotationally-symmetric metric (13.2).

The solutions due to Wahlquist and Herlt, and the interior Schwarzschild solution (16.18) are the only perfect fluid solutions (with rigid rotation) for which the Hamilton–Jacobi equation for null geodesics is separable (Bonanos 1976). The only stationary, axisymmetric and conformally flat perfect fluid solution obeying (19.13) is the interior Schwarzschild solution (16.18) (Collinson 1976b).

Jackson (1970) applied the 'complex trick' (§21.1.4) to spherically-symmetric perfect fluids, in particular to the interior Schwarzschild solution, and obtained an interior NUT metric. Herrera and Jimenez (1982) and Drake and Szekeres (1998) discussed also a generalization which applies the method to spherically-symmetric metrics, including fluid cases, but no new solutions of the types in this book have been obtained. For those types the only cases we know of where such a trick applies are those covered by Talbot (1969) and Quevedo (1992).

21.2.4 Perfect fluid solutions with differential rotation

For differential rotation ($\Omega_{,a} \neq 0$), the field equations for the metric (21.41), with $\sqrt{2}\zeta = \rho + \mathrm{i}\, z$, take the form

$$W_{,\zeta\bar{\zeta}} = \kappa_0 p \mathrm{e}^{2k-2U} W, \tag{21.62a}$$

$$\begin{aligned} &U_{,\zeta\bar{\zeta}} + \tfrac{1}{2}W^{-1}[U_{,\zeta}W_{,\bar{\zeta}} + U_{,\bar{\zeta}}W_{,\zeta}] + \tfrac{1}{2}W^{-2}\mathrm{e}^{4U}A_{,\zeta}A_{,\bar{\zeta}} \\ &\quad = \tfrac{1}{2}\mathrm{e}^{2k-2U}\left[p - \tfrac{1}{2}(\mu+p)H^{-1}\mathrm{e}^{2U}\left\{(1+\Omega A)^2 + W^2\Omega^2\mathrm{e}^{-4U}\right\}\right], \end{aligned} \tag{21.62b}$$

$$\begin{aligned} &A_{,\zeta\bar{\zeta}} - \tfrac{1}{2}W^{-1}(A_{,\zeta}W_{,\bar{\zeta}} + A_{,\bar{\zeta}}W_{,\zeta}) + 2(U_{,\zeta}A_{,\bar{\zeta}} + U_{,\bar{\zeta}}A_{,\zeta}) \\ &\quad = \kappa_0 W^2\mathrm{e}^{2k-4U}(\mu+p)\Omega(1+A\Omega)H^{-1}, \end{aligned} \tag{21.62c}$$

$$\begin{aligned} &v_{,\zeta\bar{\zeta}} + U_{,\zeta}U_{,\bar{\zeta}} + \tfrac{1}{4}W^{-2}\mathrm{e}^{-4U}A_{,\zeta}A_{,\bar{\zeta}} \\ &\quad = -\kappa_0 H^{-1}\mathrm{e}^{2k}\left[\mu\Omega W\mathrm{e}^{-4U} + p(1+\Omega A)^2\right], \end{aligned} \tag{21.62d}$$

$$2W_{,\zeta}k_{\zeta} = W_{,\zeta\zeta} + 2W(U_{\zeta})^2 - \tfrac{1}{2}W^{-1}\mathrm{e}^{4U}(A_{,\zeta})^2, \tag{21.62e}$$

with v defined by

$$\mathrm{e}^{-2v} = 2\mathrm{e}^{2k}W_{,\zeta\bar{\zeta}}. \tag{21.63}$$

Equation (21.62e) determines k via a line integral from U, A and W.

An immediate consequence of the integrability condition $T^{ab}{}_{;a} = 0$ and

$$\dot{u}_n = \frac{H_{,n}}{2H} - \frac{1}{2H}\frac{\partial H}{\partial \Omega}\Omega_{,n}, \quad H \equiv W^2\Omega^2\mathrm{e}^{-2U} - (1+A\Omega)^2\mathrm{e}^{2U} \tag{21.64}$$

is the equation

$$\mathrm{d}p = -(\mu+p)u_n\mathrm{d}u^n = -\frac{\mu+p}{2H}\left(\mathrm{d}H - \frac{\partial H}{\partial \Omega}\mathrm{d}\Omega\right). \tag{21.65}$$

It shows that the pressure p (which trivially is a function of only two variables) is a function of H and Ω even if these two functions are functionally dependent or constant. If this function $p(H,\Omega)$ is known, one can compute the mass density from

$$\mu + p = -2H\partial p/\partial H. \tag{21.66}$$

Equation (21.65) also shows that if we think of H as a function of U, W, A, Ω, then p does not depend on Ω, and if $\mu + p$ is non-zero, then

p has to satisfy

$$\frac{\partial p}{\partial \Omega} + \frac{\partial p}{\partial H}\frac{\partial H}{\partial \Omega} = 0. \tag{21.67}$$

The function $p(H,\Omega)$ can be prescribed arbitrarily as long as it satisfies this condition.

As a consequence of (21.65) one has

$$\begin{aligned} \frac{\partial p}{\partial U} &= -2\frac{\partial p}{\partial H}\left[(1+A\Omega)^2 \mathrm{e}^{2U} + W^2\Omega^2\mathrm{e}^{-2U}\right], \\ \frac{\partial p}{\partial A} &= -2\frac{\partial p}{\partial H}\mathrm{e}^{2U}\Omega(1+A\Omega), \quad \frac{\partial p}{\partial W} = 2\frac{\partial p}{\partial H}W\Omega^2\mathrm{e}^{-2U}, \end{aligned} \tag{21.68}$$

where $\partial p/\partial H$ can be replaced by $-(\mu+p)/2H$.

The field equations (21.62) can be derived from the Lagrangian

$$\begin{aligned} L = {} & 4WU_{,\zeta}U_{,\overline{\zeta}} - \mathrm{e}^{4U}W^{-1}A_{,\zeta}A_{,\overline{\zeta}} - 2(W_{,\zeta}v_{,\overline{\zeta}} + W_{,\overline{\zeta}}v_{,\zeta}) \\ & -4\kappa_0 p(H,\Omega)W\mathrm{e}^{2v-2U}W_{,\zeta}W_{,\overline{\zeta}}, \end{aligned} \tag{21.69}$$

where v can formally be treated as an independent variable (Neugebauer and Herlt 1984, Kramer 1988c). Variation with respect to W, U, A and v yields (21.62a), (21.62b), (21.62c) and (21.62d), respectively. That is to say, the field equations are just the equations of minimal surfaces in a potential space with coordinates (U, A, W, v) and line element

$$\mathrm{d}S^2 = 4W\mathrm{d}U^2 - \mathrm{e}^{4U}W^{-1}\mathrm{d}A^2 - 4\mathrm{d}W\mathrm{d}v - 4\kappa_0 p(H,\Omega)W\mathrm{e}^{2v-2U}\mathrm{d}W^2, \tag{21.70}$$

cp. Chapter 10. The symmetries of this potential space and its relation to space-time symmetries have been studied by Stephani and Grosso (1989) and Grosso and Stephani (1990).

Compared with the case of rigid rotation, the additional degree of freedom inherent in the rotation $\Omega(\zeta,\overline{\zeta})$ should lead to a plenitude of solutions. But surprisingly few differentially rotating solutions have been found so far, and for most of them some ordinary differential equations remained unsolved.

A (Petrov type I) solution with vanishing vorticity ω^n and an equation of state $\mu = p$ was found by Chinea and González-Romero (1990, 1992). Using the pressure p as one of the coordinates, it reads

$$\begin{aligned} \mathrm{d}s^2 = {} & \frac{bTT_{,r}}{rp}\left[\frac{k}{2} - \frac{r^4p^2}{4}\right]^{-1}\left[\left(\frac{1}{r^2} + \frac{k}{2T^2} - \frac{r^4p^2}{4T^2}\right)\mathrm{d}r^2 + \frac{\mathrm{d}r\,\mathrm{d}p}{rp} + \frac{\mathrm{d}p^2}{4p^2}\right] \\ & + r^2p\,(\mathrm{d}\varphi - \Omega\mathrm{d}t)^2 - a^2p^{-1}\mathrm{d}t^2, \quad \Omega(r) = a\int\frac{r\mathrm{d}r}{T(r)}, \end{aligned} \tag{21.71a}$$

where a, b and k are arbitrary constants, and $T(r)$ is a solution of

$$T_{,rr} + \left(\frac{1}{r} + \frac{kr}{T^2}\right) T_{,r} = 0. \tag{21.71b}$$

A solution admitting a homothetic vector $\mathbf{H}$ has been found by Hermann (1983) as

$$\begin{aligned} ds^2 &= e^{2z}\Big(dr^2 + r^2dz^2/s^2 \\ &\quad + r^{2(1+\sigma)/\sigma}e^{2\sigma z}\big\{[\tfrac{1}{2} - 2\lambda^2(\ln r)^2]d\varphi^2 - 4\lambda^2 \ln r d\varphi dt - 2\lambda^2 dt^2\big\}\Big), \\ \mathbf{H} &= \partial_z - \sigma(\varphi\partial_\varphi + t\partial_t), \quad s^2 = (1 + 2\sigma - 2\sigma^2)/(\sigma - 2)\sigma^3, \\ u^a &= \sqrt{-\sigma}e^{-(1+\sigma)z}r^{-(1+\sigma)/\sigma}\left(0, 0, 1, \sigma(\sigma - 2)/2(1 - \sigma) - \ln r\right), \\ \kappa_0 p &= 2\frac{(1-\sigma)(1+\sigma)^3}{(\sigma-2)\sigma^3}e^{-2z}r^{-2} = \frac{1+\sigma}{1-\sigma}\kappa_0\mu, \quad \lambda^2 = \frac{(\sigma-1)^2(1+\sigma)^2}{(\sigma-2)\sigma^3}. \end{aligned} \tag{21.72}$$

A solution with non-vanishing shear and vorticity, and with $\mu = 2p$, has been given by Chinea (1993) using an adapted tetrad formalism as

$$\begin{aligned} ds^2 &= (\boldsymbol{\omega}^1)^2 + (\boldsymbol{\omega}^2)^2 + (\boldsymbol{\omega}^3)^2 - (\boldsymbol{\omega}^4)^2, \\ \boldsymbol{\omega}^1 &= \frac{1}{\sqrt{2a\kappa_0 p_0}}e^{3v/4-\sigma^2/2}\left[-f(\sigma)dv + \frac{\sigma d\sigma}{f(\sigma)}\right], \quad \boldsymbol{\omega}^2 = \frac{1}{\sqrt{2\kappa_0 p_0}}\frac{e^{3v/4}}{\sigma}dv, \\ \boldsymbol{\omega}^3 &= -le^{v/2}\sin(S-\lambda)dt/\sqrt{7} - ne^{v/2}\cos(S-\nu)d\varphi/\sqrt{7}, \\ \boldsymbol{\omega}^4 &= u_a dx^a = le^{v/2}\cos(S-\lambda)dt + ne^{v/2}\cos(S-\nu)d\varphi, \end{aligned} \tag{21.73}$$

$$S(\sigma) = \frac{\sqrt{7}}{4\sqrt{2}}\int\frac{d\sigma}{f(\sigma)}, \quad f(\sigma) = \sqrt{1 - a\sigma^{-2}e\sigma^2},$$

$$p = p_0 e^{-3v/2}, \quad \Omega = -l\sin(S-\lambda)/n\sin(S-\nu),$$

$$\mathbf{u} = \frac{e^{-v/2}}{\sin(\nu-\lambda)}\left[\frac{1}{l}\sin(S-\nu)\partial_t - \frac{1}{n}\sin(S-\lambda)\partial_\varphi\right],$$

where a, λ, ν, l, n are arbitrary constants.

Solutions with an equation of state $p = p(\mu)$ have been studied by González-Romero (1994).

The general type D solutions with zero magnetic Weyl tensor and a four-velocity spanned by the two Weyl eigenvectors have been found (up to ordinary differential equations) by Senovilla (1992) and further discussed by Mars and Senovilla (1996). The line element has the form

$$ds^2 = N^{-2}(z)\left[dz^2 + \frac{dx^2}{hm+s^2} + \frac{hm+s^2}{m}d\varphi^2 - m\left(dt + \frac{s}{m}d\varphi\right)^2\right],$$

$$N'' = \varepsilon N, \quad \varepsilon = 0, \pm 1, \quad \Omega = -m''/s'', \tag{21.74a}$$

$$\kappa_0\mu = \kappa_0 p - 6(N'^2 - \varepsilon N^2), \quad \kappa_0(\mu + p) = N^2(z)(h'm' + s'^2 + 4\varepsilon)/2,$$

where the three functions $h(x)$, $m(x)$ and $s(x)$ have to satisfy the two differential equations

$$h''m'' + s''^2 = 0, \quad (hm + s^2)'' + 4\varepsilon = h'm' + s'^2. \tag{21.74b}$$

This class admits a conformal Killing vector $\zeta = \partial_z$, cp. Mars and Senovilla (1994), and contains the rigidly rotating solutions (21.61) as the special case $s'' = 0 = m''$. Only very special explicit solutions of (21.74) are known, where h, m and s are either powers of x (García D. 1994) or a product of powers of x with $(x^{\sqrt{1/10}} + \alpha x^{-\sqrt{1/10}})$ (Senovilla 1992).

22

Groups G_2I on spacelike orbits: cylindrical symmetry

22.1 General remarks

The metrics to be considered admit two spacelike commuting Killing vectors ($\boldsymbol{\eta} = \partial_\varphi$ and $\boldsymbol{\zeta} = \partial_z$) which generate an Abelian group G_2. Unless there is a third Killing vector (see §22.2), we assume the existence of 2-surfaces orthogonal to the group orbits; this orthogonal transitivity may be a consequence of the field equations, cp. §19.2. We start with the line element

$$\mathrm{d}s^2 = \mathrm{e}^{-2U}\left(\gamma_{MN}\mathrm{d}x^M\mathrm{d}x^N + W^2\mathrm{d}\varphi^2\right) + \mathrm{e}^{2U}\left(\mathrm{d}z + A\mathrm{d}\varphi\right)^2, \tag{22.1}$$

which is independent of φ and z, cp. §17.1 (if the field admits the reflection symmetry $\varphi \to -\varphi$, $z \to -z$, one can put $A = 0$ in (22.1) and the two commuting Killing vectors are hypersurface-orthogonal). This line element can formally be obtained from the corresponding stationary axisymmetric line element (19.17) by the complex substitution

$$t \to \mathrm{i}\,z, \quad z \to \mathrm{i}\,t, \quad A \to \mathrm{i}\,A. \tag{22.2}$$

Solutions may be mapped this way if the new field functions can be made real by e.g. analytic continuation of the parameters. The indefinite 2-metric γ_{MN} in (22.1) can always be chosen as

$$\gamma_{MN}\mathrm{d}x^M\mathrm{d}x^N = \mathrm{e}^{2k}(\mathrm{d}\rho^2 - \mathrm{d}t^2) \tag{22.3}$$

(Kompaneets 1958, Jordan *et al.* 1960).

Metrics of this kind are often called '*cylindrically-symmetric*'. Although we have adapted the notation to this custom, and shall occasionally use it, we will not be bound by its consequences. To admit an interpretation in terms of cylindrical symmetry, there should be an axis where

$\eta_a\eta^a = e^{-2U}W^2 + A^2e^{2U}$ vanishes, and the metric should be regular at this axis. For many of the known solutions there is no axis, or the metric is not regular on it; nevertheless the metric may be appropriate to describe the outer (vacuum) region of e.g. a rotating cylindrical source. Furthermore, the identification (e.g. $\varphi + 2\pi \to \varphi$) which is necessary to get the correct topology on the orbit will restrict the possible linear transformations (17.7) so that the normal forms discussed in §17.1 cannot be obtained globally (see MacCallum (1998) for references and further discussion). An extended discussion of the definition of axial (and cylindrical) symmetry can be found in Carot *et al.* (1999).

In the following we shall neglect these global considerations and classify the solutions only with respect to their local properties. We start in the following section with the subclass which admits an Abelian G_3 on T_3, treat then the vacuum solutions in §22.3, the Einstein–Maxwell and pure radiation fields in §22.4, the perfect fluid and dust solutions in Chapter 23, and finally the colliding waves in Chapter 25. Of course, generation techniques can be and have been applied also to the case of cylindrical symmetry, cp. Chapter 34.

Note that plane symmetry (§§15.4, 15.5) can be treated as a special case of (22.1): put $A = 0$ and $W^2 = e^{4U}$ and replace (e^U, φ, z) by (Y, x, y) in (22.1) to obtain the metric (15.10).

22.2 Stationary cylindrically-symmetric fields

Metrics which admit an Abelian group G_3I acting on timelike orbits T_3 are called *stationary cylindrically-symmetric*. The three Killing vectors are $\boldsymbol{\xi} = \partial_t$, $\boldsymbol{\eta} = \partial_\varphi$, $\boldsymbol{\zeta} = \partial_z$, and the metrics are special hypersurface-homogeneous space-times (cp. Chapter 13 and the methods described in §13.2). They can be obtained either as special cases of stationary axisymmetric fields (with Killing vectors $\boldsymbol{\xi}, \boldsymbol{\eta}$) or of cylindrically-symmetric fields (with Killing vectors $\boldsymbol{\eta}$, $\boldsymbol{\zeta}$) by demanding a third symmetry (and assuming orthogonal transitivity for the first two). Because of the assumption of orthogonal transitivity of two of the Killing vectors, at least one of the three Killing vectors is hypersurface-orthogonal. Accordingly, one can start from either of the forms

$$ds^2 = f^{-1}\left[e^{2k}\left(d\rho^2 + dz^2\right) + W^2 d\varphi^2\right] - f\,(dt + A d\varphi)^2, \qquad (22.4a)$$

$$ds^2 = f^{-1}\left[e^{2k}\left(d\rho^2 - dt^2\right) + W^2 d\varphi^2\right] + f\,(dz + A d\varphi)^2, \qquad (22.4b)$$

with all metric functions depending only on ρ. The two metrics (22.4) are not equivalent, but related by the complex substitution (22.2): they overlap when all three Killing vectors are (locally) hypersurface-orthogonal.

Neglecting topological considerations, the labels of the coordinates φ and z may be interchanged in metric (22.4a).

Vacuum solutions

For vacuum, and with the metric (22.4a), we get from (17.10) the equation $W_{,\rho\rho} = 0$. For the solutions with $W = \text{const}$ the Killing vector ∂_t is null; they are special *pp*-waves

$$\mathrm{d}s^2 = \mathrm{e}^{a\rho}\left[\mathrm{d}\rho^2 + \mathrm{d}z^2\right] + \rho\,\mathrm{d}\varphi^2 + 2\mathrm{d}\varphi\,\mathrm{d}t. \tag{22.5}$$

The general stationary cylindrically-symmetric vacuum solution with $W = \rho$ is given by

$$\begin{aligned} \mathrm{d}s^2 &= f^{-1}\left[\mathrm{e}^{2k}\left(\mathrm{d}\rho^2 + \mathrm{d}z^2\right) + \rho^2\mathrm{d}\varphi^2\right] - f\left(\mathrm{d}t + A\mathrm{d}\varphi\right)^2, \\ f &= \rho\left(a_1\rho^n + a_2\rho^{-n}\right), \quad n^2 a_1 a_2 = -C^2, \\ A &= \frac{C}{n a_2}\frac{\rho^n}{a_1\rho^n + a_2\rho^{-n}} + B, \quad f^{-1}\mathrm{e}^{2k} = \rho^{(n^2-1)/2} \end{aligned} \tag{22.6}$$

and its counterpart for the metric (22.4b). The constant n is real or imaginary; the other constants have to be chosen such that the metric becomes real. The metric coefficients satisfy a linear relation (17.6) and can thus be simplified as discussed in §17.1.2. For imaginary n the solutions belong to the Lewis class (§20.4). The regularity condition (19.3) on the axis of symmetry allows only flat space-time (Davies and Caplan 1971). For real n they are locally static (Frehland 1971), belong to the Weyl class (§20.2), and can be written as

$$\mathrm{d}s^2 = \rho^{-2m}\left[\rho^{2m^2}\left(\mathrm{d}\rho^2 + \mathrm{d}z^2\right) + \rho^2\mathrm{d}\varphi^2\right] - \rho^{2m}\mathrm{d}t^2 \tag{22.7}$$

(Levi-Civita 1917a). These Kasner solutions (cp. (13.51)) are flat for $m = 0, 1$, of Petrov type D for $m = 1/2,\ 2,\ -1$, and contain as a special case the Petrov solution (12.14) (Bonnor 1979a).

Λ-term solutions

The solutions of $R_{ik} = \Lambda g_{ik}$ are very similar to the vacuum solutions. The metric coefficients again satisfy a linear relation, and the solutions can be given in a form closely related to the Kasner solutions as

$$\begin{aligned} &\mathrm{d}s^2 = \mathrm{d}r^2 + G(r)^{2/3}\Big\{\sum_j \varepsilon_j \exp\left[2(p_j - \tfrac{1}{3})U(r)\right]\left(\mathrm{d}x^j\right)^2\Big\}, \quad \varepsilon = \pm 1, \\ &G'' = -3\Lambda G, \quad G'^2 + 3\Lambda G^2 = \eta = 0,\ \pm 1, \\ &U' = G^{-1}, \quad \sum_j p_j = 1, \quad \sum_j p_j^2 = (2\eta + 1)/3, \end{aligned} \tag{22.8}$$

see Krasiński (1975a) and the discussion given in MacCallum and Santos (1998). The solutions can be used as an exterior for the Gödel metric (12.26) (Bonnor *et al.* 1998). For $\Lambda = 0$ we regain the vacuum case (22.6), in rescaled radial coordinates, as $G = r = \mathrm{e}^U$.

Einstein–Maxwell fields

It is usually assumed that the metric can be written in one of the forms (22.4), that the electromagnetic field shares the symmetry of the metric and that the electric and magnetic fields are orthogonal to the orbits of the two-dimensional orthogonally transitive group, i.e. that the vector potentials have the forms

$$A_i \mathrm{d}x^i = P(\rho)\mathrm{d}t + Q(\rho)\mathrm{d}\varphi, \tag{22.9a}$$

$$A_i \mathrm{d}x^i = P(\rho)\mathrm{d}z + Q(\rho)\mathrm{d}\varphi, \tag{22.9b}$$

respectively. Condition (17.11) is satisfied, and we have $W_{,\rho\rho} = 0$.

For the static metrics (22.4*b*), with $W = \rho$, one obtains the differential equations

$$(\rho f'/f)' = c^2\rho/f^2 - 2p^2 f/\rho, \tag{22.10a}$$

$$A' = \rho c f^{-2}, \quad P' = -fp/\rho, \quad Q' = -AP', \tag{22.10b}$$

$$k' = (Q' + AP')^2 f/\rho + \rho P'^2/f + A'^2 f^2/4\rho + \rho f'^2/4f^2 \tag{22.10c}$$

(Chitre *et al.* 1975, MacCallum 1983). Equation (22.10*a*) for f is the third of the Painlevé equations. Non-transcendental solutions arise for special values of the constants. For $p = 0$ one gets a vacuum solution. For $c = 0$, the metric and 4-potential are given by

$$\begin{aligned} &\mathrm{d}s^2 = \rho^{2m^2} G^2(\mathrm{d}\rho^2 - \mathrm{d}t^2) + \rho^2 G^2 \mathrm{d}\varphi^2 + G^{-2}\mathrm{d}z^2, \\ &G = a_1\rho^m + a_2\rho^{-m}, \quad A_i\mathrm{d}x^i = (\rho G'/bG)\mathrm{d}z, \quad b^2 = 4a_1a_2m^2 > 0, \end{aligned} \tag{22.11}$$

(Bonnor 1953), where the constants b, m^2 (real) and a_1, a_2 have to be chosen so that metric and potential are real; the case $m = 1/2$ had already been given by Mukherji (1938). This solution can be interpreted as containing a magnetic field along the z-direction or, with z and φ exchanged, along the φ-direction. For a vanishing right-hand side of (22.10*c*) the result is the solution

$$\begin{aligned} \mathrm{d}s^2 &= \rho^{-4/9}\exp(a^2\rho^{2/3})(\mathrm{d}\rho^2 - \mathrm{d}t^2) + \rho^{4/3}\mathrm{d}\varphi^2 \\ &\quad + \rho^{2/3}(\mathrm{d}z + a\rho^{2/3}\mathrm{d}\varphi)^2, \\ A_i\mathrm{d}x^i &= a\rho^{2/3}\mathrm{d}z/\sqrt{2} + a^2\rho^{4/3}\mathrm{d}\varphi/\sqrt{8}, \quad a = \text{ const}, \end{aligned} \tag{22.12}$$

due to Chitre *et al.* (1975).

The Melvin solution (Bonnor (1954), rediscovered by Melvin (1964))

$$\mathrm{d}s^2 = (1 + B_0^2\rho^2/4)^2(\mathrm{d}\rho^2 + \mathrm{d}z^2 - \mathrm{d}t^2) + (1 + B_0^2\rho^2/4)^{-2}\rho^2\mathrm{d}\varphi^2 \quad (22.13)$$

is the special case $m = 1$ of (22.11) and of (31.58). The gravitational field (22.13) originates in a 'uniform' magnetic field B_0 along the z-axis.

For the metric (22.4a), two classes of solutions arise which lead to a Painlevé equation similar to (22.10a), see Van den Bergh and Wils (1983) and MacCallum (1983) for details. Again several non-transcendental cases exist. The analogue of (22.11) is

$$\begin{gathered}\mathrm{d}s^2 = \rho^{2m^2}G^2(\mathrm{d}\rho^2 + \mathrm{d}z^2) + \rho^2G^2\mathrm{d}\varphi^2 - G^{-2}\mathrm{d}t^2,\\ G = a_1\rho^m + a_2\rho^{-m}, \quad A_i\mathrm{d}x^i = (\rho G'/bG)\mathrm{d}t, \quad b^2 = -4a_1a_2m^2 > 0,\end{gathered} \quad (22.14)$$

(Raychaudhuri 1960). This solution contains a radial electric field which depends only on ρ (so that the Maxwell field admits the additional symmetry of a rotation in the φ–z-plane not shared by the metric, see e.g. Li and Liang (1989)). The Mukherji (1938) solution describing the gravitational field of a charged line-mass is contained in (22.14) for $m = 1/2$. The metric

$$\begin{aligned}\mathrm{d}s^2 &= \rho^{-4/9}\exp(-a^2\rho^{2/3})(\mathrm{d}\rho^2 + \mathrm{d}z^2) + \rho^{4/3}\mathrm{d}\varphi^2\\ &\quad - \rho^{2/3}(\mathrm{d}t + a\rho^{2/3}\mathrm{d}\varphi)^2,\\ A_i\mathrm{d}x^i &= a\rho^{2/3}\mathrm{d}t/\sqrt{2} + a^2\rho^{4/3}\mathrm{d}\varphi/\sqrt{8}, \quad a = \text{ const},\end{aligned} \quad (22.15)$$

(Islam 1983b, Van den Bergh and Wils 1983) corresponds to (22.12). There is one more static solution

$$\begin{aligned}\mathrm{d}s^2 &= a^2(\ln c\rho)^2\left[\mathrm{d}z^2 + \mathrm{d}\rho^2 + \rho^2\mathrm{d}\varphi^2\right] - a^{-2}(\ln c\rho)^{-2}\mathrm{d}t^2,\\ A_i\mathrm{d}x^i &= \mathrm{d}t/a\ln c\rho\end{aligned} \quad (22.16)$$

(Raychaudhuri 1960). A Petrov type II solution with a null electromagnetic field,

$$\begin{aligned}\mathrm{d}s^2 &= (\mathrm{d}\rho^2 + \mathrm{d}z^2)/\sqrt{\rho} + f^{-1}\rho^2\mathrm{d}\varphi^2 - f(\mathrm{d}t - \rho\,\mathrm{d}\varphi/f)^2,\\ A_i\mathrm{d}x^i &= -b\rho\,\mathrm{d}t, \quad f = 4b^2\rho^2 + c\rho\ln\rho, \quad b, c = \text{ const},\end{aligned} \quad (22.17)$$

has been found by Datta and Raychaudhuri (1968). Here the Killing vector $\boldsymbol{\eta} = \partial_\varphi$ is null and non-twisting, and the vacuum counterpart is a van Stockum solution (20.32).

Einstein–Maxwell fields which do *not* share the symmetry of the metric can occur in either of the two cases (22.4). Instead of (22.9), the 4-potential now has the form

$$A_i\mathrm{d}x^i = P(\rho)\cos kz\,\mathrm{d}t + (AP - \rho f^{-1}P'/k)\cos kz\,\mathrm{d}\varphi \quad (22.18)$$

for (22.4*a*), or an analogous form with t and z interchanged for (22.4*b*). Some explicit solutions have been found, see McIntosh (1978b), Wils and Van den Bergh (1985) and MacCallum and Van den Bergh (1985).

Dust solutions

The stationary cylindrically-symmetric dust solutions are contained in the general stationary axisymmetric dust solution (the Winicour solution) discussed in §21.2. The specialization to cylindrical symmetry can be found in Vishveshwara and Winicour (1977), and the solutions have been discussed also by King (1974). Maitra (1966) gave a special solution for non-rigidly rotating dust (with $\Theta = 0$, but $\sigma \neq 0$). For rigid rotation, specializing the van Stockum class (21.52) to cylindrical symmetry we get the metric

$$\begin{aligned} \mathrm{d}s^2 &= \mathrm{e}^{-a^2\rho^2}(\mathrm{d}\rho^2 + \mathrm{d}z^2) + \rho^2\mathrm{d}\varphi^2 - (\mathrm{d}t + a\rho^2\mathrm{d}\varphi)^2, \\ \kappa_0\mu &= 4a^2\mathrm{e}^{a^2\rho^2}, \quad \mathbf{u} = \partial_t. \end{aligned} \tag{22.19}$$

(Lanczos 1924). The solution can be matched to the vacuum solution (22.6); depending on a and the radius R of the cylinder, the external field may be locally static (for $aR < 1/2$), though the dust rotates (Tipler 1974, Bonnor 1980).

Static perfect fluid solutions

Assuming that all three Killing vectors are hypersurface-orthogonal, several simple forms of the field equations have been given which can be the starting point in the search for solutions.

Taking the form

$$\mathrm{d}s^2 = \mathrm{e}^{2\lambda(\rho)}\mathrm{d}\rho^2 + \mathrm{e}^{2\beta(\rho)}\mathrm{d}\varphi^2 + \mathrm{e}^{2\chi(\rho)}\mathrm{d}z^2 - \mathrm{e}^{2\delta(\rho)}\mathrm{d}t^2 \tag{22.20}$$

of the line element with $\lambda = 0$ and $\beta \neq \chi$, and introducing the new functions $\tau(\rho) = \beta + \chi + \delta$, and $w(\rho) = \exp(3\delta)$, Evans (1977) reduced the problem to a second-order linear differential equation for w,

$$w'' - \tau' w' + 3(\tau'' + \mathrm{e}^{-2\tau})w = 0. \tag{22.21}$$

The function $\tau(\rho)$ can be prescribed, and the metric, and p and μ, can be determined from (22.21) and from

$$\begin{aligned} \lambda = 1, &\quad 4\kappa_0 p = (\tau + 3\delta)'(\tau - \delta)' - \mathrm{e}^{-2\tau}, \\ (\beta - \chi)' = \mathrm{e}^{-\tau}, &\quad \kappa_0(5p - \mu) = 2(\mathrm{e}^{\tau})''\mathrm{e}^{-\tau}. \end{aligned} \tag{22.22}$$

Bronnikov (1979) used $\lambda = \beta + \chi + \delta$ as a coordinate condition. With $\alpha = \beta + \delta$, the field equations then lead to

$$\chi = \beta + a\rho + a_1, \quad \kappa_0 p = \alpha''\mathrm{e}^{-2\lambda}/2, \quad \kappa_0\mu = \kappa_0 p - 2\beta''\mathrm{e}^{-2\lambda} \tag{22.23a}$$

and the two equivalent equations

$$\alpha'' - (2a + 4\beta')\alpha' + 2\beta'^2 = 0, \tag{22.23b}$$

$$\alpha'' - 2a\alpha' - 2\alpha'^2 + 4\chi'^2 = 0. \tag{22.23c}$$

Both can be solved by quadratures: one can prescribe $\beta(\rho)$ and solve a linear first-order differential equation (22.23*b*) for α', or prescribe $\alpha(\rho)$ and solve (22.23*b*) for β' or (22.23*c*) for χ'. Plane symmetry corresponds to $a = 0$.

Philbin (1996) chose coordinates with $\lambda = \chi$ and arrived at

$$\beta'^2 + 2\alpha' - a_0\alpha' \mathrm{e}^{-\alpha} + (\alpha'' + \alpha'^2)/2 = 0, \quad \alpha = \beta + \delta, \tag{22.24}$$

which again can be solved by a quadrature if $\alpha(\rho)$ is prescribed.

Kramer (1988a) proposed taking the metric (22.4*a*) and using $x \equiv \frac{1}{2}\ln f$ as a new independent variable. Introducing

$$F(x) = \frac{\mu + 3p}{2p}, \quad y = \frac{\mathrm{d}k}{\mathrm{d}x} = \dot{k}, \quad z = \frac{1}{W}\frac{\mathrm{d}W}{\mathrm{d}x} = \frac{\dot{W}}{W}, \tag{22.25}$$

the field equations read

$$\dot{y} = (1 - yz)(Fy - 2), \quad \dot{z} = (1 - yz)(Fz - 2). \tag{22.26}$$

Using one or other of the above equations, several exact solutions have been constructed, e.g. for (22.21) and $\tau'' = -\mathrm{e}^{-2\tau}$ (Evans 1977), for (22.24) and $\mathrm{e}^{\alpha+\delta} = \rho + c_1\rho^3 + c_2\rho^5$ (Philbin 1996), for (22.26) and $y = (az - b)/(dz - c)$ (Haggag and Desokey 1996), for (22.26) and $z = (dy^2 + cy + b)/(y - a)$ (Haggag 1999). By making other ad hoc assumptions, solutions have also been found by Davidson (1989b, 1990a, 1990b), and Narain (1988).

For an equation of state $p = (\gamma - 1)\mu$, special cases have been given by Evans (1977) ($5p = \mu$) and by Teixeira *et al.* (1977a) ($3p = \mu$). The general solution was found by Bronnikov (1979) as

$$\mathrm{d}s^2 = \mathrm{e}^{2[\beta+\chi+\delta]}\mathrm{d}\rho^2 + \mathrm{e}^{2\beta}\mathrm{d}\varphi^2 + \mathrm{e}^{2\chi}\mathrm{d}z^2 - \mathrm{e}^{2\delta}\mathrm{d}t^2,$$
$$\beta = \frac{\gamma - 2}{4(\gamma - 1)}\alpha(\rho) + b\rho, \quad \chi = \beta + a\rho, \quad \delta = \frac{3\gamma - 2}{4(\gamma - 1)}\alpha(\rho) - b\rho, \tag{22.27a}$$

(some integration constants have been set zero by choice of coordinates), where $\alpha(\rho)$ satisfies the differential equation

$$\alpha'' + A\alpha'^2 - B\alpha' + 2b^2 = 0, \quad A = \frac{7\gamma^2 + 20\gamma - 12}{8(\gamma - 1)^2}, \quad B = 2a + b\frac{3\gamma - 2}{\gamma - 1}, \tag{22.27b}$$

which can easily be solved in terms of elementary functions. The solutions with $a = 0$, $\beta = \chi$ and α linear in ρ admit an additional homothetic vector (Debever and Kamran 1982). In the notation of Kramer (1988a), see also Haggag (1989), the solutions regular at the axis $\rho = 0$ are given by

$$\begin{aligned} \mathrm{d}s^2 &= F^m \mathrm{d}\rho^2 + F^n(F\mathrm{d}z^2 + \rho^2\mathrm{d}\varphi^2) - F^s\mathrm{d}t^2, \\ \kappa_0 p &= \frac{16(\gamma-1)^2\beta}{(2-\gamma)(7\gamma-6)}F^{-m-1}, \quad F = 1 - \beta\rho^2, \\ n &= -\frac{3\gamma-2}{7\gamma-6}, \quad m = -1 - \frac{2n\gamma}{\gamma-2}, \quad s = -4n\frac{\gamma-1}{\gamma-2}. \end{aligned} \tag{22.28}$$

If the four-velocity $\boldsymbol{u} = w^{-1}(\rho)\partial_t$ is hypersurface-orthogonal, but the two spacelike Killing vectors are *not*, one can start from a metric of the form (22.4*b*), i.e. from

$$\mathrm{d}s^2 = \frac{(\mathrm{d}y + \rho\mathrm{d}\varphi)^2}{h^2(\rho)} + \frac{E(\rho)}{h^2(\rho)}\mathrm{d}\varphi^2 + \frac{\mathrm{e}^{-b(\rho)}\mathrm{d}\rho^2}{h^2(\rho)E(\rho)} - w^2(\rho)\mathrm{d}t^2. \tag{22.29}$$

The field equations then give (Stephani 1998)

$$E(\rho) = -\rho^2 + a\rho + c, \quad w^2(\rho) = \mathrm{e}^{-b(\rho)}h^2(\rho), \tag{22.30}$$

where $h(\rho)$ and $b(\rho)$ have to obey

$$h^2b'' + 4hh'b' - 4h'^2 = 0. \tag{22.31}$$

The solutions of this differential equation can be given in either of the two forms

$$b = \int b'\mathrm{d}\rho, \quad b' = 4h^{-4}\int h^2h'^2\mathrm{d}\rho, \quad \text{or} \quad \ln h = \tfrac{1}{2}b \pm \tfrac{1}{2}\int\sqrt{b'' + b'^2}\mathrm{d}\rho. \tag{22.32}$$

So one can prescribe $h(\rho)$ or $b(\rho)$, and determine the second function from (22.32). Pressure p and energy density μ are given by

$$\kappa_0 p = -\tfrac{1}{4}\mathrm{e}^b h^2\left[1 + (b'E)'\right], \quad \kappa_0\mu = \mathrm{e}^b h^2\left[2(Eh'/h)' - \tfrac{1}{4}(Eb')' + \tfrac{3}{4}\right]. \tag{22.33}$$

The special case $h = \rho^{1/3}$, $b = \ln\rho^{4/3}$ admits a Killing tensor (Papacostas 1988). For the metrics (22.29)–(22.30) there is a linear relation between the metric coefficients of the y–φ-part, cp. §17.1.2; for $c + a^2/4 > 0$ there are two hypersurface-orthogonal Killing vectors and the solutions can be transformed into the form (22.20).

The metrics (36.22) and (36.23) also admit an Abelian G_3 on T_3.

Stationary perfect fluids

Perfect fluids with a metric of the form (22.4*a*), where all functions depend only on ρ, and a four-velocity proportional to $\partial/\partial t$, are *rigidly rotating*, cp. §21.2.3. The general metric for this case can be written as

$$\mathrm{d}s^2 = \frac{\mathrm{d}\rho^2}{EBh} + \frac{E\mathrm{d}\varphi^2}{h^2} + \frac{1}{B}h^3\mathrm{d}z^2 - \frac{(\mathrm{d}t + \rho\mathrm{d}\varphi)^2}{h^2}, \tag{22.34}$$

$$B(\rho) = h^5(\rho)E^{-1}(x)\exp\left(\int(\rho+c)E^{-1}(\rho)\mathrm{d}\rho\right), \quad \boldsymbol{u} = h\partial/\partial t,$$

where the functions $E(\rho)$ and $f = h^3(\rho)$ have to satisfy the (linear) differential equation

$$E^2 f'' + [(\rho+c)E - EE']\,f' - \tfrac{3}{4}\left[EE'' - E'^2 + (\rho+c)E' - E\right]f = 0 \tag{22.35}$$

(Krasiński 1974, 1975b, 1978). By a coordinate transformation, c can be made zero, but a non-zero c may facilitate the recognition of solutions. The pressure and mass density can be calculated from

$$\begin{aligned} \kappa_0 p &= \textstyle\int Bh'(1 - E''/2)\mathrm{d}\rho + p_0 \\ &= Bh^{-5}E^{-1}\left[\tfrac{1}{4}f^2\left\{E - E'(\rho+c) + E'^2\right\}\right. \\ &\qquad \left. + \tfrac{1}{3}Ef'\left\{f(\rho+c) + Ef' - 2E'f\right\}\right], \\ \kappa_0(\mu+p) &= Bh(1 - E''/2). \end{aligned} \tag{22.36}$$

Introducing a new function $g(\rho)$ by

$$g(\rho) = \left[E'(\rho) - \rho - c\right]E^{-1}(\rho), \tag{22.37}$$

the differential equation (22.35) can be written as

$$f'' - gf' - \tfrac{3}{4}g'f = 0 \tag{22.38}$$

and solved by a series of quadratures

$$g = \tfrac{4}{3}f^{-4/3}\int f''f^{1/3}\mathrm{d}\rho, \quad E = \exp\left(\int g\mathrm{d}\rho\right)\int\frac{(\rho+c)\mathrm{d}\rho}{\exp\left(\int g\mathrm{d}\rho\right)}. \tag{22.39}$$

So one can either prescribe the four-velocity (the function f) and use (22.39) to obtain the metric function E, or prescribe E and solve (22.35) for f.

For $E(\rho) = c_1\rho^2 + 2c_2\rho + c_3$, (22.35) gives f in terms of hypergeometric functions. For $f = \rho^n$, the choice $g \sim \rho^{-1}$ leads to simple expressions for E; included here is the self-similar solution given by Nilsson and Uggla

(1997a). The solution with $E = \rho^2/2+2c\rho+c_3$, $h = \rho^{1/3}$ admits a Killing tensor (Papacostas 1988). The choice $E = \rho(\rho + c\ln\rho)$, $h^2 = c_1\rho$ gives the subcase $\varepsilon = 0$, $N = 1$ of (21.61). Solutions with an equation of state $p = (\gamma-1)\mu$ have been discussed by Davidson (1996, 1997, 1999); they are contained as $E = a\rho^2 + c\rho$, $f = (1+a\rho/c)^{b/2}$, with $a(b^2-6b+b) = 3-2b$ and $\gamma = 1 + ab/(5ab + 6 - 12a)$. A special solution with an equation of state $\mu + p =$ const has been found by Kramer (1985); it corresponds to $f = \rho^{1/2}$, $E = \rho^2 + a\rho e^{-1/\rho}$, $c = 1$. Of the two solutions found by Sklavenites (1999), the first corresponds to $f = \rho^n$. Davidson (2000) gave a metric with an equation of state $\mu + 3p =$ const (a special case of the Wahlquist solution (21.57)); it has $f \sim (\rho - a)^{3/2}$.

Equations (22.34)–(22.39) can easily be rewritten if a new coordinate $\widetilde{\rho} = \widetilde{\rho}(\rho)$ is introduced. This was used by García D. and Kramer (1997) to find a rigidly rotating solution.

For *differential rotation*, only a few special solutions are known. Under a restriction, which in coordinates (22.4a) amounts to $W^2 + e^{2k} = f^2A^2$, García D. and Kramer (1997) found several explicit classes of solutions in terms of confluent hypergeometric functions. Davidson (1994) took the form (17.5) of the metric (with $\varepsilon = 1$) and assumed that all metric functions are powers of the 'radial' coordinate z. The solutions admitting an additional homothetic vector have been given by Debever and Kamran (1982).

22.3 Vacuum fields

To get the vacuum field equations for the metric (22.1), we take it in the form

$$ds^2 = e^{-2U}\left[e^{2k}(d\rho^2 - dt^2) + W^2 d\varphi^2\right] + e^{2U}(dz + A d\varphi)^2 . \qquad (22.40)$$

Using the corresponding equations (19.43) for the stationary axisymmetric vacuum solutions and making the complex substitution (22.2), i.e. $t \to iz$, $z \to it$, $A \to iA$, we obtain

$$2W\left(WU_{,M}\right)^{;M} = e^{4U}A_{,M}A^{,M}, \quad \left(W^{-1}e^{4U}A_{,M}\right)^{;M} = 0, \qquad (22.41)$$

and k is to be determined (for $W_{,M}W^{,M} \neq 0$) from

$$\begin{aligned} k_{,\rho} = \left[W_{,\rho}^2 - W_{,t}^2\right]^{-1}&\left[\tfrac{1}{4}e^{4U}\left(W_{,\rho}(A_{,\rho}^2 + A_{,t}^2) - 2W_{,t}A_{,t}A_{,\rho}\right)/W\right.\\ &\left.+W\left(W_{,\rho}(U_{,\rho}^2 + U_{,t}^2) - 2W_{,t}U_{,\rho}U_{,t}\right)\right] + \tfrac{1}{2}[\ln(W_{,\rho}^2 - W_{,t}^2)]_{,\rho},\\ k_{,t} = \left[W_{,\rho}^2 - W_{,t}^2\right]^{-1}&\left[\tfrac{1}{4}e^{4U}\left(2W_{,\rho}A_{,\rho}A_{,t} - W_{,t}(A_{,\rho}^2 + A_{,t}^2)\right)/W\right.\\ &\left.+W\left(2W_{,\rho}U_{,t}U_{,\rho} - W_{,t}(U_{,\rho}^2 + U_{,t}^2)\right)\right] + \tfrac{1}{2}[\ln(W_{,\rho}^2 - W_{,t}^2)]_{,t}. \end{aligned} \qquad (22.42)$$

Because of (17.10), W has to satisfy the wave equation

$$W_{,\rho\rho} - W_{,tt} = 0. \tag{22.43}$$

For its general solution $W = f(t-\rho) + g(t+\rho)$, the gradient of W can be spacelike, timelike, or null. For non-constant W, one can (locally) adjust the coordinates by

$$W = \begin{cases} \rho \\ t \\ u = t - \rho \end{cases} \quad \text{for} \quad W_{,\rho}{}^2 - W_{,t}{}^2 \begin{cases} > 0 \\ < 0 \\ = 0 \end{cases}. \tag{22.44}$$

For diagonal metrics ($A = 0$), in each of the three cases the field equations (22.41) reduce to a linear equation for U which can be solved e.g. by separation or by an integral representation (cp. Carmeli *et al.* (1981)).

If we choose $W = \rho$, the field equations read

$$\begin{aligned} U_{,\rho\rho} + \rho^{-1}U_{,\rho} - U_{,tt} &= \tfrac{1}{2}\rho^{-2}\mathrm{e}^{4U}\left(A_{,\rho}{}^2 - A_{,t}{}^2\right), \\ A_{,\rho\rho} - \rho^{-1}A_{,\rho} - A_{,tt} &= 4\left(A_{,t}U_{,t} - A_{,\rho}U_{,\rho}\right), \end{aligned} \tag{22.45}$$

where k is determined by

$$\begin{aligned} k_{,\rho} &= \rho\left(U_{,\rho}{}^2 + U_{,t}{}^2\right) + \tfrac{1}{4}\rho^{-1}\mathrm{e}^{4U}\left(A_{,t}{}^2 + A_{,\rho}{}^2\right), \\ k_{,t} &= 2\rho U_{,\rho}U_{,t} + \tfrac{1}{2}\rho^{-1}\mathrm{e}^{4U}A_{,\rho}A_{,t}. \end{aligned} \tag{22.46}$$

The integrability conditions for (22.46) are satisfied by virtue of (22.45). The substitution (22.2) takes (real) stationary axisymmetric solutions into (in general complex) cylindrically-symmetric solutions. In some cases one can obtain real counterparts by analytic continuation of the parameters. No systematic investigation of this problem has yet been made. Examples are the counterpart of the Kerr solution (Piran *et al.* 1986) and of the Tomimatsu-Sato solutions (Papadopoulos and Xanthopoulos 1990).

Similar results can be obtained for $W = t$. The case $W = \text{const}$ gives flat space-time, as can be easily seen from (25.41). $W = u$ implies the existence of a covariantly constant null Killing vector; these metrics are special plane waves. For many applications, in particular for colliding plane waves, null coordinates with $W = f(u) + g(v)$ are appropriate. Because of the many papers concerned with these colliding plane waves, a chapter (Chapter 25) has been devoted to them.

Also if for a solution the sign of $W_{,\rho}{}^2 - W_{,t}{}^2$ varies, a choice of W different from the normal forms (22.44) can be appropriate. So $W = \sin\rho \sin t$, $A = 0$ (Gowdy 1971, 1975) gives the Gowdy universes in the form

$$\begin{aligned} &\mathrm{d}s^2 = \mathrm{e}^{-2U}\left[\mathrm{e}^{2k}(\mathrm{d}\rho^2 - \mathrm{d}t^2) + \sin^2\rho \sin^2 t\, \mathrm{d}\varphi^2\right] + \mathrm{e}^{2U}\mathrm{d}z^2, \\ &\sin t(U_{,\rho}\sin\rho)_{,\rho} - \sin\rho(U_{,t}\sin t)_{,t} = 0, \end{aligned} \tag{22.47}$$

and a line integral for k. Standard separation of the equation for U gives $U = \sum c_n h_n(\cos\rho)h_n(\cos t)$, where the h_n are the Legendre functions (first and second kind). This form of the solution suggests considering ρ as a periodic coordinate also; in the case of spherical topology of the space sections, ρ, z and φ are interpreted as generalized Euler angle coordinates. The resulting space-time metrics satisfy the regularity conditions at $\rho = 0 = \pi$ and the matching conditions across $W_{,\rho}{}^2 - W_{,t}{}^2 = \cos^2\rho - \cos^2 t = 0$ by an appropriate choice of the c_n and have initial and final collapse singularities at $t = 0$ and $t = \pi$. Another possible topology is that of a 3-handle $S^1 \otimes S^2$, see e.g. Hanquin and Demaret (1983). The analogous cases (and topologies) belonging to $W = f(\rho)h(t)$, with $(f, h) =$ (sinh, cosh, exp), are discussed in Hanquin (1984) and in Hewitt (1991a). For a discussion of (other) spatially compact space-times admitting two spatial Killing vectors see Tanimoto (1998).

As in the stationary axisymmetric case, powerful generation methods are available for the construction of solutions, see Chapter 10 for details and solutions. Here we shall only give some examples.

The best-known subcases are the *Einstein–Rosen waves* (so that sometimes the whole class treated in this section is called *generalized Einstein–Rosen waves*). They are the counterpart of the *static* axisymmetric solutions (Weyl's class, §20.2) and are characterized by the existence of two *hypersurface-orthogonal* spacelike Killing vectors (so that one can put $A=0$) and a spacelike gradient of W ($W=\rho$). All solutions of this class can be obtained from the cylindrical wave equation and a line integral,

$$\rho^{-1}(\rho U_{,\rho})_{\rho} - U_{,tt} = 0, \quad k = \int \left[\rho\left(U_{,\rho}{}^2 + U_{,t}{}^2\right)\mathrm{d}\rho + 2\rho U_{,\rho}U_{,t}\mathrm{d}t\right]. \quad (22.48)$$

The complex substitution $t \to \mathrm{i}z$, $z \to \mathrm{i}t$ leading from (20.3) (Weyl's class) to (22.48) was first mentioned by Beck (1925). The interpretation of the solutions of (22.48) as cylindrical gravitational waves is due to Einstein and Rosen (1937). Because of the cylindrical symmetry, the Einstein–Rosen waves do not describe the exterior fields of bounded radiating sources. The solutions of the class (22.48) are of Petrov type *I* or *II* (Petrov 1966, p. 447). Superposing a cylindrical wave and flat space-time, Marder (1969) constructed a spherical-fronted pulse wave. For an analogous method for static Weyl solutions, see §17.2. Chandrasekhar (1986) and Chandrasekhar and Ferrari (1987) gave different forms of the field equations, discussed their relation to the Ernst equation (19.39) and constructed some special wave packet solutions.

Solutions with $W = t$, or with a t-dependent W, occur naturally as subcases of the vacuum Bianchi type solutions (§13.3) if their symmetry group has an Abelian subgroup G_2. The relations of the Einstein–Rosen waves to the vacuum Bianchi models are discussed in Jantzen (1980a), Carmeli and Charach (1980) and Hanquin and Demaret (1984); the last paper also contains some Einstein–Rosen waves with an additional homothetic vector, see also Van den Bergh (1988a).

A special solution with $W = t$ has been given by Cox and Kinnersley (1979) as

$$\begin{aligned} ds^2 &= t^{-1/2}\left[e^{2k}(d\rho^2 - dt^2) + t^2 d\varphi^2\right] + \sqrt{t}(dz + A d\varphi)^2, \\ k &= \tfrac{1}{16}\ln t - \tfrac{1}{2}\textstyle\int [A'(u)]^2\, du, \quad u \equiv \rho - t, \end{aligned} \tag{22.49}$$

where $A(\rho - t)$ is an arbitrary function.

A special standing wave solution was discussed by Halilsoy (1988b); it is given by (22.40) with

$$\begin{aligned} e^{-2U} &= \cosh^2\alpha\, e^{-2CJ_0(\rho)\cos t} + \sinh^2\alpha\, e^{2CJ_0(\rho)\cos t}, \\ A &= -2C\sinh(2\alpha)\rho J_1(\rho)\sin t, \\ k &= \tfrac{1}{2}C^2\left[\rho^2(J_0(\rho)^2 + J_1(\rho)^2) - 2\rho J_0(\rho)J_1(\rho)\cos^2 t\right], \end{aligned} \tag{22.50}$$

where the $J_i(\rho)$ are Bessel functions.

A real solution of (22.41) which seems not to have a real stationary axisymmetric counterpart was given by Papapetrou (1966). It can be written in terms of a null coordinate u as

$$\begin{aligned} &\gamma_{MN} dx^M dx^N = e^{2k}\left[G(u)v\, du^2 - du\, dv\right], \quad A^2 = v^{-1}, \\ &G(u) = \dot{F}(u)/F(u), \quad W = (u - v^{-1})F(u); \quad e^{4U} = WF(u). \end{aligned} \tag{22.51}$$

This vacuum solution contains an arbitrary function $F(u)$. The conformal factor e^{2k} was calculated by Reuss (1968).

A one-soliton solution (Belinskii and Zakharov 1978, Tomimatsu 1989) is given by

$$\begin{aligned} e^{2U} &= \frac{w(a^2+1)}{(a^2+w^2)}, \quad A = \frac{a\rho(1-w^2)}{w(a^2+1)}, \quad e^{2k} = \frac{b\sqrt{\rho}}{\sqrt{t^2-\rho^2}}, \\ w &= \left[t - \sqrt{t^2-\rho^2}\right]/\rho, \quad a, b = \text{ const.} \end{aligned} \tag{22.52}$$

Solutions where the metric functions are products of functions depending only on ρ or t (and with $A = 0$) have been found by Harris and Zund (1976) and Stein-Schabes (1986). Fischer (1980) found similarity solutions of (22.45) for which A and e^{2U}/ρ are functions only of ρ/t (corresponding to a homothetic vector $\mathbf{H} = \rho\partial_\rho + t\partial_t$) or of $\rho^2 - t^2$.

Algebraically general solutions with real ρ and σ and $\Psi_0 = -3\sigma^2/r$ admit two spacelike commuting Killing vectors, and the solutions can be constructed (up to an ordinary differential equation) using the Newman–Penrose formalism (Bilge 1990).

22.4 Einstein–Maxwell and pure radiation fields

As in the vacuum case, most of the known Einstein–Maxwell solutions have been found by using generation methods. So we refer the reader to Chapters 10 and 34 for methods and solutions, and to Chapter 25 for solutions representing colliding waves.

In the following we shall discuss the counterpart of the Einstein–Rosen waves, i.e. metrics of the form

$$ds^2 = e^{-2U}\left[e^{2k}(d\rho^2 - dt^2) + W^2 d\varphi^2\right] + e^{2U}dz^2, \tag{22.53}$$

where all metric functions depend only on ρ and t, and one can choose $W = \rho$ (assuming $W_{,\rho}^2 - W_{,t}^2 > 0$); they are related by a complex substitution to the static subcase of the stationary axisymmetric solutions considered in §19.4. For the electromagnetic counterpart of the Gowdy universes (§22.3) see Charach (1979).

The existence of the Killing vector $\boldsymbol{\zeta} = \partial_z$ enables us to introduce (real) scalar potentials Θ and η (cp. §18.4),

$$\sqrt{\kappa_0/2}\,\zeta^a F^*_{ab} = \Theta_{,b} - i\eta_{,b}. \tag{22.54}$$

If Θ and η are functions of ρ and t, then the non-zero components of the field tensor F_{mn} with respect to the metric (22.53) are

$$\begin{aligned} \sqrt{\kappa_0/2}F_{z\rho} &= \Theta_{,\rho}, & \sqrt{\kappa_0/2}F_{zt} &= \Theta_{,t}, \\ \sqrt{\kappa_0/2}F_{\varphi\rho} &= We^{-2U}\eta_{,t}, & \sqrt{\kappa_0/2}F_{\varphi t} &= We^{-2U}\eta_{,\rho}. \end{aligned} \tag{22.55}$$

It follows from (22.55) that $T^\rho_\rho + T^t_t = 0$, so that (22.43) is satisfied and we can choose $W = \rho$. By means of the substitution $t \to iz, z \to it$ we get from the field equations (19.28)–(19.35)

$$U_{,\rho\rho} + \rho^{-1}U_{,\rho} - U_{,tt} = -e^{-2U}\left(\Theta_{,\rho}^2 - \Theta_{,t}^2 + \eta_{,\rho}^2 - \eta_{,t}^2\right), \tag{22.56a}$$

$$\Theta_{,\rho\rho} + \rho^{-1}\Theta_{,\rho} - \Theta_{,tt} = 2\left(U_{,\rho}\Theta_{,\rho} - U_{,t}\Theta_{,t}\right), \tag{22.56b}$$

$$\eta_{,\rho\rho} + \rho^{-1}\eta_{,\rho} - \eta_{,tt} = 2\left(U_{,\rho}\eta_{,\rho} - U_{,t}\eta_{,t}\right), \tag{22.56c}$$

$$k_{,\rho} = \rho\left(U_{,\rho}^2 + U_{,t}^2\right) + \rho e^{-2U}\left(\Theta_{,\rho}^2 + \Theta_{,t}^2 + \eta_{,\rho}^2 + \eta_{,t}^2\right), \qquad (22.56d)$$

$$k_{,t} = 2\rho U_{,\rho}U_{,t} + 2\rho e^{-2U}\left(\Theta_{,\rho}\Theta_{,t} + \eta_{,\rho}\eta_{,t}\right), \qquad (22.56e)$$

$$\eta_{,t}\Theta_{,\rho} = \eta_{,\rho}\Theta_{,t}. \qquad (22.56f)$$

From the last equation we conclude that either one of the potentials Θ or η is a constant (and can be made zero by a gauge transformation) or the potentials are mutually dependent, $\eta = \eta(\Theta)$. The first case is contained as a special case in (22.59) and (22.60) below. For $\eta = \eta(\Theta)$, equations (22.56*b*) and (22.56*c*) are consistent only if

$$\left(\Theta_{,\rho}^2 - \Theta_{,t}^2\right) d^2\eta/d\Theta^2 = 0. \qquad (22.57)$$

For electromagnetic *null fields*, the first factor in (22.57) vanishes,

$$F_{ab}F^{*ab} = 0 \Rightarrow \Theta_{,\rho}^2 - \Theta_{,t}^2 = 0. \qquad (22.58)$$

In this case, the potentials Θ and η are arbitrary functions of $u = (t - \rho)/\sqrt{2}$ or $v = (t+\rho)\sqrt{2}$. The general solutions of the field equations (Misra and Radhakrishna 1962) are

$$ds^2 = \rho^{-1/2}\exp\left[-2^{3/2}\int\left(\dot{\Theta}^2 + \dot{\eta}^2\right)du\right]\left(d\rho^2 - dt^2\right) + \rho\left(d\varphi^2 + dz^2\right) \qquad (22.59)$$

for $\Theta = \Theta(u)$, $\eta = \eta(u)$; and a similar solution for $\Theta = \Theta(v)$, $\eta = \eta(v)$.

For electromagnetic non-null fields, (22.57) demands that η is a linear function of Θ,

$$\Theta = \Psi\cos\alpha, \quad \eta = \Psi\sin\alpha, \quad \alpha = \text{const}. \qquad (22.60)$$

In terms of the potential Ψ introduced in (22.60), the field equations (22.56*a*)–(22.56*c*) read

$$\begin{aligned} U_{,\rho\rho} + \rho^{-1}U_{,\rho} - U_{,tt} &= e^{-2U}\left(\Psi_{,t}^2 - \Psi_{,\rho}^2\right), \\ \Psi_{,\rho\rho} + \rho^{-1}\Psi_{,\rho} - \Psi_{,tt} &= 2\left(U_{,\rho}\Psi_{,\rho} - U_{,t}\Psi_{,t}\right). \end{aligned} \qquad (22.61)$$

Up to the complex replacement $t \to iz, z \to it$, these equations have exactly the same structure as the equations (19.40) for stationary axisymmetric vacuum fields.

Theorem 22.1 *From a stationary axisymmetric vacuum solution* (U, ω) *(in Weyl's coordinates) one obtains a corresponding cylindrically-symmetric Einstein–Maxwell field* (U, Ψ) *by the substitution*

$$t \to iz, \quad z \to it, \quad 2U \to U, \quad \omega \to \Psi. \qquad (22.62)$$

For instance, the solution (Radhakrishna 1963)

$$W = \rho, \quad \Psi = at, \quad e^U = ab^{-1}\rho\cosh(b\ln\rho) \tag{22.63}$$

belongs to the class related to the Lewis class (§20.4) by (22.62).

Liang (1995) used the Newman–Penrose formalism and studied the case $\Phi_1 = 0$, cp. §7.2. The resulting Petrov type II metric and the potential are given by

$$\begin{aligned} ds^2 &= \left[g(v) - \kappa_0 a^2u^2\right]^{-1/2} du\, dv \\ &\quad + \left[g(v) - \kappa_0 a^2 u^2\right]\left(e^{-\chi(v)}dz^2 + e^{\chi(v)}d\varphi^2\right), \\ \Psi &= au\sqrt{\kappa_0}e^{-\chi(v)/2}, \end{aligned} \tag{22.64a}$$

where the functions $g(v)$ and $\chi(v)$ have to satisfy

$$2g'' + \chi'^2 g = 0. \tag{22.64b}$$

The generation methods outlined in Chapter 34 apply also when two spacelike Killing vectors exist. Harrison (1965), Misra (1962, 1966) and Singatullin (1973) generated cylindrically-symmetric wave solutions of the Einstein–Maxwell equations from vacuum solutions.

The ansatz (Harrison 1965)

$$U = U(\eta), \quad \Psi = \ln\rho - \int Q(\eta)d\eta, \quad \cosh\eta = t/\rho \tag{22.65}$$

reduces (22.61) to a system of two ordinary first-order differential equations

$$Q' = 2U'(Q + \cosh\eta), \quad Q^2 + e^{2U}\left(U'^2 + C\right) = 1, \quad C = \text{const.} \tag{22.66}$$

Harrison (1965) discussed the special solution $C = 0$, $U = 0$, $Q = -1$.

Einstein–Maxwell fields with $\det F_{nm} = 0$ (and $W \neq \rho$) have been constructed by Roy and Tripathi (1972). Solutions for the general metrics (22.3) involving Painlevé transcendents have been found by Wils (1989c), and special solutions with a purely radial electric field and $A \neq 0$ by Roy and Prakash (1977). A solution due to Singh *et al.* (1965) is

$$\begin{aligned} ds^2 &= t^2\rho^{-2/3}(d\rho^2 - dt^2) + \rho^{2/3}t^{1\pm\sqrt{5}}d\varphi^2 + \rho^{2/3}t^{1\mp\sqrt{5}}dz^2, \\ \sqrt{\kappa_0}F_{14} &= \tfrac{1}{3}t\sqrt{2}\rho^{-4/3}. \end{aligned} \tag{22.67}$$

Starting with non-isotropic coordinates, Demaret and Henneaux (1983) found a solution admitting a homothetic vector.

Pure radiation fields with null vector $(1,1,0,0)$ obey, in the metric (22.53) with $W=\rho$, the equations (Krishna Rao 1964)

$$U_{,\rho\rho}+\rho^{-1}U_{,\rho}-U_{,tt}=0, \quad k_{,\rho}+k_{,t}=\rho(U_{,\rho}+U_{,t})^2. \tag{22.68}$$

All other field equations are then satisfied. Hence for every Einstein–Rosen wave (U,k_0) one can generate a pure radiation field (U,k) by

$$k=k_0+f(t-\rho). \tag{22.69}$$

This result can be generalized to the metric (22.1) with $A\neq 0$ (Krishna Rao 1970). Some other pure radiation solutions have been found by Krori and Barua (1976).

The pure radiation field

$$\mathrm{d}s^2=\mathrm{e}^{k(t-\rho)}\left(\mathrm{d}\rho^2-\mathrm{d}t^2\right)+\rho^2\mathrm{d}\varphi^2+\mathrm{d}z^2 \tag{22.70}$$

generated from flat space-time ($U=0=k_0$) has a Weyl tensor of type N (Krishna Rao 1963).

Looking at the general Einstein–Maxwell null field (22.59) we see that a pure radiation field generated by (22.69) satisfies the Maxwell equations only if $\mathrm{e}^{2U}=\rho$.

23
Inhomogeneous perfect fluid solutions with symmetry

In this chapter we cover those solutions containing a perfect fluid, and admitting at least an H_1 and at most an H_3, which are not discussed elsewhere. Most of the known solutions admit a G_2I acting on spacelike orbits, and can be considered to be cosmologies. Vacuum and Einstein–Maxwell solutions with a G_2 on S_2 in which the gradient of the W of (17.4) is timelike may also *ipso facto* be called cosmological. In this book, they and vacua with a G_1 are covered by Chapters 17–22, 25 and 34.

Solutions with a G_r, $r \geq 3$, are discussed in Chapters 13–16: see the tables in §13.5. Relations between them, in vacuum, Einstein–Maxwell and stiff fluid cases, arise from applying generating techniques when the G_3 contains a G_2I (see §10.11, Chapter 34 and, e.g., Kitchingham (1986)). Stationary axisymmetric fluid solutions appear in Chapter 21.

Theorem 10.2 enables one to generate an infinity of solutions with a G_2I on S_2 and equation of state $p = \mu$ from vacuum solutions. Vacua and stiff fluids with a G_2I on S_2 obtainable using the methods of Chapters 10 and 34 have been surveyed by e.g. Carmeli *et al.* (1981), Krasiński (1997) and Belinski and Verdaguer (2001). Any stiff fluid solution with a G_2I on S_2 should be examined to see if it arises from a vacuum using Theorem 10.2 (in which case it is in a sense trivial); in general, we give only references to the cases obtained by that method (in §10.11) and to other solutions that could have been so obtained (in §23.3), but a few of special interest are given explicitly below. Some other special equations of state also allow techniques for generating perfect fluid solutions (see §10.11).

In the case of an irrotational fluid, an extensive classification scheme was given and applied by Wainwright (1979, 1981) but has not been widely used since many of the classes do not contain any known solutions.

A number of the solutions below admit homotheties, and some were explicitly found from this property. Proper homothety generators for them are given in the tables in §11.4 rather than here.

23.1 Solutions with a maximal H_3 on S_3

Qualitative studies, and classifications, of possible solutions with an H_3 on S_3 or T_3 have been given, including treatment by some of the methods described in §13.2 for solutions with a G_3 (see e.g. Wu (1981), Hewitt *et al.* (1988), Hewitt and Wainwright (1990), Hewitt *et al.* (1991), Uggla (1992), Koutras (1992b), Carot *et al.* (1994), Carot and Sintes (1997), Kerr (1998)). Since all such solutions admit a G_2, they may be subcases of known solutions with a maximal G_2. In the qualitative study of solutions some vacuum solutions and solutions with an H_4 on a V_3 (and a G_3 on S_2; see Chapter 15) arise as critical points of the dynamical systems. These and some further perfect fluid solutions with maximal H_3 are given elsewhere in this book: see the tables in §11.4. In particular (23.36) and the special cases of (23.14), (23.32) and (23.38) appear in §23.3 rather than here because they are special cases or subfamilies of separable G_2 solutions.

Wainwright *et al.* (1979) used Theorem 10.2 to generate stiff fluid solutions from generalizations of vacua with a G_3. McIntosh (1978a) noted that most of the solutions so obtained, and the initial vacua, admit an H_3 on S_3; we give them now. (For other known vacua with an H_3, see §11.4 and the tables therein and §§21.1.5 and 22.4; other stiff fluid solutions with an H_3 appear in Wainwright and Marshman (1979), Carot *et al.* (1994) and Carot and Sintes (1997).)

Starting by generalizing the Bianchi *II* vacuum solution (13.55), Wainwright *et al.* (1979) obtained

$$\begin{aligned} &\mathrm{d}s^2 = A^2(-\mathrm{d}t^2 + \mathrm{d}x^2) + t[B(\mathrm{d}y + 2ncx\,\mathrm{d}z)^2 + B^{-1}\mathrm{d}z^2], \\ &A^2 = t^{a^2+\frac{1}{2}n^2-n}(1+c^2t^{2n})\mathrm{e}^{\frac{1}{2}b^2t^2-2abx}, \quad B = t^{n-1}(1+c^2t^{2n})^{-1}, \end{aligned} \tag{23.1}$$

with constants a, b, c and n, in which $\kappa_0\mu = (a^2 - b^2t^2)/2t^2A^2$; in the obvious tetrad, the four-velocity $\boldsymbol{u}$ obeys $u^1/u^4 = bt/a$, so the flow is tilted relative to the orbits of the H_3.

The tilted inhomogeneous $\gamma = 2$ fluid generalization of the G_3VI_h vacuum solution (13.57) is (Wainwright *et al.* 1979)

$$\mathrm{d}s^2 = g_{uu}(-\mathrm{d}u^2 + \mathrm{d}x^2) + \mathrm{e}^{2x}\sinh 2u[W\mathrm{e}^{2nx}\mathrm{d}y^2 + \mathrm{d}z^2/W\mathrm{e}^{2nx}], \tag{23.2a}$$

$$\begin{aligned} &g_{uu} = c^2(\sinh 2u)^{\frac{1}{2}(m^2+n^2-1)+a^2+b^2}(\tanh u)^{2ab+mn}\mathrm{e}^{-[m^2-n^2-3+2(a^2-b^2)]x}, \\ &W = (\sinh 2u)^n(\tanh u)^m, \quad \kappa_0\mu = 2(a^2+b^2+2ab\cosh 2u)/g_{uu}\sinh^2 2u, \end{aligned} \tag{23.2b}$$

where a, b, c, m and n are constants. Here the four-velocity obeys $u^1 = -bu^4 \sinh 2u/(a+b\cosh 2u)$, so this solution is tilted if $b \neq 0$. The case $n = b = 0$, $m^2 + 2a^2 = 1$ is a Petrov type D solution found by Allnutt (1980), while $n = m = 0$ gives (15.82). Baillie and Madsen (1985) obtained the special case $m = \sqrt{3}/F$, $n = 1/F - 1$, $a^2 = 3b^2 = 3(F^2 - 1)/F^2$, by applying Theorem 10.1 to (13.59).

Similarly, generalizing the G_3VI_0 solution (13.58), which is a limit of (13.57), led to the stiff fluid solution (Wainwright *et al.* 1979)

$$\begin{aligned} &\mathrm{d}s^2 = A^2(-\mathrm{d}t^2 + \mathrm{d}x^2) + t(B\,\mathrm{d}y^2 + B^{-1}\mathrm{d}z^2), \ \kappa_0\mu = (a^2 - b^2t^2)/2A^2t^2, \\ &A^2 = t^{a^2+\frac{1}{2}(m^2-1)} \exp\left[\left(n^2 + \tfrac{1}{2}b^2\right)t^2 + 2(nm - ab)x\right], B = t^m e^{2nx}, \end{aligned} \tag{23.3}$$

where a, b, m and n are constants, which is a corresponding limit of (23.2). It is tilted if $b \neq 0$ (in the obvious tetrad, the velocity obeys $u^1 = btu^4/a$).

There is also a $p = \mu$ solution found from the Bianchi type VII_h vacuum solution (13.62), which is given by replacing $G^2(\xi)$ in (13.62), $\epsilon = 1$, with

$$G^2 = (\sinh 2\xi)^{a^2+b^2-3/8}(\tanh\xi)^{2ab} e^{m^2\xi-(m^2+2a^2-2b^2-11/4)w}, \tag{23.4}$$

and Φ by $m\xi$ where $mA = 1$ so $h = 1/m^2$ (Wainwright *et al.* 1979). The energy density and the velocity in the obvious tetrad are given by

$$\kappa_0\mu = 2(a^2 + b^2 + 2ab\cosh 2\xi)/k^2G^2(\sinh 2\xi)^2 \tag{23.5}$$

and $u^1 = -bu^4 \sinh 2\xi/(a + b\cosh 2\xi)$. If $b \neq 0$, this solution is tilted.

Finally, among the $p = \mu$ solutions given by Wainwright *et al.* (1979), there are two with an H_3 contained in a family of comoving solutions which in general have only a G_2; they were pointed out by McIntosh (1978a) and Collins (1991). The relevant metrics take the form

$$\mathrm{d}s^2 = t^{2(q-1)}F(u)(-\mathrm{d}t^2 + \mathrm{d}x^2) + \sqrt{t}[\mathrm{d}y + W(u)\,\mathrm{d}z]^2 + t^{3/2}\mathrm{d}z^2, \tag{23.6}$$

where $u = t - x$ and q is constant. One has $F = e^{-u}$ and $W = u$. The other has $F = u^{-2p}$ and $W = 2\sqrt{2pu}$, where p is constant and $pu > 0$; the homothety is an isometry if $p = q$. For both solutions, $\kappa_0\mu = (q - 13/16)/t^2 g_{xx}$.

Among the comoving separable solutions with a G_2 in Wainwright class $A(ii)$ (see §23.3) studied by Wils (1991) are three stiff fluid solutions with an H_3 on S_3. The first two can be given as follows:

$$\begin{aligned} \mathrm{d}s^2 = \sinh t \Big[& e^{-6x}\mathrm{d}y^2 + e^{4x}(\mathrm{d}z + \sqrt{10}e^{-x}\cosh t\,\mathrm{d}x)^2 \\ &+ e^{2x}\sinh^2 t(-\mathrm{d}t^2 + 6\,\mathrm{d}x^2)\Big], \quad \kappa_0\mu = \tfrac{7}{4}e^{-2x}/\sinh^5 t\,; \end{aligned} \tag{23.7}$$

$$ds^2 = G^2(-dt^2 + 24\,dx^2) + H^2\left[f\,dy^2 + (dz + be^{-2(1+a)x}\cosh t\,dx)^2/f\right],$$
$$b^2 = 8(5 - 2a - a^2),\quad H^2 = e^{-2(1+a)x}\sinh t,\quad f = e^{2(a-5)x}(\sinh t)^a, \tag{23.8}$$
$$G^2 = e^{4(1-2a)x}(\sinh t)^{3-a},\quad \kappa_0\mu = \tfrac{1}{4}(7 - 2a - a^2)e^{(8a-4)x}(\sinh t)^{a-5},$$

where a is a constant such that $b^2 > 0$. If $a = \frac{1}{2}$, (23.8) is spatially homogeneous with a group of Bianchi type $VI_{-1/9}$, while if $a = 1$ it is the member $a^2 = \varepsilon = 1$, $b = c = m = 0$ of the class (33.5) and has Petrov type II. The third, in which the H_3 is timelike or spacelike depending on position, is the special case $\delta = 0$ of (23.51).

23.2 Solutions with a maximal H_3 on T_3

The metric (23.14) with $T = e^{at}$, $\epsilon = 1$, which admits an H_3 on T_3 and has a γ-law fluid flowing tangent to the orbits, gives the line-element considered by Hewitt *et al.* (1988, 1991). By studying symmetries of the dynamical system formed by the field equations for this metric, using methods and parametrizations closely analogous to those for the solutions (14.37)–(14.39), Uggla (1992) independently found the relevant specializations of the families given in §23.3.1 which allow $\epsilon = 1$. For the $2m = 1 + n$ and $n = 0$ cases, these are just (23.20) and (23.21) with $T = e^{at}$ and $\epsilon = 1$: the solution given by Hewitt *et al.* (1991) is a further specialization of (23.20), and (15.84), which has an H_4, is a special case $P = 1$ of the family with $n = 0$ whose general solution is (23.21).

The remaining cases, $n^2 = (3m-2)^2/(2m-1)$ and $n^2 = 1/(5-4m)$, were given only up to integrations. The coordinates in (23.14) were chosen so that $H = cG/WV$, $G^q = W^{1/k_1}V^{1/k_2}$ and $P = W^{-(n+q)/k_1 q}V^{(q-n)/k_2 q}$, where c is a constant, $n^2 - 4m + 3 = q^2$, and W and V are functions of x. For the family with $n = -(3m-2)/\sqrt{2m-1}$, one finds $k_1 = k_2 = 1/q$, $c = 2m$, $\alpha = -2n/(q+n)$ and

$$W = Ae^x + Be^{-x},\quad V = |W'|\left(C + \int [W^\alpha/W'|W'|]dx\right). \tag{23.9}$$

The case $B = 0$ (or $A = 0$) gives (23.17). For the family with $n = -1/\sqrt{5-4m}$, $k_1 = (n-q)/2(m-1)$, $k_2 = 2/q$, $\alpha + 1 = 2/k_1 q$, $c^2 = m^2(5-4m)/(4m-3)$ and

$$W = \sin x,\quad V = \cos x\left(C - \int [\sin^\alpha x/\cos^2 x]dx\right). \tag{23.10}$$

These solutions do not overlap with (23.18) unless $\mu < 0$.

Studying perfect fluid solutions with an H_3, Carot and Sintes (1997) found, in addition to stiff fluids and solutions with an H_4 or greater symmetry, two solutions with fluid flow tilted with respect to the H_3 orbit. One was a special case of the subcase of (23.14)–(23.15) with $P = 1$, $T = t$, $G = x^c$. The other is given by

$$ds^2 = e^{nu}(-A\,dt^2 + B\,dx^2 + e^{-u/2}[e^{-u/2+\lambda v}dy^2 + e^{u/2-\lambda v}dz^2]), \tag{23.11a}$$

$$\ln A = nv + a\exp(\tfrac{1}{2} - n)v + b, \quad (1+2\lambda)A = (1-2\lambda)B, \tag{23.11b}$$

$$\kappa_0(\mu - p)(1-2\lambda)g_{xx} = \lambda(2n-1)^2, \tag{23.11c}$$

$$\kappa_0(\mu + p)(1-2\lambda)g_{xx}A = 2\lambda(2n-1)A_{,v}, \tag{23.11d}$$

where $u = t + x$, $v = t - x$, and λ, n, a and b are constants with $n^2 + \lambda^2 = \frac{1}{2}$ and $0 < \lambda < \frac{1}{2}$.

Bogoyavlensky and Moschetti (1982) found as a stiff fluid solution with a maximal H_3 on T_3 and G_2 on T_2 the case $a = -1$, $N = e^z$ of (18.66). A class of such solutions is considered in Kolassis and Griffiths (1996). The solutions (23.46), given below since they are a special family of separable G_2 solutions, admit an H_3VI on a T_3 to which the fluid four-velocity is tangent. Other perfect fluid solutions with an H_3 on T_3 are given by (21.61) with $\mu = p$, (21.72), and (35.77).

23.3 Solutions with a G_2 on S_2

In almost all known fluid solutions with a G_2 on S_2 the fluid flow is orthogonal to the group orbits: metric (33.14) is an exception. When the flow is also irrotational (which follows necessarily if the group's action is orthogonally transitive) Wainwright (1981) and Sintes (1996) have given classifications. For the Abelian case G_2I one may have (Wainwright 1981):

Class A: G not orthogonally transitive;
  Class A(i): no hypersurface-orthogonal Killing vector;
  Class A(ii): one hypersurface-orthogonal Killing vector;
Class B: G orthogonally transitive;
  Class B(i): no hypersurface-orthogonal Killing vector;
  Class B(ii): two perpendicular hypersurface-orthogonal Killing vectors.

The G_2II possibilities can be similarly split (Sintes 1996), but for perfect fluid matter content all the orthogonally transitive metrics admit at least a G_3 on V_2 (Bugalho 1987, Van den Bergh 1988d), so these cases are covered by Chapters 15 and 16. Van den Bergh (1988d) shows the existence of metrics (with an H_3) where G is not orthogonally transitive.

A metric form covering all cases can be given as (Vera 1998b)

$$ds^2 = -F_0^2 dt^2 + F_1^2 dx^2 + F_2[F_3^2(e^{-az}dy + W_1 dz + W_2 dx)^2 + (dz + W_3 dx)^2/F_3^2], \tag{23.12}$$

where the F_i and W_i are functions of t and x, $a = 0$ for a G_2I, and $a = 1$ for a G_2II; $\boldsymbol{\xi} = \partial_z$ and $\boldsymbol{\eta} = \partial_y$ are the Killing vectors.

As mentioned at the start of this chapter, for the special equation of state $\mu = p$, and a four-velocity orthogonal to the orbit of an orthogonally transitive group G_2I (Class B), an infinity of solutions can be constructed by using generation methods, see §10.11 for details and examples. Solutions which were found by other methods, but could have been generated from appropriate vacuum solutions, are given in Patel (1973a), Bronnikov (1980), Roy and Narain (1981), Argüeso and Sanz (1985), Van den Bergh (1988c), Davidson (1992, 1993b) cp. (23.15), Agnew and Goode (1994), Carot *et al.* (1994), Mars (1995), Carot and Sintes (1997), Fernandez-Jambrina (1997), Mars and Senovilla (1997) and Lozanovski and McIntosh (1999). Solutions which cannot be so generated since – in the notation of Theorem 10.2 – $W_{,n}W^{,n}$ vanishes, can be found in Charach and Malin (1979), Roy and Narain (1981), Agnew and Goode (1994) and Carot *et al.* (1994). (Some of the solutions which can be generated contain special subcases or subspaces where $W_{,n}W^{,n} = 0$.) The solution discussed in Lozanovski and McIntosh (1999) is unusual in having, for suitable observers, a purely magnetic Weyl tensor, while one of the solutions in Mars and Senovilla (1997) provides an example of a non-diagonal separable singularityfree cosmology.

A number of authors have found other fluid solutions, usually not with a γ-law equation of state, by ansätze of separability, with or without the assumption that the fluid is comoving in the coordinates of (23.12); see the following subsections. An example elsewhere in this book is (36.36).

23.3.1 Diagonal metrics

The simplest cases, which we treat first, are the diagonal metrics with a G_2I, Class $B(ii)$, where $W_1 = W_2 = W_3 = 0$ and F_3 is not constant in (23.12). Mars and Wolf (1997) noted that the exchange $x \leftrightarrow t$, with appropriate choices of region, relates solutions, and used this to reduce the number of cases they needed to consider; cp. Senovilla and Vera (1998).

Separable comoving solutions For a Class $B(ii)$ comoving perfect fluid solution, assuming the functions F_i are (multiplicatively) separable gives three cases (Hewitt and Wainwright 1990, Vera 1998b), in two of which there can be a γ-law equation of state with $\gamma \neq 2$. (All the $p = \mu$ solutions separable in comoving coordinates have been determined by Agnew and Goode (1994), and include solutions found in Wainwright *et al.* (1979), Goode (1980) and Davidson (1992, 1993b); see also below and in §10.11.)

The FIRST CASE is where the F_3 of (23.12) is independent of x, so the three-surfaces $t =$ const are conformally flat. These metrics are

$$ds^2 = S^{2m}C^{2m-2}(-dt^2 + dx^2) + SC^{\alpha}(T^n dy^2 + dz^2/T^n) \qquad (23.13a)$$

where $S = \sinh 2qt$, $C = \cosh(2qx/\alpha)$, $T = \tanh qt$, q, m and n are constant, and $\alpha = (2m-2)/(1-2m)$ (Wainwright and Goode 1980). One can also replace sinh by cosh and vice versa, or replace hyperbolic functions by trigonometric functions, though the corresponding solutions generally violate the energy conditions. For (23.13a) the energy density and pressure are given by

$$\tfrac{1}{2}(1-m)\kappa_0(\mu - p) = \frac{(3-4m)q^2}{(SC)^{2m}} = \left[\frac{q^2(1+4m-n^2)}{C^{2(m-1)}S^{2(1+m)}} - \kappa_0 p\right]. \qquad (23.13b)$$

If $n^2 = 4m+1$ in (23.13) we have a γ-law equation of state with $\gamma = 2m/(m+1)$, while if $m = 3/4$ and $n^2 < 4$ we have $p = \mu > 0$.

The subcase $m = 0$ is a Petrov type D solution found by Allnutt (1980) with $p = \mu - 6q^2/\kappa_0$. It has another characterization, arising from the study (Czapor and Coley 1995) of solutions of Class $B(ii)$ admitting in addition an inheriting conformal Killing vector (see §35.4.4). The relevant metric forms in general have a linearly separable conformal factor $r(t) + s(x)$ multiplying a separable metric and do not admit a barotropic equation of state; only (23.13a) with $m = 0$ has a maximal G_2 and admits a barotropic equation of state (Vera 1998a). The limit where $m = 0$, $C = e^{qx}$ is the subcase $n = b = 0$, $m^2 + 2a^2 = 1$ of (23.3).

The SECOND CASE depends on one function T of t and takes the form

$$ds^2 = T^{2m}G^{1-2m}P^n(-dt^2 + H^2dx^2) + TG(T^nPdy^2 + dz^2/T^nP), \quad (23.14a)$$

where m and n are constants, and G, P and H are functions of x (Ruiz and Senovilla 1992). If $m = 1$ the field equations lead to $p = \mu$ solutions, so let $m \neq 1$. Then there are three subcases labelled by $\epsilon = -1$, 0 and 1:

$$T = \begin{cases} A\cosh at + B\sinh at, & \epsilon = 1, \\ At + B, & \epsilon = 0, \\ A\cos at + B\sin at, & \epsilon = -1, \end{cases} \qquad (23.14b)$$

where a, A and B are constants. The energy density and pressure obey

$$\kappa_0 p - \rho = \kappa_0 \mu - \frac{m+1}{m-1}\rho = \frac{4m+1-n^2}{4T^{2m}F^2}\left[\frac{\dot{T}^2}{T^2} - \epsilon a^2\right], \tag{23.14c}$$

$$H^2 T^{2m} F^2 \rho \equiv \frac{nG'P'}{2GP} - \frac{P'^2}{4P^2} + \frac{(3-4m)G'^2}{4G^2} + \frac{(4m-3-n^2)\epsilon a^2 H^2}{4},$$

where $F^2 = G^{1-2m}P^n$. There are two differential equations linking G, H and P, so in general one function, corresponding to a choice of the coordinate x, can be chosen freely. Four families were examined in detail by Ruiz and Senovilla (1992) and are given below. Each contains solutions with an H_3 mentioned in §23.2. The fluid has a γ-law equation of state with $\gamma = 2m/(m+1)$ if $n^2 = 4m+1$ (these cases were considered by Van den Bergh and Skea (1992)) or if $T = e^{at}$, $\epsilon = 1$, which implies a homothety. The latter case contains the solutions found by Uggla (1992), see §23.2, and also the special solutions without restrictions on n and m found by Kamani and Mansouri (1996) by taking $H^2G^{1-2m}P^n = 1$, $P = e^{qx}$, $G = e^{sx}$ for constants q and s.

The *first family* is given by $\epsilon = 0$. For general m and n,

$$P^{k+\ell} = C^{(3-4m)/(1-2m)} + r, \quad G = P^k C^{2(1-m)/(2m-1)},$$
$$k+\ell = -\frac{\sqrt{n^2+3-4m}}{1-m}, \quad k = \frac{n+\sqrt{n^2+3-4m}}{4m-3}, \quad H = sC'/P^\ell, \tag{23.15}$$

where r and s are arbitrary constants, and $C(x)$ is an arbitrary function. The $A = 0$ limit gives the static cylindrically-symmetric solutions (22.27). The case $C = (1+x^2)^q$, $k = \ell = -N = 1$, $m = (3q+2)/2(2q+1)$, $n = (q+1)/(2q+1)$ was found by Davidson (1992) and includes a $\gamma = 4/3$ solution (Davidson 1991) for $q = -2/5$. The case $P = 1$, $T = t$, $G = x^c$, for some constant c, admits an H_3 which may act on T_3 or S_3 and with respect to which the fluid is tilted. The formulae above have zero or infinite exponents for three choices of m, which thus must be treated separately: they are

$$m = (n^2+3)/4, \quad P = \exp(ax^{2n^2/(1+n^2)}),$$
$$G = P^{1/n}(ax)^{(1-n^2)/(1+n^2)}, \quad H = P^{1/n}; \tag{23.16a}$$

$$m = \tfrac{3}{4}, \quad P^{4n} = ax + r, \quad G = e^{-ax}P^{1/2n}, \quad H = e^{-ax}P^{1/2n-4n}; \tag{23.16b}$$

$$m = \tfrac{1}{2}, \quad P^{k+\ell} = C + r, \quad G = P^k/C, \quad H = sC'/CP^\ell, \tag{23.16c}$$

where C, k, ℓ, r and s are as in (23.15). The special case $P = 1$, $T = t$, $G = x$ was given by Roy and Narain (1983) and $m = 3/4$, $n = 0$, $H = 1$, $G = x$, which has $p = \mu$, was given by Patel (1973a).

The *second family* is for $\epsilon = 1$, $n^2 = (3m-2)^2/(2m-1)$, $m > \frac{1}{2}$. The general solutions involve hypergeometric functions, cp. (23.9). A special case with elementary functions is given by $H = 1$ with

$$P = f^{\sqrt{2m-1}} \exp\left[\frac{m-1}{2\sqrt{2m-1}} ax\right], \quad G = f \exp\left[-\frac{m-1}{2(2m-1)} ax\right],$$
$$f = A_1 \exp\left[\frac{5m-3}{2(2m-1)} ax\right] - A_2 \exp\left[-\frac{5m-3}{2(2m-1)} ax\right], \tag{23.17}$$

where A_1 and A_2 are arbitrary constants.

The *third family* has $\epsilon = \pm 1$, $n^2 = 1/(5-4m) > 0$; cp. (23.10). Particular solutions are given by $H = 1$ and

$$P = f^{\sqrt{5-4m}}, \quad G = f f'^{2(1-m)/(2m-1)}, \quad f'' = \frac{2m-1}{5-4m} \epsilon a^2 f, \tag{23.18}$$

where $f(x)$ can easily be found explicitly once the parameters have been chosen. There is a special case for $m = \frac{1}{2}$: with constants A_1 and A_2,

$$H = 1, \quad P = C^{\sqrt{3}}, \quad G = C e^{\epsilon a^2 C^2 / 6 A_1^2}, \quad C \equiv A_1 x + A_2. \tag{23.19}$$

Finally, Ruiz and Senovilla (1992) gave the *fourth family* $2m = 1 + n$, $n \neq \frac{1}{2}$, which can be written with $H = 1$ and

$$G = CC', \quad P = GC^{2n-2}, \quad C'^2 = \epsilon q C^2 + r - s C^{2-4n}, \tag{23.20}$$

where $C = C(x)$ and q, r, and s are constants. The case $n = 3$ gives radiation solutions (Ruiz and Senovilla 1992, Van den Bergh and Skea 1992), including those of Feinstein and Senovilla (1989b) and Senovilla (1990), the latter providing an important example of a cosmology without a singularity, and cases with a G_4 on S_3 (see §14.3).

A *further family* with $n = 0$ can be given, generalizing the work of Uggla (1992) for the case with $T = e^{at}$, in the form

$$G = x^{r/b}, \quad P = e^s, \quad s = \int \frac{dx}{bx\sqrt{F}}, \quad r^2 = 1/(4m-3), \quad H = cG/z\sqrt{F},$$
$$F = -1 + 2kx^{2r/b} - x^{1/r(m-1)b}, \quad b^2 c^2 = 2kr^2 m^2, \tag{23.21}$$

where t, y and z have been scaled so that $am = 1$ if ϵ is nonzero. s can be given explicitly in terms of elliptic functions if $(0,\ 2r,\ 1/r(m-1))$ are affine to a subset of (0, 1, 2, 3, 4). The special case with $m = 0$, $\epsilon = 1 = P$, which can be written using $G = H^{-2} = \sinh^2(ax/2)$, was given by Tariq and Tupper (1992); (15.84), with a maximal H_4, is another special case.

The THIRD CASE (Vera 1998b) is characterized by the form

$$ds^2 = T^{2m}F^2(-dt^2 + dx^2) + G(T^n P dy^2 + dz^2/T^n P), \tag{23.22}$$

where $T = T(t)$ and F, G and P are functions of x, as before. It then turns out that $\ln T$ is in general quadratic in t, and the remaining functions can be determined from quadratures once the separation constants are known. The stiff fluid example

$$P = G = x,\ F = e^{ax^2},\ T = e^{bt},\ m = n = 1, \tag{23.23}$$

where a and b are constants and $\kappa_0\mu = (a - b^2/4)/e^{ax^2+bt}$, was found by Krori and Nandy (1984) using Theorem 10.1, has an H_3 on T_3 and is included in the diagonal separable solutions of Agnew and Goode (1994), as $d_2 = \frac{1}{2}$ in Case M12.

Partially separable comoving solutions Mars (1995) considered comoving fluid solutions in which $P = P(x)$ in

$$ds^2 = F(x,t)\left[-\frac{dt^2}{M(t)} + \frac{dx^2}{N(x)}\right] + G(x,t)\left(P\,dy^2 + \frac{dz^2}{P}\right). \tag{23.24}$$

This class is invariantly defined by constancy of $\boldsymbol{\xi}\cdot\boldsymbol{\xi}/\boldsymbol{\eta}\cdot\boldsymbol{\eta}$ along the fluid flow, where $\boldsymbol{\xi}$ and $\boldsymbol{\eta}$ are the Killing vectors as in (23.12). After extensive calculation five families were found, of which one is contained in (23.14), one is a $p = \mu$ family obtainable via Theorem 10.2, and two of the remainder disobey the dominant energy condition, leaving only the case

$$F = \exp(at + \tfrac{3}{2}au + c^2e^{2au}),\quad u = t + x,\quad G = e^{a(t-x)},\quad M = 1 + e^{-2at},$$
$$N = 1 - e^{-6ax},\quad P = \exp\left(\int 2ace^{ax}dx/\sqrt{N}\right),\quad a,\ c\ \text{const}, \tag{23.25}$$
$$\mu = p + 4a^2e^{-6ax}/\kappa_0 F,\quad \kappa_0 p = a^2(2c^2e^{2at-4ax} + e^{-6ax} + 3e^{-2at})/2F.$$

Roy and Prasad (1989, 1991) found a number of comoving perfect fluid solutions from an ansatz with partial separation

$$ds^2 = e^{\alpha(x,t)}(dx^2 - dt^2) + e^{\beta(t)+\gamma(t)+2x}dy^2 + e^{\beta(t)-\gamma(t)+2qx}dz^2, \tag{23.26a}$$

where q is constant, which includes homogeneous metrics of types G_3VI_0 and G_3VI_h for $\alpha_x = 0$. The perfect fluid conditions give three equations between α, β and γ. Their solutions have matter content given by

$$\begin{aligned}\kappa_0\mu &= \kappa_0 p + (\ddot{\beta} + (\dot{\beta})^2 - (1+q)^2)/\mathrm{e}^{\alpha} \\ &= (2(1+q)\alpha_{,x} + 2\dot{\alpha}\dot{\beta} + \dot{\beta}^2 - \dot{\gamma}^2 - (1+q+q^2))/2\mathrm{e}^{\alpha}. \quad (23.26b)\end{aligned}$$

In the general case, the solutions obey

$$\alpha = \beta + (1-q)\gamma/(1+q) + \xi, \quad \xi = n\exp(m(1+q)x - \beta) + l, \quad (23.27a)$$

where β is given by one of

$$\beta = (1+q)t, \quad \text{or} \quad \mathrm{e}^{(1-m)\beta} = \begin{cases} k\sinh((1+q)(1-m)t)/(1+q) \\ k\cosh((1+q)(1-m)t)/(1+q) \end{cases}, \quad (23.27b)$$

and γ is obtained from

$$\gamma = (1-h^2)\int\left(\mathrm{e}^{-\beta}\int \mathrm{e}^{\beta}\mathrm{d}t\right)\mathrm{d}t + s\int \mathrm{e}^{-\beta}\mathrm{d}t, \quad (23.27c)$$

k, l, m, n and s being constants. There are a number of special subcases. If $m=0$, $\xi = n[(1+q)x - \beta] + l$ while if $m=1$, $\beta = wt$ for some constant w and then γ can be given explicitly; this has a limiting case in which $w=0$. If $1+q=0$, we have the cases

$$\begin{aligned}&\beta = (w\ln t)/(w+1), \quad \gamma = wnt^{1/(w+1)}, \\ &\qquad \alpha = \beta + \frac{2(w+1)}{2w+1}t^2 + r + (2nx+s)t^{1/(w+1)}\end{aligned} \quad (23.28)$$

$$\beta = wt, \quad \gamma = -n\mathrm{e}^{-wt}, \quad \alpha = r + (w^2+4)t/w + (2nx+sk)\mathrm{e}^{-wt}, \quad (23.29)$$

where w, n, r, and s are constants. If $1+q=0=\dot{\beta}$, $\mu = p < 0$.

Another simple partially separable comoving solution was found by Senin (1982) as a metric on a torus:

$$\mathrm{d}s^2 = a^2(-\mathrm{d}t^2 + \mathrm{d}x^2 + \sin^2 x\mathrm{d}y^2) + (a\cos x + b)\mathrm{d}z^2, \quad (23.30)$$

where $a = c\sin t$, $b = t$, for some constant c.

Solutions separable in non-comoving coordinates Solutions which are separable in non-comoving coordinates, but not in the comoving diagonal coordinates which always exist in Class $B(ii)$, were investigated by Senovilla and Vera (1998). Having chosen coordinates so that $F_0 = F_1$, the F_i of (23.12) were written as $F_i = \exp[T_i(t) + X_i(x)]$, and the quantities M_i were defined as the quadratic products of the T_i together with the functions of t arising in $R_{00} + R_{22}$ and $R_{11} - R_{22}$ (similar functions of

x were named N_j). The possible cases are then distinguished initially by the numbers (m, n) of linearly independent T_i and $\dot{M}_i$ respectively (the number q of linearly independent N_i' obeys $q \leq 8 - n$). Only the cases $m = 1$ (Senovilla and Vera 1998) and $m = n = 2$ (Vera 1998b) have been analysed in any detail, although not exhaustively; (23.41) gives an additional solution with $m = 1$, for example, or, by $t \leftrightarrow x$, with $m = n = 2$. In these metrics

$$\begin{aligned} F_1^2\kappa_0(\mu - p) &= \ddot{T}_2 + \dot{T}_2^{\,2} - X_2'' - X_2'^2, \\ 4F_1^2\kappa_0 p &= [2(2X_1 + X_2)'' + X_2'^2 + 4X_3'^2] - [2(2T_1 + T_2)^{\cdot\cdot} + \dot{T}_2^{\,2} + 4\dot{T}_3^{\,2}]. \end{aligned} \tag{23.31}$$

From $m = 1$, neglecting solutions with $p = \mu$ or with extra symmetry but using $x \leftrightarrow t$ if necessary to meet the dominant energy condition, five families were found. One was (35.77) and its counterpart with $x \leftrightarrow t$. Another, with $T_1 = aT(t)$, $X_1 = rx$, $T_2 = bT(t)$, $X_2 = 0 = T_3$, $X_3 = qx$, is known only up to a differential equation for T. In the following forms for the remaining three families, a, b, c, r, q and s are constants. The first family has $T_1 = (a + b/2)t$, $X_1 = sX_2$, $T_2 = bt$, $T_3 = \varepsilon brq^2 t$, $X_3 = rX_2$, where $\varepsilon = 1$, 0 or -1 and

$$\frac{2a}{b(1+2s)} = \varepsilon q^2 = \frac{4s^2 - 4s - 1 + r^2}{(2s+1)^2 - r^2}, \quad G'^2 = b^2q^2(1 + \varepsilon G^2), \tag{23.32}$$

$G \equiv \exp X_2$, and the tetrad components of the four-velocity can be found from $4G^2 g_{xx}\kappa_0 (u^1)^2 = b^2q^2(1 + 4s - r^2)(1 + \varepsilon G^2)$, the sign of u_1u_4 being given by R_{14}. When $\varepsilon = 0$ there is a homothety. The second family have $T_1 = at$, $X_1 = -(b^2 + a^2)X_2/2b^2$, $T_2 = 0 = X_3$, $T_3 = bt$ and

$$\exp X_2 = (\cos qx)^{r^2}, \quad r^2 = \frac{b^2 - a^2}{b^2 + a^2}, \quad q^2 = \frac{8b^2(a^2 + b^2)}{(b^2 - a^2)(a^2 + 3b^2)}. \tag{23.33}$$

The last family is given by $F_2 = \cos qx \cosh qt$, $T_1 = X_1 = 0$, $X_3 = bX_2$, $T_3 = -bT_2$, and a corresponding solution with cosh replaced by sinh.

In the case $m = n = 2$, the systems of equations for the many possibilities were given but only the following three families of metrics were given completely. The first family have $T_2 = rst/c$, $X_2 = sx$, $T_3 = ct$, $X_3 = rx$, $X_1 = qx$, $T_1 = at + b\exp(mt)$, where

$$c^2m^2 = s^2(r^2 - c^2), \quad sca = -(mc - rs)q, \quad 4sq = 4c^2 + s^2. \tag{23.34}$$

The second family are dust solutions which can be written

$$\begin{aligned} \mathrm{d}s^2 &= -2\mathrm{d}u\,\mathrm{d}v + F^2\mathrm{d}y^2 + H^2\mathrm{d}z^2, \quad H = (ku)^{1-b} + s(kv)^{-q+1/2}, \\ F &= r(ku)^{q+1/2} + (kv)^b, \quad \kappa_0\mu = \frac{2b(b-1)k(rsu^{q+b}v - v^{q+b}\sqrt{uv})}{u^{b+1/2}v^{q+3/2}FH}, \end{aligned} \tag{23.35}$$

where $r = \pm 1$, $s = 0$ or ± 1, b and k are constants and $4q^2 = 1 + 4b - 4b^2$. Here the velocity 1-form is $-(v\,\mathrm{d}u + u\,\mathrm{d}v)/\sqrt{2uv}$.

Finally there is a third family with a γ-law equation of state (Senovilla and Vera 2001), $\gamma = (2k+1)/(k+1)$,

$$\begin{aligned} &\mathrm{d}s^2 = F^2(-\mathrm{d}t^2 + \mathrm{d}x^2) + \mathrm{e}^{a(t+x)}G(P\mathrm{d}y^2 + \mathrm{d}z^2/P), \quad F = \mathrm{e}^{mt+cx}G^k, \\ &G = 1 - \mathrm{e}^{-(a+b)t}, \quad P = (\mathrm{e}^{b(t+x)}G)^{2\ell}, \quad 4\ell^2 = 4k+1, \\ &0 = (ak + c - m - \tfrac{1}{2}b)^2 - \tfrac{1}{4}b^2(2k+1)^2 + c(a + b - 2c + 2m), \\ &a^2 + 4b^2\ell^2 = 2a(c+m), \quad \kappa_0 p = k(a^2 - b^2)/F^2(\mathrm{e}^{(a+b)t} - 1). \end{aligned} \tag{23.36}$$

The constants a, b and k obey $a^2 > b^2$, $a + b > 0$ and $k \geq -\frac{1}{4}$. If $k = -\frac{1}{4}$, there is a G_3 on S_2. In general, (23.36) admits an H_3 if $c \neq 0$, while if $c = 0$ it admits a G_3 and includes special LRS solutions and subcases of (14.32) and (14.39), as well as the Kasner solution and *pp*-wave limits. The $\gamma = 4/3$ case can be interpreted as waves superposed on (14.32).

The solutions (23.35) suggested ansätze leading to further dust solutions (Senovilla and Vera 1997, Vera 1998b). As well as special Szekeres solutions (see §33.3.2), the following two cases were obtained.

$$\mathrm{d}s^2 = -\mathrm{d}t^2 + \mathrm{d}x^2 + (\sin au + \mathrm{e}^{-av})\mathrm{d}y^2 + \mathrm{e}^{2au}\mathrm{d}z^2, \tag{23.37}$$

with comoving dust with density $\kappa_0\mu = -4a^2\mathrm{e}^{-av}/(\sin au + \mathrm{e}^{-av})$; here and in the next case $u = t - x$, $v = t + x$.

$$\begin{aligned} &\mathrm{d}s^2 = -\mathrm{d}t^2 + \mathrm{d}x^2 + [\sqrt{au}F(u) + (av)^b]^2\mathrm{d}y^2 + (au)^{2(1-b)}\mathrm{d}z^2, \\ &F = \begin{cases} c_1(au)^q + c_2/(au)^q & \text{if } (i)\ b \in (b_-,\, b_+), \\ c_1 - c_2 \ln au & \text{if } (ii)\ b = b_\pm, \\ c_1 \cos(q \ln au + c_2) & \text{if } (iii)\ b \notin [b_-,\, b_+], \end{cases} \\ &\kappa_0\mu = -4b(b-1)(au)^b/uv(\sqrt{au}F(u) + (av)^b), \end{aligned} \tag{23.38}$$

where the c_i are constants, $b_\pm = \frac{1}{2}(1 \pm \sqrt{2})$ and $4q^2 = |1 + 4b - 4b^2|$. Here the fluid velocity 1-form is $(-t\,\mathrm{d}t + x\,\mathrm{d}x)/\sqrt{uv}$. The special case $F = 0$ is included in (14.33). There is an H_3 if $c_2 = 0$ in cases (i) or (ii) (or, equivalently by $q \to -q$, $c_1 = 0$ in case (i)). The two solutions (23.37)–(23.38) provide examples with ${B^{ab}}_{;b} = 0 \neq B^{ab}$, cp. Lesame *et al.* (1996).

An assumption of partial separability for metrics with $\mu = p + \text{const}$ in non-comoving coordinates was studied by Xanthopoulos (1987).

Solutions with a conformal motion Carot *et al.* (1996) and Mars and Wolf (1997) studied solutions of Class $B(ii)$ with an additional conformal motion (cp. §35.4), the orbits being non-null; Mars and Wolf (1997) assumed that the G_2I was an invariant subgroup in the three-parameter conformal group. Carot *et al.* (1996) gave some families only up to differential equations or systems, and found the following explicit solutions. First, in comoving coordinates,

$$ds^2 = S^{-2}[-dt^2 + dx^2 + (\cosh x)^{2m}(e^{-2t}dy^2 + \sinh^2 x\, dz^2)], \qquad (23.39a)$$

$$S = b\cosh ku(e^{-u/2}\cosh x)^{m+1}, \quad u \equiv t + \ln\cosh x, \quad 4k^2 = 1 + m^2, \qquad (23.39b)$$

$$\kappa_0\mu = \frac{3Q^2 - m}{S^2\cosh^2 x}, \quad \kappa_0 p = \frac{-Q(2m+3Q)}{S^2\cosh^2 x}, \quad Q \equiv (\ln S)_{,u} + 1; \; b,\, m, \text{ const.}$$

Mars and Wolf (1997) found the analogous solution with $\cosh x$ and $\sinh x$ exchanged, and one with $\sinh ku$ and $\cosh ku$ also exchanged.

Secondly, using the comoving coordinates of Mars and Wolf (1997),

$$ds^2 = S^{-2}(-dt^2 + dx^2/x^2 + x^{b(b+1)}\cosh^{1-b}t\, dy^2 + x^{b(b-1)}\cosh^{b+1}t\, dz^2),$$
$$S = x^k + s_0|\sinh t|^k, \quad 2k = 1 + b^2, \quad b,\, s_0 \text{ const}, \qquad (23.40)$$
$$\kappa_0\mu = A + B, \; A \equiv \tfrac{1}{4}(b^2-1)S^2/(\cosh t)^2, \; B \equiv \tfrac{3}{4}(b^2+1)^2(\sinh t)^{b^2-1},$$
$$\kappa_0 p = A + C, \; 3(b^2+1)C \equiv B[2(b^2-1)(x/\sinh t)^k - b^2 - 5],$$

together with the similar metric with $\cosh t$ and $\sinh t$ interchanged. If $b = 1$ this is conformally flat and if $b = 0$ it is Petrov type D. If $s_0 \to 0$, this is a member of the class (23.26).

The further solutions given by Carot *et al.* (1996) are: the subcase $m = 0$ of (23.13a), due to Allnutt (1980); a tilted fluid solution separable in non-comoving coordinates which, with constants a, b and c and in the notation of (23.31), is given by $2T_1 = T_2 = a - (\ln|\sinh(2bct)|)/c$, $T_3 = 0$,

$$e^{X_1} = (\cosh bx)^{(1-2c)}, \quad e^{X_2} = \sinh 2bx, \quad e^{2X_3} = \tanh bx, \qquad (23.41)$$

(here the sinh in T_2 can be replaced by cosh or exp); a counterpart of (23.41) with $t \leftrightarrow x$; and a solution conformal to a decomposable space-time (35.29) with $E = 0$, which therefore is of Petrov type D and has a six-parameter conformal group,

$$ds^2 = \frac{a^2}{F^2}\left[-dt^2/(1-t^2) + dx^2/(1+x^2) + (1-t^2)dy^2 + (1+x^2)dz^2\right],$$
$$\kappa_0\mu = 3q_0(F-1) - p_0, \quad \kappa_0 p = (3-5F)q_0 + p_0, \quad F = 1 + b\sqrt{tx}, \qquad (23.42)$$
$$p_0 \equiv (3-2F)F/a^2, \quad q_0 \equiv (F-1)(x^2-t^2)/4a^2x^2t^2, \quad a,\, b \text{ const.}$$

Mars and Wolf (1997) gave an additional solution analogous to (23.42):

$$ds^2 = S(v)^{-2}e^{-(1+a)x}(-dt^2 + a\,dx^2/q + e^{-2t}dy^2 + e^{2ax}q\,dz^2), \quad (23.43a)$$

$$w_{,v} = \frac{w(a+w)}{(w+1)}, \quad S(w) = \frac{b|w-1||w+a|^{(1-a)/2a}}{|w|^{(1+a)/2a}}, \quad \begin{matrix} a,\ b \ \text{const}, \\ q = 2e^{-ax}\cosh(ax), \end{matrix} \quad (23.43b)$$

$$\mu = \frac{3(a+1)(1+w)(\mu+p)}{2(a-1)(w-1)} + \frac{e^{(a+1)x}S^2(aw+2a+2w+1)}{\kappa_0(w-1)},$$

where $v = t + x$ and $2\kappa_0(\mu+p)a(1-w^2)g_{tt} = (a - qw^2)(a^2-1)$.

In the case where the conformal group acts on null surfaces $v =$const and is maximal the only solution known explicitly (Sintes *et al.* 1998) is

$$ds^2 = f^2(-2e^{(a+1)v}du\,dv + e^{-u+av}dy^2 + e^{-au+v}dz^2), \quad x \equiv u - v,$$

$$f = cx^{a/(a^2+1)}e^{(a+1)x/4}/\sqrt{x}, \quad \kappa_0(\mu - p) = 4(a^2 - a + 1)Q, \quad (23.44)$$

$$\kappa_0(\mu+p) = \left\{[(a^2+1)x/2]^2 - (a+1)^2\right\}Q, \quad a,\ c \ \text{const},$$

where $Q \equiv -(a-1)^2/e^{(a+1)v}x^2f^2(a^2+1)^2$. If $a = -1$ this admits a G_3.

23.3.2 Non-diagonal solutions with orthogonal transitivity

For the non-diagonal Class $B(i)$ ($a = W_2 = W_3 = 0$ and W_1 not constant in (23.12)) many of the known solutions have $p = \mu$ and are included in the references listed above. Mars and Wolf (1997) showed there are no solutions in Class $B(i)$ obeying their assumptions (just described above).

Mars and Senovilla (1997) defined separability for Class $B(i)$ as separability of the expressions for $\boldsymbol{\xi}\cdot\boldsymbol{\eta}$, $\boldsymbol{\xi}\cdot\boldsymbol{\xi}$ and $\sqrt{(\boldsymbol{\xi}\cdot\boldsymbol{\xi})(\boldsymbol{\eta}\cdot\boldsymbol{\eta}) - (\boldsymbol{\xi}\cdot\boldsymbol{\eta})^2}$. The general separable metric has the form

$$ds^2 = T_f^2F^2(-dt^2 + dx^2) + T_gG[(dy + T_wW dz)^2/T_pP + T_pP dz^2], \quad (23.45)$$

where the T_i are functions of t only and F, G, P and W are functions of x only. A detailed study of possible comoving fluids leads to several metric forms which are not diagonalizable and have a maximal G_2. Of these some turn out to be vacuum or stiff fluid and we omit those here. The stiff fluids with $\mu > 0$ found can all be obtained from vacua using Theorem 10.2. The first of the remaining cases is

$$T_f = T_p^q, \quad T_g = T_p^b, \quad T_p = T_w = e^{at}, \quad (23.46)$$

which gives a γ-law fluid with $\gamma = 2q/(q+b)$. The functions F, G, P and W are only known up to a system of differential equations implying

$\kappa_0\mu = (a^2b^2 - G_{xx}/G)/\gamma P^2 T_g^2$. This class generalizes the subclass of (23.14) in which $T = \mathrm{e}^{at}$, and has as a limit (14.33). It admits a homothety. The dust solutions within this class are the most general with these properties.

A second class can be written with $W = 1$ and, using $X = G^2$,

$$T_f = T_g^\ell, \quad T_g = T_p = \dot{T}_w/b, \quad \dot{T}_g^2 - mT_g^2 = T_0, \quad P = \sqrt{X}(X/Q_{,X})^{\ell-1},$$
$$\mathrm{d}x^2 = \frac{\mathrm{d}X^2}{4X(mX+Q)}, \quad F^2 = a^2 Q_{,X}^{1-\ell}, \quad \kappa_0\mu g_{tt} = \left(\frac{\ell T_0}{T_g^2} - \tfrac{1}{2}(\ell-1)Q_{,X}\right), \tag{23.47}$$

where $\kappa_0(\mu - p)g_{tt} = Q_{,X}$ and a, b, ℓ, T_0 and m are constants; $Q(X)$ obeys a non-linear second-order differential equation. This can be regarded as a generalization of (23.14) with $n = 1$. The equations for T_g and T_w are trivial to integrate once the sign of m is chosen. An explicit example is

$$Q = cX^p, \quad p = (4\ell - 3)/2(\ell - 1), \quad b^2 = -m(cp)^{2-2\ell}, \tag{23.48}$$

so that $m = -k^2 < 0$, $T_g = \sin kt$, $T_w = \cos kt$. Its $\ell = 0$ limit is the de Sitter space-time. The last of these classes is given by $T_w = 1 = T_g$ and

$$T_f^2 = T_p = \exp(b^2 \sin m\tau - a^2), \; G = P = \cos mx = W_x/km,$$
$$\mathrm{d}t^2 = \frac{b^2\cos^2 2m\tau(1 - \mathrm{e}^{-2b^2})\mathrm{d}\tau^2}{2 - 2\mathrm{e}^{-2b^2\sin m\tau} - 2(1-\mathrm{e}^{-2b^2})\sin^2 m\tau}, \quad a^2 = \frac{b^2}{1 - \mathrm{e}^{-2b^2}}, \tag{23.49}$$
$$\mu = p + m^2/T_f^2F^2\kappa_0, \quad \kappa_0 p = m^2(\ln Q - Q)/T_f^2F^2, \quad F = \mathrm{e}^{c\sin mx},$$

where b, c, $k = \sqrt{2}a\mathrm{e}^{-a^2}$ and m are constants. This satisfies the dominant energy condition but not the strong energy condition. The similar solution with hyperbolic functions of x has $\mu < 0$ in some regions.

23.3.3 Solutions without orthogonal transitivity

Studying separable Class $A(ii)$ metrics ($W_1 = W_2 = 0 \neq W_{3,t}$ in (23.12)), Wils (1991) found some stiff fluid solutions. Two, (23.7)–(23.8), always have an H_3. A third, which is of Petrov type D, is

$$\mathrm{d}s^2 = F^4(-\mathrm{d}t^2 + \mathrm{d}x^2)/C^6 + F[\mathrm{d}y^2/F + F(\mathrm{d}z - 2\cos t\,\mathrm{d}x/C^5)^2], \tag{23.50}$$

where $F = SC$, $S = \sin t$, $C^3 = \sin 3x$, and $\kappa_0\mu = 2C^2/S^6$. The last is

$$\mathrm{d}s^2 = t^{4+5c/6}f^{2+c}(-\mathrm{d}t^2 + \mathrm{d}x^2)/x^{2/3} + t^{2+5c/6}x^{2/3}f^{2+c}(\mathrm{d}z + \tfrac{1}{3}t^2\mathrm{d}x/x^{5/3})^2$$
$$+t^{-5c/6}f^{-c}\mathrm{d}y^2, \quad \kappa_0\mu = \tfrac{23}{24}x^{2/3}/f^{3+c}t^{6+5c/6}, \tag{23.51}$$

where $c^2 = 6$ and $f = \delta x^{2/3} + 1$ for an arbitrary constant δ.

Only two stiff fluid solutions, not obtainable via Theorem 10.2, were found explicitly in the study of separable solutions of Class $A(i)$ by Van den Bergh *et al.* (1991). First, we have (23.12) with $\epsilon = -1$, $a = 0$, and

$$F_0 = F_1 = \mathrm{e}^{K+k}, \quad F_2 = \mathrm{e}^{2s}, \quad F_3 = \mathrm{e}^{-(K+s)}, \quad W_2 = 0, \tag{23.52a}$$

where k and s are functions of x and

$$\begin{aligned} s'' &= -\,2s'^2 - c^2\mathrm{e}^{2(k-4s)},\ k'' = 2s'k' + 4c^2\mathrm{e}^{2(k-4s)}, \\ \mathrm{e}^{2K} &= \frac{a}{2b}\cosh 2bt,\ W_1 = a\int \mathrm{e}^{2s}\mathrm{d}x,\ W_3 = 2c(t-t_0)\mathrm{e}^{2(k-3s)}, \end{aligned} \tag{23.52b}$$

a, b, c and t_0 being constants; these solutions have $\kappa_0\mu = (2s'k' + c^2\mathrm{e}^{2(k-4s)} - b^2)\mathrm{e}^{-2(k+K)}$. The other solution given was

$$\begin{aligned} \mathrm{d}s^2 = \cosh(\sqrt{6}x)[&\sinh^4 t(-\mathrm{d}t^2 + \mathrm{d}x^2) + 2\sinh^2 t(\mathrm{d}z + \cosh t\,\mathrm{d}x)^2] \\ &+ 12(\mathrm{d}y + \cosh t\,\mathrm{d}z + \tfrac{1}{2}\cosh^2 t\,\mathrm{d}x)^2/\cosh(\sqrt{6}x), \end{aligned} \tag{23.53}$$

with $\kappa_0\mu = 2/\cosh(\sqrt{6}x)\sinh^6 t$.

23.4 Solutions with a G_1 or a H_2

Solutions with a maximal G_1 are rare. Most of them have been found as a by-product in the search for other classes of solutions (like the generalized Kerr–Schild metrics (32.99), the algebraically special solutions (33.12) and the metrics (35.78) admitting a group of conformal motions), or they are subcases of solutions which in general do not admit any symmetry (like the algebraically special solutions (33.11), (33.40) and (33.44), the metrics (36.30) with conformally flat slices, and the conformally flat solutions (37.39) and (37.45)). Another was found by Papadopoulos and Sanz (1985) but has $p < 0$.

Kolassis and Griffiths (1996) made a thorough search for perfect fluids admitting an orthogonally transitive H_2 when the four-velocity lies either in the orbit of the group or orthogonal to it. If the orbits are timelike (and the four-velocity lies in the orbit), it turns out that the symmetry group is at least an H_3; if the surfaces orthogonal to the orbits are spaces of constant curvature, the group is an H_5. If the orbits are spacelike (and the four-velocity orthogonal to the orbit), the fluid is necessarily stiff ($\mu = p$) and the metric is diagonal, but a third-order partial differential equation remains to be solved; the known solutions admit an H_3.

24

Groups on null orbits. Plane waves

24.1 Introduction

In classifying space-times according to the group orbits in Chapters 11–22, we postponed the case of null orbits; they will be the subject of this chapter. All space-times considered here satisfy the condition $R_{ab}k^a k^b = 0$.

A null surface N_m is geometrically characterized by the existence of a unique null direction $\boldsymbol{k}$ tangent to N_m at any point of N_m. The null congruence $\boldsymbol{k}$ is restricted by the existence of a group of motions acting transitively in N_m.

The groups G_r, $r \geq 4$, on N_3 have at least one subgroup G_3 (Theorems 8.5, 8.6 and Petrov (1966), p.179), which may act on N_3, N_2 or S_2. (A G_4 on N_3 cannot contain G_3 on T_2 since the N_3 contains no T_2.) For G_3 on S_2, one obtains special cases of the metric (15.4) admitting either a group G_3 on N_3 or a null Killing vector (see Barnes (1973a)). For G_3 on N_2, the metric also admits a null Killing vector (Petrov 1966, p.154, Barnes 1979).

Thus we need only consider here the groups G_3 on N_3 (§24.2), G_2 on N_2 (§24.3), and G_1 on N_1 (§24.4). As we study the case of null Killing vectors (G_1 on N_1) separately, we can also restrict ourselves to groups G_3 on N_3 and G_2 on N_2 generated by *non-null* Killing vectors. It will be shown that in these cases, independent of the group structure, there is always a non-expanding, non-twisting and shearfree null congruence $\boldsymbol{k}$.

None of the space-times with a G_4 on N_3 is compatible with the types of energy-momentum tensors considered in this book (see §5.2) (Lauten and Ray 1977), and all space-times admitting a G_3 on N_3 or a G_2 on N_2 (generated by non-null Killing vectors) are algebraically special and belong to Kundt's class (Chapter 31).

24.2 Groups G_3 on N_3

In this section we study space-times V_4 for which the orbits of a group G_3 are null hypersurfaces N_3 parametrized by u (u being constant in each N_3). The null vector $k_a = -u_{,a}$ is orthogonal to all vectors tangent to N_3, and therefore it is orthogonal to the three independent Killing vectors $\boldsymbol{\xi}_A$ $(A = 1, \dots, 3)$,

$$\xi^a_A k_a = 0 = k_a k^a, \quad \xi_{A(a;b)} = 0, \quad k_a = -u_{,a}, \quad k^a = c^A(x)\xi^a_A. \tag{24.1}$$

From these relations it follows immediately that the null congruence $\boldsymbol{k}$ and tensors obtained from it by covariant differentiation have zero Lie derivatives with respect to $\boldsymbol{\xi}_A$, e.g.

$$k^a{}_{;b}\,\xi^b_A - k^b\,\xi^a_{A;b} = 0, \quad k^c{}_{;ca}\xi^a_A = 0. \tag{24.2}$$

Now we use (6.33),

$$R_{ab}k^a k^b = k^b{}_{;ab}k^a = -k_{a;b}k^{a;b} = -2(\sigma\bar{\sigma} + \Theta^2) \leq 0, \tag{24.3}$$

where σ and Θ denote the shear and the expansion, respectively. By continuity, the energy conditions (5.18)

$$T_{ab}u^a u^b \geq 0, \quad T_{ab}T^a{}_c u^b u^c \leq 0, \quad u^a u_a < 0 \tag{24.4}$$

must still be true if we replace the timelike vector $\boldsymbol{u}$ by a null vector $\boldsymbol{k}$:

$$R_{ab}k^a k^b \geq 0, \quad T_{ab}T^a{}_c k^b k^c \leq 0, \quad k^a k_a = 0. \tag{24.5}$$

Comparison of (24.3) with (24.5) leads to $\sigma = 0 = \Theta$, i.e.

$$k_{a;b} = 2k_{(a}p_{b)}, \quad p_a k^a = 0, \quad R_{ab}k^a k^b = 0. \tag{24.6}$$

With (24.6), the second energy condition (24.4) reads

$$R_{ab}R^a{}_c k^b k^c = k^b{}_{;ab}k^{c;a}{}_{;c} \leq 0, \tag{24.7}$$

whereas (24.6) leads to

$$k^b{}_{;ab}k^{c;a}{}_{;c} = 2\left|R_{ab}m^a k^b\right|^2 \geq 0. \tag{24.8}$$

Comparison of the two inequalities gives

$$R_{ab}k^a k^b = 0 = R_{ab}m^a k^b \quad \Leftrightarrow \quad k_{[c}R_{a]b}k^b = 0. \tag{24.9}$$

Theorem 24.1 *If the energy-momentum tensor of a space-time with a* G_3 *on* N_3 *satisfies the energy conditions* (24.5)*, then the non-twisting (and geodesic) null congruence* $\boldsymbol{k}$ *is non-expanding and shearfree and an eigendirection of the Ricci tensor* (Kramer 1980).

Of the types of energy-momentum tensor considered in this book, only vacuum fields and Einstein–Maxwell and pure radiation fields are compatible with a group of motions acting on N_3; perfect fluids (with $\mu+p\neq 0$) are excluded by the condition $R_{ab}k^ak^b = 0$. For the compatible space-times $R_{ab}m^am^b = 0$ follows from (24.9), so that according to Theorem 7.1 they are algebraically special, $\boldsymbol{k}$ being the repeated principal null direction of the Weyl tensor.

Using the Newman–Penrose formalism (Chapter 7), we now inspect the Einstein–Maxwell equations. For the spin coefficients and tetrad components we have

$$\begin{aligned} &\rho = \sigma = \kappa = 0, \quad \varepsilon + \bar{\varepsilon} = 0, \quad \tau = \bar{\alpha} + \beta, \\ &\Phi_{00} = \Phi_{01} = \Phi_{02} = 0, \quad \Psi_0 = \Psi_1 = 0, \quad R = 0 \end{aligned} \tag{24.10}$$

($\boldsymbol{k}$ is a gradient!). We choose the null tetrad so that its Lie derivatives with respect to $\boldsymbol{\xi_A}$ vanish (invariant basis in N_3). The tetrad vectors $\boldsymbol{k}$, $\boldsymbol{m}$, $\overline{\boldsymbol{m}}$ are linear combinations (with non-constant coefficients) of the Killing vectors $\boldsymbol{\xi_A}$. Therefore, the intrinsic derivatives D, δ, $\bar{\delta}$ applied to spin coefficients and tetrad components of the field tensors give zero. With these simplifications and (24.10), the Newman–Penrose equations (7.21c), (7.21p), and the Maxwell equation (7.24) read

$$\tau\varepsilon = 0, \quad \tau\beta = 0, \quad \tau\Phi_1 = 0. \tag{24.11}$$

If we assume $\tau \neq 0$, it follows that $\varepsilon = \beta = \Phi_1 = 0$, and (7.21$l$) and (7.21$q$),

$$\Psi_2 = \alpha\bar{\alpha} + \beta\bar{\beta} - 2\alpha\beta + \Phi_{11} = \tau\bar{\tau}, \quad \Psi_2 = \tau(\bar{\beta} - \alpha - \bar{\tau}) = -2\tau\bar{\tau} \tag{24.12}$$

give contradictory results. Hence $\tau = 0$ must hold;

$$\tau \equiv -k_{a;b}m^al^b = p_am^a = 0, \quad k_{a;b} = A(u)k_ak_b \tag{24.13}$$

(note $\mathcal{L}_\xi k_{a;b} = 0$), i.e. there exists a covariantly constant null vector parallel to $\boldsymbol{k}$ (§6.1). For pure radiation fields ($\Phi_{11} = 0$), the same conclusion can be drawn.

Theorem 24.2 *Vacuum, Einstein–Maxwell and pure radiation fields admitting a group of motions acting transitively on N_3 are plane waves* (*see* §24.5).

24.3 Groups G_2 on N_2

Space-times admitting a null Killing vector will be treated separately in the next section, so we suppose that the group G_2 acting on two-dimensional null surfaces N_2 does not contain a null Killing vector. By supposition, the two spacelike Killing vectors $\boldsymbol{\xi}$ and $\boldsymbol{\eta}$ span a null surface N_2, i.e.

$$\xi_{[a}\eta_{b]}\xi^a\eta^b = 0. \tag{24.14}$$

At any point of N_2 we have a unique null direction tangent to N_2:

$$\boldsymbol{k} = \boldsymbol{\xi} - \Omega\boldsymbol{\eta}, \quad \Omega \equiv (\eta^c\eta_c)^{-1}\xi^b\eta_b \tag{24.15}$$

(cp. the similar expression after (19.19) in the case of stationary axisymmetric fields). The null vector field $\boldsymbol{k}$ is orthogonal to the two Killing vectors,

$$k^a\xi_a = 0 = k^a\eta_a, \tag{24.16}$$

and has the properties

$$\begin{aligned} &k^a{}_{;ab}k^b = 0, \quad k_{a;b}k^b = 0, \quad \omega = \mathrm{i}\, k_{[a;b]}\overline{m}^a m^b = 0, \\ &\sqrt{2}m^a = (\eta^c\eta_c)^{-1/2}\eta^a + \mathrm{i}\, q^a, \quad q^a\eta_a = 0 = q^a k_a, \quad q^a q_a = 1, \end{aligned} \tag{24.17}$$

which can be verified with the aid of the Killing equations, the commutator relation for $\boldsymbol{\xi}$ and $\boldsymbol{\eta}$, and the formula (24.14). Equations (24.17) hold for both the group structures G_2I and G_2II (§8.2).

We insert the information (24.17) on $\boldsymbol{k}$ into (6.33) and conclude, again from (24.3) together with (24.5), that the null vector field (24.15) is non-expanding and shearfree, $\Theta = \sigma = 0$. (The function Ω in (24.15) obeys the relations $\Omega_{,a}k^a = 0 = \Omega_{,a}m^a$.) From conditions (24.5) it follows that $\boldsymbol{k}$ is a Ricci eigenvector. So we can conclude again from Theorem 7.1 that if a vacuum, Einstein–Maxwell or pure radiation field admits a group of motions G_2 on N_2, then it is algebraically special.

Theorem 24.3 *In vacuum, Einstein–Maxwell or pure radiation fields admitting a group G_2 on N_2 there is a non-expanding, non-twisting, and shearfree null congruence and the metric can be transformed into* (cp. §31.2)

$$\begin{aligned} \mathrm{d}s^2 = 2P^{-2}\mathrm{d}\zeta\mathrm{d}\bar{\zeta} - 2\mathrm{d}u(\mathrm{d}v + W\mathrm{d}\zeta + \overline{W}\mathrm{d}\bar{\zeta} + H\mathrm{d}u), \\ P, H \text{ real}, \quad W \text{ complex}, \quad P_{,v} = 0, \quad W_{,vv} = 0. \end{aligned} \tag{24.18}$$

The gauge transformation

$$u' = h(u), \quad v' = v/h_{,u} + g(\zeta, \bar{\zeta}, u), \quad \zeta' = \zeta'(\zeta, u) \tag{24.19}$$

preserving the form of the line element (24.18) can be used to bring the Killing vectors $\boldsymbol{\xi}$ and $\boldsymbol{\eta}$ satisfying (24.14) into the simple form

$$\boldsymbol{\eta} = \partial_x, \quad \boldsymbol{\xi} = u\partial_x + S(u, y)\partial_v, \quad \sqrt{2}\zeta = x + \mathrm{i}y. \tag{24.20}$$

These Killing vectors necessarily commute. The metric (24.18) is independent of x, and the rest of the Killing equations imply $W_{,v} \neq 0$. The type D vacuum solution (31.41) is also contained here

The *general Einstein–Maxwell type D solutions* with a G_2 on null orbits, where both eigenvectors of the Weyl and Maxwell tensor are aligned and geodesic and shearfree, divide into two classes. They have either one null eigenvector with non-zero expansion and twist and are then given by

$$\begin{aligned} ds^2 &= \frac{(x^2+y^2)dx^2}{2nx-(e^2+g^2)} + \frac{2nx-(e^2+g^2)}{x^2+y^2}[du - y^2 dv]^2, \\ &\quad + 2dy[du + x^2 dv], \\ \Phi_{11} &= -(e^2+g^2)/2(x^2+y^2)^2, \end{aligned} \tag{24.21}$$

or both null eigenvectors are expansion- and twist-free and the solutions are given by

$$\begin{aligned} ds^2 &= \frac{K(x)}{x^2+l^2}[dv + 2ly du]^2 + \frac{x^2+l^2}{K(x)}dx^2 + 2(x^2+l^2)dy du, \\ K(x) &= 2nx - (e^2+g^2) - \Lambda\left(2l^2x^2 + x^4/3 - l^4\right), \\ \Phi_{11} &= -(e^2+g^2)/2(x^2+l^2) \end{aligned} \tag{24.22}$$

(García D. and Salazar I. 1983). If admissible, the cosmological constant Λ has been included. Solution (24.21) is a special case of the solutions given by Leroy (1978) and Debever (1971). Solution (24.22) contains the vacuum solutions found by Bampi and Cianci (1979), see also Joly *et al.* (1992). Both classes can be obtained by limiting procedures (contractions) from the type D solutions with a G_2 on non-null orbits (García D. and Salazar I. 1983, García D. and Plebański 1982b), cp. §21.1.2.

24.4 Null Killing vectors (G_1 on N_1)

The important relation (6.33),

$$\Theta_{,a}k^a - \omega^2 + \Theta^2 + \sigma\bar{\sigma} = -\tfrac{1}{2}R_{ab}k^a k^b, \tag{24.23}$$

applied to a null Killing vector $\boldsymbol{k}$ yields

$$R_{ab}k^a k^b = 2\omega^2. \tag{24.24}$$

Obviously, for vacuum solutions, and for Einstein–Maxwell and pure radiation fields for which $\boldsymbol{k}$ is an eigenvector of R_{ab}, the null Killing vector is always twistfree ($\omega = 0$) (Debney 1972). Perfect fluid solutions cannot admit a non-twisting null Killing vector, except if $\mu + p = 0$. The algebraically special perfect fluid solutions with (twisting) null Killing vectors are treated in §33.2; they admit an Abelian group G_2. The Gödel universe (12.26) admits the two twisting null Killing vectors $\partial_y \pm \partial_t$.

If there is a Killing vector with a fixed point as well as the null Killing vector, then the two commute or space-time admits a third Killing vector (Szabados 1987).

24.4.1 Non-twisting null Killing vector

For the case of a non-twisting null Killing vector, the field equations can be completely solved, i.e. reduced to a system of two-dimensional Poisson equations (Kramer (1977); for the vacuum case see Dautcourt (1964)). Starting from

$$k_{(a;b)} = 0, \quad k_a k^a = 0, \quad k_{[a} k_{b;c]} = 0 \tag{24.25}$$

we obtain the relations

$$k_a = -w u_{,a}, \quad k_{a;b} = w_{,[a} u_{,b]}, \quad w_{,a} k^a = 0. \tag{24.26}$$

We introduce coordinates $x^i = (x, y, v, u)$, similar to those of (24.18), such that the components of the null Killing vectors are given by

$$k^i = \delta^i_3, \quad k_i = -w \delta_i{}^4, \tag{24.27}$$

and that the coordinates x and y label the points of the spacelike 2-surfaces V_2 orthogonal to $\boldsymbol{k}$ (Kundt 1961). Because of its independence of v, the metric can be transformed into

$$\mathrm{d}s^2 = P^{-2}(\mathrm{d}x^2 + \mathrm{d}y^2) - 2\mathrm{d}u(w\,\mathrm{d}v - m\,\mathrm{d}x + H\mathrm{d}u), \quad g_{ij,v} = 0 \tag{24.28}$$

(see §31.2).

One can show that conditions (24.25) are incompatible with the field equations for Einstein–Maxwell non-null fields. Hence we can restrict ourselves to *electromagnetic null fields*

$$F_{ab} = 2A_{[b,a]} = 2r_{[a} k_{b]}, \quad r_a k^a = 0, \tag{24.29}$$

and to vacuum fields. The null Killing vector $\boldsymbol{k}$ is a Ricci eigenvector with zero eigenvalue,

$$R^a{}_b k^b = k^{b;a}{}_{;b} = (w^{[b} u^{,a]})_{;b} = 0. \tag{24.30}$$

Calculation of the divergence in (24.30) with respect to the metric (24.28) shows that the function w satisfies the potential equation V_2,

$$w_{,xx} + w_{,yy} = 0. \tag{24.31}$$

From (24.29) it follows that only F_{14} and F_{24} are non-zero. Maxwell equations then guarantee the integrability of the system $A_{4,1} = F_{14}$, $A_{4,2} = F_{24}$, so that the vector potential A_a can always be gauged to have the form

$$A_a = \Psi(x, y, u)u_{,a}. \tag{24.32}$$

The metric and Maxwell field are independent of v, so the Lie derivative of the field tensor $F_{ab} = 2\Psi_{,[a}u_{,b]}$ with respect to $\boldsymbol{k}$ vanishes. In virtue of the Maxwell equations, the real scalar potential defined in (24.32) satisfies the potential equation in V_2,

$$\Psi_{,xx} + \Psi_{,yy} = 0. \tag{24.33}$$

For the complex self-dual field tensor F^*_{ab} we obtain

$$F^*_{ab} = 2F_{,[a}u_{,b]}, \quad F = F(\zeta, u). \tag{24.34}$$

By a conformal transformation in the ζ–$\overline{\zeta}$-space, the solution of the potential equation (24.31) can be made (I) $w = 1$ or (II) $w = x$. For the energy-momentum tensors we are interested in, the two cases (I) and (II) include all space-times admitting a non-twisting null Killing vector.

Case (I): ($w = 1$) In this case, $\boldsymbol{k}$ is a constant vector field, $k_{a;b} = 0$, which is an invariant characterization of the *pp*-waves (§24.5). In the final form of the metric

$$\begin{aligned} ds^2 &= 2d\zeta d\overline{\zeta} - 2du\,dv - 2H(\zeta, \overline{\zeta}, u)du^2, \\ H &= \kappa_0 F\overline{F} + f + \overline{f}, \quad f = f(\zeta, u), \quad F = F(\zeta, u), \end{aligned} \tag{24.35}$$

the functions f and F are analytic in ζ and depend arbitrarily on u.

Case (II): ($w = x$) The Einstein equations and the coordinate transformations preserving the form of the metric (24.28) lead to

$$P^2 = x^{1/2}, \quad m = 0. \tag{24.36}$$

With $M = x^{-1}H$ the metric then reads

$$ds^2 = x^{-1/2}(dx^2 + dy^2) - 2x du\,[dv + M(x, y, u)du] \tag{24.37a}$$

and the remaining Einstein–Maxwell equations are

$$\Psi_{,xx} + \Psi_{,yy} = 0, \quad (xM_{,x})_{,x} + xM_{,yy} = \kappa_0({\Psi_{,x}}^2 + {\Psi_{,y}}^2). \tag{24.37b}$$

For a given potential function Ψ, one has to solve an inhomogeneous linear differential equation for M (which has the form of Poisson's equation in cylindrical polar coordinates for axisymmetric solutions). For *pure radiation fields* not necessarily obeying the Maxwell equations, M is an

Table 24.1. Metrics $ds^2 = x^{-1/2}(dx^2 + dy^2) - 2xdu\,[dv + M(x, y, u)du]$ with more than one symmetry

N_o	$M(x, y, u)$	Group	Orbit	Killing vectors $\boldsymbol{\xi}_A$
1	$P(x, u)$	G_2I	N_2	$\partial_v,\ \partial_y$
2	$P(x, ay - mu)$	G_2I	T_2	$\partial_v,\ m\partial_y + a\partial_u$
3	$P(x, u) - yB'(u)$	G_2I	T_2	$\partial_v,\ \partial_y + B(u)\partial_v$
4	$P(x, y - m\ln u)/u^2$	G_2II	T_2	$\partial_v,\ m\partial_y + u\partial_u - v\partial_v$
5	$f(x)\exp(-2ay)$	G_3V	T_3	$\partial_v,\ \partial_u,\ \partial_y + a(u\partial_u - v\partial_v)$
6	$f(x) + ay$	G_3II	T_3	$\partial_v,\ \partial_u,\ \partial_y - au\partial_v,$
7	$f(x)$	G_3I	T_3	$\partial_v,\ \partial_u,\ \partial_y,$
8	0	G_4	T_3	$\partial_v,\ \partial_u,\ \partial_y,\ u\partial_u - v\partial_v$

arbitrary function satisfying $(xM_{,x})_{,x} + xM_{,yy} > 0$. The metrics (24.37$a$) are of Petrov type II or D. Contained here are: the van Stockum solutions (20.32) for $\Psi = 0 = M_{,u}$; the (type D) static $BIII$ metric of Table 18.2 for $M = 0$; the vacuum solution given by Kundu (1979), cp. also Islam (1979); and the metric (22.17).

Any additional Killing vector of the metric (24.37a) has to be of the form $\xi^n = m\partial_y + (a_1u + a_2)\partial_u + [-a_1v + B(u)]\partial_v$, and metric and Maxwell field have to satisfy $(a_1u + a_2)M_{,u} + 2a_1M + mM_{,y} + B'(u) = 0$ and $m\Psi_{,y} + (a_1u + a_2)\Psi_{,u} + a_1\Psi = F(u)$. For the possible symmetry types, see Table 24.1; note the gauge transformations $(v, M) \to (v + h(u), M - h'(u))$. In cases 1, 3 and 5 – 8 of Table 24.1, the general solution of the Einstein–Maxwell equations can easily be given.

24.4.2 Twisting null Killing vector

As already mentioned above, in the case of a twisting null Killing vector $\boldsymbol{k}$ there has to be a non-vanishing Einstein–Maxwell or pure radiation field. The only known solution of this type is algebraically special (a special charged vacuum metric in the sense of Theorem 30.1) which admits $\boldsymbol{k} = \partial_u$ as a null Killing vector; note that this Killing vector is *not* the repeated principal null direction ∂_v (also named $\boldsymbol{k}$ in Chapters 29–30) of the metrics we are now dealing with.

As Theorem 30.1 shows, ∂_u is a null vector only if $H = 0$, and the addition of a charge (of a term $\kappa_0\Phi_1^0\overline{\Phi}{}_1^0$) can make H vanish only for $P = P_0 = \text{const}$, $m = 0$, $K = 0$; the field equations (30.19) then imply $M = M_0 = \text{const}$. So we are dealing with the special case $K = 0$ of the twisting vacuum solutions considered in §29.2.5. If we now insert the appropriate charge $\Phi_1^0(\overline{\zeta})$, the final result (found by Lukács *et al.* (1981) in a different way) is the metric

$$\begin{aligned} ds^2 &= 2(r^2+\Sigma^2)d\zeta d\overline{\zeta} - 2\left[du + Ld\zeta + \overline{L}d\overline{\zeta}\right]\left[dr + Wd\zeta + \overline{W}d\overline{\zeta}\right], \\ L &= i\tfrac{1}{2}M_0\zeta\overline{\zeta}^2, \quad \Sigma = M_0\zeta\overline{\zeta}, \quad W = iM_0\overline{\zeta}, \end{aligned} \tag{24.38a}$$

where the Maxwell field is given by

$$\sqrt{\kappa_0}\Phi_1 = \frac{M_0\overline{\zeta}}{(r+i\Sigma)^2}, \quad \sqrt{\kappa_0}\Phi_2 = \frac{M_0(r-i\Sigma)}{(r+i\Sigma)^3}. \tag{24.38b}$$

M_0 can be gauged to be 0 or 1. The solution is axially symmetric; it is flat for $M_0 = 0$.

24.5 The plane-fronted gravitational waves with parallel rays (*pp*-waves)

In this section we want to investigate *space-times admitting a (covariantly) constant null vector field* $\boldsymbol{k}$,

$$k_{a;b} = 0. \tag{24.39}$$

Such space-times are called plane-fronted gravitational waves with parallel rays (*pp*-waves). They were discovered by Brinkmann (1925) and subsequently rediscovered by several authors. Reviews of gravitational waves were given by Ehlers and Kundt (1962), Jordan *et al.* (1960), Takeno (1961), Zakharov (1972), and Schimming (1974).

The *pp*-waves belong to the wider class of solutions admitting a non-expanding, shear- and twist-free null congruence (Kundt's class, see Chapter 31). Obviously, the *pp*-waves always admit a null Killing vector and thus form a subclass of the fields treated in the previous section.

Condition (24.39) implies (see §24.4 and (35.2)) that electromagnetic non-null fields, perfect fluids and Λ-term solutions cannot occur and that the metric of vacuum, Einstein–Maxwell null and pure radiation fields can be written in the form

$$ds^2 = 2d\zeta d\overline{\zeta} - 2du dv - 2H du^2, \quad H = H(\zeta, \overline{\zeta}, u), \tag{24.40}$$

which is preserved under coordinate transformations

$$\begin{aligned} \zeta' &= e^{i\alpha}(\zeta + h(u)), \quad v' = a(v + \dot{h}(u)\overline{\zeta} + \dot{\overline{h}}(u)\zeta + g(u)), \\ u' &= (u+u_0)/a, \quad H' = a^2(H - \ddot{h}(u)\overline{\zeta} - \ddot{\overline{h}}(u)\zeta + \dot{h}(u)\dot{\overline{h}}(u) - \dot{g}(u)) \end{aligned} \tag{24.41}$$

(α, a, u_0 real constants; $g(u)$ real, $h(u)$ complex). The metric (24.40) is of Kerr-Schild type (see §32.1). We use the complex null tetrad

$$\boldsymbol{m} = \partial_\zeta, \quad \overline{\boldsymbol{m}} = \partial_{\bar\zeta}, \quad \boldsymbol{l} = \partial_u - H\partial_v, \quad \boldsymbol{k} = \partial_v, \tag{24.42}$$

and compute the Ricci and Weyl tensors:

$$R_{ab} = 2H_{,\zeta\bar\zeta}k_a k_b, \quad \tfrac{1}{2}C^*{}_{abcd} = \Psi_4 V_{ab}V_{cd}, \quad V_{ab} = 2k_{[a}m_{b]}, \quad \Psi_4 = H_{,\bar\zeta\bar\zeta}. \tag{24.43}$$

Space-times satisfying (24.39) are either Petrov type N (with the multiple principal null direction $\boldsymbol{k}$) or conformally flat (if $H_{,\bar\zeta\bar\zeta} = 0$). The null bivector $V_{ab} = 2k_{[a}m_{b]}$ is constant, $V_{ab;c} = 0$, so the *pp*-waves are complex recurrent, $C^*{}_{abcd;e} = C^*{}_{abcd}(\ln\Psi_4)_{,e}$ (see §35.2). V_{ab} is determined by the metric up to a complex constant coefficient.

If the metric (24.40) is a solution of the Einstein–Maxwell equations, the function H and the electromagnetic (null) field are given by (see (24.34), (24.35))

$$H = f(\zeta, u) + \bar f(\bar\zeta, u) + \kappa_0 F(\zeta, u)\overline{F}(\bar\zeta, u), \quad F^*_{ab} = 2k_{[a}F_{,b]}. \tag{24.44}$$

The functions f and F are analytic in ζ and depend arbitrarily on the retarded time coordinate u. For pure radiation fields, H is restricted only by $H_{,\zeta\bar\zeta} > 0$.

In general, the group G_1 on N_1 generated by the Killing vector $\boldsymbol{k} = \partial_v$ is the maximal group of motions, but larger groups exist for various special choices of H. For non-flat space-times (24.40), the Killing equations imply

$$\begin{aligned} \boldsymbol{\xi} = {} & [\mathrm{i}\, b\zeta + \beta(u)]\,\partial_\zeta + [-\mathrm{i}\, b\bar\zeta + \bar\beta(u)]\,\partial_{\bar\zeta} + (cu + d)\partial_u \\ & + \left[-cv + \dot{\bar\beta}(u)\zeta + \dot\beta(u)\bar\zeta + a(u)\right]\partial_v, \\ (\boldsymbol{\xi}H) & + 2cH + \ddot{\bar\beta}(u)\zeta + \ddot\beta(u)\bar\zeta + \dot a(u) = 0 \end{aligned} \tag{24.45}$$

(b, c, d real constants; $a(u)$ real, $\beta(u)$ complex). Ehlers and Kundt (1962) investigated the *vacuum pp-waves* and found all possible forms of $H = f(\zeta, u) + \bar f(\bar\zeta, u)$ and $\boldsymbol{\xi}$ that are compatible with (24.45). Table 24.2 summarizes the results. Note that, in each case H and $\boldsymbol{\xi}_A$ are determined up to the transformations (24.41). The *pp*-waves with $f = f(\zeta)$ and $F = F(\zeta)$ independent of u admit an Abelian group G_2 (Killing vectors ∂_v, ∂_u) on T_2 (see Table 24.2). These solutions correspond to the excluded case $W = 1$ in the metric (19.17) of stationary axisymmetric fields: the Einstein–Maxwell equations (19.28)–(19.32) are solved by $W = 1$, $\mathcal{E} = \mathcal{E}(\zeta)$, $\Phi = \Phi(\zeta)$. Hoffman (1969b) considered this case for vacuum fields.

The possible group structures for a general line element of the form (24.40) of a *pp*-wave (vacuum or not) have been determined by Sippel

Table 24.2. Symmetry classes of vacuum *pp*-waves

$\mathrm{d}s^2 = 2\mathrm{d}\zeta\mathrm{d}\bar{\zeta} - 2\mathrm{d}u\,\mathrm{d}v - 2[f(\zeta,u)+\bar{f}(\bar{\zeta},u)]\mathrm{d}u^2$ (κ, α, a real constants; $A(u)$ and $A(\zeta u^{\mathrm{i}\kappa})$ complex).

$f(\zeta,u)$	Group	Orbits	Killing vectors $\boldsymbol{\xi}_A$
$f(\zeta,u)$	G_1	N_1	∂_v
$u^{-2}A(\zeta u^{\mathrm{i}\kappa})$	G_2	T_2	$\partial_v,\ u\partial_u - v\partial_v - \mathrm{i}\kappa(\zeta\partial_\zeta - \bar{\zeta}\partial_{\bar{\zeta}})$
$f(\zeta e^{\mathrm{i}\kappa u})$	G_2	T_2	$\partial_v,\ \partial_u - \mathrm{i}\kappa(\zeta\partial_\zeta - \bar{\zeta}\partial_{\bar{\zeta}})$
$A(u)\ln(\zeta)$	G_2	N_2	$\partial_v,\ \mathrm{i}(\zeta\partial_\zeta - \bar{\zeta}\partial_{\bar{\zeta}}) + 2\int \mathrm{Im}\,A(u)\mathrm{d}u\ \partial_v$
$au^{-2}\ln\zeta$	G_3	T_3	$\partial_v,\ \mathrm{i}(\zeta\partial_\zeta - \bar{\zeta}\partial_{\bar{\zeta}}),\ u\partial_u - v\partial_v$
$\ln\zeta$	G_3	T_3	$\partial_v,\ \mathrm{i}(\zeta\partial_\zeta - \bar{\zeta}\partial_{\bar{\zeta}}),\ \partial_u$
$e^{2\kappa\zeta}$	G_3	T_3	$\partial_v,\ \partial_u,\ \partial_\zeta + \partial_{\bar{\zeta}} - \kappa(u\partial_u - v\partial_v)$
$e^{\mathrm{i}\alpha}\zeta^{2\mathrm{i}\kappa}$	G_3	T_3	$\partial_v,\ \partial_u,\ \mathrm{i}(\zeta\partial_\zeta - \bar{\zeta}\partial_{\bar{\zeta}}) + \kappa(u\partial_u - v\partial_v)$
$A(u)\zeta^2$	G_5	N_3	$\partial_v,\ \beta\partial_\zeta + \bar{\beta}\partial_{\bar{\zeta}} + \zeta\dot{\bar{\beta}}\partial_v + \bar{\zeta}\dot{\beta}\partial_v,$ $\ddot{\beta} + 2\overline{A}(u)\bar{\beta} = 0$
$au^{2\mathrm{i}\kappa-2}\zeta^2$	G_6	V_4	add $u\partial_u - v\partial_v - \mathrm{i}\kappa(\zeta\partial_\zeta - \bar{\zeta}\partial_{\bar{\zeta}})$
$e^{2\mathrm{i}\kappa u}\zeta^2$	G_6	V_4	add $\partial_u - \mathrm{i}\kappa(\zeta\partial_\zeta - \bar{\zeta}\partial_{\bar{\zeta}})$

and Goenner (1986). Since the Ricci tensor components (24.43) are *linear* in H, waves with a distributional profile may occur, e.g. when boosting a Schwarzschild metric to its singular limit; in that case, additional symmetry groups are possible (Aichelburg and Balasin 1997). For conformal symmetries of *pp*-waves see Maartens and Maharaj (1991).

Comparing the formulae $C^*{}_{abcd} = 2\Psi_4 V_{ab}V_{cd}$ and $F^*{}_{ab} = 2\Phi_2 V_{ab}$ we expect that the Weyl tensor component Ψ_4 of a vacuum *pp*-wave and the field tensor component Φ_2 of an electromagnetic wave permit analogous physical interpretations. Writing $\Psi_4 = Ae^{\mathrm{i}\Theta}$, $A > 0$, one calls A the amplitude and associates Θ with the plane of polarization at each space-time point (Ehlers and Kundt 1962). Vacuum *pp*-waves for which Θ is constant are called *linearly polarized.*

In *Einstein–Maxwell* theory, *plane waves*, first considered by Baldwin and Jeffery (1926), are defined to be *pp*-waves in which $\Psi_{4,\bar{\zeta}} = 0 = \Phi_{2,\bar{\zeta}}$. The metric is then given by

$$\mathrm{d}s^2 = 2\mathrm{d}\zeta\mathrm{d}\bar{\zeta} - 2\mathrm{d}u\,\mathrm{d}v - 2\left[A(u)\zeta^2 + \overline{A}(u)\bar{\zeta}^2 + B(u)\zeta\bar{\zeta}\right]\mathrm{d}u^2 \qquad (24.46)$$

($A(u)$ complex, $B(u)$ real); a linear function of ζ and $\bar{\zeta}$ in H can be removed by (24.41). Plane waves admit a group G_5 with an Abelian subgroup G_3 on null hypersurfaces N_3. The electromagnetic term $B(u)\zeta\bar{\zeta}$ in (24.46) does not alter the form of the Killing vectors as given in Table 24.2 for $f(\zeta,u) = A(u)\zeta^2$, but the equation for $\beta(u)$ is now $\ddot{\beta} + 2\overline{A}(u)\bar{\beta} + B(u)\beta = 0$. The four integration constants in the solution of this differential equation give rise to four independent Killing vectors. Plane waves admit a G_6 on V_4 if either $A(u) = A_0 e^{2i\kappa u}$, $B(u) = B_0$ or $A(u) = A_0 u^{2i\kappa-2}$, $B(u) = B_0 u^{-2}$ (A_0, B_0 real constants), see also §12.2.

A *pure radiation* solution which is not an Einstein–Maxwell null field is given by

$$\mathrm{d}s^2 = \mathrm{d}x^2 + \mathrm{d}y^2 + 2\mathrm{d}u\,\mathrm{d}v + k\exp\left[2\left(ax - by\right)\right]\mathrm{d}u^2, \quad a, b = \text{const} \tag{24.47}$$

(Sippel and Goenner (1986), see also Steele (1990)).

The metric (24.40) can be cast in the form

$$\mathrm{d}s^2 = g_{MN}(u)\mathrm{d}x^M\mathrm{d}x^N - 2\mathrm{d}u\,\mathrm{d}v', \quad M, N = 1, 2, \tag{24.48}$$

with the aid of the coordinate transformation

$$\zeta = \alpha_M x^M, \quad v = v' + \tfrac{1}{4}\dot{g}_{MN}x^M x^N, \quad g_{MN}(u) = 2\bar{\alpha}_{(M}\alpha_{N)},$$

$$u' = u, \quad \mathrm{Re}\left[\bar{\alpha}_{(M}\ddot{\alpha}_{N)} + 2A(u)\alpha_M\alpha_N + B(u)\bar{\alpha}_M\alpha_N\right] = 0 \tag{24.49}$$

($\alpha_M = \alpha_M(u)$ complex). The calculation of the Ricci tensor in the coordinate system (24.48) yields

$$R_{ab} = -\left(\tfrac{1}{2}g^{MN}\ddot{g}_{MN} + \tfrac{1}{4}\dot{g}^{MN}\dot{g}_{MN}\right)k_a k_b. \tag{24.50}$$

Linearly polarized plane gravitational waves have $\overline{A}(u) = \text{const}\cdot A(u)$ in the metric (24.46), and $g_{12}(u) = 0$ in (24.48). The plane wave solution (Brdička 1951),

$$\mathrm{d}s^2 = (1 - \sin\omega u)\mathrm{d}x^2 + (1 + \sin\omega u)\mathrm{d}y^2 - 2\mathrm{d}u\,\mathrm{d}v', \tag{24.51}$$

ω being a real constant, is a conformally flat Einstein–Maxwell field with constant electromagnetic null field, cp. (37.105).

Plane waves can be interpreted as gravitational fields at great distances from finite radiating bodies. Peres (1960) and Bonnor (1969) considered parallel light beams as the sources of plane waves.

25

Collision of plane waves

25.1 General features of the collision problem

Colliding plane waves are a particular case of the gravitational fields admitting two commuting spacelike Killing vectors. Solutions with this property are also treated in Chapter 22 (vacuum, Einstein–Maxwell and pure radiation fields) and in Chapter 23 (perfect fluid solutions). Here we confine ourselves to the subclass of the gravitational fields with G_2 on S_2 that allow a physical interpretation as colliding plane waves. In this chapter we can give only a short review on the topic of colliding plane waves and refer the reader to the monograph by Griffiths (1991) for a more detailed presentation of this subject.

The typical scenario is illustrated in Fig. 25.1. Two plane gravitational waves as treated in Chapter 24 move in opposite directions, and collide (head-on collision). The incoming waves (in regions II and III) determine, at least for smooth wavefronts, the data on the null surfaces $u = 0$, $v > 0$ and $v = 0$, $u > 0$, and therefore, via a characteristic initial value problem, the field in the interaction region IV in Fig. 25.1. Region I represents the situation before the collision; the corresponding background field is often assumed to be flat space-time.

To treat the problem analytically, a crucial simplifying assumption is made: there exists an orthogonally transitive Abelian group G_2 acting on spacelike 2-surfaces, even in the interaction region. Then one can cover the global situation in all four regions of Fig. 25.1 by the space-time metric (17.4) with $\varepsilon = -1$.

Usually one does not solve the characteristic Cauchy problem on the null surfaces $u = 0$, $v > 0$ and $v = 0$, $u > 0$. Instead one treats the collision process backwards, starting with an exact solution in region IV and then matching to plane wave solutions at these null surfaces, if possible. When

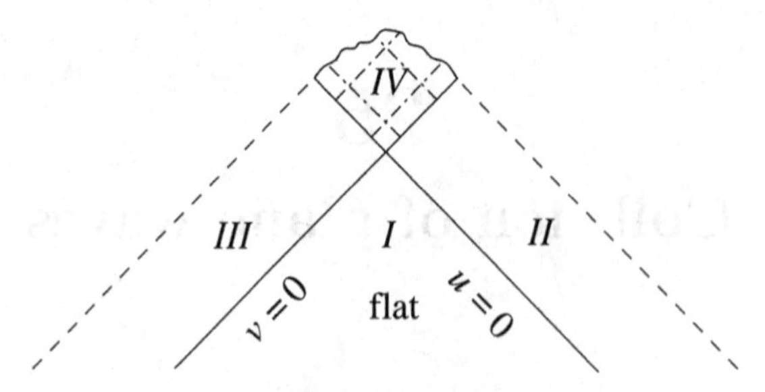

Fig. 25.1. Colliding plane waves

doing this, the so-called *colliding wave condition* has to be satisfied. This condition plays a similar role in collision processes to that which the asymptotic flatness condition does in stationary axisymmetric space-times (where the G_2 acts on timelike orbits).

In the given solution which is a candidate for a colliding wave solution in IV, one has to replace $u \to u\Theta(u)$ and $v \to v\Theta(v)$, where Θ denotes the Heaviside step function

$$\Theta(u) = \begin{cases} 1 & u \geq 0 \\ 0 & u < 0 \end{cases} . \tag{25.1}$$

Suppose one finds that for this substitution (Khan and Penrose 1971) the solution goes over at the boundary into plane wave solutions. Then this property guarantees that the metric is at least C^0 at the null surfaces $u = 0$, $v > 0$ and $v = 0$, $u > 0$, and the solution in IV we start with can be interpreted as a colliding wave solution.

Calculating the curvature invariants at the boundary surfaces between regions II and IV, and III and IV, these quantities can be smooth or have either a leading singularity term of the type of a δ-function (for impulsive waves) or of a step function (for shock waves).

Some known exact solutions also have a curvature singularity due to the mutual focusing of the colliding waves. Solutions with an event horizon across which the metric can be analytically extended have also been found. The amplitudes of the incoming waves affect the time from the instant of collision ($u = 0 = v$) to the creation of the singularity (Ferrari 1988). The asymptotics near the singularity has been discussed by Yurtsever (1989) with the result that a generalization of the Kasner solution (13.53), with space-dependent coefficients p_1, p_2, p_3, is approached.

Although the situation depicted in Fig. 25.1 is rather special, the study of that simple model provides us with valuable insight into the non-linear phenomenon of the collision of waves. The assumption of plane waves implies constant magnitude and infinite extent in all directions in the planes. These unrealistic properties of the collision models may account for the occurrence of singularities caused by mutual focusing.

In a series of papers Hauser and Ernst (1989a, 1989b, 1990, 1991) considered the initial value problem for colliding plane waves and introduced a corresponding homogeneous Hilbert problem.

25.2 The vacuum field equations

For space-times with an orthogonally transitive Abelian G_2 on S_2, Chandrasekhar and Ferrari (1984) reduced the vacuum field equations to a single differential equation for a complex potential and its complex conjugate. It turns out that this equation has exactly the same form as the Ernst equation (19.39) for stationary axisymmetric vacuum fields. To derive the Ernst equation for colliding wave solutions we start with the form of the line element due to Szekeres (1972)

$$ds^2 = -2e^M du dv + e^{-U}[\cosh w(e^V dx^2 + e^{-V} dy^2) - 2\sinh w\, dx dy], \tag{25.2}$$

where the metric functions M, U, V and w depend on the null coordinates u and v only. The colliding waves are said to have *collinear polarization* when the function w in (25.2) can be gauged to be zero so that the metric (25.2) is *diagonal*. This special case has its counterpart in the *static* axisymmetric solutions.

In the metric (25.2), Einstein's field equations read (subscripts denote partial derivatives)

$$U_{uv} = U_u U_v, \tag{25.3a}$$

$$2U_{vv} = U_v^2 + w_v^2 + V_v^2 \cosh^2 w + 2U_v M_v, \tag{25.3b}$$

$$2U_{uu} = U_u^2 + w_u^2 + V_u^2 \cosh^2 w + 2U_u M_u, \tag{25.3c}$$

$$2V_{uv} = U_u V_v + U_v V_u + 2\left(V_u w_v + V_v w_u\right)\tanh w, \tag{25.3d}$$

$$2w_{uv} = U_u w_v + U_v w_u + 2V_u V_v \sinh w \cosh w, \tag{25.3e}$$

$$2M_{uv} = U_u U_v - w_v w_u - V_u V_v \cosh^2 w. \tag{25.3f}$$

The substitution

$$e^V = (\chi^2 + \omega^2)^{-1/2}, \quad \sinh w = \omega/\chi \tag{25.4}$$

takes the Szekeres line element (25.2) into the form

$$ds^2 = -2e^M du dv + e^{-U}\left[\chi dy^2 + \chi^{-1}\left(dx - \omega dy\right)^2\right], \tag{25.5}$$

which is obviously equivalent to (17.4), the null coordinates u and v in (25.5) being related to z and t in (17.4) according to $u = (z-t)/\sqrt{2}$, $v = (z+t)/\sqrt{2}$.

The integration of (25.3a) gives (17.12),

$$W = \mathrm{e}^{-U} = f(u) + g(v). \tag{25.6}$$

(Note that U in Chapter 17 has a different meaning.)

Chandrasekhar and Ferrari (1984) introduced the complex function

$$Z = \chi + \mathrm{i}\omega, \tag{25.7}$$

which satisfies, because of the field equations (25.3d)–(25.3e), the equation

$$(Z + \overline{Z})\mathrm{e}^{U}[(Z_u \mathrm{e}^{-U})_v + (Z_v \mathrm{e}^{-U})_u] = 4Z_u Z_v, \tag{25.8}$$

which has exactly the form of the Ernst equation (19.39). It should be mentioned that we are not led to a pair of real equations for $\chi + \omega$ and $\chi - \omega$ as one could expect from the stationary axisymmetric case. Contrary to the original Ernst potential $\mathcal{E} = \Gamma$, the complex potential Z is here formed by the metric function, not by the dual quantities as in (18.34). However, the combination (25.20) below also satisfies the Ernst equation. Therefore the reduction of the field equations to the Ernst equations (25.8) and (25.21) enables one to apply the generation techniques to the colliding wave problem in two different ways.

From the potential Z one gets V and w by

$$\mathrm{e}^{2V} = (Z\overline{Z})^{-1}, \quad \sinh w = -\mathrm{i}\,(Z - \overline{Z})/(Z + \overline{Z}). \tag{25.9}$$

For any solution (V, w), the remaining function M can be determined from (25.3b)–(25.3c) by a line integral, and the last field equation (25.3f) is automatically satisfied.

In the 2-spaces orthogonal to the group orbits one can introduce coordinates η and μ according to

$$\eta = u\sqrt{1-v^2} + v\sqrt{1-u^2}, \quad \mu = u\sqrt{1-v^2} - v\sqrt{1-u^2}, \tag{25.10}$$

which leads to the inverse transformation formulae

$$W = 1 - u^2 - v^2 = \sqrt{(1-\eta^2)(1-\mu^2)}, \quad u^2 - v^2 = \eta\mu. \tag{25.11}$$

In terms of the new coordinates, and with the choice $f = \frac{1}{2} - u^2$, $g = \frac{1}{2} - v^2$, in (25.6), the metric (25.5) takes the form

$$\begin{aligned} \mathrm{d}s^2 = {} & \mathrm{e}^{N(\eta,\mu)}\left[\mathrm{d}\eta^2/(1-\eta^2) - \mathrm{d}\mu^2/(1-\mu^2)\right] \\ & + (1-\eta^2)^{1/2}(1-\mu^2)^{1/2}\left[\chi \mathrm{d}y^2 + \chi^{-1}(\mathrm{d}x - \omega \mathrm{d}y)^2\right]. \end{aligned} \tag{25.12}$$

It is sometimes useful to generalize the transformations (25.10) between (u, v) and (η, μ) to

$$f(u) = \tfrac{1}{2}\cos(\Theta + \lambda), \quad g(v) = \tfrac{1}{2}\cos(\Theta - \lambda), \quad \eta = \sin\Theta, \quad \mu = \sin\lambda. \tag{25.13}$$

This transformation implies

$$f + g = \sqrt{1-\eta^2}\sqrt{1-\mu^2}, \quad f - g = -\eta\mu, \tag{25.14}$$

and η and μ can be expressed in terms of f and g by the relations

$$\begin{aligned} \eta &= \sqrt{\tfrac{1}{2} - f}\sqrt{\tfrac{1}{2} + g} + \sqrt{\tfrac{1}{2} - g}\sqrt{\tfrac{1}{2} + f}, \\ \mu &= \sqrt{\tfrac{1}{2} - f}\sqrt{\tfrac{1}{2} + g} - \sqrt{\tfrac{1}{2} - g}\sqrt{\tfrac{1}{2} + f}. \end{aligned} \tag{25.15}$$

For $f = \frac{1}{2} - u^2$, $g = \frac{1}{2} - v^2$, one regains (25.10).

The junction conditions (see §3.8) imply that the functions V, M and w in the Szekeres metric (25.3) are continuous and U has to be smooth across the null boundaries (see Griffiths (1991)). Hence the functions $f(u)$ and $g(v)$ must be at least C^1. In the various regions of Fig. 25.1 one can assume

$$\begin{aligned} I: \mathrm{e}^{-U} &= 1 \\ II: \mathrm{e}^{-U} &= \tfrac{1}{2} + f(u), \quad f = \tfrac{1}{2} - u^{n_1}\Theta(u), \quad n_1 \geq 2, \\ III: \mathrm{e}^{-U} &= \tfrac{1}{2} + g(v), \quad g = \tfrac{1}{2} - v^{n_2}\Theta(u), \quad n_2 \geq 2, \\ IV: \mathrm{e}^{-U} &= f(u) + g(v). \end{aligned} \tag{25.16}$$

f and g are monotonically decreasing functions in II and III, respectively. Then the junction conditions for V and w do not impose additional conditions and the continuity of M eventually implies (for vacuum fields)

$$\begin{aligned} \lim_{u\to 0}\,(V_u^2\cosh^2 w + w_u^2)u^{2-n_1} &= 2n_1(n_1 - 1), \\ \lim_{v\to 0}\,(V_v^2\cosh^2 w + w_v^2)v^{2-n_2} &= 2n_2(n_2 - 1). \end{aligned} \tag{25.17}$$

Ernst *et al.* (1987) introduced the parameters

$$k = |Z_v(0,0)/2\,\mathrm{Re}\,Z(0,0)|\,, \qquad l = |Z_u(0,0)/2\,\mathrm{Re}\,Z(0,0)| \tag{25.18}$$

and proved that, when $W = 1 - u^2 - v^2$, the metric can be joined across the $u = 0$ and $v = 0$ null hypersurfaces to incident plane wave solutions if and only if $k = l = 1$.

With the scalar potentials Ψ and ϕ defined by

$$\begin{aligned} \Psi &= \sqrt{(1-\eta^2)(1-\mu^2)}\chi^{-1}, \\ \phi_\mu &= (1-\eta^2)\chi^{-2}\omega_\eta, \quad \phi_\eta = (1-\mu^2)\chi^{-2}\omega_\mu, \end{aligned} \tag{25.19}$$

one can form another Ernst potential,

$$\widetilde{Z} = \Psi + \mathrm{i}\,\phi. \tag{25.20}$$

It also obeys the Ernst equation which, in the metric (25.12), reads

$$\mathrm{Re}\,\widetilde{Z}\left\{\left[(1-\mu^2)\widetilde{Z}_\mu\right]_{,\mu} - \left[(1-\eta^2)\widetilde{Z}_\eta\right]_{,\eta}\right\} = (1-\mu^2)\widetilde{Z}_\mu^2 - (1-\eta^2)\widetilde{Z}_\eta^2. \tag{25.21}$$

The two complex potentials Z and $\widetilde{Z}$ satisfy the same Ernst equation, al- though they are defined in different ways. Therefore, taking a given Z associated with a seed space-time and identifying $\widetilde{Z}$ with Z one obtains in general another (non-diagonal) metric. In the case of two commuting *spacelike* Killing vectors this solution-generating discrete mapping preserves the reality of the metrics. The operation can be combined with other generation methods mentioned in Chapter 34. For details and applications, see Griffiths and Alekseev (1998).

25.3 Vacuum solutions with collinear polarization

For collinear polarization, the function w in (25.3) or, equivalently, ω in (25.5) and (25.12), can be put equal to zero; the metric has *diagonal* form. The solutions given in §25.4 contain also solutions with collinear polarization as particular cases. Not all of them are listed here.

The studies of colliding waves were initiated by the pioneering work by Khan and Penrose (1971). Their solution can be written, in terms of double null coordinates, as

$$\begin{aligned} \mathrm{d}s^2 = &-2\left(1-|\Gamma|^2\right)\left[(1-u^2-v^2)(1-u^2)(1-v^2)\right]^{-\frac{1}{2}}\mathrm{d}u\mathrm{d}v \\ &+(1-u^2-v^2)(1-|\Gamma|^2)^{-1}\left|(1-\Gamma)\mathrm{d}x + \mathrm{i}\,(1+\Gamma)\mathrm{d}y\right|^2, \end{aligned} \tag{25.22}$$

$$\Gamma = u\sqrt{1-v^2} + v\sqrt{1-u^2}. \tag{25.23}$$

Expressed in terms of the coordinates η and μ, the Khan–Penrose solution has the form

$$\mathrm{d}s^2 = \frac{1-\eta^2}{\sqrt{W}}\left(\frac{\mathrm{d}\mu^2}{1-\mu^2} - \frac{\mathrm{d}\eta^2}{1-\eta^2}\right) + W\left(\frac{1-\eta}{1+\eta}\mathrm{d}x^2 + \frac{1+\eta}{1-\eta}\mathrm{d}y^2\right). \tag{25.24}$$

This solution describes the gravitational field of two colliding impulsive waves. There is a spacelike curvature singularity at the hypersurface $W = 0$ $(u^2+v^2=1)$ as indicated in Fig. 25.1.

The function Γ is related to the Ernst potential Z by

$$Z = (1+\Gamma)/(1-\Gamma), \tag{25.25}$$

and relation (25.23) shows that $\Gamma = \eta$ with η given in (25.10). That means that the Khan–Penrose solution is just the counterpart of the Schwarzschild solution because the Ernst potentials Z and $\mathcal{E}$ of these two solutions coincide.

A three-parameter vacuum solution was given by Tsoubelis and Wang (1989). The metric can be written in the form

$$\mathrm{d}s^2 = -2\mathrm{e}^M \mathrm{d}u\mathrm{d}v + W(\mathrm{e}^V \mathrm{d}x^2 + \mathrm{e}^{-V}\mathrm{d}y^2), \quad f = \tfrac{1}{2} - u^{2n}, \quad g = \tfrac{1}{2} - v^{2m},$$

$$\begin{aligned} W &= f + g = 1 - u^{2n} - v^{2m} = (1-\eta^2)^{\frac{1}{2}}(1-\mu^2)^{\frac{1}{2}}, \\ V &= a\ln(1-\eta^2)(1-\mu^2) + \delta_1 \ln\frac{1-\eta}{1+\eta} + \delta_2 \ln\frac{1-\mu}{1+\mu}, \end{aligned} \tag{25.26}$$

$$\mathrm{e}^M = \frac{[(1-u^{2n})^{\frac{1}{2}}(1-v^{2m})^{\frac{1}{2}} - u^m v^n](1-\eta)^{\rho_+}(1-\mu)^{\sigma_+}}{(1-u^{2n})^{1-\frac{1}{2m}}(1-v^{2m})^{1-\frac{1}{2n}}(1+\eta)^{\rho_-}(1+\mu)^{\sigma_-}},$$

where η and μ are defined in (25.15) and the constants are related by

$$\begin{aligned} \rho_\pm &= \pm(a \pm \delta_1)^2 \mp \tfrac{1}{4}, \; n = 1/\left[1 - (\delta_1 + \delta_2)^2\right], \\ \sigma_\pm &= \pm(a \pm \delta_2)^2 \mp \tfrac{1}{4}, \; m = 1/\left[1 - (\delta_1 - \delta_2)^2\right]. \end{aligned} \tag{25.27}$$

The choice of the free parameters determines the nature of the wavefronts (smooth, shock or impulsive). The parameters δ_1, δ_2 and a can be arbitrarily chosen. For $(a + \delta_1)^2 = \frac{1}{4}$, no curvature singularity develops for $W \to 0$, and the solution is of Petrov type D in this limit and can be analytically extended across $W = 0$. The solution (25.26)–(25.27) contains as special cases the Szekeres (1970, 1972) solution ($a = 0$) and the solution given by Ferrari and Ibañez (1987a) ($\delta_1 = 1$, $\delta_2 = 0$).

A focusing effect is also absent in the solution due to Feinstein and Ibañez (1989). This solution is characterized by

$$\begin{aligned} &f(u) = \tfrac{1}{2} - (\alpha u)^n, \quad g(v) = \tfrac{1}{2} - (\beta v)^m, \\ &V = d_1 \mathrm{arccosh}(z+1)/t + d_2 \mathrm{arccosh}(1-z)/t, \\ &t = f + g, \quad z = f - g, \quad d_1^2 = 2 - 2/m, \quad d_2^2 = 2 - 2/n, \end{aligned} \tag{25.28}$$

V being a solution to the linear equation

$$V_{tt} + V_t/t - V_{zz} = 0. \tag{25.29}$$

The analogy of the colliding wave solutions with collinear polarizations on one hand and Weyl's class (§20.2) on the other was used by Ferrari and Ibañez (1987a) and Griffiths (1987) to study the counterpart of the

Erez–Rosen solution (with a gravitational quadrupole moment q). The Ernst potential $Z = \mathrm{e}^V$ is given by

$$V = a \ln\left(\frac{1+\eta}{1-\eta}\right) + q(3\mu^2 - 1)\left[\tfrac{1}{4}(3\eta^2 - 1)\ln\left(\frac{1-\eta}{1+\eta}\right) - \tfrac{1}{2}\eta\right] \quad (25.30)$$

and determines the metric in the form (25.2) with $\omega = 0$. The case $q = -2$ gives continuous tetrad components $\Psi_0, \ldots, \Psi_4$ of the Weyl tensor at $u = 0$ (Ferrari and Ibañez 1986).

Centrella and Matzner (1982) considered colliding waves in expanding cosmologies and found that only the big-bang singularity arises.

Alekseev and Griffiths (1996) discussed diagonal metrics with non-planar wavefronts. The waves are generated by strange sources in finite regions of space-time whereas the sources of plane waves are at infinity.

Halilsoy (1990a) treated an arbitrary number of successive impulsive waves. Ferrari and Ibañez (1987b) used the inverse scattering method to obtain a solution in the interactive region of colliding plane waves. The metric was given by

$$\begin{aligned} \mathrm{d}s^2 = {} & W^{(n^2-1)/2}(1-\eta)^{1+n}(1+\eta)^{1-n}[\mathrm{d}\mu^2/(1-\mu^2) - \mathrm{d}\eta^2/(1-\eta^2)] \\ & + W^{1+n}(1-\eta)\mathrm{d}y^2/(1+\eta) + W^{1-n}(1+\eta)\mathrm{d}x^2/(1-\eta), \end{aligned} \quad (25.31)$$

Further examples of solutions with collinear polarization (diagonal metrics) were discussed by Chandrasekhar (1988) and Griffiths *et al.* (1993). Colliding axisymmetric vacuum *pp*-waves with $M = 0$ and $\omega = 0$ in the metric (25.5) were treated by Ivanov (1998).

25.4 Vacuum solutions with non-collinear polarization

We begin this survey with the generalization of the Khan–Penrose solution (25.22)–(25.23). If Γ is taken as

$$\Gamma = \mathrm{e}^{\mathrm{i}\,\alpha} u\sqrt{1-v^2} + \mathrm{e}^{-\mathrm{i}\,\alpha} v\sqrt{1-u^2} \quad (25.32)$$

or, equivalently, as

$$\Gamma = p\eta + \mathrm{i}\, q\mu, \quad p^2 + q^2 = 1, \quad (25.33)$$

one obtains the Nutku–Halil solution (Nutku and Halil 1977) which describes colliding impulsive gravitational waves with non-collinear polarization. This solution corresponds to the Kerr solution in space-times with one timelike and one spacelike Killing vector. For a detailed discussion of the Nutku–Halil solution, see Chandrasekhar and Ferrari (1984).

Chandrasekhar and Xanthopoulos (1986a) constructed a solution starting with the fact that $\widetilde{Z}$ as defined in (25.20) satisfies the Ernst equation (25.21). Therefore

$$\widetilde{Z} = (1+\widetilde{\Gamma})/(1-\widetilde{\Gamma}), \quad \widetilde{\Gamma} = p\eta + \mathrm{i}\, q\mu, \quad p^2+q^2 = 1, \tag{25.34}$$

is again a solution which can be considered as another analogue of the Kerr solution. The metric is explicitly given by

$$\begin{aligned} \mathrm{d}s^2 &= X\left(\frac{\mathrm{d}\mu^2}{1-\mu^2} - \frac{\mathrm{d}\eta^2}{1-\eta^2}\right) + (1-\eta^2)(1-\mu^2)\frac{X}{Y}\mathrm{d}y^2 + \frac{Y}{X}(\mathrm{d}x - \omega \mathrm{d}y)^2, \\ X &= (1-p\eta)^2 + q^2\mu^2, \quad Y = 1 - p^2\eta^2 - q^2\mu^2, \\ \omega &= -2q(1-\mu^2)(1-p\eta)/pY + \text{ const}, \quad p^2+q^2 = 1. \end{aligned} \tag{25.35}$$

This solution is locally isometric to the Kerr solution (in the interior of the ergosphere). The spacelike curvature singularity at $W = 0$ which occurred in all other colliding wave solutions known at that time is here replaced by a null surface which acts as an event horizon and the solution can be extended beyond this horizon; a timelike curvature singularity arises. The solution (25.35) represents the collision of impulsive waves accompanied by shock waves. Chandrasekhar and Xanthopoulos (1986a) gave a thorough study of the metric (25.35).

Hoenselaers and Ernst (1990) used the Kerr metric in another range of coordinates and obtained a solution which is closely related to (25.35) but is completely free of curvature singularities, and the analytic extension reveals the asymptotically flat branch of the exterior Kerr metric. The counterparts of the Schwarzschild solution and the NUT solution (20.28) are extended across the horizon in Ferrari and Ibañez (1988).

The ansatz (Halilsoy 1988c)

$$Z = (1-\Gamma)/(1+\Gamma), \quad \Gamma = \Omega(\psi)\mathrm{e}^{\mathrm{i}\beta(\psi)}, \quad W = 1 - u^{n_1} - v^{n_2}, \tag{25.36}$$

where Ψ is a solution of the Euler–Darboux equation

$$[(1-\mu^2)\psi_\mu]_\mu - [(1-\eta^2)\psi_\eta]_\eta = 0, \tag{25.37}$$

leads to a three-parameter family of exact solutions describing the collision of shock waves for $n_1 > 2$, $n_2 > 2$. The separation of the scalar wave equation, in one solution of this class, was investigated in Halilsoy (1990c).

Ferrari *et al.* (1987b) used soliton methods (Chapter 10) and constructed a non-diagonal two-soliton solution which describes the collision of shock waves, each supporting an impulsive wave with the same wave front. For special values of the parameters, the solution reduces to the Nutku–Halil solution (25.22) with (25.32). In some cases a horizon can be created.

Ernst *et al.* (1987, 1988) and Ferrari *et al.* (1987a) took the diagonal Ferrari and Ibañez (1987b) solution (25.31) – which contains the Khan–Penrose solution (25.24) for $n = 0$ and the Chandrasekhar–Xanthopoulos solution (25.35) (with $q = 0$) for $n = 1$ – as the seed metric for a generation procedure. They derived a slight generalization of the solution of the Ferrari–Ibáñez–Bruni class (Ferrari *et al.* 1987a, 1987b) mentioned above.

By means of the powerful techniques which are described in the papers by Hauser and Ernst cited at the end of §25.1, Li *et al.* (1991a) derived a further generalization of the Ferrari–Ibáñez–Bruni class. This generalization follows from an arbitrary diagonal seed metric. For certain values of the parameters, the curvature singularities at $W = 0$ are avoidable. These cases are studied in Li *et al.* (1991b).

Li *et al.* (1991d) used the metric (25.26)–(25.27) as the seed for generating non-impulsive waves with non-collinear polarization and found that for discrete values of the parameters a horizon is formed.

Starting from the separable solution

$$\begin{aligned}\psi = &\sum_{n=0}^{\infty} \{a_n P_n(\eta) P_n(\mu) + q_n P_n(\mu) Q_n(\eta) + p_n P_n(\eta) Q_n(\mu)\} \\ &+ c \ln\left[(1-\eta^2)(1-\mu^2)\right], \quad c, a_n, p_n, q_n = \text{const},\end{aligned} \tag{25.38}$$

of the Euler–Darboux equation (25.37), where P_n and Q_n are the Legendre functions of the first and second kind, Breton B. *et al.* (1992) generated and discussed a class of solutions with the Ernst potential

$$Z = \mathrm{e}^{2\psi} \frac{\eta(1-AB) + \mathrm{i}\,\mu(A+B) - (1+\mathrm{i}\,A)(1-\mathrm{i}\,B)}{\eta(1-AB) + \mathrm{i}\,\mu(A+B) + (1+\mathrm{i}\,A)(1-\mathrm{i}\,B)}. \tag{25.39}$$

The functions $A = A(\mu, \eta)$ and $B = B(\mu, \eta)$ in this expression can be determined by straightforward integration from the equations

$$\begin{aligned}(\ln A)_\eta &= 2\left[(\eta\mu - 1)\psi_\eta + (1-\mu^2)\psi_\mu\right]/(\eta - \mu), \\ (\ln A)_\mu &= 2\left[(1-\eta^2)\psi_\eta + (\eta\mu - 1)\psi_\mu\right]/(\eta - \mu), \\ (\ln B)_\eta &= -2\left[(\eta\mu + 1)\psi_\eta + (1-\mu^2)\psi_\mu\right]/(\eta + \mu), \\ (\ln B)_\mu &= -2\left[(1-\eta^2)\psi_\eta + (1+\eta\mu)\psi_\mu\right]/(\eta + \mu),\end{aligned} \tag{25.40}$$

the integrability conditions being satisfied because ψ is a solution to (25.37). The boundary conditions (25.18) were also analysed there. The particular case with $\psi = V$ as in (25.26) has been investigated by Hassan *et al.* (1990); the gravitational waves can have different wavefronts and variable polarization, see also Li (1989b).

Panov (1979c) applied the Geroch transformation (10.30) to the Szekeres metric, i.e. to (25.27) with $a = 0$, to obtain a solution with non-collinear polarization.

The application of the inverse scattering method (Chapter 10) to the class of solutions (25.27) has been treated by Tsoubelis and Wang (1992); special members of this new class are studied.

Wang (1991) generated, using the soliton technique of Belinskii and Zakharov (Chapter 34), a five-parameter class of colliding plane gravitational waves which includes many of the known solutions, e.g. the Tsoubelis and Wang (1989) metrics (25.26)–(25.27), the Nutku and Halil (1977) solutions (25.22), (25.33) (with NUT parameter), the Chandrasekhar and Xanthopoulos (1986a) metric (25.35) (with NUT parameter) and the solutions given by Szekeres (1970, 1972), Ferrari and Ibañez (1987a) and Ernst *et al.* (1987).

The colliding plane wave solutions given so far have metrics (25.12), where the 2-metrics of the spaces orthogonal to the group orbits have, up to a conformal factor, the form $\mathrm{d}\eta^2/P(\eta) - \mathrm{d}\mu^2/Q(\mu)$, and the polynomials $P(\eta)$ and $Q(\mu)$ are at most of the second degree in their respective variables. Breton *et al.* (1998) considered the more general case when $P(\eta)$ and $Q(\mu)$ are fourth degree polynomials. The resulting colliding plane wave solutions can be expressed in terms of Jacobi functions. The limits to lower degree polynomials are studied and an explicit solution of the Einstein–Maxwell equations is given in this paper.

25.5 Einstein–Maxwell fields

In this section we summarize the work done on the collision of electromagnetic and gravitational waves. To treat this nonlinear collision problem all authors have looked for solutions of the Einstein–Maxwell equations admitting an orthogonally transitive Abelian group G_2 acting on *spacelike* 2-surfaces S_2. The general remarks in §25.1 also apply when electromagnetic waves are included. One can again start with the line element (25.2) or (25.5). The Einstein–Maxwell equations consist of (i) the set of the Einstein equations (25.3) supplemented by the source terms $\Phi_{\alpha\beta} = \kappa_0 \Phi_\alpha \bar{\Phi}_\beta, \alpha, \beta = 0, 1, 2$, cp. (7.29),

$$U_{uv} = U_u U_v, \tag{25.41a}$$

$$2U_{vv} = U_v^2 + w_v^2 + V_v^2 \cosh^2 w + 2U_v M_v + 2\kappa_0 \Phi_0 \bar{\Phi}_0, \tag{25.41b}$$

$$2U_{uu} = U_u^2 + w_u^2 + V_u^2 \cosh^2 w + 2U_u M_u + 2\kappa_0 \Phi_2 \bar{\Phi}_2, \tag{25.41c}$$

$$2V_{uv} = U_u V_v + U_v V_u + 2\left(V_u w_v + V_v w_u\right)\tanh w \\ +\kappa_0\left(\Phi_0\bar{\Phi}_2 + \Phi_2\bar{\Phi}_0\right)\operatorname{sech} w, \tag{25.41d}$$

$$2w_{uv} = U_u w_v + U_v w_u + 2V_u V_v \sinh w \cosh w \\ +\mathrm{i}\kappa_0\left(\Phi_0\bar{\Phi}_2 - \Phi_2\bar{\Phi}_0\right), \tag{25.41e}$$

$$2M_{uv} = U_u U_v - w_v w_u - V_u V_v \cosh^2 w, \tag{25.41f}$$

and (ii) the Maxwell equations

$$\Phi_{0,u} = \tfrac{1}{2}(U_u - \mathrm{i}V_u \sinh w)\Phi_0 + \tfrac{1}{2}(\mathrm{i}w_v - V_v \cosh w)\Phi_2, \tag{25.42a}$$

$$\Phi_{2,v} = \tfrac{1}{2}(U_v + \mathrm{i}V_v \sinh w)\Phi_2 - \tfrac{1}{2}(\mathrm{i}w_u + V_u \cosh w)\Phi_0, \tag{25.42b}$$

$$\Phi_1 = 0. \tag{25.42c}$$

Griffiths (1983) derived a Petrov type *II* solution to these Einstein–Maxwell equations assuming $V = V(u)$, $w = w(u)$. With $\mathrm{e}^{-U} = f(u) + g(v)$ this solution is given by

$$\Phi_0 = \tfrac{1}{2}g_v(f+g)^{-1/2}(\tfrac{1}{2}-g)^{-1/2}\mathrm{e}^{\mathrm{i}\Theta}, \quad \Theta_u = -\tfrac{1}{2}V_u \sinh w, \tag{25.43a}$$

$$\Phi_2 = \tfrac{1}{2}(V_u \cosh w + \mathrm{i}w_u)(f+g)^{-1/2}(\tfrac{1}{2}-g)^{1/2}\mathrm{e}^{\mathrm{i}\Theta}, \tag{25.43b}$$

$$\mathrm{e}^M = g_v(f+g)^{-1/2}(\tfrac{1}{2}+f)^{1/2}(\tfrac{1}{2}-g)^{-1/2}. \tag{25.43c}$$

The three arbitrary functions $f(u), V(u), w(u)$ determine the incoming gravitational wave, and $g(v)$ determines the incoming electromagnetic wave which is scattered in the interaction region, where both Φ_0 and Φ_2 are non-zero.

Assuming the existence of a shearfree geodesic null congruence orthogonal to the group orbits, Kuang *et al.* (1999), with the aid of the Newman–Penrose formalism, reduced the field equations (25.41) and (25.42) for non-null Maxwell fields to one relation between the functions $f = f(u)$, $V = V(u)$, $w = w(u)$. For prescribed V and w, this relation has the form of a linear second-order ordinary differential equation of Sturm–Liouville type for f; it reads

$$f_{uu} = \tfrac{1}{2}(\cosh^2 w V_u^2 + w_u^2)f. \tag{25.44}$$

Kuang *et al.* (1999) discussed particular solutions which describe the scattering between gravitational and electromagnetic waves.

The first example of an Einstein–Maxwell field describing a wave collision was given by Bell and Szekeres (1974). The metric in the interaction region can be written in the remarkably simple form

$$ds^2 = -2du\,dv + \cos^2(au - bv)dx^2 + \cos^2(au + bv)dy^2 \tag{25.45}$$

which is just the Bertotti–Robinson solution (12.16) in a slightly changed coordinate system. The Bell–Szekeres solution, together with its extension with the Khan–Penrose substitution (§25.1), describes the collision of two electromagnetic shock waves whose polarization vectors are aligned. This collision process generates impulsive gravitational waves; the non-zero curvature tetrad components are

$$\Phi_2 = a\Theta(u), \qquad \Psi_4 = -a\tan(bv)\delta(u)\Theta(v), \tag{25.46}$$

$$\Phi_0 = b\Theta(u), \qquad \Psi_0 = -b\tan(au)\delta(v)\Theta(u). \tag{25.47}$$

Gürses and Halilsoy (1982) and Al-Badawi and Halilsoy (1999) generalized the Bell–Szekeres solution to the case where a finite number of incoming electromagnetic shock waves collide; Griffiths (1985) and Halilsoy (1988a) found generalizations of the Szekeres solution to include non-aligned polarization.

The crucial point for the progress in constructing new exact colliding wave solutions of the Einstein–Maxwell equations was the fact that the field equations reduce to the same Ernst equations for two complex potentials $\mathcal{Z}$ and $\mathcal{H}$ as they do in the case of stationary axisymmetric Einstein–Maxwell fields (Chapter 19) for the potentials $\mathcal{E}$ and Φ. This relationship was discovered and applied by Chandrasekhar and Xanthopoulos (1985a). One can introduce the real functions Ψ as in (25.19) and ϕ according to

$$\phi_\mu = (1-\eta^2)\chi^{-2}\omega_\eta + \mathrm{i}(\mathcal{H}\bar{\mathcal{H}}_\mu - \bar{\mathcal{H}}\mathcal{H}_\mu), \tag{25.48}$$

$$\phi_\eta = (1-\mu^2)\chi^{-2}\omega_\mu + \mathrm{i}(\mathcal{H}\bar{\mathcal{H}}_\eta - \bar{\mathcal{H}}\mathcal{H}_\eta), \tag{25.49}$$

where the coordinates μ and η (partial derivatives are again denoted by subscripts) and the functions χ and ω refer to the metric (25.12). The potential $\mathcal{H}$ is related to the electromagnetic field tensor,

$$\partial_b\mathcal{H} = \mathcal{H}_b = \sqrt{\kappa_0/2}\,\xi^a F^*_{ab}, \ \xi = \xi^a\partial_a = \partial_x. \tag{25.50}$$

The Maxwell equations imply that $\mathcal{H}_b$ is a gradient such that the complex potential $\mathcal{H}$ exists. The complex potential $\mathcal{Z}$ defined by

$$\mathcal{Z} = \Psi + \mathrm{i}\phi + |\mathcal{H}|^2 \tag{25.51}$$

generalizes the potential $\tilde{Z}$ defined for vacuum fields in (25.20). In terms of $\mathcal{Z}$ and $\mathcal{H}$, the reduced field equations read

$$\begin{aligned}(\mathrm{Re}\mathcal{Z} - |\mathcal{H}|^2)\Delta\mathcal{Z} &= (\nabla\mathcal{Z})^2 - 2\mathcal{H}(\nabla\mathcal{Z})(\nabla\mathcal{H}),\\ (\mathrm{Re}\mathcal{Z} - |\mathcal{H}|^2)\Delta\mathcal{H} &= (\nabla\mathcal{Z})(\nabla\mathcal{H}) - 2\mathcal{H}(\nabla\mathcal{H})^2.\end{aligned} \tag{25.52}$$

The differential operators ∇ and Δ refer to a flat 3-space which, in terms of the coordinates η and μ, has the metric

$$d\sigma^2 = d\mu^2/(1-\mu^2) - d\eta^2/(1-\eta^2) + (1-\mu^2)(1-\eta^2)d\varphi^2. \tag{25.53}$$

The field equations (25.52) have exactly the same form as the Ernst equations (19.28)–(19.29). Therefore there is a one-to-one correspondence between stationary axisymmetric fields and colliding plane wave solutions in the Einstein–Maxwell theory. This circumstance has been widely used by several authors to construct new colliding plane wave solutions from known solutions for stationary axisymmetric fields, which in some space-time regions admit two *spacelike* Killing vectors (within the ergosphere). All generation techniques developed originally for stationary axisymmetric solutions can be applied in the case of colliding plane waves, too.

Chandrasekhar and Xanthopoulos (1987a) considered the ansatz

$$\mathcal{Z} = (1+\mathcal{G})/(1-\mathcal{G}),\ \mathcal{G} = p\eta + iq\mu,\ p^2+q^2=1,\ \mathcal{H} = Q(\mathcal{Z}+1), \tag{25.54}$$

where Q is a real constant. This choice corresponds to the Kerr–Newman solution (21.24) and generalizes ansatz (25.34). The resulting metric can be obtained from (25.35) simply by replacing the functions X and ω according to

$$\begin{aligned} X &\to X_E = \alpha^{-2}[(1-\alpha p\eta)^2 + \alpha^2 q^2\mu^2], \\ \omega &\to \omega_E = -q(1-\mu^2)(1+\alpha^2-2\alpha p\eta)/\alpha^2 pY + \text{const}\,, \end{aligned} \tag{25.55}$$

where $\alpha = \sqrt{1-4Q^2}$. For $\alpha = 1$ one regains (25.35). The derived solution is locally isometric to the Kerr–Newman solution (in the space-time regions where the Abelian group G_2 acts on spacelike orbits). The extension of the metric beyond the horizon and the interpretation of the extended space-time as a colliding plane wave solution were given in detail by Chandrasekhar and Xanthopoulos (1987a). Independently, Halilsoy (1989) considered this solution.

Starting with the Bell–Szekeres solution (25.46) as the seed metric and applying an Ehlers transformation (Chapter 10), Chandrasekhar and Xanthopoulos (1987b) derived another Petrov type D solution which can again be written in the form (25.35), but with the substitutions

$$\begin{aligned} X &\to \Pi = (1-\beta q\mu)^2 + \beta^2(1+p\eta)^2, \\ \omega &\to \omega + 2Y^{-1}\left[\beta\mu p(1-\eta^2) + \beta^2 q(1+p\eta)(1-\mu^2)/p\right]. \end{aligned} \tag{25.56}$$

The conformally flat Bell–Szekeres solution is contained as the limit $\beta = 0$, $p = 1$. A second solution given in that paper provides a generalization of the Bell–Szekeres solution in the same way as the axisymmetric distorted static black-hole solutions are a generalization of the Schwarzschild solution. The discussion of these Einstein–Maxwell fields shows that horizons form and timelike singularities develop.

Another approach uses the generalization of the complex vacuum potential $Z = \chi + \mathrm{i}\omega$ defined in (25.7), which also satisfies the Ernst equation, to the Einstein–Maxwell case. This gives rise to another correspondence between stationary axisymmetric and colliding plane wave solutions. As mentioned at the beginning of §25.4, the Kerr solution can be considered as the counterpart of the Nutku–Halil solution. The Kerr–Newman solution is then the analogue of a solution given by Chandrasekhar and Xanthopoulos (1985a). In this paper the same procedure which leads from Kerr to Kerr–Newman was applied to the Nutku–Halil solution. For the metric functions in the line element (25.12) one gets

$$\begin{aligned} &\mathrm{e}^N = \Omega^2\Delta(1-\eta^2)^{-1/4}(1-\mu^2)^{-1/4}, \quad \chi = \Omega^2\Delta\Sigma^{-1}, \\ &\omega = \tfrac{1}{2}q\mu\{(1+\alpha)^2 + (1-\alpha)^2p^{-2}[q^2(1-\mu^2)^2 + (3-\mu^2)\Sigma]\}/\alpha^2\Sigma + \text{const}, \end{aligned} \tag{25.57}$$

where $p^2 + q^2 = 1$, α is a real parameter restricted to the range $0 \leq \alpha \leq 1$, and the abbreviations

$$\begin{aligned} \Delta &= 1 - p^2\eta^2 - q^2\mu^2, \quad \Sigma = (1-p\eta)^2 + q^2\mu^2, \\ \Omega^2 &= \alpha^{-2}\left[(1-\alpha)\sqrt{1-\mu^2}\sqrt{1-\eta^2}\,\Sigma/\Delta + 1 + \alpha\right]^2 \\ &\quad + \alpha^{-2}p^{-2}\Delta^{-2}[(1-\alpha)2q(1-\mu^2)(1-p\eta)]^2 \end{aligned} \tag{25.58}$$

are used. For $\alpha = 1$, this metric reduces to the Nutku–Halil solution (25.22) with (25.33). The solution that is obtained describes the collision between two plane impulsive gravitational waves, each supporting an electromagnetic shock wave.

The method presented in Chandrasekhar and Xanthopoulos (1985a) has been applied to particular seed metrics by Halilsoy (1990b, 1993b).

Papacostas and Xanthopoulos (1989) investigated the Petrov type D Einstein–Maxwell field

$$\begin{aligned} \mathrm{d}s^2 &= (t^2+z^2)^{-1}\left[E^2(\mathrm{d}y - z^2\,\mathrm{d}x)^2 + H^2(\mathrm{d}y + t^2\mathrm{d}x)^2\right] \\ &\quad + (t^2+z^2)(\mathrm{d}z^2/H^2 - \mathrm{d}t^2/E^2), \\ E^2 &= -\tfrac{1}{2}at^2 + bt + c, \qquad H^2 = -\tfrac{1}{2}az^2 + mz + n \end{aligned} \tag{25.59}$$

(a, b, c, m, n parameters) which corresponds to the special case (21.16) of the solution due to Plebański and Demiański (1976). By means of the transformation

$$at - b = \beta\eta, \; az - m = \alpha\mu \tag{25.60}$$

one can introduce the coordinates μ and η and then with (25.10) go over to double null coordinates u and v. The solution describes the interaction of (impulsive or shock) gravitational and (shock) electromagnetic waves and the formation of a horizon. It turns out that for a particular range of the free parameters no curvature singularities occur in the extended space-time. The extension of the Khan–Penrose substitution $u \to u\Theta(u)$, $v \to v\Theta(v)$ (§25.1) to the Einstein–Maxwell case can be applied to calculate the 4-potentials of the incident waves from the metrics (25.59) in the interaction region (Dagotto *et al.* 1991). Another example of a solution describing the collision between a purely gravitational wave and an electromagnetic wave has been presented by Hogan *et al.* (1998).

The application of the Harrison transformation (34.12) to the Papacostas–Xanthopoulos solution (25.59) leads to a more general Einstein–Maxwell field which is of Petrov type I (Yanez *et al.* 1995).

With the aid of a Harrison transformation other Einstein–Maxwell fields have also been derived: García D. (1989) generalized the vacuum Ferrari–Ibáñez solution (25.31) to include electromagnetic waves, and García D. (1988) generated an electromagnetic generalization of the vacuum Nutku–Halil solution (25.22) with (25.33). The Ernst potential used is formed by taking a linear combination of the two Killing vectors ∂_x and ∂_y of the Nutku–Halil seed metric.

Li and Ernst (1989) applied Ehlers and Harrison transformations to the vacuum seed solutions in Ernst *et al.* (1987, 1988), see also Li *et al.* (1991c).

The solution given by Gürtug (1995) is an electromagnetic generalization of the diagonal vacuum metric (25.26) with $\delta_2 = 0$ and with an additional quadrupole term (as in (25.30)) in the metric function V. The resulting class of solutions contains the Khan–Penrose ($q = 0, \delta_1 = 1$) as well as the Bell–Szekeres solutions as special cases and partly overlaps with the class presented by García D. (1989, 1990).

The Bonnor transformation as given in Theorem 34.4 clearly has its counterpart in the context of wave collision. Diagonal metrics (corresponding to collinear polarization) with electromagnetic potential $A = A_x(z, t)$,

$$ds^2 = e^{2(\gamma-\psi)}(dz^2 - dt^2) + t(t^{-1}e^{2\psi}dx^2 + te^{-2\psi}dy^2), \tag{25.61}$$

can be mapped into non-diagonal vacuum metrics

$$ds^2 = e^{\alpha}(dz^2 - dt^2) + t[\chi dy^2 + \chi^{-1}(dx - \omega dy)^2] \tag{25.62}$$

and vice versa, by means of the *real* substitution

$$\chi \longleftrightarrow \mathrm{e}^{\psi}, \; \omega \longleftrightarrow A, \; \mathrm{e}^{\alpha} \longleftrightarrow \mathrm{e}^{\gamma} t^{-1/2} \tag{25.63}$$

(Breton B. 1995). The original Bonnor transformation, for stationary axisymmetric fields, is *complex*.

The construction of Panov (1979a) is in fact an application of a mapping which corresponds to the transition from Weyl's vacuum class (§20.2) to Weyl's electrovac case (§21.1) and leads from the Szekeres solution to an electrostatic generalization.

25.6 Stiff perfect fluids and pure radiation

25.6.1 Stiff perfect fluids

The key point for including perfect fluids when collision phenomena are treated is the result that the field equations for perfect fluids with the equation of state $\mu = p$ (stiff matter) and with irrotational four-velocity reduce to

$$R_{ab} = 2\sigma_{,a}\sigma_{,b}, \; u_a = (-\sigma_{,b}\sigma^{,b})^{-1/2}\sigma_{,a}, \; \kappa_0 p = \kappa_0 \mu = -\sigma_{,c}\sigma^{,c} \tag{25.64}$$

(Tabensky and Taub (1973), see also §10.11). There exists a scalar potential σ sharing the space-time isometries ($\sigma_{,a}\xi^{,a} = 0 = \sigma_{,a}\eta^{,a}$). These field equations are equivalent to the Einstein equations for a real scalar field. Moreover, solutions to (25.64) can be generated from vacuum solutions by means of Theorem 10.2 due to Wainwright *et al.* (1979). According to this theorem, in order to construct a solution with non-zero gradient $\sigma_{,a}$ one has to modify the vacuum solution as given, say, in the Szekeres form (25.5) only by the substitution

$$M \to M + \Omega, \; \Omega_{,a} = 2W(W_{,b}W^{,b})^{-1}\sigma_{,c}(2W^{,c}\sigma_{,a} - \sigma^{,c}W_{,a}), \tag{25.65}$$

where $W = \mathrm{e}^{-U}$. The Bianchi identities imply $\sigma^{,a}{}_{;a} = 0$, i.e. σ satisfies, in double null coordinates, the linear equation

$$2\sigma_{uv} + W^{-1}(W_u\sigma_v + W_v\sigma_u) = 0. \tag{25.66}$$

Investigating the collision problem in the presence of a stiff perfect fluid, Chandrasekhar and Xanthopoulos (1985b) introduced another potential $\theta = \theta(u,v)$ which is related to $\sigma = \sigma(u,v)$ according to $\theta_u = -W\sigma_u$, $\theta_v = W\sigma_v$, and satisfies the linear equation

$$2\theta_{uv} - W^{-1}(W_u\theta_v + W_v\theta_u) = 0. \tag{25.67}$$

Assuming a diagonal form of the line element,

$$ds^2 = -2e^M du\, dv + W(e^V dx^2 + e^{-V} dy^2), \tag{25.68}$$

the metric function V satisfies the same linear equation as σ,

$$2V_{uv} + W^{-1}(W_u V_v + W_v V_u) = 0. \tag{25.69}$$

Bičák and Griffiths (1994) fixed $W = f(u)+g(v) = 1+u+v$ and calculated σ for a Friedman–Robertson–Walker (FRW) open ($\varepsilon = -1$) model (§14.2) with a stiff equation of state ($p = \mu$),

$$\sigma = \frac{\sqrt{3}}{2}\ln\left(\frac{\sqrt{c+1+v} - \sqrt{c-u}}{\sqrt{c+1+v} + \sqrt{c-u}}\right), \tag{25.70}$$

identified plane surfaces in this background (the group G_3 of Bianchi type VII contains an Abelian subgroup G_2) and determined the metric function V from (25.69) and appropriate boundary conditions at the wave fronts $u = 0$ and $v = 0$. The solutions V to this characteristic initial value problem can be expressed in terms of hypergeometric functions. In the interaction region of the colliding plane gravitational waves in the FRW background, the function V is the sum of two terms, $V = V_1 + V_2$, where

$$V_1 = c_{n_1} u^{n_1}(v+1)^{-1/2} F(\tfrac{1}{2}, \tfrac{1}{2} + n_1; 1 + n_1; -u/[v+1]), \quad n_1 \geq 1 \tag{25.71}$$

(Griffiths 1993a) and V_2 is obtained from V_1 by replacing $u \longleftrightarrow v$ and $n_1 \rightarrow n_2$, $c_{n_1} \rightarrow c_{n_2}$. In the interaction region the space-time is algebraically general (Petrov type I); the gravitational waves are scattered in the background. The remaining metric function M can be determined from a line integral,

$$M = M_0 + \Omega, \quad \Omega_u = \tfrac{1}{2}(1+u+v){V_u}^2, \quad \Omega_v = \tfrac{1}{2}(1+u+v){V_v}^2, \tag{25.72}$$

where M_0 refers to the background metric and is given by

$$e^{M_0} = (1+u+v)[(c-u)(c+1+v)]^{-3/2} + \text{const.} \tag{25.73}$$

The gravitational waves in this expanding background slow down the rate of expansion, but, contrary to the case with Minkowski background, future spacelike singularities do not arise after the collision; only the big-bang singularity ($u + v + 1 = 0$) occurs. The same is true in a spatially flat ($\varepsilon = 0$) FRW background (Griffiths 1993a). However, in a closed ($\varepsilon = 1$) FRW background with $p = \mu$ the collision of gravitational waves leads to future spacelike curvature singularities (Feinstein and Griffiths 1994). Gravitational waves propagating into FRW universes were investigated also in Bičák and Griffiths (1996) and Alekseev and Griffiths (1995).

Chandrasekhar and Xanthopoulos (1985b) investigated the collision of impulsive gravitational waves coupled with stiff fluid motions. The vacuum seed metric for the transition (25.65) was taken to be the Nutku–Halil metric (25.22) with (25.33) and particular solutions to (25.67) for the potential θ were obtained by separation in the variables $r = u^2 - v^2$ and $s = 1 - u^2 - v^2$. The result of Theorem 10.2 mentioned above was rediscovered and applied in this paper. Generalizations of the particular solutions given by Chandrasekhar and Xanthopoulos (1985b) can be found in Griffiths and Ashby (1993). For solutions with stiff matter, see also Chandrasekhar and Xanthopoulos (1985c).

25.6.2 Pure radiation (null dust)

Chandrasekhar and Xanthopoulos (1986b) considered colliding impulsive gravitational waves when the space-time region after the instant of collision is filled with a mixture of pure radiation fields (null dusts) moving in different directions. Such a solution can be associated with any solution of the vacuum equations simply by multiplying the metric coefficient $g_{uv} = -e^M$ in (25.5) by the factor $\exp[F_1(u)+F_2(v)]$, with arbitrary (real) functions $F_1(u)$ and $F_2(v)$. This possibility of generating pure radiation fields from vacuum solutions corresponds to the substitution (25.65) for stiff matter. Pure radiation fields have been generated from the Nutku–Halil solution which was also taken as the seed solution in Chandrasekhar and Xanthopoulos (1985b). The solutions for a perfect fluid with the equation of state $p = \mu$ (stiff matter) and the superposition of two beams of pure radiation turned out to be different in the interaction region. However, the extension to the regions before the instant of collision by means of the Khan–Penrose substitution (§25.1) surprisingly led to the same solution in both cases. A similar situation arises even if the Riemann tensor is free of any kind of discontinuity (Tsoubelis and Wang 1991).

The problem of plane wave collision with incident pure radiation has been studied by several authors, and similar ambiguities were found. The difficulty is that the characteristic initial value problem for the gravitational field does not fix the matter field equations to be used in the interaction region. Without them, the gravitational Cauchy data fixed by the incident beams, and any shock or impulsive wave which is added, may be consistent with several matter-filled interaction regions. In a series of papers Taub (1988, 1990, 1991) treated impulsive plane gravitational waves accompanied by null dust and classified the different types of energy-momentum tensor (interacting null dusts, stiff matter, anisotropic perfect fluid, scalar fields) which can occur in the interaction region. For special values of the parameters in the Szekeres family (25.26) ($a = 0$) one gets solutions which describe shells of null dust and impulsive or shock gravitational waves (Tsoubelis 1989). They contain the particular solution due to Babala (1987), in which the interaction region is locally flat.

The conclusion of Chandrasekhar and Xanthopoulos (1987a) was that null dust can be converted into stiff matter. Feinstein *et al.* (1989) argued that admissible field equations for the underlying physics of the matter should lead to appropriate boundary conditions at the null surfaces bounding the interaction region, and thus to an unambiguous answer for the combined gravitational and matter system: this is related to uniqueness for the characteristic initial value problem. It is true for e.g. Maxwell fields and scalar fields of the usual type, although one can construct macroscopic models whose governing equations do not have this property (Hayward 1990).

Models of colliding shells accompanied by gravitational waves of null dust can be derived from the vacuum solution (25.26) (Tsoubelis and Wang 1990). Cosmological aspects of the interaction of null fluids are discussed by Letelier and Wang (1993). With the aid of the inverse scattering method and the Kaluza–Klein dimension reduction, Cruzate *et al.* (1988) were able to find solutions describing the collision of solitons in a FRW background with the equation of state $p = (\gamma - 1)\mu$. Ferrari and Ibañez (1989) used the same techniques and obtained, in the interaction region of null fields, a solution with the source term of an anisotropic fluid.

Feinstein and Senovilla (1989a) found an exact solution for the collision of a variably polarized wave with (across the null boundaries) an arbitrarily smooth wave front and a shell of null dust followed by a plane gravitational wave with constant polarization. In the Szekeres form (25.5) of the metric, the solution is given by

$$\begin{aligned} ds^2 &= -2f_u g_v W^{-3/8} e^{R(v)} du dv + W^{1/2}[W dy^2 + (dx - \omega(v) dy)^2], \\ W &= f(u) + g(v), \quad {\omega_v}^2 = 2R_v g_v, \end{aligned} \tag{25.74}$$

where $f(u)$ and two of the three functions $g(v)$, $R(v)$, $\omega(v)$ can be freely chosen.

Part III

Algebraically special solutions

26

The various classes of algebraically special solutions. Some algebraically general solutions

26.1 Solutions of Petrov type II, D, III or N

Many of the known solutions of Einstein's equations are algebraically special, i.e. of Petrov type II, D, III, N or O. One common technique for finding such metrics (except those of type O) explicitly is to start by considering a null congruence defined by a repeated principal null direction (cp. §§4.3, 7.6). The present section, in particular Table 26.1, gives a brief survey of the different subcases which naturally arise in this approach. As well as being algebraically special,

$$\Psi_0 = 0 = \Psi_1, \tag{26.1}$$

most of the known solutions of types II, D, III and N have in common that the multiple null eigenvector $\boldsymbol{k}$ of the Weyl tensor is geodesic and shearfree,

$$\kappa = 0 = \sigma, \tag{26.2}$$

and the Ricci tensor satisfies

$$R_{ab}k^a k^b = R_{ab}k^a m^b = R_{ab}m^a m^b = 0 \tag{26.3}$$

(for the definition of the complex null vector $\boldsymbol{m}$ see §3.2). Note that because of the Goldberg–Sachs Theorem and its generalizations (§7.6), and Theorem 7.4 and its corollary, these three sets of assumptions are not

Table 26.1. Subcases of the algebraically special (not conformally flat) solutions

Given are the corresponding chapter numbers for solutions obeying (26.1)–(26.3).

	$\omega = 0$		$\omega \neq 0$	Kerr–Schild
	$\Theta \neq 0$ (Robinson–Trautman)	$\Theta = 0$ (Kundt)		(both $\omega = 0$ and $\omega \neq 0$)
Vacuum	27, 28	31	27, 29	32
Einstein–Maxwell and pure radiation	27, 28	31	27, 30	32

independent of each other. In particular, *all* algebraically special vacuum solutions and *all* Einstein–Maxwell null fields obey (26.1)–(26.3).

The bulk of the material presented in the chapters on algebraically special solutions is concerned with solutions which satisfy the assumptions (26.1)–(26.3). The further subdivision of these solutions depends on the complex divergence ρ of the vector field $\boldsymbol{k}$,

$$\rho = -(\Theta + \mathrm{i}\omega), \tag{26.4}$$

see Table 26.1.

Let us now have a look at those algebraically special solutions which are *not* covered by the assumptions (26.2)–(26.3), i.e. at solutions which are of Petrov type II, D, III or N, but violate (26.2) or (26.3) or both.

For EINSTEIN–MAXWELL-FIELDS with a NON-NULL ELECTROMAGNETIC FIELD, we have to distinguish between two cases: either the two null eigenvectors of the Maxwell field are both distinct from the multiple null eigenvector(s) of the Weyl tensor (*non-aligned case*), or at least one of them is parallel to a repeated Weyl null eigenvector (*aligned case*).

In the non-aligned case, (26.3) are not satisfied. The general solution of this kind is not known. Griffiths (1986) found a class of solutions with $\kappa = \sigma = \omega = \Phi_1 = 0$, $\boldsymbol{k} = \partial_r$, $\Phi_0 = a = \text{const}\ (\neq 0)$. It reads

$$\begin{aligned} \mathrm{d}s^2 = &-2P\mathrm{d}u\mathrm{d}r - 2[PP_{,\zeta\bar\zeta} - am(u)P^{-1}\cot ar]\mathrm{d}u^2 \\ &+ 2\left|(B_{,\bar\zeta}\sin ar + P_{,\bar\zeta}\cos ar)\mathrm{d}u - \sin ar\,\mathrm{d}\zeta/a\right|^2, \end{aligned} \tag{26.5a}$$

where the real functions $P(\zeta, \overline{\zeta}, u)$ and $B(\zeta, \overline{\zeta}, u)$ have to satisfy

$$
\begin{aligned}
P^2\partial_\zeta\partial_{\overline{\zeta}}P^2\partial_\zeta\partial_{\overline{\zeta}}\ln P - 3m(\ln P)_{,u} - m_{,u} - 2a^2m^2P^{-2} \\
+ 3amP^{-1}[P_{,\zeta}B_{,\overline{\zeta}} + P_{,\overline{\zeta}}B_{,\zeta}] - P^2B_{,\zeta\zeta}B_{,\overline{\zeta}\overline{\zeta}} &= 0, \quad (26.5b) \\
P^2B_{,\zeta\overline{\zeta}} &= am.
\end{aligned}
$$

In the general case, it is of type II (and one can put $a = 1$ and $m = 1$). For $m = 0$ it contains the type III metrics studied by Cahen and Spelkens (1967); here B is an arbitrary harmonic function. For $B = 0$ ($\Rightarrow m = 0$) it contains the general non-aligned type N metrics found by Cahen and Leroy (1965, 1966) and Szekeres (1966b). As given by Szekeres, it reads

$$
\begin{aligned}
\mathrm{d}s^2 &= \tfrac{1}{2}\cos^2 ar(\mathrm{d}x^2 + \mathrm{d}y^2) - 2b(2r + a^{-1}\sin 2ar)\mathrm{d}u\,\mathrm{d}x \\
&\quad -4\mathrm{d}u\,\mathrm{d}r + 4a^{-2}(2b^2\sin^2 ar - 2\mathrm{e}^{2u} - raa_{,u})\mathrm{d}u^2, \qquad (26.6) \\
a(x,u) &= g(u)\,\mathrm{cosech}\,[\mathrm{e}^u x + f(u)], \quad b(x,u) = -\mathrm{e}^u\coth[\mathrm{e}^u x + f(u)],
\end{aligned}
$$

with $k_n = u_{,n}$. The limit $a = 0$ of (26.5) leads to the Robinson–Trautman metrics (Chapter 28).

In the aligned case, (26.3) is satisfied by definition. As shown in §7.6, the Bianchi identities yield

$$\sigma(3\Psi_2 + 2\kappa_0\Phi_1\overline{\Phi}_1) = 0 = \kappa(3\Psi_2 - 2\kappa_0\Phi_1\overline{\Phi}_1), \qquad (26.7)$$

so equation (26.2) can be violated only for Petrov type II (or D) solutions with a special constant ratio between Ψ_2 and $\Phi_1\overline{\Phi}_1$, and κ *or* σ must be non-zero (note that because of (6.33) $\kappa = \rho = 0$ would induce $\sigma = 0$). The case $\kappa = 0$, $\sigma \neq 0$, has been excluded by Kozarzewski (1965), so only $\kappa \neq 0$, $\sigma = 0$, remains to be studied. If the two null eigenvectors of a type D solution are aligned with the eigenvectors of the Maxwell tensor, then they must both be geodesic and shearfree; this is not true if a cosmological constant Λ is admitted (see García D. and Plebański (1982a) and Plebański and Hacyan (1979), where also some solutions are given).

For pure radiation fields, one again has to distinguish between the aligned and the non-aligned cases. So far, only the aligned case has been treated (the general type N and type III metrics are necessarily aligned: Plebański (1972), Urbantke (1975), Wils (1989a)). Due to the theorems given in §7.6, all aligned type II, D or III solutions have a geodesic and shearfree $\boldsymbol{k}$. The only algebraically special, aligned, pure radiation fields *not* covered by (26.1)–(26.2) are therefore of type N. Among the Weyl and

Ricci tensor components only Ψ_4 and $R_{33} = \kappa_0\Phi^2 = 2\Phi_{22}$, respectively, are different from zero for these type N fields. The Bianchi identities (7.32f), (7.32i) and (7.32j) then give $\kappa = 0$ and

$$\rho\Phi_{22} = \sigma\Psi_4, \tag{26.8}$$

and (3.25) and the field equations yield

$$\begin{aligned} \mathrm{d}\boldsymbol{\Gamma}_{41} + \boldsymbol{\Gamma}_{41} \wedge (\boldsymbol{\Gamma}_{21} + \boldsymbol{\Gamma}_{43}) &= 0, \\ \mathrm{d}(\boldsymbol{\Gamma}_{21} + \boldsymbol{\Gamma}_{43}) + 2(\boldsymbol{\Gamma}_{32} \wedge \boldsymbol{\Gamma}_{41}) &= 0, \\ \mathrm{d}\boldsymbol{\Gamma}_{32} - \boldsymbol{\Gamma}_{32} \wedge (\boldsymbol{\Gamma}_{21} + \boldsymbol{\Gamma}_{43}) &= \boldsymbol{\omega}^3 \wedge (\Phi_{22}\boldsymbol{\omega}^1 + \Psi_4\boldsymbol{\omega}^2). \end{aligned} \tag{26.9}$$

(From (26.8) we see that solutions with $\sigma = 0$ belong to Kundt's class, see Chapter 31.)

A special example of a type N field with $\sigma \neq 0$ is given by (37.12). An investigation of the type N field equations (26.9) with $\sigma \neq 0$ has been carried out by Plebański (1972). In detail, his results are as follows. After a suitable choice of the null tetrad $(\boldsymbol{m}, \overline{\boldsymbol{m}}, \boldsymbol{k}, \boldsymbol{l})$, $\boldsymbol{k}$ being unchanged, (26.9) gives

$$\boldsymbol{\Gamma}_{21} = 0, \quad \boldsymbol{\Gamma}_{34} = 0, \quad \mathrm{d}\boldsymbol{\Gamma}_{41} = 0, \quad \boldsymbol{\Gamma}_{32} \wedge \mathrm{d}\boldsymbol{\Gamma}_{32} = 0 \tag{26.10}$$

(this is true also for $\sigma = 0$). Further integration depends on whether $\boldsymbol{\Gamma}_{41}$ is zero, real (and non-zero) or complex. *For* $\boldsymbol{\Gamma}_{41} = 0$ ($\Rightarrow \rho = \sigma = \tau = 0$), we regain the *pp*-waves of §24.5. *For* $\boldsymbol{\Gamma}_{41}$ *real (and non-zero)*, the solutions with non-zero shear σ are (in real coordinates y, u, r, v)

$$\begin{aligned} \mathrm{d}s^2 &= 2\boldsymbol{\omega}^1\boldsymbol{\omega}^2 - 2\boldsymbol{\omega}^3\boldsymbol{\omega}^4, \\ \boldsymbol{\omega}^1 &= \mathrm{d}F + (v + \overline{f})\,\mathrm{d}r = \overline{\boldsymbol{\omega}}^2, \quad \boldsymbol{\omega}^3 = \mathrm{d}u + (F + \overline{F})\,\mathrm{d}r, \\ \boldsymbol{\omega}^4 &= \mathrm{d}v + (g + F f_{,u} + \overline{F}\,\overline{f}_{,u})\mathrm{d}r, \\ F &\equiv (-g_{,u} + \mathrm{i}y)\,/2f_{,uu}, \qquad f_{,uu} \neq 0. \end{aligned} \tag{26.11a}$$

They contain two disposable functions $f(u,r)$ (complex) and $g(u,r)$ (real). The corresponding radiation field is characterized by

$$\begin{aligned} \boldsymbol{\Gamma}_{41} &= -\mathrm{d}r, \quad \sigma = \overline{\rho} = -Af_{,uu}, \quad \Phi_{22} = -Af_{,uu}\overline{f}_{,uu}, \\ A^{-1} &= 2\,\mathrm{Re}\,\left\{f_{,uu}\left[-(F + \overline{F})\,F_{,u} + F_{,r} + v + \overline{f}\right]\right\}. \end{aligned} \tag{26.11b}$$

For $\boldsymbol{\Gamma}_{41}$ *complex and*

$$|\,\rho\,| = |\,\sigma\,|, \quad |\,\Phi_{22}\,| = |\,\Psi_4\,|, \tag{26.12}$$

the solutions are (in coordinates $\zeta, \overline{\zeta}, r, v$)

$$\begin{aligned} ds^2 &= 2\omega^1\omega^2 - 2\omega^3\omega^4, \\ \omega^1 &= dF + v d\zeta + \overline{h} d\overline{\zeta} = \overline{\omega}^2, \quad \omega^3 = \overline{F} d\zeta + F d\overline{\zeta}, \\ \omega^4 &= dv - (Ff + h_{,\overline{\zeta}})d\zeta - (\overline{F}\,\overline{f} + \overline{h}_{,\zeta})d\overline{\zeta}, \\ F &= -f_{,\overline{\zeta}}^{-1}(r + h_{,\overline{\zeta}\,\overline{\zeta}} - f\overline{h}), \quad f_{,\overline{\zeta}} \neq 0. \end{aligned} \tag{26.13a}$$

They contain two arbitrary complex functions $f(\zeta, \overline{\zeta})$ and $h(\zeta, \overline{\zeta})$. The corresponding radiation field is characterized by (26.12) and

$$\begin{aligned} &\boldsymbol{\Gamma}_{41} = -d\overline{\zeta}, \quad \rho = -A\overline{F}\,\overline{f}_{,\zeta}, \quad \phi_{22} = -Af_{,\overline{\zeta}}\overline{f}_{,\zeta}, \\ &\sigma = A\overline{F}f_{,\overline{\zeta}}, \quad A^{-1} = 2\mathrm{Re}\left[F(f_{,\overline{\zeta}}v + F_{,\zeta}) - \overline{f}_{,\zeta}(h - \overline{F}_{,\zeta})\right]. \end{aligned} \tag{26.13b}$$

For complex $\boldsymbol{\Gamma}_{41}$ *not satisfying* (26.12), Plebański succeeded in reducing the field equations to the differential equation

$$\begin{aligned} &\mathrm{Im}\left[\overline{h}X - h_{,\overline{\zeta}\,\overline{\zeta}} - PX^2_{,\overline{\zeta}}/X_{,u}\right] = 0, \quad X_{,u} \neq 0, \\ &X \equiv -(h_u + P_{,\zeta\zeta})P^{-1}. \end{aligned} \tag{26.14a}$$

If a solution $P(\zeta, \overline{\zeta}, u)$, $h(\zeta, \overline{\zeta}, u)$ of (26.14a) is known (P can be gauged to 1), then the metric can be obtained from

$$\begin{aligned} &ds^2 = 2\omega^1\omega^2 - 2\omega^3\omega^4, \\ &\omega^1 = d(P\overline{L}) - P_{,\overline{\zeta}}du + r d\zeta + \overline{h} d\overline{\zeta} = \overline{\omega}^2, \\ &\omega^3 = P(du + L d\zeta + \overline{L} d\overline{\zeta}), \quad \overline{L} = X_{,\overline{\zeta}}/X_{,u} \Leftrightarrow \overline{\partial} X = 0, \\ &\omega^4 = dr + P_{,\zeta\overline{\zeta}}du - (h_{,\overline{\zeta}} + P\overline{L}X)d\zeta - (\overline{h}_{,\zeta} + PL\overline{X})d\overline{\zeta}, \end{aligned} \tag{26.14b}$$

with

$$\begin{aligned} &\rho = -A\left[r + \partial(P\overline{L}) + LP_{,\overline{\zeta}}\right], \quad \sigma = A\left[h + (P\partial L) + LP_{,\zeta}\right], \\ &\partial = \partial_\zeta - L\partial_u, \; \Phi_{22} = -\overline{\sigma}\overline{X}_{,u}/P, \; \Psi_4 = \rho\overline{X}_{,u}/P, \; A = \rho^2 - \sigma^2 \end{aligned} \tag{26.14c}$$

($\Phi_{22} > 0$ has to be guaranteed). The twisting type N solutions are included (for $\sigma = 0$) in (26.14a)–(26.14c).

The general solution of (26.14a) in the case of real $h_{,u}$ (which implies real $\partial\overline{L}$ and hence real ρ) has been given by Plebański as

$$\begin{aligned} &h(u, \zeta, \overline{\zeta}) = (F - sF_{,s}) + (G - sG_{,s}) - P_{,\zeta\zeta}, \quad \sqrt{2}\zeta = x + iy, \\ &F = F(x, s), \quad G = G(y, s), \quad P = P(x, y), \\ &u = F_{,s} + G_{,s} + P \quad \Rightarrow s = s(u, x, y), \quad u_{,s} \neq 0, \end{aligned} \tag{26.15}$$

with real functions F, G and P. This solution includes the special solution found by Köhler and Walker (1975), which, in our notation, is given by $h = u^2(1+\zeta\bar{\zeta})^{-1}$.

A special class of solutions of (26.14a) can be constructed from the ansatz

$$X = (a+\mathrm{i}b)\,(\zeta+\bar{\zeta})^{-2}(u+\mathrm{i})^{-1}, \quad h = (\zeta+\bar{\zeta})^n\left[(u+\mathrm{i})G(u)+K(u)\right], \tag{26.16}$$

(Stephani 1980). For $a = 0$, the real function $K(u)$ is to be calculated from $G(u)$ according to

$$Kbu^2 = 4(1+u^2)^2 G_{,u} - (1+u^2)\left[b - n(n-1)\right]G, \tag{26.17a}$$

and $G(t)$, $t \equiv u^2$, has to satisfy

$$16t(1+t)G'' + \left[(20-8n)t - 8\right]G' + \left[b + n(n-1)(t-1)\right](1+t)^{-1}G = 0. \tag{26.17b}$$

For $b = 2n(n-1)$, the last equation is a hypergeometric differential equation; Hauser's vacuum solution (29.72) is obtained by putting $b = n = 3/2$. For $b = 0$, (26.16) leads to

$$aK = (1+u^2)n(n-1)G, \quad (1+u^2)(2+4n)G_{,u} - \left[a + 2n(n-1)u\right]G = 0. \tag{26.18}$$

For real ρ, the remaining field equation (26.14a) was also given by Köhler and Walker (1975).

PERFECT FLUID (or DUST) solutions necessarily violate $R_{ab}k^ak^b = 0$ (if $\mu + p \neq 0$). The cases where the rest of the assumptions (26.1)–(26.3) are satisfied are treated in §§33.1 and 33.2.

26.2 Petrov type D solutions

Petrov type D solutions have been intensively studied. Although we have met them in the preceding chapters, or will meet them in the following ones, we want to summarize a few of the known results here.

All vacuum solutions, and all Einstein–Maxwell solutions with a non-null double aligned electromagnetic field (both null eigenvectors are aligned and geodesic and shearfree) and a cosmological constant (which may be set zero), are known. They all admit at least an Abelian G_2 of motions which is orthogonally transitive (Debever and McLenaghan 1981, Czapor and McLenaghan 1982). They can be written in a single expression for the metric, irrespective whether the twist ω and/or the complex divergence ρ of the null vector $\boldsymbol{k}$ is zero or not and whether the group orbit is non-null or not, see Debever *et al.* (1982, 1984), García D. (1984), and the references given there. The metrics with a non-null orbit are given in §21.1.2, and those with a null orbit in §24.3.

An example of an Einstein–Maxwell type D solution where only one null eigenvector is aligned is the metric (30.36).

For perfect fluids (or dust), only special type D solutions are known. Among them are the spherically-symmetric solutions of Chapters 15 and 16, and the solutions discussed in §33.3.

26.3 Conformally flat solutions

By definition, the metric of a conformally flat space-time can be written as

$$\mathrm{d}s^2 = \mathrm{e}^{2U(x,y,z,t)}(\mathrm{d}x^2 + \mathrm{d}y^2 + \mathrm{d}z^2 - \mathrm{d}t^2). \tag{26.19}$$

An equivalent coordinate-independent definition makes use of the fact that the Weyl tensor C_{abcd} (3.50) vanishes if and only if space-time is conformally flat: a metric is conformally flat exactly if

$$R_{abcd} = \tfrac{1}{2}\left(g_{ac}R_{bd} + g_{bd}R_{ac} - g_{ad}R_{bc} - g_{bc}R_{ad}\right) - \tfrac{1}{6}R\left(g_{ac}g_{bd} - g_{ad}g_{bc}\right). \tag{26.20}$$

Being zero, the Weyl tensor does not define any null vector that can be used in constructing metrics and solutions, so other techniques have to be applied.

Equation (26.20) shows that conformally flat vacuum solutions ($R_{ab} = 0 = R$) are flat. All conformally flat solutions with a perfect fluid, an electromagnetic field or a pure radiation field are known. As most of them have been found by applying the techniques of embedding, we shall give a more detailed treatment of this subject in Chapter 37. Here we shall summarize only the main results.

Conformally flat perfect fluid solutions are either generalized interior Schwarzschild solutions (37.39) or generalized Friedmann solutions (37.45), the only dust solutions being the Friedmann models and the only stationary solution the static interior Schwarzschild solution (16.18). Conformally flat Einstein–Maxwell fields are either the Bertotti–Robinson metric (37.98)–(37.99) (with a non-null electromagnetic field), or they are special plane waves (37.104)–(37.105) (with a null electromagnetic field). Conformally flat pure radiation fields are either contained in the null electromagnetic fields (37.104) or are given by (37.106).

26.4 Algebraically general vacuum solutions with geodesic and non-twisting rays

The multiple principal null congruence of an algebraically special vacuum solution is geodesic and shearfree (cp. §26.1). Although *vacuum solutions with a geodesic but shearing* ($\kappa = 0$, $\sigma \neq 0$) *null congruence* are in general non-degenerate Petrov type I, we shall list some of them here, because the method of constructing the (non-twisting) solutions with this property is very similar to that for the Robinson–Trautman class outlined in Chapter 27 (but, in virtue of $\sigma \neq 0$, the calculations using the Newman–Penrose formalism are more lengthy).

All vacuum metrics where the null congruence with $\kappa = 0$, $\omega = 0$, $\sigma \neq 0$, $\Theta \neq 0$ *is a principal null congruence* ($\Psi_0 = 0$) were obtained by Newman and Tamburino (1962); for details see also Carmeli (1977), p. 244. Metrics with $\sigma \neq 0$, $\rho = 0$, are forbidden by the vacuum field equations, and solutions with $\sigma \neq 0$, $\omega \neq 0$, are possible only if $\rho\bar{\rho} = \sigma\bar{\sigma}$ (Unti and Torrence 1966). The twisting case has not been completely solved.

There are two classes of Newman–Tamburino solutions:

$\rho^2 \neq \sigma\bar{\sigma} \neq 0$ (*spherical class,* $x^1 + \mathrm{i}x^2 = x + \mathrm{i}y = \zeta\sqrt{2},\ x^3 = r,\ x^4 = u$)

$$
\begin{aligned}
g^{11} &= \frac{2(2\zeta\bar{\zeta})^{3/2}}{(r+a)^2}, \quad g^{22} = \frac{2(2\zeta\bar{\zeta})^{3/2}}{(r-a)^2}, \quad g^{34} = -1, \\
g^{12} &= g^{14} = g^{24} = g^{44} = 0, \quad R^2 = r^2 - a^2, \quad A = bu + c, \\
g^{13} &= 4A^2(2\zeta\bar{\zeta})^{3/2}x\left(\frac{r-a}{R^4} + \frac{r-2a}{2a^2R^2} - \frac{L}{2a^3}\right), \quad L = \frac{1}{2}\ln\frac{r+a}{r-a}, \\
g^{23} &= 4A^2(2\zeta\bar{\zeta})^{3/2}y\left(\frac{r+a}{R^4} + \frac{r+2a}{2a^2R^2} - \frac{L}{2a^3}\right), \quad a = A(2\zeta\bar{\zeta})^{1/2}, \\
g^{33} &= \frac{4A^2r^2(2\zeta\bar{\zeta})^{3/2}}{R^4} - \frac{4Ar^3(\zeta^2 + \bar{\zeta}^2)}{R^4} + \frac{2r^2(2\zeta\bar{\zeta})^{1/2}}{R^2} - \frac{2rL}{A};
\end{aligned}
\tag{26.21}
$$

$\rho^2 = \sigma\bar{\sigma}$ (*cylindrical class*)

$$
\begin{aligned}
\mathrm{d}s^2 &= 2\omega^1\omega^2 - 2\omega^3\omega^4, \quad \omega^2 = \bar{\omega}^1, \quad \omega^3 = \mathrm{d}u, \\
\omega^4 &= \mathrm{d}r - \left[2b^2\mathrm{cn}^2(bx) + \frac{c + b^2\ln[r^2\mathrm{cn}^4(bx)]}{2\mathrm{cn}^2(bx)}\right]\mathrm{d}u, \\
\omega^1 &= r\,\mathrm{d}x/2 + 4Y\mathrm{d}u + \mathrm{i}\,\mathrm{cn}(bx)\,(\mathrm{d}y + 8buY\mathrm{d}x + 2b\ln r\,\mathrm{d}u)\,/\sqrt{2}, \\
Y &= \pm\frac{b\,(1 - \mathrm{cn}^4(bx))^{1/2}}{2\sqrt{2}\mathrm{cn}(bx)},
\end{aligned}
\tag{26.22}
$$

where b and c are real constants and $\mathrm{cn}(bx)$ is an elliptic function of modulus $k = 1/\sqrt{2}$. The metric

$$
\mathrm{d}s^2 = r^2\mathrm{d}x^2 + x^2\mathrm{d}y^2 - \frac{4r}{x}\mathrm{d}u\,\mathrm{d}x - 2\mathrm{d}u\,\mathrm{d}r + x^{-2}\left[c + \ln(r^2x^4)\right]\mathrm{d}u^2 \tag{26.23}
$$

can be obtained from (26.22) by a limiting procedure.

The Newman–Tamburino solutions do not contain arbitrary functions of time; the Robinson–Trautman solutions are not included. The metric (26.21) admits at most one Killing vector ($\boldsymbol{\xi} = \partial_u$ if $b = 0$); there is only one Killing vector ($\boldsymbol{\xi} = \partial_y$) for the metric (26.22), whereas (26.23) admits an Abelian G_2, the ignorable coordinates being y and u (Collinson and French 1967).

Vacuum metrics where the null congruence with $\kappa = 0$, $\omega = 0$, $\sigma \neq 0$, $\Theta \neq 0$ *is not a principal null congruence* ($\Psi_0 \neq 0$) have been considered by Bilge (1989) under the assumption $D(\sigma/\rho) = 0$. It turns out that

$$a = \sigma/\rho = \text{ const (real)}, \quad \Psi_0 = a(\sigma^2 - \rho^2). \tag{26.24}$$

For $a \neq 0, \pm 1, \pm 2, \pm 1/2$, the Newman–Penrose equations could be solved. In coordinates $x^1 = x$, $x^2 = y$, $x^3 = r$, $x^4 = u$, the solutions are of the form

$$\begin{aligned} g^{34} &= -1, \quad g^{nm} = (\xi^n \overline{\xi}^m + \overline{\xi}^n \xi^m), \quad n, m = 1, 2, \\ g^{33} &= -2U, \quad g^{3i} = -X^i, \ i = 1, 2, \end{aligned} \tag{26.25a}$$

with

$$\begin{aligned} \xi^1 &= r^{-a^+}, \quad \xi^2 = \mathrm{i} r^{-a^-}, \quad X^1 = a^+ x \psi'(u), \\ X^2 &= a^- y \psi', \quad U = -r\psi' + (1 + a^2) r^{1-c} \mu_0, \\ a^\pm &= (1 \pm a)/(1 + a^2), \quad c = 2/(1 + a^2), \\ \mu_0 &= \begin{cases} 1 \quad \text{for } \psi' = 0, \ a^- \neq 1/2, \\ \exp\left[-(c+1)\psi\right] \quad \text{for } \psi' \neq 0, a^- \neq 1/2, \\ y \quad \text{for } \psi' = 0, \ a^- = 1/2 \text{ (special sol. only)}, \end{cases} \end{aligned} \tag{26.25b}$$

(for $\mu_0 = 1$ we have a Kasner-type metric with three Killing vectors and one homothetic vector).

Among the solutions of the type (26.25) are the Kóta *et al.* (1982) solutions (for $\psi' = 0$, $a^- = 1/2$) and type N solutions (for $\psi' = 0$, $a^- = 0$). A study of the case $D(\sigma/\overline{\sigma}) = 0$ can be found in Bilge (1991). Algebraically general solutions with real ρ and σ and $\Psi_0 = -3\sigma^2/r$ admit two spacelike commuting Killing vectors, see Bilge (1990).

27

The line element for metrics with $\kappa = \sigma = 0 = R_{11} = R_{14} = R_{44}$, $\Theta + \mathrm{i}\omega \neq 0$

27.1 The line element in the case with twisting rays ($\omega \neq 0$)

27.1.1 The choice of the null tetrad

In this chapter we shall deal with space-times which admit a geodesic non-shearing but diverging null congruence $\boldsymbol{k}$,

$$\kappa = \sigma = 0, \quad \rho = -(\Theta + i\omega) \neq 0. \tag{27.1}$$

In addition, the Ricci tensor components picked out by the null congruence $\boldsymbol{k}$ and the (complex) tetrad vector $\boldsymbol{m}$ are assumed to satisfy

$$R_{11} = R_{14} = R_{44} = 0. \tag{27.2}$$

Throughout this and the following sections, the null tetrad $(\boldsymbol{m}, \overline{\boldsymbol{m}}, \boldsymbol{l}, \boldsymbol{k})$ and the Newman–Penrose formalism will be used without further warning (see Chapters 3 and 7). All numerical indices are tetrad indices, e.g. $R_{14} = R_{ab}m^a k^b$; derivatives with respect to coordinates will be abbreviated by explicit use of the coordinate, e.g. $H_{,r} \equiv \partial H/\partial r \equiv \partial_r H$.

Due to the Goldberg–Sachs theorem (Theorem 7.1), all solutions satisfying (27.1) and (27.2) are algebraically special. Note that, because of (6.34), $\Theta = 0 = \sigma$ would imply $\omega = 0$, so that all metrics being discussed here necessarily have an expanding ($\Theta \neq 0$) null congruence $\boldsymbol{k}$.

Following Debney *et al.* (1969), we shall now evaluate (27.1)–(27.2) in order to find a preferred tetrad as well as an appropriate coordinate frame. We start with a tetrad $(\boldsymbol{m}', \overline{\boldsymbol{m}}', \boldsymbol{l}', \boldsymbol{k}')$ in which only the direction of $\boldsymbol{k}'$ is fixed. Conditions (27.1) and (27.2) remain unchanged under the set of tetrad transformations (3.15)–(3.17), i.e. under

$$\boldsymbol{k}' = \boldsymbol{k}, \qquad \boldsymbol{m}' = \boldsymbol{m} + B\boldsymbol{k}, \quad \boldsymbol{l}' = \boldsymbol{l} + B\overline{\boldsymbol{m}} + \overline{B}\boldsymbol{m} + B\overline{B}\boldsymbol{k}, \tag{27.3}$$

$$k' = k, \qquad m' = e^{iC} m, \qquad l' = l, \tag{27.4}$$

$$k' = Ak, \qquad m' = m, \qquad l' = A^{-1} l, \tag{27.5}$$

so we may use these transformations to simplify the connection forms.

Under (27.3), the spin coefficient $\tau = -\Gamma_{143} = -k_{a;b} m^a l^b$ transforms to $\tau' = \tau + B\rho$. Because $\rho \neq 0$, τ can always be made zero, and from now on we have

$$\kappa = \tau = \sigma = 0 = \Gamma_{144} = \Gamma_{143} = \Gamma_{141}, \tag{27.6}$$

i.e. the connection form $\boldsymbol{\Gamma}_{14}$ is simply

$$\boldsymbol{\Gamma}_{14} = -\rho\boldsymbol{\omega}^2 = -\boldsymbol{\Gamma}_{41}, \quad \Gamma_{142} = -\rho. \tag{27.7}$$

The tetrad components of the Ricci tensor $R_{ab} = R_{cadb}(m^c \overline{m}^d + \overline{m}^c m^d - k^c l^d - k^d l^c)$ which are zero because of (27.2), can be written in terms of the curvature tensor as

$$R_{44} = 2R_{1424} = 0, \;\; R_{41} = R_{1421} + R_{1434} = 0, \;\; R_{11} = 2R_{1431} = 0. \tag{27.8}$$

As already stated, the space-times under consideration are algebraically special, i.e. Ψ_0 and Ψ_1 vanish:

$$\Psi_0 = R_{1441} = 0, \quad 2\Psi_1 = 2R_{1434} - R_{14} = R_{1434} - R_{1421} = 0. \tag{27.9}$$

Equations (27.8) and (27.9) show that of the tetrad components R_{14cd} of the curvature tensor only R_{1432} survives. Thus the relation (3.25) between connection and curvature here reads

$$d\boldsymbol{\Gamma}_{41} + \boldsymbol{\Gamma}_{41} \wedge (\boldsymbol{\Gamma}_{21} + \boldsymbol{\Gamma}_{43}) = R_{4123}\boldsymbol{\omega}^2 \wedge \boldsymbol{\omega}^3. \tag{27.10}$$

From (27.7) and (27.10) we get

$$\boldsymbol{\Gamma}_{41} \wedge d\boldsymbol{\Gamma}_{41} = 0, \tag{27.11}$$

which is the integrability condition for the existence of a complex function $\overline{\zeta}$ such that $P\boldsymbol{\Gamma}_{41} = -d\overline{\zeta}$, see (2.44). A rotation (27.4) with $\rho = \rho'$, $\boldsymbol{\Gamma}'_{41} = e^{iC}\boldsymbol{\Gamma}_{41}$ can be used to make the function P real, and by a suitable transformation (27.5), $\boldsymbol{\Gamma}'_{41} = A\boldsymbol{\Gamma}_{41}$, we shall get $P_{,i}k^i = P_{|4} = 0$. (By means of (27.5) we could arrive at $P = 1$, but at the moment we shall not use this special gauge.) Equation (27.7) then reads

$$\boldsymbol{\omega}^2 = \overline{\boldsymbol{\omega}}^1 = -d\overline{\zeta}/P\rho = \boldsymbol{\Gamma}_{41}/\rho, \quad P = \overline{P}, \quad P_{|4} = 0. \tag{27.12}$$

Using (2.74) and (27.12), we can compute $d\boldsymbol{\omega}^2$ as

$$d\boldsymbol{\omega}^2 = \Gamma^2{}_{bc}\boldsymbol{\omega}^b \wedge \boldsymbol{\omega}^c = -(\ln P\rho)_{|b}\boldsymbol{\omega}^b \wedge \boldsymbol{\omega}^2 \tag{27.13}$$

and evaluate (27.10), (27.12) and (27.13) and their complex conjugates. The results are

$$\Gamma_{131} = \Gamma_{134} = \Gamma_{124} = \Gamma_{434} = 0 = \lambda = \pi = \epsilon, \tag{27.14}$$

$$\Gamma_{431} = (\ln \rho)_{|1}, \quad \Gamma_{121} = (\ln P\rho)_{|1}, \quad \Gamma_{123} - \Gamma_{132} = (\ln P\rho)_{|3}, \tag{27.15}$$

$$\rho_{|4} = \rho^2, \tag{27.16}$$

$$R_{4123} = \rho(\ln P)_{|3} + \rho(\Gamma_{213} + \Gamma_{433}). \tag{27.17}$$

Equations (27.6) and (27.14) give $\Gamma_{ab4} = 0$, i.e. we have chosen the tetrad $(\boldsymbol{m}, \overline{\boldsymbol{m}}, \boldsymbol{l}, \boldsymbol{k})$ to be parallelly propagated along the rays.

27.1.2 The coordinate frame

We shall now choose the coordinate frame as follows. Because of (27.12), the function ζ and its complex conjugate $\overline{\zeta}$ will serve as two (spacelike) coordinates. The third, real, coordinate is the affine parameter r along the rays,

$$k^i = \partial x^i / \partial r = (0, 0, 1, 0). \tag{27.18}$$

Because of (2.17) and (3.11) this implies

$$\mathrm{d}r = WP\overline{\rho}\boldsymbol{\omega}^1 + \overline{W}P\rho\boldsymbol{\omega}^2 - H\boldsymbol{\omega}^3 + \boldsymbol{\omega}^4, \tag{27.19}$$

where W is complex and H real. (The coefficients of $\boldsymbol{\omega}^1$ and $\boldsymbol{\omega}^2$ are chosen so that (27.22) will take a simple form.) The fourth, real, coordinate u is introduced by

$$u_{|3} = u_{,n} l^n = 1, \quad u_{|4} = u_{,n} k^n = 0. \tag{27.20}$$

Equations (27.6) and (27.14) imply that the system (27.20) is integrable: from (7.6a) we get $(\Delta D - D\Delta)u = (\gamma + \overline{\gamma})Du = 0$. Equation (27.20) gives

$$\mathrm{d}u = \overline{\rho}PL\boldsymbol{\omega}^1 + \rho P\overline{L}\boldsymbol{\omega}^2 + \boldsymbol{\omega}^3. \tag{27.21}$$

We have thus introduced all four coordinates, the result being

$$\begin{aligned}
\boldsymbol{\omega}^1 &= -\mathrm{d}\zeta/P\overline{\rho} = \overline{m}_n \mathrm{d}x^n, \quad \boldsymbol{\omega}^2 = -\mathrm{d}\overline{\zeta}/P\rho = m_n \mathrm{d}x^n, \\
\boldsymbol{\omega}^3 &= \mathrm{d}u + L\mathrm{d}\zeta + \overline{L}\mathrm{d}\overline{\zeta} = -k_n \mathrm{d}x^n, \\
\boldsymbol{\omega}^4 &= \mathrm{d}r + W\mathrm{d}\zeta + \overline{W}\mathrm{d}\overline{\zeta} + H\boldsymbol{\omega}^3 = -l_n \mathrm{d}x^n.
\end{aligned} \tag{27.22}$$

The complex functions ρ, L, W and the real functions P, H are not independent of each other and cannot be chosen arbitrarily, since the metric has to satisfy the field equations (27.10) and the conditions (27.6). We get the restrictions in question by comparing the previous results (27.6) and (27.14)–(27.16) for the Γ_{4bc} (obtained from the field equations (27.10) and from (27.6)) with those from the calculation of $\mathrm{d}\boldsymbol{\omega}^3 = -\Gamma_{4bc}\boldsymbol{\omega}^b \wedge \boldsymbol{\omega}^c$. The result is

$$\rho_{|4} = \rho^2, \quad P_{|4} = 0, \quad L_{|4} = 0, \tag{27.23}$$

$$\rho P\overline{L}_{|1} - \overline{\rho} P L_{|2} = \rho - \overline{\rho}, \quad \overline{\rho} P L_{|3} = (\ln \rho)_{|1}. \tag{27.24}$$

The evaluation of $\mathrm{d}\boldsymbol{\omega}^4 = -\Gamma_{3bc}\boldsymbol{\omega}^b \wedge \boldsymbol{\omega}^c$ gives an expression for $\Gamma_{314} - \Gamma_{341}$, which, because of (27.14), (27.15) and (27.24), yields the condition

$$\overline{\rho} P W_{|4} = -(\ln \rho)_{|1}. \tag{27.25}$$

(Evaluation of the remaining terms, which we shall postpone, would give those Γ_{3bc} not yet determined.)

Equations (27.24)–(27.25) are easy to integrate. The result is

$$\begin{aligned} \rho^{-1} &= -(r + r^0 + \mathrm{i}\Sigma), & 2\mathrm{i}\Sigma &= P^2(\overline{\partial} L - \partial \overline{L}), \\ W &= \rho^{-1} L_{,u} + \partial(r^0 + \mathrm{i}\Sigma), & \partial &\equiv \partial_\zeta - L\partial_u. \end{aligned} \tag{27.26}$$

Equations (27.26) show that all metric functions can be given in terms of the coordinate r, the real functions $r^0(\zeta, \overline{\zeta}, u)$, $P(\zeta, \overline{\zeta}, u)$ and $H(\zeta, \overline{\zeta}, r, u)$, and the complex function $L(\zeta, \overline{\zeta}, u)$. We can now summarize the main result in the following theorem:

Theorem 27.1 *A space-time admits a geodesic, shearfree and diverging ($\rho \neq 0$) null congruence $\boldsymbol{k}$ and satisfies $R_{11} = R_{14} = R_{44} = 0$ if and only if the metric can be written in the form*

$$\begin{aligned} \mathrm{d}s^2 &= 2\omega^1\omega^2 - 2\omega^3\omega^4 \\ &= \frac{2\mathrm{d}\zeta \mathrm{d}\overline{\zeta}}{P^2 \rho\overline{\rho}} - 2[\mathrm{d}u + L\mathrm{d}\zeta + \overline{L}\mathrm{d}\overline{\zeta}] \\ &\qquad \times \left[\mathrm{d}r + W\mathrm{d}\zeta + \overline{W}\mathrm{d}\overline{\zeta} + H\left\{\mathrm{d}u + L\mathrm{d}\zeta + \overline{L}\mathrm{d}\overline{\zeta}\right\}\right], \end{aligned} \tag{27.27}$$

$$\begin{aligned} m^i &= (-P\overline{\rho}, 0, PW\overline{\rho}, P\overline{\rho}L), \quad \overline{m}^i = (0, -P\rho, P\rho\overline{W}, P\rho\overline{L}), \\ l^i &= (0, 0, -H, 1), \quad k^i = (0, 0, 1, 0), \end{aligned}$$

where the complex functions ρ, W, L and the real function P are subject to (27.26) (Robinson and Trautman 1962, Debney *et al.* 1969, Talbot 1969, Robinson *et al.* 1969a, Lind 1974).

In (27.27), r is the affine parameter along the rays, u is a retarded time, and ζ and $\overline{\zeta}$ are (spacelike) coordinates on the 2-surfaces, r, u = constant. In spherical coordinates, one would have $\zeta = \sqrt{2}\exp(\mathrm{i}\varphi)\cot\vartheta/2$.

27.1.3 Admissible tetrad and coordinate transformations

The vectors $\boldsymbol{l}$ and $\boldsymbol{k}$ are fixed up to a Lorentz transformation (27.5), which will preserve (27.18) and (27.20) only in combination with a coordinate transformation. The metric (27.27) is, therefore, invariant under

$$\boldsymbol{k}' = \boldsymbol{k}F_{,u}, \quad \boldsymbol{l} = \boldsymbol{l}'F_{,u}, \quad u' = F(u,\zeta,\overline{\zeta}), \quad r = r'F_{,u}, \quad F_{,u} > 0. \tag{27.28}$$

This transformation will induce a transformation of the metric functions,

$$\rho' = \rho F_{,u}, \quad P = P'F_{,u}. \tag{27.29}$$

The vectors $\boldsymbol{m}$ (and $\overline{\boldsymbol{m}}$) are fixed up to a rotation (27.4), which keeps the metric invariant and P real if combined with a coordinate transformation $\zeta' = \zeta'(\zeta)$ (ζ' analytic in ζ):

$$\boldsymbol{m}' = e^{iC}\boldsymbol{m}, \quad \zeta' = \zeta'(\zeta), \quad e^{iC} = \left(\frac{d\overline{\zeta}'/d\overline{\zeta}}{d\zeta'/d\zeta}\right)^{1/2}, \quad P' = P\left|\frac{d\zeta'}{d\zeta}\right|. \tag{27.30}$$

The possible transformations involving only the coordinates and not the tetrad are (i) changes of the origin of the affine parameter r

$$r' = r + f(u,\zeta,\overline{\zeta}), \tag{27.31}$$

f being constant along the rays, and (ii) the transformations

$$u' = u + g(\zeta,\overline{\zeta}). \tag{27.32}$$

These two types of transformations are just the degrees of freedom inherent in the respective definitions (27.18) and (27.20) of the coordinates r and u.

27.2 The line element in the case with non-twisting rays ($\omega = 0$)

If ω vanishes,

$$\rho = \overline{\rho} = -\Theta \quad \Leftrightarrow \quad \omega = 0, \tag{27.33}$$

then the line element (27.27) can be further simplified. For $\omega = 0$, the vector field $\boldsymbol{k}$ is normal, i.e. proportional to a gradient, and thus a transformation $\boldsymbol{k}' = A\boldsymbol{k}$ will lead to

$$\boldsymbol{\omega}^3 = -k_i dx^i = du. \tag{27.34}$$

(The same result, $L = 0$, could be achieved by starting from (27.22). Because of (27.26) and (27.33), $\boldsymbol{\omega}^3 \wedge d\boldsymbol{\omega}^3$ is zero, and a transformation (27.28) gives $L = 0$.)

With $L = 0 = \Sigma$, (27.36) gives $\rho^{-1} = r + r^0(u, \zeta, \overline{\zeta})$; but by a change (27.31) of the affine parameter origin we can always make r^0 vanish. Equations (27.26) then read

$$r = -\rho^{-1} = \Theta^{-1}, \quad L = W = \Sigma = r^0 = 0, \tag{27.35}$$

and instead of (27.22) we get

$$\omega^1 = rP^{-1}\mathrm{d}\zeta, \quad \omega^2 = rP^{-1}\mathrm{d}\overline{\zeta}, \quad \omega^3 = \mathrm{d}u, \quad \omega^4 = \mathrm{d}r + H\mathrm{d}u. \tag{27.36}$$

Theorem 27.2 *A space-time admits a geodesic, shearfree, twistfree and diverging ($\rho = \overline{\rho} = -r^{-1}$) null congruence $\boldsymbol{k}$, and satisfies $R_{11} = R_{14} = R_{44} = 0$, exactly if the metric can be written in the form*

$$\begin{aligned} \mathrm{d}s^2 &= 2\omega^1\omega^2 - 2\omega^3\omega^4 \\ &= 2r^2P^{-2}\mathrm{d}\zeta\mathrm{d}\overline{\zeta} - 2\mathrm{d}u\,\mathrm{d}r - 2H\mathrm{d}u^2, \quad P_{,r} = 0, \\ m^i &= (P/r, 0, 0, 0), \quad \overline{m}^i = (0, P/r, 0, 0), \\ l^i &= (0, 0, -H, 1), \quad k^i = (0, 0, 1, 0). \end{aligned} \tag{27.37}$$

This line element is preserved under the transformations

$$\begin{aligned} \zeta' &= f(\zeta), & u' &= F(u), & r &= r'F_{,u}, \\ \boldsymbol{m}' &= \left(\overline{f}'/f'\right)^{1/2}\boldsymbol{m}, & \boldsymbol{k}' &= \boldsymbol{k}F_{,u}, & \boldsymbol{l} &= \boldsymbol{l}'F_{,u}, \\ H' &= (H + r'F_{,uu})/F^2_{,u}, & P' &= F^{-1}_{,u}\,|f'|\,P. \end{aligned} \tag{27.38}$$

Solutions which satisfy the conditions of the above theorem are called Robinson–Trautman solutions (Robinson and Trautman 1962).

28
Robinson–Trautman solutions

28.1 Robinson–Trautman vacuum solutions

28.1.1 The field equations and their solutions

By definition, Robinson–Trautman solutions are solutions meeting the requirements of Theorem 27.2, i.e. they are algebraically special solutions satisfying $\kappa = \sigma = \omega = 0$, $R_{44} = R_{41} = R_{11} = 0$. These assumptions have the advantage that the remaining field equations $R_{12} = R_{13} = R_{33} = R_{34} = 0$ can be partially solved and reduced to a single fourth order differential equation.

To perform this integration, we start from the metric (27.37) and calculate the connection forms of the 1-forms (27.36):

$$\begin{aligned} \boldsymbol{\Gamma}_{14} &= -\boldsymbol{\omega}^2/r, \quad \boldsymbol{\Gamma}_{32} = \boldsymbol{\omega}^1\left[(\ln P)_{,u} + H/r\right] - \boldsymbol{\omega}^3 P H_{,\overline{\zeta}}/r, \\ \boldsymbol{\Gamma}_{21} + \boldsymbol{\Gamma}_{43} &= -\boldsymbol{\omega}^1 P_{,\zeta}/r + \boldsymbol{\omega}^2 P_{,\overline{\zeta}}/r + \boldsymbol{\omega}^3 H_{,r}. \end{aligned} \tag{28.1}$$

The surviving tetrad components of the Ricci tensor are

$$R_{12} = R_{1212} - 2R_{1423} = \frac{2P^2}{r^2}(\ln P)_{,\zeta\overline{\zeta}} - \frac{4}{r}\frac{P_{,u}}{P} - \frac{(2Hr)_{,r}}{r^2}, \tag{28.2}$$

$$R_{13} = R_{2113} + R_{4313} = P\left[H_{,r\zeta} + (\ln P)_{,u\zeta}\right]/r, \tag{28.3}$$

$$\tfrac{1}{2}R_{33} = R_{3132} = \frac{H_{,u}}{r} + \frac{P^2}{r^2}H_{,\zeta\overline{\zeta}} + \frac{P_{,u}}{P}H_{,r} + P\left(\frac{P_{,u}}{P^2}\right)_{,u} - \frac{2H}{rP}P_{,u}, \tag{28.4}$$

$$R_{34} = R_{3434} + 2R_{1423} = (r^2 H_{,r})_{,r}/r^2 + 2(\ln P)_{,u}/r. \tag{28.5}$$

We can now evaluate the vacuum field equations. $R_{12} = 0$ immediately gives

$$2H = \Delta \ln P - 2r(\ln P)_{,u} - 2m/r, \quad \Delta \equiv 2P^2\partial_\zeta\partial_{\overline{\zeta}}, \tag{28.6}$$

where m is an arbitrary function of integration (independent of r). $R_{34} = 0$ is then satisfied identically. $R_{13} = 0 = R_{23}$ tells us that m is a function of u alone. The last equation $R_{33} = 0$ leads to

$$\Delta\Delta \ln P + 12m(\ln P)_{,u} - 4m_{,u} = 0. \qquad (28.7)$$

Theorem 28.1 *The general vacuum solution admitting a geodesic, shearfree, twistfree but diverging null congruence is the Robinson–Trautman metric* (Robinson and Trautman 1962)

$$ds^2 = \frac{2r^2}{P^2(u,\zeta,\bar\zeta)}d\zeta d\bar\zeta - 2du\,dr - \left[\Delta \ln P - 2r(\ln P)_{,u} - \frac{2m(u)}{r}\right] du^2,$$

$$\Delta\Delta(\ln P) + 12m(\ln P)_{,u} - 4m_{,u} = 0, \quad \Delta \equiv 2P^2\partial_\zeta\partial_{\bar\zeta}. \qquad (28.8)$$

This line element is invariant with respect to the transformations (27.38), *i.e. to*

$$u' = F(u), \quad r = r'F_{,u}, \quad \zeta' = f(\zeta), \quad P' = P\,|f_{,\zeta}|\,F_{,u}^{-1}, \quad m' = mF_{,u}^{-3}. \qquad (28.9)$$

They can be used e.g. to give a non-zero m the values ± 1.

In the Robinson–Trautman line element (28.8), r is the affine parameter along the rays of the repeated null eigenvector ($r_{,i}$ is not necessarily space-like!), and u is a retarded time. The surfaces $r, u = \text{const}$ may be thought of as distorted spheres (if they are closed); the solutions (28.8) are therefore often referred to as describing spherical gravitational radiation. Of course, no exact spherical gravitational waves exist, since spherical symmetry would imply $\Delta \ln P = K(u)$, and, in the gauge $m = 1$, (28.8) shows that the metric is then static (for $m = 0$, see below under type N). In some special cases, e.g. in the static Schwarzschild metric contained here (see below under type D), the parameter m has the physical meaning of the system's mass. Using a Lyapunov-functional argument, it can be shown (see Lukács *et al.* (1984), and e.g. Chruściel (1991) for results and further references) that for rather general initial values on $u = \text{const}$ and ζ–$\bar\zeta$- surfaces diffeomorphic to a sphere, the Robinson–Trautman solutions radiate and then settle down to the static Schwarzschild solution. For axisymmetric Robinson–Trautman solutions, Hoenselaers and Perjés (1993) found that, for almost all initial conditions, the final state of the solution is the C-metric.

For the Robinson–Trautman metric (28.8), the surviving components of the Weyl (curvature) tensor are

$$\Psi_2 = -mr^{-3}, \quad 2\Psi_3 = -r^{-2}P(\Delta \ln P)_{,\bar\zeta},$$
$$\Psi_4 = r^{-2}\left[P^2\left\{\tfrac{1}{2}\Delta \ln P - r(\ln P)_{,u}\right\}_{,\bar\zeta}\right]_{,\bar\zeta}. \qquad (28.10)$$

Table 28.1. The Petrov types of the Robinson–Trautman vacuum solutions

	$6m(P^2H_{,\zeta})_{,\zeta} + rP^2\,[(\Delta \ln P)_{,\zeta}]^2$	
	$\neq 0$	$= 0$
$m \neq 0$	II	D
$m = 0$	III	N, O

The conditions for the different Petrov types have been collected in Table 28.1 (with $2H = \Delta \ln P - 2r(\ln P)_{,u} - 2m/r$).

28.1.2 Special cases and explicit solutions

Type N solutions, $\Psi_2 = \Psi_3 = 0$

Type N solutions are characterized by $m = 0$ and

$$\Delta \ln P = K(u). \tag{28.11}$$

By a transformation $u' = F(u)$, the Gaussian curvature K of the 2-surface $2\mathrm{d}\zeta\mathrm{d}\overline{\zeta}/P^2$ could be normalized to $K = 0, \pm 1$. A special solution of (28.11) is

$$P = \alpha(u)\zeta\overline{\zeta} + \beta(u)\zeta + \overline{\beta}(u)\overline{\zeta} + \delta(u), \quad K = 2(\alpha\delta - \beta\overline{\beta}), \tag{28.12}$$

which gives flat space-time, as Ψ_4 vanishes.

For constant u, (28.12) is the general solution of (28.11). For varying u, the respective ζ–$\overline{\zeta}$-surfaces of constant curvature may be thought of as mapped onto each other, the mapping having an arbitrary u-dependence. Thus the general solution can be generated by the substitution $\zeta \to \zeta'(u, \zeta)$, $\mathrm{d}\zeta \to \mathrm{d}\zeta'$, $P \to P|\mathrm{d}\zeta'/\mathrm{d}\zeta|$ in (28.12) and the line element (28.8), where ζ' is an arbitrary function (analytic in ζ). In general, this substitution does not correspond to a coordinate transformation (28.9). For constant u, it represents a mapping of the ζ–$\overline{\zeta}$-plane into itself. The only one-to-one mapping of this kind is the substitution $\zeta' = (a\zeta + b)/(c\zeta + d)$, which leaves (28.12) form-invariant and thus again gives a flat four-dimensional space-time. To get a non-flat type N solution, one has to make a more general substitution, but this necessarily gives rise to at least one singular point ζ and thus to a singular line in three-dimensional $(r, \zeta, \overline{\zeta})$-space.

The appearance of singular lines (pipes) is a common feature of many of the known Robinson–Trautman solutions, which makes them inadequate for realistically describing the radiation field outside an isolated source.

For a type N shock wave, see Nutku (1991).

Type III solutions, $\Psi_2 = 0$

Type III solutions are characterized by $m = 0$ and

$$\Delta\Delta \ln P = 0, \quad (\Delta \ln P)_{,\zeta} \neq 0. \tag{28.13}$$

These conditions immediately give

$$\Delta \ln P = K = -3\left[f(\zeta, u) + \overline{f}(\overline{\zeta}, u)\right], \quad f_{,\zeta} \neq 0. \tag{28.14}$$

If f is independent of u, then by a coordinate transformation (28.9), $f(\zeta) \to \zeta$, which leaves K invariant, we arrive at

$$\Delta \ln P = K = -3(\zeta + \overline{\zeta}). \tag{28.15}$$

The only known solution of this differential equation is

$$P = (\zeta + \overline{\zeta})^{3/2} \tag{28.16}$$

(Robinson and Trautman 1962). This solution is a Bianchi type VI metric, see §13.3 and (13.64).

If f depends on ζ *and* u, then for the purpose of solving the differential equation, we may take u as a parameter in (28.14) and make the substitution $f(\zeta, u) \to \zeta$ which again leads to (28.15). As in the type N case, this substitution is *not* an allowed coordinate transformation.

Theorem 28.2 *To get the most general type III diverging non-twisting vacuum solution one has to* (a) *solve* (28.15) *for* P *and* (b) *then make the substitution* $\zeta \to f(\zeta, u)$, $P \to P\,|\mathrm{d}f/\mathrm{d}\zeta|$ (Foster and Newman 1967, Robinson 1975).

A simple example of a metric obtained by this method is

$$P = (\zeta + \overline{\zeta} + u)^{3/2}. \tag{28.17}$$

For all type III solutions, because of (28.13), the Gaussian curvature K of the ζ–$\overline{\zeta}$-surfaces is a non-constant analytic function and must have singular points.

Type D solutions, $3\Psi_2\Psi_4 = 2\Psi_3^2$, $\Psi_2 \neq 0$

The vector $\boldsymbol{l}$ used so far is not necessarily an eigenvector of the Weyl tensor. Thus a type D solution need not have $\Psi_3 = \Psi_4 = 0$; but a null rotation (27.3) which makes $\Psi'_3 = 0$ must always exist. This is guaranteed exactly if $3\Psi_2\Psi_4 = 2\Psi_3^2$, see §9.3.1, and the terms with different powers in r entering this condition eventually split it into the two equations

$$P^2\left[(\Delta \ln P)_{,\zeta}\right]^2 = 6m\left[P^2(\ln P)_{,u\zeta}\right]_{,\zeta}, \tag{28.18}$$

$$\left[P^2(\Delta \ln P)_{,\zeta}\right]_{,\zeta} = 0, \tag{28.19}$$

which have to be satisfied in addition to the field equations (28.8). Equations (28.19), (28.18) and the field equations are then (in a gauge with $m = \text{const}$) equivalent to

$$P^2(\Delta \ln P)_{,\overline{\zeta}} = h(u,\zeta), \quad 3m\partial_u P^{-2} = \partial_\zeta(hP^{-2}), \tag{28.20}$$

h being analytic in ζ. The integrability condition for this system turns out (via the Bianchi identities (7.32*d*)) to be $h_{,u} = 0$ (McIntosh, private communication 1985).

If h is zero, then $K = \Delta \ln P$ is a constant and P is independent of u. The corresponding metrics are, with $K = 0, \pm 1$,

$$ds^2 = 2r^2 d\zeta\, d\overline{\zeta}\left(1 + \tfrac{1}{2}K\zeta\overline{\zeta}\right)^{-2} - 2du\, dr - (K - 2m/r)\, du^2. \tag{28.21}$$

For $K = 1$, this is the Schwarzschild metric (15.23). The case $K = 0$ is a special Kasner metric (13.51).

If h is non-zero, then one can perform a coordinate transformation (28.9), with $hf' = \text{const}$, to achieve $h = 3\sqrt{2}$. In the gauge $m = 1$, (28.20) then shows that P is a function of $u + (\zeta + \overline{\zeta})/\sqrt{2} = u + x$ only and has to satisfy

$$P^2\, \partial_x(P^2 \partial_x \partial_x \ln P) = 6, \quad P = P(x+u), \quad \zeta + \overline{\zeta} = x\sqrt{2}. \tag{28.22}$$

This differential equation can be solved by introducing a new variable η via

$$d\eta/ds = P^{-2}, \quad s \equiv x + u. \tag{28.23}$$

The solution and the corresponding line element, best written in coordinates η, y, r, u, are

$$ds^2 = r^2\left[P^2(\eta)d\eta^2 + \frac{dy^2}{P^2(\eta)} - 2d\eta\, du + \frac{du^2}{P^2(\eta - r^{-1})}\right] - 2du\, dr, \tag{28.24}$$

$$P^{-2}(\eta) = -2\eta^3 + b\eta + c.$$

This is the static C-metric (given in Table 18.2 in different coordinates). The special case $b = 0 = c$ of it is (in a different gauge) the metric

$$P = (\zeta + \overline{\zeta} + u)^{3/4}, \quad m = 1/4 \tag{28.25}$$

(Collinson and French 1967).

Type II solutions

In the general case, the field equations (28.8) show that, for a fixed $u = u_0$, the real function $P(\zeta, \overline{\zeta}, u_0)$ can be prescribed arbitrarily. Explicit type *II* solutions can be obtained from any type *III* solution (28.15) by allowing $m = \text{const} \neq 0$. An example is (Robinson and Trautman 1962)

$$ds^2 = 2r^2(\zeta + \overline{\zeta})^{-3}\mathrm{d}\zeta\mathrm{d}\overline{\zeta} - 2\mathrm{d}u\,\mathrm{d}r + \Big[3(\zeta + \overline{\zeta}) + 2m/r\Big]\,\mathrm{d}u^2. \quad (28.26)$$

28.2 Robinson–Trautman Einstein–Maxwell fields

28.2.1 Line element and field equations

In this section, we will consider Einstein–Maxwell fields which are algebraically special, the repeated null eigenvector of the Weyl tensor being normal, geodesic and shearfree, and an eigenvector of the Maxwell tensor (*aligned case*). The conditions of Theorem (27.2) are satisfied, and we can start with the Robinson–Trautman line element (27.37), i.e. with

$$ds^2 = 2r^2P^{-2}\mathrm{d}\zeta\,\mathrm{d}\overline{\zeta} - 2\mathrm{d}u\,\mathrm{d}r - 2H\mathrm{d}u^2, \quad P_{,r} = 0. \quad (28.27)$$

By assumption, $\boldsymbol{k}$ is an eigenvector of the Maxwell tensor, and thus the components (7.26)–(7.28) vanish except for

$$\Phi_1 = \tfrac{1}{2}F_{ab}(k^al^b + \overline{m}^am^b), \quad \Phi_2 = F_{ab}\overline{m}^al^b \quad (28.28)$$

(note that Φ_2 could be made zero by performing a null rotation, see (3.15) and (3.42*b*)). The Maxwell equations (7.22)–(7.25) here read

$$r\partial_r\Phi_1 = -2\Phi_1, \quad \partial_\zeta\Phi_1 = 0, \quad (28.29)$$

$$r\partial_r\Phi_2 = -\Phi_2 + P\partial_{\overline{\zeta}}\Phi_1, \quad (28.30)$$

$$P\partial_\zeta\Phi_2 + r(H\partial_r - \partial_u)\Phi_1 = -2(H + r\partial_u \ln P)\Phi_1 + P_{,\zeta}\Phi_2. \quad (28.31)$$

Equations (28.29) and (28.30) – in that order – are integrated by

$$\Phi_1 = \overline{Q}(u, \overline{\zeta})/2r^2, \quad \Phi_2 = -P\overline{Q}_{,\overline{\zeta}}/2r^2 + Ph(u, \zeta, \overline{\zeta})/r. \quad (28.32)$$

$Q(\zeta, u)$ and $h(\zeta, \overline{\zeta}, u)$ are complex functions of their respective arguments and because of (28.31) are restricted by

$$h_{,\zeta} = \partial_u(\overline{Q}/2P^2). \quad (28.33)$$

We can now attack the remaining field equations

$$R_{12} = R_{34} = 2\kappa_0\Phi_1\overline{\Phi}_1, \tag{28.34}$$

$$R_{13} = 2\kappa_0\Phi_1\overline{\Phi}_2, \quad R_{33} = 2\kappa_0\Phi_2\overline{\Phi}_2, \tag{28.35}$$

cp. (7.10)–(7.15) and (7.29). The expressions for R_{12}, R_{13}, R_{33} and R_{34} in terms of the metric function can be taken from (28.2)–(28.5).

Equation (28.34) gives the r-dependence of the function H as

$$2H = \Delta \ln P - 2r(\ln P)_{,u} - 2m(\zeta,\overline{\zeta},u)/r + \kappa_0 Q\overline{Q}/2r^2, \tag{28.36}$$

with $\Delta \equiv 2P^2\partial_\zeta\partial_{\overline{\zeta}}$, and (28.35) yields the differential equations given below in (28.37*d*)–(28.37*e*).

Theorem 28.3 *If an Einstein–Maxwell field admits a geodesic, shearfree, diverging but normal null vector field* $\boldsymbol{k}$ *which is an eigenvector of the Maxwell and Weyl tensors, then the metric is algebraically special and can be written in the form*

$$\begin{aligned} \mathrm{d}s^2 &= 2r^2P^{-2}(\zeta,\overline{\zeta},u)\mathrm{d}\zeta\,\mathrm{d}\overline{\zeta} - 2\mathrm{d}u\,\mathrm{d}r \\ &\quad - \left[\Delta \ln P - 2r(\ln P)_{,u} - \frac{2}{r}m(\zeta,\overline{\zeta},u) + \frac{\kappa_0}{2r^2}Q(\zeta,u)\overline{Q}(\overline{\zeta},u)\right]\mathrm{d}u^2, \end{aligned} \tag{28.37a}$$

and the electromagnetic field is given by

$$\Phi_1 = \overline{Q}/2r^2, \quad \Phi_2 = -P\overline{Q}_{,\overline{\zeta}}/2r^2 + Ph(\zeta,\overline{\zeta},u)/r. \tag{28.37b}$$

P and m are real functions, h is complex and Q analytic in ζ. The four functions have to obey

$$\Delta\Delta \ln P + 12m(\ln P)_{,u} - 4m_{,u} = 4\kappa_0 P^2 h\overline{h}, \tag{28.37c}$$

$$Q\overline{Q}_{,u} - \overline{Q}Q_{,u} = 2P^2(\overline{h}\,\overline{Q}_{,\overline{\zeta}} - hQ_{,\zeta}), \tag{28.37d}$$

$$h_{,\zeta} = (\overline{Q}/2P^2)_{,u}, \quad m_{,\zeta} = \kappa_0\overline{h}\,\overline{Q}. \tag{28.37e}$$

(Equation (28.37*d*) is in fact a consequence of (28.37*e*) and the reality of m.) Like all metrics of the Robinson–Trautman class (27.37), the line element is preserved under the transformations (28.9) with the appropriate change $Q' = F_{,u}^{-2}Q$, $h' = F_{,u}^{-1}\overline{f}'^{-1}h$. In addition to that, the Einstein and Maxwell field equations are invariant with respect to $Q' = \mathrm{e}^{\mathrm{i}a}Q$, $h' = \mathrm{e}^{-\mathrm{i}a}h$, $a = \text{const}$.

To have a rough idea of the physical meaning of the quantities appearing in (28.37*a*)–(28.37*b*) one may think of m and Q as representing, respectively, mass and (electric plus magnetic) charge and of h as an electromagnetic pure radiation field. For $Q = 0$, the Maxwell field is a null field.

The surviving components of the Weyl tensor of the Robinson–Trautman solutions (28.37a)–(28.37e) are given by

$$\Psi_2 = -\frac{m}{r^3} + \frac{\kappa_0 Q\overline{Q}}{2r^4}, \quad \Psi_3 = -\frac{P}{2r^2}(\Delta \ln P)_{,\overline{\zeta}} + \frac{3P}{2r^3} m_{,\overline{\zeta}} - \frac{\kappa_0 P}{2r^4} Q\overline{Q}_{,\overline{\zeta}},$$
$$\Psi_4 = \frac{1}{r^2}\left[P^2\left(\frac{1}{2}\Delta \ln P - r\frac{P_{,u}}{P} - \frac{m}{r} + \frac{\kappa_0 Q\overline{Q}}{4r^2}\right)_{,\overline{\zeta}}\right]_{,\overline{\zeta}}. \tag{28.38}$$

28.2.2 Solutions of type III, N and O

$\Psi_2 = 0$ implies $m = Q = 0$; the Maxwell field is necessarily a null field. For type III, because of (28.37e), h is analytic in $\overline{\zeta}$ and therefore obeys

$$\Delta \ln(h\overline{h}) = 0, \tag{28.39}$$

which restricts the solutions of the only remaining field equation

$$\partial_\zeta \partial_{\overline{\zeta}} \Delta \ln P = 2\kappa_0 h\overline{h}, \quad h\overline{h} \neq 0. \tag{28.40}$$

Special type III solutions are

$$P = l(u)k(\zeta)\overline{k}(\overline{\zeta})(1 + \tfrac{1}{2}\zeta\overline{\zeta}), \; \sqrt{\kappa_0}h = \sqrt{2}\, l(u)\overline{k}(\overline{\zeta})\overline{k}'(\overline{\zeta}), \tag{28.41}$$

(Bartrum 1967) and

$$P = a\cosh[l(\zeta + \overline{\zeta}) + n)\mathrm{e}^{b(\zeta+\overline{\zeta})+c}/lb, \quad \sqrt{\kappa_0}h = 2a\mathrm{e}^{b\zeta+c}, \tag{28.42}$$

(Ivanov 1977), where a, b, c, l, n are arbitrary functions of u.

To get type N or O solutions, we have to require, in addition to $\Psi_2 = 0$, the property $\Psi_3 = 0 = (\Delta \ln P)_{,\zeta}$. Together with the field equations (28.40) this yields $h = 0$.

Theorem 28.4 *There are no non-vacuum Einstein–Maxwell fields of the Robinson–Trautman class which belong to Petrov type N or O.*

28.2.3 Solutions of type D

All Robinson–Trautman Einstein–Maxwell fields of type D are known (Cahen and Sengier 1967, Debever 1971, Leroy 1976). Following Leroy, we shall give an exhaustive list of them, without presenting the proofs (for a single expression for the metric of all type D Einstein–Maxwell fields see §21.1; cp. also §26.2). The starting point is the field equations (28.37c)–(28.37e) together with the type D condition $3\Psi_2\Psi_4 = 2\Psi_3^2$. This last is split into five equations by equating to zero the coefficients of the various powers of r. Several cases have to be distinguished.

If $Q = 0$, then Φ_1 vanishes and the electromagnetic field is a null field. The field equation (28.37e) gives $m = m(u)$, and it turns out that the metric of the type D spaces with $Q = 0$ is (Robinson and Trautman 1962)

$$\begin{aligned} \mathrm{d}s^2 &= r^2(\mathrm{d}x^2 + \mathrm{d}y^2) - 2\mathrm{d}u\,\mathrm{d}r + 2m(u)\mathrm{d}u^2/r, \\ m(u) &= -\kappa_0 \textstyle\int h(u)\overline{h}(u)\mathrm{d}u, \quad \Phi_1 = 0, \quad \Phi_2 = h(u)/r. \end{aligned} \tag{28.43}$$

If $Q \neq 0$, $Q_{,\zeta} = 0$, then because of (28.37d) Q can be transformed to a complex constant. For $h = 0$, the corresponding type D solutions are

$$\begin{aligned} \mathrm{d}s^2 &= \frac{2r^2\mathrm{d}\zeta\mathrm{d}\overline{\zeta}}{(1 + K\zeta\overline{\zeta}/2)^2} - 2\mathrm{d}u\,\mathrm{d}r - \left(K - \frac{2m}{r} + \frac{\kappa_0 Q\overline{Q}}{2r^2}\right)\mathrm{d}u^2, \\ \Phi_1 &= \overline{Q}/2r^2, \quad \Phi_2 = 0, \quad K = 0, \pm 1. \end{aligned} \tag{28.44}$$

Q (complex) and m (real) are arbitrary constants. For $K = 1$, these are the Reissner–Weyl solutions (15.21). Obviously, (28.44) is the charged counterpart of the vacuum metric (28.21).

For non-zero h, we can set $\kappa_0 Q\overline{Q} = 1$ by a suitable transformation of u. Then m becomes a function of $s = x + u$ alone. Taking m as a new variable, the corresponding type D solution can be written as

$$\begin{aligned} \mathrm{d}s^2 &= r^2\left[P^2(m)\mathrm{d}m + P^{-2}(m)\mathrm{d}y^2 - 2\,\mathrm{d}m\,\mathrm{d}u\right] \\ &\quad + r^2 P^{-2}(m - r^{-1})\mathrm{d}u^2 - 2\,\mathrm{d}r\,\mathrm{d}u, \\ P^{-2}(m) &= -m^4/2 + am^2 + bm + c, \\ \sqrt{\kappa_0}\,\Phi_1 &= \mathrm{e}^{\mathrm{i}\,q}/2r^2, \quad \sqrt{\kappa_0}\,\Phi_2 = \mathrm{e}^{\mathrm{i}\,q}/\left[rP(m)\right]. \end{aligned} \tag{28.45}$$

This is the charged C-metric, cp. (28.24) and (21.22); here a, b, c and q are real constants.

It can be shown (Leroy 1976) that the solutions (28.44) and (28.45) are exactly those for which *both* double eigenvectors of the Weyl tensor are also eigenvectors of the Maxwell field.

If $Q_{,\zeta} \neq 0$, then the type D solutions turn out to be

$$\begin{aligned} ds^2 &= r^2 e^{-x}(dx^2 + dy^2) - 2du\,dr \\ &\quad -\kappa_0 \left[e^{3x}(au+b)^2/2r^2 + e^{2x}a(au+b)/r\right] du^2, \\ \Phi_1 &= e^{3(x-i\,y)/2}(au+b)/2r^2, \\ \Phi_2 &= -e^{2x-3i\,y/2}\left[3(au+b)/2r + ae^{-x}\right]/r\sqrt{2}, \end{aligned} \tag{28.46}$$

a and b being real constants.

28.2.4 Type II solutions

Only some special cases of Petrov type II non-twisting Einstein–Maxwell fields have been treated in detail. They refer to special properties of the Maxwell field or to simple functional structures of the metric functions. In particular, the cases $h = 0, Q = 0$ and $\Delta \ln P = \text{const}$ have been studied (but the complete solution was not always found).

If h vanishes, then the field equations (28.37c)–(28.37e) give (with real q and P_0)

$$m = m(u), \quad Q = q^2(u)f(\zeta), \tag{28.47}$$

$$\Delta\Delta \ln P_0 + 12m(\ln q)_{,u} - 4m_{,u} = 0, \quad P = q(u)P_0(\zeta,\overline{\zeta}). \tag{28.48}$$

Comparing these equations with the *vacuum* Robinson–Trautman field equations (28.8), we see that the following theorem holds:

Theorem 28.5 *If*

$$ds_0^2 = 2r^2 d\zeta\, d\overline{\zeta}/P^2 - 2du\,dr - [\Delta \ln P - 2rP_{,u}/P - 2m(u)/r]du^2 \tag{28.49}$$

is a vacuum solution (flat or non-flat) such that P satisfies (28.48), *then*

$$\begin{aligned} ds^2 &= ds_0^2 - \kappa_0 q^4(u) f(\zeta)\overline{f}(\overline{\zeta})du^2/2r^2, \\ \Phi_1 &= q^2(u)\overline{f}(\overline{\zeta})/2r^2, \quad \Phi_2 = q^3(u)\overline{f}_{,\overline{\zeta}}/2r^2 \end{aligned} \tag{28.50}$$

is an Einstein–Maxwell field (a charged vacuum solution). $q(u)$ is real, $f(\zeta)$ analytic, and by a coordinate transformation $q = 1$ can be achieved. With this choice of coordinates, the field equations (28.48) *give*

$$\Delta\Delta \ln P_0 = k = \text{const}, \quad 4m = ku. \tag{28.51}$$

Examples of vacuum solutions satisfying (28.48) are (28.16), (28.21) and (28.25).

If the Maxwell field is null ($Q = 0$), the field equations (28.37*c*)–(28.37*e*) give $m = m(u)$, $h = h(\overline{\zeta}, u)$ and

$$\Delta\Delta \ln P + 12m(\ln P)_{,u} - 4m_{,u} = 4\kappa_0 P^2 h\overline{h}. \tag{28.52}$$

Besides the type *III* solution (28.41), and the type *D* solution (28.43), a known solution of this equation is (Bartrum 1967)

$$\begin{aligned} P &= f(\zeta)\overline{f}(\overline{\zeta})\,(1+\zeta\overline{\zeta}/2), \quad m = \text{ const}, \\ h &= \sqrt{2/\kappa_0}\,\overline{f}(\overline{\zeta})\overline{f}'(\overline{\zeta})\mathrm{e}^{\mathrm{i}\varphi(u)}, \quad \varphi \text{ real}. \end{aligned} \tag{28.53}$$

All type II solutions which (after a suitable choice of the coordinate ζ) *obey*

$$\Delta \ln P = 0, \quad Q = q(u)\mathrm{e}^{\zeta/\sqrt{2}}, \quad q \text{ real}, \tag{28.54}$$

are explicitly known. Because of the field equations (28.37*c*)–(28.37*e*) and the assumption (28.54), m is a function of u and x alone. It turns out that two cases can occur. If $3mq_{,u} - 2qm_{,u}$ is zero, then the solution (after a transformation $\zeta \to 4\zeta$) reads

$$\begin{aligned} \mathrm{d}s^2 &= Au\mathrm{e}^{-x}r^2(\mathrm{d}x^2+\mathrm{d}y^2) - 2\mathrm{d}u\,\mathrm{d}r - \left[\frac{r}{u} + \frac{2\kappa_0 A}{3r}\mathrm{e}^{3x} + \frac{\kappa_0}{2r^2}\mathrm{e}^{4x}\right]\mathrm{d}u^2, \\ \Phi_1 &= \frac{\mathrm{e}^{2(x-\mathrm{i}\,y)}}{2r^2}, \quad \Phi_2 = -\frac{\sqrt{2Au}}{r^2}\mathrm{e}^{(5x-4\mathrm{i}\,y)/2} - \frac{\sqrt{2Au}}{2r}\mathrm{e}^{(3x-4\mathrm{i}\,y)/2}. \end{aligned} \tag{28.55}$$

If $3mq_{,u} - 2qm_{,u}$ is non-zero, then the metric and Maxwell field have the forms (Leroy 1976)

$$\begin{aligned} \mathrm{d}s^2 &= \frac{r^2}{P^2}(\mathrm{d}x^2+\mathrm{d}y^2) - 2\mathrm{d}u\,\mathrm{d}r - \left(\frac{\kappa_0 q^2}{2r^2}\mathrm{e}^{x} - \frac{2m}{r}\right)\mathrm{d}u^2, \\ \Phi_1 &= \frac{q}{2r^2}\mathrm{e}^{(x-\mathrm{i}\,y)/2}, \quad \Phi_2 = -\frac{Pq\mathrm{e}^{(x-\mathrm{i}\,y)/2}}{2r^2\sqrt{2}} + \frac{Pm_{,x}}{\kappa_0 rq\sqrt{2}}\mathrm{e}^{-(x+\mathrm{i}\,y)/2}, \end{aligned} \tag{28.56a}$$

where P, m and q are given by

$$P = 1, \quad m = m_0 + 2\varepsilon\kappa_0 u^{-1}q_0^2\mathrm{e}^{x}, \quad q = q_0, \quad \varepsilon = 0, 1, \tag{28.56b}$$

$$\text{or} \quad P = \mathrm{e}^{x/2}, \quad m = m_0 - \tfrac{1}{2}\kappa_0 a x - \tfrac{1}{8}\kappa_0 a \ln\left|q^2\right|, \quad q^2 = au + b, \tag{28.56c}$$

$$\text{or} \quad P = \mathrm{e}^{-x/2}, \quad m = m_0 - \frac{a^2 u}{2\kappa_0 q_0^2} - a\mathrm{e}^{x}, \quad q = q_0, \tag{28.56d}$$

$$\text{or}\quad P = \mathrm{e}^{-x/2}, \quad q = q_0 u^{1/3},$$
$$m = m_0 + \left(\frac{3}{2}\frac{a^2}{\kappa_0 q^4} - \frac{a\mathrm{e}^x}{q_0} + \frac{\kappa_0}{6} q_0^2 \mathrm{e}^{2x}\right) u^{-1/3}, \tag{28.56e}$$

$$\text{or}\quad P = \mathrm{e}^{fx}, \quad m = m_0 + \frac{2\kappa_0 a \mathrm{e}^{x(1-2f)}}{(1-2f)(1-10f)}(au+b)^{\frac{1-2f}{10f-1}},$$
$$q = (au+b)^{4f/(10f-1)}, \quad f(2f-1)(2f+1)(10f-1) \neq 0. \tag{28.56f}$$

q_0, m_0, a, b, f are real constants. In all these solutions, $P = 1$ or $q = 1$ could be achieved by coordinate transformations (28.9). They all admit at least one Killing vector $\boldsymbol{\xi} = \partial_y$. For special values of the constants, the solutions may be more degenerate than type II, e.g. (28.56f) with $m_0 = 0$, $f = 1/6$ gives the type D metric (28.46) in somewhat changed coordinates.

All type II solutions which obey

$$\Delta \ln P = 0, \quad P_{,u} = 0 \tag{28.57}$$

have been found by Bajer and Kowalczyński (1985). Those which are not contained in (28.54) are

$$\mathrm{d}s^2 = 2r^2 \mathrm{d}\zeta \mathrm{d}\overline{\zeta} - 2\mathrm{d}r\,\mathrm{d}u + \{[2m_0 + 2\kappa_0 A(\zeta,\overline{\zeta},u)]/r - \kappa_0 Q\overline{Q}/2r^2\}\mathrm{d}u^2,$$
$$\Phi_1 = \overline{Q}/2r^2, \quad \Phi_2 = -\overline{Q}_{,\overline{\zeta}}/2r^2 + A_{,\overline{\zeta}}/Qr, \tag{28.58a}$$

with

$$A = \tfrac{1}{12} Q_0 \overline{Q}_0 (\zeta + \overline{\zeta})^2 u^{-1/3} - 2\mathrm{i}\, b\sqrt{Q_0 \overline{Q}_0}(\zeta - \overline{\zeta}) - 3b^2 u^{1/3},$$
$$Q = Q_0 u^{1/3}, \tag{28.58b}$$

$$\text{or}\quad A = a\sqrt{Q_0 \overline{Q}_0}(\zeta + \overline{\zeta}) - a^2 u, \quad Q = Q_0, \tag{28.58c}$$

$$\text{or}\quad A = -ab(\zeta^{-2} + \overline{\zeta}^{-2}) - a^2 u (\zeta\overline{\zeta})^{-2}, \quad Q = 2au\zeta^{-3} + 2b\zeta^{-1}, \tag{28.58d}$$

with constants a, b (real) and Q_0 (complex).

All metrics of the form $\Delta \ln P = \varepsilon = \pm 1$, $P_{,u} = 0$, i.e.

$$\mathrm{d}s^2 = \frac{2r^2 \mathrm{d}\zeta \mathrm{d}\overline{\zeta}}{\left(1 + \frac{1}{2}\varepsilon\zeta\overline{\zeta}\right)^2} - 2\mathrm{d}u\,\mathrm{d}r + \left(\frac{2m_0 + 2\kappa_0 A}{r} - \varepsilon - \frac{\kappa_0 Q\overline{Q}}{2r^2}\right)\mathrm{d}u^2 \tag{28.59}$$

have been found by Kowalczyński (1978, 1985). Some are covered by Theorem 28.5. The remaining solutions are

$$Q = Q_0\zeta\sqrt{u}, \quad A = Q_0\overline{Q}_0\left[\ln(1+\varepsilon\zeta\overline{\zeta}/2) - (\ln u)/4\right],$$
$$h = \overline{h} = \varepsilon Q_0/\left[(2+\varepsilon\zeta\overline{\zeta})\sqrt{u}\right], \quad \varepsilon = \pm 1, \tag{28.60}$$

and

$$2A = Q_0\overline{Q}_0\{\ln[1-\zeta\overline{\zeta}/2] - \ln[(1+\zeta/\sqrt{2})(1+\overline{\zeta}/\sqrt{2})] - (\ln u)/4\},$$
$$Q = Q_0\sqrt{u}, \quad h = \frac{Q_0}{\sqrt{2u}}\frac{1+\zeta/\sqrt{2}}{(2-\zeta\overline{\zeta})(1+\overline{\zeta}/\sqrt{2})}, \quad \varepsilon = -1. \tag{28.61}$$

Like most of the solutions with a ζ–$\overline{\zeta}$ -space of constant negative curvature, the solution (28.61) can be interpreted in terms of a tachyon's world line.

If $P_{,u} = Q_{,u} = h_{,u} = 0$, then the field equations (28.37*a*)–(28.37*e*) can partially be integrated and give (with real constants a_0, b_0 and m_0 and the complex constant ψ_0)

$$Q = b_0\beta(\zeta) + \overline{\psi_0}, \quad h = -\overline{\beta}'(\overline{\zeta}),$$
$$m = a_0 u - 2\kappa_0 b_0 \beta\overline{\beta} - 4\kappa_0 \mathrm{Re}\,(\psi_0\beta) + m_0, \tag{28.62}$$

the only surviving equation being

$$P^2\partial_\zeta\partial_{\overline{\zeta}}P^2\partial_\zeta\partial_{\overline{\zeta}}\ln P = a_0 + \kappa_0 P^2\beta'(\zeta)\overline{\beta}'(\overline{\zeta}) \tag{28.63}$$

(Herlt and Stephani 1984). For $a_0 \neq 0$, the only known solution is

$$P = 1, \quad \beta = \beta_0\zeta, \quad a_0 = -\kappa_0\beta_0\overline{\beta}_0. \tag{28.64}$$

For $a_0 = 0$, (28.63) can be integrated twice and yields

$$P^2\partial_\zeta\partial_{\overline{\zeta}}\ln P = \kappa_0\beta(\zeta)\overline{\beta}(\overline{\zeta}) + \varphi(\zeta) + \overline{\varphi}(\overline{\zeta}). \tag{28.65}$$

Some Einstein–Maxwell fields belonging to this class are

$$P = (\zeta + \overline{\zeta} - \tfrac{2}{3}\kappa_0\beta\overline{\beta})^{3/2}, \quad \beta = \beta_0\zeta, \tag{28.66}$$

$$P = (2/3\kappa_0 c_0)^{1/2}\,(\kappa_0\zeta\overline{\zeta} + c_0)^{3/2}, \quad \beta = \zeta, \tag{28.67}$$

$$P = (2\kappa_0\zeta\overline{\zeta})^{1/2}\,[\delta'(\zeta)\overline{\delta}'(\overline{\zeta})]^{-1/2}\,[1 + \delta(\zeta)\overline{\delta}(\overline{\zeta})/2], \quad \beta = \zeta. \tag{28.68}$$

The solutions (28.66) and (28.68) generalize (28.16) and (28.53), respectively.

28.3 Robinson–Trautman pure radiation fields

Pure radiation fields

$$T_{mn} = \Phi^2 k_n k_m \tag{28.69}$$

are very similar to electromagnetic null fields. In both cases, the condition $T^{mn}{}_{;n} = 0$ together with $k_{m;n}k^n = 0$ and $k^n{}_{;n} = 2/r$ gives the structure

$$\Phi^2 = n^2(\zeta, \overline{\zeta}, u)/r^2, \tag{28.70}$$

but in the Maxwell case $n^2 = 2h\overline{h}P^2$ is subject to the restriction $h_{,\zeta} = 0$.

Starting from (27.37), the calculations run in close analogy with those for the Einstein–Maxwell fields and lead to

Theorem 28.6 *All algebraically special solutions with pure radiation fields, the common null eigenvector* $\mathbf{k}$ *of the radiation field and the Weyl tensor being geodesic, shearfree, normal but diverging, are given by* (28.69)–(28.70) *and*

$$\begin{aligned} \mathrm{d}s^2 &= \frac{2r^2\mathrm{d}\zeta\mathrm{d}\overline{\zeta}}{P^2(u,\zeta,\overline{\zeta})} - 2\mathrm{d}u\,\mathrm{d}r - [\Delta\ln P - 2r(\ln P)_{,u} - 2m(u)/r]\mathrm{d}u^2, \\ k^i &= (0,0,1,0), \quad \Delta \equiv 2P^2\partial_\zeta\partial_{\overline{\zeta}}, \end{aligned} \tag{28.71a}$$

$$\Delta\Delta\ln P + 12m(\ln P)_{,u} - 4m_{,u} = 2\kappa_0 n^2(\zeta,\overline{\zeta},u). \tag{28.71b}$$

Equation (28.71b) shows that the functions P and m can be prescribed almost arbitrarily, the limitation being that the left-hand side must be positive; so the construction of solutions is very easy.

The components of the Weyl tensor are the same functions of P and m as given in the vacuum case by (28.10). One sees that there are no type N or O ($\Psi_2 = \Psi_3 = 0$) non-vacuum solutions. Type III solutions are characterized by $m = 0$, $(\Delta\ln P)_{,\zeta} \neq 0$, $\Delta\Delta\ln P > 0$. Examples of type III solutions are

$$P = x^a = [(\zeta+\overline{\zeta})\sqrt{2}]^a, \quad 1 < a < 1.5\ . \tag{28.72}$$

All type D solutions are known (Frolov and Khlebnikov 1975). They are generalizations either of the vacuum solutions (28.21) or of the static C-metric (28.24).

In the first case, $K = \Delta\ln P$ is a function only of u, and thus P is given by

$$P = \alpha(u)\zeta\overline{\zeta} + \beta(u)\zeta + \overline{\beta}(u)\overline{\zeta} + \delta(u), \quad K = 2(\alpha\delta - \beta\overline{\beta}), \tag{28.73}$$

$m(u)$ is arbitrary, and $n(\zeta,\overline{\zeta},u)$ can be calculated from (28.71b). For positive K, (28.73) gives the Kinnersley rocket (cp. (32.20)), i.e. a particle which accelerates by emitting pure radiation. If α, β and δ are constant and K is positive, we get the Vaidya solution (15.20). In general, the functions α, β, δ may be interpreted in terms of the acceleration of a 'particle' moving along a spacelike, timelike or null world line (Newman and Unti 1963, Frolov and Khlebnikov 1975, Taub 1976, Frolov 1977).

In the second case, the metric is given by

$$\begin{aligned} \mathrm{d}s^2 = {} & r^2P^2(\eta)\mathrm{d}\eta^2 + r^2P^{-2}(\eta)\mathrm{d}y^2 - 2r^2m^{-1}(u)\mathrm{d}u\,\mathrm{d}\eta - 2\mathrm{d}u\,\mathrm{d}r \\ & + r^2m^{-2}(u)P^{-2}(\eta - m/r)\mathrm{d}u^2, \end{aligned} \tag{28.74}$$

$$P^{-2}(\eta) = -2\eta^3 + b\eta + d$$

(in the coefficient of $\mathrm{d}u^2$, the *argument* of P is $\eta - m/r$). Here b and d are constants, and $m(u)$ is an arbitrary function, from which the radiation field can be calculated by means of (28.71*b*) as

$$-2m_{,u} = \kappa_0 n^2. \tag{28.75}$$

Robinson–Trautman pure radiation solutions have been discussed in the context of gravitational radiation, see e.g. Bonnor (1996). Radiation solutions with $n = n(u)$ approach the Vaidya solution for infinite retarded time u (Bičák and Perjés 1987).

28.4 Robinson–Trautman solutions with a cosmological constant Λ

If we assume an energy-momentum tensor of the form

$$\kappa_0 T_{mn} = -\Lambda g_{mn}, \quad \Lambda = \text{ const}, \tag{28.76}$$

or add this term to the energy-momentum tensor of the Maxwell field or that of pure radiation, then the field equations can be solved in a way similar to the case $\Lambda = 0$. The final result of the simple calculations can be stated as

Theorem 28.7 *If*

$$\mathrm{d}s^2 = 2r^2P^{-2}\mathrm{d}\zeta\mathrm{d}\bar{\zeta} - 2\mathrm{d}u\,\mathrm{d}r - 2H\mathrm{d}u^2 \tag{28.77}$$

is a vacuum, Einstein–Maxwell or pure radiation solution without *the cosmological constant* Λ, *then*

$$\mathrm{d}s^2 = 2r^2P^{-2}\mathrm{d}\zeta\mathrm{d}\bar{\zeta} - 2\mathrm{d}u\,\mathrm{d}r - (2H - \Lambda r^2/3)\mathrm{d}u^2 \tag{28.78}$$

is the corresponding solution including *the cosmological constant.*

Type N solutions with Λ have been studied by García D. and Plebański (1981), by Bičák and Podolsky (1999), and by Podolsky and Griffiths (1999).

29
Twisting vacuum solutions

In the preceding chapter, we treated the *non-twisting* solutions (of the Robinson–Trautman class) in some detail, giving or indicating nearly all proofs, and showing how the variety of known special solutions fits into the framework of the canonical form of the metric and field equations.

It is impossible to present the solutions with *twisting* degenerate eigenrays in this detailed manner, because it would nearly fill an extra volume. We share this problem with most of the authors writing on the subject. The necessity of presenting the complicated calculations in a compressed form makes some of the papers almost unreadable, and it is sometimes a formidable task merely to check the calculations. What we will do here and in the following chapters is to show why, how and how far the integration procedure of the field equations works, and what classes of solutions are known.

29.1 Twisting vacuum solutions – the field equations

29.1.1 The structure of the field equations

To get a better understanding of the structure of the vacuum field equations, we follow Sachs (1962) in dividing them into three sets:

six main equations: $$R_{11} = R_{12} = R_{14} = R_{44} = 0 \quad (29.1)$$

one trivial equation: $$R_{34} = 0 \quad (29.2)$$

three supplementary conditions: $$R_{13} = R_{33} = 0 \quad (29.3)$$

(the indices refer to the tetrad (3.8)). The reason for this splitting is the following property (which can be proved by application of the Bianchi identities): if $\boldsymbol{k}$ is a geodesic and diverging ($k^a{}_{;a} \neq 0$) null congruence, then (a) the trivial equation is satisfied identically and (b) the supplementary conditions hold for all values of the affine parameter r of the null congruence if they hold for a fixed r.

In the context of algebraically special solutions, the most remarkable property of the main equations is that they can be integrated completely, giving the r-dependence of all metric functions in terms of simple rational functions of r. The supplementary conditions then turn out to be differential equations for those constitutive parts of the metric which depend on the three remaining coordinates ζ, $\overline{\zeta}$ and u. In general, these differential equations cannot be solved; they play the role of (and are often called) *the* field equations for algebraically special vacuum solutions.

29.1.2 The integration of the main equations

We have shown in Theorem 27.1 that a space-time admits a geodesic, shearfree and diverging null congruence $\boldsymbol{k}$ (twisting or not) and satisfies $R_{44} = R_{41} = R_{11} = 0$ exactly if the metric can be written in the form

$$\begin{aligned} \mathrm{d}s^2 &= 2\boldsymbol{\omega}^1\boldsymbol{\omega}^2 - 2\boldsymbol{\omega}^3\boldsymbol{\omega}^4, \quad \boldsymbol{\omega}^1 = \overline{m}_n \mathrm{d}x^n = -\mathrm{d}\zeta/P\overline{\rho} = \overline{\boldsymbol{\omega}}^2, \\ \boldsymbol{\omega}^3 &= -k_n \mathrm{d}x^n = \mathrm{d}u + L\mathrm{d}\zeta + \overline{L}\mathrm{d}\overline{\zeta}, \\ \boldsymbol{\omega}^4 &= -l_n \mathrm{d}x^n = \mathrm{d}r + W\mathrm{d}\zeta + \overline{W}\mathrm{d}\overline{\zeta} + H\boldsymbol{\omega}^3, \end{aligned} \tag{29.4}$$

with

$$\begin{aligned} m^i &= e_1^i = (-P\overline{\rho},\, 0,\, PW\overline{\rho}, PL\overline{\rho}), \quad \overline{m}^i = e_2^i, \\ l^i &= e_3^i = (0,\, 0, -H,\, 1), \quad k^i = e_4^i = (0, 0, 1, 0), \end{aligned} \tag{29.5}$$

$$\begin{aligned} \rho^{-1} &= -(r + r^0 + \mathrm{i}\Sigma), \quad W = \rho^{-1}L_{,u} + \partial(r^0 + \mathrm{i}\Sigma), \\ 2\mathrm{i}\Sigma &= P^2(\overline{\partial}L - \partial\overline{L}), \quad \partial \equiv \partial_\zeta - L\partial_u, \quad \overline{\partial} \equiv \partial_{\overline{\zeta}} - \overline{L}\partial_u. \end{aligned} \tag{29.6}$$

P, r^0 (both real) and L (complex) are arbitrary functions of $\zeta, \overline{\zeta}$ and u, and H (real) is a function of all four coordinates.

Of the main equations (29.1) only $R_{12} = 0$ remains to be integrated. We shall do that now, thus obtaining the r-dependence of the function H.

To get the components R_{12} and R_{34} of the Ricci tensor, we start from the connection forms $\boldsymbol{\Gamma}_{ab} = -\boldsymbol{\Gamma}_{ba} = \Gamma_{abc}\omega^c$ of the metric (29.4) and evaluate (3.25c), using the expression (27.17) for R_{4123}. We obtain

$$R_{34} = \rho^2 \left[\rho^{-2}(\Gamma_{213} + \Gamma_{433})\right]_{|4} - 2\rho(\ln P)_{|3}, \tag{29.7}$$

$$\begin{aligned} R_{12} = {} & (\rho + \overline{\rho})(\Gamma_{213} + \Gamma_{433}) + 2\rho\left[\Gamma_{321} + (\ln P)_{|3}\right] + (\ln P)_{|12} \\ & -(\ln P)_{|1}(\ln P\overline{\rho})_{|2} + (\ln P\overline{\rho}^2)_{|21} - (\ln P\overline{\rho}^2)_{|2}(\ln P\rho)_{|1}, \end{aligned} \tag{29.8}$$

with

$$\Gamma_{213} = \mathrm{i}\ \mathrm{Im}[P\rho\overline{W}_{|1} + H\rho + (\ln\overline{\rho})_{|3}], \quad \Gamma_{433} = H_{|4}, \tag{29.9}$$

$$\Gamma_{321} = -\mathrm{i}\ \mathrm{Im}[P\rho\overline{W}_{|1} + H\rho] + \mathrm{Re}\ [(\ln P\overline{\rho})_{|3}]. \tag{29.10}$$

We start with $R_{34} = 0$ (which is in fact a consequence of the main equations, including $R_{12} = 0$) and integrate it by

$$\Gamma_{213} + \Gamma_{433} = -(\ln P)_{,u} + (m + \mathrm{i}M)\rho^2, \quad (m + \mathrm{i}M)_{|4} = 0, \tag{29.11}$$

m and M being real functions of integration. The real part of (29.11) yields

$$H = -(r+r^0)(\ln P)_{,u} + \mathrm{Re}\ [(m + \mathrm{i}M)\rho] + r^0{}_{,u} + K/2, \quad K_{|4} = 0. \tag{29.12}$$

Inserting both results into $R_{12} = 0$ and making repeated use of (27.24) and (29.6), we obtain K in terms of P and L. The result can be summarized as follows:

Theorem 29.1 *A space-time admits a geodesic, shearfree and diverging null congruence* $\boldsymbol{k}$ *and satisfies* $R_{44} = R_{14} = R_{11} = R_{12} = R_{34} = 0$ *exactly if the metric can be written as*

$$\begin{aligned} &\mathrm{d}s^2 = 2\boldsymbol{\omega}^1\boldsymbol{\omega}^2 - 2\boldsymbol{\omega}^3\boldsymbol{\omega}^4, \quad \boldsymbol{\omega}^1 = -\mathrm{d}\zeta/P\overline{\rho} = \overline{\boldsymbol{\omega}}^2, \\ &\boldsymbol{\omega}^3 = \mathrm{d}u + L\mathrm{d}\zeta + \overline{L}\mathrm{d}\overline{\zeta}, \quad \boldsymbol{\omega}^4 = \mathrm{d}r + W\mathrm{d}\zeta + \overline{W}\mathrm{d}\overline{\zeta} + H\boldsymbol{\omega}^3, \end{aligned} \tag{29.13a}$$

the metric functions satisfying

$$\rho^{-1} = -(r + r^0 + \mathrm{i}\Sigma), \quad 2i\Sigma = P^2(\overline{\partial}L - \partial\overline{L}), \tag{29.13b}$$

$$W = L_{,u}/\rho + \partial(r^0 + \mathrm{i}\Sigma), \quad \partial \equiv \partial_\zeta - L\partial_u, \tag{29.13c}$$

$$H = \tfrac{1}{2}K - (r + r^0)(\ln P)_{,u} - \frac{m(r + r^0) + M\Sigma}{(r + r^0)^2 + \Sigma^2} + r^0_{,u}, \tag{29.13d}$$

$$K = 2P^2\mathrm{Re}\left[\partial(\overline{\partial}\ln P - \overline{L}_{,u})\right], \tag{29.13e}$$

$$M = \Sigma K + P^2\mathrm{Re}\left[\partial\overline{\partial}\Sigma - 2\overline{L}_{,u}\partial\Sigma - \Sigma\partial_u\partial\overline{L}\right] \tag{29.13f}$$

(Kerr 1963a, Debney *et al.* 1969, Robinson *et al.* 1969a, Trim and Wainwright 1974).

The line element (29.13) shows a remarkably simple r-dependence. Furthermore, all remaining functions of $\zeta, \overline{\zeta}$ and u are given in terms of the complex function L and the real functions r^0, P and m. As r^0 and P can be removed by coordinate transformations (see §29.1.4), L (complex) and m (real) can be considered to represent the metric.

29.1.3 The remaining field equations

We will now see what conditions the remaining field equations, i.e. the supplementary conditions $R_{13} = R_{33} = 0$, impose on the functions L, m and r^0. The metric being given by (29.13a), it is a matter of straightforward calculation to compute the curvature tensor and to formulate the field equations $R_{13} = R_{33} = 0$ in terms of the above mentioned functions.

For R_{13} we obtain the expression

$$R_{13} = \rho^{-1}[\rho(\Gamma_{213} + \Gamma_{433})]_{|1} + (\ln P)_{|13} - (\ln P)_{|1}(\ln P)_{|3}, \tag{29.14}$$

and because of (29.11) and (27.24) $R_{13} = 0$ immediately yields

$$\partial(m + \mathrm{i}M) = 3(m + \mathrm{i}M)L_{,u}. \tag{29.15}$$

The calculation of $R_{33} = 0$ is rather lengthy, and we therefore will not give any details. The final consequence proves to be (Kerr 1963a, Debney *et al.* 1969, Robinson *et al.* 1969a, Trim and Wainwright 1974)

$$\begin{aligned} [P^{-3}(m + \mathrm{i}M)]_{,u} &= P[\partial + 2(\partial \ln P - L_{,u})]\partial I, \\ I &\equiv \overline{\partial}(\overline{\partial} \ln P - \overline{L}_{,u}) + (\overline{\partial} \ln P - \overline{L}_{,u})^2. \end{aligned} \tag{29.16}$$

As M is already given in terms of P and L by (29.13f), the four real equations (29.15)–(29.16) form a system of partial differential equations for P, m and L. Note that because of the definition $\partial = \partial_\zeta - L\partial_u$, the function L appears in the differential operator too!

By use of the commutator relations

$$\partial\overline{\partial} - \overline{\partial}\partial = (\overline{\partial}L - \partial\overline{L})\partial_u, \quad \partial\partial_u - \partial_u\partial = L_{,u}\partial_u, \tag{29.17}$$

the field equations, as well as the definitions (29.13c)–(29.13f) of the metric functions M, W and H, can be put into rather different forms. A formal simplification can be achieved by the introduction of a real 'potential' V for the function P by means of

$$V_{,u} = P \tag{29.18}$$

(Robinson and Robinson 1969). From (29.18) it follows that

$$I = \overline{\partial}(\overline{\partial} \ln P - \overline{L}_{,u}) + (\overline{\partial} \ln P - \overline{L}_{,u})^2 = P^{-1}(\overline{\partial}\,\overline{\partial}V)_{,u}, \tag{29.19}$$

and the system of equations (29.13f), (29.15) and (29.16) reads

$$[P^{-3}(m + \mathrm{i}M) - \partial\partial\overline{\partial}\,\overline{\partial}V]_{,u} = -P^{-1}(\partial\partial V)_{,u}(\overline{\partial}\,\overline{\partial}V)_{,u}, \tag{29.20a}$$

$$\partial(m + \mathrm{i}M) = 3(m + \mathrm{i}M)L_{,u}, \tag{29.20b}$$

$$P^{-3}M = \mathrm{Im}\, \partial\partial\overline{\partial}\,\overline{\partial}V. \tag{29.20c}$$

In the special gauge $P = 1, V = u \Rightarrow \partial V = -L$, these equations take the form given by Kerr (1963a), i.e.

$$[(m + \mathrm{i}M) + \partial\partial\overline{\partial}\,\overline{L}]_{,u} = -(\partial L)_{,u}(\overline{\partial}\,\overline{L})_{,u}, \tag{29.21a}$$

$$\partial(m + \mathrm{i}M) = 3(m + \mathrm{i}M)L_{,u}, \tag{29.21b}$$

$$M = \mathrm{Im}\,\overline{\partial}\,\overline{\partial}\partial L. \tag{29.21c}$$

We have given the field equations in three different forms, as (29.21), as (29.20), and as (29.13*f*), (29.15) and (29.16), which we will use as alternatives. As usual, we have listed the definition (29.13*f*) or (29.21*c*) or (29.20*c*) of M in terms of L and P among the field equations. It makes the structure of the equations more transparent, and in the process of integration one sometimes tries to solve the other two equations first and impose the definition of M as an additional constraint; as can be seen by comparing e.g. (29.20*a*) and (29.20*c*), this definition is in fact a first integral since the right-hand side of (29.20*a*) is automatically real. If a solution m, L, P of the field equations is known, the metric can be determined from (29.13).

If $L = 0$, M and Σ and W vanish also (in the gauge $r^0 = 0$); because of (29.20*b*) and its complex conjugate, m is a function only of u, and (29.16) turns into the field equation (28.7) of the Robinson–Trautman vacuum metrics.

The surviving components of the Weyl tensor have the form (Trim and Wainwright 1974, Weir and Kerr 1977)

$$\begin{aligned} \Psi_2 &= (m + \mathrm{i}M)\rho^3, \quad \Psi_3 = -P^3\rho^2\partial I + O(\rho^3), \\ \Psi_4 &= P^2\rho\partial_u I + O(\rho^2), \\ I &\equiv \overline{\partial}(\overline{\partial}\ln P - \overline{L}_{,u}) + (\overline{\partial}\ln P - \overline{L}_{,u})^2 = P^{-1}(\overline{\partial}\,\overline{\partial}V)_{,u}. \end{aligned} \tag{29.22}$$

The terms of higher order in ρ occurring in Ψ_3 or Ψ_4 vanish identically if $\Psi_2 = 0$ or $\Psi_2 = \Psi_3 = 0$, respectively. It can be inferred from (29.22) that a solution is flat exactly if $m + \mathrm{i}M = 0 = \partial I = \partial_u I$.

For the case of a non-zero cosmological constant Λ, the field equations (in Newman–Penrose form) have been given by Timofeev (1996), and some simple solutions can be found in Kaigorodov and Timofeev (1996).

29.1.4 Coordinate freedom and transformation properties

We shall now have a look at the possible coordinate transformations and the transformation properties of the metric functions and the field equations. As shown in §27.1.3, there are essentially three types of coordinate transformations.

The first is the freedom (27.31) in the choice of the affine parameter origin $r^0(\zeta,\bar{\zeta},u)$. As r^0 does not enter into those functions L, P, M and m which appear in the field equations (29.13b)–(29.13f), the choice of r^0 is more a matter of convenience than of practical value for the integration procedure. Most authors take

$$r^0(\zeta,\bar{\zeta},u) = 0, \tag{29.23}$$

and we shall stick to that gauge from now on.

The remaining degrees of freedom

$$\zeta' = f(\zeta), \quad u' = F(\zeta,\bar{\zeta},u), \quad r' = rF_{,u}^{-1} \tag{29.24}$$

are more important; they change L, $m+\mathrm{i}M$ and P and will often be used in the integration procedure for the field equations. Together with the appropriate changes (27.28), (27.30) of the tetrad, they give the transformation laws

$$\omega^{1'} = (f'\bar{f}'^{-1})^{1/2}\omega^1, \quad \omega^{3'} = F_{,u}\omega^3, \quad \omega^{4'} = F_{,u}^{-1}\omega^4, \tag{29.25}$$

which imply

$$\begin{aligned} &\rho' = F_{,u}\rho, \quad \Sigma = F_{,u}\Sigma', \quad (m+\mathrm{i}M)' = F_{,u}^{-3}(m+\mathrm{i}M), \\ &L' = f'^{-1}(LF_{,u} - F_{,\zeta}), \quad \partial' = f'^{-1}\partial, \quad P' = F_{,u}^{-1}\,|f'|\,P, \\ &(\partial \ln P - L_{,u})' = f'^{-1}(\partial \ln P - L_{,u} + f''/2f'), \\ &I' = \bar{f}'^{-2}(I + \bar{f}'''/2\bar{f}' - 3\bar{f}''^2/4\bar{f}'^2), \quad V' = |f'|\,V. \end{aligned} \tag{29.26}$$

As is to be expected, the field equations are either explicitly invariant or can easily be written in an invariant way. A list of (other) invariants can be found in Robinson *et al.* (1969a); among them are $r\Sigma^{-1}$, $\rho^2 K$, and $(m+\mathrm{i}M)\rho^3$.

If we examine the definition $V_{,u} = P$ of the function V in the light of the transformation laws, we recognize that V is exactly that transformation F ($f' = 1$) which transforms an arbitrary P into $P' = 1$. So the simplification of the field equations by introducing V is on an equal footing with the simplification by $P = 1$ and the Kerr form (29.21).

If L can be written as $L = F_{,\zeta}/F_{,u}$, with a *real* function F, then L' can be made zero: this characterizes the Robinson–Trautman class, see above.

29.2 Some general classes of solutions

29.2.1 Characterization of the known classes of solutions

The field equations of the algebraically special vacuum solutions become less complicated if a special dependence of the metric functions on the

Table 29.1. The possible types of two-variable twisting vacuum metrics

Type	L	P	$m+\mathrm{i}M$	I
KV	$L(x,u)$	$P(x,u)$	$\mu(x,u)$	$I(x,u)$
KV	$L(\zeta,\overline{\zeta})$	$P(\zeta,\overline{\zeta})$	$\mu(\zeta,\overline{\zeta})$	$I(\zeta,\overline{\zeta})$
HV	$L(x,u)$	$\Pi(x,u)\mathrm{e}^{y}$	$\mu(x,u)\mathrm{e}^{4y}$	$I(x,u)$
HV	$L(\zeta,\overline{\zeta})$	$\Pi(\zeta,\overline{\zeta})\mathrm{e}^{u}$	$\mu(\zeta,\overline{\zeta})\mathrm{e}^{4u}$	$I(\zeta,\overline{\zeta})$

coordinates is assumed. So far, this has been done in three different (but overlapping) ways, leading to all known solutions.

The first obvious way is to start from the Lie symmetries (cp. §10.2)

$$\begin{aligned}\mathbf{X} = A(\zeta)\partial_\zeta + \left[au + G(\zeta,\overline{\zeta},u)\right]\partial_u + P\left[(A' + \overline{A}')/2 - G_{,u}\right]\partial_P \\ + \left[L(G_{,u} + a - A') - G_{,\zeta}\right]\partial_L + (m + \mathrm{i}M)(a - 3G_{,u})\partial_{m+\mathrm{i}M} \\ + \text{complex conjugate}\end{aligned} \tag{29.27}$$

(A complex, G and a real, $\zeta = x + \mathrm{i}y$) of the field equations and to consider solutions which admit either a Killing vector (KV, $a = 0$), or a homothetic vector (HV, $a \neq 0$). The metric functions then depend on at most two coordinates. This leads to four different cases (Halford 1980, Stephani 1984), see Table 29.1.

From these cases, only that of line 2 ($X = \partial_u$ is a Killing vector) can be treated in some generality, see §§29.2.3 and 29.2.5. In the other three cases, an additional symmetry (Killing or homothetic vector) has to be assumed, see §§29.2.6 and 29.3; cp. also Chapter 38.

A second way, more specific to the algebraically special solutions, is to assume a rather special dependence on u of some combinations of the metric functions appearing in the field equations. To make this dependence explicit, we have to restrict the possible transformations (29.24) of u, which we will do by

$$P_{,u} = 0 \Leftrightarrow u' = uk(\zeta,\overline{\zeta}) + h(\zeta,\overline{\zeta}). \tag{29.28}$$

The main assumptions are now (Robinson and Robinson 1969) that $L_{,u} - \partial \ln P$ and $m + \mathrm{i}M$ are independent of u. In view of the field equations (29.16) and the gauge $P_{,u} = 0$, these assumptions are equivalent to

$$L_{,u} - (\ln P)_{,\zeta} = G(\zeta,\overline{\zeta}), \tag{29.29}$$

$$(\partial_\zeta - 2G)\partial_\zeta I = 0, \quad I \equiv -\partial_\zeta\overline{G} + \overline{G}^2. \tag{29.30}$$

Table 29.2. Twisting algebraically special vacuum solutions

$P, m+\mathrm{i}M$ and $L_{,u} - (\ln P)_{,\zeta} = G$ are assumed to be independent of u.

	$\partial_\zeta I \neq 0$	$\partial_\zeta I = 0 = I$
$L_{,u} \neq 0$	Equations only: (29.34)–(29.37)	Flat background. All solutions are known and given by (29.53), (29.54), (29.58)
$L_{,u} = 0$	Equations: (29.41)–(29.44) Background: non-twisting type III. Solutions (not exhaustive): (29.45)–(29.50), (38.6)	All solutions are known and given by (29.60)–(29.61)

The main idea then runs as follows. Assume we are able to solve the differential equation (29.30), which is in fact one of the field equations, and have found a solution G. It then turns out that the second field equation (29.20*b*) can be solved in the sense that $m+\mathrm{i}M$ is given in terms of G, P and arbitrary functions or constants of integration. The third (and last) field equation (29.20*c*) can now be reduced to a linear homogeneous differential equation for a real function, or can be solved explicitly. G is sometimes called the 'background', because $m+\mathrm{i}M = 0$ and $L = [G + (\ln P)_{,\zeta}]u$ also give algebraically special solutions. Explicit solutions have been found for several cases, see Table 29.2 and the following subsections. It is possible to reduce and simplify the field equations in a similar way even if $m+\mathrm{i}M$ depends on u (so that (29.30) is no longer true). But no solution is known so far which satisfies (29.29) but not (29.30).

A third way of characterizing the known solutions starts from the observation that the field equation (29.20*b*), i.e. $\partial(m+\mathrm{i}M) = 3(m+\mathrm{i}M)L_{,u}$ can be integrated by introducing a complex function Φ via

$$m + \mathrm{i}M = \Phi^3_{,u} \tag{29.31}$$

(Stephani 1983a). In terms of Φ, the field equation (29.20*b*) then reads $\partial(\Phi_{,u}) = L_{,u}\Phi_{,u}$, and because of the commutator relation (29.17) it can be written as

$$\partial_u \partial \Phi = 0. \tag{29.32}$$

This equation is integrated by $\partial\Phi = \varphi(\zeta, \overline{\zeta})$, and since (29.31) defines Φ only up to an additive function of ζ and $\overline{\zeta}$, φ can be set zero, and $\partial\Phi = 0$ can then be read as giving L in terms of Φ:

$$L = \Phi_{,\zeta}/\Phi_{,u}. \tag{29.33}$$

For three particular forms of Φ, the field equations have been simplified or solved:

(i) $\Phi = \Phi\left[V A(\zeta,\overline{\zeta}) + B(\zeta,\overline{\zeta})\right]$, see §29.2.2. No solution of this class is known which does not belong to (ii).

(ii) $\Phi = V A(\zeta,\overline{\zeta}) + B(\zeta,\overline{\zeta})$, see §§29.2.3 and 29.2.4. This class contains all known solutions with $m + \mathrm{i}M \neq 0$ and a Killing vector $\boldsymbol{\xi} = \partial_u$.

(iii) $\Phi = A(V) + \zeta$, see §29.2.6.

29.2.2 The case $\partial_\zeta I = \partial_\zeta(\overline{G}^2 - \partial_{\overline{\zeta}}\overline{G}) \neq 0$

The condition (29.30) is a differential equation for the complex function $G(\zeta,\overline{\zeta})$, which in full reads

$$(\partial_\zeta - 2G)\partial_\zeta(\overline{G}^2 - \partial_{\overline{\zeta}}\overline{G}) = 0. \tag{29.34}$$

If this condition is satisfied, then, due to the field equations (29.16), the 'mass aspect' $m + \mathrm{i}M$ is independent of u; we shall write it as $m + \mathrm{i}M = 2P^3 A(\zeta,\overline{\zeta})\left[\partial_\zeta I\right]^{3/2}$. Substituting this expression into (29.20b) and eliminating $L_{,u}$ by means of (29.29), one sees that A is a function only of $\overline{\zeta}$:

$$m + \mathrm{i}M = 2P^3 A(\overline{\zeta})\left[\partial_\zeta(\overline{G}^2 - \partial_{\overline{\zeta}}\overline{G})\right]^{3/2}. \tag{29.35}$$

Now only the third and last field equation (29.20c) remains to be solved. If we make the ansatz

$$\begin{aligned} L = {} & (G + \partial_\zeta \ln P)u - A(\overline{\zeta})P^{-1}\left[\partial_\zeta(\overline{G}^2 - \partial_{\overline{\zeta}}\overline{G})\right]^{1/2} \\ & + P^{-1}(G + \partial_\zeta)\left[\Phi + \mathrm{i}\Psi\right] \end{aligned} \tag{29.36}$$

for L, then this field equation shows that $\Phi(\zeta,\overline{\zeta})$ is an arbitrary (real) function and that the real function $\Psi(\zeta,\overline{\zeta})$ has to obey

$$\begin{aligned} \Psi_{,\zeta\zeta\overline{\zeta}\overline{\zeta}} - \left[(\overline{G}^2 - \overline{G}_{,\overline{\zeta}})\Psi\right]_{,\zeta\zeta} - \left[(G^2 - G_{,\zeta})\Psi\right]_{,\overline{\zeta}\overline{\zeta}} & \\ + (G^2 - G_{,\zeta})(\overline{G}^2 - \overline{G}_{,\overline{\zeta}})\Psi = 0. & \end{aligned} \tag{29.37}$$

The four equations (29.34)–(29.37) show how to construct an explicit solution: one has to solve (29.34) for $G(\zeta,\overline{\zeta})$ and then (29.37) for $\Psi(\zeta,\overline{\zeta})$. The functions $P(\zeta,\overline{\zeta})$ and $\Phi(\zeta,\overline{\zeta})$ (both real) and $A(\overline{\zeta})$ are disposable; these functions being chosen, L and $m + \mathrm{i}M$ can be determined from (29.36) and (29.35), respectively. The full metric is then obtainable from (29.13); in general, it admits no Killing vector.

Unfortunately, no explicit solution is known in the general case. All known solutions belong to subcases, with either $L_{,u} = 0$ or $\partial_\zeta(\overline{G}^2 - \partial_{\overline{\zeta}}\overline{G}) = 0$.

29.2.3 The case $\partial_\zeta I = \partial_\zeta(\overline{G}^2 - \partial_\zeta \overline{G}) \neq 0,\ L_{,u} = 0$

If the function $G(\zeta,\overline{\zeta})$ in (29.29) can be written as the derivative with respect to ζ of a real function, then we can choose the coordinate u (i.e. the function P) so that $L_{,u} = 0$ holds. We then have

$$G(\zeta,\overline{\zeta}) = -(\ln P)_{,\zeta}. \tag{29.38}$$

The coordinate u is now fixed up to a change of its origin ($k = 1$ in (29.28)). All metric functions are independent of u; $\boldsymbol{\xi} = \partial_u$ is a Killing vector.

The 'background equation' (29.34) now reads

$$\Delta\Delta \ln P \equiv 4P^2 \partial_\zeta \partial_{\overline{\zeta}} P^2 \partial_\zeta \partial_{\overline{\zeta}} \ln P = 0, \tag{29.39}$$

and the restriction $\partial_\zeta I \neq 0$ can be written as

$$2P^2 \partial_\zeta I = (\Delta \ln P)_{,\overline{\zeta}} \neq 0. \tag{29.40}$$

Surprisingly, equation (29.39) is exactly the field equation for non-twisting diverging vacuum solutions of type *III* and *N*, and (29.40) ensures that the solution is of type *III*, cp. (28.13). So the background for all solutions with $\partial_\zeta I \neq 0,\ L_{,u} = 0 = P_{,u}$ is an arbitrary type *III* Robinson–Trautman vacuum solution in which the real function P obeys

$$\Delta \ln P(\zeta,\overline{\zeta}) = -3(\zeta + \overline{\zeta}), \tag{29.41}$$

cp. (28.15). The corresponding twisting solutions can be generated as follows (Robinson 1975).

We start with an arbitrary solution P of (29.41). Because of (29.35)–(29.36) and (29.40)–(29.41), $m + \mathrm{i}M$ does not depend on ζ,

$$m + \mathrm{i}M = -3\mathrm{i}A(\overline{\zeta})\sqrt{3/2}, \tag{29.42}$$

and L can be written as

$$L = (m + \mathrm{i}M)/3P^2 + \partial_\zeta \left[(\Phi + \mathrm{i}\Psi)/P\right], \tag{29.43}$$

where Ψ obeys

$$\Psi_{,\zeta\zeta\overline{\zeta}\overline{\zeta}} - (\Psi P_{,\overline{\zeta}\overline{\zeta}}/P)_{,\zeta\zeta} - (\Psi P_{,\zeta\zeta}/P)_{,\overline{\zeta}\overline{\zeta}} + \Psi P_{,\zeta\zeta} P_{,\overline{\zeta}\overline{\zeta}}/P^2 = 0. \tag{29.44}$$

Equation (29.44) has to be solved, and then P, $m + \mathrm{i}M$ and L represent the metric. $\Phi(\zeta,\overline{\zeta})$ and $A(\overline{\zeta})$ are disposable functions, but Φ can be eliminated by a coordinate transformation $u' = u + h(\zeta,\overline{\zeta})$.

A simple solution of this kind is

$$L = (m + \mathrm{i}M)/3P^2 + a/P^2, \quad a = \text{ const (real)}. \tag{29.45}$$

A large class of solutions can be constructed from the only known solution

$$P = (\zeta + \overline{\zeta})^{3/2} = (x\sqrt{2})^{3/2} \tag{29.46}$$

of the background equation (29.41). If we choose the function Φ in (29.43) so that $(\Phi/P)_{,y} = (\Psi/P)_{,x}$ is satisfied, then L takes the form

$$L = (m + \mathrm{i}M)/3P^2 + l(x, y), \tag{29.47}$$

in which the real function l is defined by

$$l_{,y} \equiv w = 2^{-3/2}(\Psi/P)_{,\zeta\overline{\zeta}}. \tag{29.48}$$

Instead of (29.44), we then have to solve

$$w_{,xx} + w_{,yy} + 6w_{,x}/x + 3w/x^2 = 0. \tag{29.49}$$

The solution of this linear differential equation can be found by standard separation: the most general solution is a superposition of

$$\begin{aligned} w_0 &= x^{(\pm\sqrt{13}-5)/2}(a_0 + b_0 y), \\ w_s &= x^{-5/2} J_{\pm\sqrt{13}/2}(xs)\left[a_s \mathrm{e}^{sy} + b_s \mathrm{e}^{-sy}\right], \quad s \neq 0, \end{aligned} \tag{29.50}$$

$J_n(sx)$ being a Bessel function. Equations (29.13), (29.42), (29.46)–(29.48) and (29.50) exhibit the class of solutions. Special cases have been published earlier (Robinson and Robinson 1969, Held 1974b); see. also (38.6).

The field equations (29.13*f*) and (29.15)–(29.16) show that for any solution with $L_{,u} = 0$ (and, of course, $P_{,u} = 0$) a real constant can be added to m. This property can be used to generate type *II* solutions from type *III* solutions ((29.46)–(29.50) with $m + \mathrm{i}M = 0$). This idea has been discussed in the context of the generalized Kerr–Schild transformations §32.5.2 by Fels and Held (1989).

29.2.4 The case $I = 0$

In the previous two sections we dealt with the case $\partial_\zeta I \neq 0$. For $\partial_\zeta I = 0$, i.e. for $I = I(\overline{\zeta})$, the transformation law (29.26) implies that the coordinate ζ can be chosen so that I vanishes. We shall now take that gauge and discuss solutions that obey

$$P_{,u} = 0, \quad L_{,u} - \partial_\zeta \ln P = G(\zeta, \overline{\zeta}), \quad I = \overline{G}^2 - \partial_{\overline{\zeta}}\overline{G} = 0. \tag{29.51}$$

The coordinate transformations that preserve (29.51) are

$$\zeta' = (p_1\zeta + q_1)/(p_2\zeta + q_2), \quad u' = uk(\zeta,\overline{\zeta}) + h(\zeta,\overline{\zeta}). \tag{29.52}$$

Because of the form (29.22) of the Weyl tensor components, $m + \mathrm{i}M = 0$, together with $I = 0$, would give flat space-time. So we have to assume $m + \mathrm{i}M \neq 0$; that excludes Petrov types III, N and O.

The integration procedure of the field equations runs as follows. As the function $G = \partial_\zeta \ln P - L_{,u}$ has an inhomogeneous transformation law because of (29.26), G can always be made non-zero. The differential equations (29.51) are then integrated by

$$G(\zeta,\overline{\zeta}) = -[\zeta + g(\overline{\zeta})]^{-1}, \quad L(\zeta,\overline{\zeta},u) = (G + \partial_\zeta \ln P)u + P^{-1}Gl(\zeta,\overline{\zeta}). \tag{29.53}$$

Having thus evaluated the conditions (29.51), we can now attack the three field equations.

The first field equation (29.16) shows that $m + \mathrm{i}M$ does not depend on u, and can therefore be written as $m + \mathrm{i}M = 2P^3G^3A(\zeta,\overline{\zeta})$. The second field equation (29.20b) then yields $A = A(\overline{\zeta})$, i.e.

$$m + \mathrm{i}M = 2P^3G^3A(\overline{\zeta}). \tag{29.54}$$

With these results, the last field equation (29.20c) can be written in the form

$$\mathrm{Im}\left[\partial_\zeta\partial_\zeta(AG - \overline{\partial}\,\overline{\partial}V)\right] = 0 = (AG - \overline{\partial}\,\overline{\partial}V)_{,\zeta\zeta} - (\overline{A}\,\overline{G} - \partial\partial V)_{,\overline{\zeta}\overline{\zeta}}. \tag{29.55}$$

This is exactly the condition for the existence of a real function $B(\zeta,\overline{\zeta})$ such that

$$AG - \overline{\partial}\,\overline{\partial}V = B_{,\overline{\zeta}\overline{\zeta}}. \tag{29.56}$$

Because of $V_{,u} = P$ and $P_{,u} = 0$, V has the structure $V = Pu + v(\zeta,\overline{\zeta})$, with an arbitrary real function v. Inserting this result into (29.56), and using (29.54), we obtain

$$AG + \overline{G}\,\overline{l}_{,\overline{\zeta}} = B_{,\overline{\zeta}\overline{\zeta}} + v_{,\overline{\zeta}\overline{\zeta}}. \tag{29.57}$$

If we choose $v = -B$, then (29.57) is solved by

$$l(\zeta,\overline{\zeta}) = -\int \overline{A}(\zeta)\overline{G}(\overline{\zeta},\zeta)G^{-1}(\zeta,\overline{\zeta})\mathrm{d}\zeta + \overline{l}_1(\overline{\zeta}). \tag{29.58}$$

To get the explicit form of the metric, one has to prescribe the functions $P(\zeta,\overline{\zeta})$, $A(\overline{\zeta})$, $g(\overline{\zeta})$ and $\overline{l}_1(\overline{\zeta})$. The formulae (29.53), (29.54) and (29.58) will give the functions L and $m + \mathrm{i}M$, which together with P give all the information needed to construct the full metric according to (29.13). As P can be made equal to 1 by a coordinate transformation (29.52), the above vacuum solutions contain three disposable analytic functions. In general they will not admit any Killing vector. Physically the class of algebraically special diverging vacuum solutions that satisfy the conditions (29.51) can be characterized (Trim and Wainwright 1974) as the only solutions which are non-radiative in the sense that the Weyl tensor asymptotically (for large r) behaves as

$$C_{abcd} = II_{abcd}/r^3 + O(r^4). \tag{29.59}$$

29.2.5 The case $I = 0 = L_{,u}$

If we add $L_{,u} = 0$ to the conditions (29.51), then the metric becomes independent of u; $\boldsymbol{\xi} = \partial_u$ is a Killing vector. It can be shown (Trim and Wainwright 1974) that instead of $L_{,u} = 0$ the equivalent condition $\Sigma_{,u} = 0$ may be imposed. Like $I = 0$, the condition $L_{,u} = 0$ is not invariant (strictly speaking, the condition is that $L_{,u}$ can be made zero). So the coordinate transformations are now restricted to (29.52) with constant k.

Because of (29.51), $L_{,u} = 0$ yields $G = -(\ln P)_{,\zeta}$, which is compatible with $\overline{I} = G^2 - G_{,\zeta} = 0$ only if $P_{,\zeta\zeta} = 0$, i.e. with real P only if

$$P = \alpha\zeta\overline{\zeta} + \beta\zeta + \overline{\beta}\,\overline{\zeta} + \delta. \tag{29.60}$$

Equation (29.60) shows that the 2-space $2\mathrm{d}\zeta\mathrm{d}\overline{\zeta}/P^2$ has constant curvature $K = 2(\alpha\delta - \beta\overline{\beta})$.

Instead of (29.53), (29.54) and (29.58) the constitutive parts of the metric are now given by (29.60) and

$$m + \mathrm{i}M = \overline{Z}(\overline{\zeta}), \quad L = -\frac{1}{2P^2}\int \frac{Z(\zeta)\,\mathrm{d}\zeta}{(\alpha\zeta + \overline{\beta})^2} + \frac{\overline{l}_1(\overline{\zeta})}{P^2}, \tag{29.61}$$

with $Z = -\overline{A}/(\alpha\zeta + \overline{\beta})^3$ (note that we have chosen a gauge with $G = -(\alpha\overline{\zeta} + \beta)/P \neq 0$, so that in the case $K = 0$ we have to stick to a P with α or $\beta \neq 0$). The solutions contain the disposable functions $\overline{Z}(\overline{\zeta})$ and $\overline{l}_1(\overline{\zeta})$. The value of K can be transformed to $0, \pm 1$, and L may be simplified by a transformation $u' = u + h(\zeta, \overline{\zeta})$, $L' = L - h_{,\zeta}$, e.g. so that L does not contain terms in m.

A subcase of these solutions is the class $m + \mathrm{i}M = \mathrm{const}$. It contains some of the well-known type D solutions such as Kerr and NUT (see below in §29.5) as well as the Kerr and Debney (1970)/Demiański (1972) four-parameter (m, M, a, c) solution, which corresponds to

$$L = -P^{-2}[2\mathrm{i}M/\zeta + \mathrm{i}\overline{\zeta}(M + a) + \tfrac{1}{4}\mathrm{i}c\overline{\zeta}\ln(\overline{\zeta}/\sqrt{2})]. \tag{29.62}$$

For $m = \mathrm{const}$, $M = 0$ the solutions (29.60)–(29.61) specialize to the Kerr–Schild class of vacuum solutions (see §32.2), which can be characterized by $\partial\partial V = 0$, or, in the gauge $P = 1$, by $\partial L = 0$. For a more direct approach to the field equations in the case $I = 0 = L_{,u}$ see §30.7.4.

29.2.6 Solutions independent of ζ and $\overline{\zeta}$

Following Weir and Kerr (1977), we choose coordinates such that $P = 1$ and assume that in this system the metric functions do not depend on ζ and $\overline{\zeta}$. The field equation (29.21*b*) then reads

$$-L\partial_u(m + \mathrm{i}M) = 3(m + \mathrm{i}M)L_{,u} \tag{29.63}$$

(because $\partial = -L\partial_u$) and is integrated by

$$m + \mathrm{i}M = \mu_0 L^{-3}(u), \tag{29.64}$$

μ_0 being a complex constant. The second field equation (29.16) can be written as

$$\partial_u(L^2\partial I) = 3\mu_0 L^{-3}L_{,u}, \quad \partial I \equiv -L\partial_u\partial_u\overline{L}\,\overline{L}_{,u}, \tag{29.65}$$

and yields

$$-2\partial I = 2L\partial_u\partial_u\overline{L}\,\overline{L}_{,u} = 3\mu_0 L^{-4} + \nu_0 L^{-2}, \tag{29.66}$$

ν_0 being another complex constant.

This last equation and the third field equation $M = \mathrm{Im}\,\overline{\partial}\,\overline{\partial}\partial L$ give a system of three real differential equations for the complex function $L(u)$. This system can be simplified if we introduce a new real variable w by

$$L = g^{-1/2}(u)e^{-\mathrm{i}\varphi(u)}, \quad \mathrm{d}w = g^{3/2}\mathrm{d}u, \quad \partial_u = g^{3/2}\partial_w. \tag{29.67}$$

In terms of g and φ, the remaining field equations then read

$$\varphi'g'' - \varphi''g' + (2\varphi'^3 - \varphi''')g = -\mathrm{Im}\,\mu_0 \mathrm{e}^{3\mathrm{i}\varphi}, \tag{29.68a}$$

$$(2\varphi'^3 + \varphi''')g^2 = -\mathrm{Im}\,\nu_0 \mathrm{e}^{\mathrm{i}\varphi}, \tag{29.68b}$$

$$g''' + 6g'\varphi'^2 + 12g\varphi'\varphi'' = -\mathrm{Re}\,(3\mu_0\mathrm{e}^{3\mathrm{i}\varphi} + \nu_0 g^{-1}\mathrm{e}^{\mathrm{i}\varphi}), \tag{29.69}$$

where g', φ', etc. denote the derivatives with respect to w.

The solutions $g = \mathrm{const}$ ($\Rightarrow \varphi = \mathrm{const}$ and $\mu_0 = 0 = \nu_0$) give $m + \mathrm{i}M = 0 = \partial I = \partial_u I$ and are therefore flat space-times, cp. (29.22).

Solutions with $\varphi' = 0$ ($\Rightarrow \varphi = 0$ by a ζ–$\overline{\zeta}$-rotation) have zero twist ($\Sigma = 0$), belong to the Robinson–Trautman class and can easily be transformed to a coordinate system with $L = 0 = W$, $P = P(\zeta, \overline{\zeta}, u)$. Examples of non-twisting solutions of this kind are the C-metric (28.24) ($\Leftrightarrow \nu_0 = 0$), and the type III metric (28.17) ($\Leftrightarrow \mu_0 = 0$).

If $\varphi' \neq 0$, the third differential equation (29.69) is a consequence of the first two and need not be considered. The only known solution of this kind (twisting and independent of ζ and $\overline{\zeta}$) is the general solution in the case $\nu_0 = 0$. With $\nu_0 = 0$, the differential equation (29.68*b*) is integrated by $e^{-2i\varphi}(\varphi'^2 - i\varphi'') = 1/2$, and the complete solution of the system (29.68*a*)–(29.68*b*) is (Kinnersley 1969b)

$$e^{i\varphi} = \mathrm{dn}(w) + i\,\mathrm{sn}(w)/\sqrt{2}, \quad \sqrt{2}\varphi' = \mathrm{cn}\, w = \sqrt{\cos 2\varphi}, \qquad g(w) = \mathrm{Re}\, \alpha_0 e^{2i\varphi} - 4\,\mathrm{Im}\, \mu_0 \varphi' e^{-i\varphi}, \tag{29.70}$$

where the functions dn, sn and cn are the Jacobian elliptic functions of modulus $1/\sqrt{2}$, and α_0 and μ_0 are complex constants. This solution is the twisting generalization of the C-metric (28.24).

It can be shown (Valiente Kroon 2000) that the two Killing vectors of the solutions have the structure of a boost-rotation symmetry (assuming the existence of a local $\mathcal{I}$), cp. §17.2.

29.3 Solutions of type N ($\Psi_2 = 0 = \Psi_3$)

Due to the structure (29.22) of the components of the Weyl tensor, solutions of Petrov type N are characterized by $m + iM = 0$, $\partial I = 0$, $\partial_u I \neq 0$. It turns out that the type N conditions together with the field equations (29.20*c*) yield

$$\partial I = \partial[P^{-1}(\overline{\partial}\,\overline{\partial} V)_{,u}] = 0, \quad \partial_u I \neq 0, \quad \mathrm{Im}\, \partial\partial\overline{\partial}\,\overline{\partial} V = 0, \tag{29.71}$$

the remaining field equations (29.20*a*)–(29.20*b*) being automatically satisfied. Note that no solution of the classes studied in the preceding sections can be of type N, because all such solutions violate $\partial_u I \neq 0$ if satisfying the rest of (29.71).

So far only one (one-parameter) class of solutions of type N has been found, the Hauser solution (Hauser 1974, 1978). It reads in our notation

$$P = (\zeta + \overline{\zeta})^{3/2} f(w), \quad w \equiv u/(\zeta + \overline{\zeta})^2, \qquad L = 2i(\zeta + \overline{\zeta}), \quad I = 3/[(\zeta + \overline{\zeta})^2 - iu], \tag{29.72a}$$

where $f(w)$ is a solution of the hypergeometric differential equation

$$16(1 + w^2) f'' + 3f = 0. \tag{29.72b}$$

The Hauser solution obviously admits a Killing vector $\xi = \mathrm{i}\partial_{(\zeta-\bar{\zeta})}$; McIntosh found that it also admits a (special) homothetic vector (Halford 1979). It is not asymptotically flat and does not describe the radiation field of an isolated source. It contains one essential parameter, which can be chosen to be the ratio of two appropriately chosen solutions of the hypergeometric differential equation (29.72*b*).

Much effort has been invested in seeking other explicit type N solutions, but so far without any success except an understanding of the difficulties of the problem. Even if one assumes the existence of a Killing vector and a second (general) homothetic or Killing vector, only the Hauser case can be solved, although it could be shown (Herlt 1986) that with these assumptions (due to the existence of two Lie point symmetries) the type N field equations can be reduced to a third-order ordinary differential equation for a real function.

Held (1999) argued that the GHP-formalism can in principle be used to construct an explicit type N solution with one Killing vector.

29.4 Solutions of type III ($\Psi_2 = 0, \Psi_3 \neq 0$)

Solutions of type III are characterized by

$$m + \mathrm{i}M = 0, \quad \partial I \neq 0, \tag{29.73}$$

cp. (29.22). All twisting solutions known so far are subcases of the general class treated in §29.2.3, i.e. in addition to (29.73) they all satisfy $P_{,u} = 0 = \Delta\Delta \ln P$, $L_{,u} = 0$, and can therefore be generated from non-twisting type III solutions. These known solutions (Robinson and Robinson 1969, Held 1974b, Robinson 1975) are given by (29.46)–(29.50), with, of course, $m + \mathrm{i}M = 0$; they are twisting exactly if $\bar{L}_{,\zeta} \neq L_{,\bar{\zeta}}$.

29.5 Solutions of type D ($3\Psi_2\Psi_4 = 2\Psi_3^2$, $\Psi_2 \neq 0$)

All vacuum solutions of Petrov type D are known (Kinnersley 1969b). They were found by applying the Newman–Penrose formalism, the main difference from the method so far outlined in this chapter being that *both* vectors $\boldsymbol{k}$ and $\boldsymbol{l}$ were chosen to be eigenvectors of the Weyl tensor. A systematic approach on the basis of the canonical tetrad and coordinate system used in this chapter can be found in Weir and Kerr (1977).

From our point of view, the diverging type D solutions divide into two classes, the first covering all solutions of Kinnersley's Classes I and II and the second Kinnersley's Class III (twisting C-metric).

In detail, the *first class* is contained in the subcase $m + \mathrm{i}M = \text{const}$ of the solutions with $I = 0 = L_{,u}$ discussed in §29.2.5. These type D solutions are given by (29.60)–(29.61) with $l_1''' = 0$, i.e. they satisfy

$$P = \alpha\zeta\bar{\zeta} + \beta\zeta + \bar{\beta}\bar{\zeta} + \delta, \quad K = 2(\alpha\delta - \beta\bar{\beta}) = 0, \pm 1, \quad m + \mathrm{i}M = \text{ const},$$
$$L = -\frac{m - \mathrm{i}M}{2P^2} \int \frac{\mathrm{d}\zeta}{(\alpha\zeta + \bar{\beta})^2} - \frac{1}{P^2}(\lambda_0 + \lambda_1\bar{\zeta} + \lambda_2\bar{\zeta}^2). \tag{29.74}$$

Among the constants α, δ, m, M (all real) and β, λ_0, λ_1, λ_2 (all complex) only three real numbers (including m and M) are essential, i.e. cannot be removed by a coordinate transformation.

For non-zero curvature, $K = \pm 1$, the metric can be written as

$$\begin{aligned} \mathrm{d}s^2 &= 2(r^2 + \Sigma^2)P^{-2}\mathrm{d}\zeta\mathrm{d}\bar{\zeta} - 2(\mathrm{d}u + L\mathrm{d}\zeta + \bar{L}\mathrm{d}\bar{\zeta}) \\ &\quad \times \left[\mathrm{d}r + W\mathrm{d}\zeta + \bar{W}\mathrm{d}\bar{\zeta} + H(\mathrm{d}u + L\mathrm{d}\zeta + \bar{L}\mathrm{d}\bar{\zeta})\right], \\ P &= 1 + K\zeta\bar{\zeta}/2, \quad L = -P^{-2}[2\mathrm{i}M/\zeta + \mathrm{i}\bar{\zeta}(M + a)], \quad K = \pm 1, \qquad (29.75) \\ H &= \frac{K}{2} - \frac{mr + M\Sigma}{r^2 + \Sigma^2}, \quad W = \frac{Ka\zeta}{P^2}, \quad \Sigma = KM - a\frac{1 - K\zeta\bar{\zeta}/2}{1 + K\zeta\bar{\zeta}/2}, \end{aligned}$$

($\lambda_1 \equiv \mathrm{i}(a+M)$; in L the m-term has been transformed to zero). The Kerr solution (Kerr 1963a) with $M = 0$, $K = 1$, the Schwarzschild solution $M = 0 = a$ and the NUT solutions (Newman *et al.* 1963) $a = 0$ are special cases. The parameters m, M and a are called the mass, the NUT parameter and the Kerr parameter, respectively.

The *second class* of type D diverging vacuum solutions are those for which the metric does not depend on ζ and $\bar{\zeta}$, cp. §29.2.6. It turns out (Weir and Kerr 1977) that the type D condition exactly reduces to $\nu_0 = 0$. The metric is therefore (29.13) with $P = 1$ and $m + \mathrm{i}M$ and L given by (29.64), (29.67) and (29.70).

Obviously, the coordinate frame best adapted to type D metrics is one which gives a single expression for the metric of *all* diverging vacuum type D metrics. This frame is provided by the metric (21.11), where the vacuum solutions are contained as the case $e = g = \Lambda = 0$. Note that the m of (21.11) is related to, but not always equal to, the mass parameter m used elsewhere in this chapter; in the case of the twisting C-metric, the m of (21.11) corresponds to Re μ_0 in (29.64). A detailed discussion of the type D metric, its various subcases, and its generalization to Einstein–Maxwell fields is given in §21.1.

The comparatively simple form (21.11) of the general diverging type D metric indicates that the canonical frame (29.13), which in principle covers all diverging algebraically special vacuum metrics, is not the best frame for type D. It is an open question whether different coordinates could also facilitate the integration of the field equations in the other Petrov cases.

29.6 Solutions of type II

The type II solutions explicitly known belong to the classes treated in §§29.2.3–29.2.5 and its subclasses (e.g. Kerr–Schild). They contain at most three arbitrary analytic functions.

30

Twisting Einstein–Maxwell and pure radiation fields

30.1 The structure of the Einstein–Maxwell field equations

In the subsequent sections of this chapter, we are looking for exact solutions of the coupled system

$$2R_{ab} = \kappa_0 F^{*c}_{a}\overline{F^{*}_{bc}}, \tag{30.1}$$

$$F^{*ab}{}_{;b} = (F^{ab} + \mathrm{i}\widetilde{F}^{ab})_{;b} = 0, \tag{30.2}$$

of Einstein–Maxwell equations in the case when the Weyl tensor possesses a (multiple) null eigenvector $\boldsymbol{k}$ which is twisting, geodesic and shearfree, and is also an eigenvector of the Maxwell tensor F_{ab} (aligned case). The latter condition implies that (a) of the tetrad components (7.26)–(7.28) of the Maxwell tensor only

$$\Phi_1 = \tfrac{1}{2}F_{ab}(k^a l^b + \overline{m}^a m^b), \quad \Phi_2 = F_{ab}\overline{m}^a l^b, \tag{30.3}$$

can be non-zero (Φ_0 must vanish), and (b) therefore

$$R_{11} = R_{14} = R_{44} = 0 \tag{30.4}$$

holds.

The conditions of Theorem 27.1 being satisfied, we can integrate (30.4) by (27.26)–(27.27), i.e. by

$$\begin{aligned} \mathrm{d}s^2 &= 2\boldsymbol{\omega}^1\boldsymbol{\omega}^2 - 2\boldsymbol{\omega}^3\boldsymbol{\omega}^4, \\ \boldsymbol{\omega}^1 &= -\mathrm{d}\zeta/P\overline{\rho} = \overline{\boldsymbol{\omega}}^2, \quad \boldsymbol{\omega}^3 = \mathrm{d}u + L\mathrm{d}\zeta + \overline{L}\mathrm{d}\overline{\zeta}, \\ \boldsymbol{\omega}^4 &= \mathrm{d}r + W\mathrm{d}\zeta + \overline{W}\mathrm{d}\overline{\zeta} + H\boldsymbol{\omega}^3, \end{aligned} \tag{30.5}$$

with (in the gauge $r^0 = 0$)

$$\begin{aligned} \rho^{-1} &= -(r + \mathrm{i}\Sigma), \qquad 2\mathrm{i}\,\Sigma = P^2(\overline{\partial}L - \partial\overline{L}), \\ W &= \rho^{-1}L_{,u} + \mathrm{i}\,\partial\Sigma, \qquad \partial \equiv \partial_\zeta - L\partial_u. \end{aligned} \tag{30.6}$$

All metric functions are given in terms of r, $P(\zeta,\overline{\zeta},u)$, $L(\zeta,\overline{\zeta},u)$ and $H(\zeta,\overline{\zeta},r,u)$ (for more details see Chapter 27).

The metric (30.5)–(30.6) is of course subject to the rest of Einstein's equations (30.1),

$$R_{12} = R_{34} = 2\kappa_0\Phi_1\overline{\Phi}_1, \tag{30.7}$$

$$R_{13} = 2\kappa_0\Phi_1\overline{\Phi}_2, \quad R_{33} = 2\kappa_0\Phi_2\overline{\Phi}_2, \tag{30.8}$$

and Maxwell's equations (30.2) have to be solved too. In close analogy with the vacuum case (§29.1), the calculations naturally divide into two steps. In the first step, the radial dependence of the metric (i.e. in addition to (30.6) that of the function H) and of the Maxwell field will be completely determined. In the second step, the remaining field equations (both Einstein and Maxwell) will be reduced to a system of partial differential equations for the as yet undetermined functions of $\zeta,\overline{\zeta},u$ which enter into the metric and the electromagnetic field.

30.2 Determination of the radial dependence of the metric and the Maxwell field

In the metric (30.5), the spin coefficients $\tau,\lambda,\pi,\kappa,\varepsilon$ and σ vanish (cp. §27.1.1) and Φ_0 is zero by assumption. So the first part (7.22)–(7.23) of the Maxwell equations reads

$$\partial_r\Phi_1 = 2\rho\Phi_1, \quad \partial_r\Phi_2 = \rho\Phi_2 - P\rho\overline{\partial}\Phi_1 + P\rho\overline{W}\partial_r\Phi_1. \tag{30.9}$$

Using (30.6), these equations can be integrated and yield

$$\Phi_1 = \rho^2\Phi_1^0(\zeta,\overline{\zeta},u), \tag{30.10a}$$

$$\Phi_2 = \rho\Phi_2^0(\zeta,\overline{\zeta},u) + \rho^2 P(2\overline{L}_{,u}\Phi_1^0 - \overline{\partial}\Phi_1^0) + 2\mathrm{i}\rho^3 P(\Sigma\overline{L}_{,u} - \overline{\partial}\Sigma)\Phi_1^0. \tag{30.10b}$$

As in the non-twisting case, one may think of Φ_1^0 and Φ_2^0 as representing the field of charges and a pure radiation field, respectively.

We can now evaluate the field equations (30.7). Substituting the expressions (29.7) and (30.10a) for R_{34} and Φ_1, we obtain

$$\rho^2\left[\rho^{-2}(\Gamma_{213} + \Gamma_{433})\right]_{|4} = 2\rho(\ln P)_{,u} + 2\kappa_0\rho^2\overline{\rho}^2\Phi_1^0\overline{\Phi}{}_1^0. \tag{30.11}$$

This equation yields

$$\Gamma_{213} + \Gamma_{433} = -(\ln P)_{,u} + 2\kappa_0\overline{\rho}\rho^2\Phi_1^0\overline{\Phi}{}_1^0 + (m + \mathrm{i}M)\rho^2 \tag{30.12}$$

(because $\rho_{|4} = \rho^2$), and, as Γ_{213} is purely imaginary and Γ_{433} equals $H_{|4}$, we can determine H,

$$H = K/2 - r(\ln P)_{,u} - (mr + M\Sigma - \kappa_0\Phi_1^0\overline{\Phi}{}_1^0)/(r^2 + \Sigma^2), \qquad (30.13)$$

the real functions m, M and K being independent of r. After some calculations it then turns out that the term proportional to $\Phi_1^0\overline{\Phi}{}_1^0$ does not enter into those equations which as a consequence of the field equations (30.7) give M and K in terms of P and L. So we can take M and K from (29.13); they are the same functionals as in the vacuum case.

Theorem 30.1 *An (algebraically special) Einstein–Maxwell field admits a diverging, geodesic and shearfree null congruence* $\boldsymbol{k}$ *and satisfies* $R_{11} = R_{14} = R_{44} = 0$ *(aligned case),* $R_{12} = R_{34} = 2\kappa_0\Phi_1\overline{\Phi}_1$ *and the radial part (7.22)–(7.23) of Maxwell's equations exactly if the metric and the electromagnetic field can be given by*

$$\begin{aligned}
ds^2 &= 2d\zeta d\overline{\zeta}/(P^2\rho\overline{\rho}) - 2(du + Ld\zeta + \overline{L}d\overline{\zeta})\Big[dr + Wd\zeta + \overline{W}d\overline{\zeta} \\
&\qquad\qquad + H(du + Ld\zeta + \overline{L}d\overline{\zeta})\Big], \\
H &= K/2 - r(\ln P)_{,u} - (mr + M\Sigma - \kappa_0\Phi_1^0\overline{\Phi}{}_1^0)/(r^2 + \Sigma^2), \\
\rho^{-1} &= -(r + i\Sigma), \quad 2i\Sigma = P^2(\overline{\partial}L - \partial\overline{L}), \qquad\qquad (30.14) \\
W &= \rho^{-1}L_{,u} + i\partial\Sigma, \quad K = 2P^2\,\mathrm{Re}\left[\partial(\overline{\partial}\ln P - \overline{L}_{,u})\right], \\
M &= \Sigma K + P^2\,\mathrm{Re}\left[\partial\overline{\partial}\Sigma - 2\overline{L}_{,u}\partial\Sigma - \Sigma\partial_u\partial\overline{L}\right],
\end{aligned}$$

and

$$\begin{aligned}
\Phi_1 &= \rho^2\Phi_1^0, \\
\Phi_2 &= \rho\Phi_2^0 + \rho^2 P(2\overline{L}_{,u} - \overline{\partial})\Phi_1^0 + 2i\rho^3 P(\Sigma\overline{L}_{,u} - \overline{\partial}\Sigma)\Phi_1^0
\end{aligned} \qquad (30.15)$$

(Robinson *et al.* 1969b, Trim and Wainwright 1974).

Equations (30.14)–(30.15) show that all field functions can be constructed from the real functions P and m and the complex functions L, Φ_1^0, Φ_2^1, which are all independent of r. These (so far arbitrary) functions of $\zeta, \overline{\zeta}$ and u are of course subject to the remaining field equations, both Einstein and Maxwell, which we shall now formulate. Concerning the coordinate freedom and the transformation properties of the various functions we refer the reader to §29.1.4. The transformation properties given there are completed by

$$\Phi_1^{0\prime} = F_{,u}^{-2}\Phi_1^0, \quad \Phi_2^{0\prime} = (f'/\overline{f}')^{1/2}F_{,u}^{-2}\Phi_2^0. \qquad (30.16)$$

30.3 The remaining field equations

The second part (7.24)–(7.25) of Maxwell's equations, which has to be satisfied in addition to (30.9), reads

$$\delta\Phi_1 = 0, \quad \delta\Phi_2 - \Delta\Phi_1 - 2\mu\Phi_1 + 2\beta\Phi_2 = 0. \tag{30.17}$$

Taking $\mu = -\Gamma_{321}$ from (29.10), and substituting $2\beta = -(\ln P)_{|1}$ and expressions (30.14)–(30.15) for the metric and the Maxwell field, a straightforward calculation yields the simple equations

$$\begin{aligned} (\partial - 2L_{,u})\Phi_1^0 &= 0, \\ (\partial - L_{,u})(P^{-1}\Phi_2^0) + (P^{-2}\Phi_1^0)_{,u} &= 0. \end{aligned} \tag{30.18}$$

The two Einstein equations (30.8) not yet taken into account can be simplified in a way analogous to that in the vacuum case. The final result (Robinson *et al.* 1969b, Lind 1974, Trim and Wainwright 1974) reads

$$P(3L_{,u} - \partial)(m + \mathrm{i}M) = 2\kappa_0\Phi_1^0\overline{\Phi}{}_2^0, \tag{30.19a}$$

$$\begin{aligned} &P^4(\partial - 2L_{,u} + 2\partial\ln P)\partial\left[\overline{\partial}(\overline{\partial}\ln P - \overline{L}_{,u}) + (\overline{\partial}\ln P - \overline{L}_{,u})^2\right] \\ &\quad -P^3\left[P^{-3}(m + \mathrm{i}M)\right]_{,u} = \kappa_0\Phi_2^0\overline{\Phi}{}_2^0, \end{aligned} \tag{30.19b}$$

$$P^{-3}M = \mathrm{Im}\,(\partial\partial\overline{\partial}\,\overline{\partial}V), \quad V_{,u} \equiv P. \tag{30.19c}$$

The five equations (30.18)–(30.19*c*) form a system of partial differential equations for the functions P, m (real) and L, Φ_1^0, Φ_2^0 (complex). If a solution has been found, then the full metric and the Maxwell field can be obtained from (30.3), (30.14) and (30.15). Different forms of the field equations may easily be derived from (29.20) and (29.21).

Equations (30.18)–(30.19) generalize the vacuum equations (29.15)–(29.16) as well as the Einstein–Maxwell equations (28.37) of the non-twisting case. To achieve conformity with the notation in the non-twisting case, we have to put

$$\Phi_1^0 = \overline{Q}/2, \quad \Phi_2^0 = -Ph. \tag{30.20}$$

The detailed expressions for the non-zero components of the Weyl tensor can be found in Trim and Wainwright (1974). Here we mention only that they have the structure

$$\begin{aligned} \Psi_2 &= (m + \mathrm{i}M)\rho^3 + \kappa_0 Q\overline{Q}\rho^3\overline{\rho}/2, \\ \Psi_3 &= -P^3\rho^2\partial I + O(\rho^3), \quad \Psi_4 = P^2\rho\,\partial_u I + O(\rho^2), \\ I &\equiv \overline{\partial}(\overline{\partial}\ln P - \overline{L}_{,u}) + (\overline{\partial}\ln P - \overline{L}_{,u})^2 = P^{-1}(\overline{\partial}\,\overline{\partial}V)_{,u}, \end{aligned} \tag{30.21}$$

where the terms of higher order in ρ occurring in Ψ_3 or Ψ_4 vanish identically if Ψ_2 or Ψ_2 and Ψ_3, respectively, vanish.

30.4 Charged vacuum metrics

By inspecting the field equations (30.19a)–(30.19c) one immediately sees that they reduce to the vacuum case if no free radiation field is present, i.e. if

$$\Phi_2^0 = 0, \quad \Phi_1^0 = \overline{Q}(\zeta, \overline{\zeta}, u)/2 \tag{30.22}$$

holds. In that case, Maxwell's equations (30.18) yield

$$\overline{Q} = q(\zeta, \overline{\zeta}) P^2, \tag{30.23}$$

and (in the gauge $P_{,u} = 0$)

$$[\partial_\zeta - 2(L_{,u} - \partial_\zeta \ln P)]\, q = 0. \tag{30.24}$$

As q and P do not depend on u, the same is true for $L_{,u} - \partial_\zeta \ln P$, i.e. as a consequence of Maxwell's equations we obtain

$$L_{,u} - \partial \ln P = G(\zeta, \overline{\zeta}), \quad P_{,u} = 0. \tag{30.25}$$

These conditions coincides with the assumption (29.29) we made in the vacuum case to get special classes of solutions, see §29.2.1. We thus have proved the following generalization of Theorem 28.5 (on Robinson–Trautman Einstein–Maxwell fields):

Theorem 30.2 *All Einstein–Maxwell fields (aligned case) admitting a diverging, geodesic and shearfree null congruence with a non-radiative* ($\Phi_2^0 = 0$) *Maxwell field are given by*

$$\begin{aligned} \mathrm{d}s^2 &= \mathrm{d}s_0^2 - \tfrac{1}{2}\kappa_0 \overline{Q} Q \rho\overline{\rho}(\mathrm{d}u + L\mathrm{d}\zeta + \overline{L}\mathrm{d}\overline{\zeta})^2, \\ \overline{Q} &= \alpha(\overline{\zeta}) P^2 \exp\left(2 \textstyle\int G(\zeta, \overline{\zeta})\mathrm{d}\zeta\right), \quad \Phi_1 = \overline{Q}\rho^2/2, \\ \Phi_2 &= \tfrac{1}{2} P \rho^2 (2L_{,u} - \partial_{\overline{\zeta}})\overline{Q} + \mathrm{i}\, P\rho^3 \overline{Q}(\Sigma \overline{L}_{,u} - \overline{\partial}\Sigma), \end{aligned} \tag{30.26}$$

where $\mathrm{d}s_0^2$ *is an algebraically special* vacuum *metric subject to* (30.25)*, and* $\alpha(\overline{\zeta})$ *is a disposable function* (Robinson *et al.* 1969b, Trim and Wainwright 1974).

In §29.2 we gave a survey of all explicitly known vacuum solutions which satisfy the above conditions (and, moreover, $(m + \mathrm{i}M)_{,u} = 0$). For the subcases treated there, the complex (electric plus magnetic) charge Q is given (as a solution of (30.24)) as

$$\overline{Q}(\overline{\zeta}) = \alpha(\overline{\zeta}) \qquad \text{for } L_{,u} = 0, \tag{30.27}$$

$$\overline{Q}(\zeta,\overline{\zeta}) = \alpha(\overline{\zeta})P^2G^2 \quad \text{for} \quad I = 0. \tag{30.28}$$

Special cases included here are the Reissner–Weyl solution (15.21), the charged NUT and Kerr–Schild solutions etc. Note that the charged C-metric (21.22) and its twisting generalization are *not* covered by Theorem 30.2 as in both cases the Maxwell field is radiative ($\Phi_2^0 \neq 0$).

The Einstein–Maxwell fields covered by (30.28) together with (30.26) are exactly those fields which are non-radiative in the sense that the Weyl tensor and the Maxwell tensor do not contain terms with r^{-n}, where $n < 3$ and $n < 2$, respectively (Trim and Wainwright 1974). Solutions which furthermore are regular and stationary must be type D (Lind 1975a, 1975b, Held 1976a).

30.5 A class of radiative Einstein–Maxwell fields ($\Phi_2^0 \neq 0$)

Radiative Einstein–Maxwell fields have been found in the restricted case

$$P_{,u} = L_{,u} = \Phi^0_{1,u} = \Phi^0_{2,u} = 0 \tag{30.29}$$

(Herlt and Stephani 1984). If we perform a coordinate transformation $u' = u + h(\zeta,\overline{\zeta})$ to achieve

$$L = \mathrm{i}B(\zeta,\overline{\zeta})_{,\zeta} \tag{30.30}$$

with a real function $B(\zeta,\overline{\zeta})$, then (30.18)–(30.19$a$) and parts of the remaining Einstein equations yield

$$\begin{aligned} \Phi_1^0 &= \overline{\psi}(\overline{\zeta}), \quad \Phi_2^0 = P(\zeta,\overline{\zeta})\overline{\beta}'(\overline{\zeta}), \quad \beta' \neq 0, \\ m + \mathrm{i}M &= a\left[u + \mathrm{i}B(\zeta,\overline{\zeta})\right] - 2\kappa_0\overline{\psi}(\overline{\zeta})\beta(\zeta) + \overline{\alpha}(\overline{\zeta}), \end{aligned} \tag{30.31}$$

with a = const (real). The functions α, β, ψ, B and P are subject to the remaining two field equations (30.19b)–(30.19c). The first of these equations reads

$$P^2\partial_\zeta\partial_{\overline{\zeta}}P^2\partial_\zeta\partial_{\overline{\zeta}}\ln P = a + \kappa_0 P^2\beta'(\zeta)\overline{\beta}'(\overline{\zeta}). \tag{30.32}$$

Following Robinson and Robinson (1969) in their treatment of the vacuum case $\alpha = 0 = \beta'$, we call this equation the 'background equation'; it generalizes (29.39). Any solution P, β of this equation together with

$$\begin{aligned} L &= 0 = M, \quad \Phi_1^0 = b_0\overline{\beta} + \psi_0, \quad \Phi_2^0 = P\overline{\beta}', \\ m &= au - 2\kappa_0 b_0 \beta\overline{\beta} - 4\kappa_0 \mathrm{Re}\,(\psi_0\beta) + m_0, \end{aligned} \tag{30.33}$$

gives a non-twisting Einstein–Maxwell field, in which one can hope to insert twist ($L \neq 0, M \neq 0$) and an independent additional charge Φ_1^0 by solving the last surviving equation (30.19c), with L and $m + \mathrm{i}M$ given by (30.30)–(30.31). The additional charge ψ_0 can always be switched on or off.

In §28.2, we have listed the known classes of solution of the background equation (30.32). In each case twisting Einstein–Maxwell fields can be found, which generalize known twisting vacuum solutions, e.g. (29.46)–(29.50). In general, they are of Petrov type II or III. We refer the reader to the original paper (Herlt and Stephani 1984), but want to mention one special case given by

$$\begin{aligned} &P = 1, \quad \Phi_1^0 = \overline{\psi}(\overline{\zeta}), \quad \Phi_2^0 = b, \quad bL = \overline{\zeta}\psi' - \overline{\psi} + \mathrm{i}\hat{B}(\zeta, \overline{\zeta})_{,\zeta}, \\ &m + \mathrm{i}M = -\kappa_0 b^2[u + \mathrm{i}\hat{B}] - 2\kappa_0 b\,\mathrm{Re}\,(\zeta\overline{\psi}), \end{aligned} \tag{30.34}$$

where $\hat{B}(\zeta, \overline{\zeta})$ is a real function satisfying

$$\hat{B}_{,\zeta\zeta\overline{\zeta}\overline{\zeta}} = -b^2\hat{B} \tag{30.35}$$

(whose general solution can be constructed, cp. the discussion of (30.66)). The existence of a constant electromagnetic pure radiation field implies in m a term linear in u, i.e. a constant rate of loss of the mass m of the system.

The field equations can also be simplified under the assumptions $P_{,u} = (m + \mathrm{i}\,M)_{,u} = \Phi^0_{1,u} = \Phi^0_{2,u} = 0$, L linear in u (Nurowski and Tafel 1992), but no solutions have been found for $L_{,u} \neq 0$.

30.6 Remarks concerning solutions of the different Petrov types

All type D Einstein–Maxwell fields for which both null eigenvectors of the Maxwell field are multiple eigenvectors of the Weyl tensor are known. They depend on at most six arbitrary parameters; for a detailed discussion, see §21.1.2. The charged Kerr–NUT metrics covered by Theorem 30.2 and the charged and twisting C-metric are included here. Besides these solutions, one more type D solution can be generated by means of

Theorem 30.2, namely (Leroy 1978)

$$\begin{aligned} ds^2 &= 2d\zeta d\overline{\zeta}/\rho\overline{\rho} - 2(du + Ld\zeta + \overline{L}d\overline{\zeta}) \\ &\quad \times \left[dr + \tfrac{1}{4}\kappa_0 Q\overline{Q}\rho\overline{\rho}(du + Ld\zeta + \overline{L}d\overline{\zeta})\right], \qquad (30.36) \\ L &= i\Sigma\overline{\zeta}, \quad \overline{Q} = \overline{\zeta}^{-3}e^{ib}, \quad b, \Sigma = \text{const.} \end{aligned}$$

This metric is the twisting generalization of the Einstein–Maxwell field (28.46) in the subcase $m = 0$; the corresponding vacuum solution is flat. For (30.36), only one eigenvector of the Maxwell field coincides with an eigenvector of the Weyl tensor.

Comparing the list of known twisting (and diverging) type D Einstein–Maxwell fields with the exhaustive list of type D twisting vacuum or non-twisting Einstein–Maxwell solutions, one sees that for each non-twisting or vacuum solution a charged and twisting counterpart exists. Any as yet unknown type D Einstein–Maxwell field cannot be generated from vacuum solutions by simply adding a charge in the sense of Theorem 30.2; i.e. they must be radiative. As shown by Debever *et al.* (1989), purely radiative Einstein–Maxwell fields ($\Phi_0 = 0 = \Phi_1$) of type D have zero twist.

Concerning type N (or O) Einstein–Maxwell fields, an inspection of the structure (30.21) of the Weyl tensor shows that $\Psi_2 = \Psi_3 = 0$ implies $\Phi_1^0 = 0 = m + iM$ and $\partial I = 0$. But under these assumptions, the field equation (30.19*b*) yields $\Phi_2^0 = 0$; there is no Maxwell field at all. As all aligned type N (or O) Einstein–Maxwell fields must have $\kappa = \sigma = 0$, see §7.6, we thus obtain

Theorem 30.3 *There are no non-vacuum diverging Einstein–Maxwell fields (aligned case) of Petrov type N or O.*

As a consequence of this theorem, type N Einstein–Maxwell fields are either non-aligned (i.e. the repeated null eigenvector of the Weyl tensor is not an eigenvector of the Maxwell field), see §26.1 and (26.6), or they are aligned and non-diverging, see Chapter 31.

A special type II solution can be obtained by charging a vacuum solution. It is given (in a gauge with $r^0 \neq 0$, cp. §29.1.4) by

$$\begin{aligned} ds^2 &= 2d\zeta d\overline{\zeta}/\rho\overline{\rho} - 2(du + Ld\zeta + \overline{L}d\overline{\zeta})[dr \\ &\qquad + \kappa_0 \Phi_1^0 \overline{\Phi}_1^0 \rho\overline{\rho}(du + Ld\zeta + \overline{L}d\overline{\zeta})], \qquad (30.37) \\ L &= ia\overline{\zeta} + \alpha e^{\overline{\zeta}}, \quad \rho^{-1} = -(r + L_{,\overline{\zeta}}), \quad \Phi_1^0 = e^{\zeta}/2. \end{aligned}$$

It is the only type II solution which is double aligned, i.e. both null eigenvectors of the Maxwell tensor are eigenvectors of the Weyl tensor, only $\boldsymbol{k}$ being a repeated null eigenvector (Leroy 1979). The corresponding vacuum solution is flat.

It can be shown that no *regular* diverging (aligned) type *III* Einstein–Maxwell field exists, the only regular diverging, geodesic and shearfree non-radiating fields being the Kerr–Newman solutions (Lind 1975a).

30.7 Pure radiation fields

30.7.1 The field equations

Pure radiation fields

$$T_{mn} = \Phi^2 k_n k_m, \tag{30.38}$$

$\boldsymbol{k}$ being a geodesic and shearfree multiple eigenvector of the Weyl tensor, are similar to electromagnetic null fields ($\Phi_0 = 0 = \Phi_1$) insofar as (because of $T^{mn}{}_{;n} = 0$, $k_{m;n}k^n = 0$, $k^n{}_{;n} = -(\rho + \overline{\rho})$) the factor Φ^2 has the same ρ-dependence

$$\Phi^2 = n^2(\zeta, \overline{\zeta}, u)\rho\overline{\rho} \tag{30.39}$$

as the corresponding expression $2\Phi_2\overline{\Phi}_2$ in the electromagnetic case. The difference is that there $n^2 = 2P^2h\overline{h}$ is subject to the additional restriction $(\partial - L_{,u})h = 0$. Note that in the Maxwell case a null field is *necessarily* geodesic and shearfree and the metric must be algebraically special (§7.6), whereas here we have to assume these properties.

As a consequence of this similarity, the metric of a pure radiation field has exactly the form (30.14) of an Einstein–Maxwell field, with, of course, $\Phi_1^0 = 0$ (so that n^2 does not explicitly appear in the metric), and the field equations read

$$(3L_{,u} - \partial)(m + \mathrm{i}M) = 0, \tag{30.40}$$

$$M = P^3 \mathrm{Im}\, \partial\overline{\partial}\,\overline{\partial}V = \Sigma K + P^2 \mathrm{Re}\left[\partial\overline{\partial}\Sigma - 2\overline{L}_{,u}\partial\Sigma - \Sigma\partial_u\partial\overline{L}\right] \tag{30.41}$$

and

$$\begin{gathered} P^4(\partial - 2L_{,u} + 2\partial\ln P)\partial I - P^3\left[P^{-3}(m + \mathrm{i}M)\right]_{,u} = \kappa_0 n^2(\zeta, \overline{\zeta}, u)/2, \\ I \equiv \overline{\partial}(\overline{\partial}\ln P - \overline{L}_{,u}) + (\overline{\partial}\ln P - \overline{L}_{,u})^2. \end{gathered} \tag{30.42}$$

As $n^2(\zeta, \overline{\zeta}, u)$ is an arbitrary (positive) function, (30.42) is in fact only the definition of n^2, and not an equation which needs to be integrated. The actual field equations are (30.40)–(30.41), which are the same as in the vacuum case. Accordingly, one can try to generate radiation field solutions from vacuum metrics (possibly flat) by simply changing one or more metric functions in such a way that (30.40)–(30.41) remain satisfied, but (30.42) gives a non-zero n^2. Of course, the direct approach to solving the field equations is also possible. We shall discuss the different approaches now in turn; in most of them a special dependence of the metric functions on u is assumed. For the pure radiation fields of the Kerr–Schild class see §32.4.

Concerning the Petrov types of aligned pure radiation fields, only types II and III are admitted: for type N, $m+\mathrm{i}M$ and ∂I must vanish, but then n^2 vanishes too, and type D is also not permitted (Wils 1990).

30.7.2 Generating pure radiation fields from vacuum by changing P

In dealing with the coupled system (30.40)–(30.41) of the field equations it simplifies the task if we confine ourselves to metrics which satisfy (30.40) identically by $L_{,u} = 0 = M$, $m =$ const. If we start from such a vacuum (or pure radiation) metric $(L^0, P^0, m^0, M^0 = 0)$ and change only the function P by making the ansatz $P = P^0 A(\zeta, \overline{\zeta}, u)$, then the remaining field equation (30.41), i.e. $M = 0$, reduces to a single *linear* partial differential equation for A (Stephani 1979):

Theorem 30.4 *If* (L^0, m^0 = const, M^0 = 0, P^0) *is an algebraically special vacuum solution satisfying* $L^0_{,u} = 0$, *then* (L^0, m^0 = const, M^0 = 0, $P = P^0 A(\zeta, \overline{\zeta}, u)$) *represents a pure radiation field exactly if the real function A obeys*

$$\partial(\Sigma^0 \overline{\partial} A) + \overline{\partial}(\Sigma^0 \partial A) = 0,$$
$$2\mathrm{i}\,\Sigma^0 \equiv (P^0)^2(\overline{\partial} L^0 - \partial \overline{L}^0), \quad \partial \equiv \partial_\zeta - L^0 \partial_u. \tag{30.43}$$

This new solution is twisting if the original one is, and it is non-vacuum if

$$(\partial + 2\partial \ln P)\partial \left[\overline{\partial}\,\overline{\partial} \ln P + (\overline{\partial} \ln P)^2\right] + 3mP_{,u}P^{-4} > 0. \tag{30.44}$$

For $P_{,u} \neq 0$, m can always be chosen so that the inequality (30.44) holds at least for some region of space-time.

If we look for vacuum solutions satisfying the above conditions, we fall back on the two classes discussed in §§29.2.3 and 29.2.5. In both cases not only L^0 but also P^0 and Σ^0 are independent of u.

In the first case ($L_{,u} = 0, \partial_\zeta I \neq 0$) the vacuum solutions to start from are included in (29.46)–(29.50). So far no non-vacuum solutions with positive n^2 have been found.

In the second case ($L_{,u} = 0 = I$), the starting point is the Kerr–Schild class

$$L^0 = l_1(\overline{\zeta})(P^0)^{-2}, \quad P^0 = \alpha\zeta\overline{\zeta} + \beta\zeta + \overline{\beta}\overline{\zeta} + \delta, \quad m^0 = \text{ const}, \quad M^0 = 0, \tag{30.45}$$

of vacuum solutions. Explicit solutions of the resulting differential equation for A have been found in several subcases.

If A is a function only of u, then we find

$$\begin{aligned} &A(u) = a_1 u, \quad L^0 = -\mathrm{i}a^0\overline{\zeta}/(P^0)^2, \quad a^0 \text{ real}, \\ &P^0 = 1 + K\zeta\overline{\zeta}/2, \quad K = 0, \pm 1, \end{aligned} \tag{30.46}$$

and

$$A(u) = a_1 u, \quad L^0 = (b_1\overline{\zeta}^2 + b_2\overline{\zeta} + b_3)^{1/2}, \quad P^0 = 1, \quad b_1 \text{ real}, \tag{30.47}$$

as radiating solutions. In both cases, the intensity of the radiation field is given by

$$\kappa_0 n^2 = 6m^0/u. \tag{30.48}$$

In particular, (30.46) with $K = +1$ is a radiating Kerr metric (asymptotically flat) first given by Kramer (1972); it belongs to the class of Kerr–Schild metrics. If for this solution we make a coordinate transformation $u' = F(u) = a_1 u^2/2$, which transforms P back into P^0, then (30.46) transforms into

$$\begin{aligned} &L = -\mathrm{i}\, a^0(2a_1 u)^{1/2}\overline{\zeta}/(P^0)^2, \quad m = m^0(2a_1 u)^{-3/2}, \\ &P^0 = 1 + \zeta\overline{\zeta}/2, \quad K = 1, \end{aligned} \tag{30.49}$$

which differs from the Kerr metric exactly by a (special) time- (u-)dependence of the parameters m and $a = a^0(2a_1 u)^{1/2}$.

If we assume axial symmetry (Σ^0 and A dependent only on $\zeta\overline{\zeta}$ and u), then we have

$$\begin{aligned} &(\Sigma^0\zeta\overline{\zeta}A')' + \Sigma^0 a^2\zeta\overline{\zeta}(P^0)^{-4}\ddot{A} = 0, \quad A = A(\zeta\overline{\zeta}, u), \\ &L^0 = -\mathrm{i}\, a\overline{\zeta}/(P^0)^2, \quad P^0 = 1 + K\zeta\overline{\zeta}/2. \end{aligned} \tag{30.50}$$

For $K = 0$, the general solution is a superposition (with different α) of

$$\begin{aligned} &A(\zeta\overline{\zeta}, u) = (a_1 \mathrm{e}^{\alpha u} + a_2 \mathrm{e}^{-\alpha u})J_0(a\alpha\zeta\overline{\zeta}) \quad \alpha \neq 0, \\ &A(\zeta\overline{\zeta}, u) = (a_1 u + a_2)(a_3 \ln\zeta\overline{\zeta} + a_4) \quad \alpha = 0, \end{aligned} \tag{30.51}$$

J_0 being a Bessel function; a particular integral is given by $A = (\zeta^2\overline{\zeta}^2 + u^2)^{-1/2}$(Ivanov 1999). For $K = \pm 1$, (30.50) can be separated (see e.g. Kramer and Hähner (1995)), but except for (30.46) only the solution

$$A(\zeta\overline{\zeta}) = a_1 \left[\ln \zeta\overline{\zeta} - 2\ln(1 - K\zeta\overline{\zeta}/2)\right] + a_2 \tag{30.52}$$

has been found.

30.7.3 Generating pure radiation fields from vacuum by changing m

If we let P, L and M stay fixed and change only m,

$$m = m^0 + B(\zeta, \overline{\zeta}, u), \quad B \text{ real}, \tag{30.53}$$

then (30.41) remains valid and the field equations reduce to

$$(3L_{,u} - \partial)B = 0 = (3\overline{L}_{,u} - \overline{\partial})B \tag{30.54}$$

(Hughston 1971). The new solutions would be non-vacuum exactly if $(P^{-3}B)_{,u}$ were non-zero.

For twisting solutions ($\overline{\partial}L - \partial\overline{L} \neq 0$), the integrability condition of the system (30.54) imposes severe restrictions on the vacuum metric. From (30.54) and the commutator relations (29.17) one gets

$$(\partial\overline{\partial} - \overline{\partial}\partial)B = 3B\partial_u(\partial\overline{L} - \overline{\partial}L) = (\overline{\partial}L - \partial\overline{L})\partial_u B, \tag{30.55}$$

i.e. the real function B has the form

$$B(\zeta, \overline{\zeta}, u) = b(\zeta, \overline{\zeta})(\overline{\partial}L - \partial\overline{L})^{-3}, \tag{30.56}$$

and because of (30.54) the function $b(\zeta, \overline{\zeta})$ has to satisfy

$$(\ln b)_{,\zeta} = 3L_{,u} + 3\partial\ln(\overline{\partial}L - \partial\overline{L}), \tag{30.57}$$

the main implication being that the right-hand side must be independent of u.

Radiation fields have been generated from two of the known classes of vacuum solutions. For $L_{,u} = 0$ (and non-zero twist), because of (30.55) and (30.54) B is independent of u and therefore constant; but to get non-zero radiation P must depend on u. The only vacuum solution we gave in this gauge is the Hauser solution (29.72). The corresponding radiation solution is (Stephani 1980)

$$m = \text{ const}, \quad P(\zeta + \overline{\zeta}) = f\left[u/(\zeta + \overline{\zeta})^2\right], \quad L = 2\mathrm{i}(\zeta + \overline{\zeta}). \tag{30.58}$$

For $L_{,u} \neq 0$, the solutions of §29.2.4 have also been used for the generation procedure. Here P is independent of u, and L and Σ are linear in u. If we choose the origin of u so that Σ is proportional to u, then (30.57) implies that L is also proportional to u. It turns out that the only class of pure radiation solutions we can generate is

$$\begin{aligned} L &= \left(\partial_\zeta \ln P - [\zeta + \overline{g}(\overline{\zeta})]^{-1}\right) u, \quad P_{,u} = 0, \quad M = 0, \\ m &= B = bu^{-3}, \quad \kappa_0 n^2 = -6bu^{-4}, \end{aligned} \tag{30.59}$$

where $g(\zeta)$ is a disposable function and b is a negative constant.

30.7.4 Some special classes of pure radiation fields

Pure radiation fields have also been found in some other restricted cases.

The first case is that P and L are independent of u and (therefore) m is linear in u:

$$P_{,u} = 0 = L_{,u}, \quad m^0 = m^0(\zeta, \overline{\zeta}) + au, \quad a = \text{const.} \tag{30.60}$$

For $L_{,u} = 0$, we can use a gauge with

$$L = \mathrm{i}B(\zeta, \overline{\zeta}), \quad B \text{ real}, \tag{30.61}$$

where B is defined only up to a gauge $\hat{B} = B + v$, $v_{,\zeta\overline{\zeta}} = 0$. The field equation (30.40) then reads $(m + \mathrm{i}M)_{,\zeta} = aB_{,\zeta}$ and is integrated by

$$m = m^0 + au, \quad M = aB, \quad m^0, a \text{ const (real)} \tag{30.62}$$

(a constant of integration being incorporated in B), and the last field equation (30.41) reads (with $\Sigma = P^2 B_{,\zeta\overline{\zeta}}$)

$$P^2 \left(P^2 B_{,\zeta\overline{\zeta}}\right)_{,\zeta\overline{\zeta}} + 2P^4 B_{,\zeta\overline{\zeta}} (\ln P)_{,\zeta\overline{\zeta}} = aB, \quad a \neq 0. \tag{30.63}$$

This is a single partial differential equation for the two unknown functions P and B. To find solutions, we can choose one of them appropriately so that we can solve the differential equation then arising for the second function. For example, we can take P from one of the corresponding vacuum solutions of §§29.2.3, 29.2.5, thus trying to generalize these solutions to the pure radiation case; the vacuum solutions (with $M = 0$) are then contained for $a = 0$. We shall do that now for the three forms of P occurring.

If we take P from the vacuum solutions

$$\begin{aligned} P &= 1 + K\zeta\overline{\zeta}/2, \quad K = \pm 1, \quad m^0 + \mathrm{i}M^0 = 2K\overline{\zeta}\,\overline{g}_{,\overline{\zeta}} \\ \mathrm{i}B^0 &= \frac{\overline{h}(\overline{\zeta})}{P} - \frac{h(\zeta)}{P} - \int \frac{g(\zeta)}{P^2} \mathrm{d}\zeta + \int \frac{\overline{g}(\overline{\zeta})}{P^2} \mathrm{d}\overline{\zeta}, \end{aligned} \tag{30.64}$$

(i.e. from (29.60)–(29.61) given in a different gauge), or from the vacuum metrics

$$\begin{aligned} P = 1, \quad m^0 + \mathrm{i}M^0 &= 2\overline{g}(\overline{\zeta})_{,\overline{\zeta}\overline{\zeta}}, \\ \mathrm{i}B^0 &= \zeta\overline{h}(\overline{\zeta}) - \overline{\zeta}h(\zeta) + \tfrac{1}{2}\zeta^2\overline{g}(\overline{\zeta}) - \tfrac{1}{2}\overline{\zeta}^2 g(\zeta), \end{aligned} \tag{30.65}$$

then the field equation (30.63) gives

$$P^2\left(P^2B_{,\zeta\overline{\zeta}}\right)_{,\zeta\overline{\zeta}} + KP^2B_{,\zeta\overline{\zeta}} = aB, \quad a < 0 \tag{30.66}$$

(for $P = 1$, this is exactly (30.35)). Introducing the functions

$$S^{\pm} = aB + \lambda_{\pm}\Sigma, \quad \Sigma = P^2B_{,\zeta\overline{\zeta}}, \quad \lambda_{\pm} = -\tfrac{1}{2}K \pm \sqrt{a + K^2/4}, \tag{30.67}$$

this fourth-order differential equation can be split into

$$P^2S^{\pm}_{,\zeta\overline{\zeta}} = \lambda_{\pm}S^{\pm}. \tag{30.68}$$

Since the operator $P^2\partial_\zeta\partial_{\overline{\zeta}}$ is the Laplacian on the corresponding space of constant curvature, the general solution to (30.68) is known and can be constructed by standard separation techniques leading to spherical harmonics etc. The solutions of this class have been given in the axisymmetric case by Patel (1978) and Akabari *et al.* (1980) (note that the au of their solutions can be set to zero), and the solutions admitting at least a H_2 by Grundland and Tafel (1993). These solutions generalize some type D vacuum metrics.

If we take P from the vacuum solutions (29.46)–(29.50) as

$$P = (\zeta + \overline{\zeta})^{3/2}, \tag{30.69}$$

then we have to solve

$$(\zeta + \overline{\zeta})^3\left[(\zeta + \overline{\zeta})^3B_{,\zeta\overline{\zeta}}\right]_{,\zeta\overline{\zeta}} - 3(\zeta + \overline{\zeta})^4B_{,\zeta\overline{\zeta}} = aB. \tag{30.70}$$

No solution of this equation has been found so far.

The second case is that L is independent of u, and $m + \mathrm{i}M$ is constant,

$$L = \mathrm{i}B(\zeta, \overline{\zeta})_{,\zeta}, \quad m + \mathrm{i}M = \text{ const}, \quad P = P(\zeta, \overline{\zeta}, u). \tag{30.71}$$

The only field equation to be solved then reads, with $\partial = \partial_\zeta - \mathrm{i}B_{,\zeta}\partial_u$,

$$P^2\mathrm{Re}\left[\partial\overline{\partial}(P^2B_{,\zeta\overline{\zeta}})\right] + 2P^4\mathrm{Re}\left[\partial\overline{\partial}\ln P\right] = M = \text{ const}. \tag{30.72}$$

As in the first case, this is a single differential equation for the two real functions P and B, and one can get solutions by suitably prescribing one of them. So one may take e.g.

$$B = (\zeta + \overline{\zeta})^{\sigma}. \tag{30.73}$$

For $M = 0$, and with the ansatz $P = P(u, B)$, one gets (Stephani 1980)

$$4\sigma^2 B^2(P_{,BB} + P_{,uu}) + 4\sigma(2\sigma - 3)BP_{,B} + (\sigma - 2)(\sigma - 3)P = 0. \quad (30.74)$$

Solutions can be found by standard separation methods or by taking $P = B^\nu f(s)$, $s = u/B$, which leads to the hypergeometric equation

$$\begin{aligned} &4\sigma^2(s^2 + 1)f'' + 4\sigma(3 - 2\sigma\nu)sf' \\ &\quad + [4\sigma\nu(\sigma\nu + \sigma - 3) + (\sigma - 2)(\sigma - 3)]\, f = 0. \end{aligned} \quad (30.75)$$

This solution generalizes the Hauser solution (29.72), which is contained for $m = 0$, $\sigma = 2$, $\nu = 3/4$. A large class of solutions to (30.74) has been given by Tafel *et al.* (1991), and solutions with $M = 0$ admitting a Killing vector $\mathrm{i}(\partial_\zeta - \partial_{\overline{\zeta}})$ by Lewandowski *et al.* (1991).

The third case is that all metric functions are independent of u. They have to satisfy $m + \mathrm{i}M = \text{const}$, $L = \mathrm{i}B(\zeta, \overline{\zeta})_{,\zeta}$ and

$$P^2 \left(P^2 B_{,\zeta\overline{\zeta}}\right)_{,\zeta\overline{\zeta}} + 2P^4 B_{,\zeta\overline{\zeta}}(\ln P)_{,\zeta\overline{\zeta}} = M. \quad (30.76)$$

Solutions which admit at least an H_3 of homothetic motions or have $M = 0$ were found by Lewandowski and Nurowski (1990) and Grundland and Tafel (1993) (note that solutions 3.15 and 3.17 of Grundland and Tafel (1993) contain mistakes).

31
Non-diverging solutions (Kundt's class)

31.1 Introduction

In Chapters 27–30 we dealt with those algebraically special solutions for which $\boldsymbol{k}$, the multiple principal null direction of the Weyl tensor, is diverging ($\rho \neq 0$) and shearfree. Here we treat the non-diverging case, i.e. we assume $\rho = -(\Theta + \mathrm{i}\,\omega) = 0$ throughout this chapter. Since physically reasonable energy-momentum tensors have to satisfy the energy condition $T_{ab}k^a k^b \geq 0$ (§5.3), one sees from (6.33), i.e. from $\Theta_{,a}k^a - \omega^2 + \Theta^2 + \sigma\bar{\sigma} = -R_{ab}k^a k^b/2 \leq 0$, that $(\Theta + \mathrm{i}\,\omega) = 0$ implies

$$\sigma = 0 = R_{ab}k^a k^b. \tag{31.1}$$

Thus the non-twisting (and therefore geodesic) and non-expanding null congruence must be shearfree, and $R_{ab}k^a k^b = 0$ implies that the relations

$$R_{ab}k^a k^b = R_{ab}k^a m^b = R_{ab}m^a m^b = 0 \tag{31.2}$$

are satisfied for vacuum, Einstein–Maxwell and pure radiation fields. Hence, by Theorem 7.1, these space-times are algebraically special. Einstein-Maxwell and pure radiation fields are aligned, i.e. they have a common eigendirection $\boldsymbol{k}$ of the Weyl and Ricci tensor. Perfect fluid solutions violate $R_{ab}k^a k^b = 0$ unless $p + \mu = 0$.

31.2 The line element for metrics with $\Theta + \mathrm{i}\omega = 0$

The non-twisting null vector field may be chosen to be a gradient field, and coordinates u and v are then naturally introduced by

$$\boldsymbol{e}_4 = k^i \partial_i = \partial_v, \quad \boldsymbol{\omega}^3 = -k_i \mathrm{d}x^i = \mathrm{d}u. \tag{31.3}$$

As coordinates in the null hypersurfaces u = const we use the affine parameter v and two spacelike coordinates x^1, x^2. With this choice, the line element has the form

$$\mathrm{d}s^2 = g_{AB}\mathrm{d}x^A\mathrm{d}x^B - 2\mathrm{d}u(\mathrm{d}v + m_A\mathrm{d}x^A + H\mathrm{d}u), \quad A, B = 1, 2 \quad (31.4)$$

(Kundt (1961); for a detailed study of these metrics see also Kundt and Trümper (1962)). The spin coefficients ρ and σ are then

$$\begin{aligned}\rho &= -k_{a;b}m^a\overline{m}^b = -\tfrac{1}{2}g_{ab,v}m^a\overline{m}^b = -\tfrac{1}{2}g_{AB,v}m^A\overline{m}^B/2,\\ \sigma &= -k_{a;b}m^a m^b = -\tfrac{1}{2}g_{ab,v}m^a m^b = -\tfrac{1}{2}g_{AB,v}m^A m^B.\end{aligned} \quad (31.5)$$

Hence $\rho = \sigma = 0$ leads to $g_{AB,v} = 0$. Performing a coordinate transformation $x^{A'} = x^{A'}(x^B, u)$, and with complex coordinates $\sqrt{2}\,\zeta = (x^1 + \mathrm{i}\,x^2)$ and $\sqrt{2}\,\overline{\zeta} = (x^1 - \mathrm{i}\,x^2)$, one can write the line element in the form

$$\mathrm{d}s^2 = 2P^{-2}\mathrm{d}\zeta\mathrm{d}\overline{\zeta} - 2\mathrm{d}u\left(\mathrm{d}v + W\mathrm{d}\zeta + \overline{W}\mathrm{d}\overline{\zeta} + H\mathrm{d}u\right), \quad P_{,v} = 0, \quad (31.6)$$

or equivalently as

$$\begin{aligned}&\mathrm{d}s^2 = 2\omega^1\omega^2 - 2\omega^3\omega^4,\\ &\omega^1 = \mathrm{d}\zeta/P = \overline{\omega}^2, \quad \omega^3 = \mathrm{d}u, \quad \omega^4 = \mathrm{d}v + W\mathrm{d}\zeta + \overline{W}\mathrm{d}\overline{\zeta} + H\mathrm{d}u,\end{aligned} \quad (31.7)$$

with real P and H and complex W. This choice of 1-forms is very similar to the choice (27.22), with $L = 0$, in the case of diverging algebraically special metrics.

In order to calculate the Riemann tensor we found it helpful to first perform a null rotation (27.3) and instead of (31.7) to use the basis of 1-forms

$$\omega^1 = \overline{\omega}^2 = \mathrm{d}\zeta/P - P\overline{W}\mathrm{d}u, \quad \omega^3 = \mathrm{d}u, \quad \omega^4 = \mathrm{d}v + (H + P^2W\overline{W})\mathrm{d}u, \quad (31.8)$$

and corresponding tetrad vectors

$$\boldsymbol{e}_1 = \overline{\boldsymbol{e}}_2 = P\partial_\zeta, \;\; \boldsymbol{e}_3 = \partial_u + P^2(\overline{W}\partial_\zeta + W\partial_{\overline{\zeta}}) - (H + P^2W\overline{W})\partial_v, \;\; \boldsymbol{e}_4 = \partial_v. \quad (31.9)$$

The coordinate transformations preserving the form (31.6) of the metric and the associated transformations of the metric functions P, H and W are

$$\begin{aligned}\text{(i)}\quad &\zeta' = f(\zeta, u), \quad P'^2 = P^2 f_{,\zeta}\overline{f}_{,\overline{\zeta}}, \quad W' = W/f_{,\zeta} + \overline{f}_{,u}/(P^2 f_{,\zeta}\overline{f}_{,\overline{\zeta}}),\\ &H' = H - (f_{,u}\overline{f}_{,u}/P^2 + Wf_{,u}\overline{f}_{,\overline{\zeta}} + \overline{W}\overline{f}_{,u}f_{,\zeta})/(f_{,\zeta}\overline{f}_{,\overline{\zeta}}),\end{aligned} \quad (31.10a)$$

(ii) $v' = v + g(\zeta, \overline{\zeta}, u), \quad P' = P, \quad W' = W - g_{,\zeta}, \quad H' = H - g_{,u},$ (31.10b)

(iii) $u' = h(u), \quad v' = v/h_{,u},$

$$P' = P, \; W' = W/h_{,u}, \; H' = (H + vh_{,uu}/h_{,u})/(h_{,u})^2. \tag{31.10c}$$

In order to maintain the form (31.9) of the null tetrad, the effect of (31.10) must be compensated by the following tetrad rotations ($\boldsymbol{e}_2 = \overline{\boldsymbol{e}}_1$)

(i) $\boldsymbol{e}_1' = (\overline{f}_{,\overline{\zeta}}/f_{,\zeta})^{1/2}\boldsymbol{e}_1, \quad \boldsymbol{e}_3' = \boldsymbol{e}_3, \quad \boldsymbol{e}_4' = \boldsymbol{e}_4,$ (31.11a)

(ii) $\boldsymbol{e}_1' = \boldsymbol{e}_1 - Pg_{,\zeta}\,\boldsymbol{e}_4, \quad \boldsymbol{e}_4' = \boldsymbol{e}_4,$

$$\boldsymbol{e}_3' = \boldsymbol{e}_3 - Pg_{,\overline{\zeta}}\,\boldsymbol{e}_1 - Pg_{,\zeta}\,\boldsymbol{e}_2 + P^2 g_{,\zeta}g_{,\overline{\zeta}}\,\boldsymbol{e}_4, \tag{31.11b}$$

(iii) $\boldsymbol{e}_1' = \boldsymbol{e}_1, \quad \boldsymbol{e}_3' = \boldsymbol{e}_3/h_{,u}, \quad \boldsymbol{e}_4' = h_{,u}\,\boldsymbol{e}_4.$ (31.11c)

If $(\ln P)_{,\zeta\overline{\zeta}} = 0$, then one can always transform P to $P = 1$ by means of (31.10a). The condition $W_{,v} = 0$ is invariant under the transformations (31.10), and so characterizes a special subclass of metrics.

The 2-surfaces $u, v =$ const with metric $2\mathrm{d}\zeta\mathrm{d}\overline{\zeta}/P^2$ are called *wave surfaces*. The vector fields $\boldsymbol{e}_1 = P\partial_\zeta$ and $\boldsymbol{e}_2 = P\partial_{\overline{\zeta}}$ are surface-forming, i.e. their commutator is a linear combination of themselves (see (6.12)). They are tangent to the wave surfaces, whereas the vector fields $\boldsymbol{e}_1 = P(\partial_\zeta - W\partial_v)$ and $\boldsymbol{e}_2 = P(\partial_{\overline{\zeta}} - \overline{W}\partial_v)$ associated with the basis (31.7) are not.

The existence of (spacelike) 2-surfaces orthogonal to $\boldsymbol{k}$ implies $\omega = 0$, since

$$0 = k_a(m^b\overline{m}^a{}_{;b} - \overline{m}^b m^a{}_{;b}) = 2k_{a;b}m^{[a}\overline{m}^{b]} = 2\mathrm{i}\,\omega \tag{31.12}$$

(Kundt 1961). Conversely we have seen that $\omega = 0$ implies that 2-surfaces (wave surfaces) orthogonal to $\boldsymbol{k}$ exist.

The space-time geometry uniquely determines the null congruence $\boldsymbol{k}$ and the wave surfaces. Therefore the Gaussian curvature

$$K = 2P^2(\ln P)_{,\zeta\overline{\zeta}} = \Delta(\ln P) \tag{31.13}$$

of the wave surfaces is a space-time invariant.

31.3 The Ricci tensor components

In this section we list the tetrad components of the Ricci tensor with respect to the basis (31.8). As in Chapters 27–30, we shall use the conventions that numerical indices are tetrad indices, and partial derivatives are denoted by a comma.

In the basis (31.8), the independent connection forms are

$$\begin{aligned}
\boldsymbol{\Gamma}_{14} &= \tfrac{1}{2}PW_{,v}\boldsymbol{\omega}^3, \\
\boldsymbol{\Gamma}_{21}+\boldsymbol{\Gamma}_{43} &= \left[(H+P^2W\overline{W})_{,v} - \tfrac{1}{2}(P^2\overline{W})_{,\zeta} + \tfrac{1}{2}(P^2W)_{,\overline{\zeta}}\right]\boldsymbol{\omega}^3 \\
&\quad -(P_{,\zeta} - \tfrac{1}{2}PW_{,v})\boldsymbol{\omega}^1 + (P_{,\overline{\zeta}} + \tfrac{1}{2}P\overline{W}_{,v})\boldsymbol{\omega}^2, \qquad (31.14) \\
\boldsymbol{\Gamma}_{32} &= \left[-\tfrac{1}{2}P^2(\overline{W}_{,\zeta} + W_{,\overline{\zeta}}) + (\ln P)_{,u}\right]\boldsymbol{\omega}^1 - (P^2\overline{W})_{,\overline{\zeta}}\boldsymbol{\omega}^2 \\
&\quad -P(H+P^2W\overline{W})_{,\overline{\zeta}}\boldsymbol{\omega}^3 - \tfrac{1}{2}P\overline{W}_{,v}\boldsymbol{\omega}^4.
\end{aligned}$$

Inserting (31.14) into the second Cartan equations (3.25) we obtain the following expressions for the tetrad components of the Ricci tensor:

$$R_{44}=0, \quad R_{41} = -\tfrac{1}{2}PW_{,vv}, \quad R_{11} = (P^2W_{,v})_{,\zeta} - \tfrac{1}{2}P^2(W_{,v})^2, \qquad (31.15a)$$

$$R_{12} = \Delta\ln P + \tfrac{1}{2}P^2(\overline{W}_{,v\zeta} + W_{,v\overline{\zeta}} - W_{,v}\overline{W}_{,v}), \quad \Delta = 2P^2\partial_\zeta\partial_{\overline{\zeta}}, \qquad (31.15b)$$

$$R_{34} = H_{,vv} - \tfrac{1}{2}P^2(\overline{W}_{,v\zeta} + W_{,v\overline{\zeta}} - 2W_{,v}\overline{W}_{,v}) + P^2(W_{,vv}\overline{W} + W\overline{W}_{,vv}), \qquad (31.15c)$$

$$\begin{aligned}
R_{31} &= P(P^2W)_{,\zeta\overline{\zeta}} - 2P_{,\overline{\zeta}}(P^2W)_{,\zeta} + PH_{,v\zeta} + P_{,u}W_{,v} \\
&\quad -\tfrac{1}{2}PW_{,uv} + \tfrac{1}{2}P\left[(P^2W_{,v})_{,\zeta}\overline{W} + (P^2\overline{W}_{,v})_{,\zeta}W\right] \qquad (31.15d) \\
&\quad +\tfrac{1}{2}P^3\left[(W\overline{W}_{,v})_{,\zeta} - (WW_{,v})_{,\overline{\zeta}}\right] + \tfrac{1}{2}SPW_{,vv} - \mu_{,\zeta}P,
\end{aligned}$$

$$\begin{aligned}
R_{33} &= 2P^2S_{,\zeta\overline{\zeta}} + P^2\overline{W}_{,v}S_{,\zeta} + P^2W_{,v}S_{,\overline{\zeta}} - 2(P^2\overline{W})_{,\overline{\zeta}}(P^2W)_{,\zeta} \\
&\quad -2\left[\mu_{,\zeta}P^2\overline{W} + \mu_{,\overline{\zeta}}P^2W + \mu_{,u} - \mu_{,v}S + \mu S_{,v} + \mu^2\right], \qquad (31.15e)
\end{aligned}$$

with $\mu \equiv \tfrac{1}{2}P^2(\overline{W}_{,\zeta} + W_{,\overline{\zeta}}) - (\ln P)_{,u}$, $\quad S \equiv H + P^2W\overline{W}$.

31.4 The structure of the vacuum and Einstein–Maxwell equations

In this section we shall consider the Einstein–Maxwell equations including, of course, the vacuum field equations as a special case.

As mentioned in the introduction, the existence of a non-expanding and non-twisting null congruence $\boldsymbol{k}$ implies that $\boldsymbol{k}$ is also an eigendirection of the Maxwell tensor, i.e. that the tetrad components $\Phi_0 = F_{41}$ must be zero. With $\Phi_0 = 0$, the Einstein equations are

$$R_{11} = R_{14} = R_{44} = 0, \qquad (31.16a)$$

$$R_{12} = R_{34} = 2\kappa_0\Phi_1\overline{\Phi}_1, \qquad (31.16b)$$

$$R_{31} = 2\kappa_0\Phi_1\overline{\Phi}_2, \quad R_{33} = 2\kappa_0\Phi_2\overline{\Phi}_2, \qquad (31.16c)$$

and the (sourcefree) Maxwell equations (7.22)–(7.25) read

$$\Phi_{1,v} = 0, \qquad \Phi_{1,\zeta} = W_{,v}\Phi_1, \tag{31.17a}$$

$$\Phi_{2,v} = P(\Phi_{1,\overline{\zeta}} + \overline{W}_{,v}\Phi_1), \tag{31.17b}$$

$$P\Phi_{2,\zeta} = \Phi_{1,u} + P^2\left[(\overline{W}\Phi_1)_{,\zeta} + (W\Phi_1)_{,\overline{\zeta}}\right] - 2(\ln P)_{,u}\Phi_1 + P_{,\zeta}\Phi_2. \tag{31.17c}$$

Using $R_{11} = 0$, (31.17a) can be integrated, giving

$$\begin{aligned} \Phi_1 &= \Phi_1^0 = P^4\,(W_{,v})^2\,\overline{F}(\overline{\zeta},u) \quad \text{for} \quad W_{,v} \neq 0, \\ \Phi_1 &= \Phi_1^0 = \overline{F}(\overline{\zeta},u) \quad\quad\quad\quad\; \text{for} \quad W_{,v} = 0. \end{aligned} \tag{31.18}$$

For a *null field* ($\Phi_0 = 0 = \Phi_1$) the general solution of the Maxwell equations (31.17) is

$$\Phi_2 = \Phi_2^0 = P\overline{g}(\overline{\zeta},u). \tag{31.19}$$

We now study the structure of the Einstein–Maxwell equations (31.16)–(31.17). From $R_{14} = 0$ one infers that W is a function linear in v,

$$W = W_{,v}(\zeta,\overline{\zeta},u)v + W^0(\zeta,\overline{\zeta},u). \tag{31.20}$$

The equations

$$\begin{aligned} R_{11} &= (P^2 W_{,v})_{,\zeta} - \tfrac{1}{2}P^2\,(W_{,v})^2 = 0, \quad \Phi_{1,\zeta} = W_{,v}\Phi_1, \\ R_{12} &= \Delta \ln P + \tfrac{1}{2}P^2(\overline{W}_{,v\zeta} + W_{,v\overline{\zeta}} - W_{,v}\overline{W}_{,v}) = 2\kappa_0\Phi_1\overline{\Phi}_1, \end{aligned} \tag{31.21}$$

form a system of simultaneous differential equations for the v-independent functions P, $W_{,v}$ and Φ_1. No general solution is known, even for the vacuum case.

Because P, $W_{,v}$ and Φ_1 do not depend on v, the equation $R_{34} = 2\kappa_0\Phi_1\overline{\Phi}_1$ tells us that H is a quadratic function of v,

$$\begin{aligned} H &= \tfrac{1}{2}H_{,vv}(\zeta,\overline{\zeta},u)v^2 + G^0(\zeta,\overline{\zeta},u)v + H^0(\zeta,\overline{\zeta},u), \\ H_{,vv} &= \tfrac{1}{2}P^2(\overline{W}_{,v\zeta} + W_{,v\overline{\zeta}} - 2W_{,v}\overline{W}_{,v}) + 2\kappa_0\Phi_1\overline{\Phi}_1, \end{aligned} \tag{31.22}$$

and from the Maxwell equation (31.17b) it follows that Φ_2 is linear in v,

$$\Phi_2 = P\left(\Phi_{1,\overline{\zeta}} + \overline{W}_{,v}\Phi_1\right)v + \Phi_2^0(\zeta,\overline{\zeta},u). \tag{31.23}$$

The (complex) field equation $R_{31} = 2\kappa_0\Phi_1\overline{\Phi}_2$ and the Maxwell equation (31.17c) contain parts linear in v and parts independent of v. If the Einstein equations (31.16a)–(31.16b) and the Maxwell equations (31.17a)–(31.17b) are satisfied, then it follows from the Bianchi identity (7.32j) that the v-dependent parts of the equation $R_{31} = 2\kappa_0\Phi_1\overline{\Phi}_2$ and the Maxwell equation (31.17c) are identically satisfied. The v-independent parts of these equations, namely

$$\begin{aligned}R^0_{31} &= P(P^2W^0)_{,\zeta\overline{\zeta}} + P\left[(\ln P)_{,u} - \tfrac{1}{2}P^2\overline{W}^0_{,\zeta} - \tfrac{1}{2}P^2W^0_{,\overline{\zeta}}\right]_{,\zeta}\\ &\quad+\tfrac{1}{2}P\left[(P^2W_{,v})_{,\zeta}\overline{W}^0 + (P^2\overline{W}_{,v})_{,\zeta}W^0\right] - 2P_{,\overline{\zeta}}(P^2W^0)_{,\zeta}\\ &\quad+PG^0_{,\zeta} + \tfrac{1}{2}P^3\left[(W^0\overline{W}_{,v})_{,\zeta} - (W^0W_{,v})_{,\overline{\zeta}}\right] + P_{,u}W_{,v} - \tfrac{1}{2}PW_{,vu}\\ &= 2\kappa_0\Phi_1\overline{\Phi}^0_2,\end{aligned} \tag{31.24a}$$

$$P\Phi^0_{2,\zeta} = \Phi_{1,u} + P^2\left[(\overline{W}^0\Phi_1)_{,\zeta} + (W^0\Phi_1)_{,\overline{\zeta}}\right] - 2(\ln P)_{,u}\Phi_1 + P_{,\zeta}\Phi^0_2, \tag{31.24b}$$

determine the functions W^0, G^0, Φ^0_2 once P, $W_{,v}$ and Φ_1 are known. Equations (31.24) are linear in W^0, G^0, Φ^0_2 and their derivatives.

The last Einstein equation, $R_{33} = 2\kappa_0\Phi_2\overline{\Phi}_2$, contains terms up to second order in v. Provided that the other Einstein–Maxwell equations are already satisfied, the Bianchi identity (7.32k) gives $(R_{33} - 2\kappa_0\Phi_2\overline{\Phi}_2)_{,v} = 0$, i.e. the v-dependent part of the last field equation is identically satisfied. Finally, the v-independent part

$$\begin{aligned}R^0_{33} &= 2P^2H^0_{,\zeta\overline{\zeta}} + P^2(\overline{W}_{,v}H^0)_{,\zeta} + P^2(W_{,v}H^0)_{,\overline{\zeta}} + (\text{known function})\\ &= 2\kappa_0\Phi^0_2\overline{\Phi}^0_2\end{aligned} \tag{31.25}$$

is a linear partial differential equation of the second order for the remaining function H^0 (the 'known function' being obtainable from the expression (31.15e) for R_{33} in terms of functions already determined by the other field equations).

The metric and Maxwell field thus have the form

$$\begin{aligned}\mathrm{d}s^2 &= 2P^{-2}\mathrm{d}\zeta\mathrm{d}\overline{\zeta} - 2\mathrm{d}u\Big\{\mathrm{d}v + \left[vW_{,v} + W^0\right]\mathrm{d}\zeta + \left[v\overline{W}_{,v} + \overline{W}^0\right]\mathrm{d}\overline{\zeta}\\ &\qquad+\left[\tfrac{1}{2}v^2P^2(W_{,v\zeta} + \overline{W}_{,v\overline{\zeta}} - 2W_{,v}\overline{W}_{,v} + 2\kappa_0\Phi^0_1\overline{\Phi}^0_1) + vG^0 + H^0\right]\mathrm{d}u\Big\},\\ \Phi_1 &= \Phi^0_1,\quad \Phi_2 = vP(\Phi^0_{1,\overline{\zeta}} + \overline{W}_{,v}\Phi^0_1) + \Phi^0_2,\end{aligned} \tag{31.26}$$

and the Einstein–Maxwell equations split into the set (31.21) determining P, $W_{,v}$ and Φ^0_1, the set (31.24) determining W^0, G^0 and Φ^0_2 and (31.25) determining H^0. It should be emphasized that the functions W^0, G^0, Φ^0_2 and H^0 do not occur in the field equations (31.21). This remarkable fact gives us the possibility of constructing new solutions:

Theorem 31.1 *From a known solution ('background metric') one can generate other solutions with the same P, $W_{,v}$ and Φ_1, by choosing new functions W^0, G^0, Φ_2^0 satisfying the linear equations* (31.24) *and by then choosing a new function H^0 satisfying the linear equation* (31.25).

To conclude this section we give the tetrad components Ψ_2 and Ψ_3 of the Weyl tensor for Einstein–Maxwell (and pure radiation) fields:

$$\Psi_2 = \tfrac{1}{2}P^2(\overline{W}_{,v\zeta} - \tfrac{1}{2}W_{,v}\overline{W}_{,v}),$$

$$\begin{aligned}\Psi_3 = {} & P\left(\partial_{\overline{\zeta}} + \tfrac{1}{2}\overline{W}_{,v}\right)\left[\tfrac{1}{2}P^2\left(\overline{W}_{,\zeta} + W_{,\overline{\zeta}}\right) - (\ln P)_{,u}\right] \\ & -P(P^2\overline{W})_{,\zeta\overline{\zeta}} + 2P_{,\zeta}(P^2\overline{W})_{,\overline{\zeta}} - \tfrac{1}{2}PW_{,v}(P^2\overline{W})_{,\overline{\zeta}} + \tfrac{1}{2}R_{32}.\end{aligned} \tag{31.27}$$

For $W = 0$, the last expression reduces to

$$\Psi_3 = \tfrac{1}{2}P\left[H_{,v} - (\ln P)_{,u}\right]_{,\overline{\zeta}}. \tag{31.28}$$

31.5 Vacuum solutions

31.5.1 Solutions of types III and N

The vacuum solutions of types *III* and *N* in Kundt's class are completely known (Kundt 1961; for the subcase $W_{,v} = 0$ see also Pandya and Vaidya 1961). The field equation $R_{12} = 0$ and the type *III* condition $\Psi_2 = 0$ give $(\ln P)_{,\zeta\overline{\zeta}} = 0$, so that one can use a coordinate transformation (31.10*a*) to make $P = 1$. The field equation $R_{34} = 0$ determines $H_{,vv}$ as

$$H_{,vv} = -\tfrac{1}{2}W_{,v}\overline{W}_{,v}. \tag{31.29}$$

The field equation $R_{11} = 0$ and the condition $\Psi_2 = 0$ lead to

$$W_{,v} = -2n_{,\zeta}, \quad \overline{n} = n, \quad (\mathrm{e}^n)_{,\zeta\zeta} = 0 = (\mathrm{e}^n)_{,\zeta\overline{\zeta}}. \tag{31.30}$$

Two cases can occur: either $W_{,v}$ vanishes, or it can be transformed to $W_{,v} = -2/(\zeta + \overline{\zeta})$ by the transformations (31.10*a*) and (31.10*c*).

In the case $W_{,v} = 0$, and with $P = 1$, $R_{31} = 0$ gives

$$R_{31} = \left(H_{,v} + \tfrac{1}{2}W_{,\overline{\zeta}} - \tfrac{1}{2}\overline{W}_{,\zeta}\right)_{,\zeta} = 0. \tag{31.31}$$

This equation implies $H_{,v} + \frac{1}{2}(W_{,\overline{\zeta}} - \overline{W}_{,\zeta}) = \overline{f}(\overline{\zeta}, u)$. Because $H_{,v}$ is real, it follows that

$$W_{,\overline{\zeta}} - \overline{W}_{,\zeta} = \overline{f}(\overline{\zeta}, u) - f(\zeta, u), \tag{31.32}$$

which implies

$$W = \int \overline{f}(\overline{\zeta}, u)\mathrm{d}\overline{\zeta} - g_{,\zeta}, \quad g = g(\zeta, \overline{\zeta}, u) \text{ real}, \tag{31.33}$$

and so finally, using the admissible coordinate transformation (31.10*b*), one obtains $W = W(\overline{\zeta}, u)$. Thus the type *III* and *N* solutions with $W_{,v} = 0$ are given by

$$\begin{aligned}
&\mathrm{d}s^2 = 2\mathrm{d}\zeta\,\mathrm{d}\overline{\zeta} - 2\mathrm{d}u(\mathrm{d}v + W\mathrm{d}\zeta + \overline{W}\mathrm{d}\overline{\zeta} + H\mathrm{d}u), \\
&W = W(\overline{\zeta}, u), \quad H = \tfrac{1}{2}\left(W_{,\overline{\zeta}} + \overline{W}_{,\zeta}\right) v + H^0, \\
&H^0_{,\zeta\overline{\zeta}} - \mathrm{Re}\left[W^2_{,\overline{\zeta}} + WW_{,\overline{\zeta}\overline{\zeta}} + W_{,\overline{\zeta}u}\right] = 0,
\end{aligned} \tag{31.34}$$

where $\overline{W}(\zeta, u)$ is a disposable function. In general, these solutions are of Petrov type *III*. They are of type *N* if $\Psi_3 = 0 = W_{,\overline{\zeta}\overline{\zeta}}$ (see (31.27)), but if W is a linear function of $\overline{\zeta}$, we can use the remaining freedom in the coordinate transformation (31.10) to make W zero. The solutions (31.34) with $W = 0$ are the plane-fronted waves studied in detail in §24.5.

The case $W_{,v} = -2/(\zeta + \overline{\zeta})$ can be solved in a similar way. With $P = 1$ and $W = -2v/(\zeta + \overline{\zeta}) + W^0$ the equation $R^0_{31} = 0$ reads

$$\left[\tfrac{1}{2}\left(W^0_{,\overline{\zeta}} - \overline{W}^0_{,\zeta}\right) + G^0 - \frac{W^0 + \overline{W}^0}{\zeta + \overline{\zeta}}\right]_{,\zeta} = -\frac{1}{\zeta + \overline{\zeta}}\left(W^0_{,\overline{\zeta}} + \overline{W}^0_{,\zeta}\right). \tag{31.35}$$

By means of a coordinate transformation (31.10*b*) which induces

$$W^{0\prime} = W^0 - g_{,\zeta} + 2g/(\zeta + \overline{\zeta}) \tag{31.36}$$

one can arrange that the right-hand side of (31.35) vanishes. Then $(W^0_{,\overline{\zeta}} - \overline{W}^0_{,\zeta})$ is the imaginary part of an analytic function. One obtains

$$W^0 = \int \overline{f}(\overline{\zeta}, u)\mathrm{d}\overline{\zeta} - \overline{\zeta} f(\zeta, u) + h(\zeta, u), \tag{31.37}$$

and the remaining freedom in the transformation (31.10*b*) can be used to put $W^0 = W^0(\zeta, u)$. Thus the class *III* and *N* solutions with $W_{,v} \neq 0$ are given by

$$\begin{aligned}
&\mathrm{d}s^2 = 2\mathrm{d}\zeta\mathrm{d}\overline{\zeta} - 2\mathrm{d}u(\mathrm{d}v + W\mathrm{d}\zeta + \overline{W}\mathrm{d}\overline{\zeta} + H\mathrm{d}u), \\
&W = W^0(\zeta, u) - \frac{2v}{\zeta + \overline{\zeta}}, \quad H = H^0 + v\frac{W^0 + \overline{W}^0}{\zeta + \overline{\zeta}} - \frac{v^2}{(\zeta + \overline{\zeta})^2}, \\
&(\zeta + \overline{\zeta})\left[(H^0 + W^0\overline{W}^0)/(\zeta + \overline{\zeta})\right]_{,\zeta\overline{\zeta}} = W^0_{,\zeta}\overline{W}^0_{,\overline{\zeta}}
\end{aligned} \tag{31.38}$$

(Cahen and Spelkens 1967), where the function $W^0(\zeta, u)$ is disposable. In general, these solutions are of Petrov type *III*; they are of type *N* if $\Psi_3 = 0 = W^0_{,\zeta}$. In this case, W^0 can be transformed to zero.

Theorem 31.2 *The classes* (31.34) *and* (31.38) *exhaust all type III and N vacuum solutions with* $\rho = 0$. *They are characterized by* $\Delta \ln P = 0$, *i.e. by plane wave surfaces.*

These two classes were originally given by Kundt (1961), who chose W^0 to be real. For the class (31.38), W^0 is then given by

$$W^0 = \overline{W}^0 = \varphi(\zeta + \overline{\zeta})^{-1}, \quad \varphi_{,\zeta\overline{\zeta}} = 0. \tag{31.39}$$

The type *III* solution,

$$\mathrm{d}s^2 = x(v - \mathrm{e}^x)\mathrm{d}u^2 - 2\mathrm{d}u\mathrm{d}v + \mathrm{e}^x(\mathrm{d}x^2 + \mathrm{e}^{-2u}\mathrm{d}z^2), \tag{31.40}$$

given by Petrov (1962), and its generalizations (Harris and Zund 1975, Kaigorodov 1967) are particular members of the class (31.34) written in other coordinate systems. The metric (31.40) admits a non-Abelian group G_2. Solutions that admit one non-null Killing vector were found by Hoenselaers (1978a), see also McIntosh and Arianrhod (1990b).

31.5.2 Solutions of types D and II

The *type D* vacuum solutions are completely known (Kinnersley 1969b, Carter 1968b). The non-diverging ($\rho = 0$) solutions admit a group of motions G_4 on T_3 and are given in §§13.3.1 and 18.6.2. In our present notation this class can be written as

$$\begin{aligned}
&\mathrm{d}s^2 = 2\mathrm{d}\zeta\mathrm{d}\overline{\zeta}/P^2 - 2\mathrm{d}u(\mathrm{d}v + W\mathrm{d}\zeta + \overline{W}\mathrm{d}\overline{\zeta} + H\mathrm{d}u),\\
&\sqrt{2}\zeta = x + \mathrm{i}y, \quad P^2\mathrm{d}z = \mathrm{d}x, \quad P^2 = \frac{z^2 + l^2}{k(z^2 - l^2) + 2mz},\\
&W = -\frac{\sqrt{2}v}{P^2(z - \mathrm{i}l)}, \quad H = -\left[\frac{k}{2(z^2 + l^2)} + \frac{2l^2}{P^2(z^2 + l^2)^2}\right]v^2,
\end{aligned} \tag{31.41}$$

with constant m, l and $k = 0, \pm 1$. Note that the solutions can only be given implicitly ($P^2\mathrm{d}z = \mathrm{d}x$) if one sticks to an isotropic form of the wave surface metric, and that only the terms with the highest possible powers of v occur in the metric functions W and H.

The *type II* solutions are only partially known; some unrecognized ones may be among solutions found by other methods. If a vacuum field admits a null Killing vector $\boldsymbol{\xi} = \mathrm{e}^{-\sigma}\boldsymbol{k}$, then the space-time clearly belongs to

Kundt's class (see Chapter 24). In virtue of the Killing equations for the metric (31.6) the real function σ has to obey the relations

$$\sigma_{,\zeta} = W_{,v}, \quad \sigma_{,u} = H_{,v}, \quad \sigma_{,v} = 0. \tag{31.42}$$

The corresponding integrability conditions restrict the metric functions in such a way that one can construct the general solution of Einstein's field equations. We have treated these space-times in §24.4.

If there is a non-null Killing vector ∂_φ, then (in adapted coordinates, with $v = \widetilde{v}e^{-2U}$ and r and φ related to the real and imaginary part of ζ) the metric can be written as

$$\begin{aligned} ds^2 = {} & 2e^{2U}(d\varphi + w_2 du)^2 - e^{-2U} du d\widetilde{v} \\ & +2e^{-2U}\left[(dr + w_1\widetilde{v}du)^2 + (m_1\widetilde{v}^2 + m_2\widetilde{v} + m_3)du^2\right], \end{aligned} \tag{31.43}$$

where all metric functions are functions of u and r and obey

$$\begin{aligned} & w_{1,rr} + 6w_{1,r}w_1 + 4w_1^3 = 0, \\ & U_{,r}^2 = -w_1^2 - w_{1,r}, \quad w_{2,rr} + 4U_{,r}w_{2,r} = 0, \end{aligned} \tag{31.44a}$$

$$\begin{aligned} & m_1 = (w_{1,r} - 2w_1^2)/4, \quad m_{2,r} = w_{1,u} - 2U_{,r}U_{,u}, \\ & (m_{3,r} - 2m_3 w_1)_{,r} = -4U_{,u}^2 - e^{-4U}w_{2,r}^2. \end{aligned} \tag{31.44b}$$

Equations (31.44*a*) can be solved and give

$$\begin{aligned} 2U &= \ln|(r - h_1 - h_2)/(r - h_1 + h_2)| + h_3, \\ w_1 &= (r - h_1)/\left[-h_2^2 + (r - h_1)^2\right], \\ w_2 &= h_4\left[r + 4h_2 \ln|r - h_1 - h_2|\right] + 4h_2^2/(h_1 + h_2 - r), \end{aligned} \tag{31.45}$$

where the h_i are arbitrary functions of u, and (31.44*b*) then gives the m_i (Hoenselaers 1978a, McIntosh and Arianrhod 1990b). The subcases $w_1 = 1/(r - h_1)$ and $w_1 = 0$ are of Petrov types *III* and *N* (*pp*-waves), respectively.

If an algebraically special vacuum field admits a hypersurface-normal spacelike Killing vector orthogonal to $\boldsymbol{k}$, then $\boldsymbol{k}$ necessarily has zero twist. When, in addition, the expansion of $\boldsymbol{k}$ vanishes, the solutions are of Kundt's class (in general, type *II*); they are the subcase $w_2 = 0$ of (31.45) (Kramer and Neugebauer 1968a).

Using Theorem 31.1, one can generate type *II* solutions from the type *D* solution (31.41). If one changes only the metric function H, to $H + H^0$ say, then H^0 has to satisfy the equation

$$2H^0_{,\zeta\bar{\zeta}} + (\overline{W}_{,v}H^0)_{,\zeta} + (W_{,v}H^0)_{,\bar{\zeta}} = 0 \tag{31.46}$$

($W_{,v}$ as in (31.41)). Starting from (31.41) with parameters $k = 0 = l$, $m = 1$, i.e. from (24.37*a*), the vacuum solutions with a null Killing vector, satisfying (24.37*b*) with $\Psi = 0$ and $H^0 = Mx$, can be generated this way. They contain the van Stockum class (20.32) and the solution with an additional spacelike Killing vector given by Bampi and Cianci (1979), see also Joly *et al.* (1992). A different example is a solution with two spacelike Killing vectors also due to Bampi and Cianci (1979). It is likely that Theorem 31.1 also leads to other solutions not contained in previously known classes.

The vacuum metrics (17.30) admitting one Killing vector also belong here.

31.6 Einstein–Maxwell null fields and pure radiation fields

For Einstein–Maxwell null fields (aligned case, $\Phi_0 = 0 = \Phi_1$) and for pure radiation fields the field equations have the same form as in the vacuum case, except that $R^0_{33} = 0$ has to be replaced by

$$R^0_{33} = \kappa_0 \Phi^2(\zeta, \overline{\zeta}, u), \tag{31.47}$$

cp. (31.25). Φ^2 is positive for pure radiation, and for Einstein–Maxwell fields it has the structure

$$\Phi^2 = 2\Phi_2\overline{\Phi}_2 = 2P^2 g(\zeta, u)\overline{g}(\overline{\zeta}, u), \tag{31.48}$$

see (31.19). (With the aid of a coordinate transformation (31.10*a*) one can set $\Phi^2 = P^2$.) The function H^0, which occurs in (31.47) but in no other field equation, is essentially disposable for pure radiation solutions so long as $\Phi^2 > 0$. For Einstein–Maxwell fields, H^0 must be chosen so that (31.48) holds. Some examples of Einstein–Maxwell null fields are given in Wyman and Trollope (1965).

Pure radiation solutions with one non-null Killing vector can be obtained from the vacuum solutions (31.44)–(31.45) by adding a function $g(r, u)$ to m_3; to obtain Einstein–Maxwell null fields, conditions (Maxwell equations) have to be imposed on g (Hoenselaers and Skea 1989, McIntosh and Arianrhod 1990b).

No Einstein–Maxwell null fields of type D exist (Van den Bergh 1989). A type D pure radiation field can be generated from the type D vacuum solution (31.41) by adding a suitable H^0 in the sense of Theorem 31.1 (Wils and Van den Bergh 1990): one has to take (31.41) with $l = 0$ and $H^0(\zeta, \overline{\zeta}, u) = A(u)z + B(u)z^2$, which leads to $\kappa_0\Phi^2 = 2mAz^{-2}$.

All pure radiation fields of Petrov type III or more special can be obtained from the vacuum solutions (31.34) and (31.38) by omitting the differential equation for H^0. The *pp*-waves (§24.5) are well-known examples. The conformally flat Einstein–Maxwell and pure radiation solutions are given in §37.5.3.

The Einstein–Maxwell fields of Kundt's class which are also Kerr–Schild metrics (§32.3) are of Petrov type N, and the electromagnetic field is null (Debney 1973, 1974).

31.7 Einstein–Maxwell non-null fields

The non-diverging ($\rho = 0$) *type N (and type O)* Einstein–Maxwell non-null fields (aligned case) are completely known (Cahen and Leroy 1965, 1966, Szekeres 1966b). This class of solutions is given by setting

$$\begin{aligned}
&2P^2(\ln P)_{,\zeta\overline{\zeta}} = K = \text{ const } (\neq 0), \quad H = \tfrac{1}{2}Kv^2 + G^0(u)v + H^0, \\
&W = -F(\zeta,\overline{\zeta},u)_{,\zeta}, \quad F_{,\zeta\overline{\zeta}} + KFP^{-2} = 0, \quad K(F+\overline{F}) = 2(\ln P)_{,u}, \quad (31.49) \\
&\Phi_0 = 0 = \Phi_2, \quad \sqrt{2\kappa_0}\,\Phi_1 = \sqrt{K}, \quad H^0_{,\zeta\overline{\zeta}} + K^2P^{-2}F\overline{F} = KW\overline{W},
\end{aligned}$$

in the metric (31.6). The only conformally flat solution (the Bertotti–Robinson solution, cp. (12.16) and (35.35)),

$$\mathrm{d}s^2 = 2\mathrm{d}\zeta\mathrm{d}\overline{\zeta}(1 + K\zeta\overline{\zeta}/2)^{-2} - 2\mathrm{d}u\mathrm{d}v - Kv^2\mathrm{d}u^2 \quad (31.50)$$

is contained in (31.49) as the special case $F = G^0 = H^0 = 0$.

For type III and $W_{,v} \neq 0$, by substituting P^2 from the field equation $R_{11} = 0$ into the type III condition $\Psi_2 = 0$, one obtains

$$[\ln(W_{,v}P^2/\overline{W}_{,v})]_{,\zeta} = 0. \quad (31.51)$$

This equation implies

$$P^2W_{,v} = \overline{W}_{,v}\overline{f}(\overline{\zeta},u) = W_{,v}\overline{f}(\overline{\zeta},u)f(\zeta,u)P^{-2} \quad \Rightarrow \quad (\ln P)_{,\zeta\overline{\zeta}} = 0. \quad (31.52)$$

Then the field equation $R_{12} = 2\kappa_0\Phi_1\overline{\Phi}_1$ would lead to the contradiction $\Phi_1 = 0$:

Theorem 31.3 *In the non-null case* ($\Phi_1 \neq 0$)*, there are no type III Einstein–Maxwell fields with* $W_{,v} \neq 0$.

For type III and $W_{,v} = 0$, one gets (from $R_{12} = 2\kappa_0\Phi_1\overline{\Phi}_1$ and $\Phi_1 = \overline{F}(\overline{\zeta},u)$) the differential equation (Liouville equation)

$$P^2(\ln P)_{,\zeta\overline{\zeta}} = \kappa_0\left|F(\zeta,u)\right|^2, \quad (31.53)$$

which has the general solution

$$P^2 = \kappa_0 |F|^2 (1 + f\overline{f})^2 (f_{,\zeta}\overline{f}_{,\overline{\zeta}})^{-1}, \quad f = f(\zeta, u), \tag{31.54}$$

containing an arbitrary function $f(\zeta, u)$.

Hacyan and Plebański (1975) investigated the subcase $W = 0$. The functions Φ_2^0, G^0, H^0, in

$$\Phi_1 = \overline{F}(\overline{\zeta}, u), \quad \Phi_2 = P\overline{F}_{,\overline{\zeta}}v + \Phi_2^0, \quad H = \kappa_0 F\overline{F}v^2 + G^0 v + H^0, \tag{31.55}$$

with P given by (31.54), are then subject to the rest of the Einstein–Maxwell equations, i.e. to

$$\begin{aligned} &(\Phi_2^0/P)_{,\zeta} = \left(\overline{F}P^{-2}\right)_{,u}, \quad [G^0 + (\ln P)_{,u}]_{,\zeta} = 2\kappa_0 \overline{F}\,\overline{\Phi}_2^0/P, \\ &P^2 H^0_{,\zeta\overline{\zeta}} + G^0(\ln P)_{,u} + (\ln P)_{,uu} - (\ln P)_{,u}^2 = \kappa_0 \Phi_2^0 \overline{\Phi}_2^0. \end{aligned} \tag{31.56}$$

An obvious particular solution is given by

$$\begin{aligned} &ds^2 = 2d\zeta d\overline{\zeta}(1 + f\overline{f})^{-2} - 2du[dv + (f'\overline{f}'v^2 + H^0)du], \quad f = f(\zeta), \\ &\sqrt{\kappa_0}\,\Phi_1 = \overline{f}' = \sqrt{\kappa_0}\,\overline{F}, \quad \sqrt{\kappa_0}\,\Phi_2 = (1 + f\overline{f})\overline{f}''v, \quad H^0_{,\zeta\overline{\zeta}} = 0. \end{aligned} \tag{31.57}$$

From (31.28), this metric is of Petrov type *III* if $f'' \neq 0$. The electromagnetic field is determined up to a constant duality rotation. Other particular solutions of (31.56) are given in Hacyan and Plebański (1975).

Solutions with $W \neq 0$, $P_{,u} = 0 = G^0 = \Phi_2$ and constant $\kappa_0 \Phi_1 \overline{\Phi}_1$ (and cosmological constant Λ) have been constructed by Khlebnikov (1986).

It is a general feature of the type *III* solutions (aligned case) that the second eigendirection of the electromagnetic non-null field is not parallel to the single principal null direction of the Weyl tensor.

A class of *type D* Einstein–Maxwell fields can be obtained from the general type *D* vacuum class (31.41) by simply modifying the function P^2 to

$$P^2 = (z^2 + l^2)[k(z^2 - l^2) + 2mz - e^2]^{-1}, \tag{31.58}$$

the rest of the metric remaining unchanged. The tetrad components of the electromagnetic field tensor are then

$$\Phi_0 = 0, \quad \sqrt{2\kappa_0}\,\Phi_1 = \frac{e}{(z - \mathrm{i}\,l)^2}, \quad \sqrt{\kappa_0}\,\Phi_2 = \frac{-2ezv}{P(z - \mathrm{i}\,l)^2(z^2 + l^2)}. \tag{31.59}$$

The two null eigendirections of the Maxwell field coincide with the eigendirections of the type *D* Weyl tensor. Like their uncharged counterparts

(31.41), these solutions admit a group of motions G_4 on T_3 (cp. (13.48)). Unlike the vacuum case this class does not exhaust all type D solutions: it does not, for instance, contain an Einstein–Maxwell field given by Kowalczyński and Plebański (1977) in the form

$$\begin{aligned} \mathrm{d}s^2 &= 2x^{-2}[(\mathrm{d}x/A)^2 + A^2\mathrm{d}y^2 + (\mathrm{d}z/B)^2 - B^2\mathrm{d}t^2], \\ A^2 &= ax^2 + cx^3 - 2e^2x^4, \quad B^2 = b - az^2, \quad a, b, c, e = \text{const.} \end{aligned} \tag{31.60}$$

The Einstein–Maxwell type D solutions (double aligned) of Kundt's class have been determined by Plebański (1979). They are given by

$$\begin{aligned} \mathrm{d}s^2 &= \frac{x^2+l^2}{K(x)}\mathrm{d}x^2 + \frac{K(x)}{x^2+l^2}\left[\mathrm{d}\sigma + \frac{v\mathrm{d}u - u\mathrm{d}v}{1-\varepsilon uv}\right] + 2\frac{(x^2+l^2)}{(1-\varepsilon uv)^2}\mathrm{d}u\mathrm{d}v, \\ K(x) &= 2nx - (e^2+g^2) - 2\varepsilon(x^2 - l^2), \\ \Phi_{11} &= (e^2+g^2)/2(x^2+l^2)^2, \quad e, g, l, \varepsilon, n = \text{const}, \end{aligned} \tag{31.61}$$

and contain the solutions (24.21)–(24.22) – with $\Lambda = 0$ – with a group G_2 on null orbits.

As in the vacuum case, Theorem 31.1 can be used to generate new solutions from known ones.

31.8 Solutions including a cosmological constant Λ

If we want to include (add) an energy-momentum tensor $-\Lambda g_{ik}$, cp. (5.4), then we have to replace the field equations (31.16b) by

$$R_{12} = 2\kappa_0\Phi_1\overline{\Phi}_1 + \Lambda, \quad R_{34} = 2\kappa_0\Phi_1\overline{\Phi}_1 - \Lambda, \tag{31.62}$$

the rest of the field equations remaining unchanged. No general theorem saying how to incorporate Λ into a vacuum or Einstein–Maxwell field is available, but solutions have been found in a number of subcases.

Starting from type III and N vacuum solutions, one has $\Delta \ln P = \Lambda$, i.e. $P = 1 + \Lambda\zeta\overline{\zeta}/2$ instead of $P = 1$, and $H_{,vv} = -\frac{1}{2}P^2W_{,v}\overline{W}_{,v} - \Lambda$ instead of (31.29). The type N solutions falling into this class have been given by Ozsváth *et al.* (1985) (see also García D. and Plebański (1981) and Bičák and Podolsky (1999)), and the solutions with $W_{,v} = 0$ by Lewandowski (1992).

Metrics for which the multiple null eigenvector $\boldsymbol{k}$ is recurrent, $k_{a;b} = k_a p_b$, have been studied by Leroy and McLenaghan (1973).

Some special type II solutions with *constant* $\Phi_1\overline{\Phi}_1$ have been constructed by García D. and Alvarez C. (1984) and Khlebnikov (1986).

The hypersurface-homogeneous Einstein spaces of Kundt's class have been found by MacCallum and Siklos (1992).

Type D Einstein–Maxwell fields (double aligned) with Λ can be obtained from (31.61) by simply adding the term $-\Lambda(x^4/3 + 2l^2x^2 - l^4)$ in the expression for $K(x)$ (Plebański 1979).

32
Kerr–Schild metrics

32.1 General properties of Kerr–Schild metrics

32.1.1 The origin of the Kerr–Schild–Trautman ansatz

In general relativity, the field equations are often simplified by considering null vector fields. One important example is the Kerr–Schild metrical ansatz (Kerr and Schild 1965a, 1965b), which is given by

$$g_{ab} = \eta_{ab} - 2Sk_a k_b, \tag{32.1}$$

where η_{ab} is the Minkowski metric, S is a scalar function and k_a is a null vector with respect to both metrics g_{ab} and η_{ab}, so that we have

$$g_{ab}k^a k^b = \eta_{ab}k^a k^b = 0, \quad g^{ab} = \eta^{ab} + 2Sk^a k^b. \tag{32.2}$$

The ansatz (32.1)–(32.2) was first studied by Trautman (1962). His idea was that a gravitational wave should have the ability to propagate information, and that this can be achieved if both the covariant and the contravariant components of the metric tensor depend linearly on the same function S of the coordinates.

32.1.2 The Ricci tensor, Riemann tensor and Petrov type

Kerr–Schild metrics have been studied by several authors, using either the Newman–Penrose formalism or coordinate methods. The calculations are greatly simplified by the fact the 4-vector $\boldsymbol{k}$ is null.

First we note that the Christoffel symbols and the determinant of the metric tensor (32.1)–(32.2) satisfy the conditions

$$\Gamma^c_{ef}k^e k^f = 0, \quad \Gamma^c_{ef}k_c k^f = 0, \quad (-g)^{1/2} = 1. \tag{32.3}$$

These relations imply that $\boldsymbol{k}$ has the following properties:

$$k_{a;c}k^c = k_{a,c}k^c, \quad k^a{}_{;c}k^c = k^a{}_{,c}k^c, \tag{32.4}$$

$$\Gamma^c_{ab}k_c = (Sk_ak_b)_{;c}k^c; \quad \Gamma^c_{ab}k^a = -(Sk^ck_b)_{;a}k^a. \tag{32.5}$$

(A semicolon denotes the covariant derivative with respect to g_{ab}.) $\boldsymbol{k}$ is geodesic with respect to g_{ab} if and only if it is geodesic with respect to the flat metric η_{ab}, and it has the same expansion with respect to both metrics,

$$2\Theta = k^c{}_{;c} = k^c{}_{,c}. \tag{32.6}$$

The definition of the Ricci tensor together with (32.3)–(32.5) lead to the equation

$$R_{bd}k^bk^d = 2Sg^{bd}(k_{d;a}k^a)(k_{b;c}k^c) = \kappa_0 T_{bd}k^bk^d. \tag{32.7}$$

This gives rise to

Theorem 32.1 *The null vector of a Kerr–Schild metric is geodesic if and only if the energy-momentum tensor obeys the condition*

$$T_{ab}k^bk^d = 0. \tag{32.8}$$

We proceed further with energy-momentum tensors of the type (32.8). The geodesic null vector $\boldsymbol{k}$, $k^a{}_{;b}k^b = 0$, has the same twist and shear with respect to both metrics,

$$2\omega^2 = k_{[a;b]}k^{a;b} = k_{[a,b]}k^{a,b}; \quad 2\Theta^2 + 2\sigma\bar{\sigma} = k_{(a;b)}k^{a;b} = k_{(a,b)}k^{a,b}. \tag{32.9}$$

The Ricci tensor has the simple structure ($D \equiv k^i\partial_i$)

$$R_{bd} = (Sk_bk_d)^{;a}{}_{;a} - (Sk^ak_d)_{;ab} - (Sk^ak_b)_{;ad} + 2S(D^2S)k_bk_d. \tag{32.10}$$

It obeys the eigenvalue equation

$$R_{cd}k^d = -[4\omega^2 S + (k^aS_{,b}k^b)_{;a}]k_c. \tag{32.11}$$

By means of the field equations we are led to

Theorem 32.2 *The geodesic null vector* $\mathbf{k}$ *of a Kerr–Schild metric is an eigenvector of the energy-momentum tensor.*

Further we get, from the definition of the Riemann tensor,

$$k^ck^aR_{abcd} = (D^2S)k_bk_d. \tag{32.12}$$

A straightforward calculation leads to the main result of this section:

Theorem 32.3 *The geodesic null vector* $\mathbf{k}$ *of a Kerr–Schild space-time obeying* (32.8) *is a multiple principal null direction of the Weyl tensor;*

thus the space-time is algebraically special,

$$T_{ab}k^a k^b = 0 \quad \Longleftrightarrow \quad k^a{}_{;b}k^b = 0 \quad \Longleftrightarrow \quad k^c k^a C_{abcd} = H k_b k_d, \quad (32.13)$$

with $H = D^2 S - R/6$ (Gürses and Gürsey 1975).

32.1.3 Field equations and the energy-momentum tensor

The field equations of a Kerr–Schild space-time possessing a geodesic null vector $\boldsymbol{k}$ take the 'linear' form (Gürses and Gürsey 1975)

$$\kappa_0 T^d{}_b = \tfrac{1}{2}\eta_{cb}(\eta^{ae}g^{cd} - \eta^{ca}g^{de} - \eta^{da}g^{ce} + \eta^{cd}g^{ae})_{,ae}. \quad (32.14)$$

This result follows from (32.4), (32.10). Every Kerr–Schild solution is a solution of the linear field equations (32.14). The reverse is not true, because the Kerr–Schild conditions (32.1)–(32.2) have to be fulfilled.

In terms of a complex null tetrad $\{\boldsymbol{e}_a\} = (\boldsymbol{m}, \overline{\boldsymbol{m}}, \boldsymbol{l}, \boldsymbol{k})$ the tetrad components S_{44} and S_{41} of the traceless part of the Ricci tensor vanish because of Theorem 32.2. Usually one considers Kerr–Schild space-times satisfying the additional restriction

$$S_{ab}m^a m^b = S_{11} = 0. \quad (32.15)$$

With this assumption, among the energy-momentum tensor types considered in this book only

$$\kappa_0 T_{ab} = -\tfrac{1}{4}R g_{ab} + \lambda_1 k_a k_b + \lambda_2(m_a \overline{m}_b + \overline{m}_a m_b + k_a l_b + k_b l_a) \quad (32.16)$$

survives. The form (32.16) includes electromagnetic non-null fields ($\lambda_1 = R = 0$) and electromagnetic null fields and pure radiation fields ($\lambda_2 = R = 0$). Perfect fluid distributions cannot occur.

32.1.4 A geometrical interpretation of the Kerr–Schild ansatz

Newman and Unti (1963) introduced a coordinate system $\{x^a\}$ attached to an arbitrary particle world line $y^a(u)$ in flat space-time. Kinnersley and Walker (1970), and Bonnor and Vaidya (1972) have used this coordinate system to find accelerated particle solutions in general relativity. Here we confine the investigation to Kerr–Schild metrics with a geodesic, shearfree and twistfree null congruence, following closely the paper of Bonnor and Vaidya (1972).

Let $y^a(u)$ be a particle world line in Minkowski space and $\lambda^a \equiv \mathrm{d}y^a/\mathrm{d}u$ be its unit tangent vector; u is the proper time of the particle. We can extend the definition of u off the world line by drawing the forward null cone at $Q(y^a)$,

$$\eta_{ab}(x^a - y^a)(x^b - y^b) = 0, \quad (32.17)$$

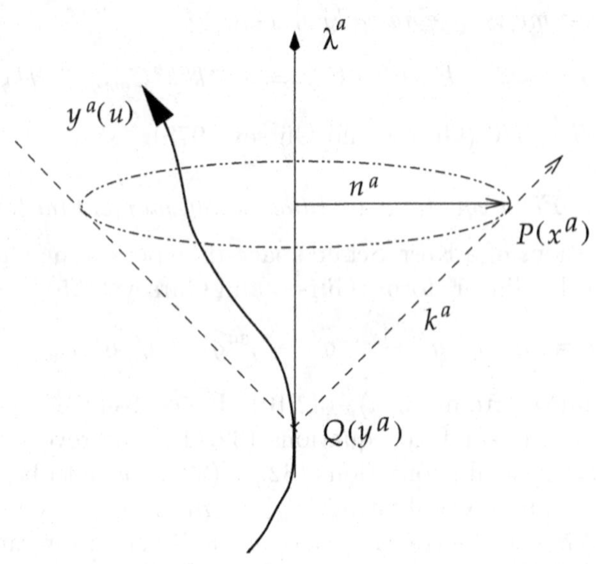

Fig. 32.1. A geodesic and shearfree null vector $\boldsymbol{k}$ and an arbitrary particle world line in flat space-time

and assigning the corresponding value of u to all points $P(x^a)$ on this null cone, see Fig. 32.1. Similarly, we can extend the definition of λ^a by $\lambda^a(P) = \lambda^a(Q)$. We identify the null vector

$$k^a = r^{-1}(x^a - y^a) \tag{32.18}$$

with the null vector in the Kerr–Schild ansatz (32.1)–(32.2) and get

$$g^{ab} = \eta^{ab} + 2Sr^{-2}(x^a - y^a)(x^b - y^b). \tag{32.19}$$

Here we have introduced the retarded distance $r \equiv \lambda_a(x^a - y^a)$. The null congruence given by (32.18) is shearfree, geodesic and twistfree with respect to both the Minkowski metric η_{ab} and the Kerr–Schild metric g_{ab}. In order to find the energy momentum tensor of the Kerr–Schild metric (32.19), we need the identities $k_{a,b} = r^{-1}[\eta_{ab} - k_a\lambda_b - k_b\lambda_a - k_a k_b(1 + rk_c\dot{\lambda}^c)]$ and $r_{,b} = k_b(1 + rk^c\dot{\lambda}_c) + \lambda_b$. (The dot denotes $\mathrm{d}/\mathrm{d}u$ and $k_c\dot{\lambda}^c$ may be interpreted as the acceleration of the particle at Q in the direction n^a, see Fig. 32.1.)

The particular scalar function $S = m(u)r^{-1} - e^2(u)r^{-2}/2$ leads by means of (32.14)–(32.19) to the energy-momentum tensor

$$\begin{aligned} \kappa_0 T_b{}^d &= -\frac{e^2}{r^4} g_b^d + \frac{2e^2}{r^4}(k^d\lambda_b + k_b\lambda^d) + \Phi^2 k^d k_b, \\ \Phi^2 &\equiv \frac{2e^2}{r^4} + \frac{4}{r}\left(\frac{m}{r} - \frac{e^2}{2r^2}\right)\dot{\lambda}_c k^c - \frac{2}{r}\left(\frac{m_{,u}}{r} - \frac{ee_{,u}}{r^2}\right). \end{aligned} \tag{32.20}$$

The Kerr–Schild solutions corresponding to (32.20) contain: (i) the Schwarzschild solution (15.19) $e = 0$, $m = \text{const}$, $\dot{\lambda}^c k_c = 0$, (ii) Vaidya's shining star solution (15.20) $e = 0$, $m = m(u)$, $\dot{\lambda}^c k_c = 0$, (iii) the Reissner–Nordström solution (15.21) $e = \text{const}$, $m = \text{const}$, $\dot{\lambda}^c k_c = 0$, (iv) Kinnersley's photon rocket (28.73) $e = 0$, $m = m(u)$, $\dot{\lambda}^c k_c \neq 0$ (Kinnersley 1969a), and (v) the general case $e = e(u)$, $m = m(u)$, $\dot{\lambda}^c k_c \neq 0$.

32.1.5 The Newman–Penrose formalism for shearfree and geodesic Kerr–Schild metrics

In solving the field equations, the Newman–Penrose formalism is more convenient than the direct procedure outlined in §32.1.2. Throughout this section the vector $\boldsymbol{k}$ in the null tetrad $(\boldsymbol{m}, \overline{\boldsymbol{m}}, \boldsymbol{l}, \boldsymbol{k})$ is the null vector of the Kerr–Schild metric (32.1)–(32.2); all numerical indices are tetrad indices.

Following Debever (1974), we deal only with those Kerr–Schild space-times which admit a geodesic, diverging and shearfree null congruence with tangent vector $\boldsymbol{k}$,

$$\kappa = \sigma = 0; \quad \Gamma_{142} = -\rho = \Theta + \mathrm{i}\omega \neq 0, \tag{32.21}$$

and have a Ricci tensor satisfying the relations

$$R_{44} = R_{41} = R_{11} = 0. \tag{32.22}$$

(Ricci tensors of this type are general enough for most of the applications of the Kerr–Schild ansatz.)

As in §27.1, we can infer from (32.21)–(32.22) and $\Psi_0 = \Psi_1 = 0$ (Theorem 32.3) that

$$\boldsymbol{\Gamma}_{41} = \rho\boldsymbol{\omega}^2 + \tau\boldsymbol{\omega}^3 \tag{32.23}$$

can be written as

$$\boldsymbol{\Gamma}_{41} = \mathrm{d}\overline{Y}. \tag{32.24}$$

The difference from §27.1 is that τ may be non-zero, and P is transformed to unity. Comparing (32.23) and (32.24) we see that

$$D\overline{Y} = 0, \quad \delta\overline{Y} = 0, \quad \overline{\delta}\overline{Y} = \rho \neq 0, \quad \Delta\overline{Y} = \tau \tag{32.25}$$

holds. Using the commutation relations and the Newman–Penrose equations, we find that ε and β vanish. A null rotation $\boldsymbol{k}' = \boldsymbol{k}$, $\boldsymbol{m}' = \boldsymbol{m} + B\boldsymbol{k}$, $\boldsymbol{l}' = \boldsymbol{l} + B\overline{\boldsymbol{m}} + \overline{B}\boldsymbol{m} + B\overline{B}\boldsymbol{k}$, preserves the relations $\kappa = \sigma = \varepsilon = \beta = 0$ and can be used to make α vanish, which because of $\alpha' = \alpha + \overline{B}\rho$ and $\rho \neq 0$ is always possible. As a consequence of the Newman–Penrose equations, the coefficient π vanishes too. Thus we have

$$\kappa = \sigma = \varepsilon = \beta = \alpha = \pi = 0. \tag{32.26}$$

At this stage, Debever (1974) proves

Theorem 32.4 *A space-time is a special Kerr–Schild space-time (defined by the properties (32.1)–(32.2) and (32.21)–(32.22)) if and only if its Newman–Penrose coefficients can be transformed to*

$$\begin{pmatrix} \kappa & \tau & \sigma & \rho \\ \pi & \nu & \mu & \lambda \\ \varepsilon & \gamma & \beta & \alpha \end{pmatrix} = \begin{pmatrix} 0 & \Delta\overline{Y} & 0 & \delta\overline{Y} \\ 0 & \nu & \mu & 0 \\ 0 & \gamma & 0 & 0 \end{pmatrix}, \quad S = \frac{\mu}{\rho} = \frac{\bar{\mu}}{\bar{\rho}} \tag{32.27}$$

with respect to a suitably chosen null tetrad.

The details of the proof are given by Debever (1974). To show that (32.27) is sufficient, one first determines the remaining Newman–Penrose coefficients by means of the Newman–Penrose equations and the commutation relations, see §§7.1, 7.3. The result reads

$$\gamma = 2^{-1}(DS + S(\rho - \bar{\rho})), \quad \nu = \bar{\delta}S - S\Delta Y, \quad \mu = S\rho. \tag{32.28}$$

Then, solving the first Cartan equations (2.76)

$$\begin{aligned} \mathrm{d}\boldsymbol{\omega}^1 &= Y_{,a}\boldsymbol{\omega}^a \wedge (\boldsymbol{\omega}^4 - S\boldsymbol{\omega}^3) = \mathrm{d}\overline{\boldsymbol{\omega}}^2, \\ \mathrm{d}\boldsymbol{\omega}^3 &= \overline{Y}_{,a}\boldsymbol{\omega}^a \wedge \boldsymbol{\omega}^1 + Y_{,a}\boldsymbol{\omega}^a \wedge \boldsymbol{\omega}^2, \\ \mathrm{d}\boldsymbol{\omega}^4 &= SY_{,a}\boldsymbol{\omega}^a \wedge \boldsymbol{\omega}^2 + S\overline{Y}_{,a}\boldsymbol{\omega}^a \wedge \boldsymbol{\omega}^1 + S_{,a}\boldsymbol{\omega}^a \wedge \boldsymbol{\omega}^3, \end{aligned} \tag{32.29}$$

one obtains the dual basis

$$\begin{aligned} &\boldsymbol{\omega}^1 = \mathrm{d}\zeta + Y\mathrm{d}v, \quad \boldsymbol{\omega}^2 = \mathrm{d}\bar{\zeta} + \overline{Y}\mathrm{d}v, \\ &\boldsymbol{\omega}^3 = \overline{Y}\mathrm{d}\zeta + Y\mathrm{d}\bar{\zeta} + Y\overline{Y}\mathrm{d}v + \mathrm{d}u, \quad \boldsymbol{\omega}^4 = S\boldsymbol{\omega}^3 + \mathrm{d}v. \end{aligned} \tag{32.30}$$

The associated metric

$$\mathrm{d}s^2 = 2(\mathrm{d}\zeta\,\mathrm{d}\bar{\zeta} - \mathrm{d}u\,\mathrm{d}v) - 2S(\overline{Y}\mathrm{d}\zeta + Y\mathrm{d}\bar{\zeta} + Y\overline{Y}\mathrm{d}v + \mathrm{d}u)^2 \tag{32.31}$$

is clearly of the Kerr–Schild type.

If we start with a line element (32.31), then the null congruence

$$\boldsymbol{\omega}^3 = -k_a\mathrm{d}x^a = \overline{Y}\mathrm{d}\zeta + Y\mathrm{d}\bar{\zeta} + Y\overline{Y}\mathrm{d}v + \mathrm{d}u \tag{32.32}$$

will be geodesic and shearfree only if Y obeys (32.25). Because of

$$\begin{aligned} D &= -(Y\partial_\zeta + \overline{Y}\partial_{\bar{\zeta}} - \partial_v - Y\overline{Y}\partial_u), \quad \delta = \partial_\zeta - \overline{Y}\partial_u, \\ \Delta &= S(Y\partial_\zeta + \overline{Y}\partial_{\bar{\zeta}} - \partial_v - Y\overline{Y}\partial_u) + \partial_u, \end{aligned} \tag{32.33}$$

these conditions read

$$Y_{,\bar{\zeta}} - YY_{,u} = 0, \quad Y_{,v} - YY_{,\zeta} = 0. \tag{32.34}$$

They show that Y satisfies an equation

$$F(Y, \bar{\zeta}Y + u, vY + \zeta) = 0, \tag{32.35}$$

where F is an arbitrary function analytic in the three complex variables Y, $\bar{\zeta}Y + u$ and $vY + \zeta$:

Theorem 32.5 (Kerr's theorem) *Any analytic, geodesic and shearfree null congruence in Minkowski space is given by* $\boldsymbol{\omega}^3 = \mathrm{d}v$ *or by*

$$\boldsymbol{\omega}^3 = \overline{Y}\mathrm{d}\zeta + Y\mathrm{d}\bar{\zeta} + Y\overline{Y}\mathrm{d}v + \mathrm{d}u, \tag{32.36}$$

where $Y(u, v, \zeta, \bar{\zeta})$ *is defined implicitly by* (32.35) *with arbitrary* F (Cox and Flaherty 1976).

The null congruence (32.32) will be geodesic but not necessarily shear-free if

$$(Y_{,\bar{\zeta}} - YY_{,u})\overline{Y} - (Y_{,v} - YY_{,\zeta})Y = 0 \tag{32.37}$$

is satisfied (Cox and Flaherty 1976). The property $\kappa = 0$ (or $\kappa = 0 = \sigma$) holds in both Minkowski space and Kerr–Schild space-time, see above.

To get explicitly all Kerr–Schild metrics one has to solve the Newman–Penrose equations. Using the basis (32.30) we find that the following Newman–Penrose equations are already fulfilled:

$$D\rho = \rho^2; \quad D\tau = \tau\rho; \quad \delta\tau = \tau^2; \quad \delta\rho = (\rho - \bar{\rho})\tau; \quad \bar{\delta}\tau - \Delta\rho = \tau\bar{\tau} + S\rho^2. \tag{32.38}$$

The Newman–Penrose equations one really has to solve read

$$\begin{aligned} &D^2 S = 2(\rho + \bar{\rho})DS - 2S\rho\bar{\rho} - \tfrac{1}{2}R, \\ &(\rho + \bar{\rho})DS - S(\rho^2 + \bar{\rho}^2) = \tfrac{1}{4}R + 2\Phi_{11}, \end{aligned} \tag{32.39}$$

$$\delta\gamma = \tau(\mu + \gamma) + \Phi_{12}, \quad \delta\nu - \Delta\mu - \mu^2 - (\gamma + \bar{\gamma})\mu - \tau\nu = \Phi_{22}. \tag{32.40}$$

The remaining Newman–Penrose equations determine the non-vanishing components of the Weyl tensor, namely

$$\Psi_2 = \mu\rho + \gamma(\rho - \bar{\rho}) + \tfrac{1}{24}R + \Phi_{11}, \quad \Psi_3 = \bar{\delta}\gamma + \rho\nu - \bar{\tau}\gamma, \quad \Psi_4 = \bar{\delta}\nu - \bar{\tau}\nu. \tag{32.41}$$

The Newman–Penrose equations (32.39) can be solved immediately in the case of vanishing Ricci scalar ($R = 0$). The solution reads

$$S = \tfrac{1}{2}\widetilde{M}(\rho + \bar{\rho}) + \widetilde{B}\rho\bar{\rho}, \quad \Phi_{11} = \widetilde{B}\rho^2\bar{\rho}^2, \quad D\widetilde{M} = D\widetilde{B} = 0, \tag{32.42}$$

$\widetilde{M}$ and $\widetilde{B}$ being real functions. So for $R = 0$, only (32.40) remains to be solved.

Table 32.1. Kerr–Schild space-times

	Vacuum	Einstein–Maxwell	Pure radiation	Perf. fluid
$\rho \neq 0$	A (32.43)	S (32.59)	S (32.78) (32.80)	$\nexists$
$\rho = 0$	A Th. 32.6	A (32.71)	A (32.71)	$\nexists$

In the following sections we shall consider the different types of the energy-momentum tensor which may occur. The main results are collected in Table 32.1.

32.2 Kerr–Schild vacuum fields

32.2.1 The case $\rho = -(\Theta + \mathrm{i}\omega) \neq 0$

The general Kerr–Schild solution of Einstein's vacuum field equations was found by Kerr and Schild (1965a, 1965b). The remaining field equations (32.40) require an intricate calculation to get the final result. For details see Debney *et al.* (1969) and §32.3.1. The solutions are

$$\begin{aligned} \mathrm{d}s^2 &= 2(\mathrm{d}\zeta\,\mathrm{d}\bar{\zeta} - \mathrm{d}u\,\mathrm{d}v) + \widetilde{m}P^{-3}(\rho + \bar{\rho})[\mathrm{d}u + \overline{Y}\mathrm{d}\zeta + Y\mathrm{d}\bar{\zeta} + Y\overline{Y}\mathrm{d}v]^2, \\ P &= pY\overline{Y} + qY + \bar{q}\overline{Y} + c, \quad \widetilde{M} = \widetilde{m}P^{-3}, \end{aligned} \tag{32.43}$$

where $\widetilde{m}$, p, c (real) and q (complex) are constants. The function Y is given by the implicit relation

$$\Phi(Y) + (qY + c)(\zeta + Yv) - (pY + \bar{q})(u + Y\bar{\zeta}) = 0, \tag{32.44}$$

where $\Phi(Y)$ is an arbitrary analytic function of the complex variable Y. The complex expansion of the null vector $\boldsymbol{k}$ is given by

$$\bar{\rho} = -P[\{\Phi + (qY + c)(\zeta + Yv) - (pY + \bar{q})(u + Y\bar{\zeta})\}_{,Y}]^{-1}. \tag{32.45}$$

The solutions have the following properties:

(i) They are all algebraically special, $\boldsymbol{k}$ being a multiple principal null direction of the Weyl tensor. Thus $\boldsymbol{k}$ is shearfree and geodesic.

(ii) They are of Petrov types *II* or *D*; types *III* and *N* cannot occur. For $\Phi = 0$ ($\leftrightarrow \rho = \bar{\rho}$) and $\Phi = \mathrm{i}aY$ they are of type *D* (McIntosh and Hickman 1988).

(iii) They all admit at least a one-parameter group of motions with Killing vector

$$\boldsymbol{\xi} = c\partial_u + q\partial_{\bar{\zeta}} + \bar{q}\partial_\zeta + p\partial_v, \tag{32.46}$$

which is simultaneously a Killing vector of flat space-time. The solutions can be simplified by performing a Lorentz transformation. One can thus assume that if $\eta_{ab}\xi^a\xi^b < 0$, $P\sqrt{2} = 1+Y\overline{Y}$; if $\eta_{ab}\xi^a\xi^b > 0$, $P\sqrt{2} = 1-Y\overline{Y}$; and if $\eta_{ab}\xi^a\xi^b = 0$, $P = 1$.
(iv) The points $(P\rho^{-1}) = 0$ are singularities of the Riemannian space. In general Y is a multivalued function with these singularities as branch points. The only solution whose singularities are confined to a bounded region is the Kerr metric (Kerr and Wilson 1979).
(v) For a timelike Killing vector $\boldsymbol{\xi}$, the particular case $\Phi = -\mathrm{i}aY$ leads to the Kerr solution (20.34). Its Kerr–Schild form reads (with r given by $(x^2+y^2)(r^2+a^2)^{-1} + z^2r^{-2} = 1$)

$$\begin{aligned} \mathrm{d}s^2 = \mathrm{d}x^2 + \mathrm{d}y^2 + \mathrm{d}z^2 - \mathrm{d}t^2 + \frac{2mr^3}{r^4+a^2z^2}\Big[\mathrm{d}t + \frac{z}{r}\mathrm{d}z \\ + \frac{r}{r^2+a^2}(x\,\mathrm{d}x + y\,\mathrm{d}y) - \frac{a}{r^2+a^2}(x\,\mathrm{d}y - y\,\mathrm{d}x)\Big]^2, \end{aligned} \tag{32.47}$$

the double Debever–Penrose null vector being

$$\omega^3 = 2^{1/2}\frac{r}{r+z}\left[\mathrm{d}t + \frac{z}{r}\mathrm{d}z + \frac{r(x\,\mathrm{d}x + y\,\mathrm{d}y)}{r^2+a^2} - \frac{a(x\,\mathrm{d}y - y\,\mathrm{d}x)}{r^2+a^2}\right] = -k_i\mathrm{d}x^i. \tag{32.48}$$

For a different characterization of the Kerr–Schild vacuum solutions see §§29.2.5 and 32.4.1.

32.2.2 The case $\rho = -(\Theta + \mathrm{i}\omega) = 0$

The non-expanding and non-twisting solutions have been treated in Chapter 31. The corresponding Kerr–Schild metrics have been considered by Trautman (1962), Urbantke (1972), Debney (1973) and McIntosh (private communication). The result is

Theorem 32.6 *The Kerr–Schild vacuum fields with a non-expanding and non-twisting null congruence* $\mathbf{k}$ *are necessarily of Petrov type N. They are the subcases* $W^0 = 0$ *of* (31.34) *and* (31.38).

32.3 Kerr–Schild Einstein–Maxwell fields

32.3.1 The case $\rho = -(\Theta + \mathrm{i}\omega) \neq 0$

Electromagnetic solutions of the Kerr–Schild type were studied by Debney *et al.* (1969), again assuming the null vector $\boldsymbol{k}$ to be geodesic. By Theorem 32.1, the energy-momentum tensor of the electromagnetic field fulfils the conditions

$$k_{a;b}k^b = 0 \quad \leftrightarrow \quad T_{ab}k^ak^b = 0 \quad \leftrightarrow \quad F_{ab}k^a = \lambda k_b. \tag{32.49}$$

Hence $\boldsymbol{k}$ is a principal null direction of F_{ab} if and only if it is geodesic. The space-time is algebraically special with $\boldsymbol{k}$ as the multiple principal null direction. The energy-momentum tensor obeys (32.22), and the Bianchi identities imply that the null vector $\boldsymbol{k}$ is shearfree. So all results of §32.1.5 are true for the electromagnetic case too. See also Chapter 30!

The result (32.49) implies

$$\Phi_1 = 2^{-1}F_{ab}(k^a l^b + \overline{m}^a m^b), \quad \Phi_2 = F_{ab}\overline{m}^a l^b, \quad \Phi_0 = 0, \tag{32.50}$$

for the tetrad components of the Maxwell tensor. The field equations therefore read

$$R_{12} = R_{34} = 2\kappa_0\Phi_1\overline{\Phi}_1, \quad R_{13} = 2\kappa_0\Phi_1\overline{\Phi}_2, \quad R_{33} = 2\kappa_0\Phi_2\overline{\Phi}_2, \tag{32.51}$$

and Maxwell's equations (7.22)–(7.25) are

$$D\Phi_1 - 2\rho\Phi_1 = 0, \qquad \delta\Phi_1 - 2\tau\Phi_1 = 0, \tag{32.52}$$

$$\bar{\delta}\Phi_1 - D\Phi_2 + \rho\Phi_2 = 0, \quad -\delta\Phi_2 + \Delta\Phi_1 + 2S\rho\Phi_1 + \tau\Phi_2 = 0. \tag{32.53}$$

Using (32.38) and the commutation relations (7.6a)–(7.6d), the field equations (32.51)–(32.53) can be partially integrated. The result is

$$\Phi_1 = \rho^2\Phi_1^0, \quad \Phi_2 = \rho\Phi_2^0 + \bar{\delta}(\rho\Phi_1^0), \quad D\Phi_1^0 = D\Phi_2^0 = 0, \tag{32.54}$$

$$S = \tfrac{1}{2}\widetilde{M}(\rho + \bar{\rho}) + \kappa_0\Phi_1^0\overline{\Phi}_1^0\rho\bar{\rho}, \quad D\widetilde{M} = 0. \tag{32.55}$$

In terms of the four unknown functions Y, $\widetilde{M}$, Φ_1^0 and Φ_2^0, the remaining Maxwell equations and Einstein equations are found to be

$$\begin{aligned} &D\Phi_1^0 = D\Phi_2^0 = 0, \quad \delta\Phi_1^0 - 2\bar{\rho}\rho^{-1}\tau\Phi_1^0 = 0, \\ &\Delta\Phi_1^0 - \tau\rho^{-1}\bar{\delta}\Phi_1^0 - \bar{\tau}\bar{\rho}^{-1}\delta\Phi_1^0 + \tau\rho^{-1}\Phi_2^0 - \bar{\rho}^{-1}\delta\Phi_2^0 = 0 \end{aligned} \tag{32.56}$$

and

$$\begin{aligned} &D\widetilde{M} = DY = \bar{\delta}Y = 0, \quad \delta\widetilde{M} - 3\widetilde{M}\bar{\rho}\rho^{-1}\tau = 2\kappa_0\bar{\rho}\Phi_1^0\overline{\Phi}_2^0, \\ &\Delta\widetilde{M} - \bar{\tau}\bar{\rho}^{-1}\delta\widetilde{M} - \tau\rho^{-1}\bar{\delta}\widetilde{M} = -\kappa_0\Phi_2^0\overline{\Phi}_2^0. \end{aligned} \tag{32.57}$$

No general solution of this system is yet known. Debney *et al.* (1969) have completely solved the case in which the electromagnetic field is further restricted by the condition

$$\Phi_2^0 = 0, \tag{32.58}$$

i.e. these Kerr–Schild metrics are a subcase of the charged vacuum solutions, cp. §30.4. The solutions then are

$$\begin{aligned} ds^2 = {} & 2(d\zeta\, d\bar{\zeta} - du\, dv) + P^{-3}\left[\widetilde{m}(\rho + \bar{\rho}) - \tfrac{1}{2}\kappa_0\Psi\overline{\Psi}P^{-1}\rho\bar{\rho}\right] \\ & \times [\overline{Y}d\zeta + Y d\bar{\zeta} + Y\overline{Y}dv + du]^2. \end{aligned} \tag{32.59}$$

As in the vacuum case, Y, ρ and P are given by (32.43)–(32.45), and Φ and Ψ are arbitrary (analytic) functions of Y. The solutions admit the Killing vector (32.46). $\Psi = 0$ gives the vacuum solution (32.43).

The particular choice

$$P = 2^{-1/2}(1 + Y\overline{Y}), \quad \Phi = -\mathrm{i}aY, \quad \Psi = e, \tag{32.60}$$

leads to the field of a rotating charged body first obtained by Newman *et al.* (1965) (cp. §21.1.3). Using the special null coordinates $\sqrt{2}\zeta = x + \mathrm{i}y$, $\sqrt{2}\overline{\zeta} = x - \mathrm{i}y$, $\sqrt{2}u = t - z$, $\sqrt{2}v = t + z$ the solution and the electromagnetic field are found to be

$$\begin{aligned} \mathrm{d}s^2 = \mathrm{d}x^2 + \mathrm{d}y^2 + \mathrm{d}z^2 - \mathrm{d}t^2 + \frac{2\widetilde{m}r^3 - e^2r^2}{r^4 + a^2z^2}\Bigg[\mathrm{d}t + \frac{z}{r}\mathrm{d}z \\ + \frac{r}{r^2+a^2}(x\,\mathrm{d}x + y\,\mathrm{d}y) - \frac{a}{r^2+a^2}(x\,\mathrm{d}y - y\,\mathrm{d}x)\Bigg]^2, \end{aligned} \tag{32.61}$$

$$(F_{xt} - \mathrm{i}F_{yz}, F_{yt} - \mathrm{i}F_{zx}, F_{zt} - \mathrm{i}F_{xy}) = er^3(r^2 + \mathrm{i}az)^{-3}(x, y, z + \mathrm{i}a),$$

where the function r is defined by the implicit relation

$$(x^2 + y^2)(r^2 + a^2)^{-1} + z^2r^{-2} = 1. \tag{32.62}$$

Taking $e = 0$ in (32.61), we recover the Kerr solution (32.47).

32.3.2 The case $\rho = -(\Theta + \mathrm{i}\omega) = 0$

The non-diverging solutions were considered in detail in Chapter 31 (Kundt's class). In §31.1 it was shown that $\rho = 0$ implies that $\boldsymbol{k}$ is both a multiple principal null direction of the Weyl tensor and the electromagnetic field tensor and that $\boldsymbol{k}$ is shearfree.

Expansionfree electromagnetic solutions of the Kerr–Schild class were treated first by Białas (1963), and later by Debney (1974), again assuming $\boldsymbol{k}$ to be a geodesic null congruence.

Choosing the basis (32.30), the only non-zero connection coefficients are given by

$$\Gamma_{233} = \overline{\delta}S - S\Delta Y, \quad \Gamma_{143} = -\Delta\overline{Y}, \quad \Gamma_{433} = DS, \tag{32.63}$$

and the vector fields $(\boldsymbol{e}_1, \boldsymbol{e}_2, \boldsymbol{e}_3, \boldsymbol{e}_4)$ obey the commutation relations $[\boldsymbol{e}_1, \boldsymbol{e}_2] = [\boldsymbol{e}_1, \boldsymbol{e}_4] = [\boldsymbol{e}_2, \boldsymbol{e}_4] = 0$. The non-zero Newman–Penrose coefficients and the non-vanishing components of the Weyl and Ricci tensors read

$$\begin{aligned} 2\gamma = DS, \quad \nu = \overline{\delta}S - S\Delta Y, \quad \tau = \Delta\overline{Y}, \\ R_{33} = 2\kappa_0\Phi_2\overline{\Phi}_2, \quad \Psi_4 = \overline{\delta}\overline{\delta}S - 2\overline{\delta}S\Delta Y. \end{aligned} \tag{32.64}$$

The only non-zero component of the electromagnetic field tensor is Φ_2.

Theorem 32.7 *All sourcefree Einstein–Maxwell fields of the Kerr–Schild class with an expansionfree (geodesic) null congruence* $\boldsymbol{k}$ *are of Petrov type N (or O).* $\boldsymbol{k}$ *is both a shearfree quadruple principal null direction of the Weyl tensor and a principal null direction of the Maxwell field. The electromagnetic field is null* (Debney 1974).

These results reduce the Einstein–Maxwell equations for expansionfree Kerr–Schild fields to the equations

$$\begin{aligned} &D^2S = 0,\ \ D\bar{\delta}S - \Delta Y DS = 0, \quad \delta\bar{\delta}S - \delta S\Delta Y - \bar{\delta}S\Delta\overline{Y} = \kappa_0\Phi_2\overline{\Phi}_2, \\ &D\Phi_2 = 0,\ \ \bar{\delta}\Phi_2 - \Delta Y\overline{\Phi}_2 = 0, \quad \delta Y = \bar{\delta}Y = DY = 0. \end{aligned} \tag{32.65}$$

In solving the field equations new coordinates z, $\overline{z}$, u', v' introduced by

$$z = \zeta + Yv, \quad \overline{z} = \bar{\zeta} + \overline{Y}v, \quad u' = u + \overline{Y}\zeta + Y\bar{\zeta} + Y\overline{Y}v, \quad v' = v \tag{32.66}$$

are more convenient than the old set $\zeta, \bar{\zeta}, u, v$. In terms of the new coordinates the dual basis (32.30) and tetrad basis read

$$\begin{aligned} &\boldsymbol{\omega}^1 = \overline{\boldsymbol{\omega}}^2 = \mathrm{d}z - r^{-1}v\bar{\tau}\mathrm{d}u', \quad \boldsymbol{\omega}^3 = r^{-1}\mathrm{d}u', \quad \boldsymbol{\omega}^4 = r^{-1}S\mathrm{d}u' + \mathrm{d}v, \\ &r = [1 - \dot{\overline{Y}}z - \dot{Y}\overline{z}]^{-1}, \quad r\mathrm{d}\overline{Y}/\mathrm{d}u' = r\dot{\overline{Y}} = \tau; \end{aligned} \tag{32.67}$$

$$\boldsymbol{e}_1 = \partial_z,\ \boldsymbol{e}_2 = \partial_{\overline{z}},\ \boldsymbol{e}_3 = v[\bar{\tau}\partial_z + \tau\partial_{\overline{z}}] + r\partial_{u'} - S\partial_v,\ \boldsymbol{e}_4 = \partial_v. \tag{32.68}$$

The field equations then are, with $Y = Y(u')$,

$$S_{,vv} = 0, \quad \overline{\Phi}_{2,v} = 0, \quad S_{,\overline{z}v} - \bar{\tau}S_{,v} = S_{,zv} - S_{,v}\tau = 0, \tag{32.69a}$$

$$\overline{\Phi}_{2,\overline{z}} - \tau\overline{\Phi}_2 = 0, \quad S_{,z\overline{z}} - S_{,z}\bar{\tau} - S_{,\overline{z}}\tau = \kappa_0\Phi_2\overline{\Phi}_2. \tag{32.69b}$$

From (32.69a) we get

$$\overline{\Phi}_2 = \overline{\Phi}_2(z, \overline{z}, u'), \quad S = vra(u') + g(z, \overline{z}, u'), \tag{32.70}$$

where $a(u')$ and $g(z, \overline{z}, u')$ are real functions. By a coordinate transformation one can set $a(u') = 0$ (Debney 1974). Then the electromagnetic Kerr–Schild solutions admitting a geodesic, shearfree and expansionfree null congruence become

$$\mathrm{d}s^2 = 2(\mathrm{d}z - v\dot{Y}\mathrm{d}u')(\mathrm{d}\overline{z} - v\dot{\overline{Y}}\mathrm{d}u') - 2r^{-1}(gr^{-1}\mathrm{d}u' + \mathrm{d}v)\mathrm{d}u'. \tag{32.71}$$

The remaining field equations (32.69b) restrict the electromagnetic field and the real function $g(z, \overline{z}, u')$ by the equations

$$(r^{-1}\overline{\Phi}_2)_{,\overline{z}} = 0; \quad (r^{-1}g)_{,z\overline{z}} = \kappa_0 r^{-1}\Phi_2\overline{\Phi}_2. \tag{32.72}$$

The vector $\boldsymbol{k}$ is a null Killing vector (see §24.4) if and only if the rotation coefficient $\tau = r\dot{\overline{Y}}$ vanishes. In this case the null vector is covariantly constant; all solutions of this type are plane-fronted gravitational waves (see §24.5). To get all solutions in the general case $\tau \neq 0$ means solving the two-dimensional Poisson equation (32.72).

32.4 Kerr–Schild pure radiation fields

For Kerr–Schild metrics with the energy-momentum tensor of pure radiation, $T_{mn} = \Phi^2 k_m k_n$, the geodesic null vector $\boldsymbol{k}$ need not be shearfree. Three cases can be distinguished, depending on whether ρ and σ are zero or not.

32.4.1 The case $\rho \neq 0$, $\sigma = 0$

For $\sigma = 0$, $\boldsymbol{k}$ is a multiple eigenvector of the Weyl tensor and the Kerr–Schild metrics belong to the class of twisting algebraically special metrics considered in Chapters 29 and 30. To see the connection between the metric used there (see e.g. (29.13) or (30.14)) and the Kerr–Schild form (32.1), one can start from (32.27) with (32.32) being satisfied. Taking coordinates $(Y, \overline{Y}, v, w)$ with

$$w = u + \zeta\overline{Y} + \overline{\zeta}Y + Y\overline{Y}v \tag{32.73}$$

and introducing a new basis with

$$\begin{aligned} \boldsymbol{\omega}^{1'} &= -(\boldsymbol{\omega}^1 + B\boldsymbol{\omega}^3) = -\overline{\rho}^{-1}\mathrm{d}Y, \quad \boldsymbol{\omega}^{2'} = \overline{\boldsymbol{\omega}}^1, \\ \boldsymbol{\omega}^{4'} &= \overline{B}\boldsymbol{\omega}^1 + B\boldsymbol{\omega}^2 + B\overline{B}\boldsymbol{\omega}^3 + \boldsymbol{\omega}^4, \quad \boldsymbol{\omega}^{3'} = \boldsymbol{\omega}^3, \\ B &= \overline{L}_{,w}, \quad L \equiv -(\overline{\zeta} + v\overline{Y}), \end{aligned} \tag{32.74}$$

one arrives at the form (29.13) of the line element, by identifying $(Y, \overline{Y}, v, w)$ here with $(\zeta, \overline{\zeta}, r, u)$ there and by setting $P = 1$.

In the new variables, (32.35), i.e. $F(Y, \overline{\zeta}Y + u, vY + \zeta) = 0$, reads

$$F(Y, w + \overline{L}Y, -\overline{L}) = 0. \tag{32.75}$$

It is equivalent to the statement that the function $L(Y, \overline{Y}, w)$ obeys $\partial L \equiv L_{,Y} - LL_{,w} = 0$. Leaving now the gauge $P = 1$ we can state:

Theorem 32.8 *There exist coordinates in which Kerr–Schild vacuum and pure radiation metrics with a geodesic, shearfree but diverging null vector $\boldsymbol{k}$ are the subcase $\partial\partial V = 0$ ($\partial L = 0$ in the gauge $P = 1$) of the algebraically special metrics* (29.13) (Debney *et al.* 1969).

Because of $\partial\partial V = 0$, M is necessarily zero, and since $m = 0$ would lead to flat space-time, we can gauge m to $m = 1$ which induces $L_{,u} = 0$ (we are sticking now to the coordinates $(\zeta, \overline{\zeta}, r, u)$ used throughout Chapter 30). To integrate $\partial\partial V = 0$, we introduce as in (29.33) a function Φ by $L = \Phi_{,\zeta}/\Phi_{,u}$ so that we have

$$M = 0, \quad m = 1, \quad \Phi = u + \mathrm{i}\, h(\zeta, \overline{\zeta}), \quad L = \mathrm{i}\, h_{,\zeta}, \quad h(\zeta, \overline{\zeta}) \text{ real.} \tag{32.76}$$

Since because of this definition of Φ we have $\partial u/\partial\zeta\,|_{\Phi,\overline{\zeta}=\text{ const}} = -\Phi_{,\zeta}/\Phi_{,u}$, we can write the operator ∂ as

$$\partial = \partial_\zeta - L\partial_u = \partial_\zeta\,|_{\Phi,\overline{\zeta}=\text{ const}}\cdot \tag{32.77}$$

The Kerr–Schild condition $\partial\partial V = 0$ is then integrated by $V = F_1(\Phi, \overline{\zeta}) + \zeta F_2(\Phi, \overline{\zeta})$ with arbitrary functions F_i, i.e. we have (Stephani 1983a)

$$\begin{aligned} V &= F_1\left[u + \mathrm{i}\, h(\zeta, \overline{\zeta}), \overline{\zeta}\right] + \zeta F_2\left[u + \mathrm{i}\, h(\zeta, \overline{\zeta}), \overline{\zeta}\right], \\ m &= 1, \quad M = 0, \quad L = \mathrm{i}\, h_{,\zeta}, \quad P = V_{,u}, \quad V, h \text{ real,} \end{aligned} \tag{32.78}$$

and the last remaining field equation (30.42) reads

$$6P_{,u}/P = \kappa_0 n^2(\zeta, \overline{\zeta}, u). \tag{32.79}$$

The closed form (32.78) looks rather promising, but one quickly realizes that it is not easy to find functions h, F_i that make V real. Except in the non-twisting case $h = 0$, the dependence on u (i.e. the radiation rate) will be rather restricted. In the vacuum case ($n^2 = 0$ in (32.79)) V and therefore also the F_i must be linear in u, and the reality condition then uniquely defines h.

In the pure radiation case (32.79) is just the definition of n^2. Only a few solutions are known; most of them have been found not by using (32.78) but by directly attacking the condition $\partial\partial V = 0$. In the axisymmetric case, the complete solution was first found by Herlt (1980), using a formalism developed by Vaidya (1973, 1974). In the present notation, it is given by

$$\begin{aligned} V &= \tfrac{1}{2}u^2 g_1 + g_3, \quad h = -g_2/g_1, \quad g_i = a_i + b_i\zeta\overline{\zeta}, \\ V &= (g_1 g_2)^{1/2}\cos au, \quad h = (\ln g_1/g_2)/2a, \\ V &= \left(b_1 \mathrm{e}^{au} + b_2 \mathrm{e}^{-au}\right)(g_0\overline{g}_0)^{1/2}, \quad h = \mathrm{i}\,(\ln g_0/\overline{g}_0)/2a, \end{aligned} \tag{32.80}$$

with constants a_i, b_i that are real except a_0, b_0. Included here is the radiating Kerr metric given by Vaidya and Patel (1973). None of these solutions admits an interpretation as an Einstein–Maxwell null field.

An example of a non-axisymmetric solution can be obtained if in the last of the above examples one inserts $g_0 = A(\overline{\zeta}) + \zeta B(\overline{\zeta})$ with arbitrary functions A and B (Stephani 1983a). Another example is the photon rocket (28.73), cp. also (32.20).

32.4.2 The case $\sigma \neq 0$

In the aligned case (32.8), space-time is algebraically special because of Theorem 32.3, and the theorems given in §7.6 show that it must be of type N. So the Kerr–Schild solutions in question are contained in the solutions discussed in §26.1. To extract them from the classes considered there, we have to incorporate the Kerr–Schild condition (32.1), i.e. we have to demand that the type N solution has non-zero shear and can be written as

$$\begin{aligned} ds^2 &= 2\omega^1\omega^2 - 2\omega^3\omega^4, \\ \omega^1 &= d\zeta + Y dv = \overline{\omega}^2, \quad \omega^4 = S\omega^3 + dv, \\ \omega^3 &= -k_a dx^a = \overline{Y} d\zeta + Y d\overline{\zeta} + Y\overline{Y} dv + du. \end{aligned} \tag{32.81}$$

Starting with (32.81), and using $\Psi_0 = \Psi_1 = \Psi_2 = \Psi_3 = 0$ and the field equations $R_{ab} = \kappa_0 \delta^3_a \delta^3_b \Phi^2$, the Newman–Penrose equations show that the null congruence $\boldsymbol{k}$ must be geodesic and twistfree,

$$\kappa = \rho - \bar{\rho} = 0, \quad \rho^2 = \sigma\bar{\sigma}, \tag{32.82}$$

and the spin coefficients ρ, σ and τ have to satisfy

$$\delta(\rho\tau\sigma^{-1} + \tau) = \bar{\delta}(\rho\bar{\tau}\sigma^{-1} + \tau) \tag{32.83}$$

(Urbantke 1975). All type N pure radiation fields satisfying (32.82) and $\sigma \neq 0$ are known and contained in (26.11) and (26.13), but only special (Kerr–Schild) metrics have been determined which obey in addition (32.83), cp. Urbantke (1975). A special example is the cylindrically symmetric solution (22.70), cp. Patel (1973b).

32.4.3 The case $\rho = \sigma = 0$

Kerr–Schild pure radiation solutions possessing a geodesic, expansionfree and shearfree null congruence belong to Petrov type N, the vector $\boldsymbol{k}$ being the quadruple principal null direction. All solutions of this class are given by (32.71) with arbitrary functions $Y(u')$ and (real) $g(z, \bar{z}, u')$.

32.5 Generalizations of the Kerr–Schild ansatz

32.5.1 General properties and results

The original Kerr–Schild ansatz relates a flat space-time to a non-flat one via (32.1). In generalizing this idea one considers two spaces $\widetilde{V}_4$ and V_4 with metrics related by

$$\widetilde{g}_{ab} = g_{ab} - 2Sk_a k_b, \quad \widetilde{g}^{ab} = g^{ab} + 2Sk^a k^b, \quad (-\widetilde{g})^{1/2} = (-g)^{1/2}, \tag{32.84}$$

where *both* g_{ab} and $\tilde{g}_{ab}$ are non-flat, and $\boldsymbol{k}$ is a null vector with respect to both metrics. Using the abbreviation

$$\begin{aligned} D^a_{bc} &= \widetilde{\Gamma}^{\,a}_{bc} - \Gamma^a_{bc} \\ &= 2Sk^a k^n (Sk_b k_c)_{;n} - g^{an}\left[(Sk_n k_b)_{;c} + (Sk_n k_c)_{;b} - (Sk_b k_c)_{;n}\right], \end{aligned} \tag{32.85}$$

one obtains for the Ricci tensors

$$\widetilde{R}_{ma} = R_{ma} + D^s_{ma;s} - D^s_{na} D^n_{ms}, \tag{32.86}$$

where the covariant derivative is taken with respect to g_{ab}.

The null tetrads of the two metrics being related by

$$\widetilde{m}^a = m^a, \quad \widetilde{l}^a = l^a - Sk^a, \quad \widetilde{k}^a = k^a, \tag{32.87}$$

one can easily derive the transformation properties of the Newman–Penrose coefficients. They read

$$\begin{aligned} &\widetilde{\kappa} = \kappa, \quad \widetilde{\sigma} = \sigma, \quad \widetilde{\rho} = \rho, \quad \widetilde{\varepsilon} = \varepsilon, \quad \widetilde{\tau} = \tau, \quad \widetilde{\pi} = \pi, \\ &\widetilde{\lambda} = \lambda + \overline{\sigma} S, \quad \widetilde{\mu} = \mu + \rho S, \quad \widetilde{\beta} = \beta + \tfrac{1}{2}\kappa S, \quad \widetilde{\alpha} = \alpha + \tfrac{1}{2}\overline{\kappa} S, \\ &\widetilde{\gamma} = \gamma + \tfrac{1}{2}(2\overline{\varepsilon} + \rho - \overline{\rho} + D)S, \quad \widetilde{\nu} = \nu + (2\alpha + 2\overline{\beta} - \pi - \overline{\tau} + \overline{\delta})S + \overline{\kappa} S^2, \end{aligned} \tag{32.88}$$

see e.g. Dozmorov (1971a), Thompson (1966), Xanthopoulos (1983c) and Bilge and Gürses (1983). For the Weyl and Ricci tensor components one obtains

$$\begin{aligned} &\widetilde{\Phi}_{00} = \Phi_{00} + 2\kappa\overline{\kappa} S, \quad \widetilde{\Psi}_0 = \Psi_0 + 2\kappa^2 S, \\ &\widetilde{\Psi}_1 = \Psi_1 + \tfrac{1}{2}\kappa DS - \tfrac{1}{2}\overline{\kappa}\sigma S + \tfrac{1}{2}(\varepsilon + 3\overline{\varepsilon} + 3\rho - 2\overline{\rho} + D)\kappa S \end{aligned} \tag{32.89}$$

and other lengthy expressions given e.g. in Bilge and Gürses (1983) (beware of misprints!).

The following properties of the generalized Kerr–Schild transformation can easily be read off:

(i) If $\widetilde{V}_4$ and V_4 are vacuum space-times, then $\boldsymbol{k}$ is geodesic with respect to both (Thompson 1966).

(ii) If $\boldsymbol{k}$ is geodesic, then the mixed components

$$\widetilde{R}^a_b = R^a_b + 2Sk^a k^i R_{bi} - g^{ai} g^{mn}\left[(Sk_n k_b)_{;i} + (Sk_n k_i)_{;b} - (Sk_b k_i)_{;n}\right]_{;m} \tag{32.90}$$

are linear in S (Taub 1981). This property has also been used and discussed by Xanthopoulos (1986). The linearity of the field equations permits the occurrence of shock waves (δ-functions).

(iii) If $\widetilde{V}_4$ and V_4 are vacuum space-times, and V_4 is algebraically special with $\boldsymbol{k}$ as a repeated null eigenvector, then the same is true for $\widetilde{V}_4$ (Thompson 1966).
(iv) If $\boldsymbol{k}$ is geodesic and if V_4 is algebraically special with $\boldsymbol{k}$ as a repeated null eigenvector ($\kappa = \Psi_0 = \Psi_1 = 0$), then $\widetilde{V}_4$ has the same properties (Bilge and Gürses 1983).
(v) If $\boldsymbol{k}$ is a principal null direction of both Weyl tensors, then it is geodesic (Bilge and Gürses 1983).
(vi) If the tetrad $(\boldsymbol{m}, \overline{\boldsymbol{m}}, \boldsymbol{l}, \boldsymbol{k})$ is parallelly propagated along $\boldsymbol{k}$ in V_4, then the same holds in $\widetilde{V}_4$ (Bilge and Gürses 1983).

When using the generalized Kerr–Schild transformation (32.84) in the search for solutions, both metrics – g_{ab} and $\widetilde{g}_{ab}$ – have to be restricted. We shall discuss now the different choices investigated so far.

32.5.2 Non-flat vacuum to vacuum

As shown above, the null congruence $\boldsymbol{k}$ is necessarily geodesic, but its shear may or may not vanish. Gergely and Perjés (1993, 1994a, 1994b, 1994c) showed that the vacuum solutions with *non-vanishing shear* ($\sigma \neq 0$) are the Kóta–Perjés metrics (18.48)–(18.50) generated from a type N vacuum metric. In general, they have non-vanishing twist ($\omega \neq 0$), but the metrics with $\omega = 0$ are contained as a limiting case (Kupeli 1988b, Gergely and Perjés 1994c): they are the Kasner metrics (13.53)–(13.51), obtained from a plane wave V_4 (Kóta and Perjés 1972).

For a general metric with *vanishing shear* (which is algebraically special) it may happen that the non-flat background metric g_{ab} is already of the Kerr–Schild type,

$$\widetilde{g}_{ab} = g_{ab} - 2S_2 k_a k_b = \eta_{ab} - 2S_1 k_a k_b - 2S_2 k_a k_b. \tag{32.91}$$

An example for this is the Schwarzschild metric, for which S_1 introduces a mass into Minkowski space, and S_2 only changes the mass parameter. These cases will be considered trivial.

To get non-trivial solutions with non-zero expansion ($\rho \neq 0$), one can start from the line element (29.4), see §29.1 for further details. Because of its definition, a generalized Kerr–Schild transformation is a transformation which adds a function S to H such that P, L and W are not changed and that $\widetilde{H} = H + S$ is again a solution of the field equations. An inspection of (29.13) reveals that the only admissible change of H is the addition of a mass term, $\widetilde{m} = m + m_S$, $S = m_S \mathrm{Re}\,\rho = m_S r/(r^2 + \Sigma^2)$, cp. also Dozmorov (1971b) and Talbot (1969). The field equations then imply

$$(P^{-3} m_S)_{,u} = 0, \quad \partial m_S = m_S L_{,u}. \tag{32.92}$$

If we now take a gauge with $m_S = \text{const}$, (32.92) gives $L_{,u} = 0 = P_{,u}$. For the background field this implies $M_{,u} = 0$, $\partial(m + \mathrm{i}M) = 0$ and therefore

$$(\partial\overline{\partial} - \overline{\partial}\partial)m = (\overline{\partial}L - \partial\overline{L})m_{,u} = 2\mathrm{i}M_{,\zeta\overline{\zeta}}. \tag{32.93}$$

So for non-vanishing twist ($\overline{\partial}L - \partial\overline{L} \neq 0$), m is at most linear in u,

$$m = au + \beta(\zeta, \overline{\zeta}), \quad a = \text{const}, \tag{32.94}$$

and the same result (with $\beta = \text{const}$) can be obtained for zero twist directly from the Robinson–Trautman field equation (28.8).

If a vanishes, all metric functions are independent of u, $\boldsymbol{\xi} = \partial_u$ is a Killing vector; this class has been considered by Fels and Held (1989) using the GHP formalism. The twisting metrics with these properties have been considered in §§29.2.3, 29.2.5, and the non-twisting in §28.1, equation (28.26); the solutions are known only for a few subcases, among them the Kerr–Schild metrics with $\partial\partial V = 0$.

If a is non-zero, then the Kerr–Schild transformation amounts to $\widetilde{m} = au + \beta(\zeta, \overline{\zeta}) + m_S$; this class was found by Kerr (1998). But then a coordinate transformation $u' = u - m_S/a$ eliminates m_S: the Kerr–Schild transformation is equivalent to a trivial coordinate transformation.

Twisting type N solutions of the generalized Kerr–Schild form do not exist (Xanthopoulos 1983b).

Similarly, for $\rho = 0$ a generalized Kerr–Schild transformation amounts to a transformation $\widetilde{H} = H + S$ that leaves P and W fixed and that thus is a special case (W^0 fixed, $\Phi_1 = 0 = \Phi_2$) of the method described in Theorem 31.1, see §§31.1, 31.4–31.5. We write $\widetilde{G}^0 = G^0 + G^0_S$ and $\widetilde{H}^0 = H^0 + H^0_S$; then because of (31.24$a$) and (31.25) the equations

$$G^0_{S,\zeta} = 0, \quad 2P^2 H^0_{S,\zeta\overline{\zeta}} + P^2\left(\overline{W}_{,v} H^0_S\right)_{,\zeta} + P^2\left(W_{,v} H^0_S\right)_{,\overline{\zeta}} = 0 \tag{32.95}$$

have to be satisfied. The general solution is not known, except for the type III and N cases (31.38) and (31.34) with $[H^0_S/(\zeta + \overline{\zeta})]_{,\zeta\overline{\zeta}} = 0$ and $H^0_{S,\zeta\overline{\zeta}} = 0$.

32.5.3 Vacuum to electrovac

Only special cases have been considered so far. If $\boldsymbol{k}$ is geodesic and shear-free, then both metrics are algebraically special and $\boldsymbol{k}$ is also an eigenvector of the Maxwell field (aligned case). As in the vacuum to vacuum case, one can go through the Einstein–Maxwell equations in the relevant Chapters 30, 31 and 24 and try to detect such Kerr–Schild transformations.

Examples are the charged vacuum metrics of Theorem 30.2 (a special case has been discussed by Bhatt and Vaidya (1991)) and the addition of a Maxwell field (F or Ψ, respectively) to the solutions (24.35) and (24.37a) admitting a null Killing vector, cp. also Gavrilina (1983) and Garfinkle (1991). For a geodesic but shearing $\boldsymbol{k}$, Kupeli (1988a) found an example of a generalized Kerr–Schild transformation leading to

$$\begin{aligned} \mathrm{d}\tilde{s}^2 &= f^{-2}(r^{2a^+}\mathrm{d}x^2 + r^{2a^-}\mathrm{d}y^2) - 2\mathrm{d}u\mathrm{d}r + 2r^{-s}f^2\mathrm{e}^{\alpha x+\beta y-\gamma u}\mathrm{d}u^2, \\ f &= (\eta + s^{-1}r^{-s})^{-1}, \quad a^+ = 1/2, \quad a^- = (2\pm 2\sqrt{2})^{-1}, \\ s &= (2+\sqrt{2})/(2+2\sqrt{2}), \quad \eta = \text{const.} \end{aligned} \tag{32.96}$$

It admits the two commuting spacelike Killing vectors, $\partial_x - (\alpha/\gamma)\partial_u$ and $\partial_y - (\beta/\gamma)\partial_u$, and starts from a special *pp*-wave (where ∂_u is covariantly constant).

32.5.4 Perfect fluid to perfect fluid

If $\boldsymbol{k}$ is geodesic (and perhaps shearfree) in flat space-time, then it remains so under any conformal transformation; this makes conformally flat space-times

$$\mathrm{d}s^2 = g_{ab}\mathrm{d}x^a\mathrm{d}x^b = \phi^2(2\mathrm{d}\zeta\mathrm{d}\bar{\zeta} - 2\mathrm{d}u\,\mathrm{d}v) \tag{32.97}$$

good candidates for parent metrics g_{ab}: taking a null tetrad

$$\begin{aligned} \boldsymbol{\omega}^1 &= \phi(\mathrm{d}\zeta + Y\mathrm{d}v), \quad \boldsymbol{\omega}^2 = \phi(\mathrm{d}\bar{\zeta} + \overline{Y}\mathrm{d}v), \\ \boldsymbol{\omega}^3 &= \overline{Y}\mathrm{d}\zeta + Y\mathrm{d}\bar{\zeta} + Y\overline{Y}\mathrm{d}v + \mathrm{d}u, \quad \boldsymbol{\omega}^4 = \phi^2\mathrm{d}v, \end{aligned} \tag{32.98}$$

one can use the results of §32.1 (Taub 1981).

Starting with a conformally flat perfect fluid solution (cp. §37.4.2), several classes of solutions have been found. If $\boldsymbol{k}$ is shearing but non-twisting ($\sigma \neq 0$, $\rho - \bar{\rho} = 0$), then one can take a Friedmann metric in the form

$$\begin{aligned} \mathrm{d}s^2 &= -2G\mathrm{d}t\,\mathrm{d}u + 2G^2M\mathrm{d}u^2 + t^{1-c}\left(\mathrm{d}x + G_{,x}t^c\mathrm{d}u/c\right)^2 \\ &\quad + t^{1+c}\left(\mathrm{d}y - G_{,y}t^{-c}\mathrm{d}u/c\right)^2, \\ M(t) &= 2t(at^c + bt^{-c}), \quad G(x,y) = \sin(2c\sqrt{a}\,x)\,h(y), \\ \boldsymbol{k} &= \partial_t, \quad h_{,yy} = -(2c)^2bh, \quad \sigma = c/2t \neq 0, \quad a,b,c = \text{ const,} \end{aligned} \tag{32.99a}$$

and perform a Kerr–Schild transformation (32.84) with

$$S(x,y) = \frac{\text{const}}{G(x,y)}\left[\frac{\sin(2c\sqrt{a}\,x)}{h(y)}\right]^{1/c}, \quad k_a\mathrm{d}x^a = -G\mathrm{d}u, \tag{32.99b}$$

(Martín and Senovilla 1986, Senovilla and Sopuerta 1994). The resulting perfect fluid solutions are of Petrov type D and admit a Killing vector ∂_u (and a second Killing vector ∂_y if $b = 0$, $h = 1$, leading to a diagonal non-separable metric). The four-velocity $\widetilde{u}_a$, $\widetilde{u}_a \mathrm{d}x^a = -\mathrm{d}t\,[2M(t) + 2S(x,y)]^{-1/2}$, does *not* lie in the plane spanned by the two null eigenvectors of the Weyl tensor. If $\boldsymbol{k}$ is non-shearing and non-twisting ($\sigma = 0 = \rho - \overline{\rho}$), several classes have been constructed by Martín and Senovilla (1986, 1988). Among them are the type D solutions (33.11) and several static spherically symmetric perfect fluids. Sopuerta (1998b) studied the case of a non-shearing but expanding $\boldsymbol{k}$, with a rigidly rotating perfect fluid, and found a class of new solutions (of Petrov types D and II).

All solutions (33.9) with $H^0 \neq 0$ are of the generalized Kerr–Schild form (32.84).

All static spherically-symmetric space-times can be written in the generalized Kerr–Schild form

$$\mathrm{d}\widetilde{s}^2 = \mathrm{e}^{2U(r)}\mathrm{d}s_0^2 - 2S(r)(\mathrm{d}t \pm \mathrm{d}r)^2, \tag{32.100}$$

where $\mathrm{d}s_0^2$ is a Minkowski space (Mitskievic and Horský 1996), but the first part is not necessarily a perfect fluid metric.

Starting with a perfect fluid LRS metric

$$\mathrm{d}s^2 = -\mathrm{d}t^2 + A^2(t)\mathrm{d}x^2 + B^2(t)\left[\mathrm{d}y^2 + \Sigma^2(y,k)\mathrm{d}z^2\right], \tag{32.101}$$

see §13.1, one can make the transformation (Senovilla 1987a)

$$\mathrm{d}\widetilde{s}^2 = \mathrm{d}s^2 + A^2(t)f(t)(\mathrm{d}x \pm \mathrm{d}t/A)^2, \quad f(t) = \left(\frac{B}{A}\right)^2\left[c_1\int\frac{A}{B^4}\mathrm{d}t + c_2\right]. \tag{32.102}$$

The new metric is again of the LRS class, with pressure and mass density being related by

$$\widetilde{\mu} + \widetilde{p} = (\mu + p)(1 + f). \tag{32.103}$$

Patel and Vaidya (1983) showed that a generalized Kerr–Schild transformation

$$\mathrm{d}\widetilde{s}^2 = a^2\left[\mathrm{d}x^2 + \mathrm{d}y^2 + \tfrac{1}{2}\mathrm{e}^{2x}\mathrm{d}z^2 - (\mathrm{d}t + \mathrm{e}^x\mathrm{d}z)^2\right] - 2S(y)(\mathrm{d}y - \mathrm{e}^x\mathrm{d}z - \mathrm{d}t)^2, \tag{32.104}$$

$S(y) = c\sin(y\sqrt{2})$, $c =$ const, $\kappa_0 p = (a^2 - 2S)/2a^4$, $\kappa_0(\mu + 3p) = 2/a^2$, relates the Gödel solution (12.26) – with $S = 0$ – to a subclass of the Wahlquist solution (21.57) which admits four Killing vectors and is thus contained in (13.2), and in a similar way the Einstein cosmos (12.24) to the subclass of the Wahlquist solution (21.58) given by Vaidya (1977), cp. also Taub (1981).

33
Algebraically special perfect fluid solutions

Most of the algebraically special perfect fluid solutions admit symmetries, and have been found exploiting these symmetries. In this chapter we want to present some methods of characterizing and constructing solutions which do not rely on symmetries, and to indicate in which of the other chapters algebraically special perfect fluid solutions can be found. For a detailed discusssion of these solutions see also Krasiński (1997).

33.1 Generalized Robinson–Trautman solutions

Generalized Robinson–Trautman solutions are characterized by the following set of assumptions:
(i) The multiple null eigenvector $\boldsymbol{k}$ of the Weyl tensor is geodesic, shear-free, and twistfree but expanding

$$\Psi_0 = \Psi_1 = 0, \quad \kappa = \sigma = \omega = 0, \quad \rho = \bar{\rho} \neq 0. \tag{33.1}$$

(ii) The energy-momentum tensor is that of a perfect fluid,

$$R_{ab} = \kappa_0(\mu + p)u_a u_b + \kappa_0(\mu - p)g_{ab}/2. \tag{33.2}$$

(iii) The four-velocity $\boldsymbol{u}$ of the fluid obeys

$$u_{[a;b}u_{c]} = 0, \quad k_{[c}k_{a];b}u^b = 0. \tag{33.3}$$

Introducing the null tetrad $(\boldsymbol{m}, \overline{\boldsymbol{m}}, \boldsymbol{l}, \boldsymbol{k})$, one sees that (33.1) and (33.3) imply that τ can be made zero by choice of $\boldsymbol{l}$ and that then $\boldsymbol{u}$ lies in the plane spanned by $\boldsymbol{l}$ and $\boldsymbol{k}$. With this choice, (33.2) yields

$$R_{11} = R_{14} = R_{13} = 0. \tag{33.4}$$

From now on, the calculations run in close analogy with those for the Robinson–Trautman solutions in Chapters 27 and 28. We will present here only the main results, all due to Wainwright (1974).

As a first step, one can infer from (33.1) and (33.4) that coordinates $\zeta, \bar{\zeta}, r, u$ can be introduced such that the metric takes the form

$$\begin{aligned} ds^2 &= 2\chi^2(r,u)P^{-2}(\zeta,\bar{\zeta},u)d\zeta\, d\bar{\zeta} - 2du\, dr - 2H(\zeta,\bar{\zeta},r,u)du^2, \\ \rho &= -\partial_r \ln\chi; \quad H = -r\partial_u \ln P - H^0(\zeta,\bar{\zeta},u) - S(r,u), \end{aligned} \tag{33.5}$$

the (irrotational) four-velocity $\boldsymbol{u}$ having the components

$$B\sqrt{2}\, u_i = (0, 0, -1, -H - B^2). \tag{33.6}$$

As a second step, the so far unknown real functions χ, P, H^0, S and B should be determined from the rest of the equations (33.2)–(33.3). Two of the field equations (33.2) can be regarded as giving the energy density μ and the pressure p in terms of the metric functions, so only two of the equations (33.2) need to be considered. These ensure the correct algebraic structure for R_{ab} (i.e. the isotropy of pressure). It eventually turns out that three different cases occur.

The *first case* is characterized by the Newman–Penrose coefficient ν being zero, i.e. by

$$\partial_u\partial_\zeta \ln P = 0, \quad \partial_\zeta H^0 = 0, \quad K = 2P^2\partial_\zeta\partial_{\bar{\zeta}} \ln P = \text{ const.} \tag{33.7}$$

Obviously, these solutions admit a G_3 acting on the 2-spaces $r =$ const, $u =$ const, of constant curvature K, and are therefore of Petrov type D or O. The corresponding metrics (including the spherically-symmetric solutions) are discussed in detail in Chapters 15 and 16.

In the *second case* ($\nu \neq 0$, χ^2 quadratic in r, $a \neq 0$) the metric is given by (33.5)–(33.6) with

$$\begin{aligned} &\chi^2 = \varepsilon(r^2 - a^2), \quad H^0 = -\varepsilon K/2, \quad B^2 = -H + c\chi^2, \quad \varepsilon = \pm 1, \\ &S = \chi^2(b - 3m\int \chi^{-4}dr) + \varepsilon a^2 c, \quad K = 2P^2\partial_\zeta\partial_{\bar{\zeta}} \ln P, \end{aligned} \tag{33.8a}$$

where the real constants a, b, c and m and the function P have to satisfy

$$\begin{aligned} &cm = 0, \quad c(c + 2b) = 0, \\ &P^2\partial_\zeta\partial_{\bar{\zeta}}K + a^2P^2\partial_u\partial_u(P^{-2}) + 6\varepsilon\, m\, \partial_u \ln P + 2ca^2(K - 2ca^2) = 0. \end{aligned} \tag{33.8b}$$

In the *third case* ($\nu \neq 0$, χ^2 linear in r), the metric is given by (33.5)–(33.6) with

$$\chi^2 = \varepsilon r, \quad S = \varepsilon ra + \varepsilon br \ln \varepsilon r, \quad B^2 = -H, \tag{33.9a}$$

where P and H^0 have to satisfy

$$2P^2\partial_\zeta\partial_{\bar{\zeta}} \ln P - \varepsilon\partial_u \ln P = b, \quad 2\partial_\zeta\partial_{\bar{\zeta}}H^0 + \varepsilon\partial_u(H^0P^{-2}) = 0. \tag{33.9b}$$

The expressions for μ, p and the non-vanishing components Ψ_2, Ψ_3 and Ψ_4 of the Weyl tensor in the two cases (33.8) and (33.9) are given in Wainwright's paper. They show that no dust solutions and no solutions of type *III* or *N* are possible.

Besides the Robinson–Trautman vacuum solutions (see Chapter 28), which are contained in (33.8) as the subcase $a = b = 0$ ($\Rightarrow \varepsilon = 1$, $S = m/r$), only a few classes of solutions to (33.8) or (33.9) have been found. All of them rely on a similarity reduction (Rainer and Stephani 1999).

If the ζ–$\bar{\zeta}$-space has a constant curvature $K = K(u)$, one gets

$$P^2 = h^{-1}(u)\left(1 + k\zeta\bar{\zeta}/2\right)^2, \quad K(u) = k/h(u), \quad k = 0, \pm 1, \tag{33.10}$$

and an ordinary differential equation for $h(u)$ which can easily be solved (Rainer and Stephani 1999); the metrics necessarily admit a G_3 on V_2 (for (33.9): if H_0 is chosen appropriately). Included here is the subcase $\partial_u P = b = 0$ of (33.9) discussed by Wainwright (1974). With the choice $\varepsilon = 1$ $(r > 0)$ and $S = 2r$, it reads

$$\begin{aligned} ds^2 &= 2r d\zeta\, d\bar{\zeta} - 2du\, dr + 2[H^0(\zeta, \bar{\zeta}, u) + 2r]du^2, \\ \kappa_0\mu &= 3r^{-1} + H^0 r^{-2}/2, \quad \kappa_0 p = -r^{-1} + H^0 r^{-2}/2, \\ \sqrt{2}u_a &= -(H^0 + 2r)^{-1/2}\delta_a^3, \quad (2\partial_\zeta\partial_{\bar{\zeta}} + \partial_u)H^0 = 0. \end{aligned} \tag{33.11}$$

If $\partial_\zeta H^0 = 0$, we return to (33.7). Non-trivial ($\partial_\zeta H^0 \neq 0$) real solutions of the linear (heat conduction) equation for H_0 can easily be constructed. The solutions with $\partial_u H^0 = 0$ have been discussed in the context of the generalized Kerr–Schild transformation, see §32.5.4.

The only known solutions with $2P^2\partial_\zeta\partial_{\bar{\zeta}}K = 0$, $K_{,\zeta} \neq 0$ are

$$\begin{aligned} ds^2 &= 2\varepsilon(r^2 - 1)h(u)\left(\zeta + \bar{\zeta}\right)^{-3} d\zeta d\bar{\zeta} - 2du dr - 2H du^2, \\ 2H &= r\frac{h'(u)}{h(u)} - \frac{3\varepsilon(\zeta + \bar{\zeta})}{h(u)} + 2\varepsilon(r^2 - 1)\left[b - 3m\int \frac{dr}{(r^2 - 1)^2}\right], \\ h &= Au + B \quad \text{for} \quad m = 0, \qquad h = Ae^{3\varepsilon m u/a^2} + B \quad \text{for} \quad m \neq 0, \end{aligned} \tag{33.12}$$

cp. (28.16).

For $2P^2\partial_\zeta\partial_{\bar{\zeta}}K \neq 0$, assuming $m = 0$ and $P = P(\zeta + \bar{\zeta})$, and in the gauge $a = 1$, Kramer (1984b) found the special solution of (33.8*b*)

$$K = 2c\left[w(\zeta + \bar{\zeta}) + 3\right], \quad P^{-2} = -e^{w+2}w/2c, \quad w' = we^{1+w/2}\sqrt{2}. \tag{33.13}$$

Taking w and $\varphi = \mathrm{e}\,(\zeta - \overline{\zeta})\,/\,\mathrm{i}\,\sqrt{2}$ as new coordinates, the resulting Petrov type II metric (with Killing vectors ∂_u and ∂_φ and an equation of state $\kappa_0(\mu - p) = -6\varepsilon c$) is

$$\mathrm{d}s^2 = \frac{(1-r^2)}{2\varepsilon c}\left[\frac{\mathrm{d}w^2}{4w} + w\mathrm{e}^w\mathrm{d}\varphi^2\right] - 2\mathrm{d}u\,\mathrm{d}r - \varepsilon c(r^2 + 2w + 3)\mathrm{d}u^2. \quad (33.14)$$

Other known solutions of this class are given by (33.8a) and

$$\begin{aligned} P^2(z) &= \left(\mathrm{e}^{2Az} + 2B\mathrm{e}^{Az} + B^2 - A^2\right)\mathrm{e}^{-Az}/4A^2, \\ z &= \zeta + \overline{\zeta} + u, \; c = 0 = m \end{aligned} \quad (33.15)$$

(Drauschke 1996) or

$$\begin{aligned} P^2(z) &= \mathrm{e}^{3\varepsilon m u}\left(\mathrm{e}^{2Az} + 2B\mathrm{e}^{Az} + B^2 - A^2\right)\mathrm{e}^{-Az}/4A^2, \\ z &= \zeta + \overline{\zeta} + 2\mathrm{e}^{-3\varepsilon m u}/3\varepsilon m, \; c = 0, m \neq 0, \end{aligned} \quad (33.16)$$

and by (33.9) and

$$P^2 = \exp\,(Az) + \varepsilon/2, \; z = \zeta + \overline{\zeta} + u, \; b = 0, \quad (33.17)$$

$$P^2 = \mathrm{e}^{-2\varepsilon b u}\,[\exp(Az) - 1]\,/A; \; z = \zeta + \overline{\zeta} + \varepsilon\mathrm{e}^{-2\varepsilon b u}/b, \; b \neq 0, \quad (33.18)$$

$$P^2 = \tfrac{1}{2}[B\mathrm{e}^{-4\varepsilon b u} - b(\zeta + \overline{\zeta})^2], \quad b \neq 0 \quad (33.19)$$

(Rainer and Stephani 1999).

Looking for solutions which satisfy the above conditions (i)–(iii) *except* $u_{[a;b}u_{c]} = 0$, Bonnor and Davidson (1985) considered metrics of the form

$$\mathrm{d}s^2 = 2\chi^2(r)P^{-2}(\zeta,\overline{\zeta})\mathrm{d}\zeta\,\mathrm{d}\overline{\zeta} - 2\mathrm{d}r\,\mathrm{d}u - 2[H_0(\zeta,\overline{\zeta}) + S(r)]\mathrm{d}u^2, \quad (33.20)$$

with a (non-expanding, shearfree but rotating) four-velocity parallel to the Killing vector ∂_u. From the resulting field equations

$$\partial_\zeta\partial_{\overline{\zeta}}H_0 = 0, \quad P^2\partial_\zeta\partial_{\overline{\zeta}}\ln P = (H_0 + S)(\chi'^2 - \chi\chi'') - S''\chi^2/2, \quad (33.21)$$

they found an equation of state $\kappa_0(3p + \mu) = \mathrm{const} = 4n$ and the special solutions

$$\begin{aligned} S &= n[1 - (r + m)\cot r], & \chi &= \sqrt{a}\sin r, \\ P^2 &= a\sqrt{2}(\zeta + \overline{\zeta})^3, & H_0 &= -3(\zeta + \overline{\zeta})/\sqrt{2}. \end{aligned} \quad (33.22)$$

Algebraically special solutions with a geodesic and twistfree, but expanding and *shearing*, multiple null eigenvector have been considered in detail by Oleson (1972). In coordinates with $\boldsymbol{k} = \partial_u$ and four-velocity $u_a = Bt_{,a}$, he gave as an explicit solution

$$\begin{aligned} \mathrm{d}s^2 &= t^{3/2}(\mathrm{d}x + 3a_1x\mathrm{d}u)^2 + t^{1/2}\left(\mathrm{d}y - [m(u)x + a_1y]\mathrm{d}u\right)^2 \\ &\quad -2\mathrm{d}t\mathrm{d}u - 2\left(5a_1t - \tfrac{1}{2}\mathrm{e}^{4a_1u}\int m^2\mathrm{e}^{-4a_1u}\mathrm{d}u\right)\mathrm{d}u^2, \end{aligned} \tag{33.23}$$

cp. also (33.48).

33.2 Solutions with a geodesic, shearfree, non-expanding multiple null eigenvector

In this section, we assume that the multiple null eigenvector $\boldsymbol{k}$ satisfies

$$\Psi_0 = \Psi_1 = 0, \quad \kappa = \sigma = \Theta = 0, \quad \rho = -\bar{\rho} \neq 0. \tag{33.24}$$

Note that, because of (6.33), $\kappa = \sigma = \rho = 0$ implies $\mu + p = 0$. We will exclude this case here, i.e. in what follows $\boldsymbol{k}$ is necessarily twisting ($\omega \neq 0$). We shall again present only the main results, all due to Wainwright (1970).

As a first step in solving the field equations, one uses (33.24) and parts of (33.2) to introduce a suitable tetrad and coordinate system. These are found to be

$$\begin{aligned} \mathrm{d}s^2 &= 2P^{-2}\mathrm{d}\zeta\mathrm{d}\bar{\zeta} - 2[\mathrm{d}u + L\mathrm{d}\zeta + \overline{L}\mathrm{d}\bar{\zeta}] \\ &\quad \times[\mathrm{d}r + W\mathrm{d}\zeta + \overline{W}\mathrm{d}\bar{\zeta} + H\{\mathrm{d}u + L\mathrm{d}\zeta + \overline{L}\mathrm{d}\bar{\zeta}\}], \\ m^i &= P(-1, 0, W, L), \quad \overline{m}^i = P(0, -1, \overline{W}, \overline{L}), \\ l^i &= (0, 0, -H, 1), \quad k^i = (0, 0, 1, 0), \quad \rho = \mathrm{i}, \end{aligned} \tag{33.25}$$

where $L(\zeta, \bar{\zeta})$ is purely imaginary and determined by the real function $P(\zeta, \bar{\zeta})$ as

$$\partial_x L = -\partial_x \overline{L} = \mathrm{i}P^{-2}\sqrt{2}, \quad x\sqrt{2} \equiv \zeta + \bar{\zeta}. \tag{33.26}$$

The coordinates and the tetrad (33.25) are of the Robinson–Trautman type (27.27) if one inserts into (27.27) the value $\rho = \mathrm{i}$ and makes P in (27.12) imaginary instead of real (for the remaining coordinate and tetrad freedom see §§27.1.3 and 29.1.4). Equation (33.26) corresponds to the first part of (27.26), the second part being invalid here. The four-velocity $\boldsymbol{u}$ turns out to have the form

$$u^a\sqrt{2} = Bk^a + B^{-1}l^a, \quad B^2 = \kappa_0(\mu + p)/4. \tag{33.27}$$

As a second step, one has to solve the remaining field equations to obtain explicit expressions for the so far unknown functions $W(\zeta, \bar{\zeta}, u)$ and $H(\zeta, \bar{\zeta}, r, u)$. Three different cases can occur.

The *first case* is characterized by $W = 0$, $\partial_r\partial_r(H + \kappa_0 p/2) \neq 0$. These solutions are the case $IIIa$, $\alpha + E = \tau = 0$ of the locally rotationally-symmetric solutions of Stewart and Ellis (1968).

The *second case* is characterized by $k_{a;b} + k_{b;a} = 0$. It includes all algebraically special (not conformally flat) perfect fluid solutions whose multiple null eigenvector $\boldsymbol{k}$ is a (twisting) Killing vector. $W(\zeta, \bar{\zeta})$, chosen to be purely imaginary, and $H(\zeta, \bar{\zeta})$ are given by

$$\partial_x W = -\partial_x \overline{W} = \mathrm{i}\kappa_0 p P^{-2}\sqrt{2}, \quad 2H = -\kappa_0 p, \tag{33.28}$$

and $P(\zeta, \bar{\zeta})$, $p(\zeta, \bar{\zeta})$ and $\mu(\zeta, \bar{\zeta})$ have to obey

$$4P^2\partial_\zeta\partial_{\bar{\zeta}} \ln P = \kappa_0(\mu - 5p), \quad 8P^2\partial_\zeta\partial_{\bar{\zeta}} p = \kappa_0(p - \mu)(\mu + 3p). \tag{33.29}$$

To get the explicit metric, one has to solve (33.29) for P, p and μ (one function can be prescribed) and then to determine H, L and W from (33.28) and (33.26). Besides $\boldsymbol{k}$, the metric admits a second Killing vector $\boldsymbol{\xi} = \partial_u$. Unless space-time is conformally flat, the four-velocity $\boldsymbol{u}$ has non-zero rotation.

Particular solutions include the Gödel universe (12.26), the Einstein universe (12.24) and the Petrov type II solution

$$\begin{aligned} &\mu = p = x, \ P^2 = \tfrac{4}{3}\kappa_0 x^3, \ 2H = -\kappa_0 x, \\ &W = -\mathrm{i}\tfrac{3}{4}\sqrt{2}/x, \ L = -\mathrm{i}\tfrac{3}{8}\sqrt{2}/\kappa_0 x^2, \end{aligned} \tag{33.30}$$

which admits a G_3 simply transitive on $x =$ const, see §13.4.

The *third case* is characterized by $\partial_r(H + \kappa_0 p/2) = 0$, $\partial_r H \neq 0$. The only non-trivial field equation remaining to be solved is

$$\begin{aligned} &P^2\partial_\zeta\partial_{\bar{\zeta}} K = \kappa_0 m(\kappa_0 m + K), \\ &K \equiv 2P^2\partial_\zeta\partial_{\bar{\zeta}} \ln P, \quad 2m \equiv \mu + 3p = \text{const}. \end{aligned} \tag{33.31}$$

Having found a solution $P(\zeta, \bar{\zeta})$ of this equation, one can obtain the other metric functions and the pressure p from (33.26) and the following set of equations

$$\begin{aligned} &\partial_x W(\zeta, \bar{\zeta}) = -\mathrm{i}(2\kappa_0 m + K)/4P^2\sqrt{2}, \quad W = -\overline{W} \\ &\partial_\zeta \ln R(\zeta, \bar{\zeta}) = 2\mathrm{i}W, \quad 2H(\zeta, \bar{\zeta}, r) = \kappa_0(m - p), \\ &8\kappa_0 p(\zeta, \bar{\zeta}, r) = 4(R\mathrm{e}^{2\mathrm{i}r} + \overline{R}\mathrm{e}^{-2\mathrm{i}r}) + 2\kappa_0 m - K. \end{aligned} \tag{33.32}$$

These metrics admit a Killing vector collinear with the four-velocity

$$u^i = (0, 0, 0, [\kappa_0(m - p)]^{-1/2}), \tag{33.33}$$

which is, therefore, expansion- and shear-free (rigid motion). Only Petrov types II and D can occur; the metrics of type D are subcases or limits of the Wahlquist solution (21.57).

Special solutions (of Petrov type D) are

$$K + \kappa_0 m = 0 \quad \Rightarrow \quad W = 0, \quad R = R(\bar{\zeta}), \tag{33.34}$$

which for $R =$ const contains a solution admitting a G_4 on $r =$ const, cp. §13.4, and the solution (Kramer 1984c)

$$K = -\kappa_0 m \left[w(\zeta + \bar{\zeta}) + 3 \right], \quad P^{-2} = e^{w+2} w / \kappa_0 m, \quad w' = w e^{1+w/2}\sqrt{2}, \tag{33.35}$$

see also (33.13), where the same functions appear in a different solution.

33.3 Type D solutions

Most of the known algebraically special perfect fluid solutions are of Petrov type D, but, as the following review shows, only a minor subset of all such type D solutions is known.

In the type D solutions, the two multiple null eigenvectors $\boldsymbol{l}$ and $\boldsymbol{k}$ define a preferred 2-space Σ. The four-velocity $\boldsymbol{u}$ may or may not lie in Σ.

For $u_{[a}k_b l_{c]} \neq 0$, most of the known solutions admit an Abelian G_2 of motions, cp. Chapters 21 and 23, and many correspond to the interior of a rigidly rotating body. Examples with a G_2 are the Wahlquist solution (21.57) and its limits, see §21.2.3, the solutions due to Mars and Senovilla (1994) and Mars and Wolf (1997) (admitting an additional conformal motion), and a solution belonging to the Kerr–Schild class (Martín and Senovilla 1986), see §32.5.4. An example of a solution with only a G_1 (but two conformal symmetries) is the metric (35.78) due to Koutras and Mars (1997).

For

$$u_{[a}k_b l_{c]} = 0, \tag{33.36}$$

all known solutions have the property that the rotation $\omega^a = \varepsilon^{abcd} u_b u_{c;d}$ and shear σ_{ab} of the velocity field furthermore obey

$$\omega_{[a}k_b l_{c]} = 0, \quad k^d \sigma_{d[a}k_b l_{c]} = 0. \tag{33.37}$$

Following Wainwright (1977b), we classify all solutions satisfying (33.36) and (33.37) according to the acceleration $\dot{\boldsymbol{u}}$ and the Newman–Penrose coefficients κ, ν, σ and λ (it can be shown that for the solutions in question $\kappa = 0$ ($\sigma = 0$) if and only if $\nu = 0$ ($\lambda = 0$)). For none of the subcases is the complete list of solutions known.

Solutions which fulfil (33.36) and have an equation of state $p = p(\mu)$ and a zero magnetic part of the Weyl tensor, satisfy either $\mathrm{d}p/\mathrm{d}\mu = 0, 1$ or admit at least three Killing vectors (Carminati and Wainwright 1985). An example with an equation of state $p = \mu + \text{const}$ is the metric (21.74).

Many of the solutions with special subspaces considered in Chapter 36 are of type D; the classification with respect to the above given classes has not yet been performed for all of them.

33.3.1 Solutions with $\kappa = \nu = 0$

The subcase $\sigma = \lambda = 0$, $\dot{u}_{[a}k_b l_{c]} = 0$ contains all the locally rotationally symmetric perfect fluid solutions considered by Stewart and Ellis (1968), see Wainwright (1977a). The type D solutions covered by §§33.1, 33.2 also belong to this subcase. The spherically-symmetric perfect fluid and dust solutions and the Kantowski–Sachs models (see Chapters 15 and 16 and §14.3) are important examples.

The subcase $\sigma = \lambda = 0$, $\dot{u}_{[a}k_b l_{c]} \neq 0$ contains a solution found by Barnes (1973b); it can be obtained from the solution (18.64) by allowing A and B to be t-dependent.

The subcase $\sigma \neq 0$, $\lambda \neq 0$, $\dot{u}_{[a}k_b l_{c]} = 0$ contains the Ozsváth solution (12.29), admitting a G_4 on V_4, the Bianchi type I solution (14.28b) and the Allnutt solution (case $m = 0$ of (23.13)), admitting an Abelian G_2.

No solutions are known for which σ, λ and $\dot{u}_{[a}k_b l_{c]}$ are all non-zero.

33.3.2 Solutions with $\kappa \neq 0$, $\nu \neq 0$

No solutions are known for which κ, ν, σ and λ are all non-zero. The static Barnes solutions (18.65) and (18.66) and their generalizations (21.61) to a rigidly rotating fluid are examples with $\sigma = \lambda = 0$, but $\dot{u}_{[a}k_b l_{c]} \neq 0$.

A large variety of explicitly known solutions occurs in the subcase $\sigma = \lambda = \dot{u}_{[a}k_b l_{c]} = 0$. They all have a metric of the form

$$\mathrm{d}s^2 = 2\mathrm{e}^{2b}\mathrm{d}\zeta\mathrm{d}\bar{\zeta} + \mathrm{e}^{2a}\mathrm{d}r^2 - \mathrm{d}t^2, \quad \boldsymbol{u} = \partial_t, \tag{33.38}$$

and have been considered by Szekeres (1975) (for dust), Szafron (1977), Tomimura (1977), Szafron and Wainwright (1977) and Wainwright (1977a). To ensure $\kappa \neq 0$, $\sigma \neq 0$, $a_{,\zeta}$ must be non-zero.

Two cases have to be distinguished, $b_{,r} \neq 0$ and $b_{,r} = 0$. In both cases, the mass density μ can be computed from

$$\kappa_0(\mu + 3p) = -2(\ddot{a} + \dot{a}^2 + 2\ddot{b} + 3\dot{b}^2). \tag{33.39}$$

If b depends on r, then the metric must have the form

$$\mathrm{d}s^2 = \Phi^2(r,t)\left[2P^{-2}(\zeta,\bar{\zeta},r)\mathrm{d}\zeta\,\mathrm{d}\bar{\zeta} + (\partial_r \ln\left\{\Phi P^{-1}\right\})^2\mathrm{d}r^2 - \mathrm{d}t^2\right], \tag{33.40a}$$

where the ζ–$\bar{\zeta}$-space is a space of constant curvature $K(r)$,

$$P(\zeta,\bar{\zeta},r) = \alpha(r)\zeta\bar{\zeta}+\beta(r)\zeta+\bar{\beta}(r)\bar{\zeta}+\delta(r), \quad K(r) = 2(\alpha\delta-\beta\bar{\beta}), \tag{33.40b}$$

and the function $\Phi(r,t)$ is a solution of the ordinary (Friedmann-type) differential equation

$$2\Phi\ddot{\Phi} + \dot{\Phi}^2 + \kappa_0 p(t)\Phi^2 = 1 - K(r). \tag{33.40c}$$

To get an explicit solution, one has to prescribe the pressure $p(t)$ and the functions $\alpha(r)$, $\delta(r)$ (both real) and $\beta(r)$ (complex), and then to solve (33.40*c*). For dust ($p = 0$), or a constant $p = p_0$ (Covarrubias 1984), the differential equation (33.40*c*) is integrated by

$$\dot{\Phi}^2 - 2m(r)\Phi^{-1} + \kappa_0 p_0\phi^2/3 = 1 - K(r). \tag{33.41}$$

In the dust case (Szekeres 1975), the solutions of (33.41) are of the form (15.37). In general, these dust solutions contain five arbitrary functions of r, and admit no Killing vector (Bonnor *et al.* 1977). Surprisingly, they can be matched to the spherically-symmetric, static, exterior Schwarzschild solution (Bonnor 1976). In the case of a non-zero p_0, (33.41) can be integrated in terms of elliptic functions (Barrow and Stein-Schabes 1984).

For a perfect fluid (with $p \neq 0$), solutions of (33.40*c*) with

$$\kappa_0 p(t) = \varphi^2(r)\Phi^{-2}(r,t) \tag{33.42}$$

can easily be obtained from the dust solutions with $1-K(r) = \text{const}\ \varphi^2(r)$, $m = \text{const}\ \varphi^3$.

For $K = 1$, the solutions of (33.40*c*) are

$$K = 1, \quad \Phi = [g(r) + h(r)f(t)]^{2/3}\dot{f}^{-1/3}, \quad 3\kappa_0 p(t) = 2\,\dddot{f}/\dot{f} - 3(\ddot{f}/\dot{f})^2 \tag{33.43}$$

(Szafron 1977, Bona *et al.* 1987a).

If b does not depend on r, then the metric must have the form

$$\begin{aligned} \mathrm{d}s^2 = -\mathrm{d}t^2 + \Phi^2(t)\Big[2P^{-2}(\zeta,\bar{\zeta})\mathrm{d}\zeta\mathrm{d}\bar{\zeta} + \\ + \Big\{A(r,t) + P^{-1}\left(U(r)\zeta\bar{\zeta} + V(r)\zeta + \overline{V}(r)\bar{\zeta}\right)\Big\}^2\mathrm{d}r^2\Big], \end{aligned} \tag{33.44a}$$

with

$$P = 1 + k\zeta\bar{\zeta}/2, \quad k = 0, \pm 1, \tag{33.44b}$$

$$2\Phi\ddot{\Phi} + \dot{\Phi}^2 + \kappa_0 p(t)\Phi^2 = -k, \tag{33.44c}$$

$$\ddot{A}\Phi^2 + 3\dot{A}\dot{\Phi}\Phi - Ak = U. \tag{33.44d}$$

To get an explicit solution, one has to specify the pressure $p(t)$ (or $\Phi(t)$) and the functions $U(r)$ (real) and $V(r)$ (complex), and then to determine $\Phi(t)$ (or $p(t)$) from (33.44c) and $A(r,t)$ from (33.44d). The energy density μ can be obtained from (33.39). Equation (33.44c) has the same structure as (33.40c), so the remarks on the integrable cases apply also here.

For dust, the general solution of (33.44c)–(33.44d) is given in Szekeres (1975), see also Bonnor and Tomimura (1976), and the perfect fluid case $\kappa_0 p(t) = \mathrm{const}\ \Phi^{-2}$ has been solved by Tomimura (1977). Some other perfect fluid solutions, corresponding to the choice $k = 0$, $p(t) = \mathrm{const}\ t^{-2}$, have been constructed by Szafron and Wainwright (1977); the solution (37.50)–(37.51) of embedding class one is also included here. Introducing a new time variable τ by $\mathrm{d}t = \Phi^3(\tau)\mathrm{d}\tau$, (33.44$d$) can be written as $\partial^2 A/\partial\tau^2 = (Ak + U)\Phi^4$, which for $k = 0$ and prescribed $\Phi(t)$ can be solved by quadratures (Stephani 1987); the special case $U = 0$ was also solved by Bona *et al.* (1987a). For $k = 0$ and $p = \mathrm{const}$, the solution was given by Barrow and Stein-Schabes (1984), for $k = 0$ and $\kappa_0 p = 1/t^4$ it is the metric (37.50) of embedding class one.

33.4 Type *III* and type *N* solutions

The only known type *III* perfect fluid solution was given by Allnutt (1981) as

$$\begin{aligned} \mathrm{d}s^2 = {} & 2\mathrm{e}^{3t+4x}\mathrm{d}x^2 + (3\mathrm{e}^{t+2(x-y)} + \mathrm{e}^{3t+6x}/a)\mathrm{d}u^2 \\ & + a\mathrm{e}^t(\mathrm{e}^{2y}-1)^{-1}\mathrm{d}y^2 - 2\mathrm{e}^{2t+3x-y}\mathrm{d}u(\mathrm{d}t + 3\mathrm{d}x + \mathrm{d}y) \end{aligned} \tag{33.45}$$

($a = \mathrm{const}$). The (geodesic) repeated principal null vector $\boldsymbol{k} = \mathrm{e}^{-2t}\partial_t$ and the four-velocity $u_a\mathrm{d}x^a = (\mathrm{e}^{-t}/a - \tfrac{3}{2}\mathrm{e}^{-3t-3x})^{-1/2}\mathrm{d}t$ are irrotational, but shearing and expanding. The energy density μ and pressure p are given by

$$\kappa_0\mu = \tfrac{27}{4}\mathrm{e}^{-t}/a - \tfrac{9}{8}\mathrm{e}^{-3t-4x}, \quad \kappa_0 p = -\tfrac{21}{4}\mathrm{e}^{-t}/a - \tfrac{9}{8}\mathrm{e}^{-3t-4x}; \tag{33.46}$$

both are positive for $\mathrm{e}^{2y} > 1$ and $\mathrm{e}^{2t+4x} > 3a/2$. The only Killing vector is ∂_u.

The following theorems restrict the properties of possible type *III* and *N* solutions ($\mu + p = 0$ is always excluded).

Theorem 33.1 *Any Petrov type III shearfree perfect fluid solution with an equation of state $p = p(\mu)$ is rotating and has zero expansion* (Carminati and Cyganowski 1997).

Theorem 33.2 *No type N perfect fluid (or dust) solutions with zero acceleration* ($\dot{u}_a = 0$) *exist.* (Oleson 1972; cp. also Kundt and Trümper 1962 and Szekeres 1966b).

Theorem 33.3 *Any Petrov type N shearfree perfect fluid space-time in which the fluid satisfies an equation of state* $p = p(\mu)$ *is stationary and admits an Abelian* G_3 (Carminati 1988).

These solutions are a subcase of the rigidly rotating cylindrically-symmetric solutions (22.34).

Theorem 33.4 *The principal null congruence of a type N perfect fluid solution is geodesic if and only if the fluid is irrotational and pressure and energy density are related by*

$$\mu = p + A(t), \quad u_a = (-2H)^{-1/2} t_{,a}. \tag{33.47}$$

If it is geodesic, then it is twistfree but has non-zero shear and expansion (Kundt and Trümper 1962, Oleson 1971).

Oleson (1971, 1972) determined all metrics covered by Theorem 33.4, i.e. all type N perfect fluid solutions with a geodesic and twistfree but shearing and expanding principal null congruence. He gave the following two classes of solutions.

Class *I:* $\rho^{-1} = -2t$, $\sigma^{-1} = -4t$

$$\begin{aligned}
\mathrm{d}s^2 &= t^{3/2}(\mathrm{d}x - 2t^{-1/2}G_{,x}\mathrm{d}u)^2 + t^{1/2}(\mathrm{d}y + 2t^{1/2}G_{,y}\mathrm{d}u)^2 \\
&\quad -2G\mathrm{d}t\mathrm{d}u - 2G^2H\mathrm{d}u^2, \\
H(t,u) &= -2[a^2(u)t^{1/2} + b^2t^{3/2}], \quad G(x,y,u) = g(x,u)h(y,u), \\
&g_{,xx} + a^2(u)g = 0, \quad h_{,yy} + b^2h = 0, \quad b = \mathrm{const}, \\
4p &= 3t^{-3/2}(a^2 - 7b^2t), \quad \mu = p + 12b^2t^{-1/2}.
\end{aligned} \tag{33.48}$$

Class *II:* $\rho^{-1} = (1-t^2)/t$, $\sigma^{-1} = 2(t^2-1)$,

$$\begin{aligned}
\mathrm{d}s^2 &= \varepsilon(1-t^2)R^{-1}(\mathrm{d}x - \varepsilon RG_{,x}\mathrm{d}u)^2 + \varepsilon(1-t^2)R(\mathrm{d}y + \varepsilon R^{-1}G_{,y}\mathrm{d}u)^2 \\
&\quad -2G\mathrm{d}t\mathrm{d}u - 2G^2H\mathrm{d}u^2, \quad (\varepsilon,\lambda) = (1,1) \text{ or } (-1,\pm1),
\end{aligned}$$

$$
\begin{aligned}
H(t,u) &= -\varepsilon[\varepsilon(1-t^2)]^{1/2}[\lambda a^2(u) - b^2(t+1)], \quad b = \text{const},\\
R(t) &= [\varepsilon(1-t)/(1+t)]^{1/2}, \quad G(x,y,u) = g(x,u)h(y,u), \qquad (33.49)\\
g_{,xx} &+ \lambda a^2(u)g = 0, \quad h_{,yy} + [\varepsilon(a^2 - 2\lambda)b^2]h = 0,\\
2p &= 3\varepsilon[\varepsilon(1-t^2)]^{-3/2}[b^2(4t^3 - 5t - 1) + \lambda a^2],\\
\mu &= p + 12b^2t[\varepsilon(1-t^2)]^{-1/2}.
\end{aligned}
$$

For both classes, the fluid has non-zero shear, acceleration and expansion provided that $a'(u) \neq 0$; $a = \text{const}$ gives conformally flat solutions. Killing vectors (maximum number two) are possible only in Class I for metrics with $b = 0$.

Part IV
Special methods

34
Application of generation techniques to general relativity

In this chapter we shall apply some of the methods outlined in Chapter 10 to the Einstein or Einstein–Maxwell equations with two Killing vectors. Maison (1978, 1979) was the first to show that these equations, which for vacuum reduce to the Ernst equation, actually are an integrable system. Not all of the methods will be discussed in detail as was mentioned in Chapter 1. In particular we shall utilize harmonic maps (§10.8), 'exponentiate' solutions of the linearized equations (§10.5) and study Bäcklund transformations (§10.6) and the homogeneous Hilbert problem (§10.7). The applications of other generation methods are summarized at the end of this chapter. For reviews of some topics to be discussed below see Gürses (1984), Harrison (1986), Guo *et al.* (1983b) and Kordas (1999).

34.1 Methods using harmonic maps (potential space symmetries)

34.1.1 Electrovacuum fields with one Killing vector

In the present section we shall consider space-times admitting a *non-null* Killing vector field ξ, i.e.

$$\xi_{(a;b)} = 0, \qquad F = \xi^a \xi_a \neq 0, \tag{34.1}$$

and use the methods outlined in §10.8. For a null Killing vector see Julia and Nicolai (1995). To recapitulate the essential results from Chapter 18, it is convenient to introduce a three-dimensional metric γ_{ab} defined by

$$\gamma_{ab} = |F|(g_{ab} - F^{-1}\xi_a\xi_b), \qquad \gamma = \det(\gamma_{ab}) \tag{34.2}$$

(Neugebauer 1969). The field equations assume the form

$$\hat{R}_{ab} = \tfrac{1}{2}F^{-2}(\mathcal{E}_{,(a} + 2\overline{\Phi}\Phi_{,(a})(\mathcal{E}_{,b)}2\Phi\overline{\Phi}_{,b)}) + 2F^{-1}\Phi_{,(a}\overline{\Phi}_{,b)}, \tag{34.3a}$$

$$0 = F\mathcal{E}_{,a}^{;a} + \gamma^{ab}\mathcal{E}_{,a}(\mathcal{E}_{,b} + 2\overline{\Phi}\Phi_{,b}), \tag{34.3b}$$

$$0 = F\Phi_{,a}^{;a} + \gamma^{ab}\Phi_{,a}(\mathcal{E}_{,b} + 2\overline{\Phi}\Phi_{,b}), \quad F = -\mathrm{Re}\,\mathcal{E} - \overline{\Phi}\Phi. \tag{34.3c}$$

By inspection we have the following:

Theorem 34.1 *For sourcefree Einstein–Maxwell fields admitting a non-null Killing vector ξ, there exists a set $\{\Phi, \mathcal{E}, \gamma_{ab}\}$ such that the Einstein–Maxwell equations* (34.3) *follow from a variational principle with the Lagrangian*

$$L = \sqrt{\gamma}[\hat{R} + \tfrac{1}{2}F^{-2}\gamma^{ab}(\mathcal{E}_{,a} + \overline{\Phi}\Phi_{,a})(\overline{\mathcal{E}}_{,b} + \Phi\overline{\Phi}_{,b}) + 2F^{-1}\gamma^{ab}\Phi_{,a}\overline{\Phi}_{,b}] \tag{34.4}$$

($\hat{R}$ denotes the curvature scalar with respect to γ_{ab}), i.e. (34.3) *are*

$$\delta L/\delta\gamma_{ab} = 0, \qquad \delta L/\delta\Phi = 0, \qquad \delta L/\delta\mathcal{E} = 0. \tag{34.5}$$

The second term of this Lagrangian is precisely of the form (10.8) and the methods discussed there can be applied. The curvature scalar in the Lagrangian relates the two complex – or equivalently four real – fields Φ and $\mathcal{E}$ to the three-dimensional metric γ_{ab}. For the definitions of the two complex scalar potentials Φ and $\mathcal{E}$ the reader is referred to (18.31) and (18.35).

From a given solution $(\Phi, \mathcal{E}, \gamma_{ab})$ of the Einstein–Maxwell equations (34.3), the quantities F, ξ_a, the space-time metric and the Maxwell tensor can be reconstructed – in that order – via (cp. §18.2)

$$\begin{aligned} -F &= \tfrac{1}{2}(\mathcal{E} + \mathcal{E}^*) + \Phi\overline{\Phi}, & K^*_{ab} &= 2F^{-1}(\xi_{[a}\mathcal{E}_{,b]})^*, \\ g_{ab} &= |F|^{-1}\gamma_{ab} + F^{-1}\xi_a\xi_b, & \sqrt{\kappa_0/2}F^*_{ab} &= 2F^{-1}(\xi_{[a}\Phi_{,b})^*. \end{aligned} \tag{34.6}$$

Taking Φ and $\mathcal{E}$, respectively their real and imaginary parts, as coordinates in a four-dimensional potential space, its metric G_{AB} from §10.8 has according to (34.4) the form

$$\mathrm{d}S^2 = \tfrac{1}{2}F^{-2}|\mathrm{d}\mathcal{E} + 2\overline{\Phi}\,\mathrm{d}\Phi|^2 + 2F^{-1}\mathrm{d}\Phi\,d\overline{\Phi}, \tag{34.7}$$

where F is given in terms of the complex potentials by the first of expressions (34.6). For stationary fields ($F < 0$) the signature of this metric is 0 whereas it is 4 if the Killing vector in space-time is spacelike ($F > 0$).

According to §10.8 one now has to solve (10.86) to find the affine collineations of the potential space (34.7). It turns out that they are all, in fact, Killing symmetries. As the potential space is not of constant curvature, the maximal dimension of its group of motions G_R is $R \leq 8$, cp. Theorem 8.17. It turns out that there are precisely eight Killing vectors and the associated finite transformations, the integrals of (10.87), are

$$\mathcal{E}' = \alpha\bar{\alpha}\mathcal{E}, \ \Phi' = \alpha\Phi, \tag{34.8a}$$

$$\mathcal{E}' = \mathcal{E} + \mathrm{i}b, \ \Phi' = \Phi, \tag{34.8b}$$

$$\mathcal{E}' = \mathcal{E}(1 + \mathrm{i}c\mathcal{E})^{-1}, \ \Phi' = \Phi(1 + \mathrm{i}c\mathcal{E})^{-1}, \tag{34.8c}$$

$$\mathcal{E}' = \mathcal{E} - 2\bar{\beta}\Phi - \beta\bar{\beta}, \ \Phi' = \Phi + \beta, \tag{34.8d}$$

$$\mathcal{E}' = \mathcal{E}(1 - 2\bar{\gamma}\Phi - \gamma\bar{\gamma}\mathcal{E})^{-1}, \ \Phi' = (\Phi + \gamma\mathcal{E})(1 - 2\bar{\gamma}\Phi - \gamma\bar{\gamma}\mathcal{E})^{-1} \tag{34.8e}$$

(Neugebauer and Kramer 1969). The complex constants α, β and γ and the real constants b and c are the eight real parameters of the isometry group G_8. Because the potential space admits a G_8, its Ricci tensor R_{AB} is proportional to the metric G_{AB} (Egorov 1955).

The Einstein–Maxwell equations (34.5) are invariant under an arbitrary combination of the transformations (34.8). Hence we have shown

Theorem 34.2 *Given a solution $(\Phi, \mathcal{E}, \gamma_{ab})$ of the Einstein–Maxwell equations with a non-null Killing field, then any set $(\Phi', \mathcal{E}', \gamma_{ab})$ obtained by application of an arbitrary sequence of transformations* (34.8) *is also a solution.*

It should be noted that $\gamma'_{ab} = \gamma_{ab}$; only the potentials are transformed.

The transformations (34.8b) and (34.8d) are simply gauge transformations of the potentials; neither the space-time metric nor the Maxwell field are changed. The transformation (34.8a) is a duality rotation

$$F^{*}_{ab}{}' = \sqrt{\alpha/\bar{\alpha}} F^{*}_{ab} \tag{34.9}$$

combined with a rescaling of the coordinate x^n along the Killing vector $\xi = \partial_n$. Only the transformations (34.8c) and (34.8e) affect space-time and Maxwell field in a non-trivial manner.

The product of the transformations (34.8b), (34.8c) and (34.8a) with $c = b^{-1}$ and $\alpha = \mathrm{i}b^{-1}$, in the limit $b \to \infty$, leads to the *inversion*

$$\mathcal{E}' = \mathcal{E}^{-1}, \qquad \Phi' = \mathcal{E}^{-1}\Phi, \tag{34.10}$$

which maps the gauge transformations (34.8*b*) and (34.8*d*) into the non-trivial transformations (34.8*c*) and (34.8*d*), respectively.

We shall list now some earlier results. Some transformations discovered before 1969 are special cases of the general transformations (34.8).

Buchdahl (1954) found a transformation corresponding to the inversion (34.10) for static vacuum fields, i.e. $\Phi = 0$ and $\mathcal{E} = \mathrm{e}^{2U}$, cp. (10.116). This takes the simple form $U' = -U$.

The transformation (34.8*c*) applied to vacuum fields, ($\Phi = 0$), i.e.

$$\mathcal{E}' = \mathcal{E}(1 + \mathrm{i}c\mathcal{E})^{-1}, \tag{34.11}$$

is known under the name 'Ehlers transformation' (1957). It maps static into stationary solutions which in general, however, suffer from a NUT-like singularity (see (20.28)).

The 'charging' transformation, (34.8*e*) with $\Phi = 0$, found by Harrison (1968), maps vacuum solutions into electrovac solutions

$$\mathcal{E}' = \mathcal{E}(1 - \gamma\bar{\gamma}\mathcal{E})^{-1}, \qquad \Phi' = \gamma\mathcal{E}(1 - \gamma\bar{\gamma}\mathcal{E})^{-1}. \tag{34.12}$$

Harrison (1965, 1968) postulated a functional dependence between the gravitational and electromagnetic potential, i.e. a relation $\Phi = \Phi(\mathcal{E})$, but did not investigate all possibilities. A linear relation between Φ and $\mathcal{E}$ has also been found by Woolley (1973) who started with the assumption that the bivectors F^*_{ab} and K^*_{ab} in (34.6) determine the same geometry. In view of the Rainich formulation (§5.4) this led to proportionality for the bivectors and consequently to proportionality for the potentials.

Asymptotically flat Einstein–Maxwell fields generated from vacuum solutions via (34.12) exhibit a typical feature: the sources of the gravitational and electromagnetic field have a rather similar structure. This follows from the analysis of the far-field behaviour (see e.g. Kramer *et al.* (1972)).

34.1.2 The group SU(2,1)

In this section we shall study the transformations (34.8) or, equivalently, the symmetry group of the metric (34.7), in more detail. The equations (34.8) give a *non-linear* representation of the group G_8 in question. A *linear* representation will be obtained in what follows.

Following Kinnersley (1973) we introduce a vector Y in a three-dimensional complex vector space and express the potentials $\mathcal{E}$ and Φ in terms of its components as

$$Y^\mu = (u, v, w), \quad \mathcal{E} = (u - w)/(u + w), \qquad \Phi = v/(u + w). \tag{34.13}$$

We introduce a metric $\eta_{\mu\nu} = \mathrm{diag}(1, 1, -1)$ and write the norm of Y as

$$|Y|^2 = \eta_{\mu\nu}\overline{Y}^{\mu}Y^{\nu} = \overline{Y}^{\mu}Y_{\mu} = \overline{Y}_{\mu}Y^{\mu} = u\bar{u} + v\bar{v} - w\bar{w} \equiv A. \qquad (34.14)$$

The Lagrangian (34.4) takes the form

$$L = \sqrt{\gamma}[\hat{R} + 2A^{-2}\gamma^{ab}(\overline{Y}^{\mu}Y_{\mu,a}Y^{\nu}\overline{Y}_{\nu,b} - A\overline{Y}^{\mu}_{,a}Y_{\mu,b})]. \qquad (34.15)$$

By disregarding the redundancy of the description (34.13), i.e. treating the Y^{μ} as independent variables which is via the variational principle tantamount to adding a suitable equation for one of the components of Y^{μ}, the field equations assume the forms

$$\delta L/\delta Y^{\mu} = 0: \quad AY^{\mu;a}_{,a} = 2\overline{Y}^{\nu}\gamma^{ab}Y^{\mu}_{,a}Y_{\nu,b}, \qquad (34.16a)$$

$$\delta L/\delta\gamma_{ab} = 0: \quad R_{ab} = A^{-2}\overline{V}_{\mu(a}V^{\mu}_{b)}, \quad V^{\mu}_{a} = \epsilon^{\mu\nu\rho}Y_{\nu}Y_{\rho,a}. \qquad (34.16b)$$

A linear homogeneous transformation

$$Y^{\mu\prime} = U^{\mu}_{\nu}Y^{\nu} \qquad (34.17)$$

is said to be pseudounitary if it preserves the norm (34.14). In matrix notation, a pseudounitary matrix U satisfies the relation

$$U^{+}\eta U = \eta, \qquad \eta = \mathrm{diag}(1, 1, -1). \qquad (34.18)$$

Equations (34.16) are invariant under the transformations (34.17)–(34.18). The significant potentials $\mathcal{E}$ and Φ are determined by the ratios of u, v and w. Hence a common factor in those functions is irrelevant and we restrict ourselves to unimodular transformations, i.e. the group $SU(2,1)$.

Theorem 34.3 *The group of symmetry transformations of the Einstein–Maxwell equations with a non-null Killing vector is the group* $SU(2,1)$ (Kinnersley 1973).

The group $SU(2,1)$ has eight independent generators. With $U = 1 - \mathrm{i}X$ they can be written as

$$X = \begin{pmatrix} a_1 & b & c \\ \bar{b} & a_2 & d \\ -\bar{c} & -\bar{d} & a_3 \end{pmatrix}, \qquad \begin{array}{l} a_1, a_2, a_3 \text{ real} \\ a_1 + a_2 + a_3 = 0 \\ b, c, d \text{ complex.} \end{array} \qquad (34.19)$$

These transformations can also be derived as Lie symmetries of (34.3*b*)–(34.3*c*), see Leibowitz and Meinhardt (1978).

We shall now consider subgroups of $SU(2,1)$ and two-dimensional subspaces of the potential space. For stationary Einstein–Maxwell fields, each two-dimensional subspace of the potential space (34.7) must be

Table 34.1. The subspaces of the potential space for stationary Einstein–Maxwell fields, and the corresponding subgroups of $SU(2,1)$

Subspace	Metric $\mathrm{d}S^2$	Curvature and signature	Subgroup of $SU(2,1)$ invariant	Potentials $\mathcal{E}$ and Φ
$\Phi=0$ $(v=0)$ or $\mathcal{E}=-1$ $(u=0)$	$\frac{2\mathrm{d}\xi\mathrm{d}\overline{\xi}}{(1-\xi\overline{\xi})^2}$ $\frac{2\mathrm{d}\Phi\mathrm{d}\overline{\Phi}}{(1-\Phi\overline{\Phi})^2}$	$K=-2$ $(++)$	$SU(1,1)$ $\overline{u}u-\overline{w}w$ $SU(1,1)$ $\overline{v}v-\overline{w}w$	$\mathcal{E}=\frac{1-\xi}{1+\xi}$
$\mathcal{E}=+1$ $(w=0)$	$\frac{-2\mathrm{d}\Phi\mathrm{d}\overline{\Phi}}{(1+\Phi\overline{\Phi})^2}$	$K=-2$ $(--)$	$SU(2)$ $\overline{u}u+\overline{v}v$	
$\mathcal{E}=\overline{\mathcal{E}}$, $\Phi=\overline{\Phi}$, $(\overline{u}=u,$ $\overline{v}=v,$ $\overline{w}=w)$	$\frac{8\mathrm{d}\zeta\mathrm{d}\eta}{(1-\zeta\eta)^2}$	$K=-\frac{1}{2}$ $(+-)$	$O(2,1)$ $u^2+v^2-w^2$	$\mathcal{E}=\frac{(1-\zeta)(1-\eta)}{(1+\zeta)(1+\eta)}$ $\Phi=\frac{\zeta-\eta}{(1+\zeta)(1+\eta)}$
$\mathcal{E}=0$ $(u=w)$	0		$\overline{v}v$	

either a null surface or a space of constant curvature (Neugebauer, private communication, Matos and Plebański 1994). There are four *inequivalent,* modulo $SU(2,1)$, cases:

(1) $\Phi=0$ or $\mathcal{E}=-1$, vacuum;
(2) $\mathcal{E}=1$;
(3) $\mathcal{E}=\overline{\mathcal{E}}$, $\Phi=\overline{\Phi}$, electrostatic fields;
(4) $\mathcal{E}=0$, conformastationary fields, cp. §18.7.

The space-times in the second class become flat when the electromagnetic field is switched off.

In Table 34.1 we have listed the metrics, signatures and Gaussian curvatures of the corresponding 2-spaces. The exceptional case (4) is a null surface in the potential space. All subspaces are planes in the variables u, v and w (see also Tanabe (1977)).

The isometry groups of the 2-spaces are three-parameter subgroups of $SU(2,1)$. These subgroups have been classified by Montgomery *et al.* (1969) up to conjugation, i.e. the generators of the inequivalent subgroups are determined up to $SU(2,1)$ transformations. Hence the inequivalent

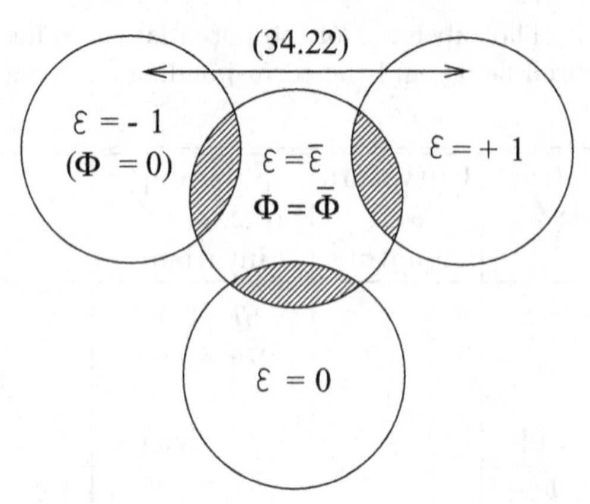

Fig. 34.1. The inequivalent classes of stationary Einstein–Maxwell fields. The double arrow indicates a relation by a complex substitution.

classes given by the representatives in Table 34.1 can be generalized by the application of arbitrary invariance transformations (34.11).

Figure 34.1 is to be understood as follows. The interior of each circle indicates the class of Einstein–Maxwell fields generated from the fields indicated by the symmetry operations of $SU(2,1)$. For instance, the class $\mathcal{E} = -1$ discovered by Demiański (1976) is equivalent to the class of vacuum fields $\Phi = 0$; the two cases are linked by elements of $SU(2,1)$. It is impossible to pass from one class to another by a $SU(2,1)$ transformation if the subspace of the potential space is truly two-dimensional.

The overlapping regions, shaded in Fig. 34.1, represent those Einstein–Maxwell fields for which the potentials depend on each other. In those cases, the 2-spaces degenerate into one-dimensional subspaces of the potential space, i.e. geodesics, cp. (10.91). The invariance transformations map each shaded region into itself. Electrostatic fields with only one independent potential belong to one of the classes $\mathcal{E} = \pm 1, 0$. The Reissner–Nordström solution (15.21) is an example: it belongs to the classes $\mathcal{E} = \text{sign}(e^2 - m^2)$; the three branches of the solution are inequivalent with respect to $SU(2,1)$ transformations which leave the quantity $e^2 - m^2$ invariant. The subclass $\mathcal{E} = 1$ contains (12.21) and the fields of massless charges constructed by Tanabe (1978). As can be inferred from Table 34.1, electrostatic fields are closely related to vacuum fields. This relation is borne out by the following:

Theorem 34.4 *From any stationary axisymmetric vacuum solution* ($\mathcal{E} = \mathrm{e}^{2U} + \mathrm{i}\psi, k$) *one obtains a static Einstein–Maxwell field* ($\mathrm{e}^{2U'}, \Phi', k'$) *or vice versa by the substitutions*

$$\mathrm{e}^{2U'} = \mathcal{E}\overline{\mathcal{E}}, \qquad \Phi' = \mathrm{i}\psi, \qquad k' = 4k. \tag{34.20}$$

The new solution will be purely electric if it is possible to continue the parameters analytically so that U' remains real and ψ becomes purely imaginary. As it stands the solution is purely magnetic (Bonnor 1961).

(Cp. also Fischer (1979) and for applications Hauser and Ernst (1978) and Hoenselaers (1982b).)

Finally we note that for asymptotically flat stationary Einstein–Maxwell fields generated from the corresponding vacuum fields by means of invariance transformations, the gyromagnetic factor is equal to that of an electron, $g = 2m$, $\mu = eJ/m$ (dipole moment μ, angular momentum J) (Reina and Treves 1975).

34.1.3 Complex invariance transformations

Allowing a *complex* space-time metric and complex fields but maintaining real space-time coordinates, we need four complex potentials $\mathcal{E}_1, \mathcal{E}_2, \Phi_1, \Phi_2$. In the potential space metric (34.7) we have to substitute $\mathcal{E} \to \mathcal{E}_1, \overline{\mathcal{E}} \to \mathcal{E}_2, \Phi \to \Phi_1, \overline{\Phi} \to \Phi_2$. After this complexification the isometry group of the potential space has eight complex parameters. The potentials and the parameters in the invariance transformations (34.8) are now considered to be independent of their complex conjugates. For instance, the complexified version of (34.8e) reads

$$\begin{aligned} \mathcal{E}_1' &= \frac{\mathcal{E}_1}{1 - 2\gamma_2\Phi_1 - \gamma_1\gamma_2\mathcal{E}_1}, \qquad \Phi_1' = \frac{\Phi_1 + \gamma_1\mathcal{E}_1}{1 - 2\gamma_2\Phi_1 - \gamma_1\gamma_2\mathcal{E}_1}, \\ \mathcal{E}_2' &= \frac{\mathcal{E}_2}{1 - 2\gamma_1\Phi_2 - \gamma_1\gamma_2\mathcal{E}_1}, \qquad \Phi_2' = \frac{\Phi_2 + \gamma_2\mathcal{E}_2}{1 - 2\gamma_1\Phi_2 - \gamma_1\gamma_2\mathcal{E}_2}. \end{aligned} \tag{34.21}$$

Here, the replacements $\gamma \to \gamma_1$, $\bar{\gamma} \to \gamma_2$ have been made; γ_1 and γ_2 are complex parameters. In some cases, complex symmetry transformations can be used to generate real solutions from complex ones (see Herlt (1978b)).

The two cases $\Phi = 0$ and $\mathcal{E} = 1$ have the same Gaussian curvature and are related by the complex transformation

$$\Phi_1' = \mathrm{i}\xi_1, \qquad \Phi_2' = \mathrm{i}\xi_2. \tag{34.22}$$

This correspondence is indicated by the $\leftrightarrow$ in Fig. 34.1. Starting with a real vacuum solution, we obtain a complex electromagnetic solution $\Phi_2 = -\overline{\Phi}_1$, which can be converted into a real solution by complex continuation of the parameters. For instance, the potentials Φ and ξ for the Reissner–Nordström solution (15.18) are $\Phi = e/r$ and $\xi = (1 - \mathcal{E})/(1 + \mathcal{E}) = m/r$. The complex transformation (34.22) and the substitution $\mathrm{i}m \to e$ take the Schwarzschild solution into the Reissner–Nordström solution with vanishing mass parameter. There is a one-to-one correspondence between stationary vacuum fields with sources characterized by masses and angular momenta and stationary Einstein–Maxwell fields with purely electromagnetic sources, i.e. charges and currents. The complex invariance transformation can change the sign of $e^2 - m^2$.

The class $\mathcal{E} = 0$ cannot be transformed into $\mathcal{E} = \pm 1$ even by complex invariance transformations.

34.1.4 Stationary axisymmetric vacuum fields

In this section we assume that space-time admits two commuting Killing vectors the trajectories of which form two-dimensional orbits, cp. Chapter 19 and Kramer *et al.* (1972). We also specialize our discussion to the case in which one of the Killing vectors is time-like. We shall indicate in §34.9 where changes occur when both Killing vectors are space-like. The space-time metric is written in Lewis–Papapetrou form as

$$\mathrm{d}s^2 = \mathrm{e}^{2K}(\mathrm{d}x^2 + \mathrm{d}y^2) + g_{MN}\,\mathrm{d}x^M\,\mathrm{d}x^N \tag{34.23}$$

such that all functions appearing depend only on x and y. The ignorable coordinates x^M $(M = 3, 4)$ are chosen such that

$$x^M = (\varphi, t), \quad \xi = \partial_t, \quad \eta = \partial_\varphi, \qquad g_{MN} = \begin{pmatrix} \eta_\alpha\eta^\alpha & \xi_\alpha\eta^\alpha \\ \xi_\alpha\eta^\alpha & \xi_\alpha\xi^\alpha \end{pmatrix}. \tag{34.24}$$

A reparametrisation yields the metric in the form (19.17) with (19.20)

$$\mathrm{d}s^2 = \mathrm{e}^{-2U}[\mathrm{e}^{2k}(\mathrm{d}x^2 + \mathrm{d}y^2) + W^2\mathrm{d}\varphi^2] - \mathrm{e}^{2U}(\mathrm{d}t + A\mathrm{d}\varphi)^2, \tag{34.25}$$

in which case, with $W^2 = -\det(g_{MN})$, g_{MN} and K are given by

$$g = (g_{MN}) = \begin{pmatrix} W^2\mathrm{e}^{-2U} - A^2\mathrm{e}^{2U} & -A\mathrm{e}^{2U} \\ -A\mathrm{e}^{2U} & -\mathrm{e}^{2U} \end{pmatrix}, \qquad K = k - U. \tag{34.26}$$

In this section we shall for brevity use the derivative operators

$$\nabla = (\partial_x, \partial_y), \quad \widetilde{\nabla} = (\partial_y, -\partial_x), \quad \partial = \partial_x + \mathrm{i}\partial_y, \quad \partial^* = \partial_x - \mathrm{i}\,\partial_y. \tag{34.27}$$

It can be shown (Hoenselaers 1976) that the Ricci scalar, the Lagrangian for vacuum fields, becomes

$$-L = \sqrt{-g}R = 2W\nabla^2 K + 2\nabla^2 W - \left[(\nabla g_{34})^2 - \nabla g_{33}\nabla g_{44}\right]/2W. \tag{34.28}$$

After rewriting the second derivatives as divergence terms – which can be omitted – the Lagrangian can be written as

$$L = 2\nabla K\nabla W + \left[(\nabla g_{34})^2 - \nabla g_{33}\nabla g_{44}\right]/2W. \tag{34.29}$$

This Lagrangian is again of the form (10.83). It has a five-parameter affine symmetry group since it is invariant under the transformations

$$g' = A^{-1}gA, \qquad K' = K + c, \tag{34.30}$$

with constant matrix A and constant c. The Killing vectors in this group correspond to the constant c and matrices A with $\det(A) = 1$. It will be seen later that the important transformations are the latter ones. The overall scaling of g is a homothety. The transformations generated by A are equivalent to a linear coordinate transformation in the φ–t-plane.

As the background space is two-dimensional we can use any linear combination of the Killing vectors of (34.29) to effect a Legendre transformation (10.97). It turns out (Hoenselaers 1979a) that there are three linear combinations such that the Legendre-transformed Lagrangian (10.98) admits more than one Killing vector. In one particular case, the parametrisation (34.26) casts the Lagrangian (34.29) into the form

$$L = 2\nabla k \nabla W - \tfrac{1}{2}[4W(\nabla U)^2 - \mathrm{e}^{4U}(\nabla A)^2/W]. \tag{34.31}$$

The equation for $A, \delta L/\delta A = 0$, is the integrability condition for the existence of another function ψ,

$$\nabla[W^{-1}\mathrm{e}^{4U}\nabla A] = 0 \quad \Leftrightarrow \quad W\widetilde{\nabla}\psi = \mathrm{e}^{4U}\nabla A. \tag{34.32}$$

Using ψ instead of A, the transformed Lagrangian becomes

$$L' = 2\nabla k \nabla W - \tfrac{1}{2}W[4(\nabla U)^2 + \mathrm{e}^{-4U}(\nabla\psi)^2]. \tag{34.33}$$

Introducing the Ernst (1968a) potential

$$\mathcal{E} = \mathrm{e}^{2U} + \mathrm{i}\psi, \tag{34.34}$$

this assumes the familiar form

$$L' = 2\nabla k \nabla W - 2W\nabla\mathcal{E}\,\nabla\overline{\mathcal{E}}\;(\mathcal{E}+\overline{\mathcal{E}})^{-2}, \tag{34.35}$$

from which the Ernst equation (19.39) and equations (19.42), the first of which corresponds to (10.93), can be derived.

The potential space metrics associated with the Lagrangians (34.33) resp. (34.35) again admit four Killing vectors and one homothety. With $\mathcal{E}$ given by (34.13), the important transformations are generated by the matrix X from (34.19) with $b = d = a_2 = 0$ and $a_3 = -a_1$. The other ones are $k' = k + \text{const}$ and $W' = \text{const}\, W$.

As there are several Killing vectors one can try to iterate the process of Legendre-transforming the Lagrangian. Again there are three linear combinations of the Killing vectors such that the transformed Lagrangian

L'' admits more than one symmetry. Moreover, it turns out in each case that L'' has exactly the same form as L. The functions, however, are not the same; they are related by equations of the form (34.32). At each step a finite invariance transformation of the form (34.8) can be used to generate a new solution of the field equations (Hoenselaers 1978b). None of these solutions, however, is asymptotically flat.

To summarize: The metric associated with the Lagrangian (34.29) admits three non-trivial (in the sense that L' admits more than one Killing vector) linear combinations of its Killing vectors. Each of them can be used to derive an L'. Each of the three Lagrangians L' has the same form as (34.33) and the metrics associated with them admit three non-trivial linear combinations of the Killing vectors, one of which leads back to L. There are thus six inequivalent (in the sense that they are related to the original Lagrangian by equations of the form (34.32) with different right-hand sides) Lagrangians L'' all of which are of the form (34.29). There are thus twelve Lagrangians L'''. The process of Legendre-transforming the Lagrangians can thus be continued ad infinitum. The group of transformations yielding new solutions of the stationary vacuum equations is thus infinite-dimensional (the same can be shown by analogous methods for the Einstein–Maxwell equations).

Asymptotically flat solutions can only be obtained if the infinity of potentials ψ is considered. We shall see in §34.3 how a convenient way of doing so can be found.

That the Lagrangians (34.29) and (34.35) admit the same invariance group suggests that they can be cast into the same form. To this end we introduce new potentials and find

$$\begin{aligned} E_\pm &= W\mathrm{e}^{-2U} \pm A, \qquad k = k' + U - (\ln W)/4, \\ L &= 2\nabla\tilde{k}\nabla W - 2W(E_+ + E_-)^{-2}\nabla E_+ \nabla E_-. \end{aligned} \tag{34.36}$$

Hence we can formulate (cp. Theorem 19.3).

Theorem 34.5 *From a given stationary axisymmetric vacuum solution* $(\mathcal{E} = \mathrm{e}^{2U} + \mathrm{i}\psi, k)$ *one gets a new solution* (U', A', k') *by the substitution*

$$U' = -U + (\ln W)/2, \quad A' = \mathrm{i}\psi, \quad k' = k - U + (\ln W)/4. \tag{34.37}$$

The new solution will be real if it is possible to continue the parameters analytically such that U *remains real and* ψ *becomes purely imaginary* (Kramer and Neugebauer 1968b).

To derive a set of equations analogous to (10.101) we note that the metrics associated with both Lagrangians (34.35) and (34.36) admit four Killing vectors and one homothety. The group acting on the $\mathcal{E}(E)$ potential is $SU(1,1)$ which has two-dimensional subgroups. We have four-dimensional simply transitive subgroups and thus can utilize Theorem 8.19 to write the field equations in the form (10.101). Using the derivative operators ∂ and ∂^* defined in (34.27) and the functions

$$M^1 = \partial k, \quad M^2 = W^{-1}\partial W, \quad M^3 = (\mathcal{E}+\overline{\mathcal{E}})^{-1}\partial\mathcal{E}, \quad M^4 = (\mathcal{E}+\overline{\mathcal{E}})^{-1}\partial\overline{\mathcal{E}}, \tag{34.38}$$

we get as field equations (Kramer and Neugebauer 1984, Hoenselaers 1988)

$$\partial^* M^1 + \tfrac{1}{2}(M^3 M^{*3} + M^4 M^{*4}) = 0, \tag{34.39a}$$

$$\partial^* M^2 + M^2 M^{*2} = 0, \tag{34.39b}$$

$$\partial^* M^3 + \tfrac{1}{2}(M^2 M^{*3} + M^3 M^{*2}) - M^3 M^{*3} + M^3 M^{*4} = 0, \tag{34.39c}$$

$$\partial^* M^4 + \tfrac{1}{2}(M^2 M^{*4} + M^4 M^{*2}) + M^4 M^{*3} - M^4 M^{*4} = 0, \tag{34.39d}$$

and the corresponding starred versions. N.b. the '$*$' operation is not to be confused with complex conjugation; it only changes the derivative operator ∂. The frame vectors e^α_A from (10.100) and the rotation coefficients (10.101) can be inferred from (34.38) and (34.39), respectively.

It should be noted that (34.39) remain the same if we replace $\mathcal{E} \to E_+$ and $\overline{\mathcal{E}} \to E_-$ in (34.38). In this case, the star operation is the same as complex conjugation. Equations (34.39) have been the starting point in Neugebauer's Bäcklund transformations (Neugebauer 1979), cp. §34.4.

Finally we note that a similar treatment is possible for Einstein–Maxwell fields with *two* commuting Killing vectors. In particular, using the variables introduced in (34.13), the interchange $\Pi : u \leftrightarrow v$ – which is a subcase of (34.19) – and a coordinate transformation $\mathcal{R}$ (34.30), Herlt (1979) (cp. also Clément (2000)) has generated the Kerr solution from the Schwarzschild solution, i.e. symbolically, *Kerr* $= \Pi^{-1}\mathcal{R}\,\Pi$ *Schwarzschild*.

The invariance transformations (34.8) or their subcases and the Bonnor transformation (34.20) have been applied frequently to vacuum and electrovac solutions, see Table 34.2.

For further information about applications to the case of colliding plane waves or, in general, two space-like Killing vectors we refer the reader to the book by Griffiths (1991).

34.2 Prolongation structure for the Ernst equation

In this section we shall study the prolongation structure for the Ernst equation. This was first investigated by Harrison (1978). Here, however, we shall not adhere to the original formulation, rather we shall use (34.39*b*)–(34.39*d*) as a starting point (the equations for M^1 decouple from the other ones). According to the procedures outlined in §10.4.3 we write

Table 34.2. Generation by potential space transformations

E, H and B refer to the Ehlers, Harrison and Bonnor transformations (34.11), (34.12), and (34.20), respectively. 'Transf.' means Transformation.

Seed metric	Transf.	Author(s)
Reissner–Nordström	E	Neugebauer and Kramer (1969)
Minkowski space	$SU(2,1)$	Hauser and Ernst (1979c)
Tomimatsu–Sato sol.	H	Ernst (1973), Wang (1974)
	$B, SU(2,1)$	Önengüt and Serdaroğlu (1975)
Erez–Rosen metric	H	Panov (1979b)
Curzon	$SU(2,1)$	Hernandez P. *et al.* (1993)
Type D vacuum, Kerr–Newman, Carter	H	García D. and Breton B. (1985, 1989), Breton B. and García D. (1986a, 1986b)
	E, H	Demiański and Newman (1966), Ernst and Wild (1976), Diaz (1985)
	$B, SU(2,1)$	Bonnor (1966), Kramer and Neugebauer (1969)
Weyl solutions	H	Salazar I. (1986)
	H	Patel and Trivedi (1975)
Stationary cylindrically-symmetric solution	H	Iyer and Vishveshwara (1983)
	H	Chamorro *et al.* (1991, 1993)
	H	Gutsunaev and Manko (1989)
	H	Quevedo and Mashhoon (1990)
	$SU(2,1)$	Krisch (1983)
	$SU(2,1)$	Denisova and Manko (1992)
Melvin solution	E	Garfinkle and Melvin (1994)
	H	Li and Ernst (1989)
Static electrovac	$SU(2,1)$	Kramer (1987)

the equations in differential forms using complex coordinates ζ and $\overline{\zeta}$ as

$$\begin{aligned}
&\mathrm{d}M^2\wedge\mathrm{d}\zeta = M^2M^{*2}\mathrm{d}\zeta\wedge\mathrm{d}\overline{\zeta}, \qquad \mathrm{d}M^{*2}\wedge\mathrm{d}\overline{\zeta} = -M^2M^{*2}\mathrm{d}\zeta\wedge\mathrm{d}\overline{\zeta},\\
&\mathrm{d}M^3\wedge\mathrm{d}\zeta = -[M^3M^{*3} - M^3M^{*4} - \tfrac{1}{2}(M^2M^{*3} + M^{*2}M^3)]\mathrm{d}\zeta\wedge\mathrm{d}\overline{\zeta},\\
&\mathrm{d}M^{*3}\wedge\mathrm{d}\overline{\zeta} = [M^3M^{*3} - M^{*3}M^4\\
&\qquad\qquad -\tfrac{1}{2}(M^2M^{*3} + M^{*2}M^3)]\mathrm{d}\zeta\wedge\mathrm{d}\overline{\zeta},\\
&\mathrm{d}M^4\wedge\mathrm{d}\zeta = -[M^4M^{*4} - M^4M^{*3} - \tfrac{1}{2}(M^2M^{*4} + M^{*2}M^4)]\mathrm{d}\zeta\wedge\mathrm{d}\overline{\zeta},\\
&\mathrm{d}M^{*4}\wedge\mathrm{d}\overline{\zeta} = [M^4M^{*4} - M^{*4}M^3 - \tfrac{1}{2}(M^2M^{*4} + M^{*2}M^4)]\mathrm{d}\zeta\wedge\mathrm{d}\overline{\zeta}.
\end{aligned} \tag{34.40}$$

As in (10.51) we introduce an as yet unspecified number of new variables, the pseudopotentials y, by

$$dy = F(M^i, M^{*i}, y)\, d\zeta + G(M^i, M^{*i}, y)\, d\bar{\zeta}, \quad (i = 2, 3, 4). \qquad (34.41)$$

(ζ and $\bar{\zeta}$ do not appear in (34.40) and we therefore neglect the dependence on them in F and G. We also suppress the index on y, F and G, the last two being vector fields with respect to y.) The following calculations are somewhat long albeit straightforward and we shall thus only describe them. The integrability condition $ddy = 0$ yields an equation which contains expressions proportional to $dM^i \wedge d\zeta$, $dM^{*i} \wedge d\bar{\zeta}$, $dM^{*i} \wedge d\zeta$ and $dM^i \wedge d\bar{\zeta}$. The former two expressions can be replaced by (34.40) to yield terms proportional to $d\zeta \wedge d\bar{\zeta}$, while the latter two arising from $\partial_{M^{*i}} F$ and $\partial_{M^i} G$ cannot be replaced. As these terms have to vanish, we conclude that F is independent of M^{*i} while G does not depend on M^i. Moreover, the two terms $\partial_y F\, dy \wedge d\zeta + \partial_y G\, dy \wedge d\bar{\zeta}$ yield via (34.41) a term $(F\partial_y G - G\partial_y F)\, d\zeta \wedge d\bar{\zeta} = [F, G]\, d\zeta \wedge d\bar{\zeta}$. The resulting equation is proportional to $d\zeta \wedge d\bar{\zeta}$ and is linear in M^i and M^{*i}. By repeated differentiation it can be shown that all second derivatives of F and G with respect to M^i resp. M^{*i} commute; it suffices to assume F and G to be linear (for a more general treatment see Finley and McIver (1995)). With (for later use we replaced the letter 'X' of (10.56) by an 'A')

$$\begin{aligned} F &= M^2 A_1 + M^3 A_2 + M^4 A_3, \\ G &= M^{*2} A_4 + M^{*3} A_5 + M^{*4} A_6, \end{aligned} \qquad (34.42)$$

where the A_ks depend on the ys only, we get after sorting with respect to M^i and M^{*i} the commutator relations

$$\begin{aligned} &[A_1, A_4] = A_1 - A_4, \quad [A_1, A_6] = \tfrac{1}{2}(A_4 - A_1), \quad [A_3, A_4] = \tfrac{1}{2}(A_4 - A_1), \\ &[A_1, A_5] = A_5 - A_1, \quad [A_3, A_6] = A_6 - A_3, \quad [A_2, A_4] = A_4 - A_2, \qquad (34.43) \\ &[A_2, A_5] = A_2 - A_5, \quad [A_2, A_6] = \tfrac{1}{2}(A_5 - A_2), \quad [A_3, A_5] = \tfrac{1}{2}(A_5 - A_2). \end{aligned}$$

The problem is now to determine that free algebra generated by the above commutators which still 'remembers' the original equations (34.40), cp. the remarks below equation (10.57).

In terms of the semidirect product $\mathcal{A}_1^1 \otimes \mathcal{V}$ of the loop algebra of $SU(1,1)$ and the Virasoro algebra, i.e.

$$\begin{aligned} &[X_i, Y_k] = Y_{i+k}, \quad [X_i, Z_k] = -Z_{i+k}, \quad [Y_i, Z_k] = 2X_{i+k}, \\ &[X_i, X_k] = 0, \quad \text{and } Y, Z, \quad [V_i, X_k] = kX_{i+k} \quad \text{and } Y, Z, \qquad (34.44) \\ &[V_i, V_k] = (k-1)V_{i+k}, \end{aligned}$$

the basic generators $A_1, ..., A_6$ are given by

$$\begin{aligned} A_1 &= -X_0 + Y_1, \quad A_2 = X_0 + Z_1, \; A_3 = \tfrac{1}{2}(V_2 - V_0), \\ A_4 &= -X_0 + Y_{-1}, \; A_5 = X_0 + Z_{-1}, \; A_6 = \tfrac{1}{2}(V_0 - V_{-2}), \end{aligned} \tag{34.45}$$

cp. Guo *et al.* (1982). The loop algebra of $SU(1,1)$ can be represented by setting $X_i = \lambda^i X_0$ etc., thereby introducing the spectral parameter λ. However, due to the presence of the Virasoro algebra this spectral parameter cannot be constant, rather it is another pseudopotential. Using a one-dimensional non-linear representation for $SU(1,1)$, viz.

$$X_0 = y\partial_y, \quad Y_0 = y^2\partial_y, \quad Z_0 = \partial_y, \tag{34.46}$$

we find for (34.41) with two pseudopotentials y and λ

$$\begin{aligned} \mathrm{d}y &= [(\lambda y^2 - y)M^3 + (\lambda - y)M^4]\mathrm{d}\zeta \\ &\quad + [(y^2\lambda^{-1})M^{*3} + (\lambda^{-1} + y)M^{*4}]\mathrm{d}\bar{\zeta}, \\ \mathrm{d}\lambda &= \tfrac{1}{2}\lambda(\lambda^2 - 1)M^2 + \tfrac{1}{2}(\lambda - \lambda^{-1})M^2. \end{aligned} \tag{34.47}$$

In terms of a given solution M^i, M^{*i} and the pseudopotentials y and λ calculated from it, a new solution is given by a linear ansatz, viz.

$$\begin{aligned} &\widetilde{M}^2 = u(\lambda)M^2, \quad \widetilde{M}^{*2} = v(\lambda)M^2, \\ &\widetilde{M}^3 = f_i M^i, \quad \widetilde{M}^4 = g_i M^i, \quad \widetilde{M}^{*3} = f_i^* M^{*i}, \quad \widetilde{M}^{*4} = g_i^* M^{*i}. \end{aligned} \tag{34.48}$$

In the course of the ensuing calculations one finds that $u = v^{-1}$ equals either λ^2 or 1. The original choice in Harrison (1978) was $u = v = 1$. The six functions $f_i(y, \lambda)$ etc. can be determined algebraically. Harrison (1983) has given an exhaustive list. He has also analysed the group structure of the transformations. We just quote one example:

$$\begin{aligned} &\widetilde{M}^2 = \lambda^2 M^2, \quad \widetilde{M}^3 = y\lambda M^3, \quad \widetilde{M}^4 = y^{-1}\lambda M^4 \\ &\widetilde{M}^{*2} = \lambda^{-2}M^{*2}, \; \widetilde{M}^{*3} = y\lambda^{-1}M^{*3}, \; \widetilde{M}^{*4} = y^{-1}\lambda^{-1}M^{*4} \end{aligned} \tag{34.49}$$

(Neugebauer 1979, Kramer and Neugebauer 1981, 1984).

34.3 The linearized equations, the Kinnersley–Chitre B group and the Hoenselaers–Kinnersley–Xanthopoulos transformations

34.3.1 The field equations

In this section we shall derive a linear problem for the stationary axisymmetric vacuum equations of the form (10.58) from first principles

(Kinnersley 1977, Kinnersley and Chitre 1977, 1978a, Chitre 1980). To this end we use a matrix f related to g from (34.29) by

$$f = g\epsilon = \begin{pmatrix} A\mathrm{e}^{2U} & W^2\mathrm{e}^{-2U} - A^2\mathrm{e}^{2U} \\ \mathrm{e}^{2U} & -A\mathrm{e}^{2U} \end{pmatrix}, \quad \epsilon = \begin{pmatrix} 0 & 1 \\ -1 & 0 \end{pmatrix}. \tag{34.50}$$

It satisfies

$$f^2 = W^2\mathbf{1}, \qquad \epsilon f^T \epsilon = f, \qquad \det f = -W^2. \tag{34.51}$$

The important part of the field equations, i.e. that part which involves second derivatives of U, A and W, can be written as

$$\nabla(W^{-1} f \nabla f) = 0. \tag{34.52}$$

This implies the existence of a matrix Ω defined by

$$\tilde{\nabla}\Omega = -W^{-1} f \nabla f \Leftrightarrow \tilde{\nabla} f = W^{-1} f \nabla \Omega. \tag{34.53}$$

Ω generalizes the imaginary part of the Ernst potential, ψ. We have

$$\mathrm{Tr}\,\Omega = 2V, \quad \tilde{\nabla} W = \nabla V. \tag{34.54}$$

The complex matrix $H = f + \mathrm{i}\,\omega$ satisfies

$$\nabla H = \mathrm{i}\, W^{-1} f \tilde{\nabla} H. \tag{34.55}$$

H generalizes the Ernst potential (34.37); indeed, $\mathcal{E}$ is the lower left element. In (34.58) it will be seen that H is closely related to an instance of the matrix $H(\lambda)$ in (10.58). It can now be shown that there exists a matrix $F(\lambda)$, the analogue of $\Phi(\lambda)$ in (10.58), such that

$$\begin{aligned} &[1 - \mathrm{i}\lambda(H + \epsilon H^+ \epsilon)]\nabla F(\lambda) = \mathrm{i}\lambda \nabla H F(\lambda), \\ &F(0) = -\mathrm{i}\mathbf{1}, \qquad \partial_\lambda F(\lambda)|_{\lambda=0} = H. \end{aligned} \tag{34.56}$$

To prove this, one operates with $\tilde{\nabla}$ on this equation and uses (34.55) and the algebraic properties of H. The matrix $F(\lambda)$ has the following properties

$$\begin{aligned} &\nabla F(\lambda) = \mathrm{i} W^{-1} f \tilde{\nabla} F(\lambda), \quad [1 - \mathrm{i}\lambda(H + \epsilon H^+ \epsilon)] F(\lambda) + S(\lambda)\overline{F}(\lambda) = 0, \\ &\epsilon F^+(\lambda)\epsilon F(\lambda) = S^{-1}(\lambda), \quad S(\lambda) \equiv (4\lambda^2 W^2 + (1 - 2\lambda V)^2)^{1/2}. \end{aligned} \tag{34.57}$$

These relations can now be used to solve (34.56) for ∇F, thereby casting it into the form (10.58). One finds

$$\nabla F(\lambda) = \mathrm{i}\lambda S(\lambda)^{-2}[(1 - 2\lambda V)\nabla H - 2\mathrm{i}\lambda W \tilde{\nabla} H] F(\lambda) \tag{34.58}$$

(for the details of these calculations see the original papers of Kinnersley and Chitre or Hoenselaers 1984).

Now we have cast the relevant equations into the form (10.58) and we also know that the solution of the linearized equations can be written in the form (10.63). Before we proceed to integrate (10.64) we shall digress for historical reasons and examine (10.67) and (10.68) in more detail.

Here, the matrix function $G(\lambda, \mu)$ as defined in (10.65) satisfies

$$\nabla G(\lambda, \mu) = \epsilon F^{+}(\lambda)\epsilon \nabla F(\mu). \tag{34.59}$$

For static solutions, i.e. solutions of Weyl's class, cp. §20.2, the matrix $F(\lambda)$ can be calculated without much ado. Indeed, if f is given by

$$f = A^{-1}RA, \qquad A = \mathrm{diag}\,(\mathrm{e}^{U}, \mathrm{e}^{-U}), \qquad R = \begin{pmatrix} 0 & W^2 \\ 1 & 0 \end{pmatrix}, \tag{34.60}$$

then

$$\begin{aligned} &F(\lambda) = A^{-1}Y(\lambda)B(\lambda), \qquad B(\lambda) = \mathrm{diag}\,(\mathrm{e}^{\beta(\lambda)}, \mathrm{e}^{-\beta(\lambda)}), \\ &Y(\lambda) = \tfrac{1}{2}S^{-1}(\lambda)\begin{pmatrix} -\mathrm{i}\,(1-2\lambda V + S(\lambda)) & -2+2\lambda V + S(\lambda) \\ 2\lambda & -2\mathrm{i} \end{pmatrix}, \\ &\nabla(W\nabla U) = 0, \quad S(\lambda)\nabla\beta(\lambda) = (1-2\lambda V)\nabla U - 2\lambda W\tilde{\nabla}U, \quad \beta(0) = U. \end{aligned} \tag{34.61}$$

34.3.2 Infinitesimal transformations and transformations preserving Minkowski space

In this subsection we shall investigate (10.67). We recall that the matrices N_{nm} are defined as the expansion coefficients of

$$G(\lambda, \mu) = \frac{1}{\lambda - \mu}[-\lambda \mathbf{1} + \mu F^{-1}(\lambda)F(\mu)] = \sum_{m,n=0}^{\infty} N_{nm}\lambda^n\mu^m \tag{34.62}$$

and transform as

$$\dot{N}_{nm} = \alpha_k N_{n+k,m} - N_{n,m+k}\alpha_k - \sum_{s=1}^{k} N_{ns}\alpha_k N_{k-s,m}. \tag{34.63}$$

Note that, due to (34.56), $G(0, \mu) = \mathrm{i}F(\mu)$. The lower left element of H, H_{21}, is the Ernst potential and from the explicit expression (34.61) with $U = \beta = 0$ one can verify by direct calculation that

$$\begin{aligned} &\alpha_{k,21}: \quad \mathcal{E} \to 1 - \mathrm{i}\epsilon(2r)^{k+1}P_{k+1}(\cos\varphi), \\ &\alpha_{k,11}: \quad \mathcal{E} \to 1 - \epsilon(2r)^{k}P_{k}(\cos\varphi), \\ &\alpha_{k,12}: \quad \mathcal{E} \to 1 - \mathrm{i}\epsilon(2r)^{k-1}P_{k-1}(\cos\varphi), \end{aligned} \qquad \begin{aligned} r^2 &= W^2 + V^2, \\ \tan\varphi &= V/W, \end{aligned} \tag{34.64}$$

with the obvious notation for the components of α_k. The weak gravitational fields generated in this manner contain static and stationary multipole moments of all orders but of the 'wrong' variety. They are all inner solutions; the sources which produce them are located at infinity.

Some of the infinitesimal transformations can be exponentiated. One set of those are transformations with only $\alpha_{k,12} \neq 0$; for details see Kinnersley and Chitre (1978a). They are equivalent to $SU(1,1)$ transformations in the chain of Lagrangians derived from the repeated application of (10.98) (Hoenselaers 1978b).

The form of (34.64) suggests that one considers transformations generated by

$$\beta_k = \alpha_{k,12} - \alpha_{k+2,21}; \tag{34.65}$$

they form an Abelian subgroup called B. The infinitesimal transformations leave Minkowski space invariant. Kinnersley and Chitre (1977) have shown that this holds also for the finite transformations. These transformations also preserve asymptotic flatness (in the sense of $\mathcal{E} \to \infty$ for large r). The action of B on the potentials can be exponentiated for the Zipoy–Voorhees solutions with integer δ: Minkowski space ($\delta = 0$) remains unchanged, the Schwarzschild solution ($\delta = 1$) yields the Kerr solution, while $\delta = 2$ gives the Kinnersley and Chitre (1978b) solution (20.41).

34.3.3 The Hoenselaers–Kinnersley–Xanthopoulos transformation

We now turn to the integration of (10.64). We write it in the abbreviated form (10.63), i.e.

$$\partial_\epsilon F(\lambda) = \lambda(\lambda - \sigma)^{-1}\left[F(\sigma)\alpha F^{-1}(\sigma)F(\lambda) - F(\lambda)\alpha\right]. \tag{34.66}$$

We are dealing with 2×2 matrices and α is tracefree. It has only been possible to integrate these equations if the constant matrix α is nilpotent,

$$\alpha^2 = 0. \tag{34.67}$$

The main line of attack is to solve for terms involving $F(\sigma)$ first. By taking the limit $\lambda \to \sigma$ we find

$$\partial_\epsilon F(\sigma) = \sigma[F(\sigma)\alpha F^{-1}(\sigma)F'(\sigma) - F'(\sigma)\alpha], \tag{34.68}$$

where the prime denotes the derivative of F with respect to its first argument, i.e. $F'(\lambda) = \partial_\lambda F(\lambda)$. Due to the nilpotence of α, we note that

$$\alpha\partial_\epsilon\left[F^{-1}(\sigma)F(\lambda)\right]\alpha = \alpha F^{-1}(\sigma)F'(\sigma)\alpha F^{-1}(\sigma)F(\lambda)\alpha. \tag{34.69}$$

As α is nilpotent, we have, for any matrix X, $\alpha X \alpha = \alpha\,\mathrm{Tr}(X\alpha) = \alpha\mathrm{Tr}(\alpha X)$. To exploit this identity in solving (34.69), we define

$$\Theta(\lambda,\sigma) = \frac{\lambda}{\lambda-\sigma}\mathrm{Tr}\left[F^{-1}(\sigma)F(\lambda)\alpha\right] \Rightarrow \Theta(\sigma,\sigma) = \sigma\mathrm{Tr}\left[F^{-1}(\sigma)F'(\sigma)\alpha\right]. \tag{34.70}$$

The equation for $\Theta(\sigma,\sigma)$ and its solution become

$$\partial_\epsilon\Theta(\sigma,\sigma) = \Theta(\sigma,\sigma)^2 \Rightarrow \Theta(\sigma,\sigma) = [1-\epsilon\Theta_0(\sigma,\sigma)]^{-1}\Theta_0(\sigma,\sigma). \tag{34.71}$$

The subscript '0' refers to the corresponding quantity for the seed solution. From (34.68) we find

$$\begin{aligned}\partial_\epsilon\left(F(\sigma)\alpha F^{-1}(\sigma)\right) &= F(\sigma)\alpha F^{-1}(\sigma)\Theta(\sigma,\sigma),\\ \Rightarrow F(\sigma)\alpha F^{-1}(\sigma) &= [1-\epsilon\Theta_0(\sigma,\sigma)]^{-2}F_0(\sigma)\alpha F_0^{-1}(\sigma).\end{aligned} \tag{34.72}$$

This can now be inserted into (34.66) which can be solved for $F(\lambda)$. The solution is

$$F(\lambda) = \left[\mathbf{1} + \frac{\epsilon\lambda}{\lambda-\sigma}\frac{F_0(\sigma)\alpha F_0^{-1}(\sigma)}{1-\epsilon\Theta_0(\sigma,\sigma)}\right]F_0(\lambda)\left(\mathbf{1}-\frac{\epsilon\lambda\alpha}{\lambda-\sigma}\right). \tag{34.73}$$

Recall that the real part of H contains the metric functions and that the lower left element of H is the $\mathcal{E}$ potential. H can be calculated from (34.56). Note that the quantities σ, ϵ and the components of α appear as parameters in the final solution.

The transformation can be iterated. The result can conveniently be written in matrix notation by introducing the vectors and matrices

$$\Theta = \left(\Theta_0(\sigma_i,\sigma_k)\epsilon_k\right), \quad K = \left(F_0(\sigma_i)\alpha\epsilon_i\right), \quad L = \left(F_0^{-1}(\sigma_i)\sigma_i^{-1}\right) \tag{34.74}$$

$(i,k = 1,\ldots,N)$. One finds

$$H = H_0 + \mathrm{i}K^T(1-\Theta)^{-1}L. \tag{34.75}$$

N.b. the components of the vectors K and L are themselves matrices; but the transposition does not affect the components of K.

The Ernst potential $\mathcal{E}$ is the lower left element of H. If the original metric is static, i.e. $F_0(\lambda)$ is given by (34.61), and α is chosen as $\alpha = \begin{pmatrix}0&1\\0&0\end{pmatrix}$, one finds (Dietz 1983b), dropping the subscript '0' and denoting $z_i = \sigma_i/2$,

$$\begin{aligned}&\mathcal{E} = \mathrm{e}^{2U}D_-D_+^{-1}, \qquad D_\pm = \det(\delta_{ik}+\gamma_\pm(z_i,z_k)\epsilon_k),\\ &\gamma_\pm(\eta,\zeta) = \mathrm{i}\mathrm{e}^{2\beta(\zeta)}S(\zeta)^{-1}([S(\eta)-S(\zeta)]/[\eta-\zeta]\pm 1),\\ &S(\zeta) = (W^2-(\zeta-V)^2)^{1/2}, \qquad S(\zeta)\nabla\beta(\zeta) = (\zeta-V)\nabla U - W\tilde{\nabla}U.\end{aligned} \tag{34.76}$$

Table 34.3. Applications of the HKX method

Seed	References
Minkowski	Hoenselaers (1981), Dietz and Hoenselaers (1982b) Hoenselaers and Dietz (1984)
Curzon	Hernandez P. *et al.* (1993)
Weyl	Dietz and Hoenselaers (1982a), Dietz (1983a, 1984b) Castejon-Amenedo and Manko (1990b) Quevedo and Mashhoon (1991) Xanthopoulos (1981), Hernandez P. *et al.* (1993)
Erez–Rosen	Quevedo (1986a, 1986b)
Zipoy–Voorhees	Dietz (1984a)

These transformations were first derived – albeit in a different form – by Hoenselaers *et al.* (1979) and are known as HKX transformations.

The special solution which we shall mention as an example of the HKX transformations is the following (Dietz and Hoenselaers 1982c, Hoenselaers 1983) (other examples are listed in Table 34.3). One starts with a superposition of two Curzon particles (20.5). One then performs two HKX transformations (34.76) and chooses the parameters so that $z_1 = -z_2 = \sigma$. The solution depends on four parameters, $m, \epsilon_1, \epsilon_2$ and σ. The last, however, can be scaled to unity. We shall not give the metric functions as they are rather lengthy, rather we shall describe the principal features of this solution, depicted in Fig. 34.2.

In the standard prolate spheroidal coordinates x and y (cp. (20.7)), the coordinate axis $\rho = 0$ splits into three parts: I: $y = 1$, II: $x = 1$, III: $y = -1$. The solution has rather complicated singularities at $x = 1, y = \pm 1$. For $\rho = 0$ to be an axis in space-time, i.e. a set of fixed points of the action of the ∂_φ Killing vector, we need $A(\rho = 0) = 0$. It follows from (34.32) that A at $\rho = 0$ is at most a step function of z, i.e. it is constant on the three parts mentioned above. It is also defined up to an additive constant which can be chosen such that $A(III) = 0$. The condition that the solution be asymptotically flat, i.e. possess no NUT-like singularity, yields one condition on the parameters, $\epsilon = \epsilon_1 = -\epsilon_2$. The condition of the existence of an axis between the objects, i.e. $A(II) = 0$, is a condition on the remaining parameters m and ϵ.

The metric function k, the conformal factor of the $\rho - z$ part of the metric, cp. (19.31), is by (19.42) also at most a step function of z at $\rho = 0$. It is also only defined up to an additive constant which can be chosen such that $k(I) = k(III) = 0$; the latter relation follows from the behaviour of the Ernst potential at infinity. For the axis II to be regular, i.e. to

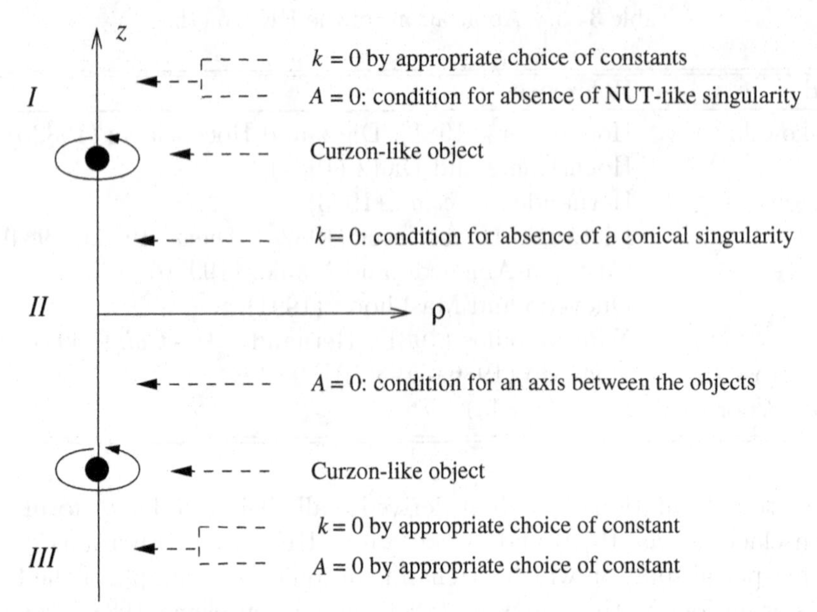

Fig. 34.2. The configuration of two Curzon-like objects in balance

be such that one can introduce locally Cartesian coordinates, one needs $k(II) = 0$. This is yet another condition on the remaining parameters. It can be shown that the two conditions $A(II) = k(II) = 0$ can be satisfied for particular numerical values of ϵ and m. It can also be shown that the solution is free of singularities except those located at $x = 1, y = \pm 1$; it is invariant under $y \to -y$.

The solution with the particular choice of parameters just defined describes two objects (whatever they are) that can be enclosed in convex regions surrounded by vacuum gravitational field. The two objects are balanced against their gravitational attraction by relativistic interaction of angular momentum.

34.4 Bäcklund transformations

In this section we follow Neugebauer (1979) (see also Neugebauer (1981), Kramer and Neugebauer (1984), Neugebauer and Kramer (1985)) and adapt (34.39c)–(34.39d) to the polynomial Bäcklund transformations outlined in §10.6. To this end we use Weyl's canonical coordinates and introduce a coordinate ζ related to them by

$$\zeta = \rho + \mathrm{i}z. \tag{34.77}$$

Note that now $M^2 = \rho^{-1}\rho_{,\zeta}$. One can reformulate (34.39c)–(34.39d) as a linear problem analogous to (10.58); however, due to the explicit coordinate dependence, the 'parameter' is no longer constant, rather it is given by

$$\Lambda = \sqrt{(K - \mathrm{i}\zeta)/(K + \mathrm{i}\zeta)}. \tag{34.78}$$

Using this Λ the linear problem becomes

$$\mathrm{d}\Phi = \left[\begin{pmatrix} M^4 & \Lambda M^4 \\ \Lambda M^3 & M^3 \end{pmatrix} \mathrm{d}\zeta + \begin{pmatrix} M^{*4} & \Lambda^{-1} M^{*4} \\ \Lambda^{-1} M^{*3} & M^{*3} \end{pmatrix} \mathrm{d}\overline{\zeta}\right] \Phi. \tag{34.79}$$

We have (cp. (34.38))

$$M^3 = \tfrac{1}{2}\mathrm{e}^{-2U}\mathcal{E}_{,\zeta},\ M^{*3} = \tfrac{1}{2}\mathrm{e}^{-2U}\mathcal{E}_{,\overline{\zeta}},\ M^4 = \tfrac{1}{2}\mathrm{e}^{-2U}\overline{\mathcal{E}}_{,\zeta},\ M^{*4} = \tfrac{1}{2}\mathrm{e}^{-2U}\overline{\mathcal{E}}_{,\overline{\zeta}} \tag{34.80}$$

(Ernst picture). We note that the matrix $\Phi(\Lambda)$ at $\Lambda = 1$, i.e. in the limit $K \to \infty$, is given by

$$\Phi(1) = \begin{pmatrix} \overline{\mathcal{E}} & 1 \\ \mathcal{E} & -1 \end{pmatrix}. \tag{34.81}$$

The matrix corresponding to $H(\lambda)$ in (10.58) is not tracefree, another manifestation that the independent variables appear explicitly in the equation. Nevertheless, we seek a new solution of (34.79) in the form

$$\Phi = qP\Phi_0,\ P = \sum_{s=0}^{n} P_s \Lambda^s \tag{34.82}$$

(Meinel *et al.* 1991), where the matrices P_s are independent of Λ. The scalar function q is determined by the conditions

$$q_{,\overline{\zeta}} = 0,\ (\Lambda^n q)_{,\zeta} = 0, \tag{34.83}$$

which are chosen such that the ansatz (34.82) inserted into (34.79) allows the comparison of coefficients of ascending resp. descending powers of Λ. The determinant of P is a polynomial in Λ of degree $2n$, i.e.

$$\det(P) = \beta \prod_{i=1}^{2n} (\Lambda - \Lambda_i), \tag{34.84}$$

where $\beta(\zeta, \overline{\zeta})$ is independent of Λ. The particular choice of Λ guarantees that $\mathrm{d}[\det(\Phi)] = \mathrm{Tr}(\Phi^{-1}\mathrm{d}\Phi) = 0$ at $\Lambda = \Lambda_i := \Lambda(K_i)$. It turns out that, in order to preserve the particular algebraic structure (34.79), $\det(P)$ is an even polynomial of Λ such that we have essentially n zeros Λ_i.

According to (10.78) we have to solve

$$P(\Lambda_i)\,\Phi_0(\Lambda_i)\,\mathbf{p}^{(i)} = 0, \quad i = 1, \ldots, n \tag{34.85}$$

and it can be shown that the vectors $\mathbf{p}^{(i)}$ are constant. An appropriate form for the solution of (34.83) is

$$q = \prod_{i=1}^{n} \sqrt{(1-\Lambda_i^2)/(\Lambda^2-\Lambda_i^2)}. \tag{34.86}$$

For the solution of (34.85) it is useful to introduce the quantities

$$\begin{pmatrix} l_i \\ m_i \end{pmatrix} = \Phi_0\left(\Lambda_i\right) \mathbf{p}^{(i)}. \tag{34.87}$$

With $P(-1) = \mathbf{1}$ one finds that the new Ernst potential can be expressed in terms of the Vandermonde-like determinant

$$\Theta(\Lambda, l, m) = \begin{vmatrix} m & \Lambda l & \Lambda^2 m & \Lambda^3 l & \cdots & \Lambda^n l \\ m_1 & \Lambda_1 l_1 & \Lambda_1^2 m_1 & \Lambda_1^3 l_1 & \cdots & \Lambda_1^n l_1 \\ m_2 & \Lambda_2 l_2 & \Lambda_2^2 m_2 & \Lambda_2^3 l_2 & \cdots & \Lambda_2^n l_2 \\ \vdots & \vdots & \vdots & \vdots & & \vdots \\ m_n & \Lambda_n l_n & \Lambda_n^2 m_n & \Lambda_n^3 l_{2n} & \cdots & \Lambda_n^n l_n \end{vmatrix}. \tag{34.88}$$

The new solution of Ernst's equation is given by

$$\mathcal{E} = \Theta(-1, -\overline{\mathcal{E}}_0, \mathcal{E}_0)/\Theta(-1, 1, 1). \tag{34.89}$$

This expression involving $n \times n$ determinants can, for even n, be rewritten as (Kramer and Neugebauer 1980, Yamazaki 1984)

$$\mathcal{E} = \mathcal{E}_0 \det\left(\frac{\alpha_p r_p - \alpha_q r_q}{K_p - K_q} - \frac{\overline{\mathcal{E}}_0}{\mathcal{E}_0}\right) \Big/ \det\left(\frac{\alpha_p r_p - \alpha_q r_q}{K_p - K_q} - 1\right), \quad \alpha_k = \frac{l_k}{m_k},$$
$$r_k^2 = (K_k - \mathrm{i}\overline{\zeta})(K_k + \mathrm{i}\zeta), \quad p = 1, 3, \ldots n-1, \quad q = 2, 4, \ldots n. \tag{34.90}$$

For the new Ernst potential (34.90) to yield a real gravitational field, the α_k have to be restricted by one of the two following expressions:

$$\alpha_k \overline{\alpha}_k = 1 \text{ or } \alpha_k \overline{\alpha}_m = 1 \ (k \neq m). \tag{34.91}$$

In terms of the parameters K_k this implies that they are either real or come in complex conjugate pairs, i.e.

$$K_k = \overline{K}_k \quad \text{or} \quad K_k = \overline{K}_{k+1}. \tag{34.92}$$

It should be noted that the α_k satisfy the Riccati equations

$$\mathrm{d}\alpha_k = (\alpha_k + \Lambda_k)\left(M^4 - \alpha_k M^3\right)\mathrm{d}\zeta + \left(\alpha_k + \Lambda_k^{-1}\right)\left(M^{*4} - \alpha_k M^{*3}\right)\mathrm{d}\overline{\zeta}. \tag{34.93}$$

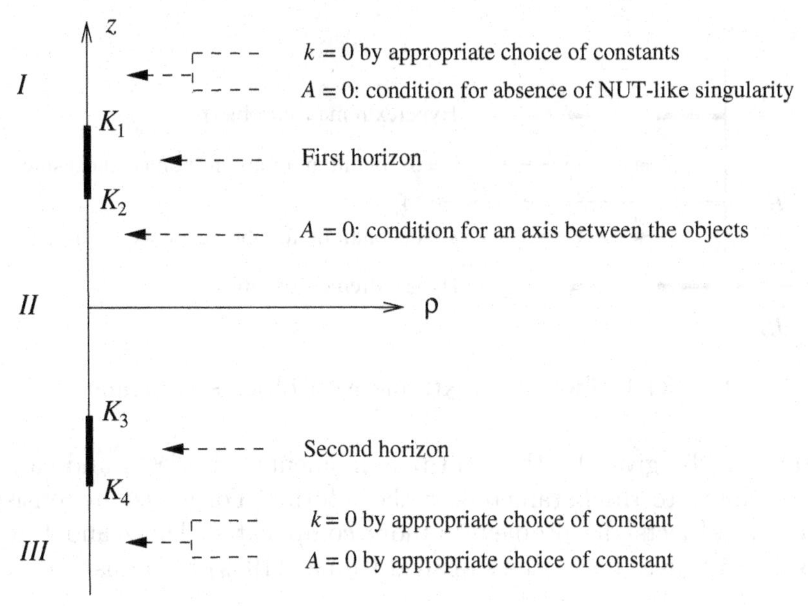

Fig. 34.3. Two rotating black holes in Weyl coordinates

For a given Ernst potential $\mathcal{E}_0$ (seed solution) one calculates the M^k from (34.80), solves (34.93) for the α_k and inserts the result into (34.90) to obtain the new Ernst potential. The most prominent example of a solution generated by this method results from the application of two Bäcklund transformations to Minkowski space (Kramer and Neugebauer 1980). It is the so-called double-Kerr solution. In terms of the determinants (34.90) it can be written as (Yamazaki 1983a, 1983b)

$$\mathcal{E} = \frac{J_-}{J_+}, \; J_\pm = \begin{vmatrix} \dfrac{S_1 - S_2}{K_1 - K_2} \pm 1 & \dfrac{S_1 - S_4}{K_1 - K_4} \pm 1 \\ \dfrac{S_3 - S_2}{K_3 - K_2} \pm 1 & \dfrac{S_3 - S_4}{K_3 - K_4} \pm 1 \end{vmatrix}, \tag{34.94}$$

$$S_k = \mathrm{e}^{\mathrm{i}\omega_k}\sqrt{\rho^2 \; + (K_k - z)^2}, \quad \omega_k = \mathrm{const.}$$

For real K it describes the superposition of two Kerr black holes. The horizons are located at $\rho = 0$, $K_1 \geq z \geq K_2$, $K_3 \geq z \geq k_4$, see Fig. 34.3.

This solution has been analysed extensively (e.g. Tomimatsu and Sato (1981), Kihara *et al.* (1983), Dietz and Hoenselaers (1985), Kramer (1986b), Dietz (1988)). The first problem is to give some physical meaning to the parameters appearing in the solution. In an asymptotically flat system with only one constituent the mass, angular momentum etc. can

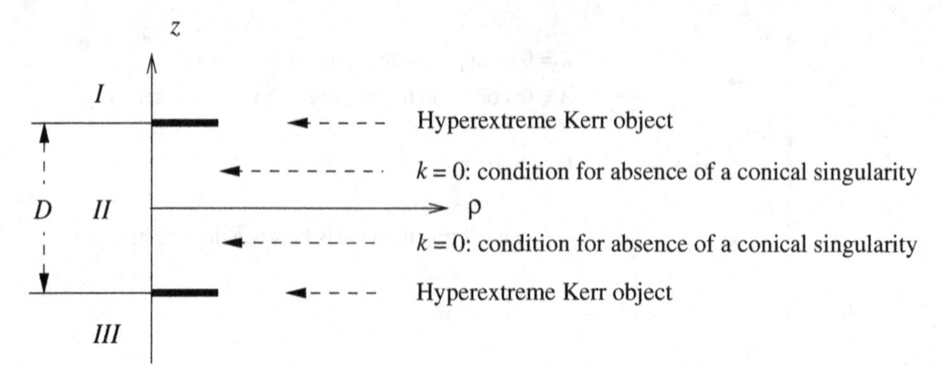

Fig. 34.4. Two hyperextreme Kerr objects in balance

be taken to be given by the multipole moments, cp. §18.8, and can easily be related to the parameters of the solution. For a system consisting of several objects the problem is more complicated. Dietz and Hoenselaers (1985) have chosen the so-called Komar (1959) integrals

$$M = -\frac{1}{4\pi}\int_\Sigma \xi_{\alpha;\beta}\mathrm{d}\Sigma^{\alpha\beta}, \quad J = \frac{1}{8\pi}\int_\Sigma \eta_{\alpha;\beta}\mathrm{d}\Sigma^{\alpha\beta} \tag{34.95}$$

as definitions for the masses and angular momenta of the individual objects. Σ is a closed 2-surface surrounding the object and $\xi = \partial_t$, $\eta = \partial_\varphi$. It can be shown that the integrals are independent of the surface, i.e. $M(\Sigma_1) = M(\Sigma_2)$, if and only if $R_{\alpha\beta} \equiv 0$ between Σ_1 and Σ_2.

Again, as was the case for the solution described in Fig. 34.2, the condition $k = 0$ on the middle part of the axis is a rather simple one. The difficult ones are the condition of asymptotic flatness, i.e. $A = 0$ at I, and the condition of the existence of an axis between the objects, i.e. $A = 0$ at II – without which the interpretation of the solution as describing two Kerr objects is not tenable. It can be shown analytically that balance cannot be achieved for positive Komar masses (Manko and Ruiz 2001).

The situation changes if one considers hyperextreme Kerr objects (Fig. 34.4). In the symmetric case, $M_1 = M_2 = M$, $J_1 = J_2 = J > 0$ (without loss of generality), it can be shown that the three conditions reduce to

$$D = M\left[Q\left(1+\sqrt{1-2/Q}\right)-2\right], \quad Q = JM^{-2}. \tag{34.96}$$

The reality of D requires $Q > 2$. Loosely speaking, a black hole is too small to store the angular momentum required for balance. It should be noted that the limit $K_1 \to \infty$ combined with an appropriate scaling of $\mathcal{E}$ yields the rotating version of the C-metric (21.11) (Hoenselaers

1985a). Ernst (see Hauser and Ernst 1979a) has used data specified on the symmetry axis to determine the parameters K_i of Neugebauer's Bäcklund transformation.

34.5 The Belinski–Zakharov technique

In this section we shall introduce the main features of the Belinski–Zakharov technique (Belinskii and Zakharov 1978, 1979); for reviews, see Miccichè (1999) and Belinski and Verdaguer (2001). Belinski and Zakharov started with the matrix $\mathbf{g}$ representing the φ–t-part of the metric, i.e.

$$\mathbf{g} = W\mathrm{e}^{-U}\begin{pmatrix} -1 & -A \\ -A & \mathrm{e}^{2U} + A^2 \end{pmatrix}, \tag{34.97}$$

and introduced the matrices

$$\mathbf{A} = -W\mathbf{g}_{,\zeta}\mathbf{g}^{-1}, \quad \mathbf{B} = W\mathbf{g}_{,\bar{\zeta}}\mathbf{g}^{-1} \tag{34.98}$$

(ζ given by (34.77)). The important part of the field equations, i.e. the part involving the φ–t-components of the metric, becomes

$$\mathbf{A}_{,\bar{\zeta}} - \mathbf{B}_{,\zeta} = 0. \tag{34.99}$$

Apart from this equation, there are integrability conditions following from (34.98). They are

$$W(\mathbf{A}_{,\bar{\zeta}} + \mathbf{B}_{,\zeta}) + [\mathbf{A}, \mathbf{B}] - W_{,\bar{\zeta}}\mathbf{A} - W_{,\zeta}\mathbf{B} = 0. \tag{34.100}$$

We are thus interested in two matrices $\mathbf{A}$ and $\mathbf{B}$ satisfying (34.99) and (34.100).

Belinski and Zakharov introduced the following operators

$$\mathcal{D}_1 = \partial_\zeta - \frac{2\lambda}{\lambda - W}W_{,\zeta}\partial_\lambda, \quad \mathcal{D}_2 = \partial_{\bar{\zeta}} + \frac{2\lambda}{\lambda + W}W_{,\bar{\zeta}}\partial_\lambda. \tag{34.101}$$

It can be shown by direct calculation that the condition that the $\mathcal{D}$s commute is equivalent to the equation for W, i.e.

$$[\mathcal{D}_1, \mathcal{D}_2] = 0 \quad \Longleftrightarrow \quad W_{,\zeta\bar{\zeta}} = 0. \tag{34.102}$$

Moreover, the $\mathcal{D}$s are invariant under the rescaling

$$\lambda \longrightarrow \lambda' = W^2\lambda^{-1}. \tag{34.103}$$

A linear pair, analogous to (10.58) albeit using two derivative operators in place of the exterior derivative, is introduced by

$$\mathcal{D}_1\Phi = \frac{1}{\lambda - W}\mathbf{A}\Phi, \quad \mathcal{D}_2\Phi = \frac{1}{\lambda + W}\mathbf{B}\Phi, \tag{34.104}$$

where Φ is a complex matrix depending on the so-called spectral parameter λ. The field equations (34.99) and the integrability conditions (34.100) are equivalent to the integrability condition of the system (34.104). In particular we have

$$\Phi(\lambda = 0) = \mathbf{g}. \tag{34.105}$$

In analogy to (34.82) we use the dressing ansatz, i.e. we look for a new solution in the form

$$\Phi(\lambda) = P(\lambda)\Phi_0(\lambda), \tag{34.106}$$

where, as usual, a subscript '0' indicates quantities pertaining to the seed metric. This ansatz implies by (34.105)

$$\mathcal{D}_1 P = \frac{1}{\lambda - W}(\mathbf{A}P - P\mathbf{A}_0), \quad \mathcal{D}_2 P = \frac{1}{\lambda + W}(\mathbf{B}P - P\mathbf{B}_0). \tag{34.107}$$

Not all solutions of (34.107) are acceptable. For instance, the reality of $\mathbf{g}$ imposes the condition

$$\overline{P}(\overline{\lambda}) = P(\lambda), \quad \overline{\Phi}(\overline{\lambda}) = \Phi(\lambda). \tag{34.108}$$

Moreover, we can only accept a symmetric metric $\mathbf{g}$. To this end we consider a new matrix $P'(\lambda)$ defined by

$$P'(\lambda) = \mathbf{g}P(\lambda')\mathbf{g}_0^{-1} \tag{34.109}$$

(λ' as given by (34.103)). By using the invariance of the $\mathcal{D}$s under (34.103), it can be shown that P' also satisfies (34.107) if $\mathbf{g}$ is symmetric. Indeed, the symmetry of $\mathbf{g}$ implies $\mathbf{g}\mathbf{A}^{\dagger-1}\mathbf{g}^{-1} = \mathbf{A}$, which in turn ensures that P' is also a solution. Consequently the symmetry of $\mathbf{g}$ is assured if there exists a scalar function $h(\lambda)$ such that

$$P'(\lambda) = h(\lambda)P(\lambda). \tag{34.110}$$

In general such a function will be different from 1. The new $\mathbf{g}$ is given by

$$\mathbf{g} = h(\lambda)P(\lambda')\mathbf{g}_0 P^T(\lambda). \tag{34.111}$$

To write this in the form

$$\mathbf{g} = P'(\lambda = 0)\mathbf{g}_0 \tag{34.112}$$

we need to impose the condition

$$\lim_{\lambda \to 0} h(\lambda)P(\lambda') = 1. \tag{34.113}$$

Finally, one requires $\det(\mathbf{g}) = -W^2$. This implies for P

$$\det\left[P(\lambda = 0)\right] = 1. \tag{34.114}$$

In fact, any solution of the linear pair (34.104) is defined up to a scaling $\mathbf{g} \to h(\lambda = 0)^{-1}\mathbf{g}$. Thus, given any solution $\mathbf{g}$ of (34.105) the physical metric $\mathbf{g}_{ph}$ ($\det \mathbf{g}_{ph} = -W^2$) can be obtained by

$$\mathbf{g}_{ph} = W(\det \mathbf{g})^{1/2}\mathbf{g}, \tag{34.115}$$

in which case we have $h = (\det \mathbf{g})^{1/2}W^{-1}$.

In order to construct a solution explicitly, we now introduce some assumptions on the pole structure of the matrix P in the complex λ-plane. Here, we shall assume that P and P^{-1} have singularities in λ and that these are simple poles, i.e. both P and P^{-1} are meromorphic. Of course, the poles will depend on W and V; note that – whenever possible – we have suppressed the coordinate dependence of the various quantities. This is analogous to prescribing the zeros of the dressing matrix P in §34.4. Let us assume that $P(\lambda)$ is not invertible at a number of points ν_k ($k = 1, \ldots, n$) and that these are simple poles for P^{-1}. Then it can be shown that P has simple poles at $\mu_k = W^2\nu_k^{-1}$.

From (34.108) it can be concluded that the poles of P are either real or come in complex conjugate pairs. Thus the general forms of P and P^{-1} are

$$P = \sum_{k=1}^{n} \left[(\lambda - \mu_k)R_k + (\lambda - \overline{\mu}_k)^{-1}\overline{R}_k\right], \tag{34.116a}$$

$$P^{-1} = \sum_{k=1}^{n} \left[(\lambda - \nu_k)^{-1}Q_k + (\lambda - \overline{\nu}_k)\overline{Q}_k\right], \tag{34.116b}$$

where the matrices R_k and Q_k are related through the condition $PP^{-1} = 1$. By inserting (34.116) into (34.107) we find an equation for μ_k the solution of which is

$$\mu_k = w_k \pm \sqrt{(w_k - V)^2 + W^2}. \tag{34.117}$$

The R_k cannot be chosen arbitrarily, rather they are given in terms of vectors n_k and m_k by

$$R_k = n_k m_k^T. \tag{34.118}$$

For a given seed metric $\mathbf{g}_0$ one obtains, at least in principle, Φ_0 from (34.104). Then one defines matrices M_k and with their help the vectors m_k by

$$M_k = \Phi_0(\lambda = \mu_k), \quad m_k = M_k^T \kappa_k, \tag{34.119}$$

Table 34.4. Applications of the Belinski–Zakharov method

The Killing vector types are AS (axisymmetric stationary), CY (cylindrical symmetry), CO (cosmological symmetry), BR (boost-rotation symmetry).

Seed	Type	Result	
Minkowski	AS	Vac.	Alekseev (1981), Gruszczak (1981)
			Azuma *et al.* (1993)
			Letelier and Oliveira (1998a)
	CY	Vac.	Tomimatsu (1989)
	BR	Vac.	Jantzen (1983)
Euclidean	AS	Vac.	Verdaguer (1982)
Weyl	AS	Vac.	Letelier (1982, 1985a, 1985c)
			Tomimatsu (1984)
			Carot and Verdaguer (1989)
Kerr	AS	Vac.	Tomimatsu (1980)
v. Stockum	AS	Vac.	Letelier (1986)
Kasner	CO	Vac.	Belinskii and Fargion (1980b)
			Carr and Verdaguer (1983)
			Ibañez and Verdaguer (1983, 1985, 1986)
			Das (1985), Griffiths and Miccichè (1999)
	CY	Vac.	Cespedes and Verdaguer (1987)
Bianchi II	CO	Vac.	Belinskii and Francaviglia (1982, 1984)
vacuum		Vac.	Bradley *et al.* (1991)
Einstein–	CY	Vac.	Letelier (1985b)
Rosen			Fustero and Verdaguer (1986)
			Oliver and Verdaguer (1989)
Non-diagonal	AS	Vac.	Das and Chaudhuri (1991)
Minkowski	AS	EM	Alekseev (1980), Wang *et al.* (1983b)
	CY	EM	Dagotto *et al.* (1993)

where κ_k are arbitrary complex vectors. The n_k are now constructed via

$$n_l = \sum_{k=1}^{2n} N_k \mu_k^{-1} \Gamma_{kl}^{-1}, \quad N_k = \mathbf{g}_0 m_k, \quad \Gamma_{kl} = \left(\mu_k \mu_l - W^2\right) m_k \mathbf{g}_0 m_l. \tag{34.120}$$

Finally the new physical metric is given by

$$\mathbf{g} = \prod_{k=1}^{2n} \left(\mu_k W^{-1}\right) \sum_{k,l=1}^{2n} \left(\mathbf{g}_0 - \mu_k^{-1} \mu_l^{-1} \Gamma_{kl}^{-1} N_k N_l^T\right). \tag{34.121}$$

The Belinski–Zakharov formulation of the field equations gives also rise to an infinite number of conserved quantities, see Wu *et al.* (1983). For poles of higher order see Gleiser *et al.* (1988b).

34.6 The Riemann–Hilbert problem

34.6.1 Some general remarks

It was recognized quite early that the linear problem equivalent to the field equations in the stationary axisymmetric case admits a formulation in terms of a Riemann–Hilbert problem (or of integral equations), in which the values of the Ernst potential on the axis of symmetry quite naturally occur, see e.g. Belinskii and Zakharov (1978), Hauser and Ernst (1979a, 1979b, 1980a, 1980b), Sibgatullin (1984, 1991), Manko and Sibgatullin (1993) and Alekseev (1980, 1985). For a detailed discussion of the mathematics involved cp. Hauser (1984). In most of the applications solutions were constructed which have been – or could have been – quite easily found by using the more direct approaches discussed in the previous sections. An example of a more intricate application is the Neugebauer–Meinel solution discussed below. For other formulations and applications of the Riemann–Hilbert problem see also Guo *et al.* (1983a), Wang *et al.* (1983a, 1984) and Nagatomo (1989).

Hauser (1980) and Hauser and Ernst (1981) used a Riemann–Hilbert formulation (referring to the linear system (34.58), cp. §10.7) to prove a conjecture due to Geroch, namely that any stationary axisymmetric solution which is regular in an open neighbourhood of at least one point on the axis can, at least in principle, be generated from flat space by application of the Geroch group, the algebra of which is given in (34.62) (that a solution is uniquely determined by the values of the Ernst potential on the axis follows from standard theorems on elliptic differential equations). Some first applications of this formulation are discussed in Ernst (1984) and Guo (1984a, 1984b). The analogous formulations for the colliding plane waves (including a generalized Geroch conjecture) have been discussed by Ernst *et al.* (1988) and in a series of papers by Hauser and Ernst, see e.g. Hauser and Ernst (1991, 2001).

The method proposed by Alekseev has been applied on several examples, see e.g. Alekseev and García D. (1996) (where also a detailed presentation of the method is given) and Miccichè and Griffiths (2000) and the references given therein.

Sibgatullin's method has been used in many applications by Sibgatullin himself and by Manko and coworkers, see e.g. Il'ichev and Sibgatullin (1985), Manko and Sibgatullin (1992), Aguirregabiria *et al.* (1993a),

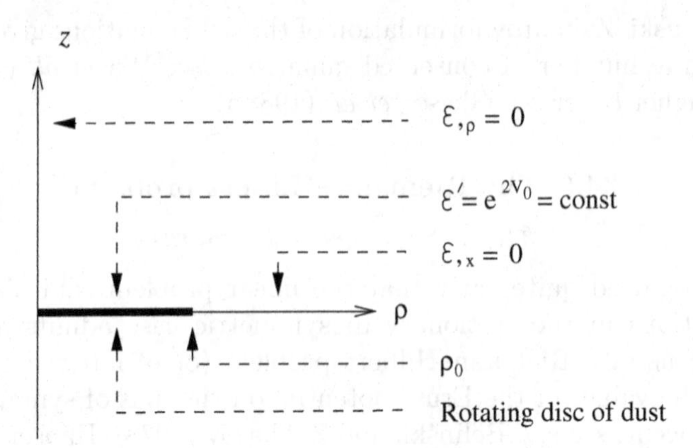

Fig. 34.5. The parameters and boundary conditions for the rotating disc of dust

Manko *et al.* (1995), Breton B. and Manko (1995), Manko and Ruiz (1998), Manko (1999) and Manko *et al.* (1999) and the references given therein. Many of these papers deal with the equilibrium problem of n bodies on the axis, including some solutions not readily given by other methods.

34.6.2 The Neugebauer–Meinel rotating disc solution

Neugebauer and Meinel (1993, 1994, 1995) and Neugebauer *et al.* (1996) have constructed and used a Riemann–Hilbert problem pertaining to the linear problem (34.79) to find the exact solution for a rigidly rotating disc of dust. The boundary data for such a configuration are shown in Fig. 34.5.

First, (34.79) are solved on the coordinate axis $\rho = 0$ and the equatorial plane $z = 0$ where they reduce to ordinary differential equations. In particular, the Ernst potential on the axis, $\mathcal{E}(\rho = 0, z)$, is given in terms of a solution, $\beta(x)$, of a linear integral equation the details of which can be found in the original papers. The Ernst potential can, according to (34.81), be read off from the matrix $\Phi(\Lambda, \rho, z)$ evaluated at $\Lambda = 1$. For arbitrary fixed values of ρ and z the matrix $\Phi(\lambda)$ is regular everywhere in the complex Λ-plane except on the curve

$$\Gamma: \ \Lambda = \sqrt{(\mathrm{i}z + \rho_0 x - \rho)/(\mathrm{i}z + \rho_0 x + \rho)}, \qquad -1 \leq x \leq 1, \ (\mathrm{Re}\Lambda > 0 \text{ for } z > 0). \tag{34.122}$$

On Γ, Φ jumps in a well-defined way, i.e.

$$[\Phi(\Lambda)]_+ = A(x)\,[\Phi(\Lambda)]_- - B(x)\Phi(-\Lambda). \tag{34.123}$$

The jump coefficients A and B – the scattering data – are algebraic functions of $\beta(x)$ and x. This is a matrix Riemann–Hilbert problem and can be reformulated as an integral equation. The approach via the axis data can also be used to derive the Kerr solution (Neugebauer 2000).

The solution for the rigidly rotating disc of dust depends on two parameters, the radius ρ_0 and the angular velocity Ω. It exists for $\mu = 2\Omega^2\rho_0^2 \mathrm{e}^{-2V_0} \leq \mu_0 = 4.62966\ldots$. The surface potential e^{-2V_0} is an implicit function of μ. For $\mu = \mu_0$ the solution reduces to the extreme Kerr solution.

The solution of the Riemann–Hilbert problem leads to a class of Ernst potentials involving elliptic and hyperelliptic functions (Meinel and Neugebauer 1996). Let, for given n, K_i, $i = 1, \ldots, n$, be arbitrary complex constants and

$$W = \left[(K + \mathrm{i}\zeta)\,(K - \mathrm{i}\zeta) \prod_{i=1}^{n} (K - K_i)(K - \overline{K}_i)\right]^{1/2}. \tag{34.124}$$

Additional quantities $K^{(m)}, m = 1, \ldots, n$, are defined as solutions of the so-called Jacobi inversion problem

$$\sum_{m=1}^{n} \int_{K_m}^{K^{(m)}} \frac{K^j dK}{W} = u_j, \quad j = 0, \ldots, n-1, \tag{34.125}$$

where the u_j are given recursively by

$$\Delta u_0 = 0, \quad \mathrm{i}\partial_\zeta u_j = \tfrac{1}{2} u_{j-1} + \zeta \partial_\zeta u_{j-1}, \; j = 1, \ldots, n. \tag{34.126}$$

All u_j are thus solutions of Laplace's equation. Finally, the Ernst potential is

$$\mathcal{E} = \exp\left(\sum_{m=1}^{n} \int_{K_m}^{K^{(m)}} \frac{K^n dK}{W} - u_n\right) \tag{34.127}$$

This Ernst potential can also be expressed in terms of theta functions (Neugebauer *et al.* 1996).

For other applications using hyperelliptic functions see Korotkin (1988, 1993), Klein and Richter (1997, 1999) and the references given therein. The relation between the different approaches is discussed in Korotkin (1997) and Meinel and Neugebauer (1997).

34.7 Other approaches

In this section we shall mention several other approaches to solving Ernst's equation.

Chinea (1983, 1984), Harrison (1980, 1983), Cosgrove (1981), Omote and Wadati (1981a, 1981b) and Nakamura (1979, 1987) have constructed various Bäcklund transformations distinct from the one discussed here in detail. Hirota's method has been applied by Masuda *et al.* (1998) and Nakamura and Ohta (1991). Twistor methods have been used by Ward (1983) and Woodhouse and Mason (1988). The infinitesimal transformation for the system (34.56) discovered by Hou and Li (1988) and Li (1988, 1989a) and studied further by Kinnersley (1990, 1991) is a generalization of the original Geroch transformations; the resulting algebra turns out to be a semidirect product of the original Kinnersley–Chitre transformations (34.63) and a Virasoro algebra operating on the fields W and V (Wu and Ge 1983). For yet other approaches see Dodd and Morris (1982), Dodd *et al.* (1984), Zhong (1985) and Korotkin and Nicolai (1994).

Relations between various solution-generating methods have been studied in detail by Cosgrove (1980, 1982a, 1982b).

34.8 Einstein–Maxwell fields

All solution-generating methods that have been discussed above and that pertain to vacuum solutions can be generalized to Einstein–Maxwell fields. In particular, Kinnersley (1980) has generalized the approach of §34.3 to include electrovac solutions. Bäcklund transformations have been found and used by Kramer and Neugebauer (1981, 1984), Neugebauer and Kramer (1983), and Kramer (1984a). Guo and Ernst (1982) and Ernst (1994) have generalized the double Kerr solution to the double Kerr-Newman solution. Alekseev's technique has been applied frequently to Einstein–Maxwell fields (Alekseev 1983, 1990). Another Bäcklund transformation has been found by Omote *et al.* (1980). Most solution-generating techniques can generate only the hyperextreme Kerr–Newman solution, i.e. $m^2 < a^2 + e^2$, whereas Sibgatullin's method also allows the generation of the underextreme solution from flat space. (An n-body solution containing such objects was given in Ruiz *et al.* (1995).) For a relation of the Einstein–Maxwell equations to the non-linear σ-model see Mazur (1983).

34.9 The case of two space-like Killing vectors

Having concentrated so far mostly on the stationary axisymmetric case we shall in this section describe briefly how to translate the main equations into the case where both Killing vectors are spacelike. The metric now assumes the form

$$ds^2 = e^{-2U}[e^{2k}(dz^2 - dt^2) + W^2 dx^2] + e^{2U}(dy + A dx)^2, \quad (34.128)$$

cp. (17.4) and (22.40), and the Killing vectors are $\xi = \partial_x$ and $\eta = \partial_y$. The field equations are to be derived from the Lagrangian given in (34.31) with the difference that here $\nabla f \, \nabla g = \partial_z f \, \partial_z g - \partial_t f \, \partial_t g$, i.e. the metric for manipulating the ∇ operator is diag$(1, -1)$. Equation (34.27) translates into

$$\nabla = (\partial_z, \partial_t), \quad \widetilde{\nabla} = (\partial_t, \partial_z), \quad \partial = \partial_z + \partial_t, \quad \partial^* = \partial_z - \partial_t \quad (34.129)$$

and the transformed Lagrangian analogous to (34.33) becomes

$$L' = 2\nabla k \nabla W - \tfrac{1}{2} W \left[4(\nabla U)^2 - e^{-4U}(\nabla \psi)^2\right]. \quad (34.130)$$

N.b. (34.32) remains unchanged. Now two real Ernst potentials are introduced by

$$\mathcal{E} = e^{2U} + \psi, \quad \mathcal{E}^* = e^{2U} - \psi, \quad (34.131)$$

and the Lagrangian becomes

$$L' = 2\nabla k \nabla W - 2W \nabla \mathcal{E} \, \nabla \mathcal{E}^* \, (\mathcal{E} + \mathcal{E}^*)^{-2}. \quad (34.132)$$

This is the same function as (34.36) and consequently every solution of the corresponding field equations gives rise to two different space-times.

The rule for translating formulae from the stationary axisymmetric situation into the one with two spacelike Killing vectors is that in every occurrence of the Ernst potential $\mathcal{E}$ or the derivative operator ∂ an 'i' should be replaced by a '$\pm$', the conjugation operation being $\pm \rightarrow \mp$. By this replacement the complex Ernst potential $\mathcal{E}$ splits into two real potentials, i.e. $\mathcal{E} = e^{2U} + i\psi \rightarrow e^{2U} \pm \psi$, cp. (34.131). Note that (34.39) can be used without change after a reinterpretation of the symbols M^k and ∂.

The field equation for the metric function W becomes

$$\nabla^2 W = W_{,zz} - W_{,tt} = 0, \quad (34.133)$$

and in contrast to the stationary case there are the possibilities of ∇W being spacelike, timelike, null or even zero. The first case is commonly referred to as 'cylindrical waves' (§22.3), the second one as 'cosmological solutions' or 'colliding waves' (Chapter 25) and the third case gives rise to plane waves (§22.3). The fourth case implies that space-time is flat for vacuum (§24.5). For a detailed discussion of colliding waves the reader is referred to Griffiths (1991).

Bäcklund transformations for the case of two space-like Killing vectors have been given by various authors; indeed, the Belinski–Zakharov technique discussed above and the methods developed by Alekseev were first formulated for this case. In addition to the papers quoted above we mention Papanicolaou (1979) and Fokas *et al.* (1999).

35
Special vector and tensor fields

35.1 Space-times that admit constant vector and tensor fields

35.1.1 Constant vector fields

Because of the definition of the curvature tensor, the very existence of a constant vector field $\boldsymbol{a}$,

$$a_{b;c} = 0, \tag{35.1}$$

imposes severe conditions on the curvature tensor and the metric: $\boldsymbol{a}$ is (proportional to) a constant vector field if and only if it satisfies

$$a_b R^b{}_{cde} = 0 \tag{35.2}$$

and all equations obtained by repeated differentiation of (35.2).

Equation (35.2) shows that a four-dimensional space admitting *four* (independent) *constant vectors* is necessarily flat. The constancy of the metric g_{ab} and the existence of *three constant vectors* imply the existence of a fourth constant vector (which completes the system of the three) via (3.2), and therefore the space is again flat.

If *two constant vectors* $\boldsymbol{a}$ and $\boldsymbol{b}$ exist (and are linearly independent), it follows from (35.2) by considering a tetrad representation of the curvature tensor that this tensor can be given in terms of a simple bivector A_{cd}:

$$R_{cdef} = A_{cd}A_{ef}, \quad A_{cd}a^c = 0 = A_{cd}b^c. \tag{35.3}$$

Two different cases occur, depending on whether $\boldsymbol{a}$ and $\boldsymbol{b}$ can both be chosen to be non-null, in which case

$$A_{cd} = p_c q_d - q_c p_d, \quad a_c b^c = 0, \quad p_c q^c = 0, \quad (p_a p^a)(q_b q^b) \neq 0, \tag{35.4}$$

or not, in which case one of them is necessarily a null vector, say $\boldsymbol{b} = \boldsymbol{k}$, and their product vanishes,

$$A_{ab} = p_a k_b - k_a p_b, \; p_a p^a \neq 0, \; a_c a^c \neq 0, \; k_a k^a = a_a k^a = k_a p^a = 0. \tag{35.5}$$

In the first case (35.4), the metric can be transformed into

$$\begin{aligned} ds^2 &= g_{AB}(x^1,x^2)dx^A dx^B + \varepsilon_1(dx^3)^2 + \varepsilon_2(dx^4)^2, \\ a^c &= \delta^c_4, \quad b^c = \delta^c_3, \quad A,B = 1,2, \quad \varepsilon_1,\varepsilon_2 = \pm 1. \end{aligned} \tag{35.6}$$

Because of (35.3), the energy-momentum tensor T_{cd} is proportional to $(a_c a_d/a^2 + b_c b_d/b^2)$. In the second case (35.5), we get

$$ds^2 = g_{AB}(x^1,x^2)dx^A dx^B + 2dx^1dx^3 + (dx^4)^2, \; a^c = \delta^c_4, \; k^c = \delta^c_3, \tag{35.7}$$

the energy momentum tensor $\kappa_0 T_{ab} = p^2 k_a k_b$ being that of a pure radiation field. (For both cases, see Eisenhart (1949), p. 162, and Takeno and Kitamura (1968).)

Finally, if the space-time admits *one constant vector* $a^c = \delta^c_4$, this vector is either non-null, the metric being

$$ds^2 = g_{\alpha\beta}(x^\nu)dx^\alpha dx^\beta + \varepsilon(dx^4)^2, \quad \varepsilon = \pm 1, \quad \alpha,\beta,\nu = 1,2,3, \tag{35.8}$$

or the vector $a^c = \delta^c_4$ is null,

$$ds^2 = g_{\alpha\beta}(x^\nu)dx^\alpha dx^\beta - 2dx^3dx^4, \quad \alpha,\beta,\nu = 1,2,3. \tag{35.9}$$

The existence of a non-null constant vector $a^c = \delta^c_4$ implies $R_{4knm} = 0$ and $\overset{4}{R}_{\alpha\beta\gamma\delta} = \overset{3}{R}_{\alpha\beta\gamma\delta}$. For vacuum solutions $R_{ab} = 0$ this yields $\overset{3}{R}_{\alpha\beta} = 0$, which is equivalent to $\overset{3}{R}_{\alpha\beta\gamma\delta} = 0$: if a vacuum solution admits a constant vector field, this vector is a null vector or space-time is flat. For perfect fluid solutions it immediately follows that for a timelike constant vector the equation of state is $\mu + 3p = 0$ (and the solution is the Einstein universe, (12.24) or (37.40), and for a space-like constant vector $\mu = p$. In the latter case the general solution is not known; particular cases are the Gödel solution (12.26), the special Kasner solution (13.53) with $p_3 = 0$ and some solutions with a G_2 (Coley and Tupper 1991).

35.1.2 Constant tensor fields

Decomposable space-times A four-dimensional space-time is *decomposable* if it is the product of a three- and a one-dimensional space

$$ds^2 = g_{\alpha\beta}(x^\nu)dx^\alpha dx^\beta + \varepsilon(dx^4)^2, \quad \varepsilon = \pm 1, \tag{35.10}$$

or the product of two two-dimensional spaces

$$ds^2 = g_{AB}(x^C)dx^A dx^B + g_{MN}(x^P)dx^M dx^N. \tag{35.11}$$

Decomposable space-times can be characterized (Petrov 1966, p. 398) by the existence of a symmetric tensor h_{mn} which is idempotent and constant,

$$h_{ab} = h_{ba}, \quad h_{ab}h^b{}_c = h_{ac}, \quad h_{ab;c} = 0. \tag{35.12}$$

This tensor can be used to split the metric tensor into two parts,

$$g_{ab} = {}_{(1)}g_{ab} + {}_{(2)}g_{ab} \ = h_{ab} + (g_{ab} - h_{ab}), \tag{35.13}$$

both satisfying (35.12). The rank of the matrix h_{ab} is three (one) or two, respectively.

The metric (35.10) is that of (35.8), admitting a constant non-null vector. In the metric (35.11), the only surviving components of the curvature tensor are R_{1212} and R_{3434}, and the Ricci tensor obeys

$$R^1_1 = R^2_2, \quad R^3_3 = R^4_4, \quad R^b_a = 0 \text{ otherwise.} \tag{35.14}$$

Correspondingly, of the energy-momentum tensors considered in this book, only that of a Maxwell field is possible, and for this one should have $R = 0$, i.e. a conformally flat space-time (35.35).

Space-times conformal to decomposable space-times, the conformal factor depending only on one of the two coordinate sets (*warped space-times*) were considered by Carot and da Costa (1993).

Constant symmetric tensors of order two It can be shown (Hall 1991) that the existence of a constant symmetric tensor of order two implies that a constant vector exists, or space-time is $(2+2)$-decomposable.

Constant non-null bivectors A constant non-null bivector F_{ab} (which need not be the actual electromagnetic field, but trivially satisfies the Maxwell equations) implies the existence of a constant self-dual bivector F^*_{ab}, which because of (5.11) has the form

$$F^*_{ab} = AW_{ab}, \quad A = \text{const.} \tag{35.15}$$

W_{ab} is constant, and so is the tensor

$$2h_{ab} = g_{ab} - W_{ac}\overline{W}^c{}_b = 2(g_{ab} + k_a l_b + k_b l_a), \tag{35.16}$$

which obeys (35.12). Consequently, if a V_4 admits a constant non-null bivector, it is decomposable and its metric can be written in the form (35.11) (Debever and Cahen 1960).

Conversely, if a V_4 is decomposable, the constant constituent parts of the metric being

$${}_{(1)}g_{ab} = 2m_{(a}\overline{m}_{b)}, \quad {}_{(2)}g_{ab} = -2k_{(a}l_{b)}, \tag{35.17}$$

then $(m_a\overline{m}_b)_{;c}$, $(k_a l_b)_{;c}$ and $W_{ab} = 2m_{[a}\overline{m}_{b]} + 2l_{[a}k_{b]}$ are constant too.

Constant null bivectors A null bivector F_{ab} can be written in the form

$$F_{ab} = p_a k_b - p_b k_a, \quad p^n p_n = 1, \quad k^a k_a = 0, \quad p^n k_n = 0. \tag{35.18}$$

If F_{ab} is constant, so is $F_{ab}F_c{}^b = k_a k_c$, which implies $k_{a;b} = 0$. It can be shown (Ehlers and Kundt 1962) that a V_4 admits a constant null bivector if and only if it is a *pp*-wave (§24.5)

$$\mathrm{d}s^2 = \mathrm{d}x^2 + \mathrm{d}y^2 - 2\mathrm{d}u\,\mathrm{d}v - 2H(x,y,u)\mathrm{d}u^2, \quad k_a = -u_{,a}. \tag{35.19}$$

35.2 Complex recurrent, conformally recurrent, recurrent and symmetric spaces

35.2.1 The definitions

A *complex recurrent* space-time V_4 is a space for which the self-dual Weyl tensor (3.53) satisfies the condition

$$C^*_{abcd;e} = C^*_{abcd} K_e. \tag{35.20}$$

Generally, the recurrence vector K_e is complex; if it is real, then the space is a *conformally recurrent* space

$$C_{abcd;e} = C_{abcd} K_e, \tag{35.21}$$

and if K_e is zero, one gets a *conformally symmetric* space.

A *recurrent space* is a space in which the Riemann tensor satisfies

$$R_{abcd;e} = R_{abcd} K_e. \tag{35.22}$$

For a recurrent space, the identities $R_{abcd;[mn]} + R_{cdmn;[ab]} + R_{mnab;[cd]} = 0$ yield $K_e = K_{,e}$. If instead of (35.22) only

$$R_{ac;e} = R_{ac} K_e \tag{35.23}$$

holds, the space is *Ricci recurrent* (for which see Hall 1976b).

A recurrent space is said to be *symmetric* if K_e vanishes,

$$R_{abcd;e} = 0. \tag{35.24}$$

Obviously, each recurrent (symmetric) space-time is conformally recurrent (conformally symmetric) and hence a complex recurrent space too.

Using the canonical forms (§4.2) of the self-dual Weyl tensor C^*_{abcd} and evaluating (35.20), it can easily be shown that there are no complex recurrent space-times of Petrov types I, II and III. Therefore, we have only to deal with types D, N and O. We shall list the main results in the following, only indicating the ideas of the proofs which can be found in Sciama (1961), Kaigorodov (1971), McLenaghan and Leroy (1972) and the references given there.

35.2.2 Space-times of Petrov type D

The canonical form (cp. Table 4.2)

$$C^*_{abcd} = 2\Psi_2(V_{ab}U_{cd} + U_{ab}V_{cd} + W_{ab}W_{cd}) \tag{35.25}$$

of a type D Weyl tensor is compatible with (35.20) only if

$$\Psi_{2,e} = \Psi_2 K_e, \quad U_{ab;e}W^{ab} = V_{ab;e}W^{ab} = 0 \tag{35.26}$$

holds. Equation (35.26) implies $W_{ab} = \text{const}$, which means that a type D complex recurrent space-time is necessarily decomposable and a product (35.11) of two two-dimensional spaces.

Because of (35.20) and (35.26), $C^*_{abcd;[ef]} = 0$ holds. Inserting (35.25) and using the decomposition (3.45) of the curvature tensor one obtains

$$\begin{aligned} &\Psi_2 = -R/12, \quad 4E_{abcd} = E\overline{W}_{ab}W_{cd}, \\ &4R_{ab} = -(R+E)(l_a k_b + k_a l_b) + (R-E)(m_a\overline{m}_b + \overline{m}_a m_b). \end{aligned} \tag{35.27}$$

One sees that the recurrence vector $K_e = \Psi_{2,e}/\Psi_2$ is real; all complex recurrent space-times of type D are *conformally recurrent*. Furthermore, (35.27) shows that the scalar curvatures of the $(\boldsymbol{m}, \overline{\boldsymbol{m}})$- and the $(\boldsymbol{k}, \boldsymbol{l})$-spaces are given by

$$_{(1)}R(x^1, x^2) = (R-E)/4, \quad _{(2)}R(x^3, x^4) = (R+E)/4. \tag{35.28}$$

If the space is *recurrent*, with K_e non-zero, then $R^2 = E^2$ holds, i.e. the space-time is a product of a flat and a curved two-dimensional space.

For a conformally symmetric space-time, R is constant, and the full set of Bianchi identities yields $E = \text{const}$: the space is symmetric and has the line element (12.8),

$$ds^2 = \frac{2d\zeta d\overline{\zeta}}{\left[1 + (R-E)\zeta\overline{\zeta}/8\right]^2} - \frac{2du\, dv}{[1 - (R+E)uv/8]^2}. \tag{35.29}$$

35.2.3 Space-times of type N

Inserting the canonical form $C^*_{abcd} = -4V_{ab}V_{cd}$ of a Weyl tensor of type N into (35.20), one obtains $k_{a;b} = k_a p_b$, the (eigen-) null vector $\boldsymbol{k}$ is recurrent. Conversely, these two relations yield (35.20): a space-time of type N is complex recurrent exactly if it contains a recurrent null vector.

The metric of these complex recurrent spaces of type N is

$$\begin{aligned} ds^2 &= 2k^{-2}(1+\varepsilon\zeta\overline{\zeta})^{-2}(d\zeta + b\,du)(d\overline{\zeta} + \overline{b}\,du) \\ &\quad -2du\left[dv + du(\varepsilon k^2 v^2 + lv + H)\right], \\ l &= (b_{,\zeta} + \overline{b}_{,\overline{\zeta}})/2 - \varepsilon(1+\zeta\overline{\zeta})^{-1}(\overline{\zeta}b + \zeta\overline{b}), \quad b = b(u,\zeta), \\ H &= \overline{H} = H(u,\zeta,\overline{\zeta}), \quad k^2 = 1 + \varepsilon^2 K^2(u), \quad \varepsilon = 0, \pm 1. \end{aligned} \tag{35.30}$$

For a conformally recurrent space, this line element specializes to (35.19) with $H = f(u,x) + g(u,y)$, and the conformally symmetric and the symmetric spaces have (35.19) with $H = x^2 - y^2 + \text{const}(x^2 + y^2)$.

Among all these spaces of type N, the only vacuum solutions are the *pp*-waves (§24.5), which can be characterized by being complex recurrent vacuum solutions (Ehlers and Kundt 1962).

35.2.4 Space-times of type O

Starting from the definition (35.22) of a *recurrent* space, one easily obtains

$$R_{,a} = RK_a, \quad 2S_{ab;[cd]} = S_{eb}R^e{}_{acd} + S_{ae}R^e{}_{bcd} = 0. \tag{35.31}$$

The Weyl tensor being zero, from this equation follows

$$S_{ac}S^c{}_b - \tfrac{1}{4}g_{ab}S_{cd}S^{cd} + \tfrac{1}{6}RS_{ab} = 0. \tag{35.32}$$

Inserting into (35.32) the canonical forms (§5.1) of the traceless tensor S_{ab} and substituting the results back into (35.23), one sees that the following cases can occur.

Space-times with a non-vanishing recurrence vector K_a must be plane waves

$$\begin{aligned} ds^2 &= dx^2 + dy^2 - 2du\,dv - \tfrac{1}{2}\kappa_0\Phi^2(u)(x^2+y^2)du^2, \\ K_a &= \Phi_{,a}/2\Phi, \quad k_{a;b} = 0, \quad R_{ab} = \kappa_0\Phi^2 k_a k_b. \end{aligned} \tag{35.33}$$

Symmetric space-times ($K_a = 0$) are either spaces of constant curvature (§8.5) or they possess a constant timelike or spacelike vector field orthogonal to a three-dimensional space of constant curvature,

$$ds^2 = \frac{dx^2 + dy^2 + \varepsilon dw^2}{\left[1 + R(x^2+y^2+\varepsilon w^2)/24\right]^2} - \varepsilon dz^2, \quad \varepsilon = \pm 1, \tag{35.34}$$

or they are a product of two 2-spaces of constant curvature

$$ds^2 = 2d\zeta d\overline{\zeta}\left[1 + \lambda\zeta\overline{\zeta}\right]^{-2} - 2du\,dv\left[1 + \lambda uv\right]^{-2} \tag{35.35}$$

(which is a Bertotti–Robinson metric (12.18), see also Kasner (1921)), or they are plane waves (35.33) with $\Phi = \text{const}$, $K_a = 0$.

35.3 Killing tensors of order two and Killing–Yano tensors

35.3.1 The basic definitions

A Killing tensor of order m is a symmetric tensor $K_{a_1 \cdots a_m}$, which satisfies

$$K_{(a_1 \cdots a_m;c)} = 0. \tag{35.36}$$

Killing tensors are generalizations of Killing vectors ($m = 1$). Since not much is known on solutions with higher-order Killing tensors, we will confine ourselves to *Killing tensors of order two*, which in accordance with (35.36) obey

$$K_{ab} = K_{ba}, \quad K_{(ab;c)} = 0. \tag{35.37}$$

Splitting the Killing tensor into its traceless part P_{ab} and the trace K, $K_{ab} = P_{ab} + K g_{ab}/4$, one easily gets

$$4P^a{}_{b;a} = -3K_{,b} \tag{35.38}$$

and

$$P_{(ab;c)} - \tfrac{1}{3} g_{(ab} P^d{}_{c);d} = 0. \tag{35.39}$$

A symmetric traceless tensor P_{ab} is said to be a *conformal Killing tensor* if it satisfies (35.39). If, in addition to (35.39), $P^a{}_{b;a}$ is a gradient, then K_{ab} constructed from P_{ab} is a Killing tensor.

Trivial examples of Killing tensors are g_{ab} and all products

$$K_{ab} = \xi_{(a}\eta_{b)} \tag{35.40}$$

of Killing vectors ξ_a, η_b (not necessarily different), and linear combinations of these with constant coefficients. Killing tensors *not* admitting this type of representation are referred to as non-trivial or non-redundant or irreducible.

Killing tensors of order two (not necessarily trivial) may also be obtained from (special) homothetic or conformal Killing vectors by constructions similar to (35.40) (Koutras 1992a). For example

$$K_{ab} = \xi_{(a}\eta_{b)} - 2\varphi\, g_{ab} \tag{35.41}$$

is a Killing tensor if the vectors ξ_a, η_a obey

$$\xi_{a;b} + \xi_{b;a} = 2\Phi g_{ab}, \quad \eta_{a;b} + \eta_{b;a} = 2\Psi g_{ab}, \quad \Phi\eta_a + \Psi\xi_a = \varphi_{,a}. \tag{35.42}$$

Condition (35.42) is satisfied e.g. if $\xi_a = \eta_a = \varphi_{,a}/2\Phi$ (ξ_a is a hypersurface-orthogonal conformal vector), or if $\Phi = 0$ ($\xi_a = \varphi_{,a}/\Psi$ is a special hypersurface-orthogonal Killing vector).

A *Killing–Yano tensor* is a skew-symmetric tensor $a_{bc\cdots de}$ satisfying

$$a_{bc\cdots de;f} + a_{bc\cdots df;e} = 0, \tag{35.43}$$

which is also a generalization of the Killing equation. At first glance this seems to be a quite different type of tensor, but it can easily be shown that (symmetric) Killing tensors can be constructed from Killing–Yano tensors. In four dimensions, the maximal order of a Killing–Yano tensor is four. In the case of Killing–Yano tensors $a_b = \xi_b$ and $a_{bcde} =$const ε_{bcde}, the corresponding Killing tensors $\xi_a \xi_b$ and $\varepsilon_{bcde}\varepsilon^{cde}{}_a = -6g_{ab}$ are trivial. For second-order tensors $\underset{A}{a}_{bc}$ the corresponding Killing tensors are

$$K_{nm} = \underset{A}{a}_{ni}\,\underset{B}{a}{}^i{}_m + \underset{B}{a}_{ni}\,\underset{A}{a}{}^i{}_m \tag{35.44}$$

and linear combinations of these. A third-order Killing–Yano tensor a_{bcd} can be replaced by the vector $a_e = \varepsilon_e{}^{bcd}a_{bcd}$, which because of (35.43) satisfies

$$4a_{e;i} = g_{ei}a^c{}_{;c}, \tag{35.45}$$

i.e. a_n is a special conformal motion which (because of the Ricci identities) is hypersurface orthogonal. If $\underset{A}{a}_b$ and $\underset{B}{a}_c$ are two such vectors, then

$$K_{bc} = \underset{A}{a}_b\,\underset{B}{a}_c + \underset{A}{a}_c\,\underset{B}{a}_b - 2\,\underset{A}{a}{}^d\,\underset{B}{a}_d\,g_{bc} \tag{35.46}$$

and linear combinations of such terms are symmetric Killing tensors.

35.3.2 First integrals, separability and Killing or Killing–Yano tensors

From the more physical point of view, the interest in Killing tensors originated in their connection with quadratic first integrals of geodesic motion and the separability of various partial differential equations, e.g. the Hamilton–Jacobi equation.

Let t^a be the tangent vector of an affinely parametrized geodesic,

$$\mathrm{D}t^a/\mathrm{D}\lambda = t^a{}_{;b}t^b = 0. \tag{35.47}$$

Then, because of (35.37),

$$\mathrm{d}K_{ab}t^a t^b/\mathrm{d}\lambda = K_{ab;c}t^a t^b t^c = 0 \tag{35.48}$$

holds, so $K_{ab}t^a t^b$ is a quadratic first integral of motion. For a conformal Killing vector P_{ab}, satisfying the weaker condition (35.39), the analogous statement is true only for null geodesics.

Skew-symmetric Killing–Yano tensors a_{bc} induce the existence of a vector $a_{bc}t^b$, which because of (35.43) is parallelly propagated along any geodesic, and to each vector $\boldsymbol{a}$ satisfying (35.45) corresponds a skew-symmetric tensor $a_c t_b - a_b t_c$ which is constant along the geodesic. Each of these cases gives rise to quadratic first integrals, arising, respectively, from the invariants of the vector and the tensor, in accordance with (35.44), (35.45) and (35.48).

For the connection between Killing tensors and the separability of partial differential equations, we refer the reader to the literature, see e.g. Kamran (1988) and Benenti (1997) and the references given there. We mention only the result of Woodhouse (1975) that the separable coordinates for the Hamilton–Jacobi equation in a Lorentzian manifold are adapted either to a Killing vector or to an eigenvector of a (symmetric) Killing tensor of order two. If the field equations can be cast into a Hamiltonian formulation, which may be possible if there is a G_r acting on V_3, then the Killing tensor of the corresponding Jacobi metric can be used in the search for solutions (Uggla *et al.* 1995b), cp. also §13.2.

For the connection between Killing tensors and solutions of the equation of geodesic deviation, see Caviglia *et al.* (1982).

35.3.3 Theorems on Killing and Killing–Yano tensors in four-dimensional space-times

In analogy with the treatment of Killing vectors and groups of motions, the final goal of studying Killing and Killing–Yano tensors is (a) to find all such tensors of a given space-time and/or (b) to classify space-times with respect to the nontrivial Killing tensors and Killing–Yano tensors they admit. Problem (a) can be attacked using computer programs, see Wolf (1998). For problem (b) we shall list some of the few known results.

Theorems and results on Killing tensors K_{ab}

Theorem 35.1 *A four-dimensional space-time admits at most 50 linearly independent Killing tensors of order two. The maximum number of 50 is attained if and only if the space is of constant curvature; in that case, all 50 Killing tensors are reducible* (Hauser and Malhiot 1975a, 1975b).

Theorem 35.2 *Every type D vacuum solution, the Weyl tensor being*

$$C^*_{abmn} = 2\Psi_2(V_{ab}U_{mn} + U_{ab}V_{mn} + W_{ab}W_{mn}), \tag{35.49}$$

admits a conformal Killing tensor

$$\begin{aligned} P_{ab} &= -\tfrac{1}{2}\Phi\overline{\Phi}W_{ac}\overline{W}^{c}{}_{b} = (A^2+B^2)\left[k_a l_b + l_a k_b + \tfrac{1}{2}g_{ab}\right], \\ \Phi &= A + \mathrm{i}B = \mathrm{const}(\Psi_2)^{-1/3}. \end{aligned} \tag{35.50}$$

The conformal Killing tensor is irreducible provided the space-time admits fewer than four Killing vectors (Walker and Penrose 1970).

Theorem 35.3 *Every type D vacuum solution, with the exception of the C-metric and its generalization (case III of Kinnersley's* (1969b *classification), admits a Killing tensor*

$$K_{ab} = (A^2+B^2)(l_a k_b + k_a l_b) + B^2 g_{ab}, \quad A+\mathrm{i}B = \mathrm{const}(\Psi_2)^{-1/3}, \tag{35.51}$$

the complex constant being adjusted such that $DA = \Delta A = \delta B = 0$ (Walker and Penrose 1970, Hughston *et al.* 1972, Hughston and Sommers 1973).

Theorem 35.4 *If a space-time admits a Killing tensor of Segre characteristic* [(11)(1,1)],

$$K_{ab} = A^2(l_a k_b + k_a l_b) + B^2(\overline{m}_a m_b + m_a \overline{m}_b), \tag{35.52}$$

then its two eigenvectors $\boldsymbol{k}$ *and* $\boldsymbol{l}$ *are both shearfree and geodesic and, therefore, principal null vectors of the Weyl tensor if* $R_{11} = R_{14} = R_{44} = 0$, *cp. Theorem 7.1. Furthermore, the functions* A *and* B *have to satisfy*

$$\begin{aligned} DA = \Delta A = \delta B = 0, \qquad & \delta A^2 = (\overline{\pi} - \tau)(A^2 + B^2), \\ DB^2 = -(\rho + \overline{\rho})(A^2 + B^2), \quad & \Delta B^2 = (\mu + \overline{\mu})(A^2 + B^2). \end{aligned} \tag{35.53}$$

If the space-time is a vacuum solution, then (35.51) *holds.*

The metrics corresponding to this type of Killing tensor have been determined by Hauser and Malhiot (1978) and Papacostas (1988).

Theorem 35.5 *If a space-time admits a Killing tensor* (35.52) *with non-constant* A *and* B *and* $A^{;a}A_{,a}B^{;b}B_{,b} \neq 0$, *then* $\Phi_{01} = \mathrm{e}\Phi_{21}$ *is a necessary and sufficient condition for the existence of an Abelian group which acts on spacelike* ($\mathrm{e} = -1$) *or timelike* ($\mathrm{e} = 1$) *orbits. For* $\mathrm{e} = 1$, *the only perfect fluid solution compatible with such a Killing tensor is the Wahlquist solution* (21.57) *with a* G_2 *on timelike orbits* (Papacostas 1988).

Perfect fluid space-times known to admit a Killing tensor (35.52) not satisfying the conditions of Theorem 35.5 are e.g. the interior Schwarzschild solution (16.18), the Kantowski–Sachs solutions, the Gödel solution (12.26) and a solution of the Wahlquist class (21.57) with a vanishing Simon tensor found by Kramer (Papacostas 1988). They all have at least a G_2. For Killing tensors in static spherically-symmetric metrics see Hauser and Malhiot (1974), in *pp*-waves Cosgrove (1978b).

Theorems and results on Killing–Yano tensors

Theorem 35.6 *If a space-time admits a non-degenerate second-order Killing–Yano tensor, then this tensor can be written as*

$$a_{bc} = A(l_b k_c - k_b l_c) + \mathrm{i}B(m_b \overline{m}_c - \overline{m}_b m_c), \tag{35.54}$$

the Weyl tensor is of type D (or 0), the degenerate, geodesic and shearfree null eigenvectors being $\boldsymbol{l}$ *and* $\boldsymbol{k}$, *the Ricci tensor satisfies* $R_{bc}a^c{}_d + R_{dc}a^c{}_b = 0$, *and the real functions A and B have to obey*

$$\begin{aligned} DA = \Delta A = \delta B = 0, \\ D(A+\mathrm{i}B) = -\rho(A+\mathrm{i}B),\ \Delta(A+\mathrm{i}B) = \mu(A+\mathrm{i}B), \\ \delta(A+\mathrm{i}B) = -\tau(A+\mathrm{i}B),\ \overline{\delta}(A+\mathrm{i}B) = \pi(A+\mathrm{i}B) \end{aligned} \tag{35.55}$$

(Collinson 1974, 1976a, Stephani 1978).

Theorem 35.7 *All type D vacuum solutions that admit a Killing tensor* K_{bc} *(see Theorem* 35.3*), also admit a Killing–Yano tensor* a_{bc}*, the two tensors being connected by* (35.44) (Collinson 1976a, Stephani 1978).

Theorem 35.8 *If a space-time admits a degenerate Killing–Yano tensor* a_{bc}*, then this tensor can be written as*

$$a_{bc} = k_b p_c - p_b k_c, \tag{35.56}$$

$\boldsymbol{k}$ *is a (null) Killing vector, the Weyl tensor is of type N (or O),* $\boldsymbol{k}$ *being the multiple eigenvector, and the Ricci tensor has to satisfy* $R_{bc}a^c{}_d + R_{dc}a^c{}_b = 0$ (Collinson 1974, Stephani 1978).

Theorem 35.9 *If a vacuum space-time admits a third-order Killing–Yano tensor, i.e. a vector* $\boldsymbol{a}$ *satisfying* (35.45)*, then* $\boldsymbol{a}$ *is a constant null vector or space-time is flat* (Collinson 1974).

Theorem 35.10 *If a space-time admits at least two independent Killing–Yano tensors, then* (a) *it is of constant curvature and admits ten independent Killing–Yano tensors, or* (b) *it is not an Einstein space but contains a 3-space of constant curvature and admits exactly four independent Killing–Yano tensors, or* (c) *space-time is decomposable* (35.11) *and admits exactly two independent Killing–Yano tensors, or* (d) *space-time admits a constant null vector and exactly two independent Killing–Yano tensors. For non-flat vacuum fields, only this last case applies* (Hall 1987).

Normal forms of metrics admitting second-order Killing–Yano tensors have been constructed by Dietz and Rüdiger (1981, 1982). Papacostas (1985) discussed Killing–Yano tensors for type D Einstein–Maxwell fields with two Killing vectors, for the interior Schwarzschild solution (16.18), and for Kantowski–Sachs solutions. For other examples see Ibohal (1997).

35.4 Collineations and conformal motions

35.4.1 The basic definitions

Besides Killing vectors (motions)

$$\mathcal{L}_\xi g_{nm} = \xi_{n;m} + \xi_{m;n} = 0 \tag{35.57}$$

and homothetic vectors (homothetic motions)

$$\mathcal{L}_\xi g_{nm} = 2a g_{nm}, \quad a = \text{const}, \tag{35.58}$$

which originate in the symmetries of Einstein's equations in the process of a similarity reduction (cp. §10.2.3), there are several other vector fields which originate in the symmetries of e.g. the equation of geodesic motion (so that their existence puts severe restrictions upon space-time) and which can also be used to characterize space-times. These are

$$\text{conformal motions:} \quad \mathcal{L}_\xi g_{nm} = 2\phi(x) g_{nm}, \tag{35.59}$$

$$\text{projective collineations:} \quad \mathcal{L}_\xi \Gamma^i_{jk} = \delta^i_j \varphi_{,k} + \delta^i_k \varphi_{,j}, \tag{35.60}$$

$$\text{affine collineations:} \quad \mathcal{L}_\xi \Gamma^i_{jk} = 0, \tag{35.61}$$

$$\text{Ricci collineations:} \quad \mathcal{L}_\xi R_{nm} = 0, \tag{35.62}$$

$$\text{curvature collineations:} \quad \mathcal{L}_\xi R^a{}_{bnm} = 0. \tag{35.63}$$

Conformal motions (Petrov 1966) preserve angles between two directions at a point and map null geodesics into null geodesics. *Projective collineations* map geodesics into geodesics; *affine collineations* preserve, in addition, the affine parameters on geodesics (Katzin and Levine 1972). Obviously, motions, affine collineations and homothetic motions are automatically curvature collineations. The connection between those vector fields and geodesic first integrals is discussed e.g. in Katzin and Levine (1981). A collection of formulae on the incorporation of all these symmetries into the field equations can be found in Zafiris (1997).

For perfect fluids, the notion of a homothetic vector has been generalized to that of a *similarity vector* (*kinematic self-similarity*) satisfying

$$\mathcal{L}_\xi (g_{mn} + u_n u_m) = 2a(g_{mn} + u_n u_m), \quad \mathcal{L}_\xi u_n = \alpha u_n, \quad a, \alpha = \text{ const}, \tag{35.64}$$

which for $\alpha = a$ is equivalent to (35.58). For applications see Carter and Henriksen (1989), Coley (1997a), Benoit and Coley (1998), and Sintes (1998), and for a further generalization Collins and Lang (1987).

We will discuss in some detail only the proper curvature collineations and the conformal motions. Definition (35.61) implies that proper affine collineations satisfy $\xi_{a;nm} = R_{anm}{}^i\xi_i$, and if they exist then space-time admits a constant symmetric tensor (see above in §35.1) which is not a multiple of the metric (Hall and da Costa 1988).

35.4.2 Proper curvature collineations

Curvature collineations were introduced by Katzin *et al.* (1969). Proper curvature collineations (i.e. those which are not affine collineations or even conformal motions) are rare. They are almost always special conformal motions (McIntosh 1980).

In the vacuum case, they can only occur in non-twisting type N metrics (Collinson 1970). For the *pp*-waves (§24.5), they were studied by Aichelburg (1970), for the Robinson–Trautman metrics (§28.1) by Vaz and Collinson (1983) and Vaz (1986), and for the non-expanding metrics (Kundt's class §31.5) by Halford *et al.* (1980).

In the Einstein–Maxwell and pure radiation cases, they can occur only if the Maxwell field is null and the Weyl tensor of type N or O (Tariq and Tupper 1977). An example is the metric (22.70), cp. Singh *et al.* (1978). For non-null Einstein–Maxwell fields, all curvature collineations are homothetic motions (McIntosh 1979).

In the perfect fluid case, proper curvature collineations occur only in the conformally flat solutions with a $\mu + 3p = 0$ equation of state (i.e. in the Einstein universe (37.40) and in a subcase of (37.45)) or in type D solutions with $\mu = p$ which admit a spacelike constant vector field (Carot and da Costa 1991).

35.4.3 General theorems on conformal motions

Conformal motions (or conformal Killing vectors) are defined by (35.59) or, equivalently, by

$$\xi_{a;b} = \phi g_{ab} + f_{ab}, \quad f_{ab} = -f_{ba}. \tag{35.65}$$

They are called *proper* if $\phi_{,a} \neq 0$, and *special* if $\phi_{;ab} = 0$. The integrability conditions of the system (35.65) turn out to be

$$\mathcal{L}_\xi C^i{}_{abc} = 0 \tag{35.66}$$

and all equations obtained from this by repeated differentiation (Eisenhart 1949, Defrise 1969, Collinson 1989). As a consequence of the defining equation one has

$$\mathcal{L}_\xi R_{ab} = 2\phi_{;ab} - \phi^{;c}{}_{;c} g_{ab}. \tag{35.67}$$

If a space-time admits several conformal motions $\boldsymbol{\xi}_A$, then they form a group and satisfy

$$\mathcal{L}_{\xi_A} g_{ab} = 2\phi_A g_{ab}, \quad [\boldsymbol{\xi}_A, \boldsymbol{\xi}_B] = C^D{}_{AB}\boldsymbol{\xi}_D, \quad \mathcal{L}_{\xi_A}\phi_B - \mathcal{L}_{\xi_B}\phi_A = C^D{}_{AB}\phi_D. \tag{35.68}$$

Theorem 35.11 *A space of dimension $n\,(>2)$ admits at most $(n+1)(n+2)/2$ independent conformal motions, not all of them being proper. If this maximum number is attained, space-time is conformally flat. If space-time is not conformally flat, then the maximal order of the group of conformal Killing vectors is 7 for Petrov type N, 6 for type D, and 5 for type III* (Eisenhart 1949, Hall and Steele 1991).

For flat space-time, and in Cartesian coordinates, the general solution to (35.65) is

$$\xi^a = A^a + p^a{}_b x^b + Bx^a + 2B_n x^n x^a - B^a x^n x_n, \quad \phi = B + 2B_n x^n, \tag{35.69}$$

where A^a, $p_{ab} = -p_{ba}$, B, B^a are constant (see e.g. Choquet-Bruhat *et al.* (1991)); among these conformal motions only those with $B^a \neq 0$ are proper. If only these B^a are non-zero, the corresponding finite transformations are

$$\widetilde{x}^a = (x^a - \lambda B^a x_n x^n)/(1 - 2\lambda B_n x^n + \lambda^2 B^n B_n), \quad \lambda = \text{const.} \tag{35.70}$$

If two spaces are connected by a conformal transformation $\hat{g}_{ab} = \mathrm{e}^{2U} g_{ab}$, cp. §3.7, and ξ^a_A is a conformal motion of g_{ab}, then it is a conformal motion of $\hat{g}_{ab}$, too,

$$\mathcal{L}_{\xi_A}\hat{g}_{ab} = 2\hat{\phi}_A \hat{g}_{ab}, \quad \hat{\phi}_A = \phi_A + U_{,n}\xi^n_A. \tag{35.71}$$

So (35.69) are also the conformal motions of any conformally flat space-time.

Relation (35.71) indicates that it may be possible to construct for a space-time with (proper) conformal vectors a conformally related space-time in which these vectors are Killing vectors: one has to find a function $U(x^n)$ such that all $\hat{\phi}_A$ are zero, i.e. such that

$$\xi^n_A U_{,n} = -\phi_A \tag{35.72}$$

holds. This will not always be possible: if there is a conformal motion with a fixed point P such that $\xi^n(P) = 0$, but $\phi(P) \neq 0$, then $\xi^n U_{,n} + \phi = 0$ cannot be solved. Taking such exceptions into account, one finds:

Theorem 35.12 *If a space-time g_{ab} admits a group of conformal motions, and the Petrov type, the dimension and the nature (spacelike, etc.)*

of the orbits is the same at each point, and there are no local non-globalizable conformal motions, then there exists a conformally related metric $\hat{g}_{ab} = e^{2U} g_{ab}$ *such that the group is a group of motions (if the Petrov type is not O, and the space-time is not conformally related to*

$$ds^2 = dx^2 + dy^2 - 2dudv - 2\left[\alpha(u)x^2 + \beta(u)y^2 + \gamma(u)xy\right] du^2, \quad (35.73)$$

i.e. to a special pp-wave), or is a group of homothetic motions (if the space-time is conformally a pp-wave). For any Petrov type, $\hat{g}_{ab}$ *can be chosen so that* $\hat{\phi}_{A;ab} = 0$ (Bilyalov 1964, Defrise-Carter 1975, Hall 1990a, Hall and Steele 1991).

If a space-time admits a *special* conformal Killing vector $\phi_{;ab} = 0$, (i.e. a constant vector field $\phi_{,a}$) with $\phi_{,a}\phi^{,a} \neq 0$, then its metric is of the form (35.8). It can be shown that the 3-space orthogonal to $\phi_{,a}$ – in which the projection of $\psi_a = f_{ab}\phi^b$ is a homothetic vector – does not admit a group of Killing vectors acting *transitively* on it; a detailed study of the still possible cases G_r, $r \leq 3$, has been given (Carot 1990, Hall 1990b).

From the above results one may guess that proper conformal vectors are particularly to be expected in those space-times which are either conformally flat or contain conformally flat subspaces. Homogeneous space-times (Chapter 12) of Petrov types *I*, *II* or *D* do not admit proper conformal vectors (Hall 2000).

35.4.4 Non-conformally flat solutions admitting proper conformal motions

The only vacuum solutions which admit a proper ($\phi_{,n} \neq 0$) conformal motion are special *pp*-waves (§24.5); up to the addition of Killing or homothetic vectors this conformal Killing vector is uniquely determined and special, $\phi_{;ab} = 0$ (Eardley *et al.* 1986, Hall 1990a). The vacuum solutions with a cosmological constant, $G_{ab} = \Lambda g_{ab}$, which admit a conformal motion with $\phi \neq 0$, are conformally flat (Garfinkle and Tian 1987).

In the case of a general non-zero energy-momentum tensor T_{ab}, the constituent parts of it (e.g. pressure p, four-velocity u^a) in general do *not* inherit the conformal symmetry, i.e. relations such as $\mathcal{L}_\xi p = -2\phi p$, $\mathcal{L}_\xi u^a = -\phi u^a$ etc. in general do *not* follow from $\mathcal{L}_\xi g_{ab} = 2\phi g_{ab}$ – in contrast to the case of a homothetic motion (§11.3) (Coley and Tupper 1990a, Saridakis and Tsamparlis 1991). The deeper reason for this non-inheritance is that conformal motions are not based on symmetries of the field equations, cp. §10.3.

Perfect fluids do not admit *special* conformal motions (Coley and Tupper 1989). If a perfect fluid with an equation of state $p = p(\mu)$, $p + \mu \neq 0$, admits a proper conformal motion parallel to the four-velocity, it is conformally flat (Coley 1991); if it is rotating and shearfree and has a zero magnetic part of the Weyl tensor, it admits (besides at least two Killing vectors) a conformal motion (Collins 1984).

Spherically-symmetric perfect fluids with a conformal vector orthogonal to the four-velocity and the orbit of the group, where the fluid inherits the symmetry in the sense discussed above, are either static, with

$$\mathrm{d}s^2 = [\mathrm{d}r^2 + (2n^2-1)^{-1}r^2(\mathrm{d}\vartheta^2 + \sin^2\vartheta\mathrm{d}\varphi^2) - \mathrm{d}t^2]/[ar^{1+n} + br^{1-n}], \quad (35.74)$$

or of the form (15.72)–(15.74), where the plane- and pseudospherically-symmetric counterparts are also included (Coley and Tupper 1990b, Coley and Czapor 1991, Kitamura 1994, 1995a, 1995b). Spherically-symmetric perfect fluids with an equation of state $p = p(\mu)$, $\mu + p \neq 0$ and a proper *arbitrary* inheriting conformal vector are contained in (15.72) or are conformally flat Friedmann universes (Coley and Czapor 1991). An example with a conformal motion *not* satisfying $\mathcal{L}_{\xi}p = -2\phi p$ is the metric (15.86). The conformal motions in a general static spherically-symmetric space-time have been determined by Maartens *et al.* (1995).

The only static perfect fluid solutions for which the proper conformal vector $\boldsymbol{\xi}$ commutes with $\boldsymbol{\eta} = \partial/\partial t$ are (18.66) or are spherically- (pseudospherically-) symmetric, *not* inheriting, and given by (Bona and Coll 1991)

$$\begin{aligned} \mathrm{d}s^2 &= 2[k + \Lambda r^2/3]^{-1}\mathrm{d}r^2 + r^2[\mathrm{d}x^2 + \Sigma^2(x,k)\mathrm{d}y^2] - r^2\mathrm{d}t^2, \\ \Sigma(x,k) &= \sin x, \ \sinh x \ \text{ for } \ k = 1, -1, \qquad (35.75) \\ \kappa_0 p &= (\Lambda + k/r^2)/2 = \kappa_0\mu - \Lambda, \quad \boldsymbol{\xi} = r[k + \Lambda r^2/3]^{1/2}\partial/\partial r. \end{aligned}$$

If a plane-symmetric solution with an equation of state $p = p(\mu)$, $\mu + p \neq 0$ admits an inheriting proper conformal motion, it is either contained in (15.72), or it is the static solution

$$\mathrm{d}s^2 = \left[\mathrm{d}x^2 + x^c(\mathrm{d}y^2 + \mathrm{d}z^2) - \mathrm{d}t^2\right](Ax^a + Bx^b)^{-2}, \quad (35.76)$$

with $b = a$, $c = 4a(1-a)/(1-2a)$, or $b = (a-1)/(2a-1)$, $c = 4a$ (Coley and Czapor 1992).

Dust solutions with a G_3 acting on S_2 and a proper conformal vector commuting with the motions of the G_3 are conformally flat (Bona 1988a).

If a perfect fluid solution with an Abelian orthogonally transitive G_2 acting on a non-null V_2 (see §17.1) admits one proper conformal motion, the orbits of the resulting three-dimensional conformal group can be non-null or null. For non-null orbits, Carot *et al.* (1996) considered the case of a general V_2 when the metric is diagonal. Mars and Wolf (1997) found all solutions when the V_2 is spacelike and the motions are an invariant subgroup of a non-Abelian conformal group (and no extra symmetry is present). The (four) metrics are necessarily diagonal, cp. also §23.3.1, and one of them,

$$\begin{aligned} \mathrm{d}s^2 &= t^a x^{c-2}[\mathrm{d}x^2 + t^{1-a-b}\mathrm{d}y^2 + t^{1-a+b}\mathrm{d}z^2 - \mathrm{d}t^2], \\ c &= 2a^2/(b^2+a^2-2a-1) \neq 0, \quad a,b = \text{const}, \\ u_n &= -\text{sign}\,(c) t^{a/2} x^{c/2-1}\left[a^2x^2 - t^2\right]^{-1/2}(t,0,0,2a/c)\,, \end{aligned} \tag{35.77}$$

admits besides the conformal Killing vector ∂_x (which commutes with the two Killing vectors ∂_y and ∂_z) a homothetic vector $\boldsymbol{\zeta} = 2t\partial_t + 2x\partial_x + (1+a+b)y\partial_y + (1+a-b)z\partial_z$. The related metric, with x and t interchanged in the metric functions, contains a metric found by Bray (1971). Czapor and Coley (1995) considered metrics with a spacelike V_2 and an inheriting conformal vector, cp. also Vera (1998a); the only perfect fluid diagonal space-time with an equation of state $p = p(\mu)$ satisfying $\mu > 0$, $\mu + p > 0$ and admitting a proper conformal vector is the Allnutt solution $m = 0$ of (23.13). The case of null orbits (and a spacelike V_2) has been studied by Sintes *et al.* (1998).

For stationary axially-symmetric perfect fluids (i.e. with a G_2 acting on a *timelike surface*) with one additional conformal vector, the following results are known: If the resulting three-dimensional Lie algebra of $\boldsymbol{\xi} = \partial/\partial t$, $\boldsymbol{\eta} = \partial/\partial\varphi$, and the conformal motion $\boldsymbol{\zeta}$ is Abelian, then the solutions (of type D, equation of state $p = \mu +$ const, in general with differential rotation) are (21.74) and a metric of Herlt's class (the twist vector is a gradient, see §21.2) with a conformally flat 3-space (Kramer 1992, 1990, Mars and Senovilla 1994). All non-static rigidly rotating solutions with a proper conformal vector are contained here (Kramer and Carot 1991).

If for a non-rotating perfect fluid (with $\mu + p > 0$) there is an Abelian group of one Killing vector ∂_x and two conformal Killing vectors $\boldsymbol{\eta}$ and $\boldsymbol{\zeta}$ acting on a spacelike hypersurface, then in coordinates with $\boldsymbol{\eta} = \partial_y$ and $\boldsymbol{\zeta} = \partial_x$ the non-conformally flat solutions are of the form

$$\begin{aligned} \mathrm{d}s^2 &= N^{\frac{1-\alpha}{\alpha}}(y) Q^{\frac{-1-\alpha}{\alpha}}(z)[A(t)\mathrm{d}x^2 + t^{1+\alpha}\mathrm{d}y^2 + t^{1-\alpha}\mathrm{d}z^2 - A(t)\mathrm{d}t^2], \\ A(t) &= a_0 + a_1 t^{1-\alpha} + a_2 t^{1+\alpha}, \quad \alpha\ (\neq 0),\ a_i = \text{const}, \\ N'^2 &= \alpha^2(a_1N^2 - b_1)q, \qquad Q'^2 = \alpha^2(a_2Q^2 - b_2), \quad b_i = \text{const}, \\ u_a\mathrm{d}x^a &= N^{\frac{1-\alpha}{2\alpha}}(y) Q^{\frac{-1-\alpha}{2\alpha}}(z)\,\mathrm{d}(tN^{1/\alpha}Q^{-1/\alpha}) \end{aligned} \tag{35.78}$$

(Koutras and Mars 1997). They are of Petrov type D and generalize the solutions (32.98)–(32.99) found by Senovilla and Sopuerta (1994) using a generalized Kerr–Schild transformation. Subcases may admit more than one Killing vector, e.g. for $a_i = 0$ a non-Abelian G_2.

For axial symmetry and exactly one conformal symmetry, the two generators commute, so that in axially-symmetric *and stationary* space-times with one conformal Killing vector the axial Killing vector commutes with the other two (Mars and Senovilla 1993a).

If a pure radiation field, $R_{ab} = \Phi^2 k_a k_b$, $\boldsymbol{k}$ being a geodesic and shearfree null vector, admits a proper conformal Killing vector, then the twist ω of $\boldsymbol{k}$ vanishes (Lewandowski 1990) and the Petrov type is either N (with $k_{a;b} = 0$) or III (with $\phi_{,a} = \lambda k_a$, $k_{a;b} = \nu k_a k_b$). In the latter case, the solutions belong to the class (31.34); an example is

$$\begin{aligned} \mathrm{d}s^2 &= 2\mathrm{d}\zeta \mathrm{d}\overline{\zeta} - 2\mathrm{d}u(\mathrm{d}v + W\mathrm{d}\zeta + \overline{W}\mathrm{d}\overline{\zeta} + [u^{-4}(\zeta+\overline{\zeta})v - W\overline{W}]\mathrm{d}u), \\ W &= u^{-4}\overline{\zeta}^2 + u^{-2} + u^{-1}, \quad \boldsymbol{\xi} = u^3\partial_u - u^2 v\partial_v + u^2(\zeta\partial_\zeta + \overline{\zeta}\partial_{\overline{\zeta}}) \end{aligned} \tag{35.79}$$

(Hall and Carot 1994). The Petrov type N case (the *pp*-waves) has been studied by Maartens and Maharaj (1991); an example with a non-special conformal Killing vector is

$$\begin{aligned} \mathrm{d}s^2 &= \mathrm{d}x^2 + \mathrm{d}y^2 - 2\mathrm{d}u\,\mathrm{d}v - 2H\mathrm{d}u^2, \\ H &= u^{-3}h(u^{-3/2}x, u^{-3/2}y) - 3u^{-2}(x^2+y^2)/8, \\ \boldsymbol{\xi} &= \tfrac{2}{3}u^3\partial_u + u(x^2+y^2)\partial_v + u^2(x\partial_x + y\partial_y), \quad \phi = u^2. \end{aligned} \tag{35.80}$$

Methods of finding the conformal Killing vector of space-times conformal to a decomposable space-time (35.11) have been discussed by Tupper (1996).

36

Solutions with special subspaces

When studying exact solutions, subspaces of space-time occur naturally and frequently. Trivially in any chosen coordinate system there are subspaces in which one or two of the coordinates are constant; these subspaces may have some significance if the coordinates they are attached to have. More significantly, subspaces arise as group orbits of groups of motions.

In this chapter we shall discuss a third idea, namely to look for (three-dimensional) subspaces which admit intrinsic symmetries or have some other special properties which are *not* shared by the full space-time. This idea was formulated by Collins (1979) in an explicit way, but had been implicitly used earlier. It has been applied also to the space of Killing trajectories of a timelike Killing vector; in particular the case where this space (which in general is not a subspace of space-time) is conformally flat has been discussed, see e.g. Perjés *et al.* (1984) and §18.7.

36.1 The basic formulae

We parametrize the hypersurfaces we are interested in by the (spacelike or timelike) coordinate x^4, and denote their normal unit vector by

$$n_a = (0,0,0,\varepsilon N), \quad n^a = (-N^\alpha/N, 1/N), \quad n^a n_a = \varepsilon = \pm 1, \\ a,b,... = 1,...,4, \quad \alpha,\beta,... = 1,...,3. \tag{36.1}$$

The space-time metric then reads

$$\mathrm{d}s^2 = g_{\alpha\beta}(\mathrm{d}x^\alpha + N^\alpha \mathrm{d}x^4)(\mathrm{d}x^\beta + N^\beta \mathrm{d}x^4) + \varepsilon(N\mathrm{d}x^4)^2 \tag{36.2}$$

and has

$$g^{ab} = \begin{pmatrix} \overset{3}{g}{}^{\alpha\beta} + \varepsilon N^\alpha N^\beta/N^2 & -\varepsilon N^\alpha/N^2 \\ -\varepsilon N^\beta/N^2 & \varepsilon/N^2 \end{pmatrix}. \tag{36.3}$$

Here and in the following equations, Greek indices are raised and lowered by $\overset{3}{g}{}^{\alpha\beta}$ and $g_{\alpha\beta}$, i.e.

$$N_\alpha = g_{\alpha\beta}N^\beta, \quad g_{\alpha\beta}\overset{3}{g}{}^{\alpha\nu} = \delta^\nu_\beta. \tag{36.4}$$

By coordinate transformations $x^{\alpha'} = x^{a'}(x^\nu, x^4)$, $x^{4'} = x^{4'}(x^4)$ one could achieve $N^\alpha = 0$. Applying standard projection techniques (see e.g. Stephani (1996)), we introduce the tensor of exterior curvature K_{ab} by

$$K_{ab} = -n_{a;b} + \varepsilon\dot{n}_a n_b, \quad \dot{n}_a \equiv n_{a;b}n^b. \tag{36.5}$$

It is symmetric ($K_{ab} = K_{ba}$) and orthogonal to the normal vector ($K_{ab}n^a = 0$). Its 3-space components are given by

$$K_{\alpha\beta} = \varepsilon N\Gamma^4_{\alpha\beta} = (N_{\alpha||\beta} + N_{\beta||\alpha} - g_{\alpha\beta,4})/2N, \tag{36.6}$$

where the double stroke $||$ denotes the covariant derivative with respect to the 3-space metric $g_{\alpha\beta}$. The (four-dimensional) Riemann tensor components can then be expressed as follows:

$$R_{\alpha\beta\mu\nu} = \overset{3}{R}_{\alpha\beta\mu v} + \varepsilon(K_{\beta\mu}K_{\alpha v} - K_{\beta v}K_{\alpha\mu}), \tag{36.7a}$$

$$R^a{}_{\beta\mu\nu}n_a = K_{\beta\nu||\mu} - K_{\beta\mu||\nu}, \tag{36.7b}$$

$$R^a{}_{\beta m\nu}n_a n^m = K_{\beta\mu}K^\mu{}_\nu + \mathcal{L}_n K_{\beta\nu} - \varepsilon\dot{n}_\beta\dot{n}_\nu + \dot{n}_{(\beta;\nu)} \tag{36.7c}$$

(if space-time is flat, we regain from (36.7a)–(36.7b) the Gauss and Codazzi equations in the form (37.23)–(37.24) used in embedding, with Ω_{ab} replaced by K_{ab}). Einstein's field equations read

$$\begin{aligned} R_{\alpha\beta} &= \overset{3}{R}_{\alpha\beta} + 2\varepsilon K_{\alpha\nu}K^\nu{}_\beta - \varepsilon K^\nu_\nu K_{\alpha\beta} + \mathcal{L}_n K_{\alpha\beta} + \dot{n}_{(\alpha;\beta)} \\ &= \overset{3}{R}_{\alpha\beta} + \varepsilon g_{\nu(\beta}K^\nu{}_{\alpha),4}/N - \varepsilon K_{\alpha\beta||\nu}N^\nu/N - \varepsilon K^\nu_\nu K_{\alpha\beta} \\ &\quad + \varepsilon K^\nu{}_\beta N_{[\alpha,\nu]}/N + \varepsilon K^\nu{}_\alpha N_{[\beta,\nu]}/N - \varepsilon N_{,\alpha||\beta}/N \\ &= \kappa_0(T_{\alpha\beta} - Tg_{\alpha\beta}/2), \end{aligned} \tag{36.8}$$

$$R^4{}_\alpha = \varepsilon(K^\beta_{\beta||\alpha} - K^\beta{}_{\alpha||\beta})/N = \kappa_0 T^4{}_\alpha, \tag{36.9}$$

$$R - 2\overset{3}{g}{}^{\alpha\beta}R_{\alpha\beta} = \varepsilon(K^\alpha_\alpha K^\beta_\beta - K_{\alpha\beta}K^{\alpha\beta}) = 2\kappa_0(T - \overset{3}{g}{}^{\alpha\beta}T_{\alpha\beta}). \tag{36.10}$$

The strategy for constructing solutions is to assume special properties of the 3-space (with metric $g_{\alpha\beta}$), of the exterior curvature $K_{\alpha\beta}$, and – for non-vacuum solutions – of the direction of the four-velocity u^a of a perfect fluid source with respect to the slicing (these ideas have not yet been applied to electrovac solutions, probably because an electromagnetic field is more naturally related to a (2+2)- than to a (3+1)-foliation of space-time). So far only some simple cases have been investigated in a systematic way.

36.2 Solutions with flat three-dimensional slices

36.2.1 Vacuum solutions

By inspection of the solutions given in this book, one may easily find examples falling into this class, e.g. all metrics admitting a G_3I on V_3. A less obvious example is the exterior Schwarzschild solution, where this property of the subspaces t = const was already observed by Lemaître (1933) when introducing the coordinates

$$\mathrm{d}s^2 = 2m\mathrm{d}R^2/r + r^2(\mathrm{d}\vartheta^2 + \sin^2\vartheta\mathrm{d}\varphi^2) - \mathrm{d}t^2, \quad R = t + \tfrac{1}{3}\sqrt{2}r^{3/2}/\sqrt{m}. \tag{36.11}$$

For a more systematic study one should consider the different structures of the exterior curvature $K_{\alpha\beta}$.

For $K_{\alpha\beta} = 0$, in coordinates with $N_\alpha = 0$, one immediately sees from (36.7) that space-time is flat: non-flat vacuum solutions do not exist (Verma and Roy 1956). The same is true for $K_{\alpha\beta} = \lambda(x^\nu, x^4)g_{\alpha\beta}$ where (36.8)–(36.9) lead to $\lambda = 0$ (Stephani and Wolf 1985).

For $K_{\alpha\beta} = Ah_\alpha h_\beta$, $h_\alpha h^\alpha = 0$, the vacuum solutions with flat slices are special *pp*-waves or special type N solutions of Kundt's class (Wolf 1986a). The general case (arbitrary $K_{\alpha\beta}$) has not been solved. But since some Petrov type III solutions of Kundt's class (Wolf 1986a) and the Petrov type I solution (12.14) also admit flat slices, one sees that all Petrov types can occur, and that these solutions may have high symmetry or no symmetry at all.

The general vacuum solution admitting flat slices is not known.

36.2.2 Perfect fluid and dust solutions

Whereas in the case of vacuum solutions the above listed solutions admitting flat slices were known before, the search for non-vacuum solutions with such slices led to new solutions.

For $K_{\alpha\beta} = 0$ and in the gauge $N_\alpha = 0$, (36.6) implies that the 3-metric $g_{\alpha\beta}$ is independent of x^4. Since the 3-space is flat, its metric can be written as

$$\mathrm{d}s^2 = \varepsilon N^2(x^4, x^\alpha)(\mathrm{d}x^4)^2 + \eta_{\alpha\beta}\mathrm{d}x^\alpha\mathrm{d}x^\beta, \quad \eta_{\alpha\beta} = \ \mathrm{diag}\ (1, 1, -\varepsilon). \tag{36.12}$$

From $T_{\alpha\beta} = (\mu + p)u_\alpha u_\beta + pg_{\alpha\beta}$ and the field equation (36.9) one sees that (since $\mu + p \neq 0$) the four-velocity has to be orthogonal or parallel to the slices, and (36.10) then shows that the first case is ruled out: one necessarily has $\varepsilon = +1$, $p = 0$. The resulting dust solutions are given by

$$N = -\tfrac{1}{2}M\left[(u_\alpha x^\alpha)^2 + \tfrac{1}{2}\eta_{\alpha\beta}x^\alpha x^\beta\right] + g_a x^\alpha + h, \quad \mu = M/N, \tag{36.13}$$

where u_α, M, g_α and h are arbitrary functions of x^4 restricted only $u_\alpha u^\alpha = -1$ and $\mu > 0$, and

$$\begin{aligned} N &= M \ln \sqrt{T} + g_\alpha x^\alpha + h, \quad u_\alpha = T_{,\alpha}, \\ T^2 &= -\eta_{\alpha\beta}(x^\alpha - f^\alpha)(x^\beta - f^\beta), \quad \mu = M/NT^2, \end{aligned} \tag{36.14}$$

where M, f^α, g_α and h are arbitrary functions of x^4 (Stephani 1982, 1987). These solutions are of Petrov type D. In general the four-velocity has non-zero rotation, and the metric does not admit a Killing vector. The non-rotating solutions ($u_{\alpha,4} = 0$) contained here belong to the class of the Szekeres dust solutions, see §33.3.2.

A Petrov type N pure radiation field $T_{mn} = \Phi^2 k_m k_n$ is contained in the metric (36.12) as

$$\begin{aligned} &N = A(k_\alpha x^\alpha, x^4) + a_\alpha(x^4)x^\alpha + b(x^4), \quad \varepsilon = 1, \\ &k^\alpha k_\alpha = 0, \quad k_\alpha = k_\alpha(x^4), \quad \Phi^2 = -\partial^2 A/\partial(k_\alpha x^\alpha)^2 \end{aligned} \tag{36.15}$$

(Stephani 1982). Its multiple null eigenvector $k_m = (k_\alpha, 0)$ is shearing.

For $K_{\alpha\beta} = \lambda(x^n, x^4)g_{\alpha\beta}$, $\lambda \neq 0$, and in coordinates with $N_\alpha = 0$, equation (36.6) shows that the 3-metric $g_{\alpha\beta}$ depends on x^4 only through an overall factor,

$$\mathrm{d}s^2 = V^{-2}(x^\nu, x^4)\left[\eta_{\alpha\beta}\mathrm{d}x^\alpha \mathrm{d}x^\beta + \varepsilon M^2(\mathrm{d}x^4)^2\right], \quad \eta_{\alpha\beta} = \mathrm{diag}(1, 1, -\varepsilon), \tag{36.16}$$

with $\lambda M = V_{,4}$. To yield a flat 3-space, the function V has to satisfy

$$VV_{,\alpha\beta} = \eta_{\alpha\beta}\eta^{\nu\mu}\left(2V_{,\nu}V_{,\mu} - VV_{,\nu\mu}\right), \tag{36.17}$$

i.e. it has to be in one of the forms

$$\begin{aligned} V &= A\eta_{\nu\mu}(x^\nu - f^\nu)(x^\mu - f^\mu) \\ \text{or } V &= B_\nu x^\nu + C, \quad \eta^{\nu\mu}B_\nu B_\mu = 0, \end{aligned} \tag{36.18}$$

where the functions A, B_ν, C and f^ν are (up to now) arbitrary functions of x^4.

If λ is a function only of x^4, then the solutions are either (for $\varepsilon = +1$, $\lambda =$ const, $\mu + p = 0$) the de Sitter universe, or (for $\varepsilon = -1$)

$$\begin{aligned} \mathrm{d}s^2 &= V^{-2}\left[\mathrm{d}x^2 + \mathrm{d}y^2 + \mathrm{d}z^2 - V_{,4}^2(\mathrm{d}x^4)^2/\lambda^2\right], \\ V &= A\left[(x - f_1)^2 + (y - f_2)^2 + (z - f_3)^2\right], \end{aligned} \tag{36.19}$$

where A, f_α are arbitrary functions of x^4. These solutions generalize the 3-flat Friedmann models and are contained in (37.45) as a limiting case; in general they admit no Killing vector.

If λ depends on the x^α, then the solutions are given by

$$\mathrm{d}s^2 = (x^4)^{-2}\left[\eta_{\alpha\beta}\mathrm{d}x^\alpha\,\mathrm{d}x^\beta + \varepsilon n^2(w,x^4)(\mathrm{d}x^4)^2\right], \quad \eta_{\alpha\beta} = \mathrm{diag}(1,1,-\varepsilon),$$

$$\varepsilon = \pm 1, \quad w = \eta_{\alpha\beta}x^\alpha x^\beta \quad \text{or} \quad w = m_\alpha x^\alpha,\ m_\alpha = \text{ const}, \quad {}' \equiv \partial/\partial w,$$

$$\kappa_0\mu = (x^4)^2 w_{,\alpha}w^{,\alpha}n''/n - 3\varepsilon/n^2, \quad \kappa_0 p = 3\varepsilon/n^2 + 4\varepsilon n'^2/n^3 n'', \quad (36.20a)$$

$$u_\alpha = w_{,\alpha}[-n''/\kappa_0(\mu+p)n]^{1/2}, \quad (u^4)^2 = -\frac{\varepsilon(x^4)^2}{n^2} + \frac{\varepsilon w_{,\alpha}w^{,\alpha}n''(x^4)^2}{\kappa_0(\mu+p)n^3},$$

(Stephani and Wolf 1985), where n is a solution of

$$\varepsilon(x^4)^2\left[w_{,\alpha}w^{,\alpha}n''/n + (w_{,\alpha}w^{,\alpha})'n'/n\right] - 4n'^2/n^3 n'' + 2n_{,4}x^4/n^3 = 0. \quad (36.20b)$$

The metrics (36.20) admit a G_3 which leaves w invariant. For $\varepsilon = -1$ and $w = \eta_{\alpha\beta}x^\alpha x^\beta$ they are spherically-symmetric (and given in non-comoving coordinates); in that case the new variable $s = \varepsilon w/(x^4)^2$ leads to

$$[s\ddot{n}/n + \dot{n}/n]\,[1 - \dot{n}/\ddot{n}n^2] + n_{,4}x^4/2n^3 = 0, \quad \dot{n} \equiv \partial n/\partial s. \quad (36.21)$$

Special solutions of this equation can easily be constructed. For flat slices in spherically-symmetric space-times see also below.

For $K_{\alpha\beta||\nu} = 0$, $K_{\alpha\beta} \neq \lambda g_{\alpha\beta}$, the field equation (36.9) implies that (for $\mu + p \neq 0$) either $u_\alpha = 0$ and $\varepsilon = -1$, or $u^4 = 0$ and $\varepsilon = +1$. Taking coordinates with $g_{\alpha\beta} = g_{\alpha\beta}(x^4)$, one has $K_{\alpha\beta} = K_{\alpha\beta}(x^4)$. Depending on the algebraic structure of the 3-tensor $K_{\alpha\beta}$ and the timelike or spacelike character of the slices, 15 subcases occur which are listed and discussed in Wolf (1986b). Among them are the following classes.

(1) The Bianchi type I models with a G_3 on S_3 or T_3. Here the field equation can be completely solved up to algebraic conditions ensuring the reality of the metric functions and the four-velocity components. The metrics with a G_3 on T_3 are

$$\begin{aligned}
\mathrm{d}s^2 &= \mathrm{d}x^2 - [\mathrm{d}t + (at - hy + fz)\mathrm{d}x]^2 + [\mathrm{d}y + (by - ht + gz)\mathrm{d}x]^2 \\
&\quad + [\mathrm{d}z + (ft - gy + cz)\mathrm{d}x]^2, \\
u_t^2 &= A^2/\kappa_0(\mu+p), \quad u_y^2 = B^2/\kappa_0(\mu+p), \quad u_z^2 = C^2/\kappa_0(\mu+p), \\
h &= AB/(b-a), \quad g = BC/(c-b), \quad f = CA/(a-c), \qquad (36.22) \\
A &= (b+c)_{,x} - b^2 - c^2 + a(b+c), \quad \kappa_0(\mu+p) = A^2 - B^2 - C^2, \\
B &= -(a+c)_{,x} + a^2 + c^2 - b(a+c), \quad \kappa_0 p = ab + bc + ca, \\
C &= -(a+b)_{,x} + a^2 + b^2 - c(a+b),
\end{aligned}$$

where $a(x)$, $b(x)$, and $c(x)$ are arbitrary functions, and

$$\begin{aligned} ds^2 &= dw^2 + 2\,(dx + N_2 dw)\,[b dx/a^2 + dy + (N_1 - bN_2/a^2)dw] \\ &\quad +(dz + N_3 dw)^2, \\ N_1 &= (1 + (b/a^2)_{,w})x + gy - sz, \quad N_2 = wz - gx - a^2 y, \\ N_3 &= sx - wy + cz, \quad \mu = 5p + 2[\,K^{\alpha}_{\alpha,w} - (K^{\alpha}_{\alpha})^2\,]/\kappa_0, \end{aligned} \tag{36.23}$$

where the functions a, g, s, w and c can be given in terms of the arbitrary functions $K^{\alpha}_{\alpha}(w)$, $b(w)$ and $p(w)$ which can be chosen such that reasonable equations of state etc. arise, see Wolf (1986b) for details. Both metrics have Petrov type I and a non-diverging, shearing and rotating four-velocity.

(2) The Gödel solution.

(3) Three classes of Petrov type D solutions with timelike slices, an equation of state $\mu = p$, a purely electric Weyl tensor and $u^{[a}k^b l^{n]} = 0$, see §33.3. They have at least either a G_2 on T_2 (and a four-velocity with $\Theta = \sigma = \dot{u} = 0$, $\omega \neq 0$) or a G_1 on S_1 (with Θ, ω, σ, $\dot{u}$ all nonzero) and are given by

$$\begin{aligned} ds^2 &= (F_{,t} + G_{,t})^2 dx^2 - [dt + a(F+G)dx]^2 \\ &\quad + \left[dy + \sqrt{ab}(F-G)dx\right]^2 + dz^2, \\ F &= \left[t + \sqrt{by/a}\right] \frac{(a-b)_{,x}(a+b)}{2(a-b)(a^2+b^2)} + \sum_{\pm} n_{\pm} \left(\frac{a-b}{a(a^2+b^2)}\right)^{1/2} \times \\ &\quad \times \exp\left\{\pm\left[a(a^2+b^2)/(a-b)\right]^{1/2}\left[t + (by/a)^{1/2}\right]\right\}, \\ G &= \left[t + (by/a)^{1/2}\right] \{a_{,x}(a-3b) - b_{,a}(b-3a)\}/2(a^2+b^2)(a-b), \\ u_t^2 &= (ab)_{,x}/\left[2n(a+b)\right](F_{,t} + G_{,t}) + b/(b-a), \\ u_x^2 &= -u_t a(F+G) + u_y\sqrt{ab}(F-G), \quad \kappa_0\mu = \kappa_0 p = ab, \\ u_y^2 &= (ab)_{,x}/\left[2n(a+b)\right](F_{,t} + G_{,t}) + a/(b-a), \quad u_z = 0, \end{aligned} \tag{36.24}$$

where $n = $ const, $a(x)$ and $b(x)$ have to satisfy $ab(a-b) = n(a+b)$, $ab > 0$, and $n_+(x)$ and $n_-(x)$ are arbitrary functions which are real if $b > a$ and complex conjugate if $b < a$.

For spherically-symmetric space-times, by means of a coordinate transformation $T = T(t, r)$, the general spherically-symmetric line element

$$ds^2 = r^2 d\Omega^2 + e^{2\lambda(r,T)} dr^2 - e^{2\nu(r,T)} dT^2, \quad d\Omega^2 \equiv d\vartheta^2 + \sin^2\vartheta\, d\varphi^2, \tag{36.25}$$

is transformed into

$$\mathrm{d}s^2 = \left[\mathrm{e}^{2\lambda} - \mathrm{e}^{2\nu}T_{,r}^2\right]\mathrm{d}r^2 + r^2\mathrm{d}\Omega^2 - \mathrm{e}^{2\nu}\left[2T_{,r}T_{,t}\mathrm{d}r + T_{,t}^2\mathrm{d}t\right]\mathrm{d}t. \quad (36.26)$$

Obviously the slices $t = \text{const}$ are flat if

$$(\partial T/\partial r)^2 = \mathrm{e}^{-2\nu(r,T)}\left[\mathrm{e}^{2\lambda(r,T)} - 1\right] \quad (36.27)$$

holds, i.e. only if $\mathrm{e}^{2\lambda(r,T)} \geq 1$, and then there are two distinct families of slices (Stephani and Wolf 1985). For static solutions this condition is satisfied if the mass function $m(r)$ defined by (16.5) is positive. It is also satisfied for the general open ($\varepsilon = -1$) Robertson–Walker metric (12.8).

If the flat slices are comoving, then for spherically-symmetric space-times the four-velocity is geodesic and the solutions are the subcase $\varepsilon = 0$ of (15.66)–(15.71) (Bona *et al.* 1987b). The subcases of the Szekeres–Szafron classes of solutions (§33.3.2) which have flat comoving slices are (33.43) and the subcase $k = 0$, $U = 0$ of (33.44*a*) (Berger *et al.* 1977, Bona *et al.* 1987a). Also the plane-symmetric fluids (15.80) and the plane-symmetric dust solutions – (15.43) with $k = 0$ – have flat slices.

36.3 Perfect fluid solutions with conformally flat slices

For vacuum solutions, no systematic search has been carried out, and also among the perfect fluid solutions only some restricted cases have been investigated. We shall first consider those solutions (or theorems on them) which have been found by *asking* for metrics with conformally flat slices, and then give a list of (further) solutions where this property has been detected in hindsight.

Verma and Roy (1956) considered metrics of the form

$$\mathrm{d}s^2 = \varphi(x, y, z, t)[\mathrm{d}x^2 + \mathrm{d}y^2 + \mathrm{d}z^2] - \mathrm{d}t^2. \quad (36.28)$$

They found as solutions the subcase x_0, y_0, $z_0 = \text{const}$ of (37.45).

All solutions with $K_{\alpha\beta} = 0$,

$$\mathrm{d}s^2 = N^2(x^4, x^\alpha)(\mathrm{d}x^4)^2 + g_{\alpha\beta}(x^\nu)\mathrm{d}x^\alpha\mathrm{d}x^\beta, \quad (36.29)$$

where $\overset{3}{R}_{\alpha\beta}$ has at most two different eigenvalues and has the four-velocity u^α as an eigenvector, have been found by Stephani (1987) and Barnes (1999). They are necessarily of Petrov type D or 0. There are two cases.

If the 3-spaces are of constant curvature, then one obtains either the dust solutions (36.13)–(36.14) (with a flat 3-space), or the rotating perfect fluid solutions

$$\begin{aligned} ds^2 &= N^2(dx^4)^2 + \eta_{\alpha\beta}dx^\alpha dx^\beta(1-pr^2/12)^{-2}, \\ N &= -Mf(\tau) + \left[h(1+pr^2/12) + h_\alpha x^\alpha + 3M/p\right](1-pr^2/12)^{-1}, \\ f(\tau) &= \frac{3}{4}\frac{\tau}{p}\ln\frac{1+\tau}{1-\tau}, \quad \tau = \left[g\left(1+\tfrac{1}{12}pr^2\right) + g_\alpha x^\alpha\right]\left(1-\tfrac{1}{12}pr^2\right)^{-1}, \\ r^2 &= \eta_{\alpha\beta}x^\alpha x^\beta, \quad \eta_{\alpha\beta} = \mathrm{diag}\,(1,1,-1), \quad \eta^{\alpha\beta}g_\alpha g_\beta = p(g^2-1)/3, \\ u_\alpha &= T_{,\alpha}, \quad \cos\left(\tfrac{1}{3}T\sqrt{3p}\right) = \tau, \quad \mu = -p + M/N\sin^2\left(\tfrac{1}{3}T\sqrt{3p}\right), \end{aligned} \tag{36.30}$$

where the pressure p is an arbitrary positive constant and M, h, h_α, g_α are arbitrary functions of x^4. For negative p (corresponding to a cosmological constant Λ), similar solutions occur.

If the 3-spaces are not of constant curvature, then the fluid is non-rotating and the metric must be of the form

$$ds^2 = N^2(x^4,\zeta,\overline{\zeta},t)(dx^4)^2 + 2e^{2b(\zeta,\overline{\zeta},t)}d\zeta\, d\overline{\zeta} - e^{2a(\zeta,\overline{\zeta},t)}dt^2. \tag{36.31}$$

Contained here are subcases of the generalized interior Schwarzschild solution (37.39), subcases of the generalized Friedmann solutions (37.45), and the Szekeres–Tomimura class (33.44*a*)–(33.44*d*) (note that here the ζ–$\overline{\zeta}$–t-space is conformally flat; for the ζ–$\overline{\zeta}$–x^4-space see below).

Dust metrics with comoving constant curvature slices (Bona and Palou 1992) are either the Szekeres dust solutions (contained in (33.38)) or they are of Bianchi type I (with zero curvature) or V (contained in (13.32)).

Properties of metrics with a S_3 of constant curvature have been discussed by Bona and Coll (1985a).

Metrics with conformally flat comoving slices have been studied by Szafron and Collins (1979).

The following solutions are known to admit conformally flat subspaces:
(i) Comoving: The Szekeres metrics (33.38) (slices t = const!) (Berger *et al.* 1977, Szafron and Collins 1979), the Wainwright–Goode solutions (23.13) (Wainwright and Goode 1980), the dust solutions (15.38) with a G_3 on V_2.
(ii) Non-comoving: Spherically-symmetric metrics which admit isotropic coordinates (16.22).

36.4 Solutions with other intrinsic symmetries

Martinez and Sanz (1985) studied metrics of the form

$$\mathrm{d}s^2 = B(t,r)\mathrm{d}r^2 + C(t,r)\left[\mathrm{d}\vartheta^2 + M^2(\vartheta)\mathrm{d}\varphi^2\right] - A(t,r,\vartheta,\varphi)\mathrm{d}t^2, \quad (36.32)$$

which on $t=$ const have (at least) one Killing vector ∂_φ.

Vacuum solutions of this form have either a G_3 on S_2 (the ϑ–φ-space has constant curvature) or are given by the special case

$$\mathrm{d}s^2 = t^{-2}\mathrm{d}r^2 + a^2t^2(\mathrm{e}^{2x^2}\mathrm{d}x^2 + x^2\mathrm{d}y^2) - \mathrm{e}^{2x^2}\mathrm{d}t^2, \quad a = \mathrm{const} \quad (36.33)$$

of (17.13) (admitting a G_2).

Perfect fluid solutions with four-velocity orthogonal to $t=$ const contain the conformally flat solutions (37.39) and (for $A_\varphi = 0$) the two rigidly rotating solutions

$$\begin{aligned} \mathrm{d}s^2 &= \mathrm{d}r^2 + \mathrm{d}\vartheta^2 + F(\vartheta)^2_{,\vartheta}\mathrm{d}\varphi^2 - F^2\mathrm{d}t^2, \\ F_{,\vartheta\vartheta} &= b/F, \quad \kappa_0\mu = \kappa_0 p = bF^{-2}, \quad b = \text{ const} \end{aligned} \quad (36.34)$$

(admitting a G_3) and

$$\begin{aligned} &\mathrm{d}s^2 = \mathrm{d}r^2/(\lambda + br^2) + r^2(\mathrm{d}\vartheta^2 + F(\vartheta)^2_{,\vartheta}\mathrm{d}\varphi^2) - r^2F^2\mathrm{d}t^2, \\ &FF_{,\vartheta\vartheta} = c - \lambda F^2, \quad \kappa_0\mu = c/r^2F^2 - 3b = \kappa_0 p - 6b, \quad \lambda, b, c = \text{ const} \end{aligned} \quad (36.35)$$

(admitting a G_2), and the metric

$$\begin{aligned} &\mathrm{d}s^2 = t^{-2}\mathrm{d}r^2 + t^2(\mathrm{d}\vartheta^2 + F(\vartheta)^2_{,\vartheta}\mathrm{d}\varphi^2) - (b\ln t + c)^{-1}F^2\mathrm{d}t^2, \\ &F_{,\vartheta\vartheta} = aF^{-1} + bF^{-1}\ln F, \quad a, b, c = \text{ const}, \\ &\kappa_0 p = (tF)^{-2}\left[a + b\ln F - c - b\ln t\right] = \kappa_0\mu + b(tF)^{-2} \end{aligned} \quad (36.36)$$

(admitting a G_2).

In generalizing (36.32), Argüeso and Sanz (1985) considered the metrics

$$\mathrm{d}s^2 = B(t,r)\mathrm{d}r^2 + C(t,r)\mathrm{d}\vartheta^2 + H(t,r)M^2(\vartheta)\mathrm{d}\varphi^2 - A(t,r,\vartheta,\varphi)\mathrm{d}t^2, \quad (36.37)$$

($H \neq C$), again with one symmetry (∂_φ) on $t =$ const. If $A_{,\varphi} \neq 0$, the solutions are either static degenerate vacuum B-metrics (with Λ) or the conformally flat perfect fluid (37.39). If $A_{,\varphi} = 0$, only the subcase $C = C(t)$, $H = H(t)$ has been considered. It leads to solutions with a G_2I and – for perfect fluids – to an equation of state $\mu = p$, see §23.1.

Space-times with a G_3 transitive on a S_3, where the S_3 admits an intrinsic G_4, have been studied by Szafron (1981), see also McManus (1995).

37

Local isometric embedding of four-dimensional Riemannian manifolds

37.1 The why of embedding

It is a well-known theorem of differential geometry (Eisenhart 1949) that one can regard every (analytic) four-dimensional space-time V_4 (at least locally) as a subspace of a flat pseudo-Euclidean space E_N of $N \leq 10$ dimensions. If we choose Cartesian coordinates y^A in describing E_N,

$$\overset{(N)}{\mathrm{d}s}{}^2 = \sum_{A=1}^{N} e_A(\mathrm{d}y^A)^2 = \eta_{AB}\,\mathrm{d}y^A\mathrm{d}y^B, \quad e_A = \pm 1, \tag{37.1}$$

then the subspace V_4 (coordinates x^a) will be given by the parametric representation

$$y^A = y^A(x^a) \tag{37.2}$$

and the metric of this subspace as induced by (37.1) is

$$\begin{aligned} &\mathrm{d}s^2 = g_{ab}\,\mathrm{d}x^a\mathrm{d}x^b = \eta_{AB}\,y^A{}_{,a}\,y^B{}_{,b}\mathrm{d}x^a\mathrm{d}x^b, \\ &A, B, \ldots = 1, \ldots, N, \quad a, b, \ldots = 1, \ldots, 4. \end{aligned} \tag{37.3}$$

If these three equations (37.1)–(37.3) hold, they describe a local isometric embedding of V_4 into E_N. The *minimum* number of extra dimensions is called the *embedding class p* (or just class) of the V_4 in question, $0 \leq p \leq 6$.

Attempts have been made to give a physical meaning to the flat embedding space, or to use it as an auxiliary space for visualizing or deriving physical properties of the embedded space-time. In the context of this book, our point of view is a more pragmatic one. The invariance of the embedding class gives rise to a classification scheme of all solutions of Einstein's field equations according to their respective embedding class. From a mathematical point of view, this classification scheme is on an equal footing with the classifications with respect to groups of motions or to the Petrov types, and it will give a refinement of both these schemes. Moreover, there is some hope of obtaining exact solutions by the method of embedding, at least for some simple cases of low embedding class, solutions which are not readily available by other methods. We are *not* interested in the embedding itself; the functions $y^A(x^a)$ will not be determined or given here. A large number of explicit embeddings can be found in Rosen (1965) and Collinson (1968b). Other aspects of the embedding problem are discussed in Goenner (1980).

Nearly all work summarized in this chapter deals with *local* embedding only, i.e. the embedding of an open and simply connected neighbourhood of a point of the given V_4. In contrast, the *global* embedding of a V_4 can be considered. It may give a deeper insight into the geometrical properties of space-time. In fact, the maximal analytic extension of the Schwarzschild solution was found by the method of embedding (Fronsdal 1959). The number of extra dimensions needed for the embedding can be considerably higher than that for local embedding; only upper limits are known: a compact (non-compact) space-time V_4 is at most of embedding class $p = 46$ ($p = 87$). For theorems and results on global embeddings, see Friedman (1965), Penrose (1965), Clarke (1970), Greene (1970). No systematic analysis of global embedding of exact solutions has yet been done.

37.2 The basic formulae governing embedding

To get a deeper insight into the geometrical properties of the embedding described by (37.1)–(37.3), we introduce an N-leg at every point of V_4 and consider the change of this N-leg along V_4, i.e. we consider the covariant derivative with respect to the coordinates x^n of V_4.

The N-leg in question consists of four vectors $y^A_{,a}$ (vectors in E_N, $a = 1, \ldots, 4$) tangent to V_4 and p unit vectors $n^{\alpha A}$ ($\alpha = 1, \ldots, p$) orthogonal to V_4 and to each other,

$$\eta_{AB} n^{\alpha A} n^{\beta B} = e^\alpha \delta^{\alpha\beta}, \quad \eta_{AB} n^{\alpha A} y^B{}_{,a} = 0, \quad e^\alpha = \pm 1. \tag{37.4}$$

(In these and the following formulae, summation over Greek indices takes place only if explicitly indicated.)

The covariant derivatives (covariant with respect to coordinates x^n and metric g_{ab}) of the basic vectors $n^{\alpha A}$ and $y^B_{,a}$ are vectors and tensors (respectively) in V_4, but again vectors in the embedding space E_N and are, therefore, linear combinations of the basic vectors. Starting from the metric (37.3), we get

$$g_{ab;c} = \eta_{AB} \left(y^A{}_{,a;c} y^B{}_{,b} + y^A{}_{,a} y^B{}_{,b;c} \right) = 0, \tag{37.5}$$

and subtracting from this the expressions obtained by substituting (cab) and (acb) for (abc), we conclude that

$$\eta_{AB}\, y^A{}_{,c}\, y^B{}_{,a;b} = 0 \tag{37.6}$$

holds. Equation (37.6) tells us that $y^B{}_{,a;b}$ is a vector orthogonal to V_4, which can therefore be expressed as a linear combination of the normal vectors $n^{\alpha B}$,

$$y^B{}_{,a;b} = \sum_\alpha e^\alpha \Omega^\alpha{}_{ab} n^{\alpha B}, \quad \Omega^\alpha{}_{ab} = \Omega^\alpha{}_{ba}. \tag{37.7}$$

The p symmetric tensors $\Omega^\alpha{}_{ab}$ (tensors in V_4) defined by this equation are generalizations to higher dimensions of the tensor of the second fundamental form used in the theory of hypersurfaces.

In a similar way we conclude from (37.4) and (37.7) that

$$n^{\alpha A}{}_{,a} = -\Omega^\alpha{}_{ab} g^{bc} y^A{}_{,c} + \sum_\beta e^\beta t^{\beta\alpha}{}_a n^{\beta A}, \quad t^{\beta\alpha}{}_b + t^{\alpha\beta}{}_b = 0 \tag{37.8}$$

holds. The $p(p-1)/2$ vectors $t^{\beta\alpha}{}_b$ (vectors in V_4) defined by (37.8) are sometimes called torsion vectors.

The $\Omega^\alpha{}_{ab}$ and the vectors $t^{\beta\alpha}{}_b$ cannot be prescribed arbitrarily. They have to satisfy the conditions of integrability of the system (37.7)–(37.8), which turn out to be

$$R_{abcd} = \sum_\alpha e^\alpha (\Omega^\alpha{}_{ab}\Omega^\alpha{}_{bd} - \Omega^\alpha{}_{ad}\Omega^\alpha{}_{bc}) \quad \text{(Gauss)}, \tag{37.9}$$

$$\Omega^\alpha{}_{ab;c} - \Omega^\alpha{}_{ac;b} = \sum_\beta e^\beta (t^{\beta\alpha}{}_c \Omega^\beta{}_{ab} - t^{\beta\alpha}{}_b \Omega^\beta{}_{ac}) \quad \text{(Codazzi)}, \tag{37.10}$$

$$t^{\beta\alpha}{}_{a;b} - t^{\beta\alpha}{}_{b;a} = \sum_\nu e^\nu (t^{\nu\beta}{}_b t^{\nu\alpha}{}_a - t^{\nu\beta}{}_a t^{\nu\beta}{}_b) + g^{cd}(\Omega^\beta{}_{cb}\Omega^\alpha{}_{da} - \Omega^\beta{}_{ca}\Omega^\alpha{}_{db}) \quad \text{(Ricci)}. \tag{37.11}$$

Equations (37.9)–(37.11) are the most important equations of embedding theory. They are written entirely in terms of the V_4 in question, using only vector and tensor fields on V_4.

If a V_4 is of embedding class p, then it must admit p tensor fields $\Omega^\alpha{}_{ab}$ and $p(p-1)/2$ vector fields $t^{\alpha\beta}{}_a$ which satisfy (37.9)–(37.11), the constants $e^\alpha = \pm 1$ being suitably chosen; and if p is the minimum number enabling (37.9)–(37.11) to be satisfied, then the V_4 is of embedding class p. For $p > 1$, the tensors $\Omega^\alpha{}_{ab}$ and the vectors $t^{\alpha\beta}{}_a$ are not uniquely defined by the embedding because of the possibility of performing a rotation (pseudo-rotation) at each point of V_4 of the basic unit vectors $n^{\alpha A}$ orthogonal to V_4. These degrees of freedom may be used to simplify (37.9)–(37.11) in special cases, see §37.5.

If one inserts the curvature tensor (37.9) into the Bianchi identities $R_{ab[cd;e]} = 0$, one gets identities involving the derivatives of the tensors $\Omega^{\alpha}{}_{ab}$. It turns out (Gupta and Goel 1975, Goenner 1977) that as a consequence of these identities parts of the Codazzi equations (37.10) are automatically satisfied. In this sense, the Gauss equations (37.9) and the Codazzi equations (37.10) are not completely independent of each other. In some exceptional cases, only the Gauss equations need to be satisfied to ensure the embedding property of a given V_4.

37.3 Some theorems on local isometric embedding

In the general case, no practical way of solving the Gauss–Codazzi–Ricci equations (37.9)–(37.11) is known, either in the sense of determining all solutions of Einstein's equations of a given embedding class or in the sense of determining the embedding class of a given metric. Progress made so far concentrates on three points, namely: (a) solutions of embedding classes one and two, (b) explicit embedding of certain metrics or classes of metrics and (c) the connections between embedding class and other properties of the metric, e.g. special vector and tensor fields, groups of motions. We shall postpone the discussion of (a) and (b) to later sections and deal here only with (c).

37.3.1 General theorems

If we denote by $V_n(s,t)$ a Riemannian space with s spacelike and t timelike directions, and by $E_N(S,T)$ the pseudo-Euclidean embedding space, then we have:

Theorem 37.1 *Any analytic Riemann manifold $V_n(s,t)$ can be (locally) isometrically embedded in $E_N(S,T)$ with $s+t=n$, $S+T=N$, $n \le N \le n(n+1)/2$, $s \le S$, $t \le T$* (Eisenhart 1949, Friedman 1965).

Correspondingly, the embedding class p of a V_n is at most $n(n-1)/2$. For space-time $n=4$, and $p \le 6$. If the embedding space is assumed to be Ricci-flat (instead of flat), then for any n only one additional dimension is needed (Campbell 1926).

If we start from the embedding (37.3), with $N = n+p$, of a V_n with metric g_{ab} into a flat $(p+n)$-dimensional space, then relations

$$\begin{aligned} z^A &= \mathrm{e}^U y^A, \quad z^{N+1} = \mathrm{e}^U(\eta_{AB}\, y^A y^B - 1/4), \\ z^{N+2} &= \mathrm{e}^U(\eta_{AB}\, y^A y^B + 1/4), \end{aligned} \tag{37.12}$$

give an explicit embedding of a $\widehat{V}_n$ with metric $\hat{g}_{ab} = \mathrm{e}^{2U} g_{ab}$ into a flat $(n+p+2)$-dimensional space.

Theorem 37.2 *If two spaces are conformally related, then their respective embedding classes differ by at most two. In particular, the embedding class of conformally flat spaces is at most two.*

An example of a conformally flat space of embedding class one is the Robertson–Walker metric

$$\mathrm{d}s^2 = f^2(t)[\mathrm{d}\varphi^2 + \sin^2\varphi(\mathrm{d}\psi^2 + \sin^2\psi\,\mathrm{d}\alpha^2)] - \mathrm{d}t^2, \qquad (37.13)$$

the embedding being given by

$$\begin{aligned} y^1 &= f(t)\cos\varphi, \quad y^2 = f(t)\sin\varphi\cos\psi, \\ y^3 + \mathrm{i}y^4 &= f(t)\sin\varphi\sin\psi\,\mathrm{e}^{\mathrm{i}\alpha}, \quad y^5 = \int\sqrt{f'^2(t)+1}\,\mathrm{d}t, \\ \mathrm{d}s^2 &= (\mathrm{d}y^1)^2 + (\mathrm{d}y^2)^2 + (\mathrm{d}y^3)^2 + (\mathrm{d}y^4)^2 - (\mathrm{d}y^5)^2. \end{aligned} \qquad (37.14)$$

37.3.2 Vector and tensor fields and embedding class

The existence of special vector and tensor fields in a V_4 may reduce its embedding class below the maximum value $p = 6$.

As an example we consider a V_4 which admits a non-null vector field $\boldsymbol{v}$ satisfying

$$\mathcal{L}_{\boldsymbol{v}} h_{ab} = \mathcal{L}_{\boldsymbol{v}}(g_{ab} - v_a v_b/v^2) = 0, \quad h_{ab}v^a = 0. \qquad (37.15)$$

In the frame of reference defined by $v^i = (0,0,0,v^4)$, condition (37.15) implies $h_{4i} = 0$, $h_{ji,4} = 0$, i.e.

$$\mathrm{d}s^2 = h_{\alpha\beta}(x^\nu)\mathrm{d}x^\alpha\mathrm{d}x^\beta + \varepsilon v^{-2}(v_i\mathrm{d}x^i)^2, \quad \alpha,\beta = 1,2,3, \quad \varepsilon = \pm 1. \quad (37.16)$$

(In the case of a timelike unit vector field, $v^a/v = u^a$, the field would be called shear- and expansionfree; it describes a rigid congruence.)

If, in addition, v_a/v is a gradient, $v_i\mathrm{d}x^i/v = \mathrm{d}x^4$, then the metric admits a (covariantly) constant vector field, cp. (35.8). As the $h_{\alpha\beta}$ part of the metric is a V_3 and can be embedded in at most six dimensions, we get

Theorem 37.3 *If a V_4 admits a non-null constant vector field, then its embedding class is $p \leq 3$.*

By inspection of the metric (35.9) admitting a constant null vector field, and using the representation $\mathrm{d}x^3\mathrm{d}x^4 = \mathrm{d}u^2 - \mathrm{d}v^2$, one gets in a similar way

Theorem 37.4 *If a V_4 admits a constant null vector field, then its embedding class is $p \leq 4$.*

As every V_2 can be embedded in a E_3, in the case of the metrics (35.6) and (35.7) which possess two constant vector fields we get

Theorem 37.5 *If a V_4 admits two constant non-null vector fields, then its embedding class is $p \leq 1$.*

Theorem 37.6 *If a V_4 admits a constant null vector field and a constant non-null vector field orthogonal to each other, then its embedding class is $p \leq 2$.*

If in (37.16) v_a/v is proportional to a gradient, the metric can be written as

$$\mathrm{d}s^2 = h_{\alpha\beta}(x^\gamma)\mathrm{d}x^\alpha \mathrm{d}x^\beta + \varepsilon f^2(x^i)(\mathrm{d}x^4)^2. \qquad (37.17)$$

With respect to the embedding class, two different cases may occur. For $f_{,4} = 0$, the metric

$$\mathrm{d}s^2 = h_{\alpha\beta}(x^\nu)\mathrm{d}x^\alpha \mathrm{d}x^\beta + \varepsilon f^2(x^\nu)(\mathrm{d}x^4)^2, \quad \alpha, \beta, \nu = 1, 2, 3, \qquad (37.18)$$

is that of a space-time admitting a normal Killing vector field $\xi^i = (0, 0, 0, \xi^4)$. By introducing (Szekeres 1966a)

$$u + \mathrm{i}v = f\mathrm{e}^{\mathrm{i}x^4}, \quad \varepsilon(\mathrm{d}u^2 + \mathrm{d}v^2) = \varepsilon[\mathrm{d}f^2 + f^2(\mathrm{d}x^4)^2], \qquad (37.19)$$

we see that the problem has been reduced to finding an embedding of the three-dimensional metric $h_{\alpha\beta}\mathrm{d}x^\alpha \mathrm{d}x^\beta - \varepsilon \mathrm{d}f^2$, which can be done in at most six dimensions:

Theorem 37.7 *If a V_4 admits a non-null normal Killing vector field, then its embedding class is $p \leq 4$.*

In the general case $f_{,4} \neq 0$, the $f^2(\mathrm{d}x^4)^2$ part of the metric (37.17) needs, according to Theorem 37.2, at most three dimensions for embedding, which add to the six dimensions of the $h_{\alpha\beta}\mathrm{d}x^\alpha \mathrm{d}x^\beta$ part. Hence we get

Theorem 37.8 *If a V_4 admits a normal non-null vector field $\boldsymbol{v}$, satisfying* (37.15), *then its embedding class is $p \leq 5$.*

Besides metrics satisfying (37.15), we may consider metrics characterized by the existence of a vector field $v_a = (x^4)_{,a}$ with

$$\begin{aligned} &v_{a;b} = \Theta(g_{ab} - \varepsilon v_a v_b)/3, \quad v_a v^a = \varepsilon \neq 0, \\ &\mathrm{d}s^2 = f^2(x^a) h_{\alpha\beta}(x^\nu)\mathrm{d}x^\alpha \mathrm{d}x^\beta + \varepsilon(\mathrm{d}x^4)^2, \quad \alpha, \beta, \nu = 1, 2, 3. \end{aligned} \qquad (37.20)$$

For the embedding of the three-dimensional space with metric $h_{\alpha\beta}$ we need at most three additional dimensions; application of Theorem 37.2 then leads to

Theorem 37.9 *If a V_4 admits a normal non-null vector field $\boldsymbol{v}$ satisfying* (37.20), *then its embedding class is $p \leq 5$.*

Table 37.1. Upper limits for the embedding class p of various metrics admitting groups

Group →	G_{10}	G_7	G_6		G_3	A_3	A_2
Orbit ↓			$G_6 \supset G_5$	$G_6 \not\supset G_5$			
V_4	$p \leq 1$	$p \leq 2$	$p \leq 4$	$p \leq 2$			
S_3, T_3				$p \leq 1$		$p \leq 3$	
N_3			$p \leq 2$			$p \leq 3$	
S_2, T_2					Flat: $p \leq 3$ Non-flat: $p \leq 2$		$p \leq 4$

37.3.3 Groups of motions and embedding class

If groups of motions of a V_4 induce the existence of subspaces of low embedding class, e.g. of subspaces of constant curvature, then the existence of the group will induce a low class for V_4 too. The details will depend on the order r of the groups G_r (A_r, if Abelian) and on the dimensions of their orbits (S = spacelike, T = timelike, N = null). Some results (Goenner 1973) are given in Table 37.1. We see that, to a certain extent, high symmetry induces low embedding class. The converse is not true; metrics of class one without symmetry are known, see e.g. (37.45).

To give an example of the method of reasoning, we consider an arbitrary spherically-symmetric line element

$$ds^2 = b^2(r,t)(d\theta^2 + \sin^2\theta d\varphi^2) + a^2(r,t)dr^2 - c^2(r,t)dt^2. \tag{37.21}$$

If we put

$$\begin{aligned} &y^1 = b\cos\theta, \quad y^2 = b\sin\theta\cos\varphi, \quad y^3 = b\sin\theta\sin\varphi, \\ &(dy^1)^2 + (dy^2)^2 + (dy^3)^2 = b^2(d\theta^2 + \sin^2\theta d\varphi^2) + (db)^2, \end{aligned} \tag{37.22}$$

and take into account that the 2-space with metric $a^2(r,t)dr^2 - c^2(r,t)dt^2 - [db(r,t)]^2$ can be embedded into E_3, we see that spherically-symmetric space-times are of class $p \leq 2$ ($p = 1$ is possible for special functions a, b, c, see Karmarkar (1948) and Ikeda *et al.* (1963)).

If one compares the theorems given in this section (and especially the methods of the proofs outlined or mentioned) with the general theory of embedding sketched in the preceding section, one may feel that the two sections are disconnected: no use has been made of the Gauss–Codazzi–Ricci equations. The reason for this incoherence is the fact, already stated above, that no systematic treatment of these equations has been carried out, with the exception of metrics of class one or two. The following sections will be devoted to these metrics.

37.4 Exact solutions of embedding class one

37.4.1 The Gauss and Codazzi equations and the possible types of Ω_{ab}

Application of the general theory outlined in §37.2 to class one yields: a V_4 is of class one if and only if there is a symmetric tensor Ω_{ab} satisfying

$$R_{abcd} = e(\Omega_{ac}\Omega_{bd} - \Omega_{ad}\Omega_{bc}), \quad e = \pm 1 \quad \text{(Gauss)}, \tag{37.23}$$

$$\Omega_{ab;c} = \Omega_{ac;b} \quad \text{(Codazzi)}, \tag{37.24}$$

$e = \pm 1$ being suitably chosen.

If $\Omega^{-1}{}_{ab}$ exists, then the Codazzi equations are a consequence of the Gauss equations and the Bianchi identities (cp. Goenner 1977):

Theorem 37.10 *If there is a non-singular symmetric tensor Ω_{ab} satisfying* (37.23)*, then space-time is of embedding class $p = 1$.*

The Gauss equations (37.23) and the field equations yield

$$\kappa_0(T_{ab} - \tfrac{1}{2}g_{ab}T^c{}_c) = R_{ab} = e(\Omega_{ab}\Omega^c{}_c - \Omega_{ac}\Omega^c{}_b). \tag{37.25}$$

Due to the algebraic simplicity of this equation, all possible tensors Ω_{ab} which correspond to an energy-momentum tensor of a perfect fluid or Maxwell type can be determined. The calculations are straightforward, starting with a suitable tetrad representation of T_{ab} and Ω_{ab}. If Ω_{ab} is known, the Petrov type can easily be obtained from (37.23). Four different cases occur, namely (Stephani 1967b):

Perfect fluid metrics, Petrov type O (conformally flat)

$$\begin{aligned} T_{ab} &= (\mu + p)u_a u_b + p g_{ab}, \quad \Omega_{ab} = A u_a u_b + C g_{ab}, \\ \kappa_o \mu &= 3C^2 > 0, \quad \kappa_0 p = C(2A - 3C), \quad e = +1. \end{aligned} \tag{37.26}$$

Perfect fluid metrics, Petrov type D

$$\begin{aligned} &T_{ab} = (\mu + p)u_a u_b + p g_{ab}, \quad \kappa_o \mu = e(3C + 2A)C, \quad \kappa_0 p = eC^2, \\ &\Omega_{ab} = 2C u_a u_b + C g_{ab} + A v_a v_b, \quad v_a v^a = 1, \quad u_a v^a = 0, \quad AC \neq 0. \end{aligned} \tag{37.27}$$

Pure radiation fields, Petrov type N

$$\begin{aligned} T_{ab} &= \Phi^2 k_a k_b, \quad \Omega_{ab} = A k_a k_b + C z_a z_b, \quad k_a k^a = 0, \\ \kappa_0 \Phi^2 &= eAC, \quad z_a z^a = 1, \quad z^a k_a = 0, \end{aligned} \tag{37.28}$$

and

$$\begin{aligned} T_{ab} &= \Phi^2 k_a k_b, \quad \Omega_{ab} = B(k_a z_b + z_z k_b), \quad k_a k^a = 0, \\ \kappa_0 \Phi^2 &= -eB^2, \quad z_a z^a = 1, \quad z^a k_a = 0. \end{aligned} \tag{37.29}$$

The corresponding space-times will be determined in the following sections. In each of the four cases the functions A, B, C and the vector fields $\boldsymbol{u}$, $\boldsymbol{v}$, $\boldsymbol{k}$, $\boldsymbol{z}$ have to be chosen so that the Gauss and Codazzi equations (37.23) and (37.24) are satisfied. To get the metric, one uses the Codazzi equations (and parts of the Gauss equations) to find preferred vector fields, adjusts the coordinates to these vector fields and tries to solve the remaining Gauss equations. In the case of an electromagnetic null field, the Maxwell equations have to be satisfied too.

Concerning the remaining types of energy-momentum tensor, the question is answered by

Theorem 37.11 *There are no embedding class one solutions of the Einstein–Maxwell equations with a non-null electromagnetic field* (Collinson 1968a), *and no embedding class one vacuum solutions.*

37.4.2 Conformally flat perfect fluid solutions of embedding class one

The problem we have to solve is the following: find all metrics with curvature tensor (37.23) and Ω_{ab} given by (37.26), i.e. all metrics with curvature tensor

$$R_{abcd} = C^2(g_{ac}g_{bd} - g_{ad}g_{bc}) + CA(g_{ac}u_b u_d + g_{bd}u_a u_c - g_{ad}u_b u_c - g_{bc}u_a u_d). \tag{37.30}$$

(Because of Theorem 37.10, the Codazzi equations are a consequence of the Bianchi identities for $C \neq A$; $C = A$ is either (if $\Theta = 0$) the Einstein universe (37.40) or (if $\Theta \neq 0$) a special case of (37.45).) For $C = 0$, space-time is flat.

An almost trivial solution of (37.30) is

$$A = 0, \quad \kappa_0 T_{ab} = -3C^2 g_{ab}, \quad C^2 = \text{const}, \tag{37.31}$$

which corresponds to a space of constant curvature (de Sitter space). Allowing for negative μ ($C^2 < 0$), we have thus shown that the spaces of constant curvature are of embedding class one (or zero, if they are flat). From now on we assume $A \neq 0$.

Starting from (37.30), the Bianchi identities give us

$$u_{a;b} = -\dot{u}_a u_b + \Theta h_{ab}/3, \quad h_{ab} \equiv g_{ab} + u_a u_b; \tag{37.32}$$

the velocity field $\boldsymbol{u}$ is normal and shearfree. In a comoving frame of reference $u_i = (0,0,0,u_4)$ the metric takes the form

$$ds^2 = h_{\mu\nu} dx^\mu dx^\nu - (u_4)^2 dt^2, \quad \mu,\nu = 1,2,3. \tag{37.33}$$

Furthermore, the Bianchi identities yield

$$C = C(t), \quad \partial_t C = \Theta A u_4/3, \quad \Theta = \Theta(t), \quad A_{,a} = \dot{A} u_a - A\dot{u}_a. \tag{37.34}$$

The calculations now depend on whether the velocity field is expansionfree ($\Theta = 0$) or not ($\Theta \neq 0$).

Expansionfree solutions If Θ vanishes, (37.32) and (37.34) lead to

$$C = \text{const}, \quad \partial_t h_{\mu\nu} = 0, \quad A = u^4 f(t). \tag{37.35}$$

Because of (37.35), the spatial part of the Gauss equations (37.30) reads

$$\overset{3}{R}_{\mu\nu\sigma\tau} = \overset{4}{R}_{\mu\nu\sigma\tau} = C^2(h_{\mu\sigma}h_{\nu\tau} - h_{\mu\tau}h_{\sigma\nu}), \tag{37.36}$$

so the 3-space $h_{\mu\nu}$ is a space of constant curvature,

$$ds^2 = dr^2/(1 - C^2r^2) + r^2(d\theta^2 + \sin^2\theta d\varphi^2) - (u_4)^2 dt^2. \tag{37.37}$$

The remaining Gauss equations

$$C(A - C)(u_4)^2 h_{\sigma\tau} = -u_4(u_{\sigma;4;\tau} - u_{\sigma;\tau;4} = u_4(u_{4,\sigma\tau} - \Gamma^\nu{}_{\sigma\tau} u_{4,\nu}) \tag{37.38}$$

are a system of differential equations for the function u_4 which can be completely integrated. The result (Stephani 1967b, Kramer *et al.* 1972) is the metric

$$\begin{aligned} ds^2 &= dr^2/[1 - C^2r^2] + r^2(d\theta^2 + \sin^2\theta d\varphi^2) - (u_4)^2 dt^2, \\ u_4 &= r f_1(t) \sin\theta \sin\varphi + r f_2(t) \sin\theta \cos\varphi + r f_3(t) \cos\theta \\ &\quad + f_4(t)\sqrt{1 - C^2r^2} - C^{-1}, \end{aligned} \tag{37.39}$$

$$\kappa_0 \mu = 3C^2 = \text{const}, \quad \kappa_0 p = -\kappa_0 \mu + 2Cu^4, \quad A = u^4 (\neq \text{const}).$$

We see that all these expansionfree conformally flat solutions are generalizations of the interior Schwarzschild solution (16.18) ($f_1 = f_2 = f_3 = 0$,

$f_4 = \text{const}$), containing four arbitrary functions of time. In the general case, the metric admits no Killing vector at all.

No dust solutions ($p = 0$) are included in the class (37.39): $p = 0$, $\mu \neq 0$ leads to $2A = 3C = \text{const}$, and $\dot{\boldsymbol{u}}$ vanishes because of (37.34), i.e. $\boldsymbol{u}$ is covariantly constant. This is compatible with the form (37.30) of the curvature tensor only if $A = C = 0$; i.e. space-time must be flat.

The case $A = C$ is the Einstein universe

$$\mathrm{d}s^2 = \mathrm{d}r^2/(1 - C^2r^2) + r^2(\mathrm{d}\theta^2 + \sin^2\theta \mathrm{d}\varphi^2) - \mathrm{d}t^2. \tag{37.40}$$

Solutions with non-vanishing expansion For $\Theta \neq 0$, we get from (37.32)–(37.34) the system

$$3\partial_t h_{\mu\nu} = -2\Theta u_4 h_{\mu\nu}, \quad A_{,\mu} = Au^4 u_{4,\mu}, \quad 3\partial_t C = \Theta A u_4, \tag{37.41}$$

which is integrated by

$$\begin{aligned} h_{\mu\nu} &= \overline{h}_{\mu\nu}(x^\sigma)V^{-2}(x^\sigma, t), \quad \partial_t V = V\Theta u_4/3, \quad A = u^4 f(t), \\ \Theta &= \Theta(t), \qquad C = C(t), \quad \partial_t C = -\Theta f/3. \end{aligned} \tag{37.42}$$

The spatial part $h_{\mu\nu}$ of the metric tensor is time-dependent; hence we have to calculate the curvature tensor of the hypersurface $t = \text{const}$ by means of

$$\overset{3}{R}_{\mu\nu\sigma\tau} = \overset{4}{R}_{\mu\nu\sigma\tau} - \tfrac{1}{4}[(\partial_t h_{\mu\sigma})(\partial_t h_{\nu\tau}) - (\partial_t h_{\mu\tau})(\partial_t h_{\sigma\nu})](u^4)^2. \tag{37.43}$$

Using (37.41) and (37.30), we get from this representation

$$\overset{3}{R}_{\mu\nu\sigma\tau} = (C^2 - \Theta^2/9)(h_{\mu\sigma}h_{\nu\tau} - h_{\mu\tau}h_{\nu\sigma}); \tag{37.44}$$

the hypersurfaces $t = \text{const}$ are spaces of constant curvature, but with a time-dependent metric. Transforming the metric $h_{\mu\nu}$ for $t = 0$ into the canonical form $h_{\mu\nu} = V^{-2}\delta_{\mu\nu}$, we find that, due to (37.41), this form is preserved in time, and the complete metric takes the form

$$\begin{aligned} &\mathrm{d}s^2 = V^{-2}(\mathrm{d}x^2 + \mathrm{d}y^2 + \mathrm{d}z^2) - (3V_{,4}/V)^2\Theta^{-2}(t)\mathrm{d}t^2, \\ &\kappa_0\mu = 3C^2(t), \quad \kappa_0 p = -\kappa_0\mu + 2CC_{,4}V/V_{,4} = C(2A - 3C) \\ &V = V_0(t) + \frac{C^2(t) - \frac{1}{9}\Theta^2(t)}{4V_0(t)}\left\{[x - x_0(t)]^2 + [y - y_0(t)]^2 + [z - z_0(t)]^2\right\}, \end{aligned} \tag{37.45}$$

(Stephani universe, Stephani 1967b). The remaining Gauss equations being satisfied, the metrics (37.45) are all of class one, with arbitrary functions $x_0(t)$, $y_0(t)$, $z_0(t)$, $C(t)$, $\Theta(t)$ and $V_0(t)$. These solutions are generalizations of the Robertson–Walker cosmological models (§14.2).

Dust solutions of this class are obtained by putting $2A = 3C$. This implies $u_4 = -1$ and $3\partial_t V = -V\Theta$, which exactly characterizes the Friedmann dust models. A detailed study of the various cases in which the conformally flat metrics (37.39) and (37.45) admit one or several Killing vectors was given by Barnes and Rowlingson (1990) and Barnes (1998); see also Bona and Coll (1985b) and Seixas (1992b).

37.4.3 Type D perfect fluid solutions of embedding class one

The type D solutions of embedding class one are characterized by (37.27), i.e. by

$$\begin{aligned} \Omega_{ab} &= 2Cu_a u_b + Cg_{ab} + Av_a v_b, \quad AC \neq 0, \\ \kappa_0 T_{ab} &= e2C(A+2C)u_a u_b + eC^2 g_{ab}, \end{aligned} \tag{37.46}$$

$$R_{abcd} = e(\Omega_{ac}\Omega_{bd} - \Omega_{ac}\Omega_{bc}). \tag{37.47}$$

For $A + 2C = 0$, the Codazzi equations (37.24) give $u_{a;b} = v_a p_b$, $v_{a;b} = u_a p_b$. The Ricci identities $u_{a;bc} - u_{a;cb} = u^d R_{dabc}$, together with the Gauss equations (37.47), then imply $C = 0$; i.e. space-time is flat. We can therefore confine the discussion to the case $2C + A \neq 0$.

As with the conformally flat solutions, two different cases occur, depending now on whether the acceleration $\dot{u}_a = u_{a;b}u^b$ vanishes or not. The two cases will be treated separately.

Solutions with vanishing acceleration After some lengthy but straightforward calculations, the Codazzi equations and parts of the Gauss equations (via the Ricci identity for u_a) give the expressions

$$\begin{aligned} u_{a;b} &= a_1 v_a v_b + a_2(g_{ab} - v_a v_b + u_a u_b), \quad C_{,a} = 2Ca_2 u_a, \\ v_{a;b} &= a_1 u_a v_b + p_a v_b, \quad p_a u^a = 0 = p_a v^a, \\ A_{,a} &= [(2C+A)a_1 - 2Ca_2]u_a + a_3 v_a + Ap_a, \end{aligned} \tag{37.48}$$

for the derivatives of u_a, v_a, A and C. In adapted coordinates $u_i = (0, 0, 0, -1)$, $v_i = (0, 0, V, 0)$, the metric then reads

$$\mathrm{d}s^2 = F^2(t)\left[\mathrm{d}x^2 + H^2(x,y)\mathrm{d}y^2\right] + V^2(x,y,z,t)\mathrm{d}z^2 - \mathrm{d}t^2; \tag{37.49}$$

it belongs to the Szekeres class (33.44a). The system of embedding equations, i.e. (37.48) and the Gauss equations, can be completely integrated. It turns out that the x–y-space is of constant curvature.

If this space is flat, then only $e = 1$ is possible, and we get the metric

$$\begin{aligned} \mathrm{d}s^2 &= t(\mathrm{d}r^2 + r^2\mathrm{d}\varphi^2) + V^2\mathrm{d}z^2 - \mathrm{d}t^2, \quad G_1^2 + G_5^2 \neq 0, \\ V(r,\varphi,z,t) &= t\sqrt{t}G_1(z) + \sqrt{t}[G_2(z)r\cos\varphi + G_3(z)r\sin\varphi \\ &\quad + \tfrac{3}{4}G_1(z)r^2 + G_4(z)] + G_5(z), \\ \kappa_0 p &= 1/4t^2, \quad \kappa_0\mu = 3\kappa_0 p - [4t\sqrt{t}G_1(z) + G_5(z)]/2Vt^2, \end{aligned} \tag{37.50}$$

(Stephani 1968b). This metric contains five arbitrary functions $G_i(z)$. It is a subcase of the type D metrics (33.44). The subclass $G_1 = G_2 = G_3 = 0$ is a metric of plane symmetry.

If the x–y-space is non-flat, one finally gets

$$\begin{aligned} \mathrm{d}s^2 &= F^2(t)\left[\mathrm{d}r^2/(1+\varepsilon r^2) + r^2\mathrm{d}\varphi^2\right] + V^2(r,z,\varphi,t)\mathrm{d}z^2 - \mathrm{d}t^2, \\ V &= G_1(z)\int F^{-1}\mathrm{d}t + G_2(z) + F\,[G_3(z)r\cos\varphi \\ &\quad + G_4(z)r\sin\varphi + \varepsilon G_5(z)\sqrt{1+\varepsilon r^2}\Big], \\ \kappa_0\mu &= 3\kappa_0 p + \frac{2}{VF^4}\left[G_1(z)\left\{\varepsilon tF + b\int\frac{\mathrm{d}t}{F}\right\} + bG_2(z)\right], \quad \varepsilon = \pm 1, \\ \kappa_0 p &= -bF^{-4}, \quad F^2 = \varepsilon(t^2 + b), \quad b = \mathrm{const}, \quad G_1^2 + G_2^2 \neq 0 \end{aligned} \tag{37.51}$$

(Stephani 1968b, Barnes 1973c). The metrics (37.50) and (37.51) cover all class one type D perfect fluid solutions without acceleration.

Solutions with acceleration Evaluation of the Codazzi equations (37.24) for the tensor Ω_{ab} given in (37.46) and partial use of the Ricci identities give

$$\begin{aligned} u_{a;b} &= a_1 v_a u_b + a_2 v_a v_b + a_3(g_{ab} - v_a v_b + u_a u_b), \quad a_1 \neq 0, \\ v_{a;b} &= a_1 u_a u_b + a_2 u_a v_b + (2C + A)A^{-1}a_1(g_{ab} - v_a v_b + u_a u_b), \\ C_{,b} &= 2Ca_3 u_b + (2C + A)a_1 v_b, \quad A_{,b} = [(2C + A)a_2 - 2Ca_3 u_b] + a_4 v_b. \end{aligned} \tag{37.52}$$

The preferred vector field $\mathbf{v}$ is parallel to the acceleration $\dot{\mathbf{u}}$, and both fields are normal; the space orthogonal to them turns out (via the Gauss equations) to be of constant curvature. So we are led to introduce a coordinate system

$$\begin{aligned} \mathrm{d}s^2 &= Y^2(r,t)\left[\mathrm{d}\rho^2/(1-k\rho^2) + \rho^2\mathrm{d}\varphi^2\right] + \mathrm{e}^{2\lambda(r,t)}\mathrm{d}r^2 - \mathrm{e}^{2\nu(r,t)}\mathrm{d}t^2, \\ u^i &= (0,0,0,-\mathrm{e}^{\nu}), \quad v_i = (0,0,\mathrm{e}^{\lambda},0), \quad k = 0, \pm 1. \end{aligned} \tag{37.53}$$

These metrics obviously admit a G_3 on V_2, cp. Chapter 15. Equations (37.52) furthermore imply

$$C = h(r)Y^{-2}(r,t), \quad Y' \equiv \partial Y/\partial r \neq 0, \quad h' \neq 0. \tag{37.54}$$

The function $h(r)$ could be gauged to $h = r$.

If Y depends only on r ($\dot{Y} = 0$), then the metric is necessarily static and thus also shearfree; conversely, if the shear vanishes, the metric is static. The resulting metric is (Kohler and Chao 1965, Stephani 1967b)

$$\begin{gathered} \mathrm{d}s^2 = k\frac{(a+2br^2)}{a+br^2}\mathrm{d}r^2 + r^2\left(\frac{\mathrm{d}\rho^2}{1-k\rho^2} + \rho^2\mathrm{d}\varphi^2\right) - (a+br^2)\mathrm{d}t^2, \\ \kappa_0 p = \frac{kb}{a+2br^2}, \quad \kappa_0\mu = kb\frac{3a+2br^2}{(a+2br^2)^2}, \quad a,\, b = \text{ const}, \quad k = \pm 1. \end{gathered} \tag{37.55}$$

If $Y' \neq 0$, the Gauss and Codazzi equations (37.47) and (37.52) yield

$$\begin{gathered} \mathrm{e}^{2\lambda}(-\ddot{\lambda} - \dot{\lambda}^2 + \dot{\lambda}\dot{\nu}) + \mathrm{e}^{2\nu}(\nu'' + \nu'^2 - \nu'\lambda') = \mathrm{e}^{2\lambda+2\nu}\frac{h}{Y^2}\left(\frac{h'}{Y'Y} - \frac{h}{Y^2}\right), \\ Y^2(k - Y'^2\mathrm{e}^{-2\lambda} + \dot{Y}^2\mathrm{e}^{-2\nu}) = eh^2(r), \\ -\dot{Y}' + Y'\dot{\lambda} + \dot{Y}\nu' = 0, \quad \nu' = 2hY'^2Y^{-2}/h' - Y'Y^{-1}, \\ h'(r) \neq 0, \quad \kappa_0 p = eh^2(r)Y^{-4}, \quad \kappa_0\mu = -\kappa_0 p + 2eh'hY^{-3}/Y', \quad e = \pm 1. \end{gathered} \tag{37.56}$$

The general solution to these equations is not known.

If one assumes an equation of state $\mu = \mu(p)$, then it turns out that only $\mu = p$ is possible, and the resulting metrics are (Stephani 1968a)

$$\begin{gathered} \mathrm{d}s^2 = \tfrac{1}{2}r^2\left[k + a_1\mathrm{e}^t + a_2\mathrm{e}^{-t}\right]\left[\mathrm{d}\rho^2/(1-k\rho^2) + \rho^2\mathrm{d}\varphi^2\right] + \mathrm{d}r^2 - \tfrac{1}{4}r^2\mathrm{d}t^2, \\ \kappa_0 p = \kappa_0\mu = (k^2 - 4a_1a_2)r^{-2}(k + a_1\mathrm{e}^t + a_2\mathrm{e}^{-t})^{-2}, \quad k = 0,\, \pm 1. \end{gathered} \tag{37.57}$$

They are a special case ($F(t) = 0$, $H = c\mathrm{e}^{-r/2}$, r rescaled) of (15.72)–(15.74), admit a homothetic vector (Collins and Lang 1987), and include a spherically-symmetric solution given by Gutman and Bespal'ko (1967) and its plane-symmetric counterpart (Kitamura 1989).

By starting from an explicit embedding, Gupta and Gupta (1986) found spherically-symmetric solutions (in non-comoving coordinates) which belong to the class (37.56). They are given by

$$\mathrm{d}s^2 = \mathrm{d}r^2 + r^2\mathrm{d}\Omega^2 - \mathrm{d}t^2 + \mathrm{d}U^2, \tag{37.58}$$

$$U = C^{-1/2)} \ln \left[2\sqrt{C}\sqrt{C\,(r^2+g(t))^2 - 4r^2 - 4g(t)} + 2C[r^2+g(t)] - 4\right],$$

$$g(t) = A^{-2}\cosh^2(At), \quad -A^{-2}\sinh^2(At), \quad -A^{-2}\cos^2(At), \tag{37.59}$$

$$\kappa_0 p = 4\left[C(r^2+g)^2 - 4A^2g^2\right]^{-1}, \quad \kappa_0\mu = 3p - p^2(r^2+g)r^2C(\ddot{g}+2)/4.$$

In a similar way, special plane-symmetric solutions were found by Gupta and Sharma (1996a, 1996b) and Bhutani and Singh (1998).

37.4.4 Pure radiation field solutions of embedding class one

As mentioned above, a pure radiation field $T_{ab} = \Phi^2 k_a k_b$, $k_a k^a = 0$, can be of embedding class one only for either of the two forms

$$\Omega_{ab} = Ak_a k_b + Cz_a z_b, \quad \kappa_0\Phi^2 = eAC > 0, \tag{37.60}$$

$$\Omega_{ab} = B(k_a z_b + z_a k_b), \quad \kappa_0\Phi^2 = -eB^2 > 0, \tag{37.61}$$

of Ω_{ab} ($k_a z^a = 0$, $z_a z^a = 1$). In both cases the Gauss equations give

$$R_{abcd} = \kappa_0\Phi^2(k_a k_c z_b z_d + z_a z_c k_b k_d - k_a k_d z_b z_c - z_a z_d k_b k_c), \tag{37.62}$$

and because of

$$C_{abcd}k^d = 0 = R_{ab}k^b \tag{37.63}$$

these solutions are of Petrov type N.

For the case (37.61), the null vector field $\boldsymbol{k}$ turns out to be geodesic, shearfree, normal and non-diverging, so that the metric necessarily belongs to Kundt's class (see Chapter 31). For the case (37.60), only metrics belonging to Kundt's class have been studied, although only a pure radiation field satisfying Maxwell equations must have a $\boldsymbol{k}$ with those properties. So from now on we assume that the metric is of Kundt's class (it is not known whether there are type N, embedding class one, solutions of the form studied in §26.1).

The case $\Omega_{ab} = Ak_a k_b + Cz_a z_b$ Here the Codazzi and Gauss equations show that (in a suitable gauge) the null vector $\boldsymbol{k}$ is constant. As shown in §24.5, such pure radiation fields of type N can be transformed to

$$\begin{aligned} &\mathrm{d}s^2 = \mathrm{d}x^2 + \mathrm{d}y^2 - 2\mathrm{d}u\,\mathrm{d}v - 2H(x,y,u)\mathrm{d}u^2, \\ &\kappa_0\Phi^2 = H_{,11} + H_{,22}, \quad z_i = (z_1, z_2, 0, z_4), \quad k_i = (0,0,0,-1). \end{aligned} \tag{37.64}$$

Table 37.2. Embedding class one solutions

T_{ab}	Petrov type	Ω_{ab}	Kinematical classification		Metrics	Type of metric	General solution known
			—		(8.33)	de Sitter	Yes
	O		$\omega_{ab} = 0$	$\Theta = 0$	(37.39)	Generalized interior Schwarzschild	Yes
Perfect fluid:		$Au_a u_b + Cg_{ab}$	$\sigma_{ab} = 0$	$\Theta \neq 0$	(37.45)	Generalized Friedmann	Yes
				$\dot{u}_a = 0$	(37.50), (37.51)	Inhomogeneous cosmological models	Yes
$pg_{ab} + (\mu + p)u_a u_b$	D	$2Cu_a u_b + Cg_{ab}$ $+Av_a v_b$	$\omega_{ab} = 0$	$\dot{u}_a \neq 0$ $\sigma_{ab} = 0$	(37.55)	Static, spherical symmetry	Yes
				$\dot{u}_a \neq 0$ $\sigma_{ab} \neq 0$	(37.57), (37.59)	G_3 on V_2	No
Pure radiation:		$Ak_a k_b + Cz_a z_b$	$k_{a;b} = 0$		(37.65)	Einstein–Maxwell plane wave	No
$\Phi^2 k_a k_b$	N	$B(k_a z_b + z_a k_b)$	$k_{a;b} = 0$		(37.66)–(37.67)	Einstein–Maxwell Plane wave	Yes
			$\Theta = \sigma = \omega = \kappa = 0$		(37.68)	Plane wave, no Maxwell field	Yes

The general solution of the Gauss–Codazzi equations is not known. A special solution and its electromagnetic field are given by

$$\begin{aligned} &\mathrm{d}s^2 = \mathrm{d}r^2 + r^2\mathrm{d}\varphi^2 - 2\mathrm{d}u\,\mathrm{d}v - 2[r\alpha(u) + \beta(u)]\mathrm{d}u^2, \quad \kappa_0\Phi^2 = \alpha r^{-1}, \\ &\quad k_i = (0,0,0,-1), \quad z_i = (0,r,0,0), \quad eA = \alpha(u), \quad C = r^{-1}, \\ &F_{ab} = k_a p_b - k_b p_a, \quad \sqrt{r}\,p_i = \sqrt{\alpha}\,(\cos\psi,\ -r\sin\psi,\ 0,\ 0), \\ &\quad \psi = \varphi/2 + \delta(u). \end{aligned} \tag{37.65}$$

Another special solution is given in Ludwig (1999).

The case $\Omega_{ab} = B(k_a z_b + z_a k_b)$. Two different types belong to this class, depending on whether the null vector $\boldsymbol{k}$ is constant or not.

If $\boldsymbol{k}$ is constant, then the Gauss–Codazzi equations imply

$$\begin{aligned} \mathrm{d}s^2 &= \mathrm{d}x^2 + \mathrm{d}y^2 - 2\mathrm{d}u\,\mathrm{d}v - [\alpha(u)x + \beta(u)y]^2\mathrm{d}u^2, \\ \kappa_0\Phi^2 &= \alpha^2 + \beta^2, \quad e = -1 \end{aligned} \tag{37.66}$$

(Collinson 1968a). The corresponding electromagnetic null field is

$$F_{ab} = k_a p_b - p_a k_b, \quad p_i = (\alpha\cos\varphi,\ \beta\sin\varphi,\ 0,\ 0), \quad \varphi = \varphi(u). \tag{37.67}$$

If $\boldsymbol{k}$ is not constant, then the metric is necessarily of the form (Collinson 1968a, Ludwig 1999)

$$\begin{aligned} \mathrm{d}s^2 = -2\Big\{ [\alpha(u) + y\gamma(u)]^2 + \beta(u)xy + \delta(u)x - v^2/(2x^2)\Big\}\mathrm{d}u^2 + \mathrm{d}x^2 \\ + \mathrm{d}y^2 + 4v\mathrm{d}u\mathrm{d}x/x - 2\mathrm{d}u\,\mathrm{d}v, \quad \kappa_0\Phi^2 = 2(\alpha + y\gamma)^2/x^2 + 2\gamma^2. \end{aligned} \tag{37.68}$$

No Einstein–Maxwell field with metric (37.68) exists, because the condition of integrability $\Delta\ln\Phi^2 = 0$ cannot be satisfied.

37.5 Exact solutions of embedding class two

37.5.1 The Gauss–Codazzi–Ricci equations

According to the general theory outlined in §37.2, a V_4 is of embedding class two if and only if there exist two symmetric tensors Ω_{ab} $(= \Omega^1{}_{ab})$, Λ_{ab} $(= \Omega^2{}_{ab})$, and a vector t_a satisfying

$$R_{abcd} = e_1(\Omega_{ac}\Omega_{bd} - \Omega_{ad}\Omega_{bc}) + e_2(\Lambda_{ac}\Lambda_{bd} - \Lambda_{ad}\Lambda_{bc}), \tag{37.69}$$

$$\begin{aligned} \Omega_{ab;c} - \Omega_{ac;b} &= e_2(t_c\Lambda_{ab} - t_b\Lambda_{ac}), \\ \Lambda_{ab;c} - \Lambda_{ac;b} &= -e_1(t_c\Omega_{ab} - t_b\Omega_{ac}), \end{aligned} \tag{37.70}$$

$$t_{a;b} - t_{b;a} = \Omega_{ac}\Lambda^c{}_b - \Lambda_{ac}\Omega^c{}_b. \tag{37.71}$$

The tensors Ω_{ab}, Λ_{ab} and t_a are not uniquely determined. If a set $(\Omega_{ab}, \Lambda_{ab}, t_a, e_1, e_2)$ satisfies the embedding equations (37.69)–(37.71), then so does $(\overline{\Omega}_{ab}, \overline{\Lambda}_{ab}, \bar{t}_a, \bar{e}_1, \bar{e}_2)$ given by

$$\begin{aligned} \overline{\Omega}_{ab} &= A\Omega_{ab} + B\Lambda_{ab}, \quad \overline{\Lambda}_{ab} = C\Omega_{ab} + D\Lambda_{ab}, \\ \bar{t}_a &= (AD - BC)t_a + e_1 CA_{,a} + e_2 DB_{,a}, \end{aligned} \tag{37.72a}$$

if the functions A, B, C, D satisfy

$$\begin{aligned} e_1 &= \bar{e}_1 A^2 + \bar{e}_2 C^2, \ e_2 = \bar{e}_1 B^2 + \bar{e}_2 D^2, \ e_1 AC + e_2 BD = 0, \\ \bar{e}_1 &= e_1 A^2 + e_2 B^2, \ \bar{e}_2 = e_1 C^2 + e_2 D^2, \ \bar{e}_1 AB + \bar{e}_2 CD = 0. \end{aligned} \tag{37.72b}$$

This transformation corresponds to a rotation (pseudorotation) in the two-dimensional space orthogonal to the V_4 in the embedding space E_6; it can be used to simplify Ω_{ab} and Λ_{ab}, the curvature tensor being given.

Some simple purely algebraic conditions can be derived from the embedding equations. Using the property

$$\varepsilon^{abcd}\Omega_{an}\Omega_{bm}\Omega_{cp}\Omega_{dq} = \varepsilon_{nmpq}\Omega/g \tag{37.73}$$

of the ε-tensor (Ω being the determinant of Ω_{ab}), one obtains (Yakupov 1968a, 1968b)

$$\varepsilon^{abcd}\varepsilon^{nmrs}\varepsilon^{pqik}R_{abnm}R_{cdpq}R_{rsik} = 0. \tag{37.74}$$

Combining the Gauss and Ricci equations, one gets (Matsumoto 1950)

$$-\tfrac{1}{2}e_1 e_2 \varepsilon^{idmn}R^{ab}{}_{cd}R_{mnab} = \varepsilon^{idmn}(t_{c;d} - t_{d;c})(t_{m;n} - t_{n;m}). \tag{37.75}$$

Because of the skew symmetry of the tensors involved, the only information contained in (37.75) is

$$\varepsilon^{cdmn}R_{abcd}R^{ab}{}_{mn} = -e_1 e_2 8\varepsilon^{cdmn}t_{c;d}t_{m;n}. \tag{37.76}$$

As the curvature tensor and the Weyl tensor always satisfy

$$\varepsilon^{edmn}R_{abcd}R^{ab}{}_{mn} = \varepsilon^{edmn}C_{abcd}C^{ab}{}_{mn}, \tag{37.77}$$

the curvature tensor may be replaced in (37.76) by the Weyl tensor (Goenner 1973).

Theorem 37.12 *If a V_4 is of embedding class two, it necessarily satisfies* (37.74) *and* (37.76).

A thorough investigation of the embedding class two space-times has been carried out only for vacuum solutions, conformally flat solutions, and Petrov type D Einstein spaces. We shall summarize the results in the following sections, see also Table 37.3.

37.5.2 Vacuum solutions of embedding class two

In addition to the algebraic conditions (37.74) and (37.76) imposed on the curvature tensor R_{abcd} and on $t_{c;d}$, another simple condition can be derived in the vacuum case from the Gauss–Codazzi–Ricci equations (Yakupov 1968a, 1968b). Calculating the derivatives $\Omega_{a[b;c;i]}$ and $\Lambda_{a[b;c;i]}$ either by means of the Codazzi and Ricci equations or by means of the Ricci identity, one gets

$$2e_2\Lambda_{a[b}t_{c;d]} = R_{ea[bc}\Omega^e_{d]}, \quad -2e_1\Omega_{a[b}t_{c;d]} = R_{ea[bc}\Lambda^e{}_{d]}, \tag{37.78}$$

and after summation over a, c and insertion of $R_{nm} = 0$,

$$\begin{aligned} &\Omega^a_a(t_{c;b} - t_{b;c}) + \Omega^a_b(t_{a;c} - t_{c;a}) - \Omega^a_c(t_{a;b} - t_{b;a}) = 0, \\ &\Lambda^a_a(t_{c;b} - t_{b;c}) + \Lambda^a_b(t_{a;c} - t_{c;a}) - \Lambda^a_c(t_{a;b} - t_{b;a}) = 0. \end{aligned} \tag{37.79}$$

Multiplication of these equations with Ω^c_d and Λ^c_d, respectively, and antisymmetrization leads to

$$\begin{aligned} -(\Omega^a_a)^2(t_{b;d} - t_{d;b}) + (t_{a;c} - t_{c;a})(\Omega^a_b\Omega^c_d - \Omega^a_d\Omega^c_b) \\ -\Omega^a_c\Omega^c_d(t_{a;b} - t_{b;a}) + \Omega^a_c\Omega^c_d(t_{a;d} - t_{d;a}) = 0 \end{aligned} \tag{37.80}$$

and an equivalent equation for Λ_{ab}. Making use of

$$R_{ab} = e_1(\Omega^c_c\Omega_{ab} - \Omega^c_a\Omega_{cb}) + e_2(\Lambda^c_c\Lambda_{ab} - \Lambda_{ac}\Lambda^c_b) = 0, \tag{37.81}$$

and of the Gauss equations (37.69), one finally gets from (37.80)

Theorem 37.13 *A vacuum space-time of embedding class two necessarily satisfies*

$$R^{ac}{}_{bd}t_{a;c} = 0. \tag{37.82}$$

If one evaluates the Gauss equations (37.69) in the Newman–Penrose formalism and partially uses the Codazzi equations (37.70), one gets

Theorem 37.14 *If an algebraically special vacuum solution is of embedding class two, then its multiple null eigenvectors are normal (and geodesic and shearfree); if it is non-degenerate (Type II or III), then the null congruence has zero divergence too* (Collinson 1966).

Starting with the normal form (Table 4.2) of the curvature tensor for each Petrov type, the purely algebraic relations (37.74), (37.76) and (37.82) can be used to get further information about the algebraic structure of the curvature tensor and the tensor $t_{a;b}$, which will simplify the evaluation of the Gauss–Codazzi–Ricci equations.

Following this line of investigation, Yakupov (1973) arrived at the following results (published without proof):

Theorem 37.15 *There are no embedding class two vacuum solutions of Petrov type III* (Yakupov 1973).

Theorem 37.16 *In all embedding class two vacuum solutions* $\mathbf{t}$ *is a gradient, i.e.* Ω_{ab} *and* Λ_{ab} *commute,* cf. (37.71) (Yakupov 1973).

Starting from Theorem 37.16, Hodgkinson (1984) determined the possible algebraic structures of Ω_{ab} and Λ_{ab} for all Petrov types, see also Van den Bergh (1996a). Using these lists, all type D and type N solutions could be found (for Petrov types I and II, the problem is still open).

The type D vacuum solutions of embedding class two are exactly the Robinson–Trautman solutions (28.21) and the subclass $l = 0$ of the solutions (31.41) of Kundt's class, or, equivalently, the subclass $l = 0$ of (13.48) and their counterpart from (13.17), or, again equivalently, the static degenerate metrics of class A and B of Table 18.2 and the type D subclass of the Kasner solutions (13.53)–(13.51) (Van den Bergh 1996b, Hodgkinson 1987). The Schwarzschild solution is included here. They all have a G_4 on V_3 and were known to have embedding class two long since (Rosen 1965, Collinson 1968b).

The Petrov type D Einstein spaces of embedding class two are either products of two 2-spaces of constant curvature as in (12.8), or they are generalizations to $\Lambda \neq 0$ of the embedding class two vacuum solutions (as the Kottler solution of Table 15.1 generalizes the Schwarzschild metric) (Hodgkinson 2000).

The type N vacuum solutions of embedding class two are exactly the pp-waves (§24.5) and the subcase

$$\begin{aligned} ds^2 &= dx^2 + dy^2 - 2du\left[dv - 2v dx/x - v^2 du/2x^2\right] \\ &\quad + \left[a(u)x\ln(x^2+y^2) + b(u)xy + c(u)x^2 + f(u)x\right]du^2 \end{aligned} \tag{37.83}$$

of the solution (31.38) of Kundt's class (Van den Bergh 1996b).

37.5.3 Conformally flat solutions

By definition, conformally flat metrics can always be transformed to

$$ds^2 = e^{2U(x,y,z,t)}[dx^2 + dy^2 + dz^2 - dt^2]. \tag{37.84}$$

They can be characterized by the property that their Weyl tensor C_{abcd} vanishes, which in view of (3.50) is equivalent to

$$R_{abcd} = \tfrac{1}{2}(g_{ac}R_{bd} + R_{ac}g_{bd} - g_{ad}R_{bc} - R_{ad}g_{bc}) - \tfrac{1}{6}R(g_{ac}g_{bd} - g_{ad}g_{bc}). \tag{37.85}$$

One easily sees from (37.85) that non-flat vacuum solutions cannot exist. Even if a space-time is known to satisfy (37.85), the explicit transformation of its metric into the form (37.84) can be rather difficult to find.

As stated in Theorem 37.2, all conformally flat metrics are at most of embedding class two. In this section, we will use the technique of embedding to determine all conformally flat gravitational fields created by a perfect fluid or an electromagnetic field (Stephani 1967a). The conformally flat Einstein–Maxwell fields have also been found by other techniques (Cahen and Leroy 1966, McLenaghan *et al.* 1975).

Conformally flat solutions with a perfect fluid

In the case of a perfect fluid

$$T_{ab} = (\mu + p)u_a u_b + p g_{ab}, \tag{37.86}$$

the condition (37.85) of zero Weyl tensor reads

$$\begin{aligned} 6R_{abcd} = {} & 3\kappa_0(\mu + p)(g_{ac}u_b u_d + u_a u_c g_{bd} - g_{ad}u_b u_c - u_a u_d g_{bc}) \\ & +2\kappa_0\mu(g_{ac}g_{bd} - g_{ad}g_{bc}). \end{aligned} \tag{37.87}$$

Comparing (37.86) with (37.30) together with (37.26), we see that the curvature tensor (37.87) can be written as

$$R_{abcd} = \Omega_{ac}\Omega_{bd} - \Omega_{ad}\Omega_{bc}, \tag{37.88}$$

$$\Omega_{ab} = A u_a u_b + C g_{ab}, \quad \kappa_0\mu = 3C^2, \quad \kappa_0 p = 2CA - 3C^2. \tag{37.89}$$

Equation (37.88), which is exactly the Gauss equation for a embedding class one space-time, indicates that the conformally flat perfect fluid solutions may be of class one. To prove this conjecture, we have to show that the Codazzi equations $\Omega_{ab;c} = \Omega_{ac;b}$ are satisfied too.

If $A = C$, i.e. $\mu = -3p$, then the curvature tensor is simply

$$R_{abcd} = \kappa_0\mu(h_{ac}h_{bd} - h_{ad}h_{bc})/3, \quad h_{ab} = g_{ab} + u_a u_b, \tag{37.90}$$

and the Bianchi identities yield

$$3u_{a;b} = \Theta h_{ab}, \quad 3\mu_{,a} = 2\mu\Theta u_a. \tag{37.91}$$

Because of these equations, Ω_{ab} satisfies the Codazzi equations and this implies that space-time is of embedding class one. In detail, for $\Theta = 0$ the corresponding solution is the Einstein universe (37.40), and for $\Theta \neq 0$ the solution is a special case of the Friedmann universes (37.45).

If $A \neq C$ and $C \neq 0$, the tensor Ω_{ab} is non-singular, and because of Theorem 37.10 the Codazzi equations are satisfied and space-time is of class one. We summarize the results in

Theorem 37.17 *All conformally flat perfect fluid solutions ($\mu \neq 0$) are of embedding class one, and are therefore all contained either in the generalized Schwarzschild type metrics* (37.39) *or in the generalized Friedmann type metrics* (37.45) (Stephani 1967a).

All conformally flat solutions with $\mu = 0$ (Barnes 1973c) are of embedding class two, because if μ was zero in (37.26), this would imply $C = 0$ and space-time would be flat.

Conformally flat solutions with electromagnetic non-null fields

In the case of a non-null field (5.13), i.e.

$$T_{ab} = \Phi^2 \overset{1}{g}_{ab} - \Phi^2 \overset{2}{g}_{ab} \tag{37.92}$$

$$g_{ab} = \overset{1}{g}_{ab} + \overset{2}{g}_{ab} = (x_a x_b + y_a y_b) - (z_a z_b - u_a u_b), \tag{37.93}$$

the curvature tensor has the form

$$R_{abcd} = \kappa_0 \Phi^2 \left[\left(\overset{1}{g}_{ac} \overset{1}{g}_{bd} - \overset{1}{g}_{ad} \overset{1}{g}_{bc} \right) - \left(\overset{2}{g}_{ac} \overset{2}{g}_{bd} - \overset{2}{g}_{ad} \overset{2}{g}_{bc} \right) \right]. \tag{37.94}$$

If we start with a tetrad representation of Ω_{ab} and Λ_{ab}, use the identities

$$\begin{aligned} &\varepsilon^{abcd}[(R_{abnm}\Omega_{cp}\Omega_{dq} - \Omega_{cq}\Omega_{dp}) + (\Omega_{an}\Omega_{bm} - \Omega_{am}\Omega_{bn})R_{cdpq}] = 0, \\ &\varepsilon^{abcd}[(R_{abnm}\Lambda_{cp}\Lambda_{dq} - \Lambda_{cq}\Lambda_{dp}) + (\Lambda_{an}\Lambda_{bm} - \Lambda_{am}\Lambda_{bn})R_{cdpq}] = 0 \end{aligned} \tag{37.95}$$

(n, m, p, q not all different), which follow from the Gauss equations (37.69), together with (37.73) and (37.94), and perform some suitable gauge transformations (37.72), a lengthy but straightforward calculation yields

$$\begin{aligned} \Lambda_{ab} &= G u_a u_b + H(u_a z_b + z_a u_b) + K z_a z_b, \\ \Omega_{ab} &= E x_a x_b + F y_a y_b, \quad e_1 EF = e_2(GK - H^2) = \kappa_0 \Phi^2. \end{aligned} \tag{37.96}$$

The Codazzi and Ricci equations (37.70)–(37.71) then tell us that t_a vanishes and

$$\overset{1}{g}_{ab;c} = 0 = \overset{2}{g}_{ab;c}, \quad \Phi_{,c} = 0 \quad \Rightarrow \quad R_{abcd;e} = 0 \tag{37.97}$$

holds; the curvature tensor is constant, and the solution in question is a symmetric, decomposable, conformally flat space-time. As shown in §35.2, the only metric with these properties is the Bertotti-Robinson metric (35.35), i.e.

$$\mathrm{d}s^2 = \mathrm{d}x^2 + \cos^2(\sqrt{\kappa_0}\Phi x)\,\mathrm{d}y^2 + \cos^2(\sqrt{\kappa_0}\Phi t)\,\mathrm{d}z^2 - \mathrm{d}t^2. \tag{37.98}$$

The corresponding Maxwell field (determined only up to a duality rotation) is constant too, and can be written as

$$F_{ab} = \Phi\sqrt{2}(u_a z_b - z_a u_b), \quad u_i = -\delta_i^4, \quad z_i = \delta_i^3 \cos\left(\sqrt{\kappa_0}\Phi t\right). \qquad (37.99)$$

Theorem 37.18 *The only conformally flat solution with non-null electromagnetic field is the metric* (37.98)–(37.99), *which is the product of two-dimensional spaces of constant curvature. Both the curvature tensor and the electromagnetic field tensor are constant* (Cahen and Leroy 1966, Stephani 1967a; see also Singh and Roy 1966).

Conformally flat solutions with pure radiation or electromagnetic null fields.

In the case of

$$T_{ab} = \Phi^2 k_a k_b \qquad (37.100)$$

the curvature tensor of a conformally flat metric is given by

$$2R_{abcd} = \kappa_0\Phi^2(k_a k_c g_{bd} + g_{ac} k_b k_d - k_a k_d g_{bc} - g_{ad} k_b k_c). \qquad (37.101)$$

One can now use a tetrad representation of Ω_{ab} and Λ_{ab} and evaluate the Gauss equations along the lines indicated above. The result is that only the following two structures are admissible:

$$\begin{aligned} &\Omega_{ab} = Ck_a k_b + D(x_a x_b + y_a y_b), \quad 2e_1 CD = \kappa_0\Phi^2, \quad k^a x_a = 0 = k^a y_a, \\ &\Lambda_{ab} = F x_a x_b + G y_a y_b, \quad e_1 D^2 + e_2 FG = 0, \quad x^a x_a = 1 = y^a y_a, \end{aligned} \qquad (37.102)$$

$$\begin{aligned} &\Omega_{ab} = Ck_a k_b + D(k_a y_b + y_a k_b) + E y_a y_b, \quad -2e_2 F^2 = \kappa_0\Phi^2, \\ &\Lambda_{ab} = F(k_a x_b + x_a k_b), \quad 2e_1(EC - D^2) = \kappa_0\Phi^2. \end{aligned} \qquad (37.103)$$

In both cases the null eigenvector is expansion- and shear-free, and the corresponding metrics belong to Kundt's class (Chapter 31). So to proceed further, one should start from the metrics (31.34) and (31.38), omit the differential equation for H^0 (cp. §31.6) and impose the condition of conformal flatness.

If the metric is of the form (31.34), then conformal flatness implies that metric and Maxwell field have the form

$$\mathrm{d}s^2 = \mathrm{d}x^2 + \mathrm{d}y^2 - 2\mathrm{d}u\,\mathrm{d}v - \kappa_o(x^2 + y^2)\Phi^2(u)\mathrm{d}u^2/2, \qquad (37.104)$$

$$F_{ab} = \Phi(u)\left(k_a p_b - p_a k_b\right), \quad p_a = (\cos\varphi, \sin\varphi, 0, 0), \quad \varphi = \varphi(u). \qquad (37.105)$$

Table 37.3. Metrics known to be of embedding class two

Type of metric	Metrics	Reference (for embedding)
Vacuum type D with G_4 on V_3	(28.21) (31.41), $l = 0$	Rosen (1965) Collinson (1968b)
Some type D vacuum solutions and Einstein spaces		§37.5.2
All conformally flat Einstein–Maxwell fields	(37.98)–(37.99) (37.104)–(37.105)	§37.5.3
All pp-waves (vacuum or not)$^+$	$\mathrm{d}x^2 + \mathrm{d}y^2 - 2\mathrm{d}u\mathrm{d}v$ $-2H\mathrm{d}u^2$	Collinson (1968b)
Melvin universe (Geon)	(22.13)	Collinson (1968b)
All metrics with a G_3 on a non-flat S_2 or T_2 $^+$	Chapters 15, 16	§37.3.3

$^+$: Some of them have embedding class one.

If the metric is of the form (31.38), then conformal flatness implies

$$\mathrm{d}s^2 = \left\{v^2/x^2 - x\left[(x^2+y^2)h_1(u) + xh_2(u) + yh_3(u) + h_4(u)\right]\right\}\mathrm{d}u^2 \\ + \mathrm{d}x^2 + \mathrm{d}y^2 - 2\mathrm{d}u\,\mathrm{d}v + 4v\,\mathrm{d}u\,\mathrm{d}x/x, \quad \kappa_0\Phi^2 = xh_1(u) \tag{37.106}$$

(Wils 1989a, Edgar and Ludwig 1997a). No Maxwell field exists for this case. There is a homothety if the $h_i(u)$ are specialized to appropriate powers of u.

Theorem 37.19 *The only conformally flat solutions with a null electromagnetic field are the special plane waves* (37.104)–(37.105) *with a constant null eigenvector* $\mathbf{k}$ (McLenaghan *et al.* 1975). *The conformally flat pure radiation fields which cannot be interpreted as a Maxwell field are given by* (37.106) (Wils 1989a, Edgar and Ludwig 1997a).

37.6 Exact solutions of embedding class $p > 2$

So far no systematic research has been done to find solutions of embedding class greater than two. We can only give some metrics where the embedding class p is known either from general theorems as given in §37.3 or from the explicit embedding. These metrics are:

$p = 3$: (i) Static axisymmetric vacuum solutions (Weyl's class, see §20.2 (Szekeres 1966a, Collinson 1968b). Some of them have class $p = 2$. (ii) The Petrov type *III* vacuum metric (31.40) with a G_2 (Collinson 1968b). (iii) The Gödel cosmos (12.26) (Collinson 1968b).

$p = 4$: All Robinson–Trautman solutions (Chapter 28) (Collinson 1968b).

Part V
Tables

38
The interconnections between the main classification schemes

38.1 Introduction

As already pointed out in Chapter 1, the solutions of Einstein's field equations could be (and have been) classified according to (at least) four main classification schemes, namely with respect to symmetry groups, Petrov types, energy-momentum tensors, and special vector and tensor fields. Whereas the first two schemes have been used in extenso in this book, the others played only a secondary role, and the connections between Petrov types and groups of motions were also treated only occasionally.

This last chapter is devoted to the interconnection of the first three of the classification schemes mentioned above. It consists mainly of tables. §38.2 gives the (far from complete) classification of the algebraically special solutions in terms of symmetry groups. §38.3 contains tables, wherein the solutions (and their status of existence and/or knowledge) are tabulated by combinations of energy-momentum tensors, Petrov types and groups of motion. In the tables the following symbols are used:

S: some special solutions are known A: all solutions are known
∄: does not exist

Th., Ch. and Tab. are abbreviations for 'Theorem', 'Chapter' and 'Table' respectively.

For perfect fluid solutions, the connection between the kinematical properties of the four-velocity (see §6.1) and groups of motions was discussed e.g. by Ehlers (1961) and Wainwright (1979).

The reader who is interested in an introductory survey of the solutions as classified by some invariant properties should consult (besides the subject index and the table of contents):

Table 11.1	for metrics with isometries,
Tables 11.2–11.4	for metrics with homotheties
Table 12.1	for homogeneous solutions,
Tables 13.2–13.4	for hypersurface-homogeneous solutions
§§26.1–26.4	for algebraically special solutions,
Table 32.1	for Kerr-Schild solutions,
Table 37.2	for solutions of embeddding class one,
Table 37.3	for solutions of embedding class two.

38.2 The connection between Petrov types and groups of motions

In Parts II and III of this book, two methods for the invariant classification of gravitational fields, namely groups of motions and Petrov types, were treated quite independently. We have seen that many solutions admitting an isometry group are algebraically special and vice versa. Occasionally, if it was known to us, we mentioned the Petrov type of a solution classified according to the underlying group structure or referred to the group of motions of a solution of a certain Petrov type. In this section and in Tables 38.3–38.10 we want to collect the known results on the connection between these two invariant classifications.

If one knows the group of motions and asks for the possible Petrov types, the following facts impose some restrictions.

(i) The existence of an *isotropy group* I_s (see §11.2) implies that the Weyl tensor is degenerate, i.e. the Petrov type is N, D or O. In particular, a group G_3 on non-null orbits V_2 implies type D or O (Theorem 15.1).

(ii) The *static* solutions are of type I, D or O (§18.6.1).

(iii) The *stationary axisymmetric* vacuum solutions cannot be of type III (see §19.5). For all admissible Petrov types, the subclasses admitting a group G_r, $r \geq 3$, were determined by Collinson and Dodd (1971). Those of type II are given by (20.32).

The *hypersurface-homogeneous* (Chapter 13) algebraically special Einstein spaces ($R_{ab} = \Lambda g_{ab}$) were determined by Siklos (1981). Apart from special plane waves (§24.5) and the Λ-term solutions of (12.8), (13.48), (13.65) and (13.67), Siklos obtained the Petrov type III

Table 38.1. The algebraically special, diverging vacuum solutions of maximum mobility

Petrov type	Twist	Maximal group	solutions
N	$\omega \neq 0$	G_2	
	$\omega = 0$	G_2	S Table 38.2
III	$\omega \neq 0$	G_2	S (38.4), $m + \mathrm{i}M = 0$
	$\omega = 0$	G_3	A (38.1)
II	$\omega \neq 0$	G_2	S (29.62), (38.4)–(38.6)
	$\omega = 0$	G_2	S Table 38.2
D	$\omega \neq 0$	G_2	A §29.5
	$\omega = 0$	G_4	A (28.21)

Robinson–Trautman solution (28.16), i.e.

$$\mathrm{d}s^2 = r^2 x^{-3}(\mathrm{d}x^2 + \mathrm{d}y^2) - 2\mathrm{d}u\mathrm{d}r + \tfrac{3}{2}x\mathrm{d}u^2, \tag{38.1}$$

where the Killing vectors ∂_y, ∂_u and $2(x\partial_x + y\partial_y) + r\partial_r - u\partial_u$ generate a group G_3VI_h, $h = -1/9$, its $\Lambda \neq 0$ generalization (13.64) and the homogeneous non-diverging solutions (12.34) and (12.35). These latter solutions belong to Kundt's class (Chapter 31) and are, in the metric form (31.6), given by ($\lambda \equiv \sqrt{|\Lambda|}$, $\Lambda < 0$)

$$\mathrm{d}s^2 = 2(\lambda x)^{-2}(\mathrm{d}x^2 + \mathrm{d}y^2) - 2\mathrm{d}u(\mathrm{d}v + 2v\mathrm{d}x/x + x\mathrm{d}u), \tag{38.2}$$

$$\mathrm{d}s^2 = 2\frac{(\mathrm{d}x^2 + \mathrm{d}y^2)}{(\lambda x)^2} - 2\mathrm{d}u\left[\mathrm{d}v + 2v\mathrm{d}x/x + \frac{4x}{3\lambda}\mathrm{d}y + 2x^4\right]. \tag{38.3}$$

They were found by Kaigorodov (1962) to be the type N and III Einstein spaces of maximum mobility (maximum order of G_r): the groups of motions are G_5 for (38.2) and G_4 for (38.3); in both cases there is a subgroup G_3VI_h, $h = -1/9$.

We now start with a specified Petrov type and ask for the groups of motions. For the various Petrov types of the algebraically special, diverging ($\rho \neq 0$) *vacuum* fields, the dimension of the maximal group G_r and the corresponding solutions are given in Table 38.1. The restrictions on the dimension of the maximal group were obtained by Kerr and Debney (1970).

The type D vacuum solutions admit either a G_2 or a G_4. The same holds true for the type D Einstein–Maxwell fields (21.11) and (30.36). Kerr and Debney (1970) determined all algebraically special, diverging, vacuum solutions admitting a group G_r, $r > 2$, and some solutions admitting a

G_2. These latter solutions are: (i) the Demianski solution (29.62), (ii) the solution

$$m + \mathrm{i}M = m_0 + \mathrm{i}M_0 = \text{const}, \quad \mathrm{P}^2 = (x\sqrt{2})^3 = (\zeta + \bar{\zeta})^3,$$
$$L = \tfrac{1}{6}\mathrm{i}M_0(2)^{-1/2}x^{-3}\left[C_1 x^{\sqrt{13}/2} + C_2 x^{-\sqrt{13}/2}\right] \tag{38.4}$$

$(C_1, C_2$ being real constants), which is a special member of (29.50) to which it is related by a transformation (29.26) with $f = \zeta$, $F_{,u} = 1$, and (iii) the solution

$$P = 1,\ \mathrm{i}M - m = 2A(1+\alpha)\zeta^\alpha,\ L = A\bar{\zeta}^2\zeta^{1+\alpha} + B\bar{\zeta}\zeta^{\alpha/3}, \tag{38.5}$$

where $\mathrm{Re}\,\alpha = -3$, and A and B are complex constants. The metrics (38.4) and (38.5) respectively admit groups G_2I and G_2II and belong to the classes (29.46)–(29.50) and (29.60)–(29.61). A vacuum solution of the class (29.46)–(29.50), with a G_2II, was found by Lun (1978): it reads

$$m + \mathrm{i}M = (m_0 + \mathrm{i}M_0)\bar{\zeta}^{-3/2}, \quad P^2 = (\zeta + \bar{\zeta})^3, \quad s \equiv y/x,$$
$$L = x^{-3/2}\left[A\left(s + \sqrt{1+s^2}\right)^{\sqrt{13}/2}\left(s - \tfrac{1}{2}\sqrt{13}\sqrt{1+s^2}\right)\right.$$
$$\left. + B\left(s + \sqrt{1+s^2}\right)^{-\sqrt{13}/2}\left(s + \tfrac{1}{2}\sqrt{13}\sqrt{1+s^2}\right)\right] + \frac{m + \mathrm{i}M}{3P^2} \tag{38.6}$$

(m_0, M_0, A, B real constants). The twisting type N vacuum solutions admit at most a (non-Abelian) G_2 (Stephani and Herlt 1985).

The classes (29.46)–(29.50) and (29.60)–(29.61) cover all algebraically special diverging vacuum solutions which admit a G_1 generated by an asymptotically timelike Killing vector field $\xi = \partial_u$ (Held 1976a, 1976b, Zenk and Das 1978). The algebraically special vacuum solutions ($\rho \neq 0$) with an orthogonally transitive G_2I (cp. §8.6) are of Petrov type D (Weir and Kerr 1977).

The groups of motion of the Robinson–Trautman vacuum solutions (Chapter 28) were systematically analysed by Collinson and French (1967) using a null tetrad formulation of the Killing equations. The results are given in Table 38.2. To the authors' knowledge, the groups of motions of the algebraically special diverging *non-empty* spaces have not been systematically investigated and the same is true for the symmetries of the (non-diverging) solutions of Kundt's classs (Chapter 31) except for the null Killing vector case and the *pp*-waves, see §§24.4–24.5 and Tables 24.1–24.2).

Tables 38.3–38.5 give the solutions listed in this book for which both the Petrov type and the symmetries are known. Special cases of some solutions may admit a higher-dimensional group or/and the Petrov type may be

Table 38.2. Robinson–Trautman vacuum solutions admitting two or more Killing vectors

The gauge $m = \text{const}$ is used. Numbers in brackets [ ] refer to equations in Collinson and French (1967). Type *III* solutions with G_2 exist for all three forms of P, but only (28.17) is explicitly known.

	$P(\zeta,\bar\zeta,u)$	*II*	*D*	*III*	*N*
G_4	$1+K\zeta\bar\zeta/2$	$\nexists$	A (28.21)	$\nexists$	$\nexists$
G_3	$(\zeta+\bar\zeta)^{3/2}$, $m=0$	$\nexists$	$\nexists$	A (38.1)	$\nexists$
	$P(\zeta,\bar\zeta)$	S (28.26)	$\nexists$		$\nexists$
G_2	$P(\zeta+\bar\zeta,u)$	S (28.25)	A (28.24)	S (28.17)	A [6.17]
	$P(\zeta+u,\bar\zeta+u)$	$\nexists$	$\nexists$		A [6.20]
	$P_{,\zeta}\neq P_{,\bar\zeta}$				–[6.22]

Table 38.3. Petrov types versus groups on orbits V_4

	G_7	G_6	G_5	G_4
I	$\nexists$ §11.2	$\nexists$ §11.2	$\nexists$ §11.2	(12.14), (12.21) (12.27)–(12.32)
D	$\nexists$ §11.2	(12.8)	(12.26)	
II	$\nexists$ §11.2	$\nexists$ §11.2	$\nexists$ §11.2	(12.29)
N	$\nexists$ §11.2	(12.6), (12.12)	(12.34)	
III	$\nexists$ §11.2	$\nexists$ §11.2	$\nexists$ §11.2	(12.35)
0	(12.24) (12.37)	(12.7), $A=0$ (12.16)		

more special. A variety of solutions are not contained in these Tables because the group or the Petrov type is not known, but these solutions are partly covered by Tables 38.6–38.10.

38.3 Tables

The tables refer to the equation numbers of the solutions given in this book and/or the relevant chapters or sections. As a rule, reference to a chapter or a section includes all (or most of the) solutions given therein. The symbols $\nexists$ (= non-existence), A (= All), S (= Some) are used as explained at the end of §38.1.

Table 38.4. Petrov types versus groups on non-null orbits V_3

	G_6	G_4	G_3
I	$\nexists$ §11.2	$\nexists$ §11.2	§§13.3.4, 13.4, 14.4 §§22.2; (13.53)–(13.62)
D	$\nexists$ §11.2	§§13.4, 14.3, 15.4 §16.1; (13.48), (15.78) in Tab. 18.1, (22.13) (28.44), (31.41) (31.58), (33.34)	(18.65), (28.43)
II	$\nexists$ §11.2	$\nexists$ §11.2	(13.65), (22.17), (33.30)
N	$\nexists$ §11.2		Tab. 24.2, (13.67)
III	$\nexists$ §11.2	$\nexists$ §11.2	(13.64), (13.67), (28.16) (38.1)
0	§14.2	(16.18)	

Table 38.5. Petrov types versus groups on non-null orbits V_2 and V_1

	G_3 on V_2	G_2 on V_2	G_1
I	$\nexists$ §11.2	Ch. 17, 20, 21, 23, 34 §§22.4; (20.16), (22.48) (21.59)–(21.60), (25.71)	§17.3; (18.73) (26.21) $b=0$, (26.22)
D	Th. 15.1 Ch. 15 §16.2 (28.43) (37.50)	§§18.6.2, 21.1.2, 29.5 (18.64), (18.66), (21.24) (21.57), (21.61), (23.43) in: (23.2), (23.13), (23.40) (23.50), in (25.26) (25.56), (25.59) (28.24), (28.45), (30.36) (31.60)–(31.61), (33.9a)	
II	$\nexists$ Th. 15.1	§29.2.6; (18.50), (20.32) (22.48), (22.64), in (23.8) (25.43), (28.26), (29.62) (33.14), (33.28) (38.4)–(38.6)	§§29.2.3, 29.2.5 32.2; (18.48)–(18.49) (28.56), (31.43) (30.51)–(30.52), (32.80) (33.31), (33.43)
N	$\nexists$ Th. 15.1	Tab. 24.2	(29.72)
III	$\nexists$ §11.2	(28.17), (31.40)	
0	§37.5.3		

Table 38.6. Energy-momentum tensors versus groups on orbits V_4 (with $\mathcal{L}_\xi F_{ab} = 0$ for the Maxwell field)

	G_7	G_6	G_5	G_4
Vacuum	∄ Th. 12.1	A (12.12), $a = 0$	∄ Th. 12.1	A (12.14)
Einstein–Maxwell non-null	∄ Th. 12.3	A (12.16)	∄ Th. 12.3	∄ Th. 12.3
Einstein–Maxwell null	∄ Th. 12.1	A (12.12)	∄ Th. 12.1	∄ §13.1
Pure radiation	A (12.37)	A (12.12)		
Λ-term	∄ §12.5	A (12.8)	A (12.34)	A (12.35)
Perfect fluid	A (12.24)	∄ Th. 12.4	A (12.26)	A (12.27)–(12.32)

Table 38.7. Energy-momentum tensors versus groups on non-null orbits V_3

	G_6	G_4	G_3
Vacuum	∄ §13.1	A §§13.3.1, 15.4 in Table 18.2 (28.21), (31.41)	S §13.3.2; (22.5)–(22.7) (38.1), Table 24.2
Einstein–Maxwell non-null	∄ §13.1	A §§13.3.1, 15.4 (28.44), (31.58)	S §13.3.4 (22.11)–(22.16)
Einstein–Maxwell null	∄ §13.1		(22.17)
Pure radiation	∄ §13.1		
Λ-term	∄ §13.1	A §13.3.1	S §13.3.2; (22.8)
Perfect fluid	S §14.2	S §§14.3, 16.1 (15.78) (35.75)–(35.79)	S §§13.4, 14.4; (18.65) (22.19), (22.21)–(22.24) (22.27)–(22.33) (22.34)–(22.39), (33.30) (36.22)–(36.23)

Table 38.8. Energy-momentum tensors versus groups on non-null orbits V_2 and V_1

	G_3 on V_2	G_2 on V_2	G_1
Vacuum	A Th. 15.5	S Ch. 20, 34; §§17.1, 17.2, 22.3, 25.3, 25.4, 29.5; Tab. 24.2 (18.50), (26.23) (28.17), (29.62) (28.24)–(28.25) (31.40), (36.33) (38.4)–(38.6)	S §§10.3.2, 17.3 in 29.2, 34.1.1 32.2.1 (18.48)–(18.49) (26.21) $b = 0$, (26.22), (29.72) (31.43
Einstein–Maxwell non-null	A Th. 15.5	S Ch. 34, §§21.1, 22.4, 25.5 Th. 22.1; (28.45) (30.36), (32.96) (31.60)–(31.61)	S §34.1.1; (18.73) (28.56)
Einstein–Maxwell null	A (28.43)	S (22.59)	
Pure radiation	A Tab. 15.1	S §25.6.2 (22.68)–(22.70)	S (30.46), (30.51) (30.52), (32.80)
Λ-term	A Th. 15.5		S
Perfect fluid	S §§15.5, 15.6, 16.2 (15.80) (33.10) (36.24) (37.53)	S Th. 10.2, Ch. 23 §§10.11.1, 21.2 25.6.1; (18.64), (18.66), (33.13) (33.28)–(33.35) (35.77)	S §§10.3.2, 23.4 (33.12), (33.20) (33.32), (35.78) (33.45)–(33.46)

Table 38.9. Algebraically special vacuum, Einstein–Maxwell and pure radiation fields (non-aligned or with $\kappa\bar{\kappa}+\sigma\bar{\sigma}\neq 0$)

No conformally flat solutions of these types exist.

			II		D		III		N
Vacuum	$\rho\neq\bar{\rho}$	S	Ch. 29 (29.45)–(29.50) (29.53)–(29.58) (29.60)–(29.62)	A	§§21.1, 29.2 §29.5; (21.11) (29.74)–(29.75)	S	§29.4 (29.46)–(29.50) $m+\mathrm{i}M=0$	S	§29.3 (29.72)
	$\rho=\bar{\rho}\neq 0$	S	§28.1, Th. 28.1 (28.26)	A	§28.1; (28.21) (28.24), (28.25)	S	Th. 28.2 (28.16)–(28.17)	A	§28.1
	$\rho=0$	S	§31.5.2 Th. 31.1	A	§31.5.2; (31.41)	A	§31.5.1; Th. 31.2	A	§§24.5, 31.5, (24.40)–(24.44) $(F=0)$ Th. 31.2
Aligned $\kappa\bar{\kappa}+\sigma\bar{\sigma}\neq 0$	E-M non-null	?	§26.1	?	§26.1	$\nexists$	§26.1	$\nexists$	§26.1
	E-M null	$\nexists$	Th. 7.4	$\nexists$	Th. 7.4	$\nexists$	Th. 7.4	$\nexists$	Th. 7.4
	pure rad.	$\nexists$	Th. 7.5	$\nexists$	Th. 7.5	$\nexists$	Th. 7.5	S	§26.1 (26.11)–(26.18)
Non-aligned	E-M non-null	S	§26.1 (26.5)	?		S	§26.1	A	§26.1; (26.6)
	E-M null	$\nexists$	Th. 7.4	$\nexists$	Th. 7.4	$\nexists$	Th. 7.4	$\nexists$	Th. 7.4
	pure rad.	?		?		$\nexists$	§26.1	$\nexists$	§26.1

Table 38.10. Algebraically special (non-vacuum) Einstein–Maxwell and pure radiation fields, aligned and with $\kappa\overline{\kappa} + \sigma\overline{\sigma} = 0$.

		II	*D*	*III*	*N*	0
E-M non-null	$\rho \neq \overline{\rho}$	S Th. 30.2 §§30.430.5 (30.26)–(30.28) (30.37)	S §§21.1, 30.6 (21.11), (24.21) (30.36)	S §30.5	$\nexists$ Th. 30.3	$\nexists$ Th. 30.3 Th. 37.18
	$\rho = \overline{\rho} \neq 0$	S Th. 28.3, 28.5 (28.55)–(28.68)	A §§28.2 (28.44)–(28.46)	$\nexists$ §28.2	$\nexists$ Th. 28.4	$\nexists$ Th. 28.4 Th. 37.18
	$\rho = 0$	S §31.7	S §31.7; (24.22) (31.58)–(31.61)	S §31.7 (31.57)	A §31.7 (31.49)	A Th. 37.18
E-M null	$\rho \neq \overline{\rho}$				$\nexists$ Th. 30.3	$\nexists$ Th. 30.3 Th. 37.19
	$\rho = \overline{\rho} \neq 0$	S Th. 28.3 (28.53)	A §28.2 (28.43)	S §28.2 (28.41)	$\nexists$ Th. 28.4	$\nexists$ Th. 28.4 Th. 37.19
	$\rho = 0$	§31.6 (24.37)			S §§24.5, 31.6, 32.3 (32.71)–(32.72)	A Th. 37.19
Pure rad.	$\rho \neq \overline{\rho}$	S §30.7	$\nexists$ §30.7.1, , 32.4		$\nexists$ §30.7 Th. 30.3	$\nexists$ Th. 37.19
	$\rho = \overline{\rho} \neq 0$	S Th. 28.6	A §28.3 (28.73)–(28.75)	S §28.3 (28.72)	$\nexists$ §28.3	$\nexists$ Th. 37.19
	$\rho = 0$	S §§31.6	S §31.6	A §31.6	A §31.6	A Th. 37.19

References

For brevity we have shortened the following standard journal abbreviations: *CQG* = *Class. Quant. Grav.*; *GRG* = *Gen. Rel. Grav.*; *JMP* = *J. Math. Phys.*; *PRD* = *Phys. Rev. D*;

Ablowitz, M.J. and Fokas, A.S. (1997). *Complex variables: Introduction and applications* (Cambridge University Press, Cambridge). *See* §10.7.

Adler, R.J. and Sheffield, C. (1973). Classification of space-times in general relativity. *JMP* **14**, 465. *See* Ch. 4.

Agnew, A.F. and Goode, S.W. (1994). The $p = \mu$ separable G_2 cosmologies with heat flow. *CQG* **11**, 1725. *See* §23.3.

Aguirregabiria, J.M., Chamorro, A., Manko, V.S. and Sibgatullin, N.R. (1993a). Exterior gravitational field of a magnetized spinning source possessing an arbitrary mass-quadrupole moment. *PRD* **48**, 622. *See* §34.6.

Aguirregabiria, J.M., Feinstein, A. and Ibanez, J. (1993b). Exponential scalar field universes I. The Bianchi I models. *PRD* **48**, 4662. *See* §14.4.

Aichelburg, P.C. (1970). Curvature collineations for gravitational *pp* waves. *JMP* **11**, 2458. *See* §35.4.

Aichelburg, P.C. and Balasin, H. (1997). Generalized symmetries of impulsive gravitational waves. *CQG* **14**, A31. *See* §24.5.

Aichelburg, P.C. and Sexl, R.U. (1971). On the gravitational field of a massless particle. *GRG* **2**, 203. *See* §9.5.

Akabari, R.P., Dave, U.K. and Patel, L.K. (1980). Some pure radiation fields in general relativity. *J. Aust. Math. Soc. B (Appl. Math.)* **21**, 464. *See* §30.7.

Al-Badawi, A. and Halilsoy, M. (1999). Interaction of successive electromagnetic waves in general relativity. *N. Cim. B* **114**, 253. *See* §25.5.

Alekseev, G.A. (1980). N soliton solutions of the Einstein–Maxwell equations. *JETP Lett.* **32**, 377. *See* §§34.5, 34.6.

Alekseev, G.A. (1981). On soliton solutions of the Einstein equations in a vacuum. *Sov. Phys. – Dokl.* **26**, 158. *See* §34.5.

Alekseev, G.A. (1983). Method for generating exact nonsoliton solutions for electrovacuum gravitational fields. *Sov. Phys. – Dokl.* **28**, 17. *See* §34.8.

Alekseev, G.A. (1985). The method of the inverse problem of scattering and the singular integral equations for interacting massless fields. *Sov. Phys. – Dokl.* **30**, 565. *See* §34.6.

Alekseev, G.A. (1990). Isomonodromy deformations and integrability of electrovacuum Einstein–Maxwell field equations with isometries, in *Solitons and applications*, eds. V.G. Makhankov, O.K. Pashaev and V.K. Fedyanin, page 174 (World Scientific, Singapore). *See* §34.8.

Alekseev, G.A. and García D., A. (1996). Schwarzschild black hole immersed in a homogeneous electromagnetic field. *PRD* **53**, 1853. *See* §§21.1, 34.6.

Alekseev, G.A. and Griffiths, J.B. (1995). Propagation and interaction of gravitational waves in some expanding backgrounds. *PRD* **52**, 4497. *See* §§10.11, 25.6.

Alekseev, G.A. and Griffiths, J.B. (1996). Exact solutions for gravitational waves with cylindrical, spherical or toroidal wavefronts. *CQG* **13**, 2191. *See* §25.3.

Alencar, P.S.C. and Letelier, P.S. (1986). Solitary and non-solitary waves on a Friedmann–Robertson–Walker background, in *Proceedings of the fourth Marcel Grossmann meeting on general relativity*, ed. R. Ruffini, page 885 (North-Holland, Amsterdam). *See* §10.11.

Aliev, A.N. and Gal'tsov, D.V. (1989a). Exact solutions for magnetized black holes. *Astrophys. Space Sci.* **155**, 181. *See* §21.1.

Aliev, A.N. and Gal'tsov, D.V. (1989b). 'Magnetized' black holes. *Sov. Phys. – Uspekhi* **32**, 75. *See* §21.1.

Aliev, A.N., Gal'tsov, D.V. and Sokolov, A.A. (1980). Rotating black hole in a strong magnetic field. *Sov. Phys. J.* **23**, 179. *See* §21.1.

Aliev, B.N. and Leznov, A.N. (1992a). Einstein's vacuum fields with nonAbelian group of motion G_2II. *CQG* **9**, 1261. *See* §17.1.

Aliev, B.N. and Leznov, A.N. (1992b). Exact solutions of the vacuum Einstein's equations allowing for two noncommutative Killing vectors (Type G_2II of Petrov classification). *JMP* **33**, 2567. *See* §17.1.

Allnutt, J.A. (1980). On the algebraic classification of perfect fluid solutions of Einstein's equations. Ph.D. thesis, Queen Elizabeth College London. *See* §§14.4, 23.1, 23.3.

Allnutt, J.A. (1981). A Petrov type-III perfect fluid solution of Einstein's equations. *GRG* **13**, 1017. *See* §33.4.

Åman, J.E. (1984). Computer-aided classification of geometries in general relativity; example: The Petrov type D vacuum metrics, in *Classical general relativity*, eds. W.B. Bonnor, J.N. Islam and M.A.H. MacCallum, page 1 (Cambridge University Press, Cambridge). *See* §9.2.

Åman, J.E. (2002). Classification programs for geometries in general relativity – manual for CLASSI, 4th edition. Report, Stockholm. *See* §§5.4, 9.1.

Åman, J.E., d'Inverno, R.A., Joly, G.C. and MacCallum, M.A.H. (1984). Quartic equations and classification of the Riemann tensor in general relativity, in *Proceedings of EUROSAM '84*. Lecture notes in computer science, vol. 174, ed. J. Fitch, page 47 (Springer Verlag, Berlin). *See* §9.3.

Åman, J.E., d'Inverno, R.A., Joly, G.C. and MacCallum, M.A.H. (1991). Quartic equations and classification of Riemann tensors in general relativity. *GRG* **23**, 1023. *See* §9.3.

Anderson, I.M. and Torre, C.G. (1996). Classification of local generalized symmetries for the vacuum Einstein equations. *Commun. math. phys.* **176**, 479. *See* §10.3.

Araujo, M.E., Dray, T. and Skea, J.E.F. (1992). Finding isometry groups in theory and practice. *GRG* **24**, 477. *See* §§2.11, 9.4.

Araujo, M.E. and Skea, J.E.F. (1988a). The automorphism groups for Bianchi universe models and computer-aided invariant classification of metrics. *CQG* **5**, 537. *See* §§9.4, 13.3.

Araujo, M.E. and Skea, J.E.F. (1988b). Automorphisms in action: spatially homogeneous Einstein–Maxwell plane waves. *CQG* **5**, 1073. *See* §13.3.

Argüeso, F. and Sanz, J.L. (1985). Some exact inhomogeneous solutions of Einstein's equations with symmetries on the hypersurfaces t = const. *JMP* **26**, 3118. *See* §§23.3, 36.4.

Assad, M.J.D. and Damião Soares, I. (1983). Anisotropic Bianchi types VIII and IX locally rotationally symmetric cosmologies. *PRD* **28**, 1858. *See* §14.3.

Azuma, T., Endo, M. and Koikawa, T. (1993). Exact disk solution of the Einstein equation. *Prog. Theor. Phys.* **90**, 585. *See* §34.5.

Azuma, T. and Koikawa, T. (1994). Equilibrium condition in the axisymmetric N-Reissner–Nordstrom solution. *Prog. Theor. Phys.* **92**, 1095. *See* §21.1.

Babala, D. (1987). Collision of a gravitational impulsive wave with a shell of null dust. *CQG* **4**, L89. *See* §25.6.

Bach, R. and Weyl, H. (1922). Neue Lösungen der Einsteinschen Gravitationsgleichungen. *Math. Z.* **13**, 134. *See* §20.2.

Baillie, G. and Madsen, M.S. (1985). A tilted Bianchi type-V perfect fluid solution for stiff matter. *Astrophys. Space Sci.* **115**, 413. *See* §§10.11, 23.1.

Bajer, K. and Kowalczyński, J.K. (1985). A class of solutions of the Einstein–Maxwell equations. *JMP* **26**, 1330. *See* §28.2.

Balasin, H. and Nachbagauer, H. (1996). Boosting the Kerr geometry in an arbitrary direction. *CQG* **13**, 731. *See* §9.5.

Baldwin, O.R. and Jeffery, G.B. (1926). The relativity theory of plane waves. *Proc. Roy. Soc. Lond. A* **111**, 95. *See* §24.5.

Bampi, F. and Cianci, R. (1979). Generalized axisymmetric spacetimes. *Commun. Math. Phys.* **70**, 69. *See* §§24.3, 24.4, 31.5.

Banerjee, A. (1970). Null electromagnetic fields in general relativity admitting timelike or null Killing vectors. *JMP* **11**, 51. *See* §18.6.

Banerjee, A.K. and Santos, N.O. (1984). Spatially homogeneous cosmological models. *GRG* **16**, 217. *See* §14.3.

Bardeen, J.M. (1970). A variational principle for rotating stars in general relativity. *Astrophys. J.* **162**, 71. *See* §19.3.

Barnes, A. (1972). Static perfect fluids in general relativity. *J. Phys. A* **5**, 374. *See* §§13.4, 18.6.

Barnes, A. (1973a). On Birkhoff's theorem in general relativity. *Commun. Math. Phys.* **33**, 75. *See* §§8.5, 15.4, 24.1.

Barnes, A. (1973b). On shearfree normal flows of a perfect fluid. *GRG* **4**, 105. *See* §§6.2, 16.2, 15.6, 33.3.

Barnes, A. (1973c). Space-times of embedding class one in general relativity. *GRG* **5**, 147. *See* §§37.4, 37.5.

Barnes, A. (1977). A class of algebraically general non-null Einstein–Maxwell fields. II. *J. Phys. A* **10**, 755. *See* §13.3.

Barnes, A. (1978). A class of homogeneous Einstein–Maxwell fields. *J. Phys. A* **11**, 1303. *See* §§13.2, 13.3.

Barnes, A. (1979). On space-times admitting a three-parameter isometry group with two-dimensional null orbits. *J. Phys. A* **12**, 1493. *See* §§13.1, 24.1.

Barnes, A. (1984). Shear-free flow of a perfect fluid, in *Classical general relativity*, eds. W.B. Bonnor, J.N. Islam and M.A.H. MacCallum, page 15 (Cambridge University Press, Cambridge). *See* §6.2.

Barnes, A. (1998). Symmetries of the Stephani universes. *CQG* **15**, 3061. *See* §37.4.

Barnes, A. (1999). On Stephani's rotating dust solutions. *CQG* **16**, 919. *See* §36.3.

Barnes, A. (2000). A comment on a paper by Carot et al. *CQG* **17**, 2605. *See* §19.1.

Barnes, A. and Rowlingson, R.R. (1989). Irrotational perfect fluids with a purely electric Weyl tensor. *CQG* **6**, 949. *See* §6.2.

Barnes, A. and Rowlingson, R.R. (1990). Killing vectors in conformally flat perfect fluid spacetimes. *CQG* **7**, 1721. *See* §37.4.

Barrabés, C. (1989). Singular hypersurfaces in general relativity: a unified description. *CQG* **6**, 581. *See* §3.8.

Barrabés, C. and Israel, W. (1991). Thin shells in general relativity and cosmology: the lightlike limit. *PRD* **43**, 1129. *See* §3.8.

Barrow, J.D. (1978). Quiescent cosmology. *Nature* **272**, 211. *See* §14.4.

Barrow, J.D. and Stein-Schabes, J. (1984). Inhomogeneous cosmologies with cosmological constant. *Phys. Lett. A* **103**, 315. *See* §33.3.

Bartrum, P.C. (1967). Null electromagnetic field in the form of spherical radiation. *JMP* **8**, 1464. *See* §28.2.

Basey, T.C. (1975). A note on the generation of twisting 'D' metrics from non-twisting 'D' metrics. *Z. Naturforsch.* **30a**, 1200. *See* §21.1.

Basu, A., Ganguly, S. and Ray, D. (1990). Perfect fluid in a static isotropic universe. *Int. J. Theor. Phys.* **29**, 435. *See* §18.6.

Batakis, N. and Cohen, J.M. (1972). Closed anisotropic cosmological models. *Ann. Phys.* (*USA*) **73**, 578. *See* §14.3.

Bayin, S.S. (1978). Solutions of Einstein's field equations for static fluid spheres. *PRD* **18**, 2745. *See* §16.1.

Bayin, S.S. and Krisch, J.P. (1986). Fluid sources for Bianchi I and III space-times. *JMP* **27**, 262. *See* §14.3.

Beck, G. (1925). Zur Theorie binärer Gravitationsf elder. *Z. Phys.* **33**, 713. *See* §22.3.

Beig, R. and Simon, W. (1980). Proof of a multipole conjecture due to Geroch. *Commun. Math. Phys.* **78**, 75. *See* §18.8.

Beig, R. and Simon, W. (1981). On the multipole expansion for stationary space-times. *Proc. Roy. Soc. London A* **376**, 333. *See* §18.8.

Beiglböck, W. (1964). Zur Theorie der infinitesimalen Holonomiegruppe in der Allgemeinen Relativitätstheorie. *Z. Phys.* **179**, 148. *See* §9.1.

Bel, L. (1959). Quelques remarques sur la classification de Petrov. *C. R. Acad. Sci.* (*Paris*) **248**, 2561. *See* Ch. 4.

Belinski, V.A. and Verdaguer, E. (2001). *Gravitational solitons* (Cambridge University Press, Cambridge). *See* §1.4, Ch. 23, §34.5.

Belinskii, V. (1979). Single-soliton cosmological waves. *Zh. Eks. Teor. Fiz.* **77**, 1239. *See* §10.11.

Belinskii, V. and Fargion, D. (1980a). The inverse scattering method in the theory of gravitation. *Rend. Sem. Mat. Univ. Politec. Torino* **38**, 1. *See* §10.11.

Belinskii, V. and Fargion, D. (1980b). Two-soliton waves in anisotropic cosmology. *Nuovo Cim. B* **59**, 143. *See* §34.5.

Belinskii, V. and Francaviglia, M. (1982). Solitonic gravitational waves in Bianchi II cosmologies. I. The general framework. *GRG* **14**, 213. *See* §34.5.

Belinskii, V. and Francaviglia, M. (1984). Solitonic gravitational waves in Bianchi II cosmologies. II. One solitonic perturbations. *GRG* **16**, 1189. *See* §34.5.

Belinskii, V. and Zakharov, V.E. (1978). Integration of the Einstein equations by the inverse scattering problem technique and the calculation of the exact soliton solutions. *Sov. Phys. JETP* **75**, 1953. *See* §§22.3, 34.5, 34.6.

Belinskii, V.A. and Zakharov, V.E. (1979). Stationary gravitational solitons with axial symmetry. *Sov. Phys. JETP* **77**, 3. *See* §34.5.

Bell, P. and Szekeres, P. (1972). Some properties of higher spin rest-mass zero fields in general relativity. *Int. J. Theor. Phys.* **6**, 111. *See* §§7.3, 7.6.

Bell, P. and Szekeres, P. (1974). Interacting electromagnetic shock waves in general relativity. *GRG* **5**, 275. *See* §25.5.

Benenti, S. (1997). Intrinsic characterization of the variable separation in the Hamilton–Jacobi equation. *JMP* **28**, 6578. *See* §35.3.

Benoit, P.M. and Coley, A.A. (1998). Spherically symmetric spacetimes and kinematic self-similarity. *CQG* **15**, 2397. *See* §35.4.

Berger, B.K., Eardley, D. and Olson, D.W. (1977). Note on the spacetimes of Szekeres. *PRD* **16**, 3086. *See* §§36.2, 36.3.

Bergmann, P.G., Cahen, M. and Komar, A.B. (1965). Spherically symmetric gravitational fields. *JMP* **6**, 1. *See* §15.4.

Bergstrom, L. and Goobar, A. (1999). *Cosmology and particle astrophysics* (Praxis and John Wiley and Sons, Chichester). *See* §14.1.

Bertotti, B. (1959). Uniform electromagnetic field in the theory of general relativity. *Phys. Rev.* **116**, 1331. *See* §§12.3, 21.1.

Bhatt, P.V. and Vaidya, S.K. (1991). Kerr–Schild type solutions of Einstein–Maxwell equations. *CQG* **8**, 1717. *See* §32.5.

Bhutani, O.P. and Singh, K. (1998). Generalized similarity solutions for the type D fluid in five-dimensional flat space. *JMP* **39**, 3203. *See* §37.4.

Białas, A. (1963). Electromagnetic waves in general relativity as the source of information. *Acta Phys. Polon.* **24**, 465. *See* §32.3.

Bianchi, L. (1898). Sugli spazii a tre dimensioni che ammettono un gruppo continuo di movimenti. *Soc. Ital. Sci. Mem. di Mat.* **11**, 267. *See* §8.2.

Bičák, J. (1997). Radiative spacetimes: exact approaches, in *Relativistic gravitation and gravitational radiation*, eds. J.-A. Marck and J.-P. Lasota (Cambridge University Press, Cambridge). *See* §17.2.

Bičák, J. (2000). Selected solutions of Einstein's field equations: their role in general relativity and astrophysics, in *Einstein's field equations and their physical implications. Selected essays in honour of Jürgen Ehlers.* Lecture notes in physics, vol. 540, ed. B.G. Schmidt, page 1 (Springer, Berlin). *See* Ch. 1.

Bičák, J. and Griffiths, J.B. (1994). Scattering and collision of gravitational waves in Friedmann–Robertson–Walker open universes. *PRD* **49**, 900. *See* §25.6.

Bičák, J. and Griffiths, J.B. (1996). Gravitational waves propagating into Friedmann–Robertson–Walker universes. *Ann. Phys.* (*USA*) **252**, 180. *See* §25.6.

Bičák, J., Hoenselaers, C. and Schmidt, B.G. (1983a). The solutions of the Einstein equations for uniformly accelerated particles without nodal singularities. I. Freely falling particles in external fields. *Proc. Roy. Soc. London A* **390**, 397. *See* §17.2.

Bičák, J., Hoenselaers, C. and Schmidt, B.G. (1983b). The solutions of the Einstein equations for uniformly accelerated particles without nodal singularities. II. Self-accelerating particles. *Proc. Roy. Soc. London A* **390**, 411. *See* §17.2.

Bičák, J., Lynden-Bell, D. and Katz, J. (1993). Relativistic disks as sources of static vacuum spacetimes. *PRD* **47**, 4334. *See* §20.2.

Bičák, J. and Perjés, Z. (1987). Asymptotic behaviour of Robinson–Trautman pure radiation solutions. *CQG* **4**, 595. *See* §28.3.

Bičák, J. and Podolsky, J. (1999). Gravitational waves in vacuum spacetimes with cosmological constant. I. Classification and geometrical properties of non-twisting type N solutions. *JMP* **40**, 4495. *See* §§31.8, 28.4.

Bičák, J. and Pravda, V. (1998). Curvature invariants in type N spacetime. *CQG* **15**, 1539. *See* §9.1.

Bičák, J. and Pravda, V. (1999). Spinning C metric: Radiative spacetime with accelerating, rotating black holes. *PRD* **60**, 044004. *See* §17.2.

Bičák, J. and Pravdová, A. (1998). Symmetries of asymptotically flat electrovacuum spacetimes and radiation. *JMP* **39**, 6011. *See* §17.2.

Bičák, J. and Schmidt, B.G. (1984). Isometries compatible with gravitational radiation. *JMP* **25**, 600. *See* §17.2.

Bičák, J. and Schmidt, B.G. (1989). Asymptotically flat radiative space-times with boost-rotation symmetry: the general structure. *PRD* **40**, 1827. *See* §17.2.

Bichteler, K. (1964). Äusserer Differentialkalkül für Spinorformen und Anwendung auf das allgemeine reine Gravitationsstrahlungsfeld. *Z. Phys.* **178**, 488. *See* §3.6.

Biech, T. and Das, A. (1990). On exact shearing perfect-fluid solutions of the nonstatic spherically symmetric Einstein field equations. *Can. J. Phys.* **68**, 1403. *See* §14.3.

Bilge, A.H. (1989). Non-twisting vacuum metrics with $D(\sigma/\rho) = 0$. *CQG* **6**, 823. *See* §26.4.

Bilge, A.H. (1990). An asymptotically flat vacuum solution. *GRG* **22**, 365. *See* §§22.3, 26.4.

Bilge, A.H. (1991). Integration of the radial Newman–Penrose equations in terms of ρ, σ and ξ^k. *CQG* **8**, 703. *See* §26.4.

Bilge, A.H. and Gürses, M. (1983). Generalized Kerr–Schild transformation, in *Group theoretical methods in physics. Proceedings of the XI international colloquium*, eds. M. Serdaroğlu and E. Ínönü, page 252 (Springer-Verlag, Berlin). *See* §32.5.

Bilyalov, R.F. (1964). Conformal transformation groups in gravitational fields. *Sov. Phys. – Dokl.* **8**, 878. *See* §35.4.

Birkhoff, G.D. (1923). *Relativity and modern physics* (Harvard Univ. Press, Cambridge, Mass). *See* §15.4.

Bluman, G.W. and Kumei, S. (1989). *Symmetries and differential equations.* Applied mathematical sciences, vol. 81 (Springer-Verlag, Berlin). *See* §10.2.

Bogoyavlenskii, O.I. (1980). *Methods of the qualitative theory of dynamical systems in astrophysics and gas dynamics* (Springer-Verlag, Berlin). *See* §13.2.

Bogoyavlensky, O.I. and Moschetti, G. (1982). The investigation of some self-similar solutions of Einstein's equations. *JMP* **23**, 1353. *See* §§11.4, 15.6, 23.2.

Bona, C. (1988a). Invariant conformal vectors in space-times admitting a group G_3 of motions acting on spacelike orbits S_2. *JMP* **29**, 2462. *See* §§11.4, 15.5, 35.4.

Bona, C. (1988b). A new proof of the generalized Birkhoff theorem. *JMP* **29**, 1440. *See* §15.4.

Bona, C. and Coll, B. (1985a). On the space-times admitting a synchronization of constant curvature. *JMP* **26**, 1583. *See* §36.3.

Bona, C. and Coll, B. (1985b). On the Stephani universe. *C. R. Acad. Sci.* (*Paris*) **301**, 613. *See* §37.4.

Bona, C. and Coll, B. (1991). Invariant conformal vectors in static spacetimes. *GRG* **23**, 99. *See* §35.4.

Bona, C. and Coll, B. (1992). Isometry groups of three-dimensional Riemannian metrics. *JMP* **33**, 267. *See* §9.4.

Bona, C. and Coll, B. (1994). Isometry groups of three-dimensional Lorentzian metrics. *JMP* **35**, 873. *See* §9.4.

Bona, C. and Palou, P. (1992). Dust metrics with comoving constant curvature slices. *JMP* **33**, 705. *See* §36.3.

Bona, C., Stela, J. and Palou, P. (1987a). On the generalization of Szafron solutions of Einstein field equations. *JMP* **28**, 654. *See* §§33.3, 36.2.

Bona, C., Stela, J. and Palou, P. (1987b). Perfect fluid spheres admitting flat 3-dimensional slices. *GRG* **19**, 179. *See* §§16.2, 36.2.

Bonanos, S. (1971). On the stability of the Taub universe. *Commun. math. phys.* **22**, 190. *See* §13.2.

Bonanos, S. (1976). Hamilton–Jacobi separable, axisymmetric, perfect-fluid solutions of Einstein's equations. *Commun. Math. Phys.* **49**, 53. *See* §21.2.

Bonanos, S. (1996). The 'post-Bianchi equations' and the integrability of the vacuum Einstein equations. *CQG* **13**, 2473. *See* §7.2.

Bonanos, S. (1998). A new spinor identity and the vanishing of certain Riemann tensor invariants. *GRG* **30**, 653. *See* §9.1.

Bonanos, S. and Kyriakopoulos, E. (1987). On a stationary asymptotically flat solution of the Ernst equation. *PRD* **36**, 1257. *See* §20.6.

Bonanos, S. and Kyriakopoulos, E. (1994). An algebraically general, stationary, axisymmetric, perfect fluid solution of Einstein's equations. *CQG* **11**, L23. *See* §21.2.

Bonanos, S. and Sklavenites, D. (1985). Relativistic, stationary, axisymmetric perfect fluids. I. Reduction to a system of two equations. *JMP* **26**, 2275. *See* §21.2.

Bondi, H. (1947). Spherically symmetrical models in general relativity. *Mon. Not. R. Astron. Soc.* **107**, 410. *See* §15.5.

Bonnor, W.B. (1953). Certain exact solutions of the equations of general relativity with an electrostatic field. *Proc. Phys. Soc. A* **66**, 145. *See* §22.2.

Bonnor, W.B. (1954). Static magnetic fields in general relativity. *Proc. Phys. Soc. Lond. A* **67**, 225. *See* §§21.1, 22.2.

Bonnor, W.B. (1961). Exact solutions of the Einstein–Maxwell equations. *Z. Phys.* **161**, 439. *See* §§19.5, 34.1.

Bonnor, W.B. (1966). An exact solution of the Einstein–Maxwell equations referring to a magnetic dipole. *Z. Phys.* **190**, 444. *See* §§21.1, 34.1.

Bonnor, W.B. (1969). The gravitational field of light. *Commun. Math. Phys.* **13**, 163. *See* §24.5.
Bonnor, W.B. (1976). Non-radiative solutions of Einstein's equations for dust. *Commun. Math. Phys.* **51**, 191. *See* §33.3.
Bonnor, W.B. (1977). A rotating dust cloud in general relativity. *J. Phys. A* **10**, 1673. *See* §21.2.
Bonnor, W.B. (1979a). A source for Petrov's homogeneous vacuum space-time. *Phys. Lett. A* **75**, 25. *See* §§12.2, 22.2.
Bonnor, W.B. (1979b). A three-parameter solution of the static Einstein–Maxwell equations. *J. Phys. A* **12**, 853. *See* §21.1.
Bonnor, W.B. (1980). The rigidly rotating relativistic dust cylinder. *J. Phys. A* **13**, 2121. *See* §22.2.
Bonnor, W.B. (1983). The sources of the vacuum C-metric. *GRG* **15**, 535. *See* §17.2.
Bonnor, W.B. (1985). An open recollapsing cosmological model with $\Lambda = 0$. *Mon. Not. Roy. Astron. Soc.* **217**, 597. *See* §15.5.
Bonnor, W.B. (1990). The C-metric in Bondi's coordinates. *CQG* **7**, L229. *See* §20.2.
Bonnor, W.B. (1992). Physical interpretation of vacuum solutions of Einstein's equations. I. Time-independent solutions. *GRG* **24**, 551. *See* §20.1.
Bonnor, W.B. (1996). Another photon rocket. *CQG* **13**, 277. *See* §28.3.
Bonnor, W.B. (2000). Non-radiative spacetimes. *CQG* **17**, 3935. *See* §21.1.
Bonnor, W.B. and Chamorro, A. (1990). Models of voids in the expanding universe. *Astrophys. J.* **361**, 21. *See* §15.5.
Bonnor, W.B. and Davidson, W. (1985). Petrov type *II* perfect fluid spacetimes with vorticity. *CQG* **2**, 775. *See* §33.1.
Bonnor, W.B. and Knutsen, H. (1993). Spherically symmetric perfect fluid solutions of Einstein's equations in noncomoving coordinates. *Int. J. Theor. Phys.* **32**, 1061. *See* §16.2.
Bonnor, W.B. and MacCallum, M.A.H. (1982). The Melnick–Tabensky solutions have high symmetry. *JMP* **23**, 1539. *See* §18.6.
Bonnor, W.B. and Martins, M.A.P. (1991). The interpretation of some static vacuum metrics. *CQG* **8**, 727. *See* §20.2.
Bonnor, W.B., Santos, N.O. and MacCallum, M.A.H. (1998). An exterior for the Gödel spacetime. *CQG* **15**, 357. *See* §22.2.
Bonnor, W.B., Sulaiman, A.H. and Tomimura, N. (1977). Szekeres's space-times have no Killing vectors. *GRG* **8**, 549. *See* §33.3.
Bonnor, W.B. and Swaminarayan, N.S. (1965). An exact stationary solution of Einstein's equations. *Z. Phys.* **186**, 222. *See* §17.2.
Bonnor, W.B. and Tomimura, N. (1976). Evolution of Szekeres's cosmological models. *Mon. Not. R. Astron. Soc.* **175**, 85. *See* §33.3.
Bonnor, W.B. and Vaidya, P.C. (1972). Exact solutions of the Einstein–Maxwell equations for an accelerated charge, in *General relativity* (*Papers in honour of J.L. Synge*), ed. L. O'Raifeartaigh, page 119 (Clarendon Press, Oxford). *See* §32.1.
Bonnor, W.B. and Vickers, P.A. (1981). Junction conditions in general relativity. *GRG* **13**, 29. *See* §3.8.
Boyer, R.H. and Lindquist, R.W. (1967). Maximal analytic extension of the Kerr metric. *JMP* **8**, 265. *See* §20.5.
Bradley, J.M. and Sviestins, E. (1984). Some rotating, time-dependent Bianchi type VIII cosmologies for heat flow. *GRG* **16**, 1119. *See* §14.3.
Bradley, M. (1986). Construction and invariant classification of perfect fluids in general relativity. *CQG* **3**, 317. *See* §§9.2, 9.4.
Bradley, M. (1988). Dust EPL cosmologies. *CQG* **5**, L15. *See* §14.1.
Bradley, M., Curir, A. and Francaviglia, M. (1991). Solitonic solutions on a Bianchi II background generated by SHEEP algebraic manipulation. *GRG* **23**, 1011. *See* §34.5.

Bradley, M. and Karlhede, A. (1990). On the curvature description of gravitational fields. *CQG* **7**, 449. *See* §§9.3, 9.4.

Bradley, M. and Marklund, M. (1996). Finding solutions to Einstein's equations in terms of invariant objects. *CQG* **13**, 3021. *See* §§9.4, 13.4.

Brans, C.H. (1965). Invariant approach to the geometry of spaces in general relativity. *JMP* **6**, 95. *See* §9.2.

Brans, C.H. (1975). Some restrictions on algebraically general vacuum metrics. *JMP* **16**, 1008. *See* §4.2.

Brans, C.H. (1977). Complete integrability conditions of the Einstein–Petrov equations, type *I*. *JMP* **18**, 1378. *See* §§7.3, 9.2.

Bray, M. (1971). Sur certaines solutions du problème intérieur en relativité générale. *C. R. Acad. Sci. (Paris) A* **272**, 841. *See* §§23.3, 35.4.

Bray, M. (1983). Some solutions of the Einstein system of field equations corresponding to Taub metric. *Ann. Inst. H. Poincaré A* **38**, 243. *See* §15.7.

Brdička, M. (1951). On gravitational waves. *Proc. Roy. Irish Acad. A* **54**, 137. *See* §24.5.

Breitenlohner, P. and Maison, D. (1987). On the Geroch group. *Ann. Inst. H. Poincaré* **46**, 215. *See* §10.5.

Breton, N., Garcia, A., Macias, A. and Yanez, G. (1998). Colliding plane waves in terms of Jacobi functions. *JMP* **39**, 6051. *See* §25.4.

Breton B., N. (1995). On some colliding plane wave solutions. *JMP* **36**, 1877. *See* §25.5.

Breton B., N., Feinstein, A. and Ibañez, J. (1992). Infinite-dimensional family of colliding wave solutions with variable polarization. *CQG* **9**, 2437. *See* §25.4.

Breton B., N. and García D., A. (1986a). Magnetic generalizations of the Reissner–Nordstrom class of metrics and Bertotti–Robinson solutions. *Nuovo Cim. B* **91**, 83. *See* §34.1.

Breton B., N. and García D., A. (1986b). The most general magnetized Kerr–Newman metric. *JMP* **27**, 562. *See* §34.1.

Breton B., N. and Manko, V.S. (1995). A binary system of 'antisymmetric' Kerr–Newman masses. *CQG* **12**, 1969. *See* §34.6.

Breuer, R.A. (1975). *Gravitational perturbation theory and synchrotron radiation.* Lecture notes in physics, vol. 44 (Springer-Verlag, Berlin). *See* §7.4.

Brickell, F. and Clark, R.S. (1970). *Differentiable manifolds: an introduction* (Van Nostrand Reinhold, London). *See* §8.1.

Brill, D.R. (1964). Electromagnetic fields in a homogeneous, nonisotropic universe. *Phys. Rev. B* **133**, 845. *See* §§13.3, 21.1.

Brinkmann, H.W. (1925). Einstein spaces which are mapped conformally on each other. *Math. Ann.* **18**, 119. *See* §§3.7, 24.5.

Bronnikov, K.A. (1979). Static fluid cylinders and plane layers in general relativity. *J. Phys. A* **12**, 201. *See* §22.2.

Bronnikov, K.A. (1980). Gravitational and sound waves in stiff matter. *J. Phys. A* **13**, 3455. *See* §23.3.

Bronnikov, K.A. and Kovalchuk, M.A. (1979). Properties of static fluid cylinders and plane layers in general relativity. *GRG* **11**, 343. *See* §15.7.

Buchdahl, H.A. (1954). Reciprocal static solutions of the equations $G_{\mu\gamma} = 0$. *Quart. J. Math. Oxford* **5**, 116. *See* §34.1.

Buchdahl, H.A. (1956). Reciprocal static solutions of the equations of the gravitational field. *Aust. J. Phys.* **9**, 13. *See* §§10.11, 16.1.

Buchdahl, H.A. (1959). General relativistic fluid spheres. *Phys. Rev.* **116**, 1027. *See* §16.1.

Buchdahl, H.A. (1964). A relativistic fluid sphere resembling the Emden polytrope of index 5. *Astrophys. J.* **140**, 1512. *See* §16.1.

Buchdahl, H.A. (1967). General-relativistic fluid spheres. III. A static gaseous model. *Astrophys. J.* **147**, 310. *See* §16.1.

Buchdahl, H.A. (1984). Remark on a family of static relativistic stellar models. *CQG* **1**, 301. *See* §16.1.

Buchdahl, H.A. (1985). Isotropic coordinates and Schwarzschild metric. *Int. J. Theor. Phys.* **24**, 731. *See* §15.4.

Buchdahl, H.A. and Land, W.J. (1968). The relativistic incompressible sphere. *J. Aust. Math. Soc.* **8**, 6. *See* §16.1.

Bueken, P. and Vanhecke, L. (1997). Examples of curvature homogeneous Lorentz metrics. *CQG* **14**, L93. *See* §§9.1, 9.2.

Bugalho, M.H. (1987). Orthogonality transitivity and cosmologies with a non-Abelian two-parameter isometry group. *CQG* **4**, 1043. *See* §§17.1, 23.3.

Burlankov, D.E. (1993). Analytic solutions for static spherically symmetric distribution of liquid in its gravitational self-field. *Theor. Math. Phys.* **94**, 455. *See* §16.1.

Cahen, M. (1964). On a class of homogeneous spaces in general relativity. *Bull. Acad. Roy. Belg. Cl. Sci.* **50**, 972. *See* §12.5.

Cahen, M. and Debever, R. (1965). Sur le théorème de Birkhoff. *C. R. Acad. Sci.* (*Paris*) **260**, 815. *See* §15.4.

Cahen, M., Debever, R. and Defrise, L. (1967). A complex vectorial formalism in general relativity. *J. Math. Mech.* **16**, 761. *See* §3.5.

Cahen, M. and Defrise, L. (1968). Lorentzian 4-dimensional manifolds with "local isotropy". *Commun. Math. Phys.* **11**, 56. *See* §§11.2, 12.3, 13.1, 13.3.

Cahen, M. and Leroy, J. (1965). Ondes gravitationnelles en présence d'un champ électromagnétique. *Bull. Acad. Roy. Belg. Cl. Sci.* **51**, 996. *See* §§26.1, 31.7.

Cahen, M. and Leroy, J. (1966). Exact solutions of Einstein–Maxwell equations. *J. Math. Mech.* **16**, 501. *See* §§26.1, 31.7, 37.5.

Cahen, M. and Sengier, J. (1967). Espaces de classe D admettant un champ électromagnétique. *Bull. Acad. Roy. Belg. Cl. Sci.* **53**, 801. *See* §28.2.

Cahen, M. and Spelkens, J. (1967). Espaces de type III solutions des équations de Maxwell–Einstein. *Bull. Acad. Roy. Belg. Cl. Sci.* **53**, 817. *See* §§26.1, 31.5.

Cahill, M.E. and Taub, A.H. (1971). Spherically symmetric similarity solutions of the Einstein field equations for a perfect fluid. *Commun. math. phys.* **21**, 1. *See* §11.3.

Calvert, G. and Woodhouse, N.M.J. (1996). Painlevé transcendents and Einstein's equation. *CQG* **13**, L33. *See* §20.6.

Campbell, J.E. (1926). *A course of differential geometry* (Clarendon, Oxford). *See* §37.3.

Capovilla, R. (1994). No new symmetries of the vacuum Einstein equations. *PRD* **49**, 879. *See* §10.3.

Carigi, L. and Herrera, L. (1986). Killing vectors and Maxwell collineations in general relativity. *Can. J. Phys.* **64**, 1496. *See* §11.1.

Carlson, Jr., G.T. and Safko, J.L. (1980). Canonical forms for axial symmetric space-times. *Ann. Phys.* (*USA*) **128**, 131. *See* §19.1.

Carmeli, M. (1977). *Group theory and general relativity* (McGraw-Hill, New York). *See* §26.4.

Carmeli, M. and Charach, Ch. (1980). Inhomogeneous generalizations of some Bianchi models. *Phys. Lett. A* **75**, 333. *See* §22.3.

Carmeli, M., Charach, Ch. and Feinstein, A. (1983). Inhomogeneous mixmaster universes: Some exact solutions. *Ann. Phys.* (*USA*) **150**, 392. *See* §10.11.

Carmeli, M., Charach, Ch. and Malin, S. (1981). Survey of cosmological models with gravitational, scalar and electromagnetic waves. *Phys. Rep.* **76**, 79. *See* §§10.11, 22.3, Ch. 23.

Carmeli, M. and Kaye, M. (1976). Transformation laws of the Newman–Penrose field variables. *Ann. Phys.* (*USA*) **99**, 188. *See* §7.1.

Carminati, J. (1981). An investigation of axially symmetric electrovac solutions. *GRG* **13**, 1185. *See* §21.1.

Carminati, J. (1987). Shear-free perfect fluids in general relativity. I. Petrov type *N* Weyl tensor. *JMP* **28**, 1848. *See* §6.2.

Carminati, J. (1988). Type-N, shear-free, perfect-fluid spacetimes with a barotropic equation of state. *GRG* **20**, 1239. *See* §33.4.

Carminati, J. and Cooperstock, F.I. (1983). Coordinate modelling for static axially symmetric electrovac metrics. *J. Phys. A* **16**, 3867. *See* §21.1.

Carminati, J. and Cooperstock, F.I. (1992). Herlt metrics and gravitational-electrostatic balance in general relativity. *GRG* **24**, 881. *See* §21.1.

Carminati, J. and Cyganowski, S. (1997). Shearfree perfect fluids in general relativity. IV. Petrov type *III* spacetimes. *CQG* **13**, 1167. *See* §§6.2, 33.4.

Carminati, J. and McIntosh, C.B.G. (1980). A non-static Einstein–Maxwell solution. *J. Phys. A* **13**, 953. *See* §11.4.

Carminati, J. and McLenaghan, R.G. (1991). Algebraic invariants of the Riemann tensor in a four-dimensional Lorentzian space. *JMP* **32**, 3135. *See* §9.1.

Carminati, J. and Wainwright, J. (1985). Perfect-fluid space-times with type-D Weyl tensor. *GRG* **17**, 853. *See* §33.3.

Carot, J. (1990). Exact solutions for space-times admitting nonnull special conformal Killing vectors. *GRG* **22**, 1135. *See* §35.4.

Carot, J. (2000). Some developments on axial symmetry. *CQG* **17**, 2675. *See* §19.1.

Carot, J., Coley, A.A. and Sintes, A.M. (1996). Space-times admitting a three-dimensional conformal group. *GRG* **28**, 311. *See* §§23.3, 35.4.

Carot, J. and da Costa, J. (1991). Perfect fluid spacetimes admitting curvature collineations. *GRG* **23**, 1057. *See* §35.4.

Carot, J. and da Costa, J. (1993). On the geometry of warped spacetimes. *CQG* **10**, 461. *See* §35.1.

Carot, J., Mas, L. and Sintes, A.M. (1994). Space-times admitting a three-parameter similarity group. *JMP* **35**, 3560. *See* §§23.1, 23.3.

Carot, J., Senovilla, J.M.M. and Vera, R. (1999). On the definition of cylindrical symmetry. *CQG* **16**, 3025. *See* §22.1.

Carot, J. and Sintes, A.M. (1997). Homothetic perfect fluid spacetimes. *CQG* **14**, 1183. *See* §§15.7, 23.1, 23.2, 23.3.

Carot, J. and Verdaguer, E. (1989). Generalised soliton solutions of the Weyl class. *CQG* **6**, 845. *See* §34.5.

Carr, B.J. and Coley, A.A. (1999). Self-similarity in general relativity. *CQG* **16**, R31. *See* §11.3.

Carr, B.J. and Verdaguer, E. (1983). Soliton solutions and cosmological gravitational waves. *PRD* **28**, 2995. *See* §34.5.

Cartan, È (1945). *Les systèmes différentiels extérieurs et leurs applications géométriques* (Herrmann, Paris). *See* §10.4.

Cartan, È (1946). *Leçons sur la géométrie des espaces de Riemann* (Gauthier-Villars, Paris). *See* §9.2.

Carter, B. (1968a). Hamilton–Jacobi and Schrödinger separable solutions of Einstein's equations. *Commun. Math. Phys.* **10**, 280. *See* §21.1.

Carter, B. (1968b). A new family of Einstein spaces. *Phys. Lett. A* **26**, 399. *See* §§15.4, 20.5, 21.1, 31.5.

Carter, B. (1970). The commutation property of a stationary, axisymmetric system. *Commun. Math. Phys.* **17**, 233. *See* §19.1.

Carter, B. (1972). Black hole equilibrium states, in *Black holes* (*Les Houches lectures*), eds. B. DeWitt and C. DeWitt, page 57 (Gordon and Breach, New York). *See* §§18.4, 19.2.

Carter, B. and Henriksen, R.N. (1989). A covariant characterisation of kinematic self-similarity. *Ann. de Physique* **14** (colloq. 1), 47. *See* §§11.3, 35.4.

Castejon-Amenedo, J. and Manko, V.S. (1990a). On a stationary rotating mass with an arbitrary multipole structure. *CQG* **7**, 779. *See* §21.1.

Castejon-Amenedo, J. and Manko, V.S. (1990b). Superposition of the Kerr metric with the generalized Erez–Rosen solution. *PRD* **41**, 2018. *See* §34.3.

Catenacci, R., Marzuoli, A. and Salmistraro, F. (1980). A note on Killing vectors in algebraically special vacuum space-times. *GRG* **12**, 575. *See* §8.4.

Caviglia, G., Zordan, C. and Salmistraro, F. (1982). Equation of geodesic deviation and Killing tensors. *Int. J. Theor. Phys.* **21**, 391. *See* §35.3.

Centrella, J. and Matzner, R.A. (1982). Colliding gravitational waves in expanding cosmologies. *PRD* **25**, 930. *See* §25.3.

Centrella, J.M., Shapiro, S.L., Evans, C.R., Hawley, J.F. and Teukolsky, S.A. (1986). Test-bed calculations in numerical relativity, in *Dynamical spacetimes and numerical relativity*, ed. J.M. Centrella, page 326 (Cambridge University Press, Cambridge). *See* §1.1.

Cespedes, J. and Verdaguer, E. (1987). Cosmological Einstein–Rosen metrics and generalised soliton solutions. *CQG* **4**, L7. *See* §34.5.

Chakrabarti, S.K. (1988). Spacetime with self-gravitating thick disc. *J. Astrophys. Astron.* **9**, 49. *See* §20.2.

Chamorro, A., Manko, V.S. and Denisova, T.E. (1991). New exact solution for the exterior gravitational field of a charged spinning mass. *PRD* **44**, 3147. *See* §34.1.

Chamorro, A., Manko, V.S. and Denisova, T.E. (1993). Exterior gravitational field of a charged magnetised axisymmetric mass. *Nuovo Cim. B* **108, ser.2**, 905. *See* §34.1.

Chandrasekhar, S. (1978). The Kerr metric and stationary axisymmetric gravitational fields. *Proc. Roy. Soc. Lond. A* **358**, 405. *See* §§19.5, 20.5.

Chandrasekhar, S. (1986). Cylindrical waves in general relativity. *Proc. Roy. Soc. Lond. A* **408**, 209. *See* §22.3.

Chandrasekhar, S. (1988). On Weyl's solution for space-times with two commuting Killing fields. *Proc. Roy. Soc. London A* **415**, 329. *See* §25.3.

Chandrasekhar, S. and Ferrari, V. (1984). On the Nutku–Halil solution for colliding impulsive gravitational waves. *Proc. Roy. Soc. London A* **396**, 55. *See* §§25.2, 25.4.

Chandrasekhar, S. and Ferrari, V. (1987). On the dispersion of cylindrical impulsive gravitational waves. *Proc. Roy. Soc. London A* **412**, 75. *See* §22.3.

Chandrasekhar, S. and Xanthopoulos, B.C. (1985a). On colliding waves in the Einstein–Maxwell theory. *Proc. Roy. Soc. London A* **398**, 223. *See* §25.5.

Chandrasekhar, S. and Xanthopoulos, B.C. (1985b). On the collision of impulsive gravitational waves when coupled with fluid motions. *Proc. Roy. Soc. London A* **402**, 37. *See* §25.6.

Chandrasekhar, S. and Xanthopoulos, B.C. (1985c). Some exact solutions of gravitational waves coupled with fluid motions. *Proc. Roy. Soc. London A* **402**, 205. *See* §25.6.

Chandrasekhar, S. and Xanthopoulos, B.C. (1986a). A new type of singularity created by colliding gravitational waves. *Proc. Roy. Soc. London A* **408**, 175. *See* §25.4.

Chandrasekhar, S. and Xanthopoulos, B.C. (1986b). On the collision of impulsive gravitational waves when coupled with null dust. *Proc. Roy. Soc. London A* **403**, 189. *See* §25.6.

Chandrasekhar, S. and Xanthopoulos, B.C. (1987a). The effect of sources on horizons that may develop when plane gravitational waves collide. *Proc. Roy. Soc. London A* **414**, 1. *See* §§25.5, 25.6.

Chandrasekhar, S. and Xanthopoulos, B.C. (1987b). On colliding waves that develop time-like singularities: a new class of solutions of the Einstein–Maxwell equations. *Proc. Roy. Soc. London A* **410**, 311. *See* §25.5.

Charach, Ch. (1979). Electromagnetic Gowdy universe. *PRD* **19**, 3516. *See* §22.4.

Charach, Ch. and Malin, S. (1979). Cosmological model with gravitational and scalar waves. *PRD* **19**, 1058. *See* §23.3.

Chatterjee, S. and Banerji, S. (1979). Axially symmetric stationary electrovac solutions. *GRG* **11**, 79. *See* §21.1.

Chau, Ling-Lie and Ge, Mo-Lin (1989). Kac–Moody algebra from infinitesimal Riemann–Hilbert transform. *JMP* **30**, 166. *See* §10.7.

Chaudhuri, S. and Das, K.C. (1996). On the structure and multipole moments of axially symmetric stationary metrics. *Pramana, J. Phys.* **46**, 17. *See* §18.8.

Chazy, J. (1924). Sur la champ de gravitation de deux masses fixes dans la théorie de la relativité. *Bull. Soc. Math. France* **52**, 17. *See* §20.2.

Chinea, F.J. (1983). New Bäcklund transformations and superposition principle for gravitational fields with symmetries. *Phys. Rev. Lett.* **50**, 221. *See* §34.7.

Chinea, F.J. (1984). Vector Bäcklund transformations and associated superposition principle, in *Solutions of Einstein's equations: Techniques and results.* Lecture notes in physics, vol. 205, eds. C. Hoenselaers and W. Dietz, page 55 (Springer, Berlin). *See* §34.7.

Chinea, F.J. (1993). A differentially rotating perfect fluid. *CQG* **10**, 2539. *See* §21.2.

Chinea, F.J. and González-Romero, L.M. (1990). Interior gravitational field of a stationary, axially symmetric perfect fluid in irrotational motion. *CQG* **7**, L99. *See* §21.2.

Chinea, F.J. and González-Romero, L.M. (1992). A differential form approach for rotating perfect fluids in general relativity. *CQG* **9**, 1271. *See* §§19.6, 21.2.

Chitre, D.M. (1980). Stationary, axially symmetric solutions of Einstein's equations, in *Gravitation, quanta and the universe*, eds. A.R. Prasanna, J.V. Narlikar and C.V. Vishveshwara, page 69 (Wiley Eastern Ltd., New Delhi). *See* §34.3.

Chitre, D.M., Güven, R. and Nutku, Y. (1975). Static cylindrically symmetric solutions of the Einstein–Maxwell equations. *JMP* **16**, 475. *See* §22.2.

Choquet-Bruhat, Y., DeWitt-Morette, C. and Dillard-Bleick, M. (1991). *Analysis, manifolds and physics: Basics* (North Holland Publ. Co., Amsterdam). *See* §§2.1, 35.4.

Christoffel, E.B. (1869). Über die transformation der homogenen Differentialausdrücke zweiten Grades. *Crelle's J.* **70**, 46. *See* §§9.1, 9.2.

Chruściel, P.T. (1991). Semi-global existence and convergence of solutions of the Robinson–Trautman (2-dimensional Calabi) equation. *Commun. Math. Phys.* **137**, 289. *See* §28.1.

Churchill, R.V. (1932). Canonical forms for symmetric linear vector functions in pseudo-euclidean space. *Trans. Amer. Math. Soc.* **34**, 784. *See* §5.1.

Clarke, C.J.S. (1970). On the isometric global embedding of pseudo-Riemannian manifolds. *Proc. Roy. Soc. Lond. A* **314**, 417. *See* §37.1.

Clarke, C.J.S. and Dray, T. (1987). Junction conditions for null hypersurfaces. *CQG* **4**, 265. *See* §3.8.

Clément, G. (2000). Generating rotating Einstein–Maxwell fields. *Ann. Phys. (Germany)* **9**, SI42. *See* §34.1.

Cocke, W.J. (1989). Table for constructing the spin coefficients in general relativity. *PRD* **40**, 650. *See* §7.1.

Cohn, P.M. (1957). *Lie groups* (Cambridge Univ. Press, Cambridge). *See* §8.1.

Coley, A.A. (1991). Fluid spacetimes admitting a conformal Killing vector parallel to the velocity vector. *CQG* **8**, 955. *See* §35.4.

Coley, A.A. (1997a). Kinematic self-similarity. *CQG* **14**, 87. *See* §35.4.

Coley, A.A. (1997b). Self-similarity and cosmology, in *Proceedings of the 6th Canadian conference on general relativity and relativistic astrophysics.* Fields Institute Communications, vol. 15, eds. S. Braham, J. Gegenberg and R. McKellar, page 19 (Amer. Math. Soc., Providence, Rhode Island). *See* §11.3.

Coley, A.A. and Czapor, S.R. (1991). General analysis of perfect fluid spherically symmetric ICKV spacetimes. *CQG* **8**, L147. *See* §35.4.

Coley, A.A. and Czapor, S.R. (1992). Plane symmetric spacetimes admitting inheriting conformal Killing vector fields. *CQG* **9**, 1787. *See* §§15.6, 35.4.

Coley, A.A. and Tupper, B.O.J. (1989). Special conformal Killing vector space-times and symmetry inheritance. *JMP* **30**, 2616. *See* §35.4.

Coley, A.A. and Tupper, B.O.J. (1990a). Spacetimes admitting inheriting conformal Killing vector fields. *CQG* **7**, 1961. *See* §35.4.

Coley, A.A. and Tupper, B.O.J. (1990b). Spherically symmetric spacetimes admitting inheriting conformal Killing vector fields. *CQG* **12**, 1111. *See* §§15.6, 35.4.

Coley, A.A. and Tupper, B.O.J. (1991). Fluid spacetimes admitting covariantly constant vectors and tensors. *GRG* **23**, 1113. *See* §35.1.

Coll, B. (1975). Sur l'invariance du champ électromagnétique dans un espace-temps d'Einstein–Maxwell admettant un groupe d'isométries. *C. R. Acad. Sci. (Paris) A* **280**, 1773. *See* §11.1.

Coll, B. and Ferrando, J.J. (1989). Thermodynamic perfect fluid. Its Rainich theory. *JMP* **30**, 2918. *See* §5.5.

Collins, C.B. (1971). More qualitative cosmology. *Commun. Math. Phys.* **23**, 137. *See* §§13.2, 13.3, 14.3, 14.4.

Collins, C.B. (1972). Qualitative magnetic cosmology. *Commun. Math. Phys.* **27**, 37. *See* §§13.3, 14.4.

Collins, C.B. (1976). Complex potential equations II: an application to general relativity. *Math. Proc. Camb. Phil. Soc.* **80**, 349. *See* §21.1.

Collins, C.B. (1977a). Global structure of the 'Kantowski-Sachs' cosmological models. *JMP* **18**, 2116. *See* §§8.2, 13.1.

Collins, C.B. (1977b). Static stars: Some mathematical curiosities. *JMP* **18**, 1374. *See* §11.3.

Collins, C.B. (1979). Intrinsic symmetries in general relativity. *GRG* **10**, 925. *See* Ch. 36, §36.2.

Collins, C.B. (1984). Shear-free perfect fluids with zero magnetic Weyl tensor. *JMP* **25**, 995. *See* §§6.2, 35.4.

Collins, C.B. (1985). Static relativistic perfect fluids with spherical, plane, or hyperbolic symmetry. *JMP* **26**, 2268. *See* §15.7.

Collins, C.B. (1986). Shear-free fluids in general relativity. *Can. J. Phys.* **64**, 191. *See* §6.2.

Collins, C.B. (1991). Higher symmetries in a class of cosmological models. *GRG* **23**, 321. *See* §§14.4, 23.1.

Collins, C.B., Glass, E.N. and Wilkinson, D.A. (1980). Exact spatially homogeneous cosmologies. *GRG* **12**, 805. *See* §14.3.

Collins, C.B. and Hawking, S.W. (1973). Why is the universe isotropic? *Astrophys. J.* **180**, 317. *See* §13.2.

Collins, C.B. and Lang, J.M. (1986). Singularities in self-similar spacetimes. *CQG* **3**, 1143. *See* §8.7.

Collins, C.B. and Lang, J.M. (1987). A class of self-similar space-times, and a generalisation. *CQG* **4**, 61. *See* §§11.4, 15.6, 15.7, 16.2, 35.4, 37.4.

Collins, C.B. and Stewart, J.M. (1971). Qualitative cosmology. *Mon. Not. R. Astron. Soc.* **153**, 419. *See* §14.3.

Collins, C.B. and Wainwright, J. (1983). On the role of shear in general relativistic cosmological and stellar models. *PRD* **27**, 1209. *See* §§6.2, 14.3, 15.6.

Collins, J.M. and d'Inverno, R.A. (1993). The Karlhede classification of type-D nonvacuum spacetimes. *CQG* **10**, 343. *See* §9.2.

Collins, J.M., d'Inverno, R.A. and Vickers, J.A. (1991). Upper bounds for the Karlhede classification of type D vacuum spacetimes. *CQG* **8**, L215. *See* §9.2.

Collinson, C.D. (1964). Symmetry properties of Harrison space-times. *Proc. Camb. Phil. Soc.* **60**, 259. *See* §17.3.

Collinson, C.D. (1966). Empty space-times of embedding class two. *JMP* **7**, 608. *See* §37.5.

Collinson, C.D. (1967). Empty space-times algebraically special on a given world line or hypersurface. *JMP* **8**, 1547. *See* §7.6.

Collinson, C.D. (1968a). Einstein–Maxwell fields of embedding class one. *Commun. Math. Phys.* **8**, 1. *See* §37.4.

Collinson, C.D. (1968b). Embedding of the plane-fronted waves and other space-times. *JMP* **9**, 403. *See* §§37.1, 37.5, 37.6.

Collinson, C.D. (1970). Curvature collineations in empty space-times. *JMP* **11**, 818. *See* §35.4.

Collinson, C.D. (1974). The existence of Killing tensors in empty space-times. *Tensor* **28**, 173. *See* §35.3.

Collinson, C.D. (1976a). On the relationship between Killing tensors and Killing–Yano tensors. *Int. J. Theor. Phys.* **15.**, 311. *See* §35.3.

Collinson, C.D. (1976b). The uniqueness of the Schwarzschild interior metric. *GRG* **7**, 419. *See* §21.2.

Collinson, C.D. (1989). A comment on the integrability conditions of the conformal Killing equation. *GRG* **21**, 979. *See* §35.4.

Collinson, C.D. and Dodd, R.K. (1969). Petrov classification of stationary axisymmetric empty space-time. *Nuovo Cim. B* **62**, 229. *See* §19.5.

Collinson, C.D. and Dodd, R.K. (1971). Symmetries of stationary axisymmetric empty space-times. *Nuovo Cim. B* **3**, 281. *See* §38.2.

Collinson, C.D. and French, D.C. (1967). Null tetrad approach to motions in empty space-time. *JMP* **8**, 701. *See* §§13.3, 26.4, 28.1, 38.2.

Collinson, C.D. and Vaz, E.G.L.R. (1982). Mappings of empty space-times leaving the curvature tensor invariant. *GRG* **14**, 5. *See* §9.2.

Cooperstock, F.I. and de la Cruz, V. (1979). Static and stationary solutions of the Einstein–Maxwell equations. *GRG* **10**, 681. *See* §21.1.

Coquereaux, R. and Grossmann, A. (1982). Analytic discussion of spatially closed Friedman universes with cosmological constant and radiation pressure. *Ann. Phys. (USA)* **143**, 296. *See* §14.2.

Cornish, F.H.J. and Uttley, W.J. (1995). The interpretation of the C metric. The vacuum case. *GRG* **27**, 439. *See* §17.2.

Cosgrove, C.M. (1977). New family of exact stationary axisymmetric gravitational fields generalising the Tomimatsu–Sato solutions. *J. Phys. A* **10**, 1481. *See* §20.6.

Cosgrove, C.M. (1978a). A new formulation of the field equations for the stationary axisymmetric vacuum gravitational field. I. General theory. *J. Phys. A* **11**, 2389. *See* §§19.5, 20.6.

Cosgrove, C.M. (1978b). A new formulation of the field equations for the stationary axisymmetric vacuum gravitational field. II. Separable solutions. *J. Phys. A* **11**, 2405. *See* §§20.6, 35.3.

Cosgrove, C.M. (1980). Relationship between the group-theoretic and soliton-theoretic techniques for generating stationary axisymmetric gravitational solutions. *JMP* **21**, 2417. *See* §34.7.

Cosgrove, C.M. (1981). Bäcklund transformations in the Hauser–Ernst formalism for stationary axisymmetric spacetimes. *JMP* **22**, 2624. *See* §34.7.

Cosgrove, C.M. (1982a). Continuous groups and Bäcklund transformations generating asymptotically flat solutions, in *Proceedings of the second Marcel Grossmann meeting on general relativity*, ed. R. Ruffini, page 287 (North-Holland, Amsterdam). *See* §34.7.

Cosgrove, C.M. (1982b). Relationship between the inverse scattering techniques of Belinskii–Zakharov and Hauser–Ernst in general relativity. *JMP* **23**, 615. *See* §34.7.

Covarrubias, G.M. (1984). A class of Szekeres space-times with cosmological constant. *Astrophys. Space Sci.* **103**, 401. *See* §33.3.

Cox, D. and Flaherty, E.J. (1976). A conventional proof of Kerr's theorem. *Commun. Math. Phys.* **47**, 75. *See* §32.1.

Cox, D. and Kinnersley, W. (1979). Yet another formulation of the Einstein equations for stationary axisymmetry. *JMP* **20**, 1225. *See* §§19.5, 22.3.

Crade, R.F. and Hall, G.S. (1982). Second order symmetric tensors and quadratic surfaces in general relativity. *Acta Phys. Polon. B* **13**, 405. *See* §5.1.

Crampin, M. and Pirani, F.A.E. (1986). *Applicable differential geometry.* London Mathematical Society lecture notes, vol. 59 (Cambridge University Press, Cambridge). *See* §2.11.

Cruzate, J., Diaz, M., Gleiser, R.J and Pullin, J.A. (1988). Soliton collision in cosmologies with matter. *CQG* **5**, 883. *See* §25.6.

Curry, C. and Lake, K. (1991). Vacuum solutions of Einstein's equations in double-null coordinates. *CQG* **8**, 237. *See* §15.4.

Curzon, H.E.J. (1924). Cylindrical solutions of Einstein's gravitational equations. *Proc. London Math. Soc.* **23**, 477. *See* §20.2.

Czapor, S.R. and Coley, A.A. (1995). Diagonal G_2 spacetimes admitting inheriting conformal Killing vector fields. *CQG* **12**, 1995. *See* §§14.4, 23.3, 35.4.

Czapor, S.R. and McLenaghan, R.G. (1982). Orthogonal transitivity, invertibility and null geodesic separability in type D vacuum solutions of Einstein's field equations with cosmological constant. *JMP* **23**, 2159. *See* §26.2.

da Costa, J. and Vaz, E.G.L.R. (1992). Vacuum type N space-times admitting homothetic vector fields with isolated fixed points. *GRG* **24**, 745. *See* §8.7.

Dąbrowski, M. and Stelmach, J. (1986). Analytic solutions of Friedman equation for spatially opened universes with cosmological constant and radiation pressure. *Ann. Phys. (USA)* **166**, 422. *See* §14.2.

Daftardar-Gejji, V. (1998). A generalization of Brinkmann's theorem. *GRG* **30**, 695. *See* §3.7.

Dagotto, A.D., Gleiser, R.J. and Nicasio, C.O. (1991). On the Khan–Penrose construction for colliding electro-vacuum plane waves. *CQG* **8**, 2085. *See* §25.5.

Dagotto, A.D., Gleiser, R.J. and Nicasio, C.O. (1993). Two-soliton solutions of the Einstein–Maxwell equations. *CQG* **10**, 961. *See* §34.5.

Daishev, R.A. (1984). Homogeneous solutions of Einstein's equations for perfect liquid. *Ukrayins'kyi Fiz. Zh.* **29**, 1163. *See* §13.4.

Dale, P. (1978). Axisymmetric gravitational fields: a nonlinear differential equation that admits a series of exact eigenfunction solutions. *Proc. Roy. Soc. Lond. A* **362**, 463. *See* §20.6.

Darmois, G. (1927). *Les équations de la gravitation einsteinienne.* Mémorial des sciences mathématique, part XXV (Gauthier-Villars, Paris). *See* §§3.8, 20.2.

Das, A. (1979). On the static Einstein–Maxwell field equations. *JMP* **20**, 740. *See* §18.6.

Das, K.C. (1980a). Axially symmetric solutions in general relativity. *J. Phys. A* **13**, 2985. *See* §21.1.

Das, K.C. (1980b). Electrovac solution. *J. Phys. A* **13**, 223. *See* §21.1.

Das, K.C. (1983). New sets of asymptotically flat static and stationary solutions. *PRD* **27**, 322. *See* §20.6.

Das, K.C. (1985). Odd-soliton solutions of the Einstein equations in a vacuum. *PRD* **31**, 927. *See* §34.5.

Das, K.C. and Chaudhuri, S. (1991). Soliton solution of Einstein field equations from nondiagonal seed. *Indian J. Pure Appl. Math.* **22**, 963. *See* §34.5.

Das, K.C. and Chaudhuri, S. (1993). Astrophysically significant solutions of the Einstein and Einstein–Maxwell equations from the Laplace's seed. *Pramana, J. Phys.* **40**, 277. *See* §21.1.

Datt, B. (1938). Über eine Klasse von Lösungen der Gravitationsgleichungen der Relativität. *Z. Phys.* **108**, 314. *See* §15.5.

Datta, B.K. (1965). Homogeneous nonstatic electromagnetic fields in general relativity. *Nuovo Cim.* **36**, 109. *See* §13.3.

Datta, B.K. and Raychaudhuri, A.K. (1968). Stationary electromagnetic fields in general relativity. *J. Math. Phys* **9**, 1715. *See* §22.2.

Dautcourt, G. (1964). Gravitationsfelder mit isotropem Killingvektor, in *Relativistic theories of gravitation*, ed. L. Infeld, page 300 (Pergamon Press, Oxford). *See* §24.4.

Davidson, W. (1987). A note on plane-symmetric static perfect-fluid spacetimes. *CQG* **4**, 1469. *See* §15.7.

Davidson, W. (1988). A non-stationary generalised Robinson–Trautman solution. *CQG* **5**, 147. *See* §15.7.

Davidson, W. (1989a). A plane-symmetric static solution for perfect fluid. *Nuovo Cim. B* **103**, 217. *See* §15.7.

Davidson, W. (1989b). A static cylindrically symmetric solution for perfect fluid in general relativity. *JMP* **30**, 1560. *See* §22.2.

Davidson, W. (1990a). Internal and external metrics for a perfect fluid cylinder in general relativity. *JMP* **31**, 1972. *See* §22.2.

Davidson, W. (1990b). A solution for a family of perfect fluid cylinders in general relativity. *GRG* **22**, 553. *See* §22.2.

Davidson, W. (1991). A big-bang cylindrically symmetric radiation universe. *JMP* **32**, 1560. *See* §23.3.

Davidson, W. (1992). A one-parameter family of cylindrically symmetric perfect fluid cosmologies. *GRG* **24**, 179. *See* §23.3.

Davidson, W. (1993a). A cylindrically symmetric cosmological solution for stiff fluid. *CQG* **10**, 1843. *See* §10.11.

Davidson, W. (1993b). A cylindrically symmetric solution of Einstein's equations describing gravitational collapse of stiff fluid. *JMP* **34**, 1908. *See* §23.3.

Davidson, W. (1994). Infinite perfect fluid in cylindrically symmetric steady differential rotation. *CQG* **11**, L129. *See* §22.2.

Davidson, W. (1996). A Petrov type *I* cylindrically symmetric solution for perfect fluid in steady rigid body rotation. *CQG* **13**, 283. *See* §22.2.

Davidson, W. (1997). Barotropic perfect fluid in steady cylindrically symmetric rotation. *CQG* **14**, 119. *See* §22.2.

Davidson, W. (1998). An inhomogeneous cosmology featuring a 'stiff' fluid density wave. *CQG* **15**, 2813. *See* §10.11.

Davidson, W. (1999). Barotropic perfect fluid in steady cylindrically symmetric rotation. II. *CQG* **16**, 2135. *See* §22.2.

Davidson, W. (2000). A cylindrically symmetric stationary solution of Einstein's equations describing a perfect fluid of finite radius. *CQG* **17**, 2499. *See* §§21.2, 22.2.

Davies, H. and Caplan, T.A. (1971). The space-time metric inside a rotating cylinder. *Proc. Camb. Phil. Soc.* **69**, 325. *See* §22.2.

Debever, R. (1959). Tenseur de super-énergie, tenseur de Riemann: cas singuliers. *C. R. Acad. Sci.* (*Paris*) **249**, 1744. *See* Ch. 4.

Debever, R. (1964). Le rayonnement gravitationnel: Le tenseur de Riemann en relativité générale. *Cahiers Phys.* **18**, 303. *See* Ch. 4.

Debever, R. (1965). Local tetrad in general relativity, in *Atti del convegno sulla relatività generale; problemi dell'energia e ondi gravitationali*, page 3 (Barbèra, Firenze). *See* §12.2.

Debever, R. (1966). Représentation vectorielle de la courbure en relativité générale: transformations conformes, in *Perspectives in geometry and relativity*, ed. B. Hoffman, page 96 (Indiana Univ. Press, Bloomington). *See* §3.4.

Debever, R. (1971). On type D expanding solutions of Einstein–Maxwell equations. *Bull. Soc. Math. Belgique* **23**, 360. *See* §§21.1, 24.3, 28.2.

Debever, R. (1974). Sur une classe d'espaces lorentziens. *Bull. Acad. Roy. Belg. Cl. Sci.* **60**, 998. *See* §32.1.

Debever, R. and Cahen, M. (1960). Champs électromagnétiques constants en relativité générale. *C. R. Acad. Sci. (Paris)* **251**, 1160. *See* §35.1.

Debever, R. and Cahen, M. (1961). Sur les espaces-temps, qui admettent un champ de vecteurs isotropes parallèles. *Bull. Acad. Roy. Belg. Cl. Sci.* **47**, 491. *See* §6.1.

Debever, R. and Kamran, N. (1982). Empty spaces and perfect fluids with homothetic transformation. *GRG* **14**, 637. *See* §§11.4, 22.2.

Debever, R., Kamran, N. and McLenaghan, R.G. (1982). Sur l'intégration complète des équations d'Einstein du vide et de Maxwell–Einstein, en type D. *Bull. Acad. Roy. Belg. Cl. Sci.* **68**, 592. *See* §§21.1, 26.2.

Debever, R., Kamran, N. and McLenaghan, R.G. (1984). Exhaustive integration and a single expression for the general solution of the type D vacuum and electrovac field equations with cosmological constant for a nonsingular aligned Maxwell field. *JMP* **25**, 1955. *See* §§21.1, 26.2.

Debever, R. and McLenaghan, R.G. (1981). Orthogonal transitivity, invertibility, and null geodesic separability in type D electrovac solutions of Einstein's field equations with cosmological constant. *JMP* **22**, 1711. *See* §§21.1, 26.2.

Debever, R., McLenaghan, R.G. and Tariq, N. (1979). Riemannian–Maxwellian invertible structures in general relativity. *GRG* **10**, 853. *See* §17.1.

Debever, R., Van den Bergh, N. and Leroy, J. (1989). Diverging Einstein–Maxwell null fields of Petrov type *D*. *CQG* **6**, 1373. *See* §30.6.

Debney, G. (1971). On vacuum space-times admitting a null Killing bivector. *JMP* **12**, 2372. *See* §17.3.

Debney, G. (1972). Null Killing vectors in general relativity. *Nuovo Cim. Lett.* **5**, 954. *See* §24.4.

Debney, G. (1973). Expansion-free Kerr–Schild fields. *Nuovo Cim. Lett.* **8**, 337. *See* §§31.6, 32.2.

Debney, G. (1974). Expansion-free electromagnetic solutions of the Kerr–Schild class. *JMP* **15**, 992. *See* §§31.6, 32.3.

Debney, G., Kerr, R.P. and Schild, A. (1969). Solutions of the Einstein and Einstein–Maxwell equations. *JMP* **10**, 1842. *See* §§27.1, 29.1, 32.2, 32.3.

Defrise, L. (1969). Groupes d'isotropie et groupes de stabilité conforme dans les espaces lorentziens. Thesis, Université Libre de Bruxelles. *See* §§8.1, 8.4, 11.2, 12.1, 13.1, 35.4.

Defrise-Carter, L. (1975). Conformal groups and conformally equivalent isometry groups. *Commun. math. Phys.* **40**, 271. *See* §35.4.

Delgaty, M.S.R. and Lake, K. (1998). Physical acceptibility of isolated, static, spherically symmetric, perfect fluid solutions of Einstein's equations. *Comp. Phys. Comm.* **115**, 395. *See* §16.1.

Demaret, J. and Henneaux, M. (1983). New solutions to Einstein–Maxwell equations of cosmological interest. *Phys. Lett. A* **99**, 217. *See* §22.4.

Demiański, M. (1972). New Kerr-like space-time. *Phys. Lett. A* **42**, 157. *See* §§21.1, 29.2.

Demiański, M. (1976). Method of generating stationary Einstein–Maxwell fields. *Acta Phys. Polon. B* **7**, 567. *See* §34.1.

Demiański, M. and Grishchuk, L.P. (1972). Homogeneous rotating universe with flat space. *Commun. Math. Phys.* **25**, 233. *See* §14.4.

Demiański, M. and Newman, E.T. (1966). A combined Kerr–NUT solution of the Einstein field equations. *Bull. Acad. Polon. Sci. Math. Astron. Phys.* **14**, 653. *See* §§20.5, 21.1, 34.1.
Denisova, T.E., Khakimov, S.A. and Manko, V.S. (1994). The Gutsunaev–Manko static vacuum solution. *GRG* **26**, 119. *See* §20.2.
Denisova, T.E. and Manko, V.S. (1992). Exact solution of the Einstein–Maxwell equations referring to a charged spinning mass. *CQG* **9**, L57. *See* §34.1.
Denisova, T.E., Manko, V.S. and Shorokhov, S.G. (1991). Generalization of the Kerr–Newman solution. *Sov. Phys. J.* **34**, 1050. *See* §21.1.
Diaz, A.G. (1985). Magnetic generalization of the Kerr–Newman metric. *JMP* **26**, 155. *See* §34.1.
Dietz, W. (1983a). N rank zero HKX transformations, in *Proceedings of the third Marcel Grossmann meeting on general relativity*, ed. Hu Ning, page 1053 (North-Holland, Amsterdam). *See* §34.3.
Dietz, W. (1983b). New representations of the HKX transformations by means of determinants. *GRG* **15**, 911. *See* §34.3.
Dietz, W. (1984a). HKX transformations. Some results, in *Solutions of Einstein's equations: Techniques and results.* Lecture notes in physics, vol. 205, eds. C. Hoenselaers and W. Dietz, page 85 (Springer, Berlin). *See* §34.3.
Dietz, W. (1984b). A new class of asymptotically flat stationary axisymmetric vacuum gravitational fields. *GRG* **16**, 246. *See* §34.3.
Dietz, W. (1988). New exact solutions of Einstein's field equations: gravitational force can also be repulsive! *Found. Phys.* **18**, 529. *See* §34.4.
Dietz, W. and Hoenselaers, C. (1982a). A new class of bipolar vacuum gravitational fields. *Proc. Roy. Soc. London A* **382**, 221. *See* §34.3.
Dietz, W. and Hoenselaers, C. (1982b). A new representation of the HKX-transformations. *Phys. Lett. A* **90**, 218. *See* §34.3.
Dietz, W. and Hoenselaers, C. (1982c). Stationary system of two masses kept apart by their gravitational spin-spin interaction. *Phys. Rev. Lett.* **48**, 778. *See* §34.3.
Dietz, W. and Hoenselaers, C. (1985). Two mass solutions of Einstein's vacuum equations: the double Kerr solution. *Ann. Phys. (USA)* **165**, 319. *See* §34.4.
Dietz, W. and Rüdiger, R. (1981). Space-times admitting Killing Yano tensors I. *Proc. Roy. Soc. Lond. A* **375**, 361. *See* §35.3.
Dietz, W. and Rüdiger, R. (1982). Space-times admitting Killing Yano tensors II. *Proc. Roy. Soc. Lond. A* **381**, 315. *See* §35.3.
d'Inverno, R.A. and Russell-Clark, R.A. (1971). Classification of the Harrison metrics. *JMP* **12**, 1258. *See* §§9.3, 17.3.
Dodd, R.K., Kinoulty, J. and Morris, H.C. (1984). Bäcklund transformations for the Ernst equation of general relativity, in *Advances in nonlinear waves*, ed. L. Debnath, page 254 (Pitman, Boston, MA). *See* §34.7.
Dodd, R.K. and Morris, H.C. (1982). Linear deformation problems for the Ernst equation. *JMP* **23**, 1131. *See* §34.7.
Dodd, R.K. and Morris, H.C. (1983). Some equations that give special solutions to the Ernst equation. *Proc. Roy. Irish Acad. A* **83**, 95. *See* §20.6.
Dolan, P. (1968). A singularity free solution of the Maxwell–Einstein-equation. *Commun. Math. Phys.* **9**, 161. *See* §12.3.
Dolan, P. and Kim, C.W. (1994). The wave equation for the Lanczos potential. I. *Proc. Roy. Soc. London A* **447**, 557. *See* §3.6.
Doroshkevich, A.G. (1965). Model of a universe with a uniform magnetic field (in Russian). *Astrofiz.* **1**, 255. *See* §§14.1, 14.3.
Doroshkevich, A.G., Lukash, V.N. and Novikov, I.D. (1973). Isotropisation of homogeneous cosmological models. *Sov. Phys. JETP* **37**, 739. *See* §13.3.
Dozmorov, I.M. (1971a). Solutions of the Einstein equations related by null vectors. II (in Russian). *Izv. Vys. Uch. Zav. Fiz.* **11**, 68. *See* §32.5.

Dozmorov, I.M. (1971b). Solutions of the Einstein equations related by null vectors. III (in Russian). *Izv. Vys. Uch. Zav. Fiz.* **11**, 76. *See* §32.5.

Drake, P. and Szekeres, P. (1998). An explanation of the Newman–Janis algorithm. Preprint gr-qc/9807025, Adelaide. *See* §21.1.

Drauschke, A. (1996). Untersuchung einer algebraisch speziellen Lösung der allgemeinen Relativitätstheorie für ideale Flüssigkeiten. Diploma thesis, Jena. *See* §33.1.

Dray, T. and Walker, M. (1980). On the regularity of Ernst's generalized C-metric. *Lett. Math. Phys.* , **4**, 15. *See* §17.2.

Droste, J. (1916-17). The field of a single centre in Einstein's theory of gravitation, and the motion of a particle in that field. *Kon. Akad. Wetensch. Amsterdam, Proc. Sec. Sci.* **19**, 197. *See* §15.4.

Duggal, K.L. (1978). Existence of two Killing vector fields on the space-time of general relativity. *Tensor* **32**, 318. *See* §17.1.

Durgapal, M.C. (1982). A class of new exact solutions in general relativity. *J. Phys. A* **15**, 2637. *See* §16.1.

Durgapal, M.C. and Fuloria, R.S. (1985). Analytic relativistic model for a superdense star. *GRG* **17**, 671. *See* §16.1.

Durgapal, M.C., Pande, A.K. and Phuloria, R.S. (1984). Physically realizable relativistic stellar structures. *Astrophys. Space Sci.* **102**, 49. *See* §16.1.

Dyer, C.C., McVittie, G.C. and Oattes, L.M. (1987). A class of spherically symmetric solutions with conformal Killing vectors. *GRG* **19**, 887. *See* §§11.4, 16.2.

Eardley, D., Isenberg, J., Marsden, J.E. and Moncrief, V. (1986). Homothetic and conformal symmetries of solutions to Einstein's equations. *Commun. Math. Phys.* **106**, 137. *See* §§11.3, 35.4.

Eardley, D.M. (1974). Self-similar spacetimes: geometry and dynamics. *Commun. Math. phys.* **37**, 287. *See* §§8.7, 11.4.

Eddington, A.S. (1924). A comparison of Whitehead's and Einstein's formulae. *Nature* **113**, 192. *See* §15.4.

Edgar, S.B. (1979). A new approach to Einstein–Petrov type *I* spaces. *Int. J. Theor. Phys.* **18**, 251. *See* §7.3.

Edgar, S.B. (1980). The structure of tetrad formalisms in general relativity: the general case. *GRG* **12**, 347. *See* §7.3.

Edgar, S.B. (1986). New approach to Einstein–Petrov type *I* spaces. II. A classification scheme. *Int. J. Theor. Phys.* **25**, 425. *See* §9.2.

Edgar, S.B. (1992). Integration methods within existing tetrad formalisms in general relativity. *GRG* **24**, 1267. *See* §7.3.

Edgar, S.B (1999). Four-dimensional tensor identities of low order for the Weyl and Ricci tensors. *GRG* **31**, 405. *See* §9.1.

Edgar, S.B. and Höglund, A. (1997). The Lanczos potential for the Weyl curvature tensor: existence, wave equation and algorithms. *Proc. Roy. Soc. London A* **453**, 835. *See* §3.6.

Edgar, S.B. and Ludwig, G. (1997a). All conformally flat pure radiation metrics. *CQG* **14**, L65. *See* §37.5.

Edgar, S.B. and Ludwig, G. (1997b). Integration in the GHP formalism III: Finding conformally flat radiation metrics as an example of an "optimal situation". *GRG* **29**, 1309. *See* §7.4.

Edwards, D. (1972). Exact expressions for the properties of the zero-pressure Friedmann models. *Mon. Not. R. Astron. Soc.* **159**, 51. *See* §14.2.

Egorov, I.P. (1955). Riemannian spaces V_4 of nonconstant curvature and maximum mobility (in Russian). *Dokl. Akad. Nauk SSSR* **103**, 9. *See* §34.1.

Ehlers, J. (1957). Konstruktionen und Charakterisierungen von Lösungen der Einsteinschen Gravitationsfeldgleichungen. Dissertation, Hamburg. *See* §34.1.

Ehlers, J. (1961). Beiträge zur relativistischen Mechanik kontinuierlicher Medien. *Akad. Wiss. Lit. Mainz, Abhandl. Math.-Nat. Kl.* **11**. *See* §§6.1, 38.1.

Ehlers, J. (1962). Transformations of static exterior solutions of Einstein's gravitational field equations into different solutions by means of conformal mappings. *Colloques Int. C.N.R.S.* (*Les théories relativistes de la gravitation*) **91**, 275. *See* §21.2.

Ehlers, J. (1981). Christoffel's work on the equivalence problem for Riemannian spaces and its importance for modern field theories of physics, in *E.B. Christoffel*, eds. P.L Butzer and F. Fehér, page 526 (Birkhäuser Verlag, Basel). *See* §9.2.

Ehlers, J. and Kundt, W. (1962). Exact solutions of the gravitational field equations, in *Gravitation: an introduction to current research*, ed. L. Witten, page 49 (Wiley, New York and London). *See* §§5.1, 6.1, 9.1, 18.6, 24.5, 35.1, 35.2.

Ehlers, J., Rosenblum, A., Goldberg, J.N. and Havas, P. (1976). Comments on gravitational radiation damping and energy loss in binary systems. *Astrophys. J.* **208**, L77. *See* §1.1.

Einstein, A. (1917). Kosmologische Betrachtungen zur allgemeinen Relativitätstheorie. *Sitzb. Preuss. Akad. Wiss.*, 142. *See* §12.4.

Einstein, A. and de Sitter, W. (1932). On the relation between the expansion and mean density of the universe. *Proc. Natl. Acad. Sci. U.S.* **18**, 213. *See* §14.2.

Einstein, A. and Rosen, N.J. (1937). On gravitational waves. *J. Franklin Inst.* **223**, 43. *See* §22.3.

Einstein, A. and Straus, E.G. (1945). The influence of the expansion of space on the gravitation fields surrounding the individual stars. *Rev. Mod. Phys.* **17**, 120. *See* §3.8.

Eisenhart, L.P. (1927). *Non-Riemannian geometry* (Amer. Math. Soc., Providence, Rhode Island). *See* §2.1.

Eisenhart, L.P. (1933). *Continuous groups of transformations* (Princeton Univ. Press., Princeton). *See* §§8.1, 8.4, 8.5, 8.6.

Eisenhart, L.P. (1949). *Riemannian geometry* (Princeton Univ. Press., Princeton). *See* §§3.1, 8.4, 35.1, 35.4, 37.1, 37.3.

Eleuterio, S.M. and Mendes, R.V. (1982). Equivalence and singularities: a note on computer algebra. *J. Comp. Phys.* **48**, 150. *See* §9.1.

Ellis, G.F.R. (1967). Dynamics of pressure-free matter in general relativity. *JMP* **8**, 1171. *See* §§6.2, 10.11, 11.2, 13.1, 13.4, 15.5.

Ellis, G.F.R. and King, A.R. (1974). Was the big bang a whimper? *Commun. Math. Phys.* **38**, 119. *See* §14.1.

Ellis, G.F.R. and MacCallum, M.A.H. (1969). A class of homogeneous cosmological models. *Commun. Math. Phys.* **12**, 108. *See* §§8.2, 13.2, 13.3, 14.4.

Ellis, G.F.R. and Madsen, M. (1991). Exact scalar field cosmologies. *CQG* **8**, 667. *See* §14.2.

Ellis, G.F.R. and McCarthy, P.J. (1987). Arnold deformation of Petrov types. *Ann. Phys.* (*USA*) **180**, 74. *See* §9.5.

Erez, G. and Rosen, N. (1959). The gravitational field of a particle possessing a quadripole moment. *Bull. Res. Counc. Israel* **8F**, 47. *See* §20.2.

Ernotte, P. (1980). Commutation properties of 2-parameter groups of isometries. *JMP* **21**, 954. *See* §19.1.

Ernst, F.J. (1968a). New formulation of the axially symmetric gravitational field problem. *Phys. Rev.* **167**, 1175. *See* §§19.5, 20.5, 34.1.

Ernst, F.J. (1968b). New formulation of the axially symmetric gravitational field problem II. *Phys. Rev.* **168**, 1415. *See* §18.4.

Ernst, F.J. (1973). Charged version of Tomimatsu–Sato spinning mass field. *PRD* **7**, 2520. *See* §§21.1, 34.1.

Ernst, F.J. (1976). Removal of the nodal singularity of the C-metric. *JMP* **17**, 515. *See* §§17.2, 21.1.

Ernst, F.J. (1977). A new family of solutions of the Einstein field equations. *JMP* **18**, 233. *See* §20.6.
Ernst, F.J. (1978a). Coping with different languages in the null tetrad formulation of general relativity. *JMP* **19**, 489. *See* §7.2.
Ernst, F.J. (1978b). Generalized C-metric. *JMP* **19**, 1986. *See* §§17.2, 20.2.
Ernst, F.J. (1984). The homogeneous Hilbert problem. Practical application, in *Solutions of Einstein's equations: Techniques and results.* Lecture notes in physics, vol. 205, eds. C. Hoenselaers and W. Dietz, page 176 (Springer, Berlin). *See* §34.6.
Ernst, F.J. (1994). Fully electrified Neugebauer spacetimes. *PRD* **50**, 6179. *See* §34.8.
Ernst, F.J., García D., A. and Hauser, I. (1987). Colliding gravitational plane waves with noncollinear polarization. I. *JMP* **28**, 2155. *See* §§25.2, 25.4, 25.5.
Ernst, F.J., García D., A. and Hauser, I. (1988). Colliding gravitational plane waves with noncollinear polarization. III. *JMP* **29**, 681. *See* §§25.4, 25.5, 34.6.
Ernst, F.J. and Wild, W.J. (1976). Kerr black holes in a magnetic universe. *JMP* **17**, 182. *See* §§21.1, 34.1.
Estabrook, F.B. (1976). Some old and new techniques for the practical use of exterior differential forms, in *Bäcklund transformations.* Lecture notes in mathematics, vol. 515, ed. R.M. Miura, page 136 (Springer Verlag, Berlin). *See* §10.4.
Estabrook, F.B. (1982). Moving frames and prolongation algebras. *JMP* **23**, 2071. *See* §10.4.
Estabrook, F.B. and Wahlquist, H.D. (1976). Prolongation structures of nonlinear evolution equations II. *JMP* **17**, 1293. *See* §10.4.
Estabrook, F.B., Wahlquist, H.D. and Behr, C.G. (1968). Dyadic analysis of spatially homogeneous world models. *JMP* **9**, 497. *See* §8.2.
Evans, A.B. (1977). Static fluid cylinders in general relativity. *J. Phys. A* **10**, 1303. *See* §22.2.
Fackerell, E.D. and Kerr, R.P. (1991). Einstein vacuum field equations with a single non-null Killing vector. *GRG* **23**, 861. *See* §17.3.
Faridi, A.M. (1990). Einstein–Maxwell equations and the groups of homothetic motion. *JMP* **31**, 401. *See* §11.3.
Farnsworth, D.L. (1967). Some new general relativistic dust metrics possessing isometries. *JMP* **8**, 2315. *See* §14.3.
Farnsworth, D.L. and Kerr, R.P. (1966). Homogeneous dust-filled cosmological solutions. *JMP* **7**, 1625. *See* §§8.2, 12.4.
Fee, G.J. (1979). Homogeneous spacetimes. M. Math. thesis, University of Waterloo. *See* §12.1.
Feinstein, A. and Griffiths, J.B. (1994). Colliding gravitational waves in a closed Friedmann–Robertson–Walker background. *CQG* **11**, L109. *See* §25.6.
Feinstein, A. and Ibañez, J. (1989). Curvature-singularity-free solutions for colliding plane gravitational waves with broken $u - v$ symmetry. *PRD* **39**, 470. *See* §25.3.
Feinstein, A., MacCallum, M.A.H. and Senovilla, J.M.M. (1989). On the ambiguous evolution and the production of matter in spacetimes with colliding waves. *CQG* **6**, L217. *See* §25.6.
Feinstein, A. and Senovilla, J.M.M. (1989a). Collision between variably polarized plane gravitational wave and a shell of null matter. *Phys. Lett. A* **138**, 102. *See* §25.6.
Feinstein, A. and Senovilla, J.M.M. (1989b). A new inhomogeneous cosmological perfect fluid solution with $p = \rho/3$. *CQG* **6**, L89. *See* §23.3.
Fels, M. and Held, A. (1989). Kerr–Schild rides again. *GRG* **21**, 61. *See* §§29.2, 32.5.
Fernandez-Jambrina, L. (1994). Moment density of Zipoy's dipole solution. *CQG* **11**, 1483. *See* §20.2.
Fernandez-Jambrina, L. (1997). Singularity-free cylindrical cosmological model. *CQG* **14**, 3407. *See* §23.3.

Fernandez-Jambrina, L. and Chinea, F.J. (1994). Differential form approach for stationary axisymmetric Maxwell fields in general relativity. *CQG* **11**, 1489. *See* §§19.6, 21.1.

Ferrando, J.J. and Sáez, J.A. (1997). A covariant determination of the Weyl canonical frames in Petrov type I spacetimes. *CQG* **14**, 129. *See* §9.3.

Ferrari, V. (1988). Focusing process in the collision of gravitational plane waves. *PRD* **37**, 3061. *See* §25.1.

Ferrari, V. and Ibañez, J. (1986). Collision of two plane gravitational waves, in *Proceedings of the fourth Marcel Grossmann meeting on general relativity*, ed. R. Ruffini, page 931 (North-Holland, Amsterdam). *See* §25.3.

Ferrari, V. and Ibañez, J. (1987a). A new exact solution for colliding gravitational plane waves. *GRG* **19**, 383. *See* §§25.3, 25.4.

Ferrari, V. and Ibañez, J. (1987b). On the collision of gravitational plane waves. A class of soliton solutions. *GRG* **19**, 405. *See* §§25.3, 25.4.

Ferrari, V. and Ibañez, J. (1988). Type-D solutions describing the collision of plane-fronted gravitational waves. *Proc. Roy. Soc. London A* **417**, 417. *See* §25.4.

Ferrari, V. and Ibañez, J. (1989). On the gravitational interaction of null fields. *CQG* **6**, 1805. *See* §25.6.

Ferrari, V., Ibañez, J. and Bruni, M. (1987a). Colliding gravitational waves with noncollinear polarization: a class of soliton solutions. *Phys. Lett. A* **122**, 459. *See* §25.4.

Ferrari, V., Ibañez, J. and Bruni, M. (1987b). Colliding plane gravitational waves: a class of nondiagonal soliton solutions. *PRD* **36**, 1053. *See* §25.4.

Ferrari, V. and Pendenza, P. (1990). Boosting the Kerr metric. *GRG* **22**, 1105. *See* §9.5.

Finch, M.R. (1987). The Painlevé–Gambier equation and the relativistic static fluid sphere. Ph.D. thesis, University of Sussex. *See* §16.1.

Finch, M.R. and Skea, J.E.F. (1989). A review of the relativistic static fluid sphere. Preprint, London. *See* §16.1.

Finkelstein, D. (1958). Past-future asymmetry of the gravitational field of a point particle. *Phys. Rev.* **110**, 965. *See* §15.4.

Finkelstein, R.J. (1975). The general relativistic fields of a charged rotating source. *JMP* **16**, 1271. *See* §21.1.

Finley, III, J.D. (1996). The Robinson–Trautman type III prolongation structure contains $K2$. *Commun. Math. Phys.* **178**, 375. *See* §10.4.

Finley, III, J.D. and McIver, J.K. (1995). Infinite-dimensional Estabrook–Wahlquist prolongations for the sine-Gordon equation. *JMP* **36**, 5707. *See* §10.4.

Fischer, E. (1979). An extension of the Bonnor transformation. *JMP* **20**, 2547. *See* §34.1.

Fischer, E. (1980). Similarity solutions of the Einstein and Einstein–Maxwell equations. *J. Phys. A* **13**, L81. *See* §§20.6, 22.3.

Fiser, K., Rosquist, K. and Uggla, C. (1992). Bianchi type V perfect fluid cosmologies. *GRG* **24**, 679. *See* §§13.2, 14.4.

Fitch, J.P. (1971). An algebraic manipulator. Ph.D. thesis, University of Cambridge. *See* §9.3.

Flaherty, Jr., E.J. (1980). Complex variables in relativity, in *General relativity and gravitation. One hundred years after the birth of Albert Einstein,* vol. 2, ed. A. Held, page 207 (Plenum, New York). *See* §2.2.

Flanders, H. (1963). *Differential forms with applications to the physical sciences* (Academic Press, New York). *See* §§2.1, 2.7.

Fodor, G. (2000). Generating spherically symmetric static perfect fluid solutions. Preprint gr-qc/0011040, Budapest. *See* §16.1.

Fodor, G., Hoenselaers, C. and Perjés, Z. (1989). Multipole moments of axisymmetric systems in relativity. *JMP* **30**, 2252. *See* §18.8.

Fodor, G., Marklund, M. and Perjés, Z. (1999). Axistationary fluids – a tetrad approach. *CQG* **16**, 453. *See* §19.6.

Fokas, A.S., Sung, L.-Y. and Tsoubelis, D. (1999). The inverse spectral method for colliding gravitational waves. *Math. Phys. Ann. Geom.* **1**, 313. *See* §34.9.

Foster, J. and Newman, E.T. (1967). Note on the Robinson–Trautman solutions. *JMP* **8**, 189. *See* §28.1.

Foyster, J.M. and McIntosh, C.B.G. (1972). A class of solutions of Einstein's equations which admit a 3-parameter group of isometries. *Commun. Math. Phys.* **27**, 241. *See* §15.4.

Frehland, E. (1971). The general stationary gravitational field with cylindrical symmetry. *Commun. math. Phys.* **23**, 127. *See* §22.2.

Friedman, A. (1965). Isometric embedding of Riemann manifolds into euclidean spaces. *Rev. Mod. Phys.* **37**, 201. *See* §§37.1, 37.3.

Friedmann, A. (1922). Über die Krümmung des Raumes. *Z. Phys.* **31**, 1991. *See* §14.2.

Friedmann, A. (1924). Über die Möglichkeit einer Welt mit konstanter negativer Krümmung des Raumes. *Z. Phys.* **31**, 2001. *See* §14.2.

Friedrich, H. and Rendall, A. (2000). The Cauchy problem for the Einstein equations, in *Einstein's field equations and their physical implications. Selected essays in honour of Jürgen Ehlers.* Lecture notes in physics, vol. 540, ed. B.G. Schmidt, page 127 (Springer, Berlin). *See* Ch. 9.

Frolov, V.P. (1977). Newman–Penrose formalism in general relativity (in Russian). *Problems General Rel. Theory Group Representations, Trudy Lebedev Inst., Akad. Nauk SSR* **96**, 72. *See* §§7.1, 28.3.

Frolov, V.P. and Khlebnikov, V.I. (1975). Gravitational field of radiating systems I. Twisting free type D metrics. Preprint no. 27, Lebedev Phys. Inst. Akad. Nauk. Moscow. *See* §28.3.

Fronsdal, C. (1959). Completion and embedding of the Schwarzschild solution. *Phys. Rev.* **116**, 778. *See* §37.1.

Ftaclas, C. and Cohen, J.M. (1978). Locally rotationally symmetric cosmological model containing a nonrotationally symmetric electromagnetic field. *PRD* **18**, 4373. *See* §13.3.

Fulling, S.A., King, R.C., Whybourne, B.G. and Cummins, C.J. (1992). Normal forms for tensor polynomials I. The Riemann tensor. *CQG* **9**, 1151. *See* §9.1.

Fustero, X. and Verdaguer, E. (1986). Einstein–Rosen metrics generated by the inverse scattering transform. *GRG* **18**, 1141. *See* §34.5.

Gaete, P. and Hojman, R. (1990). General exact solution for homogeneous time-dependent self-gravitating perfect fluids. *JMP* **31**, 140. *See* §14.3.

Gaffet, B. (1988). Common structure of several completely integrable non-linear equations. *J. Phys. A* **21**, 2491. *See* §10.4.

Gaffet, B. (1990). The Einstein equations with two commuting Killing vectors. *CQG* **7**, 2017. *See* §17.1.

García D., A. (1984). Electrovac type D solutions with cosmological constant. *JMP* **25**, 1951. *See* §§21.1, 26.2.

García D., A. (1988). Colliding-wave generalizations of the Nutku–Halil metric in the Einstein–Maxwell theory. *Phys. Rev. Lett.* **61**, 507. *See* §25.5.

García D., A. (1989). Colliding wave generalization of the Ferrari–Ibañez solution in the Einstein–Maxwell theory. *Phys. Lett. A* **138**, 370. *See* §25.5.

García D., A. (1990). Generation of colliding gravitational-electromagnetic waves by means of a Harrison transformation. *Theor. Math. Phys.* **83**, 434. *See* §25.5.

García D., A. (1994). A new stationary axisymmetric perfect fluid type D solution with differential rotation. *CQG* **11**, L45. *See* §21.2.

García D., A. and Alvarez C., M. (1984). Shear-free special electrovac type-II solutions with cosmological constant. *Nuovo Cim. B* **79**, 266. *See* §31.8.

García D., A. and Breton B., N. (1985). The gravitational field of a charged, magnetized, accelerating, and rotating mass. *JMP* **26**, 465. *See* §34.1.

García D., A. and Breton B., N. (1989). Magnetic generalizations of the Carter metrics. *JMP* **30**, 1310. *See* §34.1.

García D., A. and Hauser, I. (1988). Type-D rigidly rotating perfect fluid solutions. *JMP* **29**, 175. *See* §21.2.

García D., A. and Kramer, D. (1997). Stationary cylindrically symmetric gravitational fields with differentially rotating perfect fluids. *CQG* **14**, 499. *See* §22.2.

García D., A. and Macias, A. (1998). Black holes as exact solutions of the Einstein–Maxwell equations of Petrov type D, in *Black holes: Theory and observation.* Lecture notes in physics, vol. 514, eds. F.W. Hehl, C. Kiefer and R.J.K. Metzler, page 205 (Springer, Berlin). *See* §21.1.

García D., A. and Plebański, J.F. (1981). All nontwisting N's with cosmological constant. *JMP* **22**, 2655. *See* §§28.4, 31.8.

García D., A. and Plebański, J.F. (1982a). An exceptional type D shearing twisting electrovac with Λ. *JMP* **23**, 123. *See* §26.1.

García D., A. and Plebański, J.F. (1982b). Solutions of type D possessing a group with null orbits as contractions of the seven-parameter solution. *JMP* **23**, 1463. *See* §24.3.

García D., A. and Salazar I., H. (1983). All null orbit type D electrovac solutions with cosmological constant. *JMP* **24**, 2498. *See* §24.3.

Gardner, R.B. (1989). *The method of equivalence and its applications* (SIAM, Philadelphia). *See* §§7.4, 9.2.

Garfinkle, D. (1991). Cosmic-string traveling waves, in *Nonlinear problems in relativity and cosmology*, eds. J.R. Buchler, S.L. Detweiler and J.R. Ipser, page 68 (New York Acad. Sci, New York, NY). *See* §32.5.

Garfinkle, D., Glass, E.N. and Krisch, J.P. (1997). Solution generating with perfect fluids. *GRG* **29**, 467. *See* §17.3.

Garfinkle, D. and Melvin, M.A. (1994). Generalized magnetic universe solutions. *PRD* **50**, 3859. *See* §34.1.

Garfinkle, D. and Tian, Q. (1987). Spacetimes with cosmological constant and a conformal Killing field have constant curvature. *CQG* **4**, 137. *See* §35.4.

Gautreau, R. and Hoffman, R.B. (1972). Generating potential for the NUT metric in general relativity. *Phys. Lett. A* **39**, 75. *See* §20.3.

Gautreau, R. and Hoffman, R.B. (1973). The structure of the sources of Weyl-type electrovac fields in general relativity. *Nuovo Cim. B* **16**, 162. *See* §18.6.

Gautreau, R., Hoffman, R.B. and Armenti, A. (1972). Static multiparticle systems in general relativity. *Nuovo Cim. B* **7**, 71. *See* §21.1.

Gavrilina, G.A. (1983). Generalized Kerr–Schild space-times with null Killing vectors, in *10th international conference on general relativity and gravitation*, vol. 1. *Contributed papers. Classical relativity*, eds. B. Bertotti, F. De Felice and A. Pascolini, page 233 (Consiglio Nazionale delle Ricerche, Rome). *See* §32.5.

Géhéniau, J. (1957). Une classification des espaces einsteiniens. *C. R. Acad. Sci. (Paris)* **244**, 723. *See* Ch. 4.

Gergely, L.A. and Perjés, Z. (1993). Solution of the vacuum Kerr–Schild problem. *Phys. Lett. A* **181**, 345. *See* §32.5.

Gergely, L.A. and Perjés, Z. (1994a). Kerr–Schild metrics revisited. I. The ground state. *JMP* **35**, 2438. *See* §32.5.

Gergely, L.A. and Perjés, Z. (1994b). Kerr–Schild metrics revisited. II. The complete vacuum solution. *JMP* **35**, 2448. *See* §32.5.

Gergely, L.A. and Perjés, Z. (1994c). Vacuum Kerr–Schild metrics generated by nontwisting congruences. *Ann. Phys. (Germany)* **3**, 609. *See* §32.5.

Geroch, R. (1969). Limits of spacetimes. *Commun. Math. Phys.* **13**, 180. *See* §9.5.

Geroch, R. (1970a). Multipole moments. I. Flat space. *JMP* **11**, 1955. *See* §18.8.

Geroch, R. (1970b). Multipole moments. II. Curved space. *JMP* **11**, 2580. *See* §18.8.

Geroch, R. (1971). A method for generating solutions of Einstein's equations. *JMP* **12**, 918. *See* §18.1.

Geroch, R. (1972). A method for generating new solutions of Einstein's equation. II. *JMP* **13**, 394. *See* §§10.3, 10.5, 19.3.

Geroch, R., Held, A. and Penrose, R. (1973). A space-time calculus based on pairs of null directions. *JMP* **14**, 874. *See* §§7.4, 17.1.

Glass, E.N. (1975). The Weyl tensor and shear-free perfect fluids. *JMP* **16**, 2361. *See* §6.2.

Glass, E.N. and Goldman, S.P. (1978). Relativistic spherical stars reformulated. *JMP* **19**, 856. *See* §16.1.

Gleiser, R.J., Diaz, M.C. and Grosso, R. (1988a). Phase transitions in perturbed stiff fluid Friedmann–Robertson–Walker cosmological models. *CQG* **5**, 989. *See* §10.11.

Gleiser, R.J., Gonzalez, G.I. and Pullin, J.A. (1988b). Higher order poles in the Belinskii–Zakharov inverse scattering method. *Phys. Lett. A* **130**, 206. *See* §34.5.

Gleiser, R.J. and Pullin, J.A. (1989). Appell rings in general relativity. *CQG* **6**, 977. *See* §20.2.

Gödel, K. (1949). An example of a new cosmological solution of Einstein's field equation of gravitation. *Rev. Mod. Phys.* **21**, 447. *See* §12.4.

Godfrey, B.B. (1972). Horizons in Weyl metrics exhibiting extra symmetries. *GRG* **3**, 3. *See* §§11.4, 13.3.

Goenner, H. (1970). Einstein tensor and generalizations of Birkhoff's theorem. *Commun. Math. Phys.* **16**, 34. *See* §15.4.

Goenner, H. (1973). Local isometric embedding of Riemannian manifolds and Einstein's theory of gravitation. Habilitationsschrift, Göttingen. *See* §§37.3, 37.5.

Goenner, H. (1977). On the interdependence of the Gauss–Codazzi–Ricci equations of local isometric embedding. *GRG* **8**, 139. *See* §§37.2, 37.4.

Goenner, H. (1980). Local isometric embedding of Riemannian manifolds and Einstein's theory of gravitation, in *General relativity and gravitation. One hundred years after the birth of Albert Einstein,* vol. 1, ed. A. Held, page 441 (Plenum Press, New York). *See* §37.1.

Goenner, H. (1984). Killing vector fields and the Einstein–Maxwell field equations for null electromagnetic fields. *GRG* **16**, 795. *See* §11.1.

Goenner, H. and Stachel, J. (1970). Einstein tensor and 3-parameter groups of isometries with 2-dimensional orbits. *JMP* **11**, 3358. *See* §§15.1, 15.4.

Golda, Z., Heller, M. and Szydlowski, M. (1983). Structurally stable approximations to Friedmann–Lemaitre world models. *Astrophys. Space Sci.* **90**, 313. *See* §14.1.

Goldberg, J.N. and Kerr, R.P. (1961). Some applications of the infinitesimal-holonomy group to the Petrov classification of Einstein spaces. *JMP* **2**, 327. *See* §9.1.

Goldberg, J.N. and Sachs, R.K. (1962). A theorem on Petrov types. *Acta Phys. Polon., Suppl.* **22**, 13. *See* §7.6.

Goldman, S.P. (1978). Physical solutions to general-relativistic fluid spheres. *Astrophys. J.* **226**, 1079. *See* §16.1.

González-Romero, L.M. (1994). Rotating barotropes and baroclines in general relativity. *CQG* **11**, 2741. *See* §§13.4, 21.2.

Goode, S.W. (1980). Some aspects of spatially inhomogeneous cosmologies. M. Math. thesis, University of Waterloo. *See* §§15.7, 23.1, 23.3.

Götz, G. (1988). A plane-symmetric solution of Einstein's equations with perfect fluid and an equation of state $p = \gamma\rho$. *GRG* **20**, 23. *See* §15.7.

Gourgoulhon, E. and Bonazzola, S. (1993). Noncircular axisymmetric stationary spacetimes. *PRD* **48**, 2635. *See* §19.2.

Gowdy, R.H. (1971). Gravitational waves in closed universes. *Phys. Rev. Lett.* **27**, 826. *See* §22.3.

Gowdy, R.H. (1975). Closed gravitational-wave universes: analytic solutions with two-parameter symmetry. *JMP* **16**, 224. *See* §22.3.

Greene, R.E. (1970). Isometric embedding of Riemannian and pseudo-Riemannian manifolds. *Memoirs Amer. Math. Soc.* **97**. *See* §37.1.

Griffiths, J.B. (1983). Colliding plane gravitational and electromagnetic waves. *J. Phys. A* **16**, 1175. *See* §25.5.

Griffiths, J.B. (1985). On the Bell–Szekeres solution for colliding electromagnetic waves, in *Galaxies, axisymmetric systems and relativity. Essays presented to W.B. Bonnor on his 65th birthday*, ed. M.A.H. MacCallum, page 199 (Cambridge University Press, Cambridge). *See* §25.5.

Griffiths, J.B. (1986). Algebraically special, nonaligned, non-null Einstein–Maxwell fields. *GRG* **18**, 389. *See* §26.1.

Griffiths, J.B. (1987). Colliding plane gravitational waves. *CQG* **4**, 957. *See* §25.3.

Griffiths, J.B. (1991). *Colliding plane waves in general relativity* (Oxford University Press, Oxford). *See* §§1.4, 25.1, 25.2.

Griffiths, J.B. (1993a). The collision of gravitational waves in a stiff perfect fluid. *JMP* **34**, 4064. *See* §25.6.

Griffiths, J.B. (1993b). On the propagation of a gravitational wave in a stiff perfect fluid. *CQG* **10**, 975. *See* §10.11.

Griffiths, J.B. and Alekseev, G.A. (1998). Some unpolarized Gowdy cosmologies and noncolinear colliding plane wave spacetimes. *Int. J. Mod. Phys D* **7**, 237. *See* §25.2.

Griffiths, J.B. and Ashby, P.C. (1993). On stiff fluids and colliding plane gravitational waves coupled with fluid motions. *Phys. Lett. A* **184**, 12. *See* §25.6.

Griffiths, J.B., Hoenselaers, C. and Ashby, P.C. (1993). Colliding plane gravitational waves with colinear polarization. *GRG* **25**, 189. *See* §25.3.

Griffiths, J.B. and Miccichè, S. (1999). The extension of gravitational soliton solutions with real poles. *GRG* **31**, 869. *See* §34.5.

Gron, O. and Eriksen, E. (1987). A dust-filled Kantowski–Sachs universe with $\Lambda > 0$. *Phys. Lett. A* **121**, 217. *See* §14.3.

Grosser, M., Kunzinger, M., Steinbauer, R., Urbantke, H. and Vickers, J. (2000). Diffeomorphism invariant construction of nonlinear generalized functions. *Ann. Phys. (Germany)* **9**, SI173. *See* §3.8.

Grosso, R. and Stephani, H. (1990). Bianchi types of Lie point symmetries arising for differentially rotating stationary axisymmetric fluids. *Astron. Nachr.* **311**, 283. *See* §21.2.

Grundland, A.M. and Tafel, J. (1993). Group invariant solutions to the Einstein equations with pure radiation fields. *CQG* **10**, 2337. *See* §30.7.

Gruszczak, J. (1981). Two-soliton solutions of the Ernst equation. *J. Phys. A* **14**, 3247. *See* §34.5.

Guo, Dong-Sheng (1984a). Noniterative method for constructing exact solutions of Einstein equations, in *Solutions of Einstein's equations: Techniques and results.* Lecture notes in physics, vol. 205, eds. C. Hoenselaers and W. Dietz, page 186 (Springer, Berlin). *See* §34.6.

Guo, Dong-Sheng (1984b). Noniterative method for constructing many-parameter solutions of the Einstein and Einstein–Maxwell field equations. *JMP* **25**, 2284. *See* §34.6.

Guo, Dong-Sheng and Ernst, F.J. (1982). Electrovac generalization of Neugebauer's $N = 2$ solution of the Einstein vacuum field equations. *JMP* **23**, 1359. *See* §34.8.

Guo, Han-ying, Wu, Ke, Hsiang, Yan-Yu and Wang, Shi-kun (1982). On the prolongation structure of Ernst equation. *Commun. Theor. Phys.* **1**, 661. *See* §34.2.

Guo, Han-ying, Wu, Ke and Wang, Shi-kun (1983a). Prolongation structure, Bäcklund transformation and principal homogeneous Hilbert problem in general relativity. *Commun. Theor. Phys.* **2**, 883. *See* §34.6.

Guo, Han-ying, Wu, Ke and Wang, Shi-kun (1983b). Prolongation structure, Bäcklund transformation and principal homogeneous Hilbert problem in general relativity, in *Proceedings of the third Marcel Grossmann meeting on general relativity.* Pt. A., ed. Hu Ning, page 143 (North-Holland, Amsterdam). *See* Ch. 34.

Gupta, Y.K. and Goel, P. (1975). Class II analogue of T.Y. Thomas's theorem and different types of embedding of static spherically symmetric space-times. *GRG* **6**, 499. *See* §37.2.

Gupta, Y.K. and Gupta, R.S. (1986). Nonstatic analogues of Kohler–Chao solution of imbedding class one. *GRG* **18**, 641. *See* §37.4.

Gupta, Y.K. and Sharma, J.R. (1996a). Non-static non-conformally flat perfect fluid plates of embedding class one. *GRG* **28**, 1447. *See* §§15.7, 37.4.

Gupta, Y.K. and Sharma, J.R. (1996b). Similarity solutions for the type D fluid plates in 5-d flat space. *JMP* **37**, 1962. *See* §§15.7, 37.4.

Gürsel, Y. (1983). Multipole moments for stationary systems: the equivalence of the Geroch–Hansen formulation and the Thorne formulation. *GRG* **15**, 737. *See* §18.8.

Gürses, M. (1977). Some solutions of stationary, axially-symmetric gravitational field equations. *JMP* **18**, 2356. *See* §21.1.

Gürses, M. (1984). Inverse scattering, differential geometry, Einstein–Maxwell solitons and one soliton Bäcklund transformation, in *Solutions of Einstein's equations: Techniques and results.* Lecture notes in physics, vol. 205, eds. C. Hoenselaers and W. Dietz, page 199 (Springer, Berlin). *See* Ch. 34.

Gürses, M. and Gürsey, F. (1975). Lorentz covariant treatment of the Kerr–Schild geometry. *JMP* **16**, 2385. *See* §32.1.

Gürses, M. and Halilsoy, M. (1982). Interacting superposed electromagnetic shock plane waves in general relativity. *Lett. Nuovo Cim.* **34**, 588. *See* §25.5.

Gürtug, Ö. (1995). A new extension of the Ferrari–Ibañez colliding wave solution. *GRG* **27**, 651. *See* §25.5.

Gutman, I.I. and Bespal'ko, R.M. (1967). Some exact spherically symmetric solutions of Einstein's equations (in Russian). Sbornik Sovrem. Probl. Grav., Tbilisi. *See* §37.4.

Gutsunaev, T.I. and Elsgolts, S.L. (1993). A new class of solutions of the magnetostatic Einstein–Maxwell equations. *J. Experimental Theor. Phys.* **77**, 1. *See* §21.1.

Gutsunaev, T.I. and Manko, V.S. (1985). On the gravitational field of a mass possessing a multipole moment. *GRG* **17**, 1025. *See* §20.2.

Gutsunaev, T.I. and Manko, V.S. (1987). On the gravitational field of a mass possessing a magnetic dipole moment. *Phys. Lett. A* **123**, 215. *See* §21.1.

Gutsunaev, T.I. and Manko, V.S. (1988). On a family of solutions of the Einstein–Maxwell equations. *GRG* **20**, 327. *See* §21.1.

Gutsunaev, T.I. and Manko, V.S. (1989). New stationary electrovacuum generalizations of the Schwarzschild solution. *PRD* **40**, 2140. *See* §34.1.

Guzman S., A.R., Plebański, J.F. and Przanowski, M. (1991). Deformations of algebraic types of the energy-momentum tensor. *JMP* **32**, 2839. *See* §9.5.

Haager, G. and Mars, M. (1998). A self-similar inhomogeneous dust cosmology. *CQG* **15**, 1567. *See* §23.3.

Hacyan, S. and Plebański, J.F. (1975). Some type III solutions of the Einstein–Maxwell equations. *Int. J. Theor. Phys.* **14**, 319. *See* §31.7.

Haggag, S. (1989). Comment on 'cylindrically symmetric static perfect fluids'. *CQG* **6**, 945. *See* §22.2.

Haggag, S. (1999). Solutions of Kramer's equations for perfect fluid cylinders. *GRG* **31**, 1169. *See* §22.2.

Haggag, S. and Desokey, F. (1996). Perfect-fluid sources for the Levi-Civita metric. *CQG* **13**, 3221. *See* §22.2.

Hajj-Boutros, J. (1985). On hypersurface-homogeneous space-times. *JMP* **26**, 2297. *See* §14.3.
Hajj-Boutros, J. (1986). New cosmological models. *CQG* **3**, 311. *See* §14.3.
Hajj-Boutros, J. (1989). Radiation equilibrium in general relativity: general solution. *Mod. Phys. Lett. A* **4**, 427. *See* §16.1.
Hajj-Boutros, J. and Léauté, B. (1985). Solutions of Einstein's field equations with plane symmetry. *Lett. Nuovo Cim.* **43**, 140. *See* §15.6.
Hajj-Boutros, J. and Sfeila, J. (1987). Zero-curvature FRW models and Bianchi I space-time as solutions of the same equation. *Int. J. Theor. Phys.* **26**, 97. *See* §14.3.
Hajj-Boutros, J. and Sfeila, J. (1989). Radiating regime to matter dominated regime. *Phys. Lett. A* **136**, 274. *See* §14.3.
Halford, W.D. (1979). Petrov type N vacuum metrics and homothetic motions. *JMP* **20**, 1115. *See* §29.3.
Halford, W.D. (1980). Einstein spaces and homothetic motions. II. *JMP* **21**, 129. *See* §§11.4, 29.2.
Halford, W.D. and Kerr, R.P. (1980). Einstein spaces and homothetic motions. I. *JMP* **21**, 120. *See* §11.4.
Halford, W.D., McIntosh, C.B.G. and Van Leeuwen, E.H. (1980). Curvature collineations of non-expanding and twist-free vacuum type-N metrics in general relativity. *J. Phys. A* **13**, 2995. *See* §35.4.
Halilsoy, M. (1985). A class of transcendent solutions to the Ernst system. *Lett. Nuovo Cim.* **44**, 88. *See* §21.1.
Halilsoy, M. (1988a). Colliding electromagnetic shock waves in general relativity. *PRD* **37**, 2121. *See* §25.5.
Halilsoy, M. (1988b). Cross-polarized cylindrical gravitational waves of Einstein and Rosen. *Nuovo Cim. B* **102**, 563. *See* §22.3.
Halilsoy, M. (1988c). Distinct family of colliding gravitational waves in general relativity. *PRD* **38**, 2979. *See* §25.4.
Halilsoy, M. (1989). Colliding superposed waves in the Einstein–Maxwell theory. *PRD* **39**, 2172. *See* §25.5.
Halilsoy, M. (1990a). Colliding impulsive waves in succession. *PRD* **42**, 2922. *See* §25.3.
Halilsoy, M. (1990b). Large family of colliding waves in the Einstein–Maxwell theory. *JMP* **31**, 2694. *See* §25.5.
Halilsoy, M. (1990c). Separable wave equation in the space of colliding waves. *Phys. Lett. A* **151**, 205. *See* §25.4.
Halilsoy, M. (1992). New metrics for spinning spheroids in general relativity. *JMP* **33**, 4225. *See* §20.3.
Halilsoy, M. (1993a). Interpolation of the Schwarzschild and Bertotti–Robinson solutions. *GRG* **25**, 275. *See* §21.1.
Halilsoy, M. (1993b). Note on 'Large family of colliding waves in the Einstein–Maxwell theory'. *JMP* **34**, 3553. *See* §25.5.
Halilsoy, M. and Al-Badawi, A. (1998). Modified Reissner–Nordström metric in an external electrostatic field. *Nuovo Cim. B* **13**, 761. *See* §21.2.
Halilsoy, M. and El-Said, M. (1993). The physical interpretation of a Painlevé transcendent spacetime. *GRG* **25**, 81. *See* §20.6.
Halilsoy, M. and Gürtuĝ, Ö. (1994). On some properties of the nut–Curzon space-time. *Nuovo Cim. B* **109**, 963. *See* §20.3.
Hall, G.S. (1973). On the Petrov classification of gravitational fields. *J. Phys. A* **6**, 619. *See* §4.3.
Hall, G.S. (1976a). The classification of the Ricci tensor in general relativity theory. *J. Phys. A* **9**, 541. *See* §5.1.
Hall, G.S. (1976b). Ricci recurrent space times. *Phys. Lett. A* **56**, 17. *See* §35.2.

Hall, G.S. (1987). Killing–Yano tensors in general relativity. *Int. J. Theor. Phys.* **26**, 71. *See* §35.3.
Hall, G.S. (1988a). Homothetic transformations with fixed points in spacetime. *GRG* **20**, 671. *See* §§8.4, 8.7.
Hall, G.S. (1988b). Singularities and homothety groups in spacetime. *CQG* **5**, L77. *See* §8.7.
Hall, G.S. (1989). The global extension of local symmetries in general relativity. *CQG* **6**, 157. *See* §8.1.
Hall, G.S. (1990a). Conformal symmetries and fixed points in space-time. *JMP* **31**, 1198. *See* §35.4.
Hall, G.S. (1990b). Special conformal symmetries in general relativity. *GRG* **22**, 203. *See* §35.4.
Hall, G.S. (1991). Covariantly constant tensors and holonomy structure in general relativity. *JMP* **32**, 181. *See* Ch. 9.
Hall, G.S. (2000). Conformal vector fields and conformal-type collineations in space-times. *GRG* **32**, 933. *See* §35.4.
Hall, G.S. and Carot, J. (1994). Conformal symmetries in null Einstein–Maxwell fields. *CQG* **11**, 475. *See* §35.4.
Hall, G.S. and da Costa, J. (1988). Affine collineations in space-time. *JMP* **29**, 2465. *See* §35.4.
Hall, G.S. and Kay, W. (1988). Curvature structure in general relativity. *JMP* **29**, 420. *See* Ch. 9.
Hall, G.S., Morgan, T. and Perjés, Z. (1987). Three-dimensional space-times. *GRG* **19**, 1137. *See* §7.2.
Hall, G.S. and Negm, D.A. (1986). Physical structure of energy-momentum tensor in general relativity. *Int. J. Theor. Phys.* **25**, 405. *See* §5.2.
Hall, G.S. and Steele, J.D. (1990). Homothety groups in space-time. *GRG* **22**, 457. *See* §§8.7, 11.3.
Hall, G.S. and Steele, J.D. (1991). Conformal vector fields in general relativity. *JMP* **32**, 1847. *See* §35.4.
Hanquin, J.L. (1984). Vacuum inhomogeneous cosmological models, in *The big bang and Georges Lemaitre. Proceedings*, ed. A. Berger, page 83 (Dordrecht, Reidel). *See* §22.3.
Hanquin, J.L. and Demaret, J. (1983). Gowdy $S^1 \times S^2$ and S^3 inhomogeneous cosmological models. *J. Phys. A* **16**, L5. *See* §22.3.
Hanquin, J.L. and Demaret, J. (1984). Exact solutions for inhomogeneous generalisations of some vacuum Bianchi models. *CQG* **1**, 291. *See* §22.3.
Hansen, R.O. (1974). Multipole moments of stationary space-times. *JMP* **15**, 46. *See* §18.8.
Harness, R.S. (1982). Space-times homogeneous on a time-like hypersurface. *J. Phys. A* **15**, 135. *See* §13.2.
Harris, R.A. and Zund, J.D. (1975). Some Bel–Petrov type *III* solutions of the Einstein equations. *Tensor* **29**, 277. *See* §31.5.
Harris, R.A. and Zund, J.D. (1976). Two variable Kasner type solutions of Einstein's field equations. *Tensor* **30**, 255. *See* §22.3.
Harris, R.A. and Zund, J.D. (1978). Multi-variable Kasner type solutions of the source-free Einstein equations. *Tensor* **32**, 39. *See* §17.3.
Harrison, B.K. (1959). Exact three-variable solutions of the field equations of general relativity. *Phys. Rev.* **116**, 1285. *See* §17.3.
Harrison, B.K. (1965). Electromagnetic solutions of the field equations of general relativity. *Phys. Rev. B* **138**, 488. *See* §§22.4, 34.1.
Harrison, B.K. (1968). New solutions of the Einstein–Maxwell equations from old. *JMP* **9**, 1744. *See* §§18.4, 34.1.

Harrison, B.K. (1978). Bäcklund transformations for the Ernst equation of general relativity. *Phys. Rev. Lett.* **41**, 1197. *See* §34.2.

Harrison, B.K. (1980). New large family of vacuum solutions of the equations of general relativity. *PRD* **21**, 1695. *See* §34.7.

Harrison, B.K. (1983). Unification of Ernst-equation Bäcklund transformations using a modified Wahlquist–Estabrook technique. *JMP* **24**, 2178. *See* §§34.2, 34.7.

Harrison, B.K. (1984). Prolongation structures and differential forms, in *Solutions of Einstein's equations: Techniques and results.* Lecture notes in physics, vol. 205, eds. C. Hoenselaers and W. Dietz, page 26 (Springer, Berlin). *See* §10.4.

Harrison, B.K. (1986). Integrable systems in general relativity. *Lectures Appl. Math.* **23**, 123. *See* Ch. 34.

Harrison, B.K. and Estabrook, F.B. (1971). Geometric approach to invariance groups and solution of partial differential systems. *JMP* **12**, 653. *See* §10.4.

Harrison, E.R. (1967). Classification of uniform cosmological models. *Mon. Not. R. Astron. Soc.* **137**, 69. *See* §14.2.

Hartle, J.B. and Hawking, S.W. (1972). Solutions of the Einstein–Maxwell equations with many black holes. *Commun. Math. Phys.* **26**, 87. *See* §21.1.

Hartley, D. (1996). Overview of computer algebra in relativity, in *Relativity and scientific computing. Computer algebra, numerics, visualization*, eds. F.W. Hehl, R.A. Puntigam and H. Ruder, page 173 (Springer, Berlin). *See* Ch. 9.

Harvey, A. (1979). Automorphisms of the Bianchi model Lie groups. *JMP* **20**, 251. *See* §13.2.

Harvey, A. (1990). Will the real Kasner metric please stand up? *GRG* **22**, 1433. *See* §13.3.

Harvey, A. (1995). Identities of the scalars of the four-dimensional Riemannian manifold. *JMP* **36**, 356. *See* §9.1.

Harvey, A. and Tsoubelis, D. (1977). Exact Bianchi IV cosmological model. *PRD* **15**, 2734. *See* §13.3.

Hassan, Z., Feinstein, A. and Manko, V.S. (1990). Asymmetric collision between plane gravitational waves with variable polarisation. *CQG* **7**, L109. *See* §25.4.

Hauser, I. (1974). Type N gravitational field with twist. *Phys. Rev. Lett.* **33**, 1112. *See* §29.3.

Hauser, I. (1978). Type N gravitational field with twist. II. *JMP* **19**, 661. *See* §29.3.

Hauser, I. (1980). Complex plane representation of the Geroch group and a proof of a Geroch conjecture, in *Group theoretical methods in physics. Proceedings of the IXth international colloquium on group theoretical methods in physics*, ed. K.B. Wolf, page 424 (Springer-Verlag, Berlin). *See* §34.6.

Hauser, I. (1984). On the homogeneous Hilbert problem for effecting Kinnersley-Chitre transformations, in *Solutions of Einstein's equations: Techniques and results.* Lecture notes in physics, vol. 205, eds. C. Hoenselaers and W. Dietz, page 128 (Springer, Berlin). *See* §34.6.

Hauser, I. and Ernst, F.J. (1978). On the generation of new solutions of the Einstein–Maxwell field equations from electrovac spacetimes with isometries. *JMP* **19**, 1316. *See* §34.1.

Hauser, I. and Ernst, F.J. (1979a). Integral equation method for effecting Kinnersley-Chitre transformations. *PRD* **20**, 362. *See* §34.6.

Hauser, I. and Ernst, F.J. (1979b). Integral equation method for effecting Kinnersley-Chitre transformations II. *PRD* **20**, 1783. *See* §34.6.

Hauser, I. and Ernst, F.J. (1979c). SU(2,1) generation of electrovacs from Minkowski space. *JMP* **20**, 1041. *See* §34.1.

Hauser, I. and Ernst, F.J. (1980a). A homogeneous Hilbert problem for the Kinnersley-Chitre transformations. *JMP* **21**, 1126. *See* §34.6.

Hauser, I. and Ernst, F.J. (1980b). A homogeneous Hilbert problem for the Kinnersley-Chitre transformations of electrovac spacetimes. *JMP* **21**, 1418. *See* §34.6.

Hauser, I. and Ernst, F.J. (1981). Proof of a Geroch conjecture. *JMP* **22**, 1051. *See* §34.6.

Hauser, I. and Ernst, F.J. (1989a). Initial value problem for colliding gravitational plane waves. I. *JMP* **30**, 872. *See* §25.1.

Hauser, I. and Ernst, F.J. (1989b). Initial value problem for colliding gravitational plane waves. II. *JMP* **30**, 2322. *See* §25.1.

Hauser, I. and Ernst, F.J. (1990). Initial value problem for colliding gravitational plane waves. III. *JMP* **31**, 871. *See* §25.1.

Hauser, I. and Ernst, F.J. (1991). Initial value problem for colliding gravitational plane waves. IV. *JMP* **32**, 198. *See* §§25.1, 34.6.

Hauser, I. and Ernst, F.J. (2001). Proof of a generalized Geroch conjecture for the hyperbolic Ernst equation. *GRG* **33**, 195. *See* §34.6.

Hauser, I. and Malhiot, R.J. (1974). Spherically symmetric static space-times which admit stationary Killing tensors of rank two. *JMP* **15**, 816. *See* §35.3.

Hauser, I. and Malhiot, R.J. (1975a). Structural equations for Killing tensors of order two. I. *JMP* **16**, 150. *See* §35.3.

Hauser, I. and Malhiot, R.J. (1975b). Structural equations for Killing tensors of order two. II. *JMP* **16**, 1625. *See* §35.3.

Hauser, I. and Malhiot, R.J. (1978). Forms of all space-time metrics which admit [(11)(11)] Killing tensors with nonconstant eigenvalues. *JMP* **19**, 187. *See* §35.3.

Havas, P. (1992). Shear-free spherically symmetric perfect fluid solutions with conformal symmetry. *GRG* **24**, 599. *See* §§11.4, 16.2.

Hawking, S.W. and Ellis, G.F.R. (1973). *The large scale structure of space-time* (Cambridge Univ. Press, Cambridge). *See* §§5.3, 12.4.

Hayward, S.A. (1990). Transmutation of matter via colliding waves. *CQG* **7**, L29. *See* §25.6.

Heckmann, O. and Schücking, E. (1958). World models, in *La structure et l'evolution de l'univers*, page 149 (Editions Stoops, Brussels). *See* §§13.2, 14.4.

Heintzmann, H. (1969). New exact static solutions of Einstein's field equations. *Z. Phys.* **228**, 489. *See* §16.1.

Held, A. (1974a). A formalism for the investigation of algebraically special metrics. I. *Commun. Math. Phys.* **37**, 311. *See* §7.4.

Held, A. (1974b). A type-(3,1) solution to the vacuum Einstein equations. *Nuovo Cim. Lett.* **11**, 545. *See* §§29.2, 29.4.

Held, A. (1975). A formalism for the investigation of algebraically special metrics. II. *Commun. Math. Phys.* **44**, 211. *See* §7.4.

Held, A. (1976a). Killing vectors in empty space algebraically special metrics. I. *GRG* **7**, 177. *See* §§30.4, 38.2.

Held, A. (1976b). Killing vectors in empty space algebraically special metrics. II. *JMP* **17**, 39. *See* §38.2.

Held, A. (1999). The GHP formalism and a type N twisting vacuum metric with Killing vector. *GRG* **31**, 1473. *See* §29.3.

Henneaux, M. (1984). Electromagnetic fields invariant up to a duality rotation under a group of isometries. *JMP* **25**, 2276. *See* §§11.1, 13.3.

Henriksen, R.N. and Wesson, P.S. (1978). Self-similar space-times. I. Three solutions. *Astrophys. Space Sci.* **53**, 429. *See* §§11.4, 15.5, 16.1.

Hereman, W. (1996). Symbolic software for Lie symmetry analysis, in *CRC handbook of Lie group analysis of differential equations*, vol. 3, ed. N.H. Ibragimov, page 367 (CRC Press, Boca Raton, Florida). *See* §10.2.

Hergesell, G.A. (1985). Eben-symmetrische Raum-zeiten. Dissertation T.U. München, Mathematisches Institut. *See* §15.1.

Herlt, E. (1972). Über eine Klasse innerer stationärer axial-symmetrischer Lösungen der Einsteinschen Feldgleichungen mit idealem fluidem Medium. *Wiss. Zeitschr. FSU Jena, Math.-Naturw. Reihe* **21**, 19. *See* §21.2.

Herlt, E. (1978a). Remarks on stationary axially symmetric vacuum solutions in Einstein's general theory of relativity. *Wiss. Zeitschr. FSU Jena, Math.-Naturw. Reihe* **27**, 329. *See* §19.5.

Herlt, E. (1978b). Static and stationary axially symmetric gravitational fields of bounded sources. I. Solutions obtainable from the van Stockum metric. *GRG* **9**, 711. *See* §§20.6, 21.1, 34.1.

Herlt, E. (1979). Static and stationary axially symmetric gravitational fields of bounded sources. II. Solutions obtainable from Weyl's class. *GRG* **11**, 337. *See* §§20.6, 21.1, 34.1.

Herlt, E. (1980). Kerr–Schild–Vaidya fields with axial symmetry. *GRG* **12**, 1. *See* §32.4.

Herlt, E. (1986). Reduktion der Einsteinschen Feldgleichungen für Typ-N-Vakuumlösungen mit Killingvektor und homothetischer gruppe auf eine gewöhnliche Differentialgleichung 3. Ordnung für eine relle funktion. *Wiss. Zeitschr. FSU Jena, Naturw. Reihe* **35**, 735. *See* §29.3.

Herlt, E. (1988). Stationary axisymmetric perfect fluid solutions in general relativity. *GRG* **20**, 635. *See* §13.4.

Herlt, E. (1996). Spherically symmetric nonstatic perfect fluid solutions with shear. *GRG* **28**, 919. *See* §16.2.

Herlt, E. and Hermann, H.-C. (1980). A note on the equivalence of two axially symmetric stationary perfect fluid solutions of Einstein's field equation. *Experimentelle Technik d. Phys.* **28**, 97. *See* §21.2.

Herlt, E. and Stephani, H. (1984). Algebraically special, shearfree, diverging, and twisting Einstein–Maxwell fields. *CQG* **1**, 95. *See* §§28.2, 30.5.

Herlt, E. and Stephani, H. (1992). Invariance transformations of the class $y'' = f(x)y^2$ of differential equations. *JMP* **33**, 3983. *See* §16.2.

Hermann, H.-C. (1983). Eigenschaften und Konstruktion von Lösungen der Einsteinschen Gleichungen mit idealer Flüssigkeit, die eine homothetische gruppe gestatten. Ph.D. thesis, Jena. *See* §§11.4, 13.4, 21.2.

Hernandez P., J.L., Manko, V.S. and Martín, J. (1993). Some asymptotically flat generalizations of the Curzon metric. *JMP* **34**, 4760. *See* §§34.1, 34.3.

Hernandez P., J.L. and Martín, J. (1994). Monopole-quadrupole static axisymmetric solutions of Einstein field equations. *GRG* **26**, 877. *See* §20.2.

Herrera, L. and Jimenez, J. (1982). The complexification of a nonrotating sphere: an extension of the Newman–Janis algorithm. *JMP* **23**, 2339. *See* §21.1.

Herrera, L., Paiva, F.M. and Santos, N.O. (1999). The Levi-Civita spacetime as a limiting case of the γ spacetime. *JMP* **40**, 4064. *See* §9.5.

Herrera, L. and Ponce de León, J. (1985). Perfect fluid spheres admitting a one-parameter group of conformal motions. *JMP* **26**, 778. *See* §15.6.

Hewitt, C.G. (1991a). Algebraic invariant curves in cosmological dynamical systems and exact solutions. *GRG* **23**, 1363. *See* §§13.2, 22.3.

Hewitt, C.G. (1991b). An exact tilted Bianchi II cosmology. *CQG* **8**, 109. *See* §14.4.

Hewitt, C.G., Bridson, R. and Wainwright, J.A. (2001). The asymptotic regimes of tilted Bianchi II cosmologies. *GRG* **33**, 65. *See* §14.4.

Hewitt, C.G. and Wainwright, J. (1990). Orthogonally transitive G_2 cosmologies. *CQG* **7**, 2295. *See* §§23.1, 23.3.

Hewitt, C.G., Wainwright, J. and Glaum, M. (1991). Qualitative analysis of a class of inhomogeneous self-similar cosmological models: II. *CQG* **8**, 1505. *See* §§23.1, 23.2.

Hewitt, C.G., Wainwright, J. and Goode, S.W. (1988). Qualitative analysis of a class of inhomogeneous self-similar cosmological models. *CQG* **5**, 1313. *See* §§23.1, 23.2.

Hiromoto, R.E. and Ozsváth, I. (1978). On homogeneous solutions of Einstein's field equations. *GRG* **9**, 299. *See* §§12.1, 12.2.

Hirota, R. (1976). Direct method of finding exact solutions of nonlinear evolution equations, in *Bäcklund transformations*. Lecture notes in mathematics, vol. 515, ed. R.M. Miura, page 40 (Springer Verlag, Berlin). *See* §10.10.

Hiscock, W.A. (1981). On black holes in magnetic universes. *JMP* **22**, 1828. *See* §21.1.

Hodgkinson, D.E. (1984). Empty space-times of embedding class two. *GRG* **16**, 569. *See* §37.5.

Hodgkinson, D.E. (1987). Type D empty space-times of embedding class 2. *GRG* **3**, 253. *See* §37.5.

Hodgkinson, D.E. (2000). Petrov type D Einstein space-times of embedding class two. *JMP* **42**, 863. *See* §37.5.

Hoenselaers, C. (1976). On generation of solutions of Einstein's equations. *JMP* **17**, 1264. *See* §34.1.

Hoenselaers, C. (1978a). Algebraically special one Killing vector solutions of Einstein's equations. *Prog. Theor. Phys.* **60**, 747. *See* §§17.3, 31.5.

Hoenselaers, C. (1978b). A new solution of Ernst's equation. *J. Phys. A* **11**, L75. *See* §34.1.

Hoenselaers, C. (1978c). On the effect of motions on energy-momentum tensor. *Prog. Theor. Phys.* **59**, 1518. *See* §11.1.

Hoenselaers, C. (1978d). On the stationary Einstein–Maxwell equations. *JMP* **19**, 539. *See* §10.8.

Hoenselaers, C. (1979a). Symmetries of the stationary Einstein–Maxwell field equations. V. *JMP* **20**, 2526. *See* §34.1.

Hoenselaers, C. (1979b). Weyl conform tensor of the Tomimatsu–Sato $\delta = 3$ metric. *GRG* **11**, 325. *See* §20.5.

Hoenselaers, C. (1981). An axisymmetric stationary solution of Einstein's equations calculated by computer. *J. Phys. A* **14**, L427. *See* §34.3.

Hoenselaers, C. (1982a). Brute force method for solving Ernst's equation and limits of the Kinnersley-Chitre solution. *J. Phys. A* **15**, 3531. *See* §20.6.

Hoenselaers, C. (1982b). A static solution of the Einstein–Maxwell equations. *Prog. Theor. Phys.* **67**, 697. *See* §34.1.

Hoenselaers, C. (1983). Lagrangians, Bäcklund transformations and a stationary solution of the relativistic two-body problem, in *Proceedings of the third Marcel Grossmann meeting on general relativity*, ed. Hu Ning, page 221 (North-Holland, Amsterdam). *See* §34.3.

Hoenselaers, C. (1984). HKX-transformations. An introduction, in *Solutions of Einstein's equations: Techniques and results*. Lecture notes in physics, vol. 205, eds. C. Hoenselaers and W. Dietz, page 68 (Springer, Berlin). *See* §34.3.

Hoenselaers, C. (1985a). Limits of the double Kerr solution, in *Galaxies, axisymmetric systems and relativity. Essays presented to W.B.Bonnor on his 65th birthday*, ed. M.A.H. MacCallum, page 126 (Cambridge University Press, Cambridge). *See* §34.4.

Hoenselaers, C. (1985b). The sine-Gordon prolongation algebra. *Prog. Theor. Phys.* **74**, 645. *See* §10.4.

Hoenselaers, C. (1986). More prolongation structures. *Prog. Theor. Phys.* **75**, 1014. *See* §10.4.

Hoenselaers, C. (1988). Imagines concordes et functiones Lagrangii, in *Relativity today. Proceedings of the second Hungarian relativity workshop*, ed. Z. Perjés, page 15 (World Scientific, Singapore). *See* §§8.6, 10.8, 34.1.

Hoenselaers, C. (1992). One Killing-vector solutions of Einstein's equations, in *Relativity today. Proceedings of the third Hungarian workshop*, ed. Z. Perjes (Nova Science Publishers, New York). *See* §17.3.

Hoenselaers, C. (1993). Axisymmetric stationary solutions of Einstein's equations, in *Rotating objects and relativistic physics. Proceedings*, eds. F.J. Chinea and L.M. Gonzalez-Romero, page 29 (Springer, Berlin). *See* §10.5.

Hoenselaers, C. (1995). The Weyl solution for a ring in a homogeneous field. *CQG* **12**, 141. *See* §20.2.

Hoenselaers, C. (1997). Factor structure of rational electrovacuum metrics. *CQG* **14**, 2627. *See* §20.7.

Hoenselaers, C. and Dietz, W. (1984). The rank N HKX transformations: new stationary axisymmetric gravitational fields. *GRG* **16**, 71. *See* §34.3.

Hoenselaers, C. and Ernst, F.J. (1983). Remarks on the Tomimatsu–Sato metrics. *JMP* **24**, 1817. *See* §20.5.

Hoenselaers, C. and Ernst, F.J. (1990). Matching *pp* waves to the Kerr metric. *JMP* **31**, 144. *See* §25.4.

Hoenselaers, C., Kinnersley, W. and Xanthopoulos, B.C. (1979). Symmetries of the stationary Einstein–Maxwell equations. VI. Transformations which generate asymptotically flat spacetimes with arbitrary multipole moments. *JMP* **20**, 2530. *See* §34.3.

Hoenselaers, C. and Perjés, Z. (1990). Factor structure of rational vacuum metrics. *CQG* **7**, 2215. *See* §20.7.

Hoenselaers, C. and Perjés, Z. (1993). Remarks on the Robinson–Trautman solutions. *CQG* **10**, 375. *See* §28.1.

Hoenselaers, C. and Skea, J.E.F. (1989). Generating solutions of Einstein's field equations by typing mistakes. *GRG* **21**, 17. *See* §§17.3, 31.6.

Hoffman, R.B. (1969a). Stationary axially symmetric generalizations of the Weyl solution in general relativity. *Phys. Rev.* **182**, 1361. *See* §20.4.

Hoffman, R.B. (1969b). Stationary 'noncanonical' solutions of the Einstein vacuum field equations. *JMP* **10**, 953. *See* §24.5.

Hogan, P.A. (1990). McVittie's mass particle in an expanding universe and related solutions of Einstein's equations. *Astrophys. J.* **360**, 315. *See* §15.6.

Hogan, P.A., Barrabés, C. and Bressange, G.F. (1998). Colliding plane waves in Einstein–Maxwell theory. *Lett. Math. Phys.* **43**, 263. *See* §25.5.

Hojman, R. and Santamarina, J. (1984). Exact solutions of plane symmetric cosmological models. *JMP* **25**, 1973. *See* §15.7.

Hon, E. (1975). Application of REDUCE system to some problems in general relativity. M. Math. thesis, University of Waterloo. *See* §9.3.

Hori, S. (1996a). Generalization of Tomimatsu–Sato solutions. II. *Prog. Theor. Phys.* **95**, 557. *See* §20.5.

Hori, S. (1996b). Generalization of Tomimatsu–Sato solutions. IV. *Prog. Theor. Phys.* **96**, 327. *See* §20.5.

Horský, J. (1975). The gravitational field of a homogeneous plate with a non-zero cosmological constant. *Czech. J. Phys. B* **25**, 1081. *See* §15.7.

Horváth, I. and Lukács, B. (1992). Some notes on stationary vacuum solutions of the Einstein equations with shearing nongeodesic eigenrays. *Ann. Phys. (Germany)* **1**, 500. *See* §18.5.

Horváth, I., Lukásh, B. and Szello, L. (1997). All stationary vacuum solutions with geodesic shearfree rays. *Acta Phys. Slovaca* **47**, 499. *See* §18.5.

Hotta, M. and Tanaka, M. (1993). Shock-wave geometry with nonvanishing cosmological constant. *CQG* **10**, 307. *See* §9.5.

Hou, Bo-Yu and Li, W. (1988). A new method to generate the solutions of the reduced Einstein equations. *Chinese Phys.* **8**, 343. *See* §34.7.

Hsu, L. and Wainwright, J. (1986). Self-similar spatially homogeneous cosmologies: orthogonal perfect fluid and vacuum solutions. *CQG* **3**, 1105. *See* §11.4.

Hughston, L.P. (1971). Generalized Vaidya metrics. *Int. J. Theor. Phys.* **4**, 267. *See* §30.7.

Hughston, L.P. and Jacobs, K. (1970). Homogeneous electromagnetic and massive-vector-meson fields in Bianchi cosmologies. *Astrophys. J.* **160**, 147. *See* §13.3.

Hughston, L.P., Penrose, R., Sommers, P. and Walker, M. (1972). On a quadratic first integral for the charged particle orbits in the charged Kerr solution. *Commun. Math. Phys.* **27**, 303. *See* §35.3.

Hughston, L.P. and Sommers, P. (1973). Spacetimes with Killing tensors. *Commun. Math. Phys.* **32**, 147. *See* §35.3.

Ibañez, J. and Verdaguer, E. (1983). Soliton collisions in general relativity. *Phys. Rev. Lett.* **51**, 1313. *See* §34.5.

Ibañez, J. and Verdaguer, E. (1985). Multisoliton solutions to Einstein's equations. *PRD* **31**, 251. *See* §34.5.

Ibañez, J. and Verdaguer, E. (1986). Generation of gravitational waves in vacuum FRW models. *CQG* **3**, 1235. *See* §34.5.

Ibohal, N. (1997). On the relationship between Killing–Yano tensors and electromagnetic fields on curved spaces. *Astrophys. Space Sci.* **249**, 73. *See* §35.3.

Ibragimov, N.H. (1985). *Transformation groups applied to mathematical physics* (Reidel, Boston). *See* §10.2.

Ihrig, E. (1975). The uniqueness of g_{ij} in terms of $R^l{}_{ijk}$. *Int. J. Theor. Phys.* **14**, 23. *See* §9.1.

Ikeda, M., Kitamura, S. and Matsumoto, M. (1963). On the embedding of spherically symmetric space-times. *J. Math. Kyoto Univ.* **3**, 71. *See* §37.3.

Il'ichev, A.T. and Sibgatullin, N.R. (1985). Mathematical construction of multisheeted solutions in GTR. *Sov. Phys. – Dokl.* **30**, 464. *See* §34.6.

Illge, R. (1988). On potentials for several classes of spinor and tensor fields in curved space-times. *GRG* **20**, 551. *See* §3.6.

Islam, J.N. (1979). Recently found solution of Einstein's equations. *Phys. Rev. Lett.* **43**, 601. *See* §24.4.

Islam, J.N. (1983a). Closed form for van Stockum interior solution of Einstein's equations. *Phys. Lett. A* **94**, 421. *See* §21.2.

Islam, J.N. (1983b). On rotating charged dust in general relativity. V. *Proc. Roy. Soc. London A* **389**, 291. *See* §22.2.

Islam, J.N. (1985). *Rotating fields in general relativity* (Cambridge University Press, Cambridge). *See* §20.1.

Israel, W. (1966). New interpretation of the extended Schwarzschild manifold. *Phys. Rev.* **143**, 1016. *See* §15.4.

Israel, W. (1970). Differential forms in general relativity. *Commun. Dublin Inst. Adv. Stud. A* **19**. *See* §3.4.

Israel, W. and Khan, K.A. (1964). Collinear particles and Bondi dipoles in general relativity. *Nuovo Cim.* **33**, 331. *See* §20.2.

Israel, W. and Spanos, J.T.J. (1973). Equilibrium of charged spinning masses in general relativity. *Nuovo Cim. Lett.* **7**, 245. *See* §18.7.

Israel, W. and Stewart, J.M. (1980). Progress in relativistic thermodynamics and electrodynamics of continuous media, in *General relativity and gravitation. One hundred years after the birth of Albert Einstein,* vol. 2, ed. A. Held, page 491 (Plenum Press, New York). *See* §5.5.

Israel, W. and Wilson, G.A. (1972). A class of stationary electromagnetic vacuum fields. *JMP* **13**, 865. *See* §18.7.

Ivanov, B.V. (1998). Colliding axisymmetric pp-waves. *PRD* **57**, 3378. *See* §25.3.

Ivanov, B.Y. (1999). Expanding, axisymmetric, pure-radiation gravitational fields with a simple twist. *PRD* **60**, 104005. *See* §30.7.

Ivanov, G.G. (1977). An exact solution of Einstein's equation that describes spherical gravitational and electromagnetic waves (in Russian). *Grav. Teor. Otnos.* **12**, 69. *See* §28.2.

Iyer, B.R. and Vishveshwara, C.V. (1983). Magnetization of all stationary cylindrically symmetric vacuum metrics. *JMP* **24**, 1568. *See* §34.1.

Jackson, J.C. (1970). Infilling for NUT space. *JMP* **11**, 924. *See* §21.2.

Jacobs, K.C. (1968). Spatially homogeneous and Euclidean cosmological models with shear. *Astrophys. J.* **153**, 661. *See* §§14.3, 14.4.

Jacobs, K.C. (1969). Cosmologies of Bianchi type I with a uniform magnetic field. *Astrophys. J.* **155**, 379. *See* §§13.3, 14.1, 14.4.

Jantzen, R.T. (1980a). Soliton solutions of the Einstein equations generated from cosmological solutions with additional symmetry. *Nuovo Cim. B* **59**, 287. *See* §§10.11, 22.3.

Jantzen, R.T. (1980b). Variation of parameters in cosmology. *Ann. Phys.* (*USA*) **127**, 302. *See* §14.1.

Jantzen, R.T. (1983). Uniformly accelerated particle metrics obtained by soliton techniques. *GRG* **15**, 115. *See* §34.5.

Jantzen, R.T. (1984). Spatially homogeneous dynamics: A unified picture, in *Cosmology of the early universe*, eds. R. Ruffini and L.-Z. Fang, page 233 (World Scientific, Singapore). *See* §13.2.

Jantzen, R.T. (1986). Rotation in cosmology: Comments on 'Imparting rotation to a Bianchi type II space-time,' by M.J. Reboucas and J.B.S. d'Olival (J. Math. Phys. 27, 417 (1986)) and similar papers. *JMP* **27**, 2748. *See* §13.2.

Jantzen, R.T. (1988). Power law time lapse gauges. *PRD* **37**, 3472. *See* §13.2.

Jantzen, R.T. and Rosquist, K. (1986). Adapted slicings of space-times possessing simply transitive similarity groups. *JMP* **27**, 1191. *See* §13.2.

Jantzen, R.T. and Uggla, C. (1998). The kinematical role of automorphisms in the orthonormal frame approach to Bianchi cosmology. *JMP* **40**, 353. *See* §13.2.

Joly, G.C. and MacCallum, M.A.H. (1990). Computer-aided classification of the Ricci tensor in general relativity. *CQG* **7**, 541. *See* §9.3.

Joly, G.C., MacCallum, M.A.H. and Siklos, S.T.C. (1992). Comment on the 'generalized axisymmetric spacetimes' of Bampi and Cianci. *CQG* **9**, 795. *See* §§24.3, 31.5.

Jordan, P., Ehlers, J. and Kundt, W. (1960). Strenge Lösungen der Feldgleichungen der Allgemeinen Relativitätstheorie. *Akad. Wiss. Lit. Mainz, Abhandl. Math.-Nat. Kl.* **2**. *See* §§9.1, 18.1, 22.1, 24.5.

Jordan, P., Ehlers, J. and Sachs, R.K. (1961). Beiträge zur Theorie der reinen Gravitationsstrahlung. *Akad. Wiss. Lit. Mainz, Abhandl. Math.-Nat. Kl.* **1**. *See* §§4.3, 6.1.

Jordan, P. and Kundt, W. (1961). Geometrodynamik im Nullfall. *Akad. Wiss. Lit. Mainz, Abhandl. Math.-Nat. Kl.* **3**. *See* §5.4.

Joseph, V. (1966). A spatially homogeneous gravitational field. *Proc. Camb. Phil. Soc.* **62**, 87. *See* §13.3.

Julia, B. and Nicolai, H. (1995). Null Killing vector dimensional reduction and galilean geometrodynamics. *Nucl. Phys. B* **439**, 291. *See* §34.1.

Kac, V.G. (1990). *Infinite dimensional Lie algebras* (Cambridge University Press, Cambridge). *See* §10.4.

Kaigorodov, V.R. (1962). Einstein spaces of maximum mobility (in Russian). *Dokl. Akad. Nauk SSSR* **7**, 893. *See* §§12.5, 38.2.

Kaigorodov, V.R. (1967). Exact type III solutions of the field equations $R_{ij} = 0$ (in Russian). *Grav. Teor. Otnos., Univ. Kazan* **3**, 155. *See* §31.5.

Kaigorodov, V.R. (1971). Petrov classification and recurrent spaces (in Russian), in *Gravitatsiya*, page 52 (Nauk dumka, Kiev). *See* §35.2.

Kaigorodov, V.R. and Timofeev, V.N. (1996). Algebraically special solutions of the Einstein equations $R_{ij} = 6\Lambda g_{ij}$. *Grav. Cosmol.* **2**, 107. *See* §29.1.

Kaliappan, P. and Lakshmanan, M. (1981). Similarity solutions for the Ernst equations with electromagnetic fields. *JMP* **22**, 2447. *See* §21.1.

Kamani, D. and Mansouri, R. (1996). A new class of inhomogeneous cosmological solutions. Preprint, Potsdam. *See* §23.3.

Kammerer, J.B. (1966). Sur les directions principales du tenseur de courbure. *C. R. Acad. Sci. (Paris)* **263**, 533. *See* §7.5.

Kamran, N. (1988). Killing–Yano tensors and their role in separation of variables, in *Proceedings of the second Canadian conference on general relativity and relativistic astrophysics*, eds. A. Coley, C. Dyer and B.O.J. Tupper, page 43 (World Scientific, Singapore). *See* §35.3.

Kantowski, R. (1966). Some relativistic cosmological models. Ph. D. thesis, University of Texas at Austin. *See* §§13.1, 14.3.

Kantowski, R. and Sachs, R.K. (1966). Some spatially homogeneous anisotropic relativistic cosmological models. *JMP* **7**, 443. *See* §§13.1, 14.3.

Kar, S.C. (1926). Das Gravitationsfeld einer geladenen Ebene. *Phys. Zeitschr.* **XXVII**, 208. *See* §15.4.

Karlhede, A. (1980a). On a coordinate-invariant description of Riemannian manifolds. *GRG* **12**, 963. *See* §9.4.

Karlhede, A. (1980b). A review of the geometrical equivalence of metrics in general relativity. *GRG* **12**, 693. *See* §9.2.

Karlhede, A. and Åman, J.E. (1979). Progress towards a solution of the equivalence problem in general relativity, in *EUROSAM '79: Symbolic and algebraic computation.* Lecture notes in computer science, vol. 72, ed. E. Ng, page 42 (Springer, Berlin). *See* §9.2.

Karlhede, A. and Lindström, U. (1983). Finding space-time geometries without using a metric. *GRG* **15**, 597. *See* §9.4.

Karlhede, A. and MacCallum, M.A.H. (1982). On determining the isometry group of a Riemannian space. *GRG* **14**, 673. *See* §§9.4, 18.6.

Karmarkar, K.R. (1948). Gravitational metrics of spherical symmetry and class one. *Proc. Indian Acad. Sci. A* **27**, 56. *See* §37.3.

Kasner, E. (1921). Geometrical theorems on Einstein's cosmological equations. *Amer. J. Math.* **43**, 217. *See* §§13.3, 35.2.

Kasner, E. (1925). Solutions of the Einstein equations involving functions of only one variable. *Trans. A.M.S.* **27**, 155. *See* §13.3.

Kastor, D. and Traschen, J. (1993). Cosmological multi-black-hole solutions. *PRD* **47**, 5370. *See* §21.1.

Katzin, G.H. and Levine, J. (1972). Applications of Lie derivatives to symmetries, geodesic mappings, and first integrals in Riemannian spaces. *Colloquium Math.* **26**, 21. *See* §35.4.

Katzin, G.H. and Levine, J. (1981). Geodesic first integrals with explicit path-parameter dependence in Riemannian space-times. *JMP* **22**, 1878. *See* §35.4.

Katzin, G.H., Levine, J. and Davis, W.R. (1969). Curvature collineations: A fundamental symmetry property of the space-times of general relativity defined by the vanishing Lie derivative of the Riemann curvature tensor. *JMP* **10**, 617. *See* §35.4.

Kellner, A. (1975). 1-dimensionale Gravitationsfelder. Ph.D. dissertation, Inst. f. Theoret. Physik Göttingen. *See* §§13.1, 13.3, 14.4.

Kerns, R.M. and Wild, W.J. (1982a). Black hole in a gravitational field. *GRG* **14**, 1. *See* §20.2.

Kerns, R.M. and Wild, W.J. (1982b). Generalised Zipoy–Voorhees metric. *PRD* **26**, 3726. *See* §20.2.

Kerr, G.D. (1998). Algebraically special Einstein spaces: Kerr–Schild metrics and homotheties. Ph.D. thesis, Queen Mary and Westfield College, London. *See* §§7.4, 8.7, 23.1, 32.5.

Kerr, R.P. (1963a). Gravitational field of a spinning mass as an example of algebraically special metrics. *Phys. Rev. Lett.* **11**, 237. *See* §§20.5, 29.1, 29.5.

Kerr, R.P. (1963b). Scalar invariants and groups of motions in a four dimensional Einstein space. *J. Math. Mech.* **12**, 33. *See* §8.4.

Kerr, R.P. and Debney, G. (1970). Einstein spaces with symmetry groups. *JMP* **11**, 2807. *See* §§13.3, 29.2, 38.2.

Kerr, R.P. and Schild, A. (1965a). A new class of vacuum solutions of the Einstein field equations, in *Atti del convegno sulla relatività generale; problemi dell'energia e ondi gravitationali*, page 222 (Barbèra, Firenze). *See* §32.2.

Kerr, R.P. and Schild, A. (1965b). Some algebraically degenerate solutions of Einstein's gravitational field equations. *Proc. Symp. Appl. Math* **17**, 199. *See* §32.2.

Kerr, R.P. and Wilson, W.B. (1979). Singularities in the Kerr–Schild metrics. *GRG* **10**, 273. *See* §32.2.

Kerr, R.P. and Wilson, W.B. (1989). A simplified representation of stationary axially symmetric metrics, in *Proceedings of the fifth Marcel Grossmann meeting on general relativity*, Pt. A, eds. D.G. Blair and M.J. Cunningham, page 507 (World Scientific, Singapore). *See* §20.7.

Khan, K.A. and Penrose, R. (1971). Scattering of two impulsive gravitational plane waves. *Nature* **229**, 185. *See* §§25.1, 25.3.

Kharbediya, L.I. (1976). Some exact solutions of the Friedmann equations with the cosmological term (in Russian). *Astron. Zh.* (*USSR*) **53**, 1145. *See* §14.2.

Khlebnikov, V.I. (1986). Gravitational radiation in electromagnetic universes. *CQG* **3**, 169. *See* §§31.7, 31.8.

Kihara, M., Oohara, K., Sato, H. and Tomimatsu, A. (1983). Structure of the two Kerr-solution, in *Proceedings of the third Marcel Grossmann meeting on general relativity*, ed. Hu Ning, page 1049 (North-Holland, Amsterdam). *See* §34.4.

King, A.R. (1974). New types of singularity in general relativity: the general cylindrically symmetric stationary dust solution. *Commun. Math. Phys.* **38**, 157. *See* §22.2.

King, A.R. and Ellis, G.F.R. (1973). Tilted homogeneous cosmological models. *Commun. Math. Phys.* **31**, 209. *See* §§5.2, 14.1.

Kinnersley, W. (1969a). Field of an arbitrary accelerating point mass. *Phys. Rev.* **186**, 1335. *See* §32.1.

Kinnersley, W. (1969b). Type D vacuum metrics. *JMP* **10**, 1195. *See* §§7.1, 13.3, 21.1, 29.2, 29.5, 31.5, 35.3.

Kinnersley, W. (1973). Generation of stationary Einstein–Maxwell fields. *JMP* **14**, 651. *See* §34.1.

Kinnersley, W. (1975). Recent progress in exact solutions, in *Proceedings of GR7, Tel-Aviv, General relativity and gravitation*, eds. G. Shaviv and J. Rosen, page 109 (Wiley, New York and London). *See* §§1.1, 13.3.

Kinnersley, W. (1977). Symmetries of the stationary Einstein–Maxwell field equations. I. *JMP* **18**, 1529. *See* §34.3.

Kinnersley, W. (1980). Symmetries of the stationary Einstein–Maxwell field equations. VII. Charging transformations. *JMP* **21**, 2231. *See* §34.8.

Kinnersley, W. (1990). The Hou–Li transformations. *CQG* **7**, 1827. *See* §34.7.

Kinnersley, W. (1991). Symmetries of the stationary axisymmetric vacuum Einstein equations which preserve asymptotic flatness. *CQG* **8**, 1011. *See* §34.7.

Kinnersley, W. and Chitre, D.M. (1977). Symmetries of the stationary Einstein–Maxwell equations. II. *JMP* **18**, 1538. *See* §§10.5, 34.3.

Kinnersley, W. and Chitre, D.M. (1978a). Symmetries of the stationary Einstein–Maxwell field equations. III. *JMP* **19**, 1926. *See* §§10.5, 34.3.

Kinnersley, W. and Chitre, D.M. (1978b). Symmetries of the stationary Einstein–Maxwell equations. IV. Transformations which preserve asymptotic flatness. *JMP* **19**, 2037. *See* §§20.6, 34.3.

Kinnersley, W. and Kelley, E.F. (1974). Limits of the Tomimatsu–Sato gravitational field. *JMP* **15**, 2121. *See* §20.6.

Kinnersley, W. and Walker, M. (1970). Uniformly accelerated charged mass in general relativity. *PRD* **2**, 1359. *See* §§17.2, 21.1, 32.1.

Kitamura, S. (1978). A c.s. method for stationary axially symmetric space-times and an application to generalizations of Gödel's universe. *Tensor* **32**, 156. *See* §19.3.

Kitamura, S. (1989). On spherically symmetric and plane symmetric perfect fluid solutions and their imbedding of class one. *Tensor* **48**, 169. *See* §37.4.

Kitamura, S. (1994). On spherically symmetric perfect fluid solutions with shear. *CQG* **11**, 195. *See* §15.6.

Kitamura, S. (1995a). On spherically symmetric perfect fluid solutions with shear. II. *CQG* **12**, 827. *See* §§14.3, 15.6, 35.4.

Kitamura, S. (1995b). A remark on the invariant characterization of a class of exact spherically symmetric perfect fluid solutions with shear. *CQG* **12**, 1559. *See* §§15.6, 35.4.

Kitchingham, D.W. (1984). The use of generating techniques for space-times with two non-null commuting Killing vectors in vacuum and stiff perfect fluid cosmological models. *CQG* **1**, 677. *See* §10.11.

Kitchingham, D.W. (1986). The application of the homogeneous Hilbert problem of Hauser and Ernst to cosmological models with spatial axes of symmetry. *CQG* **3**, 133. *See* §10.11, Ch. 23.

Klein, C. and Richter, O. (1997). The Ernst equation and the Riemann–Hilbert problem on hyperelliptic Riemann surfaces. *J. Geom. Phys.* **24**, 53. *See* §34.6.

Klein, C. and Richter, O. (1999). Exact relativistic gravitational field of a stationary counterrotating dust disk. *Phys. Rev. Lett.* **83**, 2884. *See* §34.6.

Klein, O. (1953). On a class of spherically symmetric solutions of Einstein's gravitational equations. *Ark. Fys.* **7**, 487. *See* §16.1.

Klekowska, J. and Osinovski, M.E. (1971). Group-theoretical analysis of some type N solutions of the Einstein field equations in vacuo. Preprint ITP-73-II7E, Kiev. *See* §12.2.

Knutsen, H. (1986). Physical properties of some McVittie metrics. *Int. J. Theor. Phys.* **25**, 741. *See* §16.2.

Knutsen, H. (1995). On a class of spherically symmetric perfect fluid distributions in non-comoving coordinates. *CQG* **12**, 2817. *See* §16.2.

Kobayashi, S. and Nomizu, K. (1969). *Foundations of differential geometry* (2 vols.) Interscience Tracts in Pure and Applied Mathematics, vol. 15 (Interscience, New York). *See* §§2.1, 8.4.

Kobiske, R.A. and Parker, L. (1974). Solution of the Einstein–Maxwell equations for two unequal spinning sources in equilibrium. *PRD* **10**, 2321. *See* §21.1.

Köhler, E. and Walker, M. (1975). A remark on the generalized Goldberg–Sachs theorem. *GRG* **6**, 507. *See* §26.1.

Kohler, M. and Chao, K.L. (1965). Zentralsymmetrische statische Schwerefelder mit Räumen der Klasse I. *Z. Naturforsch.* **20a**, 1537. *See* §37.4.

Kolassis, C. (1989). Spacetimes with a two-dimensional group of isometries. *CQG* **6**, 683. *See* §17.1.

Kolassis, C. (1996). Vacuum spacetimes with a two-dimensional orthogonally transitive group of homothetic motions. *GRG* **28**, 787. *See* §11.4.

Kolassis, C. and Griffiths, J.B. (1996). Perfect fluid space-times with a two-dimensional orthogonally transitive group of homothetic motions. *GRG* **28**, 805. *See* §§21.2, 23.2, 23.4.

Kolassis, C. and Ludwig, G. (1996). Space-times with a two-dimensional group of conformal motions. *Int. J. Mod. Phys. A* **11**, 845. *See* §11.3.

Kolassis, C. and Santos, N.O. (1987). Spacetimes with a preferred null direction and a two-dimensional group of isometries: the null dust case. *CQG* **4**, 599. *See* §17.1.

Komar, A. (1959). Covariant conservation laws in general relativity. *Phys. Rev.* **113**, 934. *See* §34.4.

Kompaneets, A.S. (1958). Strong gravitational waves in vacuum (in Russian). *Zh. Eks. Teor. Fiz.* **34**, 953. *See* §22.1.

Kordas, P. (1995). Reflection-symmetric, asymptotically flat solutions of the vacuum axistationary Einstein equations. *CQG* **12**, 2037. *See* §19.5.

Kordas, P. (1999). Aspects of solution-generating techniques for space-times with two commuting Killing vectors. *GRG* **31**, 1941. *See* Ch. 34.

Korkina, M.P. (1981). Static configuration with an ultrarelativistic equation of state at the center. *Sov. Phys. J.* **24**, 468. *See* §16.1.

Korotkin, D.A. (1988). Finite-gap solutions of the stationary axisymmetric Einstein equation in vacuum. *Theor. Math. Phys.* **77**, 1018. *See* §34.6.

Korotkin, D.A. (1993). Elliptic solutions of stationary axisymmetric Einstein equations. *CQG* **10**, 2587. *See* §34.6.

Korotkin, D.A. (1997). Some remarks on finite-gap solutions of the Ernst equation. *Phys. Lett. A* **229**, 195. *See* §34.6.

Korotkin, D.A. and Nicolai, H. (1994). The Ernst equation on a Riemann surface. *Nuclear Phys. B* **429**, 229. *See* §34.7.

Kóta, J., Lukács, B. and Perjés, Z. (1982). Solutions of the spin coefficient equations with nongeodesic eigenrays, in *Proceedings of the second Marcel Grossmann meeting on general relativity*, ed. R. Ruffini, page 203 (North-Holland, Amsterdam). *See* §§21.1, 26.4.

Kóta, J. and Perjés, Z. (1972). All stationary vacuum metrics with shearing geodesic eigenrays. *JMP* **13**, 1695. *See* §§18.5, 32.5.

Kottler, F. (1918). Über die physikalischen Grundlagen der Einsteinschen Gravitationstheorie. *Ann. Phys. (Germany)* **56**, 410. *See* §15.4.

Koutras, A. (1992a). Killing tensors from conformal Killing vectors. *CQG* **9**, 1573. *See* §35.3.

Koutras, A. (1992b). Mathematical properties of homothetic space-times. Ph.D. thesis, Queen Mary and Westfield College, London. *See* §§8.7, 9.4, 11.4, 13.3, 14.4, 23.1.

Koutras, A. (1992c). A spacetime for which the Karlhede invariant classification requires the fourth covariant derivative of the Riemann tensor. *CQG* **9**, L143. *See* §9.3.

Koutras, A. and Mars, M. (1997). General non-rotating perfect fluid solution with an abelian spacelike C_3 including only one isometry. *GRG* **29**, 1075. *See* §§33.3, 35.4.

Koutras, A. and Skea, J.E.F. (1998). An algorithm for determining whether a space-time is homothetic. *Comp. Phys. Comm.* **115**, 350. *See* §§9.4, 11.3.

Kowalczyński, J.K. (1978). Charged tachyon in general relativity: can it be detected? *Phys. Lett. A* **65**, 269. *See* §28.2.

Kowalczyński, J.K. (1985). On a certain class of solutions of the Einstein–Maxwell equations. *JMP* **26**, 1743. *See* §28.2.

Kowalczyński, J.K. and Plebański, J.F. (1977). Metric and Ricci tensors for a certain class of space-times of type D. *Int. J. Theor. Phys.* **16**, 371; erratum **17**, 388 (1978) *See* §31.7.

Kozameh, C.N., Newman, E.T. and Tod, K.P. (1985). Conformal Einstein spaces. *GRG* **17**, 343. *See* §10.11.

Kozarzewski, B. (1965). Asymptotic properties of the electromagnetic and gravitational fields. *Acta Phys. Polon.* **27**, 775. *See* §26.1.

Kramer, D. (1972). Gravitational field of a rotating radiating source, in *Tezisy third Soviet gravitation conference*, page 321 (No Imprint, Erevan). *See* §30.7.

Kramer, D. (1977). Einstein–Maxwell fields with null Killing vector. *Acta Phys. Hung.* **43**, 125. *See* §24.4.

Kramer, D. (1978). Homogeneous Einstein–Maxwell fields. *Acta Phys. Hung.* **44**, 353. *See* §12.3.

Kramer, D. (1980). Space-times with a group of motions on null hypersurfaces. *J. Phys. A* **13**, L43. *See* §24.2.

Kramer, D. (1984a). Kerr solution endowed with magnetic dipole moment. *CQG* **1**, L45. *See* §34.8.

Kramer, D. (1984b). A new inhomogeneous cosmological model in general relativity. *CQG* **1**, 611. *See* §33.1.

Kramer, D. (1984c). A new solution for a rotating perfect fluid in general relativity. *CQG* **1**, L3. *See* §§21.2, 33.2.

Kramer, D. (1985). Perfect fluids with vanishing Simon tensor. *CQG* **2**, L135. *See* §§21.2, 22.2.

Kramer, D. (1986a). Rigidly rotating perfect fluids. *Astron. Nachr.* **307**, 309. *See* §21.2.

Kramer, D. (1986b). Two Kerr–NUT constituents in equilibrium. *GRG* **18**, 497. *See* §34.4.

Kramer, D. (1987). The Ernst equation in general relativity. *Czech. J. Phys. B* **37**, 350. *See* §34.1.

Kramer, D. (1988a). Cylindrically symmetric static perfect fluids. *CQG* **5**, 393. *See* §22.2.

Kramer, D. (1988b). The nonlinear evolution equation for Robinson–Trautman space-times, in *Relativity today. Proceedings of the second Hungarian relativity workshop*, ed. Z. Perjés, page 71 (World Scientific, Singapore). *See* §10.4.

Kramer, D. (1988c). Perfect fluids as minimal surfaces. *Astron. Nachr.* **309**, 267. *See* §21.2.

Kramer, D. (1990). Perfect fluids with conformal motion. *GRG* **22**, 1157. *See* §§21.2, 35.4.

Kramer, D. (1992). Rigidly rotating perfect fluids admitting a conformal Killing vector, in *Relativity today. Proceedings of the third Hungarian workshop*, ed. Z. Perjes (Nova Science Publishers, New York). *See* §35.4.

Kramer, D. and Carot, J. (1991). Conformal symmetry of perfect fluids in general relativity. *JMP* **32**, 1857. *See* §§21.2, 35.4.

Kramer, D. and Hähner, U. (1995). A rotating pure radiation field. *CQG* **12**, 2287. *See* §30.7.

Kramer, D. and Neugebauer, G. (1968a). Algebraisch spezielle Einstein-Räume mit einer Bewegungsgruppe. *Commun. Math. Phys.* **7**, 173. *See* §31.5.

Kramer, D. and Neugebauer, G. (1968b). Zu axialsymmetrischen stationären Lösungen der Einsteinschen Feldgleichungen für das Vakuum. *Commun. Math. Phys.* **10**, 132. *See* §§10.3, 19.5, 34.1.

Kramer, D. and Neugebauer, G. (1969). Eine exakte stationäre Lösung der Einstein–Maxwell-Gleichungen. *Ann. Phys. (Germany)* **24**, 59. *See* §34.1.

Kramer, D. and Neugebauer, G. (1980). The superposition of two Kerr solutions. *Phys. Lett. A* **75**, 259. *See* §34.4.

Kramer, D. and Neugebauer, G. (1981). Prolongation structure and linear eigenvalue equations for Einstein–Maxwell fields. *J. Phys. A* **14**, L333. *See* §§34.2, 34.8.

Kramer, D. and Neugebauer, G. (1984). Bäcklund transformations in general relativity, in *Solutions of Einstein's equations: Techniques and results.* Lecture notes in physics, vol. 205, eds. C. Hoenselaers and W. Dietz, page 1 (Springer, Berlin). *See* §§34.1, 34.4, 34.8.

Kramer, D., Neugebauer, G. and Stephani, H. (1972). Konstruktion und Charakterisierung von Gravitationsfelden. *Fortschr. Phys.* **20**, 1. *See* §§34.1, 37.4.

Kramer, D, Stephani, H., MacCallum, M. and Herlt, E. (1980). *Exact solutions of Einstein's field equations* (Cambridge University Press, Cambridge). *See* §1.2.

Krasiński, A. (1974). Solutions for the Einstein field equations for a rotating perfect fluid. I. Presentation of the flow-stationary and vortex-homogeneous solutions. *Acta Phys. Polon. B* **5**, 411. *See* §22.2.

Krasiński, A. (1975a). Solutions for the Einstein field equations for a rotating perfect fluid. II. Properties of the flow-stationary and vortex-homogeneous solutions. *Acta Phys. Polon. B* **6**, 223. *See* §22.2.

Krasiński, A. (1975b). Some solutions of the Einstein field equations for a rotating perfect fluid. *JMP* **16**, 125. *See* §22.2.

Krasiński, A. (1978). All flow-stationary cylindrically symmetric solutions of the Einstein field equations for a rotating isentropic perfect fluid. *Rep. Math. Phys.* **14**, 225. *See* §22.2.

Krasiński, A. (1989). Shear-free normal cosmological models. *JMP* **30**, 433. *See* §15.6.

Krasiński, A. (1997). *Inhomogeneous cosmological models* (Cambridge University Press, Cambridge). *See* §1.4, Ch. 33.

Krasiński, A. (1998a). Rotating dust solutions of Einstein's equations with 3-dimensional symmetry groups. II. One Killing field spanned on $u^{(\alpha)}$ and $w^{(\alpha)}$. *JMP* **39**, 401. *See* §13.4.

Krasiński, A. (1998b). Rotating dust solutions of Einstein's equations with 3-dimensional symmetry groups. III. All Killing fields linearly independent of $u^{(\alpha)}$ and $w^{(\alpha)}$. *JMP* **39**, 2148. *See* §13.4.

Krasiński, A., Quevedo, H. and Sussman, R.A. (1997). On the thermodynamical interpretation of perfect fluid solutions of the Einstein equations with no symmetry. *JMP* **38**, 2602. *See* §5.5.

Krisch, J.P. (1983). A one-fixed-point Killing parameter transform. *JMP* **24**, 2783. *See* §34.1.

Krisch, J.P. (1988). On the classification of vacuum zero Simon tensor solutions in relativity. *JMP* **29**, 446. *See* §18.7.

Krishna Rao, J. (1963). Type-N gravitational waves in non-empty space-time. *Curr. Sci.* **32**, 350. *See* §22.4.

Krishna Rao, J. (1964). Cylindrical waves in general relativity. *Proc. Nat. Inst. Sci. India A* **30**, 439. *See* §22.4.

Krishna Rao, J. (1970). Cylindrical symmetric null fields in general relativity. *Indian J. Pure Appl. Math.* **1**, 367. *See* §22.4.

Krori, K.D. and Barua, J. (1976). The gravitational field of a radiating cylinder. *Indian J. Pure Appl. Math.* **7**, 522. *See* §22.4.

Krori, K.D. and Goswami, D. (1992). A multipolar stationary object embedded in a gravitational field. *JMP* **33**, 1780. *See* §21.1.

Krori, K.D. and Nandy, D. (1984). Self-gravitating isotropic and anisotropic fluid distributions. *JMP* **25**, 2515. *See* §§10.11, 23.3.

Kruskal, M.D. (1960). Maximal extension of Schwarzschild metric. *Phys. Rev.* **119**, 1743. *See* §15.4.

Kuang, Z.Q., Lau, Y.K. and Wu, X.N. (1999). Einstein–Maxwell spacetime with two commuting spacelike Killing vector fields and Newman–Penrose formalism. *GRG* **31**, 1327. *See* §25.5.

Kuang, Z.Q., Li, J.Z. and Liang, C.B. (1987). Completion of plane-symmetric metrics yielded by electromagnetic fields. *GRG* **19**, 345. *See* §15.4.

Kuchowicz, B. (1966). Some new solutions of the gravitational field equations. Preprint, Warsaw. *See* §16.1.

Kuchowicz, B. (1967). Exact formulae for a general relativistic fluid sphere. *Phys. Lett. A* **25**, 419. *See* §16.1.

Kuchowicz, B. (1968a). Extensions of the external Schwarzschild solution. *Acad. Pol. Sci. Math. Astron. Phys.* **16**, 341. *See* §16.1.

Kuchowicz, B. (1968b). General relativistic fluid spheres. I. New solutions for spherically symmetric matter distribution. *Acta Phys. Polon.* **33**, 541. *See* §16.1.

Kuchowicz, B. (1968c). General relativistic fluid spheres. II. Solutions of the equation for $e^{-\lambda}$. *Acta Phys. Polon.* **34**, 131. *See* §16.1.

Kuchowicz, B. (1970). General relativistic fluid spheres. III. A simultaneous solving of two equations. *Acta Phys. Polon. B* **1**, 437. *See* §16.1.

Kuchowicz, B. (1971a). Two practical methods of solving Einstein's equations in isotropic coordinates. *Phys. Lett. A* **35A**, 223. *See* §16.1.

Kuchowicz, B. (1971b). Two relativistic matter distributions of radial symmetry. *Indian J. Pure Appl. Math.* **2**, 297. *See* §16.1.

Kuchowicz, B. (1972a). General relativistic fluid spheres. V. Non-charged static spheres of perfect fluid in isotropic coordinates. *Acta Phys. Polon. B* **3**, 209. *See* §16.1.

Kuchowicz, B. (1972b). On the integration of Einstein's equation for energy density inside a perfect fluid sphere. *Bull. Acad. Pol. Sci. Math. Astron. Phys.* **20**, 603. *See* §16.1.

Kuchowicz, B. (1973). General relatistic fluid spheres. VI. On physically meaningful static model spheres in isotropic coordinates. *Acta Phys. Polon. B* **4**, 415. *See* §16.1.

Kundt, W. (1961). The plane-fronted gravitational waves. *Z. Phys.* **163**, 77. *See* §§24.4, 31.2, 31.5.

Kundt, W. and Thompson, A. (1962). Le tenseur de Weyl et une congruence associée de géodésiques isotropes sans distorsion. *C. R. Acad. Sci. (Paris)* **254**, 4257. *See* §7.6.

Kundt, W. and Trümper, M. (1962). Beiträge zur Theorie der Gravitations-Strahlungsfelder. *Akad. Wiss. Lit. Mainz, Abhandl. Math.-Nat. Kl.* **12**. *See* §§7.6, 31.2, 33.4.

Kundt, W. and Trümper, M. (1966). Orthogonal decomposition of axi-symmetric stationary spacetimes. *Z. Phys.* **192**, 419. *See* §19.2.

Kundu, P. (1979). Class of 'noncanonical' vacuum metrics with two commuting Killing vectors. *Phys. Rev. Lett.* **42**, 416. *See* §24.4.

Kundu, P. (1981). Multipole expansion of stationary asymptotically flat vacuum metrics in general relativity. *JMP* **22**, 1236. *See* §18.8.

Kupeli, A.H. (1988a). Generalised Kerr–Schild transformation from vacuum backgrounds to Einstein–Maxwell spacetimes. *CQG* **5**, 401. *See* §32.5.

Kupeli, A.H. (1988b). Vacuum solutions admitting a geodesic null congruence with shear proportional to expansion. *JMP* **29**, 440. *See* §32.5.

Kustaanheimo, P. (1947). Some remarks concerning the connexion between two spherically symmetric relativistic metrics. *Comment. Phys. Math., Helsingf.* **13**, 8. *See* §16.2.

Kustaanheimo, P. and Qvist, B. (1948). A note on some general solutions of the Einstein field equations in a spherically symmetric world. *Comment. Phys. Math., Helsingf.* **13**, 1. *See* §§15.6, 16.1, 16.2.

Kyriakopoulos, E. (1987). Solutions of Einstein's equations for the interior of a stationary axisymmetric perfect fluid. *JMP* **28**, 2162. *See* §21.2.

Kyriakopoulos, E. (1988). A class of solutions of Einstein's equations for the interior of a rigidly rotating perfect fluid. *GRG* **20**, 427. *See* §21.2.

Kyriakopoulos, E. (1992). Interior axisymmetric stationary perfect fluid solution of Einstein's equations. *CQG* **9**, 217. *See* §21.2.

Kyriakopoulos, E. (1999). Petrov type *I*, stationary, axisymmetric, perfect fluid solution of Einstein's equations. *Mod. Phys. Lett. A* **14**, 7. *See* §21.2.

Lake, K. (1983). Remark concerning spherically symmetric nonstatic solutions to the Einstein equations in the comoving frame. *GRG* **15**, 357. *See* §16.2.

Lanczos, C. (1962). The splitting of the Riemann tensor. *Rev. Mod. Phys.* **34**, 379. *See* §3.6.

Lanczos, K. (1924). Über eine stationäre kosmologie im sinne der Einsteinschen Gravitationstheorie. *Zeitschr. f. Phys.* **21**, 73. *See* §§21.2, 22.2.

Lang, J.M. (1993). Contributions to the study of general relativistic shear-free perfect fluids. Ph.D. thesis, University of Waterloo. *See* §6.2.

Lang, J.M. and Collins, C.B. (1988). Observationally homogeneous shear-free perfect fluids. *GRG* **20**, 683. *See* §6.2.

Lauten, III, W.T. and Ray, J.R. (1977). Investigations of space-times with four-parameter groups of motions acting on null hypersurfaces. *JMP* **18**, 855. *See* §24.1.

Léauté, B. and Marcilhacy, G. (1979). On certain particular transcendental solutions of Einstein's equations. *Ann. Inst. H. Poincaré A* **31**, 363. *See* §20.6.

Léauté, B. and Marcilhacy, G. (1982). A new transcendent solution of Einstein's equations. *Phys. Lett. A* **87**, 159. *See* §20.6.

Léauté, B. and Marcilhacy, G. (1984). A new transcendent solution of Einstein–Maxwell equations. *Lett. Nuovo Cim.* **40, ser.2**, 102. *See* §21.1.

Leibovitz, C. (1969). Spherically symmetric static solutions of Einstein's equations. *Phys. Rev.* **185**, 1664. *See* §16.1.

Leibovitz, C. (1971). Time-dependent solutions of Einstein's equations. *PRD* **4**, 2949. *See* §§15.6, 16.2.

Leibowitz, E. and Meinhardt, J. (1978). Symmetry of charged rotating body metrics. *J. Phys. A* **11**, 1579. *See* §34.1.

Lemaître, G. (1927). Un univers homogène de masse constante et de rayon croissant, rendant compte de la vitesse radiale de nebuleuses extra-galactiques. *Ann. Soc. Sci. Bruxelles A* **47**, 49. *See* §14.2.

Lemaître, G. (1933). L'univers en expansion. *Ann. Soc. Sci. Bruxelles I A* **53**, 51. *See* §§14.2, 15.4, 15.5, 36.2.

Lemaître, G. (1949). Cosmological applications of relativity. *Rev. Mod. Phys.* **21**, 357. *See* §15.1.

Lemos, J.P.S. and Letelier, P.S. (1994). Exact general relativistic thin disks around black holes. *PRD* **49**, 5135. *See* §20.2.

Leroy, J. (1970). Un espace d'Einstein de type N à rayons non intégrables. *C. R. Acad. Sci. (Paris) A* **270**, 1078. *See* §13.3.

Leroy, J. (1976). Champs électromagnétiques à rayons intégrables, divergents et sans distorsion. *Bull. Acad. Roy. Belg. Cl. Sci.* **62**, 259. *See* §28.2.

Leroy, J. (1978). Sur une classe d'espaces-temps solutions des equations d'Einstein–Maxwell. *Bull. Acad. Roy. Belg. Cl. Sci.* **64**, 130. *See* §§24.3, 30.6.

Leroy, J. (1979). On the Robinson, Schild, and Strauss solution of the Einstein–Maxwell equations. *GRG* **11**, 245. *See* §30.6.

Leroy, J. and McLenaghan, R.G. (1973). Sur les espace-temps contenant un champ de vecteurs isotropes récurrents. *Bull. Acad. Roy. Belg. Cl. Sci.* **59**, 584. *See* §31.8.

Lesame, W.M., Ellis, G.F.R. and Dunsby, P.K.S. (1996). Irrotational dust with div$H = 0$. *PRD* **53**, 738. *See* §§6.2, 23.3.
Letelier, P.S. (1975). Self-gravitating fluids with cylindrical symmetry. *JMP* **16**, 1488. *See* §10.11.
Letelier, P.S. (1979). Self-gravitating fluids with cylindrical symmetry. II. *JMP* **20**, 2078. *See* §10.11.
Letelier, P.S. (1982). New two-soliton solution to the Einstein equations. *PRD* **26**, 3728. *See* §34.5.
Letelier, P.S. (1985a). On the superposition of stationary axially symmetric solutions to the vacuum Einstein equations. *CQG* **2**, 419. *See* §34.5.
Letelier, P.S. (1985b). Solitary wave solutions to the Einstein equations. *JMP* **26**, 326. *See* §34.5.
Letelier, P.S. (1985c). Static and stationary multiple soliton solutions to the Einstein equations. *JMP* **26**, 467. *See* §34.5.
Letelier, P.S. (1986). Soliton solutions to the vacuum Einstein equations obtained from a nondiagonal seed solution. *JMP* **27**, 564. *See* §34.5.
Letelier, P.S. and Oliveira, S.R. (1988). Superposition of Weyl solutions to the Einstein equations: cosmic strings and domain walls. *CQG* **5**, L47. *See* §20.2.
Letelier, P.S. and Oliveira, S.R. (1998a). Double Kerr–NUT spacetimes: spinning strings and spinning rods. *Phys. Lett. A* **57**, 6113. *See* §34.5.
Letelier, P.S. and Oliveira, S.R. (1998b). Superposition of Weyl solutions: the equilibrium forces. *CQG* **15**, 421. *See* §20.2.
Letelier, P.S. and Tabensky, R.R. (1974). The general solution to Einstein–Maxwell equations with plane symmetry. *JMP* **15**, 594. *See* §15.4.
Letelier, P.S. and Tabensky, R.R. (1975). Cylindrical self-gravitating fluids with pressure equal to energy density. *Nuovo Cim. B* **28**, 407. *See* §10.11.
Letelier, P.S. and Wang, A. (1993). On the interaction of null fluids in cosmology. *Phys. Lett. A* **182**, 220. *See* §25.6.
Letniowski, F.W. and McLenaghan, R.G. (1988). An improved algorithm for quartic equation classification and Petrov classification. *GRG* **20**, 463. *See* §9.3.
Levi-Civita, T. (1917a). ds^2 einsteiniani in campi newtoniani. *Rend. Acc. Lincei* **27**, 183. *See* §§18.6, 22.2.
Levi-Civita, T. (1917b). Realtá fisica di alcuni spazi normali del Bianchi. *Rend. R. Acad. Lincei, Cl. Sci. Fis. Mat. Nat.* **26**, 519. *See* §12.3.
Lewandowski, J. (1990). Conformal symmetries of pure radiation. *CQG* **7**, L135. *See* §35.4.
Lewandowski, J. (1992). Reduced holonomy group and Einstein equations with a cosmological constant. *CQG* **9**, L147. *See* §31.8.
Lewandowski, J. and Nurowski, P. (1990). Algebraically special twisting gravitational fields and CR structures. *CQG* **7**, 309. *See* §30.7.
Lewandowski, J., Nurowski, P. and Tafel, J. (1991). Algebraically special solutions of the Einstein equations with pure radiation fields. *CQG* **8**, 493. *See* §30.7.
Lewis, T. (1932). Some special solutions of the equations of axially symmetric gravitational fields. *Proc. Roy. Soc. Lond. A* **136**, 176. *See* §§19.3, 20.4.
Li, Jian-Zeng and Liang, Can-Bin (1989). Static 'semi-plane-symmetric' metrics yielded by plane-symmetric electromagnetic fields. *JMP* **30**, 2915. *See* §22.2.
Li, W. (1988). The complete Virasoro algebra for the stationary and axially symmetric Einstein field equations. *Phys. Lett. A* **129**, 301. *See* §34.7.
Li, W. (1989a). Generalization of the Hauser–Ernst formalism for two-dimensional reduced Einstein abelian gauge field equations. *Phys. Lett. A* **134**, 343. *See* §34.7.
Li, W. (1989b). New families of colliding gravitational plane waves with collinear polarisation. *CQG* **6**, 477. *See* §25.4.

Li, W. and Ernst, F.J. (1989). A family of electrovac colliding wave solutions of Einstein's equations. *JMP* **30**, 678. *See* §§25.5, 34.1.

Li, W., Hauser, I. and Ernst, F.J. (1991a). Colliding gravitational plane waves with noncollinear polarizations. *JMP* **32**, 723. *See* §25.4.

Li, W., Hauser, I. and Ernst, F.J. (1991b). Colliding gravitational waves with Killing–Cauchy horizons. *JMP* **32**, 1025. *See* §25.4.

Li, W., Hauser, I. and Ernst, F.J. (1991c). Colliding wave solutions of the Einstein–Maxwell field equations. *JMP* **32**, 1030. *See* §25.5.

Li, W., Hauser, I. and Ernst, F.J. (1991d). Nonimpulsive colliding gravitational waves with noncollinear polarizations. *JMP* **32**, 2478. *See* §25.4.

Li, W. and Hou, Bo-yu (1989). The Riemann–Hilbert transformation for an approach to a representation of the Virasoro group. *JMP* **30**, 1198. *See* §10.7.

Liang, Can-Bin (1995). A family of cylindrically symmetric solutions to Einstein–Maxwell equations. *GRG* **27**, 669. *See* §22.4.

Lichnerowicz, A. (1955). *Théories relativistes de la gravitation et de l'électromagnétisme* (Masson, Paris). *See* §§3.8, 18.1.

Liddle, A. and Lyth, D. (2000). *Cosmological inflation and large-scale structure* (Cambridge University Press, Cambridge). *See* §14.1.

Lifshitz, E.M. and Khalatnikov, I.M. (1963). Investigations in relativistic cosmology. *Advances Phys.* **12**, 185. *See* §13.3.

Lind, R.W. (1974). Shear-free, twisting Einstein–Maxwell metrics in the Newman–Penrose formalism. *GRG* **5**, 25. *See* §§7.1, 27.1, 30.3.

Lind, R.W. (1975a). Gravitational and electromagnetic radiation in Kerr–Maxwell spaces. *JMP* **16**, 34. *See* §§30.4, 30.6.

Lind, R.W. (1975b). Stationary Kerr–Maxwell spaces. *JMP* **16**, 39. *See* §30.4.

Lindblom, L. (1980). Some properties of static general relativistic stellar models. *JMP* **21**, 1455. *See* §18.6.

Lindblom, L. (1981). Some properties of static general relativistic stellar models. II. *JMP* **22**, 1324. *See* §18.6.

Linet, B. (1987). A charged black hole and a cosmic string in equilibrium. *CQG* **4**, L33. *See* §18.7.

Lor, J.C. and Rozoy, L. (1991). The Ricci curvatures of a Riemannian or pseudo-Riemannian manifold do not always determine its metric. (in French). *Helvetica Phys. Acta* **64**, 104. *See* §9.2.

Lorencz, K. and Sebestyén, A. (1986). On a fourth-order equation of axisymmetric stationary vacuum spacetimes, in *Proceedings of the fourth Marcel Grossmann meeting on general relativity*, ed. R. Ruffini, page 997 (North-Holland, Amsterdam, Netherlands). *See* §19.5.

Lorenz, D. (1981). Tilted electromagnetic Bianchi type I and type II cosmologies. *Phys. Lett. A* **83**, 155. *See* §14.3.

Lorenz, D. (1982a). Exact Bianchi type I solutions with a cosmological constant. *Phys. Lett. A* **92**, 118. *See* §14.4.

Lorenz, D. (1982b). On the solution for a vacuum Bianchi type-III model with a cosmological constant. *J. Phys. A* **15**, 2997. *See* §14.3.

Lorenz, D. (1983a). Exact Bianchi–Kantowski–Sachs solutions of Einstein's field equations. *J. Phys. A* **16**, 575. *See* §14.3.

Lorenz, D. (1983b). On the general vacuum solution with a cosmological constant for Bianchi type-VI_0. *Acta Phys. Polon. B* **14**, 479. *See* §13.3.

Lorenz, D. (1983c). Spatially self-similar cosmological model of Bianchi type-${}_1I$. *Astrophys. Space Sci.* **93**, 419. *See* §15.7.

Lorenz-Petzold, D. (1984). On the general vacuum and stiff matter solutions for 'diagonal' Bianchi type-VI_0 and type-VII_0 models. *Acta Phys. Polon. B* **15**, 117. *See* §§13.3, 14.4.

Lorenz-Petzold, D. (1987a). Bianchi type-VI_0 and type-VI_h perfect fluid solutions. *Astrophys. Space Sci.* **134**, 415. *See* §14.4.
Lorenz-Petzold, D. (1987b). Some new Bianchi type-I perfect fluid solutions. *Astrophys. Space Sci.* **132**, 147. *See* §14.3.
Lovelock, D. (1967). A spherically symmetric solution of the Maxwell–Einstein equations. *Commun. Math. Phys.* **5**, 257. *See* §12.3.
Lozanovski, C. and Aarons, M. (1999). Irrotational perfect fluid spacetimes with a purely magnetic Weyl tensor. *CQG* **16**, 4075. *See* §14.3.
Lozanovski, C. and McIntosh, C.B.G. (1999). Perfect fluid spacetimes with a purely magnetic Weyl tensor. *GRG* **31**, 1355. *See* §23.3.
Ludwig, G. (1969). Classification of electromagnetic and gravitational fields. *Amer. J. Math.* **37**, 1225. *See* Ch. 4.
Ludwig, G. (1970). Geometrodynamics of electromagnetic fields in the Newman–Penrose formalism. *Commun. Math. Phys.* **17**, 98. *See* §5.4.
Ludwig, G. (1988). Extended Geroch–Held–Penrose formalism. *Int. J. Theor. Phys.* **27**, 315. *See* §7.4.
Ludwig, G. (1996). The post-Bianchi identities in the GHP formalism. *Int. J. Mod. Phys. D* **5**, 407. *See* §7.3.
Ludwig, G. (1999). Type N pure radiation fields of embedding class one. *Int. J. Mod. Phys. D* **8**, 677. *See* §37.4.
Ludwig, G. and Edgar, S.B. (2000). A generalized Lie derivative and homothetic or Killing vectors in the Geroch–Held–Penrose formalism. *CQG* **17**, 1683. *See* §9.4.
Ludwig, G. and Scanlan, G. (1971). Classification of the Ricci tensor. *Commun. Math. Phys.* **20**, 291. *See* §5.1.
Lukács, B. (1973). All vacuum metrics with space-like symmetry and shearing geodesic timelike eigenrays. Report KFKI-1973-38, Centr. Res. Inst. Phys. Acad. Sci., Budapest. *See* §18.5.
Lukács, B. (1974). All stationary, rigidly rotating incoherent fluid metrics with geodesic and/or shearfree eigenrays. Report KFKI-1974-87, Centr. Res. Inst. Phys. Acad. Sci., Budapest. *See* §21.2.
Lukács, B. (1983). All vacuum metrics with space-like symmetry and shearing geodesic eigenrays. *Acta Phys. Slovaca* **33**, 225. *See* §18.5.
Lukács, B. (1985). Electrovac solutions with common shearing geodesic eigenrays. *Acta Phys. Hung.* **58**, 149. *See* §18.5.
Lukács, B. (1992). On the Bonnor counterparts of the Tomimatsu–Sato solutions. *Acta Phys. Hung.* **70**, 289. *See* §21.1.
Lukács, B. and Perjés, Z. (1973). Electrovac fields with geodesic eigenrays. *GRG* **4**, 161. *See* §18.5.
Lukács, B. and Perjés, Z. (1976). Time-dependent Maxwell fields in a stationary geometry, in *Proceedings of the first Marcel Grossmann meeting on general relativity*, ed. R. Ruffini, page 281 (North-Holland, Amsterdam). *See* §11.1.
Lukács, B., Perjés, Z., Porter, J. and Sebestyén, A. (1984). Lyapunov functional approach to radiative metrics. *GRG* **16**, 691. *See* §28.1.
Lukács, B., Perjés, Z. and Sebestyén, A. (1981). Null Killing vectors. *JMP* **22**, 1248. *See* §24.4.
Lukács, B., Perjés, Z., Sebestyén, A. and Sparling, G.A.J. (1983). Stationary vacuum fields with a conformally flat three-space. I. General theory. *GRG* **15**, 511. *See* §18.7.
Lukash, V.N. (1974). Gravitational waves that conserve the homogeneity of space (in Russian). *Zh. Eks. Teor. Fiz.* **67**, 1594. *See* §13.3.
Lun, A.W.C. (1978). A class of twisting type *II* and type *III* solutions admitting two Killing vectors. *Phys. Lett. A* **69**, 79. *See* §38.2.

Lun, A.W.C., McIntosh, C.B.G. and Singleton, D.B. (1988). Expanding algebraically degenerate vacuum spaces with homothetic motions. *GRG* **20**, 745. *See* §§11.4, 17.3.

Maartens, R., Lesame, W.M. and Ellis, G.F.R. (1998). Newtonian-like and anti-Newtonian universes. *CQG* **15**, 1005. *See* §6.2.

Maartens, R. and Maharaj, S.D. (1991). Conformal symmetries of *pp*-waves. *CQG* **8**, 503. *See* §§24.5, 35.4.

Maartens, R., Maharaj, S.D. and Tupper, B.O.J. (1995). General solution and classification of conformal motions in static spherical spacetimes. *CQG* **12**, 2577. *See* §35.4.

Maartens, R. and Nel, S.D. (1978). Decomposable differential operators in a cosmological context. *Commun. Math. Phys.* **59**, 273. *See* §§14.1, 14.3, 14.4.

MacCallum, M.A.H. (1971). A class of homogeneous cosmological models. III. Asymptotic behaviour. *Commun. Math. Phys.* **20**, 57. *See* §13.3.

MacCallum, M.A.H. (1972). On 'diagonal' Bianchi cosmologies. *Phys. Lett. A* **40**, 385. *See* §13.2.

MacCallum, M.A.H. (1973). Cosmological models from a geometric point of view, in *Cargèse lectures in physics*, vol. 6, ed. E. Schatzman, page 61 (Gordon and Breach, New York). *See* §13.2.

MacCallum, M.A.H. (1979a). Anisotropic and inhomogeneous relativistic cosmologies, in *General relativity: an Einstein centenary survey*, eds. S.W. Hawking and W. Israel, page 533 (Cambridge Univ. Press, Cambridge). *See* §13.2.

MacCallum, M.A.H. (1979b). The mathematics of anisotropic cosmologies, in *Proceedings of the first international Cracow school of cosmology*, ed. M. Demiański, page 1 (Springer-Verlag, Berlin). *See* §13.2.

MacCallum, M.A.H. (1980). Locally isotropic spacetimes with non-null homogeneous hypersurfaces, in *Essays in general relativity: a festschrift for Abraham Taub*, ed. F.J. Tipler, page 121 (Academic Press, New York). *See* §§11.2, 13.1.

MacCallum, M.A.H. (1983). Static and stationary 'cylindrically symmetric' Einstein–Maxwell fields and the solutions of Van den Bergh and Wils. *J. Phys. A* **16**, 3853. *See* §22.2.

MacCallum, M.A.H. (1990). On "generalized Kasner" metrics and the solutions of Harris and Zund. *Wiss. Zeitschr. FSU Jena, Math.-Naturw. Reihe* **39**, 102. *See* §17.3.

MacCallum, M.A.H. (1996). Computer algebra and applications in relativity and gravity, in *Recent developments in gravitation and mathematical physics: Proceedings of the first Mexican school on gravitation and mathematical physics*, eds. A. Macias, T. Matos, O. Obregon and H. Quevedo, page 3 (World Scientific, Singapore). *See* Ch. 9.

MacCallum, M.A.H. (1998). Hypersurface-orthogonal generators of an orthogonally transitive G_2I, topological identifications, and axially and cylindrically symmetric spacetimes. *GRG* **30**, 131. *See* §§17.1, 22.1.

MacCallum, M.A.H. (1999). On the classification of the real four-dimensional Lie algebras, in *On Einstein's path: Essays in honor of Engelbert Schucking*, ed. A.L. Harvey, page 299 (Springer Verlag, New York). *See* §8.2.

MacCallum, M.A.H. and Åman, J.E. (1986). Algebraically independent n-th derivatives of the Riemannian curvature spinor in a general spacetime. *CQG* **3**, 1133. *See* §9.3.

MacCallum, M.A.H. and Santos, N.O. (1998). Stationary and static cylindrically symmetric Einstein spaces of the Lewis form. *CQG* **15**, 1627. *See* §22.2.

MacCallum, M.A.H. and Siklos, S.T.C. (1980). Homogeneous and hypersurface-homogeneous algebraically special Einstein spaces. *GR9 Abstracts* **1**, 54. *See* §13.3.

MacCallum, M.A.H. and Siklos, S.T.C. (1992). Algebraically special hypersurface-homogeneous Einstein spaces in general relativity. *J. Geometry Phys.* **8**, 221. *See* §§13.2, 13.3, 31.8.

MacCallum, M.A.H. and Skea, J.E.F. (1994). SHEEP: a computer algebra system for general relativity, in *Algebraic computing in general relativity. Proceedings of the first Brazilian school on computer algebra*, vol. 2, eds. M.J. Rebouças and W.L. Roque, page 1 (Oxford University Press, Oxford). *See* Ch. 9, §§9.2, 9.4.

MacCallum, M.A.H., Stewart, J.M. and Schmidt, B.G. (1970). Anisotropic stresses in homogeneous cosmologies. *Commun. Math. Phys.* **17**, 343. *See* §13.2.

MacCallum, M.A.H. and Taub, A.H. (1972). Variational principles and spatially homogeneous universes including rotation. *Commun. Math. Phys.* **25**, 173. *See* §13.2.

MacCallum, M.A.H. and Van den Bergh, N. (1985). Non-inheritance of static symmetry by Maxwell fields, in *Galaxies, axisymmetric systems and relativity. Essays presented to W.B. Bonnor on his 65th birthday*, ed. M.A.H. MacCallum, page 138 (Cambridge University Press, Cambridge). *See* §22.2.

Maharaj, S.D., Maartens, R. and Maharaj, M.S. (1993). A note on a class of spherically symmetric solutions. *Nuovo Cim. B* **108**, 75. *See* §16.2.

Maison, D. (1978). Are the stationary, axially symmetric Einstein equations completely integrable? *Phys. Rev. Lett.* **41**, 521. *See* Ch. 34.

Maison, D. (1979). On the complete integrability of the stationary, axially symmetric Einstein equations. *JMP* **20**, 871. *See* Ch. 34.

Maitra, S.G. (1966). Stationary dust-filled cosmological solution with $\Lambda = 0$ and without closed timelike lines. *JMP* **7**, 1025. *See* §22.2.

Majumdar, S.D. (1947). A class of exact solutions of Einstein's field equations. *Phys. Rev.* **72**, 390. *See* §§18.7, 21.1.

Man, Y.-K. and MacCallum, M.A.H. (1997). A rational approach to the Prelle–Singer algorithm. *J. Symb. Comp.* **24**, 31. *See* §13.2.

Manko, O.V., Manko, V.S. and Sambria G., J.D. (1999). Remarks on the charged, magnetized Tomimatsu–Sato $\delta = 2$ solution. *GRG* **31**, 1539. *See* §34.6.

Manko, V.S. (1990). New axially symmetric solutions of the Einstein–Maxwell equations. *GRG* **22**, 799. *See* §21.1.

Manko, V.S. (1999). Generating techniques and analytically extended solutions of the Einstein–Maxwell equations. *GRG* **31**, 673. *See* §34.6.

Manko, V.S. and Khakimov, S.A. (1991). On the gravitational field of an arbitrary axisymmetric mass possessing a magnetic dipole moment. *Phys. Lett. A* **154**, 96. *See* §21.1.

Manko, V.S., Martín, J. and Ruiz, E. (1995). Six-parameter solution of the Einstein–Maxwell equations possessing equatorial symmetry. *JMP* **36**, 3063. *See* §34.6.

Manko, V.S. and Moreno, C. (1997). Extension of the parameter space in the Tomimatsu–Sato solutions. *Mod. Phys. Lett. A* **12**, 613. *See* §20.5.

Manko, V.S. and Novikov, I.D. (1992). Generalizations of the Kerr and Kerr–Newman metrics possessing an arbitrary set of mass-multipole moments. *CQG* **9**, 2477. *See* §21.1.

Manko, V.S. and Ruiz, E. (1998). Extended multi-soliton solutions of the Einstein field equations. *CQG* **15**, 2007. *See* §34.6.

Manko, V.S. and Ruiz, E. (2001). Exact solution of the double-Kerr equilibrium problem. *CQG* **18**. *See* §34.4.

Manko, V.S. and Sibgatullin, N.R. (1992). Metric of a rotating, charged, magnetised mass. *Phys. Lett. A* **168**, 343. *See* §34.6.

Manko, V.S. and Sibgatullin, N.R. (1993). Construction of exact solutions of the Einstein–Maxwell equations corresponding to a given behaviour of the Ernst potentials on the symmetry axis. *CQG* **10**, 1383. *See* §34.6.

Marder, L. (1969). Gravitational waves in general relativity. XI. Cylindrical-spherical waves. *Proc. Roy. Soc. Lond. A* **313**, 83. *See* §22.3.

Marek, J. (1968). Some solutions of Einstein's equations in general relativity. *Proc. Camb. Phil. Soc.* **64**, 167. *See* §20.6.

Mariot, L. (1954). Le champ électromagnétique singulier. *C. R. Acad. Sci. (Paris)* **238**, 2055. *See* §7.6.

Marklund, M. (1997). Invariant construction of solutions to Einstein's field equations – LRS perfect fluids I. *CQG* **14**, 1267. *See* §§9.4, 13.4, 14.3.

Marklund, M. and Bradley, M (1999). Invariant construction of solutions to Einstein's field equations – LRS perfect fluids II. *CQG* **16**, 1577. *See* §§9.4, 16.2.

Marklund, M. and Perjés, Z. (1997). Stationary rotating matter in general relativity. *JMP* **38**, 5880. *See* §19.6.

Mars, M. (1995). New non-separable diagonal cosmologies. *CQG* **12**, 2831. *See* §23.3.

Mars, M. (1999). 3+1 description of silent universes: a uniqueness result for the Petrov type *I* vacuum case. *CQG* **16**, 3245. *See* §6.2.

Mars, M. and Senovilla, J.M.M. (1993a). Axial symmetry and conformal Killing vectors. *CQG* **10**, 1633. *See* §§19.1, 35.4.

Mars, M. and Senovilla, J.M.M. (1993b). Geometry of general hypersurfaces in spacetime – junction conditions. *CQG* **10**, 1865. *See* §3.8.

Mars, M. and Senovilla, J.M.M. (1994). Stationary and axisymmetric perfect fluid solutions with conformal motion. *CQG* **11**, 3049. *See* §§21.2, 33.3, 35.4.

Mars, M. and Senovilla, J.M.M. (1996). Study of a family of stationary and axially symmetric differentially rotating perfect fluids. *PRD* **54**, 6166. *See* §21.2.

Mars, M. and Senovilla, J.M.M. (1997). Non-diagonal G_2 separable perfect-fluid spacetimes. *CQG* **14**, 205. *See* §23.3.

Mars, M. and Senovilla, J.M.M. (1998). On the construction of global models describing rotating bodies; uniqueness of the exterior gravitational field. *Mod. Phys. Lett. A* **13**, 1509. *See* §21.2.

Mars, M. and Wolf, T. (1997). G_2 perfect-fluid cosmologies with a proper conformal Killing vector. *CQG* **14**, 2303. *See* §§23.3, 33.3, 35.4.

Martín, J. and Senovilla, J.M.M. (1986). Petrov type *D* perfect-fluid solutions in generalized Kerr–Schild form. *JMP* **27**, 2209. *See* §§32.5, 33.3.

Martín P., F. and Senovilla, J.M.M. (1988). Petrov types *D* and *II* perfect-fluid solutions in generalized Kerr–Schild form. *JMP* **29**, 937. *See* §32.5.

Martinez, E. and Sanz, J.L. (1985). Space-times with intrinsic symmetries on the three-spaces $t = constant$. *JMP* **26**, 785. *See* §36.4.

Martins, M.A.P. (1996). The sources of the A and B degenerate static vacuum fields. *GRG* **28**, 1309. *See* §20.2.

Mason, D.P. and Pooe, C.A. (1987). Rotating rigid motion in general relativity. *JMP* **28**, 2705. *See* §6.2.

Masuda, T., Sasa, N. and Fukuyama, T. (1998). Neugebauer–Kramer solutions of the Ernst equation in Hirota's direct method. *J. Phys. A* **31**, 5717. *See* §34.7.

Matos, T. and Plebański, J.F. (1994). Axisymmetric stationary solutions as harmonic maps. *GRG* **26**, 477. *See* §34.1.

Matsumoto, M. (1950). Riemann spaces of class two and their algebraic characterization. *JMP (Japan)* **2**, 67. *See* §37.5.

Matte, A. (1953). Sur de nouvelles solutions oscillatoires des équations de la gravitation. *Can. J. Math.* **5**, 1. *See* §§3.5, 4.2.

May, T.L. (1975). Uniform model universe containing interacting blackbody radiation and matter with internal energy. *Astrophys. J.* **199**, 322. *See* §14.2.

Mazur, P.O. (1983). A relationship between the electrovacuum Ernst equations and nonlinear σ-model. *Acta Phys. Polon. B* **14**, 219. *See* §34.8.

McIntosh, C. (1978a). Self-similar cosmologies with equation of state $p = \mu$. *Phys. Lett. A* **69**, 1. *See* §§11.4, 23.1.

McIntosh, C.B.G. (1968). Relativistic cosmological models with both radiation and matter. *Mon. Not. R. Astron. Soc.* **140**, 461. *See* §14.2.

McIntosh, C.B.G. (1972). Cosmological models with two fluids. I. Robertson–Walker metric. *Aust. J. Phys.* **25**, 75. *See* §14.2.

McIntosh, C.B.G. (1976a). Homothetic motions in general relativity. *GRG* **7**, 199. *See* §11.3.

McIntosh, C.B.G. (1976b). Homothetic motions with null homothetic bivectors in general relativity. *GRG* **6**, 915. *See* §8.7.

McIntosh, C.B.G. (1978b). Einstein–Maxwell space-times with symmetries and with non-null electromagnetic fields. *GRG* **9**, 277. *See* §§11.1, 22.2.

McIntosh, C.B.G. (1979). Symmetry mappings in Einstein–Maxwell space-times. *GRG* **10**, 61. *See* §§8.7, 11.3, 13.3, 35.4.

McIntosh, C.B.G. (1980). Symmetries and exact solutions of Einstein's equations, in *Gravitational radiation, collapsed objects and exact solutions.* Lecture notes in physics, vol. 124, ed. C. Edwards, page 469 (Springer, Berlin). *See* §35.4.

McIntosh, C.B.G. (1992). Real Kasner and related complex 'windmill' vacuum spacetime metrics. *GRG* **24**, 757. *See* §§13.3, 20.2.

McIntosh, C.B.G. and Arianrhod, R. (1990a). 'Degenerate' nondegenerate spacetime metrics. *CQG* **7**, L213. *See* §4.3.

McIntosh, C.B.G. and Arianrhod, R. (1990b). Kundt solutions of Einstein's equations with one non-null Killing vector. *GRG* **22**, 995. *See* §§31.5, 31.6.

McIntosh, C.B.G., Arianrhod, R., Wade, S.T. and Hoenselaers, C. (1994). Electric and magnetic Weyl tensors: classification and analysis. *CQG* **11**, 1555. *See* §§4.1, 4.3, 6.2.

McIntosh, C.B.G. and Foyster, J.M. (1972). Cosmological models with two fluids. II. Conformal and conformally flat metrics. *Aust. J. Phys.* **25**, 83. *See* §14.2.

McIntosh, C.B.G. and Hickman, M.S. (1988). Single Kerr–Schild metrics: a double view. *GRG* **20**, 793. *See* §32.2.

McIntosh, C.B.G., Lun, A.W.C. and Halford, W.D. (1987). Type II and III twisting vacuum metrics and symmetries. *CQG* **4**, 117. *See* §11.4.

McIntosh, C.B.G. and Steele, J.D. (1991). All vacuum Bianchi type *I* metrics with a homothety. *CQG* **8**, 1171. *See* §11.4.

McIntosh, C.B.G. and van Leeuwen, E.H. (1982). Spacetimes admitting a vector field whose inner product with the Riemann tensor is zero. *JMP* **23**, 1149. *See* §8.7.

McLenaghan, R.G. and Leroy, J. (1972). Complex recurrent space-times. *Proc. Roy. Soc. Lond. A* **327**, 229. *See* §35.2.

McLenaghan, R.G. and Tariq, N. (1975). A new solution of the Einstein–Maxwell equations. *JMP* **16**, 11. *See* §12.3.

McLenaghan, R.G., Tariq, N. and Tupper, B.O.J. (1975). Conformally flat solutions of the Einstein–Maxwell equations for null electromagnetic fields. *JMP* **16**, 829. *See* §37.5.

McManus, D.J. (1995). Lorentzian three-metrics with degenerate Ricci tensors. *JMP* **36**, 1353. *See* §36.4.

McVittie, G.C. (1929). On Einstein's unified field theory. *Proc. Roy. Soc. Lond. A* **124**, 366. *See* §15.4.

McVittie, G.C. (1933). The mass-particle in an expanding universe. *Mon. Not. R. Astron. Soc.* **93**, 325. *See* §16.2.

McVittie, G.C. (1984). Elliptic functions in spherically symmetric solutions of Einstein's equations. *Ann. Inst. H. Poincaré* **3**, 235. *See* §16.2.

McVittie, G.C. and Wiltshire, R.J. (1975). Fluid spheres and R- and T-regions in general relativity. *Int. J. Theor. Phys.* **14**, 145. *See* §§15.1, 16.2.

McVittie, G.C. and Wiltshire, R.J. (1977). Relativistic fluid spheres and noncomoving coordinates. I. *Int. J. Theor. Phys.* **16**, 121. *See* §16.2.

Mehra, A.L. (1966). Radially symmetric distribution of matter. *J. Aust. Math. Soc.* **6**, 153. *See* §16.1.
Meinel, R. and Neugebauer, G. (1995). Asymptotically flat solutions to the Ernst equation with reflection symmetry. *CQG* **12**, 2045. *See* §19.5.
Meinel, R. and Neugebauer, G. (1996). Solutions of Einstein's field equations related to Jacobi's inversion problem. *Phys. Lett. A* **210**, 160. *See* §34.6.
Meinel, R. and Neugebauer, G. (1997). Reply to "Some remarks on finite-gap solutions of the Ernst equation" by Korotkin. *Phys. Lett. A* **229**, 220. *See* §34.6.
Meinel, R., Neugebauer, G. and Steudel, H. (1991). *Solitonen* (Akademie Verlag, Berlin). *See* §34.4.
Melnick, J. and Tabensky, R.R. (1975). Exact solutions to Einstein field equations. *JMP* **16**, 958. *See* §18.6.
Melvin, M.A. (1964). Pure magnetic and electric geons. *Phys. Lett.* **8**, 65. *See* §22.2.
Mészáros, A. (1985). Non-static fluid spheres in general relativity. *Astrophys. Space Sci.* **108**, 415. *See* §15.6.
Micciché, S. (1999). Physical properties of gravitational solitons. Ph.D. thesis, Loughborough. *See* §34.5.
Micciché, S. and Griffiths, J.B. (2000). Soliton solutions with real poles in the Alekseev formulation of the inverse-scattering method. *CQG* **17**, 1. *See* §34.6.
Michalski, H. and Wainwright, J. (1975). Killing vector fields and the Einstein–Maxwell field equations in general relativity. *GRG* **6**, 289. *See* §§11.1, 19.2.
Misner, C.W. (1968). The isotropy of the universe. *Astrophys. J.* **151**, 431. *See* §13.2.
Misner, C.W. and Sharp, D.H. (1964). Relativistic equations for adiabatic, spherically symmetric gravitational collapse. *Phys. Rev. B* **136**, 571. *See* §16.2.
Misner, C.W. and Taub, A.H. (1968). A singularity-free empty universe (in Russian). *Zh. Eks. Teor. Fiz.* **55**, 233. *See* §13.3.
Misner, C.W. and Wheeler, J.A. (1957). Classical physics as geometry: Gravitation, electromagnetism, unquantized charge, and mass as properties of empty space. *Ann. Phys. (USA)* **2**, 525. *See* §5.4.
Misra, M. (1962). A non-singular electromagnetic field. *Proc. Camb. Phil. Soc.* **58**, 711. *See* §22.4.
Misra, M. (1966). Some exact solutions of the field equations of general relativity. *JMP* **7**, 155. *See* §22.4.
Misra, M. and Radhakrishna, L. (1962). Some electromagnetic fields of cylindrical symmetry. *Proc. Nat. Inst. Sci. India A* **28**, 632. *See* §22.4.
Misra, R.M. and Narain, U. (1971). A new solution of the field equations with perfect fluid, in *Relativity and gravitation, Proceedings of the international seminar, Technion City, Israel*, eds. G.C. Kuper and A. Peres, page 253 (Gordon & Breach, London). *See* §13.4.
Misra, R.M. and Srivastava, D.C. (1973). Dynamics of fluid spheres of uniform density. *PRD* **8**, 1653. *See* §16.2.
Mitskievic, N.V. and Horský, J. (1996). Conformal Kerr–Schild ansatz for all static spherically symmetric spacetimes. *CQG* **13**, 2603. *See* §32.5.
Montgomery, W., O'Raifeartaigh, L. and Winternitz, P. (1969). Two-variable expansions of relativistic amplitudes and the subgroups of the SU(2,1) group. *Nucl. Phys. B* **11**, 39. *See* §34.1.
Morgan, L. and Morgan, T. (1969). The gravitational field of a disk. *Phys. Rev.* **183**, 1097. *See* §20.2.
Morisetti, S., Reina, C. and Treves, A. (1980). Petrov classification of vacuum axisymmetric space-times. *GRG* **12**, 619. *See* §19.5.
Morris, H.C. (1976). Prolongation structure and nonlinear evolution equations in two spatial dimensions. *JMP* **17**, 1870. *See* §10.4.
Mukherji, B.C. (1938). Two cases of exact gravitational fields with axial symmetry. *Bull. Calc. Math. Soc.* **30**, 95. *See* §22.2.

Murphy, G.L. (1973). Big-bang model without singularities. *PRD* **8**, 4231. *See* §14.1.

Muskhelishvili, N.I. (1953). *Singular integral equations* (Nordhoff, Groningen). *See* §10.7.

Mustapha, N.M., Ellis, G.F.R., van Elst, H. and Marklund, M. (2000). Partially locally rotationally symmetric perfect fluid cosmologies. *CQG* **17**, 3135. *See* §11.2.

Nagatomo, K. (1989). Explicit description of ansatz E_n for the Ernst equation in general relativity. *JMP* **30**, 1100. *See* §34.6.

Nakahara, M. (1990). *Geometry, topology and physics* (Adam Hilger, Bristol). *See* §2.1.

Nakamura, A. and Ohta, Y. (1991). Bilinear, Pfaffian and Legendre function structures of the Tomimatsu–Sato solutions of the Ernst equation in general relativity. *J. Phys. Soc. of Japan* **60**, 1835. *See* §34.7.

Nakamura, Y. (1979). On inverse scattering formula for Ernst equation. *Math. Japon.* **24**, 469. *See* §34.7.

Nakamura, Y. (1987). On a linearisation of the stationary axially symmetric Einstein equations. *CQG* **4**, 437. *See* §34.7.

Narain, S. (1988). Some static and nonstatic perfect fluid models in general relativity. *GRG* **20**, 15. *See* §22.2.

Nariai, H. (1950). On some static solutions of Einstein's gravitational field equations in a spherically symmetric case. *Sci. Rep. Tôhoku Univ., I* **34**, 160. *See* §16.1.

Nariai, H. (1951). On a new cosmological solution of Einstein's field equations of gravitation. *Sci. Rep. Tôhoku Univ., I* **35**, 62. *See* §15.4.

Narlikar, V.V. (1947). Some new results regarding spherically symmetrical fields in relativity. *Curr. Sci.* **16**, 113. *See* §15.6.

Narlikar, V.V. and Karmarkar, K.R. (1946). On a curious solution of relativistic field equations. *Curr. Sci.* **15**, 69. *See* §13.3.

Narlikar, V.V. and Moghe, D.N. (1935). Some new solutions of the differential equations for isotropy. *Phil. Mag.* **20**, 1104. *See* §16.2.

Narlikar, V.V., Patwardhan, G.K. and Vaidya, P.C. (1943). Some new relativistic distributions of radial symmetry. *Proc. Natl. Inst. Sci. India* **9**, 229. *See* §16.1.

Neugebauer, G. (1969). Untersuchungen zu Einstein–Maxwell-Feldern mit eindimensionaler Bewegungsgruppe. Habilitationsschrift, Jena. *See* §§18.7, 34.1.

Neugebauer, G. (1979). Bäcklund transformations of axially symmetric stationary gravitational fields. *J. Phys. A* **12**, L67. *See* §§34.1, 34.2.

Neugebauer, G. (1981). Relativistic gravitational fields of rotating bodies. *Phys. Lett. A* **86**, 91. *See* §34.4.

Neugebauer, G. (2000). Rotating bodies as boundary value problems. *Ann. Phys. (Germany)* **9**, 342. *See* §34.6.

Neugebauer, G. and Herlt, E. (1984). Einstein–Maxwell fields inside and outside rotating sources as minimal surfaces. *CQG* **1**, 695. *See* §21.2.

Neugebauer, G., Kleinwaechter, A. and Meinel, R. (1996). Relativistically rotating dust. *Helv. Phys. Acta* **69**, 472. *See* §34.6.

Neugebauer, G. and Kramer, D. (1969). Eine Methode zur Konstruktion stationärer Einstein–Maxwell-Felder. *Ann. Phys. (Germany)* **24**, 62. *See* §§10.8, 18.4, 21.1, 34.1.

Neugebauer, G. and Kramer, D. (1983). Einstein–Maxwell solitons. *J. Phys. A* **16**, 1927. *See* §34.8.

Neugebauer, G. and Kramer, D. (1985). Stationary axisymmetric electrovacuum fields in general relativity, in *Galaxies, axisymmetric systems and relativity. Essays presented to W.B. Bonnor on his 65th birthday*, ed. M.A.H. MacCallum, page 149 (Cambridge University Press, Cambridge). *See* §34.4.

Neugebauer, G. and Meinel, R. (1984). General N-soliton solution of the AKNS class on arbitrary background. *Phys. Lett. A* **100**, 467. *See* §10.6.

Neugebauer, G. and Meinel, R. (1993). The Einsteinian gravitational field of the rigidly rotating disk of dust. *Astrophys. J., Lett.* **414**, L97. *See* §34.6.

Neugebauer, G. and Meinel, R. (1994). General relativistic gravitational field of a rigidly rotating disk of dust: axis potential, disk metric, and surface mass density. *Phys. Rev. Lett.* **73**, 2166. *See* §34.6.

Neugebauer, G. and Meinel, R. (1995). General relativistic gravitational field of a rigidly rotating disk of dust: Solution in terms of ultraelliptic functions. *Phys. Rev. Lett.* **75**, 3046. *See* §§21.2, 34.6.

Newman, E.T., Couch, E., Chinnapared, K., Exton, A., Prakash, A. and Torrence, R.J. (1965). Metric of a rotating charged mass. *JMP* **6**, 918. *See* §§21.1, 32.3.

Newman, E.T. and Janis, A.I. (1965). Note on the Kerr spinning-particle metric. *JMP* **6**, 915. *See* §21.1.

Newman, E.T. and Penrose, R. (1962). An approach to gravitational radiation by a method of spin coefficients. *JMP* **3**, 566. *See* §§7.1, 7.2, 7.6.

Newman, E.T. and Tamburino, L.A. (1962). Empty space metrics containing hypersurface orthogonal geodesic rays. *JMP* **3**, 902. *See* §26.4.

Newman, E.T., Tamburino, L.A. and Unti, T. (1963). Empty-space generalization of the Schwarzschild metric. *JMP* **4**, 915. *See* §§13.3, 29.5.

Newman, E.T. and Unti, T.W.J. (1963). A class of null flat-space coordinate systems. *JMP* **4**, 1467. *See* §§28.3, 32.1.

Nilsson, U.S. and Uggla, C. (1997a). Rigidly rotating stationary cylindrically symmetric perfect fluid models. *CQG* **14**, 2931. *See* §22.2.

Nilsson, U.S. and Uggla, C. (1997b). Stationary Bianchi type *II* perfect fluid models. *JMP* **38**, 2616. *See* §13.4.

Nordström, G. (1918). On the energy of the gravitational field in Einstein's theory. *Proc. Kon. Ned. Akad. Wet.* **20**, 1238. *See* §§15.4, 21.1.

Novikov, I.D. (1963). On the evolution of a semiclosed universe (in Russian). *Zh. Astrof.* **40**, 772. *See* §15.4.

Novikov, I.D. (1964). R- and T-regions in a spacetime with a spherically symmetric space (in Russian). *Comm. State Sternberg Astron. Inst.* **132**, 3. *See* §14.3.

Novotný, J. and Horský, J. (1974). On the plane gravitational condensor with the positive gravitational constant. *Czech. J. Phys. B* **24**, 718. *See* §15.4.

Nurowski, P., Hughston, L.P. and Robinson, D.C. (1999). Extensions of bundles of null directions. *CQG* **16**, 255. *See* §9.4.

Nurowski, P. and Robinson, D.C. (2000). Intrinsic geometry of a null hypersurface. *CQG* **17**, 4065. *See* §9.4.

Nurowski, P. and Tafel, J. (1992). New algebraically special solutions of the Einstein–Maxwell equations. *CQG* **9**, 2069. *See* §30.5.

Nutku, Y. (1991). Spherical shock waves in general relativity. *PRD* **44**, 3164. *See* §28.1.

Nutku, Y. and Halil, M. (1977). Colliding impulsive gravitational waves. *Phys. Rev. Lett.* **39**, 1379. *See* §25.4.

O'Brien, S. and Synge, J.L. (1952). Jump conditions at discontinuities in general relativity. *Comm. Dublin Inst. Advanced Studies A* **9**, 1. *See* §3.8.

Öktem, F. (1976). On parallel null 1-planes in space-time. *Nuovo Cim. B* **34**, 169. *See* §6.1.

Oleson, M. (1971). A class of type [4] perfect fluid space-times. *JMP* **12**, 666. *See* §33.4.

Oleson, M. (1972). Algebraically special fluid space-times with shearing rays. Ph.D. thesis, University of Waterloo. *See* §§33.1, 33.4.

Oliver, G. and Verdaguer, E. (1989). A family of inhomogeneous cosmological Einstein–Rosen metrics. *JMP* **30**, 442. *See* §§10.11, 34.5.

Olver, P.J. (1986). *Applications of Lie groups to differential equations.* Graduate Texts in Mathematics, vol. 107 (Springer-Verlag, Berlin). *See* §10.2.

Olver, P.J. (1995). *Equivalence, invariants, and symmetry* (Cambridge University Press, Cambridge). *See* §§7.4, 9.2.

Omote, M., Michihiro, Y. and Wadati, M. (1980). A Bäcklund transformation of the axially symmetric stationary Einstein–Maxwell equations. *Phys. Lett. A* **79**, 141. *See* §34.8.

Omote, M. and Wadati, M. (1981a). The Bäcklund transformations and the inverse scattering method of the Ernst equation. *Progr. Theor. Phys.* **65**, 1621. *See* §34.7.

Omote, M. and Wadati, M. (1981b). Bäcklund transformations for the Ernst equation. *JMP* **22**, 961. *See* §34.7.

Önengüt, G. and Serdaroğlu, M. (1975). Two-parameter static and five-parameter stationary solutions of the Einstein–Maxwell equations. *Nuovo Cim. B* **27**, 213. *See* §34.1.

Orlyanski, O.Yu. (1997). Singularity-free static fluid spheres in general relativity. *JMP* **38**, 5301. *See* §16.1.

Ovsiannikov, L.V. (1982). *Group analysis of differential equations* (Academic Press, New York). *See* §10.2.

Ozsváth, I. (1965a). Homogeneous solutions of the Einstein–Maxwell equations. *JMP* **6**, 1255. *See* §§12.3, 12.5.

Ozsváth, I. (1965b). New homogeneous solutions of Einstein's field equations with incoherent matter. *Akad. Wiss. Lit. Mainz, Abh. Math.-Nat. Kl.* **1.** *See* §12.4.

Ozsváth, I. (1965c). New homogeneous solutions of Einstein's field equations with incoherent matter obtained by a spinor technique. *JMP* **6**, 590. *See* §12.4.

Ozsváth, I. (1966). Two rotating universes with dust and electromagnetic field, in *Perspectives in geometry and relativity*, ed. B. Hoffmann, page 245 (Indiana University Press, Bloomington). *See* §§12.5, 13.1.

Ozsváth, I. (1970). Dust-filled universes of class II and class III. *JMP* **11**, 2871. *See* §12.4.

Ozsváth, I. (1987). All homogeneous solutions of Einstein's vacuum field equations with a non vanishing cosmological term, in *Gravitation and geometry: a volume in honour of Ivor Robinson.* Monographs and textbooks in physical science, vol. 4, eds. W. Rindler and A. Trautman, page 309 (Bibliopolis, Naples). *See* §12.5.

Ozsváth, I., Robinson, I. and Roźga, K. (1985). Plane-fronted gravitational and electromagnetic waves in spaces with cosmological constant. *JMP* **26**, 1755. *See* §31.8.

Ozsváth, I. and Schücking, E. (1962). An anti-Mach metric, in *Recent developments in general relativity*, page 339 (Pergamon Press–PWN Warsaw, Oxford). *See* §12.2.

Ozsváth, I. and Schücking, E. (1969). The finite rotating universe. *Ann. Phys. (USA)* **55**, 166. *See* §12.4.

Paiva, F.M., Rebouças, M.J., Hall, G.S. and MacCallum, M.A.H. (1998). Limits of the energy-momentum tensor in general relativity. *CQG* **15**, 1031. *See* §§5.1, 9.3, 9.5.

Paiva, F.M., Rebouças, M.J. and MacCallum, M.A.H. (1993). On limits of spacetimes – a coordinate-free approach. *CQG* **10**, 1165. *See* §9.5.

Paklin, N.N. (1994). Exact regular models for static relativistic stars. *Russian Phys. J.* **37**, 836. *See* §16.1.

Pandya, I.M. and Vaidya, P.C. (1961). Wave solutions in general relativity. I. *Proc. Nat. Inst. Sci. India A* **27**, 620. *See* §31.5.

Panov, V.F. (1979a). Collision of plane electromagnetic-gravitational waves. *Sov. Phys. J.* **5**, 91. *See* §25.5.

Panov, V.F. (1979b). Cylindrical gravitational wave in Melvin's universe. *Izv. Vys. Uch. Zav., Fiz.* **12**, 24. *See* §34.1.

Panov, V.F. (1979c). Interaction of plane gravitational waves with variable polarization (in Russian). *Fiz.* **9**, 79. *See* §25.4.

Panov, V.F. (1979d). Obtaining stationary Einstein–Maxwell fields. I. *Sov. Phys. J.* **22**, 702. *See* §21.1.

Pant, D.N. (1994). Varieties of new classes of interior solutions in general relativity. *Astrophys. Space Sci.* **215**, 97. *See* §16.1.

Pant, D.N. and Sah, A. (1982). Class of solutions of Einstein's field equations for static fluid spheres. *PRD* **26**, 1254. *See* §16.1.

Pant, D.N. and Sah, A. (1985). Massive fluid spheres in general relativity. *PRD* **32**, 1358. *See* §16.1.

Papacostas, T. (1985). Space-times admitting Penrose–Floyd tensors. *GRG* **17**, 149. *See* §35.3.

Papacostas, T. (1988). Hauser–Malhiot spaces admitting a perfect fluid energy-momentum tensor. *JMP* **29**, 1445. *See* §§21.2, 22.2, 35.3.

Papacostas, T. and Xanthopoulos, B.C. (1989). Collisions of gravitational and electromagnetic waves that do not develop curvature singularities. *JMP* **30**, 97. *See* §25.5.

Papadopoulos, D. and Sanz, J.L. (1985). New inhomogeneous viscous-fluid cosmologies. *Lett. Nuovo Cim.* **42**, 215. *See* §23.4.

Papadopoulos, D. and Xanthopoulos, B.C. (1990). Tomimatsu–Sato solutions describe cosmic strings interacting with gravitational waves. *PRD* **41**, 2512. *See* §22.3.

Papanicolaou, N. (1979). Gravitational duality and Bäcklund transformations. *JMP* **20**, 2069. *See* §34.9.

Papapetrou, A. (1947). A static solution of the equations of the gravitational field for an arbitrary charge distribution. *Proc. Roy. Irish Acad. A* **51**, 191. *See* §§18.7, 21.1.

Papapetrou, A. (1953). Eine rotationssymmetrische Lösung in der Allgemeinen Relativitätstheorie. *Ann. Phys.* (*Germany*) **12**, 309. *See* §20.3.

Papapetrou, A. (1963). Quelques remarques sur les champs gravitationnels stationnaires. *C. R. Acad. Sci.* (*Paris*) **257**, 2797. *See* §18.4.

Papapetrou, A. (1966). Champs gravitationnels stationnaires à symmétrie axiale. *Ann. Inst. H. Poincaré A* **4**, 83. *See* §§19.2, 19.3, 22.3.

Papapetrou, A. (1971a). Les relations identiques entre les équations du formalisme de Newman–Penrose. *C. R. Acad. Sci.* (*Paris*) *A* **272**, 1613. *See* §§7.1, 7.3.

Papapetrou, A. (1971b). Quelques remarques sur le formalism de Newman–Penrose. *C. R. Acad. Sci.* (*Paris*) *A* **272**, 1537. *See* §§7.1, 7.3.

Parker, L., Ruffini, R. and Wilkins, D. (1973). Metric of two spinning charged sources in equilibrium. *PRD* **7**, 2874. *See* §21.1.

Pasqua, M. (1975). A solution to the Einstein–Maxwell equations admitting a group G_7 of automorphisms (in Italian). *Atti Accad. Naz. Lincei, Rend. Cl. Sci. Fis. Mat. Nat.* (*Italy*) **59**, 91. *See* §12.2.

Patel, L.K. (1973a). A cylindrically symmetric non-uniform universe filled with perfect fluid. *Curr. Sci.* **42**, 457. *See* §23.3.

Patel, L.K. (1973b). A new form of Rao's solution. *Vidya B* **16**, 71. *See* §32.4.

Patel, L.K. (1978). Radiating Demianski-type space-times. *Indian J. Pure Appl. Math.* **9**, 1019. *See* §30.7.

Patel, L.K. and Trivedi, V.M. (1975). On axially symmetric electrovac universes. *Vidya B* **18**, 127. *See* §34.1.

Patel, M.D. and Vaidya, P.C. (1983). A rotating mass embedded in an Einstein–Gödel universe. *GRG* **15**, 777. *See* §32.5.

Patera, J. and Winternitz, P. (1977). Subalgebras of real three and four-dimensional Lie algebras. *JMP* **18**, 1449. *See* §8.2.

Patnaik, S. (1970). Einstein–Maxwell fields with plane symmetry. *Proc. Camb. Phil. Soc.* **67**, 127. *See* §15.4.

Patwardhan, G.K. and Vaidya, P.C. (1943). Relativistic distributions of matter of radial symmetry. *J. Univ. Bombay* **12**, 23. *See* §16.1.

Peacock, J. (1999). *Cosmological physics* (Cambridge University Press, Cambridge). *See* §14.1.

Penrose, R. (1960). A spinor approach to general relativity. *Ann. Phys.* (*USA*) **10**, 171. *See* §3.6, Ch. 4, §9.3.
Penrose, R. (1965). A remarkable property of plane waves in general relativity. *Rev. Mod. Phys.* **37**, 215. *See* §37.1.
Penrose, R. and Rindler, W. (1984). *Spinors and space-time*, vol. I (Cambridge University Press, Cambridge). *See* §§2.2, 3.6, 7.1, 7.4.
Penrose, R. and Rindler, W. (1986). *Spinors and space-time*, vol. II (Cambridge University Press, Cambridge). *See* §§2.2, 3.6, 5.1, 7.1, 9.1, 9.3.
Peres, A. (1960). Null electromagnetic fields in general relativity theory. *Phys. Rev.* **118**, 1105. *See* §24.5.
Perjés, Z. (1970). Spinor treatment of stationary space-times. *JMP* **11**, 3383. *See* §18.5.
Perjés, Z. (1971). Solutions of the coupled Einstein–Maxwell equations representing the fields of spinning sources. *Phys. Rev. Lett.* **27**, 1668. *See* §18.7.
Perjés, Z. (1985a). An almost conformal approach to axial symmetry, in *Galaxies, axisymmetric systems and relativity*, ed. M.A.H. MacCallum, page 166 (Cambridge Univ. Press, Cambridge). *See* §19.5.
Perjés, Z. (1985b). Improved characterization of the Kerr metric, in *Proceedings of the third seminar on quantum gravity*, eds. M.A. Markov, V.A. Berezin and V.P. Frolov, page 446 (World Scientific, Singapore). *See* §18.7.
Perjés, Z. (1986a). Stationary vacuum fields with a conformally flat three-space. II. Proof of axial symmetry. *GRG* **18**, 511. *See* §18.7.
Perjés, Z. (1986b). Stationary vacuum fields with conformally flat three-space. III. Complete solution. *GRG* **18**, 531. *See* §18.7.
Perjés, Z. (1986c). Stationary vacuum space-times in Ernst coordinates, in *Proceedings of the fourth Marcel Grossmann meeting on general relativity*, ed. R. Ruffini, page 1003 (North-Holland, Amsterdam). *See* §18.7.
Perjés, Z. (1988). Ernst coordinates. *Acta Phys. Hung.* **63**, 89. *See* §18.7.
Perjés, Z. (1993). Einstein–Maxwell fields with no vacuum counterpart. *CQG* **10**, 1649. *See* §21.1.
Perjés, Z. and Kramer, D. (1996). An electrovacuum spacetime satisfying the oval equation of general relativity. *CQG* **13**, 3241. *See* §21.1.
Perjés, Z., Lukács, B., Sebestyén, A., Valentini, A. and Sparling, G.A.J. (1984). Solution of the stationary vacuum equations of relativity for conformally flat 3-spaces. *Phys. Lett. A* **100**, 405. *See* Ch. 36.
Persides, S. and Xanthopoulos, B.C. (1988). Some new stationary axisymmetric asymptotically flat space-times obtained from Painlevé transcendents. *JMP* **29**, 674. *See* §20.6.
Peters, P.C. (1979). Toroidal black holes? *JMP* **20**, 1481. *See* §20.2.
Petrov, A.Z. (1954). Classification of spaces defined by gravitational fields. *Uch. Zapiski Kazan Gos. Univ.* **144**, 55. *See* §4.2.
Petrov, A.Z. (1962). Gravitational field geometry as the geometry of automorphisms, in *Recent developments in general relativity*, page 379 (Pergamon Press–PWN Warsaw, Oxford). *See* §§12.2, 31.5.
Petrov, A.Z. (1963a). Central-symmetric gravitational fields (in Russian). *Zh. Eks. Teor. Fiz.* **44**, 1525. *See* §15.4.
Petrov, A.Z. (1963b). On Birkhoff's theorem (in Russian). *Uch. Zapiski Kazan. Gos. Univ.* **123**, 61. *See* §15.4.
Petrov, A.Z. (1966). *New methods in general relativity* (in Russian) (Nauka, Moscow). *See* §§4.2, 8.1, 8.2, 8.4, 8.5, 11.1, 12.5, 13.1, 17.1, 22.3.
Phan, D. (1993). A G_2II algebraically general twistfree vacuum spacetime. *GRG* **25**, 1009. *See* §§11.4, 17.1.
Philbin, T.G. (1996). Perfect-fluid cylinders and walls – sources for the Levi-Civita space-time. *CQG* **13**, 1217. *See* §22.2.

Piper, M.S. (1997). Computer algebra and power series in general relativity. Ph.D. thesis, Queen Mary and Westfield College. *See* §9.3.

Piran, T., Safier, P.N. and Katz, J. (1986). Cylindrical gravitational waves with two degrees of freedom: an exact solution. *PRD* **34**, 331. *See* §22.3.

Pirani, F.A.E. (1957). Invariant formulation of gravitational radiation theory. *Phys. Rev.* **105**, 1089. *See* Ch. 4.

Pirani, F.A.E. (1965). Introduction to gravitational radiation theory, in *Lectures on general relativity, Brandeis 1964*, vol. 1, page 249 (Prentice-Hall, Englewood Cliffs, New Jersey). *See* Ch. 4, §7.3.

Plebański, J.F. (1964). The algebraic structure of the tensor of matter. *Acta Phys. Polon.* **26**, 963. *See* §§5.1, 5.3.

Plebański, J.F. (1967). On conformally equivalent Riemannian spaces. Centro de investigación y estudios avanzados, IPN, Mexico. *See* §8.5.

Plebański, J.F. (1972). Solutions of Einstein equations $g_{\mu\nu} = -\rho k_\mu k_\nu$ determined by the condition that K_μ is the quadruple Debever–Penrose vector. Handwritten notes, Warszawa. *See* §26.1.

Plebański, J.F. (1975). A class of solutions of Einstein–Maxwell equations. *Ann. Phys.* (*USA*) **90**, 196. *See* §21.1.

Plebański, J.F. (1979). The nondiverging and nontwisting type D electrovac solutions with λ. *JMP* **20**, 1946. *See* §§31.7, 31.8.

Plebański, J.F. (1980). Group of diagonal empty space-times. *Phys. Rev. Lett.* **45**, 303. *See* §20.2.

Plebański, J.F. and Demiański, M. (1976). Rotating, charged and uniformly accelerated mass in general relativity. *Ann. Phys.* (*USA*) **98**, 98. *See* §§21.1, 25.5, 29.5.

Plebański, J.F. and García D., A. (1982). Multiexponent Weyl metrics. *Phys. Rev. Lett.* **48**, 1447. *See* §20.2.

Plebański, J.F. and Hacyan, S. (1979). Some exceptional electro-vac type D metrics with cosmological constant. *JMP* **20**, 1004. *See* §26.1.

Plebański, J.F. and Stachel, J. (1968). Einstein tensor and spherical symmetry. *JMP* **9**, 269. *See* §15.4.

Podolsky, J. and Griffiths, J.B. (1999). Expanding impulsive gravitational waves. *CQG* **16**, 2937. *See* §28.4.

Pollney, D., Skea, J.E.F. and D'Inverno, R.A. (2000). Classifying geometries in general relativity: I. Standard forms for symmetric spinors. *CQG* **17**, 643. *See* §9.2.

Ponce de León, J. (1988). Solutions of Einstein's equations relevant to the description of 'bubbles' in the early universe. *JMP* **29**, 2479. *See* §§11.4, 16.2.

Prasanna, A.R. (1968). A solution for an isolated charged body in isotropic coordinates. *Current Sci.* **37**, 430. *See* §15.4.

Pravda, V. (1999). Curvature invariants in type-*III* spacetimes. *CQG* **16**, 3315. *See* §9.1.

Pryse, P.V. (1993). Group invariant solutions of the Ernst equation. *CQG* **10**, 163. *See* §20.6.

Quevedo, H. (1986a). Class of stationary axisymmetric solutions of Einstein's equations in empty space. *PRD* **33**, 324. *See* §34.3.

Quevedo, H. (1986b). The Kerr–NUT metric with all mass multipole moments, in *Proceedings of the fourth Marcel Grossmann meeting on general relativity*, ed. R. Ruffini, page 1015 (North-Holland, Amsterdam). *See* §34.3.

Quevedo, H. (1989). General static axisymmetric solution of Einstein's vacuum field equations in prolate spheroidal coordinates. *PRD* **39**, 2904. *See* §20.2.

Quevedo, H. (1990). Multipole moments in general relativity. Static and stationary vacuum solutions. *Fortschr. d. Phys.* **38**, 733. *See* §20.1.

Quevedo, H. (1992). Complete transformations of the curvature tensor. *GRG* **24**, 693. *See* §21.1.

Quevedo, H. and Mashhoon, B. (1990). Exterior gravitational field of a charged rotating mass with arbitrary quadrupole moment. *Phys. Lett. A* **148**, 149. *See* §34.1.

Quevedo, H. and Mashhoon, B. (1991). Generalization of Kerr spacetime. *PRD* **43**, 3902. *See* §34.3.

Quevedo, H. and Sussman, R.A. (1995). On the thermodynamics of simple non-isentropic perfect fluids in general relativity. *CQG* **12**, 859. *See* §5.4.

Rácz, I. (1997). On Einstein's equations for space-times admitting a non-null Killing vector. *JMP* **38**, 4202. *See* §18.3.

Rácz, I. and Zsigrai, J. (1996). Generating new perfect-fluid solutions from known ones. *CQG* **13**, 2783. *See* §§10.11, 21.2.

Radhakrishna, L. (1963). Some exact non-static cylindrically symmetric electrovac universes. *Proc. Nat. Inst. Sci. India A* **29**, 588. *See* §22.4.

Rainer, A. and Stephani, H. (1999). Similarity reduction of a class of algebraically special perfect fluids. *J. Math. Phys* **40**, 897. *See* §33.1.

Rainich, G.Y. (1925). Electrodynamics in general relativity. *Trans. Amer. Math. Soc.* **27**, 106. *See* §5.4.

Ram, S. (1988). Bianchi type VI_0 space-times with perfect fluid source. *JMP* **29**, 449. *See* §14.4.

Ram, S. (1989a). Generation of LRS Bianchi type I universes filled with perfect fluids. *GRG* **21**, 697. *See* §14.3.

Ram, S. (1989b). Homogeneous Bianchi type VI_0 perfect fluid space-times. *Int. J. Theor. Phys.* **28**, 97. *See* §14.4.

Ram, S. (1989c). LRS Bianchi type I perfect fluid solutions generated from known solutions. *Int. J. Theor. Phys.* **28**, 917. *See* §14.3.

Ram, S. (1989d). Spatially homogeneous cosmological models of Bianchi type III. *JMP* **30**, 757. *See* §14.3.

Ram, S. (1990). Bianchi type V perfect-fluid space-times. *Int. J. Theor. Phys.* **29**, 901. *See* §14.4.

Ramos, M.P.M. (1998). Invariant differential operators and the Karlhede classification of type N non-vacuum solutions. *CQG* **15**, 435. *See* §9.2.

Ramos, M.P.M. and Vickers, J.A.G. (1996a). Invariant differential operators and the Karlhede classification of type N vacuum solutions. *CQG* **13**, 1589. *See* §9.2.

Ramos, M.P.M. and Vickers, J.A.G. (1996b). A spacetime calculus based on a single null direction. *CQG* **13**, 1579. *See* §§7.2, 9.2.

Ray, D. (1976). Self-gravitating fluids with cylindrical symmetry. *JMP* **17**, 1171. *See* §10.11.

Ray, J.R. and Thompson, E.L. (1975). Space-time symmetries and the complexion of the electromagnetic field. *JMP* **16**, 345. *See* §11.1.

Raychaudhuri, A.K. (1955). Relativistic cosmology, I. *Phys. Rev.* **98**, 1123. *See* §6.2.

Raychaudhuri, A.K. (1958). An anisotropic cosmological solution in general relativity. *Proc. Phys. Soc. Lond.* **72**, 263. *See* §14.4.

Raychaudhuri, A.K. (1960). Static electromagnetic fields in general relativity. *Ann. Phys. (USA)* **11**, 501. *See* §22.2.

Raychaudhuri, A.K. and Saha, S.K. (1981). Viscous fluid interpretation of electromagnetic field, I. *JMP* **23**, 2554. *See* §5.2.

Reina, C. and Treves, A. (1975). Gyromagnetic ratio of Einstein–Maxwell fields. *PRD* **11**, 3031. *See* §34.1.

Reissner, H. (1916). Über die Eigengravitation des elektrischen Feldes nach der Einsteinschen Theorie. *Ann. Phys. (Germany)* **50**, 106. *See* §§15.4, 21.1.

Reula, O. (1989). On existence and behaviour of asymptotically flat solutions to the stationary Einstein equations. *Commun. Math. Phys.* **122**, 615. *See* §18.4.

Reuss, J.-D. (1968). Un champ gravitationnel cylindrique et non stationnaire. *C. R. Acad. Sci. (Paris) A* **266**, 794. *See* §22.3.

Robertson, H.P. (1929). On the foundations of relativistic cosmology. *Proc. Nat. Acad. Sci. USA* **15**, 822. *See* §12.1.
Robertson, H.P. (1935). Kinematics and world-structure. *Astrophys. J.* **82**, 284. *See* §12.1.
Robertson, H.P. (1936). Kinematics and world-structure II. *Astrophys. J.* **83**, 187. *See* §12.1.
Robinson, I. (1959). A solution of the Einstein–Maxwell equations. *Bull. Acad. Polon. Sci. Math. Astron. Phys.* **7**, 351. *See* §§12.3, 21.1.
Robinson, I. (1961). Null electromagnetic fields. *JMP* **2**, 290. *See* §7.6.
Robinson, I. (1975). On vacuum metrics of type (3,1). *GRG* **6**, 423. *See* §§28.1, 29.2, 29.4.
Robinson, I. and Robinson, J.R. (1969). Vacuum metrics without symmetry. *Int. J. Theor. Phys.* **2**, 231. *See* §§29.1, 29.2, 29.4, 30.5.
Robinson, I., Robinson, J.R. and Zund, J.D. (1969a). Degenerate gravitational fields with twisting rays. *J. Math. Mech.* **18**, 881. *See* §§27.1, 29.1.
Robinson, I. and Schild, A. (1963). Generalization of a theorem by Goldberg and Sachs. *JMP* **4**, 484. *See* §7.6.
Robinson, I., Schild, A. and Strauss, H. (1969b). A generalized Reissner–Nordström solution. *Int. J. Theor. Phys.* **2**, 243. *See* §§30.2, 30.3, 30.4.
Robinson, I. and Trautman, A. (1962). Some spherical gravitational waves in general relativity. *Proc. Roy. Soc. Lond. A* **265**, 463. *See* §§27.1, 27.2, 28.1, 28.2.
Roque, W.L. and Ellis, G.F.R. (1985). The automorphism group and the field equations for Bianchi universes, in *Galaxies, axisymmetric systems and relativity. Essays presented to W.B. Bonnor on his 65th birthday,* ed. M.A.H. MacCallum, page 54 (Cambridge University Press, Cambridge). *See* §13.2.
Rosen, G. (1962). Symmetries of the Einstein–Maxwell equations. *JMP* **3**, 313. *See* §13.3.
Rosen, J. (1965). Embedding of various relativistic Riemann spaces in pseudo-Euclidean spaces. *Rev. Mod. Phys.* **37**, 204. *See* §§37.1, 37.5, 37.6.
Rosquist, K. (1980). Global rotation. *GRG* **12**, 649. *See* §12.4.
Rosquist, K. (1983). Exact rotating and expanding radiation-filled universe. *Phys. Lett. A* **97**, 145. *See* §14.4.
Rosquist, K. (1994). Hamiltonian approach to relativistic star models. *CQG* **12**, 1305. *See* §16.1.
Rosquist, K. and Jantzen, R.T. (1985). Exact power law solutions of the Einstein equations. *Phys. Lett. A* **107**, 29. *See* §§11.4, 14.4.
Rosquist, K. and Jantzen, R.T. (1988). Unified regularization of Bianchi cosmology. *Phys. Rep.* **166**, 89. *See* §13.2.
Rosquist, K. and Uggla, C. (1991). Killing tensors in two-dimensional space-times with applications to cosmology. *JMP* **32**, 3412. *See* §13.2.
Roy, A.R. and Rao, J.R. (1972). Reciprocal static solutions of Einstein's equation for spheres of fluid. *Indian J. Pure Appl. Phys.* **10**, 845. *See* §16.1.
Roy, S.R. and Narain, S. (1981). Some cylindrical symmetric nonstatic perfect fluid distributions in general relativity with pressure equal to density. *Int. J. Theor. Phys.* **20**, 709. *See* §23.3.
Roy, S.R. and Narain, S. (1983). Some non-static self gravitating fluids in space-times of cylindrical symmetry. *Indian J. Pure Appl. Math.* **14**, 96. *See* §23.3.
Roy, S.R. and Prakash, S. (1977). Some solutions of Einstein–Maxwell equations for cylindrically symmetric space-time with two degrees of freedom in general relativity. *Indian J. Pure Appl. Math.* **8**, 1132. *See* §22.4.
Roy, S.R. and Prasad, A. (1989). Some inhomogeneous generalizations of Bianchi type VI_h cosmological models in general relativity. *Progr. Math.* **23**, 153. *See* §23.3.
Roy, S.R. and Prasad, A. (1991). Inhomogeneous generalizations of Bianchi type VI_h models with perfect fluid. *Astrophys. Space Sci.* **181**, 61. *See* §23.3.

Roy, S.R. and Tiwari, O.P. (1982). Some anisotropic non-static perfect fluid cosmological models in general relativity. *J. Phys. A* **15**, 1747. *See* §14.4.
Roy, S.R. and Tripathi, V.N. (1972). Cylindrically symmetric electromagnetic field of the second class. *Indian J. Pure Appl. Math.* **3**, 926. *See* §22.4.
Ruban, V.A. (1971). On the nature of the singularities in the spatially homogeneous Einstein fields. Preprint no. 355, A.F. Ioffe Institute Leningrad. *See* §§13.3, 14.3.
Ruban, V.A. (1977). Dynamics of anisotropic homogeneous generalizations of the Friedmann cosmological models. *Zh. Eks. Teor. Fiz.* **45**, 629. *See* §14.3.
Ruban, V.A. (1978). The dynamics of the spatially homogeneous (axisymmetric) cosmological models. II. Preprint no. 412, Leningrad Institute of Particle Physics. *See* §§14.3, 14.4.
Ruiz, E. (1986). Harmonic coordinates for Kerr's metric. *GRG* **18**, 805. *See* §20.5.
Ruiz, E., Manko, V.S. and Martín, J. (1995). Extended N-soliton solution of the Einstein–Maxwell equations. *PRD* **51**, 4192. *See* §34.8.
Ruiz, E. and Senovilla, J.M.M. (1992). General class of inhomogeneous perfect-fluid solutions. *PRD* **45**, 1995. *See* §23.3.
Rund, H. (1976). Variational problems and Bäcklund transformations associated with the sine-Gordon and Korteweg-deVries equations and their extensions, in *Bäcklund transformations.* Lecture notes in mathematics, vol. 515, ed. R.M. Miura, page 199 (Springer Verlag, Berlin). *See* §10.9.
Ryan, M.P. and Shepley, L.C. (1975). *Homogeneous relativistic cosmologies* (Princeton Univ. Press, Princeton). *See* §§12.4, 13.2, 13.3.
Sachs, R.K. (1962). Gravitational waves in general relativity. VIII. Waves in asymptotically flat space-time. *Proc. Roy. Soc. Lond. A* **270**, 103. *See* §29.1.
Salazar I., H. (1986). Magnetized diagonal Weyl metrics. *JMP* **27**, 277. *See* §34.1.
Salazar I., H., García D., A. and Plebański, J.F. (1983). Symmetries of the nontwisting type-N solutions with cosmological constant. *JMP* **24**, 2191. *See* §11.4.
Sapar, A. (1970). Basic dependencies for cosmological models with matter and radiation (in Russian). *Astron. Zh.* (*USSR*) **47**, 503. *See* §14.2.
Saridakis, E. and Tsamparlis, M. (1991). Symmetry inheritance of conformal Killing vectors. *JMP* **32**, 1541. *See* §35.4.
Sato, H. (1982). Metric solution of a spinning mass: Kerr–Tomimatsu–Sato class solution, in *Proceedings of the second Marcel Grossmann meeting on general relativity*, ed. R. Ruffini, page 353 (North-Holland, Amsterdam). *See* §20.5.
Sato, H. (1984). Voids in expanding universe, in *General relativity and gravitation. Invited papers and discussion reports of the 10th international conference on general relativity and gravitation*, eds. B. Bertotti, F. de Felice and A. Pascolini, page 289 (Reidel, Dordrecht). *See* §15.5.
Saunders, P.T. (1967). Non-isotropic model universes. Ph.D. thesis, London. *See* §§13.3, 14.4.
Schiffer, M.M., Adler, R.J., Mark, J. and Sheffield, C. (1973). Kerr geometry as complexified Schwarzschild geometry. *JMP* **14**, 52. *See* §21.1.
Schimming, R. (1974). Riemannsche Räume mit ebenfrontiger und mit ebener Symmetrie. *Math. Nachr.* **59**, 129. *See* §24.5.
Schmidt, B.G. (1967). Isometry groups with surface-orthogonal trajectories. *Z. Naturforsch.* **22a**, 1351. *See* §8.6.
Schmidt, B.G. (1968). Riemannsche Räume mit mehrfach transitiver Isometriegruppe. Ph.D. thesis, Hamburg. *See* §§8.6, 11.2, 12.1.
Schmidt, B.G. (1971). Homogeneous Riemannian spaces and Lie algebras of Killing fields. *GRG* **2**, 105. *See* §8.6.
Schmidt, B.G. (1984). The Geroch group is a Banach Lie group, in *Solutions of Einstein's equations: Techniques and results.* Lecture notes in physics, vol. 205, eds. C. Hoenselaers and W. Dietz, page 113 (Springer, Berlin). *See* §10.5.

Schmidt, B.G. (1996). Vacuum spacetimes with toroidal null infinities. *CQG* **13**, 2811. *See* §17.2.
Schmidt, H.J. (1998). Consequences of the noncompactness of the Lorentz group. *Int. J. Theor. Phys.* **37**, 691. *See* §9.1.
Schouten, J.A. (1954). *Ricci-calculus* (Springer-Verlag, Berlin). *See* §§2.1, 3.1.
Schutz, B. (1980). *Geometrical methods of mathematical physics* (Cambridge University Press, Cambridge). *See* §2.1.
Schwarzschild, K. (1916a). Über das Gravitationsfeld eines Masenpunktes nach der Einsteinschen Theorie. *Sitz. Preuss. Akad. Wiss.*, 189. *See* §15.4.
Schwarzschild, K. (1916b). Über das Gravitationsfeldeiner Kugel aus inkompressibler Flüssigkeit nach der Einsteinschen Theorie. *Sitz. Preuss. Akad. Wiss.*, 424. *See* §16.1.
Sciama, D.W. (1961). Recurrent radiation in general relativity. *Proc. Camb. Phil. Soc.* **57**, 436. *See* §35.2.
Segre, C. (1884). Sulla teoria e sulla classificazione delle omografie in uno spazio lineare ad un numero qualcunque di dimensioni. *Memorie della R. Accademia dei Lincei, ser. 3a* **XIX**, 127. *See* §5.1.
Seixas, W. (1991). Extensions to the computer-aided classification of the Ricci tensor. *CQG* **8**, 1577. *See* §9.3.
Seixas, W. (1992a). Computer-aided classification of exact solutions. Ph.D. thesis, Queen Mary and Westfield College, London. *See* §9.4.
Seixas, W. (1992b). Killing vectors in conformally flat perfect fluids via invariant classification. *CQG* **9**, 225. *See* §§9.2, 37.4.
Sen, N. (1924). Über die Grenzbedingungen des Schwerefeldes an Unstetigkeitsflächen. *Ann. Phys. (Germany)* **73**, 365. *See* §3.8.
Sengier-Diels, J. (1974a). Espaces pseudo-riemanniens homogenènes à quatre dimensions. *Bull. Acad. Roy. Belg. Cl. Sci.* **60**, 1469. *See* §12.1.
Sengier-Diels, J. (1974b). Sur les espaces homogenes de la relativité. Ph.D. thesis, Université Libre de Bruxelles. *See* §12.1.
Senin, Y.E. (1982). Cosmology with toroidal distribution of a fluid, in *Problems of gravitation theory and elementary particle theory*, 13th issue, ed. K.P. Stanyukovich, page 107 (Energoizdat, Moscow). *See* §23.3.
Senovilla, J.M.M. (1987a). New LRS perfect-fluid cosmological models. *CQG* **4**, 1449. *See* §§14.3, 32.5.
Senovilla, J.M.M. (1987b). On Petrov type-D stationary axisymmetric rigidly rotating perfect-fluid metrics. *CQG* **4**, L115. *See* §21.2.
Senovilla, J.M.M. (1987c). Stationary axisymmetric perfect-fluid metrics with $q + 3p = const.$ *Phys. Lett. A* **123**, 211. *See* §21.2.
Senovilla, J.M.M. (1990). New class of inhomogeneous cosmological perfect-fluid solutions without big-bang singularity. *Phys. Rev. Lett.* **64**, 2219. *See* §23.3.
Senovilla, J.M.M. (1992). New family of stationary and axisymmetric perfect-fluid solutions. *CQG* **9**, L167. *See* §21.2.
Senovilla, J.M.M. (1993). Stationary and axisymmetric perfect-fluid solutions to Einstein's equations, in *Rotating objects and relativistic physics. Proceedings*, eds. F.J. Chinea and L.M. Gonzalez-Romero, page 73 (Springer, Berlin). *See* §21.2.
Senovilla, J.M.M. and Sopuerta, C.F. (1994). New G_1 and G_2 inhomogeneous cosmological models from the generalized Kerr–Schild transformation. *CQG* **11**, 2073. *See* §§32.5, 35.4.
Senovilla, J.M.M., Sopuerta, C.F. and Szekeres, P. (1998). Theorems on shear-free perfect fluids with their Newtonian analogues. *GRG* **30**, 389. *See* §6.2.
Senovilla, J.M.M. and Vera, R. (1997). Dust G_2 cosmological models. *CQG* **14**, 3481. *See* §23.3.
Senovilla, J.M.M. and Vera, R. (1998). G_2 cosmological models separable in non-comoving coordinates. *CQG* **15**, 1737. *See* §23.3.

Senovilla, J.M.M. and Vera, R. (2001). New family of inhomogeneous γ-law cosmologies: Example of gravitational waves in a homogeneous $p = \rho/3$ background. *PRD* **63**, 084008. *See* §23.3.

Shikin, I.S. (1966). Homogeneous anisotropic cosmological model with magnetic fields (in Russian). *Dokl. Akad. Nauk SSSR* **171**, 73. *See* §§14.1, 14.3.

Shikin, I.S. (1967). A uniform axisymmetrical cosmological model in the ultrarelativistic case. *Dokl. Akad. Nauk SSSR* **176**, 1048. *See* §13.2.

Shikin, I.S. (1968). Solutions of the gravitational equations for homogeneous flat anisotropic models. *Dokl. Akad. Nauk SSSR* **179**, 817. *See* §§14.3, 14.4.

Shikin, I.S. (1972). Gravitational fields with groups of motion on two-dimensional transitivity hypersurfaces in a model with matter and magnetic field. *Commun. Math. Phys.* **26**, 24. *See* §13.1.

Shikin, I.S. (1979). Evolution of plane-symmetric self-similar space-times containing perfect fluid. *GRG* **11**, 433. *See* §15.7.

Shukla, H.C. and Patel, L.K. (1977). On cylindrically symmetric cosmological models. *Vidya B* **20**, 35. *See* §14.3.

Sibgatullin, N.R. (1984). Construction of the general solution of the system of Einstein–Maxwell equations for the stationary axisymmetric case. *Sov. Phys. – Dokl.* **29**, 802. *See* §34.6.

Sibgatullin, N.R. (1991). *Oscillations and waves in strong gravitational and electromagnetic fields* (Springer Verlag, Berlin). *See* §34.6.

Siklos, S.T.C. (1976a). Singularities, invariants and cosmology. Ph.D. thesis, Cambridge. *See* §§8.2, 9.1, 11.2, 12.2, 13.2.

Siklos, S.T.C. (1976b). Two completely singularity-free NUT space-times. *Phys. Lett. A* **59**, 173. *See* §13.3.

Siklos, S.T.C. (1980). Field equations for spatially homogeneous space-times. *Phys. Lett. A* **76**, 19. *See* §13.2.

Siklos, S.T.C. (1981). Some Einstein spaces and their global properties. *J. Phys. A* **14**, 395. *See* §§12.2, 12.5, 13.2, 13.3, 38.2.

Siklos, S.T.C. (1985). Lobatchevski plane gravitational waves, in *Galaxies, axisymmetric systems and relativity. Essays presented to W.B. Bonnor on his 65th birthday*, ed. M.A.H. MacCallum, page 247 (Cambridge University Press, Cambridge). *See* §§10.11, 12.5.

Simon, W. (1984). Characterization of the Kerr metric. *GRG* **16**, 465. *See* §§18.7, 20.5.

Simon, W. (1994). A class of static perfect fluid solutions. *GRG* **26**, 97. *See* §16.1.

Singatullin, R.S. (1973). Exact wave solutions of Einstein–Maxwell equations defined by the Einstein–Rosen solutions (in Russian). *Grav. Teor. Otnos., Univ. Kazan* **9**, 67. *See* §22.4.

Singh, J.K. and Ram, S. (1995). A class of new Bianchi I perfect fluid space-times. *Astrophys. Space Sci.* **225**, 57. *See* §14.3.

Singh, K.P. and Abdussattar (1973). Plane-symmetric cosmological model. II. *J. Phys. A* **6**, 1090. *See* §14.3.

Singh, K.P. and Abdussattar (1974). A plane-symmetric universe filled with perfect fluid. *Curr. Sci.* **43**, 372. *See* §14.3.

Singh, K.P., Radhakrishna, L. and Sharan, R. (1965). Electromagnetic fields and cylindrical symmetry. *Ann. Phys.* (*USA*) **32**, 46. *See* §22.4.

Singh, K.P. and Roy, S.R. (1966). Electromagnetic behaviour in space-times conformal to some well-known empty space-times. *Proc. Nat. Inst. Sci. India* **A32**, 223. *See* §37.5.

Singh, K.P. and Singh, D.N. (1968). A plane symmetric cosmological model. *Mon. Not. R. Astron. Soc.* **140**, 453. *See* §14.3.

Singh, K.P., Singh, G. and Ram, S. (1978). Curvature collineation for the field of total radiation. *Indian J. Pure Appl. Math.* **9**, 906. *See* §35.4.

Singleton, D.B. (1990). Homothetic motions and vacuum Robinson–Trautman solutions. *GRG* **22**, 1239. *See* §§11.4, 17.3.

Sintes, A.M. (1996). Inhomogeneous cosmologies with special properties. Ph.D. thesis, Palma de Mallorca. *See* §§11.4, 23.3.

Sintes, A.M. (1998). Kinematic self-similar locally rotationally symmetric models. *CQG* **15**, 3689. *See* §35.4.

Sintes, A.M., Coley, A.A. and Carot, J. (1998). Lie groups of conformal motions acting on null orbits. *GRG* **30**, 151. *See* §§13.4, 23.3, 35.4.

Sippel, R. and Goenner, H. (1986). Symmetry classes of *pp*-waves. *GRG* **18**, 1229. *See* §§12.5, 24.5.

Sistero, R.F. (1972). Relativistic non-zero pressure cosmology. *Astrophys. Space Sci.* **17**, 150. *See* §14.2.

Skea, J.E.F. (1997). The invariant classification of conformally flat pure radiation spacetimes. *CQG* **14**, 2393. *See* §9.3.

Skea, J.E.F. (2000). A spacetime whose invariant classification requires the fifth covariant derivative of the Riemann tensor. *CQG* **17**, L69. *See* §§9.1, 9.2.

Sklavenites, D. (1985). Relativistic, stationary, axisymmetric perfect fluids. II. Solutions with vanishing magnetic Weyl tensor. *JMP* **26**, 2279. *See* §21.2.

Sklavenites, D. (1992a). Geodesic Bianchi type cosmological models. *GRG* **24**, 47. *See* §14.4.

Sklavenites, D. (1992b). Stationary and static axisymmetric perfect fluid solutions. *GRG* **24**, 935. *See* §21.2.

Sklavenites, D. (1999). Stationary perfect fluid cylinders. *CQG* **16**, 2753. *See* §22.2.

Skripkin, V.A. (1960). Point explosion in an ideal incompressible fluid in the general theory of relativity (in Russian). *Dokl. Akad. Nauk SSSR* **135**, 1072. *See* §16.2.

Sneddon, G.E. (1976). Hamiltonian cosmology: a further investigation. *J. Phys. A* **9**, 229. *See* §13.2.

Sneddon, G.E. (1999). The identities of the algebraic invariants of the four-dimensional Riemann tensor. III. *JMP* **40**, 5905. *See* §9.1.

Soh, C.W. and Mahomed, F.M. (1999). Noether symmetries of $y'' = f(x)y^n$ with applications to nonstatic spherically symmetric perfect fluid solutions. *CQG* **16**, 3553. *See* §§15.6, 16.2.

Sopuerta, C.F. (1997). New study of silent universes. *PRD* **55**, 5936. *See* §6.2.

Sopuerta, C.F. (1998a). Covariant study of a conjecture on shear-free barotropic perfect fluids. *CQG* **15**, 1043. *See* §6.2.

Sopuerta, C.F. (1998b). Stationary generalized Kerr–Schild spacetimes. *JMP* **39**, 1024. *See* §32.5.

Srinivasa Rao, K.N. and Gopala Rao, A.V. (1980). Some electrovac models of homogeneous gravitational force fields in general relativity. *JMP* **21**, 2261. *See* §18.6.

Srivastava, D.C. (1987). Exact solutions for shear-free motion of spherically symmetric perfect fluid distributions in general relativity. *CQG* **4**, 1093. *See* §16.2.

Srivastava, D.C. (1992). Exact solutions for shear-free motion of spherically symmetric charged perfect fluid distributions in general relativity. *Fortschr. Phys.* **40**, 31. *See* §16.2.

Stachel, J. (1982). Globally stationary but locally static space-times: a gravitational analog of the Aharonov–Bohm effect. *PRD* **26**, 1281. *See* §18.6.

Steele, J.D. (1990). On homogeneous pure radiation fields. *CQG* **7**, L81. *See* §§12.5, 24.5.

Steele, J.D. (1991). Simply-transitive homothety groups. *GRG* **23**, 811. *See* §11.4.

Steenrod, N.J. (1951). *The topology of fibre bundles* (Princeton University Press, Princeton). *See* §2.11.

Stein-Schabes, J. (1986). Nonstatic vacuum strings: exterior and interior solutions. *PRD* **33**, 3545. *See* §22.3.

Stephani, H. (1967a). Konform flache Gravitationsfelder. *Commun. Math. Phys.* **5**, 337. *See* §37.5.

Stephani, H. (1967b). Über Lösungen der Einsteinschen Feldgleichungen, die sich in einen fünfdimensionalen flachen Raum einbetten lassen. *Commun. Math. Phys.* **4**, 137. *See* §37.4.

Stephani, H. (1968a). Die Einbettung von Lösungen der Einsteinschen Gleichungen – eine Methode zur Gewinnung und invarianten Klassifizierung von Lösungen. Habilitationsschrift Universität Jena, Jena. *See* §37.4.

Stephani, H. (1968b). Einige Lösungen der Einsteinschen Feldgleichungen mit idealer Flüssigkeit die sich in einen fünfdimensionalen flachen Raum einbetten lassen. *Commun. Math. Phys.* **9**, 53. *See* §37.4.

Stephani, H. (1978). A note on Killing tensors. *GRG* **9**, 789. *See* §35.3.

Stephani, H. (1979). A method to generate algebraically special pure radiation field solutions from the vacuum. *J. Phys. A* **12**, 1045. *See* §30.7.

Stephani, H. (1980). Some new algebraically special radiation field solutions connected with Hauser's type N vacuum solution, in *9th international conference on general relativity and gravitation. Abstracts of contributed papers for the discussion groups*, vol. 1, ed. E. Schmutzer, page 76 (Friedrich-Schiller-Universität, Jena). *See* §§26.1, 30.7.

Stephani, H. (1982). Two simple solutions to Einstein's field equations. *GRG* **14**, 703. *See* §36.2.

Stephani, H. (1983a). Algebraically special, diverging vacuum and pure radiation fields revisited. *GRG* **15**, 173. *See* §§29.2, 32.4.

Stephani, H. (1983b). A new interior solution of Einstein's field equations for a spherically symmetric perfect fluid in shear-free motion. *J. Phys. A* **16**, 3529. *See* §16.2.

Stephani, H. (1984). Algebraically special, shearfree, diverging, and twisting vacuum and Einstein–Maxwell fields, in *Solutions of Einstein's equations: Techniques and results.* Lecture notes in physics, vol. 205, eds. C. Hoenselaers and W. Dietz, page 321 (Springer, Berlin). *See* §29.2.

Stephani, H. (1987). Some perfect fluid solutions of Einstein's field equations without symmetries. *CQG* **4**, 125. *See* §§33.3, 36.2, 36.3.

Stephani, H. (1988). Symmetries of Einstein's field equations with a perfect fluid source as examples of Lie–Bäcklund symmetries. *JMP* **29**, 1650. *See* §10.3.

Stephani, H. (1989). *Differential equations – their solutions using symmetries* (Cambridge University Press, Cambridge). *See* §10.2.

Stephani, H. (1996). *General relativity* (Cambridge University Press, Cambridge). *See* §36.1.

Stephani, H. (1998). Some static perfect fluid solutions with an Abelian G_3. *CQG* **15**, L55. *See* §22.2.

Stephani, H. and Grosso, R. (1989). Perfect fluids with 4-velocity spanned by two commuting Killing vectors. *CQG* **6**, 1673. *See* §21.2.

Stephani, H. and Herlt, E. (1985). Twisting type-N vacuum solutions with two non-commuting Killing vectors do exist. *CQG* **2**, L63. *See* §§11.4, 38.2.

Stephani, H. and Wolf, T. (1985). Perfect fluid and vacuum solutions of Einstein's field equations with flat 3-dimensional slices, in *Galaxies, axisymmetric systems and relativity. Essays presented to W.B. Bonnor on his 65th birthday*, ed. M.A.H. MacCallum, page 275 (Cambridge University Press, Cambridge). *See* §36.2.

Stephani, H. and Wolf, T. (1996). Spherically symmetric perfect fluids in shear-free motion – the symmetry approach. *CQG* **13**, 1261. *See* §16.2.

Sternberg, S. (1964). *Lectures on differential geometry* (Prentice-Hall, Englewood Cliffs, N.J.). *See* §§2.1, 2.7, 9.3.

Stewart, B.W. (1982). A generalisation of the Buchdahl transformation and perfect fluid solutions. *J. Phys. A* **15**, 1799. *See* §§10.11, 16.1.

Stewart, B.W., Witten, L. and Papadopoulos, D. (1987). Gravitational field with toroidal topology. *GRG* **19**, 827. *See* §20.2.

Stewart, J.M. (1990). *Advanced general relativity* (Cambridge University Press, Cambridge). *See* §7.1.

Stewart, J.M. and Ellis, G.F.R. (1968). Solutions of Einstein's equations for a perfect fluid which exhibit local rotational symmetry. *JMP* **9**, 1072. *See* §§10.11, 11.2, 13.1, 13.2, 13.4, 33.2, 33.3.

Stoeger, W.R., Nel, S.D., Maartens, R. and Ellis, G.F.R. (1992). The fluid-ray tetrad formulation of Einstein's field equations. *CQG* **9**, 493. *See* §7.2.

Suen, W.-M. (1986). Multipole moments for stationary, non-asymptotically-flat systems in general relativity. *PRD* **34**, 3617. *See* §18.8.

Suhonen, E. (1968). General relativistic fluid sphere at mechanical and thermal equilibrium. *Kgl. Danske Vidensk. Sels., Mat.-Fys. Medd.* **36**, 1. *See* §16.1.

Sussman, R.A. (1987). On spherically symmetric shear-free perfect fluid configurations (neutral and charged). I. *JMP* **28**, 1118. *See* §16.2.

Sussman, R.A. (1988a). On spherically symmetric shear-free perfect fluid configurations (neutral and charged). II. Equation of state and singularities. *JMP* **29**, 945. *See* §16.2.

Sussman, R.A. (1988b). On spherically symmetric shear-free perfect fluid configurations (neutral and charged). III. Global view. *JMP* **29**, 1177. *See* §16.2.

Sussman, R.A. (1989). Radial conformal Killing vectors in spherically symmetric shear-free space-times. *GRG* **21**, 1281. *See* §16.2.

Swift, S.T., d'Inverno, R.A. and Vickers, J.A. (1986). Everywhere invariant spaces of metrics and isometries. *GRG* **18**, 1093. *See* §8.4.

Synge, J.L. (1937). Relativistic hydrodynamics. *Proc. London Math. Soc.* **43**, 376. *See* §6.2.

Synge, J.L. (1960). *Relativity: The general theory* (North-Holland, Amsterdam). *See* §18.7.

Synge, J.L. (1964). The Petrov classification of gravitational fields. *Commun. Dublin Inst. Adv. Stud. A* **15**. *See* §§4.1, 4.2.

Szabados, L.B. (1987). Commutation properties of cyclic and null Killing symmetries. *JMP* **28**, 2688. *See* §§19.1, 24.4.

Szafron, D.A. (1977). Inhomogeneous cosmologies: New exact solutions and their evolution. *JMP* **18**, 1673. *See* §33.3.

Szafron, D.A. (1981). Intrinsic isometry groups in general relativity. *JMP* **22**, 543. *See* §36.4.

Szafron, D.A. and Collins, C.B. (1979). A new approach to inhomogeneous cosmologies: intrinsic symmetries. II. Conformally flat slices and an invariant classification. *JMP* **20**, 2354. *See* §36.3.

Szafron, D.A. and Wainwright, J. (1977). A class of inhomogeneous perfect fluid cosmologies. *JMP* **18**, 1668. *See* §33.3.

Szekeres, G. (1960). On the singularities of a Riemannian manifold. *Publ. Mat. Debrecen* **7**, 285. *See* §15.4.

Szekeres, P. (1965). The gravitational compass. *JMP* **6**, 1387. *See* §3.5.

Szekeres, P. (1966a). Embedding properties of general relativistic manifolds. *Nuovo Cim.* **43**, 1062. *See* §§37.3, 37.6.

Szekeres, P. (1966b). On the propagation of gravitational fields in matter. *JMP* **7**, 751. *See* §§7.6, 26.1, 31.7, 33.4.

Szekeres, P. (1968). Multipole particles in equilibrium in general relativity. *Phys. Rev.* **176**, 1446. *See* §20.2.

Szekeres, P. (1970). Colliding plane waves. *Nature* **228**, 1183. *See* §§25.3, 25.4.

Szekeres, P. (1972). Colliding plane gravitational waves. *JMP* **13**, 286. *See* §§25.2, 25.3, 25.4.

Szekeres, P. (1975). A class of inhomogeneous cosmological models. *Commun. Math. Phys.* **41**, 55. *See* §33.3.

Tabensky, R.R. and Taub, A.H. (1973). Plane symmetric self-gravitating fluids with pressure equal to energy density. *Commun. Math. Phys.* **29**, 61. *See* §§10.11, 15.7, 25.6.

Tafel, J., Nurowski, P. and Lewandowski, J. (1991). Pure radiation field solutions of the Einstein equations. *CQG* **8**, L83. *See* §30.7.

Takeno, H. (1961). The mathematical theory of plane gravitational waves in general relativity. *Sci. Rep. Res. Inst. Theor. Phys., Hiroshima Univ.* **1**. *See* §24.5.

Takeno, H. (1966). The theory of spherically symmetric space-times. *Sci. Rep. Res. Inst. Theor. Phys. Hiroshima Univ.* **5**, 4. *See* §15.1.

Takeno, H. and Kitamura, S. (1968). On the space-time admitting a parallel null vector field. *Tensor* **19**, 207. *See* §35.1.

Talbot, C.J. (1969). Newman–Penrose approach to twisting degenerate metrics. *Commun. Math. Phys.* **13**, 45. *See* §§7.1, 21.1, 27.1, 32.5.

Tanabe, Y. (1977). *SU*(2,1) symmetry of the Einstein–Maxwell fields. *Prog. Theor. Phys.* **57**, 840. *See* §34.1.

Tanabe, Y. (1978). Exact solutions of the stationary axisymmetric Einstein–Maxwell equations. *Prog. Theor. Phys.* **60**, 142. *See* §34.1.

Tanimoto, M. (1998). New varieties of Gowdy space-times. *JMP* **39**, 4891. *See* §22.3.

Tariq, N. and Tupper, B.O.J. (1975). A class of algebraically general solutions of the Einstein–Maxwell equations for non-null electromagnetic fields. *GRG* **6**, 345. *See* §§12.3, 13.2, 13.3.

Tariq, N. and Tupper, B.O.J. (1977). Curvature collineations in Einstein–Maxwell space-time and in Einstein spaces. *Tensor* **31**, 42. *See* §35.4.

Tariq, N. and Tupper, B.O.J. (1992). Conformal symmetry inheritance with cosmological constant. *JMP* **33**, 4002. *See* §§15.7, 23.3.

Taub, A.H. (1951). Empty space-times admitting a three parameter group of motions. *Ann. Math.* **53**, 472. *See* §§13.2, 13.3, 15.4.

Taub, A.H. (1956). Isentropic hydrodynamics in plane symmetric space-times. *Phys. Rev.* **103**, 454. *See* §15.7.

Taub, A.H. (1968). Restricted motions of gravitating spheres. *Ann. Inst. H. Poincaré A* **9**, 153. *See* §16.2.

Taub, A.H. (1972). Plane-symmetric similarity solutions for self-gravitating fluids, in *General relativity* (*Papers in honour of J.L. Synge*), ed. L. O'Raifeartaigh, page 133 (Clarendon Press, Oxford). *See* §15.7.

Taub, A.H. (1976). High frequency gravitational radiation in Kerr–Schild space-times. *Commun. Math. Phys.* **47**, 185. *See* §28.3.

Taub, A.H. (1980). Space-times with distribution-valued curvature tensors. *JMP* **21**, 1423. *See* §3.8.

Taub, A.H. (1981). Generalised Kerr–Schild space-times. *Ann. Phys.* (*USA*) **134**, 326. *See* §32.5.

Taub, A.H. (1988). Collision of impulsive gravitational waves followed by dust clouds. *JMP* **29**, 2622. *See* §25.6.

Taub, A.H. (1990). Interaction of null dust clouds fronted by impulsive plane gravitational waves. *JMP* **31**, 664. *See* §25.6.

Taub, A.H. (1991). Interaction of null dust clouds fronted by plane impulsive gravitational waves. II. *JMP* **32**, 1322. *See* §25.6.

Tauber, G.E. (1967). Expanding universe in conformally flat coordinates. *JMP* **8**, 118. *See* §14.2.

Teixeira, A.F.F., Wolk, I. and Som, M.M. (1977a). Exact relativistic cylindrical solution of disordered radiation. *Nuovo Cim. B* **41**, 387. *See* §22.2.

Teixeira, A.F.F., Wolk, I. and Som, M.M. (1977b). Exact relativistic solution of disordered radiation with planar symmetry. *J. Phys. A* **10**, 1679. *See* §15.7.

Thomas, T.Y (1934). *Differential invariants of generalised spaces* (Cambridge University Press, Cambridge). *See* §9.1.

Thompson, A. (1966). A class of related space-times. *Tensor* **17**, 92. *See* §32.5.

Thompson, I.H. and Whitrow, G.J. (1967). Time-dependent internal solutions for spherically symmetrical bodies in general relativity. *Mon. Not. R. Astron. Soc.* **136**, 207. *See* §16.2.

Thorne, K.S. (1967). Primordial element formation, primordial magnetic fields and the isotropy of the universe. *Astrophys. J.* **148**, 51. *See* §§14.1, 14.3.

Timofeev, V.N. (1996). Algebraically special vacuum gravitational fields with a cosmological constant. *Russ. Phys. J.* **39**, 585. *See* §29.1.

Tipler, F.J. (1974). Rotating cylinders and the possibility of global causality violation. *PRD* **9**, 2203. *See* §22.2.

Tod, K.P. (1992). On choosing coordinates to diagonalize the metric. *CQG* **9**, 1693. *See* §9.4.

Tolman, R.C. (1934a). Effect of inhomogeneity in cosmological models. *Proc. Nat. Acad. Sci. U.S.* **20**, 169. *See* §15.5.

Tolman, R.C. (1934b). *Relativity, thermodynamics and cosmology* (Oxford Univ. Press, Oxford). *See* §14.2.

Tolman, R.C. (1939). Static solutions of Einstein's field equations for spheres of fluid. *Phys. Rev.* **55**, 364. *See* §16.1.

Tomimatsu, A. (1980). On the soliton solutions of stationary axialsymmetric gravitational fields. *Progr. Theor. Phys.* **63**, 1054. *See* §34.5.

Tomimatsu, A. (1981). Axially symmetric, stationary gravitational field equations and pseudospherical surfaces. *Progr. Theor. Phys.* **66**, 526. *See* §19.5.

Tomimatsu, A. (1984). Distorted rotating black holes. *Phys. Lett. A* **103**, 374. *See* §34.5.

Tomimatsu, A. (1989). The gravitational Faraday rotation for cylindrical gravitational solitons. *GRG* **21**, 613. *See* §§22.3, 34.5.

Tomimatsu, A. and Sato, H. (1972). New exact solution for the gravitational field of a spinning mass. *Phys. Rev. Lett.* **29**, 1344. *See* §20.5.

Tomimatsu, A. and Sato, H. (1973). New series of exact solutions for gravitational fields of spinning masses. *Prog. Theor. Phys.* **50**, 95. *See* §20.5.

Tomimatsu, A. and Sato, H. (1981). Multi-soliton solutions of the Einstein equation and the Tomimatsu–Sato metric. *Prog. Theor. Phys. Supplement* **70**, 215. *See* §§20.5, 34.3.

Tomimura, N. (1977). Particular exact inhomogeneous solution with matter and pressure. *Nuovo Cim. B* **42**, 1. *See* §33.3.

Tooper, R.F. (1964). General relativistic polytropic fluid spheres. *Astrophys. J.* **140**, 434. *See* §16.1.

Torre, C.G. and Anderson, I.M. (1993). Symmetries of the Einstein equations. *Phys. Rev. Lett.* **70**, 3525. *See* §10.3.

Trautman, A. (1962). On the propagation of information by waves, in *Recent developments in general relativity*, page 459 (Pergamon Press–PWN, Oxford). *See* §§32.1, 32.2.

Trim, D.W. and Wainwright, J. (1974). Nonradiative algebraically special space-times. *JMP* **15**, 535. *See* §§29.1, 29.2, 30.2, 30.3, 30.4.

Trümper, M. (1965). On a special class of type-I gravitational fields. *JMP* **6**, 584. *See* §§4.3, 6.2.

Trümper, M. (1967). Einsteinsche Feldgleichungen für das axialsymmetrische, stationäre Gravitationsfeld im Innern einer starr rotierenden idealen Flüssigkeit. *Z. Naturforsch.* **22a**, 1347. *See* §21.2.

Tsamparlis, M. and Mason, D.P. (1983). On spacelike congruences in general relativity. *JMP* **24**, 1577. *See* §6.2.
Tseitlin, M.G. (1985). Solutions of two-dimensional Einstein equations parametrized by arbitrary functions and generated by the $O(2,1)$ sigma model. *Theor. Math. Phys.* **64**, 679. *See* §§10.11, 20.6.
Tsoubelis, D. (1989). Plane gravitational waves colliding with shells of null dust. *CQG* **6**, L117. *See* §25.6.
Tsoubelis, D. and Wang, A. (1991). On the gravitational interaction of plane symmetric clouds of null dust. *JMP* **32**, 1017. *See* §25.6.
Tsoubelis, D. and Wang, A. (1992). Head-on collision of gravitational plane waves with noncollinear polarization: a new class of analytic models. *JMP* **33**, 1054. *See* §25.4.
Tsoubelis, D. and Wang, A.Z. (1989). Asymmetric collision of gravitational plane waves: a new class of exact solutions. *GRG* **21**, 807. *See* §25.3.
Tsoubelis, D. and Wang, A.Z. (1990). Impulsive shells of null dust colliding with gravitational plane waves. *GRG* **22**, 1091. *See* §25.6.
Tupper, B.O.J. (1976). A class of algebraically general solutions of the Einstein–Maxwell equations for non-null electromagnetic fields. II. *GRG* **7**, 479. *See* §12.3.
Tupper, B.O.J. (1981). The equivalence of electromagnetic fields and viscous fluids in general relativity. *JMP* **22**, 2666. *See* §5.2.
Tupper, B.O.J. (1983). The equivalence of perfect fluid space-times and viscous magnetohydrodynamic space-times in general relativity. *GRG* **15**, 849. *See* §5.2.
Tupper, B.O.J. (1996). Conformal symmetries of conformal-reducible spacetimes with nonzero Weyl tensor. *CQG* **13**, 1679. *See* §35.4.
Uggla, C. (1990). New exact perfect fluid solutions of Einstein's equations. *CQG* **7**, L171. *See* §14.4.
Uggla, C. (1992). Inhomogeneous self-similar cosmological models. *CQG* **9**, 2287. *See* §§23.1, 23.2, 23.3.
Uggla, C., Bradley, M. and Marklund, M. (1995a). Classifying Einstein's field equations with applications to cosmology and astrophysics. *CQG* **12**, 2525. *See* §§9.4, 13.2.
Uggla, C., Jantzen, R.T. and Rosquist, K. (1995b). Exact hypersurface-homogeneous solutions in cosmology and astrophysics. *PRD* **51**, 5522. *See* §§13.2, 35.3.
Uggla, C. and Rosquist, K. (1990). New exact perfect fluid solutions of Einstein's equations. II. *CQG* **7**, L279. *See* §14.4.
Uggla, C., Rosquist, K. and Jantzen, R.T. (1990). Geometrizing the dynamics of Bianchi cosmology. *PRD* **42**, 404. *See* §13.2.
Unti, T.W.J. and Torrence, R.J. (1966). Theorem on gravitational fields with geodesic rays. *JMP* **7**, 535. *See* §26.4.
Urbantke, H. (1972). Note on Kerr–Schild type vacuum gravitational fields. *Acta Phys. Aust.* **35**, 396. *See* §32.2.
Urbantke, H. (1975). Der metrische Ansatz von Trautman, Kerr und Schild und einige seiner Anwendungen. *Acta Phys. Aust.* **41**, 1. *See* §§26.1, 32.4.
Vaidya, P.C. (1943). The external field of a radiating star in general relativity. *Curr. Sci.* **12**, 183. *See* §15.4.
Vaidya, P.C. (1968). Nonstatic analogues of Schwarzschild's interior solution in general relativity. *Phys. Rev.* **174**, 1615. *See* §16.2.
Vaidya, P.C. (1973). Some algebraically special solutions of Einstein's equations. II. *Tensor* **27**, 276. *See* §32.4.
Vaidya, P.C. (1974). A generalized Kerr–Schild solution of Einstein's equations. *Proc. Camb. Phil. Soc.* **75**, 383. *See* §32.4.
Vaidya, P.C. (1977). The Kerr metric in cosmological background. *Pramana* **8**, 512. *See* §21.2.

Vaidya, P.C. and Patel, L.K. (1973). Radiating Kerr metric. *PRD* **7**, 3590. *See* §32.4.

Vajk, J.P. (1969). Exact Robertson–Walker cosmological solutions containing relativistic fluids. *JMP* **10**, 1145. *See* §14.2.

Vajk, J.P. and Eltgroth, P.G. (1970). Spatially homogeneous anisotropic cosmological models containing relativistic fluid and magnetic field. *JMP* **11**, 2212. *See* §14.3.

Valiente Kroon, J.A. (2000). On conserved quantities, symmetries and radiative properties of peeling and non-peeling (polyhomogeneous) asymptotically flat spacetimes. Ph.D. thesis, Queen Mary and Westfield College, London. *See* §§17.2, 29.2.

Van den Bergh, N. (1986a). Conformally Ricci flat Einstein–Maxwell solutions with a null electromagnetic field. *GRG* **18**, 1105. *See* §10.11.

Van den Bergh, N. (1986b). Conformally Ricci-flat perfect fluids. *JMP* **27**, 1076. *See* §10.11.

Van den Bergh, N. (1986c). Conformally Ricci-flat spacetimes admitting a Killing vector field parallel to the gradient of the conformal scalar field. *Lett. Math. Phys.* **12**, 43. *See* §10.11.

Van den Bergh, N. (1986d). Irrotational and conformally Ricci-flat perfect fluids. *GRG* **18**, 649. *See* §§10.11, 14.4.

Van den Bergh, N. (1986e). Shearfree and conformally Ricci-flat perfect fluids. *Lett. Math. Phys.* **11**, 141. *See* §10.11.

Van den Bergh, N. (1987). Comment on conformally Ricci-flat perfect fluids of Petrov-type N. *GRG* **19**, 1131. *See* §10.11.

Van den Bergh, N. (1988a). A class of inhomogeneous cosmological models with separable metrics. *CQG* **5**, 167. *See* §22.3.

Van den Bergh, N. (1988b). Conformally Ricci-flat perfect fluids. II. *JMP* **29**. *See* §10.11.

Van den Bergh, N. (1988c). Nonrotating and nonexpanding perfect fluids. *GRG* **20**, 131. *See* §§13.4, 23.3.

Van den Bergh, N. (1988d). Perfect-fluid models admitting a non-Abelian and maximal two-parameter group of isometries. *CQG* **5**, 861. *See* §23.3.

Van den Bergh, N. (1989). Einstein–Maxwell null fields of Petrov type D. *CQG* **6**, 1871. *See* §31.6.

Van den Bergh, N. (1996a). Lorentz- and hyperrotation-invariant classification of symmetric tensors and the embedding class-2 problem. *CQG* **13**, 2817. *See* §37.5.

Van den Bergh, N. (1996b). Vacuum solutions of embedding class 2: Petrov types D and N. *CQG* **13**, 2839. *See* §37.5.

Van den Bergh, N. (1999). The shear-free perfect fluid conjecture. *CQG* **16**, 117. *See* §6.2.

Van den Bergh, N. and Skea, J.E.F. (1992). Inhomogeneous perfect fluid cosmologies. *CQG* **9**, 527. *See* §23.3.

Van den Bergh, N. and Wils, P. (1983). A class of stationary Einstein–Maxwell solutions with cylindrical symmetry. *J. Phys. A* **16**, 3843. *See* §22.2.

Van den Bergh, N. and Wils, P. (1985a). Exact solutions for nonstatic perfect fluid spheres with shear and an equation of state. *GRG* **17**, 223. *See* §16.2.

Van den Bergh, N. and Wils, P. (1985b). The rotation axis for stationary and axisymmetric space-times. *CQG* **2**, 229. *See* §19.1.

Van den Bergh, N., Wils, P. and Castagnino, M (1991). Inhomogeneous cosmological models of Wainwright class A1. *CQG* **8**, 947. *See* §23.3.

van Elst, H. and Ellis, G.F.R. (1996). The covariant approach to LRS perfect fluid spacetime geometries. *CQG* **13**, 1099. *See* §6.2.

van Elst, H, Uggla, C., Lesame, W.M., Ellis, G.F.R. and Maartens, R. (1997). Integrability of irrotational silent cosmological models. *CQG* **14**, 1151. *See* §6.2.

van Stockum, W.J. (1937). The gravitational field of a distribution of particles rotating about an axis of symmetry. *Proc. Roy. Soc. Edinburgh A* **57**, 135. *See* §§20.4, 21.2.

Vandyck, M.A.J. (1985). Some time-dependent axially symmetric metrics generalising the Weyl and the Einstein–Rosen line elements. *CQG* **2**, 241. *See* §17.3.

Vaz, E.G.L.R. (1986). Bianchi types of a three-parameter group of curvature collineations. *GRG* **18**, 1187. *See* §35.4.

Vaz, E.G.L.R. and Collinson, C.D. (1983). Curvature collineations for type-N Robinson–Trautman space-times. *GRG* **15**, 661. *See* §35.4.

Vera, R. (1998a). On "Diagonal G_2 spacetimes admitting inheriting conformal Killing vector fields". *CQG* **15**, 2037. *See* §§14.4, 23.3, 35.4.

Vera, R. (1998b). Theoretical aspects concerning separability, matching and matter contents of inhomogeneities in cosmology. Ph. D. thesis, University of Barcelona. *See* §23.3.

Verdaguer, E. (1982). Stationary axisymmetric one-soliton solutions of the Einstein equations. *J. Phys. A* **15**, 1261. *See* §34.5.

Verma, D.N. and Roy, S.R. (1956). Special metric forms and their gravitational significance. *Bull. Calc. Math. Soc.* **48**, 127. *See* §§36.2, 36.3.

Vishveshwara, C.V. and Winicour, J. (1977). Relativistically rotating dust cylinders. *JMP* **18**, 1280. *See* §§21.2, 22.2.

Vishwakarma, R.G., Abdusattar and Beesham, A. (1999). LRS Bianchi type-*I* models with a timedependent cosmological "constant ". *PRD* **60**, 063507. *See* §14.3.

Volkoff, G.M. (1939). On the equilibrium of massive spheres. *Phys. Rev.* **55**, 413. *See* §16.1.

Voorhees, B.H. (1970). Static axially symmetric gravitational fields. *PRD* **2**, 2119. *See* §20.2.

Wahlquist, H.D. (1968). Interior solution for a finite rotating body of perfect fluid. *Phys. Rev.* **172**, 1291. *See* §21.2.

Wahlquist, H.D. (1992). The problem of exact interior solutions for rotating rigid bodies in general relativity. *JMP* **33**, 304. *See* §21.2.

Wahlquist, H.D. and Estabrook, F.B. (1966). Rigid motions in Einstein spaces. *JMP* **7**, 894. *See* §6.2.

Wahlquist, H.D. and Estabrook, F.B. (1975). Prolongation structures of nonlinear evolution equations. *JMP* **16**, 1. *See* §10.4.

Wainwright, J. (1970). A class of algebraically special perfect fluid space-times. *Commun. Math. Phys.* **17**, 42. *See* §33.2.

Wainwright, J. (1974). Algebraically special fluid space-times with hypersurface-orthogonal shearfree rays. *Int. J. Theor. Phys.* **10**, 39. *See* §33.1.

Wainwright, J. (1977a). Characterization of the Szekeres inhomogeneous cosmologies as algebraically special space-times. *JMP* **18**, 672. *See* §33.3.

Wainwright, J. (1977b). Classification of the type D perfect fluid solutions of the Einstein equations. *GRG* **8**, 797. *See* §33.3.

Wainwright, J. (1979). A classification scheme for non-rotating inhomogeneous cosmologies. *J. Phys. A* **12**, 2015. *See* Ch. 23, §38.1.

Wainwright, J. (1981). Exact spatially inhomogeneous cosmologies. *J. Phys. A* **14**, 1131. *See* Ch. 23, §23.3.

Wainwright, J. (1983). A spatially homogeneous cosmological model with plane-wave singularity. *Phys. Lett. A* **99**, 301. *See* §14.4.

Wainwright, J. (1984). Power law singularities in orthogonal spatially homogeneous cosmologies. *GRG* **16**, 657. *See* §14.4.

Wainwright, J. (1985). Self-similar solutions of Einstein's equations, in *Galaxies, axisymmetric systems and relativity. Essays presented to W.B. Bonnor on his*

65th birthday, ed. M.A.H. MacCallum, page 288 (Cambridge University Press, Cambridge). *See* §§11.3, 13.4.

Wainwright, J. and Ellis, G.F.R. (1997). *Dynamical systems in cosmology* (Cambridge University Press, Cambridge). *See* §§1.4, 11.3, 13.2, 14.1, 14.4.

Wainwright, J. and Goode, S.W. (1980). Some exact inhomogeneous cosmologies with equation of state $p = \gamma\mu$. *PRD* **22**, 1906. *See* §§23.3, 36.3.

Wainwright, J., Ince, W.C.W. and Marshman, B.J. (1979). Spatially homogeneous and inhomogeneous cosmologies with equation of state $p = \mu$. *GRG* **10**, 259. *See* §§5.5, 10.11, 14.1, 14.3, 14.4, 17.1, 23.1, 23.3, 25.6.

Wainwright, J. and Marshman, B.J. (1979). Some exact cosmological models with gravitational waves. *Phys. Lett. A* **72**, 275. *See* §§10.11, 23.1.

Wainwright, J. and Yaremovicz, P.A.E. (1976a). Killing vector fields and the Einstein–Maxwell field equations with perfect fluid source. *GRG* **7**, 345. *See* §11.3.

Wainwright, J. and Yaremovicz, P.A.E. (1976b). Symmetries and the Einstein–Maxwell field equations: the null field case. *GRG* **7**, 595. *See* §§11.1, 11.3.

Walker, A.G. (1936). On Milne's theory of world-structure. *Proc. London Math. Soc.* **42**, 90. *See* §12.1.

Walker, M. and Kinnersley, W. (1972). Some remarks on a radiating solution of the Einstein–Maxwell equations, in *Methods of local and global differential geometry in general relativity.* Lecture notes in physics, vol. 14, eds. D Farnsworth, J. Fink, J. Porter and A. Thompson, page 48 (Springer-Verlag, Berlin). *See* §21.1.

Walker, M. and Penrose, R. (1970). On quadratic first integrals of the geodesic equations for type $\{2,2\}$ spacetimes. *Commun. Math. Phys.* **18**, 265. *See* §35.3.

Wang, A. (1991). The effect of polarization of colliding plane gravitational waves on focusing singularities. *Int. J. Mod. Phys. A* **6**, 2273. *See* §§25.4, 34.5.

Wang, M.Y. (1974). Class of solutions of axial-symmetric Einstein–Maxwell equations. *PRD* **9**, 1835. *See* §34.1.

Wang, Shi-kun, Guo, Han-ying and Wu, Ke (1983a). The N-fold charged Kerr family solution, in *Proceedings of the third Marcel Grossmann meeting on general relativity.* Pt. B., ed. Hu Ning, page 1039 (North-Holland, Amsterdam). *See* §34.6.

Wang, Shi-kun, Guo, Han-ying and Wu, Ke (1983b). The N-fold Kerr family and charged Kerr family solutions. *Commun. Theor. Phys.* **2**, 921. *See* §34.5.

Wang, Shi-kun, Guo, Han-ying and Wu, Ke (1984). Principal Riemann–Hilbert problem and N-fold charged Kerr solution. *CQG* **1**, 379. *See* §34.6.

Ward, J.P. (1976). Equilibrium – its connection to global integrability conditions for stationary Einstein–Maxwell fields. *Int. J. Theor. Phys.* **15**, 293. *See* §18.7.

Ward, R.S. (1983). Stationary axisymmetric space-times: a new approach. *GRG* **15**, 105. *See* §34.7.

Warner, F.W. (1971). *Foundations of differential manifolds and Lie groups* (Scott, Foresman, Glenview, Illinois). *See* §8.1.

Waylen, P.C. (1982). The general axially symmetric static solution of Einstein's vacuum field equations. *Proc. Roy. Soc. London A* **382**, 467. *See* §20.2.

Weiler, A. (1874). Über die verschiedenen Gattungen der Complexe zweiten Grades. *Math. Ann.* **VII**, 145. *See* §5.1.

Weinberg, S. (1972). *Gravitation and cosmology; principles and applications of the general theory of relativity* (Wiley, New York). *See* §§8.5, 14.1.

Weir, G.J. and Kerr, R.P. (1977). Diverging type-D metrics. *Proc. Roy. Soc. Lond. A* **355**, 31. *See* §§29.1, 29.2, 29.5, 38.2.

Weyl, H. (1917). Zur Gravitationstheorie. *Ann. Phys. (Germany)* **54**, 117. *See* §§20.2, 21.1.

Whelan, J.T. and Romano, J.D. (1999). Quasistationary binary inspiral. I. Einstein equations for the two Killing vector spacetime. *PRD* **60**, 084009. *See* §19.3.

Whiston, G.S. (1981). Four-dimensional kinks. *J. Phys. A* **14**, 2861. *See* Ch. 9.

White, A.J. and Collins, C.B. (1984). A class of shear-free perfect fluids in general relativity. I. *JMP* **25**, 332. *See* §6.2.

Whitman, P.G. (1983). Solutions to the general-relativistic Tolman–Wyman equation. *PRD* **27**, 1722. *See* §16.1.

Whittaker, J.M. (1968). An interior solution in general relativity. *Proc. Roy. Soc. Lond. A* **306**, 1. *See* §16.1.

Wils, P. (1989a). Homogeneous and conformally Ricci flat pure radiation fields. *CQG* **6**, 1243. *See* §§12.5, 26.1, 37.5.

Wils, P. (1989b). A new Painlevé solution of the Ernst equations. *Phys. Lett. A* **135**, 425. *See* §13.3.

Wils, P. (1989c). Painlevé solutions in general relativity. *CQG* **6**, 1231. *See* §22.4.

Wils, P. (1990). Aligned twisting pure radiation fields of Petrov type *D* do not exist. *CQG* **7**, 1905. *See* §30.7.

Wils, P. (1991). Inhomogeneous perfect fluid cosmologies with a non-orthogonally transitive symmetry group. *CQG* **8**, 361. *See* §§23.1, 23.3.

Wils, P. and Van den Bergh, N. (1985). A case of dual interpretation of Einstein–Maxwell fields. *GRG* **17**, 381. *See* §22.2.

Wils, P. and Van den Bergh, N. (1990). Petrov type *D* pure radiation fields of Kundt's class. *CQG* **7**, 577. *See* §31.6.

Winicour, J. (1975). All stationary axisymmetric rotating dust metrics. *JMP* **16**, 1806. *See* §21.2.

Wolf, T. (1986a). About vacuum solutions of Einstein's field equations with flat three-dimensional hypersurfaces. *JMP* **27**, 2354. *See* §36.2.

Wolf, T. (1986b). A class of perfect fluid metrics with flat three-dimensional hypersurfaces. *JMP* **27**, 2340. *See* §§13.4, 14.4, 21.2, 36.2.

Wolf, T. (1998). Structural equations for Killing tensors of arbitrary rank. *Comp. Phys. Comp.* **115**, 316. *See* §35.3.

Woodhouse, N.M.J. (1975). Killing tensors and the separation of the Hamilton–Jacobi equation. *Commun. Math. Phys.* **44**, 9. *See* §35.3.

Woodhouse, N.M.J. and Mason, L.J. (1988). The Geroch group and non-Hausdorff twistor spaces. *Nonlinearity* **1**, 73. *See* §34.7.

Woolley, M.L. (1973). On the role of the Killing tensor in the Einstein–Maxwell theory. *Commun. Math. Phys.* **33**, 135. *See* §34.1.

Wu, Yongshi and Ge, Molin (1983). A new approach to the algebraic structure in stationary axially symmetric gravity, in *Proceedings of the third Marcel Grossmann meeting on general relativity.* Pt. B., ed. Hu Ning, page 1067 (North-Holland, Amsterdam). *See* §34.7.

Wu, Yongshi, Ge, Molin and Hou, Boy (1983). Infinitely conserved currents and hidden symmetry algebra related to the Belinskii–Zakharov's formulation of gravity, in *Proceedings of the third Marcel Grossmann meeting on general relativity*, Pt. A, ed. Hu Ning, page 367 (North-Holland, Amsterdam). *See* §34.5.

Wu, Zhong Chao (1981). Self-similar cosmological models. *GRG* **13**, 625. *See* §23.1.

Wyman, M. (1946). Equations of state for radially symmetric distributions of matter. *Phys. Rev.* **70**, 396. *See* §§15.6, 16.2.

Wyman, M. (1949). Radially symmetric distributions of matter. *Phys. Rev.* **75**, 1930. *See* §16.1.

Wyman, M. (1976). Jeffery-Williams lecture 1976: Nonstatic radially symmetric distributions of matter. *Can. Math. Bull.* **19**, 343. *See* §16.2.

Wyman, M. and Trollope, R. (1965). Null fields in Einstein–Maxwell field theory. *JMP* **6**, 1965. *See* §31.6.

Xanthopoulos, B.C. (1981). Exterior spacetimes for rotating stars. *JMP* **22**, 1254. *See* §34.3.

Xanthopoulos, B.C. (1983a). Local toroidal black holes that are static and axisymmetric. *Proc. Roy. Soc. London A* **388**, 117. *See* §20.2.

Xanthopoulos, B.C. (1983b). On Petrov type N vacuum solutions of the Einstein equations. *Phys. Lett. A* **99**, 304. *See* §32.5.

Xanthopoulos, B.C. (1983c). The optical scalars in Kerr–Schild-type spacetimes. *Ann. Phys.* (*USA*) **149**, 286. *See* §32.5.

Xanthopoulos, B.C. (1986). Linear superposition of solutions of the Einstein–Maxwell equations. *CQG* **3**, 157. *See* §§10.3, 32.5.

Xanthopoulos, B.C. (1987). Perfect fluids with $\epsilon = p + constant$ equation of state in general relativity. *JMP* **18**, 905. *See* §23.3.

Xu, Xiao-hai (1987). A new procedure generating the solutions of stationary axially symmetric gravitational field equations. *Commun. Theor. Phys.* **8**, 97. *See* §20.5.

Yakupov, M.S. (1968a). Algebraic characterization of second order Einstein spaces (in Russian). *Grav. Teor. Otnos., Univ. Kazan* **4/5**, 78. *See* §37.5.

Yakupov, M.S. (1968b). On Einstein spaces of embedding class two (in Russian). *Dokl. Akad. Nauk SSSR* **180**, 1096. *See* §37.5.

Yakupov, M.S. (1973). Einstein spaces of class two (in Russian). *Grav. Teor. Otnos., Univ. Kazan* **9**, 109. *See* §37.5.

Yamazaki, M. (1980a). A new family of spinning mass solutions. *J. Phys. Soc. of Japan* **49**, 1649. *See* §20.6.

Yamazaki, M. (1980b). On a new family of spinning mass solutions. *Prog. Theor. Phys.* **64**, 861. *See* §20.6.

Yamazaki, M. (1980c). On the Kinnersley-Chitre spinning mass field. *Prog. Theor. Phys.* **63**, 1950. *See* §20.6.

Yamazaki, M. (1982). The Kerr–Tomimatsu–Sato family of spinning mass solutions, in *Proceedings of the second Marcel Grossmann meeting on general relativity*, ed. R. Ruffini, page 371 (North-Holland, Amsterdam). *See* §20.5.

Yamazaki, M. (1983a). On the Kramer–Neugebauer spinning masses solutions. *Prog. Theor. Phys.* **69**, 503. *See* §34.4.

Yamazaki, M. (1983b). Stationary line of N Kerr masses kept apart by gravitational spin-spin interaction. *Phys. Rev. Lett.* **50**, 1027. *See* §34.3.

Yamazaki, M. (1984). N Kerr particles, in *Solutions of Einstein's equations: Techniques and results.* Lecture notes in physics, vol. 205, eds. C. Hoenselaers and W. Dietz, page 311 (Springer, Berlin). *See* §34.4.

Yanez, G., Breton B., N. and García D., A. (1995). Generalization of the Papacostas–Xanthopoulos solution. *Revista Mexicana de Fisica* **41**, 480. *See* §25.5.

Yano, K. (1955). *The theory of Lie derivatives and its application* (North-Holland, Amsterdam). *See* §8.7.

York, Jr., J.W. (1971). Gravitational degrees of freedom and the initial-value problem. *Phys. Rev. Lett.* **26**, 1656. *See* §3.7.

Yurtsever, U. (1989). Singularities and horizons in the collisions of gravitational waves. *PRD* **40**, 329. *See* §25.1.

Zafiris, E. (1997). Incorporation of spacetime symmetries in Einstein's field equations. *JMP* **38**, 5854. *See* §35.4.

Zakharov, V.D. (1965). A physical characteristic of Einsteinian spaces of degenerate type II in the classification of Petrov (in Russian). *Dokl. Akad. Nauk SSSR* **161**, 563. *See* §4.2.

Zakharov, V.D. (1970). Algebraical and group theoretical methods in general relativity: Invariant Petrov type characterisation of the type of Einstein spaces (in Russian). *Probl. Teor. Grav. Elem. Chastitis* **3**, 128. *See* §4.2.

Zakharov, V.D. (1972). *Gravitational waves in Einstein's theory of gravitation* (in Russian) (Nauka, Moscow). *See* §§4.2, 24.5.

Zakhary, E. (1994). Classification of the Riemann tensor. Ph.D. thesis, Monash University. *See* §9.3.

Zakhary, E. and McIntosh, C.B.G. (1997). A complete set of Riemann invariants. *GRG* **29**, 539. *See* §9.1.

Zenk, L.G. and Das, A. (1978). An algebraically special subclass of vacuum metrics admitting a Killing motion. *JMP* **19**, 535. *See* §38.2.

Zhong, Zai-Zhe (1985). Generation of new solutions of the stationary axisymmetric Einstein equations by a double complex function method. *JMP* **26**, 2589. *See* §34.7.

Zipoy, D.M. (1966). Topology of some spheroidal metrics. *JMP* **7**, 1137. *See* §20.2.

Zund, J.D. (1986). An invariant-theoretic approach to the transition problem and the Bel–Petrov classification. *Tensor* **43**, 196. *See* §9.3.

Zund, J.D. and Brown, E. (1971). The theory of bivectors. *Tensor* **22**, 179. *See* §3.4.

Index

Printed in the United States
By Bookmasters